Molecular Surgical Pathology

Liang Cheng · George J. Netto · John N. Eble
Editors

Molecular Surgical Pathology

Second Edition

 Springer

Editors
Liang Cheng, MD, MS
Professor and Vice Chair
Director of Anatomic Pathology
Director of Molecular Pathology
Department of Pathology and Laboratory Medicine
Warren Alpert Medical School of Brown University
Lifespan Academic Medical Center
Associate Director for Shared Resources
Legorreta Cancer Center at Brown University
Providence, Rhode Island, USA

John N. Eble, MD, MBA, FRCPA, FRCPath
(deceased) Nordschow Professor and Chair
Department of Pathology and Laboratory Medicine
Indiana University School of Medicine
Indianapolis, Indiana, USA

George J. Netto, MD
Simon Flexner Professor and Chair
Department of Pathology and Laboratory Medicine
Perelman School of Medicine
at the University of Pennsylvania
Philadelphia, Pennsylvania, USA

ISBN 978-3-031-35120-4   ISBN 978-3-031-35118-1   (eBook)
https://doi.org/10.1007/978-3-031-35118-1

This Springer imprint is published by the registered company Springer Nature Switzerland AG
The registered company address is: Gewerbestrasse 11, 6330 Cham, Switzerland

This book is dedicated to the memory of our late friend and colleague, Dr. John N. Eble, a brilliant pathologist, consummate educator, and visionary leader in our field.

Liang Cheng & George J Netto

Preface

Since the first edition of this book in 2012, the field of molecular pathology has expanded exponentially. The advent of complete cancer genome sequencing, new diagnostic molecular tests, and biomarkers for targeted therapy and immunotherapy have demonstrated the importance of molecular tools both in patient care and in the practice of pathology. Furthermore, the development of many new molecular technologies has revolutionized the practice of modern pathology. The rapid growth of this field has led to an expansion of knowledge of molecular processes and many more clinical applications in our daily practice of surgical pathology. Thus, this updated second edition *Molecular Surgical Pathology* aims to serve as "quick reference" for pathologists, oncologists, geneticists, primary care physicians, and other medical professionals with an interest in this evolving field.

The second edition of *Molecular Surgical Pathology* has been vastly expanded with the addition of ten new chapters, including *Molecular Techniques, Molecular Diagnostics for Surgical Pathologists* (Chap. 1); *Bioinformatics, Digital Pathology, and Computational Pathology for the Surgical Pathologists* (Chap. 2); *Molecular Cytopathology* (Chap. 3); *Next-Generation Immunohistochemistry in the Workup of Neoplasm of Uncertain Lineage and CUP* (Chap. 4); *Molecular Pathology of Gastroesophageal Tumors* (Chap. 6); *Molecular Pathology of Salivary Gland Tumors* (Chap. 19); *Molecular Pathology of the Heart and Cardiovascular System* (Chap. 20); *Molecular Pathology of Endocrine Tumors* (Chap. 21); *Molecular Pathology of Lymphoma* (Chap. 24); and *Molecular Pathology of Leukemia* (Chap. 25). The book focuses on the practical utility of molecular techniques and molecular biomarkers for the practicing surgical pathologist. The emphasis is on the impact of molecular pathology on tumor classification, diagnosis, and differential diagnosis, as well as its implications for precision medicine and individualized patient care.

We are incredibly grateful to our contributing authors for sharing their knowledge and expertise with our readers. We also thank the dedicated and talented staff at the Springer, especially Lillie Mae Gaurano, Dhanapal Palanisamy, Vishal Anand, and Henry Rogers, who have provided invaluable support throughout the development and production of this book. We hope that the second edition *Molecular Surgical Pathology* becomes a valuable resource for all of our readers.

Providence, RI, USA
Philadelphia, PA, USA
Indianapolis, IN, USA

Liang Cheng
George J. Netto
John N. Eble

Contents

Contributors

N. Volkan Adsay Department of Pathology, Koç University School of Medicine, Istanbul, Turkey

Khaleel I. Al-Obaidy Department of Pathology and Laboratory Medicine, Henry Ford Health, Detroit, MI, USA

Zainab I. Alruwaii Department of Pathology, Regional Laboratory and Blood Bank, Eastern Province, Dammam, Saudi Arabia

Sunil S. Badve Department of Pathology and Laboratory Medicine, Emory University School of Medicine, Atlanta, GA, USA

Boris Bastian Helen Diller Family Comprehensive Cancer Center, Departments of Pathology and Laboratory Medicine, University of California at San Francisco, San Francisco, CA, USA

Andrew M. Bellizzi Department of Pathology, University of Iowa Hospitals and Clinics and Carver College of Medicine, Iowa City, IA, USA

Katharina Biermann Department of Pathology, Erasmus MC-University Medical Center Rotterdam, Rotterdam, The Netherlands

Adam L. Booth Department of Pathology, Northwestern University Feinberg School of Medicine, Chicago, IL, USA

Devon S. Chabot-Richards Department of Pathology, University of New Mexico, Albuquerque, NM, USA

Liang Cheng The Legorreta Cancer Center at Brown University, Department of Pathology and Laboratory Medicine, Warren Alpert Medical School of Brown University, Lifespan Academic Medical Center, Providence, RI, USA

Jiayu Chen Departments of Pathology, Urology and Oncology, The Sidney Kimmel Comprehensive Cancer Center at Johns Hopkins, Johns Hopkins University School of Medicine, Baltimore, MD, USA

Wei Chen Department of Pathology, The Ohio State University, Wexner Medical Center, Columbus, OH, USA

Veronica K. Y. Cheung Tissue Pathology and Diagnostic Oncology, Royal Prince Alfred Hospital, NSW Health Pathology, University of Sydney Central Clinical School, Sydney, Australia

Alanna J. Church Department of Pathology, Boston Children's Hospital and Harvard Medical School, Boston, MA, USA

Paola Dal Cin Department of Pathology, The Center for Advanced Molecular Diagnostics (CAMD), Brigham and Women's Hospital and Harvard Medical School, Boston, MA, USA

Jennifer A. Cotter Department of Pathology and Laboratory Medicine, Children's Hospital Los Angeles, Keck School of Medicine of the University of Southern California, Los Angeles, CA, USA

Angelo M. De Marzo Departments of Pathology, Urology and Oncology, The Sidney Kimmel Comprehensive Cancer Center at Johns Hopkins, Johns Hopkins University School of Medicine, Baltimore, MD, USA

Carlo De la Sancha Department of Pathology, University of California at San Francisco, San Francisco, CA, USA

Lori A. Erickson Department of Laboratory Medicine and Pathology, Mayo Clinic, Rochester, MN, USA

Nuria Eritja University of Lleida, Institute of Biomedical Research of Lleida (Irblleida), Lleida, Spain

Brian S. Finkelman Department of Pathology and Laboratory Medicine, University of Rochester Medical Center, Rochester, NY, USA

Adam S. Fisch Department of Pathology, Massachusetts General Hospital, Harvard Medical School, Boston, MA, USA

Wendy L. Frankel Department of Pathology, The Ohio State University, Wexner Medical Center, Columbus, OH, USA

Sonia Gatius Department of Pathology, Hospital Universitari Arnau de Vilanova, University of Lleida, Institute of Biomedical Research of Lleida (Irblleida), Lleida, Spain

Yesim Gökmen-Polar Department of Pathology and Laboratory Medicine, Emory University School of Medicine, Atlanta, GA, USA

Raul S. Gonzalez Department of Pathology and Laboratory Medicine, Emory University Hospital, Atlanta, GA, USA

Ruta Gupta Tissue Pathology and Diagnostic Oncology, Royal Prince Alfred Hospital, NSW Health Pathology, University of Sydney Central Clinical School, Sydney, Australia

Eyas M. Hattab Department of Pathology and Laboratory Medicine, University of Louisville School of Medicine, Louisville, KY, USA

Ralph H. Hruban Department of Pathology, The Sol Goldman Pancreatic Cancer Research Center, Johns Hopkins University, Baltimore, MD, USA

Yin P. (Rex) Hung Department of Pathology, Massachusetts General Hospital and Harvard Medical School, Boston, MA, USA

Dan Jones Department of Pathology, The Ohio State University, Wexner Medical Center, Columbus, OH, USA

Loren Joseph Department of Pathology, Beth Israel Deaconess Medical Center and Harvard Medical School, Boston, MA, USA

Clayton E. Kibler Department of Pathology, University of New Mexico, Albuquerque, NM, USA

Jae W. Lee Department of Pathology, The Sol Goldman Pancreatic Cancer Research Center, Johns Hopkins University, Baltimore, MD, USA

Neal I. Lindeman Department of Pathology and Laboratory Medicine, Weill Cornell Medicine, New York Presbyterian Hospital, New York, NY, USA

Thomas Longerich Institute of Pathology, University Hospital Heidelberg, Heidelberg, Germany

Leendert H. J. Looijenga Princess Maxima Center for Pediatric Oncology, Utrecht, The Netherlands

Ying-Chun Lo Department of Laboratory Medicine and Pathology, Mayo Clinic, Rochester, MN, USA

Kruti P. Maniar Department of Pathology and Laboratory Medicine, NorthShore , University of Chicago Pritzker School of Medicine, Chicago, IL, USA

Adrian Marino-Enriquez Department of Pathology, The Center for Advanced Molecular Diagnostics (CAMD), Brigham and Women's Hospital and Harvard Medical School, Boston, MA, USA

Xavier Matias-Guiu Department of Pathology, Hospital Universitari Arnau de Vilanova, University of Lleida, Institute of Biomedical Research of Lleida (Irblleida), Lleida, Spain

Department of Pathology, Hospital Universitari de Bellvitge, Bellvitge Institute of Biomedical Research (IDIBELL), University of Barcelona, Barcelona, Spain

Phillip D. Michaels Department of Pathology, Brigham and Women's Hospital and Harvard Medical School, Boston, MA, USA

Amar Mirza Department of Pathology, University of California San Francisco, San Francisco, CA, USA

Sambit K. Mohanty Department of Pathology and Laboratory Medicine, Advanced Medical Research Institute, Bhubaneswar, Odisha, India

The Advanced Medical Research Institute and CORE Diagnostics, Bhubaneswar, Odisha, India

William G. Nelson Departments of Pathology, Urology and Oncology, The Sidney Kimmel Comprehensive Cancer Center at Johns Hopkins, Johns Hopkins University School of Medicine, Baltimore, MD, USA

George J. Netto Department of Pathology and Laboratory Medicine, Perelman School of Medicine at the University of Pennsylvania, Philadelphia, PA, USA

Anil V. Parwani Department of Pathology, The Ohio State University, Wexner Medical Center, Columbus, OH, USA

Gordana Raca Division of Pathology and Laboratory Medicine, Department of Clinical Pathology, University of Southern California, Children's Hospital of Los Angeles, Los Angeles, CA, USA

Sinchita Roy-Chowdhuri Department of Pathology, MD Anderson Cancer Center, Houston, TX, USA

Roberto Ruiz-Cordero Department of pathology and Laboratory Medicine, University of Miami, Miami, FL, USA

Dipti P. Sajed Department of Pathology and Laboratory Medicine, David Geffen School of Medicine at UCLA, Los Angeles, CA, USA

Peter Schirmacher Institute of Pathology, University Hospital Heidelberg, Heidelberg, Germany

Karen Sfanos Departments of Pathology, Urology and Oncology, The Sidney Kimmel Comprehensive Cancer Center at Johns Hopkins, Johns Hopkins University School of Medicine, Baltimore, MD, USA

Saba Shafi Department of Pathology, The Ohio State University, Wexner Medical Center, Columbus, OH, USA

Ie-Ming Shih Departments of Pathology, Oncology, and Gynecology and Obstetrics, Johns Hopkins University, Baltimore, MD, USA

Maie A. St. John Department of Head & Neck Surgery, David Geffen School of Medicine at UCLA, Los Angeles, CA, USA

James R. Stone Department of Pathology, Massachusetts General Hospital and Harvard Medical School, Boston, MA, USA

Keith F. Stringer Division of Pathology and Laboratory Medicine, Department of Pediatrics, University of Cincinnati, Cincinnati Children's Hospital Medical Center, Cincinnati, OH, USA

Sean R. Williamson Robert J Tomsich Pathology and Laboratory Medicine Institute, Cleveland Clinic, Cleveland, OH, USA

Laura D. Wood Department of Pathology, The Sol Goldman Pancreatic Cancer Research Center, Johns Hopkins University, Baltimore, MD, USA

Srinivasan Yegnasubramanian Departments of Pathology, Urology and Oncology, The Sidney Kimmel Comprehensive Cancer Center at Johns Hopkins, Johns Hopkins University School of Medicine, Baltimore, MD, USA

Molecular Techniques/Molecular Diagnostics for Surgical Pathologists: An Overview

Loren Joseph, Gordana Raca, and Keith F. Stringer

Contents

L. Joseph (✉)
Department of Pathology, Beth Israel Deaconess Medical Center and Harvard Medical School, Boston, MA, USA
e-mail: ljjoseph@bidmc.harvard.edu

G. Raca
Division of Pathology and Laboratory Medicine, Department of Clinical Pathology, University of Southern California, Children's Hospital of Los Angeles, Los Angeles, CA, USA
e-mail: graca@chla.hospitals.edu

K. F. Stringer
Division of Pathology and Laboratory Medicine, Department of Pediatrics, University of Cincinnati, Cincinnati Children's Hospital Medical Center, Cincinnati, OH, USA
e-mail: stringk@ucmail.uc.edu

Introduction

- The microscope was not embraced by all pathologists when introduced
  - Optimal use required the development of fixation, sectioning, staining, and literature on clinical correlation
- DNA testing was eagerly anticipated in surgical pathology (except, perhaps, by insurers)
  - Initial tests were for a few single nucleotide substitutions and for a few translocations, tested by Southern blot analysis, a laborious multiday technique using radioactive reagents
  - Sanger sequencing (1975) and the polymerase chain reaction (PCR) (1990) together dramatically expanded the number of genomic variants which could be interrogated
- Next-generation sequencing (NGS) has further transformed the molecular testing landscape
  - NGS has much greater sensitivity than Sanger sequencing
  - The cost of NGS analysis has declined to the extent that, when there are clinical indications for sequencing more than a few exons, it is less expensive to perform NGS than Sanger sequencing

© The Author(s), under exclusive license to Springer Nature Switzerland AG 2023
L. Cheng et al. (eds.), *Molecular Surgical Pathology*, https://doi.org/10.1007/978-3-031-35118-1_1

"

- NGS enables sequencing entire genomes and transcriptomes, even at the single-cell level
- Some NGS methods enable the sequencing of single strands of RNA and DNA from samples without amplification

- There are multiple roles for molecular genetics in surgical pathology including tumor classification, prognosis, and therapy selection (see below)
- Morphological analysis of an H&E slide, for now, remains the most sophisticated assay available for <u>gene expression</u> in tumors
  - It has several advantages over molecular analysis
    - It is rapid and relatively inexpensive
    - It is adequately reproducible
    - It has extensive clinical correlation
    - In selected conditions, it can suggest, though never prove, some genomic variations
      - BCR–ABL translocations in most CML
      - NPM1 mutations in AML
      - Microsatellite instability in some colorectal cancers
  - Histopathology has clear limitations
    - It cannot detect the large and increasing number of genomic variations which influence prognosis and treatment selection
    - It cannot distinguish among important subsets of morphologically similar cells, although immunohistochemistry (IHC) helps
      - Traditional IHC is limited in how many markers can be tested on a given cell
      - Quantitative IHC analysis is not practical by visual inspection
      - Assessing cell–cell proximity is not practical by visual inspection
      - Techniques like cyclic immunofluorescence allow greater multiplexing and quantitation
      - Digitized imaging combined with machine learning could advance the analysis of cell–cell spatial relationships
      - The antibodies must be selected in advance, limiting the discovery
- There are three classes of analyte for molecular diagnostics
  - Chromosomes
    - Cytogenetic methods permit cell-by-cell analysis
    - Cytogenetic fluorescent in situ hybridization (FISH) permits cell-by-cell analysis with cell-level histologic localization
    - Whole-genome sequencing data can be used to construct a virtual karyotype
      - This can use 'low-pass' coverage; it does not require full sequencing of the genome

- Until recently (see Sect. "In Situ Sequencing"), this could only give an aggregate result for complex samples rather than permit a cell-by-cell analysis
  - Genome sequence (DNA) (including modifications like methylation)
    - The preponderance of molecular testing in surgical pathology, to date, has targeted sequence changes in genomic DNA from formalin-fixed paraffin-embedded (FFPE) sections
  - Transcriptome (RNA) sequence (including identification of fusions)
    - Gene expression, while a major focus of research, has not yet become a significant molecular <u>*diagnostic*</u> modality in surgical pathology (by test volume)

- Now that entire genomes can be sequenced, it remains an open question of how much should be sequenced in a given case
  - 'Complete' sequencing of 'the' human genome (really ~82%) was reported in 2003
    - The latest genome version added 220 million nucleotides to 'the' reference genome
    - The reference sequence is now estimated as 99.8% complete
    - The gene count is approximately 22,000
    - Much of the noncoding genome encodes various RNAs such as miRNA and lncRNA, some of which are suspected of having a role in cancer
  - Systematic sequencing of over 33,000 tumors by two consortia, one national, one international
    - The Cancer Genome Atlas (TCGA) (https://www.cancer.gov/about-nci/organization/ccg/research/structural-genomics/tcga, accessed 2021/09/14) analyzed 11,000 cases spanning 33 primary tumor types
    - International Cancer Genomics Consortium (ICGC) (https://dcc.icgc.org/, accessed 2021/09/14) analyzed 22,330 cases spanning 22 primary types
    - A variety of molecular methods were applied
      - All tumors were, at a minimum, subjected to NGS for a panel of several hundred genes
      - Over 2658 tumors underwent whole-genome sequencing
      - The TCGA data includes histological images
    - A meta-analysis (Martínez-Jiménez, et al, 2020) of 28,000 tumors, primarily from these consortia, identified 568 driver genes
      - 28,000 tumors spanning 66 tumor types identified 568 driver genes
      - Most driver genes are each associated with only a few tumor types
      - 10–20 of the driver genes are responsible for causing/contributing to multiple types of cancer

- Noncoding variants
  - Some have been implicated as 'drivers' (for example, in the *TERT* promoter)
  - The analysis of both sequence variants and expression levels for noncoding RNA, such lncRNA, is far from complete, notwithstanding the consortia analysis
- Clinical utility
  - Formal definitions of clinical utility do NOT include the cost of testing
    - Cost *is* included in cost-benefit analysis (CBE) and cost-effectiveness analysis (CEA)
    - Clinical utility is concerned with clinical outcome, especially survival, which has, among its benefits, simplicity of measurement
  - The clinical utility of specific molecular tests has been clearly demonstrated in numerous specific settings—such as CML (*BCR–ABL*), melanoma (*BRAF*), lung cancer (several oncogene fusions)
  - A still unsettled issue is whether all cancers should be screened with some molecular panel
    - Multiple national-level clinical trials have been completed or are in progress
    - Some of the studies excluded patients who had tumors with already known predictive molecular biomarkers (weighting the trial against finding utility for gene panels)
    - The number of known treatment-mutation associations continues to grow
    - Most trials are intended to determine the utility of molecular panels for matching patients with specific treatments, NOT with the utility of molecular markers for prognosis
  - Many payors, including the Federal Government, now cover at least one NGS panel for a patient's tumor
- Future directions
  - Single-cell analysis

Roles for Molecular Surgical Pathology

- Prognostic biomarkers
  - Prognosis is 'The likely outcome or course of a disease' (https://www.cancer.gov/publications/dictionaries/cancer-terms/def/prognosis; Accessed 2021/8/3)
  - This is the expectation for disease without mention of treatment—the 'natural' course
  - Since cancer is now seldom untreated, 'prognosis' is often used to mean the likely outcome on 'standard therapy'

- A 'prognostic biomarker' in molecular genetic pathology is a finding (a chromosomal alteration, a mutation, altered gene expression) that stratifies patients into better or worse outcome groups
  - Examples
    - FLT3 ITD mutation: This is a marker of poor prognosis in AML
    - EGFR variants in lung cancer: One of the variants is a predictor of responsiveness to EGFR inhibitors BUT is also a positive prognostic factor in untreated disease
- A prognostic biomarker can become a predictive marker
  - The presence of an FLT3 mutation is now also predictive of response to midostaurin
- Predictive biomarkers
  - A 'predictive factor' is 'A condition or finding that can be used to help predict whether a person's cancer will respond to a specific treatment' (https://www.cancer.gov/publications/dictionaries/cancer-terms/def/predictive-factor)
  - Since a positive response to treatment based on a predictive marker can lead to an altered outcome, it is common to refer to the predictive biomarker as a 'prognostic' marker; although not strictly correct usage, the meaning is often clear from the context
  - A paradigmatic example of the goal of predictive biomarkers and of 'Precision Medicine' is the finding of the *BCR–ABL* translocation in CML, followed by the development of highly effective inhibitors
  - A drug-sensitive mutation (predictive biomarker) in one tumor type does not guarantee that the presence of the same mutation in a different tumor type will predict response to treatment
    - BRAF V600 mutations in melanoma predict a beneficial response to a specific inhibitor (vemurafenib)
    - BRAF V600 mutations in colon cancer are not associated with a beneficial response
  - Although predictive molecular markers are identified in many tumor types, the finding does not, so far, usually influence initial therapy selection
    - Treatment with drugs based on molecular markers is often only considered in stage IV cancer patients after other therapies have failed
    - If a stage IV cancer patient is considered for molecularly informed therapy, a frequent issue is whether to analyze the primary tumor or any current tumor site

- ♦ Given the variety of tumors, mutations, and patient histories, a definitive general answer is not possible
- ♦ Sample availability often limits the choice
- ♦ Several studies have shown that, in the absence of therapy, primary tumors, and metastases show mostly the same driver gene mutations
  - ▪ Until recently, activated oncogene mutations are the ones most likely to have to target drugs available
  - ▪ Chemotherapy and radiation therapy often accelerate clonal divergence among the primary and subsequent metastases
- Inherited risk markers
  - Inherited mutations in some genes, such as *BRCA1*, are associated with a marked increase in the likelihood of developing cancer, especially at a young age
    - ○ The affected genes are predominantly tumor suppressors or DNA repair genes
    - ○ Mutations in oncogenes tend to be activating and act in dominant fashion
    - ○ Mutations in tumor suppressor and repair genes are 'recessive'
  - 'Two-hit hypothesis' (Knudson's Hypothesis)
    - ○ This postulates that one requires a mutation in each of the alleles of a tumor suppressor to make the cell/tissue permissive for tumor development
      - ♦ Cancer development probably requires more than the two 'hits'
    - ○ Two mutations in the same risk gene are occasionally found in sequencing a tumor
      - ♦ One variant is assumed to be inherited, and one acquired (the 'somatic' mutation)
      - ♦ The variant allele frequency (VAF) of the mutation, when compared to the percent of tumor cells in the sample, can sometimes suggest which mutation is inherited, which somatic
      - ♦ if the two mutations are positioned more than 100–200 base pairs apart, second generation sequencing cannot usually determine if they are on the same allele (or even in the same cell)
        - ▪ Single-cell methods can determine if two 'hits' are in the same cell
        - ▪ Long-read formats of third-generation NGS can potentially determine if the mutations are in the same cell and if they are on the same allele
        - ▪ If two distinct 'hits' in the same gene are present in distinct cells, it suggests genomic instability at worst or a field effect at (not very) best

- For patients with solid tumors, sequencing genomic DNA from a blood sample can help establish which mutation, if either, is inherited
  - ○ In some cases, both mutations in a single risk gene have been shown to be somatically acquired
- For patients with hematologic malignancies
  - ○ Sequencing genomic DNA from cultured skin fibroblasts can help establish which mutation, if either, is inherited
  - ○ Neither buccal swabs, oral rinse, nor skin punch biopsies are suitable as sources of germline DNA because such samples include nucleated blood cells which might be carrying the mutation in question
- For patients with a strong family history of cancer, it is appropriate to test a panel of risk genes
  - ○ The clinical utility of screening patients without a clearly positive family history, let alone no family history, is contentious
  - ○ Full gene sequencing of a panel of multiple risk genes is likely to generate one or more variants of uncertain significance, which might exacerbate patient distress
- Tumor classification
  - WHO classifications often include molecular markers
  - Gene expression
    - ○ Gene expression studies can identify subsets of a tumor type that differing in prognosis and/or response to a possible therapy
    - ○ As an early example, microarray analysis of the transcriptome led to delineation of important classes of breast cancer, confirming morphological impressions
  - Gestational trophoblastic tumors
    - ○ Classification depends on maternal and paternal contributions to genome
    - ○ Analysis is typically by 'DNA fingerprinting' of short tandem repeats
  - Cancers of unknown primary
    - ○ Classification algorithms have been proposed, some using mRNA, some miRNA
    - ○ As IHC choices expand, the incidence of unclassified tumors has declined
    - ○ A tumor so atypical that it requires molecular classification should not be assumed to respond typically for the type finally identified
- Process QA
  - Mislabeled slides, mislabeled cassettes, and wandering tissues (floaters) are rare but critical challenges in the histology laboratory
  - DNA fingerprinting (the same assay used in forensic human identification, in bone marrow engraftment

analysis, and in gestational trophoblastic tumor analysis) can help resolve many of these

- Screening
  - Monitoring for minimal residual/recurrent disease MRD by sequencing cfDNA
    - The technical ability to detect cfDNA with mutations consistent with the primary tumor is not in doubt
    - The clinical utility is, at present, uncertain in general
  - Screening the apparently healthy for malignancy by sequencing cfDNA
    - The ability to detect mutations with at least moderate specificity and sensitivity in several sample types is clearly established
    - One approach uses shifts in the cfDNA size distribution to detect any contribution from malignant cells
    - At least two molecular assays have received FDA approval
      - COLOGUARD™ (stool)
        - Tests for mutations *KRAS2* mutations, methylation of *NDRG4* and *BMP3*, *ACTB* DNA (as a control), and the presence of hemoglobin in stool
        - Approved for average-risk individuals aged 45 years and older
      - Epi proColon®
        - A qualitative test for methylation of the septin 9 promoter in cfDNA from plasma
        - Approved to screen adults of either sex, 50 years or older, defined as average risk for CRC, who have a history of not completing colonoscopy

Background Information for Molecular Genomics

Background on Genome Organization

- (Refer reader to MGP 2nd edition, chapter 1)
- Genome 3×10^9 nucleotides (haploid), ~22,000 genes
- Gene size ranges up to 2000,000 bases (the *DMD* gene)
- Coding genes account for ~5% of the total genome
- Mitochondrial genome
  - Circular, ~16,000 nucleotides
  - Some mitochondrial DNA copurifies with nuclear DNA in a typical extraction

- With disaggregated single cells, mitochondria can be separated out in a cytoplasmic prep prior to DNA extraction
- Pseudogenes
  - Pseudogenes are stretches of genomic DNA with high homology to 'real' active genes or to the mRNA for 'real' actively transcribed genes
    - The proper name for the 'KRAS' gene is KRAS2 because the original DNA sequence was a pseudogene
    - Pseudogenes are usually considered 'inert', mainly a problem for designing primers and probes specific for the 'active' gene
  - Processed pseudogenes
    - These are contiguous stretches of genome which are nearly identical to some portion of an mRNA for the presumed homologous precursor gene
      - Exons have been spliced out (as one expects for mRNA)
      - Many pseudogenes show an accumulation of mutations relative to the homologous gene
      - RNA *is* transcribed from some pseudogenes (sometimes in the opposite direction of what is expected for the homologue)
      - It has been proposed that the RNA transcribed from a pseudogene, even if not translated, can exert an effect by several mechanisms
        - Binding miRNA which might otherwise bind the homologous mRNA
        - Antisense RNA from a pseudogene could directly hybridize and block the homologous mRNA
        - As yet, there are no clinical applications for these speculations
  - Unprocessed pseudogenes
    - Unlike processed pseudogenes, these contain introns, complicating design for analysis of genomic mutations
- Gene organization
  - Exon
    - The coding portion of a gene; average size is 100–200 base pairs (bp)
    - This is suitable for the typical second-generation NGS assay
  - Intron
    - The region between the exons within a gene
    - Many translocation breakpoints occur over large variable regions in introns
  - The fusion mRNA from the translocation usually juxtaposes the same exons after splicing out the introns regardless of the intronic breakpoint

- Splicing sites
  - There are approximate consensus motifs at each end of an exon and each end of an intron which mediates splicing out the intron by the splicesome
  - The splicesome is a massive multiprotein complex
    - There are cancer-associated mutations in many of the corresponding genes
- Mutations in genes for splicesome components can lead to variants in many transcripts on RNA sequencing, which are NOT mirrored by genomic sequencing
- Promoters
  - Most often in the region immediately 5′ (upstream) of the coding region of a gene
  - 200–2000 bp long
  - Site for binding by various factors regulating transcription
  - Majority of gene promoters are rich in CpG dinucleotides
    - Methylation of gene promoters is usually at multiple CpG sites
    - Methylation is generally described as associated with decreased transcription, hence 'silencing' of a gene; it is likely the overall pattern of CpG sites methylated is more important than methylation of one or a few specific CpG dinucleotides
- Telomeres
  - A unique hexanucleotide motif is present in thousands of copies at the ends of every chromosome
  - The length of each set of repeats (not the motif) is maintained by a complex of structural proteins, a small RNA, and the enzyme telomerase
  - The number of motif repeats at each end decreases with every cell division in a nonmalignant cell
  - Although the role of telomeres in cancer is fascinating, the main diagnostic application at present is in analyzing inherited bone marrow failure
    - Telomeres are abnormally shortened in several syndromes
    - Sequencing the genes involved in telomere maintenance is most straightforward
    - Measuring the average size of telomeres can be done in two ways
      - Flow-FISH
        - A fluorescent DNA probe targeting the repeat motif binds in proportion to the number of repeats
        - The cells are analyzed with a flow cytometer
        - This can be performed on selected cell subsets
      - PCR
        - Pools of subtelomeric primers can amplify all the telomere repeats
        - The resulting size distribution gives insight into the average telomere length
- Repeats
  - There are over a dozen families of sequence motifs that are present throughout the genome in multiple copies
  - The '*Alu*' repeat is probably the best known; it is the most widespread
    - The basic repeat motif is ~350 base pairs
    - There are over 1,000,000 copies scattered throughout the genome
    - It is thought to have originated from a retrovirus which inserted into the genome
    - In any given individual several copies can 'be active' and insert randomly somewhere in the genome, usually with no effect but occasionally causing havoc
    - Diagnostically, repeats are of concern because if a PCR primer or probe inadvertently overlaps a repeat like *Alu*, it is unlikely to work
  - Short tandem repeats (STR) (Microsatellite repeats)
    - The repeat motif can be anywhere from 1 to 6 bp in length
    - They can occur in coding regions but are most common in noncoding regions
    - If the motif is a single nucleotide, the repeat region is called a 'homopolymer' or a mononucleotide repeat
    - The number of repeats in a homopolymer is typically 4–6 (96.7%)
    - Any given di-, tri-, or tetranucleotide, such as 'ACGT', can be found constituting multiple microsatellites across the genome
      - A given microsatellite can be studied by amplifying it with primers in the unique surrounding sequence
      - The number of repeats can be inferred from the length of the PCR amplicon or by sequencing
      - The two allelic copies of any given microsatellite frequently differ in the number of repeats
      - The number of repeats occurring at a given locus in a population is quite variable
      - Identifying the alleles (number of repeats) at several STR loci, typically 10–20, each on a different chromosome, gives a DNA 'fingerprint' for the individual

- ♦ Specific trinucleotide microsatellite in 30–40 specific gene loci can undergo marked expansion in the number of repeats during neonatal development
  - ▪ Most expansions are associated with neurological syndromes, including Fragile X and the spinocerebellar ataxias
- ♦ Measuring the STR at multiple loci is the basis of DNA fingerprinting and of engraftment analysis

DNA Sequence Variation

- Single nucleotide variants (SNV)
  - This is where a nucleotide differs from the nucleotide in 'the' reference sequence
  - SNV is preferred to 'mutation'
    - o 'Mutation' is potentially pejorative, and the linkage with pathology often not as tight as when an association is first discovered
    - o Single nucleotide polymorphism (SNP)
      - ♦ There is no definite cutoff for frequency in a population that distinguished an SNP from a 'mutation'
    - o 'SNV' is not meant to convey any information, positive or negative, about an association with disease
  - They are characterized in several ways
    - o <u>Synonymous</u>: There is no change in amino acid if one is encoded
      - ♦ Synonymous variants can still cause pathology
        - • Alter splicing sites
        - • Alter miRNA binding sites
        - • Alter mRNA degradation
    - o <u>Nonsynonymous</u> variants do alter the encoded amino acid
      - ♦ <u>Missense</u> substitutions encode a different amino acid
      - ♦ <u>Nonsense</u> mutations encode a 'Stop' codon
  - Noncoding variants can also cause pathology
    - o Alter RNA splicing junctions
    - o Alter promoter function by altering the binding of transcription factors
    - o Alter RNA fate by binding regulatory factors in 5′ and 3′ untranslated regions
    - o Alter miRNA binding sites
  - Insertion-Deletions (Indels)
    - o Insertions and deletions in a coding region by a number of nucleotides which is **not** divisible by three, will alter all the amino acids encoded after the position of the variant

- ♦ This is called a 'frameshift'
- o More complex variants can both delete several nucleotides and insert several nucleotides, at the same site
- o The cutoff distinguishing between indels and 'structural variants' is arbitrary but many authors use 50 bp as a cutoff
  - ♦ This is a convenient cutoff for second-generation NGS, which cannot usually encompass larger alterations within a single read
- Loss of heterozygosity (LOH)
  - o The distribution of SNVs varies widely across the genome but, on average, one occurs every 1000–4000 bp
  - o When sequencing or genotyping by microarray shows no heterozygous sites over a long swathe of DNA, it is described as LOH
    - ♦ This can result from duplication of a portion of the chromosome, replacing the other allelic version
    - ♦ This can result from the deletion of one copy of the chromosomal region
  - o Genotyping by microarray can usually distinguish the two possibilities by the signal strength
- Structural variations
  - o Translocations
  - o Deletions
  - o Insertions
  - o Inversions
  - o Gene amplification
- Epigenetic changes
  - o Methylation DNA
  - o Hydroxymethylation, other

Background for RNA Analysis

- Classes of RNA
  - rRNA (ribosomal RNA)
    - o Accounts for 90–95% of total RNA molecules/cell
  - mRNA
    - o Although there are only about 22,000 genes, alternative splicing leads to over 100,000 different transcripts
  - miRNA
    - o These are short, nucleotide RNA molecules processed from much larger precursors (pri- and pre-miRNA)
    - o Over 1000 have been identified

- o An even shorter region of the miRNA can bind the miRNA to complementary regions in an mRNA
  - ♦ The complementarity is often imperfect
  - ♦ One miRNA can bind to many different mRNA
- o The miRNA–mRNA hybrid is recognized by the multiprotein RISC complex, which can degrade the mRNA or only block its translation
- o Because of the ability of a given miRNA to bind many different mRNA and the ability of one mRNA to bind several different miRNA, predicting or demonstrating the effect of a mutation in the miRNA or in the mRNA binding site for miRNA is difficult
- o One of the better-known disease associations is in chronic lymphocytic leukemia (CLL)
  - ♦ Deletion of chr 13q14 occurs in about 50% of CLL patients
  - ♦ It is associated with a better prognosis (than without the deletion)
  - ♦ miR15A/miR16-1 are in the minimally deleted region associated with better prognosis
  - ♦ These two miRNA influence mRNA levels for CCND1, −2, −3, CDK4 and −6, among other genes, which suggests some impact on the cell cycle
  - ♦ Loss of miR15A/16-1 expression is considered a *bad* prognostic finding in mantle cell lymphoma
- – Long noncoding RNA (lncRNA)
  - o These are noncoding RNA defined as longer than 200 nucleotides and can be as large as several hundred thousand nucleotides
  - o A recent annotated database counts 77,000 distinct lncRNA, in line with other estimates (give or take 10,000)(http://www.lncrnakb.org),
  - o The role of most remains to be determined
  - o The best known is *XIST,* which is responsible for silencing one of the two X chromosomes
- – There are several other RNA classes (tRNA, circular RNA, and piwiRNA)
  - o Variants in several are associated with inherited disorders
- – Pervasive transcription
  - o RNA transcripts, short and long, arise across much of the genome
  - o Many of the RNAs have functions as RNA molecules other than encoding protein
  - o There is now some evidence that mRNA can encode 'noncanonical' reading frames, which generate some small proteins
    - ♦ Even if this 'ghost proteome' is confirmed to be reproducible and 'real', it will, for now, leave open the question of whether or not it represents 'noise'

- – Methylation of RNA
  - o Modified ribonucleotides have been reported in most types of RNA
  - o These are primarily methylations
  - o Their effect on RNA function is uncertain
- • RNA variations (potentially detectable by RNA-seq)
  - – Many genomic variants can be seen in RNA, including SNVs and indels
  - – One cannot assume both alleles are expressed comparably
  - – Alternative splicing
    - o A gene can often generate a variety of similar mRNA, differing in the combinations of exons used
    - o The distribution of alternative transcripts for a given gene can vary by the tissue
    - o RT-PCR, unless it targets from the first known exon to the last, might miss some transcripts
      - ♦ Northern blot analysis can show transcripts of different sizes at the same time, provided the probe includes an exon common to all transcript variants
    - o Some transcripts retain introns, which further uncouples mRNA and protein abundance
  - – 5′ RACE and 3′ RACE are PCR-based techniques that can amplify multiple transcripts which can then be sized by electrophoresis or sequencing
    - o The short reads of second-generation NGS make it difficult (but not impossible) to infer alternative transcript usage
    - o Third generation NGS generates long reads, potentially full length, so alternative splicing is more readily detected and quantified
  - – Transcription level ('gene expression')
    - o Generally, the higher the mRNA level, the higher the subsequent protein level
      - ♦ Correlation with amount of protein is loose
      - ♦ One can assert that if there is no mRNA there will be no protein
  - – RNA analysis for clinical management remains, at present, limited
    - o Confirmation of translocations (e.g., *BCR–ABL*)
    - o Quantitation of the *BCR–ABL* translocation
    - o Detection of RNA viruses by RT-PCR or in situ hybridization
    - o Many research studies have sought prognostic or predictive combinations of multiple gene-expression levels for numerous malignancies, including breast cancers and thyroid cancers
      - ♦ Several are available in reference laboratories for clinical use
      - ♦ Several incorporate both gene expression and the presence/absence of mutations

Overview of Molecular Methods

- Sanger sequencing and primary PCR methods (such as real-time PCR or allele-specific expression) are logistically limited to analyzing, at most, a few genes
- NGS
  - Diagnostic emphasis has remained on the identification of genomic DNA mutations
  - NGS allows sequence analysis of many genes at once, as well as the exome or the whole genome
  - NGS offers four key advantages over the single gene or hotspot testing by Sanger
    - Logistical: A single sample can be analyzed for many genes in one test
      - This is critical for small tissue-limited specimens like biopsies
    - Financial: Testing for more than two to three targets can be more cost-effective with NGS
    - Sensitivity
      - Sanger sequencing typically can detect a variant only when the VAF is ~20%
        - This corresponds to a tumor content of 40%, assuming diploidy
      - Real-time PCR sensitivity is typically in the 1–5% VAF range but can sometimes be optimized to the 0.1% VAF range
    - Quantitation
      - Sanger sequence is nonquantitative
      - Real-time PCR can be quantitative
        - The most sensitive methods typically suppress amplification of the normal sequence in a nonquantitative manner
  - NGS, by its scale, enables additional analyses including
    - Transcriptomics (RNA-Seq)
    - Methylomics (genome wide assessment of methylated sites)
  - The optimal panel(s) of genes to target by NGS is (and might remain) a lively topic of controversy
    - As sequencing and computation become faster and less expensive, it is possible that comprehensive genomic and transcriptomic sequencing will become standard
    - Arguments in favor of comprehensive sequencing
      - If you do not sequence the entire genome, you might miss something
      - If you do not sequence the entire exome, you might miss something
    - Arguments against comprehensive sequencing
      - Variants of unknown significance (VUS)
        - The broader the coverage of the genome, the more variants found
        - This violates the 'Tenth Law of the House of God', 'If you don't take a temperature, you can't find a fever'
    - Sensitivity
      - In analysis of non-mosaic inherited disorders, NGS studies aim for 30× coverage
      - Because the tumor content of a sample is usually much less than 100%, the coverage must be higher, typically at least 200×
- Single-cell analysis
  - Molecular pathology has traditionally focused on DNA from 'bulk' tumor tissue
  - Several NGS methods now enable the analysis of single cells
    - Whole genome, whole transcriptome, whole methylome, and some proteome methods have been demonstrated
  - None of the methods achieve 100% coverage (e.g., of the genome or transcriptome) for a given single cell, but the cell throughput is now so high that statistical methods can permit important inferences
  - Some studies have successfully assayed two 'omes in each cell, permitting correlations
  - This is **not**, at present, a clinical diagnostic modality
- Spatial 'omics analysis
  - Highly multiplexed FISH methods have been demonstrated
  - Several NGS methods, using molecular barcoding to computationally 'pin' millions of mRNA to their position in a tissue section, enable genomic and transcriptomic analysis
  - These are **not** (see p. 61), at present, clinical diagnostic modalities

Sample Preparation

Sample Types

- Any tissue, fresh or fixed, biopsy or resection, is potentially suitable
- Any collection of cells (from blood, urine, CSF, and body cavity) is potentially suitable
- Supernatants from body fluids (serum, urine, CSF) are potentially suitable
  - Most molecular diagnostic testing has been on genomic DNA from FFPE tissues

- Increasingly testing is performed on cytology specimens and on circulating tumor nucleic acids
- Fresh tissue
  - Fresh tissue is essential if one needs long DNA or full-length RNA
  - Fresh tissue is essential for cytogenetic karyotyping because the cells need to be able to divide
  - Fresh tissue is not essential for FISH or cytogenetic microarray (CMA)
  - Although gene expression can be analyzed with RNA from FFPE tissue, it is much more efficient with RNA from fresh tissue
- Bone
  - Many protocols for processing for morphology involve acid decalcification
  - Acid destroys DNA
  - Ethylene diamine tetra acetic acid (EDTA) incubation
    - Ethylene diamine tetra acetic acid (EDTA) chelates calcium
    - Incubation of thin bone sections/small fragments in EDTA softens the bone sufficiently for sections to be cut <u>and</u> keeps DNA intact
    - Incubation in EDTA alters the appearance of cells
- Bone marrow
  - Usually, the aspirate is sent for molecular testing
    - The adequacy of the aspirate is judged qualitatively based on the presence of spicules as well as bone marrow cells
    - Usually, several consecutive aspirates are drawn, each for a different test
      - The first aspirate goes for morphology; subsequent 'pulls' go for flow cytometry, cytogenetics, and molecular testing
      - Because of the traumatic nature of the bone marrow biopsy, even the smoothest procedure leads to hemorrhage into the marrow space
      - As a result, each succeeding aspirate is increasingly diluted by peripheral blood, reducing the sensitivity achievable by testing the aspirate
        - This can lead to discrepant results among the test methods because they are not truly using the same sample
      - In fibrotic conditions, the aspirate might contain negligible numbers of bone marrow cells
  - The bone marrow core routinely undergoes acid decalcification, which precludes molecular analysis
- Blood
  - Sample collection tube
    - Lavender top (EDTA) and yellow top (Citrate) both work well
    - Green top (heparin) tubes
      - Heparin can inhibit DNA polymerases
        - Heparinase treatment has been reported, by some, to rescue preps

- Red top (serum tubes, no additive)
  - Because the sample clots, trapping most cells along with much of their DNA, this is not ordinarily acceptable
  - DNA *can be* extracted from serum in small quantity
  - DNA in serum is not simply representative of cfDNA because it is a mixture of DNA released from cells during clotting and more 'physiological' circulating cfDNA
  - If this is the only sample available and one is testing for an inherited variant, this limited DNA *might* be satisfactory
- Cell subsets
  - Cells can be sorted by flow cytometry
  - Cells can be sorted for single surface markers using antibody-coated magnetic beads (or other solid support)
  - Circulating tumor cells (CTC)
    - Although there is unequivocal evidence that there are CTC in many classes of tumors (though not in every case), circulating tumor (derived) DNA (ctDNA) is much easier to detect
    - Some tumor types, like prostate, are associated with a much higher incidence of CTC than in other tumor types
    - In general, cell-free DNA has proven more frequently accessible than CTC
- Serum—for cell-free DNA (cfDNA)
- Body fluids
  - Urine
  - CSF
  - Pleural fluid, ascites
  - The above sample types can all be spun down to create cell pellets from which DNA/RNA can be extracted
  - Cell-free DNA can also be extracted from these fluids
- Buccal swabs and oral rinses
  - Primarily for inherited disease testing
  - Many reference laboratories will send kits to patient homes or physician offices
  - DNA yield can include substantial bacterial DNA
  - The utility is limited if one is looking for an inherited risk variant for hematopoietic neoplasms because much of the DNA can come from inflammatory cells, which might be carrying somatically acquired mutations

Sample Adequacy

- Tumor content
  - Minimum percent of tumor content needed depends on sensitivity of method for testing
  - Sensitivity depends in part on

- ○ The sequence context surrounding the nucleotide(s) of interest
- ○ The ploidy of the tumor
- ○ The specificity and binding stability of the primers
- For Sanger sequencing, the VAF usually needs to be at least 20%; this corresponds to a tumor content of at least 40%
- For real-time PCR detection of sequence variants, the typical VAF is in the 1–5% range (tumor content of 2–10%, assuming diploidy)
  - ○ Signal-to-noise is lower when testing for a single nucleotide variant than for presence or absence of a longer sequence, especially with a translocation
  - ○ Even for a homozygous position, a low-level background signal is common, perhaps a result of limited nucleotide selectivity by the polymerases
  - ○ There are methods to suppress the normal background, increasing sensitivity but losing the ability to quantify the variant
    - ♦ Pretreat the sample with a restriction enzyme that cuts the normal sequence but not the mutated sequence (provided a suitable enzyme exists)
    - ♦ Use a short 'blocking' oligo' synthesized with modified nucleotides (e.g., LNA, PNA)
      - ▪ This binds the normal sequence tightly, interfering with amplification
      - ▪ A single nucleotide mismatch (at the site of the SNV) destabilizes binding by the blocking oligo
- NGS clinical reports for genomic DNA typically site a sensitivity of 1–5% variant allele frequency
  - ○ NGS uses polymerase both for the initial amplification of the template and for the sequencing reaction
  - ○ Polymerases introduce errors, which are then perpetuated
  - ○ The introduction of unique molecular identifiers (UMI) partially mitigates this (see p. 48)
  - ○ 'Nevertheless, data scrubbing cannot universally produce high-confidence calls of mutations much below 1% abundance. The more aggressive the approach, the greater the risk of excluding true rare variants'
- Determination of tumor content
  - ○ Guesstimate
    - ♦ Visual estimation by the pathologist
    - ♦ Pathologists habitually focus on the malignant cells
    - ♦ Background cells, like lymphocytes are often small and can be underestimated, but have as much DNA as a diploid tumor cell
    - ♦ A College of American Pathologists (CAP) proficiency test is available to assess this
  - ○ Stereometry

- ♦ This is an 'old-fashioned' technique similar to counting cells intersecting with a grid in a hemocytometer
- ○ Computerized image analysis
- Minimum number cells
  - Typical input for real-time PCR is on the order of 10–100 ng/reaction
  - 10 ng double-stranded DNA corresponds to ~3000 diploid cells
  - The number of input genomes should match the claimed sensitivity
    - ○ If an assay is asserted to identify 1 cell out of 1 million, the input DNA should be at least ~3 million cells (~10 micrograms of DNA)
      - ♦ A factor of approximately '3×' allows for stochastic variation
- Enriching for tumor content (FFPE)
  - Macrodissection
    - ○ Serial sections are cut onto glass slides
    - ○ The middle slide is stained as a 'guide slide' for dissection of the other slides
    - ○ A handheld instrument, such as a scalpel, is used to scrape up tumor-rich areas on the unstained slides, to be processed for DNA
    - ○ The histology of the area used can be documented on the guide slide
    - ○ By enriching the proportion of tumor cells in the material used to extract DNA, one is enriching for mutations in the tumor—essentially increasing sensitivity of detection
  - Microdissection
    - ○ Laser capture microdissection (LCM)
      - ♦ Several instruments allow the operator to use a microscope and computer screen to select, document, and precisely recover cells from user-defined areas on slides
      - ♦ The corresponding regions on the slide are detached from the slide (cut out, in a manner of speaking, by a microscopic laser beam), dropped or catapulted into a microtube, then processed for DNA or RNA
      - ♦ The systems can work with FFPE to recover DNA or with fresh-frozen tissues for DNA and RNA
      - ♦ Frozen or FFPE sections are cut onto special slides and stained
      - ♦ Large regions can be captured en masse—such as an entire colonic crypt
      - ♦ Individual cells can be captured; the method is too laborious to use for high-throughput single-cell sequencing
    - ○ Dissecting microscope: A lower-tech approach
    - ○ Visual microscopic assessment of a stained slide without a coverslip is hard

- o Increased sensitivity of NGS has reduced the need for enrichment by LCM
- Cell selection
  - o Flow sorting
    - ♦ Primarily applicable to hematopoietic cells
    - ♦ Disaggregation of solid tumors often leads to loss of surface antigens
  - o Magnetic beads
    - ♦ Biotinylated antibodies bind to target cells
    - ♦ Avidin-coated beads are used to pull out the antibody bound cells
- Preparation of single cells from solid tumors
  - o At present, this is mainly for research
  - o Cells are disaggregated by enzymatic digestion
    - ♦ In some settings, like gastrointestinal mucosa, incubation in an EDTA-containing solution will release the mucosal epithelial cells
  - o Most single-cell studies have been transcriptome studies of unfixed cells
  - o Single-cell nuclei can be prepared from FFPE tissue
    - ♦ These nuclei are adequate for genomic sequencing
    - ♦ These nuclei contain RNA precursor transcripts which can also be analyzed by RNA-seq
- Sample Handling
  - o Genomic DNA is robust—it can be recovered from 100,000-year-old bones
  - o DNA preparation ideally should be initiated the same day the tissue is obtained
    - ♦ If tissue cannot be processed promptly, DNA preparation should be delayed by refrigeration at 4 °C
    - ♦ Physical disruption of the tissue and suspension in lysis buffer prior to storage is the surest way to ensure the stability of DNA and RNA
    - ♦ Refrigeration, especially of blood and marrow, might change the distribution of intact cells still available when DNA extraction is initiated
      - ▪ DNA preparation (see below) usually involves spinning down the cells, and then extracting DNA
      - ▪ If leukemic cells, for example, are more 'fragile' than normal blood cells and no longer intact after refrigeration, then the mutations they carry will be underrepresented in the cell pellet made to prepare DNA
      - ▪ If the cell pellet is made and suspended in lysis buffer <u>before</u> extended storage, then both DNA and RNA will be stable

- o mRNA levels vary widely in response to many variables in the cellular environment
  - ♦ Preoperative, intraoperative, and postoperative variables (meds, anesthetic time, time to processing sample) all impact RNA expression
    - ▪ Some tissues show stable mRNA profiles by NGS for at least 24 h refrigeration
  - ♦ The best approach is to homogenize the tissue as soon as possible in a lysis solution
  - ♦ There are proprietary solutions (such as *RNAlater*) in which small pieces of tissue can be stored without homogenization for extended periods
    - ▪ The tissue section must be small and thin, to permit rapid infiltration
    - ▪ Nonproprietary solutions have been described including DMSO
    - ▪ OCT
      - ✦ If the need for molecular testing is uncertain, tissue frozen in OCT or the equivalent and stored appropriately can yield high-quality DNA and RNA indefinitely
  - ♦ The measurement of a 'housekeeping' gene (see 'Real-time PCR') can provide a rough check of the adequacy of the storage process (there are no true 'housekeeping genes')

Fixatives

- Fresh (no fixation)
  - Applications
    - o Karyotyping: Requires growing dividing cells
      - ♦ Includes DNA repair syndromes like Fanconi's Anemia
    - o Flow-FISH (see 'Telomeres', p. 6)
  - Advantages of isolation of DNA or RNA from fresh cells
    - o Mean size dist. of DNA and RNA much greater than with fixation
    - o Third generation NGS can take advantage of long reads for both DNA and RNA
    - o Fixation generates sequence artifacts (these can be recognized and controlled bioinformatically if one does use FFPE-derived DNA/RNA)
- Formalin fixation
  - This is the predominant fixative in surgical pathology
  - Formalin crosslinks proteins, thereby inhibiting enzymes, including DNAses
  - Formalin inactivation of RNases might be more limited, but other processes also act to fragment RNA during FFPE processing

- – Drawbacks to formalin fixation
  - ○ Paraffin embedding requires heating
  - ○ Heating can cause DNA sequence modification
  - ○ Formalin fixation causes a low level of DNA sequence modification
    - ◆ *N*-hydroxymethyl mono-adducts on guanine, adenine, and cytosine
    - ◆ *N*-methylene crosslinks between adjacent purines in DNA
    - ◆ The formalin-induced changes are randomly distributed through the genome but generate sufficient sequence changes such that bioinformatic analysis of a large enough swathe of DNA sequence can show a pattern of mutation, a 'signature', that the DNA comes from an FFPE sample
    - ◆ Formalin-induced changes are clinically problematic in two settings
      - ▪ When there is a very low concentration of input DNA a random formalin-induced alteration could have a large VAF (true for any sequence artifact)
      - ▪ One is looking for a very low-level alteration, as in minimal residual disease
- – DNA is more fragmented in FFPE than in fresh tissue
  - ○ Chromosomes are on the order of 100,000,000 bp
  - ○ DNA from fresh tissue has a typical size of 30,000 bp
  - ○ DNA in even the best preps (other than for pulsed-field gel electrophoresis) is fragmented
    - ◆ This is attributed to mechanical shearing during routine pipetting
  - ○ DNA from FFPE has a typical average length in the 100 s of bps
    - ◆ Since the average exon length is 100–200 bp, and the typical second-generation NGS read is on the order of 100–200 bp, such degraded DNA does not preclude analysis
    - ◆ If all the DNA appears less than the needed size on gel, PCR might still generate a larger amplicon BUT it is not reliable for quantitation and might represent a contaminant
- – RNA is invariably severely degraded
  - ○ RNA can still be used for well-defined targets of PCR like some fusion transcript sites
  - ○ Detection of miRNA can be robust, and they are already short
- • Alcohol-based fixatives
  - – These are mainly used in cytology
  - – Excellent for DNA and RNA recovery
- • Hg-based fixatives
  - – These include B5 and Zenker
  - – They cause poor recovery of DNA

- • Acid based
  - – Bouin, for example, renders DNA unusable
- • Staining
  - – For molecular purposes, mainly of interest for assessing tumor content
  - – Occasionally stained slides carry the only material available for analysis
    - ○ H&E
      - ◆ DNA can each be recovered from H&E-stained slides
      - ◆ The efficiency of recovery is disputed, possibly a reflection of the nonstandardized production of hematoxylin
    - ○ Cytology stains
      - ◆ Recovery of DNA and RNA is usually excellent for diagnostic procedures

DNA/RNA Extraction Basic Steps

- • Traditional DNA prep for tissues
  - – Works for fresh tissue, and for FFPE tissue AFTER paraffin removal
  - – Considered the 'gold standard' prep; it is described in some detail because it is useful if one needs to scale up inexpensively, and it illustrates useful subprotocols
  - – Disruption in lysis buffer
    - ○ Tissues can be disrupted by one of the following methods immediately after the addition of lysis buffer:
      - ◆ Dounce homogenizer
      - ◆ Polytron homogenizer
      - ◆ Tissue shredder
      - ◆ Bead-mills
    - ○ Lysis buffer (typical composition) includes:
      - ◆ SDS: A detergent that disrupts proteins and lipids
      - ◆ Proteinase K: Preferred protease because it is active in SDS
      - ◆ EDTA. All DNases require divalent cations (usually Mg 2+); EDTA protects by chelating doubly charged cations
      - ◆ Tris buffer, pH ~8
  - – Phase separation: To remove proteins and lipids
    - ○ Add an equal volume of a phenol/chloroform/iso-amyl alcohol mix to lysate
      - ◆ Phenol and chloroform are both toxic
    - ○ Vortex vigorously and spin at on low speed
    - ○ The lysis buffer (mostly water) forms the upper layer, the phenol/CHCl3 the bottom layer
      - ◆ DNA is in solution in the aqueous phase
      - ◆ Lipids disperse in the organic bottom phase
      - ◆ Proteins, which typically have both hydrophobic and hydrophilic stretches, aggregate at the interface

- Transfer the upper aqueous phase to a fresh centrifuge tube
- Alcohol precipitates the DNA to concentrate it
  - There are numerous recipes for adding various alcohols and salts to precipitate DNA
  - Alcohols are less polar than water; alcohol displaces water from a shell around the DNA and allows cations to form ionic bonds with the negative phosphates of DNA, collapsing the strands
  - When the mixture is centrifuged at high speed, DNA forms a pellet
  - The supernatant is poured off
  - The DNA pellet can then be resuspended in the desired buffer
    - The most common buffer is 10 mM Tris (pH 8), 10 mM EDTA (TE8 buffer)
- Traditional DNA prep: Whole blood
  - The first step is the removal of red blood cells (RBC)
    - Free heme can inhibit subsequent enzymatic reactions
    - There are three approaches
      - Buffy coat preparation
        - The buffy coat is pipetted off and processed for DNA (see below)
      - Density: Gradient centrifugation
        - The blood sample is layered on top of a layer of a highly branched hydrophilic polysaccharide like Ficoll and centrifuged
        - The erythrocytes, the densest blood cells, go to the bottom of the tube
        - Mononuclear cells stop at the interface
        - More selective separations are possible, depending on the choice of gradient
        - The mononuclear cells are pipetted from the interface into a fresh centrifuge tube, rinsed in buffer several times, then processed for DNA/RNA
      - RBC lysis with ammonium chloride
        - The blood is mixed with a solution of ammonium chloride
        - All cells take in the ammonium chloride
        - Intact mononuclear cells pump out the ammonium chloride
        - Erythrocytes cannot pump out the ammonium ions but do let in water to maintain osmolality, leading to swelling and bursting within minutes
        - The sample is centrifuged, forming a pellet of the intact nucleated cells
  - DNA can then be prepared from any of the resulting cell pellets using the traditional method or one of the kitted methods (see below)

- Cell suspensions can be added to the lysis buffer directly
- This applies to any cell collection (CSF, urine, body fluid)

Preparation of DNA from Formalin Fixed Paraffin Embedded Tissue

- The number of sections needed will depend on the specimen size and tumor content
  - In general, even for needle biopsies, ten sections, cut at 5 μ, on charged, dry but unbaked slides, is sufficient
- Foundation Medicine offers the following guidelines
  - Optimal sample area: 25 mm^2; minimal: 5 mm^2
  - 10 sections at 5 μ thickness
  - Tumor content minimum is 20%, and the optimum is given as 80%
- Macrodissection
  - Stain either slide #5 or #6 to serve as a guide slide for the macrodissection
    - Do not process the guide slide for DNA
    - Outline the areas on the slide suitable for processing
    - Line up the guide slide to the unstained slide one
      - The outline of the tissue sections usually permits unambiguous alignment
      - Mark the underside of the unstained slide, indicating areas of tumor
      - Consider photodocumenting the macrodissection markings
    - Scrape the regions of interest with a razor blade or scalpel blade
      - Transfer the fragments into a suitably labeled microfuge tube
      - Paraffin fragments/flakes can perform impressive electrostatic gymnastics, so work in a clean, small, confined area
      - If the block is known to contain adequate tumor (i.e., enrichment by macrodissection is not necessary), then scrolls can be collected directly into microfuge tubes for deparaffinization
- 'Microdissection' (using a laser capture instrument or a dissecting microscope) is distinct from 'macrodissection'
  - The two have different CPT codes
  - Microdissection is reimbursed (if at all) at a higher rate
- Deparaffinizing tissue
  - Fill each microfuge tube with xylene (toxic, flammable)
    - This should be performed in a fume hood
    - There are alternatives to xylene, including:
      - Citrisolve: Supposed to be nonhepatotoxic (unlike xylene)

- o Paraffin partitions preferentially into the xylene
- Rinse the tissues in the tube with a graded series of ethanol solutions, starting at 90%, going down to 70%, to remove xylene
- Air-dry the tube to remove residual alcohol
- Resuspend the tissue in the desired buffer
- Proceed with the manual or a kitted protocol for extracting DNA from tissue

DNA Methylation

- CpG dinucleotides occur with a marked increase in frequency in the promoters of many genes
- The cytosines in many of these clusters ('CpG' islands) are methylated
- The rule-of-thumb, which has exceptions, is that highly methylated promoters are not actively transcribed
- It is unlikely that there is a set of specific CpG dinucleotides in a given promoter that must be methylated, rather it is a matter of the density of methylated dinucleotides
- Detection of which nucleotides are methylated has, until the development of third-generation NGS, required bisulfite treatment of DNA prior to sequencing
  - Bisulfite treatment converts unmethylated cytosines to uracil
  - Bisulfite treatment leaves methylated cytosines unchanged
  - Subsequent PCR and sequencing will copy the uracil as a dT, and methylated dC will be copied as dC
- Sanger sequencing of a region comparison to a reference genome sequence would allow several inferences from the sequence of a bisulfite-treated sample
  - dT where the reference showed dC, implies the dC had NOT been methylated
  - Infer that dC in the sequence from a bisulfite-treated sample, probably represented a methylated dC in the sample
  - This assumes 100% effective bisulfite treatment
  - Sanger sequencing is usually done on DNA from a bulk sample
    - o Many of the cells will differ in methylation status for a given DNA region
    - o Even if the same promoter region is heavily methylated in two cells, the specific dC nucleotides methylated could differ
- Each second-generation NGS sequencing reaction analyzes sequence from a single allele
  - This is true whether the library is made from a single cell or from a bulk tumor
  - NGS sequencing of bisulfite-treated DNA will allow exact inferences as to which positions were methylated in a given region

- The information for sequences from each region for hundreds or thousands of cells can be aggregated for an overview
- Third-generation NGS sequencing of fresh DNA allows direct detection of methylated dC (and possibly other methylations such as hydroxymethylation of dC)
- Clinical utility of methylation studies
  - Patients with malignancy, especially hematopoietic malignancy, are frequently treated with methylation inhibitors
  - Diagnostically, the role of methylation analysis is limited
    - o Methylation of the *MGMT* gene as a predictor for the use of temozolomide
    - o Methylation of the *MLH1* gene to help distinguish sporadic microsatellite instability (MSI) from Lynch syndrome

RNA Extraction

- Procedures roughly parallel those for DNA
- There are important differences with handling of DNA
  - RNA levels start changing and degrading as soon as the specimen is obtained
  - RNases are present in all cells
    - o Some cell types, like pancreatic cells, have high levels
    - o RNases, unlike DNases, do NOT require cations, so they are not blocked by EDTA
    - o RNases allegedly can renature after boiling
- Lysis buffers typically use guanidinium isothiocyanate (GITC) or guanidinium HCl together with beta-mercaptoethanol rather than SDS and proteinase K
  - GITC is a more powerful denaturant than SDS
  - GITC is potentially toxic
- There are commercial preparations as well as lab-brew recipes which combine GITC, beta-ME, phenol, and chloroform into a single solution
  - The lysis buffer is prepared at acidic pH
  - At this pH, DNA concentrates at the interface, RNA goes into the aqueous phase
- Comments
  - Standard RNA extraction protocols do NOT efficiently recover small RNAs such as miRNAs
  - There are protocols optimized for miRNA recovery
- DNA/RNA commercial kitted extraction methods
  - Commercial kitted extraction methods can be used with deparaffinized tissue and with cells, including fresh blood (albeit with different initial steps)
  - Some kits work with whole blood directly, typically on the scale of 200 µL/sample
  - There are two common approaches
    - o Spin-filter inserts (like baskets) which go inside a 1.5 mL microfuge tube

- ♦ The filter composition is variable—including glass fibers, derivatized silica, and ion exchange membranes
- ♦ The lysate is added to the tube, the tube is spun, and the DNA or RNA binds to the membrane
- ♦ The insert is transferred to a fresh microfuge tube
- ♦ Elution buffer is added, releasing the nucleotide from the membrane
- ♦ The tube is centrifuged, the DNA/RNA goes to the bottom of the tube
- ○ Microscopic beads (typically 1 μm in diameter or smaller) of various types are mixed with the lysate and bind the DNA/RNA
  - ♦ Magnetic beads
    - ▪ The beads are properly described as paramagnetic; they only behave magnetically when in an external magnetic field
    - ▪ Magnetic bead surfaces can be coated so that the surface is charged and binds DNA/RNA
    - ▪ Magnetic beads can be derivatized so that the surface carries specific oligonucleotides, which can then capture and pull out DNA or RNA with specific sequences
  - ♦ Derivatized silica beads
  - ♦ Advantages of beads over filters
    - ▪ Greater surface area for interaction than typical filter
    - ▪ More amenable to automation
    - ▪ No risk of clogging filter
  - ♦ Disadvantage of beads over filters
    - ▪ Carryover of magnetic beads can inhibit downstream reactions
- ○ Some types of beads can be used to roughly size-select DNA
  - ♦ Typically, the beads have a polystyrene core covered by magnetite, which in turn is coated with carboxyl moieties (which can bind DNA)
  - ♦ When the beads are added to a mixture of DNA, salt, and polyethylene glycol(PEG), the DNA is 'crowded' onto the bead surface
    - ▪ The percent of PEG controls the DNA size
- ○ Some filters can be used to size-select DNA
  - ♦ The filters are packed with variable formulations of matrices like Sephadex
- – Nonmanual systems
  - ○ There are numerous instruments and widely varying formats
  - ○ Vacuum manifolds
    - ♦ Facilitates running multiple samples in parallel using filter inserts
  - ○ Fully automated systems
    - ♦ Typically, can perform on-board centrifugation

- ○ These are especially well suited to using bead purification
- ○ Substantially greater reagent cost per sample than the manual method
- ○ Less labor intensive (so total cost might be less than reagent cost suggests)
- ○ Some systems require samples to be run in multiples of a minimum number of samples/run
- – CAUTION: DNA prep protocols often bring down a lot of RNA, and RNA preps often bring along a lot of DNA
- – Consider including DNAse in RNA preps, RNase in DNA preps
  - ○ Be sure to include a step for neutralizing these enzymes before downstream work
- • cfDNA extraction
  - – The common general methods/kits for DNA purification do not efficiently capture cell-free DNA because of its small average size
  - – Manipulation of samples with cells (e.g., blood) invariably releases some genomic DNA from intact cells, often at levels greater than that of 'endogenous' cfDNA
- • DNA/RNA storage
  - – DNA storage
    - ○ DNA in Tris-EDTA (TE) or Tris-Acetate-EDTA (TAE) buffer can be stored indefinitely at −20 °C
    - ○ DNA in TE or TAE buffer can be stored for months, probably years, at 4 °C
  - – RNA storage
    - ○ Unlike DNA, RNA should be kept frozen as much as possible
    - ○ Traditionally RNA is stored at −80 °C
      - ♦ It is unlikely there is much enzymatic degradation at −20 °C
  - – The concentration of DNA and RNA
    - ○ Precipitation
      - ♦ There are numerous formulations of cations and alcohols which can precipitate DNA and RNA
        - ▪ Usually involve centrifugation in a microcentrifuge, typically at 10–20,000 g for 10–20 min
      - ♦ Centrifugation should be done at 4 °C
        - ▪ Centrifugation for this length of time will heat up the centrifuge chamber
      - ♦ Some inert additives like linear polyacrylamide can improve the efficiency of precipitation
      - ♦ The resulting pellet is poorly adherent but can be rinsed several times with buffer to remove residual alcohol
      - ♦ The pellet is finally resuspended in the desired volume of the desired buffer

- ○ Spin-columns
  - ♦ As described for purification above, they can also be used to concentrate dilute DNA preparations

Quality Assessment of DNA and RNA

- There are three standard quality measures of DNA and RNA
  - – Purity: Determined by UV absorption spectrometry
  - – Integrity (average size): Determined by electrophoresis
  - – Concentration: Determined by UV absorption or dye-binding fluorometry

- UV absorption spectrum
  - – Scanning UV absorption spectrometry
    - ○ Although only three wavelengths are commonly checked for UV absorption, UV scanning spectrometers have become so small, inexpensive, and convenient that it is common to measure a swathe of the UV absorption spectrum (Fig. 1.1a, b)
    - ○ The contour of the scan provides an additional quality check (Fig. 1.1c)
    - ○ UV absorption spectrometry does NOT distinguish between intact DNA or RNA and nucleotides in solution

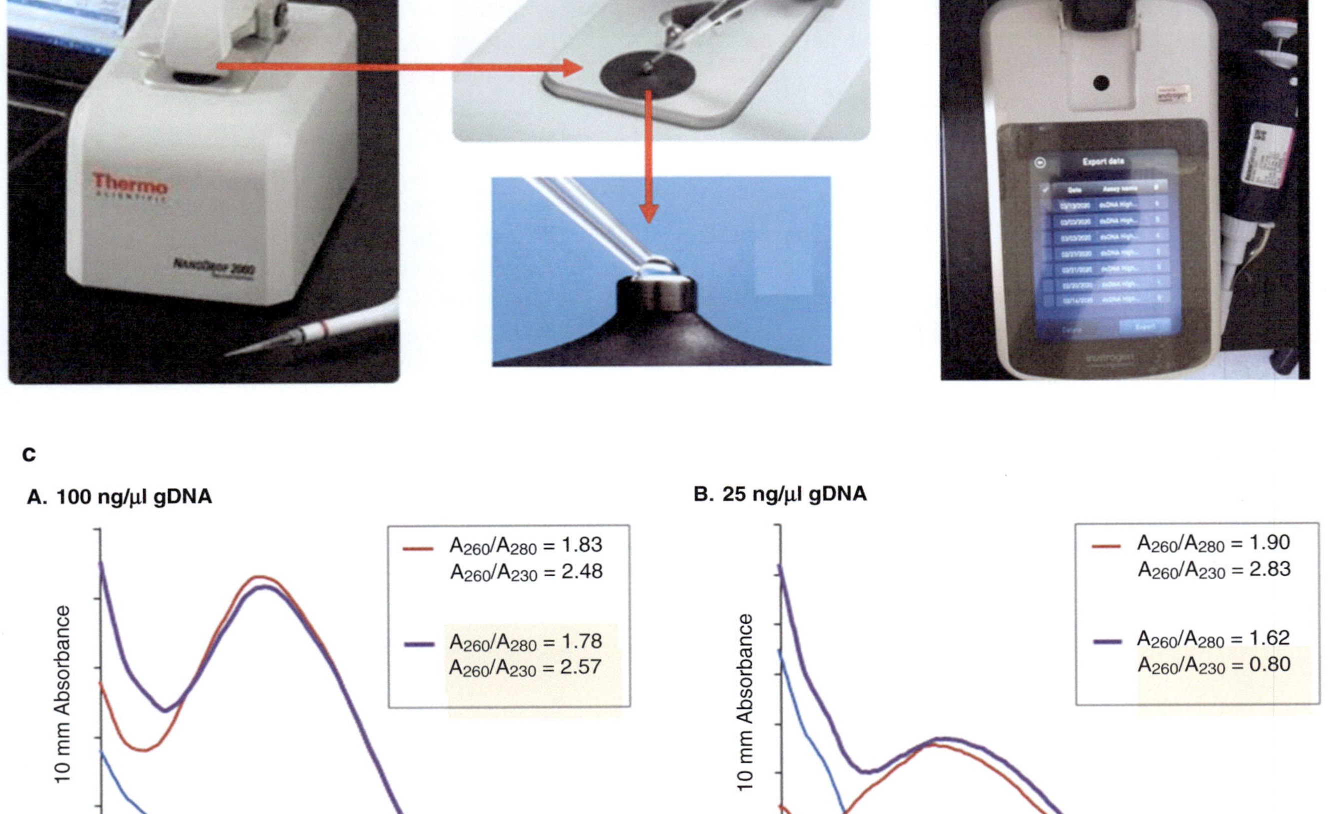

Fig. 1.1 DNA/RNA Quantitation. (**a**) Nanopore spectrophotometer. (**b**) Representative tracings from the Nanopore. (**c**) Qubit Fluorometer

- ♦ For purifying DNA/RNA, this is unlikely to matter
- ♦ One could NOT use it to measure the yield of cDNA in a reverse transcription (RT) reaction nor for the yield of an unpurified amplicon from a PCR
- A260/A280 ratio
  - ○ Deoxy- and ribodeoxynucleotides absorb well at 260 nm, weakly at 280 nm
  - ○ DNA(RNA) absorbs at 260 nm as if the nucleotides were free in solution
  - ○ Amino acids, especially aromatic amino acids, absorb more at 280 nm
  - ○ The generally accepted optimum for this ratio is 1.8 for DNA and 2.0 for RNA
  - ○ A ratio > 2.0 does NOT indicate superpure DNA/RNA, rather it is likely to be a sign that a contaminant is present (especially phenol, in older protocols)
    - ♦ Some sources attribute this to excess RNA
  - ○ **IF** the A260/A280 ratio is acceptable, the A260 can also be used to estimate the concentration of DNA or RNA using the conversion equations:
    - ♦ [DNA] = A260*50.0 *dilution = concentration in mg/mL
    - ♦ [RNA] = A260*40.0 *dilution = concentration in mg/mL
- A260/A230 ratio
  - ○ Absorption at 230 nm is contributed to by phenol, GITC, guanidinium hydrochloride, Triton, and Tween
  - ○ DNA also absorbs weakly at 230 nm
  - ○ Acceptable range is 2.3–2.4 for DNA, 2.1–2.3 for RNA
  - ○ The ratio is sensitive to salt concentrations
- Dye-binding fluorometry
  - ○ Dyes like SYBR Green fluoresce when excited at a specific wavelength, but ONLY when bound to DNA
  - ○ The DNA must be long enough to have a major groove for the dye to bind
  - ○ There are a variety of dyes, some are relatively specific to RNA
  - ○ The typical compact benchtop fluorometer dedicated to DNA/RNA measurement operates at fixed wavelengths for excitation and detection (Fig. 1.1c)
    - ♦ Compact scanning spectrofluorometers are also available
    - ♦ The more DNA present, the more dye bound, and the greater the signal
    - ♦ Two-point calibration is required for the expected concentration range

- – Discrepant concentration estimates by UV absorption and dye binding
  - ○ It is not unusual for measurements of DNA concentration to differ by four to fivefold when the two methods are compared
  - ○ Deoxynucleotides in a degraded specimen contribute to UV absorption but not to dye binding
    - ♦ RNA contributes significantly more to UV absorption than it does to DNA dye-binding fluorometry
    - ♦ The level of RNA relative to DNA is tissue dependent
      - ▪ In mononuclear blood cells, there is negligible RNA (relative to DNA)
      - ▪ In hepatocytes, there is many-fold more RNA than in lymphocytes
- Measuring the size distribution of DNA and RNA
  - – Determining size by electrophoresis can be an important QC measure (Fig. 1.2a–e)
    - ○ Many assays are sufficiently standardized and robust that it is cost-effective to run the sample (provided the concentration is adequate and the UV spectral scan shows a satisfactory curve) without first checking DNA integrity on a gel
    - ○ Gel electrophoresis can be used for troubleshooting if an assay fails
    - ○ NGS assays are sufficiently expensive electrophoresis is often used to assess the quality of the libraries
  - – DNA from fresh tissue or blood should show a band at 20,000–30,000 bp (Fig. 1.2c)
  - – DNA from fixed tissue
    - ○ Expect a broad size distribution but mostly less than 1000 bp (Fig. 1.2c)
    - ○ The longer the period of fixation, the shorter the size distribution (the greater the degradation)
  - – RNA from fresh tissue shows two bands, ~2000 bp (18S) and ~5000 (28S) bp (Fig. 1.2e)
    - ○ These represent ribosomal rRNAs
    - ○ They account for 90–95% of total RNA in a cell
    - ○ mRNA comes in sizes ranging from a few 100 bp to over 300,000 bp
      - ♦ The *DMD* gene, at 2000,000 bp gives rise to the longest *RNA* transcript
      - ♦ The mature *DMD* transcript is ~14,000 bp
      - ♦ The Titin gene is ~300,000 bp long and gives rise to an 80,781 bp mRNA
    - ○ mRNA, if seen on gel electrophoresis, forms a faint background haze
    - ○ Additional low molecular weight RNA bands can be seen occasionally
      - ♦ 7S RNA (300 bp), part of the signal recognition particle, is abundant in RBC
      - ♦ 5S RNA, another ribosome component RNA

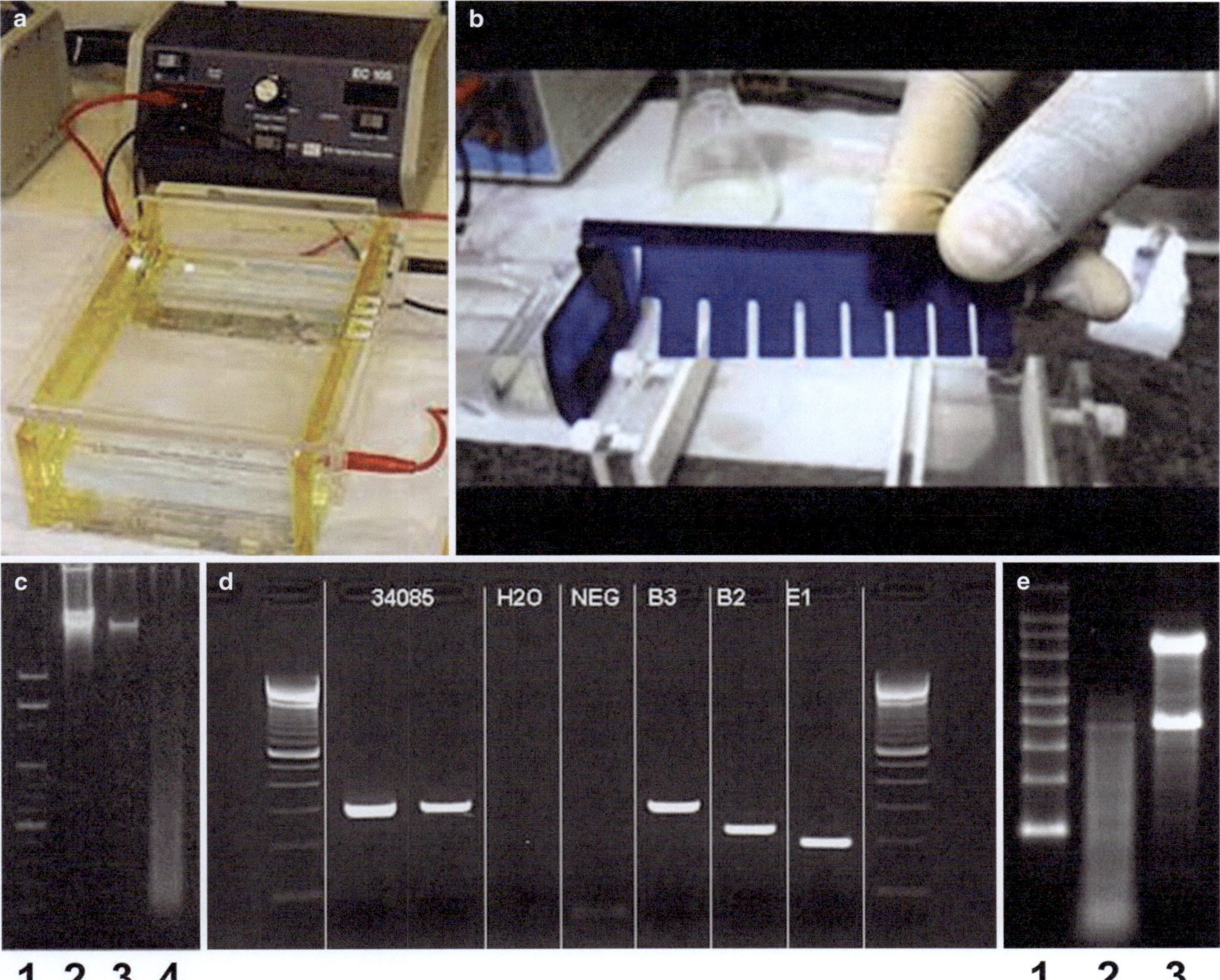

Fig. 1.2 Agarose gel electrophoresis. (**a**) Agarose gel electrophoresis gel box and power supply. (**b**) Comb for creating sample wells in a lab-prepared agarose gel. (**c**) Representative ethidium (EtBr) staining patterns for sample DNA. Lane 1—Molecular weight markers (MW); Lane 2—High MW genomic DNA; Lane 3—High MW genomic DNA, lower concentration; Lane 4—Degraded DNA, typical for FFPE. (**d**) Agarose gel showing PCR amplicons for *BCR–ABL*, lanes 2, 3, and 6–8. Source: Generously provided by Dr. Lynette M. Sholl, Department of Pathology, Mass General—Brigham Hospital, Boston, MA. (**e**) Formaldehyde-agarose gel showing EtBr pattern for RNA. Lane 1—RNA MW markers; Lane 2—Degraded RNA; Lane 3—Intact, high-quality RNA showing the 28S and 18S bands

- Degraded RNA
  - 18S and 28S are equimolar, so the longer RNA should bind more dye and give a brighter signal on a gel
  - An early sign of degradation is when the 18S band is brighter than the 28S band
  - Highly degraded RNA can present as a smear, with all staining <2000 bp and sometimes all <100 bp for FFPE RNA
  - The above observations are qualitative
    - Some microchip electrophoresis instruments provide a quantitative measure for RNA degradation (e.g., the RIN [RNA integrity number])
- With PCR amplification, it is possible to get a longer amplicon than the sample's size distribution on a gel suggests is possible
- This results from amplifying a negligible amount of long DNA/RNA present undetected on the gel and should be interpreted with caution (concern for contamination)
- General considerations for analysis of RNA
  - For most molecular techniques (e.g., PCR, sequencing), RNA analysis must first be copied into DNA by the enzyme reverse transcriptase (RT)
  - Requires a short region of double-stranded template to initiate copying

- ♦ Effect by adding short (~20 bp) synthetic DNA primers complementary to mRNA of interest
  - ○ Reverse transcriptase generates a DNA copy, growing from the 3′ toward the 5′ end, and remaining as an RNA-DNA 'hybrid' duplex
  - ○ The RNA strand is typically removed prior to subsequent steps by digestion with RNase-H, which is particularly effective on RNA-DNA hybrids
  - ○ NaOH denaturation and destruction of the RNA is an alternative
  - ○ The RT enzyme has a higher error rate than most DNA polymerases in PCR
- Although rRNA comprises 90–95% of total RNA preps, it is not efficiently copied in RT reactions, presumably due to the high level of secondary structure
- Choice of primers can influence interpretation
  - ○ Can add primers for multiple genes
  - ○ Can add multiple primers along a single mRNA
  - ○ Can add an oligo-dT primer which should prime most mRNA via the poly(A) tail
    - ♦ Poly-dT primers are not always extended across the entirety of longer mRNA, leading to a 3′ bias in the resulting sequences
    - ♦ Not all mRNA have poly(A) tails (e.g., histones)
    - ♦ This will not capture miRNA or lncRNA (unless the latter has a run of poly A)
  - ○ One can add random hexamers (or 8-mers or 12-mers), which should prime copying along most of the length of an mRNA, although not necessarily ensuring a single full-length copy
  - ○ Neither random primers nor gene-specific custom primers can ensure uniform priming of all mRNA or all exons
- Gene expression
  - ○ Microarrays were applied early in the development of the method to measuring the expression of all the genes (mRNA) in a sample at one time
    - ♦ Most gene-expression microarrays are not designed to discriminate among the various transcripts for a given gene, but exon-aware microarrays have been designed
    - ♦ Reproducibility and analysis of microarray gene-expression data were seen as prohibitively complex for clinical application
    - ♦ This is ironic in view of the later development of RNA-Seq (NGS)
    - ♦ The findings of a large microarray study of breast cancer, found to have clinical utility, inspired the development of a set of quantitative real-time PCR assays for the most significant genes (Oncotype), which is available as a clinical test

- ♦ Similar gene expression assays have been developed for specific clinical issues in several other cancers.
- ○ Gene expression continues as a major focus of research in cancer but now relies mainly on RNA-Seq, an NGS-based method
- ○ Some clinical assays offer a mixture of mutation detection and gene expression by RNA-seq (for example, Affirma) or mutation detection by DNA sequencing and gene expression by RNA-seq (for example, ThyroSeq), still others include miRNA levels

Cytogenetics

Background Information for Cytogenetics

- Most genes (and genomic regions), in most people, are present in two copies, one on each of two corresponding autosomal chromosomes (and in the two small 'pseudo-autosomal' regions on 'X' and 'Y')
- Cytogenetic variations occur in many forms and can be categorized along several axes
  - Inherited (germline) versus somatic (spontaneously acquired)
  - Microscopic (detectable in the microscope) versus submicroscopic (high-resolution microarray)
  - 'Size' of the variation
    - ○ The alteration can involve gain or loss of an entire chromosome, gain(duplication), loss (deletion), gain of multiple copies (amplification) or translocation of a subchromosomal region or segment, with the lower limit of detection dependent on the technology used for analysis
    - ○ Smallest change detectable by karyotyping is on the order of 1–10 Mb (average size of a band)
    - ○ Smallest change routinely detectable by FISH is on the order of 100,000 bp (NOT applicable to translocations)
    - ○ Smallest change detectable by microarray is ~1000 (the average probe density)
- Most, possibly all cancer cells, have cytogenetic abnormalities (the estimate depends on the method used for detection and the size range considered 'cytogenetic')

Classes of Pathogenic Cytogenetic Variation

- Ploidy refers to gain or loss of the whole complement of chromosomes (hyperdiploid, hypodiploid)
- Aneuploidy

- This is the presence of an abnormal number of chromosomes in a cell
- Aneuploidy is present in the majority but not all cancers (68% in one study of 42,000 solid tumors)
  - This does not exclude the presence of subchromosomal changes
  - Aneuploidy could render these cells susceptible to mitotic checkpoint inhibitor therapies
- Chromosome loss is much more common than chromosome copy gain for all chromosomes except 7, 12, and 20
- Aneuploidy, and especially heterogeneity of abnormal chromosome copy number, might be prognostic factors
- Aneuploidy can be assessed in cfDNA by NGS
  - This is also the basis for noninvasive pregnancy testing for chromosomal CNV
- Aneuploidy can be found at low levels in normal tissues, especially brain
- Structural variations (subchromosomal)
  - Deletions
    - Often involves loss of tumor suppressors
  - Amplifications
    - Often involves oncogenes
    - If there is an oncogene with an activating mutation in a tumor, and that oncogene shows amplification, it is usually the mutated allele that is amplified
  - Translocations
    - Most breakpoints occur over a variable intronic region
    - The fusion mRNA often stereotypically juxtaposes the same exon from each of the contributing genes, so the fusion site is precisely defined at the mRNA level
    - Many translocations can be detected by karyotyping and by FISH
      - Some translocations involve 'small' (by karyotype standards) regions of a chromosome, especially near a chromosome end, which are not readily identified by karyotype
      - If a specific translocation is sought, FISH is a rapid method that does NOT require dividing cells
        - It does require having the necessary FISH probe, which is a nontrivial consideration in view of the many known translocations)
      - Sometimes the probe will show that the gene targeted has been interrupted by a translocation but cannot identify the fusion partner
        - These can often be identified by molecular methods
          - ✦ Anchored PCR is a suitable RNA-based NGS method

          - ✦ It is closely derived from 5′ and 3′ RACE methods
    - Translocations can be detected by molecular or NGS-based methods
      - At the genomic DNA level, by 'tiling' over the intron
        - Use 'pull-down' probes 'tiled' across the reference intron to obtain amplicons from across the region of interest for sequencing of each end
        - One or more of the amplicons should have end sequences from two different chromosomes (or remote regions on the same chromosome)
      - At the RNA level by RT-PCR
        - If specific 5′ and 3′ gene partners are sought (e.g., *BCR* and *ABL*), setup PCR with appropriate primers, one from an appropriate exon in each gene
        - If only the 3′ partner gene is known or suspected (e.g., MLL) one can set up the RT reaction using one or more primers targeting the 3′ partner gene, which will extend in the 5′ direction, where the RT will eventually be copying sequence from the 5′ partner gene
      - Microarrays are not usually able to help with the identification of a translocation, but the translocation process does frequently delete modest-size regions at each breakpoint, which might be detectable by CMA
  - Inversions
    - These can range from a few thousand nucleotides to over a million
    - They can be difficult to detect by karyotyping and even by FISH
      - Examples
        - ROS1 in lung cancer
        - inv(16)(p13q22) [t(16;16)] in AML
  - Extrachromosomal chromosomal fragments
    - These are fragments of a chromosome seen on karyotyping but too small to classify
    - Once gained, many persist throughout cell division
    - They can contain active genes, inactive genes, or both
    - If the gene is active, this becomes a form of gene amplification
    - Their presence can be underestimated by cytogenetics
    - They have been proposed to be an important prognostic factor
- LOH
  - Often involves loss of a tumor suppressor
  - This is only detected by CMA (or by NGS)

- The SNP track shows a long stretch of homozygosity for SNPs
- Chromosomal instability (CIN), syndromes and neoplasia
  - There are multiple mechanisms that can result in CIN including defects associated with chromosomal segregation, the DNA damage response pathway, and telomere shortening
  - Inherited or acquired mutations in some DNA repairs genes (e.g., ATM, NBS1, multiple Fanconi genes) can lead to very heterogenous abnormal karyotypes
  - Curiously, MSI and CIN appear to be mutually exclusive
    - MSI in cancer is NOT a state which can be detected by cytogenetic methods
    - Both MSI and CIN tumors accumulate point mutations
  - Increasing the level of CIN is a potential therapeutic approach
    - PARP inhibitors interfere with ssDNA repair and appear beneficial in CIN
- Chromothripsis (from Greek for 'chromosome' and 'shattering into pieces')
  - Three major features
    - Scores or even thousands of chromosomal rearrangements within a localized (subchromosomal) region
      - Can involve several chromosomes in one cell
    - A low degree of chromosomal gain or loss for the rearranged region
    - Preservation of heterozygosity
  - Impact on treatment and prognosis uncertain
  - Found in 2–3% of all tumors
  - Found in ~25% of osteosarcoma and chordoma
- Kataegis (ancient Greek for 'thunderstorm')
  - Clusters of mutations over a region, usually several hundred basepairs long,
  - More frequent than chromothripsis; most common in breast cancer
  - Sites are often those associated with chromosomal breakpoints leading to the suggestion that it represents an effort to repair DNA damage, which might otherwise have gone on to create a breakpoint

- APOBEC enzymes are implicated in repair process
- This is best detected by NGS
- CNV
  - Some people have stable deletions (relative to some reference genome) of one or more portions of one or more subchromosomal regions
    - Although some variations might have no discernible effect on health in 'normal circumstances', it is possible that they could have an effect in the event of exposure to a stressor (infection, environment, drugs)
    - Similar considerations apply to point mutations and short indels, which create loss of function in genes
      - An early (2012) NGS estimate uncovered, on average, 100 loss-of-function mutations per person (in the study population)
      - 20 of the 100 were biallelic loss of function changes
    - CNV has been implicated in a variety of medical disorders including autism and schizophrenia
    - Sporadic CNV occurs in several organs (liver, brain) in the absence of cancer

Cytogenetic Methods

- (see MGP 2nd edition, chapter 2)
- There are three major cytogenetic methods
  - Karyotyping
  - FISH
  - CMA

Karyotyping

- Karyotyping can identify all the chromosomes in a cell by staining and viewing the size and banding pattern microscopically
- This requires live dividing cells in metaphase
- Karyotyping for surgical pathology has mostly been limited to hematopoietic neoplasms, to soft tissue tumors, and to some brain tumors

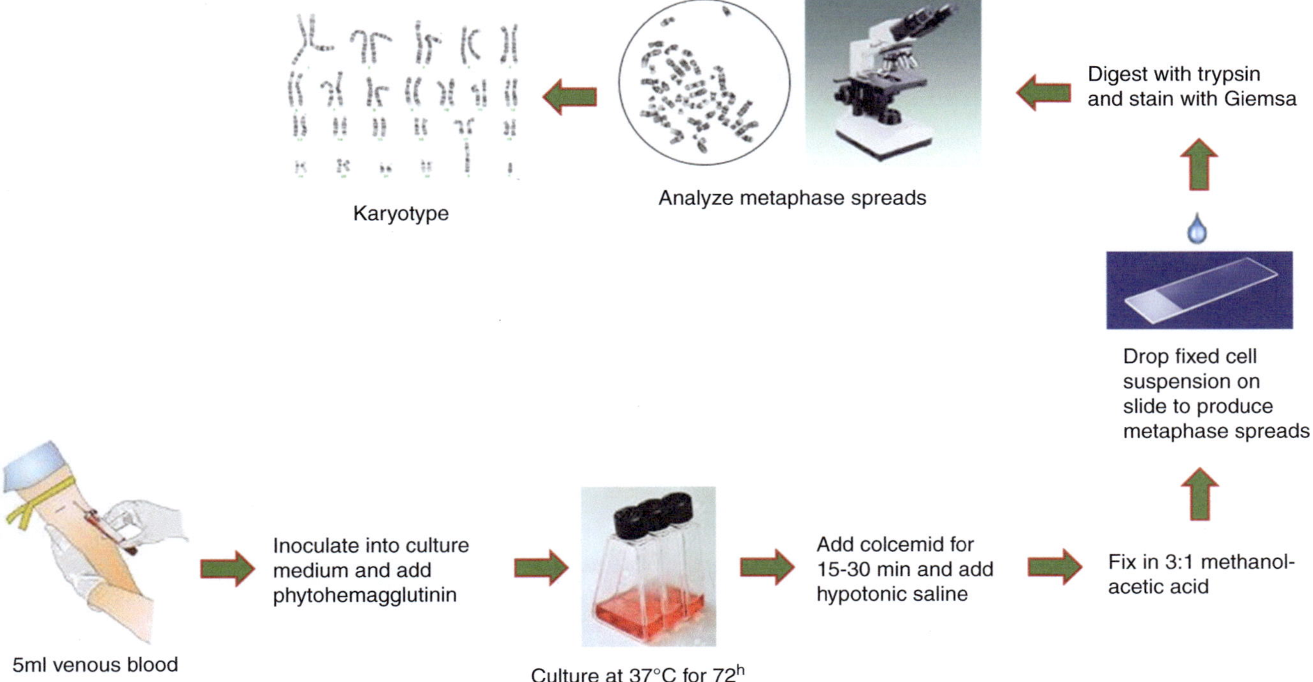

Fig. 1.3 Procedure for chromosome preparation and analysis (Karyotyping)

- Although solid tumors typically have many chromosomal abnormalities, getting the tumor cells to grow and divide is a laborious and uncertain process
- Single cells are readily obtained from blood and marrow, and from solid tumors by disaggregation, then maintained in tissue culture for 24–48 h
- The cells, from any source, which do grow are not necessarily representative
- Growth of specific cell types, especially hematopoietic, can be promoted by the addition of specific mitogens
- Outline of steps in karyotyping (Fig. 1.3)
  - Briefly grow cells in vitro
  - Swell in a hypotonic solution
  - Treat with colchicine
  - 'Drop' the cells onto a glass slide
  - Treat with trypsin to digest chromosomal proteins
  - Stain
  - View with a microscope and photograph
  - Cut out the pictures of each chromosome, sort and analyze (Fig. 1.4)
- Analysis of stained chromosomes is a pattern recognition skill, similar to surgical pathology except all the objects of interest are in grayscale and nearly identical
- The quality of the karyotype image can depend on the source as well as on the technique of the cytotechnologist; chromosomes from bone marrow aspirates are invariably less crisply stained than those from blood cells
- Software programs exist which can classify many chromosome images but require human review before reporting
- The chromosomes for a given cell stay close to the other chromosomes from that cell

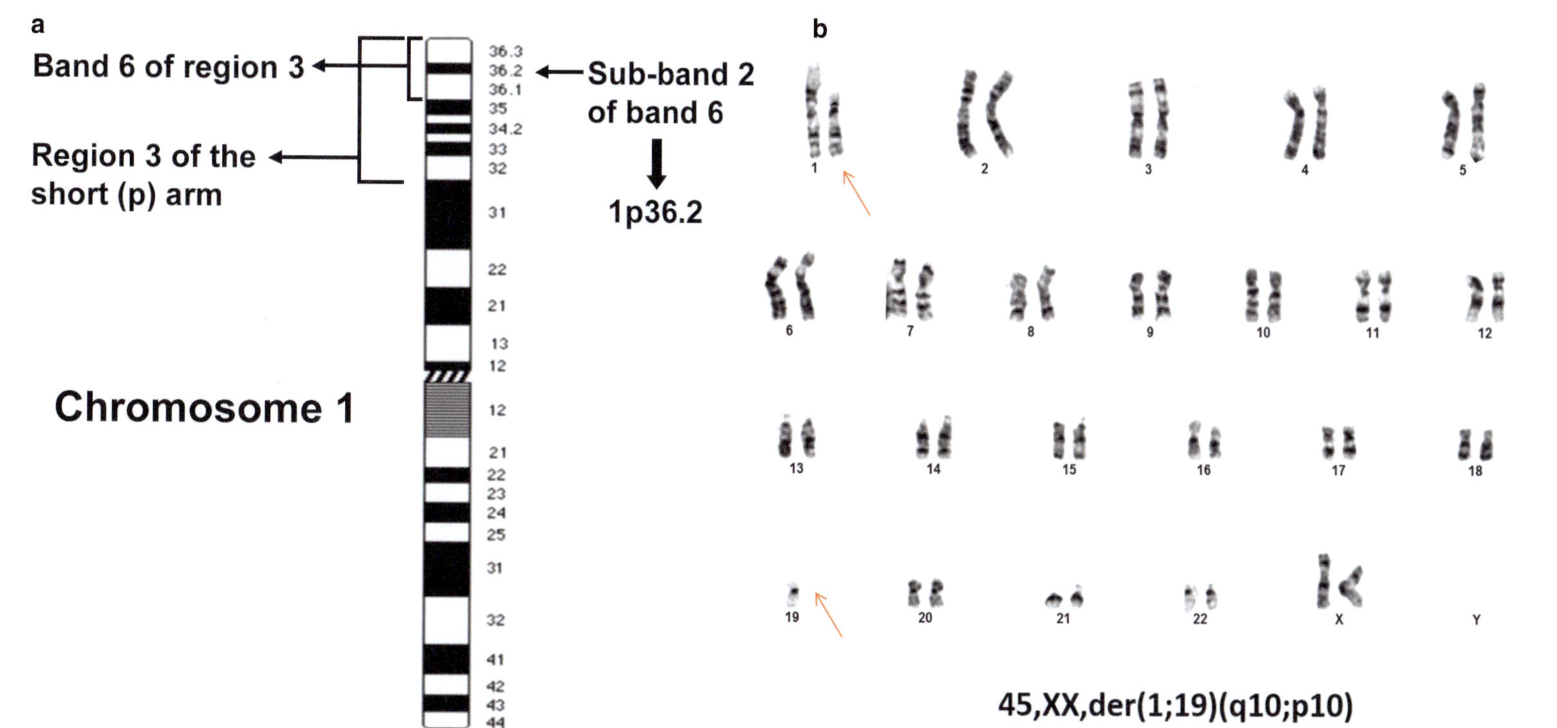

Fig. 1.4 Karyotyping. (**a**) Numbering of regions, bands, and subbands on chromosomes allows precise designation of breakpoint locations of structural abnormalities. (**b**) Karyotype showing 45,XX,der(1;19)(q10;p10). The translocation t(1;19)(q10;p10) is common in oligodendroglioma, the tumor studied here. Courtesy of Dr. Christine Bryke, Beth Israel Deaconess Medical Center. * Denotes droplets that are positive for the L858R mutation. ** Denotes droplets that show a very weak signal for the probe for the L858R mutation. These most likely represent 'spectral' cross-talk from the fluor labeling the probe for the L858 reference sequence

- This helps show if one or more abnormalities occur in the same cell
- Giemsa stain is the most common, producing light and dark bands ('G-banding')
    - The dark bands are mainly heterochromatin, the light bands are mainly euchromatin (transcriptionally active)
- A successful karyotype shows ~<u>400</u> bands total
- A typical band encompasses 5–10 million bp
- Typically, 20 cells are 'scored'
    - If there appears to be limited or variability, two or three cell karyotypes will be analyzed in detail
    - If an abnormality is found, the other cell karyotypes will be scanned for the specific abnormality
    - If a specific translocation is sought, FISH provides faster and more sensitive testing

Fluorescent In Situ Hybridization

- In FISH, a probe derived from cloned fragment of human genomic DNA, labeled with one or more fluors, is hybridized to fresh lightly fixed cells or to cells in an FFPE section on a glass slide to interrogate them for the presence of specific chromosomal regions (Fig. 1.5)
    - The probes originate from cloned fragments typically with a size on the order of 100,000–500,000 bp

- DNA polymerase, random primers, and dNTPs, including one fluorescently labeled dNTP, are used to generate DNA copies from all along the clone
    - Because most cloned genomic regions will contain one or more repeat regions (e.g., *Alu*) a subset of the probe can potentially hybridize to all chromosomes
        - This can be circumvented by including unlabeled DNA (commercially available, not from the patient in question) enriched for repeat regions
        - This will 'soak up' the labeled probe molecules, which recognize repeats
    - If the cells on the slide are from FFPE, they are permeabilized by an agent like ethanol
    - The DNA is denatured
    - The fluorescent label is applied to the slide and allowed to hybridize
    - The slide is viewed with fluorescence microscopy
        - The microscope will have objectives optimized for viewing each of the fluors if probes are multiplexed
        - Photomicrographs are taken for each fluor
        - The images can be combined with appropriate software
- FISH probes can be multiplexed to a limited extent because typical fluors overlap in their emission spectra
- With an appropriate choice of probe and label, FISH can detect chromosome copy number, gene amplification,

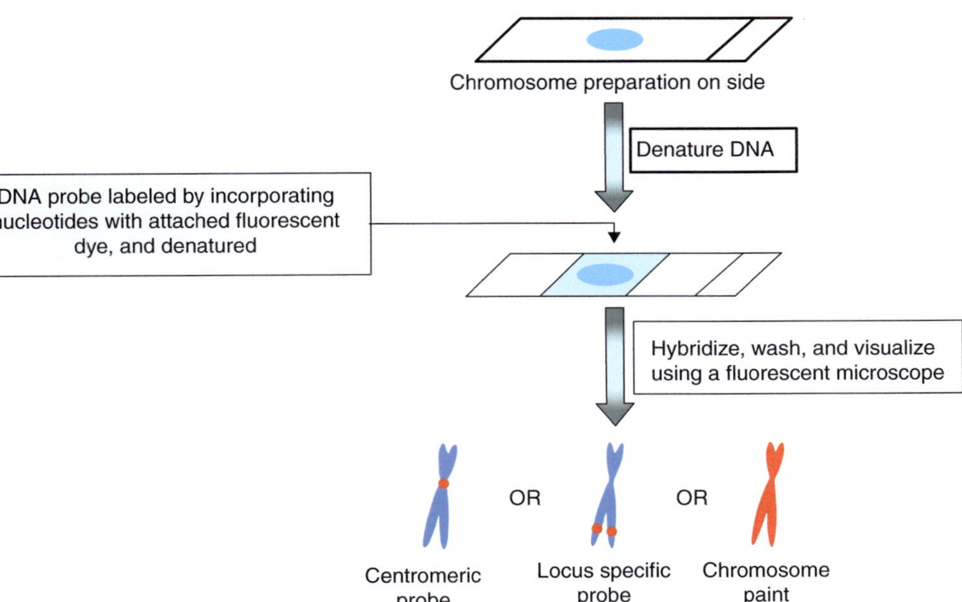

Fig. 1.5 FISH procedure. Courtesy of Dr. Christine Bryke, Beth Israel Deaconess Medical Center

gene deletion, translocations, and, in some cases, inversions
- FISH probes for chromosome enumeration are straightforward
  - X and Y probes labeled for FISH provide a rapid method to assess bone engraftment chimerism when the donor and recipient are sex-mismatched
  - A single probe, labeled with a single fluor, can be used to count the number of chromosomes carrying the gene region in a given cell
    - It cannot prove that it is on an intact version of the expected chromosome
    - It can be useful in demonstrating polysomy (more than two chromosomes carrying the targeted region)
    - Some translocations can split the region spanned by the probe, and as a result, the two resulting chromosome fragments derived can each demonstrate a FISH signal
      - Depending on the break site, one signal might be visibly stronger than the other, giving a hint about the process
      - If the two signals are the same size, it might suggest polysomy
    - Gene amplification
      - The size of the fluor signal can be many times larger than usual because of amplification of the target in place
      - Polysomy
        - There can be several bright spots, of the typical size, in multiple locations throughout the nucleus
        - This would represent polysomy—either an increase in copy number for the whole chromosome or of a subchromosomal region
      - There can be numerous bright spots of varying intensity throughout the nucleus, reminiscent of a meteor shower
        - This can occur when the targeted region is part of a marker chromosome, and some attain a very high copy number

- Break-apart probes
  - These are designed for improved detection of translocations
  - The 'probe' consists of two cDNA probes, each labeled with a different fluor and targeted at different end of the region of interest
  - With an intact chromosome, the two probes bind in proximity and the two fluors (e.g., red and green) would be seen as a single color (yellow)
  - If a translocation is present, spots of the two colors would be seen spatially apart
- FISH for monitoring response to therapy in hematologic malignancy
  - For hematopoietic neoplasms with characteristic cytogenetic abnormalities
  - Typically, 200 or 400 cells are scored
  - Different probes will have different limits of detection (on the order of 1–5%), which must be determined in each reporting laboratory
- SKY (spectral karyotyping)
  - Specially designed and combinatorically labeled probes can be used to image all the chromosomes, in their entirety at one time
  - Requires cells in metaphase, precluding use of FFPE tissue
  - Although this produces by far the most chromatically pleasing images in molecular pathology, it has not seen clinical adoption
- Common surgical pathology applications of FISH probes
  - Numerous lymphoma, leukemia, and myelodysplastic syndromes have cytogenetic abnormalities, which are required for diagnosis by WHO criteria
  - In CNS tumors
    - BRAF fusion genes are modestly frequent in gliomas
    - EGFRexon 1p/19q codeletions (Fig. 1.6)
  - Soft tissue sarcoma: There are many translocations, and many are diagnosis-specific
  - Lung cancer: ROS, RET and ALK translocations

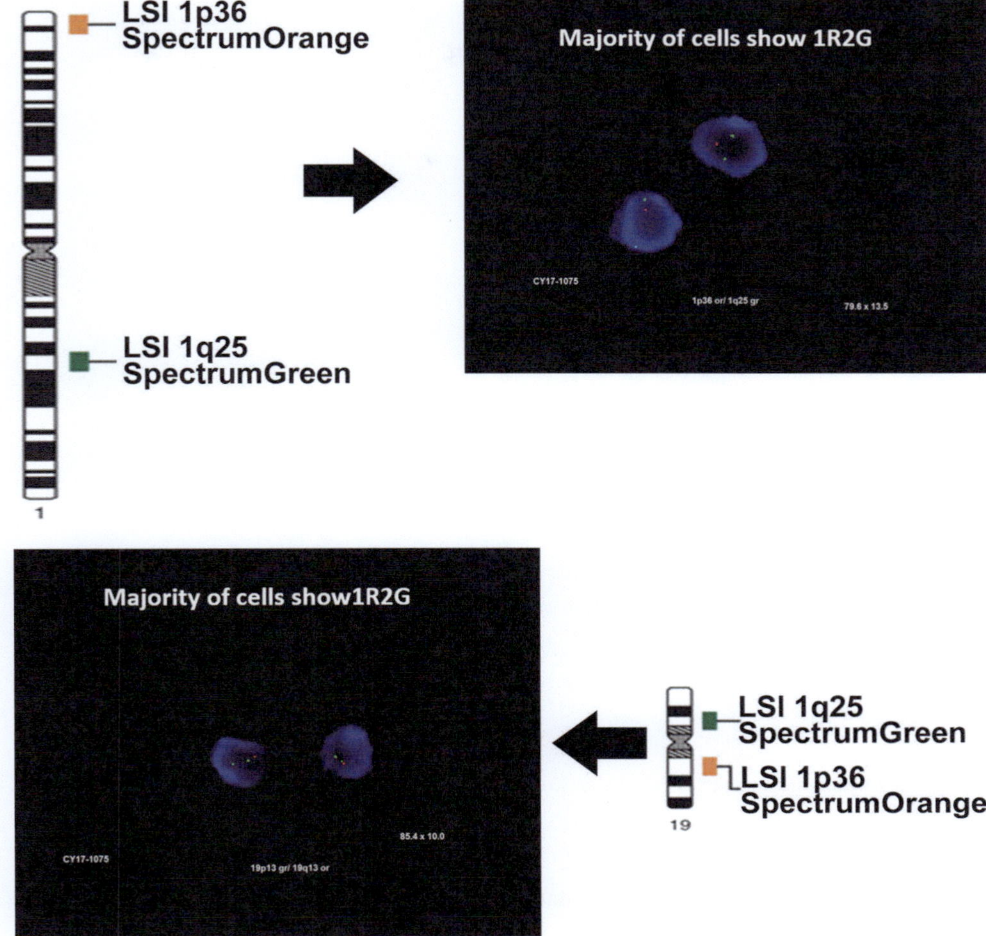

Fig. 1.6 Break-apart FISH probes for t(1;19) in an oligodendroglioma. Top panel shows FISH analysis with a dual-label probe for chromosome 1. Most cells show 1 red signal and two green signals (1R2G), indicating loss of 1p, the region of the chromosome labeled with the red fluor. The lower panel shows a similar analysis with a dual-label probe for chromosome 19. This shows a loss of 19q

Microarrays

- A microarray is a solid support that bears predetermined spatially indexed DNA oligonucleotides of known sequence, which can hybridize to either labeled DNA or labeled RNA derived from the sample(s) of interest (Fig. 1.7)
- Microarrays can be made from a variety of solid supports (glass slides, silicon chips, nylon membranes)
- The oligonucleotides immobilized on support are called 'probes'

- This caused some terminological confusion when Southern and Northern blots were still widely performed
- Oligonucleotide lengths range from ~20 to ~80 nt
- The oligos are arranged onto the array in one of two methods
  - Synthetic oligos can be deposited on a slide by an inkjet printer (or equivalent)
    - Oligonucleotides can be synthesized on a silicon chip in combination with photolithography techniques adapted from the semiconductor industry

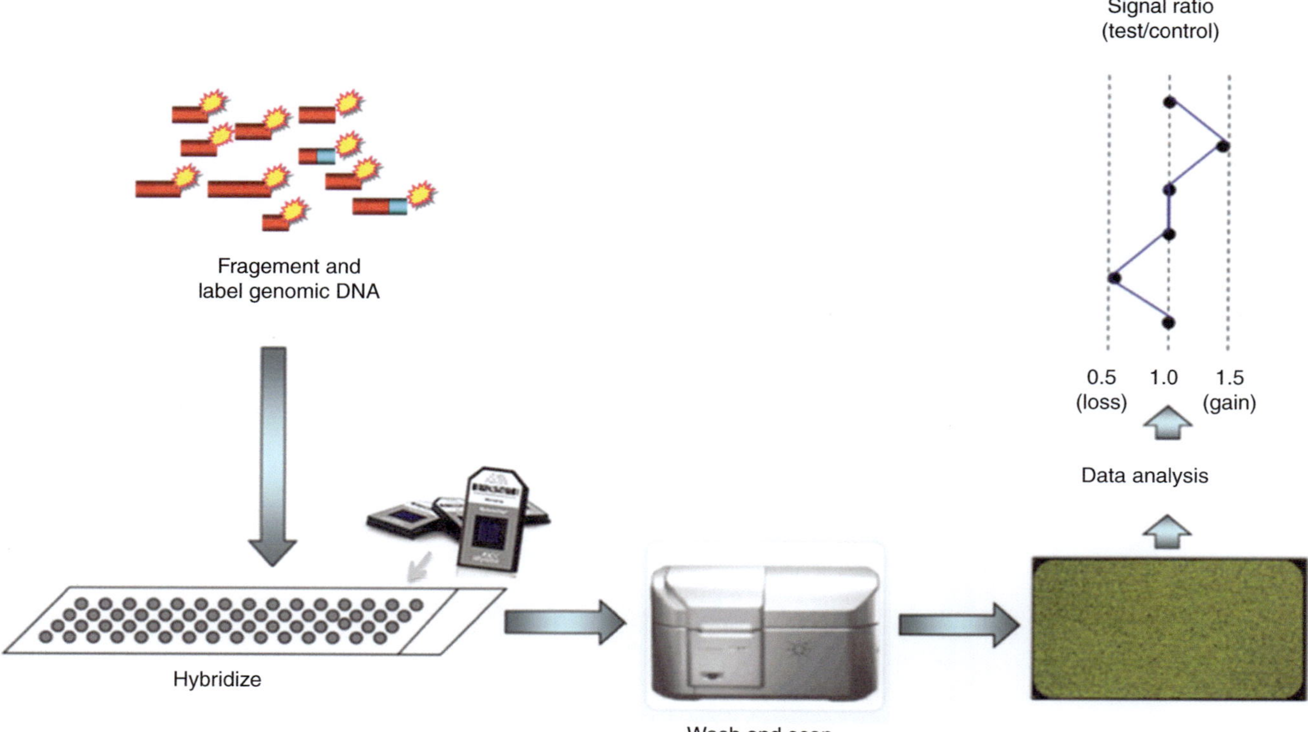

Fig. 1.7 Cytogenomic microarray procedure. Courtesy of Dr. Christine Bryke, Beth Israel Deaconess Medical Center

- Characterization of a representative cytogenomic microarray
  - 2.7 million markers spanning the genome
  - Spacing of probes, on average, 880 bases apart in genic regions, and 1700 bases apart in nongenic regions
  - 1.9 million oligos are nonpolymorphic markers for assessing gene copy number
  - 750,000 oligos are for detecting SNPs
- Long oligonucleotides are useful for detecting the presence or absence of the corresponding genomic segment in the sample
- Hybridization of long oligonucleotides is not sensitive to the presence of one or a few SNPs in the region
- Short oligonucleotides (20–25 nt) are for the detection of SNPs
  - To detect a specific SNP cytogenomic arrays will contain at least two short oligonucleotides that target that SNP
    - They are usually designed to have the SNP in the center of the probe
    - One probe will be for one allele, and the other probe will detect the other allele
  - The sample DNA is labeled with a single fluor
  - Which 'spots' lights up determines if the sample is homozygous for one allele, the other allele, heterozygous, or deleted at that position
- Depending on the design, CMA can be used to analyze
  - Chromosomes numbers, subchromosomal insertions and deletions, and gene copy numbers
  - SNPs
- Different microarray designs can be used for noncytogenetic molecular tests
  - RNA expression: Gene-expression arrays (GEA)
  - DNA sequencing (a deprecated method)
- CMA has demonstrated clinical utility
  - Neonatal genetics: CMA has been approved by the FDA for first-tier testing of intellectual disability, multiple congenital disorders, and autism
  - Hematopoietic neoplasms
    - Microarrays can help define deletions, amplifications, and breakpoints identified by karyotyping
    - It is not routinely used
  - CNS tumors
    - Some CNS tumors have characteristic alterations, such as the 1p/19q codeletion (Figs. 1.4 and 1.8)
  - CMA and FISH can each be used to confirm and deepen the results by the other method

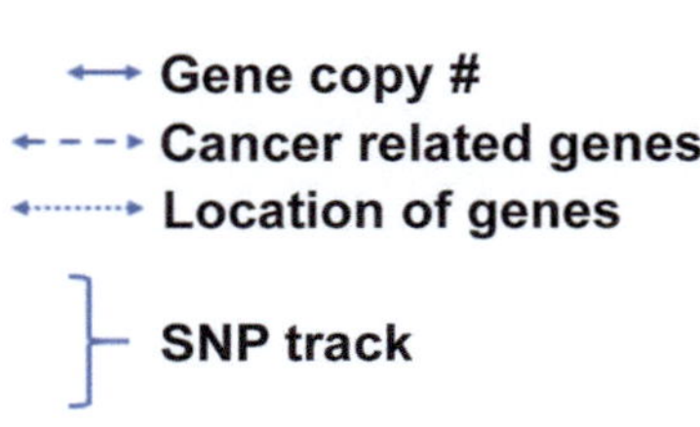

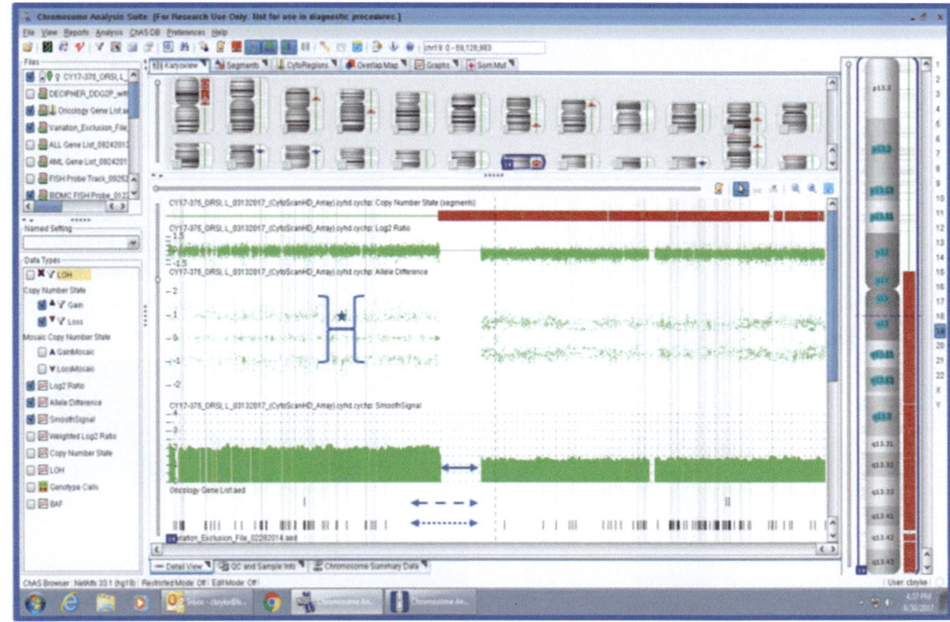

Fig. 1.8 Microarray results for an oligodendroglioma. The image shows data for chromosome 19 (highlighted in the virtual karyotype at the top of the diagram). The gene copy number tracks a diminished signal to the right of the double arrow (see legend), consistent with the deletion of that region. The SNP track to the left of the * shows three lines of clustered dots. Each represents an SNP site. One line represents homozygosity for one variant at a given position, one line represents homozygosity for the other variant, and one line represents the SNPs that are heterozygous. The SNP track to the right of the * shows two tracks; each is homozygous. There are no heterozygous SNPs because this portion of chromosome 19 has been deleted in most cells

Molecular Diagnostics Basic Methods

DNA/RNA Electrophoresis

- DNA/RNA electrophoresis background
  - DNA and RNA electrophoresis is useful as a quality control measure, as noted earlier
  - Determining size of DNA and cDNA by electrophoresis is also a means of detecting variants, although somewhat passe
    - Presence or absence of a PCR amplicon for a possible translocation in cDNA
    - Indels leading to altered PCR amplicon sizes (e.g., *FLT3 ITD*)
    - Clonality in T- and B-cells
      - Amplification of the VDJ and VJ regions leads to amplicon size distributions which can indicate the presence of a prominent clonal population
    - Presence or absence of a restriction site affected by a sequence variant leads to different size of DNA amplicons after restriction digestion
    - Presence or absence of an amplicon following allele-specific PCR
  - All DNA and RNA nucleotides are negatively charged at neutral pH due to the phosphates
  - The nucleotides are all roughly similar in weight, so the charge/mass density is identical regardless of the length of the DNA or RNA
  - All DNA and RNA polymers in the appropriate buffer migrate toward the anode
    - The speed of migration is inversely related to the length
    - Some DNA sequences can have unusual conformations which affect mobility
      - There could be complementary sequences within the DNA strand which self anneal
      - High percentage of GC nucleotides
      - Circular DNAs such as plasmids can show anomalous migration
  - For measuring DNA size, all methods use at least one sample for calibration

- A molecular weight ladder composed of multiple DNA fragments of known length (usually commercially obtained)
- The log of the length for each band can be plotted versus the distance migrated, and a straight line derived
- The size of an unknown DNA fragment can be inferred from its mobility
  - For measuring RNA, an external size standard is optional
    - High-quality RNA samples should have intact 18S and 28S RNA bands
    - These bands serve as internal size standards
- Agarose gel electrophoresis
  - The most common, versatile, and least expensive electrophoretic method
  - Agarose is derived from seaweed; it is a carbohydrate polymer that forms a sieve through which DNA molecules must wend
  - Agarose gels can be used to size DNA and RNA (on separate gels)
    - They can measure over a range of roughly 50–20,000 bp
    - Larger fragments can enter the gel but barely migrate
    - The typical gel is 0.8–1.0% agarose (by weight)
      - Although this can detect DNA down to ~50 bp, it cannot resolve small differences (e.g., 100 vs. 120 bp)
      - Gels up to 4–5%, show better resolution of small DNA bands but poor resolution of long bands
      - Polyacrylamide gels can resolve DNA under 100 bp
  - Agarose gels can be prepared in most molecular diagnostic labs
    - A typical gel tray is shown (see Fig. 1.2)
    - A plastic 'comb' is placed in a slot on a rectangular plastic template near the end with the anode (electrode post)
    - Agarose in the buffer is microwaved until dissolved
    - The molten agarose is poured into the template, where it cools
    - Once the gel has hardened, the 'comb' is pulled out, leaving indentations that will become the sample wells
    - The gel tray is transferred into a gel box
    - Buffer is poured into the gel box until the gel is just covered
    - Samples are loaded into the wells, and the lid is placed on the apparatus

- The gel box, which has a positive and a negative electrode at opposite ends, is connected to a power supply
- The DNA/RNA will migrate toward the anode
- DNA/RNA separated on the gel is detected by soaking the gel in a dye that fluoresces under UV light, but only when the dye is bound to DNA
  - Ethidium bromide (EtBr) is the most common dye
  - EtBr is commonly held to be mutagenic, possibly carcinogenic, but this appears to be, at worst, a theoretical concern
  - Agarose gels for RNA
    - Many RNA assume complex 3D structures, perhaps due to self-complementarity and facilitated by the single-stranded nature
    - Consequently, the mobility of RNA in its native state correlates poorly with mobility on gels
    - To improve the correlation, gels for RNA include a denaturing agent, usually formaldehyde, added to the gel when it is molten
    - The heated formaldehyde gives off vapors, so the gel must be poured and run in a fume hood
- Polyacrylamide gel electrophoresis
  - Polyacrylamide is a polymer that can be generated 'inhouse' in the typical molecular diagnostic laboratory
  - Polyacrylamide gels are typically formed in a vertical gel 'mold'—two glass plates separated by plastic mm thick spacers, with wells for loading at the top
  - Horizontal gels are commercially available
    - They are loaded like agarose gels
    - These are mainly used for protein work
  - This is the gel matrix used for Western (protein) blots, albeit with different buffers
- Microfluidic 'chip' electrophoresis (see Fig. 1.9)
  - A microfluidic chip electrophoresis instrument offers separations similar to those in capillary electrophoresis but on a much smaller scale
    - There are a variety of similar instruments; the Agilent TapeStation is used for illustration purposes only
    - A chip is typically flat, the size of a credit card
    - The TapeStation chip contains linear grooves, one for each sample (see Fig. 1.9)
    - The grooves contain a 'sieving matrix'
    - The sample is mixed with a fluorescent dye and loaded at one end of each groove
    - The loading volume is ~1 microliter and is performed by the instrument
    - The 'chip' has positive and negative electrodes 'printed' onto it

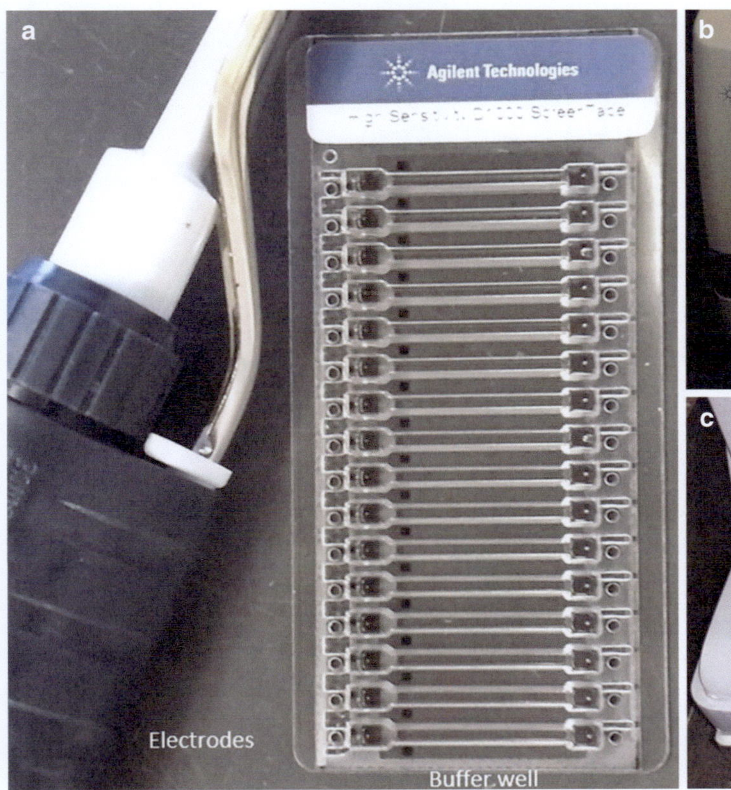

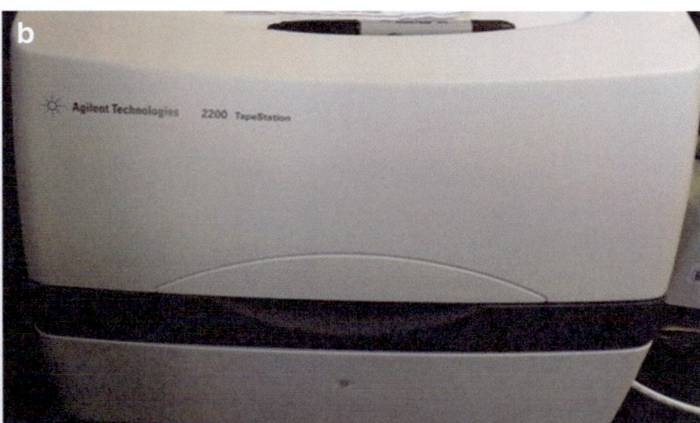

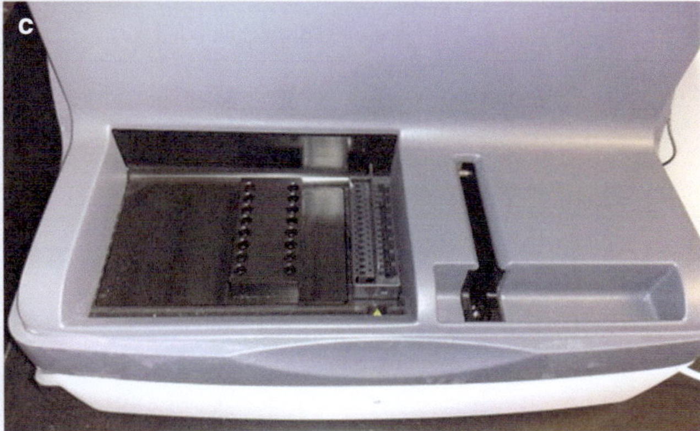

Fig. 1.9 Microfluidic electrophoresis platform. Microfluidic electrophoresis 'chip'. This is for the Agilent TapeStation. (**a**) The 'chip', is next to a pipettor for size comparison. (**b, c**) The instrument is larger than the typical agarose gel box BUT it also loads the samples onto the chip from the sample tubes, which are loaded in the middle block. Separations take a few minutes. The instrument is connected to a computer that can generate a virtual gel picture mimicking those seen in Fig. 1.2c and Fig. 1.23

- o Separations take 1–2 min
- o At the end of the separation, the chip is imaged
- o The image is presented as a virtual gel mimicking agarose gel image
- It is natural to wonder if these can be used for DNA sequencing
  - o One group reported 150 bp in 450 s, with a single bp resolution
  - o That report was in 1995, the approach has not caught on
- Capillary electrophoresis analyzers (aka Sanger sequencers) (see Fig. 1.10)
  - The capillary electrophoresis analyzer typically has 4–96 capillaries
  - Each capillary is made of fused silica; it is flexible
  - Each capillary has a pore diameter typically on the order of 20–100 nm

- The typical length of the capillary ranges from ~36 to 50 cm
  - o The longer the capillary, the greater the maximum size analyzed
- As with agarose electrophoresis, the shorter the DNA strand, the faster it moves through the capillary
- The tube contains a 'liquid' formulation of polyacrylamide which permits single nucleotide resolution
- The capillary tubes are optically clear
- A detection window in the instrument interrogates a fixed position in the capillaries by constantly illuminating it with a laser of fixed wavelength and monitoring several wavelengths for fluorescent emission
- The resulting fluorescent signals, plotted against time, generate a 'sequence chromatogram' (see Fig. 1.23)

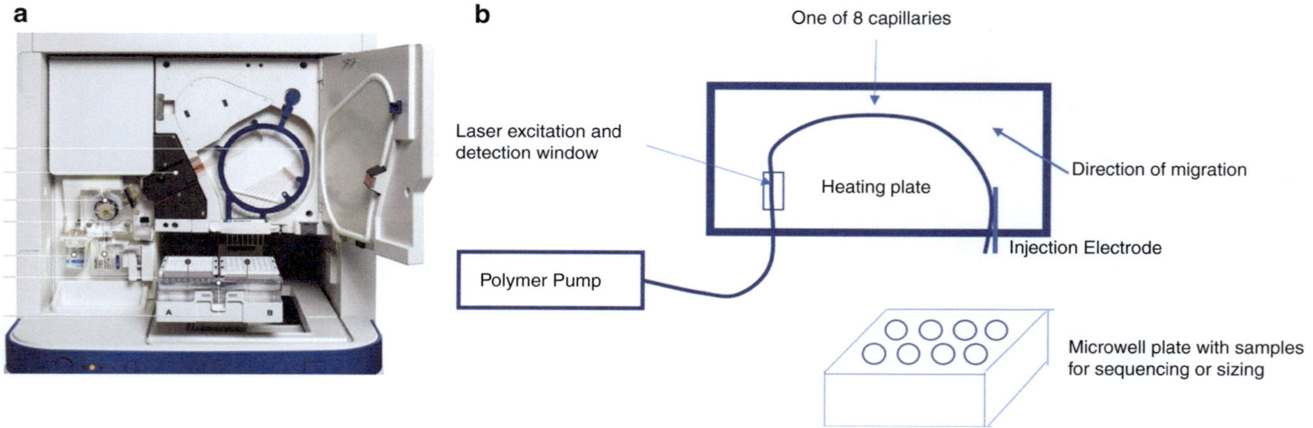

Fig. 1.10 Capillary electrophoresis analyzer. (**a**) An open view of an ABI3500 8 capillary electrophoresis analyzer (capillaries are so fine, they are barely visible, forming a loop around the more prominent blue plastic ring). (**b**) Schema for capillary electrophoresis (see text)

The Art of Detection

Target Amplification Versus Signal Amplification

- Many assays in molecular diagnostics can be classified based on whether the target is amplified, to enhance the prospect of detection, or the signal is amplified
- The polymerase chain reaction (PCR) is the archetype of a target amplification assay, but it is NOT the first target amplification assay in clinical pathology
  - Clinical microbiology has routinely used bacterial growth in culture or on plates
- A significant risk with target amplification is that the amplified target can contaminate the laboratory and affect results for other samples
- PCR is not the only target amplification format used in molecular diagnostics
  - Some methods combine it with RNA polymerase
    o RNA polymerase can make many copies from one target molecule in a single round of amplification
    o RNA molecules are less likely to cause contamination of the laboratory
    o In transcription-mediated amplification (TMA), rounds of amplification with RNA polymerase alternate with rounds of amplification with DNA polymerase, so there is still an opportunity for contamination

- Rolling circle amplification (RCA) (see below) is an isothermal technique that has found wide application both for target amplification and for signal amplification
  - Hybridization chain reaction (HCR) (see below): This uses linear amplification of DNA to generate a signal for IHC or FISH detection of nucleic acids
- Branched DNA (bDNA) (see below)
  - This is a signal-amplification assay in which a DNA oligonucleotide probe is used to detect RNA or DNA targets
  - Additional synthetic sequences on the probe are used to build up a signal, much like in IHC

Methods: PCR

- The PCR principle invariably seems obvious once it is explained (Fig. 1.11)
- PCR is an indispensable tool in many molecular methods such as NGS
- PCR itself can be used to detect the presence or absence of specific DNA sequences (variants, mutations) and to quantify them (gene copy number, gene expression level), but the capacity to multiplex is limited
- With the explosion in genomic information, there are few settings in surgical pathology in which one wants to know about the presence or absence of a single sequence variant or the expression level of a single gene

Fig. 1.11 General schema for basic PCR

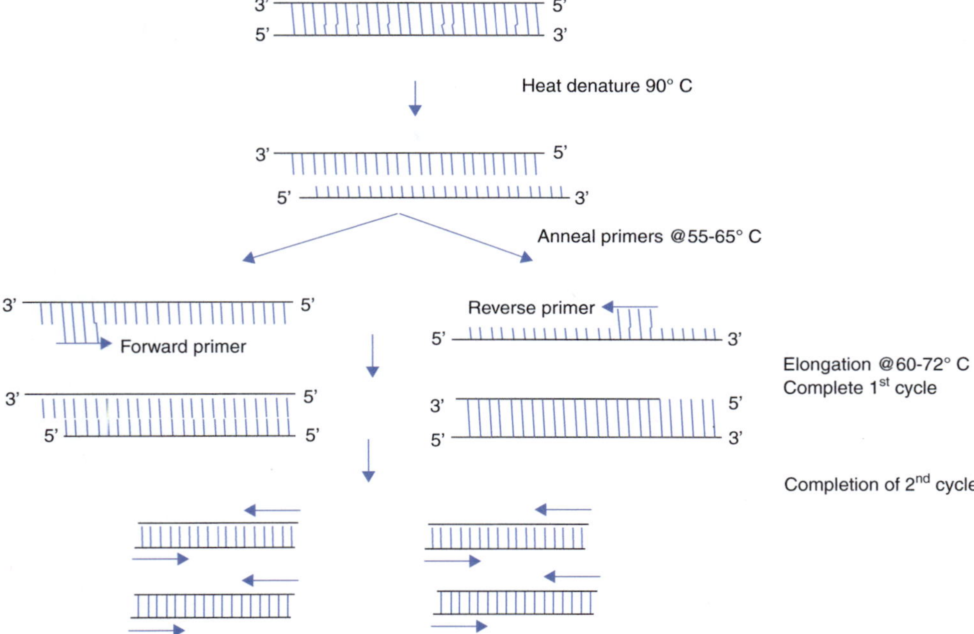

- Knowledge of PCR remains important to understanding more complex assays like NGS (which usually employ PCR)

Principles of PCR

- Genomic DNA or cDNA is heat denatured, then cooled (Fig. 1.11)
- Two synthetic DNA oligonucleotides of known sequence, typically 20–25 nucleotides in length, the 'primers', anneal (bind) at a lower temperature to the complementary sequences on the denatured DNA
- The annealed primer is extended by a DNA polymerase
- Repetitive cycles of DNA denaturation, primer annealing, and DNA synthesis initiated from the primers leads to a geometric increase in the number of DNA copies, 'amplicons', of the region bracketed
- Considerations for the PCR primers
- There are computer programs that can predict the optimal binding temperature for a given primer sequence (or, alternatively, given a sequence of interest to amplify, can identify optimal primers)
- Primers can anneal to imperfectly matched sequences
  - The longer the primer, the more tolerant it is of mismatches, except at the 3′ end
  - The lower the annealing temperature, the more tolerant of mismatches
  - A common pitfall is when a primer has a complementary overlap at its 3′ end with either the 3′ end of the other primer or with itself

- The primer concentrations are high relative to other targets
- The polymerase is quite happy to extend one primer to another
- This can generate an abundant 40–50 bp 'primer-dimer' product
- The archetypal PCR thermal cycle has three temperature phases
  - Denaturation
    - The longer the DNA and the higher the percent of GC content, the greater the temperature and the longer the incubation needed for denaturation
    - Most duplex DNA, including human genomic DNA, as well as PCR amplicons, denature fully at 95 °C well within 1 min
    - Exposure to the denaturation temperature cumulatively damages the polymerase, but for a typical PCR of 30–35 cycles with 1 min at 95 °C, typical PCR master mixes work satisfactorily
    - Polymerases are available with increased heat stability
  - Annealing phase
    - The thermocycler cools to a temperature at which the two primers can stably bind to their intended complementary target sequences
      - The optimal binding temperature is denoted Tm for the given reaction conditions (salt, buffer, etc.), the melting curve shows maximal rate of change
      - Factors that favor more specific binding
        - Higher annealing temperature

- Shorter primer sequence
- Shorter annealing time
  - Extension phase
    - In this phase, the polymerase extends the primer
      - The classic three-temperature phase PCR profile uses an extension temperature of 68 °C or 72 °C; the optimum temperature depends on the enzyme
      - As the temperature rises from the annealing temperature to the optimal extension temperature, the enzyme can still extend bound primers, albeit not as efficiently as at 72 °C (or whatever is that polymerase's optimum)
      - A typical extension rate is 1000 nucleotides/min
- Detection of amplification: There are two broad categories
  - Endpoint: The PCR reaction is completed for a predetermined number of cycles of annealing and denaturation
    - The PCR reaction product can be detected by several methods, including colorimetry, but agarose gel electrophoresis is the most common
    - This is excellent for qualitative detection: 'Present' or 'Absent'
      - Whether the sample starts with one target molecule or thousands, by the end of ~35 cycles of amplification, both give intensely bright bands on gel electrophoresis
      - Laborious workarounds existed for quantitation, such as performing PCR on a series of dilutions of the sample, to find the dilution at which no product formed

Real-Time PCR

- An output signal is read at least once <u>during</u> every cycle for every sample
- There are numerous assay formats, but all rely on generating a fluorescent light signal, detected through UV transparent optically clear reaction vessels (microfuge tubes, microtiter plates, capillary tubes)
- The signal curve over time can be used to quantify the amount of target at the beginning in the sample
  - The assay can measure gene copy number (DNA) or gene expression (mRNA)
  - The assays can be modified to detect the presence or absence of a specific single nucleotide (SNVs, mutations)
- There are three broad formats based on the signal detection method (Fig. 1.12)
  - Dye binding
  - Hybridization probes
  - Hydrolysis probes
- Amplification curve features

- All three methods generate amplification signal curves with similar features (Fig. 1.13a)
- Detection threshold
  - This is critical to the process of quantification
  - There is no one 'correct' way to choose a detection threshold
    - Even though the amplification curve appears flat at 'zero' for many early cycles, amplification occurs in the reaction
    - Even though the amplification curve appears flat at 'zero' for many early cycles, typical software tries to compress the entire amplification curve so that it fits within one computer screen
    - Expanding the signal scale axis will reveal weak signals with wide variation for some of the early cycles
    - A typical formula for determining the 'detection threshold' would be to determine the mean and SD of the first 10 or 20 cycles, then use the mean +10 SD as the detection threshold
- C_t (or C_p) is the point at which the amplification curve crosses the 'threshold' is called either the C_p (crossing point) or C_t (crossing threshold)
  - The more of the target in the starting material, the earlier a signal is detected; this corresponds to a <u>lower</u> C_t than for a sample with very little of the target (Fig. 1.13b)
- Log-phase
  - This represents the phase in which the PCR is proceeding with maximal efficiency
- Log-linear phase Usually occurs just after the Ct.
- Plateau phase
  - This is commonly assumed to be due to the exhaustion of reactants
  - It is more likely to reflect competition at the annealing phase between primers and full-length amplicon strands binding to the complementary target

PCR Contamination

- Opening PCR reaction tubes and manipulating the content (to check on a gel and setup sequencing) risks contaminating subsequent sample preps and reactions with the PCR amplicons from the earlier reaction
- PCR amplicons amplify very well compared to whole-genome DNA or cDNA
- The mechanisms by which contamination occurs are not all fully accounted
- Aerosolization is often invoked
- Contamination of pipette tips was likely a major source earlier; this has been markedly reduced by the introduction of filter pipette tips

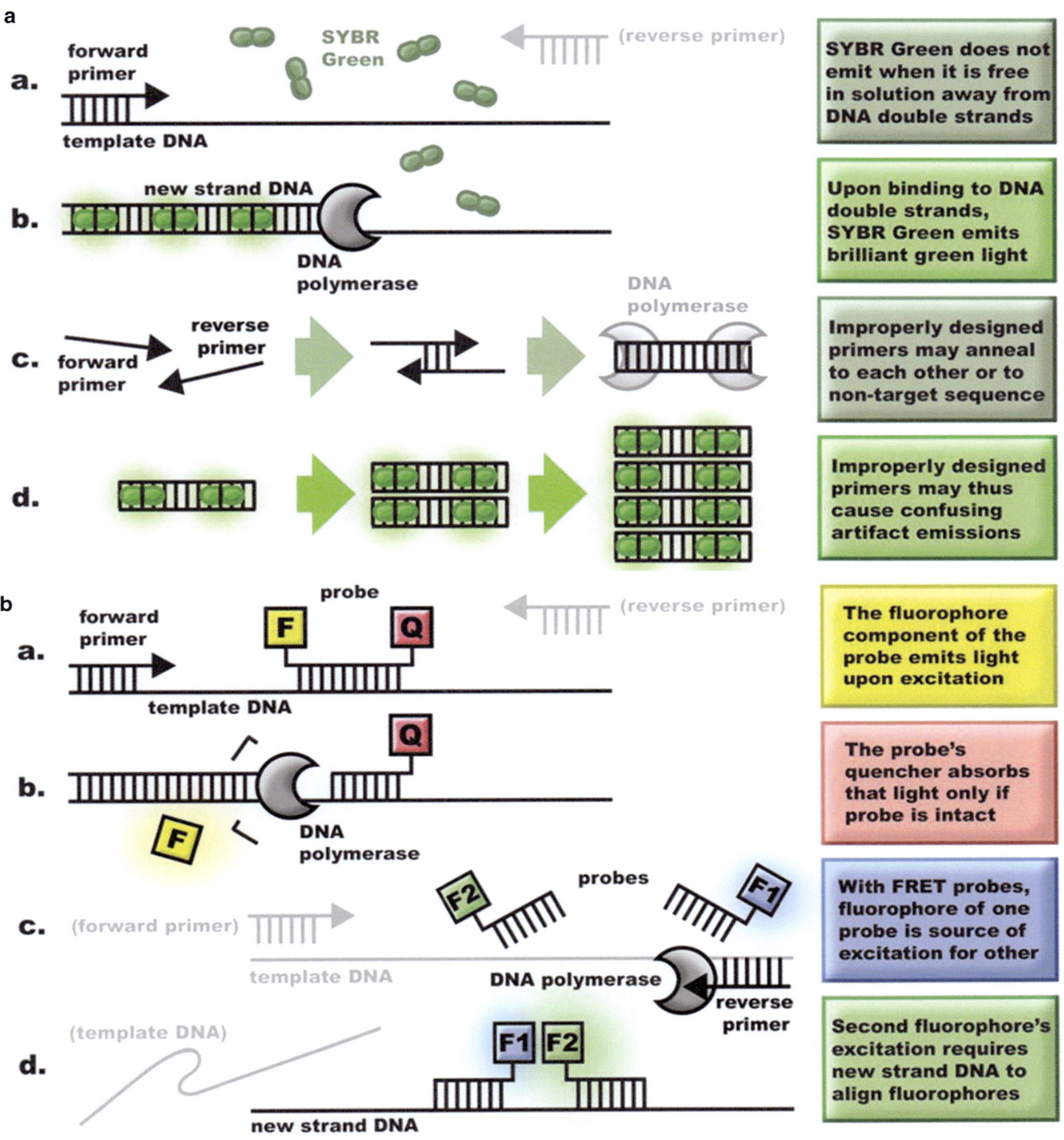

Fig. 1.12 Real-time PCR formats. (**a**) Dye-binding (SYBR green) format. (**b**) Top panels (*a,b*) Hydrolysis probe format; bottom panels (*c,d*). Hybridization probe format

- Real-time PCR, by avoiding the need to open the reaction tube and manipulate the reaction, dramatically reduces contamination
- The advent of NGS, which often employs one or more rounds of PCR, the products of which are often extensively manipulated, would seem to increase the risk
- Good Lab Practice
  - Having separate spaces for each major step of PCR is the ideal
    - Sample extraction/PCR setup/PCR analysis
  - If separate rooms are not practical, then laminar flow hoods or dead air boxes can be used successfully

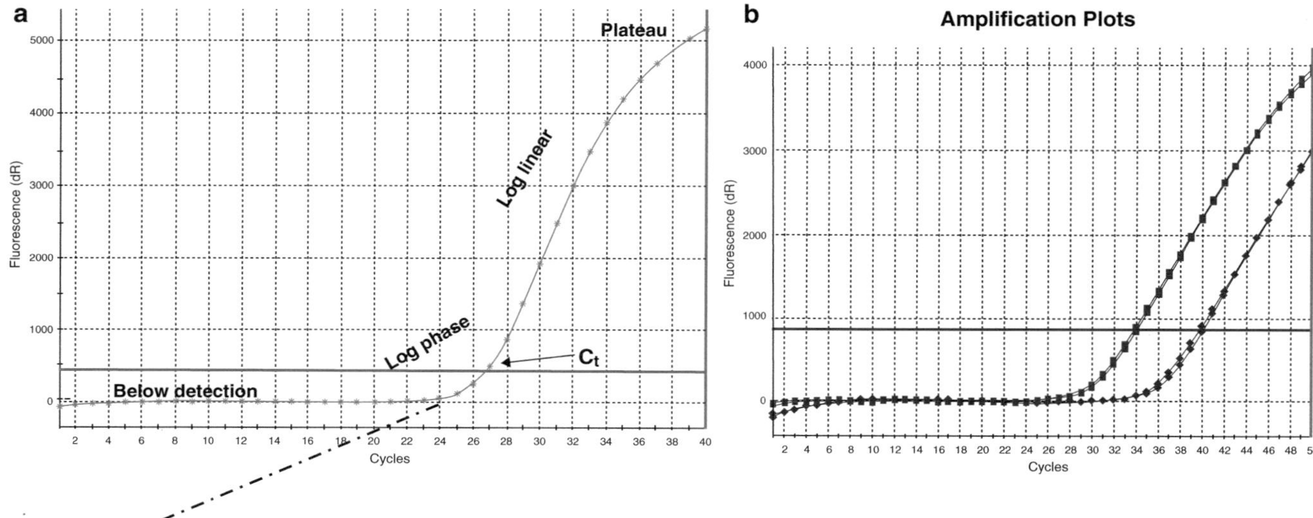

Fig. 1.13 Typical real-time PCR Amplification Curve Annotated. (**a**) The parts of the PCR amplification curve are annotated (described in text). (**b**) Shows amplification curves for two samples differing about 64-fold in starting concentration, reflected by the roughly 6 cycle difference in Ct for the curves ($2^6 = 64$, assuming 100% efficiency in each cycle)

- Unidirectional lab workflow
- Items like pipettes and microfuge tubes are never carried from the post-PCR analysis space to an earlier stage
- Disposable gowns are recommended, which are kept for use in specific rooms (areas)
- Use of Uracil N-glycosylase (UNG)
  - DNA polymerase can incorporate deoxyuridine triphosphate (dUTP), in place of dTTP
  - Substitute dUTP for a percent of dTTP in setting up the PCR
  - The master mix also includes UNG
  - No change in thermocycling conditions is required
  - The resulting amplicons will each have a few dUTP incorporated
  - When the next assay is setup, it is incubated for several minutes to give UNG time to cleave any dUTP in preexisting amplicons from earlier reactions, which might have contaminated the new samples or the reagents
- Additional comments on PCR
  - Long-range PCR
    - PCR amplification of amplicons up to 50,000 base pairs is feasible
    - This requires specialized polymerases
    - This requires considerable optimization effort
  - Two-temperature cycles
    - Using a single intermediate temperature for both annealing and for extension works well in practice
  - Temperature Ramping

  - The speed with which a thermocycler heats and cools between phases can be set, within limits, on some instruments
- Setting up PCR for multiple targets in a single run on a single instrument is difficult if each reaction has different requirements
  - Thermocyclers with independently programmable wells are available
- Slow ramp times give primers more opportunity to bind to imperfect matching sequences and then be extended
- PCR introduced errors
  - The typical Taq polymerase enzyme introduces random errors when copying
  - Because it is random, this error goes undetected with most applications using the PCR amplicon pool, such as Sanger sequencing or real-time PCR as readout
  - If the amplicon is used for NGS, the erroneous base could be read
  - One or more stages using PCR amplification are essential steps in all second-generation NGS protocols and in most for third generation NGS
  - High-fidelity DNA polymerases are available
  - They are often not as robust as the less faithful enzymes and more expensive
  - Reverse transcriptase is essential to the analysis of RNA by PCR and sequencing, has an error rate higher than that of the DNA polymerases but has received much less attention

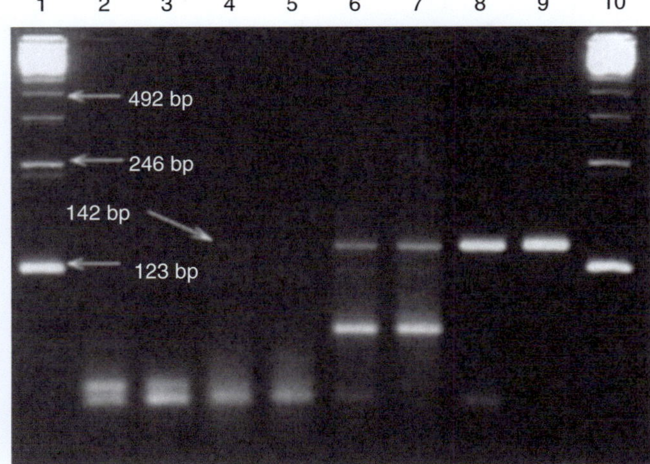

Fig. 1.14 Hot-start PCR agarose gel. Source Birch, D. E. Simplified hot-start PCR. *Nature* 381, 445–446 (1996). PCR amplifications of a 142bp product run in duplicate. Lanes 2, 3: 0 input copies, a Taq DNA polymerase, no preincubation, no hot start. Lanes 4, 5: 10 input copies, a Taq DNA polymerase, no preincubation, no hot start. Lanes 6, 7: 10 input copies, aTaq DNA polymerase, no preincubation, manual hot start. Lanes 8, 9: 10 input copies, a hotstart Taq, 9 min preincubation. Lanes 1, 10: 123bp DNA ladder

- Hot-start PCR (Fig. 1.14)
  o Primers can anneal to imperfectly matching complementary sequences
  o The lower the annealing temperature, the more extensively this happens
  o The most common wrong template is another primer (in either the same or the opposite direction) leading to 'primer-dimers'
  o Primers can also bind to imperfect matches in the genomic DNA, presumably binding to transiently single-stranded regions
  o As the temperature is raised to the extension temperature, even if this only takes a fraction of a minute, some of the imperfectly matched primers could be extended by a polymerase generating a single-stranded amplicon of the wrong target
  o If either primer binds to this errant strand in a later cycle, again because of imperfect annealing, and the polymerase extends it, the now duplex unintended amplicon has 5′ and 3′ sequences which are perfect matches for the primers and will amplify efficiently in all subsequent rounds
  o Minimizing the generation of undesired amplicons can be achieved by optimization of variables like primer sequence, annealing temperature, and buffer composition
  o Hot-start PCR refers to formats in which the initial PCR reaction cannot initiate until the DNA target is denatured

♦ After the initial denaturation, the thermocycler temperature does not drop below the annealing temperature until the reaction ends
♦ The easiest way to implement hot-start PCR is to use a modified enzyme
♦ Some polymerases are provided bound by an antibody, which causes thermally labile inhibition
♦ Some polymerases are inhibited by chemical modifications which are reversed at the denaturation temperature

Real-Time PCR Formats

- Dye-binding
  - The reaction mix contains a dye that binds duplex DNA
  - This is similar to the detection of DNA and RNA in an agarose gel
  - When excited with light of a specific wavelength, the dye molecules fluoresce, emitting light at a second wavelength, usually in the visible range, BUT ONLY IF the dye is bound to the duplex DNA
  - The more PCR product made, the more dye that binds, and the stronger the signal for that cycle
  - When the sample is heat denatured, the dye comes off, and the signal vanishes
- Hydrolysis probes
  - Often called 'TaqMan' probes
    o 'TaqMan' is a trademarked name for a commercial version
    o The name is derived from the source of the polymerase, *Thermus aquaticus*, and an early cartoon representation of the process
  - Principle of the hydrolysis probe real-time PCR format
    o The reaction includes the usual PCR elements (forward and reverse primers, polymerase, buffer) as well as a 'probe'
    o The probe is a DNA oligonucleotide, typically 20–25 nucleotides long, designed to hybridize to a sequence of interest between the two primers
    o At one end of the probe is a fluor, such as FITC
    o At the other end of the probe is a 'quencher molecule'; this molecule absorbs light at the wavelength emitted by the fluor
    o Quenching occurs by a quantum-mechanical process known as FRET (fluorescent energy transfer resonance)
    o In every PCR cycle, the entire reaction volume is irradiated during the extension phase, with UV light at wavelengths intended to excite whatever fluors are in the reaction

- o If the probe is intact, floating in the PCR reaction, the fluor absorbs the exciting light, transfers it to the quencher, and remains dark
- o In every cycle, an increasing number of probe molecules will hybridize into a strand of the PCR amplicon
- o As the polymerase extends the primer along the DNA template, it will encounter a probe molecule 'sitting on the track'
  - ♦ Assuming the correct polymerase is in use (not all polymerases have their exonuclease capability intact), the polymerase will clear the track, degrading the probe as it extends the growing DNA strand from the primer
  - ♦ The nucleotide with the fluor will now float freely in suspension, no longer in proximity to the nucleotide with a quenching moiety
  - ♦ In the next thermal cycle, when the UV light excites the reaction volume, the freed labeled nucleotide will fluoresce detectably
  - ♦ The more PCR amplicons generated, the more probes are degraded, and the greater the fluorescent signal in each successive cycle
  - ♦ Once released into solution, the fluorescent signal does not decrease
- Principle of hybridization probes
  - Often known as LightCycler probes
  - The reaction includes the usual PCR elements (forward and reverse primers, polymerase, buffer) as well as <u>TWO</u> 'probes'
  - Both probes are DNA oligonucleotides designed to bind immediately adjacent regions in the target
  - One probe has a 3′ nucleotide labeled with a fluor, which is excited by light at wavelength λ_1 and emits at λ_2
  - The other probe has a different fluor on its 5′ nucleotide
  - o This fluor is excited by light of wavelength λ_2 from the first probe)
  - o This fluor emits at wavelength λ_3
  - o The energy of the fluorescent excitation is transferred from one primer to the other, again by FRET, but instead of 'quenching' the energy by absorption, the energy is radiated at a third wavelength
  - o The real-time instrument is designed to <u>only</u> detect the third wavelength; therefore, the first probe will NOT be detected at any time even though it will be fluorescing when irradiated
  - o The second probe will only be excited AND detected when it is bound to the template, in proximity to the first probe

- – Melting curves
  - o In a positive sample, at the end of a completed PCR assay, there will be duplex PCR amplicons
  - o The final reaction is cooled, and the temperature then **slowly** raised to the denaturation temperature
    - ♦ This works for dye-binding formats and hybridization probe formats
    - ♦ The hydrolysis probe format does NOT show any melting curve
    - ♦ The light signal should show a sigmoidal curve (Fig. 1.15a)
  - o The amplicon will have a characteristic T_m, the temperature at which the melting curve shows the maximum rate of decrease
    - ♦ This point is where the first derivative of the melt curve reaches a maximum
    - ♦ Graphing the first derivative of fluorescence with respect to temperature, dF/dT gives a curve such

a

Fluorescence

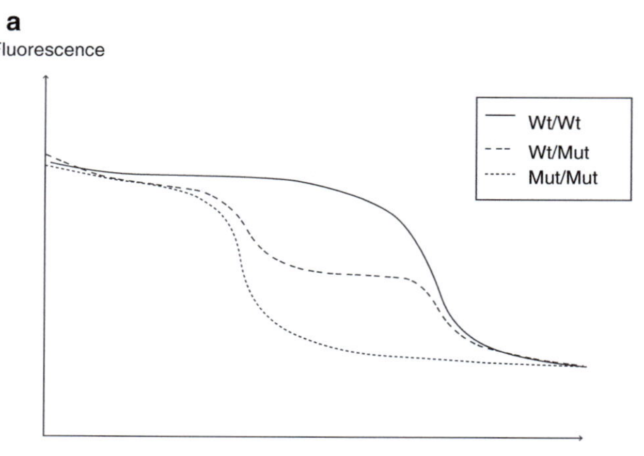

b

-dF/dT

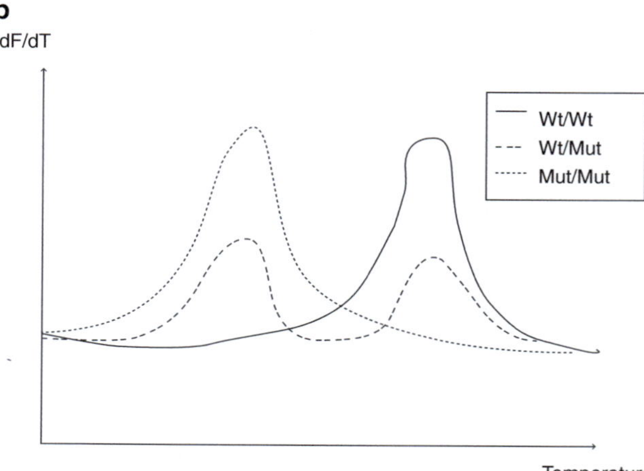

Fig. 1.15 Melt curves for real-time PCR. (**a**) Dissociation of ds DNA, (**b**) Negative first derivative of melting curve

as seen in (Fig. 1.15b); the maximum rate of change of dF/dT is much easier to spot, and the corresponding temperature discovered, than in the original melting curve

- For a target with a known Tm, this provides a cross-check that the amplicon generated is the intended one
  - If the melt curve shows two or more inflection points, it means that there are two or more amplicons present
  - The amplicons usually represent variants of the intended target with sequence variants that do not interfere with PCR amplification but which do affect duplex stability (dye binding) or probe binding
  - The longer the probe, the less sensitive it is to sequence variants in its target
  - Hybridization probe formats often use one long (anchor) probe and one short (sensor) probe; the latter intended to bind in the region where variants occur

- Data analysis for PCR
  - The C_t relates to the starting concentration of the target sequence
    - The higher the starting level, the earlier the curve rises (low C_t)
    - In the ideal PCR reaction, the amount of amplicon doubles in every cycle
    - Consider two samples that differ 10-fold in the level of gene X
      - The resulting PCR reactions should show a difference in Ct of 3.3 cycles
      - An easy way to understand this is that 3 cycles will lead to an eightfold (2^3) increase in amplicons, 4 cycles will lead to a 16-fold (2^4) increase in amplicons, 3.3 cycles ($\log_2 10$) seem about right for a 10-fold increase
      - This is strictly an arithmetic notion for the purpose of data analysis
  - Using C_t permits relative quantitation among samples
    - Even slight differences in the starting material, such as the quality of the DNA (or RNA), lead to large differences in the signal because of the many cycles of amplification
    - Gene expression itself, even with ideal controls, can vary widely
    - To deal with this variation, relative quantitation typically uses two PCR reactions, one for the gene of interest, one for a 'housekeeping' gene, then taking the ratio

- 'Housekeeping genes' are genes asserted to be expressed at a uniform level in every cell
- This is nonsense
- It is, however, a community custom
- The (relative) quantity of the gene of interest in a sample is normalized by the (relative) quantity of the housekeeping gene
- Which housekeeping gene to use is the subject of intense debate
- Vandesompele recommends using the geometric mean of ten housekeeping genes
  - With gene-expression microarrays or RNA-seq (NGS), it is much easier to use multiple control genes

- Absolute quantitation requires the use of calibration curves with defined controls
  - This still leaves open the issue of how to normalize
    - Housekeeping gene
    - Input cell number (hard to measure, especially from FFPE)
    - Input total RNA

SNV Detection by PCR

- Allele-specific primer extension (ASPE)
  - The requirement that the several nucleotides at the 3′ end match perfectly with the target so the polymerase can function can be used to design PCR assays to detect single nucleotide variants
  - In the simplest format, this requires two PCR reactions
  - For example, if you want to determine which allele is present for a biallelic SNP (T/G)
    - Setup one reaction with a primer that is complementary to the sequence right up to the site of the SNP and ends in 'A'
    - Setup a second reaction that uses an identical primer except that it ends in 'C'
    - Both PCR reactions use a common primer in the opposite direction
    - If the sample is homozygous for one base, only one reaction will give a product
    - If the sample is heterozygous, both reactions will yield products
- SNV detection by dye-binding real-time PCR
  - Achieved by melting curve analysis (see above)
- SNV detection by hybridization probe real-time PCR
  - This is also achieved by melting curve analysis
  - When several different mutations can occur in the region of the sensor probe, the Tm can often be used to distinguish them

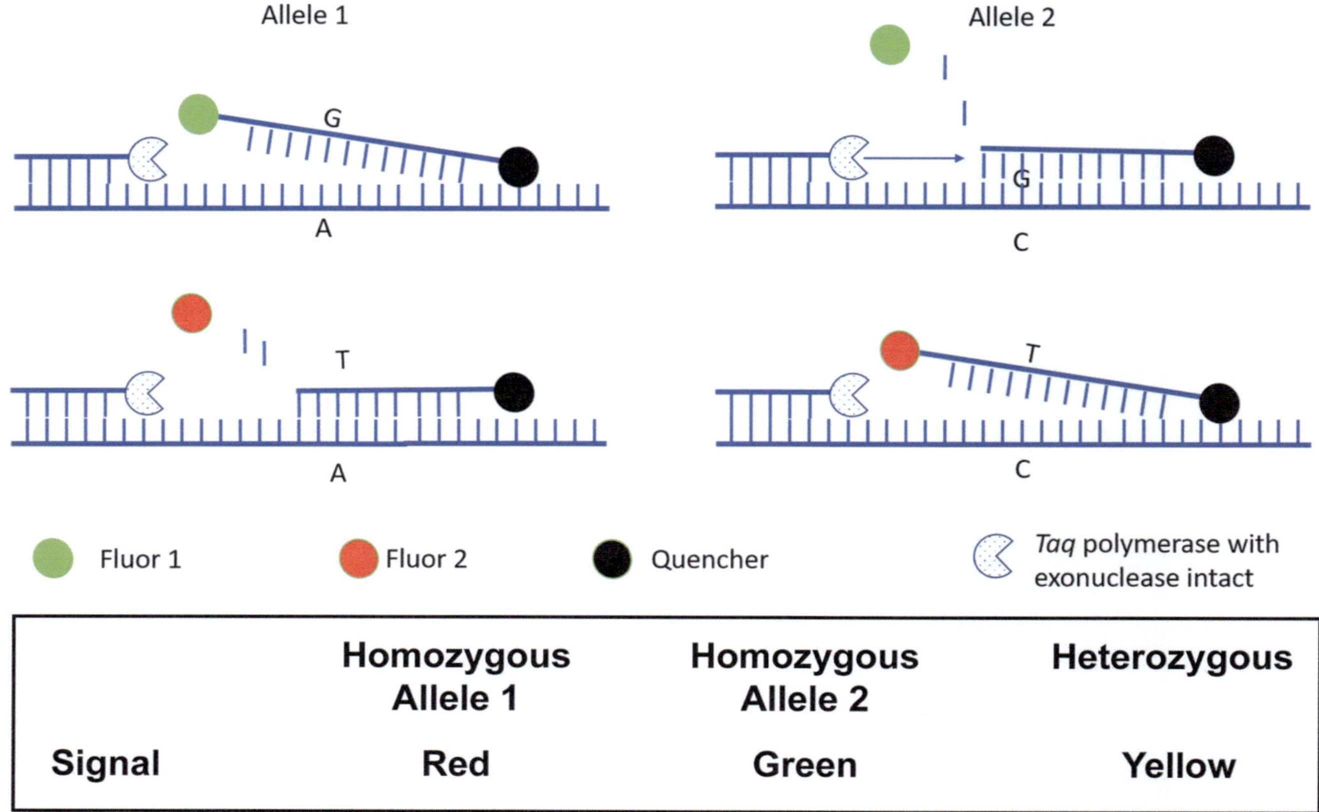

Fig. 1.16 Real-time hydrolysis probes for SNP genotyping

- SNV detection by hydrolysis probe real-time PCR (Fig. 1.16)
  - This is achieved by using two probes in the same PCR reaction
  - One is a perfect match for the reference sequence but has a single nucleotide mismatch with the variant of interest
  - One probe is a perfect match for the variant but has a single nucleotide mismatch with the variant sequence
  - The two probes have different fluors
  - If the probe is short, it will not bind stably with the mismatched target long enough for the polymerase to hydrolyze off the fluor
  - If the sample is heterozygous, both probes are hydrolyzed
  - If one variant is a mutation, this assay format can often detect it at the 1–5% variant allele frequency level

Methods: Digital Droplet PCR (ddPCR)

- It can be somewhat more sensitive than real-time PCR but it allows absolute quantitation, greater precision, and lower background without the need for calibration but all with lower throughput
- Outline of procedure (Fig. 1.17)

  - For genomic DNA, the sample is diluted and distributed into several thousand droplets such that each droplet is likely to receive at most one haploid genome equivalent of DNA, hence at most one copy of the gene of interest
    - Many droplets will not contain either allele of the gene of interest
    - Few, if any, droplets should contain two copies of the gene of interest
    - The droplets contain all the reagents needed for PCR, in effect making each droplet a PCR chamber
  - This PCR is endpoint, so a hydrolysis probe format is suitable, with two probes with distinct fluor labels, one for each variant (for example, the reference nucleotide and a possible mutation)
    - If the target is a variant that has no 'normal' variant, like the *BCR–ABL* translocation, one could include another primer—to amplify a control mRNA, as a control to ensure the reaction worked
  - At the end of the PCR assay, the droplets are streamed past a detector
    - Each droplet should show one of the two colors or no color (i.e., a copy of the gene was not distributed into that droplet)

Fig. 1.17 ddPCR Schema

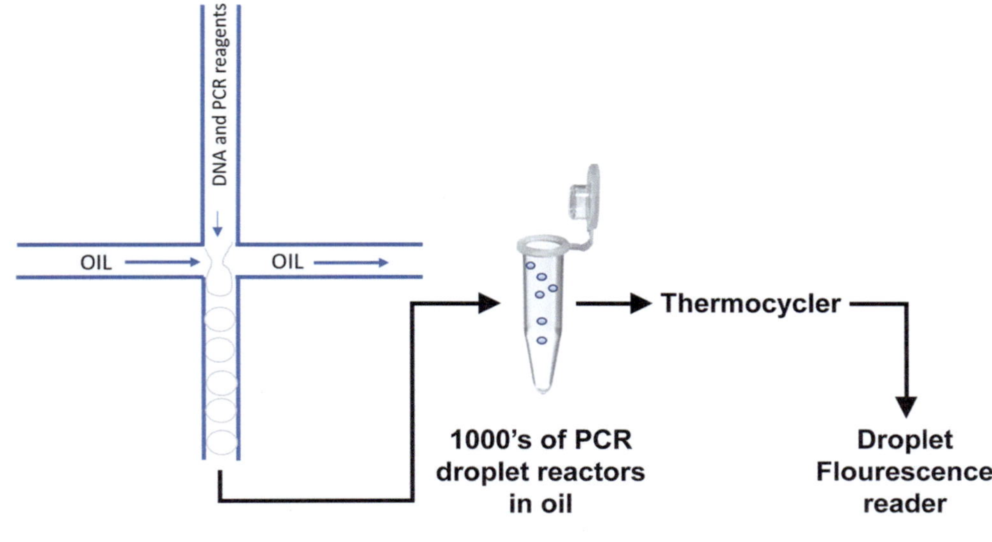

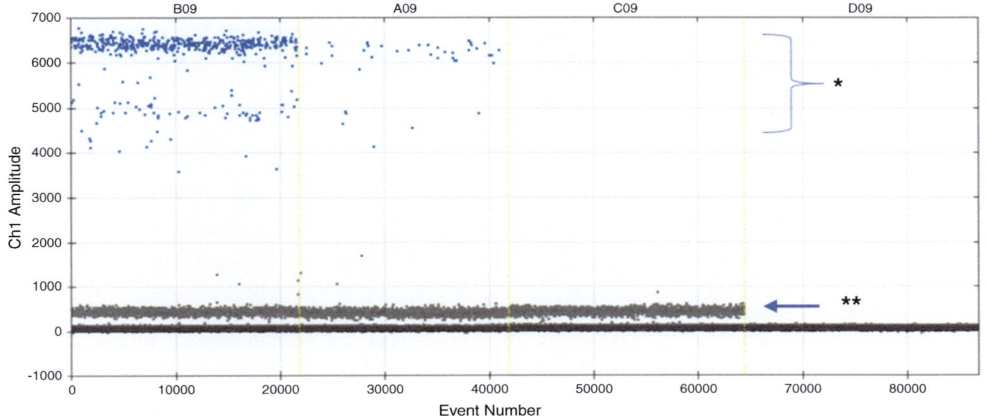

***. Denotes droplets which are positive for the L858R mutation**

**** Denotes droplets which show a very weak signal for the fluor**

Fig. 1.18 ddPCR for EGFR L858R mut controls. Source: Dr. Lynette M. Sholl, Dep't of Pathology, Mass General—Brigham Hospital, Boston, MA. The y-axis shows the fluorescence intensity for the probe specific to the L858R mutation. The x-axis represents the number of the droplet as they go through the cytometer. Very faint vertical yellow lines separate the events for three different samples. Roughly the first 20,000 events are for a strong positive L858R control. The next ~20,000 events are for the low-positive L858R control. The last ~20,000 events are for a negative control (almost no positive droplets)

- o Calculations use the Poisson distribution, the fraction of positive drops, and the dilution of the sample to derive the frequency of the variant in the starting material
- o The results can be displayed in a 'raindrop' plot (Fig. 1.18) or a two-color plot (not shown)

Methods: Rolling Circle Amplification (RCA)

- DNA 'Padlock' probes can be used to form a circular template (Fig. 1.19)

- – The end sequences of the padlock, which can be some distance apart in the probe, are designed to hybridize into adjoining sequences in a single-stranded target
- – The two ends can be covalently bound by ligase, provided they are a perfect match to the target
- – If the two ends can be ligated, this will form a closed circular template for the RCA
- – This can be used to distinguish single nucleotide variants
  - o In one reaction, a padlock is used for which the final nucleotide at one end is a complement for one SNP genotype

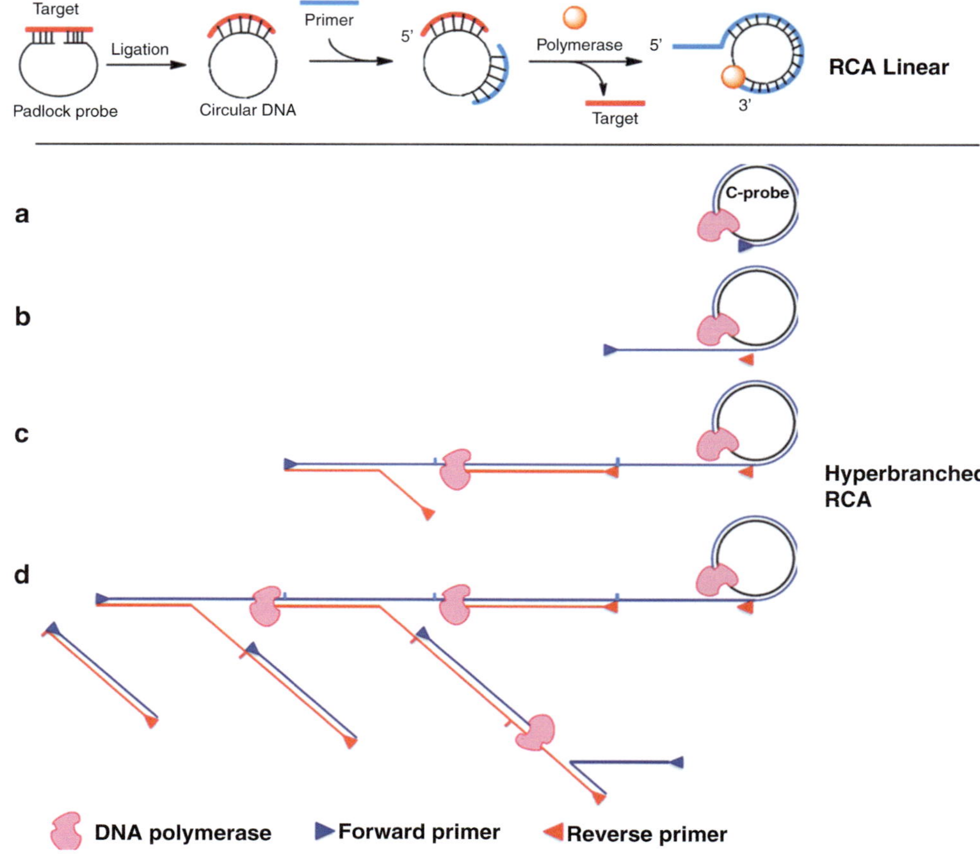

Fig. 1.19 Rolling circle amplification and padlock probes. Above the line: Padlock probes undergoing binding, ligation and rolling circle amplification. Below the line: (**a**) Forward primer binds to ligated circular probe and is extended by DNA polymerase. (**b**) Continuous amplification generates a long ssDNA, which contains binding sites for reverse primer. (**c**) Reverse primers bind to the long ssDNA and are extended by DNA polymerase. (**d**) The nascent ssDNA generated from reverse primer extension can also serve as templates for new DNA. Source: Yan, L. et al. Isothermal amplified detection of DNA and RNA. *Mol. Biosyst.* **10**, 970–1003 (2014) Figs. 13, 15. Open Source copyright

- o In a second reaction, a padlock probe is used in which the end of the probe is a complement to the other SNP genotype
- o A primer complementary to the padlock is hybridized to the padlocks, and the polymerase begins to extend it
- o If the padlock probe has been successfully ligated, the polymerase will make multiple circuits around the padlock, extending the new DNA strand
- phi29 polymerase is commonly used for RCA
- The newly synthesized strand can be used as a linearly amplified target for sequence analysis or to bind a reporter probe, specific for padlock probe of interest
- Inclusion of a reverse primer can generate a complex network of amplified strands with potentially exponential amplification
- DNA oligonucleotides can be bound to antibodies and used as reporters via the RCA
  - The number of spectrally distinct fluors which can be bound to antibodies is small
  - Using DNA oligonucleotide labels, antibodies can be more highly multiplexed,
    - o A pool of labeled antibodies is simultaneously bound to a target section

- o A circular template DNA, complementary to the oligonucleotide label for one of the antibodies, is added
- o The RCA generates a linearly amplified target which can be detected by a fluorescently labeled probe
- After imaging, the probe is stripped off
- Multiple subsequent sequential rounds of probe hybridization, RCA, and imaging can be performed, potentially with the same fluor, but coupled to a different probe each time

Methods: Hybridization Chain Reaction (HCR)

- This method is primarily applied to in situ detection of <u>single</u> molecules of mRNA (Fig. 1.20)
- Approximately 20–40 short (20 bp) oligonucleotides, designed to bind sequentially along a particular mRNA are hybridized to the tissue section
- Each of the oligonucleotides for a given gene, share an additional synthetic sequence, the 'initiator', which does NOT bind the target mRNA

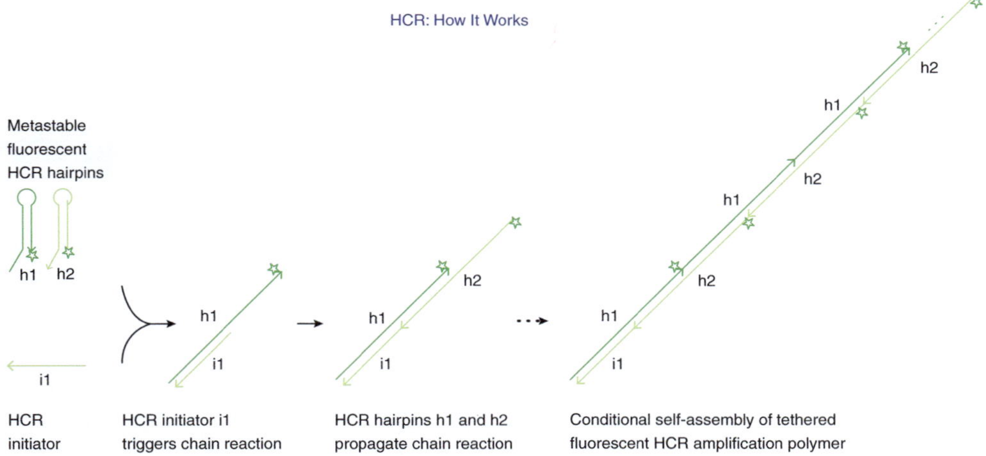

Fig. 1.20 Hybridization chain reaction. Source: Generously provided by Molecular Instruments, Inc.

- The reaction includes two short oligonucleotides, H1 and H2, with complementary but staggered sequences and a fluor
  - Each oligo has self-complementary ends so that each can form a hairpin structure at rest
  - H1 binds to the exposed sequence, the 'initiator', of oligo probes bound to the target
  - Binding opens the H1 hairpin, which can now bind to an H2 hairpin
  - The newly opened H2 oligo, which extends the complementary strand, can now bind another H1 oligo, and so forth
  - This generates linear amplification
  - HCR generates a signal sufficiently amplified that single molecules can be detected over autofluorescence

Methods: Branched DNA (bDNA)

- bDNA uses a short oligonucleotide probe to bind the target nucleic acid (DNA or RNA) in a signal-amplification format (Fig. 1.21)
  - bDNA assays are commercialized by Bayer for virus detection and quantitation in solution
  - bDNA can be used to detect sequences in tissue sections as described here
  - Its most common application at present is for FISH of RNA
- The oligonucleotide probe has an additional sequence at one end, which is NOT complementary to the target
- After the probe binds the target, the section can be rinsed to remove excess probe
- 'Preamplifier' molecules are added
  - One end of the preamplifier is complementary to the synthetic DNA tail on the probe

- The preamplifier is captured by hybridization to a bound bDNA probe
- Most of the preamplifier probably encodes a repeated motif
- Each 'preamplifier' is, in turn, bound by multiple oligonucleotide 'amplifiers'
  - Here the 'amplifier' is probably a nucleic acid polymer
    - A terminal sequence is complementary to a portion of the 'preamplifier' sequence
    - The amplifier is linked to multiple copies of an enzyme such as alkaline phosphatase, which can generate a signal with a suitable substrate (like in IHC)

Methods: Whole-Genome Amplification (WGA)

- This is a preparatory technique rather than a specific diagnostic detection technique
- It is important for genomic analysis of small quantities of DNA in general and indispensable for single-cell analysis of genomic DNA
- Early methods included PCR with degenerate oligonucleotide primers
- Process improvements included
  - The use of highly processive polymerases which can displace other newly synthesized strands
  - The use of transposases to both fragment and add transposon-tagged sequences to DNA (and cDNA)
- Problems with WGA persist
  - Incomplete coverage of the genome, which varies from cell to cell
  - Allelic dropout (missing variants)
  - Copy number variation inconsistency
  - Only optimized for intact DNA from fresh tissue

Fig. 1.21 Branched-chain DNA (RNA) Amplification (bDNA)

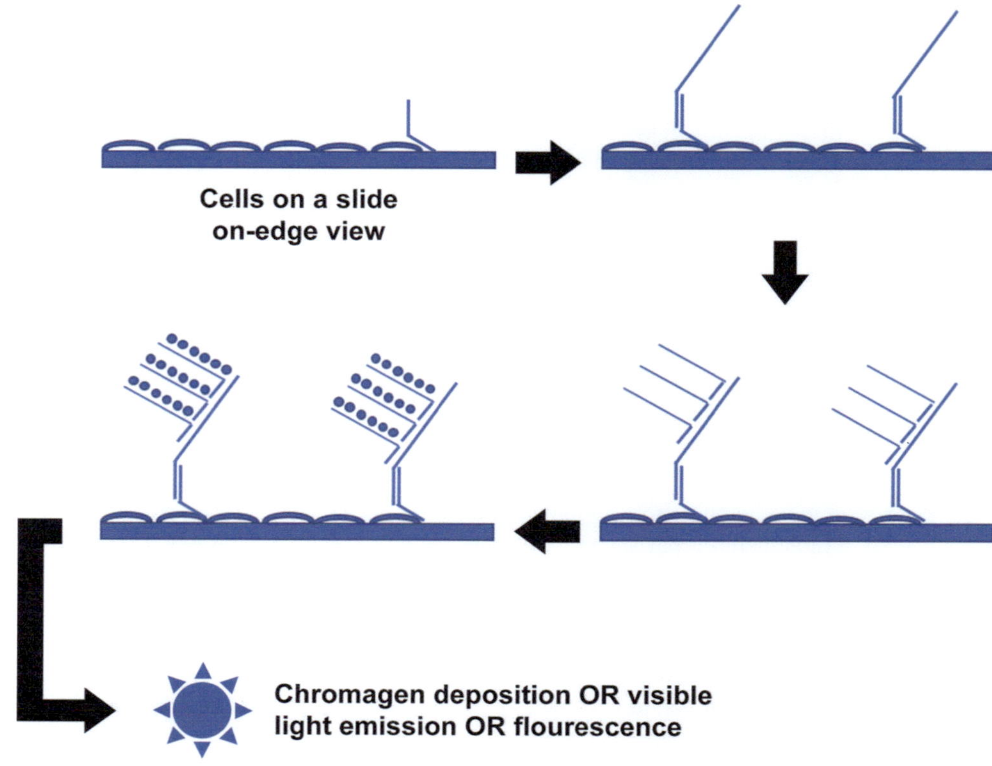

Sequencing

General Remarks

- When only a few mutations are of clinical concern, mutation-specific PCR assays could be appropriate
  - The MYD88 L265P mutation is important when considering the diagnosis of marginal zone lymphoma
  - PCR can be quantitative, which is helpful for monitoring response
    - A paradigmatic example is monitoring the level of the *BCR–ABL* transcript over six logs in chronic myelogenous leukemia
  - It is impractical to develop, validate and maintain 'off-the-shelf' assays except for the most common mutations
- Sequencing is essential for most tumors if molecular information is sought
  - Most oncogenes only have a few commonly mutated sites
  - Tumor suppressors, such as *TP53*, often lack mutation hotspots but have many low-frequency sites across hundreds or thousands of nucleotides
  - Sequencing of immunoglobulin and T-cell receptor genes and transcripts allows more precise and sensitive clonality analysis than does traditional amplicon sizing by PCR
  - The number of genes considered to have prognostic/predictive/value in any given tumor type is still growing
  - In many biopsies, the DNA yield limits the number of individual tests possible
    - Each mutation-specific reaction requires as much DNA as a single sequencing reaction if they are to have comparable sensitivity
    - Each Sanger sequencing reaction requires as much DNA as an NGS panel
  - Tumor mutation burden (TMB), which *might* prove to be an important prognostic/predictive factor, requires substantial sequencing (on the order of 500,000–1000,000 base pairs) for a reliable calculation
- Generations of sequencing technology
  - First generation is also known as Sanger sequencing
  - Second generation is NGS
    - If used without further qualification, at present, NGS refers to the second generation (see below for more detail)
  - Third generation: These are also referred to as NGS methods

- o They are distinguished by the ability to sequence individual molecules of DNA without preamplification
  - o They can also sequence amplicons
- Sanger sequencing is considered the 'gold standard' for accurate sequencing (although not infallible)
  - It has vastly lower throughput than second or third-generation NGS
  - An individual Sanger sequencing read can have a length of up to 1000 bp, which is 5- to 10-fold longer than the average second-generation NGS read
  - The analytical sensitivity of Sanger (20% VAF) is lower than that typical of second-generation NGS (1–5% VAF) or of real-time PCR (~1–5%) or ddPCR (0.1–1%) for point mutations
  - Even when sensitivity is not an issue (high percent of tumor content, or analyzing germline sequence), the cost of sequencing multiple exons of 2 or 3 genes is now more costly than sequencing with an NGS panel
- NGS (second and third generation) can sequence hundreds of genes, even entire genomes, with sensitivity similar to that of real-time PCR mutation assays while using as little DNA as that of a single Sanger sequencing reaction
- NGS (third generation)
  - One third-generation system (Nanopore) can sequence RNA directly (without a cDNA intermediate)
  - Third-generation formats (Nanopore, PacBio) can detect several DNA modifications such as methylation directly, whereas Sanger and second-generation NGS require bisulfite pretreatment of the DNA
  - The error rate of third-generation systems was initially significantly higher than for second-generation NGS and limited the sensitivity of detection to a variant allele fraction of at least 10–15%
    - o Most errors appear to be 'random'
    - o Modifications have enabled both commonly cited third-generation technologies to overcome the random error rate by increasing consensus reads
- Sequencing DNA versus RNA (cDNA)
  - Identification of mutations in genomic DNA has been, and for now, remains the focus of molecular genetic surgical pathology
  - The predominant sample type is FFPE tissue
    - o This is problematic for RNA analysis because RNA from FFPE tissue is degraded (fragmented) to a much shorter average size than the DNA
    - o Sequencing of RNA from FFPE is possible by second-generation NGS, but fresh tissue is far superior when available
  - RNA (cDNA) sequencing can identify sequence variants in coding regions

- o This presumes both alleles are expressed in the tissue sampled
- o This can provide insight into the effect of gene amplification if present
- RNA sequencing is often the best way to identify a fusion gene/transcript
  - o Breakpoints for translocations usually occur in introns
  - o Because the introns are spliced out of mRNA for fusion genes, the breakpoint sites can be quite variable but still generate the same mature fusion transcript
  - o Because the intronic (genomic) location of the breakpoint is variable, genomic sequencing requires sequencing much of the intron
    - ♦ This *can* be done by 'tiling'—'pulling down' DNA fragments spanning the intron and sequencing them (see below)
    - ♦ One of the DNA fragments sequenced will show the sequence expected for the genomic region targeted at one end and the sequence for the genomic region of the fusion partner at the other end
    - ♦ This is labor intensive
  - o To detect the fusion at the mRNA level, one can setup PCR with cDNA, a forward primer from a suitable exon of one gene, and a reverse primer from a suitable exon of the other gene
- If only one gene in a fusion is known (for example, from FISH cytogenetic analysis), it is possible to clone and sequence the unknown fusion partner

Methods: Sequencing/Sanger Sequencing

- Principle of Sanger sequencing (Fig. 1.22)
  - An oligonucleotide primer binds to a heat-denatured PCR amplicon and is extended by a DNA polymerase, incorporating nucleotides from a pool of dNTPs and fluorescently labeled dideoxynucleotides (ddNTPs), generating newly synthesized DNA chains of every length up to hundreds of base pairs
  - The new DNA molecules are sorted by size on capillary electrophoresis 'Sequencer'
  - The sequence of fluorescent nucleotides is read out as the new chains flow past a fluorescence detection window in the sequencer
- Outline of Sanger sequencing protocol
  - Duplex genomic DNA or cDNA (usually a PCR amplicon) is heat denatured in a PCR reaction mix containing
    - o An oligonucleotide primer complementary to one end

Fig. 1.22 Schema for Sanger sequencing

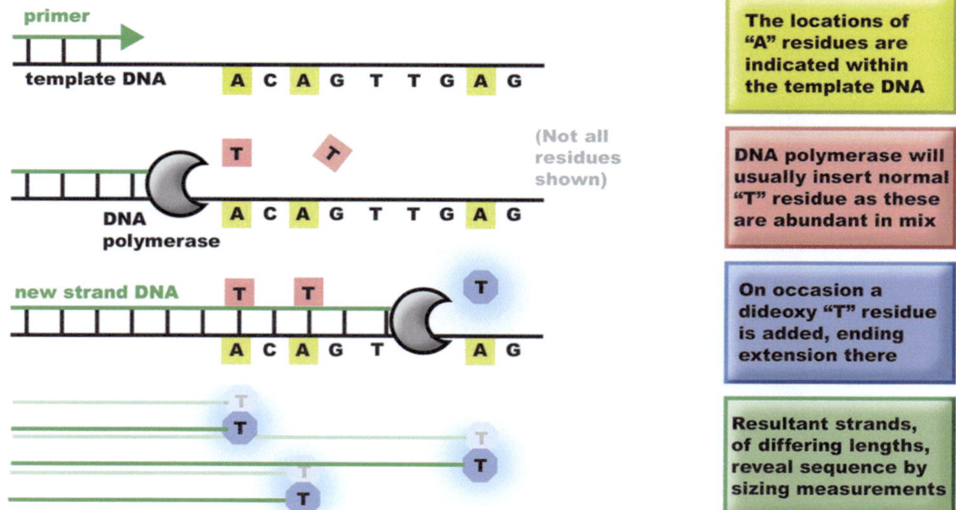

- All four dNTPs
- All four dideoxy dNTPs, each labeled with a different fluor
- Buffer
- DNA polymerase
- After the primer binds, it is extended by the DNA polymerase, incorporating nucleotides complementary to the template strand, until the first ddNTP is randomly incorporated
- When a dideoxynucleotide is incorporated, further extension stops
- In the modern format, cycle sequencing, the reaction is heated and cooled, as in PCR, generating more sequencing reaction products
- The dideoxynucleotides are randomly incorporated, so there will be newly synthesized chains at every length, from that of (primer length + 1) nucleotides to the full length of the amplicon
- Each class of dideoxynucleotide (ddATP, ddCTP, ddGTP, and ddTTP) carries a different color fluor
- Since only one ddNTP is incorporated into every synthesized DNA chain, every synthesized DNA chain has only one associated fluor
  - Suppose a sample is heterozygous, say at nucleotide number 150 in the amplicon, with either an 'A' or a 'C' (in one direction)
  - Some newly synthesized DNA strands will end with a fluorescent ddTTP, complementary to that 'A'
  - Some newly synthesized DNA strands will end with a fluorescent ddGTP, complementary to that 'C'
  - Every 150 nt long DNA molecule will end in either 'T' or 'G'

- The products of the sequencing reaction are denatured, then separated by size on a capillary electrophoresis analyzer (aka 'Sequencer')
  - A computer keeps track of the nucleotide 'colors' which pass the window, generating a DNA sequence text file
- Comments on Sanger sequencing
  - This is the archetypal example of 'sequencing by synthesis' (SBS)
  - It is customary to set up two separate sequencing reactions for each template, one with a forward primer, and one with a reverse primer, as mutual double checks
  - A single Sanger sequencing reaction, depending on reaction conditions and instruments, can generate sequences from 10 s of nucleotides in length, up to ~1000 nucleotides
    - If the amplicon is short enough, both the forward and reverse sequencing reactions will entirely cover it, and the two complementary sequences can be used to check each other
    - If one needs to sequence a larger amplicon, the amplicon can be aliquoted into multiple sequencing reactions, each with a different sequencing primer targeted along its length ('primer walking')
  - Capillary analyzers (see Fig. 1.10)
    - Sequencing reactions are typically loaded in 96-well microwell plates
    - Analyzers have arrays of capillaries (4–96)
    - Each capillary has a typical inner diameter of 0.5 mm and an associated electrode wire
    - The array is briefly dipped into the microwell plate, and the electrodes are given a negative charge, driv-

ing the negatively charged sequencing products into the capillary

- o A pump continuously fills the capillaries with a 'liquid' polyacrylamide 'POP'
- o All DNA strands, regardless of length, have roughly the same charge-to-mass ratio, so they experience the same force
  - ♦ The shorter strands move faster than the longer strands
  - ♦ All the strands of a given size, say 150 base pairs, move past the detection window nearly uniformly
    - If there is no variant in the PCR amplicon at nt 150, all the new strands, 150 base pairs in length, will end with the same fluorescent ddNTP
    - If a variant is present at position 150 in the amplicon (either a point substitution or an insertion/deletion), two colors will be detected at that position
- Analyzing the DNA sequence
  - A typical cited accuracy rate is 99.9% (one error out of 1000 nucleotides)
  - Each nucleotide call is assigned a 'quality score' (aka 'Phred') by the software used, indicating confidence in the nucleotide call
  - There are a variety of programs able to read the sequence generated and compare it to the expected sequence for the gene targeted
    - o Most sequencing software can identify point mutations well but vary in the ability to detect and/or analyze insertions deletions
    - o Review of the chromatogram by the molecular pathologist is important

- There are two abnormal chromatogram patterns to recognize
  - o Point substitutions (Fig. 1.23a)
    - ♦ Two colors will be displayed at the same nucleotide position
      - In inherited heterozygotes, where the two alleles are present in equal numbers, the two-colored peaks will be very roughly similar in fluorescent signal strength
      - With somatic mutations or inherited mosaicism, the peak heights very roughly correspond to the proportion of the variant represented
  - o Indels (Fig. 1.23b)
    - ♦ Before the polymerase reaches the site of the indel, the chromatogram shows a single color at each position
    - ♦ At the beginning of the indel, there will be two peaks with different colors, just like with a point substitution variant, BUT most of the subsequent nucleotide positions will also show two colors
      - The pattern looks like two sequence chromatograms are superimposed on each other (in fact, there are)
    - ♦ If the software does not analyze the indels, they can be worked out 'by hand' by marking off the peaks corresponding to the expected reference sequence on the chromatogram
      - The remaining peaks represent the sequence, either of the insertion or of the region after the deletion
      - After the polymerase has traversed the shorter of the two amplicons, the remaining sequence chromatogram will show only single peaks

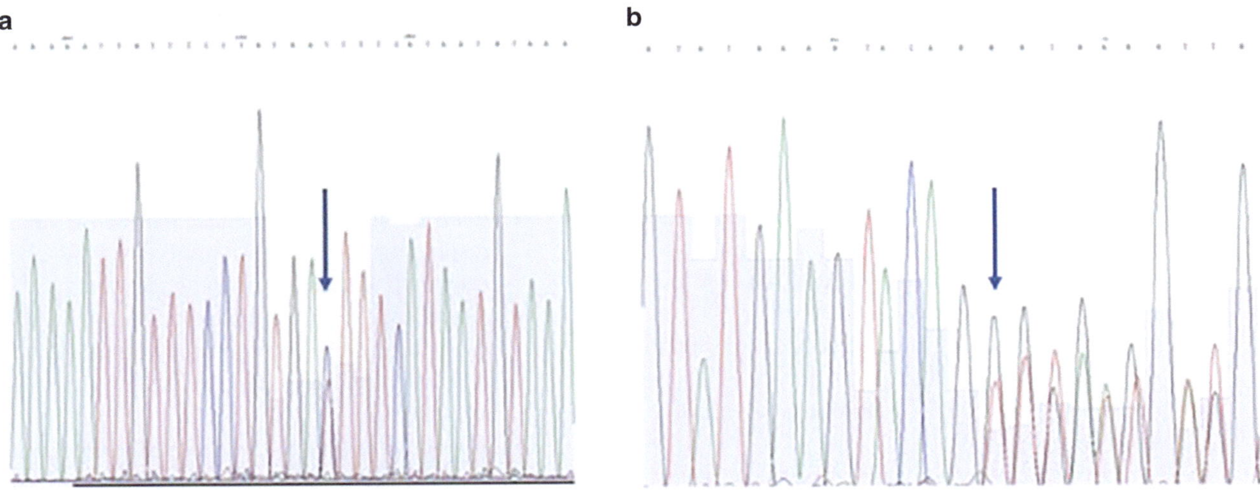

Fig. 1.23 Classic Sanger sequencing chromatogram patterns. (**a**) Arrow indicates the position of a heterozygous SNV. (**b**) Arrow indicates the position of the beginning of a deletion

- ✦ The number of single peaks at the end of the sequence chromatogram reveals the size of the indel
- Sanger sequencing (or at least the capillary analyzer) is still useful
  - For a small number of targets, if analytical sensitivity is not critical, Sanger is faster and less expensive than NGS
  - The capillary analyzer is also used for sizing DNA amplicons
    - ○ This is helpful in other molecular analyses such as identifying FLT3 internal tandem duplications and bone marrow engraftment assays
  - second generation NGS methods have reads typically under 300 bp in length
    - ○ This makes it hard to align repeat regions with the reference genome
    - ○ Longer Sanger reads can sometimes resolve this

Second-Generation Sequencing

- Principle
  - second generation NGS implements massively parallel (hundreds of thousands to millions), physically separate, sequencing reactions simultaneously in a small space (the size of a microscope slide), starting with a single DNA or cDNA fragment at each location and recorded in real-time
- Overview
  - There are two broad approaches to DNA (genomic and cDNA) preparation
    - ○ Fragment the DNA and modify the fragments for sequencing
      - ✦ Sequence all the fragments (whole-genome sequencing)
      - ✦ Alternatively, select fragments for genes of interest
        - ▪ Select thousands of fragments, even an entire exome
        - ▪ Select fewer fragments for panels for specific cancers
    - ○ Perform multiplex PCR on DNA directly to amplify the regions of interest
      - ✦ Can multiplex primers for several hundred targets in a single reaction
      - ✦ These pools can be combined after PCR amplification
      - ✦ Amplicons are then modified to enable sequencing
        - ▪ Barcoding adaptors can be added so different samples can be pooled for sequencing

- The library of DNA fragments or amplicons is diluted so that individual DNA molecules will occupy physically distinct spaces on the solid support used to capture and later sequence the DNA
  - ○ Another set of PCR cycles is used to amplify the captured fragments
  - ○ The resulting amplicons remain at the site of amplification, thereby increasing the strength of the eventual sequence signals
- Using variations on Sequencing by Synthesis, the amplicons are sequenced, with the results detected in real-time
- Outline of typical library construction procedure for fragmented DNA (Fig. 1.24)
  - Fragment genomic DNA
    - ○ There are several approaches
      - ✦ Shearing with ultrasonic vibration
        - ▪ The average size distribution can be controlled by parameters like frequency and pulse duration for ultrasonic shearing
      - ✦ Restriction enzymes
      - ✦ CRISPr-Cas9 digestion (used by PacBio for third-generation NGS)
        - ▪ Multiple gRNA pairs can target specific regions
        - ▪ Regions cut by CRISPr-Cas9 are suitable for adding adaptors
      - ✦ Transposases
        - ▪ These are retrovirus-derived enzymes that can cleave DNA and add on specific DNA sequences (which become adaptors)
  - DNA 'polishing' (of fragment ends)
    - ○ If the DNA preparation/selection method generates fragmented ends, these must be repaired ('polished') with several enzymes so the ends are suitable for the addition of adaptors
  - Adaptor addition
    - ○ Use ligation to add adaptors (partially duplex synthetic oligonucleotides) with multiple functional motifs
    - ○ The two ends of each fragment add somewhat different adaptors
      - ✦ Each adaptor has a sequence that will allow corresponding primers to amplify by PCR all the fragments in the library being prepared
      - ✦ A sequence motif complementary to the capture oligo for the chosen sequencing instrument
      - ✦ A motif that can bind the primer(s) used for sequencing
      - ✦ A barcode sequence to identify all the amplicons in the current reaction coming from the same sample ('sample index')

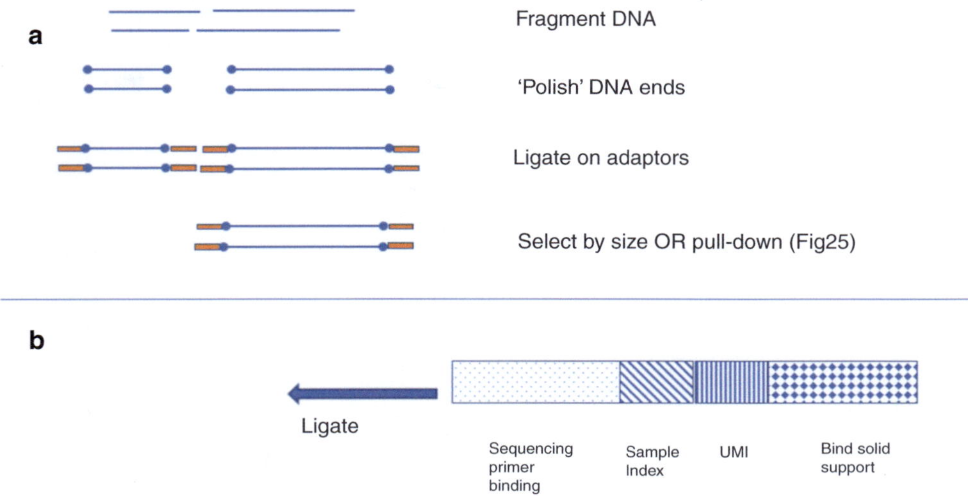

Fig. 1.24 (**a**) Steps in processing DNA to create a 'library' for 2nd generation NGS. (**b**) Simplified schematic of a multifunctional barcoded adaptor (or adaptor/primer) used for NGS library creation. *UMI* unique molecular identifier

- ♦ A random sequence barcode called a unique molecular identifier (UMI)
- ♦ Every DNA fragment (molecules) from a sample that adds an adaptor during library preparation gets an adaptor with the same sample ID barcode sequence AND a random (UMI) barcode unique to that molecule
  - ▪ Not all adaptors include UMI barcodes
  - ▪ UMI barcodes play a role in correcting the calculation of gene and transcript copy number
- – Size selection
  - ○ To optimize sequencing efficiency, the fragmented DNA can be further size selected by one of several methods, including gel electrophoresis
    - ♦ Size selection removes adaptor sequences
    - ♦ For short amplicon products it is often sufficient to purify the products using suitably derivatized paramagnetic beads
  - • The typical read length is 100–200 base pairs, so it is likely that such an amplicon could be read completely in both directions
    - ♦ It is likely that every region of interest will be multiply represented in the library, in both directions, so that reading an entire short amplicon in both directions is usually not necessary
  - ○ Fragments much longer than the read length, for example, 1000–2000 base pairs long, can also be sequenced for 150–200 bp from each end in the Illumina system
    - ♦ The sequences from the two ends will, during analysis, be aligned to a reference genome
    - ♦ If the two ends match different chromosomes or regions on the same chromosome farther apart than the expected size of the amplicon, it is evidence that a translocation has been detected

- ♦ PCR chimeras form frequently, so the detection of a fusion should be verified by an independent method
- – Selection of library components for sequencing
  - ○ For whole-genome sequencing, the library can be loaded onto a flow cell for sequencing without further selection
  - ○ Panels of targets can be pulled out of the library for sequencing (Fig. 1.25)
    - ♦ This can scale from a few exons to an entire exome
    - ♦ Fragments are hybridized with a mixture of RNA probes, each complementary to a region of interest
    - ♦ The probes are synthesized commercially
      - ▪ The user can specify the desired designs
      - ▪ Some panels are so common, such as the 'medical exome', that the entire set of RNA probes is available for order
      - ▪ Adding additional probes to a pre-existing pool does not require a redesign of the entire pool, but does require validation both that the added probe works and that it does not interfere with an existing probe
    - ♦ The RNA probes carry a biotin tag, which can be used with avidin beads to pull out the DNA-RNA hybrids
    - ♦ Because the genomic DNA fragments randomly, the fragments bound by copies of a given RNA probe will overlap but will usually have different ends
    - ♦ The RNA is destroyed after the pull-down
    - ♦ Pull-down probes are typically 70–80 nucleotides long, so binding is not prevented by the presence of a few sequence differences (e.g., SNVs) between the target sequence and the probe sequence

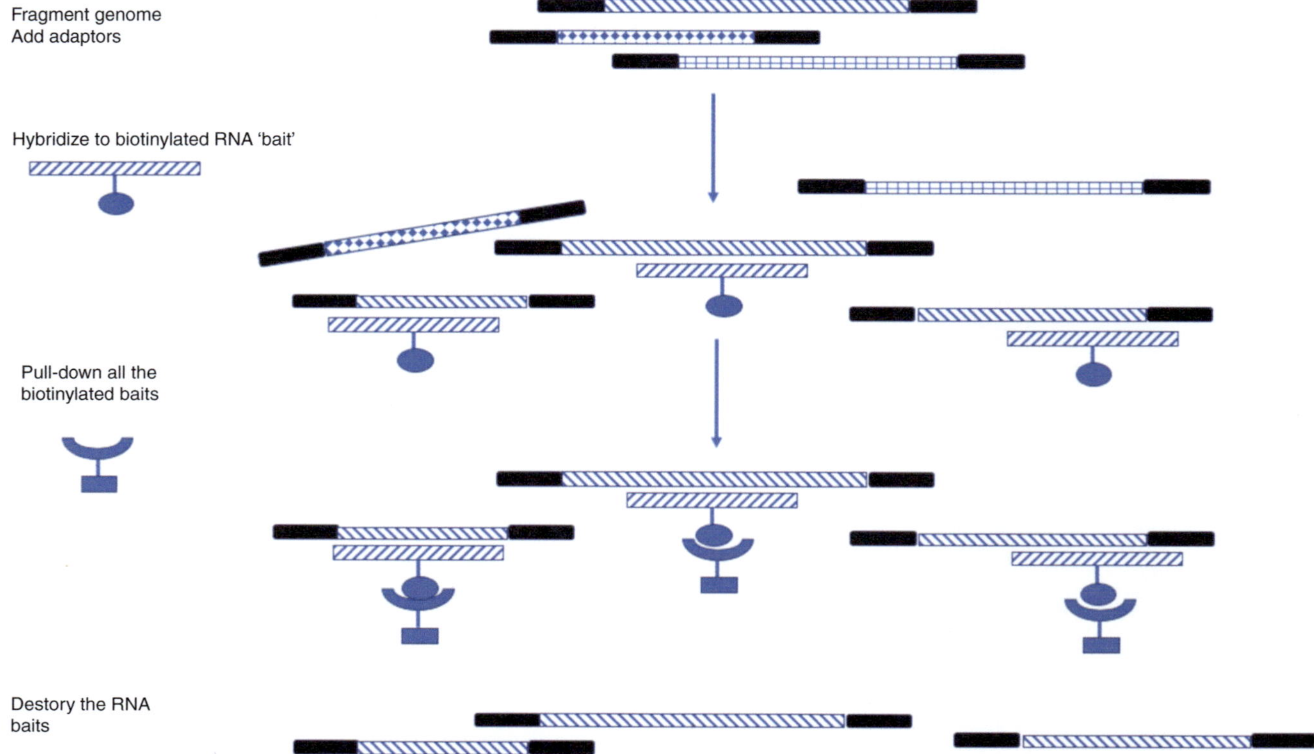

Fig. 1.25 Selection of library elements for regions of interest with pull-down baits

- ◆ The captured DNA, which has suitable motifs on each end because of library construction, can be loaded into the sequencing instrument
- Outline of typical library construction procedure for PCR reamplification
  - DNA for sequencing is generated by PCR with multiplex pools of up to 400 primer pairs directly on genomic DNA without prior fragmentation
    - ○ For genes of interest with closely related genes or pseudogenes, the primers can sometimes be designed to regions, including introns, which will allow discrimination among the genes
    - ○ The presence of an SNV (inherited or a somatic mutation) can prevent the binding of a primer, leading to the dropout of an allele
    - ○ With so many primers, it is not unusual to amplify unexpected regions of the genome by chance, but these can be filtered out bioinformatically
  - Addition of adaptors
    - ○ The gene-specific primers all have a common 5′ sequence motifs, which facilitates the addition of adaptor sequences by another round of PCR amplification with suitably structured primers
    - ○ Adaptors can include features described for adaptor ligation above
  - Size selection can be done by various methods, perhaps most easily by binding to coated paramagnetic beads

Second Generation Sequencing Platforms

- There are two systems in widespread use, and they use distinct sequencing methods
  - Illumina
  - Thermo-Fisher (aka Ion Torrent)
  - A third system, Nanoball, developed by Complete Genomics and now owned by MGI, is not covered further
    - ○ This system relies on rolling circle amplification (RCA), a method that also finds widespread application in in situ amplification for imaging (FISH) and in situ sequencing
- The second- and third-generation NGS systems continue to evolve, so the methods outlined here might well undergo further modifications, but the broad approach of each system will probably remain unchanged

Outline of a Representative Illumina Protocol

- The library is loaded onto a 'flow cell' (roughly the size of a microscope slide)
- The flow cell is uniformly covered by immobilized capture oligonucleotides
- Before loading, the library is diluted so that the average distance between molecules captured on a flow cell will be sufficiently great that the light signals from sequencing at each spot will not overlap
  - Cluster generation (Fig. 1.26a)
    - The captured DNA molecules undergo multiple rounds of PCR in situ
    - The process, 'bridge PCR', generates clusters of PCR amplicons that remain in the vicinity
    - In the first round, the capture oligonucleotide is extended, copying the hybridized DNA fragment
    - In the denaturation phase, the DNA fragment, which is not bound to the flow cell, can float away
    - The newly synthesized elongated strand, which is immobilized (through the capture oligo), can 'bend over' (forming a 'bridge') and hybridize to another capture oligo
    - The next PCR extension round synthesizes another copy of the target molecule, again tethered to the flow cell
    - During the next denaturation phase, the two 'full-length' single DNA strands, separate, but remain tethered to the flow cell
    - The PCR reaction is repeated multiple times, building up a 'cluster' of identical amplicons in the immediate region
    - After cluster generation but before sequencing, the 'reverse' strands are cleaved off by a reaction that appears to involve a dUTP in one adaptor
    - Cluster generation can be done onboard, in the sequencing instrument, or in a dedicated instrument, from which the flow cell is then transferred
  - Sequencing reaction (Fig. 1.26b)
    - The flow cell is flooded with the sequencing primer (common to all amplicons), polymerase, and four reversible terminator dNTPs (Fig. 1.26a)
      - There are several kinds of reversible terminators

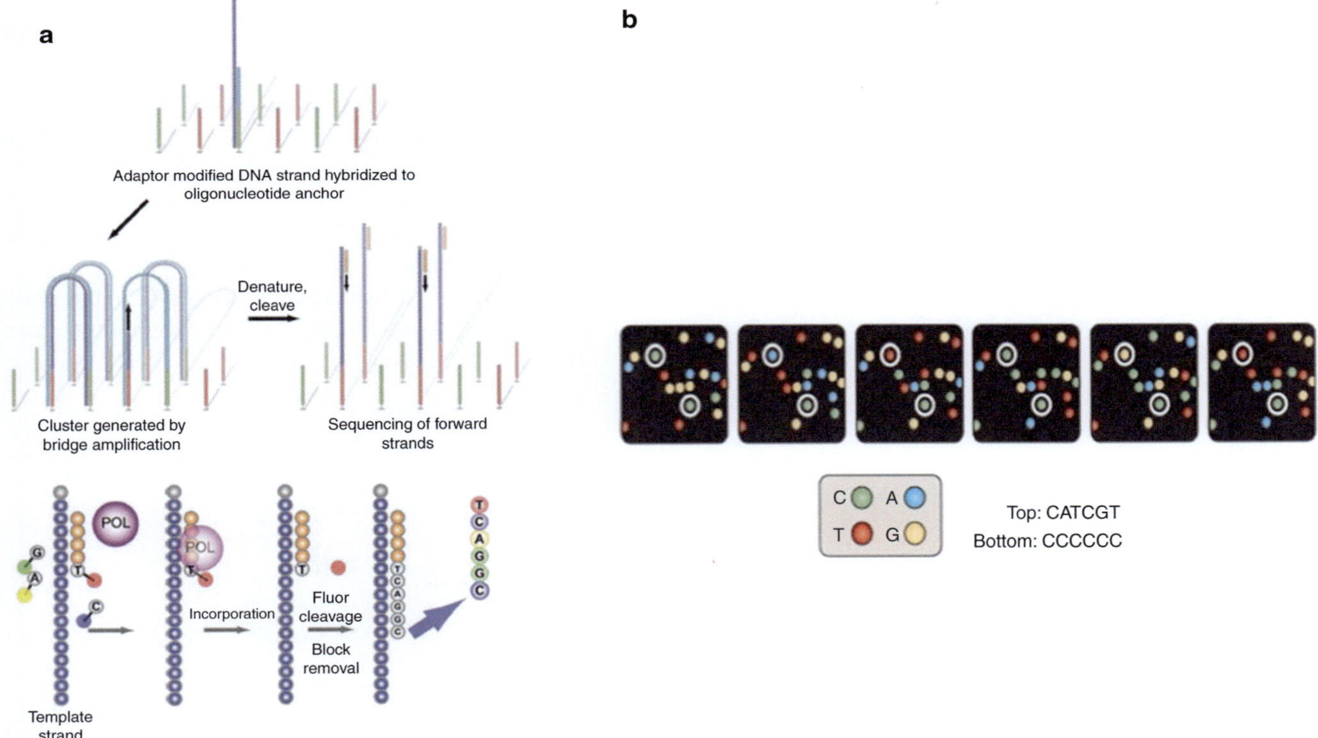

Fig. 1.26 Schema for Illumina NGS. (**a**) Cluster Generation and Sequencing. Source: Voelkerding, K. V., Dames, S. A. & Durtschi, J. D. Next-generation sequencing: from basic research to diagnostics. *Clin. Chem.* **55**, 641–658 (2009) Fig. 1.2. (**b**) Example of sequential Illumina NGS sequencing images. Source: Metzker, M. L. Sequencing technologies—the next generation. *Nat. Rev. Genet.* **11**, 31–46 (2009) from Fig. 1.2b

- • In the Illumina system, the reversal agent also cleaves the fluor which had just been incorporated, reducing the background for the next round
- o On each molecule, in each cluster, the polymerase can only extend the newly synthesized chain by a single base because the addition of a reversible terminator stops further extension
- o The slide is imaged
- o The slide is treated with a chemical that removes the 3′-O-azidomethyl group and regenerates the 3′-OH, for the next round of extension
- o The flow cell is again flooded with reagents, and an additional fluorescent reversible nucleotide is added to every growing chain
- o The cycle of extension and reversal is completed for a predetermined number of rounds
  - • Initially, most of the copies in a cluster add the same nucleotide, during a given round, generating a strong fluorescent signal
- o As more rounds proceed, not all molecules keep up
  - • This leads to 'dephasing', in which molecules in the same cluster can add different terminators, giving signals with two or more colors

Outline of a Representative Thermo-Fisher (Ion Torrent) Protocol

- • A water-in-oil emulsion is prepared
  - – It includes the following components
    - o The library of adaptor-modified PCR amplicons
    - o Microscopic solid beads (possibly dried agarose) coated with capture oligonucleotides
    - o The reagents for PCR (polymerase, unlabeled dNTPs, primers, and buffer)
    - o Mineral oil

- – The mixture is vigorously vortexed
- – Millions of vesicles of water are formed, all containing PCR reagents
- – Many but not all vesicles will also contain one bead and one amplicon
  - o Such a vesicle is, in effect, a microscopic PCR reaction chamber
- – Vesicles without a bead or without an amplicon will have no effect
- – Vesicles that contain a bead and more than one amplicon eventually give uninterpretable sequencing signals, such that results from that bead will be discarded
  - o The amount of input amplicon DNA is kept low to minimize this
- • PCR amplification of the bound library
  - – The emulsion is passed through an unusual capillary thermocycler, but the key point is to perform a PCR with uniform and rapid temperature regulation
  - – After the first denaturation, one of the amplicon strands hybridizes to a capture oligo on a bead
  - – The extension reaction copies the captured strand, creating a copy now covalently linked by the capture oligo to the bead
  - – Multiple PCR cycles lead to the decoration of the bead with additional copies of the captured strand
- • Sequencing
  - – The emulsion is 'cracked', releasing the beads
  - – The beads are loaded onto the sequencing 'chip'
  - – The chip is composed of 100,000 to millions of wells (depending on the system), each of which can accommodate at most one bead
    - o A modest number of wells are expected to be empty
    - o None of the wells can contain two beads
  - – Each well has, at its bottom, what is essentially a pH meter, wired to report the pH to the system's computer (Fig. 1.27)

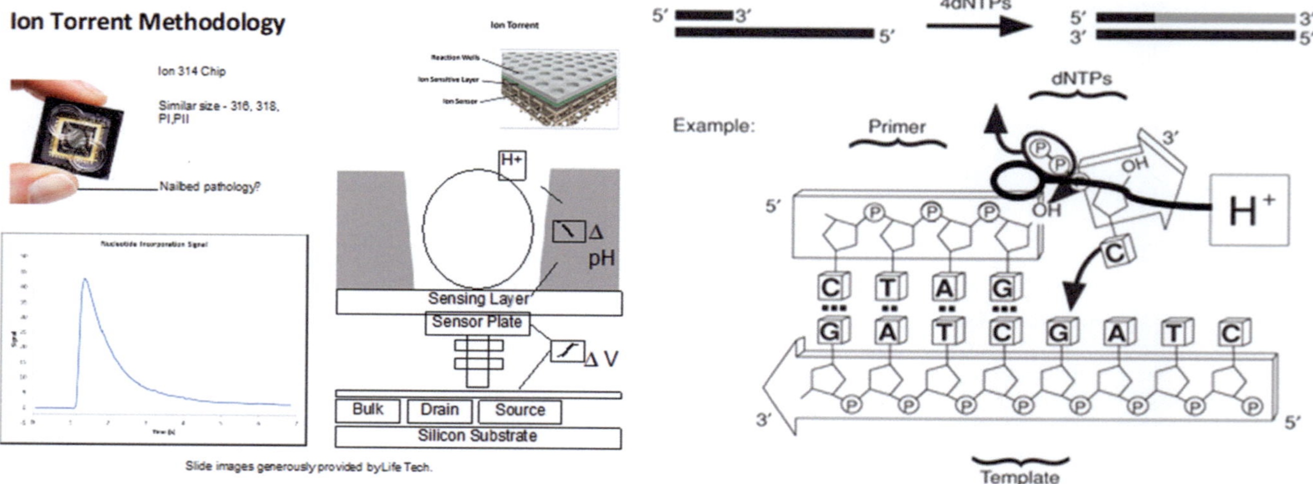

Fig. 1.27 Schema for Thermo Ion Torrent sequencing. Courtesy of Ion Torrent 2014 for a talk at an Ion Torrent seminar at U. Chicago

- The chip is flooded with buffer, a primer, DNA polymerase, and a single kind of dNTP (for now, assume it is dATP)
- If the next nucleotide on the strand being copied is a dT, then the DNA polymerase will extend the newly synthesized chain by adding a dATP
  - When the dATP is added to the new strand, a hydro-generation ion is released
  - Release of the hydrogeneration is registered by the pH meter
  - If the strand being copied has multiple consecutive dTs, then multiple dATPs will be incorporated, and the pH meter will register a higher signal
  - Depending on the sequence context, linearity of the signal is limited to perhaps 7–10 nucleotide homopolymers
  - Amplicons for which the next nucleotide to be copied is a dA, dC, or dG will not add any nucleotide
- After the addition of dATP is complete, the chip rinsed
- The chip is now flooded with PCR components and a different dNTP
- The chip is sequentially flooded with one dNTP at a time, followed by rinsing
- The cycle of adding dA/dC/dG/dT is continued for a predetermined number of cycles
- The computer keeps track of which nucleotide is added at every well in every cycle, building up the corresponding sequence
- A heat map from the author's first NGS assay indicates which wells had active sequencing (shown in orange) and which had none (Fig. 1.28)
- Data from that run—chip performance and sample variants found—are shown (Fig. 1.29)

Third-Generation Sequencing

- Like second-generation NGS, third-generation NGS implements massive parallel sequencing reactions in a small space
- There are, so far, two widely used third-generation NGS systems
  - PacBio instruments
  - Oxford Nanoporetech instruments
- Third-generation can analyze PCR amplified targets but can also, <u>unlike</u> second-generation NGS, sequence DNA directly without amplification
  - The Oxford Nanoporetech instruments can sequence fresh RNA directly (without conversion to cDNA)
  - Both systems can detect some modifications in fresh DNA, especially methylation of dC
    - Both first- and second-generation sequencing can detect methylated cytosine but require that the DNA substrate first be treated with bisulfite (see 'Sample Prep')
    - Without initial bisulfite conversion of methylated cytosines, the methylation signal is lost because of PCR amplification (which is insensitive to methylated dC)

PacBio

- The sequencing cells have up to 8,000,000 wells
- Each well has an optically clear base with a polymerase molecule affixed to it
- DNA preparation
  - Fragment into the range 250–50,000 bp
  - Ends are 'repaired'
  - Adaptors are added, which circularize the template DNA
  - The sample is denatured and diluted so that, at most, one DNA strand goes into each well
- Each well contains a primer and sequencing reagents including
  - All four fluorescently labeled nucleotides
  - No dideoxynucleotides or reversible terminators
- Once the enzyme binds the DNA-primer duplex, it starts to copy the DNA strand, incorporating the appropriate complementary nucleotide and cleaving off the fluorescent moiety, which now gives a signal detected through the base
  - Since there is only one template molecule per well, only one color should be detected at a time
  - The polymerase, immobilized on the base, cleaves the fluor near the base where it can be detected
  - After detection, the fluor drifts away
  - Because the template is circularized the enzyme can copy around the template multiple times (depending on the size and the time for sequencing)
    - The sequences from multiple reads around the template provide a consensus sequence
    - The consensus sequence markedly suppresses random polymerase errors
- An epifluorescence detector is constantly scanning the bottom of the chip, recording the incorporation of each dNTP (the color released) in each well

Oxford Nanoporetech

- The various instruments use flow cells with multiple-minute chambers
- Each chamber is divided into an upper and lower portion by a biological bilayer
- Each bilayer contains a single biological pore complex
  - There are several choices, such as alpha-hemolysin, for the pore
  - A voltage is applied to every chamber ('channel' per Oxford)

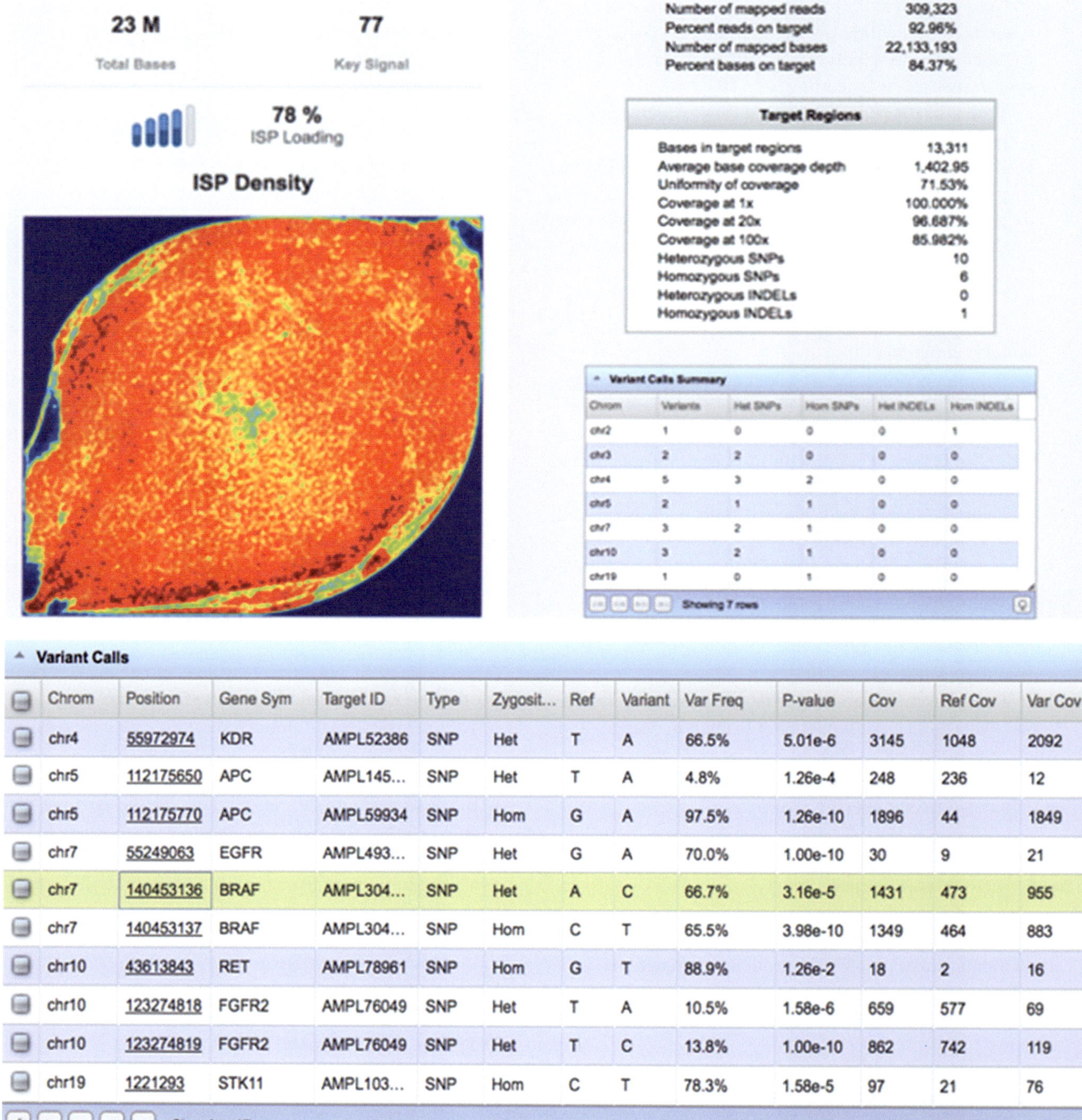

Number of mapped reads	309,323
Percent reads on target	92.96%
Number of mapped bases	22,133,193
Percent bases on target	84.37%

Target Regions

Bases in target regions	13,311
Average base coverage depth	1,402.95
Uniformity of coverage	71.53%
Coverage at 1x	100.000%
Coverage at 20x	96.687%
Coverage at 100x	85.982%
Heterozygous SNPs	10
Homozygous SNPs	6
Heterozygous INDELs	0
Homozygous INDELs	1

Variant Calls Summary

Chrom	Variants	Het SNPs	Hom SNPs	Het INDELs	Hom INDELs
chr2	1	0	0	0	1
chr3	2	2	0	0	0
chr4	5	3	2	0	0
chr5	2	1	1	0	0
chr7	3	2	1	0	0
chr10	3	2	1	0	0
chr19	1	0	1	0	0

Showing 7 rows

Variant Calls

Chrom	Position	Gene Sym	Target ID	Type	Zygosit...	Ref	Variant	Var Freq	P-value	Cov	Ref Cov	Var Cov
chr4	55972974	KDR	AMPL52386	SNP	Het	T	A	66.5%	5.01e-6	3145	1048	2092
chr5	112175650	APC	AMPL145...	SNP	Het	T	A	4.8%	1.26e-4	248	236	12
chr5	112175770	APC	AMPL59934	SNP	Hom	G	A	97.5%	1.26e-10	1896	44	1849
chr7	55249063	EGFR	AMPL493...	SNP	Het	G	A	70.0%	1.00e-10	30	9	21
chr7	140453136	BRAF	AMPL304...	SNP	Het	A	C	66.7%	3.16e-5	1431	473	955
chr7	140453137	BRAF	AMPL304...	SNP	Hom	C	T	65.5%	3.98e-10	1349	464	883
chr10	43613843	RET	AMPL78961	SNP	Hom	G	T	88.9%	1.26e-2	18	2	16
chr10	123274818	FGFR2	AMPL76049	SNP	Het	T	A	10.5%	1.58e-6	659	577	69
chr10	123274819	FGFR2	AMPL76049	SNP	Het	T	C	13.8%	1.00e-10	862	742	119
chr19	1221293	STK11	AMPL103...	SNP	Hom	C	T	78.3%	1.58e-5	97	21	76

Showing 17 rows

Fig. 1.28 Representative data from an Ion Torrent sequencing run

- Ions in the electrolyte solutions both above and below the membrane buffer the voltage drop except near the pore itself
- Ions flow through the pore, generating measurable current in each chamber
- When DNA goes through the pore, it reduces or completely blocks ion flow, as reflected by the change in current
- The pattern of change in current over time can be used to infer the sequence of DNA in the pore which caused the change in current
- DNA preparation
  - Double-stranded DNA needs to add adaptors with a 'leader sequence' and a motor protein
  - The 'leader' sequence helps the DNA fragment recognize the pore

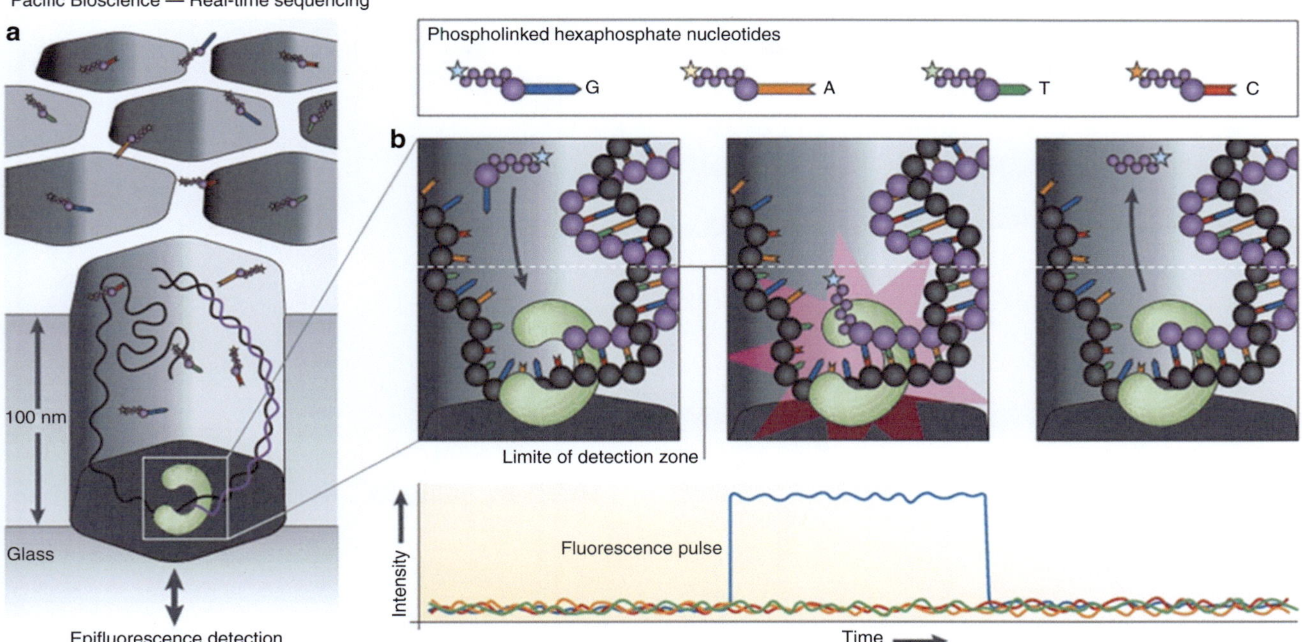

Fig. 1.29 Illustrative 3rd generation NGS - PacBio sequencing (**a**). Cutaway schematic of a single cell with a volume in femtoliters, showing a single polymerase immobilized at the base of the cell. (**b**). Top panel – Dye labelled hexaphosphate nucleotides. (There are no unlabelled nucleotides). Middle panel – Expanded view of the immobilized polymerase. The labelled nucleotide being added, is immobilized in the active site, long enough to dominate fluorescence detected in the base of the well. Bottom panel – Representative recording of fluorescence over time from a single well, showing addition of a single nucleotide. This is recorded individually from the bottom of every well. (Metzker, 2009)

- The motor protein, a helicase, unzips the double-stranded DNA so that one DNA strand can go through the pore
- How the sequence is determined
  - The pore length for one recent pore type accommodates about six nucleotides
  - A more recent pore accommodates about 12 nucleotides
  - The current reflects all the nucleotides in the pore at that moment, not a single nucleotide
  - The change in current reflects both the newest nucleotide to enter AND the most recent nucleotide to exit
  - This allows for quite a few sequence combinations which must be correlated with the changes in contour of the conductivity curve
    - Base calling uses recurrent neural nets or hidden Markov models
    - These are not *as* amenable to human review as a Sanger chromatogram
      - Although the Sanger chromatogram looks like 'real data', it is a graphic representation, with interpolation and smoothing, of a finite set of numbers, in which each data point consists of the time since injection, and the fluorescent intensity of each detector during that interval
- Operating characteristics

- Different flow cells range from 126 to 2675 channels per flow cell
  - Flow cells can be networked into larger platforms
  - The smallest flow cell, the Flongle, weighs 20 g (Fig. 1.30b, right panel)
  - The next smallest flow cell, the MinIon, with 10-fold greater capacity than the Flongle, weighs 87 g (Fig. 1.30a, left panel)
  - The flow cells are disposable
  - Lifetime is limited by damage to the pores, an irregular function of the amount of sequencing, and the time elapsed since initiating use
  - Sequencing speed is on the order of 450 bp/s
  - The longest reported sequence for a single read is 4 Mb
  - The current average (modal) accuracy claimed for single-molecule sequencing is 98%
- Additional features
  - Methylated cytosines, and other modified bases, can be recognized or suggested by distinctive changes in conductance
  - RNA can be sequenced directly
    - Full-length sequencing permits clear delineation of different isoforms
  - A '2D' method has been demonstrated which allows sequencing of both strands of a DNA duplex; however, it is currently deprecated by the company

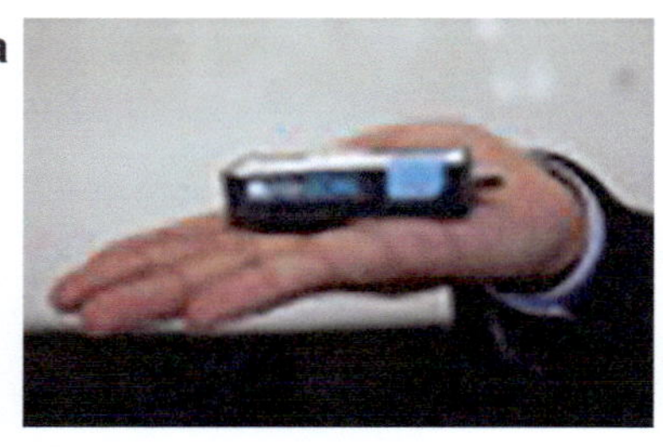

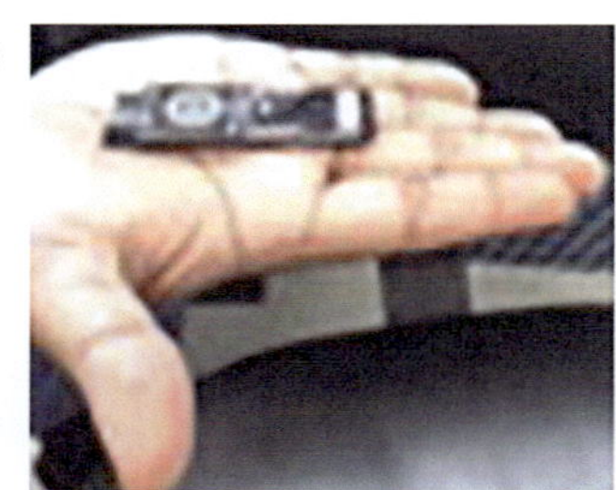

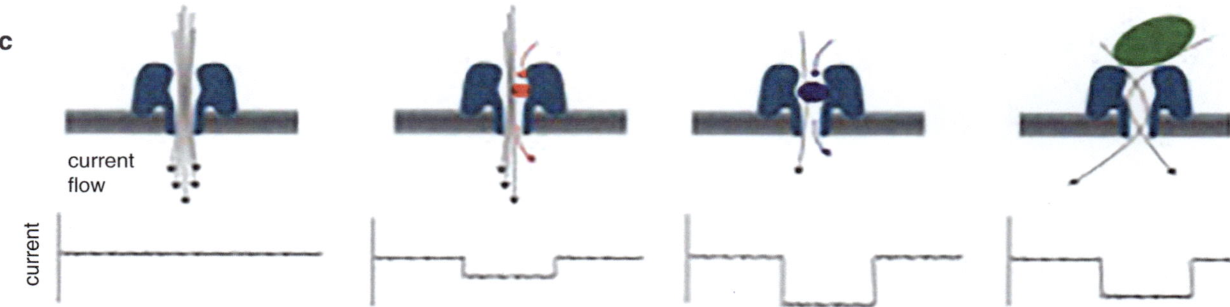

Fig. 1.30 Illustrative 3rd generation NGS - Nanopore sequencing. (**a**) Minion device, a complete sequencing instrument, with a USB port. (**b**) The 'Flongle', a lower capacity flow cell. With an adaptor it fits inside the Minion device. (**c**) Schematic of Nanopore sequencing. In each cell of the array a single pore-forming protein is embedded in a membrane. Ions in the buffer flow through the unobstructed pore generating a current measured in each cell. When a nucleotide enters the pore it restricts the ionic current flow. The degree and duration of current diminution is characteristic for different nucleotides, including modified nucleotides, and RNA

Other Applications of NGS

- RNA-Seq (transcriptomics)
  - This is the identification and enumeration of many or most of the RNAs expressed in a sample
  - Can sequence RNA of any class
    - A given protocol is not usually optimal for all classes of RNA
    - Most work is done on mRNA as a measure of gene expression
    - Sequencing miRNA is challenging because of the short size of the target
    - Sequencing lncRNA is challenging because the average sequence read in second-generation NGS is very short, and the lncRNA can be very long
  - Usually, libraries are designed to capture all the mRNA, creating a 'transcriptome', but one can design primers/probes to only capture a subset of genes
    - Most research is done with fresh tissue, but FFPE can be used, especially if a limited number of transcripts are targeted
    - RNA-seq has not been widely incorporated into molecular diagnostics, but several panels targeting a limited set of genes have found clinical application
- Methylomics
  - Sample aliquots with and without bisulfite treatment can be sequenced
  - Comparison of the methylome and the transcriptome from the same sample would facilitate analysis of the effect of methylation on gene transcription
- ATAC-seq (assay for transposase-accessible chromatin sequencing)
  - A hyperactive form of the Tn5 transposase inserts itself into open chromatin positions
  - NGS can identify the sites where the transposase is inserted
  - The process is also called 'Tagmentation'

Bioinformatics

- (Refer to Chap. 2 in the book)
- There are several broad steps in the bioinformatic analysis of NGS data
  - Calling the bases with Q scores (quality scores)
  - Aligning the bases to a reference sequence
  - Calling variants (relative to a reference sequence)
  - Assessing variant pathogenicity
- Breadth of coverage: Refers to the range of the genome/transcriptome targeted; for most purposes, it is used descriptively, but for several applications, it is important
  - Calculation of gene copy number
  - Calculation of tumor mutation burden

- Depth of coverage: Reflects how many independent reads encompass each nucleotide in the targeted region
  - The more reads, the lower the percent of variant which can be detected (greater analytic sensitivity)

Single-Cell 'Omics'

General Remarks

- 'Single-cell analysis' is not new to surgical pathology
  - Histomorphological analysis routinely integrates the appearance of individual cells with the context of the surrounding cells
  - Cytogenetic testing using karyotyping or FISH are intrinsically single-cell assays
    - FISH is limited to results for the probes used
    - Karyotyping is not limited to a few possible results but has input limits
      - The cells must be able to divide
      - At most 10–20 cells can be analyzed
  - Flow cytometry
    - Primarily used with hematopoietic cells
    - Can be used with disaggregated cells in limited measure
      - Disaggregation usually destroys surface markers
      - DNA [total] content can be measured by staining
      - Some cytoplasmic targets can be targeted by fluorescent markers, which can enter permeabilized cells
    - Results are limited to the antibodies used
    - Can interrogate thousands of cells
- In an 'omic study, the intention is to sequence ALL of one or more 'omes
  - The whole genome of each cell
    - WGA routinely amplifies 90% of a single-cell genome
  - The entire transcriptome (at least all the mRNA)
    - Whole transcriptome analysis (RNA-Seq) of single cells can quantitate 5–15% of polyadenylated transcripts within each cell
    - Although not exhaustive, this is sufficient to assign a detailed cell 'type'
  - The entire epigenome
    - The methylome: All the methylated nucleotides
    - Alternatively, all sites actively bound by regulatory proteins such as transcription factors and histones
  - The entire proteome (which is hard to define)
- A few studies have reported analysis of two or more 'omes in the same cells
- The great majority of studies have been directed at transcriptomes
  - The first single-cell 'omic study was performed in 2009, literally analyzing a single cell

- By 2017, over 1000,000 cells were analyzed in a single experiment
- Potential benefits of single-cell analysis
  - Identification of novel normal cell types
    - The premier example is the discovery of the pulmonary 'ionocyte', which is the site of pathological malfunction of the CFTR in cystic fibrosis
    - Permits unequivocal pairing of immunoglobulin heavy chains with light chains
  - Improved error correction
    - In NGS analysis of a 'bulk' sample, low-frequency mutations cannot always be distinguished from sequencing errors
    - In single-cell genomic sequencing, a polymerase error, unless it happens in the very first round of PCR, will be identifiable as an artifact
      - (Assuming diploidy) A somatic mutation and a germline variant should each have a variant allele frequency of 0% or 50% (occasionally 100%)
  - Identification of distinct cell states
    - Cells of the same broad 'type' can be in different 'states' of differentiation, of proliferation, of metabolism, of interaction with other cells
    - An important goal is to identify states (if they exist) which distinguish malignant, nonmalignant, and premalignant cells
      - This might require analysis of cells en masse
      - This might require correlation with the spatial analysis
    - Delineating 'distinct' states is a complex statistical and philosophical challenge in biology
- Cellular phylogenetics
  - Tumor heterogeneity
    - Most tumors evolve over years or decades
    - Many clones coexist in every tumor
      - Most share a small set of 'founder mutations' in oncogenes and/or tumor suppressors
      - Clones will also have accumulated distinctive sets of additional mutations, the majority of which are 'passenger mutations', presumed biologically inert but computationally useful
  - When the NGS of a bulk sample identifies multiple variants, the variant allele frequencies can be used, to a limited extent, to infer the order in which the mutations arose
    - Single-cell analysis permits more accurate and granular phylogenies
    - Whether or not mutations in two genes are present in the same or different cells is readily determined
  - Lineage/fate determination in normal tissues
    - In nonmalignant tissue, such as a normal colonic crypt, similar techniques are used to infer cell fate/lineage from stem cell to terminally differentiated

cell, along with the changes in transcriptome at each stage
  ○ Presumably, normal cells do not have many mutations, making phylogenetic tracing difficult, but mitochondrial genomes accumulate variants in normal cells (and in malignant cells) at a rate sufficient to permit statistical reconstruction of the order of development, using the mitochondrial mutations to trace lineage
- Clinical significance
  – At present single-cell analysis is primarily of research interest, but clinically relevant scenarios can be envisioned
    ○ Two mutations are present in a tumor
    ○ One variant is 'druggable', one is not
    ○ If they are in the same cell, one could target the tumor with the indicated drug

Single-Cell Protocols

- Preparation of single cells is feasible from fresh tissue, fresh-frozen, and OCT embedded
- Typical disaggregation protocol
  – Macrodissect small pieces of <u>fresh</u> tissue
  – Incubated in an enzyme cocktail (for example, collagenase, elastase)
    ○ Enzymatic digestion variably eliminates cell-surface markers
  – An alternative is to soak in EDTA
- Rigorous exclusion of damaged and of clustered cells is a critical QA step
- Spatial information is lost
  – Transcriptomic analysis (RNA-seq) can often broadly re-identify the cell type
- Cell hashing
  – After cell dissociation but before library preparation starts, one can mark the cells with oligonucleotide-tagged antibodies directed at either universal cell-surface markers or specific cell-surface markers (which must be proven during validation to survive disaggregation)
  – When DNA is prepared from the cells, it will include the oligonucleotides from whichever antibodies are bound to the surface
    ○ The oligonucleotides are barcode like
      ♦ There is a sequence complementary to a common region of all the capture barcodes
      ♦ There is a sequence motif identifying the associated antibody
    ○ When the cells are lysed early during library construction, the oligonucleotides are released

  ○ The tags will be sequenced in the general sequencing workflow
  ○ Depending on the antibodies, tags can give information about the cell type
  ○ The tags allow multiple samples to be mixed during the library preparation and sequencing, provided the antibodies for each sample have distinct tags
- Preparation of single cells from FFPE tissue
  ○ Neither enzymatic digestion nor cation chelation are feasible
  ○ Preps of intact nuclei are possible
    ♦ Can be used for genomic analysis (DNA sequencing)
    ♦ Can be used for limited transcriptomic analysis because all mRNA start as longer RNA transcripts in the nucleus
- Setting up the single-cell analysis
  ○ Manipulation of the cells or nuclei
    ♦ Early methods deposited single cells into wells of a microtiter plate, with an NGS library subsequently constructed in each well
    ♦ Microfluidic droplets
      ▪ A stream of single file cells or nuclei suspended in oil is merged with a stream of reagent droplets, each suspended in oil (similar to procedures for ddPCR and for Ion Torrent NGS)
        ✦ Each droplet contains all the reagents necessary to create an NGS library including reverse transcriptase for a transcriptome
        ✦ Reagent droplets contain barcoded oligonucleotides, often attached to a bead
      ▪ The oligonucleotide typically contains adaptor sequences which
        ✦ Attach to the bead, and, later, to the sequencing flow cell
        ✦ A barcode uniquely identifying all the oligonucleotides attached to a given bead
        ✦ For analysis of mRNA, a poly-dT end can hybridize with a poly A tail to facilitate cDNA synthesis
      ▪ Each cell combines with one reagent droplet and one bead to form a microscopic reaction chamber
      ▪ Many 'merged' droplets will lack cellular DNA or beads; they will not generate a signal
      ▪ Some merged droplets might reflect cell doublets and could be mistaken in analysis for aneuploid or hyperdiploid cells, but cell suspensions are so diluted that 90% of merged

droplets are expected to be empty, doublets rare

- Creation of the library
  - The droplets, even in the millions, occupy little space and can be pooled for reverse transcription, which occurs while maintaining droplet integrity
  - The mRNA in a given cell, which undergoes reverse transcription, generates a cDNA that incorporates a primer with a barcode uniquely identifying the cell
  - Each cDNA can *also* incorporate a UMI, uniquely identifying the specific mRNA molecule reverse transcribed; the UMI is important for quantitative analysis later in the bioinformatic process
  - At this point, the droplet contents (cDNA) can be released and pooled
- Sequencing reactions
  - Library construction is similar to that of bulk samples
  - The sequencing reactions are the same as for bulk samples
- Bioinformatic analysis
  - Base calling, quality score assignment, and alignment are similar to that for NGS for bulk tumor
  - Bioinformatic analyses are beyond the scope of this chapter
    - Cell identification
    - Cell clustering (cells of the same 'type' can be in different 'states')
    - Lineage inference/tumor evolution
    - Correlation of results for different 'omic modalities
- At present shallow sequencing of individual cells while increasing the number of cells sequenced is considered the most cost-effective approach to analysis
  - For transcriptomics, Svensson compared numerous experimental systems and concluded that calling 250,000 reads/cell would be saturating
  - Even 250,000 reads per cell will leave the count for many genes at '0'
- Several studies have reported analysis of two or more 'omes in the same cells
  - CITE-seq profiles both transcription and protein abundance
  - G&T-seq captures both genomic DNA and RNA
  - scNMT-seq combines gene-expression profiling, methylation, and chromatin accessibility
  - ASAP-seq combines chromatin accessibility, cell-surface protein abundance, and cellular lineage (inferred from mitochondrial DNA sequence variation)
  - DOGMA-seq combines ASAP-seq with RNA-seq
- Commercial systems for single-cell profiling
  - Zhang et al. compared three of the currently most popular: inDrop, Drop-seq, and 10× Chromium
  - None are for clinical use at present

Spatial 'Omics: Not Quite the Final Frontier

General Remarks

- Spatial 'omic methods generate information on a single-cell basis while retaining knowledge of the location and appearance of the cell
- So far, most spatial genomic analyses are two-dimensional
- Analysis of serial sections could, of course, give a three-dimensional tissue representation
- Several methods using ISH probes for RNA are read out by confocal microscopy, providing subcellular and three-dimensional localization (over a short vertical distance)
- Most work is on transcriptomics (RNA)
- For NGS methods, the depth of reads per cell is substantially lower than for the single-cell techniques described earlier
  - 'Anchor' genes are genes that are known (or thought) to be limited in expression to specific cell types
  - Information on anchor genes from NGS analysis of disaggregated single cells from the same (or a similar) sample can be leveraged to more fully characterize the cells and cell states identified in the sparser data of the spatial analysis by NGS or ISH
- All systems work with fresh or fresh-frozen tissue,
- A few have been shown to work with FFPE
- The field of view demonstrated for different methods varies widely, from 0.5 mm^2 to 6.5 mm × 6.5 mm
- There are many methods, but they fall into several broad approaches
  - Microdissection
  - In situ hybridization of fluorescent probes (FISH)
  - In situ sequencing
  - In situ capture followed by ex situ sequencing
- Microdissection
  - Practically this is limited to 'laser capture'
  - Can extract full-length mRNA (fresh, frozen tissues)
  - Can generate high-complexity NGS libraries
  - Cell throughput severely limited compared to other methods

In Situ Hybridization (ISH)

Single Molecule FISH (smFISH)

- This is a research tool at present
- Detection of single RNA molecules in situ has required super-resolution microscopy or confocal microscopy
- These incur high instrumentation costs
- Long scan times are required for small fields
- There are many implementations of smFISH (Fig. 1.31)
- Several methods amplify the signal to obtain better signal-to-noise

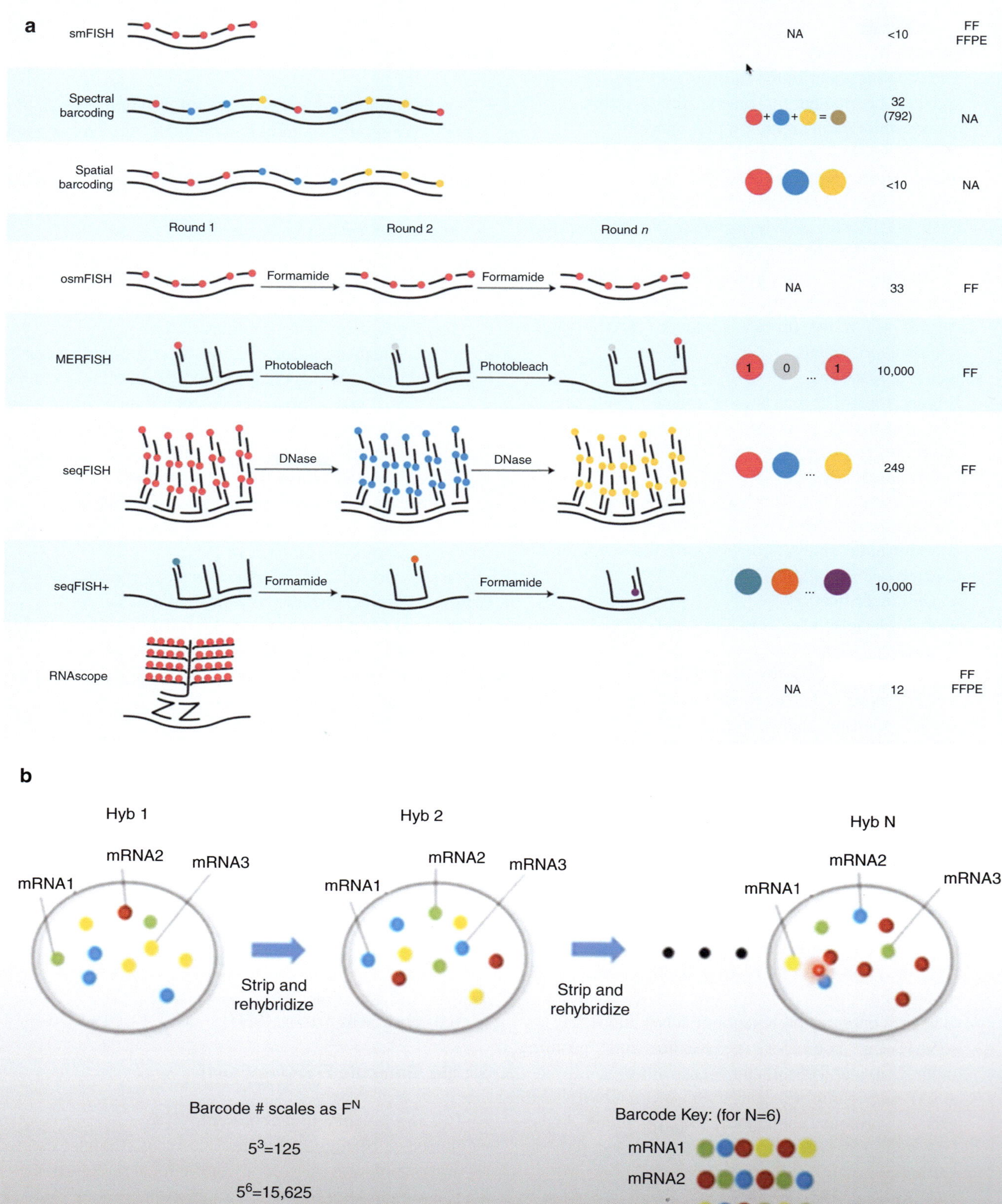

Fig. 1.31 Varieties of single molecule FISH transcriptomics. (**a**) Table of various smFISH methods. (**b**) Example of how signals are read. Source: Lewis, S. M. et al. Spatial omics and multiplexed imaging to explore cancer biology. *Nat. Methods* (2021) doi:10.1038/s41592-021-01203-6

- Branched DNA (with MERFISH)
- Hybridization chain reaction (with SeqFISH)
- Some methods can be combined with 'expansion microscopy' permitting improved subcellular resolution and improved multiplexing of fluors (decreased optical crowding)
- Highly multiplexed methods require sequential rounds of hybridization, imaging, and stripping (or photobleaching the previous fluors) and can require several days.

Representative Method: SeqFISH+

- The technique has been commercialized by the Spatial Genomics Company
- SeqFISH+ can target up to 10,000 different RNA simultaneously
- Uses confocal microscopy
- 'Primary probes' include a central ~20 nt targeting the sequence of interest
- Primary probes are NOT fluorescently labeled
- Each end of the primary probe has 'overhang' sequence motifs (a form of barcodes)
- Each motif can hybridize a fluorescently labeled probe with complementary sequence
- A set of probes with fluors are hybridized, targeting the barcodes
- The protocol uses five fluors in such a way as to generate 60 pseudocolors
- The number of pseudocolors reduces the apparent crowding so that individual mRNA can be imaged
- The spatial signal is imaged
- The probes are stripped off
- Another set of fluorescent probes are hybridized, imaged, stripped
- The cycle is repeated
- From the combinations of fluors seen at a given spot in the cell, the identity of the primary probe (and hence of the target mRNA) can be determined
- Shown to work on fresh-frozen tissue sections

Representative Method: MERFISH

- Multiplexed error-robust FISH (MERFISH) (commercialized by VizGeneration Inc)
- Can combine bDNA and multiplexed FISH probes
- Can measure at least 140 RNA species in a 40 mm² sample (~40,000 human cells) with nearly 100% detection efficiency, or 10,000 RNA species with 80% 'detection efficiency' in a smaller section that had undergone expansion microscopy
- Field of view size (220 × 220 μm) requiring 40–100 min total time scanning
- Probes

- The method of using combinatorial barcoding 5′ and 3′ extensions of the primary probes and sequential hybridizations is like that for SeqFISH(+)
- Details of probe design, error correction, and signal detection differ from those for SeqFISH(+)

In Situ Sequencing

- Several methods have shown the capability of generating short sequence reads from RNA or DNA 'in situ', on slides
- The earliest are best considered as proof-of-principle with very short reads
- More recent approaches, such as STARMAP, show improved capabilities
  - Can multiplex detection of over 1000 mRNA
- Requires hydrogel creation in cell interiors
- Can show the location of mRNA in 3D (hydrogel fixes positions of mRNAs)
- Designs primers and padlock probes to bind near each other for each target so the primer can synthesize long DNA strands via Rolling Circle Amplification ('DNA nanoballs')
  - Can sequence mRNA in situ by a ligation method with high accuracy
- The nucleotide added is fluorescent, its position can be imaged, and the accumulating sequence ascertained
  - Field size is very small
- Sequence length appears to still consist of short reads, up to six nucleotides (each addition requires a round of sequencing reaction and imaging)
- Repeated rounds of imaging degrade the growing sequence
- Signals from previous bases interfere with the signal from a newly added base
- This is primarily intended to identify the mRNA ($4^6/2 = 2048$ possible sequences)
- It is not intended primarily to detect mutations

In Situ Capture–Ex Situ Sequencing (Generic)

- A tissue section is applied to a slide that carries a patterned array of densely spaced DNA oligonucleotides with multiple barcoded domains (Fig. 1.32)
  - A 3′ oligonucleotide oligo-dT tail to capture, then prime polyadenylated mRNA for reverse transcription
  - A UMI (unique molecular identifier)
  - A 'spatial zip code' reports where on the slide the oligo was synthesized or deposited
  - Capture motifs which will later allow the resulting amplicon to be captured on a sequencer solid support
- Since the primers are randomly synthesized on the solid support, it is important to know where each spatial bar-

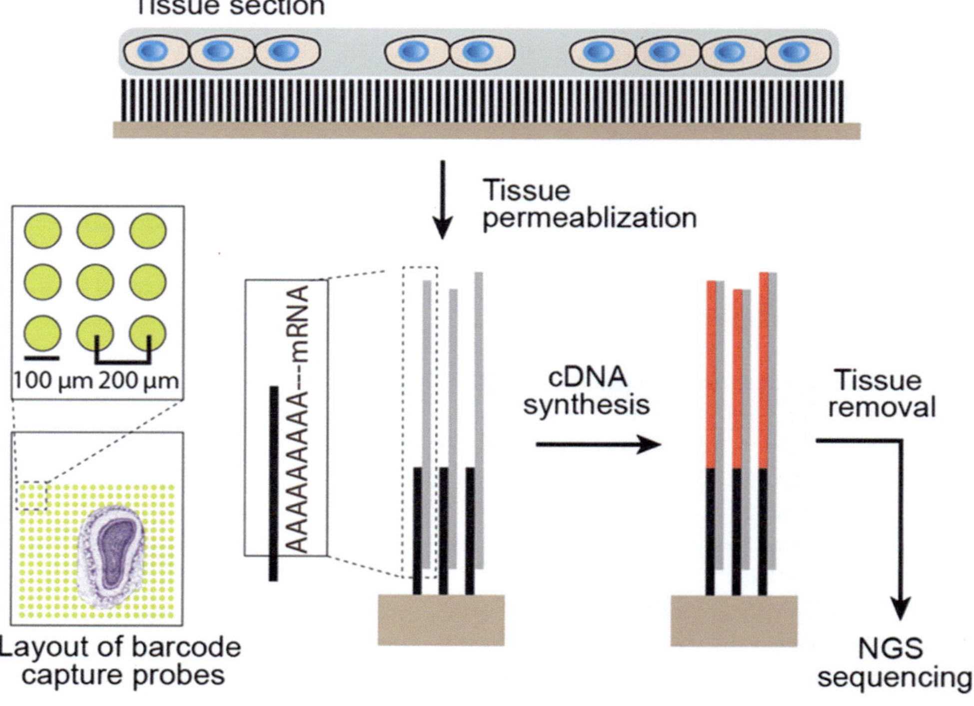

Fig. 1.32 In situ capture for ex situ sequencing with spatial barcodes. Source: Strell, C. et al. Placing RNA in context and space—methods for spatially resolved transcriptomics. *FEBS J.* **286**, 1468–1481 (2019)

code is located before a library is constructed and sequenced

- In some settings, the capture primers are synthesized in situ, so the correlation between barcode and location is known
- In some settings, the capture primers are randomly synthesized (on a bead or slide)

- After the tissue section is applied to the array, it is chemically permeabilized so that the oligonucleotide primers can penetrate and extend into the cytoplasm along with reverse transcriptase
- The primers hybridize to the poly A tails of overlying mRNA in the overlying cell; the reverse transcriptase extends the primer; in some implementations, the second strand is also synthesized in situ
- The cells are lysed, and the cDNA extracted
- The recovered cDNA is identified by sequencing in a second-generation NGS instrument
- Sequencing yields both the sequence of the mRNA hybridized and the spatial zip code, which ties the result back to a specific cell
- The number of reads for a specific mRNA in a specific cell is used to quantify that mRNA

Representative Implementation: GeoMx Digital Space Profiler

- Commercialized by Nanostring, Inc
- Combines in situ hybridization AND capture, with ex situ analysis (Fig. 1.33)

- Uses antibodies or ISH oligonucleotide probes (directed at RNA)
  - Typical panel of 40–50 antibodies
    - Four to five preselected antibodies are labeled with distinct fluors
    - These are used to visually delineate regions of interest (ROIs)
    - Most antibodies are conjugated to oligonucleotide labels but NOT to fluors
      - The oligonucleotide sequence identifies the antibody and its specificity
      - The oligonucleotides can be cleaved off by UV light
  - Typical RNA probe panel has ~900 oligonucleotide probes targeting 90 genes
    - Can use RNA probe panel targeting the entire transcriptome
- RNA panels include several fluorescently labeled RNAscope probes directed at abundant targets, to permit cell-type identification for ROI delineation
- Shown to work for sections of fresh, frozen, and FFPE tissue
- Outline of protocol
  - Incubate panel of antibodies directed at cell-surface markers
  - Image the fluorescent antibodies
  - Identify regions of interest (ROI), ~100 microns × 100 microns each
  - Using the digital image on a computer screen for guidance, UV light is directed at an ROI, cleaving the oligonucleotides from the bound antibodies

Fig. 1.33 Multiplex digital spatial profiling of proteins and RNA in fixed tissue. Source: Merritt, C. R. et al. Multiplex digital spatial profiling of proteins and RNA in fixed tissue. *Nat. Biotechnol.* **38**, 586–599 (2020)

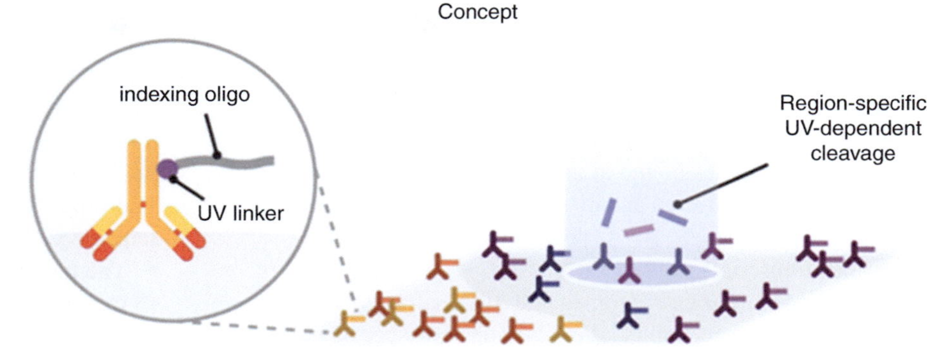

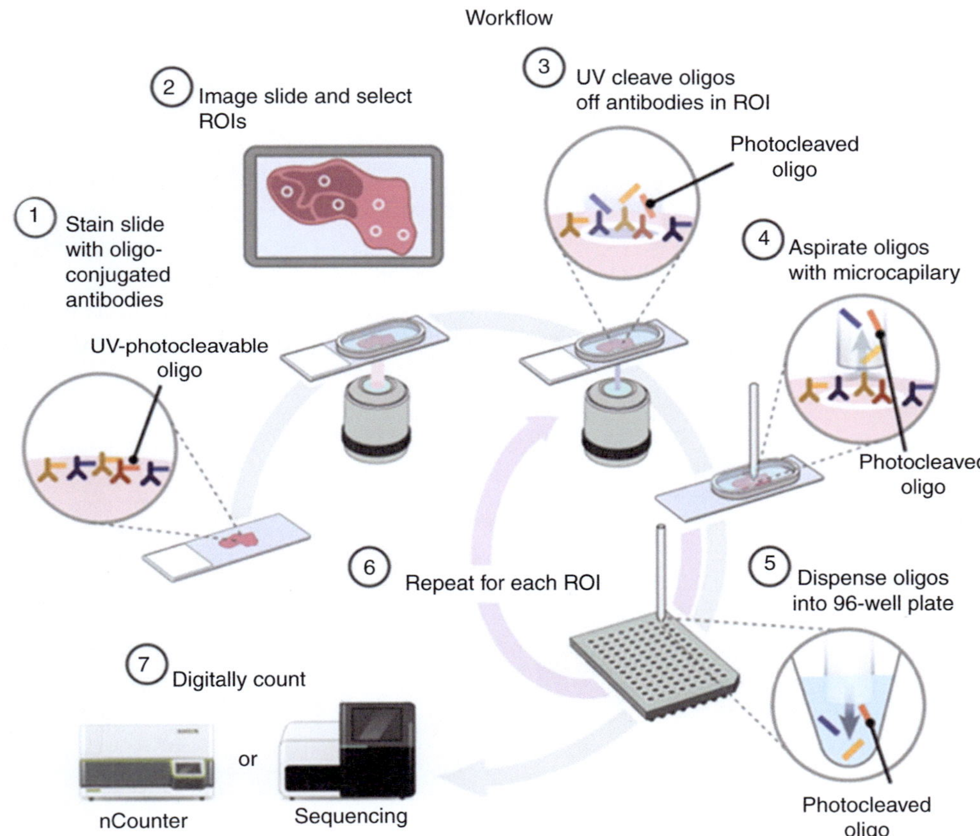

- A capillary over the ROI aspirates the released oligonucleotides, which are later analyzed on a second NanoString platform, the nCounter
  - The oligonucleotide tags are composed of combinations of short oligo sequence 'tags', which are recognized by hybridizing to labeled probes in the nCounter system (not shown here)
- Next, the section is permeabilized so the RNAscope and GeoMx probes can enter
  - The RNAscope probe(s) are targeted at one or more mRNA targets which can define the cells or region of interest (like epithelial cell markers)
- The RNAscope probes are visualized, further defining ROIs
- The GeoMx probes typically contain a 35–50 nt target-binding region and a 60–66 nucleotide cleavable index tag
- A UV light is used to irradiate the desired region of interest (based on the RNAscope pattern)
- The UV light releases unique DNA sequence tags of any GeoMx probes, which also bound in the ROI
- The solution over an ROI is aspirated
- The indicial oligos are identified by hybridization to complementary oligos on a separate instrument, the n-Counter (as were the) antibody oligo tags

- ○ In the WTA format, the oligo tags are sequenced by NGS (they are too varied to be catalogued by combinations of hybridization tags)
- – From the sequences detected by the nCounter (or by NGS), the identity, specificity, and number of antibodies or oligo probes bound at any given x,y position in an ROI can be inferred and displayed visually
- – GeoMX can detect RNA down to 600 individual mRNA transcripts per cell
- – Profiling whole transcriptomes and 100 s of antibodies, one instrument can handle 12 tissue slides per day

Representative Implementation: Seq-Scope Illumina

- Principle
  - – In situ capture, ex situ cDNA sequencing using an Illumina sequencer both for array capture and for sequencing, providing subcellular resolution
- Outline of procedure (Fig. 1.34)
  - – Creating the capture array
    - ○ The flow cell carries capture oligos as per usual for Illumina sequencing
      - ♦ In the first round of sequencing steps, before a sample is applied, the flow cell is flooded with high-density molecular index oligonucleotides (HDMI) (Fig. 1.35)

- ♦ Each HDMI oligo has a random 20–32 bp barcode and a poly-dT sequence
- ♦ The HDMI-oligos are loaded into the flow cell and amplified for cluster generation in 'traditional' Illumina sequencing, except that no sample is added
- ♦ The amplified and clustered HDMI-oligos are sequenced
  - ▪ The random HDMI barcode, which is shared by all the oligos in a cluster, becomes a spatial index for that cluster
- ♦ The enzyme *Dra*I is used to cutoff the P5 primer used to sequence the oligo
- ♦ NaOH is used to denature the oligos bound to the flow cell from complementary strands
- ♦ This exposes the next oligo primer motif, poly-dT, which can capture mRNA
- – Loading the tissue section
  - ○ A 10 µ section of OCT-embedded frozen tissue is maneuvered onto the HDMI array and fixed in paraformaldehyde
  - ○ The subsequent steps are broadly similar to those described for high-density spatial transcriptomics (HDST)
    - ♦ Synthesis of the complementary is initiated by adaptor-tagged random hexamers

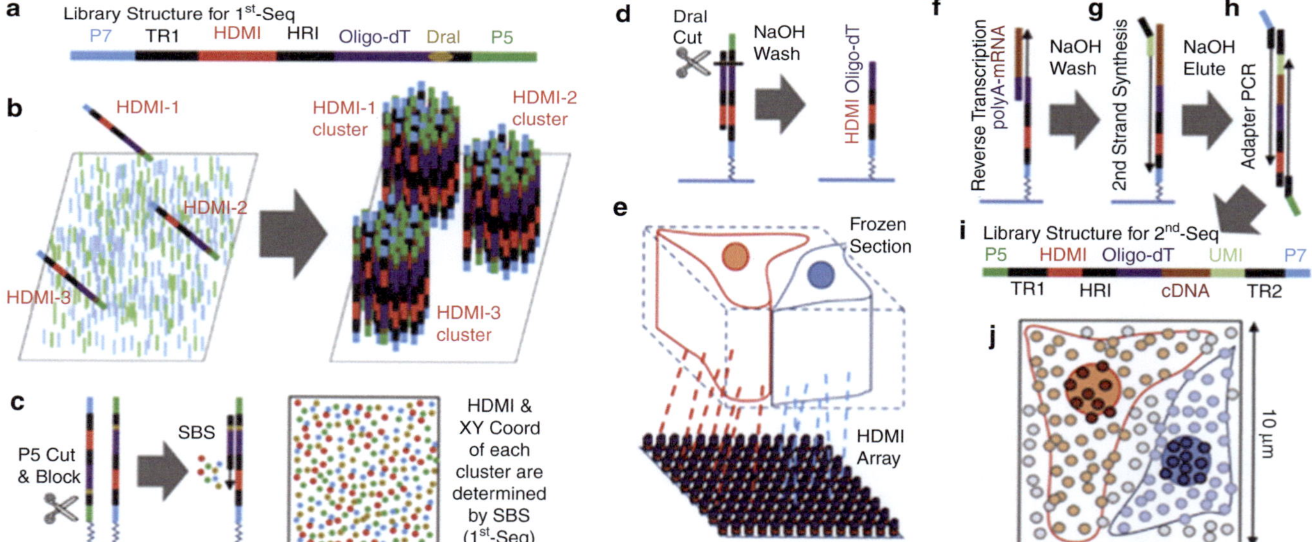

Fig. 1.34 Seq-Scope overview. (**a**) Schematic representation of barcode adaptors (see also Figs. 1.24, 1.32). (**b**) Solid-phase amplification of HDMI-oligo molecules captured on the flow cell surface (somewhat similar to bridge amplification in Illumina 2ⁿᵈ gen sequencing). (**c**) Illumina sequencing in situ determines the HDMI sequence and XY coordinates of each cluster. (**d**) Oligonucleotide clusters are modified to expose oligo-dT, the RNA-capture domain. (**e**) Frozen section applied to slide. Oligo clusters capture RNA released from overlying section. (**f**) cDNA is generated by reverse transcription. (**g**) A complementary second strand is synthesized using random priming. (**h, i**) PCR with a new adaptor generates a sequencing library containing the cDNA (and the enclosing HDMI sequences). (**j**) The cDNA library can be released from the section and set up for 2ⁿᵈ or 3ʳᵈ NGS ex situ. A sequence result will include indices identifying the sample and location for the cDNA sequence. Source: Cho, C.-S. et al. Microscopic examination of spatial transcriptome using Seq-Scope. *Cell* **184**, 3559–3572.e22 (2021)

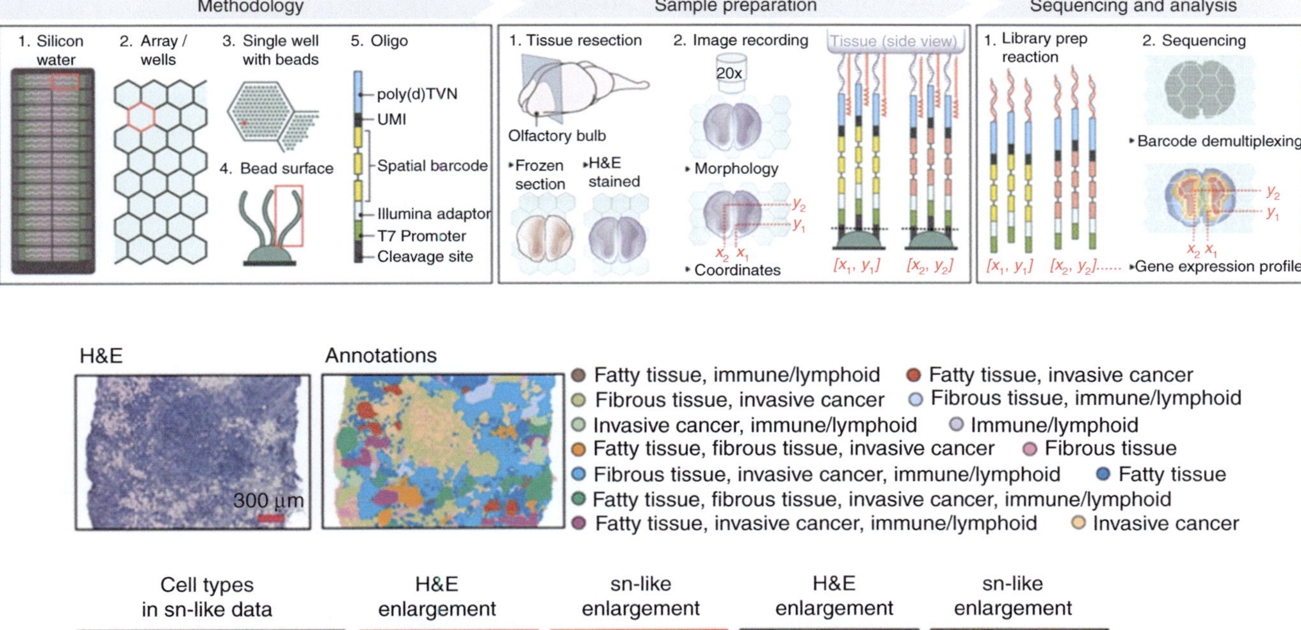

Fig. 1.35 High-density spatial transcriptomics (HDST). Source: Vickovic, S. et al. High-definition spatial transcriptomics for in situ tissue profiling. *Nat. Methods* **16**, 987–990 (2019)

♦ After the tissue is removed, the second strands are eluted and put through an Illumina sequencing reaction

- Comments
  - At present, there is a <u>single</u> reported proof-of-principle, described only for fresh tissue
  - Illumina sequencer models HiSeq2500 and NovaSeq are shown to provide 90 mm² and 800 mm² imaging areas
  - Seq-Scope has a center-to-center resolution (from one oligo cluster to the nearest oligo cluster) of 0.5–0.8 μm (0.6 μm on average), comparable to an optical microscope
  - Subcellular resolution permits discrimination of spliced and unspliced RNA, the latter usually being predominantly nuclear in location

- The capture array permits single-cell resolution
  - o Hexagonal array of >1.4 million 2 μm wells filled with randomly distributed beads
  - o Each bead contains many copies of a spatially barcoded mRNA-poly(d)T capture primers
- Before a tissue section is added, the array is 'decoded' by several rounds of hybridization with fluorescent oligonucleotides complementary to the spatial barcode motifs
- The tissue section is added, then permeabilized so mRNA can hybridize to the capture primers
- The captured mRNA is converted into cDNA, which in turn is used to prepare an NGS library
- The library is sequenced ex situ, in an NGS instrument
- The barcodes in each sequence include the spatial index, described above, which can be referred to a cellular location on the slide

Representative Implementation: HDST

- Principle
  - This is an in situ capture ex situ sequencing format

Representative Implementation: Visium (10× Genomics)

- Principle

- In situ capture from tissue sections on special glass slides with ex situ sequencing
- Glass microscope slides
  - Each contains two or four 6.5 × 6.5 mm capture arrays, each with 5000 'spots'
  - Each 'spot' has covalently bound barcoded probes to capture mRNA
- The protocol design is broadly like that in HDST (above)
- On average, mRNA from between 1 and 10 cells are captured per spot, so it is not quite routinely single-cell resolution (which also clearly depends on cell size)

In Situ Genomic (DNA) Sequencing Extended by Ex Situ Sequencing

- Principle
  - Chromosomes are spatially organized
  - This organization is thought to influence gene expression
  - The goal of this method is to determine the 3D localization of chromosomes by combining limited in situ sequencing with more extensive ex situ sequencing
- Outline of procedure
  - Create an in situ 'sequencing library'
    - Use a Tn5 transposase to randomly incorporate DNA sequencing adaptors throughout the nuclear genomes in a fixed and permeabilized tissue section
    - Circularize the fragments with oligos incorporating UMIs and amplify with Rolling Circle Amplification
  - Perform in situ sequencing of the circular fragments
    - The reads are necessarily short and intended to identify the UMIs
  - Extract the circularized fragments from the tissue and use them for standard paired-end NGS sequencing
    - Sequencing should give recover UMI sequences near the fragment ends
    - Sequencing will also give ~100 nucleotides at each end, which will usually be enough to uniquely identify the chromosomal/genomic location for the amplicons
    - The UMI sequences read in this NGS assay can be used to connect the chromosomal identification with the spatial location of the UMI sequence identified by in situ sequencing
- Comments
  - This is based on a single recent study of cells in culture and mouse embryos
  - Spatially localized several thousand genomic loci (recovered sequences roughly every 1 Mb across the genome)
  - Could distinguish parental alleles/chromosomes

Analysis

- Models for analyzing spatial relationships and their impact among cells in a tumor are much less well-developed in molecular genetics but have been an active field for IHC

Achievements of Spatial Genomics

- So far there have been no revolutionary results from spatial omics, but there have been findings of interest
- Human Cell Atlas and the Human Tumor Atlas Network consortium
  - Perhaps the most important result so far is the establishment of cell atlases of transcriptomes for every normal tissue
  - A Human Tumor Cell Atlas Network consortium, has also been established

Future Directions/Predictions

- The clinical application of second-generation NGS to the analysis of cancer tissues will continue to expand
  - The optimal test (whole genome, whole exome, and targeted panel) for a given context will remain a topic of contention
  - Analysis of fusion genes by NGS, especially in hematology, will grow more slowly but will expand
- The clinical role of third-generation NGS in cancer remains to be defined despite the advantages of long reads for both DNA and RNA from fresh tissue
- The clinical utility of the detection of cfDNA to screen for cancer and to monitor minimal residual disease remains to be defined
  - ddPCR will maintain a role for sensitive detection of frequent point mutations in cfDNA
  - NGS will predominate for a more comprehensive approach to monitoring
- Transcriptomics will remain in the realm of translational research, although some small, targeted panels have already been employed clinically
- The avant-garde methods described, 'single-cell genomics' and 'spatial 'omics', are maturing rapidly but will remain research tools
  - Practical constraints (field of view, turnaround time, bioinformatic analysis) might yield technological improvements such that it can be performed in a clinical lab
  - Clinical validation can be a years- or decades-long process
  - Knowledge of these capabilities is still useful to pathologists, especially in training

Further Reading

Alder JK, Hanumanthu VS, Strong MA, DeZern AE, Stanley SE, Takemoto CM, et al. Diagnostic utility of telomere length testing in a hospital-based setting. Proc Natl Acad Sci U S A. 2018;115(10):E2358–65.

Alon S, Goodwin DR, Sinha A, Wassie AT, Chen F, Daugharthy ER, et al. Expansion sequencing: spatially precise in situ transcriptomics in intact biological systems. Science. 2021;371(6528):eaax2656.

Berge AFL, La Berge AF. Debate as scientific practice in nineteenth-century Paris: the controversy over the microscope. Perspect Sci. 2004;12:424–53.

Birch DE. Simplified hot-start PCR. Nature. 1996;381(6581):445–6.

Burton MP, Schneider BG, Escamilla-Ponce N, Gulley ML. Comparison of histologic stains for use in PCR analysis of microdissected, paraffin-embedded tissues. BioTechniques. 1998;24(1):86–91.

Buza N, Hui P. Genotyping diagnosis of gestational trophoblastic disease: frontiers in precision medicine. Mod Pathol. 2021;34(9):1658–72.

Cardon T, Fournier I, Salzet M. Shedding light on the ghost proteome. Trends Biochem Sci. 2021;46(3):239–50.

Cho C-S, Xi J, Si Y, Park S-R, Hsu J-E, Kim M, et al. Microscopic examination of spatial transcriptome using Seq-Scope. Cell. 2021;184(13):3559–3572.e22.

Cohen JD, Li L, Wang Y, Thoburn C, Afsari B, Danilova L, et al. Detection and localization of surgically resectable cancers with a multi-analyte blood test. Science. 2018;359:926–30.

Cohen SA, Turner EH, Beightol MB, Jacobson A, Gooley TA, Salipante SJ, et al. Frequent PIK3CA mutations in colorectal and endometrial tumors with 2 or more somatic mutations in mismatch repair genes. Gastroenterology. 2016;151(3):440–447.e1.

Cooper LA, Demicco EG, Saltz JH, Powell RT, Rao A, Lazar AJ. PanCancer insights from the cancer genome atlas: the pathologist's perspective. J Pathol. 2018;244(5):512–24.

Cristiano S, Leal A, Phallen J, Fiksel J, Adleff V, Bruhm DC, et al. Genome-wide cell-free DNA fragmentation in patients with cancer. Nature. 2019;570(7761):385–9.

Davidson CJ, Zeringer E, Champion KJ, Gauthier M-P, Wang F, Boonyaratanakornkit J, et al. Improving the limit of detection for sanger sequencing: a comparison of methodologies for KRAS variant detection. BioTechniques. 2012;53(3):182–8.

Eng C-HL, Lawson M, Zhu Q, Dries R, Koulena N, Takei Y, et al. Transcriptome-scale super-resolved imaging in tissues by RNA seqFISH. Nature. 2019;568(7751):235–9.

Howat WJ, Wilson BA. Tissue fixation and the effect of molecular fixatives on downstream staining procedures. Methods. 2014;70(1):12–9.

Hu Z, Li Z, Ma Z, Curtis C. Multi-cancer analysis of clonality and the timing of systemic spread in paired primary tumors and metastases. Nat Genet. 2020;52:701–8.

Hutter C, Zenklusen JC. The cancer genome atlas: creating lasting value beyond its data. Cell. 2018;173(2):283–5.

Ivády G, Madar L, Dzsudzsák E, Koczok K, Kappelmayer J, Krulisova V, et al. Analytical parameters and validation of homopolymer detection in a pyrosequencing-based next generation sequencing system. BMC Genomics. 2018;19(1):158.

Knouse KA, Davoli T, Elledge SJ, Amon A. Aneuploidy in cancer: Seq-ing answers to old questions. Annu Rev Cancer Biol. 2017;1(1):335–54.

Lampignano R, Neumann MHD, Weber S, Kloten V, Herdean A, Voss T, et al. Multicenter evaluation of circulating cell-free DNA extraction and downstream analyses for the development of standardized (pre)analytical work flows. Clin Chem. 2019;66(1):149–60.

Lanzós A, Carlevaro-Fita J, Mularoni L, Reverter F, Palumbo E, Guigó R, et al. Discovery of cancer driver long noncoding RNAs across 1112 tumour genomes: new candidates and distinguishing features. Sci Rep. 2017;7:41544.

Levy-Lahad E, Lahad A, King M-C. Precision medicine meets public health: population screening for BRCA1 and BRCA2. J Natl Cancer Inst. 2015;107(1):420.

Lewis SM, Asselin-Labat M-L, Nguyen Q, Berthelet J, Tan X, Wimmer VC, et al. Spatial omics and multiplexed imaging to explore cancer biology. Nat Methods. 2021;18(9):997–1012.

Lhermitte B, Egele C, Weingertner N, Ambrosetti D, Dadone B, Kubiniek V, et al. Adequately defining tumor cell proportion in tissue samples for molecular testing improves interobserver reproducibility of its assessment. Virchows Arch. 2017;470(1):21–7.

Livhits MJ, Zhu CY, Kuo EJ, Nguyen DT, Kim J, Tseng C-H, et al. Effectiveness of molecular testing techniques for diagnosis of indeterminate thyroid nodules: a randomized clinical trial. JAMA Oncol. 2021;7(1):70–7.

Lowe D The Myth of ethidium bromide. In the Pipeline. 2016. https://www.science.org/content/blog-post/myth-ethidium-bromide

Ludwig LS, Lareau CA, Ulirsch JC, Christian E, Muus C, Li LH, et al. Lineage tracing in humans enabled by mitochondrial mutations and single-cell genomics. Cell. 2019;176(6):1325–1339.e22.

MacArthur DG, Balasubramanian S, Frankish A, Huang N, Morris J, Walter K, et al. A systematic survey of loss-of-function variants in human protein-coding genes. Science. 2012;335(6070):823–8.

Madissoon E, Wilbrey-Clark A, Miragaia RJ, Saeb-Parsy K, Mahbubani KT, Georgakopoulos N, et al. scRNA-seq assessment of the human lung, spleen, and esophagus tissue stability after cold preservation. Genome Biol. 2019;21(1):1.

Marberger M, McConnell JD, Fowler I, Andriole GL, Bostwick DG, Somerville MC, et al. Biopsy misidentification identified by DNA profiling in a large multicenter trial. J Clin Oncol. 2011;29(13):1744–9.

Martínez-Jiménez F, Muiños F, Sentís I, et al. A compendium of mutational cancer driver genes. Nat Rev Cancer. 2020;20(10):555–72.

Metzker ML. Sequencing technologies—the next generation. Nat Rev. Genet. 2009;11(1):31–46.

Mullis K. The unusual origin of PCR. Sci Am. 1990;262(4):56–61, 64–5.

Murciano-Goroff YR, Drilon A, Stadler ZK. The NCI-MATCH: a national, collaborative precision oncology trial for diverse tumor histologies. Cancer Cell. 2021;39(1):22–4.

Neben CL, Zimmer AD, Stedden W, van den Akker J, O'Connor R, Chan RC, et al. Multi-gene panel testing of 23,179 individuals for hereditary cancer risk identifies pathogenic variant carriers missed by current genetic testing guidelines. J Mol Diagn. 2019;21(4):646–57.

Paik S, Shak S, Tang G, Kim C, Baker J, Cronin M, et al. A multigene assay to predict recurrence of tamoxifen-treated, node-negative breast cancer. N Engl J Med. 2004;351(27):2817–26.

Payne AC, Chiang ZD, Reginato PL, Mangiameli SM, Murray EM, Yao C-C, et al. In situ genome sequencing resolves DNA sequence and structure in intact biological samples. Science. 2021;371(6532):aay3446.

Petrackova A, Vasinek M, Sedlarikova L, Dyskova T, Schneiderova P, Novosad T, et al. Standardization of sequencing coverage depth in NGS: recommendation for detection of clonal and subclonal mutations in cancer diagnostics. Front Oncol. 2019;9:851.

Quan P-L, Sauzade M, Brouzes E. dPCR: a technology review. Sensors. 2018;18(4):1271.

Rebrikov DV, Trofimov DY. Real-time PCR: a review of approaches to data analysis. Appl Biochem Microbiol. 2006;42(5):455–63.

Recondo G, Mahjoubi L, Maillard A, Loriot Y, Bigot L, Facchinetti F, et al. Feasibility and first reports of the MATCH-R repeated biopsy trial at Gustave Roussy. NPJ Prec Oncol. 2020;4(1):27.

Rheinbay E, Nielsen MM, Abascal F, Wala JA, Shapira O, Tiao G, et al. Analyses of non-coding somatic drivers in 2,658 cancer whole genomes. Nature. 2020;578(7793):102–11.

Salk JJ, Schmitt MW, Loeb LA. Enhancing the accuracy of next-generation sequencing for detecting rare and subclonal mutations. Nat Rev. Genet. 2018;19(5):269–85.

Schramm G, Bruchhaus I, Roeder T. A simple and reliable 5′—RACE approach. Nucleic Acids Res. 2000;28(22):E96.

Shem S. The house of god. New York: Dell; 1978.

Srinivasan M, Sedmak D, Jewell S. Effect of fixatives and tissue processing on the content and integrity of nucleic acids. Am J Pathol. 2002;161(6):1961–71.

Ståhl PL, Salmén F, Vickovic S, Lundmark A, Navarro JF, Magnusson J, et al. Visualization and analysis of gene expression in tissue sections by spatial transcriptomics. Science. 2016;353(6294):78–82.

Svensson V, Natarajan KN, Ly L-H, Miragaia RJ, Labalette C, Macaulay IC, et al. Power analysis of single-cell RNA-sequencing experiments. Nat Methods. 2017;14(4):381–7.

Tay Y, Rinn J, Pandolfi PP. The multilayered complexity of ceRNA crosstalk and competition. Nature. 2014;505(7483):344–52.

Thiede C, Creutzig E, Illmer T, Schaich M, Heise V, Ehninger G, et al. Rapid and sensitive typing of NPM1 mutations using LNA-mediated PCR clamping. Leukemia. 2006;20(10):1897–9.

van 't Veer LJ, Dai H, van de Vijver MJ, He YD, AAM H, Mao M, et al. Gene expression profiling predicts clinical outcome of breast cancer. Nature. 2002;415:530–6.

van Dijk E, van den Bosch T, Lenos KJ, El Makrini K, Nijman LE, van Essen HFB, et al. Chromosomal copy number heterogeneity predicts survival rates across cancers. Nat Commun. 2021;12(1):3188.

Vandesompele J, De Preter K, Pattyn F, Poppe B, Van Roy N, De Paepe A, et al. Accurate normalization of real-time quantitative RT-PCR data by geometric averaging of multiple internal control genes. Genome Biol. 2002;3(7):RESEARCH0034.

Vickovic S, Eraslan G, Salmén F, Klughammer J, Stenbeck L, Schapiro D, et al. High-definition spatial transcriptomics for in situ tissue profiling. Nat Methods. 2019;16(10):987–90.

Wang D, Eraslan B, Wieland T, Hallström B, Hopf T, Zolg DP, et al. A deep proteome and transcriptome abundance atlas of 29 healthy human tissues. Mol Syst Biol. 2019;15(2):e8503.

Watkins TBK, Lim EL, Petkovic M, Elizalde S, Birkbak NJ, Wilson GA, et al. Pervasive chromosomal instability and karyotype order in tumour evolution. Nature. 2020;587(7832):126–32.

Woolley AT, Mathies RA. Electrophoresis chips. Anal Chem. 1995;67:3676–80.

Wu S, Bafna V, Mischel PS. Extrachromosomal DNA (ecDNA) in cancer pathogenesis. Curr Opin Genet Dev. 2021;66:78–82.

Xia C, Fan J, Emanuel G, Hao J, Zhuang X. Spatial transcriptome profiling by MERFISH reveals subcellular RNA compartmentalization and cell cycle-dependent gene expression. Proc Natl Acad Sci U S A. 2019;116(39):19490–9.

Yan L, Zhou J, Zheng Y, Gamson AS, Roembke BT, Nakayama S, et al. Isothermal amplified detection of DNA and RNA. Mol BioSyst. 2014;10(5):970–1003.

Zheng Z, Liebers M, Zhelyazkova B, Cao Y, Panditi D, Lynch KD, et al. Anchored multiplex PCR for targeted next-generation sequencing. Nat Med. 2014;20(12):1479–84.

Zollinger DR, Lingle SE, Sorg K, Beechem JM, Merritt CR. GeoMxTM RNA assay: high multiplex, digital, spatial analysis of RNA in FFPE tissue. Methods Mol Biol. 2020;2148:331–45.

Bioinformatics, Digital Pathology, and Computational Pathology for Surgical Pathologists

Sambit K. Mohanty, Saba Shafi, and Anil V. Parwani

Contents

S. K. Mohanty
The Advanced Medical Research Institute and CORE Diagnostics, Bhubaneswar, Odisha, India

S. Shafi · A. V. Parwani (✉)
Department of Pathology, The Ohio State University, Wexner Medical Center, Columbus, OH, USA
e-mail: anil.parwani@osumc.edu

Introduction

- Surgical pathologists dedicate limitless hours in a structured training environment equipped with textbooks, scientific literature, principles, algorithms, and professional expertise to achieve proficiency in the diagnosis of disease and to interpret laboratory data
  - Based on their expertise and experience, they provide a report with the diagnosis, with or without prognostic factors
- With the emergence of genomics, proteomics, informatics, and associated metadata, in addition to the clinical, imaging, and laboratory data, surgical pathologists are now well equipped with managing, interpreting, and leveraging data of unmatched complexity

- **Bioinformatics** is a branch of science that refers to the management, acquisition, manipulation, and presentation of complex biological data sets, while clinical informatics is the application of information management in health care to promote safe, efficient, effective, and personalized care
- **Computational pathology (CPATH)** is the analysis of digitized pathology images with associated metadata, typically using artificial intelligence (AI) methods
  - Deep learning (DL) is a type of AI method, commonly used in CPATH that can 'learn' how to perform tasks based on examples
  - Training a typical supervised deep learning algorithm in CPATH involves large amounts of laboratory training data. Understanding the components of bioinformatics pipelines is also important for surgical pathologist
  - A few examples of bioinformatics-based computational pathology include predicting cancer outcomes from histology and genomics using convolutional networks and automated Gleason grading of prostate cancer tissue microarrays via deep learning

Bioinformatics, a Distinct Field in Pathology and Where Are We Now

- **Bioinformatics** is a blend of Computer Science, Statistics, and Biology. Several pathology laboratories, particularly the molecular pathology laboratories, have recruited specialists in dry laboratories who have skills in data management, statistics, computer programming, and data visualization that are needed for big-data applications in pathology, especially in the analyses of surgical pathology of cancers
- **Pathology Informatics (PI)** is a specialized branch of computational biology that deals with genetic and genomic information
  - Bioinformaticians/Pathology informaticians must have a strong background in population genetics, should have skills in computer programming, and must also have sufficient skills in information systems to set up and maintain computer-based algorithms required for genome-level data analyses and presentations
  - In essence, PI involves the collection, evaluation, reporting, and storage of large complex data sets derived from tests performed in clinical surgical pathology laboratories, and research laboratories to improve patient care and enhance our understanding of disease-related processes
  - PI is also ingrained in translational research tools and translates those into clinical practice required for timely, high-quality, accurate, regulated, contemporary, and safe patient diagnostic, prognostic, and predictive applications
  - Researchers with a background in PI who are entering the field of clinical laboratory practice need additional training to fit themselves into the clinical environment. Also, scientists coming from a wet laboratories background require targeted training in computational and information technologies

Pathology Bioinformaticians and Their Role

- A clinical bioinformatician manages the complete suite of information and computational technologies that tracks a sample from intake through clinical report generation and delivery
- Additionally, they are the key managers of data aggregation and data sharing. These specialists typically develop and maintain laboratory information management systems so that samples coming to the laboratories can be tracked throughout the entire testing process and result data is properly linked to the case. For example, once the raw data is produced, bioinformaticians develop, manage, and operate analyses pipelines that synthesize the results into forms comprehensible to the laboratory staff tasked with reporting the results
- Bioinformaticians maintain and develop analysis information management systems which are also used to collect and monitor performance metrics and quality control
- Sequencing database calling algorithms (primary data analysis), alignment and variant calling (secondary analysis), and variant annotation and filtering (tertiary analysis) must all fit together—ideally in a set of linked methods that minimize manual processes such as data reformatting
- All these steps require the use of complex software that needs professional supervision to manage installation, versioning and updating, knowledge of error modes, ability to manage outputs, carry out statistical analyses of validation data, and maintain quality standards
- The latter is exceptionally important because the quality metrics for sequencing are complex and failure to understand them deeply leads to increased costs, slower turnaround, excessive requirements for confirmatory testing, and even interpretive errors

Specialized Training in Bioinformatics

- When a new clinical laboratory discipline begins to coalesce, only some members are scientists who are already practicing clinical bioinformatics

- Some of these scientists have mastered bioinformatics skills by self-instruction, while other laboratories recruit research bioinformatics specialists or have new trainees enter the diagnostic laboratories with a knowledge of computational biology
- A group of scientists is effectively putting together this new discipline, and in the process, they are creating a pronounced impact on clinical diagnosis
- The next step is the recognition of the need for training standards and a process for the certification of specialists
- Individuals who have already in practice should be offered the opportunity to certify under a credentialing program
- Scientists who are already practicing in this area should be immediately eligible for certification after passing a skill-and knowledge-oriented board examination
- Two-year fellowships for new trainee scientists entering the field would be ideal to ensure that they have mastered all the skills necessary for dry laboratories clinical diagnostics
- These individuals must have training in human, population, and medical genetics and genomics
- They must master the basic principles of human disease genetics, gain exposure to the real-world problems of genetic diagnosis, and get specific training in ethical clinical practice
- These individuals need exposure and competence in the programming languages used in routine laboratory practice and systems architecture, and the algorithmic basis for primary and secondary analyses and develop an approach for tertiary analyses and reporting

Definitions and Terminologies Related to Bioinformatics and Computational Pathology

Annotation

- Indication of the position and/or outline of structures or objects within digital images, usually produced by humans using a computer mouse or drawing tablet. Annotations may have associated metadata
- Annotations can be manually generated or can be established by algorithm tools

Artificial Intelligence (AI)

- A branch of computer science that deals with the simulation of intelligent behavior in computers
- Recently, there has been a momentous drive to apply advanced AI technologies to diagnostic medicine

- The introduction of AI has provided vast new opportunities to improve health care and has introduced a new wave of heightened precision in oncologic pathology
- The impact of AI on oncologic pathology has now become apparent, and its use with respect to oral oncology is still in the nascent stage

Black Box/Glass Box

- A neural network can be perceived as a black box that lacks a clear depiction of the image features used for a decision
- However, several methods can be employed to transform it into a glass box to understand the relationship between the input parameters and the output of the network

Cloud Computing

- The practice of using a network of remote servers hosted on the internet to store, manage, and process data, rather than a local server or a personal computer

Computational Pathology (CPATH)

- A branch of pathology that involves computational analyses of a broad array of methods to analyze patient specimens for the study of disease

Convolutional Neural Network (CNN)

- A type of deep neural network particularly designed for images. It uses a kernel or filter to convolve an image, which results in features useful for differentiating images

Deep Learning

- The subset of machine learning composed of algorithms that permit software to train itself to perform tasks by exposing multilayered artificial neural networks to vast amounts of data
- Data is fed into the input layers and is sequentially processed in a hierarchical manner with increasing complexity at each layer, modeled loosely after the hierarchical organization in the brain
- Optimization functions are iteratively trained to shape the processing functions of the layers and the connections between them

Data Augmentation

- Method commonly used in deep learning to increase the training data using operations such as rotating, cropping, zooming, and image histogram-based modifications
- This provides several advantages such as promoting positional and rotational invariance, robustness to staining variability, and improves the generalizability of the classifier

Digital Pathology (DP)

- A blanket term that encompasses tools and systems to digitize pathology slides and associated metadata, their storage, review, analyses, and enabling infrastructure

Gold Standard

- The practical standard that is used to capture the 'ground truth'
- The gold standard may not always be perfectly correct, but in general, it is viewed as the best approximation

Ground Truth

- A category, quantity, or laboratories assigned to a dataset that provides guidance to an algorithm during training
- Depending on the task, the ground truth can be a patient- or slide-level characterization or can be applied to objects or regions within the image
- The ground truth is an abstract concept of the 'truth'

Image Analysis

- A method to extract typically quantifiable information from images
- Image analysis can be applied to images of histology slides, but the term itself is broader and applies to the extraction of information from any image, biomedical or not

Machine Learning (ML)

- A branch of AI in which computer software learns to perform a task by being exposed to representative data

Metadata

- In the context of DP, the term metadata describes descriptive data associated with the individual, sample, or slide
- It may include image acquisition information, patient demographic data, pathologist annotation or classification, or outcome data from treatment
- Typically, metadata are entries that allow searches in databases
- Highly complex, large, multiple-time-point associated data, such as longitudinal image data (such as radiology) or genomic data, is not usually called 'metadata'

Supervised Machine Learning

- Supervised learning is used to train a model to predict an outcome or to classify a dataset based on a laboratory associated with a data point (i.e., ground truth)
- An example of supervised machine learning includes the design of classifiers to distinguish benign from malignant regions based on manual annotations

Unsupervised Machine Learning

- Unsupervised learning seeks to identify natural divisions in a dataset without the need for ground truth, often using methods such as cluster analysis or pattern matching
- Examples of unsupervised machine learning include the identification of images with similar attributes or the clustering of tumors into subtypes

Whole-Slide Image (WSI)

- Digital representation of an entire histopathology glass slide, digitized at microscope resolution
- These whole-slide scans are typically produced using slide scanners
- Slide scan viewing software enables inspection of the image in a way that mimics the use of a traditional microscope; the image can be viewed at different magnifications

Differences Between Traditional Image Analysis and Computational Pathology

- Traditional image analysis differs from the deep learning based computational pathology in the following ways (Table 2.1)

Table 2.1 Differences between traditional image analysis and deep learning-based computational pathology

Characteristics	Image analysis	Deep learning-based computational pathology
Definition and tasks performed	It is the extraction of meaningful information from the digital images by means of digital image processing techniques. It is the automation of repetitive tasks, e.g., cell counting, quantification of immune staining areas, etc.	It is the analysis of digitized pathology images with associated metadata, typically using AI methods, and the image features are correlated with the patient metadata that assist in disease diagnosis and predict therapy.
Tuning of the parameters used	Image features are manually tuned.	Automatic learning and extraction of a large number of features by computer-based algorithms.
Algorithm testing	A few marked regions of the stained slide.	The entire stained slide.
Computer unit used	Central processing unit.	Graphics processing unit.
Number of images required in the training set	It depends on the application, usually low.	Usually, very high images are required.

Digital Pathology, Machine Learning, and Computational Pathology

- The development of slide scanners has made the process of virtualizing and digitalizing the whole glass slides possible
- DP involves the process of digitizing histopathology, immunohistochemistry, and cytology slides using whole-slide scanners and the interpretation, management, and analysis of these digitized whole-slide images using computational approaches
- The digital data of the slides can be stored in a central cloud-based space allowing for remote access to the information for manual review by a pathologist or automated review by a computer-assisted data algorithm
- This enables AI, a branch of computational biology that generates the data algorithms, to be applied to pathology
- Based on the degree of intelligence, AI can be divided into two major categories such as weak AI and strong AI
- Weak AI or artificial narrow intelligence refers to the classification of data based on a well-established statistic model that has already been trained to perform specific tasks, while strong AI or artificial general intelligence (AGI) can create a system, which can function intelli-

gently and independently by executing machine learning from any available normalized data

- Generally, ML is an AI process to allow a computer to automatically learn and improve from the data set by itself and to solve problems without being programmed during the process
- ML is an advanced branch of AGI using a large amount of initial data, training set, to build statistic algorithms to interpret and act on new data later
- Currently, various ML-based approaches have been developed and tested in pathology to assist pathologic diagnosis using the basic morphology pattern such as cancer cells, cell nuclei, cell divisions, ducts, blood vessels, etc.
- Deep learning, otherwise known as deep structured learning, is a subfield of ML that is based on artificial neural networks (ANNs) in which the statistic models are established from input training data
- Deep neural networks provide architectures for deep learning
- The ANNs can perform their own determination as to whether their interpretation or prediction is correct, resembling a biological complex neural network of the human brain
- ANNs are comprised of three functional layers of artificial neurons, known as nodes, which include an input layer, multiple hidden layers, and an output layer
- The artificial neurons are connected to each other in the ANNs, and the strength of their connections is known as weights
- The connections between artificial neurons in the ANNs are assessed using statistical methods, including clustering algorithms, K-nearest neighbor, support vector machines (SVM), and logistic regressions
- The involved artificial neurons, which are related to the output event, and their associated connections, which bear different weight, need to be trained by qualified big dataset to achieve an optimized algorithm for specific tasks
- The CNNs are a type of deep multilayer neural network particularly designed for visual image
- It employs convolutional kernels, a set of learnable filters, to build up a pooling layer that can effectively reduce the dimensions of the image data while still retaining its characteristics
- By flattening an image, removing, or reducing the dimensions, convolutional kernels act as a preprocess treatment that then allows for computer vision and machine vision models to process, analyze, and classify the digital images, or parts of the image, into known categories
- With slide scanning technology getting faster and more reliable, a larger volume of WSI data becomes available to train and validate CNN models

- In combination with clinical information, biomarkers, and multi-omics data, CPATH will become part of the new standard of care
- Computational pathology not only facilitates a more efficient pathology workflow, but also provides a more comprehensive and personalized view, enabling pathologists to address the progress of complex diseases for better patient care

Training Algorithm

Case Selection

- Patients' selection is the initial step in training the algorithm (Fig. 2.1)
- Both training set and validation set must include all sample types or variants which are related to the subject of diseases including stages, grades, histologic classification, complication, etc., to eliminate false-negative and false-positive scenarios
- Still very much a machine-driven process, algorithms have no way to recognize the variants that have not been included in the training set
- The criteria for the samples and subsequent slide selection for the learning set need to be established by experienced pathologists alongside a computational team
- Confounding variables must be isolated and removed. For example, the patients with other medical conditions which may interfere with the outcome should be eliminated
- In addition, inadequate slide preparation including blurred vision, over- or under-staining, air-bubbles, and folded tissue can produce inaccurate results and wrong algorithms

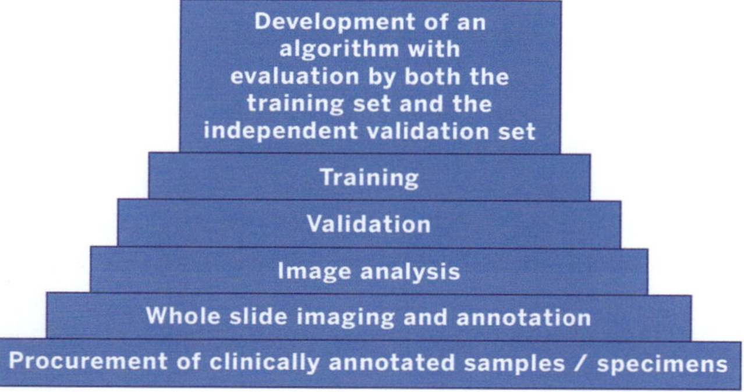

Fig. 2.1 The process of creating an algorithm is basically divided into the above steps. The initial step involves the procurement of clinically annotated samples/specimens, followed by the process of whole-slide imaging and annotation. Based on the image analysis data, an algorithm is developed and trained by both the training set and the independent validation set

- The comprehensive initial and follow-up clinical information, as well as laboratory results, should be collected and included
- The more relevant the information included, the more accurate the resulting algorithm

Whole-Slide Imaging

- Histopathology has been an integral part of the work of pathologists since the seventeenth century
- Today, histopathology largely remains a manual process in which pathologists examine glass slides using conventional brightfield microscopy
- Advances in digitization of glass slides in pathology occurred much later than the digital transformation witnessed in radiology, where digital sensors are widespread
- When histologic glass slides are digitized, they can be remotely viewed by a pathologist on a computer screen, or digitally analyzed using image analysis techniques
- At its inception, digital image analysis was predominantly used by researchers and often limited to individual fields of views, which was cumbersome and could introduce bias
- WSI allowed developments that have brought us from the application of traditional image analysis techniques on small manually selected regions of interest, to what is the current state-of-the-art in digital pathology: techniques that process the entire slide image automatically
- This allowed researchers to identify features not easily analyzed by visual evaluation alone
- Several slide scanning systems for WSI have been approved by the US Food and Drug Administration (FDA) to be used in clinical settings (Table 2.2)
- The first FDA-approved Ultra-Fast Scanner, the Philips IntelliSite Pathology Solution (PIPS), has a resolution of 0.25 μm/pixel, scanning speed of 60 s for a 15 × 15-mm scan area, and scanning capacity of 300 slides in one load
- The Aperio AT2 DX System from Leica Biosystems has 400 slide capacity for brightfield and fluorescent slides
- File sizes of digital images at applicable resolution vary depending on the scan area on the glass slides
- In general, pathology images are tremendously large, in the range of 1–3 GB per image
- Therefore, it requires a high capacity and fast digital working computer
- Furthermore, the number of slides needed to achieve a clinically accepted algorithm may vary by tissue type and diagnosis
- Campanella et al. showed that at least 10,000 slides are necessary for training to reach a good performance

Table 2.2 Various whole-slide imaging systems/devices

Name of the device	Vendor	Brightfield scanning technology	Fluorescence scanning technology	Slide loading capacity	Brightfield scanning speed	Fluorescence scanning speed	Highest throughput slides/h	Resolution (μm/pixel) × 20/×40 objective
VENTANA iScan HT	ROCHE	–	Integrated LED	160	Less than 2 min per slide	–	–	–
Aperio AT2	Leica	TDI line scan	No	400	72 s (20×)/3 min (40×)	No	50	0.50/0.25
VS200	Olympus	2/3-inch complementary metal oxide semiconductor, 3.45 × 3.45 μm pixel size	1-inch complementary metal oxide semiconductor, 3.45 × 3.45 μm pixel size	210	80 s	–	–	0.274/0.137
PathFusion	ASI	5 MP complementary metal oxide semiconductor color	–	99	–	–	–	–
TissueScope LE120	HURON digital pathology	0.75	No	120	Less than 1 min per slide	No	–	0.4/0.2
Philips IntelliSite pathology solution	Philips	Time delay and integration line scanning	–	300	60 s	–	–	0.25
3DHISTECH Pannoramic 250 FLASH III	Epredia	12 MP 12-bit camera with xenon flash illumination	Additional 4.2 MP 16-bit camera with 6-channel LED	250	35 s (20×)/1 min 35 s (40×)	5 min @ 31× 15 min @ 62×	60	0.242/0.121
ProScanner APro 5	AMOS	–	–	–	–	–	–	–
HLM8	Histo-line laboratories	–	–	–	–	–	–	–
Axiocam 208 color	ZEISS	No	1/2.1-inch Sony complementary metal oxide semiconductor, 1.85 × 1.85 μm pixel size	–	–	–	–	–
Axiocam 202 mDavidono	ZEISS	1/2.3-inch Sony complementary metal oxide semiconductor, 5.86 × 5.86 μm pixel size	No	–	–	–	–	–

Table 2.3 Basic principles for the development of supervised deep-learning algorithm development in computational pathology

Obtaining ground truth data
• Patient outcome data
• A field from the pathology report or laboratory information system
• A quantitative score assigned to the case
• Manually provided by a pathologist
• Considerations in acquisition
– Streamlined workflows
– Single common annotation tool
– Trade-off between quantity and accuracy
Good practices
• Training images must be representative of the image algorithm it is designed to be applied to
• Use a wide variety of data sources
• Use consistent preimaging steps
• Apply manual or automated image quality control processes
• Use larger and more representative training sets
• Calibrate algorithms for each laboratory prior to being used for clinical work
• Apply image preprocessing strategies such as color normalization
• Data augmentation to artificially add variation and increase (or balance) the training data
• Test developed models using a variety of test and validation sets to avoid overfitting

- The authors also observed the discrepancy in the prediction between Leica Aperio and PIPS and found that brightness, contrast, and sharpness affect the prediction performance
- Whole-slide image analysis techniques are now routinely utilized for basic and translational research, drug development, and clinical diagnostics including laboratory-developed tests and in vitro diagnostics

Traditional Image Analysis Enhanced by Machine Learning

- Traditional digital image analysis focuses on three broad categories of measurements: localization, classification, and quantification of image objects
- This method is an iterative process where typically a few parameters are manually tuned, built into an algorithm, and often tested only on a region of the slide image

- Aspects that fail a quality control review are tweaked until the algorithm performance meets predetermined analysis criteria
- ML has facilitated significant advancements within the field of image analysis, as it often allows the generation of more robust algorithms that need fewer iterative optimizations for each dataset, compared with methods where parameters are manually tuned
- Supervised ML techniques, in which an algorithm is trained using ground truth laboratories, are particularly effective in image segmentation (detection of specific objects) and classification (such as tumor diagnosis) tasks
- The ground truths may be a category or laboratories assigned to a dataset that provides guidance to an algorithm
- The capabilities of ML have dramatically expanded in the last decade due to the developments in deep learning, an approach that enables an algorithm to automatically discover relevant image features that contribute to computer vision tasks
- One of the first uses of deep learning in histopathology was the work of Ciresan et al. in the International Conference on Pattern Recognition (ICPR) challenge in 2012, which focused on fully automated recognition of mitotic figures in hematoxylin and eosin (H&E)-stained breast cancer tissue
- Using CNNs, the authors were able to generate results that far exceeded those of the competition
- These early studies applied ML to histopathology using small, manually selected regions of interest, but later research showed that these techniques could work equally well on whole-slide images
- In DP, ML-enhanced image analysis is now widely employed by researchers and implemented in several commercially available image analysis software products (Table 2.3)
- For digital slide analysis, Senaras et al. described a novel deep-learning framework, called DeepFocus, which enables the automatic identification of blurry regions in digital slides for immediate re-scan to improve image quality for pathologists and image analysis algorithms

- Janowczyk et al. presented an open-source tool called HistoQC to assess color histograms, brightness, and contrast of each slide and to identify cohort-level outliers (e.g., darker or lighter stain than other slides in the cohort)
- These methods play an essential role in the quality control of whole-slide images to standardize the quality of images in computational pathology
- Due to improvements in various smart image-recognition algorithmic discriminators, based on high-capacity deep neural network models, the pathologist can be released from extensive manual annotations for each whole-slide images at the pixel level so that they can focus on other parts of the clinical workflow
- The patch-based whole-slide images (224×224 to 256×256) have been widely used in many machine learning domains to train classifiers for diagnostic or prognostic tasks
- For example, Campanella et al. employed multiple instance learning (MIL) approaches with 'bag' and 'instance' based on convolutional neural networks and recurrent neural networks to classify the prostate cancer images of H&E slides
- Kapil et al. applied deep semi-supervised architecture and auxiliary classifier generative adversarial networks, including one generator network and one discriminator network, to automatically analyze the PD-L1 expression in immunohistochemistry slide of late-stage nonsmall cell lung cancer needle biopsies
- Barker et al. revealed an elastic net linear regression model and weighted voting system to differentiate glioblastoma and lower-grade glioma with an accuracy of 93.1%
- Several machine learning tools such as QuPath, Halo, Visiopharm, Image J are readily available for automatic quantification of various biomarkers
- Of these, QuPath is an open-source digital pathology tool that has been used for the objective assessment and scoring of several markers like Ki67, KLF4, SOX2, etc.
- QuPath performs WSI analysis through tissue and nuclei segmentation and automatically computes a series of features with various algorithms
- Briefly, stain vectors are identified using QuPath tools, then applied to ROI using the 'tissue detection' command (Fig. 2.2)
- After a step of evaluable tissue selection, excluding all nonassessable areas and regions without tumors, the 'cell detection' tool is used to segment nuclei
- If necessary, it is possible to then proceed to draw further annotations around areas of interest
- These can be processed one-by-one by running 'positive cell detection' on an annotation when it is selected, or else they can be processed all together (in parallel)
  - This has been used for the assessment of the proliferative compartment of solid tumors on H&E-stained sections as well as for an objective determination of Ki67 in cancers

Fig. 2.2 The relationship between different levels of artificial intelligence and important hurdles. The four challenges are experienced computational experts who can develop algorithms for particular issues, hardware limitations, qualified applicable data, and ethical issues

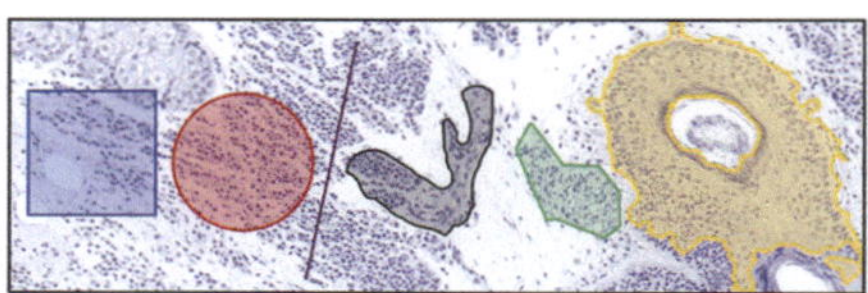

Annotations made using different drawing tools (left to right) Rectangle, Ellipse, Line, Brush, Polygon, Wand

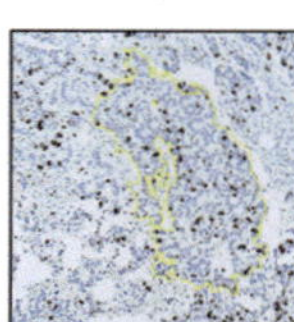

Annotations of region of interest (ROI)

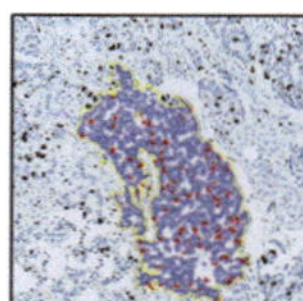

Estimating stain vectors and positive cell detection of ROI

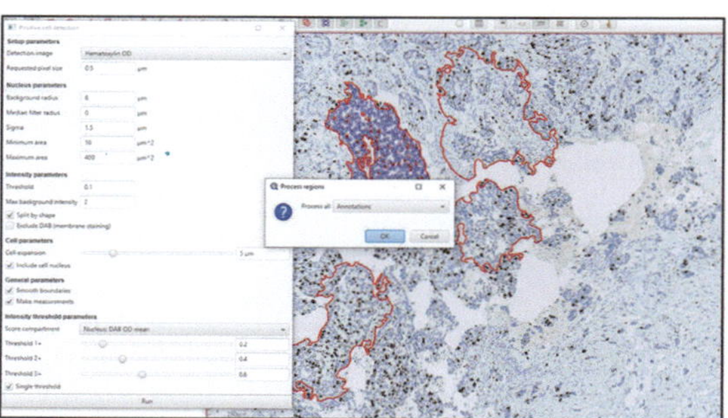

Positive cell detection in parallel for multiple annotations

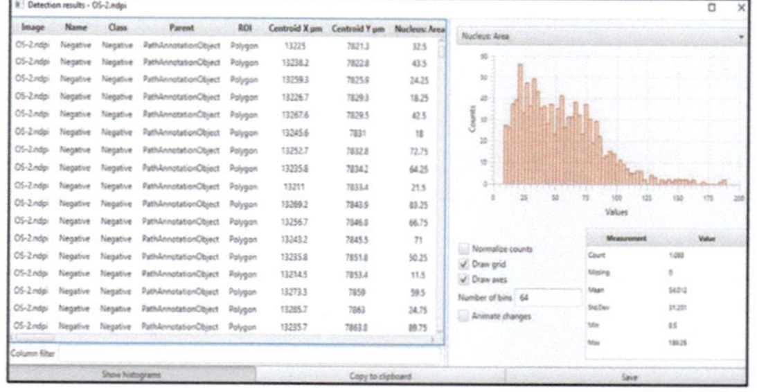

Cell detection results table

Pathologist-Centered Medical System

- Although most AI research is still focused on the detection and grading of tumors in digital pathology and radiology, computational pathology is not limited to the detection of a morphological pattern
- It can also contribute to the complex process of analysis and judgment using demographic information, digital pathology, −omics, and laboratory results
- Therefore, AI has the potential to contribute to nearly all aspects of the clinical workflow, from more accurate diagnosis to prognosis, and individualized treatment
- Multiple sources of clinical data are incorporated into mathematic models to generate diagnostic inferences and predictions, to enable physicians, patients, and laboratory personnel to make the best possible medical decisions
- For example, deep neural networks have been applied to automated biomarker assessment of breast tumor images, such as HER2, ER, and Ki67
- Hamidinekoo et al. created a novel convolutional neural network-based mammography–histology–phenotype–linking–model to connect and map the features and phenotypes between mammographic abnormalities and their histopathological representation
- Mobadersany et al. developed a genomic survival convolutional neural network model to integrate information from both histology images and genomic data to predict time-to-event outcomes and demonstrated that the prediction accuracy surpassed the current clinical paradigm for predicting the overall survival of patients diagnosed with glioma
- As electronic health record (EHR) systems enable us to collect medical data such as age, race, gender, social history, and clinic history, applying these data as independent factors of a particular disease to an appropriate mathematic algorithm becomes feasible
- These integrated data allow pathologists to gain deeper insights and to switch between different algorithms of treatment at different stages of the disease and/or for different statuses of the patient
- As the health-related apps on mobile devices and smart personal trackers become popular, direct access to continuous real-time health information, such as temperature, heart rate, respiratory rate, electrocardiogram, body mass index, blood glucose, and blood oxygen content, can be recorded into individual health data
- These data can then be incorporated into the EHR and laboratory information systems (LIS) to reintegrate into a virtualized and digitalized person, which was not possible previously and was beyond what the human brain alone can accomplish
- This new system of data-driven care requires the pathology, as a cornerstone of modern medicine, to integrate data, algorithms, and analytics to deliver high-quality and efficient care
- The combination of computational pathology and big-data mining offers the potential to create a revolutionary way of practicing evidence-based, personalized medicine

Global Pathology Service Model

- Three essential advancements happened in recent years: the possibility to store a great amount of data from network-attached storage to cloud storage, the growing speed of network from WIFI-6 to 5 G, and high-performance central processing unit (CPU) and graphics processing unit
- These technological improvements not only enhance people's daily life, but also have a great impact on medicine, especially digital and computational pathology (Fig. 2.3)
- Together with the surging development of network and information technology, these technologic improvements allow for the centralization of medical and computing resources—with the benefit of larger sample data volume for optimization of algorithms
- Furthermore, the central cloud-based AI laboratory and data bank of digital and computational pathology make the global network of computational pathology possible
- In local laboratories or centralized scanning centers, histology slides can be converted to whole-slide images and numerical data
- This data can then be transferred to the central laboratory together with EHR data and multi-omics data for further analysis
- Patients in different geographic areas around the world can benefit from more efficient and effective diagnosis, treatment, and follow-up
- In the meantime, pathologists can access the information they need to care for patients or to collaborate with specialists anytime and anywhere
- Deep-learning platforms have the potential to facilitate the discovery of more complicated or subtle connections and to help pathologists make the best clinical decisions to meet every patient's needs

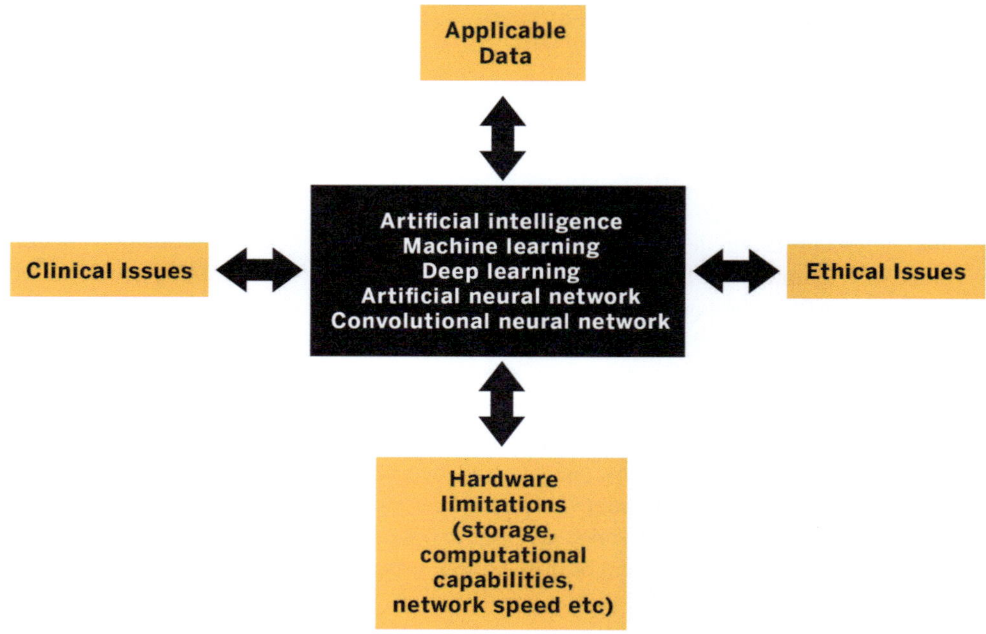

Fig. 2.3 A computational pathology team includes pathologists to identify a clinically relevant issue, data scientists to develop and train the algorithm, and engineers to support the operating environment. Subsequently, during the actual clinical practice, the pathologists play an important role in applying and monitoring the algorithm and relay feedback to the developers for optimization

Examples of Bioinformatics and AI-Driven Pathology Workflow

- Increasingly, AI detection is being applied to different subspecialties with various sample types
- Early reports on accuracy have shown to be promising and that the AI-assisted systems have the potential to classify accurately at an unprecedented scale and lay the foundation for the deployment of computational pathology in nearly all subspecialties

Prostate Cancer

- Campanella et al. validated a high-capacity deep neural network-based algorithm to analyze image classification and categorization of 44,732 whole-slide images across three different cancer types, including prostate cancer, basal cell carcinoma, and breast cancer metastases to axillary lymph nodes
- In terms of whole-slide images, they found that ×5 magnification has higher accuracy
- They trained a statistic model with MIL-based tile classifier for each tissue type and achieved area under receiver operating curve (AUC) above 0.98 for all cancer types
- Its clinical application would allow pathologists to exclude 65–75% of slides while retaining 100% sensitivity
- Wildeboer et al. discussed deep learning techniques based on different imaging sources including magnetic resonance imaging, echogenicity in ultrasound imaging, and radio density in computed tomography as computer-aided diagnostic tools for prostate cancer

- They found that the algorithm of convolutional neural network architecture performed equal or better than SVM or random forest classifiers in machine learning
- As usage of AI for diagnosing prostate cancer in biopsies are limited to individual studies, they lack validation in multinational settings
- The PANDA challenge, the largest histopathology competition to date, joined by 1290 developers, has been organized to catalyze development of reproducible AI algorithms for Gleason grading using 10,616 digitized prostate biopsies
- They validated a diverse set of submitted algorithms that reached pathologist-level performance on independent cross-continental cohorts, fully blinded to the algorithm developers
- In United States and European external validation sets, the algorithms achieved agreements of 0.862 (quadratically weighted κ, 95% confidence interval (CI), 0.840–0.884) and 0.868 (95% CI, 0.835–0.900) with expert uropathologists
- Successful generalization across different patient populations, laboratories, and reference standards, achieved by a variety of algorithmic approaches, warrants evaluating AI-based Gleason grading in prospective clinical trials

Colorectal Cancer

- Korbar et al. developed multiple deep-learning algorithms, modified version of a residual network architecture, which can accurately classify whole-slide images of five types of colorectal polyps, including hyperplastic,

sessile serrated, traditional serrated, tubular, and tubulo-villous/villous polyps
- Among 2074 images, 90% of them were used for model training and the remaining 10% of images were assigned to the validation set
- The overall accuracy for classification of colorectal polyps was 93% (confidence interval (CI) 95%, 89.0–95.9%)
- Bychkov et al. combined CNNs and recurrent neural network architectures to predict colorectal cancer outcomes based on tissue microarray (TMA) samples from 420 colorectal cancer patients
- Their results show that the AUC of deep neural network-based outcome prediction was 0.69 (hazard ratio, 2.3; CI 95%, 1.79–3.03)
- For comparison, pathology experts performed inferiorly on both TMA samples (HR, 1.67; CI 95%, 1.28–2.19; AUC, 0.58) and whole-slide level (HR, 1.65; CI 95%, 1.30–2.15; AUC, 0.57), which implied that deep neural networks could extract more prognostic information from the tissue morphology of colorectal cancer than an experienced pathologist

Breast Cancer

- Wang et al., the team of winner of competitions in the CAMELYON16 challenge, used input 256 × 256-pixel patches from positive and negative regions of the whole-slide images of breast sentinel lymph nodes to train various classification models including GoogLeNet Patch, AlexNet, VGG16, and FaceNet
- The patch classification accuracy is 98.4, 92.1, 97.9, and 96.8% separately
- Among the algorithms, GoogLeNet has the best performance and is generally faster and more stable, which achieved AUC of 0.925 for whole-slide images classification
- With the assistance of deep learning system, the accuracy of pathologist's diagnoses improved significantly as the AUC increased from 0.966 to 0.995, representing ~85% reduction of human error rate
- Furthermore, the open resource of a data set of annotated whole-slide images for CAMELYON16 and CAMELYON17 challenges enable testing of new machine learning and image analysis strategies for digital pathology

Cancer Cytopathology

- Martin et al. applied convolutional neural networks for classifying cervical cytology images into five diagnostic categories, including negative for intraepithelial lesion or malignancy, atypical squamous cells of undetermined sig-

nificance, low-grade squamous intraepithelial lesion, atypical squamous cells cannot exclude low-grade squamous intraepithelial lesion and high-grade squamous intraepithelial lesion, and achieved accuracies of 56%, 36%, 72%, 17%, and 86% separately, which implies convolutional neural networks are able to learn cytological features
- In another cytopathology study, the authors used morphometric algorithm and semantic segmentation network based on VGG-19 to classify urine cytology whole-slide images according to Paris System for Urine Cytopathology and achieved a sensitivity of 77%, false-positive rate of 30% and AUC of 0.8

Challenges and Limitations

- Despite the promises of CPATH and advancement in ML with promising results and benefits, most algorithms used in current clinical practice are limited to traditional image analysis of immunohistochemical stains, which do not employ advanced ML techniques such as deep learning
- In this section, we address the many barriers to implementing CPATH for clinical use, and potential strategies to overcome them (Fig. 2.4)

Infrastructure Considerations

- Implementation of CPATH may require a significant investment in IT infrastructure
- In general, data to be analyzed is captured as images of tissue sections, often scanned at 20× or 40× objective magnification
- In clinical practice, pathology images are commonly larger than 50,000 by 50,000 pixels
- As a benchmark, this can translate into estimated file sizes ranging from 0.5 to 4 GB for 40× images, depending on the size of the scan area and image compression type
- The large size of these images may present a problem for evaluation, storage, and inventory management
- The primary computing obstacles that users face are processor speed and memory requirements of local workstations, data storage requirements, and limitations of the network
- For CPATH to perform effectively, it is important that there are safeguards to ensure that images are fully loaded, and that the analysis algorithm is not interrupted due to insufficient bandwidth, processing power or memory
- Additional considerations when running deep-learning algorithms include, but are not limited to, the number of intended users, flexibility of the server or cloud configuration to accommodate new algorithms or caseloads, cybersecurity, and associated costs

Fig. 2.4 Cell detection by QuPath involves annotation of ROI, estimation of stain vectors, followed by positive cell detection, which gives a tabulated result and thus removes subjectivity in pathology

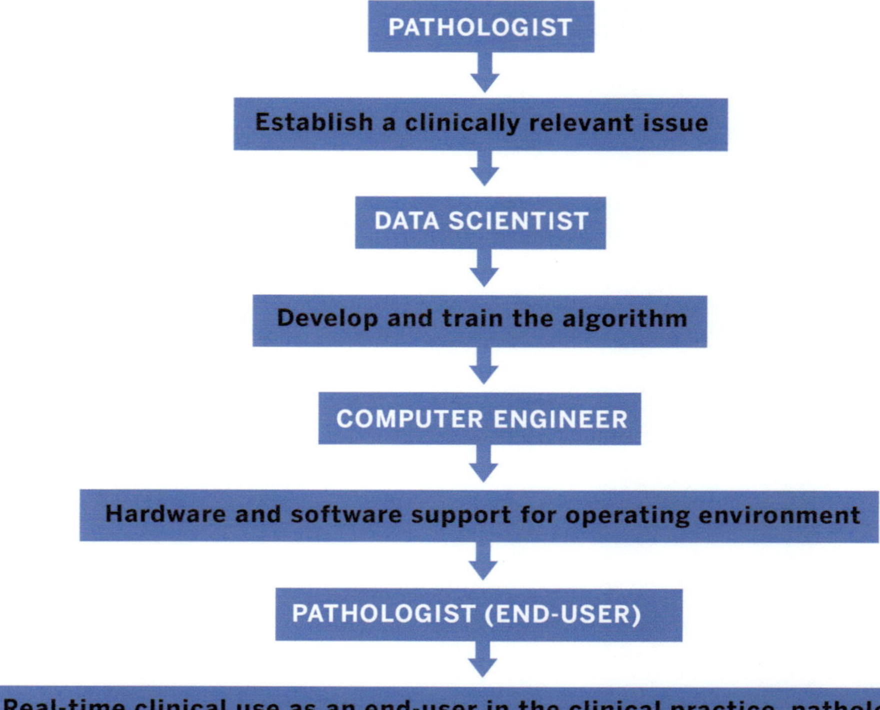

Processor Speed and Sources

- The performance of any image processing is highly dependent on processor speed
- Deep learning is best performed using graphics processing units (GPUs), which can provide significant performance enhancement over central processing units (CPUs)
- Most computers are designed to perform computations on their CPU and use the GPU simply to render graphics
- It may be necessary to purchase a more powerful GPU designed for deep learning; these are generally more expensive and tend to generate more heat
- Some laboratories may therefore elect to dedicate high-performance workstations strictly for deep learning
- However, some vendors offer the ability to perform image analysis at the server or cloud level, which may provide significantly more resources and can potentially distribute deep learning capabilities to a much larger user base

Network Limitations

- For implementations in which either data is stored remotely, or image processing is performed remotely, network bandwidth becomes an important consideration

- The large size of whole-slide images presents a potential hurdle for efficient processing in environments that lack sufficient bandwidth
- Depending on the network implementation, there are several data transfer considerations
- First, digital slide data from the whole-slide scanner must be transferred to its network storage location, which requires the file in its entirety
- Second, the digital slide must be transferred from its network storage location to the image analysis environment (which may reside locally, elsewhere on the network, or in the cloud), which can often be accomplished in a more efficient manner, since the entire image is unlikely to be analyzed at once
- Training a deep learning network on an entire slide image at full resolution is currently very challenging, so it usually operates on a smaller tiled image or patch
- Downscaling (reducing resolution) of these images is one possible approach, but this may lead to loss of discriminative details as using small, high-resolution tiles may lose tissue context
- The optimal resolution and tile size for analysis are highly case-dependent
- If only small regions of interest are to be processed, or if the processing can occur at a reduced magnification,

smaller portions of the virtual slide file need to be transferred due to the pyramid structure of most WSI file formats

Acquiring Training Data

- Deep learning is generally extremely data-hungry, especially compared with traditional image analysis where 'important features' are manually selected, as it must automatically identify these features
- For supervised learning, in addition to the raw image sets, a ground truth must be included in the dataset to provide appropriate diagnostic context
- Algorithms can then be trained to predict or characterize an image guided by the ground truth provided
- The ground truth may be derived from patient outcome data, a field extracted from the pathology report or laboratory information system (e.g., histologic grade), a quantitative score assigned to the case (e.g., molecular testing), or it can be a factor manually provided by a pathologist reviewing the case specifically to support the algorithm training
- Obtaining clinical ground truth data suitable for algorithm development is often time-consuming and challenging
- It usually takes a long time to generate enough survival data from clinical patients, and clinical data are generally locked into an unstructured format within one or more disparate electronic medical records
- The data must be manually or automatically curated before being incorporated into an algorithm
- Furthermore, training may also require manual annotations applied to the digital slide, including the designation of specific areas of interest, for instance identifying cancer from benign tissue
- Obtaining adequately annotated datasets for deep learning by a trained expert can be difficult due to the amount of time required, associated expenses, and the tedious nature of the task
- The use of streamlined workflows and a single common annotation tool with an intuitive user interface can make the task of creating and sharing manual regional annotations considerably easier
- Web-based tools may be ideal in sharing annotations between different research groups, as they avoid the need to install specific software on multiple systems
- In addition, research has shown that, for some tasks, annotations of expert observers (i.e., pathologists) may not always be necessary
- Yet there is generally a trade-off between quantity and accuracy
- Also, training images must be representative of the images that the algorithm is designed to be applied to, and appropriately 'balanced'; for example, contain approximately similar numbers of examples for different objects it is intended to identify

Data Variability

- It is important that supervised algorithms are developed using a wide variety of data sources, to handle variations more robustly when exposed to other datasets
- A pertinent consideration is to implement prospective review using retrospective data during development, and/or verification or validation
- When algorithms are developed using limited datasets supplied by only one or few pathology laboratories, the algorithms may not have incorporated all the variations and artifacts encountered across different labs, including preimaging, imaging, and postimaging steps within the WSI workflow
- This is in part because, in surgical pathology, there is currently no accepted global standard for tissue processing, staining, and slide preparation
- Even digital acquisition may introduce variability. As such, an algorithm designed to perform well on one set of WSIs may not perform equally well when generalized and used around the world by many laboratories
- This could be somewhat alleviated by implementing consistent preimaging steps, applying manual or automated image quality control processes, using larger and more representative training sets, and calibrating algorithms for each lab prior to being used for clinical work
- It is also possible to apply image preprocessing strategies such as color normalization to reduce the impact of stain and processing variability, and data augmentation to artificially add variation and increase (or balance) the training data to make it more representative of the application data
- Other best practices include testing developed models using a variety of test and validation sets to avoid overfitting, and clearly reporting characteristics of the patients used to build a model, since additional training data may be required for it to perform well on other populations
- One may consider the addition of prospective real-world data collection to monitor and optimize performance

Public Sources

- There are currently only limited publicly available datasets with annotated images and associated nonimage patient data that are required for CPATH
- This may be one of the greatest factors limiting progress in the field of CPATH

- However, some initiatives to overcome these hurdles are described below
- The Cancer Genome Atlas (TCGA) has performed comprehensive molecular profiling on approximately 10,000 cancers (National Cancer Institute, The Cancer Genome Atlas, https://cancergenome.nih.gov/)
- In addition to the collection of molecular and clinical data, TCGA has collected WSI data from a subset of its participants
- Some other examples of public digital slide datasets are the breast cancer images used for the CAMELYON competition; the Medical Image Computing and Computer Assisted Intervention Society (MICCAI) 2014 brain tumor digital pathology challenge for distinguishing brain cancer subtypes; and the Tumor Proliferation Assessment Challenge (TUPAC16), which includes hundreds of cases
- The Grand Challenge website (https://grand-challenge.org/challenges/) maintains a list of all challenges that have been organized in the field of medical image analysis
- Some of these challenges offer developers additional pathology digital datasets for CPATH
- However, care must be taken as the quality of public samples may be variable, so they should be carefully tested before use

Crowdsourcing

- An alternative for obtaining large-scale image annotations is crowdsourcing, in which this function is outsourced to an undefined and generally large group of nonexpert people in the form of an open call
- Crowdsourced image annotation has been successfully used to serve a diverse set of scientific goals, including the detection of malaria from blood smears, and estrogen receptor classification
- Compared with public sources or pathologist annotations, crowdsourcing may be cheaper and quicker, but it has the potential to introduce noise
- It is possible that this noise can be compensated by a sufficiently large body of training data, and by having multiple people annotate the same slide to achieve consensus
- But it is imperative to ensure that all annotators are taught to perform the task in the same way

Active Learning

- Active learning is considered semi-supervised learning, which may reduce the size of required training data
- In active learning, the algorithm interactively queries for expert assistance to obtain annotations for ambiguous data points

- Essentially, the algorithm uses a sampling strategy to select small sets of data iteratively for experts to label, only when it has trouble determining the outcome
- For each iteration, the classifier is updated, and then all unlabeled data are re-evaluated for their ability to further improve the classifier
- Thus, active learning offers a solution to the problem of limited data annotations in pathology by having the pathologist engage actively with the algorithm, which evolves through continuous learning

Quality Control and Reliability of the Algorithm

- It is currently difficult to establish strict quality control steps for deep learning algorithms, especially in segmentation problems, for various reasons
- A general principle in training any machine learning algorithm is to split the annotated data into 'training' and 'test' datasets and ensure that these sets are independent when assessing performance
- The algorithm should be trained on the training set and applied to the test set, then the results should be compared with the 'ground truth' associated to the test set
- However, quality control for the segmentation step may suffer from the 'gold-standard paradox'
- This paradox arises from histopathological assessments by the pathologist being considered the gold standard, but the algorithm data may in fact be more reproducible than human assessment
- This may be partially overcome by comparing the algorithm data to patient outcome, to see whether it is better able to predict outcome compared with manual pathology assessment/scoring
- Still, the best methods to determine the reliability of an algorithm applied to novel datasets are an area of active debate
- In addition, local regulations apply to legally market any clinical-grade software solution
- In the USA, such an algorithm should be developed under the Food and Drug Administration's existing Quality System Regulation (QSR, 21 CFR Part 820), and Good Machine Learning Practices (GMLP; https://www.fda.gov/media/122535/download), which are currently being discussed

Understanding Algorithms

- A principal concern with the use of deep learning is that it is very difficult to understand some of the features and neural pathways used to make decisions

- When deep learning is used to automatically extract features from an image that are directly correlated to clinical endpoints, without including a segmentation step where image objects are first extracted (see the section on correlating images to patient response), it is particularly challenging to understand why the algorithm reached its conclusions
- Artificial neural networks have accordingly been described as a 'black box'
- This has led to several concerns: difficulty in correcting an underperforming algorithm; lack of transparency, explicability, and provability for humans who may not trust how an algorithm generates reliable results; and regulatory concerns because, unlike traditional image analysis, in deep learning the image features are abstracted in a way that is very difficult for a human to understand
- In response, there have been efforts to convert deep-learning algorithms into a 'glass box' by clarifying the inputs and their relation to measured outputs, making it more interpretable by a human using a variety of techniques
- By providing information to the reviewing pathologist about the histopathologic features used by the algorithm in a particular instance, trust in the algorithm can be fostered, and synergy between pathologist and machine can be achieved that may exceed the performance of either AI or pathologist alone

Ethics

- In the new era of computation-driven decision-making processes based on AI and machine learning, computational pathology will involve more complicated interactions of massive information from clinical history, omics data, living environment to social habits
- It is very likely that the experts involved in these decision-making processes will no longer be exclusively pathologists
- Instead, the decision-making panel will include other experts such as data statisticians and bio-informaticians, which may raise ethical concerns
- A continuous massive, sensitive health data transfer among clinics, laboratories, and data banks can enable higher precision medicine but, at the same time, increases the security vulnerability
- Policies around the strict protection of patient privacy and personal data create an obstacle for computational pathology to access the health databases need to create more comprehensive training data sets
- General Data Protection Regulation was enacted in May 2018 in Europe to impose new responsibilities on organizations that process the data of European Union citizens for scientific research

- This concept highlights the proportionate approach to regulate computational pathology-related security and ethical issues while not limiting innovation unduly, which is difficult but critical

Cyber-Security

- Cyber-security concerns of CPATH primarily stem from storing large amounts of medical data in cloud-based systems that can be accessed via the Internet
- To minimize a data breach, it is prudent to decouple CPATH data (i.e., digital images) from patient data (i.e., personal identifiers such as medical record number and date of birth)
- Several cloud service providers now offer Health Insurance Portability and Accountability Act (HIPAA) compliant solutions
- In addition, the FDA has created guidelines for cyber-security (U.S. Food and Drug Administration, Postmarket Management of Cyber-security in Medical Devices, 2016)
- The European Union's general data protection regulation (GDPR) imposes similar security requirements on those who process personal data

Future Directions

- Technological innovation in health care is growing at an increasingly fast pace and has been integrated into both our daily lives, such as smart healthy tracker, and diagnostic algorithm in medical practice
- With the rapid development of digital pathology, molecular pathology, and informatics pathology, computational pathology is increasingly involved in many subspecialties such as pulmonary, renal, gastrointestinal, neurology, and gynecology pathology
- We believe the initial phase of AI will start with specific tasks such as the diagnoses of particular cancers and classification of tissue types, which require limited and simple criteria
- For example, the common subtypes and variants of benign and malignant neoplasm in prostate should be included in the training and validation to ensure the feasibility of daily pathology practice
- As a result of more data collection and more powerful computing capacity over time, the clinical applications of AI will be broader, and the number of nonspecific cases in the gray zone or with red flags classified by AI for manual review will be decreased
- The growing medical data, including genomics, proteomics, informatics, and whole-slide images, is expected

to integrate to become a data-rich pathomics and lead to rapid development and prosperity of an AI-assisted computational pathology

- Although many challenges remain, computational pathology with the deployment of digital pathology technology and statistic algorithm will continue to improve clinical workflows and collaboration among pathologist and other members of the patient care team
- The improved infrastructure of the network environment, the enhanced computing capacity, and broad integration of informatics has ushered in new horizons for both computational pathology and collaborative pattern, which makes data travel and cloud-based central laboratory and data bank to deliver better care for patients at lower costs possible

Conclusions

- In the new era of deep learning-assisted pathology, data banking, integration, and cloud laboratory are becoming an essential part of daily practice of pathology
- Furthermore, pathologists, data scientists, and industry are starting to incorporate genomics, proteomics, bioinformatics, and computer algorithms into a large amount of complex clinical information
- Through this process, computational pathology can contribute valuable insights to the diagnosis, prognosis, and, ultimately treatment of disease
- Although many technical and ethical challenges need to be addressed, computational pathology as a synergistic system will lead to an integrated workflow, enabling clinical teams to share and analyze image data in a broader platform
- Currently, deep learning has been applied to solve more and more specialized tasks in medicine
- Several studies discussed above show that algorithm assistance has the potential to not only improve the sensitivity and accuracy of the diagnoses but also improve turnaround time
- Moreover, around 75% of pathologists across 59 countries in the world are interested and excited about using AI as a diagnostic tool
- Finally, despite the challenges and obstacles, the potential of computational pathology will change and improve the current healthcare system in promising and exciting ways

Further Reading

Abels E, Pantanowitz L, Aeffner F, Zarella MD, van der Laak J, Bui MM, Vemuri VN, Parwani AV, Gibbs J, Agosto-Arroyo E, Beck AH, Kozlowski C. Computational pathology definitions, best practices, and recommendations for regulatory guidance: a white paper from the digital pathology association. J Pathol. 2019;249(3):286–94. https://doi.org/10.1002/path.5331.

Acs B, Pelekanou V, Bai Y, Martinez-Morilla S, Toki M, Leung S, Nielsen TO, Rimm DL. Ki67 reproducibility using digital image analysis: an inter-platform and inter-operator study. Lab Invest. 2019;99(1):107–17. https://doi.org/10.1038/s41374-018-0123-7.

Aeffner F, Wilson K, Martin NT, Black JC, Hendriks CLL, Bolon B, Rudmann DG, Gianani R, Koegler SR, Krueger J, Young GD. The gold standard paradox in digital image analysis: manual versus automated scoring as ground truth. Arch Pathol Lab Med. 2017;141(9):1267–75. https://doi.org/10.5858/arpa.2016-0386-RA.

Aeffner F, Zarella MD, Buchbinder N, Bui MM, Goodman MR, Hartman DJ, Lujan GM, Molani MA, Parwani AV, Lillard K, Turner OC, Vemuri VNP, Yuil-Valdes AG, Bowman D. Introduction to digital image analysis in whole-slide imaging: a white paper from the Digital Pathology Association. J Pathol Inform. 2019;10:9. https://doi.org/10.4103/jpi.jpi_82_18. Erratum in: J Pathol Inform. 2019;10:15.

Allen TC. Regulating artificial intelligence for a successful pathology future. Arch Pathol Lab Med. 2019;143(10):1175–9. https://doi.org/10.5858/arpa.2019-0229-ED.

Amgad M, Elfandy H, Hussein H, Atteya LA, Elsebaie MAT, Abo Elnasr LS, Sakr RA, Salem HSE, Ismail AF, Saad AM, Ahmed J, Elsebaie MAT, Rahman M, Ruhban IA, Elgazar NM, Alagha Y, Osman MH, Alhusseiny AM, Khalaf MM, Younes AF, Abdulkarim A, Younes DM, Gadallah AM, Elkashash AM, Fala SY, Zaki BM, Beezley J, Chittajallu DR, Manthey D, Gutman DA, Cooper LAD. Structured crowdsourcing enables convolutional segmentation of histology images. Bioinformatics. 2019;35(18):3461–7. https://doi.org/10.1093/bioinformatics/btz083.

Barker J, Hoogi A, Depeursinge A, Rubin DL. Automated classification of brain tumor type in whole-slide digital pathology images using local representative tiles. Med Image Anal. 2016;30:60–71. https://doi.org/10.1016/j.media.2015.12.002.

Bera K, Schalper KA, Rimm DL, Velcheti V, Madabhushi A. Artificial intelligence in digital pathology—new tools for diagnosis and precision oncology. Nat Rev. Clin Oncol. 2019;16(11):703–15. https://doi.org/10.1038/s41571-019-0252-y.

Bulten W, Kartasalo K, Chen PC, Ström P, Pinckaers H, Nagpal K, Cai Y, Steiner DF, van Boven H, Vink R, Hulsbergen-van de Kaa C, van der Laak J, Amin MB, Evans AJ, van der Kwast T, Allan R, Humphrey PA, Grönberg H, Samaratunga H, Delahunt B, Tsuzuki T, Häkkinen T, Egevad L, Demkin M, Dane S, Tan F, Valkonen M, Corrado GS, Peng L, Mermel CH, Ruusuvuori P, Litjens G, Eklund M, PANDA Challenge Consortium. Artificial intelligence for diagnosis and Gleason grading of prostate cancer: the PANDA challenge. Nat Med. 2022;28(1):154–63. https://doi.org/10.1038/s41591-021-01620-2.

Bychkov D, Linder N, Turkki R, Nordling S, Kovanen PE, Verrill C, Walliander M, Lundin M, Haglund C, Lundin J. Deep learning-based tissue analysis predicts outcome in colorectal cancer. Sci Rep. 2018;8(1):3395. https://doi.org/10.1038/s41598-018-21758-3.

Campanella G, Hanna MG, Geneslaw L, Miraflor A, Werneck Krauss Silva V, Busam KJ, Brogi E, Reuter VE, Klimstra DS, Fuchs TJ. Clinical-grade computational pathology using weakly supervised deep learning on whole slide images. Nat Med. 2019;25(8):1301–9. https://doi.org/10.1038/s41591-019-0508-1.

Candido Dos Reis FJ, Lynn S, Ali HR, Eccles D, Hanby A, Provenzano E, Caldas C, Howat WJ, McDuffus LA, Liu B, Daley F, Coulson P, Vyas RJ, Harris LM, Owens JM, Carton AF, McQuillan JP, Paterson AM, Hirji Z, Christie SK, Holmes AR, Schmidt MK, Garcia-Closas M, Easton DF, Bolla MK, Wang Q, Benitez J, Milne RL, Mannermaa A, Couch F, Devilee P, Tollenaar RA, Seynaeve C, Cox A, Cross SS, Blows FM, Sanders J, de Groot R, Figueroa J, Sherman M, Hooning M, Brenner H, Holleczek B, Stegmaier C, Lintott C, Pharoah PD. Crowdsourcing the general public for large scale molecular pathology studies in cancer. EBioMedicine. 2015;2(7):681–9. https://doi.org/10.1016/j.ebiom.2015.05.009.

Chang HY, Jung CK, Woo JI, Lee S, Cho J, Kim SW, Kwak TY. Artificial intelligence in pathology. J Pathol Transl Med. 2019;53(1):1–12. https://doi.org/10.4132/jptm.2018.12.16.

Chen J, Qian F, Yan W, Shen B. Translational biomedical informatics in the cloud: present and future. Biomed Res Int. 2013;2013:658925. https://doi.org/10.1155/2013/658925.

Chico V. The impact of the general data protection regulation on health research. Br Med Bull. 2018;128(1):109–18. https://doi.org/10.1093/bmb/ldy038.

Ching T, Himmelstein DS, Beaulieu-Jones BK, Kalinin AA, Do BT, Way GP, Ferrero E, Agapow PM, Zietz M, Hoffman MM, Xie W, Rosen GL, Lengerich BJ, Israeli J, Lanchantin J, Woloszynek S, Carpenter AE, Shrikumar A, Xu J, Cofer EM, Lavender CA, Turaga SC, Alexandari AM, Lu Z, Harris DJ, DeCaprio D, Qi Y, Kundaje A, Peng Y, Wiley LK, Segler MHS, Boca SM, Swamidass SJ, Huang A, Gitter A, Greene CS. Opportunities and obstacles for deep learning in biology and medicine. J R Soc Interface. 2018;15(141):20170387. https://doi.org/10.1098/rsif.2017.0387.

Cireşan DC, Giusti A, Gambardella LM, Schmidhuber J. Mitosis detection in breast cancer histology images with deep neural networks. Med Image Comput Comput Assist Interv. 2013;16(Pt 2):411–8. https://doi.org/10.1007/978-3-642-40763-5_51.

Cong L, Feng W, Yao Z, Zhou X, Xiao W. Deep learning model as a new trend in computer-aided diagnosis of tumor pathology for lung cancer. J Cancer. 2020;11(12):3615–22. https://doi.org/10.7150/jca.43268.

Evans AJ, Bauer TW, Bui MM, Cornish TC, Duncan H, Glassy EF, Hipp J, McGee RS, Murphy D, Myers C, O'Neill DG, Parwani AV, Rampy BA, Salama ME, Pantanowitz L. US Food and Drug Administration approval of whole slide imaging for primary diagnosis: a key milestone is reached and new questions are raised. Arch Pathol Lab Med. 2018;142(11):1383–7. https://doi.org/10.5858/arpa.2017-0496-CP.

Farnell DA, Huntsman D, Bashashati A. The coming 15 years in gynaecological pathology: digitisation, artificial intelligence, and new technologies. Histopathology. 2020;76(1):171–7. https://doi.org/10.1111/his.13991.

Fuchs TJ, Buhmann JM. Computational pathology: challenges and promises for tissue analysis. Comput Med Imaging Graph. 2011;35(7–8):515–30. https://doi.org/10.1016/j.compmedimag.2011.02.006.

Gabril MY, Yousef GM. Informatics for practicing anatomical pathologists: marking a new era in pathology practice. Mod Pathol. 2010;23(3):349–58. https://doi.org/10.1038/modpathol.2009.190.

George MR, Johnson KA, Gratzinger DA, Brissette MD, McCloskey CB, Conran RM, Dixon LR, Roberts CA, Rojiani AM, Shyu I, Timmons CF Jr, Hoffman RD. Will I need to move to get my first job? Geographic relocation and other trends in the pathology job market. Arch Pathol Lab Med. 2020;144(4):427–34. https://doi.org/10.5858/arpa.2019-0150-CP.

Granter SR, Beck AH, Papke DJ Jr. AlphaGo, deep learning, and the future of the human microscopist. Arch Pathol Lab Med. 2017;141(5):619–21. https://doi.org/10.5858/arpa.2016-0471-ED.

Hamidinekoo A, Denton E, Rampun A, Honnor K, Zwiggelaar R. Deep learning in mammography and breast histology, an overview and future trends. Med Image Anal. 2018;47:45–67. https://doi.org/10.1016/j.media.2018.03.006.

Hanna MG, Parwani A, Sirintrapun SJ. Whole slide imaging: technology and applications. Adv Anat Pathol. 2020;27(4):251–9. https://doi.org/10.1097/PAP.0000000000000273.

Harmon SA, Tuncer S, Sanford T, Choyke PL, Türkbey B. Artificial intelligence at the intersection of pathology and radiology in prostate cancer. Diagn Interv Radiol. 2019;25(3):183–8. https://doi.org/10.5152/dir.2019.19125.

Hassell LA, Blick KE. Training in informatics: teaching informatics in surgical pathology. Surg Pathol Clin. 2015;8(2):289–300. https://doi.org/10.1016/j.path.2015.02.008.

Hassell LA, Blick KE. Training in informatics: teaching informatics in surgical pathology. Clin Lab Med. 2016;36(1):183–97. https://doi.org/10.1016/j.cll.2015.09.014.

Hou L, Samaras D, Kurc TM, Gao Y, Davis JE, Saltz JH. Patch-based convolutional neural network for whole slide tissue image classification. Proc IEEE Comput Soc Conf Comput Vis Pattern Recognit. 2016;2016:2424–33. https://doi.org/10.1109/CVPR.2016.266.

Houssami N, Kirkpatrick-Jones G, Noguchi N, Lee CI. Artificial intelligence (AI) for the early detection of breast cancer: a scoping review to assess AI's potential in breast screening practice. Expert Rev. Med Devices. 2019;16(5):351–62. https://doi.org/10.1080/17434440.2019.1610387.

Hughes AJ, Mornin JD, Biswas SK, Beck LE, Bauer DP, Raj A, Bianco S, Gartner ZJ. Quanti.us: a tool for rapid, flexible, crowd-based annotation of images. Nat Methods. 2018;15(8):587–90. https://doi.org/10.1038/s41592-018-0069-0.

Irshad H, Montaser-Kouhsari L, Waltz G, Bucur O, Nowak JA, Dong F, Knoblauch NW, Beck AH. Crowdsourcing image annotation for nucleus detection and segmentation in computational pathology: evaluating experts, automated methods, and the crowd. Pac Symp Biocomput. 2015;2015:294–305. https://doi.org/10.1142/9789814644730_0029.

Jang HJ, Cho KO. Applications of deep learning for the analysis of medical data. Arch Pharm Res. 2019;42(6):492–504. https://doi.org/10.1007/s12272-019-01162-9.

Janowczyk A, Zuo R, Gilmore H, Feldman M, Madabhushi A. HistoQC: an open-source quality control tool for digital pathology slides. JCO Clin Cancer Inform. 2019;3:1–7. https://doi.org/10.1200/CCI.18.00157.

Jovanović J, Bagheri E. Semantic annotation in biomedicine: the current landscape. J Biomed Semantics. 2017;8(1):44. https://doi.org/10.1186/s13326-017-0153-x.

Kapil A, Meier A, Zuraw A, Steele KE, Rebelatto MC, Schmidt G, Brieu N. Deep semi supervised generative learning for automated tumor proportion scoring on NSCLC tissue needle biopsies. Sci Rep. 2018;8(1):17343. https://doi.org/10.1038/s41598-018-35501-5.

Komura D, Ishikawa S. Machine learning approaches for pathologic diagnosis. Virchows Arch. 2019;475(2):131–8. https://doi.org/10.1007/s00428-019-02594-w.

Komura D, Ishikawa S. Machine learning methods for histopathological image analysis. Comput Struct Biotechnol J. 2018;16:34–42. https://doi.org/10.1016/j.csbj.2018.01.001.

Korbar B, Olofson AM, Miraflor AP, Nicka CM, Suriawinata MA, Torresani L, Suriawinata AA, Hassanpour S. Deep learning for classification of colorectal polyps on whole-slide images. J Pathol Inform. 2017;8:30. https://doi.org/10.4103/jpi.jpi_34_17.

Krempel R, Kulkarni P, Yim A, Lang U, Habermann B, Frommolt P. Integrative analysis and machine learning on cancer genomics data using the cancer systems biology database (CancerSysDB). BMC Bioinformatics. 2018;19(1):156. https://doi.org/10.1186/s12859-018-2157-7.

Landau MS, Pantanowitz L. Artificial intelligence in cytopathology: a review of the literature and overview of commercial landscape. J Am Soc Cytopathol. 2019;8(4):230–41. https://doi.org/10.1016/j.jasc.2019.03.003.

LeCun Y, Bengio Y, Hinton G. Deep learning. Nature. 2015;521(7553):436–44. https://doi.org/10.1038/nature14539.

Levine AB, Schlosser C, Grewal J, Coope R, Jones SJM, Yip S. Rise of the machines: advances in deep learning for cancer diagnosis. Trends Cancer. 2019;5(3):157–69. https://doi.org/10.1016/j.trecan.2019.02.002.

Li C, Wang X, Liu W, Latecki LJ, Wang B, Huang J. Weakly supervised mitosis detection in breast histopathology images using concentric loss. Med Image Anal. 2019;53:165–78. https://doi.org/10.1016/j.media.2019.01.013.

Lin JC, Fan CT, Liao CC, Chen YS. Taiwan biobank: making cross-database convergence possible in the big data era. Gigascience. 2018;7(1):1–4. https://doi.org/10.1093/gigascience/gix110.

Litjens G, Bandi P, Ehteshami Bejnordi B, Geessink O, Balkenhol M, Bult P, Halilovic A, Hermsen M, van de Loo R, Vogels R, Manson QF, Stathonikos N, Baidoshvili A, van Diest P, Wauters C, van Dijk M, van der Laak J. 1399 H&E-stained sentinel lymph node sections of breast cancer patients: the CAMELYON dataset. Gigascience. 2018;7(6):giy065. https://doi.org/10.1093/gigascience/giy065.

Litjens G, Kooi T, Bejnordi BE, Setio AAA, Ciompi F, Ghafoorian M, van der Laak JAWM, van Ginneken B, Sánchez CI. A survey on deep learning in medical image analysis. Med Image Anal. 2017;42:60–88. https://doi.org/10.1016/j.media.2017.07.005.

Louis DN, Feldman M, Carter AB, Dighe AS, Pfeifer JD, Bry L, Almeida JS, Saltz J, Braun J, Tomaszewski JE, Gilbertson JR, Sinard JH, Gerber GK, Galli SJ, Golden JA, Becich MJ. Computational pathology: a path ahead. Arch Pathol Lab Med. 2016;140(1):41–50. https://doi.org/10.5858/arpa.2015-0093-SA.

Louis DN, Gerber GK, Baron JM, Bry L, Dighe AS, Getz G, Higgins JM, Kuo FC, Lane WJ, Michaelson JS, Le LP, Mermel CH, Gilbertson JR, Golden JA. Computational pathology: an emerging definition. Arch Pathol Lab Med. 2014;138(9):1133–8. https://doi.org/10.5858/arpa.2014-0034-ED.

Martin V, Kim TH, Kwon M, Kuko M, Pourhomayoun M, Martin S. A more comprehensive cervical cell classification using convolutional neural network. J Am Soc Cytopathol. 2018;7(5):S66. https://doi.org/10.1016/j.jasc.2018.06.156.

Martino F, Varricchio S, Russo D, Merolla F, Ilardi G, Mascolo M, dell'Aversana GO, Califano L, Toscano G, Pietro G, Frucci M, Brancati N, Fraggetta F, Staibano S. A machine-learning approach for the assessment of the proliferative compartment of solid tumors on hematoxylin-eosin-stained sections. Cancers. 2020;12(5):1344. https://doi.org/10.3390/cancers12051344.

Mazzanti M, Shirka E, Gjergo H, Hasimi E. Imaging, health record, and artificial intelligence: hype or hope? Curr Cardiol Rep. 2018;20(6):48. https://doi.org/10.1007/s11886-018-0990-y.

Miller DD, Brown EW. Artificial intelligence in medical practice: the question to the answer? Am J Med. 2018;131(2):129–33. https://doi.org/10.1016/j.amjmed.2017.10.035.

Mintz Y, Brodie R. Introduction to artificial intelligence in medicine. Minim Invasive Ther Allied Technol. 2019;28(2):73–81. https://doi.org/10.1080/13645706.2019.1575882.

Mobadersany P, Yousefi S, Amgad M, Gutman DA, Barnholtz-Sloan JS, Velázquez Vega JE, Brat DJ, Cooper LAD. Predicting cancer outcomes from histology and genomics using convolutional networks. Proc Natl Acad Sci U S A. 2018;115(13):E2970–9. https://doi.org/10.1073/pnas.1717139115.

Mohanty SK, Parwani AV, Crowley RS, Winters S, Becich MJ. The importance of pathology informatics in translational research. Adv Anat Pathol. 2007;14(5):320–2. https://doi.org/10.1097/PAP.0b013e3180ca8a79.

Moscatelli M, Manconi A, Pessina M, Fellegara G, Rampoldi S, Milanesi L, Casasco A, Gnocchi M. An infrastructure for precision medicine through analysis of big data. BMC Bioinformatics. 2018;19(Suppl 10):351. https://doi.org/10.1186/s12859-018-2300-5.

Murphy JFA. The general data protection regulation (GDPR). Ir Med J. 2018;111(5):747.

Nakata N. Recent technical development of artificial intelligence for diagnostic medical imaging. Jpn J Radiol. 2019;37(2):103–8. https://doi.org/10.1007/s11604-018-0804-6.

Niazi MKK, Parwani AV, Gurcan MN. Digital pathology and artificial intelligence. Lancet Oncol. 2019;20(5):e253–61. https://doi.org/10.1016/S1470-2045(19)30154-8.

Pantanowitz L, Sinard JH, Henricks WH, Fatheree LA, Carter AB, Contis L, Beckwith BA, Evans AJ, Lal A, Parwani AV, College of American Pathologists Pathology and Laboratory Quality Center. Validating whole slide imaging for diagnostic purposes in pathology: guideline from the College of American Pathologists Pathology and Laboratory Quality Center. Arch Pathol Lab Med. 2013;137(12):1710–22. https://doi.org/10.5858/arpa.2013-0093-CP.

Paparella ML, et al. Quantitative analysis of KLF4 and SOX2 expression in oral carcinomas reveals independent association with oral tongue subsite location and histological grade. Cancer Biomark. 2021;32:37–48.

Parwani AV. Next generation diagnostic pathology: use of digital pathology and artificial intelligence tools to augment a pathological diagnosis. Diagn Pathol. 2019;14(1):138. https://doi.org/10.1186/s13000-019-0921-2.

Parwani AV. Preface. Pathology informatics. Surg Pathol Clin. 2015;8(2):xi–xii. https://doi.org/10.1016/j.path.2015.04.001.

Rashidi HH, Tran NK, Betts EV, Howell LP, Green R. Artificial intelligence and machine learning in pathology: the present landscape of supervised methods. Acad Pathol. 2019;6:2374289519873088. https://doi.org/10.1177/2374289519873088.

Robertson S, Azizpour H, Smith K, Hartman J. Digital image analysis in breast pathology-from image processing techniques to artificial intelligence. Transl Res. 2018;194:19–35. https://doi.org/10.1016/j.trsl.2017.10.010.

Roux L, Racoceanu D, Loménie N, Kulikova M, Irshad H, Klossa J, Capron F, Genestie C, Le Naour G, Gurcan MN. Mitosis detection in breast cancer histological images an ICPR 2012 contest. J Pathol Inform. 2013;4:8. https://doi.org/10.4103/2153-3539.112693.

Roy S, Coldren C, Karunamurthy A, Kip NS, Klee EW, Lincoln SE, Leon A, Pullambhatla M, Temple-Smolkin RL, Voelkerding KV, Wang C, Carter AB. Standards and guidelines for validating next-generation sequencing bioinformatics pipelines: a joint recommendation of the Association for Molecular Pathology and the College of American Pathologists. J Mol Diagn. 2018;20(1):4–27. https://doi.org/10.1016/j.jmoldx.2017.11.003.

Roy S. Molecular pathology informatics. Clin Lab Med. 2016;36(1):57–66. https://doi.org/10.1016/j.cll.2015.09.007.

Roy S. Molecular pathology informatics. Surg Pathol Clin. 2015;8(2):187–94. https://doi.org/10.1016/j.path.2015.02.013.

Saltz J, Gupta R, Hou L, Kurc T, Singh P, Nguyen V, Samaras D, Shroyer KR, Zhao T, Batiste R, Van Arnam J, et al. Spatial organization and molecular correlation of tumor-infiltrating lymphocytes using deep learning on pathology images. Cell Rep. 2018;23(1):181–193.e7. https://doi.org/10.1016/j.celrep.2018.03.086.

Sarwar S, Dent A, Faust K, Richer M, Djuric U, Van Ommeren R, Diamandis P. Physician perspectives on integration of artificial intelligence into diagnostic pathology. NPJ Digit Med. 2019;2:28. https://doi.org/10.1038/s41746-019-0106-0.

Senaras C, Niazi MKK, Lozanski G, Gurcan MN. DeepFocus: detection of out-of-focus regions in whole slide digital images using deep learning. PLoS One. 2018;13(10):e0205387. https://doi.org/10.1371/journal.pone.0205387.

Shen D, Wu G, Suk HI. Deep learning in medical image analysis. Annu Rev. Biomed Eng. 2017;19:221–48. https://doi.org/10.1146/annurev-bioeng-071516-044442.

Smaïl-Tabbone M, Rance B, Section Editors for the IMIA Yearbook Section on Bioinformatics and Translational Informatics. Contributions from the 2019 literature on bioinformatics and translational informatics. Yearb Med Inform. 2020;29(1):188–92. https://doi.org/10.1055/s-0040-1702002.

Steiner DF, MacDonald R, Liu Y, Truszkowski P, Hipp JD, Gammage C, Thng F, Peng L, Stumpe MC. Impact of deep learning assistance on the histopathologic review of lymph nodes for metastatic breast cancer. Am J Surg Pathol. 2018;42(12):1636–46. https://doi.org/10.1097/PAS.0000000000001151.

Stoeklé HC, Mamzer-Bruneel MF, Frouart CH, Le Tourneau C, Laurent-Puig P, Vogt G, Hervé C. Molecular tumor boards: ethical issues in the new era of data medicine. Sci Eng Ethics. 2018;24(1):307–22. https://doi.org/10.1007/s11948-017-9880-8.

Tappeiner E, Pröll S, Hönig M, Raudaschl PF, Zaffino P, Spadea MF, Sharp GC, Schubert R, Fritscher K. Multi-organ segmentation of the head and neck area: an efficient hierarchical neural networks approach. Int J Comput Assist Radiol Surg. 2019;14(5):745–54. https://doi.org/10.1007/s11548-019-01922-4.

Tizhoosh HR, Pantanowitz L. Artificial intelligence and digital pathology: challenges and opportunities. J Pathol Inform. 2018;9:38. https://doi.org/10.4103/jpi.jpi_53_18.

Vaickus LJ, Suriawinata AA, Wei JW, Liu X. Automating the Paris system for urine cytopathology-a hybrid deep-learning and morphometric approach. Cancer Cytopathol. 2019;127(2):98–115. https://doi.org/10.1002/cncy.22099.

Wang DY, Khosla A, Gargeya R, Irshad H, Beck AH. Deep learning for identifying metastatic breast cancer. arXiv. 2016. https://arxiv.org/abs/1606.05718.

Wang S, Yang DM, Rong R, Zhan X, Fujimoto J, Liu H, Minna J, Wistuba II, Xie Y, Xiao G. Artificial intelligence in lung cancer pathology image analysis. Cancers (Basel). 2019;11(11):1673. https://doi.org/10.3390/cancers11111673.

Wen S, Kurc TM, Hou L, Saltz JH, Gupta RR, Batiste R, Zhao T, Nguyen V, Samaras D, Zhu W. Comparison of different classifiers with active learning to support quality control in nucleus segmentation in pathology images. AMIA Jt Summits Transl Sci Proc. 2017;2018:227–36.

Wildeboer RR, van Sloun RJG, Wijkstra H, Mischi M. Artificial intelligence in multiparametric prostate cancer imaging with focus on deep-learning methods. Comput Methods Prog Biomed. 2020;189:105316. https://doi.org/10.1016/j.cmpb.2020.105316.

Yagi Y. Color standardization and optimization in whole slide imaging. Diagn Pathol. 2011;6(Suppl 1):S15. https://doi.org/10.1186/1746-1596-6-S1-S15.

Zarella MD, Feldscher A. Laboratory computer performance in a digital pathology environment: outcomes from a single institution. J Pathol Inform. 2018;9:44. https://doi.org/10.4103/jpi.jpi_47_18.

Zarella MD, Quaschnick MR, Breen DE, Garcia FU. Estimation of fine-scale histologic features at low magnification. Arch Pathol Lab Med. 2018;142(11):1394–402. https://doi.org/10.5858/arpa.2017-0380-OA.

Zarella MD, Yeoh C, Breen DE, Garcia FU. An alternative reference space for H&E color normalization. PLoS One. 2017;12(3):e0174489. https://doi.org/10.1371/journal.pone.0174489.

Zhang DY. An outline of office-based bladder and prostate biopsy pathology. Hauppauge, NY: Nova Science Publishers; 2019.

Molecular Cytopathology

Roberto Ruiz-Cordero and Sinchita Roy-Chowdhuri

Contents

Introduction

- Cytopathology, the microscopic examination of cells from a wide range of body sites, is commonly used to investigate diseases
- Applying novel technologies to cytopathology samples to detect genomic alterations led to the birth of molecular cytopathology
- A growing understanding of the morphologic and genomic alterations of a variety of disease processes has propelled molecular cytopathology to the forefront of molecular diagnostics and personalized medicine
- With an increasing trend in testing limited-volume samples obtained through minimally invasive techniques in advanced-stage cancer patients, optimization and judicious triaging of all sample sources becomes crucial

- Deoxyribonucleic acid (DNA), ribonucleic acid (RNA), and more recently, their cell-free components, can be extracted from cytology specimens and used for molecular testing
- Cytology specimens, particularly nonformalin-fixed cytology specimen substrates, provide a rich source of high-quality nucleic acids
- Molecular diagnostics can be applied to cytology samples from cancer and noncancer patients to detect actionable genetic aberrations or microorganisms
- The wide range of cytology samples allows for more flexibility at the expense of increased complexity in assay validation due to wide variability in specimen preparations

Cytology Sample Collection

- Exfoliative cytology
  - Spontaneous—Cells collected after they have been spontaneously shed into a body cavity or body compartment, for example, urine, ascites, pleural effusions, and cerebrospinal fluids
  - Mechanical—Cells are manually scraped or brushed off a body surface, for example, cervicovaginal samples, bronchial washings/brushings, bronchoalveolar lavages, renal pelvic/ureteral brushings, and bile duct brushings
- Interventional cytology
  - Fine-needle aspiration—Mechanical aspiration of cells from a tissue or lesion
  - Imprint cytology—Touch imprint of a fresh piece of tissue

R. Ruiz-Cordero
Department of Pathology and Laboratory Medicine,
University of Miami, Miami, FL, USA
e-mail: rxr1314@med.miami.edu

S. Roy-Chowdhuri (✉)
Department of Pathology, MD Anderson Cancer Center,
Houston, TX, USA
e-mail: SRoy2@mdanderson.org

© The Author(s), under exclusive license to Springer Nature Switzerland AG 2023
L. Cheng et al. (eds.), *Molecular Surgical Pathology*, https://doi.org/10.1007/978-3-031-35118-1_3

Cytology Sample Preparation

- Cytology samples can be prepared using a variety of methods (Fig. 3.1)
  - Cells can be smeared on to a slide and air-dried, or they can be fixed in alcohol
    - Alcohol-fixed slides can be stained with Papanicolaou or hematoxylin and eosin stains
    - Air-dried slides can be stained with a Romanowsky-type stain (e.g., Diff-Quik or May-Greenwald-Giemsa)
    - Using cell scraping or cell lifting techniques, aspirate smears can be used to extract both DNA and RNA to perform molecular testing, with comparable results to those obtained using formalin-fixed paraffin-embedded (FFPE) tissue
    - Cytology smears can supplement the nucleic acid yield of cell block samples to decrease the rate of quantity-not-sufficient (QNS) samples
    - Aspirate smears can also be used for in situ hybridization studies such as fluorescence in situ hybridization (FISH)
  - Cells can be placed directly into different types of collection media, such as ethanol, methanol, RPMI, saline, formalin, or proprietary preservative media (e.g., CytoLyt, CytoRich Red)
    - Similarly, the needle and syringe can be rinsed into these different types of media after the preparation of aspirate smears
    - Cells in all these media and the nucleic acids that are released and remain in the supernatant (except formalin) can be used for genomic testing
      - In the case of physiological culture media such as RPMI, the viable cells collected can be cultured for karyotype testing
  - Once in media or if the sample is a body fluid, cells can be centrifuged directly on to a slide (cytospin preparation) or processed for liquid-based cytology (LBC)

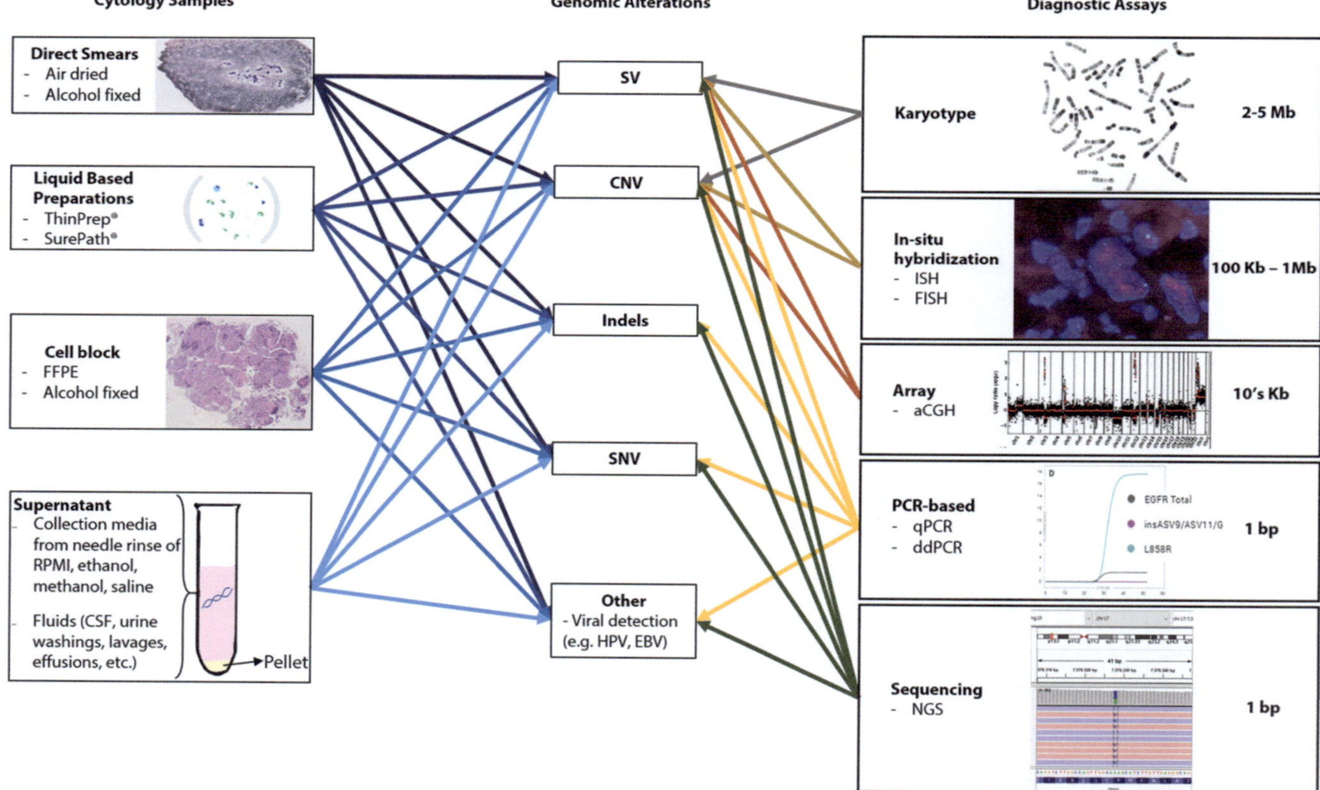

Fig. 3.1 Cytology sample types, genomic alterations, and different diagnostic assays. The left side of the image shows the different cytology sample types and how they are all suitable as a substrate to detect any of the possible genomic alterations depicted in the center of the image (blue arrows). The right side of the image shows the different types of molecular diagnostic assays and their specific resolution. The colored arrows indicate the types of genomic alterations that each diagnostic assay is capable of detecting. *FFPE* formalin-fixed paraffin-embedded, *RPMI* Roswell Park Memorial Institute Medium, *CSF* cerebrospinal fluid, *SV* structural variant, *CNV* copy number variant, *SNV* single nucleotide variant, *HPV* human papillomavirus, *EBV* Epstein-Barr virus, *ISH* in situ hybridization, *FISH* fluorescence ISH, *aCGH* array comparative genomic hybridization, *qPCR* quantitative polymerase chain reaction, *ddPCR* droplet-digital PCR, *NGS* next generation sequencing

– In LBC, cells are suspended in a proprietary preservative media rather than smeared; suspended cells are vortexed and centrifuged onto a slide resulting in a monolayer film of concentrated, evenly distributed, and fixed cells (Fig. 3.1)
  o LBC has been shown to improve specimen quality, increase sensitivity and reduce false negative rate when compared with conventional smears in specific tumor types
  o The preservative media fixes the cells and helps reduce debris, contaminating blood and mucus for a cleaner background
  o Nucleic acids are adequately preserved in LBC media for weeks to months at room temperature for downstream molecular testing
– Cells can be fixed in 10% neutral buffered formalin (preferably) and embedded in paraffin to create a cell block preparation like conventional surgical biopsies
  o There are multiple methods to produce a cell block sample
  o In the era of molecular cytopathology with high-sensitivity molecular assays, the plasma-thrombin method for cell block preparation is not appropriate because it can be a potential source of nucleic acid contamination

  o The quality of nucleic acids is inferior when cells are fixed in formalin due to crosslinking of biomolecules
  o When compared to formalin-fixed paraffin-embedded tissue, overall cellularity, and tumor fraction of fine-needle aspiration smears as well as sequencing metrics, can be significantly higher than core needle biopsies
– Supernatant—More recently, studies have shown that adequate amounts of nucleic acids can be retrieved from residual supernatant material that is commonly discarded
  o The supernatant obtained from postcentrifuged cytology samples is an underutilized resource of intact nucleic acids and their cell-free forms
  o The supernatant can be collected directly from exfoliative specimens (e.g., pleural and pericardial fluid, ascites, urine, cerebrospinal fluid, and amniotic fluid) or from different types of collection media of aspiration samples, including RPMI and LBC, to successfully extract nucleic acids for downstream molecular testing
• Table 3.1 summarizes the advantages and disadvantages of the different types of cytology samples and their potential uses for molecular testing

Table 3.1 Advantages and disadvantages of the use of different cytology sample types for molecular testing

Cytology samples	Advantages	Disadvantages	Possible uses
Direct smears	• Immediate assessment for adequacy	• Sacrificing slide from archival material (important for medicolegal issues)	• Scrape of cells for NGS or other PCR-based tests
	• High-quality nucleic acid • Whole cells with whole nuclei • Superior tumor mapping in samples with low tumor fraction	• Additional validation	• In situ hybridization for FISH
Liquid-base preparations	• Standardized processing with optimal preservation of cells and nucleic acids	• Lack of immediate assessment	• Scrape of cells for NGS or other PCR-based tests
	• Whole cells with whole nuclei • High-quality nucleic acid	• Additional validation • Nucleic acid retrieval may be variable based on preservative/fixative	• In situ hybridization for FISH
Cell block	• Ease of acquisition • Allows for multiple serial sections for immunostains and molecular testing • Standardized validation in most molecular laboratories	• Lack of immediate assessment • Possible suboptimal cellularity • Nucleic acid may be suboptimal because of formalin fixation • Partial nuclei on standard 4- to 5-micron sections • Variability in preparation methods/lack of standardization • Some methods can lead to cross-contamination	• NGS • PCR-based tests • FISH • Immunohistochemistry

(continued)

Table 3.1 (continued)

Cytology samples	Advantages	Disadvantages	Possible uses
Supernatant	• Readily available/easy to collect and repurpose	• Lack of assessment for cellularity	• NGS
	• Helps optimize all sample material as supernatant is usually discarded	• Nucleic acid retrieval may be variable based on medium	• PCR-based tests
	• High-quality nucleic acids		

NGS next generation sequencing, *FISH* fluorescence in situ hybridization, *PCR* polymerase chain reaction

Types of Molecular Alterations

- Cancer is, for the most part, a genetic disease whereby genomic alterations lead to uncontrollable cell growth and potential spread to other parts of the body
- A multitude of oncogenic genomic alterations can occur in a tumor cell:
  - Single nucleotide variants (SNV)—Substitution of a single nucleotide for another
    - Synonymous—Does not result in amino acid change
    - Missense—Results in amino acid change
    - Nonsense—Results in a stop codon leading to premature truncation of the protein
  - Insertion or deletion (Indel)—Insertions or deletions of less than 50 base pairs of DNA
    - Frameshift—The length of the indel is not a multiple of three, changing the reading frame of all remaining bases and leading to incorrect translation of the protein
    - In-frame—The length of the indel is a multiple of three without impacting translation
  - Copy number variation (CNV)—Increase or decrease in the number of copies of a particular DNA segment within the genome
    - Gain/Amplification—Increase in the production of many copies of a specific sequence of the genome
    - Loss/Deletion—When a portion of the genome is missing in one or both alleles
  - Structural variants (SV)—Greater than 50 base pairs are rearranged in a part of the genome
    - Deletion—Part of a chromosome is missing
    - Duplication—Part of a chromosome is repeated
    - Insertion—Addition of material from a different chromosome
    - Inversion—A portion of a chromosome is inverted within the same chromosome
    - Translocation—Material of two or more chromosomes is exchanged in a balanced or unbalanced way

Assays to Detect Different Molecular Alterations

- Given that entire cells are obtained and can be easily manipulated, cytology samples are a great source of nucleic acids for molecular testing
- Depending on the type of alteration to be detected, different testing methods can be used (Fig. 3.1)
- The method of choice depends not only on the type(s) of alteration(s) but also on resources, expertise and a combination of test volume, cost, and turnaround time
  - Karyotype—By arresting cells in metaphase and using G-banding techniques, all chromosomes can be analyzed to detect structural and numerical abnormalities
    - Advantages
      - Provides an overview of the entire genome
    - Limitations
      - Depends on the success of tumor cell growth
      - Laborious preparation and high-level of expertise for analysis
      - Resolution is low and certain alterations can be missed; therefore, a negative karyotype does not exclude malignancy
      - Cannot be performed on fixed specimens
  - Fluorescence in situ hybridization (FISH)—By using fluorescent probes that bind to specific parts of the genome, it is useful for the detection of CNVs and SVs
    - Advantages
      - Useful on most specimen types, capability to evaluate changes at a single cell level
      - When FISH is performed on cytology smears, disaggregated cells are easier to analyze, and more accurate signal counting can be expected because of the lack of nuclear truncation, as seen in FFPE samples
    - Limitations
      - More extensive validation when using non-FFPE material with alternative fixatives
      - Requires a high level of expertise

- ♦ Potential inability to identify invasive from non-invasive tumor cells
- ♦ Does not analyze the full set of chromosomes
- Chromosomal microarray analysis—Oligonucleotide probes are attached to fluorescent dyes that label genomic DNA through hybridization across the genome
  - ○ Advantages
    - ♦ Higher resolution
    - ♦ Provides analysis of the full set of chromosomes
    - ♦ Useful in cases of mosaicism
  - ○ Limitations
    - ♦ Labor intensive
    - ♦ Longer turnaround time
    - ♦ Requires viable cells
- Polymerase chain reaction (PCR)—Copies of nucleic acid sequences can be exponentially amplified through a series of thermal cycles
  - ○ There are several modifications of PCR methods suitable for different applications; the most common include
    - ♦ Real-time PCR
    - ♦ Reverse-transcriptase PCR
    - ♦ Multiplex PCR
    - ♦ Nested PCR
    - ♦ Long-range PCR
    - ♦ Allele-specific PCR
    - ♦ Droplet-digital PCR
  - ○ Their definition and uses are beyond the scope of this chapter
  - ○ Advantages
    - ♦ High analytical sensitivity
    - ♦ Fast turnaround time
    - ♦ Multiplexing capabilities

- ♦ Can provide quantitative measures
- ♦ Works well in cytology specimens despite staining and prolonged storage of up to 15 years
- ○ Limitations
  - ♦ Possibility of false positive results from DNA contamination or nonspecific primer amplification
  - ♦ Multiplex capabilities are limited
  - ♦ Prior sequence data is needed for primer design
- Sequencing
  - ○ Through different detection methods, the sequence of a genome can be determined using polymerase chain reactions
    - ♦ Common sequencing methods
      - ▪ Pyrosequencing
      - ▪ Sanger sequencing and massively parallel sequencing (MPS) also known as Next-generation sequencing (NGS)
  - ○ The ability to multiplex multiple samples to simultaneously look for multiple genes and genomic alterations within the same high-throughput platform has made NGS the preferred sequencing methodology
  - ○ Advantages
    - ♦ High analytic sensitivity
    - ♦ Can detect SNVs, indels, CNVs, and SVs depending on the assay design
  - ○ Limitations
    - ♦ Higher cost and level of complexity
    - ♦ Variable turnaround time
    - ♦ Requires considerable resources and expertise
    - ♦ Depending on the sequencing platform and the number of targets and samples tested, the cell requirement can range from 100 to 15,000 cells (Fig. 3.2)

Analytic Sensitivity of Assay

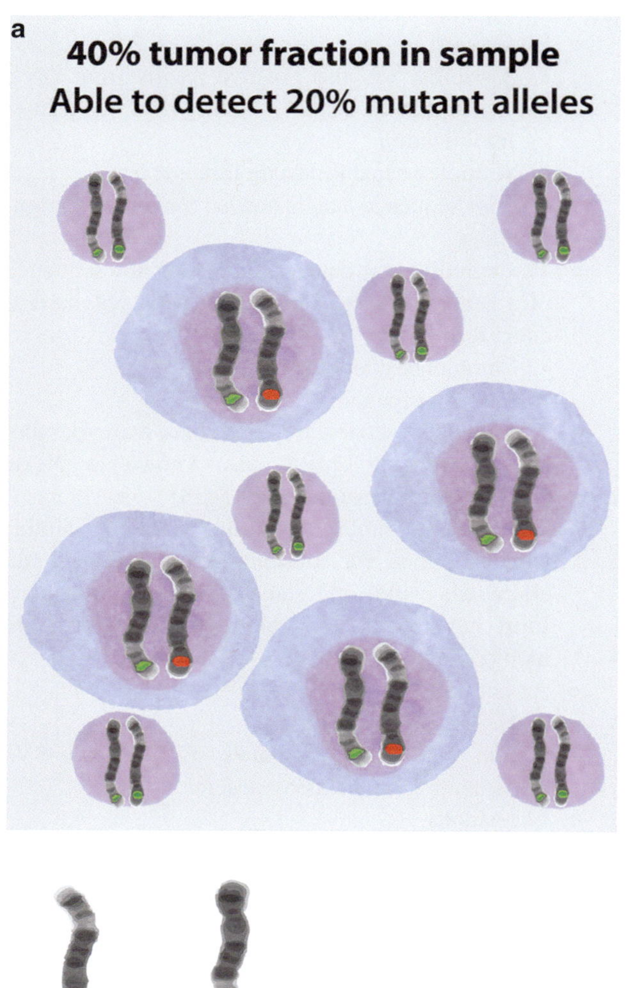

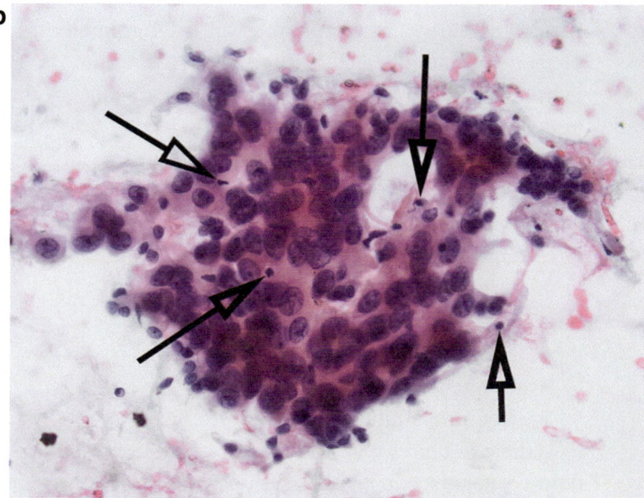

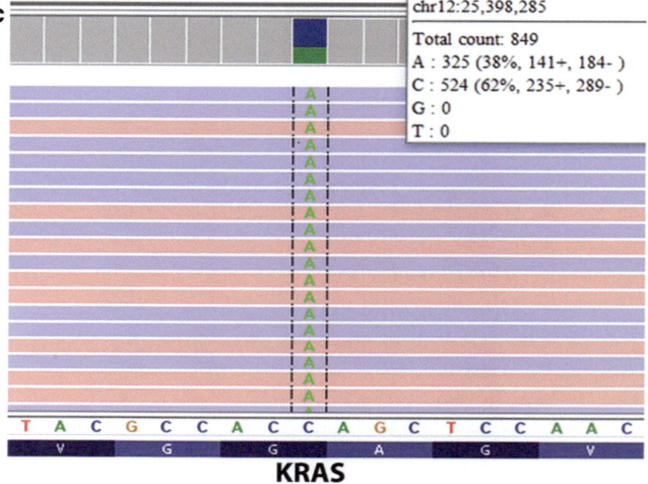

Fig. 3.2 Analytic sensitivity of a molecular diagnostic assay. Analytic sensitivity of a molecular diagnostic assay is directly proportional to the proportion of tumor cells in a sample, also known as the tumor fraction. Because tumor cells, for the most part, are heterozygous, in other words, one allele harbors the genomic alteration while the other allele is wild-type, a sample with 40% tumor fraction as shown in (**a**) with 4 large tumor cells out of a total of 10 cells, will have only 20% of the alleles mutated. For such a sample, the minimum analytic sensitivity of an assay able to detect the mutant alleles is 20%. (**b**) A case of lung adenocarcinoma showed abundant cohesive aggregates of malignant cells with admixed small lymphocytes indicated by the arrows. (**c**) Next-generation sequencing of the cytology specimen identified the hotspot p.G12D missense activating mutation in KRAS at a variant allelic frequency of 38% (total count for [a] in the white box), indicating that the tumor fraction of the sample was between 75% and 80%

Assay Validation and Molecular Testing Workflows

- The validation process is resource and time intensive as it requires rigorous laboratory practices; however, Clinical Laboratory Improvement Amendments (CLIA) and College of American Pathologists (CAP) require laboratories to diligently validate all cytopreparation types for molecular testing

- Federal and state agencies as well as organizations such as CAP, Clinical Laboratory Standards Institute (CLSI), American College of Medical Genetics (ACMG), and Association for Molecular Pathology (AMP), have issued a variety of guidelines for molecular test validations for clinical samples that are also applicable to cytology specimens
- In general, validation for laboratory-developed test (LDT) includes

- Pre-analytic phase, such as specimen stability, transport, and storage condition
- Analytic phase, such as accuracy, precision, report range, sensitivity, and specificity
- Postanalytic phase, such as cutoff criteria and data interpretation
- The wide variety of cytopreparatory methods and lack of standardization across laboratories increases the number of preanalytical variables that can affect nucleic acid yield and sequencing results; DNA yield can vary depending on the type of glass slides used (fully frosted slides yield less DNA)
  - The harvesting technique (cell lifting yields lower DNA than cell scraping)
  - The type of mounting medium (polymer-based yield higher DNA than xylene-based)
  - The medium used for cell collection (CytoLyt yields higher DNA than CytoRich Red)
  - The type of fixative and stains used (Air-dried smears may have superior long-term DNA preservation but lower DNA yields and NGS success than alcohol-fixed smears)
- The determination of minimum cellularity and tumor fraction cutoff to obtain an adequate analytic sensitivity that allows for the reproducible detection of specific genomic alterations is important for validating PCR-based assays (Fig. 3.2)
- Most PCR or RT-PCR assays requiring ~1 ng DNA and ~4 ng RNA need approximately 200 cells (6 pg DNA/cell and 20 pg RNA/cell)

- Higher input requirements require higher numbers of cells for an adequate yield
- The tumor fraction is another important factor to consider during validation as tumor cells are always in a background of normal cells such as inflammatory or stromal cells that dilute the mutant allele fraction in a sample and can result in false negative results (Fig. 3.3)
  - Different molecular assays have different analytic sensitivity (lowest limit of detection)
  - It is imperative to test and validate cytology specimens in individual laboratories
- Comprehensive validation of cytology specimens for NGS shows comparable performance with FFPE tissue samples, albeit better sequencing quality metrics and fewer DNA mass quantity failures
- Finally, a robust quality assurance (QA) policy needs to be in place to ensure that the test results are reliable; this would include
  - Competency and training of personnel
  - Adequate facilities and laboratory space
  - Assay validation with documentation as per standards and guidelines
  - Laboratory regulation and procedure manuals
  - Proficiency testing
  - Laboratory inspection programs

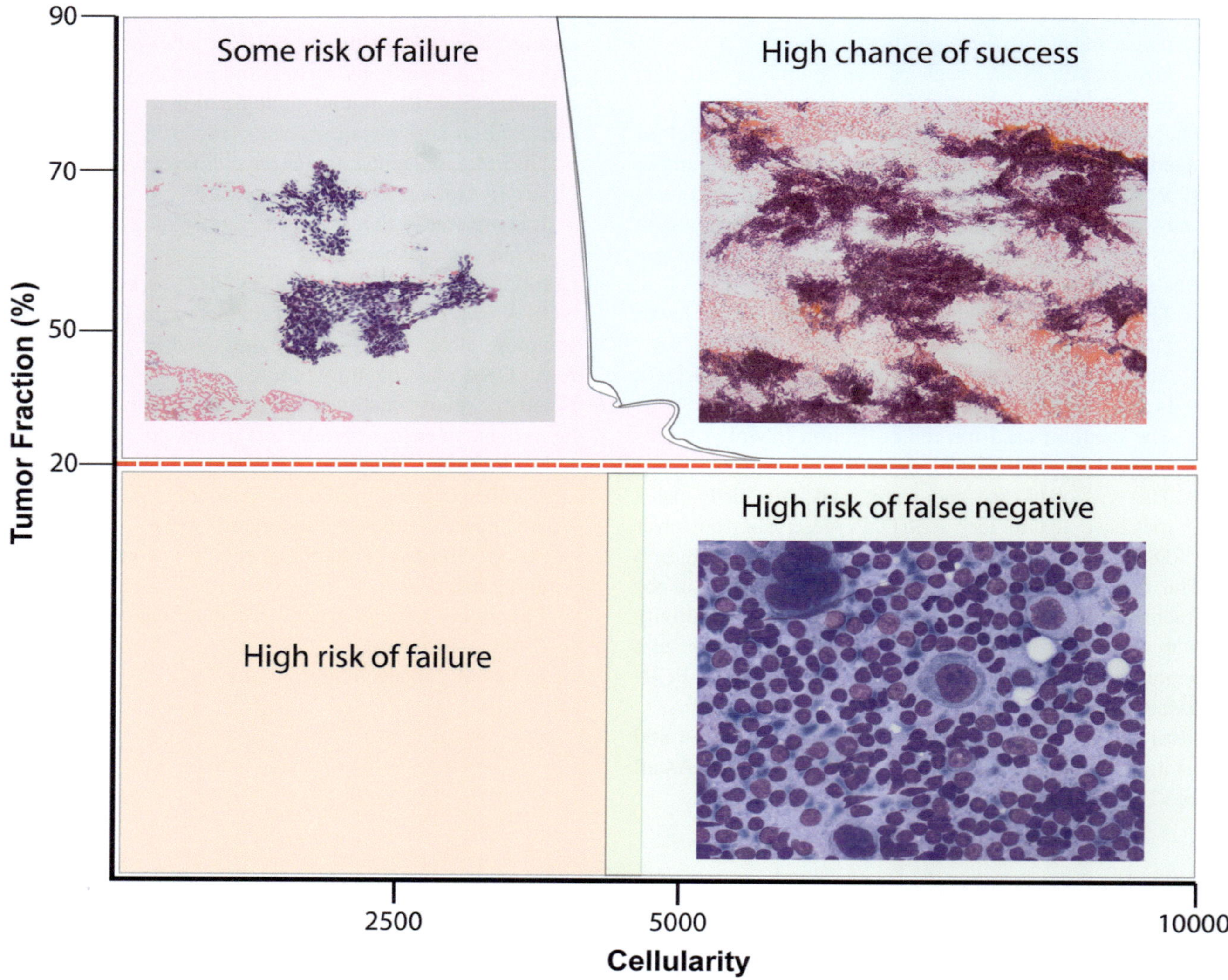

Fig. 3.3 Effect of cellularity and tumor fraction on NGS success rate. The success rate of NGS is dependent on two key factors: tumor fraction and overall cellularity. Ideal samples are those with high tumor fraction and high cellularity, as highlighted in the top right in light blue. Samples with tumor fraction above 20% (red dotted line) or higher and lower overall cellularity, as shown in the top left in pink have some risk of failure. In contrast, samples with tumor fractions below 20%, despite higher cellularity, have a high risk of false negative results. Samples with low cellularity and low tumor fraction should not be used for sequencing because of the high risk of failure. NGS, next-generation sequencing

Clinical Settings for Molecular Cytopathology

- Molecular testing is often used as an ancillary test to aid in the diagnosis of cytology specimens (Table 3.2). Some key examples include:
  - Human papillomavirus (HPV) testing applied to gynecologic cytology specimens
  - HPV testing in the diagnostic work-up of FNAs from head and neck squamous cell carcinomas
  - Various molecular assays including gene expression classifier analysis, microRNA expression profiling, and sequencing assays for detection of gene mutations and gene rearrangements in the diagnosis of indeterminate thyroid FNAs
  - FISH analysis for indeterminate urinary tract cytology specimens and pancreatobiliary brushings
  - Sequencing assay for pancreatic cyst fluid
  - FISH testing for subclassification of salivary gland neoplasms

Table 3.2 Types of molecular tests used as a diagnostic aid in cytology samples of different sites and tumor types

Specimen type	Molecular test	Molecular targets (mutations and rearrangements)
Cervicovaginal	High-risk HPV testing	HPV
Oropharyngeal SqCC	High-risk HPV testing	HPV
Thyroid FNA	GEP analysis	*BRAF, HRAS, NRAS, KRAS, TERT, NTRKs, PAX8, RET, ALK*
	MicroRNA expression profiling	
	NGS	
Urine	FISH assay	Aneuploidy
Pancreaticobiliary	FISH assay	Aneuploidy
Pancreatic cyst fluid	NGS	*KRAS, GNAS, BRAF, TP53, CDKN2A*
Soft tissue neoplasms/sarcomas	FISH assay	*EWS, ETV6, USP6, PDGF, PAX, NTRK, FUS, SS18, EWSR1,*
	RT-PCR, NGS	*MDM2, CDK4*
Salivary gland neoplasms	FISH assay	*NTRK, MYB, MAML2, EWSR1*
Hematolymphoid neoplasms	Clonality testing	*BCR-ABL1, PML-RARA, ETV6, RUNX1, SF3B1, TET2, IDH1,*
	FISH assay	*IDH2, CALR, DNMT3A, TP53, KRAS, BRAF*, 17p deletion,
	Microarray assay (e.g., array-CGH)	trisomy 12, somatic hypermutation
	Sequencing assays	
	GEP	
Nonsmall cell lung carcinoma	NGS	*EGFR, KRAS, BRAF, TP53, ALK, ROS1, RET, MET, NTRKs,*
	FISH assay	*NRG1*
Breast carcinoma	FISH assay	*ERBB2*
Gastroesophageal adenocarcinoma	FISH assay	*ERBB2*
Colorectal adenocarcinoma	NGS, PCR-based	*ERBB2, KRAS*, MSI
	FISH assay	
Gastrointestinal stromal tumor	NGS, PCR-based	*KIT, PDGFRA*
Melanoma	NGS, PCR-based	*BRAF*
Brain tumors	FISH assay	*EGFR, MDM2, IDH1, IDH2*, 1p 19q co-deletion, BRAF
	NGS, PCR-based	
Endometrial adenocarcinoma	FISH assay	*ERBB2, POLE*, MSI
	NGS	

HPV human papillomavirus, *SqCC* squamous cell carcinoma, *GEP* gene expression profiling, *FNA* fine-needle aspiration, *NGS* next-generation sequencing, *FISH* fluorescence in situ hybridization, *PCR* polymerase chain reaction, *RT-PCR* reverse-transcriptase PCR, *CGH* comparative genomic hybridization

    – FISH testing and/or NGS-based fusion assays for subclassification of soft tissue tumors and sarcomas

    – Clonality studies, FISH, and PCR-based testing for diagnosis and subclassification of hematolymphoid neoplasms

- Molecular assays are also utilized to interrogate biomarkers that provide information pertinent to prognosis and/or targeted therapeutic strategies (Table 3.1)
- Some key examples include
  - Molecular testing of nonsmall cell lung carcinomas for *EGFR, KRAS*, and *BRAF* gene mutations; *MET* exon 14 skipping mutations; and *ALK, ROS1, RET*, and *NTRK* gene rearrangements
  - *ERBB2* gene amplification testing for breast, gastroesophageal, endometrial, and colorectal carcinomas
  - Molecular testing of thyroid carcinomas for *BRAF* mutations, *RET* and *NTRK* rearrangements
  - Molecular testing of metastatic melanomas for various mutations including *BRAF, NRAS*, and *KIT* gene mutations

Further Reading

Balassanian R, Ng DL, Zante A. Stop using expired plasma for cell blocks. Cancer Cytopathol. 2019;127:737–8.

Balla A, Hampel KJ, Sharma MK, Cottrell CE, Sidiropoulos N. Comprehensive validation of cytology specimens for next-generation sequencing and clinical practice experience. J Mol Diagn. 2018;20:812–21.

Dejmek A, Zendehrokh N, Tomaszewska M, Edsjö A. Preparation of DNA from cytological material: effects of fixation, staining, and mounting medium on DNA yield and quality. Cancer Cytopathol. 2013;121:344–53.

Dudley JC, Schroers-Martin J, Lazzareschi DV, et al. Detection and surveillance of bladder cancer using urine tumor DNA. Cancer Discov. 2019;9:500–9.

Gillooly JF, Hein A, Damiani R. Nuclear DNA content varies with cell size across human cell types. Cold Spring Harb Perspect Biol. 2015;7:a019091.

Grace A, Kay E, Leader M. Liquid-based preparation in cervical cytology screening. Curr Diagn Pathol. 2001;7(2):91–5.

Guseva NV, Jaber O, Stence AA, Sompallae K, Bashir A, Sompallae R, Bossler AD, Jensen CS, Ma D. Simultaneous detection of single-nucleotide variant, deletion/insertion, and fusion in lung and thyroid carcinoma using cytology specimen and an RNA-based next-generation sequencing assay. Cancer Cytopathol. 2018;126:158–69.

Harada S, Agosto-Arroyo E, Levesque JA, Alston E, Janowski KM, Coshatt GM, Eltoum IA. Poor cell block adequacy rate for molecular testing improved with the addition of diff-quik-stained smears: need for better cell block processing. Cancer Cytopathol. 2015;123:480–7.

Heymann JJ, Yoxtheimer LM, Park HJ, et al. Preanalytic variables in quality and quantity of nucleic acids extracted from FNA specimens of thyroid gland nodules collected in CytoLyt: cellularity and storage time. Cancer Cytopathol. 2020;128:656–72.

Hoffman EA, Frey BL, Smith LM, Auble DT. Formaldehyde cross-linking: a tool for the study of chromatin complexes. J Biol Chem. 2015;290:26404–11.

Hwang DH, Garcia EP, Ducar MD, Cibas ES, Sholl LM. Next-generation sequencing of cytologic preparations: an analysis of quality metrics. Cancer Cytopathol. 2017;125:786–94.

Jennings LJ, Arcila ME, Corless C, Kamel-Reid S, Lubin IM, Pfeifer J, Temple-Smolkin RL, Voelkerding KV, Nikiforova MN. Guidelines for validation of next-generation sequencing-based oncology panels: a joint consensus recommendation of the Association for Molecular Pathology and College of American pathologists. J Mol Diagn. 2017;19:341–65.

Kanagal-Shamanna R, Portier BP, Singh RR, et al. Next-generation sequencing-based multi-gene mutation profiling of solid tumors using fine needle aspiration samples: promises and challenges for routine clinical diagnostics. Mod Pathol. 2014;27:314–27.

Larson MH, Pan W, Kim HJ, et al. A comprehensive characterization of the cell-free transcriptome reveals tissue- and subtype-specific biomarkers for cancer detection. Nat Commun. 2021;12:2357.

Liu H, Huang X, Zhang Y, Ye H, El Hamidi A, Kocjan G, Dogan A, Isaacson PG, Du M-Q. Archival fixed histologic and cytologic specimens including stained and unstained materials are amenable to RT-PCR. Diagn Mol Pathol. 2002;11:222–7.

Masumoto N, Fujii T, Ishikawa M, Mukai M, Saito M, Iwata T, Fukuchi T, Kubushiro K, Tsukazaki K, Nozawa S. Papanicolaou tests and molecular analyses using new fluid-based specimen collection technology in 3000 Japanese women. Br J Cancer. 2003;88:1883–8.

Monaco SE, Dacic S. Fluorescence in situ hybridization in cytopathology. In: Modern techniques in cytopathology, vol. 25. Basel: Karger; 2019. p. 19–33.

Monaco SE, Teot LA, Felgar RE, Surti U, Cai G. Fluorescence in situ hybridization studies on direct smears: an approach to enhance the fine-needle aspiration biopsy diagnosis of B-cell non-Hodgkin lymphomas. Cancer Cytopathol. 2009;117:338–48.

Pentsova EI, Shah RH, Tang J, et al. Evaluating cancer of the central nervous system through next-generation sequencing of cerebrospinal fluid. J Clin Oncol. 2016;34:2404–15.

Ramani NS, Chen H, Broaddus RR, et al. Utilization of cytology smears improves success rates of RNA-based next-generation sequencing gene fusion assays for clinically relevant predictive biomarkers. Cancer Cytopathol. 2021;129:374–82.

Roy-Chowdhuri S, Chen H, Singh RR, et al. Concurrent fine needle aspirations and core needle biopsies: a comparative study of substrates for next-generation sequencing in solid organ malignancies. Mod Pathol. 2017;30:499–508.

Roy-Chowdhuri S, Mehrotra M, Bolivar AM, et al. Salvaging the supernatant: next generation cytopathology for solid tumor mutation profiling. Mod Pathol. 2018;31:1036–45.

Roy-Chowdhuri S, Roy S, Pantanowitz L. Next-generation sequencing in cytopathology. In: Modern techniques in cytopathology, vol. 25. Basel: Karger; 2020. p. 34–42.

Saqi A, Balassanian R. Cell blocks: evolution, modernization, and assimilation into emerging technologies. In: Modern techniques in cytopathology, vol. 25. Basel: Karger; 2020. p. 6–18.

Soltani M, Nemati M, Maralani M, Estiar MA, Andalib S, Fardiazar Z, Sakhinia E. Cell-free fetal DNA in amniotic fluid supernatant for prenatal diagnosis. Cell Mol Biol. 2016;62:14–7.

Sung S, Sireci A, Remotti H, Hodel V, Mansukhani M, Fernandes H, Saqi A. Plasma-thrombin: potential source of DNA contamination in cell blocks. J Am Soc Cytopathol. 2018;7:S5.

Yang S-R, Mooney KL, Libiran P, et al. Targeted deep sequencing of cell-free DNA in serous body cavity fluids with malignant, suspicious, and benign cytology. Cancer Cytopathol. 2020;128:43–56.

Next-Generation Immunohistochemistry in the Workup of Neoplasm of Uncertain Lineage and CUP

Andrew M. Bellizzi

Contents

Introduction: Immunohistochemistry Complements a Morphology and Epidemiology-Based Approach to Tumor Diagnosis

- Most tumors are diagnosed by morphology alone
- Immunohistochemistry (IHC) is a useful adjunct when the line of differentiation (e.g., carcinoma, lymphoma, melanoma, and sarcoma) is not apparent on the H&E
- A judicious panel of immunostains should be applied to adjudicate a differential diagnosis generated based on morphology and epidemiology
- Principal epidemiologic considerations include age, gender, and anatomic site of disease
- The possibility of primary versus metastatic tumor should always be considered
  - When facing diagnostic uncertainty, pathologists tend to forget the range of primary tumors presenting at that anatomic site (e.g., intrahepatic cholangiocarcinoma is among the most common "carcinomas of unknown primary" presenting in the liver)
- IHC markers can be divided into screening markers and differentiation markers

A. M. Bellizzi (✉)
Department of Pathology, University of Iowa Hospitals and Clinics and Carver College of Medicine, Iowa City, IA, USA
e-mail: andrew-bellizzi@uiowa.edu

- Screening markers are sensitive for identifying the broad tumor class
    - Broad-spectrum epithelial markers: carcinoma
    - SOX10 and S-100: melanoma
    - CD45 and CD43: hematolymphoid tumors
    - SALL4: germ cell tumor
  - Differentiation markers are used to further subtype tumors (e.g., site of origin in an adenocarcinoma, confirm melanocytic differentiation in a melanoma, assign B- or T-cell lineage in a lymphoma; subtype a germ cell tumor)
- In high-grade tumors of uncertain lineage, it is generally advisable to not order differentiation markers until the broad tumor class is established
- Every biopsy of possible tumor should be considered a potential molecular pathology specimen
  - Divide specimen into at least two blocks and reserve one for molecular testing
  - Order extra unstained slides upfront to account for number expected to be ordered based on range of possible results from initial IHC panel
- Because IHC assays contain so many variable parameters, they are generally not interchangeable between laboratories (i.e., experience with a given marker can vary substantially from laboratory to laboratory)
- The information presented in this chapter reflects the central mass of the primary literature, my anecdotal experience, and, in many instances, extensive testing in my own diagnostic laboratory

Immunohistochemistry Basics

IHC Protocol

- Basic steps in the (indirect) IHC procedure include pretreatment (deparaffinization, rehydration, epitope retrieval), blocking of endogenous peroxidase, application of primary antibody, application of detection reagent, addition of chromogen, application of counterstain (typically hematoxylin), dehydration, and coverslipping
- Epitope retrieval restores "antigenicity," which is typically reduced by formalin fixation
- Epitope retrieval is used with most assays
- Epitope retrieval may be heat-induced (in buffer of high or low pH) or enzymatic
  - HIER HpH (74% of assays in my laboratory)
  - HIER LpH (14%)
  - Proteinase K (3%)
  - No retrieval (10%; S-100, several peptide hormones)
- The primary antibody confers the specificity of the signal
- Primary antibodies can be polyclonal or monoclonal

- Polyclonal antibodies are often the "first available" for a new marker; they tend to show more lot-to-lot variation
- The secondary antibody is typically directed against the Fc portion of the primary antibody
- Contemporary detection chemistry typically employs a polymer studded with secondary antibodies and enzyme (i.e., horseradish peroxidase [HRP] or alkaline phosphatase)
- The chromogen is acted upon by the enzyme, producing a colorimetric product
- For HRP-based assays, endogenous peroxidase activity must be abolished before adding the chromogen or false-positive staining will result

IHC Optimization/Validation

- IHC assays should be carefully optimized and validated before being placed into clinical service
- Optimization entails examining an assay at various conditions to achieve a favorable signal-to-noise ratio
- Conditions which are often modified in an IHC optimization include type and duration of epitope retrieval, dilution of primary antibody, duration of primary antibody incubation, use of a signal amplifying "linker," and duration of detection reagent incubation
- Nordic immunohistochemical Quality Control (NordiQC) publishes detailed platform-specific protocols for most common IHC targets (https://www.nordiqc.org/)
- Validation involves testing a set of expected positives and expected negatives to ensure that the assay performs as expected
- IHC assays should be validated in a fit-for-purpose fashion (i.e., evaluating all the intended applications for a given marker)
- Resources to guide validation include the College of American Pathologists Center Guideline: *Principles of Analytic Validation of Immunohistochemical Assays: Guideline from the College of American Pathologists Pathology and Laboratory Quality Center* and the four-part *Evolution of Quality Assurance for Clinical Immunohistochemistry in the Era of Precision Medicine* (published in *Applied Immunohistochemistry and Molecular Morphology* in 2017).

IHC Readout

- IHC readout is a characterization of the strength and localization of signal
- Because the relationship between signal and analyte concentration is typically unknown, IHC is not quantitative—instead, it is ordinal

- The intensity and extent of signal may provide additional diagnostic information (e.g., diffuse, strong CDX2-positivity in an adenocarcinoma suggests a lower gastrointestinal [GI] origin, while lesser staining is more common with tumors of pancreatobiliary, upper GI, or primary mucinous ovarian origin)
- In general, the stronger the IHC signal, the greater the diagnostic weight that should be given to the result
- For many IHC markers, weak staining in rare cells should be given little diagnostic weight (e.g., PAX8, GATA-3, and TRPS1)
- For others, similar quality staining appears to be more meaningful (e.g., TTF-1, CDX2, and islet 1)

IHC Common Pitfalls

- IHC reagents can get trapped under the edge of the tissue section, leading to nonspecific signal (aka, edge artifact) (Fig. 4.1)
- IHC signal can be greatly influenced by preanalytic factors, including delayed and incomplete fixation and decalcification
- Many of the best IHC markers are lineage-restricted transcription factors but all nuclear antigens appear to be especially adversely affected by unfavorable preanalytic conditions (Fig. 4.2)
- It is imperative to know which compartment of the cell (i.e., nucleus, cytoplasm, or membrane) a given IHC marker should highlight, and staining of another compartment does not constitute a "positive" result

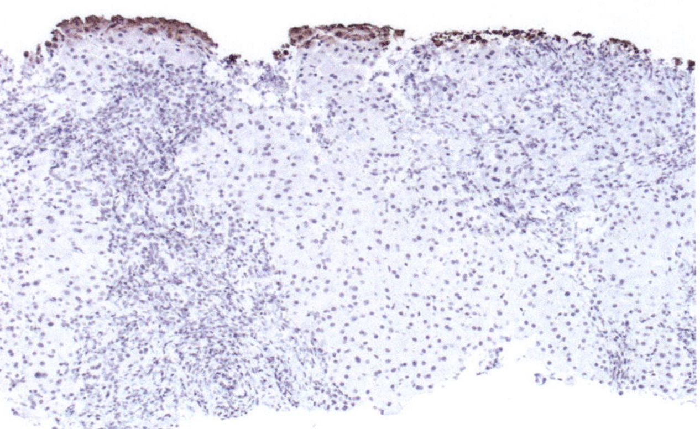

Fig. 4.1 Edge artifact: this staining artifact is attributable to reagents getting trapped under the edges of tissue (which may be more loosely adherent to the glass slide) and then not washed away during subsequent steps of the IHC procedure

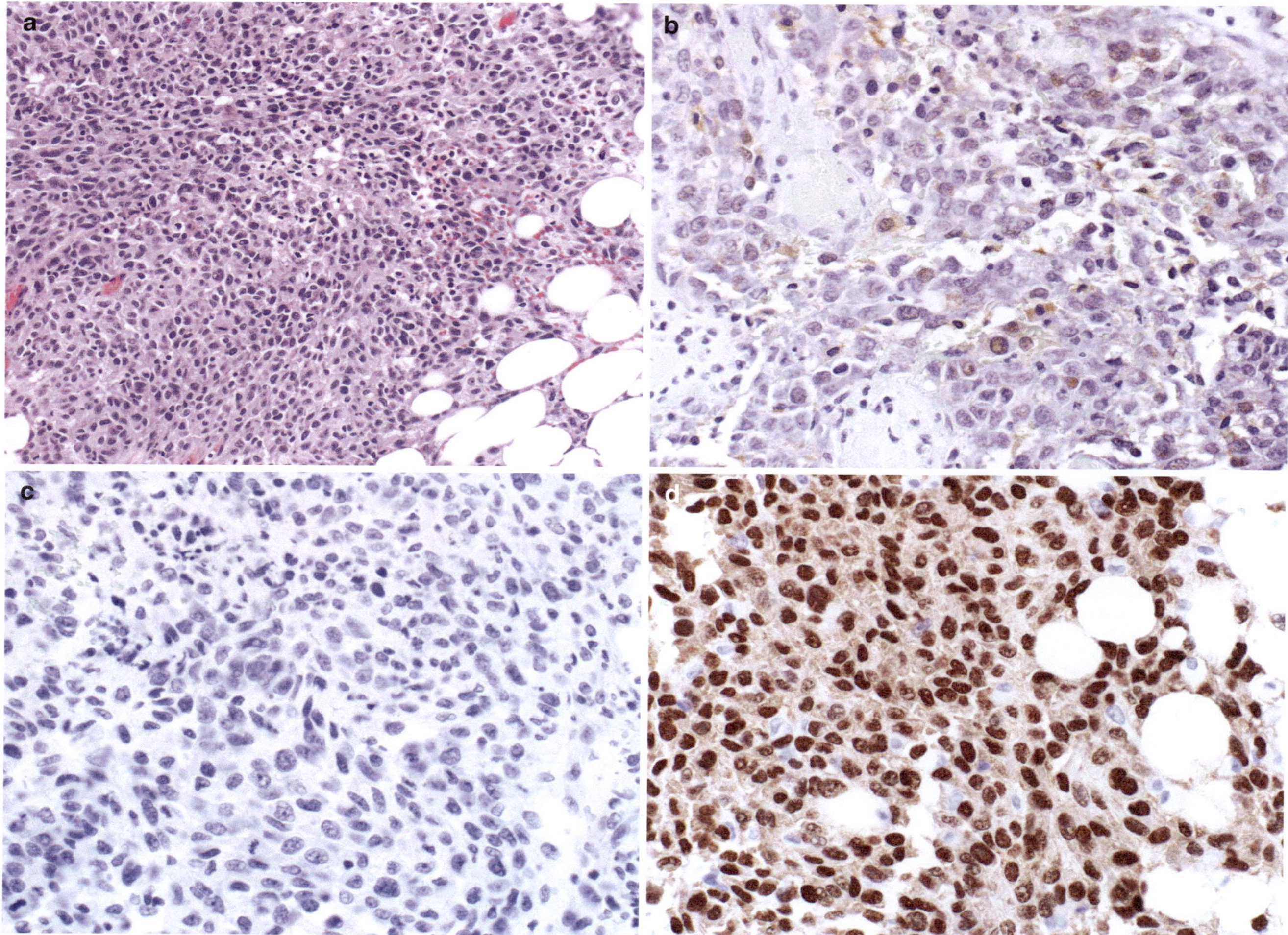

Fig. 4.2 S-100 versus SOX10 in decalcified specimen: Decalcified bone biopsy demonstrating an epithelioid malignant neoplasm (**a**) stained for S-100 (**b**) and SOX10 (**c**). Repeat SOX10 immunostain in a non-decalcified block from the same specimen (**d**). Although SOX10 typically outperforms S-100 as a marker of melanocytic, nerve sheath, and myoepithelial differentiation, like all nuclear antigens it appears susceptible to reduced staining in decalcified and poorly fixed specimens. In bone biopsies for tumor, if possible, it is recommended to separate out non-bony tissue for a non-decalcified block

Next-Generation Immunohistochemistry

- In next-generation IHC, antigenic targets are "specifically selected" rather than "empirically discovered"
- Targets are derived, especially, from the molecular genetic and developmental biology literature and include:
  - Markers identified by gene or protein expression profiling (e.g., GATA-3, MUC4, OTP, and SATB2)
  - Protein correlates of molecular genetic events (e.g., MDM2, SMARCB1, SMARCA4, Rb, SMAD4, BAP1, and MTAP)
  - Lineage-restricted transcription factors (e.g., PAX8, ERG, SALL4, SOX10, p40, SF1, and NKX3.1)
  - (Markers often fulfill more than one next-generation IHC "qualification")
- Some next-generation IHC markers, especially protein correlates of molecular genetic events, achieve near-perfect sensitivity and specificity (e.g., STAT6 in solitary fibrous tumor; SS18-SSX and SSX C-terminus in synovial sarcoma)
- Lineage-restricted transcription factors tend to be robustly expressed, independent of the differentiation status of the

Table 4.1 GATA-3 as exemplar oligospecific transcription factor—multiple diagnostic utilities

GATA-3-positive	GATA-3-negative
Cutaneous squamous cell carcinoma (80%)	Pulmonary squamous cell carcinoma
Mesothelioma (50%)	Lung adenocarcinoma
Pancreatic adenocarcinoma (33%)	Other GI adenocarcinomas
Chromophobe renal cell carcinoma (50%)	Clear cell renal cell carcinoma
Choriocarcinoma/yolk sac tumor	Embryonal carcinoma/seminoma
Paraganglioma	Neuroendocrine tumor
Breast carcinoma	
Urothelial carcinoma	

tumor, while cytoplasmic and membranous differentiation markers tend to show reduced expression in more poorly differentiated tumors

- Lineage-restricted transcription factors often demonstrate multiple diagnostic utilities (i.e., they are "oligospecific") (Table 4.1)
- Publicly available online tools (e.g., https://www.cbioportal.org/) can be used to visualize gene expression across TCGA tumor types, allowing one to infer potential diagnostic utilities (Fig. 4.3)

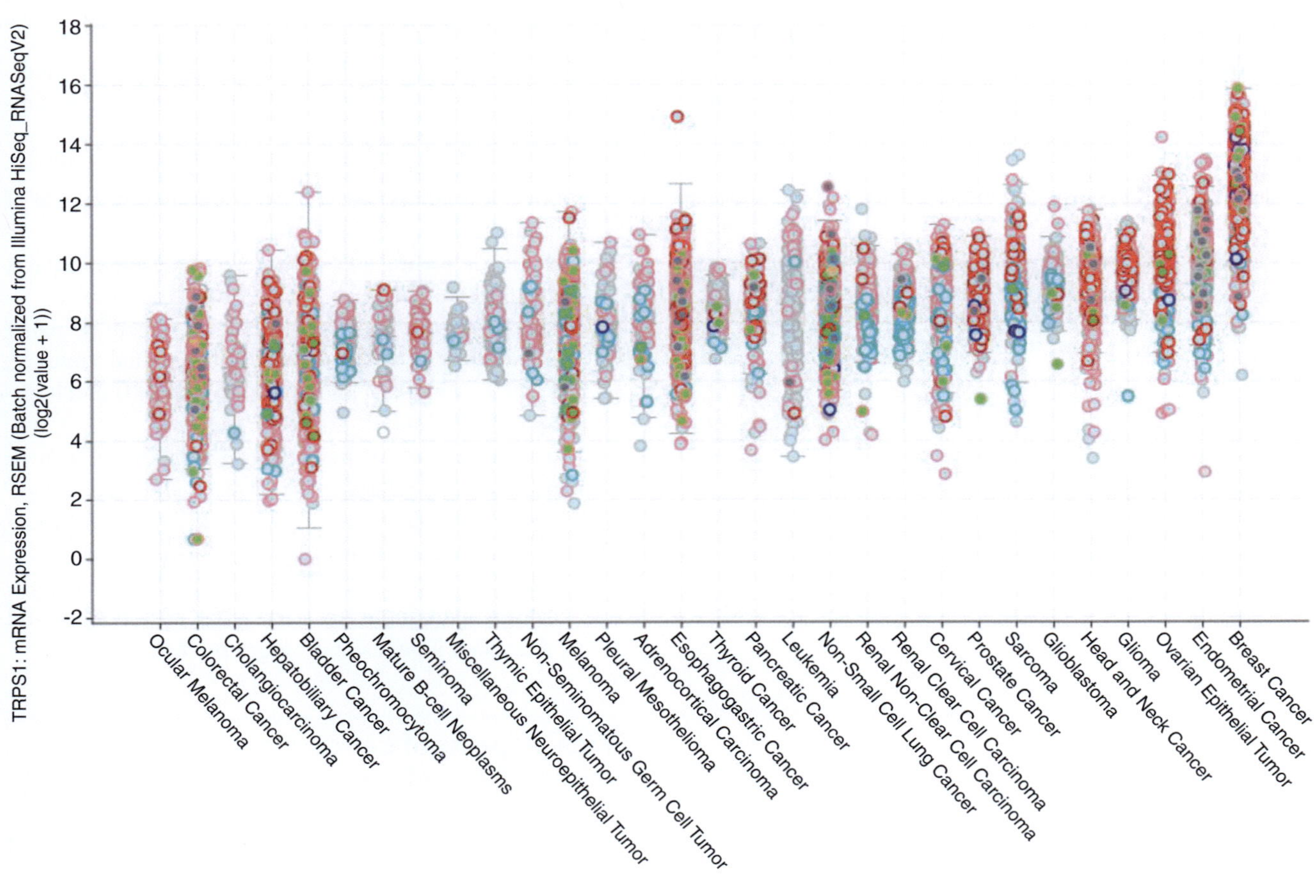

Fig. 4.3 cBioPortal Visualization of TRPS1 gene expression across TCGA tumor types: cBioPortal can be used to examine gene expression across tumor types, allowing one to infer potential diagnostic utilities and identify potential pitfalls. Although breast cancer has the highest TRPS1 expression, it overlaps with expression in endometrial cancer and ovarian epithelial tumors. High-expressing sarcomas also demonstrate overlap. Although I have found TRPS1 IHC to be a useful diagnostic adjunct, particularly in triple-negative breast cancer, I have observed strong expression in some Müllerian cancers and sarcomas (https://www.cbioportal.org/)

Cancer Epidemiology

- One should be able to generate a differential diagnosis for any biopsy of tumor given patient age, gender, and anatomic site of disease
- Tables 4.2, 4.3, and 4.4 were generated largely based on the Surveillance, Epidemiology, and End Results Program (SEER) Cancer Statistics Review (CSR) 1975–2016 (Cancer Statistics Review, 1975–2016—SEER Statistics) and present the estimated annual incidence and proportion among incident cases of (1) broad tumor class, (2) carcinoma histotype (i.e., adenocarcinoma, squamous cell carcinoma, urothelial carcinoma, neuroendocrine tumor [NET], and neuroendocrine carcinoma [NEC]), and (3) adenocarcinoma site of origin
- Based on these Tables, several conclusions can be drawn:
  - Most tumors are carcinomas (80%), followed by hematolymphoid neoplasms (10%) and melanoma (6%)

Table 4.2 Estimated annual cancer incidence stratified by broad tumor class

Tumor type	Estimated annual incidence	% of all incident cases (%)
Carcinoma	1,335,410	80
Hematolymphoid	174,250	10
Melanoma	94,810	6
Sarcoma	16,490	1
Germ cell tumor	10,422	0.6
Mesothelioma	3300	0.2
Pheochromocytoma/ paraganglioma	2608	0.2
Other and unspecified primary sites	31,810	2

SEER Cancer Statistics Review, 1975–2016 (https://seer.cancer.gov/archive/csr/1975_2016/)

Table 4.3 Estimated annual carcinoma incidence stratified by histotype

Tumor type	Estimated annual incidence	% of all incident cases (%)
Adenocarcinoma	1,046,444	77
Squamous cell carcinoma	165,900	12
Non-HPV-mediated	112,810	8
HPV-mediated	53,090	4
Urothelial carcinoma	91,544	7
Neuroendocrine	60,494	4
Neuroendocrine carcinoma	49,406	4
Neuroendocrine tumor	11,088	0.8

SEER Cancer Statistics Review, 1975–2016 (https://seer.cancer.gov/archive/csr/1975_2016/)

Table 4.4 Estimated annual adenocarcinoma incidence stratified by site of origin

Tumor type	Estimated annual incidence	% of all incident cases (%)
Breast	268,670	26
Prostate	164,690	16
Colorectum	140,250	13
Lung	124,036	12
Müllerian	84,358	8
Pancreatobiliary	72,331	7
Kidney	58,806	6
Thyroid	53,990	5
Upper GI tract	44,566	4
Hepatocellular carcinoma	34,747	3

SEER Cancer Statistics Review, 1975–2016 (https://seer.cancer.gov/archive/csr/1975_2016/)

- Outside of somatic soft tissue, retroperitoneum, mediastinum, and paratestis, sarcoma is distinctly uncommon (only 1% of all tumors)
- Outside of gonads, retroperitoneum, and mediastinum, germ cell tumor is distinctly uncommon (only 0.6% of all tumors)
- Outside of pleura and peritoneum, mesothelioma is exceedingly unlikely (only 0.2% of all tumors)
  - These results validate use of a screening panel including a broad-spectrum epithelial marker, CD45 (or CD43), and SOX10 (or S-100) in a high-grade neoplasm of uncertain lineage
- Most carcinomas, including those with a solid growth pattern, are adenocarcinoma (80% of all carcinomas)
- Two-thirds of all adenocarcinomas are of breast, prostate, colorectal, or lung origin
  - These results are accounted for in subsequent algorithms pertaining to carcinoma histotyping and adenocarcinoma site of origin prediction

Clinical Aspects of Cancer (Typically Carcinoma) of Unknown Primary

- Definition: "Cancer of unknown primary (CUP) is defined as a carcinoma or undifferentiated neoplasm for which a standardized diagnostic workup fails to identify the primary tumor responsible for metastatic seeding." (from European Society for Medical Oncology [ESMO] CUP Clinical Practice Guideline)
- Incidence: 2–9% of all tumors with a declining incidence (attributed to more widespread availability of better diagnostic tools)

- 30,620 estimated new cases in 2022 (2% of all tumors), though concern for some underreporting (from North American Association of Central Cancer Registries)
- Sites of Involvement
  - Multiple in >50% of cases with liver most common site; other common sites include lung, lymph node, bone, brain, and abdominal cavity
  - Solitary or oligometastatic (i.e., ≤5 metastases)
- Survival
  - Median survival: 8 to 12 months
  - Poor prognosis cases (80% of CUPs): 3 to 10 months
  - Good prognosis cases (20%): 12 to 36 months
  - 3-year survival superior for squamous cell carcinoma (42%) versus adenocarcinoma/undifferentiated carcinoma (3.5%)
  - CUP survival is worse than that of metastatic prostate, breast, colon, and kidney cancer; similar to that of metastatic lung and bladder cancer; and better than that of metastatic pancreas, stomach, and liver cancer
  - CUP survival is worse than that of lymphoma, melanoma, and germ-cell tumor
  - Independent poor prognostic factors include male gender, poor performance status, adenocarcinoma histology, high number of metastatic sites, and high neutrophil:lymphocyte ratio
- Prognostic groups
  - Poor prognosis
    - Male
    - Age ≥ 65
    - Poor performance status
    - Comorbidities
    - Metastasis to multiple organs
    - Malignant ascites (except due to serous carcinoma)
    - Peritoneal involvement
    - Multiple brain metastases
    - Adenocarcinoma with multiple lung/pleural or bone metastases
  - Good prognosis
    - Solitary or oligometastatic disease amenable to local excision and/or radiotherapy
    - Midline nodal disease
    - Squamous cell carcinoma in cervical lymph nodes
    - Isolated inguinal lymphadenopathy
    - Poorly differentiated NEC
    - Serous carcinoma in the peritoneum
    - Adenocarcinoma in axillary lymph node (woman)
    - Blastic bone metastasis with positive PSA (man)
- Workup
  - Standard (Minimum) Workup
    - History and physical examination, including breast, GU, pelvic, and rectal examinations
    - Basic blood work, including CBC, electrolytes, LFTs, creatinine, and calcium
    - Fecal occult blood test
    - Contrast-enhanced CT (or MRI) of neck/chest/abdomen/pelvis
    - Mammogram (woman)
  - Additional workup dictated by initial findings (e.g., endoscopy, tumor markers, urine cytology, etc.)
  - ^{18}F-FDG-PET may be especially useful in single-site/oligometastatic disease to rule out additional sites of involvement
- Pathology Workup
  - Immunohistochemistry is the diagnostic mainstay
    - One should be able to generate a differential diagnosis for any biopsy of tumor given patient age, gender, and anatomic site of disease
    - 25% of CUP patients have a personal history of malignancy, which should be the first diagnostic consideration
  - Next-Generation Sequencing
    - NGS is often pursued in this setting to identify potential therapeutic targets (including checkpoint inhibition for MSI-H and/or TMB-H tumors)
    - NGS results occasionally help inform the pathology diagnosis
    - In patients with a personal history of malignancy, it is occasionally useful to compare mutational signatures in prior and current tumor
  - Gene Expression Profiling
    - The National Comprehensive Cancer Network (NCCN) and ESMO Clinical Practice Guidelines do not endorse routine clinical use of commercially available GEP assays
    - Two recent clinical trials of GEP-driven choice of chemotherapy failed to show survival benefit (NCT01540058, PMID: 30653423)
    - Vendor of a commercial assay claims 37% improved survival for GEP-directed over empiric therapy based on a trial with no prospective control arm (i.e., comparison was to historical control)
- Therapy for widespread disease
  - Platinum-based doublet chemotherapy is the historical standard or care
  - No specific regimen can be recommended
  - Goal is to prolong survival, improve quality of life, and minimize toxicity
  - Poorly to undifferentiated adenocarcinomas are more likely to initially respond
  - For adenocarcinoma, NCCN lists of number of regimens, including those typically active in nonsmall cell lung, GI, and bladder cancers
  - For squamous cell carcinoma, many of the NCCN-listed regimens are also active in adenocarcinoma

Broad Tumor Classes and Associated Screening Markers

- Broad tumor classes include carcinoma, hematolymphoid neoplasm, melanoma, sarcoma, germ cell tumor, mesothelioma, and pheochromocytoma/paraganglioma
- Table 4.5 presents screening markers for these broad tumor classes, when these diagnostic categories should be considered (in the context of morphology, anatomic site, and results of initial IHC), and additional confirmatory markers for the noncarcinomas

- Key Pitfalls not discussed elsewhere:
  - Diffuse, strong S-100 and SOX10+ in a malignant spindle cell neoplasm points to a diagnosis of melanoma, while malignant peripheral nerve sheath tumors only show weak and often negative staining (Figs. 4.4 and 4.5)
  - Spindle cell (positive in 40–60%) and desmoplastic (positive in 10–20%) melanomas are often and usually negative, respectively, for the melanocytic differentiation markers MiTF, melan A, and HMB-45

Table 4.5 Broad tumor classes with associated screening markers

Broad tumor class	Screening markers	When to consider	Confirmatory markers
Carcinoma	**Broad-spectrum keratin**; EpCAM (i.e., MOC-31, Ber-EP4), EMA, claudin-4	Always	See additional algorithms
Hematolymphoid	**CD45** or CD43	Always; **reconsider in a "triple-negative" neoplasm**	CD45-negative lymphoma panel: CD43, CD79a, MUM1, ALK, CD30
Melanoma	**SOX10** or S-100	Always	Melan A, HMB-45, tyrosinase, BRAF V600E
Sarcoma	None	Spindle cell morphology; tumor in mediastinum, retroperitoneum, paratestis, or somatic soft tissue	Unclassified malignant neoplasm in mediastinum, retroperitoneum, paratestis: MDM2/CDK4 (DDLPS) Epithelioid neoplasm-defying typing: ERG (angiosarcoma), INI1 (epithelioid sarcoma) Last ditch effort: CD34 (rarely + in carcinoma)
Germ cell	**SALL4** or PLAP	Tumor in mediastinum, retroperitoneum, or gonads; "triple-negative" neoplasm; **keratin-positive neoplasm defying typing/site of origin assignment**	Seminoma: OCT4, KIT, D2-40 Embryonal carcinoma: OCT4, CD30, SOX2 Yolk sac tumor: Glypican-3, AFP Trophoblastic tumors: GATA-3, inhibin A, PD-L1, β-HCG
Mesothelioma	None (diagnostic consideration in keratin-positive tumor)	Tumor in the pleura or peritoneum	Calretinin, WT-1, D2-40, CK5/6, BAP1 (loss), MTAP (loss)
Pheochromocytoma/ Paraganglioma	None	Epithelioid morphology; "triple-negative" neoplasm; general neuroendocrine marker+ tumor defying site of origin assignment	GATA-3, PHOX2B, tyrosine hydroxylase, SDHB (loss)

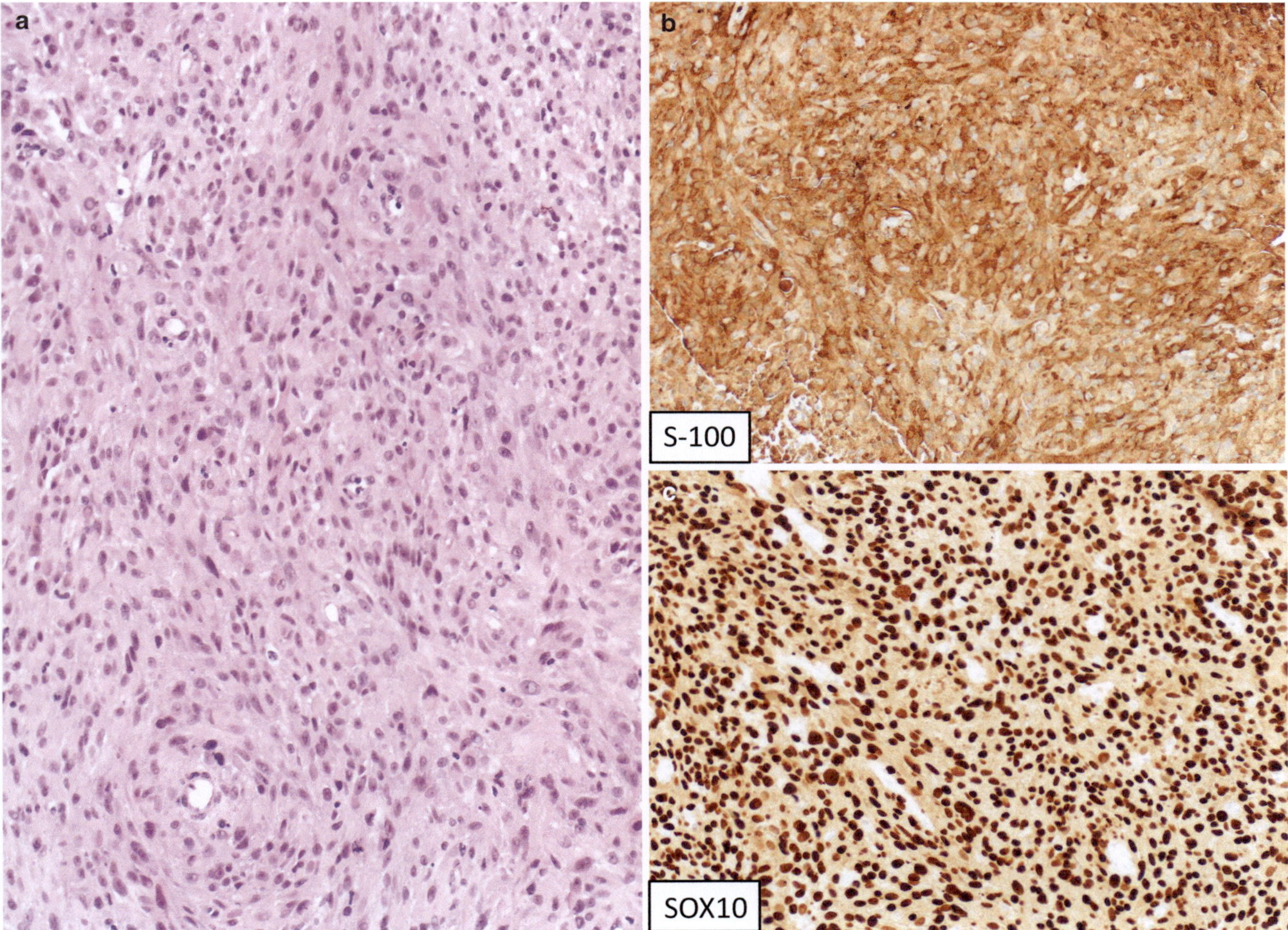

Fig. 4.4 Strong S-100 and/or SOX10-positivity in a spindle cell neoplasm supports a diagnosis of melanoma and obviates a diagnosis of malignant peripheral nerve sheath tumor: 55-year-old woman with core biopsy of an inguinal lymph node (**a**) demonstrates diffuse, strong positivity for both S-100 (**b**) and SOX10 (**c**). This case was sent in consultation for "rule out nerve sheath tumor." Spindle cell melanoma is often negative for markers of terminal melanocytic differentiation, compounding diagnostic confusion

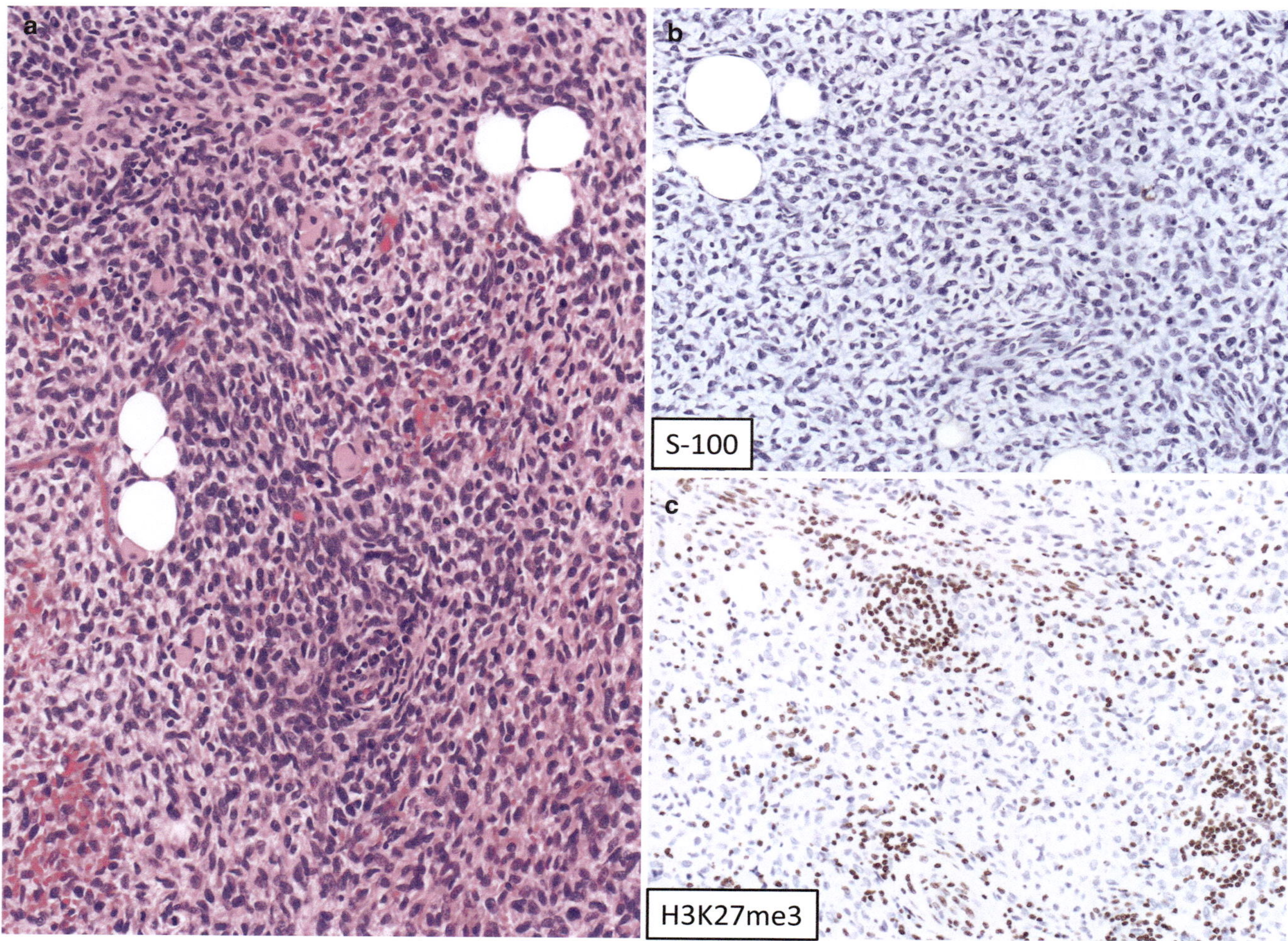

Fig. 4.5 Immunohistochemistry to support a diagnosis of malignant peripheral nerve sheath tumor: 12-year-old with NF1 and recurrent cellular, spindle cell tumor in the right psoas (**a**) is S-100 negative (**b**) but demonstrates loss of H3K27me3 staining with intact internal control (**c**)

Broad-Spectrum Epithelial Markers

- Broad-spectrum epithelial markers are used to screen for carcinoma (though tumors from several other broad tumor classes may be positive—see "noncanonical expression of screening markers")
- Broad-spectrum epithelial markers include broad-spectrum keratins, antibodies to epithelial cell adhesion molecule (EpCAM) (i.e., MOC-31, Ber-EP4), EMA, and claudin-4
- Broad-spectrum keratins should recognize low and high-molecular-weight keratins
  - Simple epithelia variably express low-molecular-weight keratins: 7, 8, 18, 19, 20
  - Stratified epithelia variably express high-molecular-weight keratins: 1–6, 9–17
- The most widely used commercially available broad-spectrum keratins are AE1/AE3 and AE1/AE3/PCK26,

though a number of other antibodies are available (Table 4.6)
  - AE1/AE3 is actually a cocktail of the monoclonal antibodies AE1 and AE3
  - Because AE1/AE3 does not recognize keratin 18 (K18), there are a few commercially available cocktails that include an additional component that recognizes K18 (e.g., AE1/AE3/5D3)
- Many laboratories have created their own "pan-keratin" cocktails, though these are not clearly superior to commercially available reagents
- Broad-spectrum keratins are commonly poorly optimized, leading to reduced sensitivity
  - The "pass rate" in 10 NordiQC assessments from 2003 to 2020 has improved from 53% to (only) 75%
  - Failure is typically attributable to inadequate antigen retrieval, too dilute primary antibody, or use of "less successful" primary antibodies

Table 4.6 Comparison of the specificity of various keratin antibody clones

Which screening keratin should I use?																			
Clone	1	2	3	4	5	6	7	8	9	10	11	12	13	14	15	16	17	18	19
AE1/AE3	X	X	X	X	X	X		X		X				X	X	X			X
MAK-6								X						X	X	X		X	X
MNF116					X	X		X									X		
CAM5.2							X	X											
KL1	X	X			X	X	X	X			X			X		X	X	X	
34βE12	X				X					X				X					

- AE1/AE3-based assays are in widest clinical use
- CAM5.2 only recognizes low and 34βE12 only recognizes high-molecular-weight keratins, so these should not be used in isolation
- It is not the number of keratins, per se, but rather the affinity that matters (e.g., CAM5.2 may be more sensitive than AE1/AE3 in HCC/RCC)
- Stratified epithelia: K1-6, 9–17
- Simple epithelia: K7, 8, 18, 19, 20

Ordóñez NG. *Hum Pathol.* 2013 Jul;44(7):1195–215

- Broad-spectrum keratin assays should show at least moderate staining in the vast majority of hepatocytes (while in the most poorly performing assays staining is confined to bile duct epithelial cells); similarly, assays should show at least moderate staining in the vast majority of proximal renal tubules and parietal epithelial cells (while in the most poorly performing assays staining is confined to the distal nephron)
- Epithelial membrane antigen (EMA) is encoded by the *MUC1* gene
    - In addition to use as a broad-spectrum epithelial marker, it is also employed in the differential diagnosis of sebaceous and squamous cell carcinoma (EMA+) versus basal cell carcinoma (EMA−) and ovarian surface epithelial tumor (EMA+) versus sex-cord stromal tumor (EMA−)
    - Key Pitfall: EMA is positive in a number of CD45 weak-to-negative hematolymphoid neoplasms
- MOC-31 and Ber-EP4 are monoclonal antibodies to epithelial cell adhesion molecule (EpCAM)
    - Utility in the adenocarcinoma (EpCAM+) versus mesothelioma (EpCAM−) differential has led to MOC-31 and Ber-EP4's undue reputation as "adenocarcinoma markers"
    - Antibodies to EpCAM may also be used as broad-spectrum epithelial markers, with frequent reactivity in neuroendocrine epithelial neoplasms (well-differentiated>poorly differentiated), urothelial carcinoma (~50%), and visceral squamous cell carcinoma (50%; staining is more frequent and more extensive in poorly differentiated tumors)
    - They are also employed in the differential diagnosis of cholangiocarcinoma (EpCAM+) versus hepatocellular carcinoma (EpCAM-; though up to 20% show at least focal staining), and basal cell carcinoma (EpCAM+) versus cutaneous squamous cell carcinoma (EpCAM−)

- Claudin-4 may be used as a broad-spectrum epithelial marker, in addition to performing similarly to EpCAM in the adenocarcinoma (claudin-4+) versus mesothelioma (claudin-4−) and cholangiocarcinoma (claudin-4+) versus hepatocellular carcinoma (claudin-4−) differentials
    - Claudin-4 may be useful to distinguish some sarcomas that often express other broad-spectrum epithelial markers from carcinomas (i.e., epithelioid sarcoma, epithelioid angiosarcoma, and epithelioid hemangioendothelioma are claudin-4−)

S-100 Versus SOX10

- S-100 and SOX10 are typically used to screen for melanocytic differentiation and are also expressed by nerve sheath and myoepithelial cells and cognate tumors
- Table 4.7 compares S-100 and SOX10 expression
- S-100 is a 24-member family of calcium-binding proteins that are expressed in a cell type-specific manner
    - Polyclonal S-100 antibodies predominantly recognize S-100B (brain), which is the dominant S-100 protein expressed by melanocytes, glia, Schwann cells, chondrocytes, adipocytes, and myoepithelial cells (SOX10 is not expressed by adipocytes or chondrocytes)
    - Most laboratories use polyclonal (rather than monoclonal) S-100 antibodies
    - Polyclonal S-100 antibodies also react (less strongly) with S-100A1 and S-100A6, which are expressed by some carcinomas
    - S-100+/SOX10− is seen in
        - Chordoma
        - Histiocytic and dendritic cell tumors
        - Some translocation-associated mesenchymal tumors (e.g., biphenotypic sinonasal sarcoma, and ossifying fibromyxoid tumor)

Table 4.7 Comparison of the performance of S-100 versus SOX10 in tumor diagnosis

S-100−, SOX10+	S-100+, SOX10+
• Some MPNSTs, melanomas, and carcinomas with myoepithelial differentiation (sensitivity issue)	• Melanoma • Nerve sheath tumors • Neoplasms with myoepithelial differentiation
S-100−, SOX10−	**S-100+, SOX10−**
• Most carcinomas, sarcomas, lymphomas • Mesothelioma • Germ cell tumor • Pheochromocytoma/paraganglioma (though both markers expressed by sustentacular cells)	• Tumors of adipocytic/chondroid lineage • Chordoma • Ossifying fibromyxoid tumor • Biphenotypic sinonasal sarcoma • Lipofibromatosis-like neural tumor • Infantile fibrosarcoma-like tumor • Rare cases of Ewing, RMS, SS • S-100+ carcinomas without myoepithelial differentiation (S-100A1 and/or S-100A6+) • S-100+ histiocytic/dendritic cell tumors (Langerhans cell histiocytosis, Rosai-Dorfman, interdigitating dendritic cell tumor (100%); histiocytic sarcoma, Erdheim-Chester, blastic plasmacytoid dendritic cell tumor (30%); follicular dendritic cell sarcoma, juvenile xanthogranuloma (occ.))

Polyclonal S-100 (S-100B >> S-100A1 >> S-100A6)
RMS rhabdomyosarcoma, *SS* synovial sarcoma

- SOX10 is a transcription factor critical to the development of the neural crest and its derivatives, including Schwann cells and melanocytes
  - SOX10 expression in the peripheral nervous system induces glial differentiation and maturation, while it is turned off in cells that become neurons and chromaffin cells
  - SOX10 expression in melanocytes induces expression of MiTF
  - SOX10+/S-100− occurs in some malignant peripheral nerve sheath tumors, carcinomas with myoepithelial differentiation, and melanomas due to the increased sensitivity of SOX10
  - SOX10 is strongly expressed by 60% of triple-negative breast cancers (a tumor type that shows myoepithelial differentiation)
  - It is expressed at a similar rate in "true" basaloid squamous cell carcinomas
  - Carcinomas that are S-100+ because of expression of S-100A1 or A6 are SOX10−
  - SOX10 is also expressed in luminal cells of serous acini and the intercalated duct and in secretory cells of the eccrine coil, as well as cognate tumors, including acinic cell carcinoma, polymorphous adenocarcinoma, cylindroma, and spiradenoma

Noncanonical Expression of Screening Markers

- Table 4.8 presents instances of "noncanonical expression of (mostly) screening markers" (i.e., expression of broad-spectrum epithelial markers by noncarcinoma, expression of melanoma markers by nonmelanoma, and expression of hematolymphoid markers by nonhematolymphoid tumors)
  - Key Pitfalls:
    - In addition to synovial sarcoma, sarcomas with epithelioid cytomorphology and leiomyosarcoma (30–40%) are often broad-spectrum keratin/EMA+
    - EMA-positivity in hematolymphoid neoplasms overlaps substantially with weak-to-negative CD45 expression (Table 4.9)
    - Aside from seminomas, germ cell tumors are usually broad-spectrum keratin/EMA+, and germ cell tumor should be considered in possible carcinomas of occult origin that defy typing/site of origin assignment, especially in the midline of the brain, mediastinum, retroperitoneum, and gonads
    - CD138 (syndecan-1) is often used as a marker of plasmacytic differentiation but is expressed by ≥40% of carcinomas

Table 4.8 Noncanonical expression of broad tumor class screening markers

Marker category	Noncanonical expressors
Broad-spectrum epithelial markers	• **Sarcomas with epithelioid cytomorphology, small round blue cell sarcomas, leiomyosarcoma (30–40% keratin and/or EMA-positive)**
	• **EMA-positivity in plasma cell neoplasms (most), ALCL (50–95%), DLBCL variants (T-cell/histiocyte-rich, ALK+, plasmablastic, primary effusion lymphoma)**, NLPHL, FDCS
	• **Up to 25% of metastatic melanomas** (keratin probably>EMA)
	• **Yolk sac tumor and choriocarcinoma are always and embryonal carcinoma is often broad-spectrum keratin/EMA-positive**; seminoma is rarely positive
Melanoma markers	• **S-100 in 10–40% of carcinomas**, especially salivary gland, breast, and cutaneous adnexal tumors (when using a polyclonal antibody)
	• **SOX10** in tumors with myoepithelial differentiation, including most **TNBC**
	• Melan A (clone A103) in adrenal cortical tumors, sex-cord stromal tumors, t(6;11) translocation renal cell carcinomas, clear cell sarcoma, PEComa
	• MiTF in cutaneous fibrohistiocytic lesions (e.g., dermatofibroma) and undifferentiated pleomorphic sarcoma
Hematolymphoid markers	• **"CD45 never lies"**
	• **CD138 (syndecan-1) expressed by ≥ 40% of carcinomas**
	• CD5/CD7 frequently expressed by GI tract tumors
	• MUM1 expressed by nearly all conventional melanomas (but typically not spindle cell or desmoplastic variants)

ALCL anaplastic large cell lymphoma, *DLBCL* diffuse large B-cell lymphoma, *NLPHL* nodular lymphocyte-predominant Hodgkin lymphoma, *FDCS* follicular dendritic cell sarcoma, *TNBC* triple negative breast cancer

Table 4.9 EMA-positivity in hematolymphoid neoplasms overlaps substantially with tumor types showing CD45 weak-to-negative staining, leading to diagnostic confusion

EMA-positive hematolymphoid neoplasms	CD45 weak-to-negative hematolymphoid neoplasms
	Lymphoblastic leukemia/lymphoma
	Classical Hodgkin lymphoma
Plasma cell neoplasm	**Plasma cell neoplasm**
Plasmablastic lymphoma	**Plasmablastic lymphoma**
Anaplastic large cell lymphoma	**Anaplastic large cell lymphoma**
ALK+ DLBCL	**ALK+ DLBCL**
Follicular dendritic cell sarcoma	**Follicular dendritic cell sarcoma**
T-cell/histiocyte-rich DLBCL	
Primary effusion lymphoma	

DLBCL diffuse large B-cell lymphoma

Immunohistochemical Approach to Small Round Blue Cell Tumor

- Table 4.10 presents an initial screening panel in a small round blue cell tumor, with the most useful markers highlighted in bold; it also lists the main differential consideration(s) given a positive result for a given marker, as well as additional differential considerations and potential pitfalls
  - CD99 must show strong, membranous positivity to support a diagnosis of Ewing sarcoma, and this quality of staining is occasionally also seen in lymphoblastic lymphoma and mesenchymal chondrosarcoma
  - Myogenin expression is stronger and more extensive in alveolar than embryonal rhabdomyosarcoma and expression in atrophic skeletal muscle is a pitfall
  - Among sarcomas, strong INSM1+ supports a diagnosis of extraskeletal myxoid chondrosarcoma
- Table 4.11 presents the frequency of diagnoses among a cohort of 41 tumors initially considered to represent possible Ewing sarcoma, in which an *EWSR1* rearrangement was not initially demonstrated, as well as associated diag-

Table 4.10 Initial screening panel in a small round blue cell tumor with potentially most useful markers highlighted in bold

Marker	Expressed by	Also expressed by
CD99	Ewing sarcoma	Lymphoblastic lymphoma, Mesenchymal chondrosarcoma
NKX2.2	Ewing sarcoma	Olfactory neuroblastoma, Mesenchymal chondrosarcoma
Desmin	Rhabdomyosarcoma, Desmoplastic small round cell tumor	Triton tumor (i.e., MPNST with rhabdomyosarcomatous differentiation)
Myogenin	Rhabdomyosarcoma (ARMS>ERMS)	Atrophic skeletal muscle
CD45	Lymphoma	
TdT	Lymphoblastic lymphoma	
INSM1 (CgA/Syn)	Neuroendocrine carcinoma, Neuroblastoma, DSRCT	Extraskeletal myxoid chondrosarcoma (INSM1)
Pan-K	Carcinoma, Desmoplastic small round cell tumor	Poorly differentiated synovial sarcoma; occasionally aberrantly expressed by sarcoma, melanoma
SOX10	Melanoma, MPNST (<50%)	Tumors with myoepithelial differentiation

ARMS alveolar rhabdomyosarcoma, *CgA* chromogranin A, *DSRCT* desmoplastic small round cell tumor, *ERMS* embryonal rhabdomyosarcoma, *MPNST* malignant peripheral nerve sheath tumor, *Syn* synaptophysin

Table 4.11 Additional diagnostic considerations in small round blue cell tumors with useful diagnostic markers not mentioned in the prior table highlighted in bold

Tumor type	Frequency (%)	Diagnostic markers
Ewing sarcoma	41	CD99, NKX2.2, *EWSR1*
CIC-rearranged	29	**WT-1**, ETV4, DUX4, *CIC*
BCOR-associated	13	**BCOR**, SATB2, *BCOR, CCNB3, YWHAE*
Neuroblastoma	8	PHOX2B, TH, GATA-3, CgA, Syn, INSM1
Malignant rhabdoid tumor	8	**SMARCB1 (INI1)**
Lymphoblastic lymphoma	4	TdT
Clear cell sarcoma	4	SOX10, *EWSR1*
Small cell carcinoma	4	INSM1, Rb
Rhabdomyosarcoma	4	Desmin, myogenin, *FOXO1* (ARMS)
DSRCT	4	Pan-K, desmin, **NSE**, WT-1 C-terminus, *EWSR1*
MPNST	4	**H3K27me3**, SOX10
PD synovial sarcoma	4	**SS18-SSX, SSX C-terminus**, TLE1, pan-K, *SS18*
GIST	4	**DOG1, KIT**
SMARCA4-deficient sarcoma	4	**SMARCA4 (BRG1)**

This table is based on findings in a cohort of small round blue cell tumors favored for Ewing sarcoma in which an *EWSR1*-rearrangement was not initially detected. Genes are italicized and may be assessed by FISH or molecular methods. *ARMS* alveolar rhabdomyosarcoma, *CgA* chromogranin A, *DSRCT* desmoplastic small round cell tumor, *GIST* gastrointestinal stromal tumor, *MPNST* malignant peripheral nerve sheath tumor, *PD* poorly differentiated, *Syn* synaptophysin
Machado I, et al. *Ann Diagn Pathol*. 2018 Jun;34:1–12

nostic markers; markers in bold are considered most useful among those not already mentioned in the previous table (taking availability into consideration); gene names are italicized, and these may be assessed by a number of methods, including FISH and anchored multiplex PCR

Immunohistochemical Approach to Carcinoma Typing

- Table 4.12 presents a conceptional framework to carcinoma typing with associated useful IHC markers
  - Most carcinomas are adenocarcinoma (77%), followed in descending order by squamous cell carcinoma (12%), urothelial carcinoma (7%), and NET/NEC (4%)
  - Sarcomatoid carcinoma (with or without recognizable heterologous elements), undifferentiated/dedifferentiated carcinoma, and mixed carcinomas of two or more distinct cell lineages are uncommon

Table 4.12 Seven different carcinoma types with useful immunohistochemical markers to support the diagnosis

Carcinoma type	Useful IHC markers (in general, expression should be extensive)
Adenocarcinoma (77%)	Site-specific transcription factors
Squamous cell carcinoma (12%)	p40 (>95%), high-molecular-weight keratin (>99%)
Urothelial carcinoma (7%)	p40 (85%), high-molecular-weight keratin (>95%), GATA-3 (80%), uroplakin II (67%), CK20 (50%)
Neuroendocrine tumor/carcinoma (4%)	Chromogranin A, synaptophysin, INSM1
Sarcomatoid carcinoma without or with recognizable heterologous elements	SMA, desmin; myogenin (if rhabdomyosarcomatous component)
Undifferentiated (pure)/dedifferentiated (concurrent differentiated component) carcinoma	SMARCA4 (BRG1), SMARCB1 (INI1), MMR, EBER, HR-HPV (rare), NUT (rare)
"Mixed" carcinomas composed of two (or more) distinct cell lineages	

- Figure 4.6 presents an algorithmic approach to carcinoma typing, including useful IHC markers
  - Carcinomas are recognized by (among other features) their cohesive nature and broad-spectrum epithelial marker-positivity
  - "Garden variety adenocarcinomas" are readily recognized as adenocarcinoma based on glandular/tubular/papillary architecture and/or mucin production
    - Site of origin assignment in garden variety adenocarcinoma is covered in subsequent algorithms
    - Key Pitfall: poorly differentiated/solid adenocarcinoma is often mistaken on the H&E for squamotransitional or neuroendocrine carcinoma
  - "Large polygonal cell adenocarcinomas" are composed of large cells with granular eosinophilic or clear cytoplasm, often with centrally placed nuclei, and tend to grow as nests or cords or show diffuse architecture; as a general rule, large polygonal cell adenocarcinomas are CK7/CK20-double negative; the principal differential includes:
    - Hepatocellular carcinoma (Hep Par 1+, glypican-3+, arginase-1+; EpCAM−)
    - Renal cell carcinoma (PAX8+)
    - Adrenal cortical carcinoma (SF1+)
      - If SF1 is not available, melan A (must use clone A103), inhibin A, and calretinin are usually (80%) positive
      - Key Pitfall: synaptophysin is positive in 60%, leading to incorrect diagnoses of NET/NEC (chromogranin A is consistently negative)
  - Squamous and urothelial carcinomas demonstrate substantial morphologic overlap and are considered together here as "squamotransitional carcinomas"

Diagnosis of Carcinoma Type

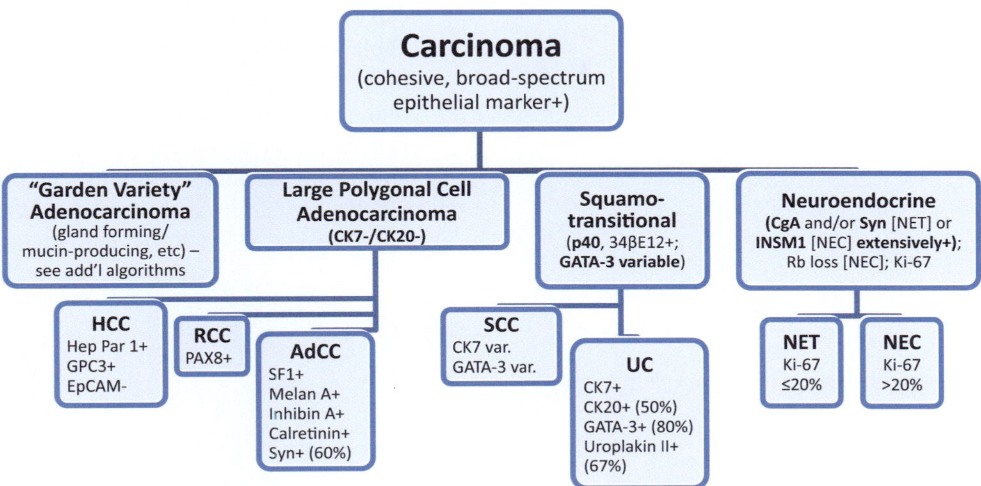

Fig. 4.6 Immunohistochemical approach to carcinoma typing. *AdCC* indicates adrenal cortical carcinoma, *CgA* chromogranin A, *GPC3* glypican-3, *HCC* hepatocellular carcinoma, *NEC* poorly differentiated neuroendocrine carcinoma, *NET* well-differentiated neuroendocrine tumor, *RCC* renal cell carcinoma, *SCC* squamous cell carcinoma, *Syn* synaptophysin, *UC* urothelial carcinoma, *var.* variable

- o High-molecular-weight keratin (e.g., clone 34βE12) is the most sensitive squamotransitional screening marker (>99%) but is only ~80% specific in the differential diagnosis with solid adenocarcinoma
- o p40 (or p63) is ≥95% sensitive for squamous cell carcinoma and 85% sensitive for urothelial carcinoma
- o GATA-3 is strongly expressed by urothelial carcinoma (80%) but is also often expressed by anogenitourinary and cutaneous squamous cell carcinoma (though expression is typically less strong in squamous cell carcinoma)
- – Neuroendocrine epithelial neoplasms include NET and NEC; NETs and some NECs are characterized by organoid architecture (i.e., nested or trabecular), and NETs and small cell NECs typically have finely granular (aka "salt and pepper") chromatin
  - o Chromogranin A and synaptophysin are typically used to support the presence of neuroendocrine differentiation in NET
  - o INSM1 is preferred to support the presence of neuroendocrine differentiation in NEC
    - ♦ INSM1 is more sensitive than the combination of chromogranin A and synaptophysin in NEC
    - ♦ INSM1 is less sensitive than chromogranin A and synaptophysin in NET
  - o Among poorly differentiated carcinoma, Rb loss also supports a diagnosis of NEC
  - o Ki-67 is essential for NET grading and is prognostic and potentially predictive in NEC

Coordinate Expression of CK7/CK20

- The low-molecular-weight keratins CK7 and CK20 are often used together to suggest the site of origin of adenocarcinomas of occult origin
- Table 4.13 presents common patterns of CK7/CK20 coordinate expression
- Some authors have questioned the continued value of CK7/CK20 assessment in our contemporary "next-generation IHC-centric" approach to site of origin assignment
- The main residual value is in identifying consistently CK7/CK20-double negative tumors, which may not have been initially considered in one's differential diagnosis (Fig. 4.7)
  - – These include prostate cancer, hepatocellular carcinoma, clear cell renal cell carcinoma, adrenal cortical carcinoma, squamous cell carcinoma (visceral primaries can be CK7+), NET (tumors of lung origin are often CK7+), visceral NEC, and yolk sac tumor
- Key Pitfalls:
  - – CK7/CK20-double negativity is often incorrectly equated with broad-spectrum-keratin-negativity when, in fact, it should call to mind the differential directly above
  - – Among the consistently CK7+ tumors (e.g., breast, pancreas), 5–10% may be CK7−
  - – Tables depicting coordinate expression of CK7/CK20 typically only take the most common variant of a given tumor into account (e.g., renal cell carcinoma is usu-

Table 4.13 Coordinate expression of CK7/CK20

Site (tumor)	CK7	CK20
Prostate, hepatocellular carcinoma, renal cell carcinoma, adrenal cortical carcinoma, squamous cell carcinoma, NET, visceral NEC, germ cell tumor (i.e., yolk sac tumor, seminoma)	–	–
Lung, breast, Müllerian, thyroid, bladder, upper GI (UGI), pancreatobiliary (PB), mucinous ovarian	+	–
Bladder, UGI, PB, mucinous ovarian, occasional colon (especially rectum), occasional lung (especially mucinous)	+	+
Colon, Merkel cell, occasional UGI	–	+

NEC neuroendocrine carcinoma, *NET* neuroendocrine tumor

ally presented as CK7/CK20-double negative, and clear cell renal cell carcinoma is, but most other renal cell carcinomas are CK7+)

- Tables depicting coordinate expression of CK7/CK20 often only present one staining pattern for a given tumor, but many tumors present at least two frequent patterns (e.g., urothelial carcinoma, lung adenocarcinoma with mucinous histology, and primary mucinous ovarian carcinoma are evenly divided between CK7+/CK20+ and CK7+/CK20− presentations)

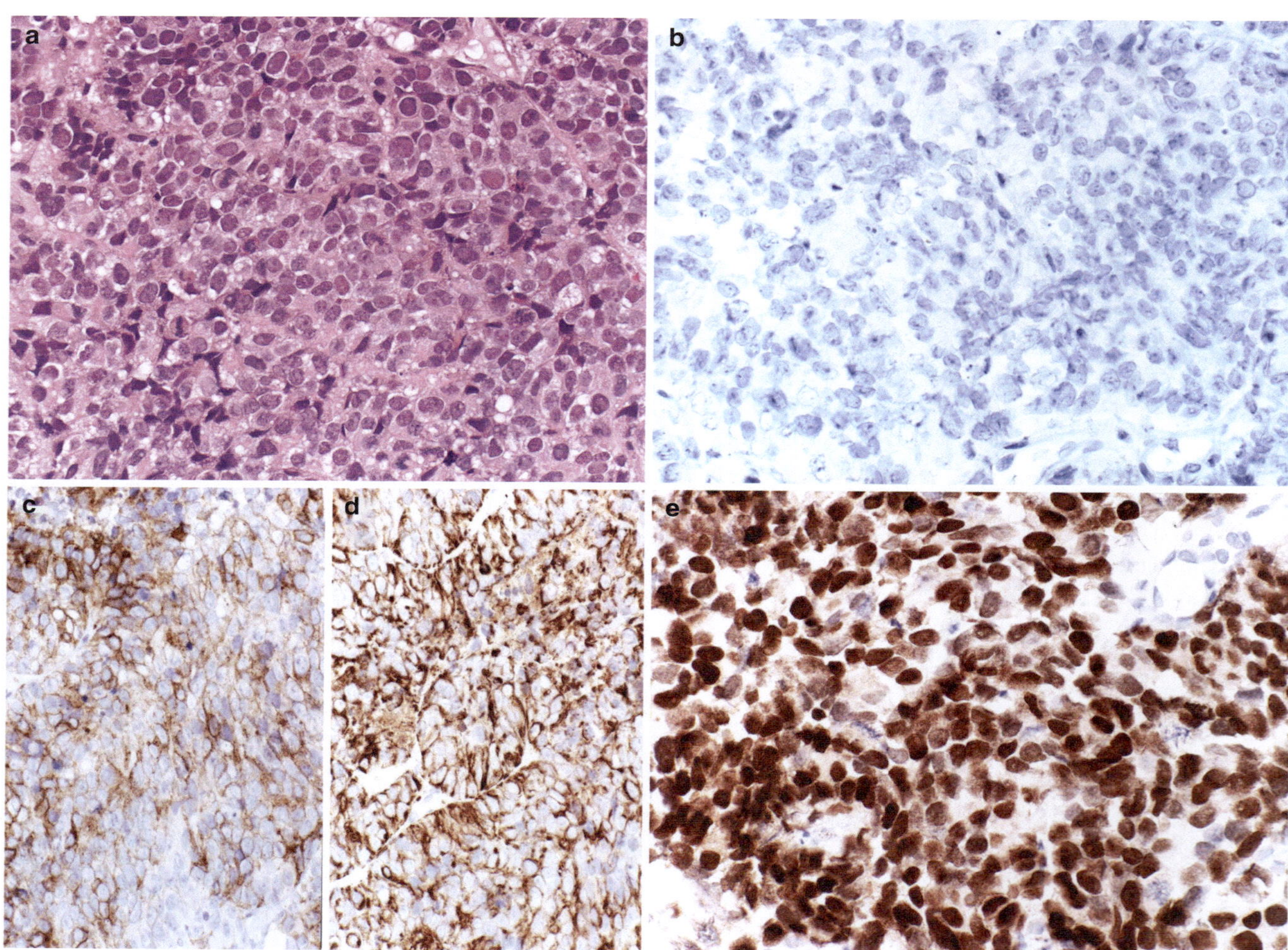

Fig. 4.7 Value of CK7/CK20 coordinate expression: 77-year-old man with tumor in the liver (**a**) disposed as nests and sheets with uniform cytomorphology, including prominent nucleoli. CK7 (depicted in **b**) and CK20-negativity was initially incorrectly equated with broad-spectrum keratin-negativity, resulting in significant diagnostic confusion, compounded by the detection of KIT-positivity (**c**). The biopsy was ultimately initially signed out as "poorly differentiated carcinoma of unknown primary" based on pan-keratin-positivity (**d**). On consultation, the combination of this morphology and CK7/CK20-double-negativity raised the possibility of prostate cancer, which was confirmed with NKX3.1 (**e**)

- Tables do not take focality of staining into account (as a general rule, I discount focal CK7 staining, and, in the absence of strong CK7 staining, I similarly discount focal CK20 staining)

Immunohistochemical Approach to "Garden Variety Adenocarcinoma" Presenting in the Liver

- Figure 4.8 presents an algorithmic immunohistochemical approach to an adenocarcinoma readily recognizable on the H&E presenting in the liver
- This algorithm is constructed based on the frequency of adenocarcinomas presenting as liver metastases of occult origin
- Some very common cancers (e.g., prostate and Müllerian) rarely present as hepatic metastasis of occult origin and, thus, are not accounted for in the initial panel
- Based on an initial panel of CK7, CK20, CDX2, and TTF-1 with the addition of GATA-3 in a woman six main patterns of reactivity are observed:
  - CK20 and/or CDX2 homogeneous (i.e., diffuse, strong staining)
    - Interpretation: Lower GI Pattern
      - Consider adding SATB2 to support colorectal or appendiceal origin
      - Recommend colonoscopy
      - Pitfall: upper GI adenocarcinomas occasionally present as lower GI pattern
      - Overstaining for CDX2 can obscure the distinction of lower and upper GI pattern
      - CK7+ is largely irrelevant in the setting of lower GI pattern (though it may be more common with rectal primaries)
  - CK20/CDX2 heterogeneous (i.e., anything less than diffuse, strong staining)
    - Interpretation: Upper GI Pattern
      - Seen with upper GI and pancreatobiliary (30%) adenocarcinomas
      - Recommend upper endoscopy and pancreas protocol CT
      - Consider adding SMAD4 to support pancreatobiliary (lost in 50%) over upper GI (lost in ≤10%) origin
  - TTF-1+ (anything more than rare cells)
    - Interpretation: Lung Primary
      - TTF-1 and napsin A are similarly sensitive for lung adenocarcinoma (≤80%)
      - Performing both simultaneously increases sensitivity to ≤90%
      - Consider performing both upfront in patients with a lung mass
  - GATA-3+ (the stronger and more extensive the staining, the more reliable the result)
    - Interpretation: Breast Primary
      - Luminal breast cancers are nearly always GATA-3+ (99%), and staining is typically diffuse and strong
      - ER-/HER2+ breast cancers are probably nearly always GATA-3+, but staining is often less strong/less extensive than in luminal cancers
      - Triple-negative breast cancers (TNBC) are usually GATA-3+, but staining is often less strong/less extensive than in luminal cancers
      - Given the performance of GATA-3 (and SOX10 and TRPS1—see below), mammaglobin and GCDFP-15 have no role in the evaluation for a possible breast primary
  - CK7+ only
    - Interpretation: Nonspecific Result
      - Consider adding:
        - Napsin A (lung)
        - PAX8 (Müllerian)
        - SOX10 (TNBC; 60% show extensive, strong staining)

Fig. 4.8 Immunohistochemical approach to "Garden Variety Adenocarcinoma" Presenting in the Liver, *AdCC* indicates adrenal cortical carcinoma, *Arg1* arginase, *GPC3* glypican-3, *HCC* hepatocellular carcinoma, *iCC* intrahepatic cholangiocarcinoma, *PB* pancreatobiliary, *RCC* renal cell carcinoma, *TNBC* triple-negative breast cancer, *UGI* upper gastrointestinal

"Garden Variety Adenocarcinoma" in the Liver

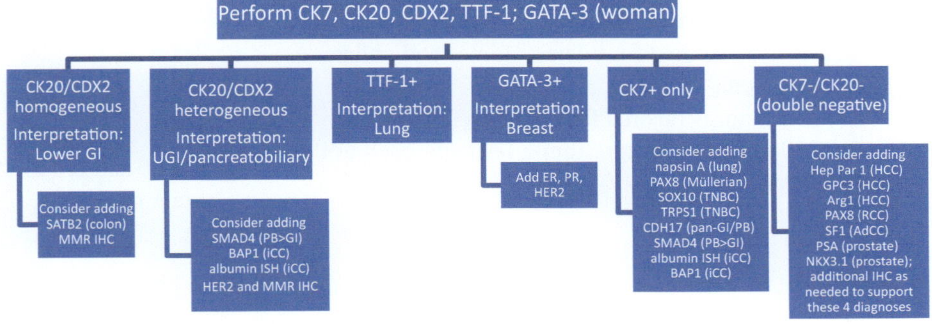

- TRPS1 (TNBC; moderate-to-strong staining is 80% sensitive, 95% specific among carcinomas; positivity is seen in some sarcomas so cannot be used alone to support diagnosis of metaplastic carcinoma with spindle cell morphology)
    - CDH17 (pan-GI/pancreatobiliary marker with incrementally greater sensitivity than CDX2)
    - SMAD4 (loss supports pancreatobiliary over upper GI origin)
    - Albumin in situ hybridization (intrahepatic cholangiocarcinoma; also expressed by hepatocellular carcinoma, though these will rarely be extensively CK7+)
    - BAP1 (lost in 25% of intrahepatic cholangiocarcinoma)
  - CK7/CK20-double negative
    - Interpretation: Nonspecific Result
      - Consider adding:
        - Hep Par 1 (hepatocellular carcinoma)
        - Glypican-3 (hepatocellular carcinoma)
        - Arginase-1 (hepatocellular carcinoma)
        - PAX8 (renal cell carcinoma)
        - SF1 (adrenal cortical carcinoma)
        - PSA (prostate cancer)
        - NKX3.1 (prostate cancer)
      - Reconsider possibility of:
        - Neuroendocrine tumor, which may show pseudoglandular architecture (chromogranin A, synaptophysin)
        - Large cell neuroendocrine carcinoma (INSM1, Rb)
        - Squamous cell carcinoma, which can be acantholytic (p40)

Immunohistochemical Approach to Primary Versus Metastatic Ovarian Tumor with Mucinous Features

- Figure 4.9 presents an algorithmic approach to the differential of primary versus metastatic ovarian tumors with mucinous features based on the gross examination
  - Metastasis is favored for bilateral tumor and unilateral tumor <13 cm
  - Primary tumor is favored for tumors ≥13 cm
- Figure 4.10 presents an algorithmic IHC approach to the same differential

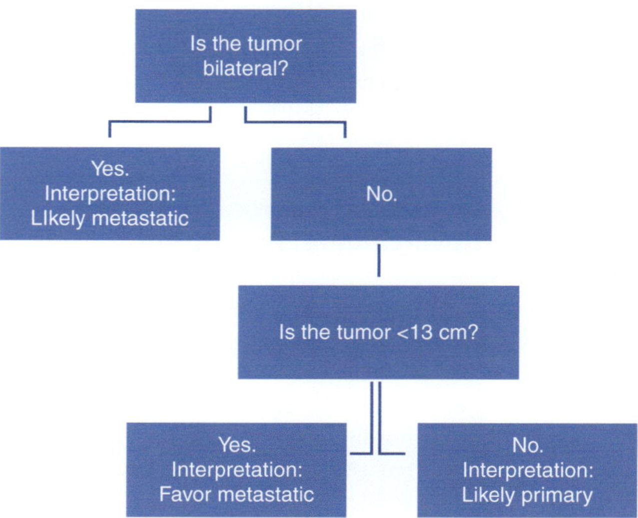

Fig. 4.9 Gross algorithm for primary ovarian surface epithelial tumor with mucinous features versus metastasis

- Based on an initial panel including CK7, CK20, CDX2, and PAX8 three main patterns are observed:
  - PAX8+ (the stronger and more extensive the staining, the more reliable the result)
    - Interpretation: Müllerian origin
      - Primary mucinous ovarian tumors are less likely to be positive than other primary surface ovarian tumors (30% in the aggregated literature) and are less extensively positive
      - Using contemporary monoclonal PAX8 antibodies, though, I have found staining to be frequent (75%) and often moderate-to-strong, multifocal
      - 95% of primary mucinous ovarian tumors are CK7+ (usually diffuse)
      - p16 and/or high-risk HPV ISH can be added if endocervical adenocarcinoma is a consideration
  - CK20/CDX2 heterogeneous (i.e., anything less than diffuse, strong staining)
    - Interpretation: Nonspecific Result (Primary or Metastatic)
      - Consider adding:
        - SMAD4 (loss supports pancreatobiliary over tubal gut origin)
        - SATB2 (lower GI origin)
        - p16 and/or high-risk HPV ISH (endocervical origin)
      - Otherwise, defer interpretation to gross algorithm
  - CK20 and/or CDX2 homogeneous (i.e., diffuse, strong staining)

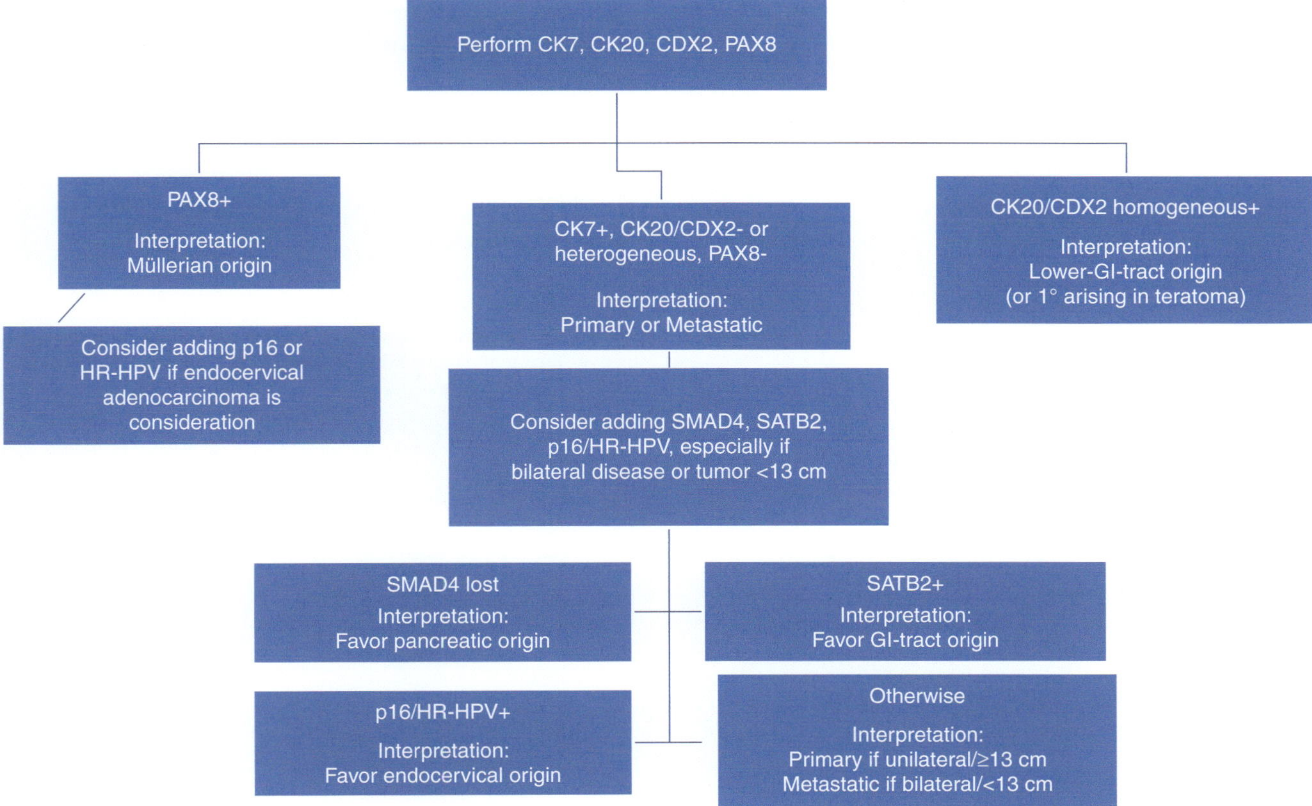

Fig. 4.10 Immunohistochemical algorithm for primary ovarian surface epithelial tumor with mucinous features versus metastasis

- o Interpretation: Lower GI Pattern
  - Consider adding SATB2 to support colorectal or appendiceal origin
  - Key Pitfall: SATB2 is expressed by rare mucinous primaries arising in association with a teratoma

Immunohistochemical Approach to Tumors Presenting at Other Sites

- Table 4.14 presents additional diagnostic considerations (i.e., beyond lung, GI, and breast cancer) and associated diagnostic markers for tumors presenting in the mediastinum, pleura, peritoneum, retroperitoneum, somatic soft tissue, and bone
  - Mediastinum: lung (regardless of TTF-1 result) is principal consideration with additional consideration to thymic epithelial neoplasms (polyclonal PAX8, p40; KIT and CD5 in thymic carcinoma), germ cell tumor (SALL4), and liposarcoma (MDM2, CDK4)
  - Pleura: mesothelioma should be excluded (calretinin, WT-1, D2-40, CK5/6, BAP1 [loss], MTAP [loss])
  - Peritoneum: mesothelioma should, again, be excluded, along with Müllerian tumors (PAX8)
  - Retroperitoneum: special consideration should be given to renal cell carcinoma (PAX8), adrenal cortical carcinoma (SF1), germ cell tumor (SALL4), and liposarcoma (MDM2, CDK4) (Fig. 4.11)
  - Somatic soft tissue: always consider sarcoma
  - Bone
    - o Blastic metastasis: prostate (PSA, NKX3.1), breast (GATA-3)
    - o Lytic metastasis: kidney (PAX8), thyroid (TTF-1/PAX8 co-expressing)
    - o Mixed metastasis: lung (TTF-1)

Table 4.14 Site-specific additional differential diagnostic considerations with associated useful markers

Site	Additional considerations
Mediastinum	Lung is the principal consideration, regardless of TTF-1 result
	Thymic neoplasm: pPAX8, p40; KIT, CD5 (latter 2 in thymic carcinoma)
	Germ cell neoplasm: SALL4
	Well- and dedifferentiated liposarcoma: MDM2/CDK4
Pleura	Mesothelioma: Calretinin, WT-1, D2-40, CK5/6, BAP1 (loss), MTAP (loss)
Peritoneum	Müllerian adenocarcinoma: PAX8
	Mesothelioma: Calretinin, WT-1, D2-40, CK5/6, BAP1 (loss), MTAP (loss)
Retroperitoneum	Renal cell carcinoma: PAX8
	Adrenal cortical carcinoma: SF1, melan A, inhibin A, calretinin, synaptophysin
	Germ cell tumor: SALL4
	Liposarcoma: MDM2/CDK4
Somatic soft tissue	Always consider sarcoma, even if keratin-positive
Bone	Blastic metastasis: Prostate (PSA, PrAP, NKX3.1), breast (GATA-3)
	Lytic metastasis: Kidney (PAX8), thyroid (thyroglobulin if TTF-1/PAX8 co-expressing)
	Mixed metastasis: Lung (TTF-1)

Immunohistochemical Approach to Distinction of Squamous and Urothelial Carcinoma

- Table 4.15 presents the sensitivity of IHC markers commonly used to support the presence of squamotransitional differentiation or used to distinguish squamous and urothelial carcinoma

Table 4.15 Immunohistochemical markers useful to support the presence of squamotransitional differentiation and to distinguish squamous and urothelial carcinoma

	Squamous cell carcinoma (%)	Urothelial carcinoma (%)
p40	≥95	85
34βE12	≥95	≥95
CK5/6	≥95	50
GATA-3	20	80
Uroplakin II	<1	70
CK7	30	≥90
CK20	<2	50

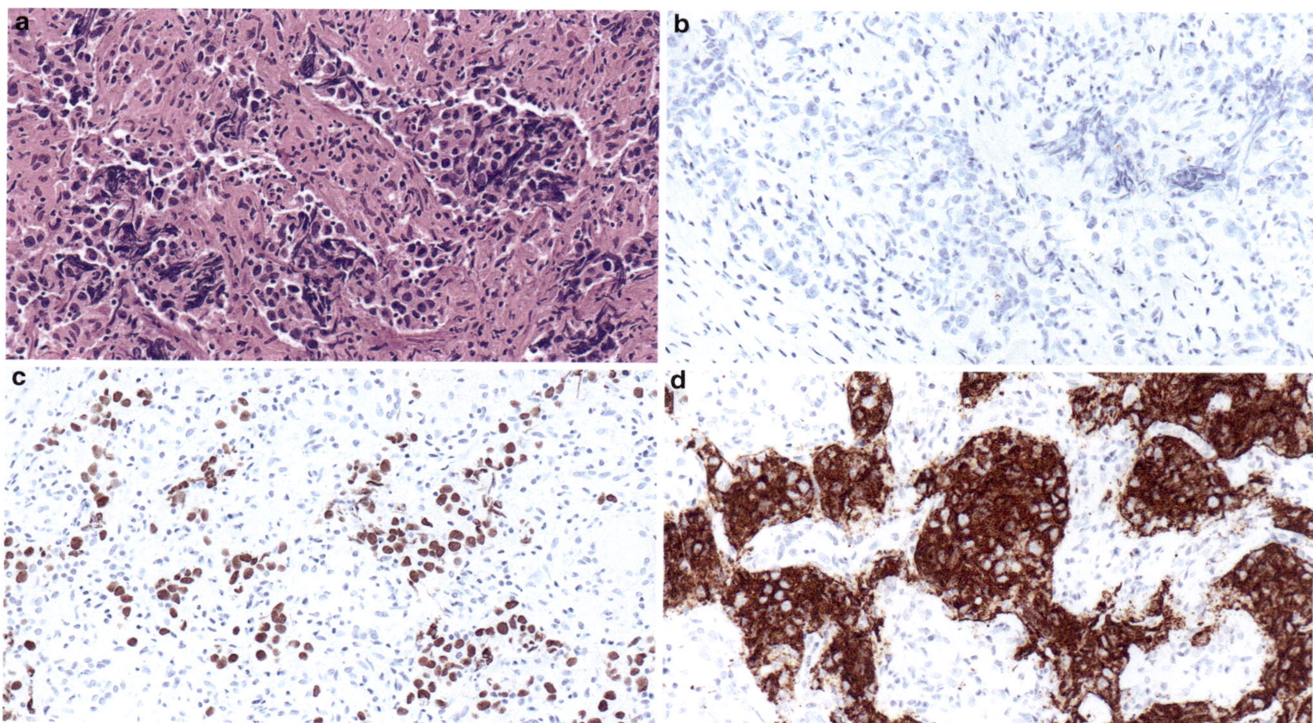

Fig. 4.11 Use of site-specific algorithm to support the diagnosis of seminoma and avoid diagnostic error: 44-year-old man with retroperitoneal mass disposed as nests of epithelioid cells in a fibroblastic stroma (**a**). An immunohistochemical panel on a prior biopsy demonstrated negativity for keratin AE1/AE3, CK5/6, S-100, SOX10, HMB-45, melan A, SMA, and desmin before the block was depleted. On repeat biopsy, AE1/AE3 (**b**) and S-100-negativity was again demonstrated, and our team was considering sending NGS and a sarcoma fusion panel. My fellow saw the case and thought, "germ cell tumor is always a diagnostic consideration in the retroperitoneum," and ordered a SALL4 (**c**). KIT-positivity (**d**) confirmed the diagnosis of seminoma

- Squamous cell and urothelial carcinoma demonstrate substantial morphologic overlap, with urothelial carcinoma often demonstrating a component of squamous differentiation
  - At potentially metastatic sites (especially liver, lymph node, bone, brain, peritoneum, and lung) urothelial carcinoma should be considered before making a diagnosis of squamous cell carcinoma
- High-molecular-weight keratin is the most sensitive squamotransitional marker, but it is less specific than p40
- p40 is less sensitive for urothelial carcinoma (85%) than squamous cell carcinoma (≥95%)
- CK5/6-negativity in a squamotransitional carcinoma supports the diagnosis of urothelial carcinoma (50%)
- Positive urothelial carcinoma markers in this differential include GATA-3 (80% sensitive), uroplakin II (70% sensitive; very specific), and CK20 (50% sensitive; very specific)
- GATA-3-positivity is occasionally seen in squamous cell carcinoma (20%), especially in tumors of cutaneous and anogenitourinary origin, though expression is typically weaker and less extensive than in urothelial carcinoma
- GATA-3-positivity has been described in lung and head and neck squamous cell carcinomas in some studies, while many authors have found tumors at these sites to be negative
- p16-positivity in a squamous cell carcinoma (i.e., moderate-to-strong staining in ≥70% of tumor cells) is a surrogate for high-risk HPV-positivity
  - This testing should be performed in oropharyngeal primaries and cervical metastasis of occult origin

  - p16-positivity in squamous cell carcinomas at other metastatic sites typically implicates high-risk HPV and is often seen in tumors of anogenital origin
  - Key Pitfall: p16 should not be used alone to support the diagnosis of squamous cell carcinoma; positivity is also seen in serous cancer and NEC

Immunohistochemical Approach to Neuroendocrine Tumor Site of Origin Assignment

- Figure 4.12 presents the algorithm used for NET site of origin assignment at the University of Iowa
- Figure 4.13 presents an alternative algorithm using more commonly available markers
- Up to 20% of NETs present as metastasis of occult origin, usually to the liver but also to other sites
- NETs always demonstrate diffuse, strong expression of chromogranin A and/or synaptophysin
- Demonstration of broad-spectrum epithelial marker-positivity (e.g., keratin AE1/AE3) is almost mandatory to distinguish NET from paraganglioma/pheochromocytoma
  - Key Pitfalls:
    - Paraganglioma/pheochromocytoma is usually islet 1+
    - NET may demonstrate S-100/SOX10+ sustentacular cells (especially tumors of lung and appendiceal origin), while paraganglioma/pheochromocytoma can lack sustentacular cells
- Positive markers for paraganglioma/pheochromocytoma in this differential diagnosis include GATA-3 (positive in

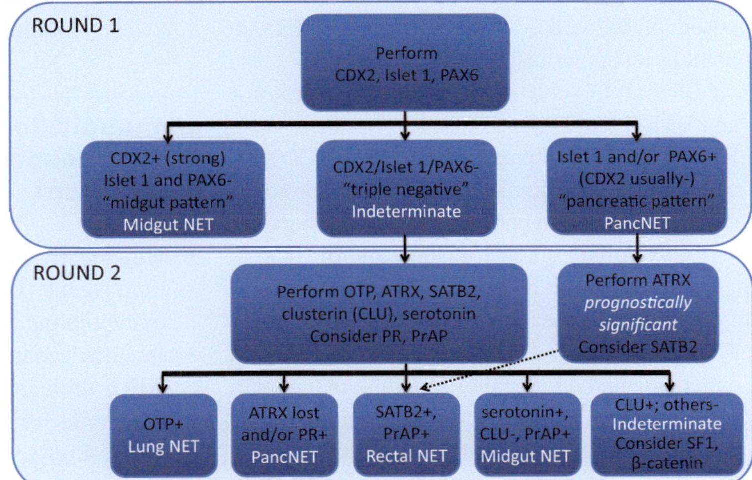

Fig. 4.12 University of Iowa immunohistochemical algorithm for well-differentiated neuroendocrine tumor site of origin assignment

Fig. 4.13 Immunohistochemical algorithm for well-differentiated neuroendocrine tumor site of origin assignment using widely available markers

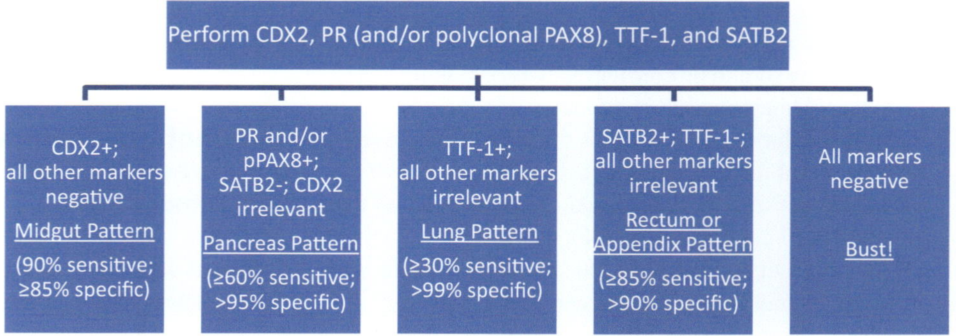

up to 90%), PHOX2B, and tyrosine hydroxylase (100% of pheochromocytomas and 40% of paragangliomas)

- SDHB IHC should be performed in paraganglioma/pheochromocytoma with loss seen in 30% of thoracoabdominal and 15% of head and neck paragangliomas and 5% of pheochromocytomas

- Most NETs of occult origin are of jejunoileal origin, followed in descending order by tumors of pancreatic, lung, and rectal origin

- Based on an initial panel of CDX2, PR (and/or polyclonal PAX8), TTF-1, and SATB2 five main patterns of reactivity are observed:
  - CDX2+ (moderate-to-strong, multifocal-to-diffuse)
    - Interpretation: Midgut Pattern
      - Jejunoileal tumors are composed of serotonin-producing EC-cells
      - Appendiceal tumors are often composed of EC-cells, but these rarely present as metastasis of occult origin
      - Ovarian insular carcinoid tumors are also composed of EC-cells, as well as some NETs of renal or testicular origin
      - Additional EC-cell markers include serotonin and VMAT1
      - Key Pitfall: 15% of pancreatic NETs are CDX2+, which may be confounding for laboratories using less sensitive pancreatic NET markers
  - PR and/or polyclonal PAX8+ (or islet 1 and/or PAX6+, if using more successful markers)/SATB2−
    - Interpretation: Pancreatic Pattern
      - These algorithms do not distinguish tumors of pancreatic and duodenal origin, though the latter rarely present as metastasis of occult origin (with the exception of gastrin-expressing duodenal tumors involving regional lymph nodes)
      - Rectal NETs usually express pancreatic NET origin markers but are typically strongly SATB2+

- TTF-1+ (or OTP+)
  - Interpretation: Lung Pattern
    - TTF-1 is only 30–40% sensitive for carcinoid tumor of lung origin
    - OTP is 2–3× more sensitive than TTF-1 without sacrificing specificity
      - OTP is 80–90% sensitive in typical and 50–60% sensitive in atypical carcinoid tumor
      - OTP-negativity in a carcinoid tumor of lung origin is prognostically adverse
      - TTF-1+/OTP− lung NETs essentially do not occur (i.e., there is no benefit to adding TTF-1 in an OTP− tumor)
  - SATB2+/lung NET marker negative (TTF-1 or OTP-)
    - Interpretation: Rectum (or Appendix) Pattern
      - SATB2+ is seen at a similar rate in rectal and appendiceal NETs, though the latter exceptionally present as metastasis of occult origin
      - SATB2+ is seen in up to 15% of lung NETs, but staining is typically weak and patchy

Immunohistochemical Approach to Neuroendocrine Carcinoma Site of Origin Assignment

- Historically, metastatic NEC has been treated with a platinum agent (i.e., cisplatin or carboplatin) + etoposide, extrapolating from small cell lung cancer data
- In some centers, large cell NEC is treated like small cell NEC, while in others, it is treated like nonneuroendocrine carcinoma arising at that anatomic site
- Increasingly, extrapulmonary visceral small cell NEC may be treated like nonneuroendocrine carcinoma arising at that anatomic site
- Merkel cell carcinoma (i.e., NEC of cutaneous origin) is very sensitive to checkpoint inhibition, and patients with

metastatic Merkel cell carcinoma may be treated with checkpoint inhibitor monotherapy

- The goal of site of origin assignment in NEC of occult origin is to distinguish tumors of cutaneous (CK20+ in ≥90%) from visceral origin (TTF-1+ in ≥90% of small cell lung carcinomas but only 40% in large cell and extra-pulmonary visceral NEC)
- In CK20/TTF-1-double negative tumors:
  - Additional Merkel cell carcinoma markers include:
    ○ neurofilament (≥70%)
    ○ Merkel cell carcinoma large T antigen (clone CM2B4; 75%; UV-light driven tumors are CM2B4−/p53-mutant pattern)
    ○ SATB2 (strong staining in 70%; less extensive staining is nonspecific)
  - An additional visceral NEC marker is ASCL1, which is only incrementally more sensitive than TTF-1
- Key Pitfall: NECs demonstrate "marked transcription factor lineage infidelity" (i.e., they tend to express multiple transcription factors independent of site of origin)
  - As such, aside from TTF-1 (visceral origin) and strong SATB2 (cutaneous origin), transcription factor IHC has little-to-no role in NEC site of origin assignment

Immunohistochemical Approach to the Distinction of NET G3 from NEC

- Up to 5% of NETs have a Ki-67 proliferation index >20%
- Although these typically arise in a background of G1/G2 NET, they may also arise de novo and can demonstrate morphology ambiguous for the distinction of NET versus large cell NEC
- The range of Ki-67 proliferation indices in NET G3 and large cell NEC is overlapping
- Biallelic inactivation of *TP53* and *RB1* is the molecular genetic hallmark of small cell lung cancer
  - Mutant-pattern p53 supports a diagnosis of NEC over NET G3
  - Rb loss supports a diagnosis of NEC over NET G3
  - Mutant-pattern p53 has been described in NET G3, but many of these cases represent heavily pretreated, pro-gressed NETs with ambiguous morphology
- Other NEC markers in this differential include CXCR4, ASCL1, and POU2F3
  - POU2F3 is an emerging NEC marker that is prefer-entially expressed in chromogranin A/synaptophy-sin weak-to-negative tumors but is only 15% sensitive
  - Key Pitfall: POU2F3 is also expressed by basaloid squamous cell carcinoma, and there is little experience outside of lung

- Other NET markers in this differential include strong SSTR2 expression and ATRX/DAXX loss (or positive ALT FISH)
  - SSTR2 is expressed by up to 30% of NEC, though expression is usually weaker and patchier than in NET
  - Key Pitfall: SSTR2 is only positive in 30–40% of lung NETs
  - ATRX/DAXX loss (or ALT FISH+) is largely limited to NETs of pancreatic origin (though it is seen in up to 40%), while NET G3 may arise at any site
- Extensive expression of organ-specific NET site of origin markers may support the diagnosis of NET G3 over NEC
  - Key Pitfalls:
    ○ Islet 1 is often strongly expressed by NEC
    ○ Strong SATB2 staining is typical of Merkel cell car-cinoma, and TTF-1+ is typical of visceral NEC

Immunohistochemical Approach to Undifferentiated/Dedifferentiated Carcinoma

- Undifferentiated malignant neoplasms demonstrate no light microscopic or immunohistochemical evidence in support of any lineage
- Undifferentiated carcinomas demonstrate no evidence of glandular, squamous, or neuroendocrine differentiation, though they are broad-spectrum epithelial marker positive and/or arise in the context of an epithelial precursor lesion
- Dedifferentiated carcinoma is an undifferentiated carci-noma arising either concurrently with a differentiated car-cinoma or in the setting of past differentiated carcinoma
- Undifferentiated/dedifferentiated carcinomas demon-strate the entire spectrum of broad-spectrum epithelial marker-positivity from rare cells positive to diffuse, strong positive
- Tumors of any lineage (i.e., carcinoma, lymphoma, mela-noma, sarcoma, etc.) can dedifferentiate
- Undifferentiated/dedifferentiated carcinomas occur in five main settings:
  - SWI/SNF subunit inactivation (mainly SMARCA4/BRG1 but occasionally SMARCB1/INI1 or other subunits)
    ○ These tumors often demonstrate rhabdoid cytomorphology
      ◆ SMARCA4-deficiency has been described in carcinomas arising at virtually every anatomic site (Fig. 4.14)
      ◆ SMARCA4-deficiency also underlies small cell carcinoma of the ovary, hypercalcemic type and sinonasal teratocarcinosarcoma
      ◆ SMARCB1-deficiency is most common among SWI/SNF-deficient undifferentiated carcinomas arising in the sinonasal tract

- ♦ SMARCB1-deficiency also occurs in diverse tumor types including
  - Malignant rhabdoid tumor (approaching 100%)
  - Atypical teratoid/rhabdoid tumor (approaching 100%)
  - Poorly differentiated chordoma (approaching 100%)
  - Epithelioid sarcoma (90%)
  - Renal medullary carcinoma (90%)
  - Epithelioid malignant peripheral nerve sheath tumor (70%)
  - Epithelioid schwannoma (40%)
  - Myoepithelial carcinoma (20–40%)
  - Extraskeletal myxoid chondrosarcoma (20%)
- ♦ Any tumor that is characteristically deficient for a particular subunit can occasionally-to-rarely substitute deficiency for any other subunit

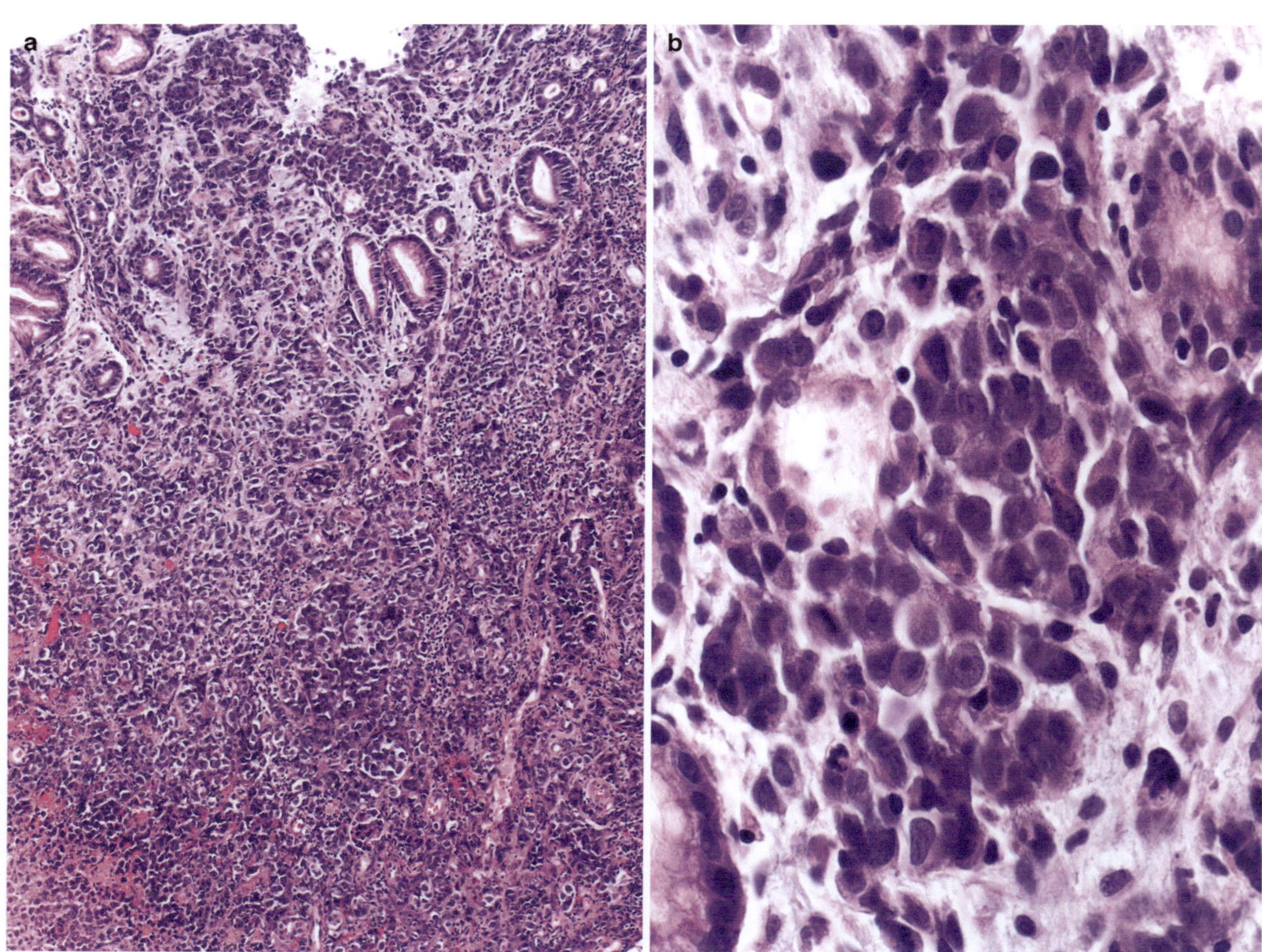

Fig. 4.14 SMARCA4-deficient undifferentiated carcinoma: 54-year-old man with gastric cardia mass. The tumor extensively infiltrates the specimen (**a**) with high-power demonstrating loosely cohesive, monotonous round cells some of which are somewhat rhabdoid-appearing (**b**). This case was referred by a former trainee to "rule out large cell neuroendocrine carcinoma" based at least in part on the synaptophysin staining seen in (**e**). EMA had been shown to be weakly positive (**c**), while pan-keratin and CAM5.2 were negative (not shown). SALL4 staining is depicted in (**d**). The tumor had also been shown to not express chromogranin A, CDX2, KIT, DOG1, SOX10, CD45, CD43, CD20, CD30, and EBV (EBER). The only stain I added was SMARCA4 (**f**), loss of which supports the diagnosis of SMARCA4-deficient undifferentiated carcinoma. Although poorly differentiated neuroendocrine carcinomas can show this limited degree of synapto-physin-positivity, this alone would be insufficient to support the diagnosis. This degree of SALL4-positivity is not uncommon in high-grade carcinomas and is, similarly, insufficient to support a diagnosis of germ cell tumor

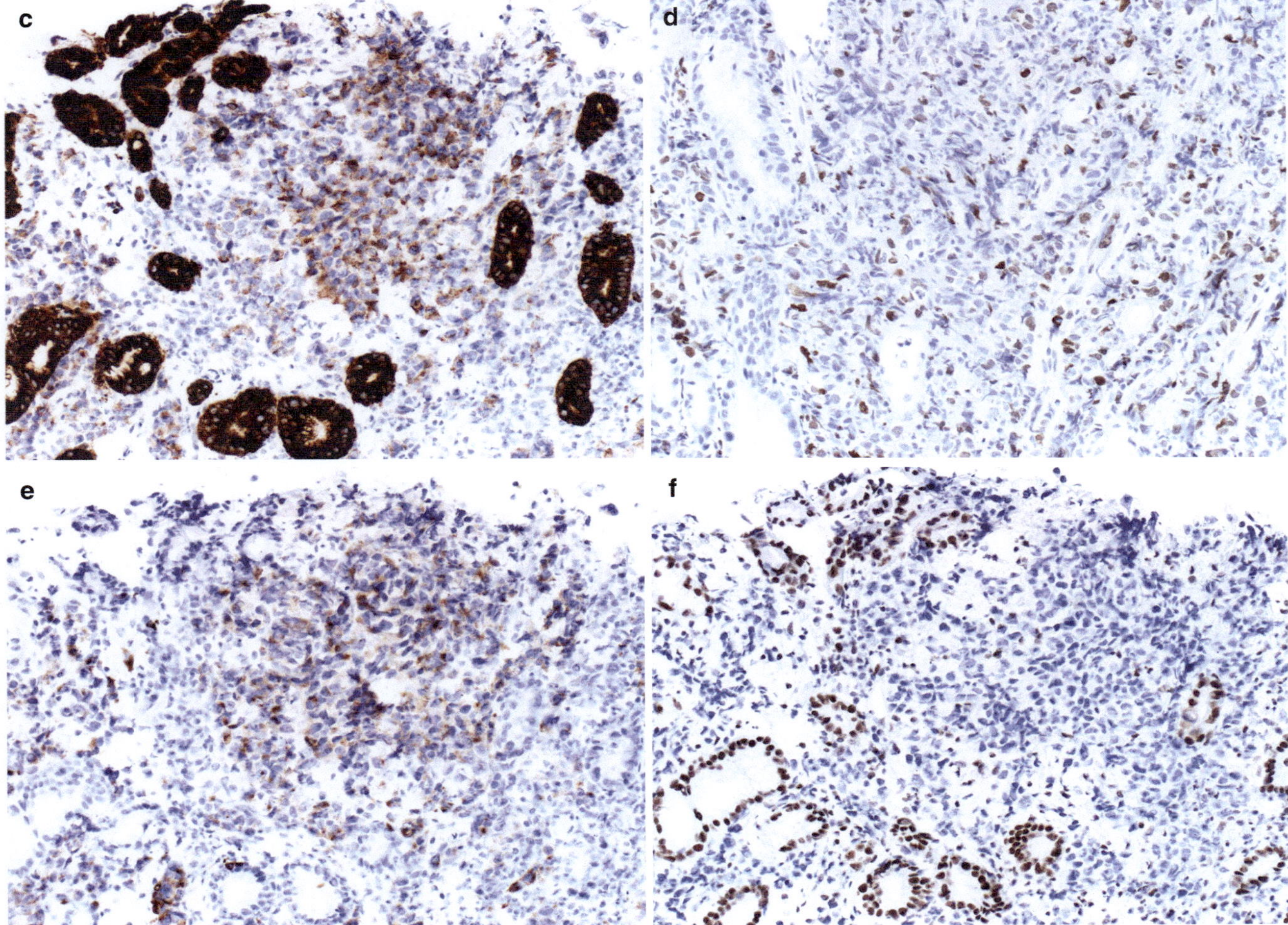

Fig. 4.14 (continued)

- ♦ Although SWI/SNF subunit deficiency appears to underpin dedifferentiation, it may also be seen at a low rate in differentiated carcinomas
- – dMMR/MSI-H
  - ○ These tumors often demonstrate prominent tumor-associated lymphoid stroma attributable to a large number of tumor neoantigens (Fig. 4.15)
- – Viral oncogenesis
  - ○ These tumors also typically demonstrate prominent tumor-associated lymphoid stroma
  - ○ Mainly due to EBV (aka, lymphoepithelial carcinoma, lymphoepithelioma-like carcinoma)
    - ♦ Tumors in the head and neck predominate in the nasopharynx and express p40
    - ♦ Tumors in the GI tract predominate in the stomach
    - ♦ Support the diagnosis with EBER ISH
  - ○ Tumors with similar morphologic features have rarely been reported in association with high-risk HPV
- – NUT carcinoma (aka NUT midline carcinoma, t(15;19) carcinoma)
  - ○ Tumors predominate in the thorax and head and neck
  - ○ May demonstrate foci of abrupt keratinization
  - ○ Support the diagnosis with NUT IHC, FISH, or molecular methods
- – Personal history of a differentiated carcinoma
  - ○ Support the diagnosis by comparing NGS results in current undifferentiated and prior differentiated carcinoma

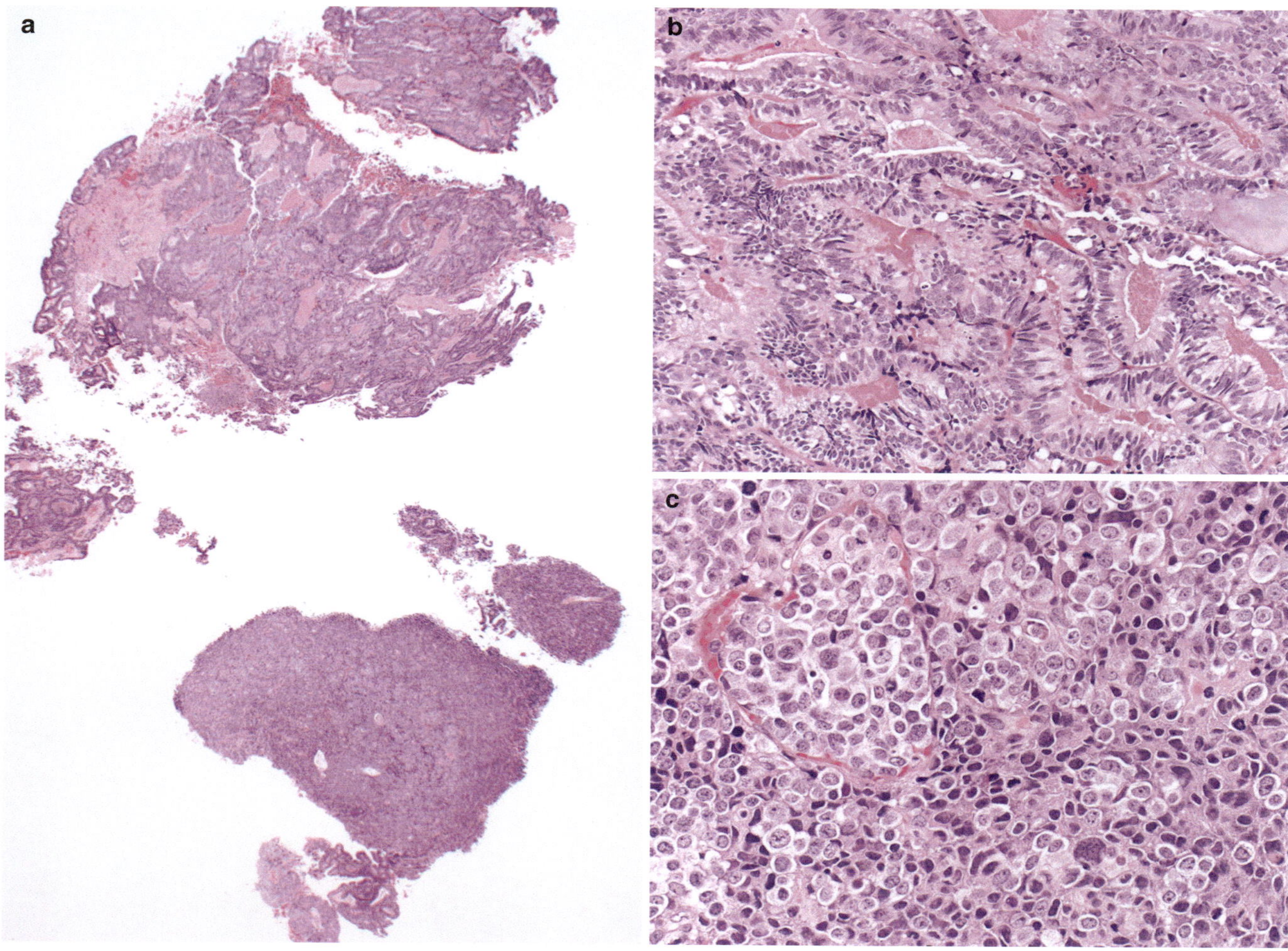

Fig. 4.15 Dedifferentiated endometrial carcinoma: 68-year-old woman presenting with heavy bleeding 25-years after her last period. Endometrial biopsy (**a**) shows concurrent low-grade (**b**) and high-grade (**c**) components. While the low-grade component shows strong keratin AE1/AE3 (**d**) and ER (**e**) staining, only scattered cells stain for keratin (**h**) and ER is entirely negative (**i**) in the high-grade component. Both components show wild-type pattern p53-staining (**f, j**). This is a dedifferentiated endometrial cancer arising in the setting of MLH1-deficiency (**g, k**). At least half of dedifferentiated endometrial cancers show dMMR/MSI-H, while about one-third are SMARCA4-deficient.

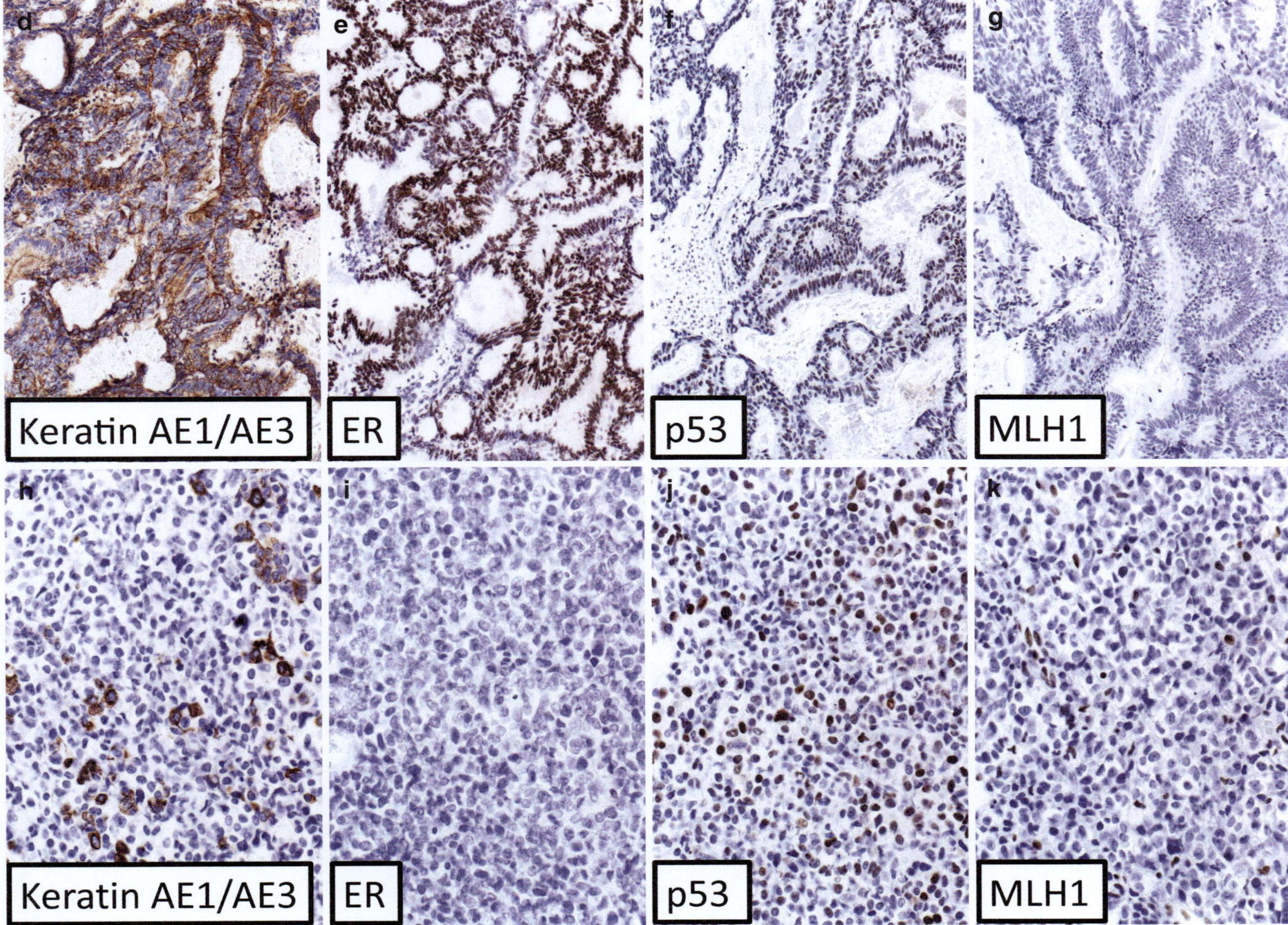

Fig. 4.15 (continued)

Immunohistochemical Approach to High-Grade Neoplasms Negative for Screening Markers

- Table 4.16 lists differential considerations in a broad-spectrum epithelial marker, SOX10 (or S-100), CD45 "triple-negative" malignant neoplasm with potentially useful additional diagnostic markers
  - Carcinoma types that tend to be broad-spectrum epithelial marker weak-to-negative include:
    - SWI/SNF-deficient carcinoma (SMARCA4 or SMARCB1 loss)
    - Sarcomatoid carcinoma (may be p40+)
    - NEC (INSM1+, Rb loss)
    - Adrenal cortical carcinoma (SF1+)
  - Mesenchymal neoplasms are often "triple-negative"
    - In the setting of an undifferentiated malignant neoplasm, CD34+ favors undifferentiated sarcoma over undifferentiated carcinoma
  - A subset of hematolymphoid neoplasms tend to be CD45−:
    - Acute lymphoblastic leukemia/lymphoma (TdT+)
    - ALK-positive large cell lymphoma (ALK+)
    - Plasma cell neoplasm (light chain restricted)
    - Classical Hodgkin lymphoma (CD30/CD15/PAX5+)
    - Plasmablastic lymphoma (light chain restricted, often EBER+)
  - Dedifferentiated melanoma (BRAF V600E+ is supportive, though nonspecific)
  - Germ cell tumor (diffuse SALL4+ is supportive)
  - Pheochromocytoma/paraganglioma (combination of general neuroendocrine marker strong-positivity and GATA-3+ is supportive)

Table 4.16 Fifteen diagnoses to consider before "busting" on a broad-spectrum keratin/SOX10 (or S-100)/CD45 (or CD43) "triple-negative" malignant neoplasm

Tumor type	Additional diagnostic markers
SWI/SNF-deficient carcinoma	SMARCA4 (BRG1), SMARCB1 (INI1)
Sarcomatoid carcinoma	Additional broad-spectrum epithelial markers, including HMW-keratin, p40
Poorly differentiated neuroendocrine ca	INSM1, Rb, TTF-1, POU2F3, CXCR4, chromogranin A, synaptophysin
Adrenal cortical carcinoma	SF1, melan A, inhibin A, calretinin, synaptophysin
Sarcoma	MDM2/CDK4; SMA, desmin; CD34 (rarely expressed by carcinoma); additional dictated by histology
Gastrointestinal stromal tumor	DOG1, KIT
Follicular dendritic cell sarcoma	CD21, CD23, CD35
Acute lymphoblastic leukemia/lymphoma	TdT, CD34, CD43
ALK-positive large-cell lymphoma	ALK, CD30
Plasma cell neoplasm (anaplastic)	CD79a, CD138, MUM1, kappa/lambda light chains
Classical Hodgkin lymphoma	CD30, CD15, PAX5
Plasmablastic lymphoma	CD79a, CD138, MUM1, EBV EBER, kappa/lambda light chains
Melanoma (dedifferentiated)	BRAF V600E
Germ cell tumor	SALL4 or PLAP
Pheochromocytoma/ paraganglioma	Chromogranin A, synaptophysin, GATA-3, PHOX2B, TH

HMW high-molecular weight, *TH* tyrosine hydroxylase

Immunohistochemistry for Hereditary Cancer Predisposition Syndromes

- Table 4.17 lists useful IHC markers and associated tumor types in hereditary cancer predisposition syndromes
- It is desirable to identify patients with hereditary cancer predisposition syndromes as they benefit from syndrome-specific surveillance and may choose to undergo prophylactic surgical procedures; it also facilitates the identification of affected family members
- Hereditary cancer predisposition syndromes are typically autosomal dominant in which affected individuals inherit one defective allele in a tumor suppressor
- Loss of function in the second allele drives oncogenesis
- Most mutations encode premature translation-termination codons (aka nonsense mutations), leading to nonsense-mediated mRNA decay and immunohistochemically detectable loss of protein expression

- These tumor suppressors are ubiquitously expressed proteins, and loss of protein expression can only be evaluated in the setting of intact internal control staining
  - Key Pitfall: delayed fixation and incomplete fixation leads to poor preservation of these antigens, and loss must only be "called" in the setting of intact internal control
- Most of these tumor suppressors are also implicated in sporadic tumors, and loss of protein expression is, thus, not tantamount to the diagnosis of a hereditary cancer predisposition syndrome
- IHC for these tumor suppressors may be useful to adjudicate "variants of uncertain significance" identified in molecular testing
- Select Specific Diseases
  - Lynch syndrome
    - MMR IHC is widely deployed to screen for Lynch syndrome and to identify dMMR tumors with a high likelihood of response to checkpoint inhibition
    - Adenocarcinomas of the endometrium and GI tract, upper urinary tract urothelial carcinoma, and sebaceous neoplasms are most frequently screened
    - Most dMMR tumors are sporadic due to *MLH1* promoter methylation leading to loss of expression of MLH1 and its dependent partner PMS2
    - Tumors with combined loss of MSH2/MSH6 or isolated loss of MSH6 or PMS2 are more likely to be Lynch syndrome
    - BRAF V600E IHC and/or *BRAF*-mutation testing can be used to identify sporadic MLH1/PMS2-deficient cancers (60% are positive and this result effectively excludes Lynch syndrome)
    - For *BRAF* wild-type colon cancers and for all other MLH1/PMS2-deficient cancers, *MLH1* promoter methylation is pursued, with the presence of methylation excluding Lynch syndrome
    - Patients with sporadic mutations in a DNA mismatch repair gene are referred to as "Lynch-like" and do not have as significant a risk of developing other Lynch-associated cancers; their family members may not require genetic counseling
  - Hereditary paraganglioma-pheochromocytoma syndrome
    - Succinate dehydrogenase (SDH) is a multi-subunit protein that participates in the Krebs cycle and electron transport chain
    - Germline mutations in *SDHA*, *SDHAF2*, *SDHB*, *SDHC*, and *SDHD* destabilize the complex and lead to loss of SDHB expression (i.e., SDHB IHC is used to screen for inactivation of any subunit)

Table 4.17 Immunohistochemistry to screen for hereditary cancer predisposition syndromes

Marker(s)	Syndrome	Associated tumors
MMR	Lynch syndrome	Adenocarcinomas of endometrium, colon, stomach, small intestine, pancreas, ovary (nonserous); upper urinary tract urothelial carcinoma; sebaceous neoplasms; glioblastoma
SDHB	Hereditary paraganglioma-pheochromocytoma syndromes	Paraganglioma, pheochromocytoma, gastrointestinal stromal tumor, renal cell carcinoma
FH/2SC	Hereditary leiomyomatosis and renal cell cancer	Cutaneous and uterine leiomyomas and renal cell carcinoma
BAP1	BAP1 tumor predisposition syndrome	Uveal melanoma, mesothelioma, renal cell carcinoma, cutaneous melanoma, melanocytic *BAP1*-mutated atypical intradermal tumor (aka BAPoma), intrahepatic cholangiocarcinoma
PRKAR1A	Carney complex	Lentigo, blue nevus, myxoma, growth hormone-positive pituitary adenoma, primary pigmented nodular adrenocortical disease, malignant melanotic nerve sheath tumor, testicular large-cell calcifying Sertoli cell tumor
SMARCB1 > SMARCA4	Rhabdoid tumor predisposition syndrome	Atypical teratoid/rhabdoid tumor, extrarenal rhabdoid tumor, renal rhabdoid tumor, small cell carcinoma of the ovary hypercalcemic type; schwannomatosis is also due to germline *SMARCB1* mutations, but patients do not develop rhabdoid tumors
Parafibromin	Hyperparathyroidism-jaw tumor syndrome	Parathyroid adenoma, parathyroid carcinoma, ossifying fibroma of the jaw
Menin	Multiple endocrine neoplasia type 1	Parathyroid tumors, pituitary adenoma, pancreatic and lung neuroendocrine tumors, adrenal cortical tumors
PTEN	Cowden syndrome	Adenocarcinoma of the thyroid, breast, kidney, endometrium, and colon; trichilemmomas; intestinal hamartomas; multiple adenomatous nodules in the thyroid

- o Germline mutations in *SDHA* (but not in other subunits) also lead to loss of SDHA expression
- o SDH-deficiency is seen in 30% of thoracoabdominal and 15% of head and neck paragangliomas and 5% of pheochromocytomas
  - ♦ These have a very high rate of germline mutation
- o SDH-deficient gastrointestinal stromal tumors occur in the stomach of young patients, demonstrate epithelioid cytomorphology and a plexiform growth pattern, and demonstrate metastatic risk greater than that predicted by NCCN criteria
  - ♦ Many are sporadic
- o The dyad of gastrointestinal stromal tumor and paraganglioma is referred to as Carney-Stratakis syndrome
- o Carney triad consists of gastrointestinal stromal tumor, pulmonary chondroma, and paraganglioma and is due to sporadic *SDHC* promoter methylation
- o Rare SDH-deficient renal cell carcinomas demonstrate solid growth, uniform cytomorphology, flocculent eosinophilic cytoplasm, and cytoplasmic vacuolation/inclusion
  - ♦ These have a very high rate of germline mutation
- – Hereditary leiomyomatosis and renal cell cancer

- o Fumarate hydratase (FH) is another Krebs cycle enzyme
- o FH-deficiency leads to accumulation of the S-(2-succino)cysteine (2SC) oncometabolite
- o FH and 2SC IHC are often performed in tandem, with FH considered more specific and 2SC more sensitive
- o FH-deficient renal cell carcinomas are extremely clinically aggressive and typically demonstrate papillary architecture and viral-inclusion-like macronucleoli with perinucleolar clearing
  - ♦ These have a very high rate of germline mutation
- o FH-deficient uterine smooth muscle tumors typically demonstrate staghorn-shaped blood vessels, alveolar-pattern edema, macronucleoli with perinucleolar clearing, and eosinophilic cytoplasmic globules
  - ♦ In a prospective cohort from UCSF, 1.4% of 2060 women with a uterine smooth muscle tumor were flagged as having a potential FH-deficient tumor
  - ♦ Among 10 of 30 women who underwent genetic testing 60% had a germline mutation
- o FH-deficiency is very common in patients with multiple cutaneous leiomyomas.

Further Reading

Agaimy A. Moving from "single gene" concept to "functionally homologous multigene complex": the SWI/SNF paradigm. Semin Diagn Pathol. 2021;38(3):165–6.

Agaimy A, Erlenbach-Wünsch K, Konukiewitz B, Schmitt AM, Rieker RJ, Vieth M, et al. ISL1 expression is not restricted to pancreatic well-differentiated neuroendocrine neoplasms, but is also commonly found in well and poorly differentiated neuroendocrine neoplasms of extrapancreatic origin. Mod Pathol. 2013;26(7):995–1003.

Ai D, Yao J, Yang F, Huo L, Chen H, Lu W, et al. TRPS1: a highly sensitive and specific marker for breast carcinoma, especially for triple-negative breast cancer. Mod Pathol. 2021;34(4):710–9.

Andrici J, Gill AJ, Hornick JL. Next generation immunohistochemistry: emerging substitutes to genetic testing? Semin Diagn Pathol. 2018;35(3):161–9.

Baine MK, Febres-Aldana CA, Chang JC, Jungbluth AA, Sethi S, Antonescu CR, et al. POU2F3 in SCLC: clinicopathologic and genomic analysis with a focus on its diagnostic utility in neuroendocrine-low SCLC. J Thorac Oncol. 2022;17(9):1109–21.

Barletta JA, Hornick JL. Succinate dehydrogenase-deficient tumors: diagnostic advances and clinical implications. Adv Anat Pathol. 2012;19(4):193–203.

Bellizzi AM. Assigning site of origin in metastatic neuroendocrine neoplasms: a clinically significant application of diagnostic immunohistochemistry. Adv Anat Pathol. 2013;20(5):285–314.

Bellizzi AM. An algorithmic immunohistochemical approach to define tumor type and assign site of origin. Adv Anat Pathol. 2020;27(3):114–63.

Bellizzi AM. Immunohistochemistry in the diagnosis and classification of neuroendocrine neoplasms: what can brown do for you? Hum Pathol. 2020;96:8–33.

Bellizzi AM. SATB2 in neuroendocrine neoplasms: strong expression is restricted to well-differentiated tumours of lower gastrointestinal tract origin and is most frequent in Merkel cell carcinoma among poorly differentiated carcinomas. Histopathology. 2020;76(2):251–64.

Bellizzi AM, Frankel WL. Colorectal cancer due to deficiency in DNA mismatch repair function: a review. Adv Anat Pathol. 2009;16(6):405–17.

Bishop JA, Teruya-Feldstein J, Westra WH, Pelosi G, Travis WD, Rekhtman N. p40 (DeltaNp63) is superior to p63 for the diagnosis of pulmonary squamous cell carcinoma. Mod Pathol. 2012;25(3):405–15.

Borrisholt M, Nielsen S, Vyberg M. Demonstration of CDX2 is highly antibody dependant. Appl Immunohistochem Mol Morphol. 2013;21(1):64–72.

Carter CS, Skala SL, Chinnaiyan AM, McHugh JB, Siddiqui J, Cao X, et al. Immunohistochemical characterization of fumarate hydratase (FH) and succinate dehydrogenase (SDH) in cutaneous leiomyomas for detection of familial Cancer syndromes. Am J Surg Pathol. 2017;41(6):801–9.

Cerami E, Gao J, Dogrusoz U, Gross BE, Sumer SO, Aksoy BA, et al. The cBio cancer genomics portal: an open platform for exploring multidimensional cancer genomics data. Cancer Discov. 2012;2(5):401–4.

Cheung CC, D'Arrigo C, Dietel M, Francis GD, Fulton R, Gilks CB, et al. Evolution of quality assurance for clinical immunohistochemistry in the era of precision medicine: part 4: tissue tools for quality assurance in Immunohistochemistry. Appl Immunohistochem Mol Morphol. 2017;25(4):227–30.

Cheung CC, D'Arrigo C, Dietel M, Francis GD, Gilks CB, Hall JA, et al. Evolution of quality assurance for clinical immunohistochemistry in the era of precision medicine: part 1: fit-for-purpose approach to classification of clinical immunohistochemistry biomarkers. Appl Immunohistochem Mol Morphol. 2017;25(1):4–11.

Chu P, Wu E, Weiss LM. Cytokeratin 7 and cytokeratin 20 expression in epithelial neoplasms: a survey of 435 cases. Mod Pathol. 2000;13(9):962–72.

Chu PG, Weiss LM. Expression of cytokeratin 5/6 in epithelial neoplasms: an immunohistochemical study of 509 cases. Mod Pathol. 2002;15(1):6–10.

Cimino-Mathews A, Subhawong AP, Elwood H, Warzecha HN, Sharma R, Park BH, et al. Neural crest transcription factor Sox10 is preferentially expressed in triple-negative and metaplastic breast carcinomas. Hum Pathol. 2013;44(6):959–65.

Copete M, Garratt J, Gilks B, Pilavdzic D, Berendt R, Bigras G, et al. Inappropriate calibration and optimisation of pan-keratin (pan-CK) and low molecular weight keratin (LMWCK) immunohistochemistry tests: Canadian immunohistochemistry quality control (CIQC) experience. J Clin Pathol. 2011;64(3):220–5.

Czeczok TW, Gailey MP, Hornick JL, Bellizzi AM. High-grade neuroendocrine carcinomas are characterized by marked transcription factor lineage infidelity: an evaluation of 36 diagnostic markers in 83 tumors. Mod Pathol. 2014;27(Suppl 2):152A.

Doyle LA, Möller E, Dal Cin P, Fletcher CD, Mertens F, Hornick JL. MUC4 is a highly sensitive and specific marker for low-grade fibromyxoid sarcoma. Am J Surg Pathol. 2011;35(5):733–41.

Drier JK, Swanson PE, Cherwitz DL, Wick MR. S100 protein immunoreactivity in poorly differentiated carcinomas. Immunohistochemical comparison with malignant melanoma. Arch Pathol Lab Med. 1987;111(5):447–52.

Fitzgibbons PL, Bradley LA, Fatheree LA, Alsabeh R, Fulton RS, Goldsmith JD, et al. Principles of analytic validation of immunohistochemical assays: guideline from the College of American Pathologists Pathology and Laboratory Quality Center. Arch Pathol Lab Med. 2014;138(11):1432–43.

Gill AJ, Hes O, Papathomas T, Šedivcová M, Tan PH, Agaimy A, et al. Succinate dehydrogenase (SDH)-deficient renal carcinoma: a morphologically distinct entity: a clinicopathologic series of 36 tumors from 27 patients. Am J Surg Pathol. 2014;38(12):1588–602.

Gloyeske NC, Woodard AH, Elishaev E, Yu J, Clark BZ, Dabbs DJ, et al. Immunohistochemical profile of breast cancer with respect to estrogen receptor and HER2 status. Appl Immunohistochem Mol Morphol. 2015;23(3):202–8.

Herpel E, Rieker RJ, Dienemann H, Muley T, Meister M, Hartmann A, et al. SMARCA4 and SMARCA2 deficiency in non-small cell lung cancer: immunohistochemical survey of 316 consecutive specimens. Ann Diagn Pathol. 2017;26:47–51.

Hornick JL. Replacing molecular genetic testing with immunohistochemistry using antibodies that recognize the protein products of gene rearrangements: "next-generation" immunohistochemistry. Am J Surg Pathol. 2021;45(4):584–6.

Howlader N, Noone AM, Krapcho M, Miller D, Brest A, Yu M, et al. SEER cancer statistics review, 1975–2016. Bethesda, MD: National Cancer Institute; 2020. https://seer.cancer.gov/archive/csr/1975_2016/.

Iwata J, Fletcher CD. Immunohistochemical detection of cytokeratin and epithelial membrane antigen in leiomyosarcoma: a systematic study of 100 cases. Pathol Int. 2000;50(1):7–14.

Krämer A, Bochtler T, Pauli C, Baciarello G, Delorme S, Hemminki K, et al. Cancer of unknown primary: ESMO clinical practice guideline for diagnosis, treatment and follow-up. Ann Oncol. 2023;34(3):228–46.

Lau HD, Chan E, Fan AC, Kunder CA, Williamson SR, Zhou M, et al. A clinicopathologic and molecular analysis of Fumarate Hydratase-deficient renal cell carcinoma in 32 patients. Am J Surg Pathol. 2020;44(1):98–110.

Le DT, Durham JN, Smith KN, Wang H, Bartlett BR, Aulakh LK, et al. Mismatch repair deficiency predicts response of solid tumors to PD-1 blockade. Science. 2017;357(6349):409–13.

Le DT, Uram JN, Wang H, Bartlett BR, Kemberling H, Eyring AD, et al. PD-1 blockade in tumors with mismatch-repair deficiency. N Engl J Med. 2015;372(26):2509–20.

Machado I, Yoshida A, Morales MGN, Abrahão-Machado LF, Navarro S, Cruz J, et al. Review with novel markers facilitates precise categorization of 41 cases of diagnostically challenging, "undifferentiated small round cell tumors". A clinicopathologic, immunophenotypic and molecular analysis. Ann Diagn Pathol. 2018;34:1–12.

Magnusson K, de Wit M, Brennan DJ, Johnson LB, McGee SF, Lundberg E, et al. SATB2 in combination with cytokeratin 20 identifies over 95% of all colorectal carcinomas. Am J Surg Pathol. 2011;35(7):937–48.

Michels S, Swanson PE, Frizzera G, Wick MR. Immunostaining for leukocyte common antigen using an amplified avidin-biotin-peroxidase complex method and paraffin sections. A study of 735 hematopoietic and nonhematopoietic human neoplasms. Arch Pathol Lab Med. 1987;111(11):1035–9.

Miettinen M, McCue PA, Sarlomo-Rikala M, Biernat W, Czapiewski P, Kopczynski J, et al. Sox10—a marker for not only schwannian and melanocytic neoplasms but also myoepithelial cell tumors of soft tissue: a systematic analysis of 5134 tumors. Am J Surg Pathol. 2015;39(6):826–35.

Miettinen M, McCue PA, Sarlomo-Rikala M, Rys J, Czapiewski P, Wazny K, et al. GATA3: a multispecific but potentially useful marker in surgical pathology: a systematic analysis of 2500 epithelial and nonepithelial tumors. Am J Surg Pathol. 2014;38(1):13–22.

Miettinen M, Wang Z, McCue PA, Sarlomo-Rikala M, Rys J, Biernat W, et al. SALL4 expression in germ cell and non-germ cell tumors: a systematic immunohistochemical study of 3215 cases. Am J Surg Pathol. 2014;38(3):410–20.

Moh M, Krings G, Ates D, Aysal A, Kim GE, Rabban JT. SATB2 expression distinguishes ovarian metastases of colorectal and Appendiceal origin from primary ovarian tumors of mucinous or Endometrioid type. Am J Surg Pathol. 2016;40(3):419–32.

Moll R, Franke WW, Schiller DL, Geiger B, Krepler R. The catalog of human cytokeratins: patterns of expression in normal epithelia, tumors and cultured cells. Cell. 1982;31(1):11–24.

NCCN. NCCN clinical practice guidelines in oncology (NCCN guidelines): occult primary (cancer of unknown primary [CUP]). 2023. https://www.nccn.org/professionals/physician_gls/pdf/occult.pdf.

Nonaka D, Papaxoinis G, Mansoor W. Diagnostic utility of orthopedia homeobox (OTP) in pulmonary carcinoid tumors. Am J Surg Pathol. 2016;40(6):738–44.

O'Connell FP, Pinkus JL, Pinkus GS. CD138 (syndecan-1), a plasma cell marker immunohistochemical profile in hematopoietic and non-hematopoietic neoplasms. Am J Clin Pathol. 2004;121(2):254–63.

Ordonez NG. Broad-spectrum immunohistochemical epithelial markers: a review. Hum Pathol. 2013;44(7):1195–215.

Pinkus GS, Etheridge CL, O'Connor EM. Are keratin proteins a better tumor marker than epithelial membrane antigen? A comparative immunohistochemical study of various paraffin-embedded neoplasms using monoclonal and polyclonal antibodies. Am J Clin Pathol. 1986;85(3):269–77.

Pinkus GS, Kurtin PJ. Epithelial membrane antigen—a diagnostic discriminant in surgical pathology: immunohistochemical profile in epithelial, mesenchymal, and hematopoietic neoplasms using paraffin sections and monoclonal antibodies. Hum Pathol. 1985;16(9):929–40.

Rabban JT, Chan E, Mak J, Zaloudek C, Garg K. Prospective detection of germline mutation of Fumarate Hydratase in women with uterine smooth muscle tumors using pathology-based screening to trigger genetic counseling for hereditary leiomyomatosis renal cell carcinoma syndrome: a 5-year single institutional experience. Am J Surg Pathol. 2019;43(5):639–55.

Riihimäki M, Thomsen H, Hemminki A, Sundquist K, Hemminki K. Comparison of survival of patients with metastases from known versus unknown primaries: survival in metastatic cancer. BMC Cancer. 2013;13:36.

Rodríguez-Soler M, Pérez-Carbonell L, Guarinos C, Zapater P, Castillejo A, Barberá VM, et al. Risk of cancer in cases of suspected lynch syndrome without germline mutation. Gastroenterology. 2013;144(5):926–32.e1; quiz e13–4.

Rooper LM, Sharma R, Li QK, Illei PB, Westra WH. INSM1 demonstrates superior performance to the individual and combined use of synaptophysin, chromogranin and CD56 for diagnosing neuroendocrine tumors of the thoracic cavity. Am J Surg Pathol. 2017;41(11):1561–9.

Schaefer IM, Agaimy A, Fletcher CD, Hornick JL. Claudin-4 expression distinguishes SWI/SNF complex-deficient undifferentiated carcinomas from sarcomas. Mod Pathol. 2017;30(4):539–48.

Torlakovic EE, Cheung CC, D'Arrigo C, Dietel M, Francis GD, Gilks CB, et al. Evolution of quality assurance for clinical immunohistochemistry in the era of precision medicine. Part 3: technical validation of immunohistochemistry (IHC) assays in clinical IHC laboratories. Appl Immunohistochem Mol Morphol. 2017;25(3):151–9.

Torlakovic EE, Cheung CC, D'Arrigo C, Dietel M, Francis GD, Gilks CB, et al. Evolution of quality assurance for clinical immunohistochemistry in the era of precision medicine—part 2: immunohistochemistry test performance characteristics. Appl Immunohistochem Mol Morphol. 2017;25(2):79–85.

Vang R, Gown AM, Barry TS, Wheeler DT, Yemelyanova A, Seidman JD, et al. Cytokeratins 7 and 20 in primary and secondary mucinous tumors of the ovary: analysis of coordinate immunohistochemical expression profiles and staining distribution in 179 cases. Am J Surg Pathol. 2006;30(9):1130–9.

Vang R, Gown AM, Wu LS, Barry TS, Wheeler DT, Yemelyanova A, et al. Immunohistochemical expression of CDX2 in primary ovarian mucinous tumors and metastatic mucinous carcinomas involving the ovary: comparison with CK20 and correlation with coordinate expression of CK7. Mod Pathol. 2006;19(11):1421–8.

Wang NP, Zee S, Zarbo RJ, Bacchi CE, Gown AM. Coordinate expression of cytokeratins 7 and 20 defined unique subsets of carcinomas. Appl Immunohistochem. 1995;3(2):99–107.

Weissinger SE, Keil P, Silvers DN, Klaus BM, Möller P, Horst BA, et al. A diagnostic algorithm to distinguish desmoplastic from spindle cell melanoma. Mod Pathol. 2014;27(4):524–34.

Yemelyanova AV, Vang R, Judson K, Wu LS, Ronnett BM. Distinction of primary and metastatic mucinous tumors involving the ovary: analysis of size and laterality data by primary site with reevaluation of an algorithm for tumor classification. Am J Surg Pathol. 2008;32(1):128–38.

Molecular Pathology of Colorectal Tumors

5

Wei Chen, Dan Jones, and Wendy L. Frankel

Contents

Introduction

Epidemiology

- Colorectal cancer (CRC) is the second most common cancer in women and the third most common cancer in men worldwide

- Incidence is rapidly increasing in many low- and middle-income countries
- Increasing incidence in young adults reported recently in Western countries
- In the United States, incidence rate dropped by ~1% each year from 2013 to 2017; however, incidence has been rising in younger people (by 2% each year in adults under age 50, 2012–2016), and younger adults present with more advanced disease

Etiology

- Common risk factors: Sedentary lifestyle, excess body fat, diet enriched with processed and red meat, alcohol, chronic bowel inflammation
- The role of smoking is controversial
- Genetic predisposition important, depending on the type of mutation (see section, Hereditary cancer syndromes)
- Rare but known risk factors: pelvic irradiation, cystic fibrosis, ureterosigmoidostomy, and acromegaly
- Factors that decrease the risk of CRC: Consumption of dietary fiber and dairy products, increased levels of physical activity, prolonged use of nonsteroidal anti-inflammatory drugs and hormone replacement therapy in women

Prognosis and Treatment

- Five-year relative survival increased significantly from 53% to 62% for colon cancer, and from 51% to 65% for rectal cancer
- Ongoing advancements in treatment (improved surgical methods and targeted therapy) and other factors (diagnostics, preoperative and postoperative treatment, and care) lead to a continuous improvement in the relative survival
- Treatment choice for metastatic right vs. left-sided CRC is different, owing to their different biology as midgut vs. hind-

W. Chen · D. Jones · W. L. Frankel (✉)
Department of Pathology, The Ohio State University,
Wexner Medical Center, Columbus, OH, USA
e-mail: wei.chen2@osumc.edu; wendy.frankel@osumc.edu

gut derivatives. Currently, for right-sided CRC, Bevacizumab should be considered. For left-sided CRC with wildtype *RAS* and *BRAF*, anti-Epidermal growth factor receptor (EGFR) agent plus chemotherapy is the first-line treatment.

Screening Guidelines

- Average risk individuals—Initial CRC screening at 45 years
  - Lowered age to initiate screening (instead of the traditional 50 years) is prompted by the recent alarming rise of CRC in people younger than 50 years
  - Average risk definition: No personal history of adenoma, sessile serrated adenoma, CRC, or inflammatory bowel disease; negative family history of CRC or confirmed advanced adenoma (high-grade dysplasia, ≥1 cm, villous histology) or advanced sessile serrated adenoma (≥1 cm, any dysplasia)
  - Recommended by American Cancer Society and the US Preventive Services Task Force
  - National Comprehensive Cancer Network (NCCN) acknowledges that there is some evidence to support a benefit for screening earlier at 45 years, and the decision may be dependent on race/ethnicity, patient preference, and resources available
- Increased risk individuals—Screening before 45 years (variable depending on the scenario)
  - Personal history of adenoma or sessile serrated adenoma, CRC, or inflammatory bowel disease
  - Positive family history: ≥1 first-degree relative with CRC at any age; second- and third-degree relatives with CRC at any age; first-degree relative with confirmed advanced adenoma(s) or advanced sessile serrated adenoma
  - Concerning symptoms (iron deficiency anemia, rectal bleeding, or a change in bowel habits)
  - Known hereditary syndrome (see section, Hereditary Cancer Syndromes)

Pathogenesis and Molecular Classifications of Colorectal Cancer

Colorectal Carcinogenesis

- Conventional adenoma-carcinoma sequence (chromosomal instability, CIN pathway, 84% of CRC)
  - Characterized by large-scale chromosomal copy-number gains/losses (e.g., entire chromosome arm and/or sub-arm) gains and losses, affecting a small group of genes

- Key genes mutated: *APC* (80%), *TP53* (60%), *KRAS* (45%), *SMAD4*, and *PIK3CA*
  - *APC*
    - Tumor suppressors, mostly inactivating mutations, are found in 80% of both adenomas and carcinomas
    - An early event in adenoma formation causes decreased degradation of β-catenin and, thus, abnormal signaling through the WNT pathway
  - *KRAS*
    - Oncoprotein, activating mutations (mostly codons 12, 13, 61, less commonly in codons 117 and 146), found in about 40% of both adenomas and carcinomas
    - *KRAS* mutations all behave similarly, dysregulating the mitogen-activated protein kinase (MAPK) pathway that drives cell cycle proliferation, and similarly less common *BRAF* activating mutations (e.g., V600E) and *NRAS* mutations
    - Mutations in these three genes are largely mutually exclusive but frequently co-occur with other growth coregulators such as *PIK3CA* or *AKT* oncogenic mutations and/or inactivating mutations or deletions of *PTEN* (Fig. 5.1)
  - *SMAD4*
    - Tumor suppressor, usually inactivated by deletion of a large part of chromosome 18q (in ~60% of CRC), often with mutation of the other allele, reducing signaling through the TGF-β inhibitory pathway
  - *TP53*
    - Tumor suppressor, deletions, and/or inactivating mutations, a later event in the adenoma-carcinoma transition in sporadic cancers, but occurring earlier in inflammatory bowel disease-associated carcinogenesis
- Microsatellite instability (MSI) (hypermutant pathway, 13% of CRC)
  - A high mutation rate (hypermutant) due to defective DNA mismatch repair (dMMR) affects many genes (see below for pathogenesis)
  - dMMR mostly due to sporadic/nonsyndromic hypermutated MSI cancers with *MLH1* promoter hypermethylation, causing loss of MLH1 expression
    - Almost all these tumors develop through the serrated pathway, with the high frequency of *BRAF* V600E and less often *KRAS* activating mutations and the CpG island methylator (CIMP) phenotype
  - dMMR in a minority due to an inherited syndrome (Lynch syndrome, LS) with one germline dMMR, or two acquired MMR mutations

Fig. 5.1 Targeting EGFR and HER2 signaling pathways in colorectal cancer. Epidermal growth factor receptor (EGFR/HER1) is a transmembrane protein receptor of the ErbB family, which also includes HER2/ErbB-2. EGFR pathway is frequently activated in colorectal cancer, through downstream RAS/MAPK and PI3K/AKT signaling pathways, leading to tumor cell proliferation and survival. Anti-EGFR antibodies (such as cetuximab and panitumumab) bind to the extracellular domain of the *EGFR*, preventing its activation. However, activating mutations of downstream modulators such as *RAS*, *BRAF*, *PI3K*, loss of function mutation in *PTEN*, or amplification/mutation of other ErbB receptors such as *HER2* may confer resistance to anti-EGFR therapy

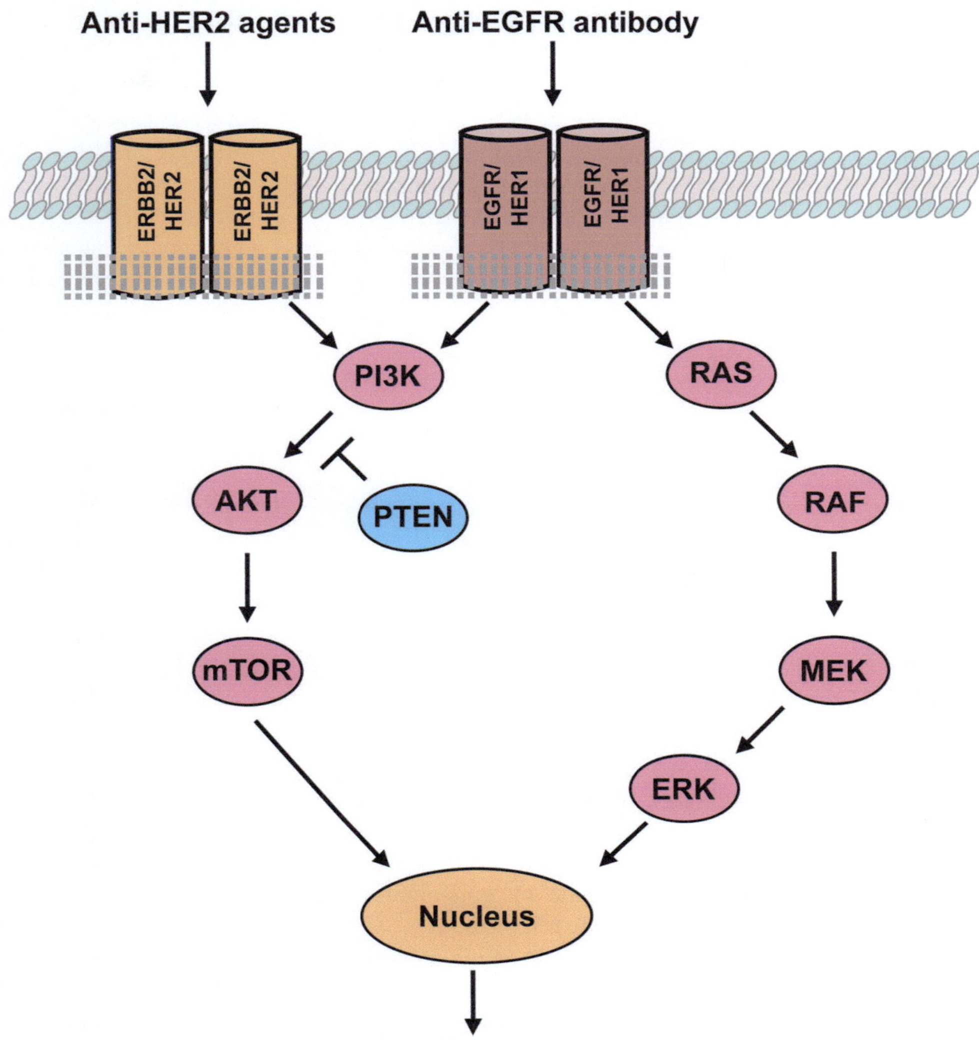

- Defective proofreading polymerase (ultramutant pathway, 3% of CRC)
  - Missense mutations in the proofreading domain of *POLE* (rarely *POLD1*) or inactivating *POLE/POLD1* mutations that confer very high mutation rates, affecting very large numbers of genes (mostly passenger mutations with few driver mutations)
  - Characteristic nucleotide base change spectrum is increased C → A transversions
- In general, right-sided colon cancer is more likely to demonstrate MSI and CpG island methylation, whereas left-sided CRC often shows chromosomal instability.

Molecular Classifications of Colorectal Cancer

DNA-Based Genomic Classification by the Cancer Genome Atlas (TCGA)
- Two groups by mutation rate
  - Hypermutated (~15% of CRC)
  - Nonhypermutated (~85% of CRC)
- Hypermutated and nonhypermutated cancers correlate with the MSI/POLE pathways and CIN pathway, respectively
- Compared with nonhypermutated group, hypermutated tumors have a higher frequency of mutations and less complex karyotypes

RNA-Based Transcriptomic Profiling by Colorectal Cancer Subtyping Consortium

- Four consensus molecular subtype (CMS) groups based on biological processes indicated by RNA expression patterns
- CMS1 (MSI-immune, 14%): Almost all hypermutated MSI cancers are in this group. The expression pattern is of strong immune activation, consistent with the histologic observation of prominent tumor-infiltrating lymphocytes in MSI tumors, explaining their responsiveness to immune checkpoint inhibitors
- CMS2 (canonical, 37%), CMS3 (metabolic, 13%), and CMS4 (mesenchymal, 23%) are all microsatellite-stable CRCs
- A residual unclassified group (mixed features, 13%) shows transition phenotype or intratumoral heterogeneity

Practical Molecular Diagnostics of Colorectal Cancer

Universal Screening Algorithm

- See Fig. 5.2
- Screening patients with CRC for LS is cost-effective and recommended by many professional organizations, and is the standard of care
- MSI by PCR and MMR expression by immunohistochemistry (IHC) are both acceptable test methods
- IHC is the more commonly reported first step in universal tumor screening and is a better option in cases with a low tumor cell percentage, intense inflammatory reaction, or no normal tissue available as required by MSI test
- If MMR deficiency is detected, an additional workup is necessary to diagnose or rule out a germline mutation
- In cases with MLH1/PMS2 loss, *BRAF* mutation analysis is an alternative to *MLH1* hypermethylation testing; however, not all *MLH1* methylated cases harbor *BRAF* mutation; therefore, at the authors' institution, *MLH1* promoter hypermethylation test is the reflex test for cases with MLH1/PMS2 loss
- Cases with dMMR other than MLH1/PMS2 loss need a referral to genetics for germline consideration and workup
- Lynch-like (Fig. 5.3)
  - Umbrella terminology for dMMR cases where no methylation or germline mutation is found
  - The most common etiology is biallelic somatic mutation or single somatic mutation with loss of heterozygosity

  - Other possible etiologies include misinterpretation of immunohistochemical results, germline alterations that are not detectable by currently available tests, and other genetic defects with MSI phenotype such as *MUTYH* mutation, *POLE* mutation, and somatic mosaicism
  - Somatic tumor testing is essential for the distinction
- Larger targeted mutation panels that combine microsatellite testing and gene mutation analyses are under investigation and promising

MMR Immunohistochemistry Interpretation and Pitfalls

- Interpretation of MMR IHC is typically straightforward
- Attention to adequate internal control staining (lymphocytes, benign epithelial and stromal cells) and comparison with the tumor important
- No universal cutoff for what is considered intact/normal staining
  - Varies: Any convincing tumor staining, 1%, 5%, and 10%
  - We use 5% cutoff at our institution
- Usual patterns
  - Normal/Intact patterns
    - Presence of all four proteins suggests microsatellite stability
    - Typically, diffuse staining in the tumor nuclei, as strong as control
    - Patchy but intact staining can occur due to uneven antibody diffusion, variable fixation, or tissue hypoxia (Fig. 5.4a, b)
  - Usual loss patterns
    - Loss of nuclear staining for any of the four proteins indicates MSI and suggests the most likely involved gene and the need for additional testing
    - Loss of MSH2 alone or loss of MSH2/MSH6 suggests mutation in *MSH2*
    - Loss of MLH1 alone or loss of MLH1/PMS2 suggests mutation or methylation in *MLH1* (Fig. 5.5)
    - Isolated loss of MSH6 suggests mutation in *MSH6*
    - Isolated loss of PMS2 suggests mutation in *PMS2* but can be seen with heterogeneous *MLH1* promoter hypermethylation or *MLH1* mutations (especially missense mutations)
- Unusual patterns
  - Tumor staining is weaker than internal background control (Fig. 5.4c)

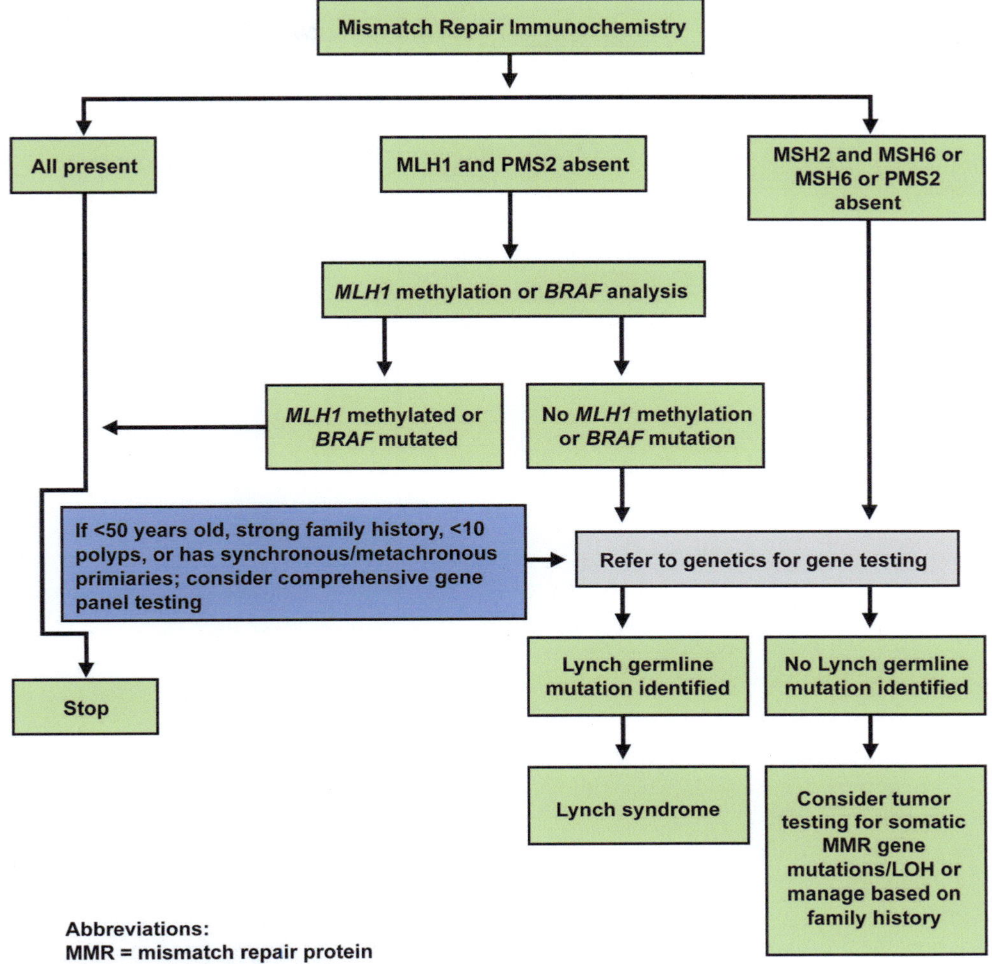

Fig. 5.2 Universal screening algorithm for mismatch repair deficiency and Lynch syndrome workup in colorectal cancer. In the universal screening algorithm, all primary colorectal cancers are immunostained for MLH1, MSH2, MSH6, and PMS2. The presence of all four proteins indicates intact expression and no further workup is necessary unless clinically suspicious (blue rectangle). When MLH1 and PMS2 are absent, *MLH1* promoter hypermethylation or *BRAF* V600E mutation analysis is performed. *MLH1* methylation or *BRAF* mutation essentially exclude Lynch syndrome. Non-*MLH1* hypermethylated cases without *BRAF* mutation, undergo germline sequencing of *MLH1* or *PMS2*; those with methylation are presumed to represent sporadic tumors and Lynch workup is not indicated. Some laboratories use *BRAF* mutational analysis first, rather than *MLH1* methylation testing after immunohistochemistry in those tumors found to have absence of MLH1 and PMS2. However, not all *MLH1* methylated cases harbor *BRAF* mutation, so negative *BRAF* mutation test may still be followed by *MLH1* methylation testing. Cases with absence of MSH2/MSH6, or isolated loss of MSH6, or isolated loss of PMS2, may represent Lynch syndrome and should undergo germline sequencing. Finally, if Lynch germline mutation is not identified in deficient MMR cases, tumor sequencing for double somatic mutation should be considered. In any patient diagnosed with colorectal cancer under the age of 50, a comprehensive germline panel should be considered

- o If such staining is confirmed on a repeat staining, it should be reported as abnormal and additional studies are warranted
- o Some of such cases can be truly dMMR
- Cytoplasmic-only staining is considered lost or absent for nuclear staining/dMMR
- Rectal tumors after neoadjuvant treatment (Fig. 5.4d)
  - o May show weak, patchy, nucleolar, or even absent MSH6 expression without dMMR
- o Testing on presurgical biopsy helpful to confirm proficient MMR
- Heterogeneous staining or loss of MSH6 expression can occasionally be seen
  - o Due to a nongermline mutation in the *MSH6* unstable coding mononucleotide tract, secondary to alternation in another MMR gene
  - o Most are seen with *MLH1* mutation or methylation, causing loss of MLH1 and PMS2 staining and MSI

Fig. 5.3 "Lynch-like": the unexplained mismatch repair deficiency cases in colorectal cancer. Mismatch repair deficient tumors without *MLH1* methylation, *BRAF* mutation, or identified germline mismatch repair gene mutations have been referred to by the waste-basket term 'Lynch-Like'. Possible causes include tumor screening error, unidentifiable mutation, somatic MMR mutations, other cancer syndrome causing somatic MMR mutations, and somatic mosaicism

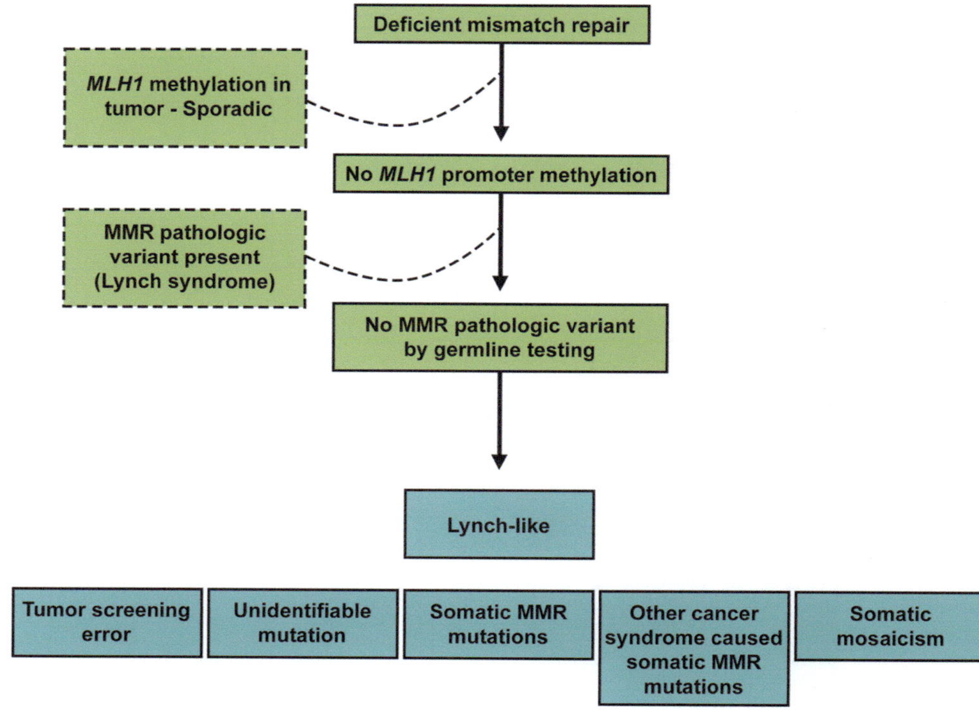

- Retained MMR staining despite dMMR
  - 3–10% of LS tumors that have dMMR with MSI show no abnormality on IHC
  - Presumably disabled protein function but retained protein antigenicity detectable by IHC, or other possible etiologies
  - For this reason, the presence of MMR staining does not unequivocally exclude the possibility of a germline mutation

Specimen Selection for CRC Molecular Testing

- Sample requirements
  - In general, a minimum of 10% tumor content is recommended
  - Increase "signal" and decrease "noise"
    - Choose the tumor block with the highest tumor content (% of tumor cells)
    - Avoid lymphoid tissue, necrosis, fat, and mucin
    - Microdissection for tumor enrichment may be performed
  - Pretreatment biopsies may be the preferred sample for rectal tumors post-neoadjuvant therapy
- Biopsy versus resection
  - For MMR IHC, biopsy specimen works equally well as resection specimen, and surgical treatment may be impacted by diagnosing LS on biopsy

- One versus all tumors
  - Finding one tumor to be microsatellite stable does not exclude the possibility that another will be unstable (LS patients may have sporadic tumors)
  - If multiple carcinomas, test all synchronous/metachronous carcinomas
- Metastasis versus primary tumor
  - High concordance rate between primary and metastatic tumors
  - Testing on metastasis is preferred for treatment-predictive targeted biomarker testing but there is no preference for dMMR testing
- Adenoma versus carcinoma
  - Carcinoma is the preferred specimen for MSI screening
  - In some situations, such as lack of cancer availability, testing on adenoma may be performed per clinical request, and the reporting of MMR status should be accompanied by a disclaimer that the presence of MMR expression in an adenoma does not exclude the possibility of LS
  - Loss of MMR staining can be identified in 70–90% LS-associated adenomas, especially in those with polyp size >10 mm, villous component, or high-grade dysplasia, as these advanced adenomas have a higher probability of having acquired the second mutation hit
- Serrated polyps

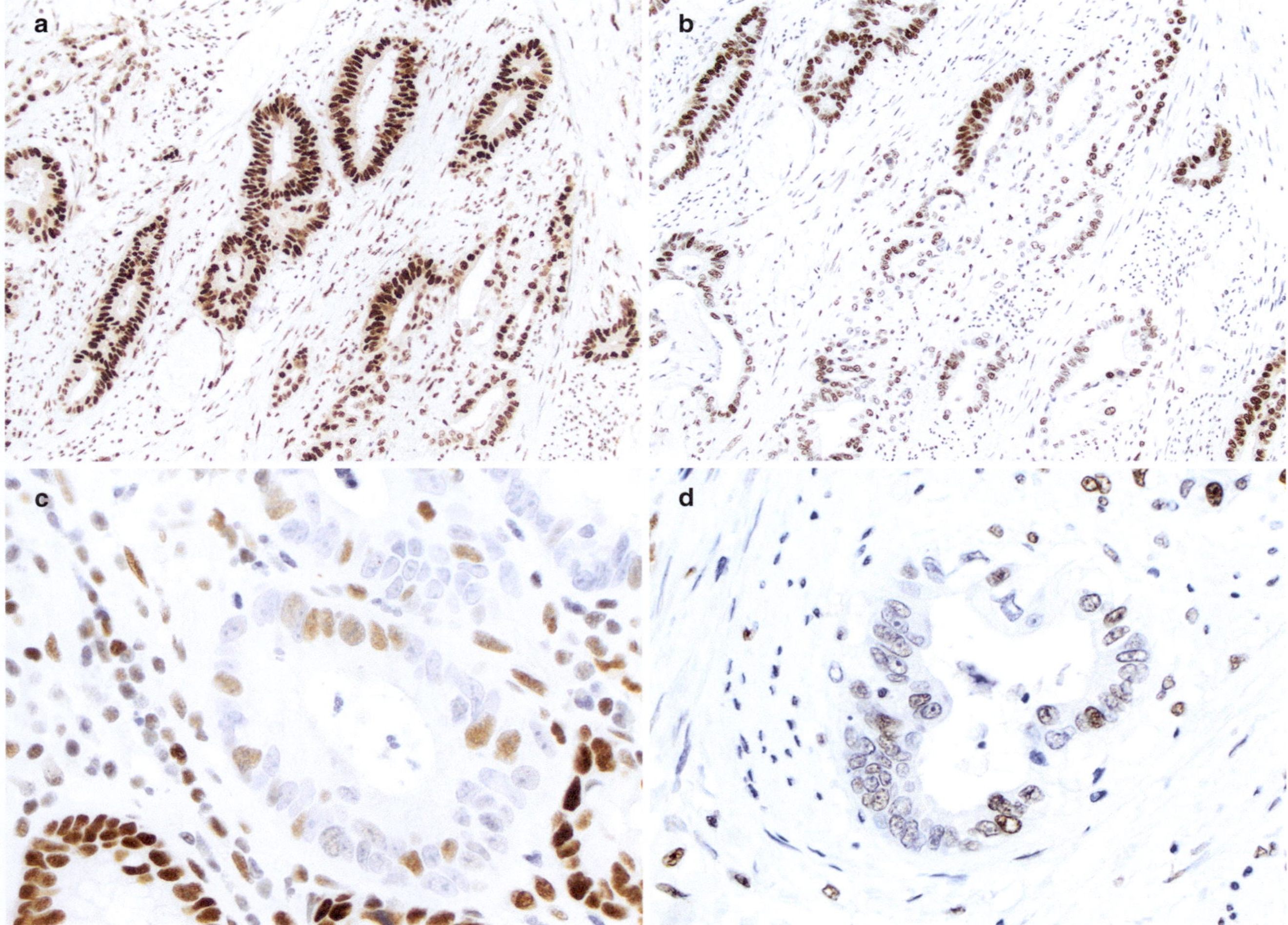

Fig. 5.4 Mismatch repair protein immunohistochemistry: Common challenges in interpretation. (**a, b**) Some variation of staining may be seen and does not need to be mentioned, as shown in two areas from the same case. (**c**) Staining of the tumor nuclei is weaker than that of the adjacent normal crypts (bottom) and stromal cell nuclei; this is considered abnormal staining pattern and additional study is warranted. (**d**) MSH6 showing weak to loss of tumor nuclear staining in the setting of post neoadjuvant chemoradiation therapy—if there is concern for loss, consider repeat on pretreatment biopsy. Notice the presence of staining in the background stromal cells (right upper corner), which serves as positive internal control

- Serrated polyps should not be used for LS screening as they are not typically precursor lesions for LS-associated CRC
- Dysplastic serrated polyps may exhibit dMMR due to *BRAF* mutation, unrelated to LS

Established Predictive Biomarkers

- The most frequent alterations of signaling pathways in CRC are activation of the WNT, MAPK, and PI3K/PTEN growth signaling pathways and dysregulation of the TGF-β and p53 inhibitory pathways. The following biomarkers are relevant for targeted therapies in these signaling pathways (Table 5.1)

RAS (*KRAS* and *NRAS*) for EGFR Antibody Immunotherapy Selection

- *RAS* genes are mutated in 45–50% of CRC
- Affect MAPK pathway of RAS/RAF/MEK/ERK, which is downstream of EGFR
- EGFR overexpression is a common targetable feature of CRC, but downstream RAS mutations are associated with lack of response to antibody-mediated EGFR blockade (Fig. 5.1)
- Extended *RAS* panel is recommended, which includes *KRAS* and *NRAS* codons 12 and 13 of exon 2, 59 and 61 of exon 3, and 117 and 146 of exon 4; mutations of these codons correlate with resistance to anti-EGFR therapy (panitumumab and cetuximab)

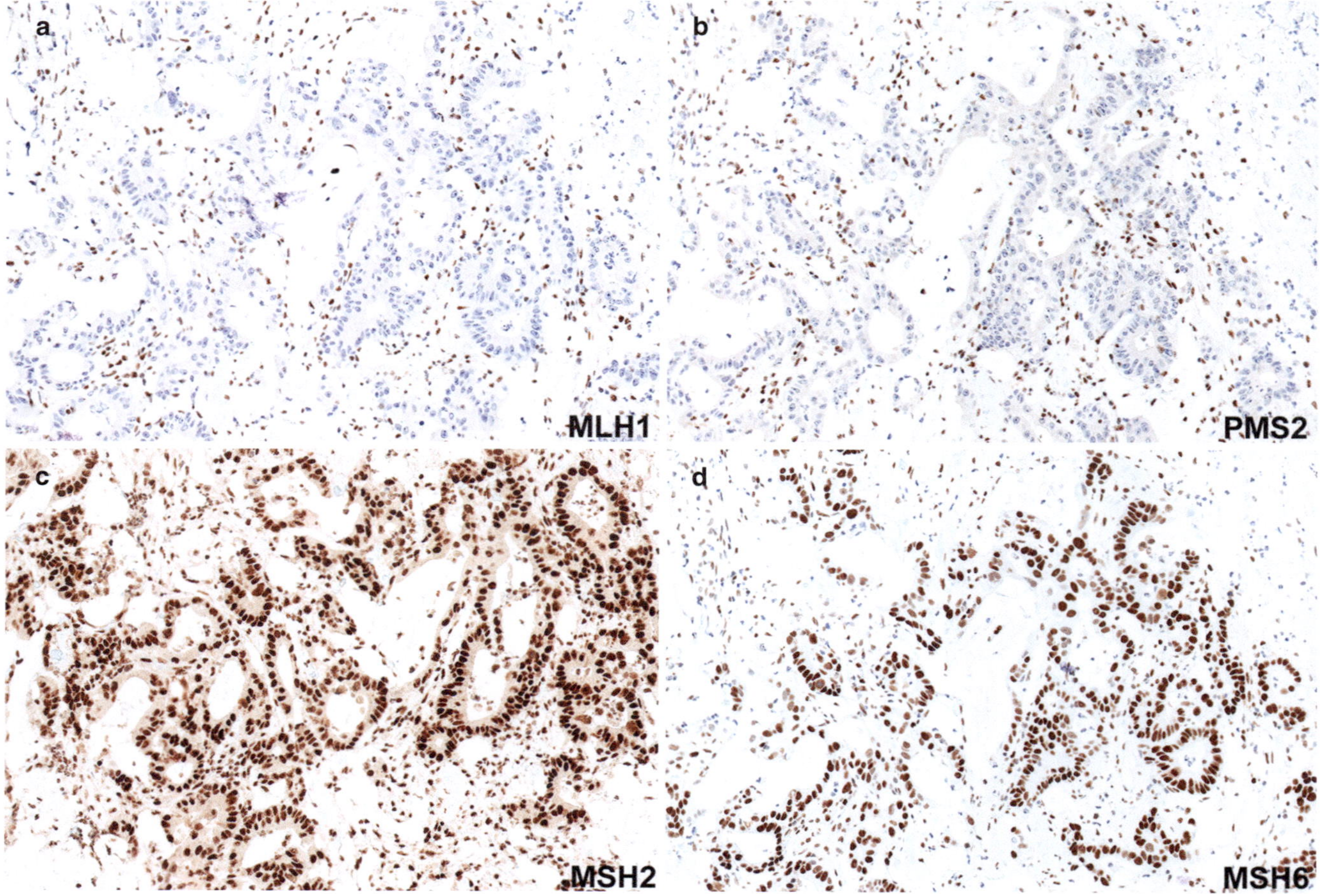

Fig. 5.5 Mismatch repair immunohistochemistry: Most common absent/lost staining pattern. The most common absent/lost staining pattern: Loss of MLH1 (**a**) and PMS2 (**b**) with intact expression of MSH2 (**c**) and MSH6 (**d**). Notice the presence of staining in the background stromal cells and lymphocytes (**a, b**), which serves as a positive internal control. *MLH1* promoter hypermethylation was detected in this case of sigmoid colon cancer

- Although initial studies showed some response in CRC with *KRAS* G13D, these cases may be no more likely to respond to EGFR inhibitors than tumors with other *KRAS* mutations

BRAF Mutation

- *BRAF* gene is mutated in ~10% of CRC
- *BRAF* is downstream of *EGFR* and *RAS*
- Detection methods
  - *BRAF* V600E-specific PCR (may lack sensitivity in some fixatives)
  - Hotspot PCR or sequencing
  - Next-generation sequencing (NGS)
- *BRAF* V600E mutation is mutually exclusive with *KRAS* mutations
- *BRAF* mutation analysis has multiple indications
  - Lynch syndrome workup: The presence of *BRAF* V600E mutation excludes LS
  - Predictive: *BRAF* V600E mutation makes response to anti-EGFR highly unlikely unless given together with a *BRAF* inhibitor
  - Prognostic: *BRAF* mutations in and around codon 600 carry an adverse prognosis

Microsatellite Instability for Therapy Selection

- MSI is present in ~15% of CRC
- LS workup: MSI is one of the screening tests
- Detection methods
  - PCR-based panel of 5–10 well-studied microsatellites (requires nontumor samples for comparison)
  - Real-time PCR-based single nucleotide polymorphisms (SNP) assays (8–10 loci)
  - NGS (typically 50–100 loci)
- Prognostic: MSI confers a good prognosis, if *BRAF* wildtype
- Predictive

Table 5.1 Predictive biomarkers in colorectal cancer

Biomarkers	Which patients to test	Uses	Use caution
Established predictive biomarkers			
KRAS/NRAS (DNA-based NGS, Liquid biopsy)	Stage IV CRC candidates for anti-EGFR therapy	• Predict resistance to anti-EGFR (HER1) therapy	• Do not test *KRAS* only
BRAF V600E mutational analysis (DNA-based NGS, Liquid biopsy)	CRC patients with loss of MLH1/PMS2 by IHC	• Evaluate germline vs. sporadic in MLH1/PMS2 absent tumors	• Do not test on serrated polyp to screen for Lynch syndrome
	Stage IV CRC patients	• Predict poor prognosis if mutated • Predict resistance to anti-EGFR therapy (controversial) • Predict response to *BRAF* V600E inhibitors (investigational)	• Caution with BRAF IHC—difficult to interpret
MMR status (IHC, PCR, NGS)	All primary CRC patients	• Predict response to anti-PD1/PD-L1 therapy (for advanced-stage patients)	
Biomarkers partially established and/or under development			
Cancer immune-related markers (PD1/PD-L1 by IHC; tumor mutation burden by NGS)	Clinical trials only	• Predict response to PD1/PD-L1 inhibitors	
HER2 overexpression/amplification (IHC, FISH, NGS)	Stage IV CRC patients with wildtype *RAS/BRAF*	• Predict resistance to anti-HER2 therapy	• HER2 IHC scoring uses HERACLES diagnostic criteria (Mod Pathol 2015; 28:1481)
NTRK fusions (IHC, FISH, DNA- or RNA-based NGS)	Stage IV CRC patients with wildtype *RAS/BRAF*	• Predict response to *NTRK*-targeted therapy	
PTEN loss (expression by IHC or deletion by FISH)	Clinical trials only	• Predict possible resistance to anti-EGFR therapy	• Do not deny anti-EGFR therapy if loss (insufficient data)
PIK3CA mutations (DNA-based NGS)	Clinical trials only	• Predict possible resistance to anti-EGFR therapy • Predict possible improved survival with aspirin use if mutated	• Do not deny anti-EGFR therapy if mutated (insufficient data)

CRC colorectal cancer, *IHC* immunohistochemistry, *GI* gastrointestinal, *MMR* mismatch repair protein, *NGS* next-generation sequencing

- MSI confers a good response to PD1/PD-L1 inhibitors
- MSI reduces the benefit of fluorouracil-based chemotherapy

Emerging and Partially Established Biomarkers

Cancer Immunotherapy
- PD1/PD-L1 expression level
  - Programmed death ligands 1 and 2 (PD-L1 and PD-L2) on tumor cells can suppress the immune response by binding to programmed cell death protein 1 (PD-1) receptor on T-effector cells
  - Many tumors (including CRC) upregulate PD-L1 (occasionally by PDL1/CD274 gene amplification), which may support immune evasion
  - There is no role of PD-L1 testing yet in CRC, except in clinical trials; eligibility for checkpoint inhibitors for

metastatic dMMR/MSI-H CRC is based on dMMR by IHC or MSI-high by PCR
- High tumor mutation burden (TMB)
  - dMMR (and *POLE*-mutated) CRCs usually contain hundreds to thousands of missense mutations that encode novel proteins ("neoantigens") that may produce anti-tumor immune responses manifested by intense tumor-infiltrating lymphocytes and therefore are often preferentially sensitive to PD-1 inhibitors
  - High TMB may be used to select patients for immunotherapy
  - TMB is calculated by counting the number of likely somatic coding mutations within a proportion of the tumor genome following an NGS study; it typically requires at least one megabase of the coding sequence for accurate study
  - Threshold for a "high" TMB score varies by NGS assay and tumor type, with the strongest association

with therapy response in the 10–15 mutations/megabase

- Given the variable associations with response, NCCN guidelines do not currently recommend TMB biomarker testing for CRC unless specified for trial enrollment

Other Biomarkers with Targeted Therapy Implications

- HER2/*ERBB2* overexpression/amplification
  - Amplification/overexpression in 3–5% of CRC (Fig. 5.6), with prevalence higher in *RAS/BRAF*-wildtype tumors (5–14%) and MMR-proficient and left-sided CRC
  - EGFR family tyrosine kinase receptor driving proliferation through the MAPK pathway (Fig. 5.1)
  - Detection methods: IHC, in situ hybridization (ISH), and NGS copy-number analysis

- *HER2*-targeted therapy often employed as secondary therapy in *HER2*-amplified CRC with no downstream RAS pathway mutations
- HERACLES diagnostic criteria for HER2 immunohistochemistry
  - o Membrane staining of tumor cells is evaluated and graded as follows
  - o Negative: No staining (0); faint (1+) segmental or granular staining with any cellularity; moderate (2+) staining with any pattern in <50% tumor cells; intense (3+) circumferential, basolateral, or lateral staining in ≤10% tumor cells
  - o Equivocal: Moderate (2+) circumferential, basolateral, or lateral staining in ≥50% tumor cells (confirmed by repeat IHC)—send for ISH
  - o Positive: Intense (3+) circumferential, basolateral, or lateral staining in ≥50% tumor cells; or intense (3+) circumferential, basolateral, or lateral staining

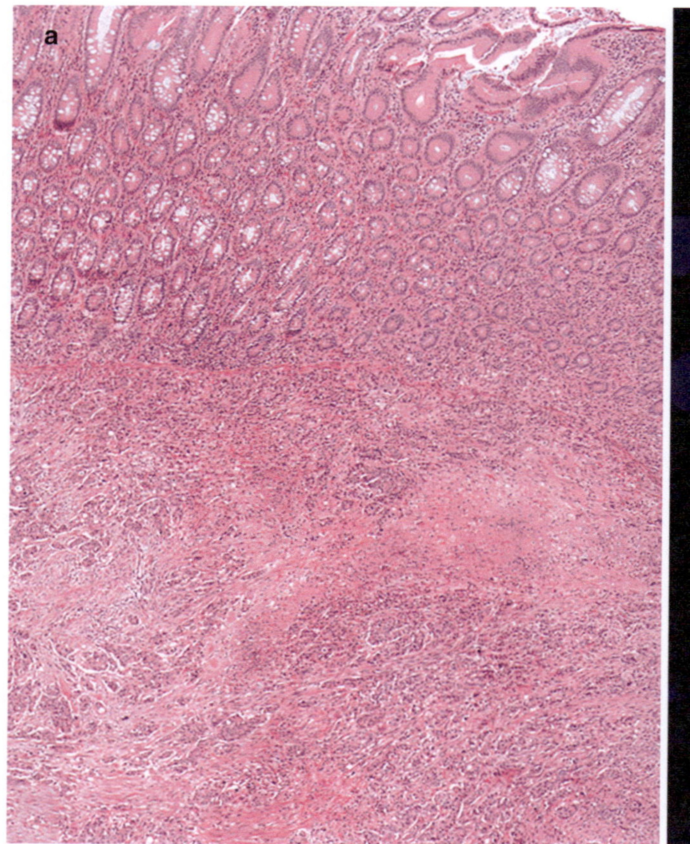
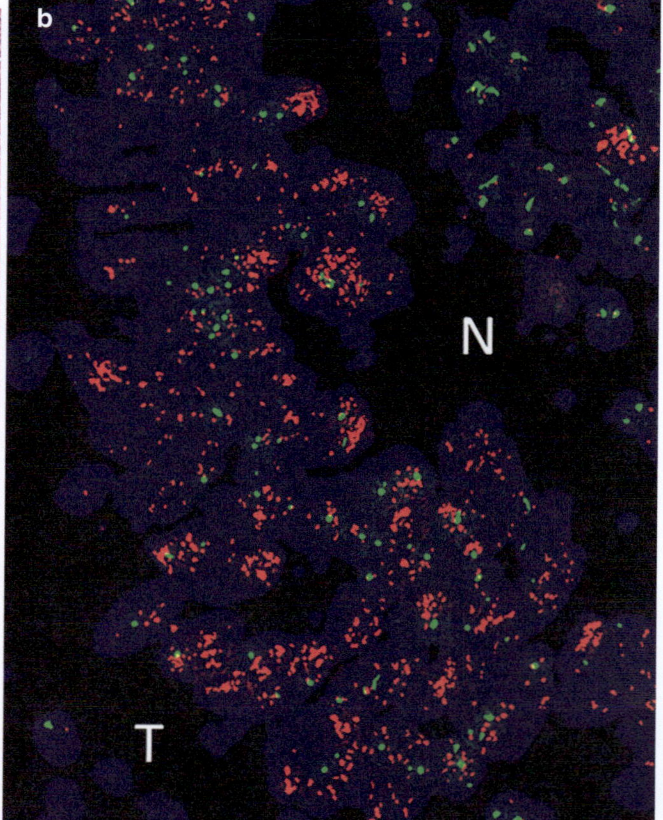

Fig. 5.6 Colorectal cancer with highly amplified *HER2* gene detected by fluorescence in situ hybridization. (**a**) H&E stained section showing poorly-differentiated colorectal adenocarcinoma underlying normal mucosa. (**b**) Fluorescence in situ hybridization showing *HER2* amplification: Green probe is a chromosome normalizer detected against the pericentromeric sequences from chromosome 17. Red probe is against HER2. T, tumor; N, normal

in >10% but <50% of tumor cells (confirmed by repeat IHC)—send for ISH

- NTRK activating gene fusions
  - 0.2–1% of CRC carry targetable *NTRK* gene fusions, usually *NTRK1*
  - Associated with large, high-stage, right-sided CRC with frequent lymphovascular invasion
  - Majority of CRC harboring *NTRK* fusions are MMR-deficient, *MLH1* hypermethylated, with no RAS pathway mutations
  - May have a poor prognosis with conventional therapies
  - Detection methods: IHC, FISH, or NGS
  - Predictive: NTRK inhibitors (larotrectinib and entrectinib) are treatment options for patients with metastatic CRC that is *NTRK* gene fusion-positive
- PTEN Loss
  - Mutated in 5–14% of CRC
  - Phosphatase that downregulates AKT/PIK3CA signaling; biallelic PTEN loss by deletion and mutation is a later-stage event in carcinogenesis
  - Detection methods: IHC (protein loss), deletion by FISH, NGS
  - Loss of PTEN associated with lack of responsiveness to EGFR antagonists
- PIK3CA Mutation
  - Mutated in 10% of CRCs, mainly in exons 9 and 20
  - PIK3CA-AKT-mTOR pathway comodulates proliferation with the EGFR/HER2 MAPK pathway
  - Detection methods: Hotspot mutation sequencing panel, mutation-specific real-time PCR or NGS panel
  - Prognostic: Mutations associated with a worse clinical outcome
  - Predictive
    - Mutations may predict a negative response to anti-EGFR therapy
    - Mutations may predict a positive response with acetylsalicylic acid therapy
- MET Dysregulation
  - Receptor tyrosine kinase can be dysregulated by amplification, activating mutations, alternate splicing (exon 14 exon skipping), and gene fusions
  - Detection methods: Comprehensive NGS panel (to detect all mechanisms)
  - Predictive: Response to MET kinase inhibitors most common in high-level amplification and exon 14 skipping mutations
- Liquid biopsy
  - Methodology: NGS analysis of the patient's peripheral blood to detect low levels of circulating mutations or gene fusions from circulating intact tumor cells, tumor mRNA in exosomes, or tumor nucleic acid fragments (cell-free nucleic acid)
  - Advantages: Minimally invasive, can profile tumor if no biopsy is available
  - Disadvantages: Some CRCs have minimal circulating nucleic acid, may detect mutations associated with other (preinvasive) neoplasms
  - May have utility as a predictive tool for therapy response or disease recurrence, or early detection
- Gene expression signatures (transcriptome)
  - Risk stratification: Several gene expression profiles (such as Oncotype DX test and the ColDx test) have been developed to predict recurrence after surgery by providing a score for recurrence in intermediate-stage CRC

Hereditary Cancer Syndromes

- For a summary and comparison of hereditary polyposis/cancer syndromes, see Table 5.2.

Lynch Syndrome

Definition

- Lynch syndrome is an autosomal dominant disorder characterized by predisposition to a wide variety of cancers, caused by constitutional pathogenic mutations affecting the DNA mismatch repair genes *MLH1*, *MSH2*, *MSH6*, and *PMS2*
- The term "hereditary non-polyposis colorectal cancer (HNPCC)" is no longer recommended as LS patients sometimes develop polyps and are at risk for cancers in addition to those arising in the colorectum
- Prevalence: 1 in 25 unselected CRC cases
- Cancer risk
  - At age 75 years, CRC risk 10–50%
  - Specific gene mutated confers different risk for CRC (*MSH2* and *MLH1* > *MSH6* > *PMS2*)
  - Cancers at other sites have different risk depending on the specific mutated gene: Gastric cancer risk increased with *MLH1* or *MSH2* pathogenic variants; Urothelial cancer risk increased with *MSH2* mutation
  - In LS patients, CRC surveillance is recommended to start at 20–25 years old (or 2–5 years prior to the earliest CRC) with *MLH1* or *MSH2* mutation but at 30–35 with *MSH6* or *PMS2* mutation due to lower risk assessment
- Localization: Depending on the gene involved, cancers occurring in LS can arise in the colorectum, stomach, small bowel, endometrium, gallbladder, hepatobiliary tract, pancreas, renal pelvis, and/or ureter, bladder, kidney, ovary, brain, or prostate

Table 5.2 Summary of hereditary polyposis/cancer syndromes

Syndrome	Molecular alterations	Inheritance pattern	Most common type(s) of polyps	Risk of colorectal cancer
Lynch syndrome	*MLH1, PMS2, MSH2, MSH6*, etc.	AD	Typically, no or few adenomas can have more	10–50% (risk at 75 years)
Constitutional mismatch repair deficiency syndrome	*MLH1, PMS2, MSH2, MSH6*, etc	AR	Adenomas	Unknown
Familial adenomatous polyposis	*APC*	AD	Adenomas (typically, >100)	100% (lifetime risk)
MUTYH-associated polyposis	*MUTYH*	AR	Adenomas (typically, 10–100), occasional serrated polyps	60–70%
Serrated polyposis syndrome	Unknown	Unknown	Serrated lesion/polyps (any subtype)	Unknown
Peutz–Jeghers polyposis syndrome	*STK11 (LKB1)*	AD	Peutz–Jeghers polyps	39%
Juvenile polyposis syndrome	*SMAD4* or *BMPR1A*	AD	Juvenile polyps	68% (risk at 60 years)
Cowden syndrome	*PTEN*	AD	Hamartomatous polyps, juvenile polyps, adenomas, hyperplastic polyps, intramucosal lipoma, among others	9%
Hereditary mixed polyposis syndrome	*GREM1*	AD	Various types (adenomas, hyperplastic polyps, inflammatory polyps, prolapse-type polyps, lymphoid aggregates)	Unknown
NTHL1-associated polyposis	*NTHL1*	AR	Adenomas	Unknown
Polymerase proofreading-associated polyposis	*POLD1, POLE*	AD	Adenomas	30–70% (risk at 65–70 years)
AXIN2-associated polyposis	*AXIN2*	AD	Adenomas	Unknown

Modified from Chapter 6 Table 6.06, *WHO Classification of Tumours of the Digestive System, vol. 1. 5th ed. Lyon: International Agency for Research on Cancer, 2019*

Pathologic Features

- Gross
  - LS-associated CRCs show similar gross features as their sporadic counterpart
- Light microscopy
  - Microsatellite unstable CRC commonly has the following features
    - Tumor-infiltrating lymphocytes
    - Crohn-like peritumoral lymphocytic reaction
    - Poor differentiation/medullary growth pattern
    - Mucinous and signet-ring cell features
  - These histological features are not specific; universal screening of CRC for LS by ancillary testing is recommended in all cases regardless of histology (see section, Universal Screening Algorithms)
- Differential diagnosis
  - Sporadic MSI tumor due to *MLH1* promoter hypermethylation
    - ~15% of non-LS, MSI colon cancers are due to sporadic somatic biallelic hypermethylation of the *MLH1* gene promoter, through the sessile serrated lesion pathway
    - ~85% of these tumors harbor *BRAF* V600E mutations, useful to distinguish them from LS cases
  - Sporadic MSI tumor due to biallelic somatic mutations in the MMR gene

  - Such tumors lack *MLH1* promoter hypermethylation or germline MMR alterations by sequencing
  - Tumor sequencing is necessary to distinguish from LS
  - Other hereditary conditions affecting DNA repair
    - MSI CRC arising in several other syndromes unrelated to LS may acquire somatic MMR mutations resulting in MSI and immunohistochemical abnormalities simulating LS. Patients often have clinical overlap with LS (personal/family histories/specific extraintestinal tumor types). Comprehensive multigene panel sequencing is diagnostic and will reveal the specific affected gene
      - *MUTYH*-associated polyposis (MAP)
        - May have sebaceous skin tumors due to recessive *MUTYH* mutations
      - Polymerase proofreading-associated polyposis (PPAP)
        - Extraintestinal tumors (e.g., endometrial adenocarcinoma) can occur
      - *NTHL1*-associated polyposis (NAP)
        - Sebaceous skin tumors have been reported

Genetic Features

- Primary cause of LS is a constitutional pathogenic mutation affecting an MMR gene (*MLH1, MSH2, MSH6*, or *PMS2*)

- MMR proteins correct single-base mismatches and insertion-deletion loops of short, repeated nucleotide sequences that occur during DNA synthesis
- MLH1/PMS2 and MSH2/MSH6 form two functional pairs. If MLH1 or MSH2 is lost, its partner becomes unstable, but not vice versa
- Some LS have mutations involving adjacent genes that affect an MMR gene, such as mutation in *EPCAM* (*TACSTD1*) affecting *MSH2* or in *LRRFIP2* affecting *MLH1*
- Rare LS have hereditary epigenetic mechanisms that manifest as abnormal DNA methylation affecting *MLH1* or *MSH2*, some of which may be caused by alterations of adjacent genes
- For dMMR to manifest, a somatic mutation or deletion must be acquired in the other allele of the gene with the germline mutation

Lynch Syndrome Variants and Related Syndromes

- Constitutional mismatch repair deficiency syndrome
  - Rare, germline biallelic mutations of MMR gene (manifested by dMMR in nontumor cells)
  - Multiple adenomas at a very young age
  - Early-onset (pediatric) colorectal, hematological, urinary tract, and brain (glioblastoma) cancers and neurofibromatosis type 1–like skin features (cafe au lait spots)
- Digenic LS
  - Mutations in more than one MMR gene
  - Unclear whether more severe than classic LS
- LS due to constitutional hypermethylation of the *MLH1* promoter
  - Usually, sporadic not heritable
  - Some have heritable chromosomal rearrangements that cause *MLH1* promoter methylation, by involving the *LRRFIP2* gene adjacent to *MLH1* on chromosome 3
- *MSH3*-associated polyposis
  - Rare, biallelic recessive inheritance of mutations in *MSH3* causes adenomatous polyposis
  - Elevated microsatellite alterations at selected tetranucleotides: Tumor arising in this setting have MSI at di-, tri-, tetra-, and pentanucleotide repeats but do not have classic mononucleotide repeats
  - Few cases reported include colorectal and duodenal adenomas, CRC, gastric cancer, and early-onset astrocytoma
- Muir–Torre syndrome
  - Concurrence of a sebaceous skin tumor (sebaceous adenoma, sebaceoma, sebaceous carcinoma, or keratoacanthoma) with any internal cancer

- Many patients with LS have such skin tumors and therefore diagnosed with Muir–Torre syndrome, but not vice versa
- Patients with *MUTYH*-associated adenomatous polyposis and *NTHL1*-associated polyposis may have sebaceous skin tumors and mimic LS
- Turcot syndrome
  - Syndrome with coexistence of a hereditary colon cancer syndrome (such as LS or familial adenomatous polyposis) and central nervous system (CNS) tumors
  - Most common primary CNS tumor in LS patients is glioblastoma multiforme, different than familial adenomatous polyposis (FAP) patients (see next section)

Familial Adenomatous Polyposis and Variants

Definition

- An autosomal dominant syndrome caused by pathogenic *APC* mutations, characterized by >100 adenomatous polyps in the colorectum, extracolonic manifestations, and desmoid tumors
- Prevalence: 1 in 8000–10,000
- Cancer risk: If colectomy is not performed, nearly 100% risk of CRC by the age of 45 years
- Localization
  - Colorectum: Hundreds of adenomas developed during adolescence
  - Gastric adenomas: 9–50% FAP patients, including adenomas and fundic gland polyps (low-grade dysplasia frequent)
  - Small bowel adenomas: 50–100% FAP patients; predilection for ampulla, nonampulla duodenum, and proximal jejunum
  - Desmoid tumors: 10% FAP patients, mostly in the small bowel mesentery, abdominal wall, or extremities
  - Less frequent extraintestinal malignancies: Hepatoblastoma and cancers of the thyroid (papillary carcinoma sometimes cribriform morular variant), biliary tree, pancreas, and CNS
  - Frequent benign extraintestinal features: Osteomas, dental abnormalities (supernumerary teeth and odontomas), and congenital hypertrophy of the retinal pigment epithelium

Clinical Features

- Colon
  - Usually >100 (and as many as several thousand) adenomatous polyps
  - Onset of colorectal adenomatous polyps in the second decade of life

- Small bowel
  - Almost all FAP patients develop duodenal adenomas; 4–10% of patients develop duodenal adenocarcinoma
  - Small bowel adenomas and cancer typically present a decade later than colonic counterparts
- Stomach
  - >60% of FAP patients develop gastric polyps—mainly benign fundic gland polyps but also adenomas
  - Gastric adenocarcinoma and proximal polyposis of the stomach is a rare subtype of FAP, featuring severe and predominant fundic gland polyposis without duodenal and colorectal polyposis
- Desmoid tumors
  - Prior surgery and certain types of *APC* mutations increase the risk of desmoid fibromatosis

Pathologic Features

- Gross
  - Resected colon carpeted by numerous polypoid or villous adenomas (Fig. 5.7)
- Light microscopy
  - Colorectal and duodenal polyps
    - Classic adenomas of varying type (tubular, tubulovillous, or villous), grade (low or high), and size are indistinguishable from sporadic adenomas
    - Monocryptal adenomas and oligocryptal adenomas (microadenomas) in otherwise normal-looking mucosa are often seen and characteristic of FAP

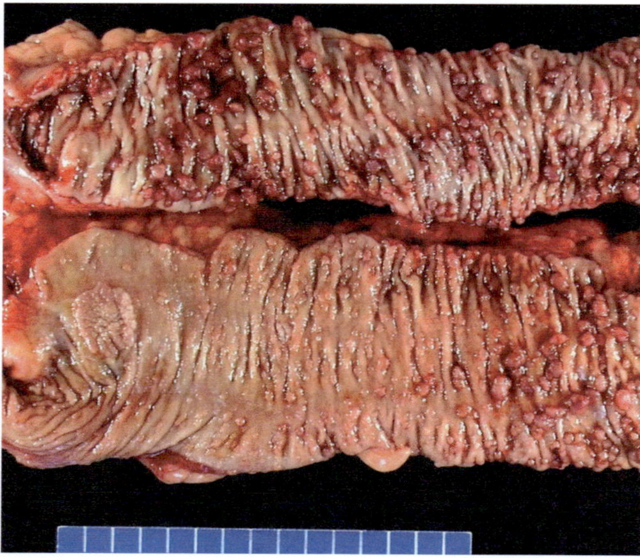

Fig. 5.7 Gross photo of colectomy specimen with familial adenomatous polyposis. Total proctocolectomy specimen from a 31-year-old patient with familial adenomatous polyposis. The entire colorectal mucosa is carpeted by adenomatous polyps of varying sizes (hundreds of polyps)

- Gastric polyps
  - 80% are fundic gland polyps, often multiple; 40% show low-grade dysplasia, but high-grade dysplasia and carcinoma rare
  - 20% are adenomas: Foveolar-type adenomas (17%), pyloric gland adenomas (3%), rarely intestinal-type adenomas
- Desmoid tumors
  - Bland, indolent, invasive fibroblastic tumors
- Differential diagnosis
  - Gold standard for FAP diagnosis is the presence of a pathogenic germline *APC* mutation and the presence of >100 colorectal adenomas. Other polyposis conditions may mimic FAP and variants, although the number of polyps is usually lower, with different pattern of inheritance and extracolonic features
  - Other colonic polyposis conditions
    - *MUTYH*-associated polyposis
    - Polymerase proofreading-associated polyposis
    - *NTHL1*-associated polyposis
    - Hereditary mixed polyposis syndrome
    - Constitutional mismatch repair deficiency syndrome
    - Multiple polyps associated with mutations in other genes (e.g., *MSH3*, *BUB1*, *AXIN2*, and *FAN1*)
    - Hamartomatous polyposis

Genetic Features

- Caused by either inheritance of a mutated *APC* gene or a new germline mutation in the same gene (up to one-third of cases are de novo)
- *APC* is a tumor suppressor gene. Loss of function (especially its binding of the AXIN and β-catenin proteins) results in impaired degradation of β-catenin with upregulation of WNT signaling
- Adenomas arise owing to somatically acquired second hit to the nonmutant *APC* gene
- Severity of disease varies with the position of the mutation in the *APC* gene
- Mutations located in or around the mutation cluster region (around codon/amino acid 1309) are associated with the highest number of adenomas (thousands—severe polyposis)
- Mutations outside this region are associated with many hundreds of adenomas
- Attenuated FAPs (<100 adenomas) harbor mutations nearer the N-terminus or within the alternatively spliced region of exon 9
- Desmoid tumors are associated with germline mutations involving codons 1310–2011 in the mid- to C-terminal portion of APC protein

FAP Variants

- Attenuated FAP
  - A variant of FAP with fewer than 100 colonic polyps (usually <30)
  - *APC* mutations that occur at the most 5′ or 3′ aspect of the gene
  - Slightly reduced risk (lifetime risk of 80%) and later onset of CRC (mean age of 56 years)
- Gardner syndrome
  - A variant of FAP characterized by extraintestinal manifestations, including osteomas, dental anomalies, desmoid tumor, epidermoid cysts, and congenital hypertrophy of the retinal pigmented epithelium
  - Mutations in *APC* causing Gardner syndrome are clustered in a region (codons 1395–1493) encoding a series of amino acid repeats responsible for the binding to β-catenin
  - Nuclear expression of β-catenin by IHC
- Turcot syndrome
  - A syndrome with the coexistence of a hereditary colon cancer syndrome (such as FAP or LS) and CNS tumors
  - CNS tumors in FAP patients are medulloblastoma > astrocytomas and ependymomas, different than LS (see prior section)

Mut Y Homolog-Associated Polyposis Syndrome

Definition

- Mut Y homolog (MUTYH)-associated polyposis (MAP) syndrome is an autosomal recessive, constitutional DNA repair disorder of base excision repair
- Prevalence: 1 in 2000
- Patient with pathogenic *MUTYH* mutations have a predilection for acquiring somatic *APC* mutations
- Adenomatosis resembles attenuated FAP phenotype as the burden of adenomas typically 10–100 (range from zero to several hundreds)
- Cancer risk: Lifetime CRC risk ~80%; increased risk of several extraintestinal neoplasm
- Localization: Colorectal and duodenal (20% of cases) polyposis; gastric adenomas and fundic gland polyps; sebaceous skin tumors

Pathologic Features

- Differential diagnosis
  - Attenuated FAP
    - Serrated polyps common in MAP but not in attenuated FAP
  - LS

Genetics Features

- Biallelic mutations in the *MUTYH* gene, located on chromosome 1p
- *MUTYH* encodes a protein in the DNA base excision repair pathway
- Monoallelic *MUTYH* mutation also confers increased cancer risks for colorectal, breast, gastric, endometrial, and liver cancers; the risk of CRC for carriers of monoallelic mutations in *MUTYH* with a first-degree relative with CRC is sufficiently high to warrant more intensive screening than for the general population

Serrated Polyposis Syndrome

Definition

- Multiple serrated polyps in the large intestine with an increased risk of CRC
- Localization: Large intestine, not the upper GI tract, small intestine, or extracolonic
- Cancer risk
  - Differs depending on the phenotype, first clinical presentation, and polyp histology
  - CRC diagnosed in 16% to 29% of patients, most before or at the time of serrated polyposis diagnosis

Clinical Features

- Clinical criteria for serrated polyposis (WHO 2019)
  - Criterion 1: ≥ 5 serrated lesions/polyps proximal to the rectum, all being ≥5 mm in size, with ≥2 being ≥10 mm in size
  - Criterion 2: > 20 serrated lesions/polyps of any size distributed throughout the large bowel, with ≥5 being proximal to the rectum
  - Any histological subtype of serrated lesion/polyp (hyperplastic polyp, sessile serrated lesion without or with dysplasia, traditional serrated adenoma, and unclassified serrated adenoma) is included in the final polyp count
  - The polyp count is cumulative over multiple colonoscopies
- 25% of patients fulfill only clinical criterion 1; 45% fulfill only clinical criterion 2; 30% have both phenotypes
- Ranging from patients that barely meet the clinical definition to patients with high polyp burden and multiple large polyps fulfilling both criteria
- Median cumulative polyp number 30–40, range 6–240 polyps
- Average age at diagnosis 50–60 years of age
- Serrated polyps less likely to bleed than conventional adenomas (fecal blood tests do not perform well)

Pathologic Features

- Gross
  - Similar to sporadic lesions
- Light microscopy
  - Any histological subtype of serrated lesion/polyp: Hyperplastic polyp, sessile serrated lesion without or with dysplasia, traditional serrated adenoma, and unclassified serrated adenoma
  - Once cytologic dysplasia develops, sessile serrated lesions may progress relatively rapidly into cancer
- Differential diagnosis
  - PTEN hamartoma tumor syndrome/Cowden syndrome and juvenile polyposis may, in rare cases, fulfill the criteria of serrated polyposis, and they almost always have other polyp types or extraintestinal features

Genetic Features

- No high-penetrance candidate genes have yet been identified
- Pathogenic germline variants of *RNF43* in 2% of patients with serrated polyposis
- May be a component of MUTYH-associated polyposis or hereditary mixed polyposis syndrome
- Similar to sporadic serrated polyps, *BRAF* mutation is found in 73%, and *KRAS* mutations are found in 8% of serrated polyps in the setting of serrated polyposis
- CRC can arise via the serrated pathway, while others likely through the conventional adenoma-carcinoma pathway

Peutz–Jeghers Polyposis Syndrome

Definition

- Peutz–Jeghers polyposis syndrome (PJS) is an autosomal dominant polyp and cancer syndrome characterized by mucocutaneous melanin pigmentation and GI hamartomatous polyposis (polyps typically number in the tens)
- Prevalence: 1 in 50,000 to 200,000 births
- Localization
  - Approximately 95% PJS patients have polyps in the small intestine
  - 25% have polyps in the colon and stomach
- Cancer risk: Overall risk of developing any cancer by the age of 70 years is 81% (gastrointestinal, breast, pancreas, reproductive organs)

Clinical Features

- Mucocutaneous pigmentation facilitates the diagnosis of asymptomatic patients in familial cases
- Abdominal pain, intestinal bleeding, anemia, and intussusception (first two decades of life)

Pathologic Features

- Gross
  - Sessile or pedunculated with thick stalk, smooth lobulated surface
- Light microscopy
  - Small bowel polyps (Fig. 5.8a, b): Arborizing bands of smooth muscle divide hyperplastic epithelium into lobules; normal lamina propria (no increased inflammation); surface commonly eroded with regenerative changes
  - Gastric or colonic polyps (Fig. 5.8c, d): Similar but less well-developed features
- Differential diagnosis
  - Sporadic hamartomatous polyp
  - Other hamartomatous polyposis syndrome (Juvenile or Cowden)
  - Filiform polyp of inflammatory bowel disease (background colitis)
  - Mucosal prolapse (stromal fibromuscular change)
  - Misplaced dysplastic epithelium in Peutz–Jeghers polyp may simulate invasive adenocarcinoma
  - Gastric Peutz–Jeghers polyps lack specific histology, difficult to distinguish from gastric juvenile polyps or sporadic hyperplastic polyps
- Diagnostic criteria for PJS
  - WHO 2019
    - $\geq$3 histologically confirmed Peutz–Jeghers polyps
    - Any number of Peutz–Jeghers polyps with a family history of PJS
    - Characteristic, prominent mucocutaneous pigmentation with a family history of PJS
    - Any number of Peutz–Jeghers polyps and characteristic, prominent mucocutaneous pigmentation
  - NCCN v1.2023
    - Two or more of the following
    - $\geq$ 2 GI Peutz–Jeghers polyps
    - Mucocutaneous hyperpigmentation of the mouth, lips, nose, eyes, genitalia, or fingers
    - Family history PJS

Genetic Features

- A germline mutation in tumor suppressor gene *STK11* (formerly called *LKB*1, chromosome 19p13.3), found in >90% PJS patients
- Disease phenotype more severe in patients with truncating mutations compared to those with missense mutations

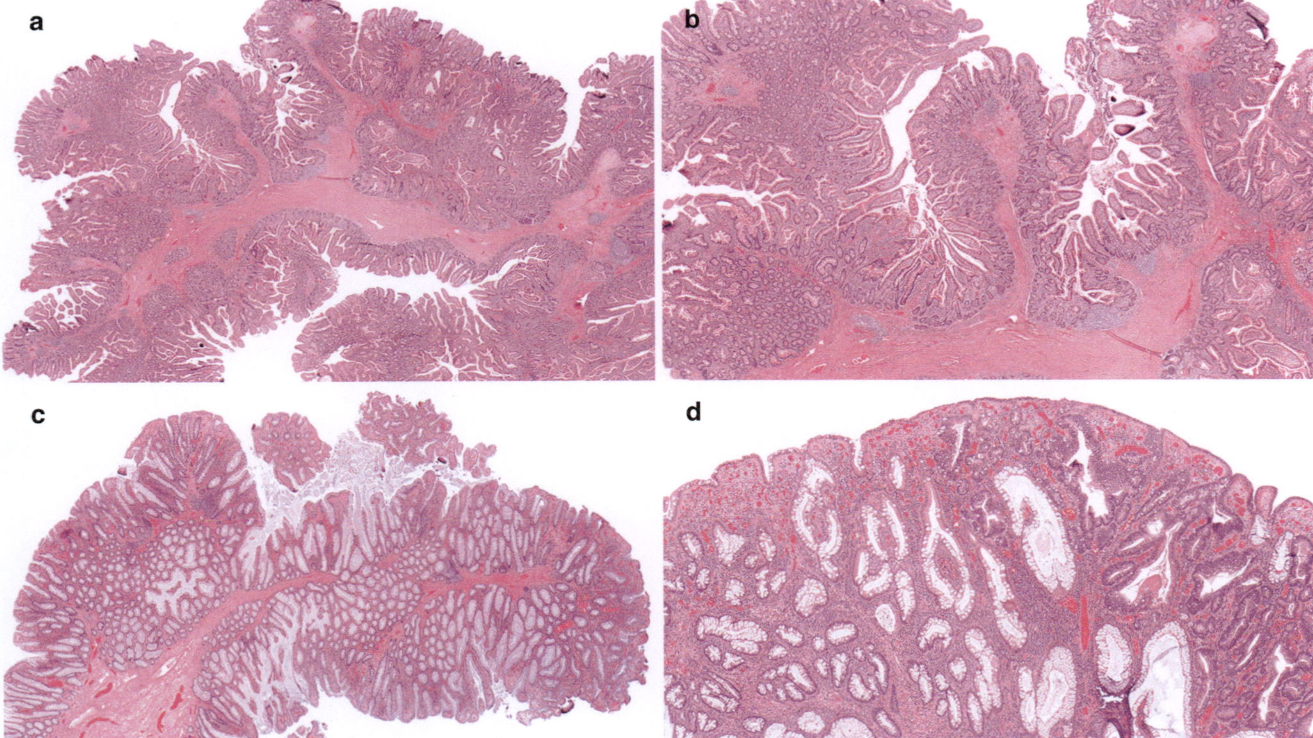

Fig. 5.8 Peutz–Jeghers polyposis syndrome: Peutz–Jeghers polyps in small bowel and colon. 50-year-old male with Peutz–Jeghers polyposis syndrome. Peutz–Jeghers polyps feature arborizing bands of smooth muscle supporting hyperplastic epithelium, which is typically better developed in the small bowel (**a, b**) than colon (**c**). Focal low-grade dysplasia (**d**, *right*) is present in the colonic Peutz–Jeghers polyp

Juvenile Polyposis Syndrome

Definition
- Autosomal dominant polyp and cancer syndrome characterized by multiple hamartomatous polyps (resembling inflammatory polyps) of the GI tract
- Prevalence: 1 in 100,000 persons
- Most common GI hamartomatous polyposis syndrome
- Localization
  - Polyps predominantly in colon (ranging in number from 1 to >100)
  - Stomach (85% of patients)
  - Small intestine (14–33% of patients)
- Cancer risks
  - 34-fold increased relative risk of CRC; risk at 60 years is 68%
  - Gastric cancer lifetime risk 10% to 30%

Clinical Features
- Gastrointestinal bleeding, prolapsed rectal polyp, the passage of tissue per rectum (autoamputated polyp), intussusception
- Extensive extracolonic polyposis in the small bowel or stomach associated with GI intussusception or systemic dysfunction (protein-losing enteropathy and malabsorption)
- Juvenile polyposis of infancy is a subtype with severe symptoms
- Congenital abnormalities and extraintestinal manifestations frequent in the nonfamilial form of the disease

Pathologic Features
- Gross
  - Pedunculated (>75%), the external surface usually smooth and erythematous due to erosion; mulberry-like atypical juvenile polyps only occur in syndromic patients
- Light microscopy
  - Stomach (Fig. 5.9a, b)
    - Irregular hyperplastic glands lined by foveolar epithelium and abundant edematous stroma
    - Dysplasia in 15% of these polyps (can show intestinal or pyloric gland differentiation)
  - Colon (Fig. 5.9c, d)
    - Eroded surface, abundant edematous stroma with inflammatory cells, cystically dilated crypts with reactive epithelium
    - Dysplasia often seen in atypical or multilobulated juvenile polyps

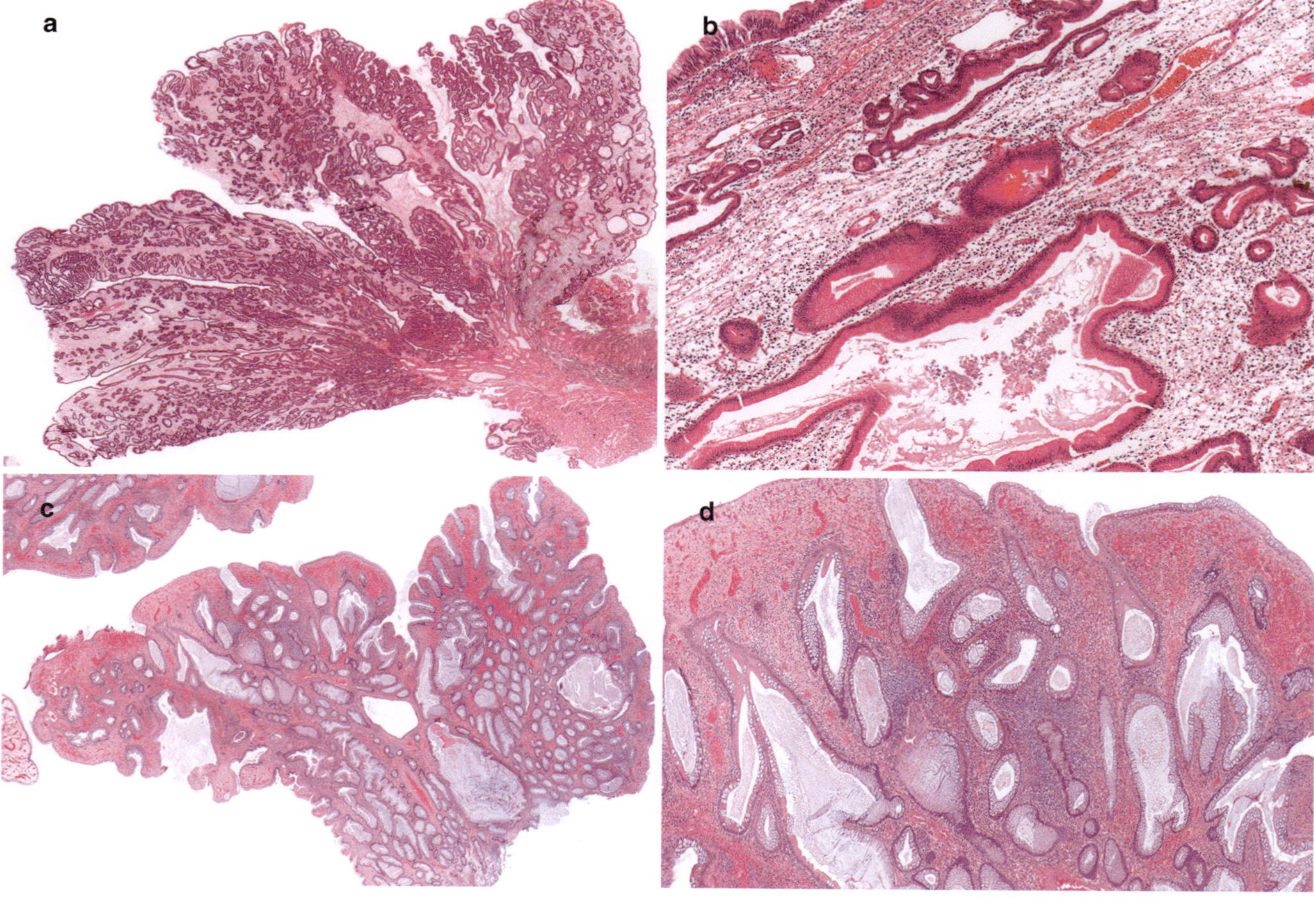

Fig. 5.9 Juvenile polyposis syndrome: Juvenile polyps in stomach and colon. (**a, b**) Gastrectomy specimen from a 39-year-old female with juvenile polyposis syndrome. This large juvenile polyp features irregular hyperplastic glands lined by foveolar epithelium, cystically dilated glands, and abundant edematous stroma. (**c, d**) Colonic juvenile polyp from a 44-year-old female with juvenile polyposis syndrome. The polyp shows focal eroded surface, abundant edematous stroma with inflammatory cells, and cystically dilated crypts with reactive epithelium

- Differential diagnosis
  - Sporadic juvenile polyps
    - Most common in rectosigmoid (54%)
    - Some have polyps proximal to splenic flexure (37%)
    - Increased crypt-to-stromal ratio may be helpful features for hereditary juvenile polyps with *SMAD4* (but not *BMPR1A*) mutation
  - Other hamartomatous polyposis syndrome (Peutz–Jeghers or Cowden)
  - Inflammatory pseudopolyp (background colitis)
  - Solitary rectal ulcer syndrome (stromal fibromuscular change)
  - Difficult to reliably distinguish between gastric hyperplastic, juvenile, and Peutz–Jeghers polyps based on histologic features alone
- Diagnostic criteria for juvenile polyposis syndrome (JPS) (WHO 2019)
  - >3–5 juvenile polyps of the colorectum
  - Juvenile polyps throughout the GI tract
  - Any number of juvenile polyps with a family history of JPS
  - Other syndromes involving hamartomatous GI polyps should be ruled out clinically or by pathological examination

Genetics

- A germline mutation in *SMAD4* (a central mediator of the TGFβ signaling) on chromosome 18q21.1 (~30% patients) or *BMPR1A* on chromosome 10q22.3 (~20% patients)
- *SMAD4* mutation is associated with more severe gastric polyposis
- *BMPR1A* (bone morphogenetic protein receptor 1A) mutation is more likely to also have cardiac defects

- <50% of patients with JPS have a family history of the disease

Cowden Syndrome

Definition

- A rare autosomal dominant disorder featuring multiple hamartomas involving organs derived from any of the three germ layers, with cancer predisposition
- Belongs to PTEN hamartoma tumor syndrome (PTHS), which also includes two other rare entities, Bannayan–Riley–Ruvalcaba syndrome and Proteus syndrome
- Prevalence: 1 in 200,000–250,000 in European populations
- Localization: Throughout GI tract; diffuse esophageal glycogenic acanthosis in >80% of Cowden syndrome patients
- Cancer risks: Breast, thyroid, endometrial, renal cell, and colon cancers (low risk, 9%); melanoma; and other cancers

Clinical Features

- Gastrointestinal polyps present in virtually all patients with Cowden syndrome
- Diffuse esophageal glycogenic acanthosis in >80% patients
- Mucocutaneous lesions (multiple facial trichilemmomas, acral keratoses, papillomatous papules, and diffuse esophageal glycogenic acanthosis are considered pathognomonic)
- Macrocephaly

Pathologic Features

- Gross
  - Similar to juvenile polyps
- Light microscopy
  - May show a mixture of histology in colon and duodenum: Ganglioneuromas (Fig. 5.10a), hamartomatous polyps (Fig. 5.10b), juvenile polyps (Fig. 5.10c, d), colonic intramucosal lipomas (Fig. 5.10e, f), adenomas, hyperplastic polyps
  - Most patients have numerous gastric polyps: Hyperplastic or hamartomatous polyps cannot be reliably distinguished; typically, without dysplasia
  - Diffuse esophageal glycogenic acanthosis in combination with colonic polyposis may be diagnostic of Cowden syndrome
- Differential diagnosis

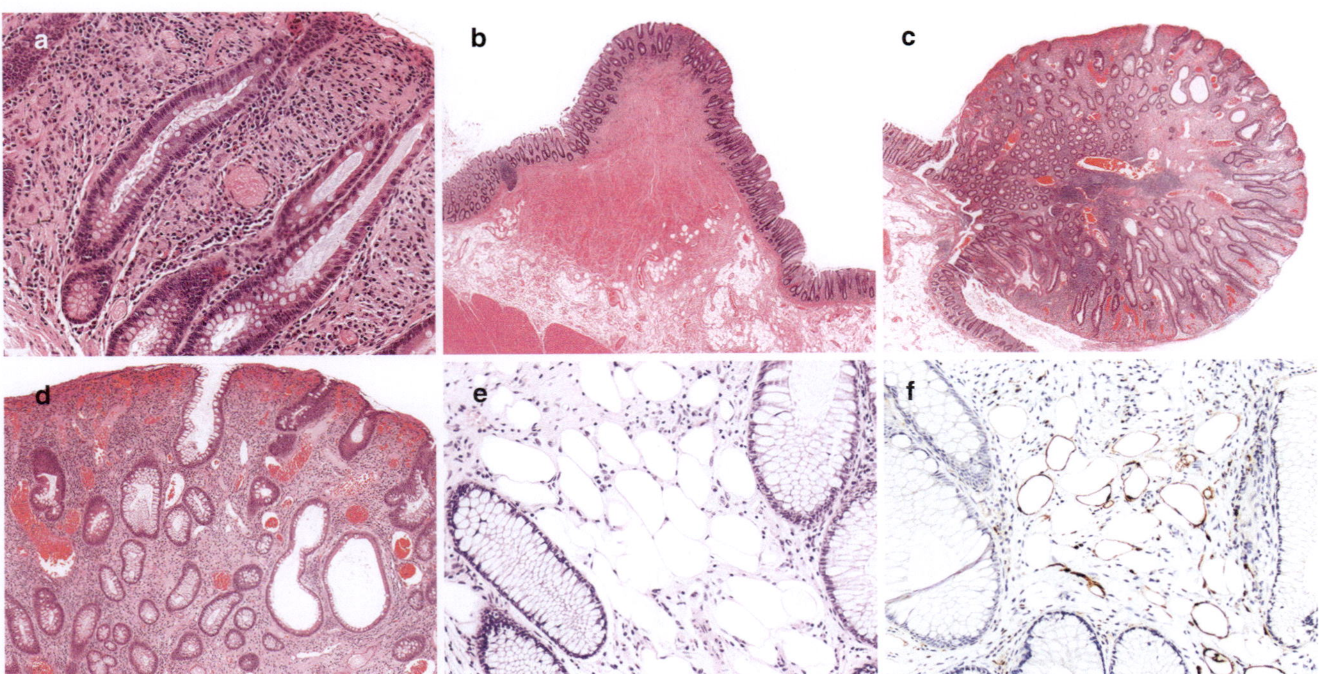

Fig. 5.10 Cowden syndrome: Colonic ganglioneuroma, hamartomatous polyp, juvenile polyp, and intramucosal lipoma. 30-year-old female with Cowden syndrome. Colectomy specimen demonstrated ganglioneuromas (**a**), hamartomatous polyps (**b**), juvenile polyps (**c, d**), and intramucosal lipomas (**e, f**). S100 immunostain (**f**) shows membranous and focal nuclear staining in the vacuolated cells, confirming adipocytes (not pseudolipomatosis)

- Juvenile polyposis syndrome (molecular testing; Juvenile polyposis has no associated mucocutaneous, breast, or thyroid lesions)
- Cronkhite-Canada syndrome (abnormal intervening nonpolypoid mucosa with atrophy)—Sporadic, multiple extraintestinal manifestations
- Inflammatory bowel disease associated pseudopolyps (abnormal intervening nonpolypoid mucosa)

Genetics Features

- Germline mutation of the tumor suppressor phosphatase and tensin homolog (*PTEN*) gene, chromosome 10q23
- Inactivation of the second copy of the gene allows deregulation of the AKT pathway

Other Polyposis Syndromes That Can Include Multiple Adenomas

Hereditary Mixed Polyposis Syndrome

Definition

- Hereditary mixed polyposis syndrome (HMPS) is an autosomal dominant polyposis syndrome characterized by a variety of colorectal polyps (adenomas, hyperplastic polyps, inflammatory polyps, prolapse-type polyps, lymphoid aggregates) with a high-risk of CRC
- HMPS shows consistent phenotypic overlap with JPS

Genetic Features

- Caused by a 40 kb upstream duplication that leads to increased and ectopic expression of the BMP antagonist *GREM1*
- A germline *BMPR1A* mutation has been identified in Chinese HMPS families
- Linkage between HMPS and chromosome 15q has also been reported

NTHL1-Associated Polyposis

Definition

- NTHL1-associated polyposis (NAP) is an autosomal recessive, constitutional DNA repair disorder of base excision repair, caused by *NTHL1* mutation
- Characterized by adenomatous polyps of the large bowel arising in adulthood, generally by the age of 50 years old
- Prevalence unknown, rarer than MAP
- Increased cancer risks: CRC, breast cancer, endometrial cancer, and skin sebaceous tumors
- Adenomas and adenocarcinomas in NAP are similar to sporadic counterparts, but they have a characteristic C → T somatic mutation spectrum

Polymerase Proofreading-Associated Polyposis

Definition

- Polymerase proofreading-associated polyposis (PPAP) is an autosomal dominant adenomatous polyposis syndrome due to mutations in the proofreading (exonuclease) domains of *POLE* and *POLD1*
- In contrast to other constitutional DNA repair disorders (such as MAP and NAP), PPAP is not recessively inherited
- Prevalence unknown, rarer than MAP
- Cancer risk: 30–70% risk at 65–70 years old
- Localization
  - Colorectum and duodenum, adenomatous polyps, and adenocarcinoma
  - Extraintestinal tumors (e.g., endometrial adenocarcinoma)

Pathologic Features

- Gross and light microscopy
  - Similar to sporadic counterparts, but with hypermutant somatic mutations
- CRCs associated with PPAP are rich in neoantigens, good targets for PD1/PDL1 immune checkpoint inhibitor immunotherapy

AXIN2-Associated Polyposis Syndrome

Definition

- AXIN2-associated polyposis syndrome is an autosomal dominant adenomatous polyposis syndrome caused by pathogenic variant in *AXIN2*
- Increased risk for CRC (exact risk unknown)
- *AXIN2* is functionally related to APC, as it is a regulator of β-catenin degradation in the WNT signaling pathway
- Localization: Colorectal and gastric polyps
- Ectodermal dysplasia (including oligodontia) is seen in some cases

Current Challenges and Emerging Trends

- The indications for and interest in expanded upfront tumor sequencing with NGS are continuing to increase in CRC
  - NGS panels that include assessment of somatic and germline mutations in the syndromic genes are most useful in CRC patients under the age of 50 or those with high suspicion for hereditary cancer syndromes
  - An NGS approach for universal LS tumor screening may eventually supplant IHC and/or PCR-MSI testing, as a recent study showed that 38.6% of those with a

hereditary cancer syndrome including 6.3% of those with LS can be missed with standard assays

- Given the limitations of TMB, improved NGS-based biomarkers for immune checkpoint inhibitor therapy are being tested that include a broader assessment of the multiple DNA repair pathways besides POLE and MMR, including base excision repair, nucleotide excision repair, homologous recombination repair, full profiling of the BRCA/Fanconi complex, nonhomologous end-joining, checkpoint regulators, and mitotic factors
- While NGS reimbursement remains a challenge, the advancement of technology and ever-decreasing cost make it promising for more widely available large panel NGS that covers all the indications above

• As combination therapies continue to evolve, the need for predictive models that combine different testing methodologies is increasing, including

- Enhanced tissue-based diagnostic used digital characterization of tumor microenvironment and multispectral immune profiling
- New classes of epigenetic markers including microRNA and genomewide methylation

• Leveraging advances in artificial intelligence and statistical methods to create more complex prognostic models that combine pathology, radiology, and molecular predictors

• Population genetic screening with a multigene panel testing could efficiently identify at-risk carriers. Germline sequencing for all patients with CRC may be cost effective and possible in the future

Further Reading

American Cancer Society. American Cancer Society: colorectal cancer facts & figs. 2020–2022. https://www.cancer.org/content/dam/cancer-org/research/cancer-facts-and-statistics/colorectal-cancer-facts-and-figures/colorectal-cancer-facts-and-figures-2020-2022.pdf. Accessed 8 Aug 2021.

Amin MB, Edge S, Greene F, et al., editors. AJCC cancer staging manual. 8th ed. New York: Springer International Publishing; 2017.

Arvai KJ, Hsu YH, Lee LA, et al. A transition zone showing highly discontinuous or alternating levels of stem cell and proliferation markers characterizes the development of PTEN-Haploinsufficient colorectal cancer. PLoS One. 2015;10(6):e0131108.

Bartley AN, Hamilton SR, Alsabeh R, et al. Template for reporting results of biomarker testing of specimens from patients with carcinoma of the colon and rectum. Arch Pathol Lab Med. 2014;138(2):166–70.

Brouwer NPM, Bos ACRK, Lemmens VEPP, et al. An overview of 25 years of incidence, treatment and outcome of colorectal cancer patients. Int J Cancer. 2018;143(11):2758–66.

Chen W, Frankel WL. A practical guide to biomarkers for the evaluation of colorectal cancer. Mod Pathol. 2019;32(Suppl 1):1–15.

Chen W, Hampel H, Pearlman R, et al. Unexpected expression of mismatch repair protein is more commonly seen with pathogenic missense than with other mutations in lynch syndrome. Hum Pathol. 2020;103:34–41.

Chen W, Swanson BJ, Frankel WL. Molecular genetics of microsatellite-unstable colorectal cancer for pathologists. Diagn Pathol. 2017;12(1):24.

Gray PN, Tsai P, Chen D, et al. TumorNext-lynch-MMR: a comprehensive next generation sequencing assay for the detection of germline and somatic mutations in genes associated with mismatch repair deficiency and lynch syndrome. Oncotarget. 2018;9(29):20304–22.

Hampel H, Frankel WL, Martin E, et al. Screening for the lynch syndrome (hereditary nonpolyposis colorectal cancer). N Engl J Med. 2005;352(18):1851–60.

Hampel H, Pearlman R, Beightol M, et al. Assessment of tumor sequencing as a replacement for lynch syndrome screening and current molecular tests for patients with colorectal cancer. JAMA Oncol. 2018;4(6):806–13.

Haraldsdottir S, Roth R, Pearlman R, Hampel H, Arnold CA, Frankel WL. Mismatch repair deficiency concordance between primary colorectal cancer and corresponding metastasis. Familial Cancer. 2016;15(2):253–60.

Lasota J, Chlopek M, Wasag B, et al. Colorectal adenocarcinomas harboring ALK fusion genes: a clinicopathologic and molecular genetic study of 12 cases and review of the literature. Am J Surg Pathol. 2020;44(9):1224–34.

Markow M, Chen W, Frankel WL. Immunohistochemical pitfalls: common mistakes in the evaluation of lynch syndrome. Surg Pathol Clin. 2017;10(4):977–1007.

Mazzoni SM, Fearon ER. AXIN1 and AXIN2 variants in gastrointestinal cancers. Cancer Lett. 2014;355(1):1–8.

National Comprehensive Cancer Network. National Comprehensive Cancer Network clinical practice guidelines in oncology (NCCN guidelines): genetic/familial high-risk assessment: colorectal (version 1.2023 - May 30, 2023). https://www.nccn.org/professionals/physician_gls/pdf/genetics_colon.pdf. Accessed July 12, 2023

National Comprehensive Cancer Network. National Comprehensive Cancer Network clinical practice guidelines in oncology (NCCN Guidelines): colorectal cancer screening (version 1.2023-May 17, 2023). https://www.nccn.org/professionals/physician_gls/pdf/colorectal_screening.pdf. Accessed July 12, 2023.

Odze RD, Goldblum JR. Surgical pathology of the GI tract, liver, biliary tract, and pancreas. 3rd ed. Philadelphia: Saunders; 2014.

Pai RK, Pai RK. A practical approach to the evaluation of gastrointestinal tract carcinomas for lynch syndrome. Am J Surg Pathol. 2016;40(4):e17–34.

Pearlman R, Frankel WL, Swanson BJ, et al. Prospective statewide study of universal screening for hereditary colorectal cancer: the Ohio colorectal cancer prevention initiative. JCO Precis Oncol. 2021;5:PO.20.00525.

Pearlman R, Frankel WL, Swanson B, et al. Prevalence and Spectrum of Germline cancer susceptibility gene mutations among patients with early-onset colorectal cancer. JAMA Oncol. 2017;3(4):464–71.

Roth RM, Haraldsdottir S, Hampel H, Arnold CA, Frankel WL. Discordant mismatch repair protein immunoreactivity in lynch syndrome-associated neoplasms: a recommendation for screening synchronous/metachronous neoplasms. Am J Clin Pathol. 2016;146(1):50–6.

Sartore-Bianchi A, Trusolino L, Martino C, et al. Dual-targeted therapy with trastuzumab and lapatinib in treatment-refractory, KRAS codon 12/13 wildtype, HER2-positive metastatic colorectal cancer (HERACLES): a proof-of-concept, multicentre, open-label, phase 2 trial. Lancet Oncol. 2016;17(6):738–46.

Sepulveda AR, Hamilton SR, Allegra CJ, et al. Molecular biomarkers for the evaluation of colorectal cancer: guideline from the American Society for Clinical Pathology, College of American Pathologists,

Association for Molecular Pathology, and American Society of Clinical Oncology. J Mol Diagn. 2017;19(2):187–225.

Song Y, Huang J, Liang D, et al. DNA damage repair gene mutations are indicative of a favorable prognosis in colorectal cancer treated with immune checkpoint inhibitors. Front Oncol. 2020;10:549777.

Valtorta EC, Martino A, Sartore-Bianchi F, et al. Assessment of a HER2 scoring system for colorectal cancer: results from a validation study. Mod Pathol. 2015;28(11):1481–91.

van Hattem WA, Langeveld D, de Leng WW, et al. Histologic variations in juvenile polyp phenotype correlate with genetic defect underlying juvenile polyposis. Am J Surg Pathol. 2011;35(4):530–6.

Vyas M, Firat C, Hechtman JF, et al. Discordant DNA mismatch repair protein status between synchronous or metachronous gastrointestinal carcinomas: frequency, patterns, and molecular etiologies. Familial Cancer. 2021;20(3):201–13.

Weren RD, Ligtenberg MJ, Geurts van Kessel A, et al. NTHL1 and MUTYH polyposis syndromes: two sides of the same coin? J Pathol. 2018;244(2):135–42.

WHO. WHO classification of tumours of the digestive system, vol. 1. 5th ed. Lyon: International Agency for Research on Cancer; 2019.

Win AK, Cleary SP, Dowty JG, Baron JA, Young JP, Buchanan DD, Southey MC, Burnett T, Parfrey PS, Green RC, Marchand LL. Cancer risks for monoallelic MUTYH mutation carriers with a family history of colorectal cancer. Int J Cancer. 2011;129(9):2256–62.

Win AK, Dowty JG, Cleary SP, Kim H, Buchanan DD, Young JP, Clendenning M, Rosty C, MacInnis RJ, Giles GG, Boussioutas A. Risk of colorectal cancer for carriers of mutations in MUTYH, with and without a family history of cancer. Gastroenterology. 2014;146(5):1208–11.

Wong NA, Gonzalez D, Salto-Tellez M, et al. RAS testing of colorectal carcinoma-a guidance document from the Association of Clinical Pathologists Molecular Pathology and Diagnostics Group. J Clin Pathol. 2014;67(9):751–7.

Molecular Pathology of Gastroesophageal Tumors

6

Adam L. Booth and Raul S. Gonzalez

Contents

Introduction

- Elucidation of molecular underpinnings in gastroesophageal neoplasms has expanded rapidly in the past decade
  - Carcinomas have been categorized based on predominant molecular characteristics
  - Key alterations have been described for numerous mesenchymal neoplasms
- Most molecular details have not translated into personalized therapy
  - Exceptions include mismatch repair and HER2 status in adenocarcinomas, and specific mutation details in gastrointestinal stromal tumors

A. L. Booth
Department of Pathology, Northwestern University Feinberg
School of Medicine, Chicago, IL, USA
e-mail: adam.booth@northwestern.edu

R. S. Gonzalez (✉)
Department of Pathology and Laboratory Medicine, Emory
University Hospital, Atlanta, GA, USA
e-mail: rsgonza@emory.edu

- Multiple familial predisposition syndromes impart an increased risk of gastric carcinoma, including polyposis and nonpolyposis conditions

Clinical and Molecular Features of Esophageal Neoplasms

Esophageal Squamous Cell Carcinoma

- Epidemiology
  - Esophageal squamous cell carcinoma (ESCC) is the most common esophageal cancer worldwide, accounting for approximately 90% of esophageal cancers and with a global incidence of 5.2 per 100,000
  - Highest incidence in Asian and Eastern countries, while rates have been decreasing in Western countries over the past four decades
  - ESCC is more common in men (69%) than women (31%)
  - Most frequently arise in the middle third of the esophagus, less so in the distal third
- Etiology
  - Risk factors include states leading to chronic irritation and inflammation
  - Alcohol consumption, chewing or smoking tobacco, diets low in fruit and vegetables (considered to be due to decreased antioxidants and vitamin deficiencies), ingestion of hot beverages, nitrosamines, and caustic injury
  - Achalasia, Plummer–Vinson syndrome, and previous mediastinal/thoracic radiation
  - Human papillomavirus has been reported as a causative risk factor but is not considered a major risk factor (in contrast to other squamous cell carcinomas)
  - Precursor lesions and dysplasia may be asymptomatic, with advanced disease presenting as dysphagia and associated anorexia
  - Genetic risk factors:

- o Nonepidermolytic palmoplantar keratoderma (tylosis): Autosomal dominant disease resulting from a mutation in *RHBDF2* on chromosome 17q25 is associated with a high frequency of ESCC
  - o Fanconi anemia: A predominantly autosomal recessive inherited disorder resulting from mutations in genes associated with DNA repair predisposing patients to bone marrow failure, congenital malformations, and cancer
  - o Alcohol consumption in individuals with *ALDH2* and *ALDH1A1* mutations
- Prognosis and treatment
  - Low-grade squamous dysplasia is routinely followed by periodic screening, while high-grade dysplasia requires prompt treatment
  - Treatment for squamous dysplasia includes endoscopic excision by endoscopic mucosal resection and endoscopic submucosal dissection or ablation by multipolar electrocoagulation, argon plasma coagulation, and radiofrequency ablation
  - Invasive ESCC prognosis depends on stage; treatment usually includes neoadjuvant chemo/radiotherapy followed by surgery
- Screening
  - Screening is currently only recommended for high-risk populations
- Pathogenesis
  - Cell-cycle controls
    - o *TP53* mutations are the most frequent, occurring as driver mutations in precursor dysplastic lesions
  - Cyclin-dependent kinase inhibitor 2A (*CDKN2A*) is a tumor suppressor gene encoding proteins P16INK4 and P14ARF
    - o P16INK4a expression is frequently reduced, most often through aberrant gene methylation and less so loss of heterozygosity or mutations; normally, P16INK4a interacts with RB1
    - o Similarly, P14ARF inhibits MDM2, preventing it from inhibiting p53; loss or reduced expression of P14ARF has a similar effect as mutations in *TP53*
  - Loss of heterozygosity in retinoblastoma 1 (*RB1*) leads to decreased or absent expression
- Cell differentiation
  - The NOTCH pathway is involved in squamous differentiation and squamous epithelial homeostasis
  - NOTCH alterations are an early event in ESCC pathogenesis, specifically *NOTCH1* and *NOTCH3*
  - NOTCH1 is involved in the balance between basal and differentiated esophageal cells
    - o In dysplasia and carcinoma, expression is reduced, hindering maturation of squamous epithelium
- Cell signaling
  - Epidermal growth factor receptor (*EGFR*) is overexpressed in precancerous lesions and the majority of ESCCs
  - *EGFR* is associated with higher pathologic stage, lymph node metastasis, and reduced overall survival
  - Downstream mutations or amplifications in *RAS* or *AKT* are seen in greater than 75% of cases
  - Mutations in phosphatidylinositol-3-kinase catalytic subunit (*PIK3CA*), which encodes the p110α protein catalytic subunit of phosphatidylinositol-3-kinase (PI3K), leading to hyperactivation of the PI3K/AKT/mTOR pathway and cell proliferation
  - *PTEN* acts as a tumor suppressor gene regulating *AKT*
    - o Loss of PTEN expression has been shown to be a negative prognostic factor
  - Transcription factor SOX2 is required for normal esophageal squamous development (known to be mutated in esophageal malformations); expression is necessary for proliferation and anchorage-independent growth
    - o Overexpression is seen in 70% of ESCC
    - o SOX2 activates the AKT/mammalian target of rapamycin complex 1 (mTORC1) signaling, promoting proliferation
- Molecular classifications
  - The Cancer Genome Atlas (TCGA): Molecular platforms, including array-based somatic copy number analysis, whole-exome sequencing, array-based DNA methylation profiling, messenger ribonucleic acid (RNA) sequencing, microRNA (miRNA) sequencing, and reverse-phase protein array (RPPA), were used to identify three molecular subtypes via integrative clustering
    - o ESCC1
      - ♦ NRF2 pathway alterations, which play a signification role in adaptation to oxidative stressors, carcinogens, and chemotherapy agents
      - ♦ Mutations in *NFE2L*; amplifications of *SOX2*, *TP63*, and *YAP1*; *VGLL4/ATG7* deletions
      - ♦ Notably, ESCC1 cases most closely resemble lung and head and neck SCCs
    - o ESCC2
      - ♦ Higher rates of mutations in *NOTCH1* and *ZNF750*; *KDM6A* and *KDM2D* inactivating alterations
      - ♦ Amplification of *CDK6*
      - ♦ Inactivating mutations in *PTEN* and *PIK3R1*
    - o ESCC3
      - ♦ PI3K pathway activation
      - ♦ Alterations of *KMT2D/MLL2* and *SMARCA4*

- Histology
  - Dysplasia: The World Health Organization (WHO) recommends the use of a two-tier grading system for the classification of dysplasia based on cytologic and architectural features
    - Low-grade: Mild cytologic atypia involving the lower half of the epithelium
    - High-grade: Any degree of severe cytologic atypia or atypia involving greater than half of the squamous epithelium
  - Squamous cell carcinoma
    - Uncontrolled growth of neoplastic squamous epithelium beyond the basement membrane
    - Three-tiered grading system based on mitotic activity, cytologic atypia, and presence of keratinization
      - Grade 1 (well-differentiated)
        - A pushing pattern of invasion composed of large, well-ordered cells with abundant eosinophilic cytoplasm
        - Low mitotic rate
        - Minimal cytologic atypia
        - Keratinization is common (keratin pearls)
      - Grade 2 (moderately differentiated)
        - Discernably less ordered
        - Intermediate cytologic atypia, mitoses
        - Keratinization without well-formed pearls
      - Grade 3 (poorly differentiated)
        - Sheets or nests of basaloid tumor cells with frequent mitoses
        - Marked nuclear atypia
        - Absent or minimal keratinization
  - Variants
    - Verrucous squamous cell carcinoma
      - Arises in the setting of inflammatory disorders, esophagitis, and chronic irritation
      - Tumor growth is slow and appears as an exophytic, warty-like projection composed of very well-differentiated tumor cells with keratinization and minimal atypia
      - Rare cases have been associated with HPV 11 and HPV 51
    - Basaloid squamous cell carcinoma
      - Basaloid cells with nested, solid, or cribriform architecture, often with central comedonecrosis
      - Notably, this type has a striking male predominance, is clinically aggressive, and is not associated with HPV
    - Spindle cell squamous cell carcinoma (sarcomatoid SCC/pseudosarcoma/carcinosarcoma)
      - Presents as a polypoid mass with obstructive symptoms
      - Tumor is biphasic, consisting of neoplastic squamous cells and spindled mesenchymal-like cells

- Components of osseous, cartilaginous, and/or muscular differentiation may also be seen

Esophageal Adenocarcinoma

- Epidemiology
  - In contrast to ESCC, the incidence of esophageal adenocarcinoma (EAC) has continued to rise in Western and developed countries for several decades, but recent data suggest it is leveling off
    - This increase is largely considered to be due to gastroesophageal reflux disease (GERD)
  - The global incidence is 0.7 cases per 100,000, rapidly increasing in many low- and middle-income countries
  - Increasing incidence in young adults reported recently in Western countries
  - In the United States, incidence rate dropped by ~1% each year from 2013 to 2017
    - Incidence has been rising in younger people (by 2% each year in adults under age 50, 2012 to 2016)
    - Younger adults present with more advanced disease
  - Almost always arises in the lower third of the esophagus at the gastroesophageal junction; rarely in the middle third in potentially heterotopic mucosa
- Etiology
  - Risk factors for GERD (obesity, smoking, and dietary factors) put patients at risk for developing EAC
  - Additional risk factors are male sex and Caucasian race
  - *Helicobacter pylori* infection is inversely associated with EAC
- Prognosis and treatment
  - Rates of progression from low-grade to high-grade glandular dysplasia range from 3–23%, with annual incidence of progression to EAC of 0.5%
  - Treatment for dysplasia may involve ablative therapy and/or endoscopic mucosal resection/endoscopic submucosal dissection
  - Neoadjuvant radio/chemotherapy precedes surgical resection followed by evaluation of residual disease
- Screening and surveillance
  - Intestinal metaplasia (Barrett esophagus if columnar mucosa of at least 1 cm in length identified on endoscopy) is identified in 7–10% of individuals with chronic GERD
  - Screening for Barrett may be employed for at-risk patients, defined as family history of EAC or Barrett esophagus or GERD with at least 1 risk factor (age > 50, obesity, tobacco use, male sex)

- American Society for Gastrointestinal Endoscopy guidelines for the screening and surveillance of Barrett esophagus
  - No dysplasia: Repeat endoscopy in 3–5 years
  - Indefinite for dysplasia: Optimize medical therapy and repeat endoscopy in 12 months
  - Low-grade dysplasia: Endoscopic eradication therapy and follow-up endoscopy every 6 months, then annually
  - High-grade dysplasia/intramucosal carcinoma
    - Endoscopic eradication therapy followed by complete eradication of the Barrett esophagus
    - Surveillance should begin every 3 months
- Molecular pathogenesis
  - *TP53* mutations are the most frequent, occurring as driver mutations in precursor dysplastic lesions
    - Aberrant p53 immunohistochemical staining appears to indicate an increased risk of progression, even for nondysplastic cases of Barrett esophagus
  - *CDKN2A* is frequently inactivated and encodes proteins P16INK4 and P14ARF
    - Additionally, it is associated with Epstein–Barr virus (EBV) infection
  - *ELMO1* and *DOCK2* mutations are present in 17% of EAC; these allow for greater cellular motility, favoring tumor invasion
  - Mutations in *PIK3CA*, resulting in hyperactivation of the PI3K/AKT/mTOR pathway
  - *PTEN* inactivating mutations or deletions
  - *ERBB2* mutations leading to HER2 amplification/overexpression
  - Amplifications in *KRAS*, *VEGF*, *EFGR*, and *IGF1R*
  - Ectopic overexpression of *GATA6* transcription factor
  - Abnormalities in chromatin-remodeling factors *ARID1A*, *SMARCA4*, and *ARID2* are present in approximately 1/5 of cases overall
  - *SMAD4* tumor suppressor mutations
  - Tumor suppressor *RUNX1* deletions
  - CpG island methylator phenotype (CIMP) leading to aberrant gene silencing
  - *JAK2* overexpression
- Molecular classifications
  - TCGA, using numerous molecular platforms as above, identified three molecular subtypes
    - Chromosomally unstable
    - Microsatellite unstable
    - EBV-positive
  - The three molecular subtypes were found to correspond to three of the four identified TCGA subtypes of gastric adenocarcinoma
    - This provided molecular support for a gradation of adenocarcinoma from the esophagus extending into

the stomach, explaining the challenging approach to staging of adenocarcinomas near the gastro-esophageal junction and cardia
- Histology
  - Metaplasia: the identification of intestinal metaplasia, defined as the presence of goblet cells, characterizes the initial histologic transformation
  - In the United States, in conjunction with endoscopically identified "salmon-colored" mucosa, the finding is diagnostic of Barrett esophagus
  - Dysplasia: the WHO recommends the use of a two-tier grading system for the classification of dysplasia based on cytologic and architectural features:
    - Low-grade: Absence of surface maturation with distinct transition from nondysplastic epithelium; cytologic atypia including hyperchromasia, greater nuclear:cytoplasmic ratio, and pseudostratification, but maintenance of nuclear polarity
    - High-grade: Features of low-grade dysplasia with loss of nuclear polarity
  - Intramucosal carcinoma: features of high-grade dysplasia with crowded or cribriform glands, prominent nucleoli, and single cells in the lamina propria, often with intraluminal necrosis
  - Esophageal adenocarcinoma: malignant neoplasm arising from epithelial cells of glandular or mucinous differentiation
    - Current staging guidelines state the tumor epicenter must be within 2 cm of the gastroesophageal junction to be considered EAC (rather than gastric adenocarcinoma)
    - Three-tiered grading system based on the quantity of gland formation
      - Grade 1 (well-differentiated): greater than 95% glands
      - Grade 2 (moderately differentiated): 50–95% glands
      - Grade 3 (poorly differentiated): less than 50% glands

Esophageal Undifferentiated Carcinoma

- Rare, recently defined carcinoma subtype with no evidence of glandular, squamous, or neuroendocrine differentiation
- Aggressive with extremely poor prognosis
- Histology shows sheets of malignant cells with an undifferentiated or rhabdoid appearance (Fig. 6.1a)
- Diagnosis can be confirmed by immunohistochemistry (IHC), which demonstrates loss of a protein associated with the SWI/SNF chromatin-remodeling complex,

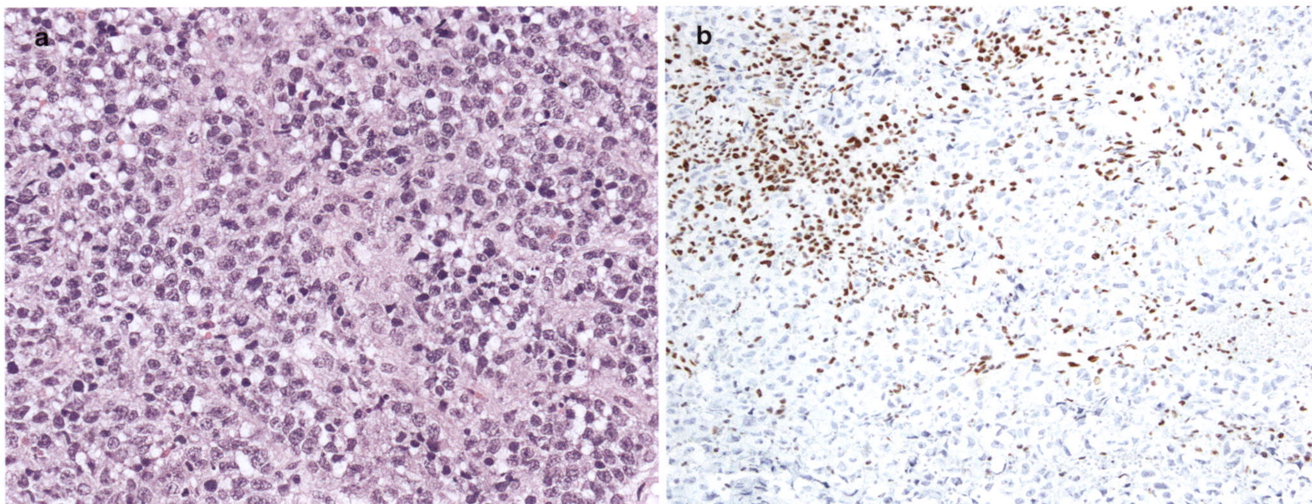

Fig. 6.1 (**a**) Undifferentiated carcinoma of the esophagus manifests as sheets of malignant cells with prominent nuclear atypia, frequent mitotic figures, and in some cases, rhabdoid cytologic features. (**b**) Aberrant SMARCA4 loss by immunohistochemistry is seen in this example, with background lymphocytes serving as a positive control (image courtesy of Dr. Amitabh Srivastava)

including SMARCB1 (INI1), SMARCA2 (BRM), or SMARCA4 (BRG1) (Fig. 6.1b)

Other Esophageal Neoplasms

- Granular cell tumor
  - Lower esophagus is the most common site for granular cell tumors in the gastrointestinal tract
  - Histology resembles granular cell tumors of soft tissue, with plump, bland lesional cells having abundant granular eosinophilic cytoplasm
    - Rare examples show nuclear atypia and patchy necrosis
  - Pseudoepitheliomatous hyperplasia of the overlying squamous epithelium is sometimes seen, potentially mimicking squamous cell carcinoma
  - IHC is positive for S100 and CD68
  - Most soft tissue cases demonstrate an inactivating mutation in *ATP6AP1* or *ATP6AP2*; specific esophageal abnormalities have not been reported
  - Numerous syndromes can predispose to multiple granular cell tumors, including neurofibromatosis type 1 (germline mutation in *NF1*), Noonan syndrome (germline mutation in *KRAS*, *PTPN11*, *RAF1*, or *SOS1*), and Noonan syndrome with multiple lentigines (also called LEOPARD syndrome; germline mutation in *MAP 2 K1* or the previously mentioned genes)
    - These syndromic cases do not necessarily result in gastrointestinal granular cell tumors
- Leiomyoma
  - Can occur anywhere in the gastrointestinal tract, though esophagus is a common site
  - Smooth muscle neoplasm formed by fascicles of bland spindle cells with prominent eosinophilic cytoplasm and positivity for smooth muscle actin and desmin by IHC
  - Benign by definition
  - Few molecular events reported
    - *FN1::ALK* fusion has been documented in one case
    - Rare familial cases of esophageal leiomyomatosis occur in the setting of *COL4A5* or *COL4A6* germline mutations (Alport syndrome)
- Atypical lipomatous tumor/well-differentiated liposarcoma
  - Uncommon neoplasm arising in proximal esophagus
  - Usually presents as a large polypoid/pedunculated mass
    - Formerly called "giant fibrovascular polyp"
  - Composed of bland-appearing fibroconnective tissue with squamous lining (Fig. 6.2a)
  - Adipocytic component may be abundant or focal, and atypia may be sparse
  - Positive for MDM2 and CDK4 by IHC
  - Positive for *MDM2* amplification by FISH (Fig. 6.2b)
  - Prognosis generally good unless dedifferentiated component present

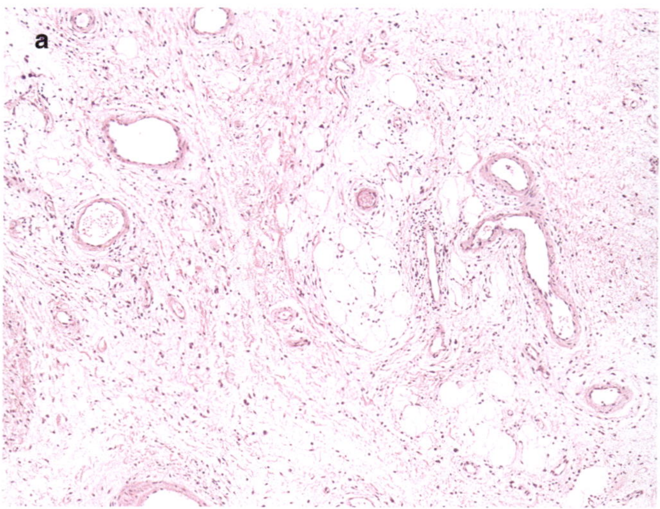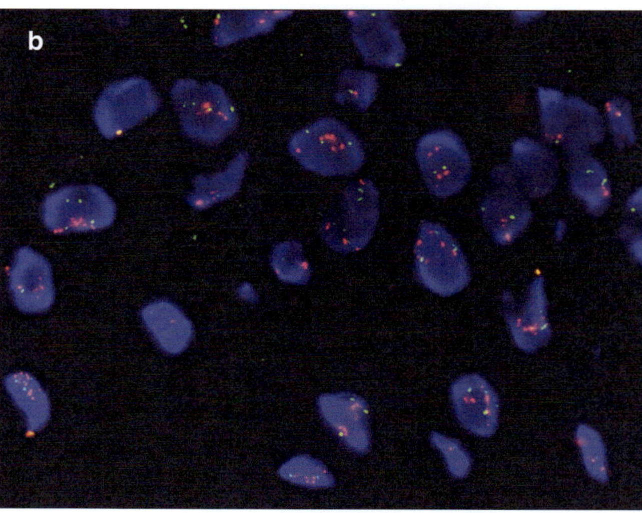

Fig. 6.2 (**a**) Atypical lipomatous tumor of the esophagus presents clinically as a giant fibrovascular polyp. Microscopically, it consists of expansive fibrovascular tissue. Adipocytes and nuclear atypia may be abundant or focal. (**b**) The diagnosis can be confirmed via fluorescence in situ hybridization showing *MDM2* amplification (red signal) (image courtesy of Dr. Madina Sukhanova)

Clinical and Molecular Features of Gastric Neoplasms

Precursor Polyps

- Fundic gland polyp with dysplasia
  - The majority of fundic gland polyps arise secondary to proton pump inhibitor use and may show *CTNNB1* exon 3 mutations
  - CpG island methylation is also common in fundic gland polyps, with no relationship to dysplastic progression or polyp origin (sporadic vs. syndromic)
  - Up to 49% of fundic gland polyps in patients with familial adenomatous polyposis show foveolar low-grade dysplasia
  - Dysplasia can rarely arise in sporadic fundic gland polyps as well
  - Dysplastic fundic gland polyps show mutations in *APC*, and dysplasia sometimes shows nuclear reactivity for beta-catenin despite wildtype *CTNNB1*
  - High-grade dysplasia is very rare in fundic gland polyps, and malignant progression is extraordinarily rare
- Hyperplastic polyp with dysplasia
  - Most gastric hyperplastic polyps arise in the setting of inflammation, including atrophic gastritis and *Helicobacter* infection (the latter causing increased epithelial cell turnover due to increased COX2 expression)
  - Risk of foveolar low-grade dysplasia is low (<5%) but increases if polyp is larger than 2 cm
  - Large hyperplastic polyps may show mutations in *KRAS* or *BRAF*, and dysplasia sometimes shows *APC* or *CTNNB1* mutations
  - Malignant progression is uncommon and appears related to *TP53* mutations in at least some cases
- Foveolar-type gastric adenoma
  - Almost always arises in patients with familial adenomatous polyposis
  - Dysplastic epithelium shows *APC* mutations and sometimes *KRAS* mutations
    - *APC* mutations in adenomas may indicate a decreased risk of malignant progression
  - Rate of malignant progression appears very low
- Intestinal-type gastric adenoma
  - Typically arises in the setting of atrophic gastritis with intestinal metaplasia
  - Dysplastic epithelium shows mutations in *ARID2*, *APC*, *ERBB2*, and *KRAS*
  - Other chromosome abnormalities include gains of 8, 9q, 11q, and 20, and losses of 5q, 6, 10, and 13
  - Microsatellite instability is occasionally present
  - Rate of malignant progression is roughly 24%
- Pyloric gland adenoma
  - Typically arises in the setting of atrophic gastritis
  - Dysplastic epithelium shows mutations in *GNAS*, *KRAS*, *CTNNB1*, and/or *SMAD4*
  - Somatic *APC* mutations are often seen in both familial adenomatous polyposis-associated polyps and sporadic polyps
    - Polyps arising in Lynch syndrome demonstrate mismatch repair loss

- – Other chromosome abnormalities include gains of 9, 11q, and 20, and losses of 5q, 6, 10, and 13q
  - – Rate of malignant progression may be up to 50%, though deep invasion is less common
- Oxyntic gland adenoma
  - – Rare adenoma more commonly seen in the East than in the West
  - – Dysplastic epithelium shows mutations in Wnt signaling pathway genes (*APC*, *AXIN1*, *AXIN2*, and *CTNNB1*)
  - – Some examples may invade submucosa and are often termed "adenocarcinoma of fundic gland type"; these may additionally show *KRAS* mutation
- Serrated adenoma
  - – Rare, understudied precursor neoplasm of the stomach with serrated architecture, eosinophilic cytoplasm, and association with invasive adenocarcinoma
  - – MLH1 expression loss by IHC and *KRAS* mutations have been described

Gastric Adenocarcinoma

- Epidemiology
  - – Despite declining incidence, gastric cancer represents the third most common cause of cancer mortality worldwide
  - – Incidence is greatest in Asia, Eastern Europe, and South America
  - – Adenocarcinoma accounts for the vast majority of gastric cancers
- Etiology
  - – Risk factors include *Helicobacter pylori* infection, smoking, EBV infection, and diets rich in salt, cured and smoked meats, pickled foods, nitrates/nitrites, and smoking
  - – Some genetic risk factors exist (see section on hereditary cancer syndromes)
  - – Diets high in antioxidants decrease risk
- Prognosis and treatment
  - – Gastric dysplasia is associated with an increased risk of progression to invasive adenocarcinoma
  - – Studies have reported an overall progression of low-grade dysplasia at 12 months in approximately 15% of cases
  - – Rates of progression of high-grade dysplasia to carcinoma at 12 months vary from 59–69%
  - – Treatment of low-grade dysplastic lesions typically involves endoscopic surveillance, whereas high-grade dysplasia requires endoscopic mucosal resection
  - – Invasive carcinoma is routinely treated with neoadjuvant therapy followed by surgical resection

- Screening
  - – Routine general screening programs have been developed in parts of the world with a higher incidence; these countries have seen a decrease in mortality because of early diagnosis
  - – In countries with lower incidence of gastric cancer, such as the United States, there is low evidence for gastric cancer screening
    - ○ National gastroenterology societies also recommend against routine surveillance of patients with gastric intestinal metaplasia
  - – Patients with intestinal metaplasia and increased risk for gastric cancer (e.g., first-degree relatives with gastric cancer, racial/ethnic minorities, immigrants from regions of high incidence of gastric cancer, extensive intestinal metaplasia, and incomplete intestinal metaplasia) may undergo surveillance every 3–5 years with ongoing discussion between the patients and their gastroenterologists regarding risks/benefits
- Molecular pathogenesis
  - – Overall similar to EAC molecular pathogenesis above, with a few additional factors
    - ○ *CDH1*, *RHOA* mutation
    - ○ *CLDN18::ARHGAP* fusion
- Molecular classifications
  - – Intrinsic: two major types based on genomic expression signatures
    - ○ Genomic-intestinal: genes associated with carbohydrate and protein metabolism (fucosyltransferase 2 (*FUT2*)) and cell adhesion (lectin (*LGALS4*); cadherin 17 (*CDH17*))
    - ○ Genomic-diffuse: functional annotations enriched in genes associated with cell proliferation (aurora kinase B (*AURKB*)) and fatty acid metabolism (ELOVL family member 5 (*ELOVL5*))
  - – Lei: three subtypes identified by unsupervised hierarchical clustering of gene expression patterns
    - ○ Proliferative: characterized by gene sets related to the cell cycle: Kyoto Encyclopedia of Genes and Genomes (KEGG) cell cycle, KEGG DNA replication, and 13 Gene Ontology (GO) gene sets, high number of *TP53* mutations, intestinal histology
    - ○ Metabolic: characterized by gene sets from several KEGG metabolism pathways and GO digestion, low *TP53* mutations, intestinal histology
    - ○ Mesenchymal: overrepresents the following gene sets: KEGG focal adhesion, KEGG extracellular-matrix–receptor interaction, and GO cell adhesion, low *TP53* mutations, diffuse histology
  - – TCGA, using numerous molecular platforms as above, identified four molecular subtypes
    - ○ Chromosomally unstable

- o Microsatellite unstable
- o Genomically stable
- o EBV-positive
- Asian Cancer Research Group (ACRG) analyzed mRNA expression, genomewide copy number microarray, and targeted sequencing, identifying four molecular subtypes
  - o Microsatellite stable/epithelial/*TP53* loss
  - o Microsatellite stable/epithelial/*TP53* intact
- o Microsatellite instability
- o Microsatellite stable/epithelial–mesenchymal transition
- Table 6.1 provides additional information on TCGA and ACRG classifications
- Histology
  - Metaplasia: gastric intestinal metaplasia defined as the presence of goblet cells replacing the normal gastric foveolar epithelium

Table 6.1 TCGA and ACRG molecular classifications of gastric carcinoma

Subtype	Clinical	Frequency (%)	Molecular alterations	Histology	Potential targets
TCGA					
Chromosomally unstable	Gastroesophageal junction/cardia	49.8	• *TP53* mutation • RTK-RAS activation • Mutations of *SMAD4* and *APC*	Intestinal type	RTKs, EGFR, VEGFA, CCNE1, CCND1, CDK6
Microsatellite unstable	Older age (median 72 years)	21.7	• Gastric-CIMP • Hypermutation in *TP53, PIK3CA, ERBB3, ARID1A* • *MLH1* silencing • Mitotic pathways activation	Gastric carcinoma with lymphoid stroma/medullary (some)	PIK3CA, ERBB2/3, EGFR, PD-L1, MLH1 silencing
Genomically stable	Younger age (median 59 years)	19.7	• *CDH1, RHOA* mutation • *CLDN18::ARHGAP* fusion • Cell adhesion, angiogenesis pathways enriched • Rare *TP53* mutations	Diffuse type (some)	RHOA, CLDN18
EBV-positive	Predominantly male Fundus/body	8.8	• EBV • CIMP • *PD-L1/2, JAK2* overexpression • Mutation in *PIK3CA, ARID1A, BCOR* • *CDKN2A* silencing • Immune cell signaling • Rare *TP53* mutations	Gastric carcinoma with lymphoid stroma/ lymphoepithelioma-like (some)	PIK3CA, JAK2, PD-L1/PD-L2
ACRG					
Microsatellite stable/ epithelial/*TP53* loss		35.7	• Highest prevalence of *TP53* and *RHOA* mutations • *APC, ARID1A, KRAS, PIK3CA, SMAD4* enriched	Intestinal type	
Microsatellite stable/ epithelial/*TP53* intact	Male Frequent EBV infection	26.3	• Frequent mutations in *ARID1A, PIK3CA, SMAD4, APC*	Intestinal type	
Microsatellite instability		22.7	• Silencing of *MLH1* • Mutations in *ARID1A, MTOR, KRAS, PIK3CA, ALK, PTEN* • Overexpression of PD-L1 • T cell infiltrate	Intestinal type	
Microsatellite stable/ epithelial–mesenchymal transition	Younger age	15.3	• Loss of *CDH1* • Loss of cellular adhesion, angiogenesis, motility	Diffuse type	

Modified from *Wang et al. Gastroenterol Res. 2019;12:275–282*
ACRG Asian Cancer Research Group, *CIMP* CpG island methylator phenotype, *EBV* Epstein–Barr virus, *TCGA* The Cancer Genome Atlas

- o Complete intestinal metaplasia includes enterocytes with a brush border and Paneth cells, alongside the goblet cells
- o Incomplete intestinal metaplasia more resembles goblet cells interspersed among foveolar-type cells with apical mucin caps
- Dysplasia: the WHO recommends the use of a two-tier grading system for the classification of dysplasia based on cytologic and architectural features
  - o Low-grade: Slight glandular crowding with normal overall shape and size, cytologic atypia including hyperchromasia, mildly increased nuclear:cytoplasmic ratio, maintenance of nuclear polarity
  - o High-grade: Variability in glandular architecture, cytologic atypia including hyperchromasia, increased nuclear:cytoplasmic ratio, loss of nuclear polarity
- Intramucosal carcinoma: features of high-grade dysplasia with crowded or cribriform glands, prominent nucleoli, single cells in the lamina propria
- Gastric adenocarcinoma: malignant neoplasm arising from epithelial cells of glandular or mucinous differentiation

- o The tumor epicenter must be greater than 2 cm from the gastroesophageal junction to be considered gastric origin for staging purposes
- o Three-tiered grading system based on the quantity of gland formation
  - ◆ Grade 1 (well-differentiated): greater than 95% glands
  - ◆ Grade 2 (moderately differentiated): 50–95% glands
  - ◆ Grade 3 (poorly differentiated): less than 50% glands
- Table 6.2 provides additional information and comparisons of various histologic classifications for gastric carcinoma
- Histology and molecular alterations do not strongly correlate for gastric adenocarcinoma, though some relationships do exist
  - o Carcinomas arising in patients with hereditary diffuse gastric cancer (e.g., *CDH1* germline mutations) are always signet ring cell carcinomas
  - o Gastric carcinoma with lymphoid stroma may arise due to EBV infection (lymphoepithelioma-like carcinoma) or microsatellite instability (medullary carcinoma) (Fig. 6.3a, b)

Table 6.2 Comparison of histologic classifications of gastric carcinoma

Laurén (1965)	Nakamura (1968)	JGCA (2017)	WHO (2019)
Intestinal	Differentiated	• Papillary: Pap	• Papillary
		• Tubular 1, well-differentiated: tub1	• Tubular, well-differentiated
		• Tubular 2, moderately differentiated: tub2	• Tubular, moderately differentiated
Indeterminate	Undifferentiated	• Poorly 1 (solid type): por1	• Tubular (solid), poorly differentiated
Diffuse	Undifferentiated	• Signet ring cell: Sig	• Poorly cohesive, signet ring cell phenotype
		• Poorly 2 (nonsolid type): por2	• Poorly cohesive, other cell types
Intestinal/diffuse/ indeterminate	Differentiated/ undifferentiated	• Mucinous	• Mucinous
Mixed	n/a	• Description according to the proportion (e.g., por2 > sig > tub2)	• Mixed
n/a	n/a	*Special types:*	*Other histological subtypes:*
		• Adenosquamous carcinoma	• Adenosquamous carcinoma
		• Squamous cell carcinoma	• Squamous cell carcinoma
		• Undifferentiated carcinoma	• Undifferentiated carcinoma
		• Carcinoma with lymphoid stroma	• Carcinoma with lymphoid stroma
		• Hepatoid adenocarcinoma	• Hepatoid carcinoma
		• Adenocarcinoma with enteroblastic differentiation	• Adenocarcinoma with enteroblastic differentiation
		• Adenocarcinoma of fundic gland type	• Adenocarcinoma of fundic gland type
			• Micropapillary adenocarcinoma

Modified from Chapter 3, Table 3.3, *WHO classification of tumours of the digestive system, vol. 1. 5th ed. Lyon: International Agency for Research on Cancer, 2019*
JCGA Japanese Gastric Cancer Association, *WHO* World Health Organization

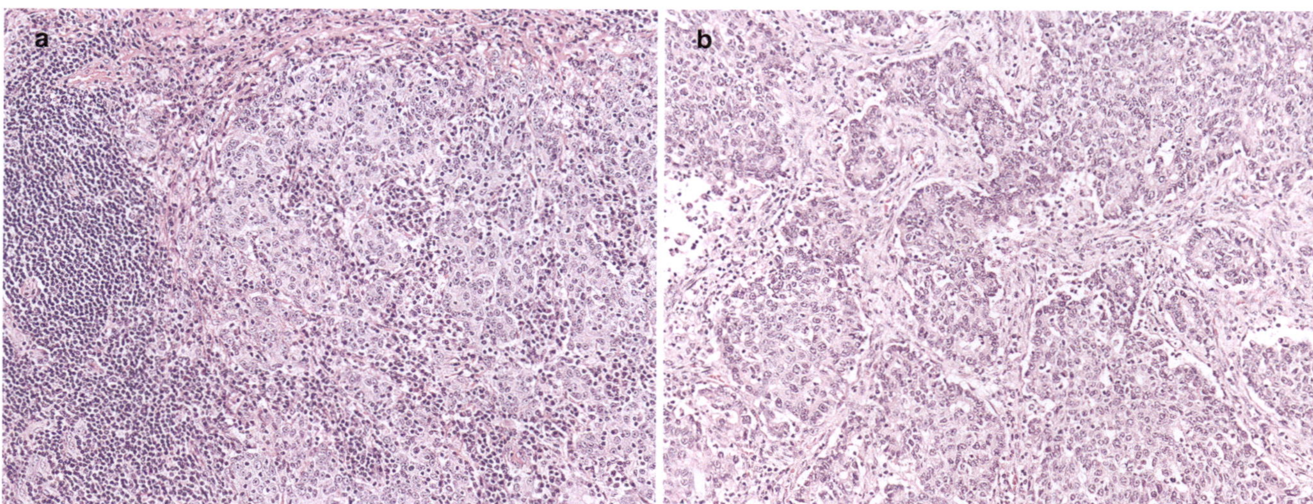

Fig. 6.3 (**a**) Gastric carcinomas with lymphoid stroma show prominent intraepithelial and tumor-adjacent lymphocytes, and lesional cells are arranged in syncytia with plentiful cytoplasm and prominent nucleoli. This case was positive for EBV by immunohistochemistry, indicating it is a lymphoepithelioma-like carcinoma. (**b**) This gastric carcinoma with lymphoid stroma was negative for EBV but showed aberrant loss of some mismatch repair proteins by immunohistochemistry, indicating it is a medullary carcinoma

Gastric Neuroendocrine Carcinoma

- Epidemiology
  - Account for fewer than 1% of gastric cancers
  - More frequent in males
- Molecular pathogenesis
  - Demonstrate higher mutation rates than conventional gastric adenocarcinomas
    - *TP53* gene mutations are the most frequent
    - Aberrations in the Rb/p16/cyclin D1 pathway
    - Hypermethylation of Ras-Association domain gene Family 1A (*RASSF1A*)
    - *KRAS* mutations have rarely been identified
    - Loss of *MLH1* and PMS2 expression and *MLH1* promoter hypermethylation have been reported
- Histology: Poorly differentiated carcinomas with evident necrosis and frequently with a Ki67 proliferation rate greater than 50% and greater than 20 mitoses per mm²; neuroendocrine nature confirmed by immunohistochemistry (e.g., synaptophysin and chromogranin)
  - Large cell: Large cells with abundant eosinophilic cytoplasm and vesicular chromatin with prominent nucleoli
  - Small cell: Cells with high nuclear:cytoplasmic ratio and hyperchromatic nuclei with indistinct chromatin
  - Mixed adenocarcinoma-neuroendocrine carcinoma
    - Features of both neuroendocrine and conventional adenocarcinomas, with each component representing at least 30% of the tumor
    - Molecular evidence suggests a monoclonal origin
    - These now fall under the broad category of mixed neuroendocrine–nonneuroendocrine neoplasms

Other Gastric Neoplasms

- Neuroendocrine tumor
  - There are multiple types of neuroendocrine tumors (NETs) that arise in the stomach, often in the setting of hypergastrinemia
  - Type 1 NETs occur in the setting of atrophic (autoimmune) gastritis with ECL cell hyperplasia
    - These NETs are often multifocal but are clinically indolent
    - *MEN1* gene abnormality (loss of heterozygosity) in 48%
  - Type 2 NETs occur in patients with MEN1 or Zollinger–Ellison syndrome, generally with a gastrinoma elsewhere
    - These NETs are more aggressive and sometimes metastasize
    - *MEN1* gene abnormality in 75%
  - Type 3 NETs are sporadic, with no known etiology
    - These NETs are the most aggressive of these types and often metastasize
    - Molecular properties remain understudied, though the extensive loss of heterozygosity of the X chromosome has been reported

- Other types have been variably described in the literature, without definitive molecular characterization
- NETs must not be confused with neuroendocrine carcinomas, which are rare and highly aggressive
- Gastrointestinal stromal tumor (GIST)
  - Most common malignant mesenchymal neoplasm of the gastrointestinal tract; slightly more than half of cases arise in the stomach
  - It may be spindled or epithelioid
  - Risk of disease progression based on tumor site, size, and mitotic rate
  - First-line nonsurgical therapy is the tyrosine kinase inhibitor imatinib (Gleevec™), though molecular characteristics may impact efficacy
  - Most GISTs have *KIT* (75%) or *PRGFRA* (10%) gain-of-function mutations, which lead to upregulation of the RAS/MAPK and PI3K/AKT/mTOR pathways
  - Most *KIT* mutations occur in exon 11
    - GISTs with *KIT* exon 9 mutations may require higher doses of imatinib
    - *KIT* exon 13, 14, 17, and 18 mutations are usually secondary mutations that confer resistance to imatinib but not to other tyrosine kinase inhibitors
    - Most *PDGFRA* mutations are D842V exon 18 point mutations, which impart significant tyrosine kinase inhibitor resistance but respond to avapritinib and can be identified by PDGFRA immunohistochemistry
  - SDH gene subunit mutations occur in most "wildtype" (non-*KIT* mutated, non-*PDGFRA* mutated) GISTs (5–10%)
    - *SDHA* mutation is more commonly seen than *SDHB*, *SDHC*, or *SDHD*

- These GISTs almost always occur in the stomach of younger patients, often female
- Classic morphology is epithelioid cells forming a multinodular tumor (Fig. 6.4a)
- SDHB IHC expression is aberrantly lost, regardless of which SDH gene subunit is mutated (Fig. 6.4b)
- These GISTs often metastasize to lymph nodes, and classic risk-stratification criteria do not hold, but the prognosis is good overall
- These GISTs are resistant to imatinib but not to other tyrosine kinase inhibitors
- *BRAF* V600E mutations occur in roughly 2% of GISTs, imparting imatinib resistance
- *KRAS* mutations have also rarely been reported in GIST
- *NF1* mutations may occur in small bowel GISTs but almost never in gastric GISTs
  - This mutation also indicates resistance to imatinib
- "Quadruple wildtype GISTs" lack mutations in *KIT*, *PDGFRA*, *BRAF/RAS/NF1*, and SDH complex genes
  - A subset of these demonstrates increased FGF4 expression secondary to a focal gain in 11q13.3 (involving *FGF3/FGF4*)
  - Others appear to harbor cryptic low-allele-fraction *KIT* mutations
- *MAX* inactivation (via homozygous deletions or hemizygous mononucleotide alterations) occurs early in GIST development
- Additional mutations accumulate in disease progression, inactivating tumor suppressor genes such as *CDKN2A*, *DMD*, *RB1*, and *TP53*
- Other genetic events suggesting poor outcome include deletions in 1p and 22q; silencing of *P16*, *RKIP*, and

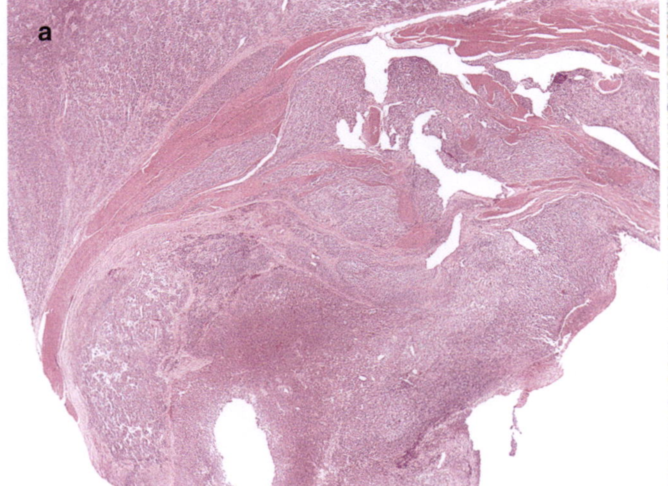

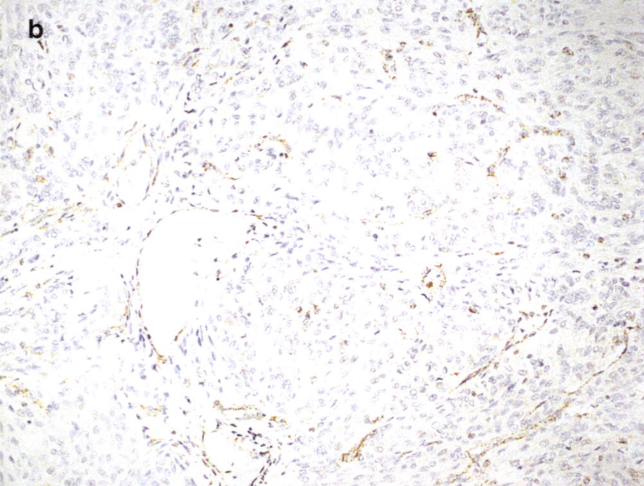

Fig. 6.4 (a) Some gastrointestinal stromal tumors harbor deficiencies in succinate dehydrogenase gene subunits. These tumors are often multinodular and epithelioid, arising in the stomach of younger patients.

(b) Immunohistochemistry for SDHB confirms loss of protein expression, with background lymphocytes and endothelial cells serving as positive internal controls (image courtesy of Dr. Anthony J Gill)

KCTD10; upregulation of *AURKA* and *SLITRK3*; and dysregulation of *ROR2*
  – NTRK gene fusions have been described in GIST, but those lesions more likely represent a unique gastrointestinal mesenchymal neoplasm mimicking GIST
  – GISTs are almost always positive for KIT and/or DOG1 by IHC regardless of molecular abnormalities
    ○ Occasional *PDGFRA*-mutant GISTs are negative for KIT
- Inflammatory fibroid polyp
  – Benign mesenchymal neoplasm most commonly arising in the gastric antrum
  – Bland spindle cell neoplasm characterized by intratumoral eosinophils, "onion-skin" concentric fibrosis around small blood vessels, and CD34 IHC positivity
  – Most gastric cases have a *PDGFRA* activating mutation in exon 18, most commonly c.2525A>T (p. D842V); exon 12 and 14 mutations can also occur, with the former more common in small intestine
  – Rare familial cases have been reported secondary to *PDGFRA* germline mutations
- Leiomyosarcoma
  – Malignant smooth muscle tumor that uncommonly arises in the gastrointestinal tract
    ○ Esophagus and stomach are both rare sites
  – May show a range of atypia, mitotic activity, and necrosis
  – Positive for smooth muscle actin and desmin by IHC
  – Extensive molecular characterization is lacking, though *RB1* and *TP53* mutations have been reported
- Gastrointestinal neuroectodermal tumor
  – Rare malignancy also known as "clear cell sarcoma-like tumor of the gastrointestinal tract"
  – Small bowel and stomach are most common sites
  – Sheets of malignant cells with prominent nucleoli and ample amphophilic cytoplasm; osteoclast-like giant cells may be seen scattered throughout the tumor (Fig. 6.5)
  – IHC is positive for S100, SOX10, and neuroendocrine markers
  – Aggressive, with a high rate of malignancy and poor prognosis
  – Always shows a fusion between *EWSR1* and either *ATF1* or *CREB1*
- Schwannoma
  – Stomach is the most common site for gastrointestinal tract schwannomas
  – There are several histologic differences compared to soft tissue schwannomas, including a peripheral lymphoid cuff, less cellular palisading, less prominent thick-walled vessels, and no capsule
    ○ Rarely, "soft tissue-type schwannomas" can occur in the gastrointestinal tract

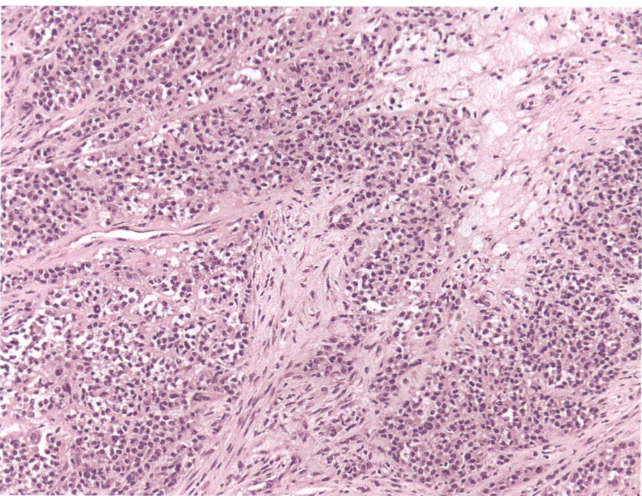

Fig. 6.5 Gastrointestinal neuroectodermal tumors are rare aggressive neoplasms that can arise in the stomach. They demonstrate sheets of epithelioid cells with a moderate nuclear:cytoplasmic ratio. This case harbored an *EWSR1::ATF1* fusion; other examples may harbor *EWSR1::CREB1* fusion

  – IHC for S100 is positive
  – Abnormalities of chromosomes 2, 18, and 22 have been reported
  – *NF1* abnormalities occur in both gastrointestinal and soft tissue schwannomas
  – *NF2* abnormalities are rare in gastrointestinal schwannomas, unlike soft tissue schwannomas
- Glomus tumor
  – Almost all gastrointestinal tract glomus tumors arise in the stomach, usually in women
  – Histology resembles their soft tissue counterparts, with monotonous cells with rounded nuclei, eosinophilic cytoplasm, and crisp cell borders; IHC for smooth muscle actin is positive
  – NOTCH genes are rearranged in many glomus tumors, most typically a *NOTCH2::MIR143* fusion
  – Rare cases may behave aggressively; these often have a *BRAF* V600E mutation
- Plexiform fibromyxoma
  – Rare gastric antral neoplasm with no particular age or sex predominance
  – Histology shows a variably cellular bland spindle cell neoplasm arranged in a plexiform/multinodular pattern, with a fibrotic/myxoid background (Fig. 6.6)
  – Roughly one-third of cases show either a *MALAT1::GLI1* fusion or *GLI1* polysomy, with strong positivity for GLI1 by IHC
- Gastroblastoma
  – Very rare biphasic gastric neoplasm, with distinct epithelial and mesenchymal components
  – Typically arises in young men

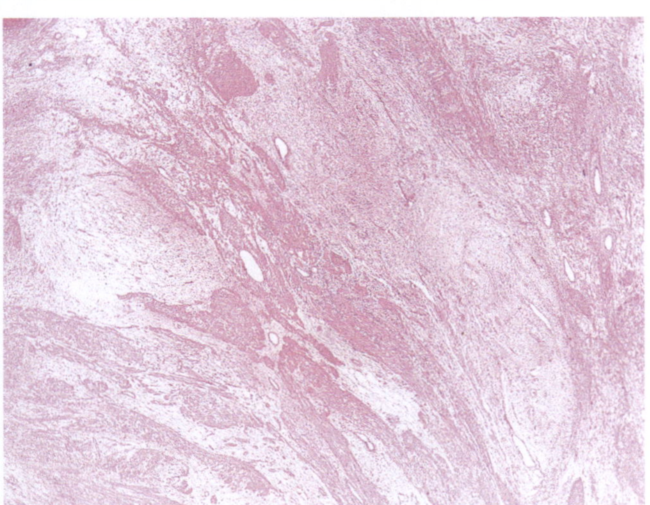

Fig. 6.6 Plexiform fibromyxoma is a rare benign neoplasm almost always arising in the gastric antrum. It characteristically displays bland spindle cells arranged in elongated plexiform nests. A sizable minority of cases contain a *MALAT1::GLI1* fusion or *GLI1* polysomy

- Has a characteristic *MALAT1::GLI1* fusion (same as plexiform fibromyxoma), detectable by FISH and other modalities
- Accordingly, strong positivity for GLI1 by IHC

Genetic Predisposition Syndromes

- See Table 6.3 for a summary of genetic predisposition syndromes relevant to upper gastrointestinal tract neoplasms
- Familial adenomatous polyposis (FAP)
  - Autosomal dominant syndrome caused by germline mutations in *APC*
    - Numerous *APC* mutations have been described; relevance to gastric disease remains unclear
  - Most striking manifestation is hundreds of adenomatous colon polyps
  - Patients are at high risk of developing dysplastic fundic gland polyps
    - Carcinoma almost never arises from these, though the phenomenon appears more common in attenuated FAP
  - Patients are also at high risk of developing foveolar-type gastric adenomas
- Gastric adenocarcinoma and proximal polyposis of the stomach (GAPPS)
  - Rare autosomal dominant syndrome caused by a germline point mutation in *APC* promoter 1B
    - All reported mutations (c.-191T>C, c.-192A>G, c.-195A>C, and c.-125delA) reduce transcription of *APC* due to altered YY1 transcription factor binding

- For this reason, essentially a unique variant of FAP
  - Patients develop innumerable carpeting polyps in the body/fundus of the stomach, including fundic gland polyps, adenomas, and hyperproliferative aberrant pits
    - Most patients have >100 polyps, though patients with >30 polyps and a first-degree relative are also considered to have GAPPS
  - Patients are also at risk of developing intestinal-type gastric adenocarcinoma
- Hereditary diffuse gastric cancer
  - Rare autosomal dominant syndrome caused by germline mutation in one of numerous genes, most commonly *CDH1*
    - Other implicated genes include *ATM*, *BRCA2*, *CTNNA1*, *MSR1*, *PALB2*, *PRSS1*, *SDHB*, and *STK11*
  - Patients are at high risk of developing multifocal diffuse (signet ring) gastric carcinoma, though age of development is variable
    - Female patients are also at increased risk for lobular breast carcinoma (~40%)
  - In situ lesions can also be seen, along with Pagetoid spread of malignant cells in overlying epithelium (Fig. 6.7)
  - Patients often undergo prophylactic total gastrectomy; the general recommendation is to submit these entirely for microscopic examination to detect subtle disease foci
- Lynch syndrome
  - Autosomal dominant syndrome caused by germline mutation in a gene encoding a mismatch repair protein (usually *MLH1*, *MSH2*, *MSH6*, or *PMS2*)
  - Most common sites of malignancy in Lynch syndrome are colon and endometrium
  - Lifetime risk of developing stomach cancer is roughly 1%, particularly in the setting of *MLH1* or *MSH2* mutations
    - Carcinomas should lack expression of certain mismatch repair proteins, depending on specific gene defect
- Juvenile polyposis syndrome
  - Autosomal dominant syndrome characterized by juvenile polyps throughout the gastrointestinal tract, particularly the colon
    - More than half of adult patients have a germline mutation in *SMAD4* or *BMPR1A*, ranging from point mutations to entire gene deletion
    - Infant patients have contiguous *BMPR1A* and *PTEN* mutations
    - Some patients appear to have *ENG* germline mutations
  - Gastric manifestations vary from occasional polyps to massive gastric polyposis, with large/confluent polyps carpeting the stomach

Table 6.3 Genetic predisposition syndromes increasing risk of gastric malignancy

Syndrome	Inheritance	Chromosome	Gene	Protein	Protein mechanism	Pathologic manifestations in stomach
Familial adenomatous polyposis	Autosomal dominant	5q22.2	*APC*	APC	• APC suppresses Wnt signaling pathway	• Fundic gland polyps with increased dysplasia risk • Foveolar-type adenomas
Gastric adenocarcinoma and proximal polyposis of the stomach	Autosomal dominant	5q22.2	*APC*	APC	• APC downregulated via mutation in promoter region 1B	• Numerous proximal stomach polyps of varying morphology • Increased risk of intestinal-type adenocarcinoma
Hereditary diffuse gastric cancer	Autosomal dominant	16q22.1	*CDH1*	E-cadherin	• E-cadherin regulates cell proliferation and cell-cell adhesions	• In situ and invasive signet ring cell carcinoma (multifocal)
		5q31.2 Various others	*CTNNA1*	α-E-catenin	• α-E-catenin anchors cadherin proteins	
Lynch syndrome	Autosomal dominant	3p22.2 2p21-p16.3 2p16.3 7p22.1	*MLH1* *MSH2* *MSH6* *PMS2* Rare others	MLH1 MSH2 MSH6 PMS2	• Mismatch repair proteins correct small errors that occur during DNA replication, which can lead to lengthened microsatellites if not repaired	• Increased risk of mismatch repair-deficient adenocarcinoma
Juvenile polyposis syndrome	Autosomal dominant	18q21.2	*SMAD4*	SMAD4	• SMAD4, part of the TGF-β pathway, helps regulate cell growth and proliferation	• Gastric juvenile polyps (sparse to massive/carpeting)
		10q23.2	*BMPR1A*	BMPR1A	• BMPR1A regulates activation of SMAD proteins	• Increased risk of adenocarcinoma
Peutz–Jeghers syndrome	Autosomal dominant	19p13.3	*STK11*	STK11	• STK11 is a tumor suppressor that regulates cell polarity and apoptosis via regulation of AMPK	• Characteristic hamartomatous polyps • Increased risk of adenocarcinoma
Cowden syndrome	Autosomal dominant	10q23.31	*PTEN*	PTEN	• PTEN negatively regulates the Akt pathway, which manages cell division and apoptosis	• Hamartomatous polyps • Little to no increased risk of adenocarcinoma
Carney–Stratakis syndrome	Autosomal dominant	1p36.13 1q23.3 11q23.1	*SDHB* *SDHC* *SDHD*	SDHB SDHC SDHD	• Succinate dehydrogenase complex involved in the citric acid cycle; dysfunction causes oxidative stress and possibly tumorigenesis	• Increased risk of SDH-deficient gastrointestinal stromal tumors

- ○ Gastric juvenile polyps can be stroma-predominant or epithelium-predominant and may be difficult to distinguish from hyperplastic polyps (Fig. 6.8)
- ○ Loss of SMAD4 expression by IHC seen in half of gastric polyps in patients with *SMAD4* germline mutation
- ○ Roughly 15% of gastric polyps develop dysplasia, though the risk is likely higher in massive disease
- ○ Rarely, polyps progress to invasive adenocarcinoma
- Peutz–Jeghers syndrome
  - – Autosomal dominant syndrome caused by germline mutations in tumor suppressor gene *STK11*
    - ○ These may be point mutations or large deletions
  - – Patients develop characteristic hamartomatous gastrointestinal polyps (mostly in small intestine, but ~25% of patients develop gastric polyps) and mucocutaneous melanin pigmentation
    - ○ Polyps demonstrate an "arborizing" network of smooth muscle bundles, with normal mucosal elements arranged into intervening nests (Fig. 6.9)
      - ♦ Features hard to discern in small polyps and in some gastric polyps regardless of size
    - ○ Polyps can develop dysplasia or carcinoma
      - ♦ Lifetime risk of gastric cancer is roughly 29%
  - – "Sporadic" (nonsyndromic) Peutz–Jeghers polyps are controversial and appear exceedingly rare if they do exist

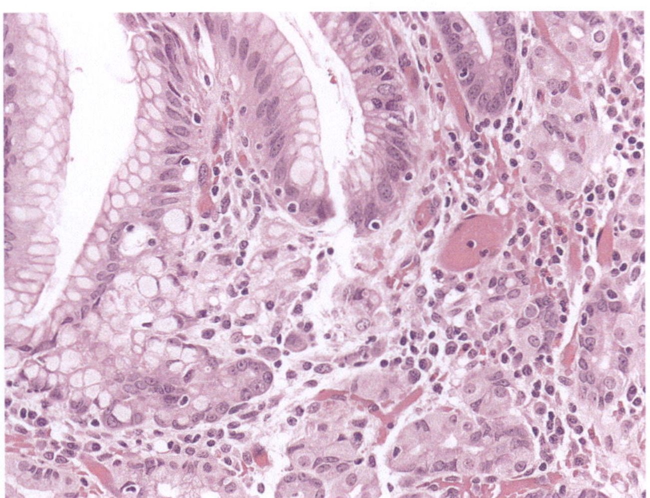

Fig. 6.7 Patients with hereditary diffuse gastric cancer (e.g., from *CDH1* germline mutations) may undergo prophylactic gastrectomy. Careful examination of such specimens can demonstrate both early superficial signet ring cell carcinomas, as well as signet ring cell carcinoma in situ (as depicted here)

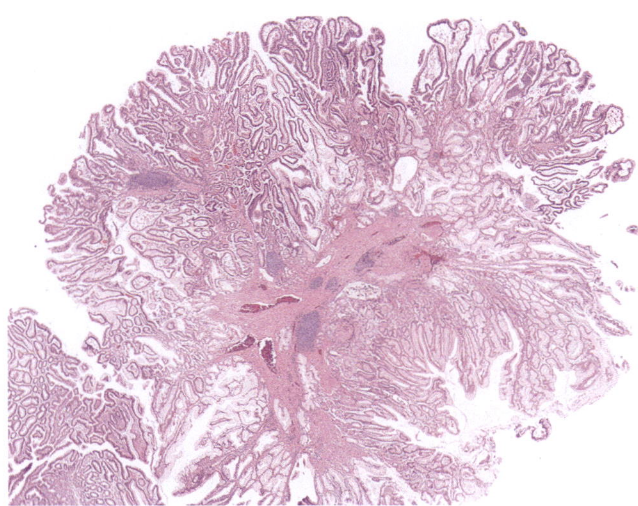

Fig. 6.9 Peutz–Jeghers syndrome is caused by *STK11* mutations and may affect the stomach. Syndromic polyps show an arborizing pattern of musculature, with benign epithelium arranged in packets along this framework. In the stomach, this arborizing pattern is often subtle or underdeveloped

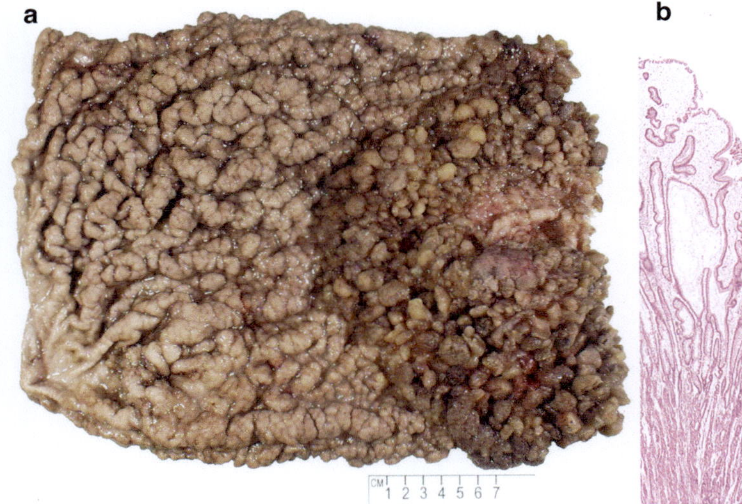

Fig. 6.8 (**a**) Juvenile polyposis syndrome may affect the stomach, in some cases leading to massive gastric involvement by polyposis. *SMAD4* is the most common germline mutation in these patients. (**b**) Microscopically, the polyps may be edematous and stroma-predominant, or tightly packed and epithelium-predominant. Dysplasia and carcinoma can occur

- Cowden syndrome
  - Autosomal dominant syndrome caused by germline mutations in *PTEN*
    - Allows for deregulation of PI3K/AKT pathway
  - Patients develop various forms of hamartomatous polyp throughout the gastrointestinal tract
    - Most patients have gastric polyps, which resemble hyperplastic polyps
    - Dysplasia does not appear to occur in these polyps
    - Most patients also have glycogenic acanthosis of the esophagus
    - Likely no increased risk of gastric cancer
- Carney–Stratakis syndrome
  - Rare autosomal dominant syndrome caused by germline mutations in *SDHA*, *SDHB*, *SDHC*, or *SDHD*
    - These genes encode subunits of the succinate dehydrogenase enzyme complex
    - Not all patients with one of these germline mutations display this syndrome

- Patients develop SDH-deficient gastric GISTs and aggressive soft tissue paragangliomas
- Distinct from the Carney triad (gastric SDH-deficient GIST, soft tissue paraganglioma, and pulmonary chondroma), which is not familial and is almost never related to a succinate dehydrogenase gene mutation

Molecular Diagnostics for Gastroesophageal Cancers

- Robust molecular testing is not always indicated and may depend on oncologist preference and individual patient scenarios
- IHC examination of upper gastrointestinal tract cancers, with molecular correlation and predictive relevance, is discussed below

- PD-L1
  - Programmed death ligand 1 (PD-L1) expression can be evaluated by IHC performed on formalin-fixed paraffin-embedded tissue
  - In upper gastrointestinal tract malignancies, the Combined Positive Score (CPS) is calculated by counting tumor cells and tumor-associated macrophages and lymphocytes with at least partial membranous staining; the sum is then divided by the total number of viable tumor cells counted and multiplied by 100 (note: at least 100 cells are required for specimen adequacy)
  - CPS score relates to patient eligibility for PD-1 inhibitors (e.g., nivolumab and pembrolizumab)
    - CPS < 1 in gastric and gastroesophageal carcinomas is considered negative (Fig. 6.10a)

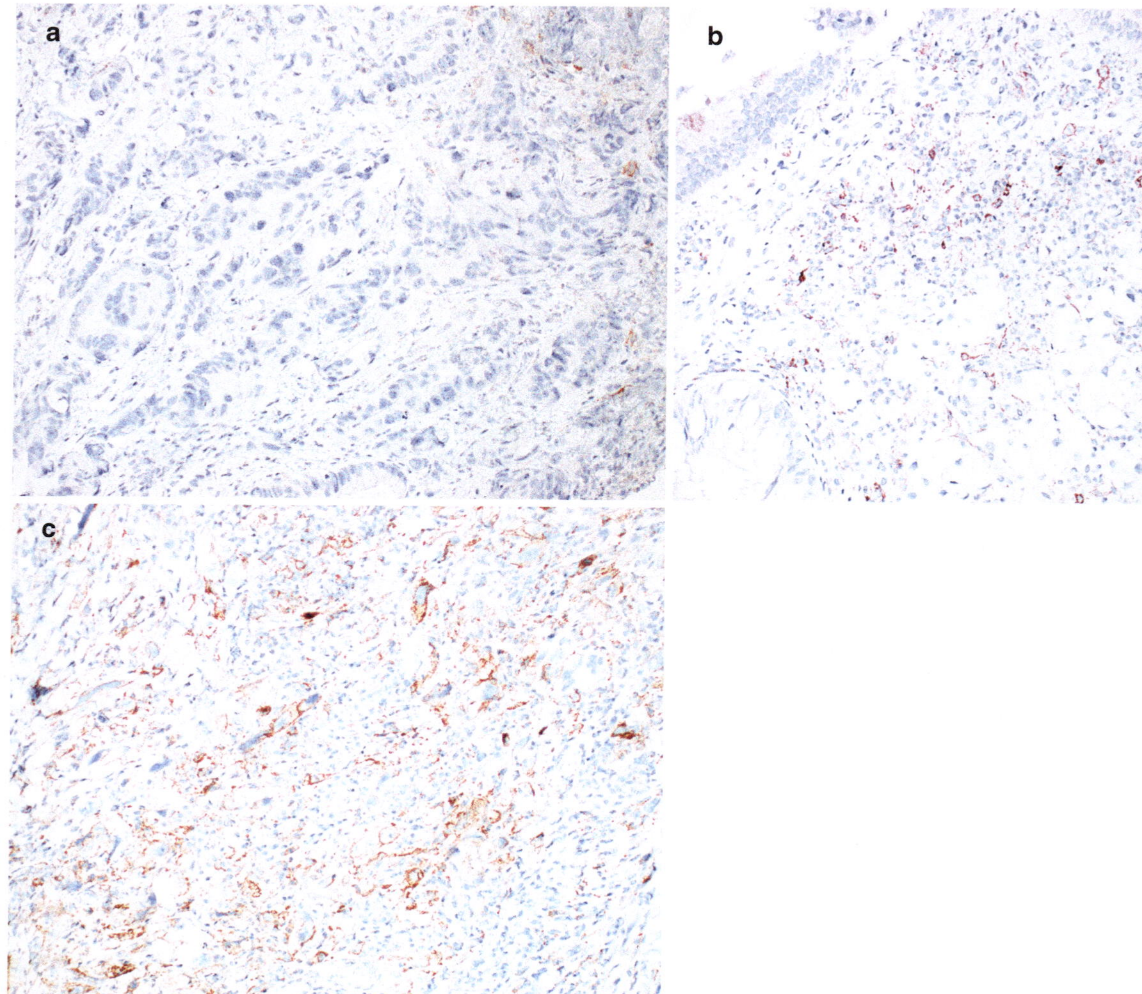

Fig. 6.10 (**a**) Esophageal adenocarcinoma demonstrating a Combined Positive Score of <1 on immunohistochemistry for PD-L1 at high magnification. (**b**) Poorly differentiated gastric carcinoma demonstrating a Combined Positive Score between 1 and 10 on immunohistochemistry for PD-L1 at high magnification. (**c**) Poorly differentiated gastric carcinoma demonstrating a Combined Positive Score of >10 on immunohistochemistry for PD-L1 at high magnification. (All images courtesy of Dr. David Escobar)

- o CPS $\geq$ 1 in gastric and gastroesophageal carcinomas is considered positive (Fig. 6.10b)
- o CPS $\geq$ 10 is the current cutoff for second-line treatment in ESCC (Fig. 6.10c)
- Trials are ongoing, and new staining cutoffs and treatment indications may be proposed in the future
- Note: Multiple antibody clones are commercially available and have different relevance to tumor response to various PD-1 inhibitors, so pathologists must be aware of which clone is being used in practice prior to reporting interpretation
- HER2
  - HER2 overexpression and *ERBB2* amplification are indications for therapy using HER2 inhibitors (e.g., trastuzumab)
  - IHC staining interpretation is specific to upper gastrointestinal tract carcinomas and should not be confused with breast or colon criteria (see Table 6.4 and colon cancer chapter for more information)
  - Initial evaluation is performed by immunohistochemistry in gastroesophageal adenocarcinomas (Fig. 6.11a–d)
  - If IHC staining is equivocal (2+), reflex to in situ hybridization and rely on results of that testing
- Mismatch repair (MMR) protein immunohistochemistry
  - Analogous to MMR staining in other organs, such as colon (see colon cancer chapter for more information)
  - Current guidelines state any nuclear staining represents intact expression (retention, proficiency) of the protein
    - o Lymphocytes and background normal mucosa provide helpful internal controls
  - Proficient (intact) pattern
    - o Presence of all four proteins suggests microsatellite stability of the tumor
  - Deficient (loss) patterns
    - o Loss of nuclear staining for any of the four proteins indicates microsatellite instability and suggests the most likely involved gene and the need for additional testing
    - o Loss of MSH2 alone or loss of MSH2/MSH6 = likely mutation in *MSH2*
    - o Loss of MLH1 alone or loss of MLH1/PMS2 = likely mutation in *MLH1* or sporadic promoter hypermethylation
    - o Isolated loss of MSH6 = likely mutation in MSH6
    - o Isolated loss of PMS2 = likely mutation in *PMS2*, but can be seen with heterogeneous *MLH1* promoter hypermethylation or *MLH1* mutations (especially missense mutations)
- Deficient patterns raise the possibility of Lynch syndrome, and additionally, these malignancies may respond to PD-1 inhibition therapy

Table 6.4 Comparison of HER2 IHC interpretation in gastroesophageal cancer and colorectal cancer

Interpretation	GEC (biopsy specimen)	GEC (surgical specimen)	CRC
Negative	No membranous reactivity in any cells	No reactivity or membranous reactivity in <10% of cells	No staining, or staining in <10% of cells
Negative	$\geq$ 5 cells with a faint or barely perceptible membranous reactivity (1+) irrespective of percentage of cells positive	Faint or barely perceptible membranous reactivity in $\geq$10% of cells (1+); cells are reactive only in part of their membrane	Faint staining (1+), any cellularity in a segmental or granular pattern
Negative			Moderate staining (2+), < 50% of cells, any pattern
Negative			Intense staining (3+), $\leq$ 10% of cells in a circumferential, basolateral, or lateral pattern
Equivocal (reflex to FISH)	$\geq$ 5 cells with a weak to moderate complete, basolateral, or lateral membranous reactivity irrespective of percentage of cells positive (2+)	Weak to moderate complete basolateral, or lateral membranous reactivity in $\geq$10% of cells (2+)	Moderate staining (2+), $\geq$ 50% of cells in a circumferential, basolateral, or lateral pattern
Positive	$\geq$ 5 cells with a strong complete basolateral, or lateral membranous reactivity irrespective of percentage of cells positive (3+)	Strong complete, basolateral or lateral membranous reactivity in $\geq$10% of cells (3+)	Intense staining (3+), > 10% of cells in a circumferential, basolateral, or lateral pattern

GEC gastroesophageal cancer, *CRC* colorectal cancer, *FISH* fluorescence in situ hybridization

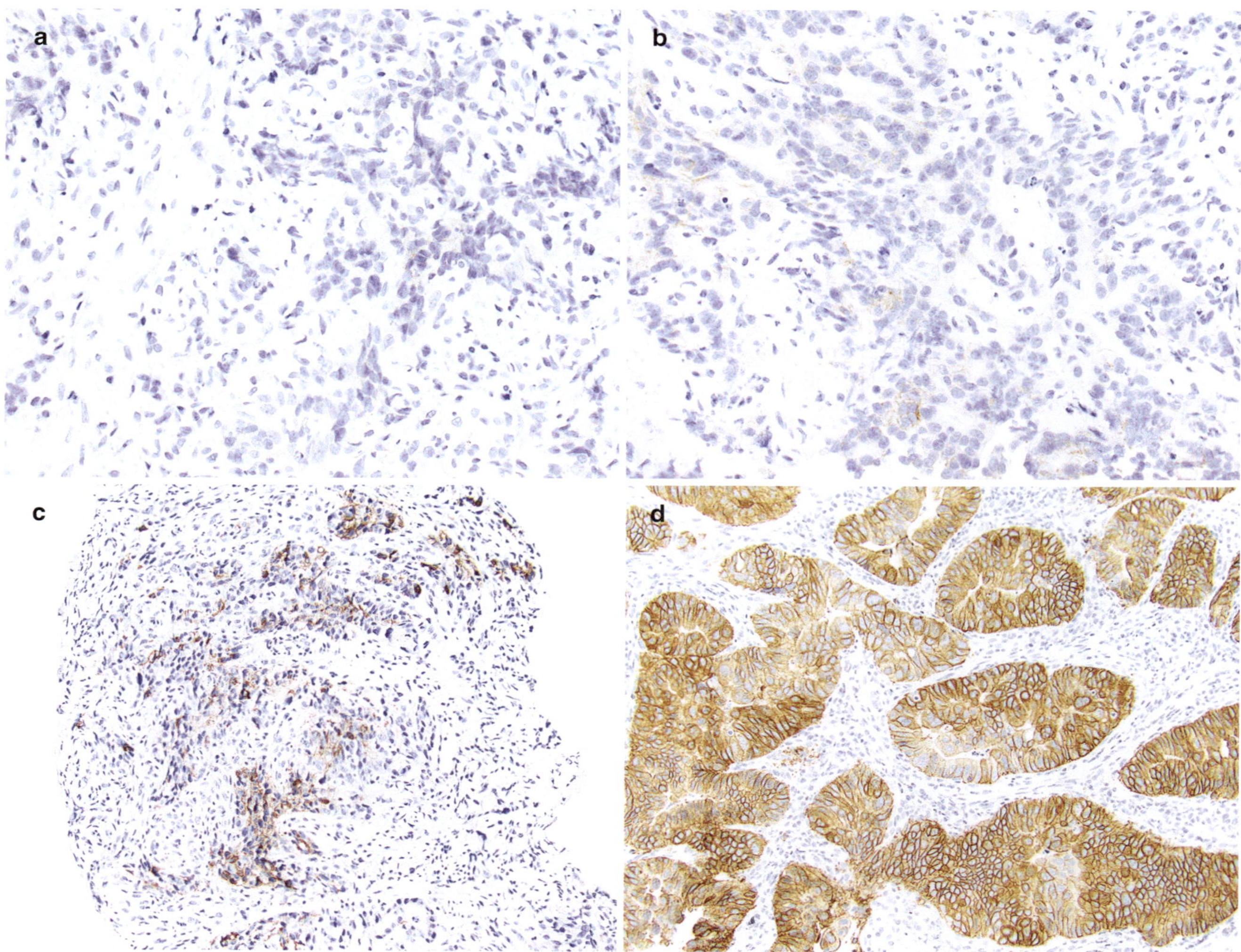

Fig. 6.11 (**a**) The results of immunohistochemistry for HER2 can help predict response to therapy in gastroesophageal carcinoma. This biopsy sample displays essentially no reactivity and is negative (0). (**b**) This biopsy sample displays faint membranous reactivity and is also negative (1+). (**c**) This biopsy sample displays moderate complete membranous reactivity staining and is equivocal (2+), meaning the case should then be assessed by fluorescence in situ hybridization for *ERBB2* amplification. (**b**) This biopsy sample displays strong complete membranous reactivity and is positive (3+)

Further Reading

Abraham SC, Montgomery EA, Singh VK, Yardley JH, Wu TT. Gastric adenomas: intestinal-type and gastric-type adenomas differ in the risk of adenocarcinoma and presence of background mucosal pathology. Am J Surg Pathol. 2002;26(10):1276–85.

Abraham SC, Nobukawa B, Giardiello FM, Hamilton SR, Wu TT. Fundic gland polyps in familial adenomatous polyposis: neoplasms with frequent somatic adenomatous polyposis coli gene alterations. Am J Pathol. 2000;157(3):747–54.

Abraham SC, Nobukawa B, Giardiello FM, Hamilton SR, Wu TT. Sporadic fundic gland polyps: common gastric polyps arising through activating mutations in the beta-catenin gene. Am J Pathol. 2001;158(3):1005–10.

Abraham SC, Park SJ, Cruz-Correa M, Houlihan PS, Half EE, Lynch PM, Wu TT. Frequent CpG Island methylation in sporadic and syndromic gastric fundic gland polyps. Am J Clin Pathol. 2004;122(5):740–6.

Abraham SC, Park SJ, Lee JH, Mugartegui L, Wu TT. Genetic alterations in gastric adenomas of intestinal and foveolar phenotypes. Mod Pathol. 2003;16(8):786–95.

Agaram NP, Wong GC, Guo T, et al. Novel V600E BRAF mutations in imatinib-naive and imatinib-resistant gastrointestinal stromal tumors. Genes Chromosomes Cancer. 2008;47(10):853–9.

Agaram NP, Zhang L, Jungbluth AA, Dickson BC, Antonescu CR. A molecular reappraisal of Glomus tumors and related Pericytic neoplasms with emphasis on NOTCH-gene fusions. Am J Surg Pathol. 2020;44(11):1556–62.

Aggarwal G, Sharma S, Zheng M, et al. Primary leiomyosarcomas of the gastrointestinal tract in the post-gastrointestinal stromal tumor era. Ann Diagn Pathol. 2012;16(6):532–40.

Aretz S, Stienen D, Uhlhaas S, et al. High proportion of large genomic STK11 deletions in Peutz-Jeghers syndrome. Hum Mutat. 2005;26(6):513–9.

Arnason T, Liang WY, Alfaro E, Kelly P, Chung DC, Odze RD, Lauwers GY. Morphology and natural history of familial adenomatous polyposis-associated dysplastic fundic gland polyps. Histopathology. 2014;65(3):353–62.

Astolfi A, Indio V, Nannini M, et al. Targeted deep sequencing uncovers cryptic KIT mutations in KIT/PDGFRA/SDH/RAS-P Wildtype GIST. Front Oncol. 2020;10:504.

Bang YJ, Van Cutsem E, Feyereislova A, et al. Trastuzumab in combination with chemotherapy versus chemotherapy alone for treatment of HER2-positive advanced gastric or gastro-oesophageal junction cancer (ToGA): a phase 3, open-label, randomised controlled trial. Lancet. 2010;376(9742):687–97.

Bartley AN, Washington MK, Ventura CB, et al. HER2 testing and clinical decision making in gastroesophageal adenocarcinoma: guideline from the College of American Pathologists, American Society for Clinical Pathology, and American Society of Clinical Oncology. Arch Pathol Lab Med. 2016;140(12):1345–63.

Bass AJ, Thorsson V, Shmulevich I, et al. Comprehensive molecular characterization of gastric adenocarcinoma. Nature. 2014;513(7517):202–9.

Bianchi LK, Burke CA, Bennett AE, Lopez R, Hasson H, Church JM. Fundic gland polyp dysplasia is common in familial adenomatous polyposis. Clin Gastroenterol Hepatol. 2008;6(2):180–5.

Boikos SA, Pappo AS, Killian JK, et al. Molecular subtypes of KIT/PDGFRA Wildtype gastrointestinal stromal tumors: a report from the National Institutes of Health Gastrointestinal Stromal Tumor Clinic. JAMA Oncol. 2016;2(7):922–8.

Boikos SA, Xekouki P, Fumagalli E, et al. Carney triad can be (rarely) associated with germline succinate dehydrogenase defects. Eur J Hum Genet. 2016;24(4):569–73.

Boni A, Lisovsky M, Dal Cin P, Rosenberg AE, Srivastava A. Atypical lipomatous tumor mimicking giant fibrovascular polyp of the esophagus: report of a case and a critical review of literature. Hum Pathol. 2013;44(6):1165–70.

Buffart TE, Carvalho B, Mons T, et al. DNA copy number profiles of gastric cancer precursor lesions. BMC Genomics. 2007;8:345.

Burkart AL, Sheridan T, Lewin M, Fenton H, Ali NJ, Montgomery E. Do sporadic Peutz-Jeghers polyps exist? Experience of a large teaching hospital. Am J Surg Pathol. 2007;31(8):1209–14.

Capelle LG, Van Grieken NC, Lingsma HF, et al. Risk and epidemiological time trends of gastric cancer in Lynch syndrome carriers in The Netherlands. Gastroenterology. 2010;138(2):487–92.

Carneiro F, Huntsman DG, Smyrk TC, et al. Model of the early development of diffuse gastric cancer in E-cadherin mutation carriers and its implications for patient screening. J Pathol. 2004;203(2):681–7.

Chlumská A, Waloschek T, Mukenšnabl P, Martínek P, Kašpírková J, Zámečník M. Pyloric gland adenoma: a histologic, immunohistochemical and molecular genetic study of 23 cases. Cesk Patol. 2015;51(3):137–43.

Codipilly DC, Qin Y, Dawsey SM, Kisiel J, Topazian M, Ahlquist D, Iyer PG. Screening for esophageal squamous cell carcinoma: recent advances. Gastrointest Endosc. 2018;88(3):413–26.

Cristescu R, Lee J, Nebozhyn M, et al. Molecular analysis of gastric cancer identifies subtypes associated with distinct clinical outcomes. Nat Med. 2015;21(5):449–56.

D'Adda T, Candidus S, Denk H, Bordi C, Höfler H. Gastric neuroendocrine neoplasms: tumour clonality and malignancy-associated large X-chromosomal deletions. J Pathol. 1999;189(3):394–401.

D'Adda T, Keller G, Bordi C, Höfler H. Loss of heterozygosity in 11q13-14 regions in gastric neuroendocrine tumors not associated with multiple endocrine neoplasia type 1 syndrome. Lab Investig. 1999;79(6):671–7.

de Boer WB, Ee H, Kumarasinghe MP. Neoplastic lesions of gastric adenocarcinoma and proximal polyposis syndrome (GAPPS) are gastric phenotype. Am J Surg Pathol. 2018;42(1):1–8.

Debelenko LV, Emmert-Buck MR, Zhuang Z, et al. The multiple endocrine neoplasia type I gene locus is involved in the pathogenesis of type II gastric carcinoids. Gastroenterology. 1997;113(3):773–81.

Delnatte C, Sanlaville D, Mougenot JF, et al. Contiguous gene deletion within chromosome arm 10q is associated with juvenile polyposis of infancy, reflecting cooperation between the BMPR1A and PTEN tumor-suppressor genes. Am J Hum Genet. 2006;78(6):1066–74.

Fukayama M, Rugge M, Washington MK. Tumours of the stomach. In: WHO Classification of Tumours Editorial Board, editor. WHO classification of tumours: digestive system tumours. 5th ed. Lyon: IARC; 2019. p. 59–110.

Giardiello FM, Brensinger JD, Tersmette AC, et al. Very high risk of cancer in familial Peutz-Jeghers syndrome. Gastroenterology. 2000;119(6):1447–53.

Girardi DM, Silva ACB, Rêgo JFM, Coudry RA, Riechelmann RP. Unraveling molecular pathways of poorly differentiated neuroendocrine carcinomas of the gastroenteropancreatic system: a systematic review. Cancer Treat Rev. 2017;56:28–35.

Gonzalez RS, Adsay V, Graham RP, et al. Massive gastric juvenile-type polyposis: a clinicopathological analysis of 22 cases. Histopathology. 2017;70(6):918–28.

Gonzalez RS, Cates JMM, Revetta F, McMahon LA, Washington K. Gastric carcinomas with lymphoid stroma: categorization and comparison with solid-type colonic carcinomas. Am J Clin Pathol. 2017;148(6):477–84.

Graham RP, Nair AA, Davila JI, et al. Gastroblastoma harbors a recurrent somatic MALAT1-GLI1 fusion gene. Mod Pathol. 2017;30(10):1443–52.

Graham RP, Yasir S, Fritchie KJ, Reid MD, Greipp PT, Folpe AL. Polypoid fibroadipose tumors of the esophagus: 'giant fibrovascular polyp' or liposarcoma? A clinicopathological and molecular cytogenetic study of 13 cases. Mod Pathol. 2018;31(2):337–42.

Gupta S, Li D, El Serag HB, et al. AGA clinical practice guidelines on management of gastric intestinal metaplasia. Gastroenterology. 2020;158(3):693–702.

Hansford S, Kaurah P, Li-Chang H, et al. Hereditary diffuse gastric cancer syndrome: CDH1 mutations and beyond. JAMA Oncol. 2015;1(1):23–32.

Hashimoto T, Ogawa R, Matsubara A, et al. Familial adenomatous polyposis-associated and sporadic pyloric gland adenomas of the upper gastrointestinal tract share common genetic features. Histopathology. 2015;67(5):689–98.

Heinrich MC, Patterson J, Beadling C, et al. Genomic aberrations in cell cycle genes predict progression of KIT-mutant gastrointestinal stromal tumors (GISTs). Clin Sarcoma Res. 2019;9:3.

Hidaka Y, Mitomi H, Saito T, et al. Alteration in the Wnt/β-catenin signaling pathway in gastric neoplasias of fundic gland (chief cell predominant) type. Hum Pathol. 2013;44(11):2438–48.

Horton RK, Ahadi M, Gill AJ, et al. SMARCA4/SMARCA2-deficient carcinoma of the esophagus and gastroesophageal junction. Am J Surg Pathol. 2021;45(3):414–20.

Huss S, Wardelmann E, Goltz D, et al. Activating PDGFRA mutations in inflammatory fibroid polyps occur in exons 12, 14 and 18 and are associated with tumour localization. Histopathology. 2012;61(1):59–68.

Ibrahim A, Chopra S. Succinate dehydrogenase-deficient gastrointestinal stromal tumors. Arch Pathol Lab Med. 2020;144(5):655–60.

Jain R, Chetty R. Gastric hyperplastic polyps: a review. Dig Dis Sci. 2009;54(9):1839–46.

Karamzadeh Dashti N, Bahrami A, Lee SJ, Jenkins SM, Rodriguez FJ, Folpe AL, Boland JM. BRAF V600E mutations occur in a subset of glomus tumors, and are associated with malignant histologic characteristics. Am J Surg Pathol. 2017;41(11):1532–41.

Kim J, Bowlby R, Mungall AJ, et al. Integrated genomic characterization of oesophageal carcinoma. Nature. 2017;541(7636):169–75.

Kim J, Braun D, Ukaegbu C, et al. Clinical factors associated with gastric cancer in individuals with lynch syndrome. Clin Gastroenterol Hepatol. 2020;18(4):830–837.e1.

Kwon MJ, Min BH, Lee SM, et al. Serrated adenoma of the stomach: a clinicopathologic, immunohistochemical, and molecular study of nine cases. Histol Histopathol. 2013;28(4):453–62.

Lasota J, Wasag B, Dansonka-Mieszkowska A, et al. Evaluation of NF2 and NF1 tumor suppressor genes in distinctive gastrointestinal nerve sheath tumors traditionally diagnosed as benign schwannomas: s study of 20 cases. Lab Investig. 2003;83(9):1361–71.

Lee JH, Abraham SC, Kim HS, et al. Inverse relationship between APC gene mutation in gastric adenomas and development of adenocarcinoma. Am J Pathol. 2002;161(2):611–8.

Lee SE, Kang SY, Cho J, et al. Pyloric gland adenoma in lynch syndrome. Am J Surg Pathol. 2014;38(6):784–92.

Lee SY, Saito T, Mitomi H, et al. Mutation spectrum in the Wnt/β-catenin signaling pathway in gastric fundic gland-associated neoplasms/polyps. Virchows Arch. 2015;467(1):27–38.

Lei Z, Tan IB, Das K, et al. Identification of molecular subtypes of gastric cancer with different responses to PI3-kinase inhibitors and 5-fluorouracil. Gastroenterology. 2013;145(3):554–65.

Levi Z, Baris HN, Kedar I, et al. Upper and lower gastrointestinal findings in PTEN mutation-positive Cowden syndrome patients participating in an active surveillance program. Clin Transl Gastroenterol. 2011;2:e5.

Li J, Woods SL, Healey S, et al. Point mutations in exon 1B of APC reveal gastric adenocarcinoma and proximal polyposis of the stomach as a familial adenomatous polyposis variant. Am J Hum Genet. 2016;98(5):830–42.

Lim CH, Cho YK, Kim SW, et al. The chronological sequence of somatic mutations in early gastric carcinogenesis inferred from multiregion sequencing of gastric adenomas. Oncotarget. 2016;7(26):39758–67.

Lin L, Bass AJ, Lockwood WW, et al. Activation of GATA binding protein 6 (GATA6) sustains oncogenic lineage-survival in esophageal adenocarcinoma. Proc Natl Acad Sci U S A. 2012;109(11):4251–6.

Liu TC, Lin MT, Montgomery EA, Singhi AD. Inflammatory fibroid polyps of the gastrointestinal tract: spectrum of clinical, morphologic, and immunohistochemistry features. Am J Surg Pathol. 2013;37(4):586–92.

Liu X, Chu KM. Molecular biomarkers for prognosis of gastrointestinal stromal tumor. Clin Transl Oncol. 2019;21(2):145–51.

Ma C, Giardiello FM, Montgomery EA. Upper tract juvenile polyps in juvenile polyposis patients: dysplasia and malignancy are associated with foveolar, intestinal, and pyloric differentiation. Am J Surg Pathol. 2014;38(12):1618–26.

Markowski AR, Markowska A, Guzinska-Ustymowicz K. Pathophysiological and clinical aspects of gastric hyperplastic polyps. World J Gastroenterol. 2016;22(40):8883–91.

Miettinen M, Dow N, Lasota J, Sobin LH. A distinctive novel epitheliomesenchymal biphasic tumor of the stomach in young adults ("gastroblastoma"): a series of 3 cases. Am J Surg Pathol. 2009;33(9):1370–7.

Miettinen M, Paal E, Lasota J, Sobin LH. Gastrointestinal glomus tumors: a clinicopathologic, immunohistochemical, and molecular genetic study of 32 cases. Am J Surg Pathol. 2002;26(3):301–11.

Nogueira AM, Carneiro F, Seruca R, Cirnes L, Veiga I, Machado JC, Sobrinho-Simões M. Microsatellite instability in hyperplastic and adenomatous polyps of the stomach. Cancer. 1999;86(9):1649–56.

Nozu K, Minamikawa S, Yamada S, et al. Characterization of contiguous gene deletions in COL4A6 and COL4A5 in alport syndrome-diffuse leiomyomatosis. J Hum Genet. 2017;62(7):733–5.

Odze RD, Lam AK, Ochiai A, Washington MK. Tumours of the oesophagus. In: WHO Classification of Tumours Editorial Board, editor. WHO classification of tumours: digestive system tumours. 5th ed. Lyon: IARC; 2019. p. 23–58.

Panagopoulos I, Gorunova L, Lund-Iversen M, Lobmaier I, Bjerkehagen B, Heim S. Recurrent fusion of the genes FN1 and ALK in gastrointestinal leiomyomas. Mod Pathol. 2016;29(11):1415–23.

Pantaleo MA, Nannini M, Corless CL, Heinrich MC. Quadruple wild-type (WT) GIST: defining the subset of GIST that lacks abnormalities of KIT, PDGFRA, SDH, or RAS signaling pathways. Cancer Med. 2015;4(1):101–3.

Papke DJ, Forgó E, Charville GW, Hornick JL. PDGFRA immunohistochemistry predicts PDGFRA mutations in gastrointestinal stromal tumors. Am J Surg Pathol. 2022;46(1):3–10.

Pareja F, Brandes AH, Basili T, et al. Loss-of-function mutations in ATP6AP1 and ATP6AP2 in granular cell tumors. Nat Commun. 2018;9(1):3533.

Pasini B, McWhinney SR, Bei T, et al. Clinical and molecular genetics of patients with the carney-stratakis syndrome and germline mutations of the genes coding for the succinate dehydrogenase subunits SDHB, SDHC, and SDHD. Eur J Hum Genet. 2008;16(1):79–88.

Qumseya B, Sultan S, Bain P, et al. ASGE guideline on screening and surveillance of Barrett's esophagus. Gastrointest Endosc. 2019;90(3):335–359.e2.

Redston M, Noffsinger A, Kim A, et al. Abnormal TP53 predicts risk of progression in patients with Barrett's esophagus regardless of a diagnosis of dysplasia. Gastroenterology. 2022;162(2):468–81.

Ricci R, Martini M, Cenci T, et al. PDGFRA-mutant syndrome. Mod Pathol. 2015;28(7):954–64.

Rogers WM, Dobo E, Norton JA, et al. Risk-reducing total gastrectomy for germline mutations in E-cadherin (CDH1): pathologic findings with clinical implications. Am J Surg Pathol. 2008;32(6):799–809.

Rosenbaum MW, Gonzalez RS. Targeted therapy for upper gastrointestinal tract cancer: current and future prospects. Histopathology. 2021;78(1):148–61.

Saab J, Estrella JS, Chen YT, Yantiss RK. Immunohistochemical and molecular features of gastric hyperplastic polyps. Adv Cytol Pathol. 2017;2(1):25–31.

Saha A, Chowdhury S, Bandyopadhyay D. Multifocal cutaneous granular cell tumor: an uncommon clinical entity. Indian J Dermatol. 2015;60(6):627–9.

Salomao M, Luna AM, Sepulveda JL, Sepulveda AR. Mutational analysis by next generation sequencing of gastric type dysplasia occurring in hyperplastic polyps of the stomach: mutations in gastric hyperplastic polyps. Exp Mol Pathol. 2015;99(3):468–73.

Savant D, Zhang Q, Yang Z. Squamous Neoplasia in the esophagus. Arch Pathol Lab Med. 2021;145(5):554–61.

Schaefer IM, Wang Y, Liang CW, et al. MAX inactivation is an early event in GIST development that regulates p16 and cell proliferation. Nat Commun. 2017;8:14674.

Spans L, Fletcher CD, Antonescu CR, Rouquette A, Coindre JM, Sciot R, Debiec-Rychter M. Recurrent MALAT1-GLI1 oncogenic fusion and GLI1 up-regulation define a subset of plexiform fibromyxoma. J Pathol. 2016;239(3):335–43.

Stachler MD, Jin RU. Molecular pathology of Gastroesophageal cancer. Surg Pathol Clin. 2021;14(3):443–53.

Stockman DL, Miettinen M, Suster S, et al. Malignant gastrointestinal neuroectodermal tumor: clinicopathologic, immunohistochemical, ultrastructural, and molecular analysis of 16 cases with a reappraisal of clear cell sarcoma-like tumors of the gastrointestinal tract. Am J Surg Pathol. 2012;36(6):857–68.

Straub SF, Drage MG, Gonzalez RS. Comparison of dysplastic fundic gland polyps in patients with and without familial adenomatous polyposis. Histopathology. 2018;72(7):1172–9.

Su HA, Yen HH, Chen CJ. An update on clinicopathological and molecular features of plexiform fibromyxoma. Can J Gastroenterol Hepatol. 2019;2019:3960920.

Sweet K, Willis J, Zhou XP, et al. Molecular classification of patients with unexplained hamartomatous and hyperplastic polyposis. JAMA. 2005;294(19):2465–73.

Szucs Z, Thway K, Fisher C, et al. Molecular subtypes of gastrointestinal stromal tumors and their prognostic and therapeutic implications. Future Oncol. 2017;13(1):93–107.

Tan IB, Ivanova T, Lim KH, et al. Intrinsic subtypes of gastric cancer, based on gene expression pattern, predict survival and respond differently to chemotherapy. Gastroenterology. 2011;141(2):476–85, 485.e1–11

Testa U, Castelli G, Pelosi E. Esophageal cancer: genomic and molecular characterization, stem cell compartment and clonal evolution. Medicines (Basel). 2017;4(3):E67.

Torbenson M, Lee JH, Cruz-Correa M, Ravich W, Rastgar K, Abraham SC, Wu TT. Sporadic fundic gland polyposis: a clinical, histological, and molecular analysis. Mod Pathol. 2002;15(7):718–23.

Urbini M, Indio V, Tarantino G, et al. Gain of FGF4 is a frequent event in KIT/PDGFRA/SDH/RAS-P WT GIST. Genes Chromosomes Cancer. 2019;58(9):636–42.

Ushiku T, Kunita A, Kuroda R, et al. Oxyntic gland neoplasm of the stomach: expanding the spectrum and proposal of terminology. Mod Pathol. 2020;33(2):206–16.

Voltaggio L, Murray R, Lasota J, Miettinen M. Gastric schwannoma: a clinicopathologic study of 51 cases and critical review of the literature. Hum Pathol. 2012;43(5):650–9.

Vyas M, Yang X, Zhang X. Gastric Hamartomatous polyps-review and update. Clin Med Insights Gastroenterol. 2016;9:3–10.

Wang Q, Liu G, Hu C. Molecular classification of gastric adenocarcinoma. Gastroenterology Res. 2019;12(6):275–82.

Wang Y, Marino-Enriquez A, Bennett RR, et al. Dystrophin is a tumor suppressor in human cancers with myogenic programs. Nat Genet. 2014;46(6):601–6.

Wani S, Qumseya B, Sultan S, et al. Endoscopic eradication therapy for patients with Barrett's esophagus-associated dysplasia and intramucosal cancer. Gastrointest Endosc. 2018;87(4):907–931.e9.

Wu CE, Tzen CY, Wang SY, Yeh CN. Clinical diagnosis of gastrointestinal stromal tumor (GIST): from the molecular genetic point of view. Cancers (Basel). 2019;11(5):E679.

Molecular Pathology of Pancreatic Tumors

Jae W. Lee, N. Volkan Adsay, Ralph H. Hruban, and Laura D. Wood

Contents

J. W. Lee · R. H. Hruban · L. D. Wood (✉)
Department of Pathology, The Sol Goldman Pancreatic Cancer
Research Center, Johns Hopkins University, Baltimore, MD, USA
e-mail: ldwood@jhmi.edu

N. V. Adsay
Department of Pathology, Koç University School of Medicine,
Istanbul, Turkey

Introduction to Normal Pancreas

- The exocrine and endocrine functions of the pancreas are performed by distinct cell types
  - Exocrine pancreas (enzymes secreted into ducts) consists of enzyme-producing acinar cells as well as mucin-secreting duct cells that transport the enzymes
  - Endocrine pancreas (hormones released into the bloodstream) consists of multiple types of hormone-secreting cells
- Histology of the pancreas parallels this functional division, with cellular components divided into four compartments: acinar, ductal, endocrine, and interstitial
  - The exocrine pancreas (acinar and ductal components) consists of clustered acinar cells that secrete digestive enzymes into small ductules, which join to form progressively larger ducts that ultimately drain into the duodenum
  - The endocrine compartment is composed of the islets of Langerhans scattered among the acini
  - These compartments are supported by interstitial stroma
- Neoplasms of the pancreas can be broadly divided based on the direction of differentiation of the neoplastic cells
- Neoplasms with distinct differentiation exhibit unique clinical, morphologic, and molecular features

© The Author(s), under exclusive license to Springer Nature Switzerland AG 2023
L. Cheng et al. (eds.), *Molecular Surgical Pathology*, https://doi.org/10.1007/978-3-031-35118-1_7

Neoplasms with Ductal Differentiation

Pancreatic Ductal Adenocarcinoma

Clinical Features

- Approximately 466,000 deaths/year in the world, 48,220 deaths/year in the United States
  - Seventh leading cause of cancer death worldwide, third leading cause of cancer death in the United States
  - Slight male predominance (male to female ratio of approximately 3.2:2.8); a disease of the elderly, as most patients are diagnosed between 60 and 80 years of age
  - Incidence of ductal adenocarcinoma is higher in African-Americans than in Caucasians; the age-adjusted incidence rate for Caucasian males in the United States is 15.4 per 100,000 and for African-American males is 17.1 per 100,000
  - Most important risk factor is cigarette smoking; other known risk factors include age, obesity, chronic pancreatitis, and diabetes mellitus
  - See Hereditary/Genetic Syndromes section for inherited disorders associated with pancreatic ductal adenocarcinoma
- Symptoms of ductal adenocarcinoma are often nonspecific, including back pain, unexplained weight loss, and jaundice
- Aggressive neoplasm that is almost uniformly fatal
  - The overall 5-year survival rate is ~10%
  - Patients who undergo surgical resection and adjuvant chemotherapy have a 5-year survival rate of nearly 30%
  - Patients with metastatic disease have a 5-year survival rate of only 3.5%
  - Resectability and stage are the most important determinants of prognosis

Pathologic Features

- Majority of ductal adenocarcinomas (60–70%) arise in the head of the pancreas
- Usually form solitary ill-defined masses
- Average size of surgically resected carcinoma in pancreatic head is 3 cm in greatest dimension, while average size of carcinoma in pancreatic tail is 5 cm in greatest dimension
- On gross cut section, ductal adenocarcinomas are firm white-yellow poorly demarcated masses that obscure the normal lobular architecture of the pancreas
  - If the carcinoma obstructs the pancreatic duct, the pancreatic duct upstream from the carcinoma may be dilated, and the upstream pancreatic parenchyma may be firm and atrophic due to obstructive chronic pancreatitis
  - If the carcinoma obstructs the distal common bile duct as it courses through the pancreas, the upstream bile duct will also be dilated
- Microscopically, pancreatic ductal adenocarcinoma is an invasive gland-forming neoplasm that elicits an intense stromal desmoplastic response
  - Histologic features of adenocarcinoma include haphazard arrangement of glands, intracytoplasmic mucin of variable types and amounts, nuclear pleomorphism, incomplete glandular lumina, luminal necrosis, neoplastic glands immediately adjacent to muscular vessels, perineural invasion, and lymphovascular invasion
  - Venous invasion is particularly common
  - These carcinomas are graded based on their degree of differentiation
    - Well-differentiated: Well-formed neoplastic glands lined by well-oriented cuboidal to columnar epithelium with an enlarged round to oval nuclei and eosinophilic cytoplasm
    - Moderately-differentiated: Irregularly shaped and small glands with poorly oriented cells, moderate nuclear pleomorphism, prominent nucleoli, and decreased mucin production
    - Poorly-differentiated: Mixture of irregular glands, solid sheets, and single neoplastic cells with marked nuclear pleomorphism, minimal mucin production, and significant mitotic activity
- In addition to the morphology described, there are several morphological variants of ductal adenocarcinoma with unique prognostic and molecular features
- Pancreata with invasive ductal adenocarcinoma also frequently harbor microscopic noninvasive epithelial proliferations within the pancreatic ducts known as pancreatic intraepithelial neoplasia (PanIN)
  - These lesions are believed to be precursors to invasive adenocarcinoma and are graded based on architectural and cytologic atypia
    - Low-grade PanIN: Flat or papillary lesions composed of uniform columnar cells with small round basal nuclei and abundant supranuclear mucin; some have moderate nuclear atypia (loss of polarity, pseudostratification, enlargement, hyperchromasia)
    - High-grade PanIN: Flat or papillary lesions with severe architectural atypia (budding, cribriforming, luminal necrosis) and nuclear atypia (enlargement, hyperchromasia, loss of orientation, and polarity, prominent nucleoli); mitoses are common

- Immunohistochemistry
  - Positive in neoplastic cells: cytokeratins (CK7, CK8, CK13, CK18, CK19), CEA, CA19–9, CA125, MUC1, MUC3, MUC4, MUC5AC
  - Variably and/or focally positive in neoplastic cells: CK20, MUC2, MUC6
  - Typically negative in neoplastic cells: vimentin, chromogranin, synaptophysin, trypsin, chymotrypsin, lipase, and BCL-10
    - SMAD4 expression is lost in approximately 55% of ductal adenocarcinomas, reflecting genetic inactivation of the *SMAD4* gene (Fig. 7.1a)
    - p53 protein expression is abnormal (null or strongly positive) in 75% of ductal carcinomas, reflecting somatic mutations in the *TP53* gene

Genetic Features

- See Table 7.1
- *KRAS* is the most frequently altered oncogene in ductal adenocarcinomas—somatic mutations in *KRAS* occur in >90% of ductal adenocarcinomas
  - These somatic mutations cluster in specific hotspots (most commonly codon 12) and activate the gene product
    - *KRAS* encodes a small GTPase that mediates downstream signaling from growth factor receptors, playing a crucial role in proliferation, survival, and differentiation; the protein encoded by *BRAF* is activated by *KRAS* and functions in the same cell signaling pathways
    - *KRAS* wildtype ductal adenocarcinomas have been reported to have microsatellite instability/defective DNA mismatch repair, kinase-fusion genes, and activation of MAPK pathway
- *p16/CDKN2A* is the most frequently altered tumor suppressor gene in ductal adenocarcinoma—loss of p16 function is seen in >90% of ductal adenocarcinomas
  - Mechanisms of functional alteration include intragenic mutation coupled with the loss of the second allele, homozygous deletion of the gene, and promoter methylation of *p16/CDKN2A*
    - Encoded by *p16/CDKN2A*, the p16 protein plays a crucial role in cell cycle regulation; p16 blocks cell cycle progression by preventing the inactivation of Rb, another crucial cell cycle regulator
- Somatic mutations in *TP53* also occur frequently in ductal adenocarcinomas, reported in 75% of cases
  - Strong diffuse nuclear immunohistochemical labeling for p53 is associated with gene mutation (Fig. 7.1b); null expression (complete loss of labeling) can also be seen

- *TP53* inactivation occurs most frequently through a small intragenic mutation, coupled with loss of the wildtype allele
    - The protein encoded by *TP53* plays a key role in the cellular stress response, inhibiting cell growth and promoting cell death in the setting of cellular stress
- Somatic inactivation of *SMAD4* occurs in approximately 55% of ductal adenocarcinomas
  - This gene is inactivated through either homozygous deletion or intragenic mutation coupled with the loss of the wildtype allele
    - The protein encoded by *SMAD4* mediates cellular signaling downstream of the transforming growth factor β (TGFβ) receptor, playing a crucial role in the regulation of proliferation, migration, and apoptosis
    - Mutations in *SMAD4* have been associated with poor prognosis
    - Immunohistochemical loss of SMAD4 protein expression is correlated with the presence of genetic inactivation and can be used as a diagnostic tool to distinguish ductal adenocarcinoma from nonneoplastic pancreatic disease (Fig. 7.1a)
    - Less frequent somatic mutations have also been reported in other members of the TGFβ signaling pathway, including *TGFBR2* and *ALK5*
- Although less prevalent, somatic mutations have also been observed in *RNF43*, *GATA6* (amplification), *ARID1A*, *GNAS*, *RREB1*, *PBRM1*, and other genes
- Specific key genes (*KRAS*, *TP53*, *SMAD4*, and *p16/CDKN2A*) are crucial components of tumorigenesis in ductal adenocarcinoma that are preserved in different tumor subgroups
  - The prevalence of alterations in *KRAS*, *TP53*, and *SMAD4* is similar in familial and sporadic ductal adenocarcinomas
  - Carcinomas from smokers contain significantly more somatic mutations than those from never smokers, but mutations in known driver genes, including *KRAS*, *TP53*, *p16/CDKN2A*, and *SMAD4*, do not differ between the two groups
- Studies of copy number alterations have revealed numerous areas of gains and losses in ductal adenocarcinomas, and cytogenetic analyses have revealed complex karyotypes
  - Some of these large chromosome alterations target the known oncogenes and tumor suppressor genes discussed above
  - Others identify genomic regions that may contain loci involved in pancreatic tumorigenesis, but further characterization and validation are necessary

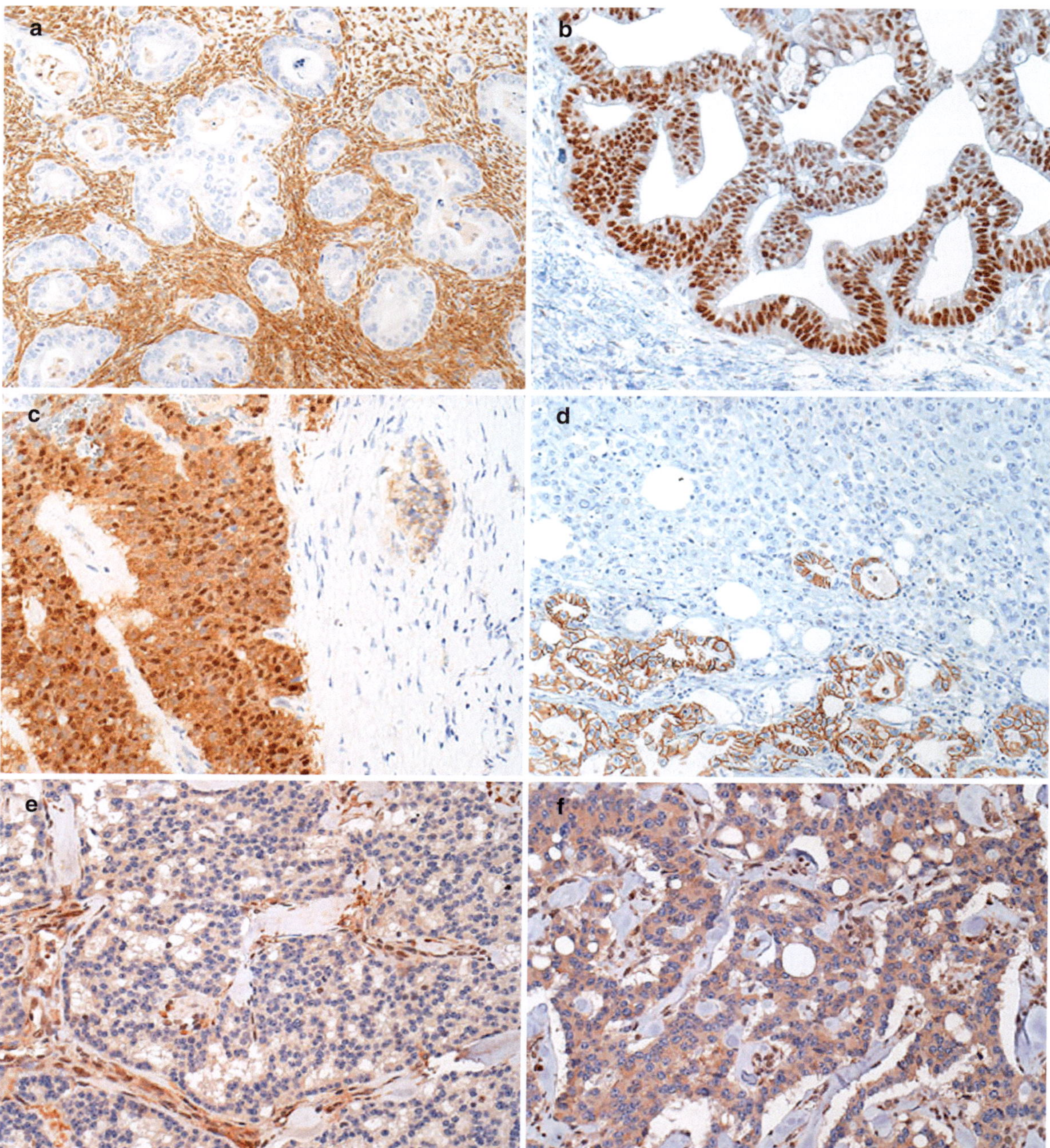

Fig. 7.1 Immunohistochemical labeling reflects genetic alterations in pancreatic neoplasms. (**a**) Pancreatic ductal adenocarcinomas frequently contain inactivating somatic mutations in *SMAD4*, leading to loss of SMAD4 protein expression in neoplastic cells. Immunohistochemical labeling for SMAD4 shows retention of nuclear labeling in the reactive stroma, while the neoplastic glands are strikingly negative. (**b**) Somatic mutations in *TP53*, leading to abnormal nuclear accumulation of p53 protein, occur in a large proportion of pancreatic ductal adenocarcinomas. Immunohistochemical labeling for p53 shows strong nuclear labeling in the neoplastic epithelium, while the reactive stroma does not label. (**c**) Solid pseudopapillary neoplasms frequently contain somatic mutations in *CTNNB1*, leading to abnormal nuclear accumulation of β-catenin protein. Immunohistochemical labeling for β-catenin shows strong nuclear and cytoplasmic labeling in the neoplasm on the *left*. On the *right*, a nonneoplastic duct has normal membranous staining for β-catenin. (**d**) In contrast to well-differentiated pancreatic ductal adenocarcinoma, undifferentiated carcinomas frequently show loss of E-cadherin protein expression. Immunohistochemical labeling of a mixed carcinoma demonstrates retention of E-cadherin expression in the well-differentiated component (*lower half*), while E-cadherin is lost in the undifferentiated component (*upper half*). (**e**, **f**) Approximately 45% of well-differentiated pancreatic neuroendocrine tumors (PanNETs) contain inactivating somatic mutations in *DAXX* or *ATRX*, leading to loss of protein expression in the neoplasm. Immunohistochemical labeling shows that although endothelial cells show strong nuclear labeling, neoplastic cells are negative for ATRX (**e**) and DAXX (**f**) in tumors with somatic mutations

Table 7.1 Somatic mutation prevalence of commonly altered genes in pancreatic neoplasms

Neoplasm	Gene	Chromosome	Alteration prevalence	Mechanisms of alteration
ACC	*APC*	5	15%	Inactivating/truncating mutation with LOH
	BRAF and *RAF1*	3 and 7	25%	Gene fusions
	CTNNB1	3	5%	Missense mutation
IOPN	*PRKACB*	1	75%	Gene fusions
	PRKACA	19	25%	Gene fusions
IPMN	*KRAS*	12	80%	Missense mutation
	RNF43	17	75%	Missense mutation or nonsense mutation with LOH
	GNAS	20	60%	Missense mutation
	p16/CDKN2A	9	Only in HGD/carcinoma	Missense mutation with LOH, homozygous deletion, promoter methylation
	TP53	17	Only in HGD/carcinoma	Missense mutation with LOH
	SMAD4	18	Only in HGD/carcinoma	Missense mutation with LOH, homozygous deletion
	KLF4	9	>50%	Hot spot mutations (K409 and S411), enriched in low-grade IPMNs
	PIK3CA	3	10%	Missense mutation
MCN	*KRAS*	12	80%	Missense mutation
	RNF43	17	40%	Missense mutation or nonsense mutation with LOH
	p16/CDKN2A	9	Only in HGD/carcinoma	Missense mutation with LOH, homozygous deletion, promoter methylation
	TP53	17	Only in HGD/carcinoma	Missense mutation with LOH
	SMAD4	18	Only in HGD/carcinoma	Missense mutation with LOH, homozygous deletion
PanNET	*MEN1*	11	45%	Missense mutation with LOH
	DAXX/ATRX	6/X	45%	Missense or nonsense mutation with LOH
	mTOR pathway	Multiple	15%	Multiple
	VHL	3	25%	Promoter methylation
PB	*CTNNB1*	3	55%	Missense mutation
	APC	5	10%	Inactivating/truncating mutation with LOH
	Unknown	11	85%	Loss of heterozygosity
PDA	*KRAS*	12	95%	Missense mutation
	p16/CDKN2A	9	95%	Missense mutation with LOH, homozygous deletion, promoter methylation
	TP53	17	75%	Missense mutation with LOH
	SMAD4	18	55%	Missense mutation with LOH, homozygous deletion
SCA	*VHL*	3	>20%	Missense mutation with LOH
SPN	*CTNNB1*	3	95%	Missense mutation

ACC acinar cell carcinoma, *IOPN* intraductal oncocytic papillary neoplasm, *IPMN* intraductal papillary mucinous neoplasm, *MCN* mucinous cystic neoplasm, *PanNET* well-differentiated pancreatic neuroendocrine tumor, *PB* pancreatoblastoma, *PDA* pancreatic ductal adenocarcinoma, *SCA* serous cystadenoma, *SPN* solid-pseudopapillary neoplasm, *HGD* high-grade dysplasia, *carcinoma* invasive carcinoma, *LOH* loss of heterozygosity

- – Chromothripsis, the occurrence of large numbers of chromosomal rearrangements localized to one or a few chromosomes, can be seen in ductal adenocarcinomas
- MicroRNAs, small noncoding RNAs that negatively regulate gene expression, are also altered in ductal adenocarcinoma
  - – Multiple microRNAs are differentially expressed in carcinomas compared to nonneoplastic pancreas and chronic pancreatitis
    - o Specific microRNA expression profiles show prognostic significance
    - o MicroRNAs may have diagnostic utility, as microRNA levels in fine needle aspirations have been correlated with ductal adenocarcinoma
- PanINs, the noninvasive precursor lesions to pancreatic cancer, acquire the molecular changes common in invasive ductal adenocarcinoma
  - – While some molecular changes reliably occur early in pancreatic tumorigenesis, others are limited to high-grade dysplasia and invasive lesions
    - o Somatic *KRAS* mutations are present even in low-grade PanINs

- o Loss of p16 expression can also be an early event, with loss in a subset of low-grade PanIN lesions; however, the prevalence of detected p16 loss increases with increasing PanIN grade
- o Telomere shortening has been reported in a large proportion of PanINs including low-grade lesions, with shortening in approximately 90% of low-grade PanINs
    - ♦ Thus, telomere shortening is one of the most frequently occurring early events in pancreatic tumorigenesis
- o Mutations in *CDKN2A*, *TP53*, and *SMAD4* are late genetic alterations that occur predominantly in invasive carcinoma and are typically absent in PanINs
- Transcriptomics studies based on gene arrays and next-generation RNA sequencing demonstrate that pancreatic ductal adenocarcinomas have distinct subtypes, including basal-like and classical subtypes
    - Basal-like subtype is associated with worse outcomes than other subtypes
    - *GATA6* amplification is more common in the classical subtype
    - Distinct subtypes may be amenable to chemotherapies directed toward subtype-specific targets
- Recent genomic, transcriptomic, and proteomic studies have revealed a complex molecular landscape of ductal adenocarcinomas
    - Four patterns of variation in chromosomal structure (stable, locally rearranged, scattered, and unstable) have been described with potential clinical utility
    - A subset of adenocarcinomas harbors multiple *KRAS* mutations, with some showing evidence of biallelic mutations
    - Adenocarcinomas with low "epithelial-mesenchymal transition" and high mTOR pathway scores are associated with a favorable prognosis
- Studies of somatic mutations in metastases revealed a period of approximately 15 years between the occurrence of the initiating mutation of ductal adenocarcinoma and the acquisition of metastatic ability, suggesting a broad time window for early detection; however, once ductal adenocarcinoma is fully developed, it may spread rapidly

Variants of Ductal Adenocarcinoma

Adenosquamous Carcinoma

- Uncommon morphologic variant of pancreatic ductal adenocarcinoma, accounting for 1–4% of malignant exocrine neoplasms

- Slight male predominance (male to female ratio of 1.5:1); mean age at diagnosis 65 years
- Aggressive neoplasm with poor prognosis, median survival of 6 months
- Microscopically, characterized by infiltrating carcinoma with both glandular and squamous differentiation (Fig. 7.2). By definition, each component must comprise at least 30% of the tumor
- Immunohistochemically, p63 and p40 are positive in squamous component
- Molecular features similar to ductal adenocarcinoma, with frequent alterations in *KRAS*, *p16/CDKN2A*, *SMAD4*, and *TP53*
- Mutations in *UPF1*, which encodes the core component of the nonsense-mediated RNA decay pathway, have been reported; however, the biological significance of mutations in *UPF1* is controversial, and these mutations may represent nonpathogenic inherited variants rather than pathogenic mutations
- Appears to fall into the basal-like subtype in transcriptomics studies
- May be more sensitive to platinum-containing chemotherapies

Colloid Carcinoma

- Uncommon morphologic variant of pancreatic ductal adenocarcinoma, accounting for 1–3% of malignant exocrine neoplasms
- Slight male predominance, mean age at diagnosis 65 years

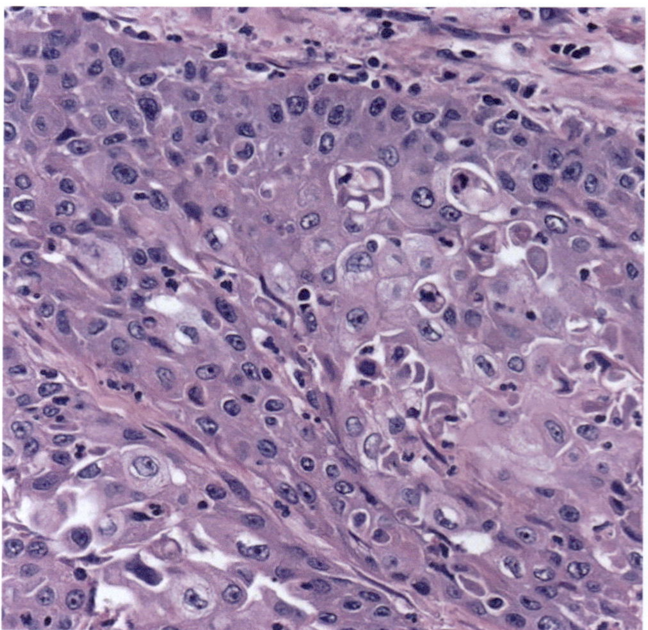

Fig. 7.2 Adenosquamous carcinoma

- Better prognosis than ductal adenocarcinoma, with a 5-year survival of approximately 55%
- Microscopically, characterized by large extracellular pools of mucin containing suspended well-differentiated neoplastic cells—almost always arises in association with intestinal-type intraductal papillary mucinous neoplasm (IPMN) (Fig. 7.3)
- By definition, the colloid component has to comprise at least 80% of the tumor
- Immunohistochemically, the neoplastic cells are positive for MUC2 and CDX2, which are frequently positive in intestinal-type IPMNs but usually negative in ductal adenocarcinoma
- Most have activating mutations in *GNAS*, an oncogene that is frequently mutated in IPMNs and associated adenocarcinomas; lower prevalence of *KRAS* mutations (approximately 30%) and *TP53* mutations (approximately 20%) compared to ductal adenocarcinoma

Hepatoid Carcinoma

- Very rare variant of pancreatic ductal adenocarcinoma, with only a few cases reported in the literature
- Too few cases have been reported to comprehensively determine clinical outcome
- Microscopically characterized by large polygonal neoplastic cells with abundant eosinophilic cytoplasm (Fig. 7.4)

- Immunohistochemically, neoplastic cells express markers of hepatocyte differentiation (HepPar1, polyclonal CEA, CD10, AFP)
- Poorly characterized at the molecular level

Medullary Carcinoma

- Uncommon morphologic variant of pancreatic ductal adenocarcinoma, accounting for <5% of ductal adenocarcinomas; for cases occurring in the periampullary region, many are proven to be of ampullary origin under careful scrutiny
- Age and sex distribution similar to that of pancreatic ductal adenocarcinoma
- More likely to report a family history of cancer than patients with pancreatic ductal adenocarcinoma
- Medullary carcinomas of the pancreas have been reported in patients with hereditary nonpolyposis colorectal cancer syndrome (HNPCC) or with synchronous colorectal adenocarcinomas (Table 7.2)
- Better prognosis than pancreatic ductal adenocarcinoma
- Microscopically, characterized by poor differentiation, pushing borders, syncytial growth, and focal to extensive necrosis (Fig. 7.5)
- Some medullary carcinomas have distinct molecular features
  - High prevalence of microsatellite instability
  - Lack somatic mutations in *KRAS*, although oncogenic *BRAF* mutations have been reported

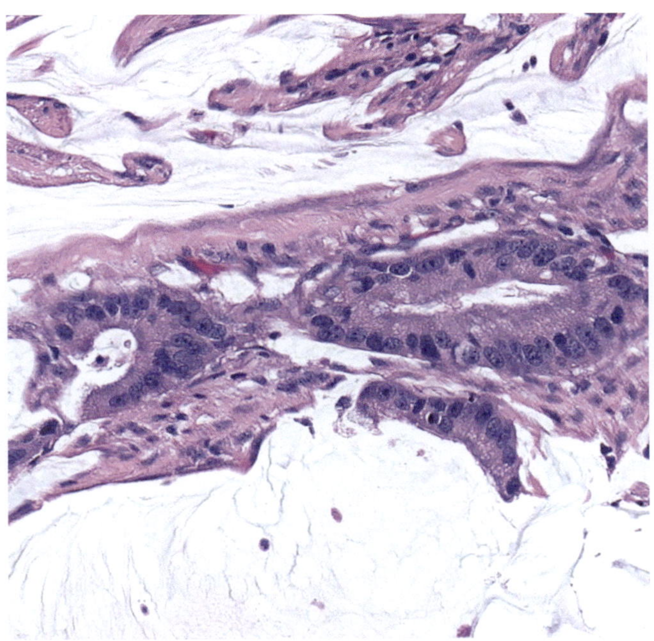

Fig. 7.3 Colloid carcinoma

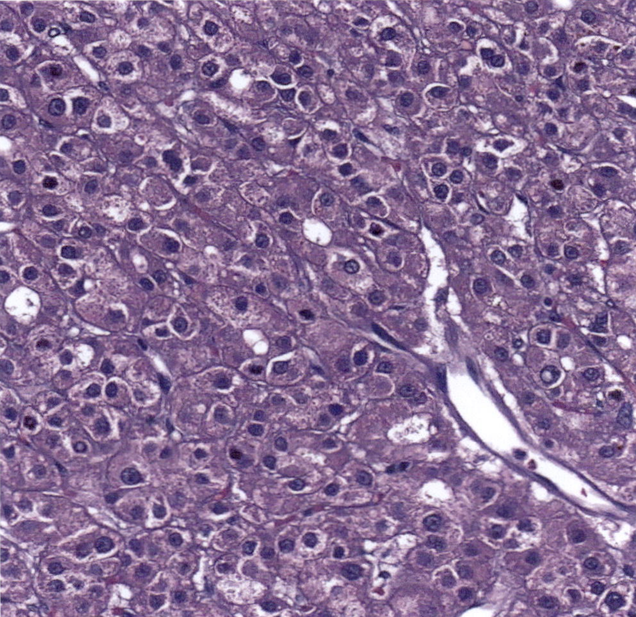

Fig. 7.4 Hepatoid carcinoma

Table 7.2 Inherited disorders with increased risk of pancreatic neoplasia

Syndrome	Gene	Chromosome	Neoplasm
Mutations in the Fanconi anemia/BRCA pathway	*BRCA2, BRCA1, PALB2 (FANCN)*	13, 16, 11	PDA
	FANCC, FANCG	Multiple	PDA
Ataxia-telangiectasia	*ATM*	11	PDA
Familial atypical multiple mole melanoma syndrome (FAMMM)	*p16/CDKN2A*	9	PDA
Peutz–Jeghers syndrome (PJS)	*STK11/LKB1*	19	PDA, IPMN
Hereditary pancreatitis	*PRSS1, SPINK1, CTRC, CFTR*	7, 5, 1, 7	PDA
Lynch syndrome/hereditary nonpolyposis colorectal cancer (HNPCC)	Multiple, including *MLH1, MSH2*	Multiple	PDA (medullary variant)
Familial adenomatous polyposis (FAP)	*APC*	5	PDA (?)
von Hippel–Lindau syndrome (VHL)	*VHL*	3	SCA, PanNET
Multiple endocrine neoplasia type 1 (MEN1)	*MEN1*	11	PanNET
Tuberous sclerosis complex (TSC)	*TSC1, TSC2*	9, 16	PanNET
Neurofibromatosis type 1 (NF1)	*NF1*	17	PanNET
Mahvash disease	*GCGR*	17	PanNET
Beckwith–Wiedemann syndrome (BWS)	Multiple	11	PB

PDA pancreatic ductal adenocarcinoma, *IPMN* intraductal papillary mucinous neoplasm, *SCA* serous cystadenoma, *PanNET* well-differentiated pancreatic neuroendocrine tumor, *PB* pancreatoblastoma, (?) weak or questionable association

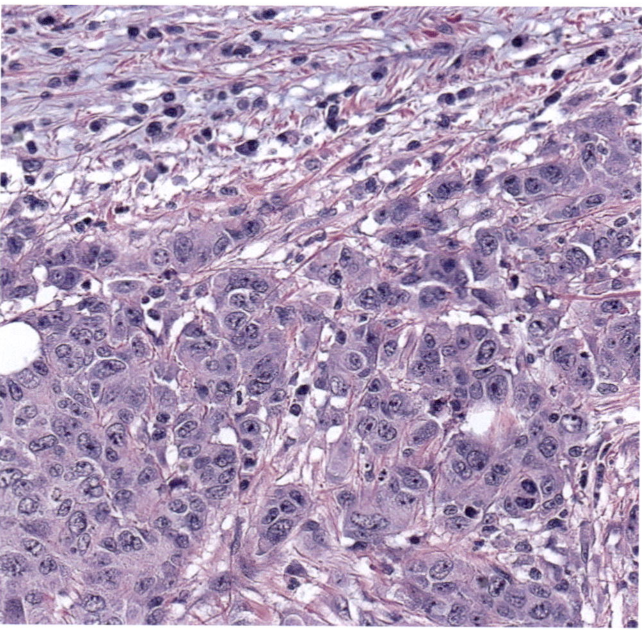

Fig. 7.5 Medullary carcinoma

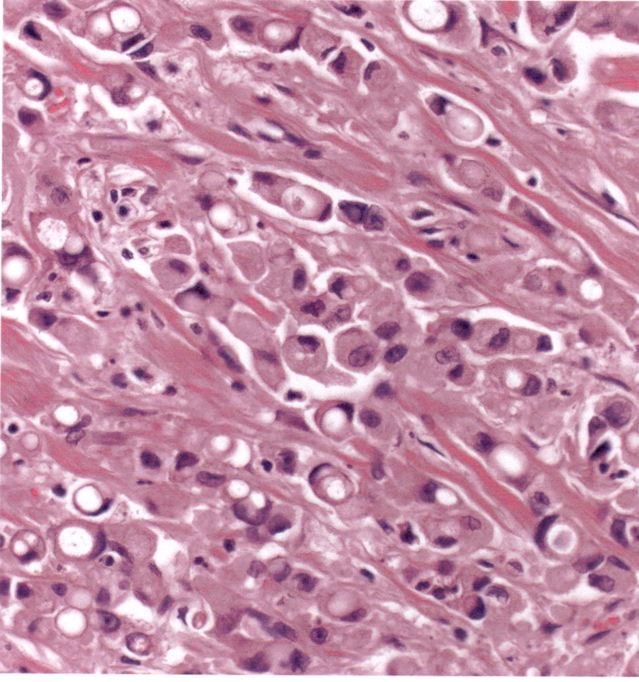

Fig. 7.6 Signet ring cell carcinoma. Image courtesy of Elizabeth D. Thompson, MD, PhD

- The diagnosis of a medullary carcinoma has therapeutic implications—microsatellite-unstable carcinomas are particularly sensitive to immune checkpoint inhibitors; these cancers may be less sensitive to fluorouracil-based chemotherapies

Signet Ring Cell Carcinoma

- Uncommon morphologic variant of pancreatic ductal adenocarcinoma, accounting for <1% of ductal adenocarcinomas

- Composed of individual cells with abundant intracellular mucin that pushes the nuclei to the side, creating a "signet ring" shape (Fig. 7.6), not to be confused with vacuolated carcinoma cells that may resemble signet ring cells
- It may have a component of ductal adenocarcinoma
- Needs to be distinguished from gastric carcinoma involving the pancreas and from lobular carcinoma of the breast metastatic to the pancreas; for cases occurring in the peri-

ampullary region, many are proven to be of ampullary origin

- Not well characterized at the molecular level

Undifferentiated Carcinoma

- Uncommon morphologic variant of pancreatic ductal adenocarcinoma, accounting for <10% of ductal adenocarcinomas
- Male predominance (male to female ratio of 3:1); average age at diagnosis 63 years
- Aggressive neoplasm with poor prognosis, with an average survival of only 5 months
- Multiple variants are being recognized and can be mixed within a single tumor:
  - Anaplastic carcinoma—mix of noncohesive pleomorphic mononuclear cells and bizarre multinucleated giant cells (Fig. 7.7)
  - Sarcomatoid carcinoma—atypical spindle cells arranged in fascicles (Fig. 7.8)
  - Rhabdoid carcinoma—distinctive epithelioid/rhabdoid cells with eosinophilic paranuclear cytoplasmic filamentous inclusions (Fig. 7.9)
    - Has heterogeneous but sometimes distinctive genetic mutations
    - Some harbor inactivating mutations in a SMARC family gene (*SMARCB1*, *SMARC2*, or *SMARC4*) and absence of *KRAS* alterations
- Immunohistochemically, the neoplastic cells express markers of epithelial differentiation (cytokeratins)

- Frequent loss of E-cadherin protein expression (Fig. 7.1d)
  - The loss of E-cadherin is a possible explanation for the neoplasm's discohesive morphology
  - Although *E-cadherin* promoter methylation has been reported in an undifferentiated carcinoma cell line, no somatic mutations in *E-cadherin* have been identified in neoplasms with loss of E-cadherin expression

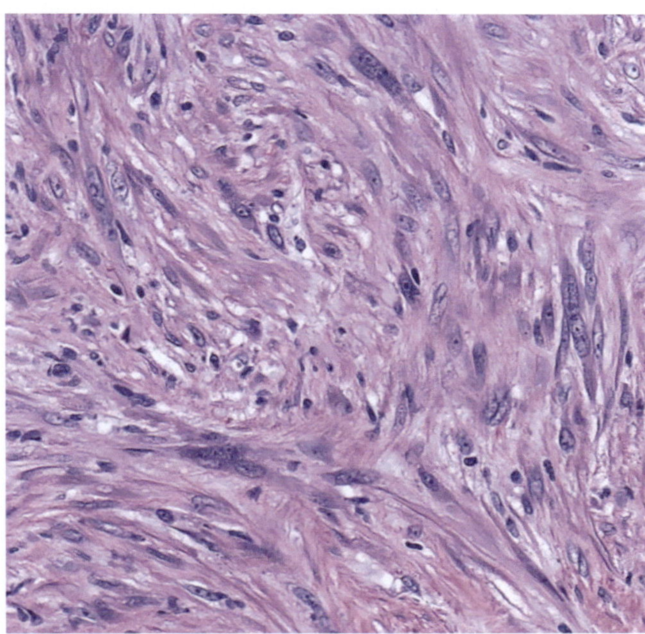

Fig. 7.8 Undifferentiated carcinoma (sarcomatoid)

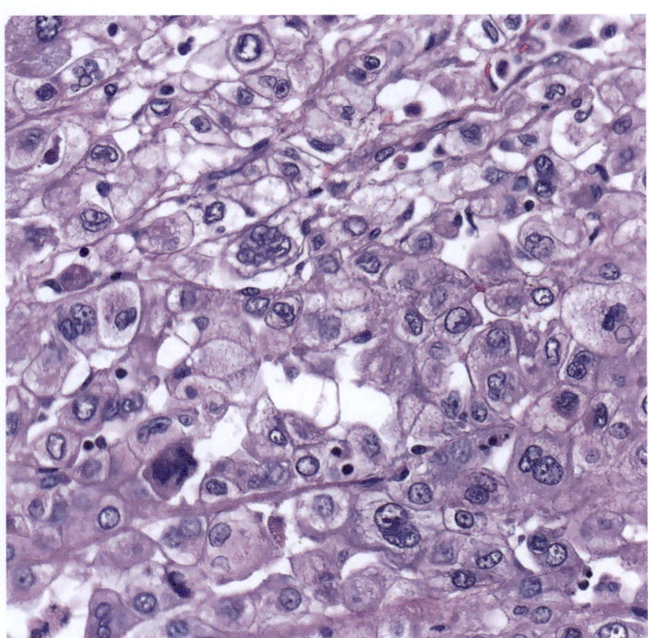

Fig. 7.7 Undifferentiated carcinoma (anaplastic)

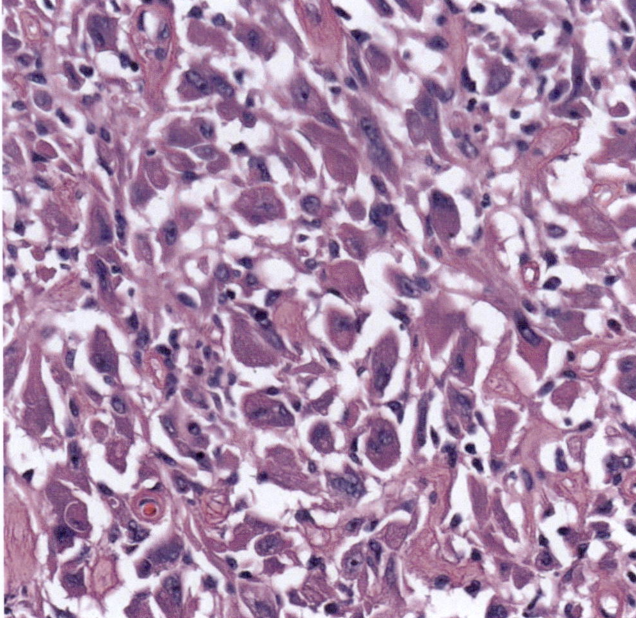

Fig. 7.9 Undifferentiated carcinoma (rhabdoid)

- Loss of E-cadherin is associated with decreased survival compared to well-, moderately-, and poorly-differentiated carcinomas with intact E-cadherin expression

Undifferentiated Carcinoma with Osteoclast-Like Giant Cells

- Uncommon morphologic variant of pancreatic ductal adenocarcinoma
- Slight female predominance in reported cases; average age at diagnosis 62 years
- Patients with pure undifferentiated carcinoma with osteoclast-like giant cells (UCOGC) have a significantly better prognosis than patients with UCOGC that is associated with an epithelial neoplasm (median overall survival of 36 vs. 15 months)
- Microscopically, composed of at least two distinct cell populations (Fig. 7.10):
  - Large multinucleated osteoclast-like giant cells with multiple round uniform nuclei
  - Undifferentiated discohesive pleomorphic mononuclear cells with marked nuclear pleomorphism and hyperchromasia
- Immunohistochemically, the osteoclast-like giant cells express histiocyte markers (CD68, CD45, and KP1), while the atypical mononuclear cells variably express markers of epithelial differentiation (cytokeratin, EMA)
- Genetic alterations in UCOGC are similar to mutations commonly present in conventional adenocarcinoma, including activating mutations in *KRAS* and inactivating mutations in *CDKN2A*, *TP53*, and *SMAD4*
- Molecular analyses have clarified the nature of the two characteristic cell types
  - While the mononuclear cells contain frequent somatic mutations in *KRAS*, these mutations are less frequently present in the osteoclast-like giant cells
    - Moreover, while pleomorphic mononuclear cells variably overexpress p53, osteoclast-like giant cells are negative
  - The molecular and immunohistochemical findings support the conclusion that while the mononuclear cells are neoplastic, the osteoclast-like giant cells are reactive
    - The presence of *KRAS* mutations in these giant cells is most likely due to phagocytosis of tumor DNA by the nonneoplastic cells

Intraductal Papillary Neoplasms

Clinical Features

- Incidence and prevalence are difficult to estimate because most are asymptomatic
  - Account for <5% of exocrine pancreatic neoplasms but most cystic neoplasms of the pancreas
  - May be more prevalent than currently appreciated—incidental pancreatic cysts are identified in approximately 3% of adults undergoing abdominal imaging, and many of these are IPMNs
- Slight male predominance (male to female ratio of 3:2); average age at diagnosis 63 years
  - Patients with noninvasive IPMNs are, on average, slightly younger than patients with an IPMN with an associated invasive carcinoma
- Frequently discovered incidentally on abdominal imaging, also can present with symptoms due to pancreatic duct obstruction (abdominal pain, weight loss, recurrent pancreatitis)
  - Some patients report a history of nonspecific symptoms and/or chronic pancreatitis for many years before diagnosis
- Approximately one-third of surgically resected IPMNs have an associated invasive adenocarcinoma
  - Evolutionary analyses based on whole exome or targeted sequencing help establish that that IPMNs are precursors of invasive adenocarcinoma
- Prognosis depends almost entirely on the presence of an associated invasive carcinoma
  - Surgical resection is curative in most, but not all, patients with a noninvasive IPMN (90–95% 5-year survival)

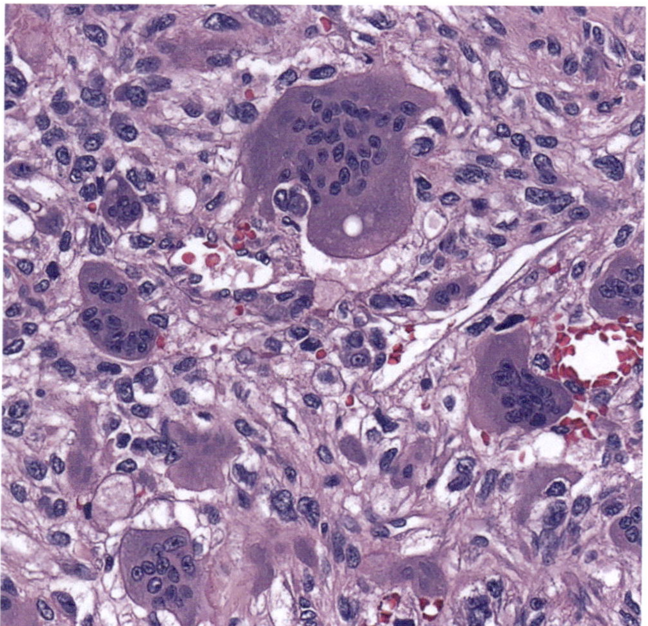

Fig. 7.10 Undifferentiated carcinoma with osteoclast-like cells

- o Some patients with "noninvasive IPMN" eventually die of metastatic carcinoma—this is believed to be due to metachronous multifocal disease
  - For patients with an associated invasive carcinoma, prognosis depends on stage, with an average 5-year survival of 40–50%, significantly higher than that of patients with pancreatic ductal adenocarcinoma not arising from an IPMN
    - o Much of this improved survival appears to be due to the lower stage at which IPMN-associated invasive carcinomas are diagnosed
- See Hereditary/Genetic Syndromes section for inherited disorders associated with IPMN

Pathologic Features
- Majority (70%) arise in the pancreatic head, with 20% in the body or tail and 10% diffusely involving the entire gland
- Multicentricity is common, occurring in up to 40% of cases
- By definition, IPMNs are >1 cm in greatest dimension, but can be quite large and involve the entire length of the pancreas
  - IPMNs with associated invasive carcinoma are usually larger than noninvasive IPMNs
- On gross cut section, the main pancreatic duct or one of its branches is dilated, containing papillary projections and thick mucin
  - The relationship of the dilated duct to the main pancreatic duct is used to classify the neoplasm:
    - o Main duct-type—neoplasm involves the main pancreatic duct
    - o Branch duct-type—neoplasm involves the secondary branches but not the main pancreatic duct
    - o Combined type—neoplasm involves the main pancreatic duct and its branches (clinical behavior similar to main duct-type)
      - ◆ Main duct IPMNs are more likely to have high-grade dysplasia or an associated invasive carcinoma than are branch duct-type IPMNs
- Microscopically, IPMNs consist of mucin-secreting columnar epithelium lining the pancreatic duct system, variably forming papillary projections
  - Multiple histologic subtypes are recognized
    - o Intestinal type: Papillae with long villous projections, lined by epithelial cells with apical mucin and cigar-shaped nuclei as well as scattered goblet cells
    - o Gastric–foveolar type: Flat epithelium or small papillae lined by epithelium with small basal nuclei, eosinophilic cytoplasm, and apical mucin; often occurs in branch duct IPMNs with low-grade dysplasia

- o Pancreatobiliary type: Complex papillary structures with bridging and cribriform architecture, lined by cuboidal epithelial cells with little extracellular mucin and round nuclei with open chromatin and variably prominent nucleoli; more likely than other types to harbor high-grade dysplasia
  - Noninvasive neoplasms are categorized based on the degree of dysplasia in their mucinous epithelium
    - o Low-grade dysplasia: Mild architectural and cytologic atypia
    - o High-grade dysplasia: Marked architectural and cytologic atypia, with irregular papillae with cribriform architecture and budding, loss of nuclear polarity, significant nuclear pleomorphism, and mitotic figures
  - Approximately one-third of resected IPMNs have an associated invasive adenocarcinoma
    - o In one-half of the cases, the carcinoma is an invasive colloid carcinoma, characterized by large pools of extracellular mucin with free-floating neoplastic epithelial cells
      - ◆ These patients have a better prognosis than those with tubular carcinoma arising in association with an IPMN
      - ◆ Colloid carcinomas in the pancreas are almost universally associated with IPMNs of the intestinal subtype
    - o In the remaining one-half of the cases, the carcinoma is a tubular carcinoma, morphologically indistinguishable from a pancreatic ductal adenocarcinoma arising without an IPMN, composed of an invasive gland-forming carcinoma with desmoplastic stroma and minimal extracellular mucin production
- Immunohistochemistry
  - Positive in neoplastic cells: Cytokeratins and CEA
  - Variably positive in neoplastic cells: Scattered basal neuroendocrine cells expressing chromogranin and synaptophysin
  - MUC expression varies based on histologic subtype
    - o Intestinal type: Positive for MUC2, CDX2, and MUC5AC, negative for MUC1
      - ◆ When they progress, they are likely to progress to invasive colloid carcinomas that are also MUC2 and CDX2 positive and MUC1 negative
    - o Gastric–foveolar type: Positive for MUC5AC, negative for MUC2 and MUC1
    - o Pancreatobiliary type: Positive for MUC1 and MUC5AC, negative for MUC2
      - ◆ When they progress, they are likely to progress to invasive ductal adenocarcinomas that are MUC1 positive and MUC2 negative

Genetic Features

- See Table 7.1
- Frequent alterations in genes commonly mutated in pancreatic ductal adenocarcinoma (*KRAS*, *TP53*, *SMAD4*, *p16/CDKN2A*)
  - Somatic mutations in *KRAS* occur frequently in IPMNs, involving up to 75% of neoplasms, with increasing mutation prevalence in neoplasms with high-grade dysplasia or associated adenocarcinoma
    - When extremely sensitive techniques are employed, 80% of IPMNs have been found to harbor *KRAS* gene mutations
      - Somatic *KRAS* mutations occur in all histologic subtypes of IPMN
      - Rare somatic mutations in *BRAF* have also been reported in IPMN
  - p53 overexpression is most prevalent in areas of high-grade dysplasia and invasive carcinoma associated with IPMNs
    - Somatic mutations of *TP53* have also been reported, and these occurred only in cases with high-grade dysplasia
  - SMAD4 expression is retained in the vast majority of noninvasive IPMNs and lost in approximately one-third of IPMN-associated invasive carcinomas, suggesting a role for SMAD4 in promoting invasion; the prevalence of SMAD4 loss in IPMN-associated invasive carcinomas is lower compared to that in pancreatic ductal adenocarcinomas not arising in association with an IPMN
  - Loss of p16 protein expression occurs in both noninvasive IPMNs and invasive carcinoma arising in association with an IPMN, but loss is much more prevalent in invasive carcinomas (100% of invasive carcinomas vs. 10% of noninvasive IPMNs in one study)
- Hypermethylation of the *p16/CDKN2A* promoter occurs in >50% of noninvasive IPMNs and IPMNs with associated adenocarcinoma
- Approximately 60% of IPMNs possess somatic mutations in *GNAS*
  - The mutations in IPMNs all occurred at a previously described oncogenic hotspot (codon 201), and in cases with an associated adenocarcinoma, *GNAS* mutations were detected in both in situ and infiltrating components
  - Mutations in *GNAS* are most prevalent in intestinal-type IPMNs
    - The protein encoded by *GNAS* couples transmembrane receptors to their downstream signaling proteins, such as adenylyl cyclase, playing a crucial role in numerous cell signaling pathways
- Up to 75% of IPMNs possess somatic mutations in *RNF43*, which encodes an E3 ubiquitin ligase
  - Most of these mutations lead to the insertion of stop codons, and there is frequent loss of heterozygosity at chromosome 17q (the location of *RNF43*), providing strong evidence that *RNF43* is a tumor suppressor gene
  - Multiple distinct mutations in *RNF43* can be present in an IPMN
  - Genetic alterations in *RNF43* are enriched in IPMNs compared to associated invasive carcinomas
- Somatic mutations in PIK3CA (some at previously described oncogenic hotspots) occur in approximately 10% of IPMNs
  - The protein encoded by *PIK3CA* is the catalytic component of crucial cell signaling kinase that regulates numerous pathways involved in growth, proliferation, and apoptosis
- Hotspot mutations in *KLF4* (K409 and S411) are present in more than 50% of IPMNs
  - *KLF4* encodes a member of the Kruppel family of transcription factors
  - Mutations in *KLF4* are more prevalent in low-grade IPMNs than in high-grade IPMNs
- Rare somatic mutations in *EGFR* and *HER2* have been described in IPMNs
- IPMNs arise from multiple independent clones
  - Sequencing studies utilizing microdissection of multiple regions of IPMNs and single cells revealed marked genetic heterogeneity in IPMNs
  - 18% of pancreatic ductal adenocarcinomas and co-occurring IPMNs have distinct mutations in cancer driver genes, suggesting that an independent clone may have given rise to the invasive carcinoma
- Promoter hypermethylation occurs in several genes (including *APC*, *E-cadherin [CDH1]*, *MLH1*, *MGMT*) in noninvasive IPMNs and IPMNs with associated adenocarcinoma, with more prevalent methylation in neoplasms with adenocarcinoma
  - Hypermethylation of multiple genes is also more prevalent in IPMNs with an associated invasive adenocarcinoma
- Large-scale chromosomal alterations have been identified in IPMNs with all levels of dysplasia, but copy number alterations are much more frequent in IPMNs with high-grade dysplasia
- Analyses of microRNA expression have revealed significantly higher expression of miR-21, miR-221, and miR-17–3p in IPMNs as compared to nonmucinous cysts
- Mutations in neoplastic cells can be detected in aspirated IPMN cyst fluid, indicating that molecular studies of cyst fluid represent a promising diagnostic tool to preoperatively classify cystic lesions in the pancreas
  - More than 95% of IPMNs contain a somatic mutation in either *KRAS* or *GNAS*, illustrating that molecular analyses are a highly sensitive assay for the identification of IPMNs

Intraductal Oncocytic Papillary Neoplasm

- Uncommon neoplasms, accounting for <10% of all surgically resected intraductal papillary neoplasms
- They form large heterogeneous partially cystic partially solid masses that may appear unresectable due to the peritumoral inflammation fibrosis spreading to neighboring sites, but virtually all are curable with complete removal
- The papillae of these neoplasms are distinct with complex arborizing patterns, and the epithelium forms architecturally complex structures with bridging and cribriform architecture; the neoplastic cells have abundant granular pink cytoplasm and round nuclei
- Immunolabeling highlights the almost universal expression of MUC5AC; one-half also express MUC1 and MUC2; immunolabeling for mitochondrial antigens, such as 111.3, is positive
- Intraductal oncocytic papillary neoplasms have recurring fusions of *ATP1B1-PRKACB*, *DNAJB1-PRKACA*, or *ATP1B1-PRKACA*
  - Mutations that are commonly identified in IPMNs, including alterations in *KRAS* and *TP53*, are absent in intraductal oncocytic papillary neoplasms

Intraductal Tubulopapillary Neoplasm

- These neoplasms tend to be solid nodular tumors
- Morphologically, they form back-to-back tubular glands with cribriform architecture, prominent comedo-like necrosis, minimal mucin production, and occasional papillary formations
- Typically, negative for MUC2 and MUC5AC, and most are positive for MUC6
- Certain chromatin remodeling genes (*MLL1*, *MLL2*, *MLL3*, *BAP1*, *PBRM1*, *EED*, and *ATRX*) were found to be mutated in one-third of cases, and more than a one-fourth harbor phosphatidylinositol 3-kinase (PI3K) pathway (*PIK3CA*, *PIK3CB*, *INPP4A*, and *PTEN*) mutations; about one-fifth of cases shows *FGFR2* fusions
  - Mutations that are commonly identified in IPMNs, including alterations in *KRAS* and *TP53*, are absent in intraductal tubulopapillary neoplasms

---

Mucinous Cystic Neoplasm

Clinical Features

- Uncommon neoplasms, accounting for <10% of all surgically resected cystic pancreatic neoplasms
- The vast majority, but not all, occur in women (female to male ratio of approximately 20:1), with a mean age at presentation of 40–50 years
- No known genetic predilection or association with genetic syndromes
- Frequently discovered incidentally on abdominal imaging, also can present with nonspecific symptoms due to compression of adjacent organs (abdominal pain or fullness)
- Associated invasive adenocarcinoma is present in about 15% of mucinous cystic neoplasms (MCNs)
  - Evolutionary analyses based on whole exome or targeted sequencing helped establish that MCNs, as well as IPMNs, are precursors of invasive adenocarcinoma
- Prognosis depends almost entirely on the presence of an associated invasive carcinoma
  - Surgical resection is curative in almost all patients with a noninvasive MCN
  - For patients with an associated invasive carcinoma, prognosis depends on stage, with an average 5-year survival of 25–50%

Pathologic Features

- Almost all (>95%) occur in body and tail of the pancreas
- Usually solitary, with a thick well-demarcated capsule
- Wide size range (2–35 cm), average 6–10 cm in greatest dimension
- On gross cut section, the neoplasm is most frequently a multiloculated thick-walled cyst with adherent thick mucin
  - Cyst within cyst appearance
  - Cysts contain mucin
  - Some locules may also contain hemorrhagic or necrotic debris admixed with mucin
  - Cyst walls can be smooth or may contain papillary excrescences or mural nodules, the latter associated with high-grade dysplasia or an associated invasive carcinoma; importantly, the cysts do not communicate with the larger ducts of the pancreas
- Microscopically, by definition, two components are present; the cysts are lined by mucin-producing columnar epithelial cells, and there is an underlying ovarian-type stroma
  - Architecturally, MCNs consist of a multiloculated cyst surrounding by a thick fibrous capsule, separating the cyst from the adjacent uninvolved pancreas
  - Cyst-lining epithelial cells are tall and columnar, with small basal nuclei and abundant apical mucin
    - Commonly observed directions of differentiation in the epithelium include pseudopyloric, gastric–foveolar, and intestinal
  - Associated with the mucinous epithelium, these neoplasms possess a characteristic ovarian-type stroma,

consisting of dense spindle cells with elongated nuclei and sparse cytoplasm

- This stroma is required for the diagnosis of MCN
  - Noninvasive neoplasms are categorized based on the degree of dysplasia in their mucinous epithelium
    - Low-grade dysplasia: Mild architectural and cytologic atypia
    - High-grade dysplasia: Severe architectural and cytologic atypia, with irregular papillae, loss of nuclear polarity, and significant nuclear pleomorphism
  - Approximately one-third of resected MCNs have an associated invasive adenocarcinoma, most commonly conventional ductal adenocarcinoma although several less common carcinoma variants have been reported
    - The transition from low-grade to high-grade dysplasia, and even to invasive carcinoma, can be abrupt and focal, necessitating extensive histologic sampling of these neoplasms to assess for an invasive carcinoma
    - Rare cases of biphasic malignant neoplasms containing both carcinomatous and high-grade spindle cell ("sarcomatous") components have been reported in association with MCNs
- Immunohistochemistry
  - Epithelial cyst lining
    - Positive in neoplastic cells: Cytokeratins and CEA
    - Variably positive in neoplastic cells: Scattered basal neuroendocrine cells positive for chromogranin and synaptophysin
    - Mostly negative or only focally positive in neoplastic cells: CDX2 and MUC2 (highlighting scattered goblet cells or basal components)
    - Differing expression of mucin markers has been associated with the transition to invasive carcinoma; noninvasive MCNs are MUC5AC-positive and MUC1-negative, while invasive carcinomas are frequently MUC5AC-positive and MUC1-positive
  - Ovarian-type stroma
    - Positive in the stromal cells: Vimentin, smooth muscle actin, desmin, calretinin, inhibin
    - Variably positive in the stromal cells: Progesterone receptor (50–75% of neoplasms), estrogen receptor (25% of neoplasms)
    - Negative in the stromal cells: S100 protein, CD34

Genetic Features

- See Table 7.1
- Frequent alterations in genes commonly mutated in pancreatic ductal adenocarcinoma (*KRAS*, *TP53*, *SMAD4*, *p16/CDKN2A*)

- Somatic *KRAS* mutations are frequently detected in MCNs, with mutation prevalence correlated with degree of dysplasia
  - Typical *KRAS* mutations have been identified in approximately 30% of neoplasms with low-grade dysplasia and approximately 80% of neoplasms with high-grade dysplasia or invasive carcinoma
    - *KRAS* mutations can be detected in cyst fluid aspirated from MCNs, suggesting that mutational analysis of cyst fluid may become an important ancillary diagnostic test to characterize cystic lesions in the pancreas, including MCNs
- p53 overexpression is limited to areas of high-grade dysplasia and invasive carcinoma in MCNs, and *TP53* mutation has been reported in neoplasms with high-grade dysplasia
- Loss of SMAD4 protein expression, a surrogate for *SMAD4* gene mutation, is associated with the transition to invasive carcinoma
  - SMAD4 protein expression is intact in virtually all noninvasive neoplasms but only a subset of invasive carcinomas is associated with MCNs
- Mutation in *p16/CDKN2A* has also been reported in a neoplasm with high-grade dysplasia; hypermethylation of the *p16/CDKN2A* promoter has been identified in approximately 15% of MCNs
- These findings suggest a model of stepwise dysplasia and carcinogenesis in MCNs in which *KRAS* mutation is an early event and loss of *SMAD4* is a late event
- Sequencing of all protein-coding genes in eight MCNs revealed frequent somatic alterations in *KRAS*, *RNF43*, and *TP53*
  - Most of the mutations in *RNF43* were nonsense substitutions, further confirming this gene's role as a tumor suppressor in mucin-producing cystic neoplasms of the pancreas
  - Studies identified an average of 16 nonsynonymous somatic mutations per MCN, fewer than in IPMN and invasive ductal adenocarcinoma
- No alterations in β-catenin, which contrasts with solid pseudopapillary neoplasms of the pancreas
- No evidence of microsatellite instability; aneuploidy has been reported in carcinomas associated with MCNs and is associated with poor prognosis
- Gene expression studies suggest different expression profiles in the epithelial and stromal components, with activation of the Notch pathway (*JAG1* and *HES1*) in the epithelial component and activation of estrogen metabolism (*STAR* and *ESR*) in the stromal component
  - The genetic or epigenetic underpinnings of these expression differences have not been fully elucidated

- MCNs have been proposed to originate from primordial germ cells that are entrapped in the dorsal pancreas during early human embryogenesis
- In rare invasive carcinomas with both carcinomatous and "sarcomatous" components arising in MCNs, the histologically distinct components contain nearly identical patterns of loss of heterozygosity
  - These findings suggest a monoclonal origin for the two components with subsequent genetic and morphologic diversion in this rare subtype

Serous Cystadenoma

Clinical Features

- Uncommon neoplasm, accounting for 1–2% of all pancreatic neoplasms, but close to one-third of all cystic neoplasms of the pancreas
- Occurs primarily in adults (mean age 60–65 years) with female predominance (female to male ratio of 7:3)
- A synchronous neoplasm, most commonly neuroendocrine tumor, is found in 13% of cases
- Serous cystadenomas (SCAs) are frequently discovered incidentally on abdominal imaging but also can present with symptoms due to abdominal mass (abdominal pain, nausea, and vomiting)
  - Jaundice is rare
- Excellent prognosis—neoplasms are slow growing (approximately 0.60 cm/year); most SCAs do not require surgery, but symptomatic SCAs can be treated with surgical resection
  - Exceedingly rare malignant variant (serous cystadenocarcinoma), defined by distant metastasis, also has favorable prognosis, with most patients alive at the time their reports were published (average follow-up 36 months)
- Patients with von Hippel–Lindau syndrome (VHL) have an increased risk of developing SCAs (see Hereditary/Genetic Syndromes section)

Pathologic Features

- Occur more frequently in the pancreatic body and tail (50–75% of cases)
- In sporadic cases, most often solitary and well-demarcated mass
  - Patients with VHL frequently develop multiple SCAs, which may involve the pancreas diffusely
- Wide size range (up to 25 cm), average 6 cm in greatest dimension

- On gross cut section, most commonly a well-circumscribed, slightly bosselated, sponge-like mass composed of numerous small thin-walled cysts filled with clear to straw-colored watery fluid
  - Often have a central scar with radiating fibrous septa
  - The most common pattern is one of innumerable small (1–3 mm) cysts, referred to as a microcystic pattern
  - However, uncommon variants exist, most of which are defined by their gross appearances, including:
    - Macrocystic (oligocystic) variant: Few or single large smooth-walled cyst(s), resembles mucinous cyst neoplasm and other megacystic lesion
    - Solid variant: Soft, fleshy pink-tan solid mass mimicking a neuroendocrine tumor
    - VHL-associated variant: Multiple lesions grossly similar to microcystic SCAs may diffusely involve the pancreas
    - Combined serous neuroendocrine neoplasm: The two components may be grossly distinct or intimately admixed; many of these patients have VHL
- Microscopically, composed of cysts lined by uniform cuboidal cells with clear glycogen-rich cytoplasm and central round nuclei
  - Most common architectural pattern (microcystic) is that of numerous small cysts lined by a single layer of cuboidal epithelial cells; micropapillae may be present in cyst lining
  - Neoplastic cells do not infiltrate adjacent pancreas, but atrophy of the adjacent pancreas is common, particularly when the neoplasm compresses the pancreatic duct
  - Cytoplasm is most often clear due to abundant glycogen, but rarely cytoplasm can have an oncocytic appearance
  - Nuclei are round and uniform, small with dense chromatin and inconspicuous nucleoli
  - Distinctive capillary network hugging the epithelium is a characteristic finding and is a feature shared with other VHL-associated clear cell neoplasms with prominent vascularity (e.g., hemangioblastomas and clear cell renal cell carcinomas)
- Immunohistochemistry
  - Positive in neoplastic cells: Cytokeratins (including AE1/AE3, CAM5.2, CK7, CK8, CK18, and CK19), inhibin, MUC6, carbonic anhydrase IX
  - Negative in neoplastic cells: CEA, chromogranin, synaptophysin, pancreatic hormones (insulin, glucagon, somatostatin), MUC2, MUC5
  - The neoplastic cells express vascular endothelial growth factor A (VEGF-A)
- Large serous cystadenocarcinomas occasionally show invasion into adjacent organs, including colon, stomach, and lymph nodes

- Serous cystadenocarcinomas are defined by the presence of distant metastases but are otherwise microscopically indistinguishable from SCAs

Genetic Features

- See Table 7.1
- Somatic inactivating mutations in *VHL* gene have been reported in up to 50% of sporadic SCAs
  - Loss of heterozygosity of chromosome 3p at the *VHL* locus also occurs in a large proportion of sporadic neoplasms
    - Neoplasms in patients with VHL syndrome also consistently show loss of heterozygosity of chromosome 3p (*VHL* gene)
- Sequencing of all protein-coding genes in eight SCAs confirmed frequent somatic alterations in *VHL* (altered in 50% of SCAs studied)
  - No additional genes with frequent somatic alterations were identified
  - These studies identified an average of 10 nonsynonymous somatic alterations per tumor, approximately one-half the number of alterations in IPMN and far fewer than in invasive ductal adenocarcinoma
- Lack alterations in genes frequently mutated in pancreatic ductal adenocarcinoma (*KRAS*, *TP53*, *SMAD4*)
- Aberrant methylation at a few loci reported in only a minority of cases
- Lack alterations in β-catenin, which is frequently mutated in solid pseudopapillary neoplasms of the pancreas
- Microsatellite instability has not been identified in sporadic neoplasms

Neoplasms with Neuroendocrine Differentiation

- Well-differentiated pancreatic neuroendocrine tumors (PanNETs) and pancreatic neuroendocrine carcinomas (PanNECs) are two distinct neuroendocrine neoplasms that can arise in the pancreas
- Although these tumors both have neuroendocrine differentiation, they are distinct morphologically, genetically, and in their clinical behavior
- PanNETs histologically have clear neuroendocrine differentiation, with an organoid growth pattern and nuclei with a "salt and pepper" chromatin
- By contrast, neuroendocrine carcinomas have the histologic appearance of a more aggressive carcinoma with coarsely clumped chromatin and prominent nucleoli

Well-Differentiated Pancreatic Neuroendocrine Tumors

Clinical Features

- Well-differentiated pancreatic neuroendocrine tumors (PanNETs) are uncommon neoplasms, accounting for 1–2% of all pancreatic neoplasms
- Can occur at any age, but are most common between 30 and 60 years (average age at diagnosis 50 years)—slight female predominance (female to male ratio of 1.15:1)
- Presenting symptoms vary based on size and type of PanNET
  - PanNETs that secrete hormones frequently present with a characteristic clinical syndrome due to hormone excess (functional PanNETs)
    - The presence of a clinical syndrome of hormone excess is required to classify a PanNET as functional
  - Small nonfunctional PanNETs are usually found incidentally on abdominal imaging or in pancreata resected for other reasons
  - Larger nonfunctional PanNETs may present with nonspecific symptoms related to a large abdominal mass (abdominal pain, nausea)
    - Jaundice infrequently occurs in patients with PanNETs
- Prognosis depends on size, grade, and stage of the PanNET
  - Small nonfunctional microadenomas (<0.5 cm) are considered clinically benign and are completely cured by surgical resection
  - PanNETs are graded based on their proliferative rate, as assessed by mitotic count or Ki-67 labeling index (see below)
    - However, all PanNETs, except microadenomas, are regarded as malignant neoplasms, with an average 5-year survival of 65%
- See Hereditary/Genetic Syndromes section for inherited disorders associated with PanNETs

Pathologic Features

- Can occur anywhere in the pancreas, with some specific localization for functional types
  - For example, gastrinomas and somatostatinomas tend to be in the duodenum
- Sporadic PanNETs are usually solitary and well-demarcated; patients with multiple endocrine neoplasia 1 (MEN1) are frequently diagnosed with multiple synchronous PanNETs
- Functional PanNETs, particularly insulinomas, are frequently small at diagnosis (<2 cm) due to early clinical detection from symptoms of hormone excess

- Nonfunctional PanNETs are generally larger (frequently >5 cm), at least partially due to later detection in absence of a specific clinical syndrome
- On gross cut section, most PanNETs are solid pink-tan masses with well-defined borders, varying in consistency from soft and fleshy to densely fibrotic
  - Infrequently, areas of hemorrhage, necrosis, and cystic degeneration can occur, particularly in larger neoplasms
  - In contrast, neuroendocrine carcinomas are usually large firm masses with ill-defined borders
- Microscopically, PanNETs consist of an organoid proliferation of uniform cells with neuroendocrine features
  - Numerous architectural patterns have been described, including trabecular, nested, and gyriform
    - Many neoplasms have a mixed architectural pattern
  - Some PanNETs have a fibrotic capsule
  - Individual cells have a distinct neuroendocrine morphology, with a moderate amount of finely granular cytoplasm, round nuclei with coarsely clumped ("salt and pepper") chromatin, and occasional nucleoli
    - Cytoplasmic hyaline globules (like those seen in solid pseudopapillary neoplasm) may rarely be present
  - Some morphologic findings are suggestive of specific functional PanNETs
    - Amyloid deposition is common in insulinomas
    - Psammoma bodies occur most commonly in somatostatinomas
  - PanNETs are graded based on proliferative rate, as assessed by mitotic count or Ki-67 labeling index
    - Grade 1 (low grade) PanNET: 0–1 mitotic figures per 10 high-power fields (HPF), Ki-67 index of 0–2%
    - Grade 2 (intermediate grade) PanNET: 2–20 mitotic figures per 10 HPF, Ki-67 index of 3–20%
    - Grade 3 (high grade) PanNET: >20 mitotic figures per 10 HPF, Ki-67 index >20%
- Immunohistochemistry
  - Positive in neoplastic cells: Synaptophysin (diffuse), chromogranin (diffuse or focal), CD56 (less specific), CD57 (less specific), cytokeratins 8 and 18
    - Functional PanNETs are frequently positive for their secreted hormone (insulin, glucagon, etc)
      - However, immunohistochemical evidence of hormone expression is not sufficient to classify a PanNET as functional in the absence of specific clinical symptoms
    - If >25% of neoplastic cells express acinar markers (trypsin, chymotrypsin, BCL10), the neoplasm should be classified as a mixed acinar neuroendocrine carcinoma

Genetic Features
- See Table 7.1
- Somatic mutations in *MEN1* occur in approximately 45% of sporadic PanNETs
  - Loss of heterozygosity at the *MEN1* locus also occurs in 30–70% of PanNETs, including some PanNETs without somatic *MEN1* mutation
  - Mutations in *MEN1* are associated with better prognosis
  - Neoplasms with *MEN1* mutation show loss of expression or aberrant localization of the MEN1 protein
    - In addition, some neoplasms without *MEN1* mutation also exhibit aberrant MEN1 protein localization, suggesting that mechanisms in addition to *MEN1* mutation contribute to MEN1 protein dysfunction in PanNETs
- Genes involved in a chromatin remodeling complex (*DAXX* and *ATRX*) are somatically mutated (including numerous inactivating mutations) in approximately 45% of sporadic PanNETs
  - These genes are part of a complex that is important for telomere maintenance, and inactivation of these genes in PanNETs is associated with the telomerase-independent telomere maintenance mechanism known as ALT (alternative lengthening of telomeres)
  - Neoplasms with somatic mutations in *DAXX* or *ATRX* show loss of protein expression by immunohistochemistry (Fig. 7.1e, f)
- Somatic mutations of genes in the cell signaling pathway of mammalian target of rapamycin (mTOR), including *PIK3CA*, *PTEN*, and *TSC2*, occur in approximately 15% of sporadic PanNETs
  - In addition to somatic mutation of the *TSC2* gene, loss of heterozygosity at the *TSC2* locus on chromosome 16p occurs in approximately one-third of PanNETs
  - The alterations in the mTOR pathway may carry clinical significance, as drugs targeting this pathway have been developed for clinical use
- Lack of alterations in genes commonly mutated in pancreatic ductal adenocarcinoma (*KRAS*, *TP53*, *SMAD4*, *p16/CDKN2A*)
  - Somatic mutations in *KRAS* and *BRAF* have not been identified in PanNETs
  - Promoter hypermethylation of *p16/CDKN2A* occurs in approximately 50% of gastrinomas, but somatic mutation or homozygous deletion of *p16/CDKN2A* has not been reported in PanNETs
- Promoter methylation and deletion of *VHL* occurs in up to 25% of sporadic PanNETs and is associated with activation of the HIF1α signaling pathway
- Sporadic PanNETs have been reported to arise in patients with germline mutations in the DNA repair genes *MUTYH*, *CHEK2*, and *BRCA2*

- Large-scale chromosomal gains and losses (including some recurrent alterations) are also present in PanNETs
  - Moreover, PanNETs with multiple chromosomal abnormalities are more likely to present with metastatic disease and possess a worse prognosis
- The prevalence of microsatellite instability in PanNETs is not known, although none of the tumors in a recent large-scale PanNET sequencing study showed defects in mismatch repair

Neuroendocrine Carcinoma

Clinical Features

- High-grade neuroendocrine carcinomas are very rare, accounting for <1% of pancreatic carcinomas
- These are extremely aggressive cancers, and most have widespread metastases at diagnosis; their median survival is less than 1 year, and a 5-year survival rate is only 15%

Pathologic Features

- Two histologic subtypes of neuroendocrine carcinoma
  - Small cell carcinoma: Highly cellular infiltrative neoplasm of small-to-medium-sized cells with scant cytoplasm, nuclear molding, and finely granular chromatin
  - Large cell carcinoma: Nested pattern of large cells with amphophilic cytoplasm and large oval nuclei with coarsely clumped chromatin and prominent nucleoli
- Neuroendocrine carcinoma may have variable or absent expression of neuroendocrine markers (chromogranin, synaptophysin)
- The Ki-67 labeling index is typically >60%, and can be as high as 90%

Genetic Features

- Inactivating mutations in *TP53* and *RB* mutations are common, and loss of *SMAD4* may also be seen
- Lack the genetic alterations seen in well-differentiated neuroendocrine tumors (*DAXX*, *ATRX*, *MEN1*, *TSC2*)

Neoplasms with Ambiguous Direction of Differentiation

Solid Pseudopapillary Neoplasm

Clinical Features

- Rare pancreatic neoplasm, accounting for <5% of all pancreatic malignancies
- Occur predominantly in young women (mean age 28 years), with a female to male ratio of 9:1
- Frequently discovered incidentally on abdominal imaging, also can present with symptoms due to abdominal mass (abdominal pain, nausea, early satiety)
  - Rarely rupture of the tumor can cause an acute abdomen
- Overall prognosis is very favorable, with 85–95% of patients cured after complete surgical resection
  - Even recurrences and metastases are frequently surgically resectable, with only rare patients dying of disease (10-year survival is >95%)
  - Stage is the best predictor of outcome in patients with solid pseudopapillary neoplasm
- No known genetic predilection or association with genetic syndromes, although a few patients with familial adenomatous polyposis have been reported to have solid pseudopapillary neoplasm

Pathologic Features

- Not localized to a specific part of the pancreas (evenly distributed throughout head, body, and tail)
- Almost all are solitary; most are grossly well-demarcated and often appear grossly encapsulated
- Wide size range (0.5–25 cm), but on average tumors are large (9–10 cm in greatest dimension)
- On gross cut section, most have both solid areas (soft white-gray to yellow) and cystic degenerative areas (irregularly shaped with friable material and hemorrhage); proportion of individual components is variable among tumors
- Microscopically, consist of uniform poorly cohesive cells supported by delicate small blood vessels; the neoplastic cells have no known counterpart in the normal pancreas
  - Architecture is often a mix of solid areas and areas with degenerative changes
    - In the degenerative areas, characteristic pseudopapillae are formed when some poorly cohesive neoplastic cells drop away, leaving a thin layer of neoplastic cells surrounding a small blood vessel
    - Cystic degeneration may also occur
  - Although grossly well-demarcated, neoplastic cells often, microscopically, delicately infiltrate into the adjacent nonneoplastic pancreas
  - Foamy macrophages, cholesterol clefts, and hemorrhage are common
  - Cytoplasm of the neoplastic cells is frequently eosinophilic, although clear or foamy change can occur
    - Vacuoles may be present, and hyaline globules can be a clue to the diagnosis
  - Nuclei are round to ovoid with stippled chromatin and frequent nuclear grooves
- Immunohistochemistry
  - Abnormal nuclear labeling with β-catenin (Fig. 7.1c)

- Diffusely positive in neoplastic cells: β-catenin (nuclear and cytoplasmic), CD10, LEF1, CD99 (dot-like), cyclin D1, TFE3, α-1 antitrypsin, progesterone receptor
- Variably positive in neoplastic cells: Synaptophysin, cytokeratin
- Negative in neoplastic cells: Chromogranin, pancreatic hormones (insulin, somatostatin, glucagon), lipase, trypsin, chymotrypsin, estrogen receptors

Genetic Features
- See Table 7.1
- Activating somatic mutations in β-catenin gene (*CTNNB1*) occur in 95% of cases; these alterations lead to abnormal nuclear accumulation of β-catenin protein in almost 100% of cases (Fig. 7.1c)
  - The β-catenin protein has diverse functions in normal cells
    - β-catenin is involved in cell adhesion through its interactions with E-cadherin
    - When not associated with E-cadherin, β-catenin is normally targeted for degradation
    - When degradation of β-catenin is inhibited by Wnt signaling, β-catenin translocates to the nucleus and leads to transcription of target genes
  - Mutations in *CTNNB1* frequently affect key phosphorylation sites, preventing β-catenin degradation
  - *CTNNB1* mutations lead to overexpression of cyclin D1 protein, a key cell cycle regulator and downstream target of β-catenin, in the majority of solid pseudopapillary neoplasms
  - *CTNNB1* mutations also are associated with loss of E-cadherin in the cell membrane, sometimes with mislocalization to the cytoplasm or nucleus, explaining the tumor's discohesive morphology
    - As a result, solid pseudopapillary neoplasms do not label with antibodies to the extracellular domain of E-cadherin, while immunolabeling with antibodies to the cytoplasmic domain of the protein produces an abnormal nuclear pattern of labeling
    - No mutations in E-cadherin have been reported
  - Activation of β-catenin signaling (with overexpression of AXIN2, TBX3, SP5, and NOTUM) and Notch signaling (with overexpression of HEY1, HEY2, and NOTCH2) pathways has been documented by gene expression profiling
  - Discovery of the genetic underpinning of this tumor (*CTNNB1* mutation) has led to a crucial diagnostic test (β-catenin immunolabeling) to distinguish solid pseudopapillary neoplasm from other solid cellular neoplasms of the pancreas (Fig. 7.1c)

- Lack alterations in genes frequently mutated in pancreatic ductal adenocarcinoma (*KRAS*, *TP53*, *SMAD4*)
  - Conversely, most pancreatic ductal adenocarcinomas and PanNETs lack mutations in *CTNNB1* and nuclear localization of the β-catenin protein

---

Neoplasms with Acinar Differentiation

Acinar Cell Carcinoma

Clinical Features
- Rare pancreatic neoplasm, accounting for <2% of all pancreatic malignancies
- Most occur in adults (mean age, 58 years), although a small proportion (5–10%) occurs in children
- Male predominance (male-to-female ratio of 3.6:1)
- Acinar cell carcinomas have been reported in patients with deleterious germline *BRCA2* gene variants
- Most frequently present with nonspecific abdominal symptoms (pain, vomiting, weight loss); jaundice is rare due to growth pattern of tumor
  - Approximately 15% develop lipase hypersecretion syndrome, characterized by subcutaneous fat necrosis, eosinophilia, and polyarthralgia; more common in metastatic disease
- Prognosis is poor but relatively better than ductal adenocarcinomas, with a 5-year survival of about 40%
  - Outcome for acinar cell carcinoma is better than for stage-matched pancreatic ductal adenocarcinoma
  - Stage is the best predictor of outcome in acinar cell carcinoma

Pathologic Features
- Occurs slightly more often in the head of the pancreas but can occur anywhere in the gland
- Usually forms solitary mass, but can be multilobulated
  - In contrast to the invasive growth of ductal adenocarcinoma, acinar cell carcinomas are frequently grossly well-circumscribed and may be encapsulated
- Usually large (average 10 cm in greatest dimension)
- Most often a solid mass with a soft, red-tan, and fleshy cut surface, but can have frank necrosis and cystic degeneration
- May invade into adjacent organs as well as into the pancreatic ductal system
- Rarely, a multicystic form can occur (acinar cell cystadenocarcinoma)
- Microscopically, composed of neoplastic cells with architectural, histologic, or immunohistochemical evidence of acinar differentiation

- Multiple possible architectural patterns, most commonly acinar (with pyramidal-shaped cells forming small lumina) or solid (with sheets of cells not forming well-defined structures), although glandular and trabecular patterns can occur
- Microscopically, invasive growth is common (vascular invasion, perineural invasion, and even invasion into the pancreatic duct system)
- Cytoplasm typically contains amphophilic or eosinophilic zymogen granules, although in some cases, these granules are not well developed and can be difficult to appreciate on routinely stained sections
- Nuclei are round to oval, characteristically with a single prominent nucleolus
- Immunohistochemistry
  - Diffusely positive in neoplastic cells: BCL10 (the antibody to BCL10 cross-reacts with carboxyl ester hydrolase), trypsin, chymotrypsin, lipase, elastase, cytokeratin (CK8 and CK18)
  - Variably positive in neoplastic cells: Chromogranin, synaptophysin, β-catenin (nuclear and cytoplasmic in a minority of neoplasms)
    - If chromogranin or synaptophysin is positive in >25% of neoplastic cells, the neoplasm should be classified as a mixed acinar neuroendocrine carcinoma

Genetic Features
- See Table 7.1
- Somatic alterations in the APC/β-catenin pathway occur in 20–25% of acinar cell carcinomas, including activating mutations in *CTNNB1* as well as truncating mutations in *APC*
- Recurrent rearrangements involving *BRAF* and *RAF1* (*CRAF*) have been identified in 25% of patients with acinar cell carcinoma
  - *SND1-BRAF* is the most prevalent fusion, which results in activation of the MAPK pathway
- Acinar cell carcinoma may harbor targetable genetic alterations, including somatic mutations in *BRCA2*, *PALB2*, *ATM*, *BAP1*, and *JAK1*, as well as *RAF* fusions
- Genetic alterations frequently observed in pancreatic ductal adenocarcinoma (*KRAS*, *TP53*, *SMAD4*) occur at a lower frequency in acinar cell carcinoma
- A subset of carcinomas has microsatellite instability
- Promoter methylation of several tumor suppressor genes has been reported in few tumors, although the functional consequences of this methylation remain to be explored
- Large-scale chromosomal losses and gains have been reported, including several chromosomal regions that are altered in multiple carcinomas

- The regions altered are different from those frequently altered in pancreatic ductal adenocarcinoma
- In addition, most acinar cell carcinomas harbor numerous chromosomal gains and losses

Pancreatoblastoma

Clinical Features
- Rare pancreatic neoplasm, accounting for very small proportion of adult pancreatic malignancies but approximately 25% of pancreatic neoplasms in the first decade of life
- Majority occur in children <10 years of age, but can occur in adults (median age in children, 2.4 years; median age in adults, 40 years)
- Slight male predominance (male to female ratio 1.3–2:1)
- Have been reported in children with the Beckwith–Wiedemann syndrome
- Often diagnosed due to palpable abdominal mass in young patients, but can also cause vague abdominal symptoms (pain, vomiting, weight loss)
  - Jaundice is uncommon
- Up to one-third of patients with pancreatoblastoma have elevated serum levels of α-fetoprotein (AFP), due to production of AFP by the tumor
- Prognosis is poor, with an overall survival of approximately 50%
  - Cure by surgery can occur in young patients without metastases, but many patients with an initially complete resection will develop local recurrence (20%) or metachronous metastasis (25%)
  - Stage is the best predictor of outcome in patients with pancreatoblastoma

Pathologic Features
- Not associated with a specific location in the pancreas (evenly distributed throughout head, body, and tail)
- Like acinar cell carcinoma, it usually forms a solitary mass but can be lobulated, usually grossly well-demarcated, and often encapsulated
- Wide size range (1.5–20 cm), but on average are large (average 10.6 cm in greatest dimension)
- On gross cut section, usually similar to acinar cell carcinomas: Solid, soft, and fleshy, with color varying from gray to tan-yellow
  - Cystic degeneration is common in patients with Beckwith–Wiedemann syndrome
- Microscopically, composed of multiple different components, including, at a minimum, cells with acinar differentiation and squamoid nests

- Acinar component is usually predominant, consisting of relatively uniform cells with round nuclei, prominent nucleoli, and granular cytoplasm, often polarized around small lumina
- Squamoid nests are a characteristic feature and are required to establish the diagnosis
  - They consist of whorled groups of plump squamoid cells with eosinophilic cytoplasm; focal keratinization can be present; the nuclei in these nests are often clear
- Subset of pancreatoblastomas exhibits a focal neuroendocrine component, consisting of uniform round cells with characteristic finely stippled chromatin
  - Neuroendocrine cells are usually diffusely scattered among cells with acinar differentiation, but may focally form solid nests or trabeculae
- Minority of pancreatoblastomas also have a ductal component, consisting of columnar cells (often mucin-producing) that surround larger lumina
- Nests of neoplastic cells of all types are separated by variably cellular stroma, which can exhibit cartilaginous or even osseous differentiation
- A primitive component of immature monotonous cells may also be present

- Immunohistochemistry
  - Acinar component: Positive for cytokeratins (CK8 and CK18), BCL10, trypsin, chymotrypsin, and lipase
  - Neuroendocrine component: Positive for chromogranin, synaptophysin, NSE; usually negative for pancreatic hormones (insulin, glucagon, somatostatin)
  - Ductal component: Positive for cytokeratins (CK7 and CK19), CEA
  - Squamoid nests: Largely negative for most immunohistochemical markers, but the optically clear nuclei contain biotin and may nonspecifically label with a variety of antibodies
    - This nonspecific labeling sometimes highlights the squamoid nests, making them easier to identify

Genetic Features
- See Table 7.1
- Frequent allelic loss of chromosome 11p (86% of cases in one study)
  - Loss of heterozygosity of 11p has also been reported in other embryonal neoplasms, such as hepatoblastoma and Wilms tumor, suggesting the possibility of a common genetic pathway in embryonal tumors
- Majority of cases have somatic alterations in the APC/β-catenin pathway, including activating mutations in *CTNNB1* as well as inactivating mutations in *APC*, leading to abnormal nuclear accumulation of β-catenin protein

- β-catenin localization can be patchy in individual tumors, with nuclear β-catenin most frequently found in squamoid nests; cyclin D1 overexpression is also most frequently present in squamoid nests
- Imprinting dysregulation of *IGF2* because of copy-neutral loss of heterozygosity, gain of paternal allele, and gain of methylation has been reported
- Lack the alterations in genes commonly mutated in pancreatic ductal adenocarcinoma (*KRAS*, *TP53*, *SMAD4*)
  - Rare loss of SMAD4 expression has been reported, but no *KRAS* gene mutations or altered p53 expression have been identified
- Several reports describe cases with complex karyotypes with multiple regions of chromosomal loss and gain
  - Trisomy 8 has been reported in a single case
  - No target genes for large-scale chromosomal alterations have been identified

Hereditary/Genetic Syndromes

- Family history of pancreatic cancer significantly increases an individual's risk of developing pancreatic cancer
- Approximately 10% of patients with pancreatic cancer have a familial basis (Table 7.2)
- Increased risk of pancreatic cancer is a feature of several genetic syndromes, but the genetic basis for most of the familial pancreatic cancer remains unknown

Mutations in the Fanconi Anemia/BRCA Pathway

- Proteins in this pathway are crucial for repair of DNA interstrand crosslinks
- Germline mutations in *BRCA2* and *BRCA1* result in increased risk of breast and ovarian cancer and account for a subset of patients with familial pancreatic ductal adenocarcinoma; importantly, some patients with *BRCA*-related familial pancreatic ductal adenocarcinoma do not report a family or personal history of breast or ovarian cancer
  - Cells with biallelic inactivation of *BRCA2* and *BRCA1* are exquisitely sensitive to therapies that target their DNA repair defect, such as mitomycin C and poly(ADP)-ribose polymerase (PARP) inhibitors
- Germline mutations in *PALB2*, also known as *FANCN*, whose protein product interacts with BRCA2, account for a subset (~3%) of familial pancreatic ductal adenocarcinomas
- Germline mutations in other Fanconi pathway genes (*FANCC*, *FANCG*) have been reported in young patients

with pancreatic ductal adenocarcinoma, but their importance in familial pancreatic ductal adenocarcinoma has not been firmly established

- The importance of *FANCA* in familial pancreatic ductal adenocarcinoma has also been investigated, but no significant contribution to cancer susceptibility could be established
- 7–10% of pancreatic ductal adenocarcinomas that lack germline mutations in the Fanconi anemia/BRCA pathway may harbor somatic mutations that confer a BRCA-like phenotype
  - Pancreatic ductal adenocarcinomas with this phenotype may be susceptible to poly(ADP)-ribose polymerase (PARP) inhibitors or other DNA damaging agents; germline sequencing is recommended for all patients with pancreatic cancer, regardless of family history, to identify targetable alterations in the Fanconi anemia/BRCA pathway

Ataxia-Telangiectasia Mutated

- Germline mutations in *ATM*, a kinase that interacts with Fanconi anemia proteins, are associated with an increased risk of pancreatic ductal adenocarcinoma
- In addition to conventional adenocarcinomas, adenosquamous and colloid carcinomas may arise in patients with germline *ATM* pathogenic variants
- Cancer with *ATM* inactivation appears to be particularly sensitive to radiation therapy

Familial Atypical Multiple Mole Melanoma Syndrome

- Germline mutations in *p16/CDKN2A* result in increased risk of both melanoma (with multiple nevi and atypical nevi) and pancreatic ductal adenocarcinoma in familial atypical multiple mole melanoma syndrome (FAMMM)

Peutz–Jeghers Syndrome

- Peutz–Jeghers syndrome (PJS) is an inherited syndrome caused by germline mutations in *STK11/LKB1* and characterized by gastrointestinal hamartomas as well as cancer predisposition
- Patients with PJS have a very high risk of pancreatic ductal adenocarcinoma, and carcinomas in PJS patients show somatic loss of the wildtype *STK11/LKB1* allele
- IPMNs have been reported in patients with PJS, a cancer predisposition syndrome associated with germline mutations in the *STK11/LKB1* gene on chromosome 19p

- In PJS, the IPMNs exhibit loss of heterozygosity at the *STK11/LKB1* gene locus, suggestive of a second somatic hit to the wildtype allele in the neoplasm
- In addition to reports of IPMNs in patients with PJS, sporadic IPMNs also undergo somatic mutation (5%) and loss of heterozygosity (25%) at the *STK11/LKB1* locus

Hereditary Pancreatitis

- Germline mutations in *PRSS1*, *SPINK1*, *CTRC*, and *CFTR* (cystic fibrosis) cause hereditary pancreatitis, characterized by continuing or relapsing pancreatic inflammatory disease
- Patients with hereditary pancreatitis have a markedly increased risk of developing pancreatic ductal adenocarcinoma, particularly if they smoke cigarettes

Hereditary Nonpolyposis Colorectal Cancer

- Germline mutations in *MSH2*, *MLH1*, *PMS1*, *PMS2*, and *MSH6/GTB* cause HNPCC (also known as Lynch syndrome), in which patients demonstrate an increased risk of carcinomas of the colon as well as other sites
- The neoplasms in these patients exhibit defects in the DNA mismatch repair machinery, resulting in microsatellite instability
- Patients with HNPCC have an increased risk of developing pancreatic ductal adenocarcinoma
  - Importantly, pancreatic ductal adenocarcinomas with microsatellite instability exhibit a distinct "medullary" morphology
  - Microsatellite instability is uncommon in tubular type (ordinary) ductal adenocarcinomas
- A few patients with Lynch syndrome (HNPCC) have been reported to have IPMNs, with loss of expression of mismatch repair proteins in the neoplastic cells of the pancreas
  - This suggests the possibility of IPMN as a rare extra-colonic manifestation of Lynch syndrome
- Cancers with microsatellite instability are often extremely responsive to treatment with immune checkpoint inhibitors

Familial Adenomatous Polyposis

- Germline mutations in *APC* result in a markedly increased risk of adenomatous colorectal polyps as well as colorectal adenocarcinoma
  - An increased risk of pancreatic ductal adenocarcinoma has also been reported in these patients; however, some

of this may reflect the increased risk of duodenal carcinomas, which frequently invade the pancreas and can mimic primary pancreatic adenocarcinoma

- Similarly, although the association is not well-established, IPMNs have been reported in patients with familial adenomatous polyposis (FAP)
- A few patients with FAP have been reported to have pancreatoblastoma

von Hippel–Lindau Syndrome

- VHL is an autosomal dominant disorder characterized by neoplasms in various organs, including hemangioblastomas of the eye and cerebellum, pheochromocytomas, and renal cell carcinomas
- VHL is caused by mutations in the *VHL* gene on chromosome 3p, which is involved in the regulation of the HIF1α pathway
- Up to 90% of patients with VHL develop SCAs of the pancreas, and these pancreatic lesions may be the first presentation in patients with VHL
- The SCAs that arise in patients with VHL are often combined serous neuroendocrine neoplasms
- Patients with VHL may present with diabetes mellitus due to diffuse involvement of the pancreas by SCAs
- 5–10% of patients with VHL also develop PanNETs, with somatic loss of wildtype allele of *VHL* in neoplastic cells

Other Hereditary/Genetic Syndromes

- PanNETs occur as features of multiple inherited syndromes
  - MEN1
    - MEN1 is caused by germline mutations in tumor suppressor gene *MEN1* on chromosome 11q
    - MEN1 is an autosomal dominant clinical syndrome characterized by neuroendocrine lesions of the parathyroid, pituitary, pancreas, duodenum (gastrinomas), and adrenal
    - Pancreas is involved in 60–70% of patients with MEN1
  - Tuberous sclerosis complex (TSC)
    - TSC is caused by germline mutations in tumor suppressor genes *TSC1* on chromosome 9q or *TSC2* on chromosome 16p
    - TSC is an autosomal dominant clinical syndrome characterized by hamartomas and rare malignant neoplasms
    - PanNETs have rarely been reported in children with TSC

- Neurofibromatosis type 1 (NF1)
  - NF1 is caused by germline mutations in tumor suppressor gene *NF1* on chromosome 17q
  - NF1 is an autosomal dominant clinical syndrome characterized most prominently by nervous system abnormalities
    - PanNETs expressing somatostatin (somatostatinomas) are a rare finding in NF1
    - Although pancreatic tumors have been reported, somatostinomas arise more frequently in the duodenum or ampulla of Vater of patients with NF1
- Mahvash disease
  - Mahvash disease is an autosomal recessive, hereditary syndrome and is caused by mutations in the glucagon receptor gene *GCGR*
- Several cases of pancreatoblastoma reported in patients with Beckwith–Wiedemann syndrome, a disorder associated with imprinting dysregulation on chromosomal 11p, leading to overgrowth of various organs and predisposition to embryonal tumors

Summary of Molecular Pathology of Pancreatic Cancer

- Neoplasms of the pancreas can be classified morphologically and immunohistochemically based on their direction of differentiation; these classifications frequently parallel distinct molecular alterations
- Ductal adenocarcinoma is the most common pancreatic neoplasm
  - The accumulation of somatic alterations in key oncogenes and tumor suppressor genes drives the transformation of normal cells through noninvasive dysplastic precursors into invasive ductal adenocarcinoma
- Neoplasms with other directions of differentiation (PanNETs, solid pseudopapillary neoplasms, acinar cell carcinomas) exhibit distinct sets of molecular alterations not shared with ductal adenocarcinoma
- Knowledge of molecular alterations has led to key diagnostic tests currently in use in pancreatic cancer, such as immunohistochemical labeling for SMAD4 in ductal adenocarcinoma and β-catenin in solid pseudopapillary neoplasms (Fig. 7.1)
- Molecular analyses of clinical samples, such as DNA sequencing of pancreatic cyst fluid to distinguish precancerous from benign cysts, show promise in augmenting current diagnostic practice to provide the best possible clinical care

Further Reading

Abraham SC, Klimstra DC, Wilentz RE, et al. Solid pseudopapillary tumors of the pancreas are genetically distinct from pancreatic ductal adenocarcinomas and almost always harbor beta-catenin mutations. Am J Pathol. 2002;160:1361–9.

Abraham SC, Wu TT, Hruban RH, et al. Genetic and immunohistochemical analysis of pancreatic acinar cell carcinoma: frequent allelic loss on chromosome 11p and alterations in the APC/beta-catenin pathway. Am J Pathol. 2002;160:953–62.

Abraham SC, Wu TT, Klimstra DC, et al. Distinctive molecular genetic alterations in sporadic and familial adenomatous polyposis-associated pancreatoblastomas: frequent alterations in the APC/beta-catenin pathway and chromosome 11p. Am J Pathol. 2001;159:1619–27.

Agaimy A, Haller F, Frohnauer J, et al. Pancreatic undifferentiated rhabdoid carcinoma: KRAS alterations and SMARCB1 expression status define two subtypes. Mod Pathol. 2014;28:248–60.

Elias KM, Tsantoulis P, Tille J-C, et al. Primordial germ cells as a potential shared cell of origin for mucinous cystic neoplasms of the pancreas and mucinous ovarian tumors. J Pathol. 2018;246:459–69.

Felsenstein M, Noë M, Masica DL, et al. IPMNs with co-occurring invasive cancers: neighbours but not always relatives. Gut. 2018;67:1652–62.

Fischer CG, Guthrie VB, Braxton AM, et al. Intraductal papillary mucinous neoplasms arise from multiple independent clones, each with distinct mutations. Gastroenterology. 2019;157:1123–37.

Fujikura K, Hosoda W, Felsenstein M, et al. Multiregion whole-exome sequencing of intraductal papillary mucinous neoplasms reveals frequent somatic *KLF4* mutations predominantly in low-grade regions. Gut. 2021;70(5):928–39. https://doi.org/10.1136/gutjnl-2020-321217.

Hruban RH, Adsay NV, Albores-Saavedra J, et al. Pancreatic intraepithelial neoplasia: a new nomenclature and classification system for pancreatic duct lesions. Am J Surg Pathol. 2001;25:579–86.

Hruban RH, Canto MI, Goggins M, et al. Update on familial pancreatic cancer. Adv Surg. 2010;44:293–311.

Hruban RH, Pittman MC, Klimstra DS. AFIP atlas of tumor pathology: tumors of the pancreas. Washington, DC: American Registry of Pathology; 2007.

Hutchings D, Jiang Z, Skaro M, et al. Histomorphology of pancreatic cancer in patients with inherited ATM serine/threonine kinase pathogenic variants. Mod Pathol. 2019;32:1806–13.

Jiao Y, Shi C, Edil BH, et al. DAXX/ATRX, MEN1, and mTOR pathway genes are frequently altered in pancreatic neuroendocrine tumors. Science. 2011;331:1199–203.

Jones S, Zhang X, Parsons DW, et al. Core signaling pathways in human pancreatic cancers revealed by global genomic analyses. Science. 2008;321:1801–6.

Kasajima A, Konukiewitz B, Schlitter AM, et al. An analysis of 130 neuroendocrine tumors G3 regarding prevalence, origin, metastasis, and diagnostic features. Virchows Arch. 2022;480:359–68.

Klimstra DS, Adsay V. Acinar neoplasms of the pancreas-a summary of 25 years of research. Semin Diagn Pathol. 2016;33:307–18.

Kuboki Y, Fischer CG, Guthrie VB, et al. Single-cell sequencing defines genetic heterogeneity in pancreatic cancer precursor lesions. J Pathol. 2019;247:347–56.

Luchini C, Pea A, Lionheart G, et al. Pancreatic undifferentiated carcinoma with osteoclast-like giant cells is genetically similar to, but clinically distinct from, conventional ductal adenocarcinoma. J Pathol. 2017;243:148–54.

Moffitt RA, Marayati R, Flate EL, et al. Virtual microdissection identifies distinct tumor- and stroma-specific subtypes of pancreatic ductal adenocarcinoma. Nat Genet. 2015;47:1168–78.

Noë M, Niknafs N, Fischer CG, et al. Genomic characterization of malignant progression in neoplastic pancreatic cysts. Nat Commun. 2020;11:4085.

The Cancer Genome Atlast Research Network. Integrated genomic characterization of pancreatic ductal adenocarcinoma. Cancer Cell. 2017;32:185–203.

Vogelstein B, Kinzler KW. Cancer genes and the pathways they control. Nat Med. 2004;10:789–99.

Vortmeyer AO, Lubensky IA, Fogt F, et al. Allelic deletion and mutation of the von Hippel-Lindau (VHL) tumor suppressor gene in pancreatic microcystic adenomas. Am J Pathol. 1997;151:951–6.

Waddell N, Pajic M, Patch A-M, et al. Whole genomes redefine the mutational landscape of pancreatic cancer. Nature. 2015;518:495–501.

WHO Classification of Tumors Editorial Board. WHO classification of tumours: digestive system. Lyon: International Agency for Research on Cancer; 2019.

Wu J, Jiao Y, Dal Molin M, et al. Whole-exome sequencing of neoplastic cysts of the pancreas reveals recurrent mutations in components of ubiquitin-dependent pathways. Proc Natl Acad Sci U S A. 2011;108:1188–93.

Wu J, Matthaei H, Maitra A, et al. Recurrent GNAS mutations define an unexpected pathway for pancreatic cyst development. Sci Transl Med. 2011;3:92ra66.

Molecular Pathology of Liver Tumors

8

Thomas Longerich and Peter Schirmacher

Contents

T. Longerich (✉) · P. Schirmacher
Institute of Pathology, University Hospital Heidelberg,
Heidelberg, Germany
e-mail: thomas.longerich@med.uni-heidelberg.de

Introduction

The different cell types of the liver provide the framework for a variety of benign and malignant tumor entities, each of which shows a more or less defined spectrum of associated molecular alterations. Among the malignant neoplasm carcinomas are prevailing, with malignant mesenchymal tumors being exceedingly rare. Hepatocellular carcinoma (HCC) is one of the most frequent tumors worldwide, and like intrahepatic cholangiocarcinoma (iCCA) shows a rising incidence. For both tumor types, specific risk factors and molecular subgroups have been identified. Hepatocellular adenoma is a paradigm entity in terms of morphomolecular subclassification, which has recently been adopted to facilitate the subtyping of some hepatocellular carcinomas. Molecular markers support the differential diagnosis of highly differentiated hepatocellular tumors, and the detection of genetic rearrangement may be helpful in subtyping mesenchymal liver tumors.

© The Author(s), under exclusive license to Springer Nature Switzerland AG 2023
L. Cheng et al. (eds.), *Molecular Surgical Pathology*, https://doi.org/10.1007/978-3-031-35118-1_8

The role of predictive (therapy-guiding) molecular markers is rising in malignant liver tumors and already has its position in iCCA; concerning HCC, this field is still emerging.

Hepatocellular Adenoma

Definition

- Hepatocellular adenoma (HCA) is a benign liver neoplasm of hepatocellular differentiation

Clinical Features

- The incidence is 3–4/100,000 people in Europe and North America but lower in Asia
- Most cases (85%) occur in women before menopause
- Etiology is well-defined
  - Endogenous or exogenous exposure to estrogenic (e.g., contraception >2 years) or androgenic (e.g., bodybuilding; androgen therapy of Fanconi anemia or acquired aplastic anemia) steroids
  - Metabolic disease (e.g., mature onset diabetes of the young [MODY] type III, glycogenosis type 1 [von Gierke disease] or type 3 [Forbes disease])
  - Familial adenomatous polyposis coli (adenomatosis polyposis coli [APC] mutation)
  - β (beta) thalassemia with iron overload
- Clinical presentation
  - Mild chronic or acute abdominal pain due to intratumoral or intraperitoneal hemorrhage (increased risk: size >5 cm, sonic hedgehog-activated HCA [shHCA])
  - Elevated levels of liver enzymes
  - Incidental finding of a liver mass by imaging
- Malignant transformation is uncommon, but varies strongly with the HCA subtype (driving factor: CTNNB1 mutation), and certain settings are associated with an increased risk (exposure to anabolic steroids, glycogenosis, vascular diseases, male gender, HNF1α(alpha)-inactivated adenoma >5 cm in patients >60 years)
- Discontinuation may result in spontaneous regression of drug-induced HCA

Pathologic Features

- HCA typically lacks a capsule (↔ progressed hepatocellular carcinoma [HCC])
- Hepatocyte-like cells arranged in plates (1–2 cells thick) and occasional pseudoglands
- Supplied by unpaired arteries (without accompanying bile duct)
- Preexisting portal tracts may be entrapped in the periphery of the lesion
- Cell size is typically mildly increased compared to the normal hepatocytes of the surrounding liver
- No or only mild nuclear atypia (except β(beta)-catenin-activated HCA variants)
- Cytoplasm can be normal, clear, or steatotic, or it can contain pigment (bile, lipofuscin)
- The neoplastic hepatocytes of HCA are usually surrounded by a regular reticulin framework (↔ HCC)
- Areas of infarction, hemorrhage or regression indicate an increased risk of spontaneous rupture
- Features associated with certain HCA-subtypes
  - HNF1α(alpha)-inactivated HCA (H-HCA): pronounced steatosis, sometimes microadenomas in background liver (particularly in adenomatosis; e.g., MODY type III diabetes)
  - Inflammatory HCA (IHCA): inflammatory infiltrates, sinusoidal dilatation, minor ductular component, sometimes intratumoral hemorrhage
  - β(beta)-catenin-activated HCA variants: cytological atypia, pseudogland formation
  - Sonic hedgehog-activated HCA: intratumoral hemorrhage

Differential Diagnosis

- Highly differentiated HCC: diffuse capillarization of sinusoids (CD34 staining), nuclear atypia, pseudogland formation, interstitial invasion, immunohistochemical marker panel positive (at least two out of three markers positive: glypican-3 [GPC3], heat shock protein 70 [HSP70], glutamine synthetase [GS])
- Focal nodular hyperplasia: central scar with abnormal vessels and radiating fibrosis, presence of ductules, IHC: map-like GS expression
- Macroregenerative nodule: comparison to surrounding liver tissue, periseptal GS expression
- Epithelioid angiomyolipoma: may resemble HCA, but HMB45, Melan A, and MITF positive

Genetic Features

- Heterogeneous, clonal lesions showing a few chromosomal alterations in up to 20% of cases
- Genomic aberrations include gains at 1p, 1q, 7p, 11q, 17q, and 20; the detection of more than two alterations in one tumor favors (early) HCC
- Morphomolecular HCA-subtypes differ with respect to clinical, genomic, pathological, and radiological features

- The genetic predisposition, risk factors, diagnostic marker, and clinicopathological characteristics of HCA subtypes are summarized in Fig. 8.1
- HNF1α(alpha)-inactivated HCA
  - HNF1α(alpha) is a transcription factor involved in hepatocellular differentiation
  - Mutational inactivation of *HNF1A* is found in 30–35% of HCA
  - Sporadic *HNF1A* mutations in 90% of cases, 10% are inherited
  - HNF1α(alpha)-inactivated HCA occur almost exclusively in women and carry a very low risk of malignant transformation even in case of adenomatosis (≥10 HCA), but progression to HCC has been observed in large HCA in older patients
  - Heterozygous *HNF1A* germline mutations are responsible for maturity-onset diabetes of the young type 3 (MODY3; an autosomal dominant type of diabetes); H-HCA in MODY3 patients carry a second somatic HNF1α(alpha)-inactivating mutation
  - Downregulation of the HNF1α(alpha) target gene fatty-acid binding protein 1 (FABP1) may contribute to the steatotic phenotype through impaired fatty-acid trafficking
  - Histology: prominent steatosis, but lack of inflammatory infiltrates and atypia, in contrast to background liver parenchyma tumor lacks FABP1 expression (Fig. 8.2)
- Inflammatory HCA
  - Inflammatory HCA (IHCA) constitutes the largest subgroup of HCA (35–40%)
  - Inflammatory HCA occurs mostly in women and is associated with obesity and fatty liver disease
  - A systemic inflammatory syndrome (e.g., elevated CRP level in serum, increased erythrocyte sedimentation rate) can be associated with inflammatory HCA (curable by tumor resection)
  - IL6-/JAK-/STAT-pathway activation is related mostly to e mutations in *IL6ST* (~60%), *FRK*, *STAT3*, *GNAS*, or *JAK1* genes (altogether accounting for ~60%) or

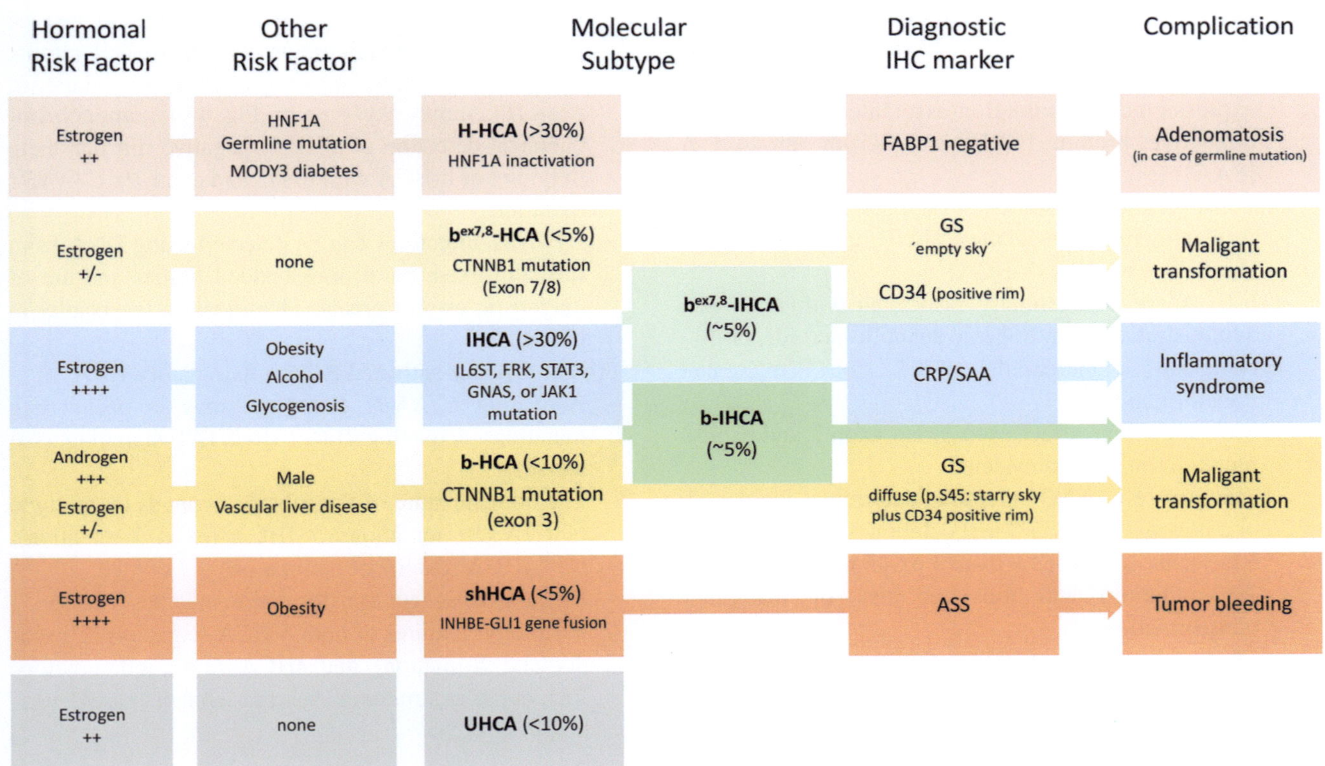

Fig. 8.1 Classification of HCA by genotype and phenotype (modified according to Nault et al., J Hepatol, 2017). ASS, argininosuccinate synthetase; b-HCA, β(beta)-catenin-activated HCA; b^ex7/8-HCA, β(beta)-catenin-activated HCA due to CTNNB1 exon 7 or exon 8 mutation; b-IHCA, β(beta)-catenin-activated inflammatory HCA; b^ex7/8-IHCA, β(beta)-catenin-activated inflammatory HCA due to CTNNB1 exon 7 or exon 8 mutation; CRP, c-reactive protein; FABP1, fatty acid binding protein1; GS, glutamine synthetase; HNF1A, hepatocyte nuclear factor 1-α(alpha); H-HCA, HNF1A-inactivated HCA; IHCA, inflammatory HCA; SAA, serum amyloid A; shHCA, sonic hedgehog-activated HCA; UHCA, unclassified HCA

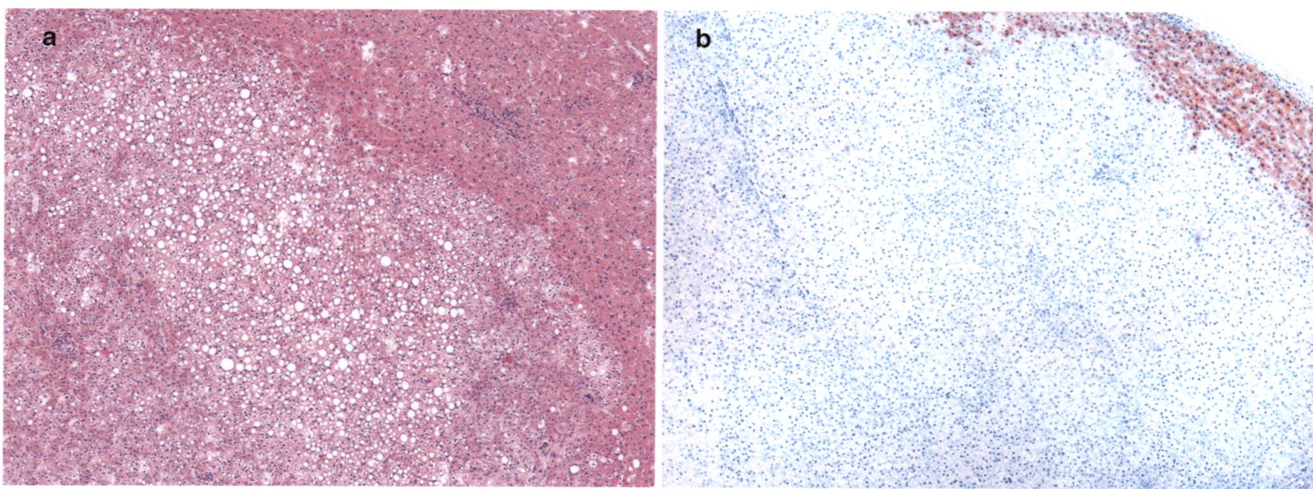

Fig. 8.2 HNF1α(alpha)-inactivated HCA. (**a**) Note the prominent fatty change and expansive growth. (**b**) The nonatypical tumor cells show loss of FAPB1 expression, while the surrounding normal liver (upper right) reveals regular diffuse positivity (original magnification 40-fold)

rarely recurrent chromosome rearrangements involving *ROS*, *FRK*, or *IL6* (~2% of cases)

- Histology: focal or diffuse inflammation and prominent vascular changes (sinusoidal dilatation, congestion, and thick-walled arteries); minor ductular proliferates (keratin [K] 7 positive) frequently detectable portal tract-like structures; increased expression of inflammation-associated proteins (e.g., c-reactive protein [CRP] and serum amyloid A [SAA]) (Fig. 8.3)

- β(beta)-Catenin-activated HCA (atypical adenoma)
  - Activating mutations of the *CTNNB1* gene is present in about 20% of HCA (b-HCA)
  - b-HCA are preferentially associated with male sex, administration of androgenic anabolic steroids, or glycogenosis; increased risk for malignant transformation
  - Most mutations affect exon 3, but exon 7 and exon 8 mutations are also prevalent
  - The type of *CTNNB1* mutation determines the degree of pathway activation (large deletion of exon 3 > exon 3 missense mutations > exon 3 p.S45 mutation > exon 7/8 mutation) and thus the risk of malignant transformation
  - Glutamine synthetase (GS) is a β(beta)-Catenin target gene and used as a surrogate marker for *CTNNB1* mutation, as nuclear β(beta)-Catenin expression is frequently lacking or may be seen only in the minority of tumor cell nuclei; thus GS is superior to β(beta)-Catenin staining in terms of subtyping
  - Histology: areas of minor to moderate nuclear atypia, pseudoglands; GS staining pattern depends on the type of *CTNNB1* mutation (exon 3: diffuse and strong, except exon 3 p.S45 showing a ´starry sky´ pattern); exon 7/8: ´empty sky´ pattern (Fig. 8.4); immunohistochemical detection of a CD34-negative rim may help in typing of b-HCA with p.S45 and exon 7/8 *CTNNB1* mutation
  - *CTNNB1* mutations can be detected using DNA from formalin-fixed, paraffin-embedded tissues and are of diagnostic use for precise classification (especially in biopsies)

- β(beta)-Catenin-activated HCA inflammatory HCA
  - Activating *CTNNB1* mutations may co-occur with mutations activating IL6-/JAK-/STAT signaling (see above)
  - The nomenclature of the subtype depends on the type of *CTNNB1* mutations: b-IHCA (exon 3 mutation), b[ex7/8]-IHCA (exon 7 or 8 mutation)
  - Risk for malignant transformation matches b-HCA
  - Histology: features of both b-HCA (e.g., atypia, pseudogland formation) and IHCA (e.g., inflammatory foci, ductular structures, vascular changes) are detectable (Fig. 8.5)

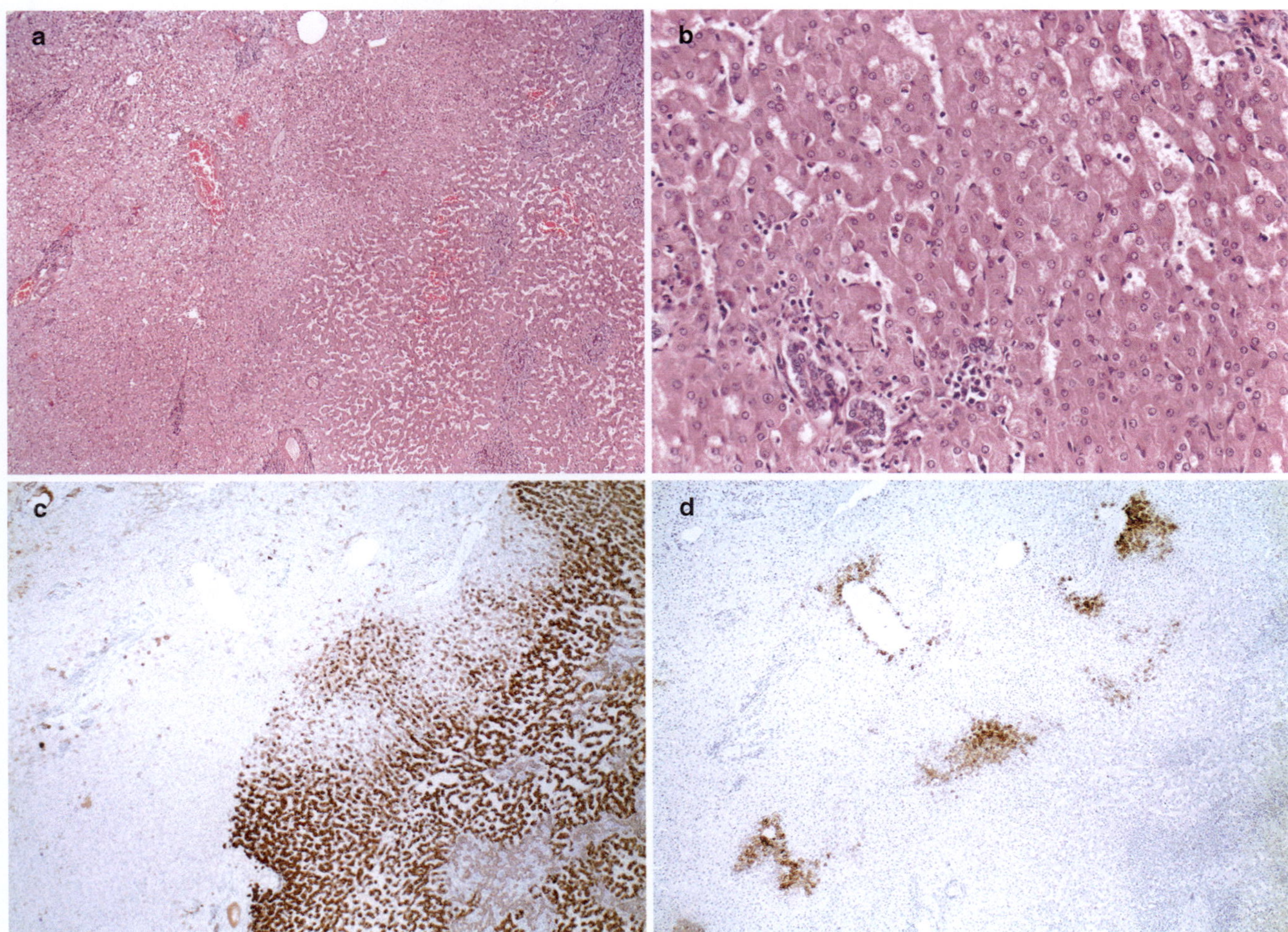

Fig. 8.3 Inflammatory HCA. (**a**) Highly differentiated hepatocellular tumor (right part) with sinusoidal dilatation showing gradual transition into the adjacent original liver parenchyma (original magnification 100-fold). In this example, there are several ductular structures. (**b**) Higher magnification reveals a lack of atypia, the presence of ductules, and small inflammatory foci (original magnification 200-fold). (**c**) The tumor cells are diffusely positive for Amyloid A, while the background liver is mainly negative. (**d**) Lack of a map-like GS staining within the focal lesion rules out the differential diagnosis of Focal Nodular Hyperplasia. Note the orthotopic GS staining in the background liver

- Sonic hedgehog-activated HCA
  - shHCA accounts for <5% of HCA and shows an association with a high risk of tumor rupture and bleeding, even in small tumors
  - Activation of sonic hedgehog signaling occurs due to a small somatic deletion of the *INHBE* gene leading to an *INHBE-GLI1* fusion gene
  - Histology: intratumoral bleeding, argininosuccinate synthetase 1 positivity (Fig. 8.6)
- Unclassified HCA
  - Less than 10% of HCA cannot be categorized into one of the abovementioned subtypes and remain molecularly unclassified (UHCA) so far
  - UHCA does not carry an increased risk for malignant transformation

Prognosis and Predictive Factors

- The risk for rupture and intraperitoneal hemorrhage increases for any HCA when the tumor diameter exceeds 5 cm and may result in hemorrhagic shock and death
- Pregnancy and shHCA are considered risk factors for symptomatic bleeding
- HCA >6 cm and β(beta)-catenin-activated HCA subtypes are at risk for malignant transformation, especially when detected in men
- HCA >5 cm in diameter requires treatment independent of the subtype due to the increased risk of rupture
- Cases of adenomatosis require individualized approaches (e.g., liver transplantation)

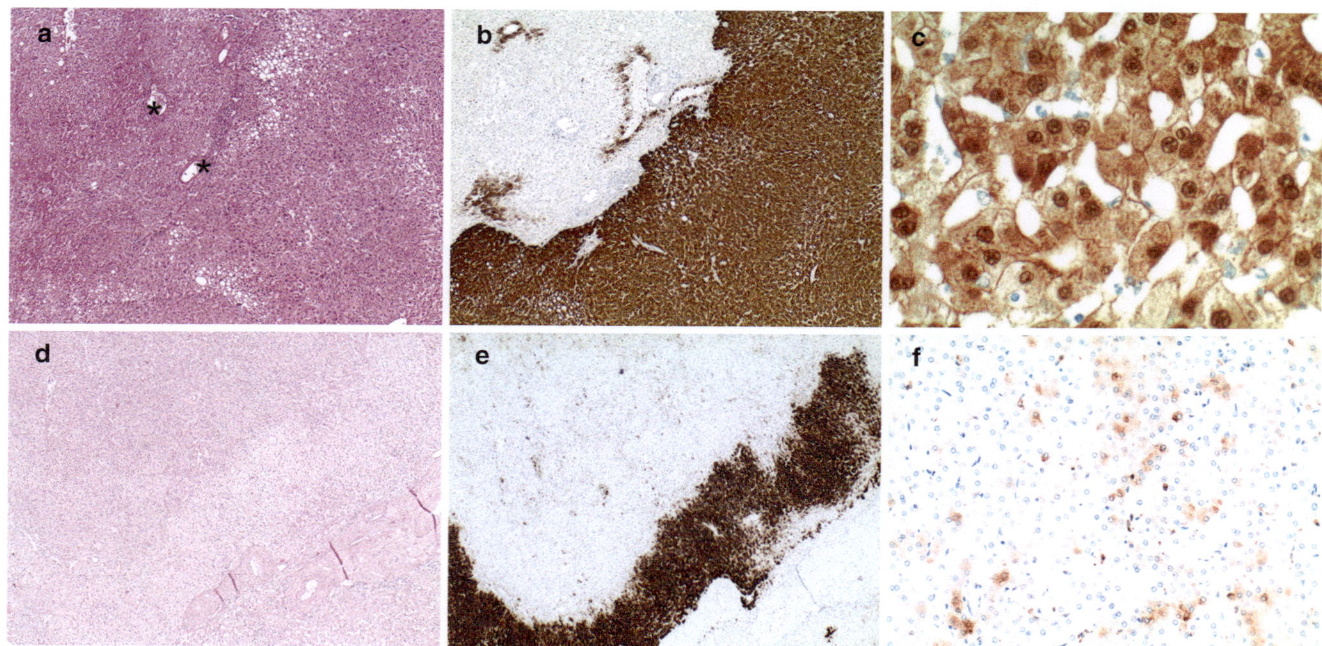

Fig. 8.4 β(beta)-catenin-activated HCA. (**a**) B-HCA (right) with exon 3 mutation (p.S33C) showing an ill-defined border and mild fatty change. Note the portal tracts (*) in the surrounding liver (Original magnification 40-fold). (**b**) Diffuse and strong expression of GS within the tumor, while the surrounding liver reveals orthotopic perivenular GS staining. (**c**) Nuclear β(beta)-catenin accumulation can be seen (Original magnification 400-fold). (**d**) B$^{ex7/8}$-HCA showing a rim of paler stained hepatocytes adjacent to dysmorphic vascular aggregates (Original magnification 40-fold). (**e**) Few hepatocytes are GS-positive in the center of the lesion, while the peripheral rim shows strong and diffuse staining. (**f**) Higher magnification (400-fold) reveals GS positivity in individual cells and groups of hepatocytes

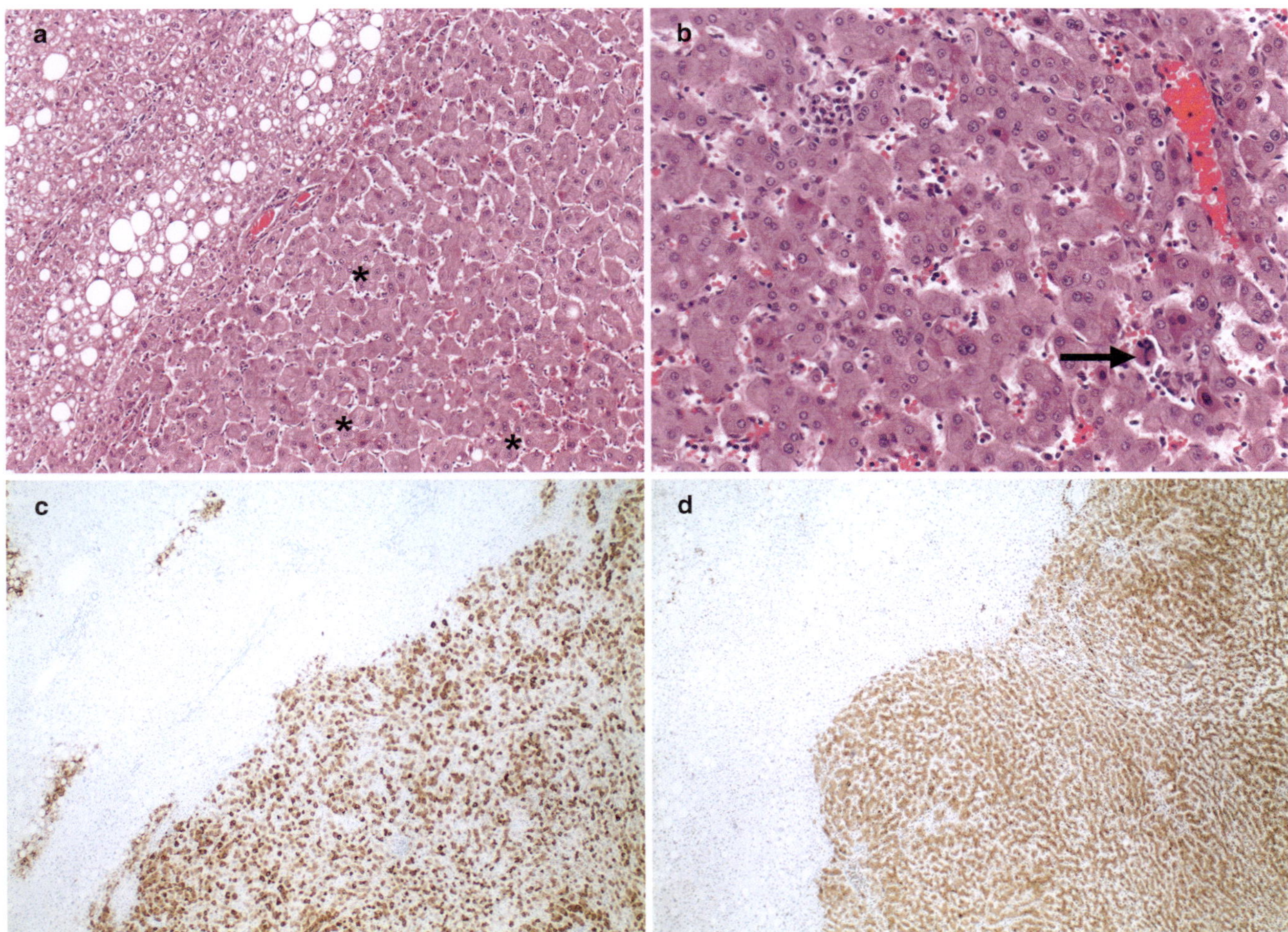

Fig. 8.5 B-IHCA: β(beta)-catenin-activated inflammatory HCA. (**a**) Highly differentiated hepatocellular tumor (lower right) showing a microtrabecular differentiation. Note there are small foci of nuclear crowding (*). The tumor is sharply demarcated from the background liver revealing moderate fatty change (original magnification 100-fold). (**b**) Higher magnification reveals mild trabecular disarray, nuclear crowding, an inflammatory focus (upper left), and apoptosis with nuclear condensation and fragmentation (arrow). Interstitial invasion was not detected (original magnification 200-fold). (**c**) The tumor cells are diffusely positive for GS, while the background liver shows orthotopic perivenular staining (original magnification 40-fold). (**d**) In addition, SAA expression is upregulated in the tumor cells

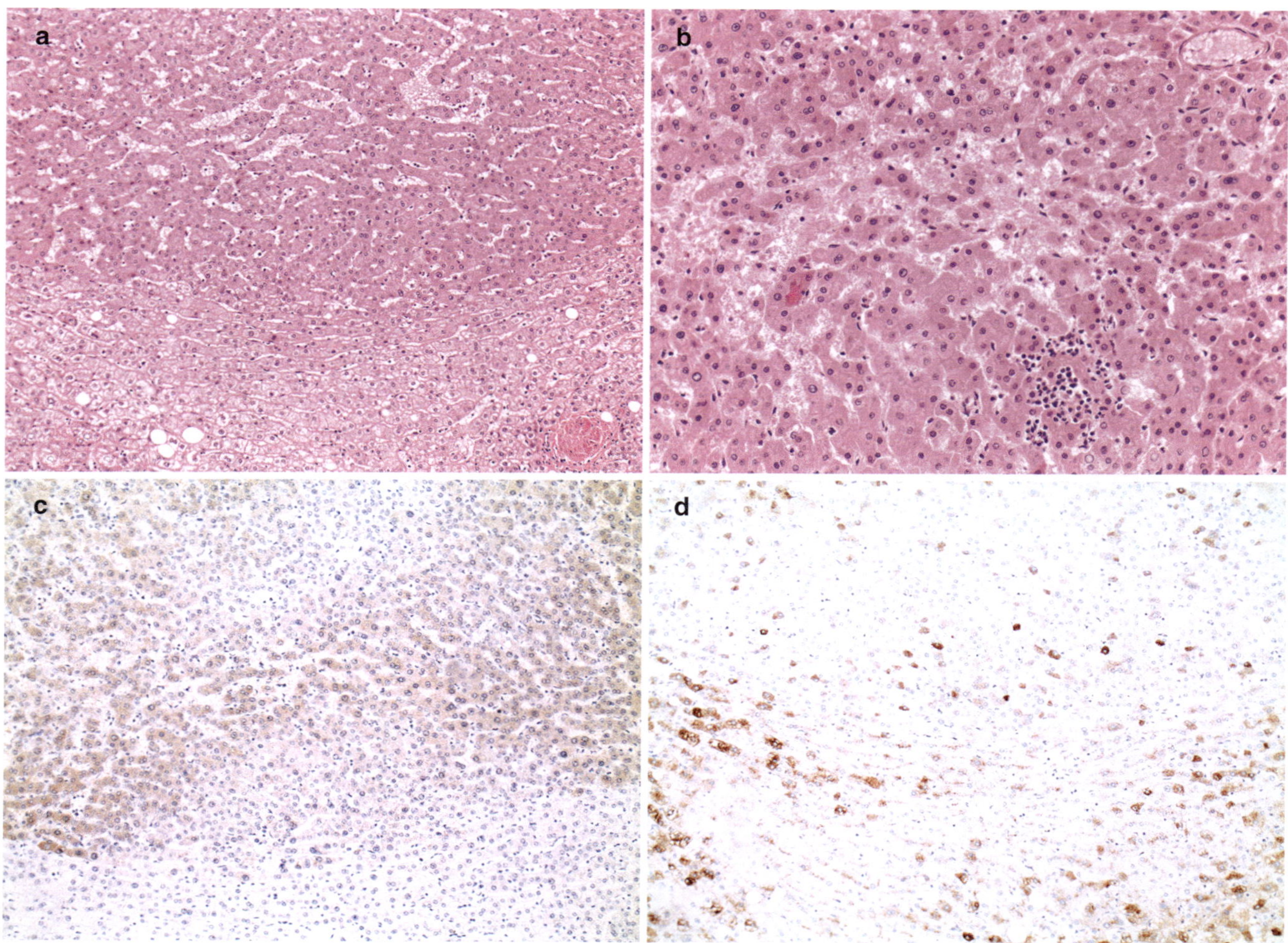

Fig. 8.6 shHCA. (**a**) Highly differentiated hepatocellular tumor (upper part) with sinusoidal dilatation showing gradual transition into the adjacent original liver parenchyma with minimal fatty change (original magnification 100-fold). (**b**) Higher magnification reveals a lack of atypia, irregular sinusoidal dilatation with microhemorrhage, and small inflammatory foci (original magnification 200-fold). (**c**) The tumor cells are diffusely positive for argininosuccinate synthetase, while the background liver is mainly negative. (**d**) There is no qualitative difference been serum amyloid A expression between lesional and nonlesional tissue, thus ruling out the differential diagnosis of inflammatory HCA

Hepatocellular Carcinoma

Definition

- Malignant primary liver tumor with hepatocellular differentiation

Clinical Features

- HCC is the sixth most frequent cancer and the fourth leading cause of cancer-related death worldwide
- HCC accounts for about 80% of primary liver cancer, with >800,000 new cases per year worldwide
- About 75% of cases occur in developing countries with endemic exposure to risk factors (e.g., Africa, China, and southeast Africa)
- Males have a significantly higher prevalence than females ranging from 2:1 to 4:1
- A defined cause of the underlying chronic liver disease can be found in more than 90% of patients
- Etiological risk factors: chronic viral hepatitis (hepatitis B virus ± hepatitis D virus, hepatitis C virus), alcoholic and nonalcoholic steatohepatitis, ingestion of mycotoxins (e.g., aflatoxin B1), several inherited diseases (e.g., genetic hemochromatosis, glycogen storage diseases, hereditary tyrosinemia, bile salt export pump [BSEP] deficiency/progressive familial intrahepatic cholestasis type 2 [PFIC2])

- Rare risk factors are: β(beta)-Catenin-activated HCA variants, α1-AT-deficiency, Wilson disease, and exposure to chemicals (e.g., vinyl chloride)
- Symptoms are either related to the tumor itself or to the advanced stage of the underlying chronic liver disease: abdominal pain, weight loss, nausea, ascites, jaundice, or (hepato)splenomegaly
- Raised α(alpha)-fetoprotein (AFP) levels >400 ng/mL or continuously rising AFP >100 ng/mL are indicative of HCC, but AFP elevation is only found in less than 50% of patients; especially early HCC (see below) are generally AFP negative

Pathologic Features

- Depending on its content of bile and fat HCC macroscopically varies in color from yellow to light tan and is often surrounded by a pseudocapsule

- The background liver frequently shows (macro- or micronodular) cirrhotic remodeling
- Four macroscopic HCC patterns are distinguished: single distinct nodule, dominant nodule with multiple satellite nodules (within 2 cm of the main nodule), diffuse (multiple small nodules separated by fibrous septa, thus mimicking cirrhosis), multiple, separated distinct nodules (suggesting metachronous development of independent primaries)
- Hepatocarcinogenesis is considered a stepwise process, in which HCC develops from premalignant lesions (e.g., dysplastic foci (<1 mm), low-grade dysplastic nodule (LGDN), high-grade dysplastic nodule (HGDN), Fig. 8.7)
- Dysplastic nodules show a clonal growth and mild cytological and structural atypia, but lack definite signs of malignancy, like severe trabecular or cellular atypia and interstitial or vascular invasion
- Compared to LGDN cell density and atypia is increased in HGDN (due to small cell change and increased nuclear to cytoplasmic ratio resulting in a nuclear crowding)

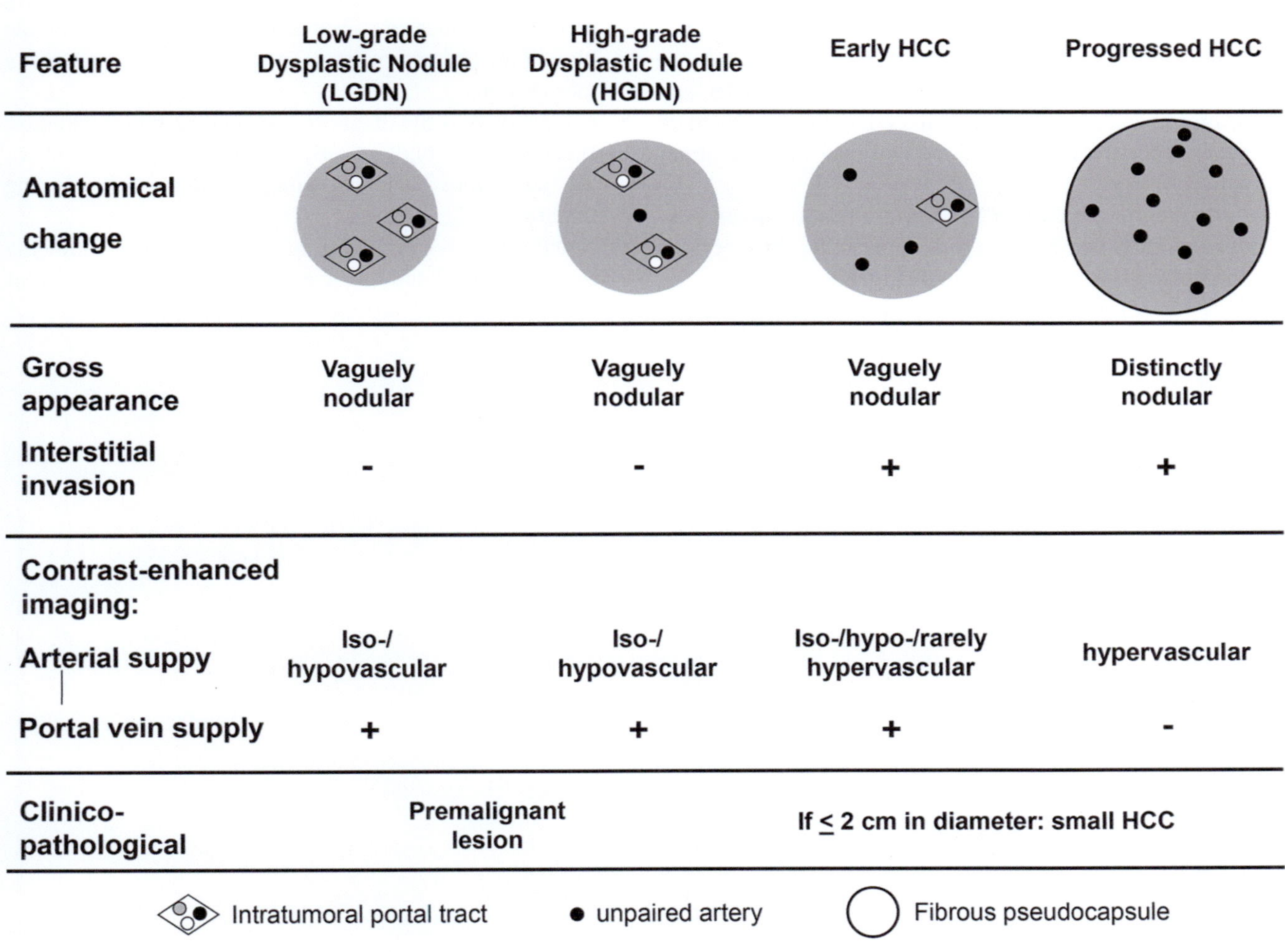

Feature	Low-grade Dysplastic Nodule (LGDN)	High-grade Dysplastic Nodule (HGDN)	Early HCC	Progressed HCC
Anatomical change				
Gross appearance	Vaguely nodular	Vaguely nodular	Vaguely nodular	Distinctly nodular
Interstitial invasion	-	-	+	+
Contrast-enhanced imaging:				
Arterial suppy	Iso-/ hypovascular	Iso-/ hypovascular	Iso-/hypo-/rarely hypervascular	hypervascular
Portal vein supply	+	+	+	-
Clinico-pathological	Premalignant lesion		If ≤ 2 cm in diameter: small HCC	

Fig. 8.7 Classification of small nodular lesions in cirrhotic liver according to the International Consensus Group for Hepatocellular Neoplasia, 2009

- Small HCC are by definition ≤2 cm; they can be further divided into early HCC (eHCC) and progressed HCC (pHCC, Fig. 8.7)
- eHCC shows a vague nodular growth and focal interstitial invasion
  - It lacks a capsule and derives its blood mainly from the portal-venous flow (and may thus be not diagnosed based on radiologic criteria)
- pHCC is encapsulated and may show a nodule-in-nodule growth
  - It is supplied by numerous unpaired arteries (causing hyperarterialization in dynamic radiological imaging) and lacks portal triads
  - Caveat: residual portal tracts may be entrapped in the periphery of the nodule
- Cytologically, HCC may be hepatoid, pleomorphic, or sarcomatoid (spindle cells)
  - Clear-cell change due to glycogen deposition or fatty change may be seen
  - Bile production, lipofuscin deposition, or cytoplasmic inclusions (Mallory-Denk-bodies, hyaline bodies, pale bodies) may be detected
- Four principal microarchitectural growth patterns are known: trabecular, solid (compact), pseudoglandular, and macrotrabecular (trabeculae being ≥10 cells thick); mixed patterns are frequently detected and may indicate tumor heterogeneity
- According to the current WHO classification about one-third of HCC can be subtyped representing distinct morphomolecular entities (Table 8.1, Fig. 8.8)

- The three-tiered WHO grading system evaluates whether tumor cells resemble mature hepatocytes (G1, well-differentiated) and is otherwise based on the degree of nuclear atypia (G2, moderately differentiated: clearly malignant, moderate atypia with occasional multinucleated cells allowed; G3, poorly differentiated: marked nuclear pleomorphism, includes HCC with sarcomatoid and anaplastic morphology)

Diagnosis and Differential Diagnosis

- Noninvasive HCC diagnosis is based on contrast-enhanced imaging studies (e.g., computed tomography, magnetic resonance tomography, and sonography) and may be corroborated by the detection of an elevated serum AFP level
- Diagnostic findings in progressed HCC are hypervascularity during arterial imaging phase followed by a rapid washout during portal-venous phase
- Early HCC cannot reliably be diagnosed using imaging techniques, since unpaired arteries (responsible for the blood flow characteristics of progressed HCC) are rare in these lesions
- Liver biopsy is the diagnostic technique with the highest specificity and high sensitivity and is recommended for all suspicious lesions without diagnostic imaging features
- Needle-tract seeding is a rare complication following biopsy of HCC and has been observed in up to 2.7% of

Table 8.1 WHO subtypes of HCC

Subtype	Frequency (%)	Clinical features	Histology	Genetic features
Steatohepatitic	5–20	May be associated with steatohepatitis	Fatty change, ballooning, inflammatory foci, Mallory-Denk bodies	IL-6/JAK/STAT activation; low frequency of CTNNB1-, TERT- and TP53-mutations
Clear cell	3–7	Unknown	>80% of tumor cells with clear-cell morphology, mild fatty change is acceptable	Unknown
Macrotrabecular massive	5	High serum AFP, poor prognosis	Macrotrabecular growth in >50% of tumor, vascular invasion common	TP53 mutation, FGF19 amplification
Scirrhous	4	May mimic cholangiocarcinoma on imaging	>50% of tumor showing a dense intratumoral fibrosis	TSC1/2 mutation; activated TGFβ signaling
Chromophob	3	Unknown	Tumor cells with chromophob cytoplasm, mainly bland nuclei, but areas with anaplasia and microcysts	Alternative lengthening of telomeres
Fibrolamellar	1	Young, no background liver disease	Large oncocytic tumor cells (K7 and CD68 positive) with prominent nucleoli, dense lamellar intratumoral fibrosis	DNAJB1-PRKACA fusion gene
Neutrophil-rich	<1	Leukocytosis, CRP- und IL-6-elevation, poor prognosis	Prominent infiltration by polymorphic granulocytes, sarcomatoid areas may be seen	G-CSF expression by tumor cells
Lymphocyte-rich	<1	Unknown	Lymphocytes > tumor cells	Unknown, not EBV-related

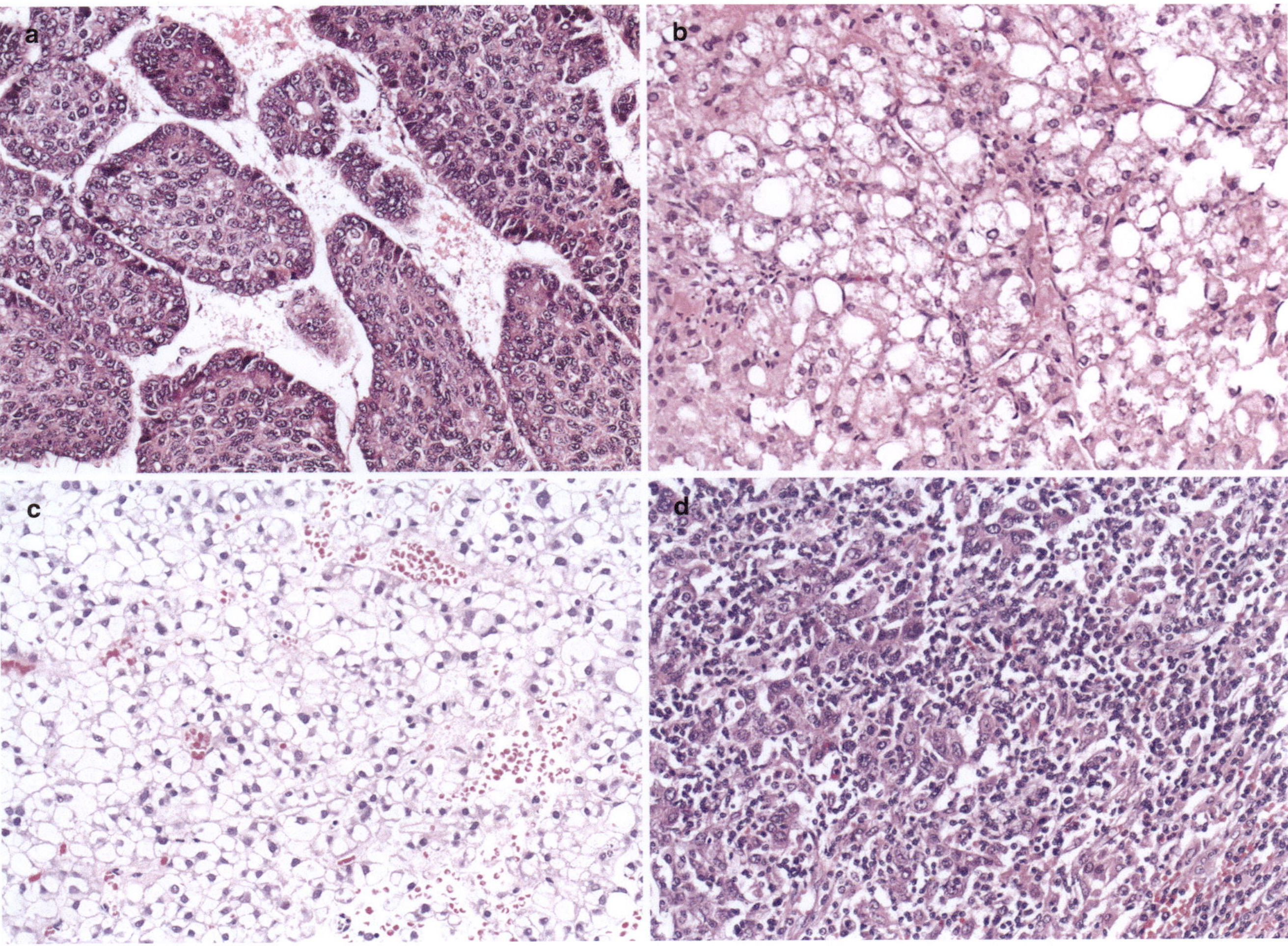

Fig. 8.8 HCC subtypes. (**a**) Macrotrabecular HCC showing more than ten cells wide trabeculae. (**b**) Steatohepatitic HCC with trabecular disarray, fatty change, hepatocellular ballooning, and satellitosis. (**c**) Clear-cell HCC consisting of solid sheets of cells with clear cytoplasm and mild fatty change. (**d**) Lymphocyte-rich HCC revealing a diffuse infiltration by lymphocytes (original magnification 200-fold)

cases. It does not infer with patients´ outcome and may be prevented by a coaxial biopsy technique

- Poorly differentiated HCC needs to be distinguished from liver metastasis
  - HCC typically (90% of cases) express carbamoyl phosphatase synthetase-1 (as detected by the Hepar-1 antibody)
  - Compared to Hepar-1, the sensitivity of Arginase-1 immunostaining to detect a hepatocellular differentiation is higher in poorly differentiated HCC
- A canalicular expression pattern using antibodies against polyclonal CEA (pCEA) and CD10 is also diagnostic of HCC
- Other markers that support HCC diagnosis include AFP, fibrinogen, whereas K7 and K19 are only positive in a fraction of HCC and may indicate an aggressive tumor biology

- Clear-cell HCC needs to be differentiated from other clear-cell tumors (e.g., kidney)
- HCC are distinguished from benign hepatocellular lesions (e.g., Focal Nodular Hyperplasia, HCA) by their architectural and cytological atypia, although the differences may be subtle in early/well-differentiated HCC
- Architectural atypia includes an irregular trabecular pattern (more than 2-cells wide cords separated by capillarized sinusoids), pseudoglandular (pseudogland formed due to abnormal and dilated bile canaliculi between tumor cells), solid growth, and focal loss of reticulin pattern
- Nodule-in-nodule growth (nodule of lower or different differentiation surrounded by a better-differentiated component) is indicative of HCC
- Well-differentiated HCC has to be differentiated from premalignant lesions (in particular HGDN and β(beta)-Catenin-activated HCA variants), which can be supported

by immunohistology: a marker panel consisting of glypi-can-3 (GPC3), glutamine synthetase (GS), and heat shock protein 70 (HSP70) has evolved as standard diagnostic procedure for this purpose

- The oncofetal protein GPC3 is frequently reactivated in early HCC
- GS is expressed by a rim of perivenular hepatocytes and in a periseptal localization in progressed fibrosis/cirrhosis
  - A strong and diffuse staining (≥10% of tumor cells) without a zonal restriction is considered positive
- Positive expression of HSP70 (≥10% of tumor cells) was reported in 80% of HCC, but only occasionally in HGDN
- Applying GPC3, GS, and HSP70 as a 3-marker panel significantly increased accuracy of HCC diagnosis (if at least two of the three markers were positive; sensitivity 60–70%, specificity >95%), particularly when diagnostic

interstitial invasion is absent in a biopsy specimen (Fig. 8.9)
- Further helpful markers include vascular markers (e.g., CD34) to detect diffuse capillarization of the tumor sinusoids; the lack of a K7- or K19-positive ductular reaction around a nodule is indicative of early stromal invasion
- Primary liver cancer showing no specific differentiation, except for an epithelial nature best on H&E morphology and immunohistochemistry is termed undifferentiated carcinoma

Genetic Features

- Global DNA hypomethylation in HCC has been associated with induction of genomic instability, activation of oncogenes, and loss of imprinting, while hypermethyl-

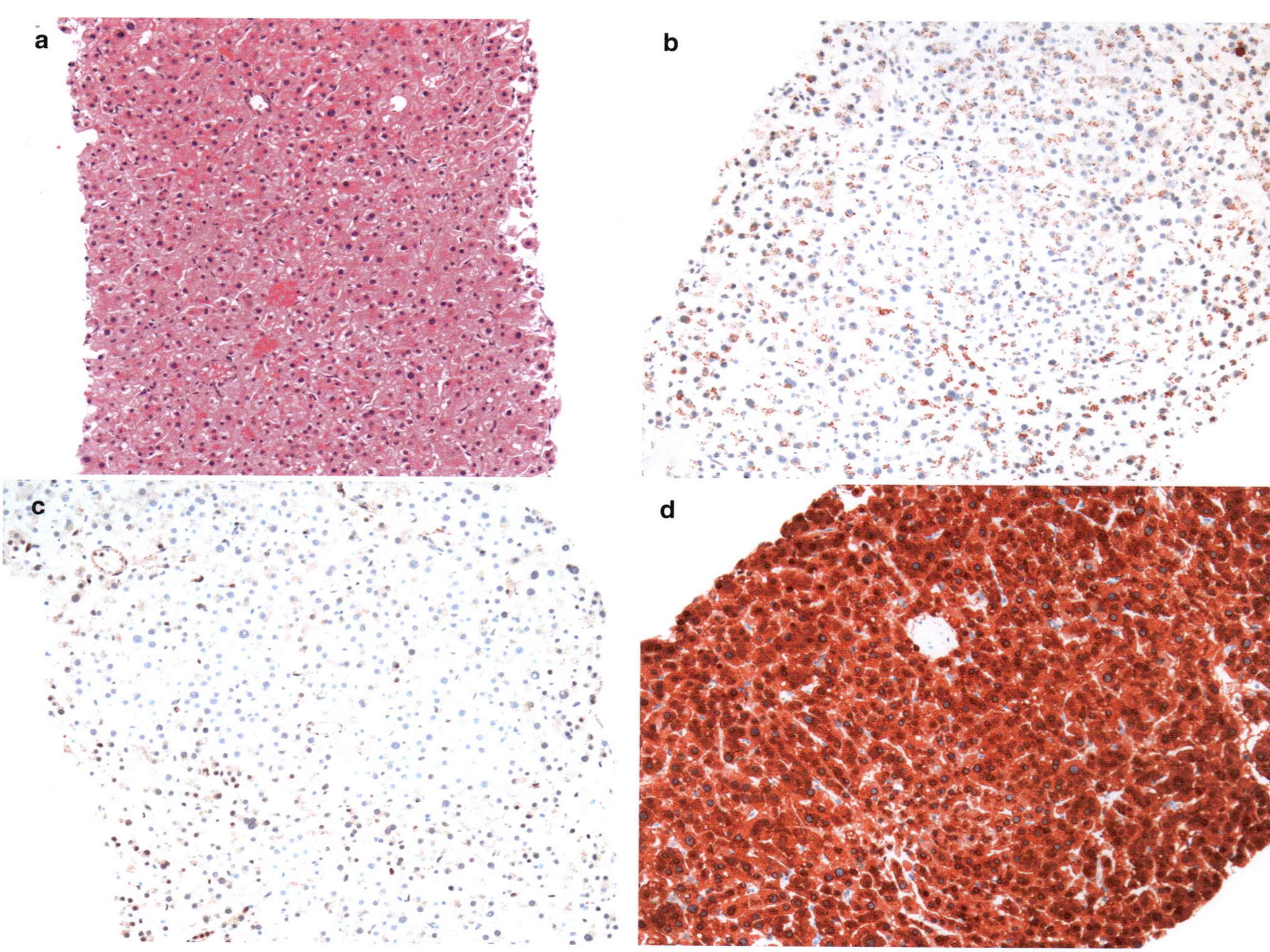

Fig. 8.9 (**a**) Early well-differentiated HCC showing areas of nuclear crowding and mild trabecular disarray. Additionally, a solitary artery can be seen. (**b**) GPC3 is expressed by most tumor cells. (**c**) Predominantly nuclear HSP70 expression in areas with nuclear crowding. (**d**) Homogeneous overexpression of GS (original magnification 100-fold)

ation of CpG islands located especially in gene regulatory sequences results in transcriptional silencing of tumor suppressor genes and target genes of polycomb repressive complexes, which may lead to a stem cell-like chromatin pattern

- HCC shows chromosomal instability with a mostly moderate number of genetic imbalances, in part responsible for the activation of oncogenes or inactivation of tumor suppressor genes
- Amplifications of genomic material frequently involve chromosomes 1q, 8q, 6p, and 17q, while losses are most prevalent at 8p, 16q, 4q, 17p, and 13q
- Distinct chromosomal imbalances (gains of 1q21–23 and 8q22–24) precede malignant transformation and may be already detectable at lower numbers in premalignant HGDN
- HCC carrying a *TP53* mutation may be more chromosomally instable compared to *CTNNB1*-mutated HCC
- In regions with high aflatoxin B1 exposure, mutations in codon 249 of TP53 are considered etiology-specific and can be detected in nearly 50% of all HCC cases
- HCC comprises a molecular heterogeneous group of tumors, which is characterized by a rather low number of frequent mutations (e.g., *TERT*, *TP53*, and *CTNNB1*) and a considerable number of low-frequency mutations, which accumulate in certain cellular pathways (Fig. 8.10)
- Based on integrative molecular profiling 3 to 6 molecular subclasses of HCC can be differentiated, some of which reveal an association with the HCC subtypes recognized by the WHO (Fig. 8.11)
- HCC belonging to the transcriptomic Boyault classes G1 to G3 are characterized by a *TP53* gene mutation, high proliferative activity, chromosomal instability and frequently show high AFP serum levels and vascular invasion suggesting a more aggressive tumor biology
- The subclass G1 is enriched for *ATM*-, *AXIN1*-, and *RPS6KA3*-mutated HCC
  - Activation of RAS/MAPK-, PI3K/AKT-, Notch-, and IGF2-signaling can be demonstrated
  - A clear-cell phenotype may be observed, and HCC belonging to G1 are likely to show markers also found in progenitor cells (e.g., K19 and EpCam expression)
- Subclass G2 has no association with *RPS6KA3* mutation and does not show progenitor cell features but is characterized by active TGFβ-signaling

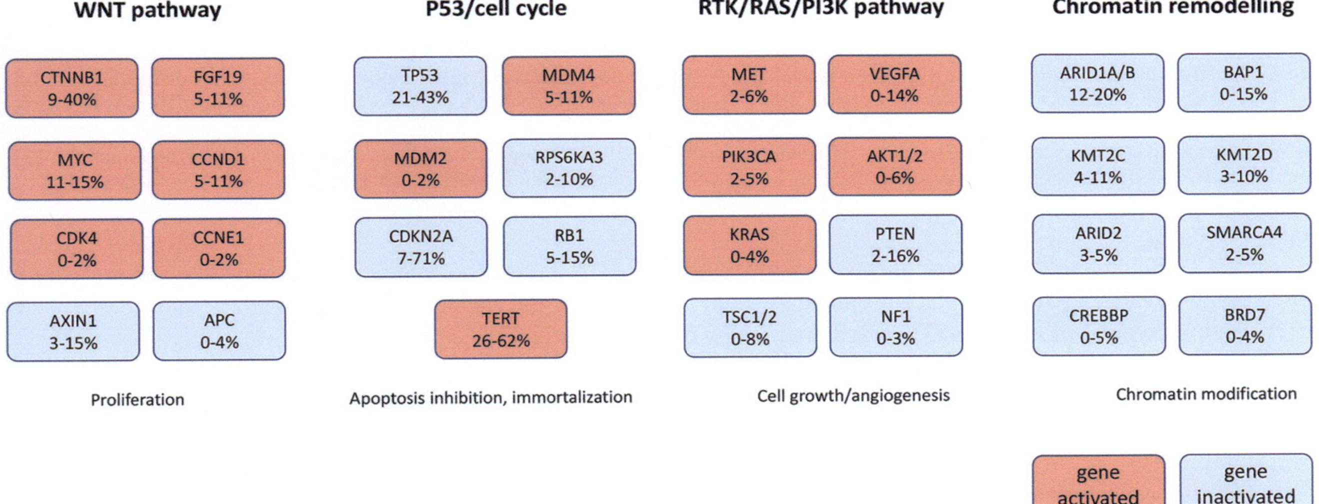

Fig. 8.10 Signaling Pathways frequently affected by somatic alterations in HCC (modified according to Cancer Genome Atlas Research Network, Cell 2017). The frequencies of alteration are indicated. AKT1/2, AKT Serine/Threonine Kinase 1/2; APC, APC Regulator of WNT Signaling Pathway; ARID1A, AT-Rich Interaction Domain 1A; ARID1B, AT-Rich Interaction Domain 1B; ARID2, AT-Rich Interaction Domain 2; BAP1, BRCA1 Associated Protein 1; BRD7, Bromodomain Containing 7; CCND1, Cyclin D1; CCNE1, Cyclin E1; CDK4, Cyclin Dependent Kinase 4; CDKN2A, Cyclin Dependent Kinase Inhibitor 2A; CREBBP, CREB Binding Protein; CTNNB1, Catenin Beta 1, β-catenin; FGF19, Fibroblast Growth Factor 19; KMT2C, Lysine Methyltransferase 2C; KMT2D, Lysine Methyltransferase 2D; KRAS, KRAS Proto-Oncogene, GTPase; MDM2, MDM2 Proto-Oncogene; MDM4, MDM4 Proto-Oncogene; MET, MET Proto-Oncogene, Receptor Tyrosine Kinase; MYC, MYC Proto-Oncogene, BHLH Transcription Factor; NF1, Neurofibromin 1; PIK3CA, Phosphatidylinositol-4,5-Bisphosphate 3-Kinase Catalytic Subunit Alpha; PTEN, Phosphatase And Tensin Homolog; RB1, RB Transcriptional Corepressor 1; RPS6KA3, Ribosomal Protein S6 Kinase A3; SMARCA4, SWI/SNF Related, Matrix Associated, Actin Dependent Regulator Of Chromatin, Subfamily A, Member 4; TERT, Telomerase Reverse Transcriptase; TP53, tumor protein 53; TSC1/2, TSC Complex Subunit 1/2; VEGFA, Vascular Endothelial Growth Factor A

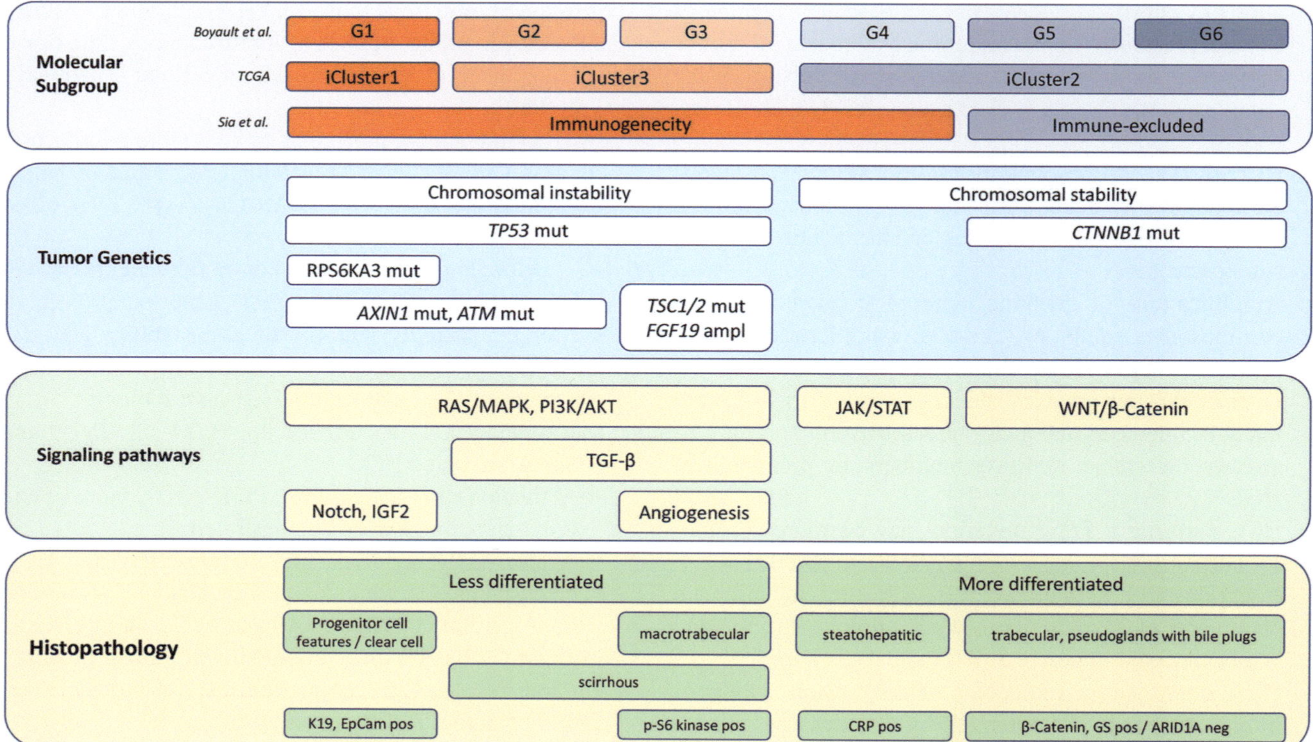

Fig. 8.11 Integration of Molecular Subgroups, tumor genetics, altered signaling pathways, and histopathology (modified according to Calderaro et al. J Hepatol 2019). Chromosomal instable HCC is characterized by TP53 mutations and active RAS/MAPK- and PI3K/AKT signaling, while chromosomal stable tumors have either activated JAK/STAT signaling and a steatohepatitic phenotype or show activation of WNT/β-catenin signaling, mainly due to CTNNB1 mutations. ampl, amplification; ARID1A, AT-Rich Interaction Domain 1A; CRP, C-Reactive Protein; CTNNB1, Catenin Beta 1, β-catenin; EpCam, Epithelial Cell Adhesion Molecule; FGF19, Fibroblast Growth Factor 19; GS, glutamine synthetase; IGF2, Insulin-Like Growth Factor 2; JAK, Janus Kinase; K19, keratin 19; MAPK, Mitogen-Activated Protein Kinase; mut, mutation; PI3K, Phosphatidylinositol-4,5-Bisphosphate 3-Kinase; p-S6 kinase, phospho-S6 kinase; RAS, RAS Proto-Oncogene, GTPase; RPS6KA3, Ribosomal Protein S6 Kinase A3; STAT, Signal Transducer And Activator Of Transcription; TGFβ, Transforming Growth Factor Beta; TP53, tumor protein 53; TSC1/2, TSC Complex Subunit 1/2; WNT, WNT Family Member

- HCC belonging to this subtype may be rich in stroma, eventually leading to a scirrhous morphology
- Alterations typical of subclass G3 include mutations of *TSC1* or *TSC2* and amplification of the *FGF19* gene locus
  - The macrotrabecular-massive HCC subtype belongs to this category as do some HCC with scirrhous morphology
  - Detection of phosphorylated S6 kinase may be used for immunohistochemical subtyping
  - Macrovascular invasion and high AFP serum levels are characteristic findings
- HCC belonging to the transcriptomic classes G4 to G6 are better differentiated, chromosomally more stable, and have a comparably lower proliferative activity
- The subclass G4 is characterized by activation of JAK-/STAT signaling, which leads to upregulation of inflammatory markers (e.g., CRP) and may show a steatohepatitic phenotype
  - G4-HCC tend to be comparably smaller and usually lack satellite nodules

- The hallmark of subclasses G5 and G6 is the presence of WNT/β-Catenin activation (mainly due to *CTNNB1* mutation)
  - HCC related to nonalcoholic steatohepatitis may mainly belong to these nonimmunogenic subclasses
  - The activation level of WNT/β-Catenin is higher in G6 than in G5, and satellite nodules are more frequently found in subclass G6

Prognosis and Predictive Factors

- The overall 5-year survival rate of HCC patients with symptomatic unresectable HCC is less than 5%
- Clinical features indicating a worse prognosis include high serum AFP level, large tumor size and/or numerous HCC nodules, coexpression of markers suggesting a progenitor cell-related phenotype (e.g., K19, EpCam, SALL4), advanced liver cirrhosis, and detection of macrovascular invasion by imaging

- Histopathological features associated with a poor outcome include high tumor grade, vascular invasion, presence of satellite nodules, advanced tumor stage, presence of liver cirrhosis
- Genetic features suggesting a worse prognosis are the presence of *FGF19* amplification, *TP53* mutation, and expression profiles leading to categorization as molecular subclass G1–3
- Long-term survival can be achieved in patients with small HCC, which can be treated by locoregional tumor ablation, liver resection, or liver transplantation (with the added value of replacing the chronically damaged liver)
- Palliative treatment includes transarterial chemoembolization, selective internal radiotherapy, and systemic treatment for advanced stage HCC
- The current first line treatment is combined atezolizumab–bevacizumab, which did not only improve survival compared to (the old standard) sorafenib treatment, but also improved the quality of patients´ life
  - Second- and third-line therapies are available, but no predictive factors have been identified so far; thus precision oncology for HCC is currently not of direct therapeutic impact

Intrahepatic Cholangiocarcinoma

Definition

- Malignant intrahepatic epithelial neoplasm with biliary differentiation arising proximal of second-order bile ducts

Clinical Features

- Second most frequent type of liver cancer (5–15% of cases, variability related to geographic region); average age at diagnosis >50 years, CA19–9 serum levels are typically elevated
- Incidence is very high in some areas of southeast Asia (due to liver fluke infestation and hepatolithiasis), but the incidence in the Western world is rising (likely due to increasing prevalence of nonalcoholic steatohepatitis)
- Biliary tree malformations (e.g., Caroli's syndrome) are a risk factor for all types of cholangiocarcinoma
- Etiological risk factors for large duct type (see below) intrahepatic cholangiocarcinoma (iCCA) include primary sclerosing cholangitis (especially in the context of concomitant ulcerative colitis), liver fluke infection, and hepatolithiasis
- Risk factors for small duct type iCCA (see below) are identical to HCC and include liver cirrhosis, chronic viral hepatitis, alcoholic and nonalcoholic steatohepatitis, and genetic hemochromatosis besides exposure to chemicals (e.g., thorium dioxide, smoking)
- Symptoms are usually nonspecific (malaise, abdominal pain, and weight loss) or depend on the tumor location, and the stage of disease as well as secondary consequences (e.g., cholangitis due to obstruction)
- Peripheral mass-forming tumors (see below) may grow undetected to large size, while tumors close to the hilum may become clinically apparent due to biliary obstruction
- Symptoms of the preexisting/predisposing disease may dominate and mask tumor related symptoms (e.g., PSC)

Pathologic Features

- Most iCCA develop in the peripheral hepatic parenchyma (so-called mass-forming type), while tumors located proximal to the hepatic hilum may macroscopically show a periductal infiltrating pattern or even an intraductal tumor growth; mixed pattern may occur
- Grading follows a three-tiered system according to standard morphological UICC criteria
- Histologically, two main subtypes are distinguished: small duct type and large duct type (Fig. 8.12)
- Small duct iCCA is morphologically similar to the adenocarcinoma component of combined HCC-CCA and characterized by small ductal components growing in a tubular, ductular, or infrequently tubulopapillary pattern, which are formed by more or less cuboidal, nonmucin-secreting tumor cells in context of a desmoplastic stroma reaction
  - By IHC, tumor cells may be positive for EMA, CD56, and CRP
- Variants of small duct type iCCA may resemble a ductal plate malformation (frequently with inspissated bile) or show cholangiocellular pattern (characterized by an antler-like anastomosing ductular phenotype)
  - Other rare subtypes are adenosquamous and squamous, mucinous, signet ring cell, clear cell, mucoepidermoid, lymphoepithelioma-like, and sarcomatous morphologies
- Large duct iCCA is morphologically similar to perihilar cholangiocarcinoma and featured by a columnar, mucin-secreting tumor cell forming ductal, papillary and/or tubular structures within a desmoplastic stroma
  - Immunohistochemically, these tumor cells are positive for MUC5A, MUC6, and S100
- As in extrahepatic locations, variants of large duct iCCA include clear-cell adenocarcinoma and variants with intestinal, mucinous, poorly cohesive (signet ring cells),

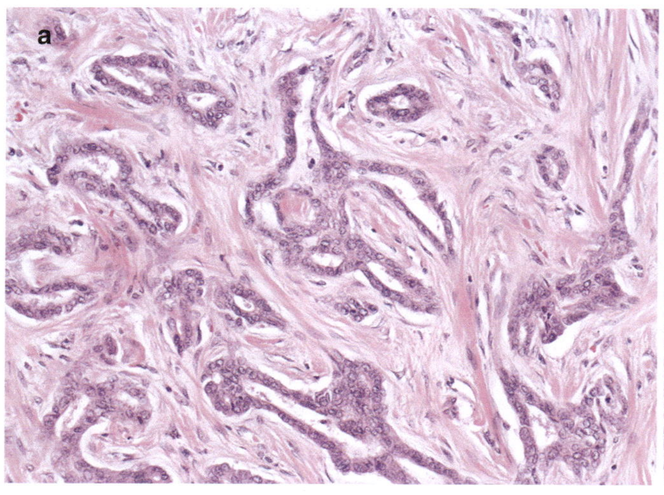

Fig. 8.12 Intrahepatic cholangiocarcinoma subtypes. (**a**) Small duct type iCCA showing irregularly branching ductular structures line by cuboidal cells with monomorphic round nuclei. (**b**) Large duct type iCCA with densely packed tubular structures lined by a columnar epithelium with mostly basally located vesicular nuclei and distinct nucleoli (original magnification 200-fold)

- squamous or adenosquamous differentiation; mixed neuroendocrine–nonneuroendocrine neoplasm (MiNEN) may be observed
- Precursor lesions of large duct iCCA include biliary intraepithelial neoplasia and intraductal papillary neoplasm; precursor lesions of small duct iCCA have not been defined so far (biliary adenofibroma?)
- Diagnostic evaluation of resection specimen includes typing, staging (TNM criteria including vascular and perineural infiltration and resection margin status), grading, and analysis of the nontumorous liver

Differential Diagnosis

- Histologically no definitive distinction from other adenocarcinomas of the pancreatobiliary system is possible; in questionable cases, metastasis has to be ruled out by immunohistochemistry and integration of the clinical and the radiological presentation
- Well-differentiated iCCA needs to be distinguished from benign biliary lesions like bile duct adenoma or biliary adenofibroma
  - Perineural invasion is indicative of an irregular branching, and the presence of nuclear atypia favors iCCA
  - Immunostaining for p53, Ki-67, EZH2, and p16 may be helpful
- Diagnosis of combined hepatocellular-cholangiocarcinoma requires the unequivocal demonstration of tumor components with hepatocellular differentiation
- Undifferentiated primary liver carcinoma lacks any further differentiation beyond its epithelial nature

Genetic Features

- While there is no universally accepted precursor lesion of small duct iCCA, there is experimental evidence that iCCA may originate from hepatocytes
- Mutations of chromatin-regulating genes (e.g., *ARID1A*, *BAP1*, and *PBRM1*) are frequently detected, and activation of NOTCH signaling in hepatocytes may result in iCCA development
- iCCA with mutations of *Isocitrate Dehydrogenase 1* or *2* (*IDH1/2*) show distinct mRNA, copy number, and DNA methylation features; they display high mitochondrial and low chromatin modifier gene expression promoting biliary differentiation via epigenetic silencing of hepatocyte nuclear factor-4α(alpha) expression
- Integrative molecular profiling identified four molecular subgroups of iCCA: extrahepatic CCA-like, *IDH* mutant, *Cyclin D1* (*CCND1*) amplificated/highly methylated, and *PBRM1* mutation or *FGFR2* translocation and methylation
- From the perspective of precision oncology, recurrent genetic alterations of (small duct) iCCA (Table 8.2) are highly relevant, as potential drug targets can be identified in more than 50% of iCCA patients including *IDH1-*, *BAP1-*, *BRCA1*, *BRCA2*, and *BRAF*[V600] mutations
- Microsatellite instability due to mutations of mismatch repair proteins encoded by *MLH1*, *MSH2*, *MSH6*, and *PMS2* occurs, but with low frequency (<1%)
- Tumors with wild-type *RAS* genes may show genetic rearrangements involving the *FGFR2*, *NRG1*, *ALK*, *ROS1*, or *NTRK1–3* genes
- *CDNK2A* inactivation (due to enhanced *CDNK2A* promoter methylation) is recurrently observed in iCCA

Table 8.2 Recurrent genetic alterations in iCCA

Molecular alteration	Frequency (%)
ARID1A mutation	5–15
BAP1 mutation	5–15
BRAF^{V600E} mutation	3–6
CDKN2A/B mutation	10–15
ELF3 mutation	1–2
ERBB2 mutation	2–3
FGFR2 translocation	15–30
IDH1/2 mutation	10–20
KRAS/NRAS mutation	10–20
MSI-H	1–2
NRG1 translocation (mostly detected in mucinous adenocarcinoma)	<1
NTRKs translocation	1–3
PBRM1 mutation	10–17
PRKACA/PRKACB translocation (IPNB-associated)	0
SMAD4 mutation	2–10
TP53 mutation	20–30

ARID1A AT-Rich Interaction Domain 1A, *BAP1* BRCA1 Associated Protein 1, *BRAF* BRD7, Bromodomain Containing 7, *CCND1* Cyclin D1, *CCNE1* Cyclin E1, *CDK4* Cyclin Dependent Kinase 4, *CDKN2A* Cyclin Dependent Kinase Inhibitor 2A, *CREBBP* CREB Binding Protein, *CTNNB1* Catenin Beta 1, β-catenin, *FGF19* Fibroblast Growth Factor 19, *KMT2C* Lysine Methyltransferase 2C, *KMT2D* Lysine Methyltransferase 2D, *KRAS* KRAS Proto-Oncogene, GTPase, *MDM2* MDM2 Proto-Oncogene, *MDM4* MDM4 Proto-Oncogene, *MET* MET Proto-Oncogene, Receptor Tyrosine Kinase, *MYC* MYC Proto-Oncogene, BHLH Transcription Factor, *NF1* Neurofibromin 1, *PIK3CA* Phosphatidylinositol-4,5-Bisphosphate 3-Kinase Catalytic Subunit Alpha, *PTEN* Phosphatase And Tensin Homolog, *RB1* RB Transcriptional Corepressor 1, *RPS6KA3* Ribosomal Protein S6 Kinase A3, *SMARCA4* SWI/SNF Related, Matrix Associated, Actin Dependent Regulator Of Chromatin, Subfamily A, Member 4, *TERT* Telomerase Reverse Transcriptase, *TP53* tumor protein 53, *TSC1/2* TSC Complex Subunit 1/2, *VEGFA* Vascular Endothelial Growth Factor A

related to primary sclerosing cholangitis, hepatolithiasis, and other risk factors as well

- Mutations of SWI/SNF components (e.g., *ARID1A*, *SMARCA4*, *SMARCB1*, and *SMARCA2*) may result in SWI/SNF deficiency and undifferentiated carcinomas (showing an aggressive rhabdoid phenotype)

Prognosis and Predictive Factors

- iCCA is an aggressive cancer entity with high mortality
- Only few patients are candidates for potentially curative surgical resection
- Small duct iCCA has a higher postoperative survival rate compared to large duct iCCA
- Intrahepatic and extrahepatic metastases, macroscopic vascular invasion, advanced TNM-stage, and noncurative resection are negative prognostic factors

- Immunohistological expression of CRP has been associated with a better prognosis, while EMA positivity indicates a poor outcome
- Mutations in *TP53* and *ARID1A* genes and amplification of the *MET* gene are independent predictors of poor prognosis
- Drugs targeting FGFR2 and NTRK1–3 rearrangements as well as IDH1 mutations have been approved for iCCA
  - Advanced iCCA patients with microsatellite instability may benefit from immune checkpoint blockade
  - Further promising candidates for precision oncology represent mutations in, e.g., *BAP1*-, *BRCA1*, *BRCA2*, *BRAF*, and genetic rearrangements like *NRG1* or *ROS1* fusions

Combined Hepatocellular-Cholangiocarcinoma

Definition

- Malignant primary liver carcinoma with unequivocal and intimately mixed components of hepatocytic and cholangiocytic differentiation (Fig. 8.13); collision tumors are excluded from this entity

Clinical Features

- Combined hepatocellular-cholangiocarcinoma (HCC-CCA) is rare, accounting for less than 5% of all primary liver carcinomas
- HCC-CCA share the etiological risk factors established for HCC and (small duct) iCCA
- Definite diagnosis by radiological imaging is impossible due to the similarities with either classical HCC or iCCA
- Radiological imaging aids in the detection of liver cirrhosis, the extent of intrahepatic disease, vascular involvement, and extraphepatic spread
- Diagnosis may be incidentally made on a liver biopsy specimen, but most diagnoses evolve from histological evaluation of surgical specimen

Pathologic Features

- The macroscopic appearance depends on the predominant component; thus detailed histological sampling of heterogeneously appearing tumor areas is necessary

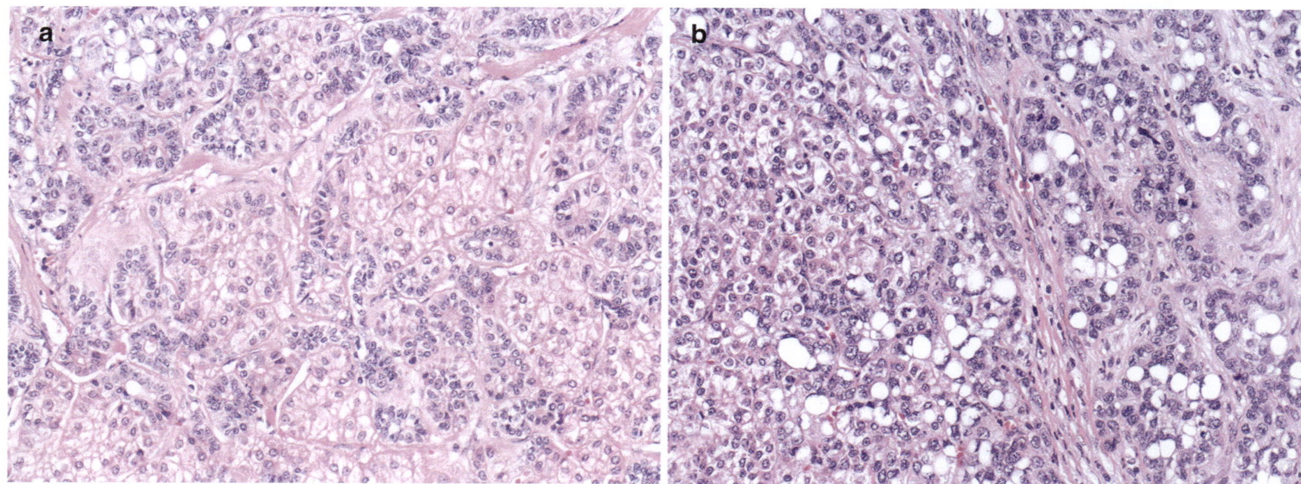

Fig. 8.13 Combined HCC-CCA. (**a**) Tumor cells with hepatocytic and cholangiocytic differentiation are deeply intermingled. (**b**) Close transition from the iCCA (right part) and the HCC (outer left) components. Interface area with presence of intermediary hepatobiliary cells (original magnification 200-fold)

- All histological pictures typical of HCC and iCCA may be observed, and the components are either close to each other or intimately intermingled
- Immunohistochemistry (as outlined above for HCC and iCCA) may support the definition of the biphenotypic differentiation of HCC-CCA

Differential Diagnosis

- Depending on the representativeness of the tissue (e.g., biopsy) available for histological evaluation HCC, iCCA, or a collision tumor of HCC and iCCA may be relevant differential diagnoses; unequivocal presence of both a hepatocellular and a cholangiocellular component establishes the HCC-CCA diagnosis regardless of the percentage of each component
- Primary liver cancer showing a monotonous appearance with microscopic features intermediate between hepatocytes and cholangiocytes should be termed intermediate cell carcinoma of the liver; an intermediate hepatobiliary phenotype being restricted to the interface of HCC and iCCA components is insufficient to diagnose intermediate cell carcinoma

Genetic Features

- The cellular origin and the pathogenesis of HCC-CCA are still unlcear and may involve cellular plasticity/ transdifferentiation and/or an origin from hepatic progenitor/stem cells

- Chromosomal aberrations observed in HCC-CCA recapitulate those previously reported in both HCC and iCCA, like gains of 1q, and 8q, and losses of 4q, 8p, 9q, 16q, and 17p
- HCC and iCCA components derive from a common clonal origin in most cases, as shown by shared chromosomal aberrations (e.g., loss of heterozygosity at chromosomes 4q, 8p, 16q, and 17p) and gene mutations (e.g., *TERT*, *TP53*) as well as the detection of clonal hepatitis B virus integration events
- HCC-CCA frequently harbor alterations known from HCC, while alterations recurrently identified in iCCA are rarely found (e.g., *ERBB2*, *KRAS*, *IDH1*, *IDH2*, *FGFR2*, and *BAP1*)
- Recurrently mutated genes include *TP53*, *TERT*, *AXIN1*, *KMT2D*, *KEAP1*, *ARID1A*, *RB1*, *PTEN*, *RPS6KA3*, *CTNNB1*, *IDH1*, *PBRM1*, *ARID2*, *BRD7*, and *CTNNB1*; recurrent copy number alterations have been described at the following gene loci: *MYC* (8q24.21), *MET* (7q31.2), *CCNE1* (19q12), *CDK6* (7q21.2), *TERT* (5p15.33), *TP53* (17p13.1), *CDKN2A* (9p21.3), *AXIN1* (16p13.3), *RB1* (13q14.2), *CCND1*, and *FGF19* (11q13)
- *TERT* alteration seems to be an early event in tumor evolution, while focal amplifications of *CCND1*, *MET*, and *ERRB2*, as well as WNT pathway alterations, evolve later during tumor progression and may be restricted to only one tumor component
- Experimental evidence from mice suggests that the tumor microenvironment epigenetically shapes the lineage commitment of primary liver cancer; genetic alterations inducing HCC in the context of an apoptotic milieu may promote iCCA development in a necroptosis-associated cytokine microenvironment

Prognosis and Predictive Factors

- The prognosis of HCC-CCA is worse than HCC and similar to iCCA with repeatedly reported 5-year survival rates of less than 25%
- HCC-CCA may frequently show tumor thrombus formation (like HCC) and lymph node metastasis (like CCA)
- Resectability is a positive prognostic factor: partial hepatectomy with hilar lymph node dissection is the best treatment option in noncirrhotic patients but may only be possible in selected patients suffering from liver cirrhosis
- Liver transplantation seems to be no choice due to the short postoperative survival time compared to HCC
- Local ablative therapy (e.g., radiofrequency ablation) may be a treatment option for local recurrences
- Due to the more aggressive biological behavior compared to HCC, iCCA protocols are mostly selected for systemic treatment of HCC-CCA
- HCC-CCA with morphological features of progenitor/stem cells are frequently positive for SALL4 and may show an even more aggressive tumor biology
- Predictive factors are not established so far

Hepatoblastoma

Definition

- Primary malignant blastomatous liver tumor recapitulating hepatic ontogenesis and consisting of various combinations of variably mature epithelial and mesenchymal elements

Clinical Features

- Hepatoblastoma (HB) is rare and accounts for 1% of pediatric malignancies
- 80–90% of HB are detected before the age of 5 years, and the median age at onset is 18 months
- Most cases arise sporadic, but HB can be associated with congenital diseases and syndromes (e.g., Beckwith–Wiedemann syndrome (BWS), familial adenomatous polyposis (FAP), trisomy 18 and others)
- HB incidence is increased up to 2000-fold in kindreds with FAP (*APC* mutation) and BWS (imprinting defect with overexpression of insulin growth factor II [IGF2])
- HB is associated with premature birth and low birth weight
- Typical symptoms are an enlarged abdomen that may be accompanied by anorexia or weight loss
  - Other unspecific symptoms, like nausea, abdominal discomfort, or pain may be present, whereas jaundice is rare (<5%)
- HB is associated with paraneoplastic hematological symptoms (e.g., anemia, thrombocytosis)
- Marked AFP elevation is found in 90% of cases; the remaining (AFP negative) cases show a more aggressive course and are associated with high-risk histological features
- In most cases (up to 85%), a single mass is detected in imaging studies; up to 50% of HB show areas of calcification or ossification
- Pulmonary metastases are present in 10–20% of patients at diagnosis
  - Other sites of metastases are bone, brain, ovaries, and eyes
- Preoperative chemotherapy is able to reduce tumor size in >80% of cases

Pathologic Features

- HB displays a variety of morphological differentiation and growth pattern that may be present to a variable extent
- Based on the International Pediatric Liver Tumors Consensus Classification, HB is classified as epithelial or mixed epithelial and mesenchymal (Fig. 8.14)
- Epithelial HB may show fetal, embryonal, small cell undifferentiated (SCUD), cholangioblastic, macrotrabecular, and mixed epithelial pattern
- The fetal pattern (Fig. 8.14a) resembles hepatocytes of the developing liver
  - Due to variable glycogen and lipid content, the cytoplasm of these hepatoblasts shows a typical light and dark pattern at scanning magnification
  - The nuclei are small and round with fine chromatin distribution
  - Immunohistologically fetal HB shows a variable AFP expression and a membranous β(beta)-catenin expression
- The embryonal pattern (Fig. 8.14b) resembles the liver during later embryonal development
  - The hepatoblasts lack visible glycogen or lipids and have a scant, dark granular cytoplasm and a large nucleus with coarse chromatin. β(beta)-catenin is expressed at the cell membrane, and in less differentiated areas, nuclear staining is seen
- The SCUD pattern (Fig. 8.14c) consists of noncohesive sheets of so-called small blue cells without discernible cytoplasm

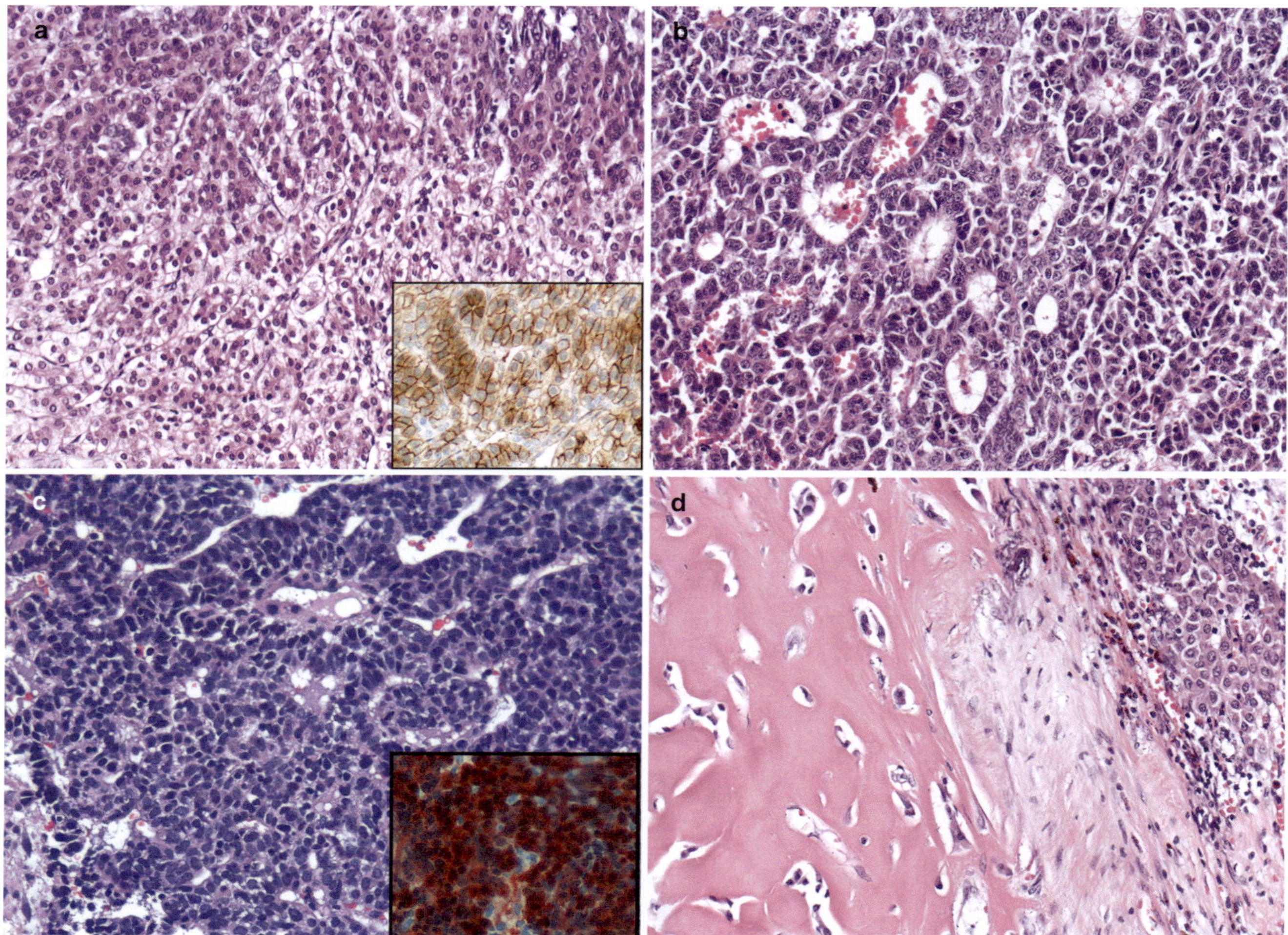

Fig. 8.14 Pattern of hepatoblastoma. (**a**) Fetal pattern showing dark (upper part) and light areas. Insert: β(beta)-catenin shows membranous expression. (**b**) Embryonal pattern consisting of hepatoblasts with dark cytoplasm and large nuclei with prominent nucleoli forming sheets and pseudorosettes. (**c**) SCUD pattern consisting of small blue tumor cells arranged in solid sheets. Insert: strong nuclear and cytoplasmic β(beta)-catenin expression. (**d**) Mesenchymal pattern with osteoid formation besides embryonal elements (upper right) (original magnification 200-fold)

- Apoptotic figures, mitosis, and necrosis are frequently noted
- SCUD pattern tumors can be either SMARCB1 (INI1) positive or negative
- INI-negative tumors show rhabdoid features and probably represent hepatic rhabdoid tumors
- SCUD represents the least differentiated pattern and is not associated with AFP elevation
- The macrotrabecular pattern consists of wide trabeculae (≥5 cells)
  - Cytologically the macrotrabeculae may be formed by embryonal or fetal hepatoblasts, pleomorphic cells, or hepatoid cells resembling macrotrabecular HCC
- The cholangioblastic pattern shows small ducts either lying within or around hepatocellular elements
- The mesenchymal pattern in mixed epithelial and mesenchymal hepatoblastoma (Fig. 8.14d) may comprise mature or immature fibrous tissue, osteoid or osteoid-like tissue, and hyaline cartilage
  - A minority of mixed hepatoblastoma may show teratoid features, whereas the remaining cases display complex heterologous tissues of all three germ layers (e.g., endodermal, neuroectodermal, and complex mesenchymal)
  - Osteoblast-like cells adjacent to osteoid may be epithelial-derived as indicated by CK8 and AFP expression and blending with epithelial components
- Independent of the subtype neoadjuvant chemotherapy may result in fibrosis, hemorrhage, and necrosis, whereas squamous metaplasia may be seen in residual viable tumor areas
  - Other treatment-induced changes include ballooning, fatty change, nuclear pleomorphism, and foreign body reaction

Differential Diagnosis

- HB must be differentiated from infantile hemangioendothelioma, mesenchymal hamartoma (benign mesenchymal liver tumor characterized by a multicystic loose connective tissue accompanied by a ductal component with ductal plate malformation and chromosomal aberrations involving 19q13.4), or embryonal sarcoma of the liver (undifferentiated mesenchymal tumor with complex karyotype, may develop from mesenchymal hamartoma and shares 19q13.4 abnormalities) that occur in the same age group
- Calcifying nested stromal epithelial tumor (CNSET), a low-grade neoplasm characterized by a nested epithelial component and a cellular myofibroblastic stroma with psammomatous calcifications
  - CNSET shows recurrent *CTNNB1* gene mutations, which is mostly sporadic, but may be associated with BWS
- Rhabdoid features may occur in SCUD HB and may require differentiation from a primary rhabdoid tumor of the liver, which lacks nuclear expression of SMARCB1/INI1
- Well-differentiated HCC is differentiated from fetal HB by the absence of light and dark areas and the presence of risk factors (e.g., conatal HBV infection and BSEP deficiency)

Genetic Features

- Alterations in the APC/β(beta)-catenin pathway play an important role in the pathogenesis of HB as demonstrated by nuclear β(beta)-catenin accumulation due to deletions in exons 3 and 4 of the *CTNNB1* gene in about 70–90% of cases, the presence of APC mutations in sporadic HB, and the association with FAP
- Although no consistent pattern of chromosomal anomalies has been detected, low number alterations include gains of 1q, 2, 4q, 7, 8q, 10, 12q, 17q, 20, X as well as losses of 1p, 2q, 4q, and 11p
- Gains of 2q, 8q, and 20q have been associated with a poor outcome
- LOH of the maternally imprinted allele at 11p15 is nearly pathognomonic for patients with BWS. This region encodes *CDKN1C*, *IGF2*, and *H19*
- Mutations of the transcription factor *NFE2L2* affect residues recognized by the KEAP1/CUL3 complex for proteasomal degradation; thus, *NFE2L2*-mutated HB are insensitive to KEAP1-mediated downregulation of NFE2L2 signaling

- *TERT* promoter mutations are rarely found (↔HCC)
- HB displays a general overexpression of several developmental genes, including the imprinted genes *IGF2* and *DLK1* as well as genes involved in the hedgehog pathway
- Based on expression profiling, several classification systems have been proposed categorizing HB into three groups (Fig. 8.15)
  - Low-risk (HB1/C1/fetal phenotype)
  - Intermediate risk (HB3/C2A/embryonal cluster 2)
  - High-risk (HB2/C2A, embryonal cluster E1)
- Clusters E1/E2 are derived from liver progenitor cells at an earlier differentiation stage (embryonal/combined histology) and harbor hypermethylation of HNF4A/CEBPA-binding regions leading to a fetal gene expression profile; due to promoter hypomethylation the *NQO1* gene is highly expressed, which induces chemoresistance and leads to stabilization of ODC1 (a key enzyme for polyamine formation and cell proliferation)
- Cluster F arises from hepatoblasts at a relatively mature stage (fetal phenotype), harbors genetic features that are opposite of those observed in clusters E1/E2

Prognosis and Predictive Factors

- Prognosis depends on tumor stage by pretreatment extent of disease (PRETEXT), posttreatment extent of tumor, age, serum AFP levels, and histological subtype (Table 8.3)
- High NFE2L2 activity and high expression of LIN28B, HMGA2, SALL4, and AFP are associated with poor outcome (cluster HB2)
  - Low WNT signaling activity and low expression of LIN28B and HMGA2 indicate a favorable prognosis, which is associated with a fetal phenotype and low mitotic activity (cluster HB1)
- Overexpression of the NFE2L2 target gene *NQO1* is associated with metastasis, vascular invasion, and poor outcome
- Poor-prognosis HB expresses high levels of AFP and SALL4, and this subgroup may also be identifiable by a four-gene signature: hydroxysteroid 17-beta dehydrogenase 6, integrin alpha 6 (belongs to cluster C1), topoisomerase 2-alpha (belongs to cluster C2A), and vimentin (belongs to cluster C2B)
- About 40% of HB are primarily resectable
- Preoperative chemotherapy may allow downstaging and secondary curative resection
- HB with pure fetal histology and low mitotic rate do not require additional chemotherapy

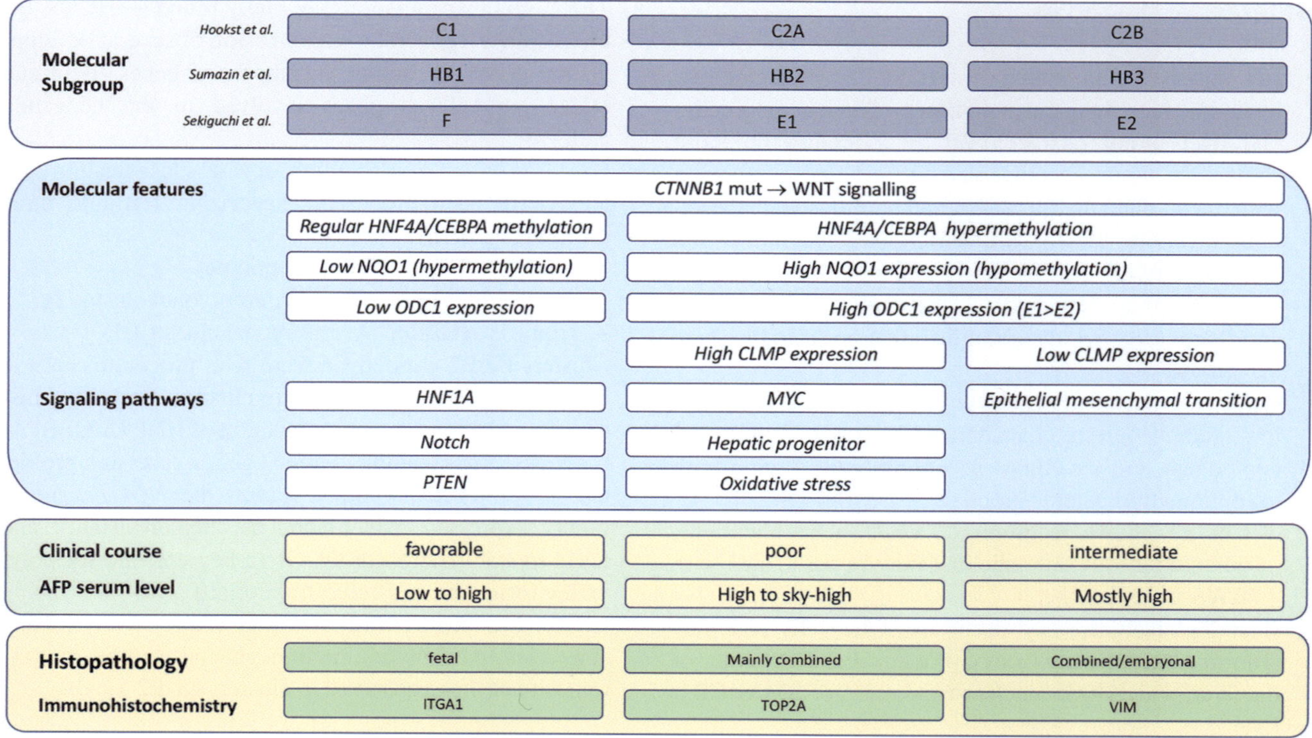

Fig. 8.15 Integration of molecular subgroups, molecular and clinic-pathological features, and histopathology of hepatoblastoma

Table 8.3 Factors affecting prognosis of hepatoblastoma patients

Parameter	Good prognosis	Bad prognosis
PRETEXT	Stage I, II	Stage IV, multifocality, unresectable vascular involvement, extrahepatic disease, tumor rupture
Posttreatment response		Poor response, progressive disease, unresectability, positive margins, relapse
Age	<1 year	>6 years
AFP serum level	Low (<1000 µg/L) Very high (>1000,000 µg/L)	
Histological pattern	Pure fetal with low mitotic activity (well differentiated)	Small cell undifferentiated pattern

- Liver transplantation is the only curative treatment option for unresectable HB and results in 10-year survival rates of 66–78%
  - Contraindication is the persistence of extrahepatic disease unresponsive to chemotherapy

Hepatic Neuroendocrine Neoplasms

Definition

- Hepatic neuroendocrine neoplasms (HNEN) are primary malignant epithelial liver tumors with morphological and immunohistological features of neuroendocrine differentiation ranging from well-differentiated neuroendocrine tumors to poorly differentiated neuroendocrine carcinoma
  - Mixed neuroendocrine–nonneuroendocrine neoplasms (MiNEN) may occur; the diagnosis requires at least 30% of each component

Clinical Features

- Neuroendocrine neoplasms often metastasize to the liver, but primary hepatic neuroendocrine tumors are extremely rare, requiring rigorous exclusion of an extrahepatic primary before accepting the diagnosis of a hepatic primary
- The clinical presentation of hepatic NENs may include nonspecific abdominal symptoms, such as abdominal discomfort or diarrhea
- Symptoms of hormone hypersecretion (e.g., Zollinger-Ellison syndrome, Cushing syndrome, and hypercalcemia) have been reported

Pathologic Features

- Hepatic NEN may arise anywhere within the liver and typically presents as a circumscribed mass-forming lesion

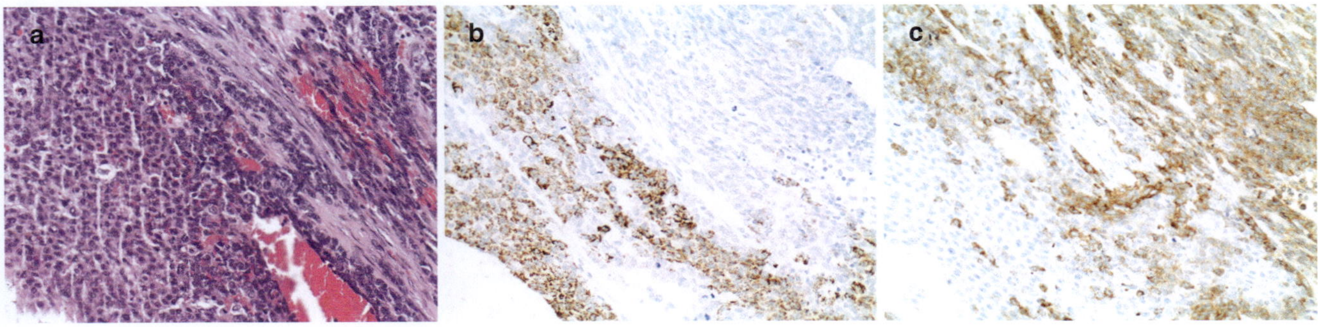

Fig. 8.16 Mixed neuroendocrine–nonneuroendocrine neoplasm of the liver. (**a**) Mixed tumor showing relatively small hepatoid cells growing in a solid patter (lower left) and a small cell component (upper right). (**b**) The hepatoid component is positive for Hepar-1, while the (**c**) small cell compartment shows synaptophysin expression (original magnification 200-fold)

- Histology may vary from a well-differentiated neuroendocrine tumor (NET G1: <2% Ki-67 labeling index, G2 2–10%, G3, >20–50%) showing an organoid growth pattern like nests, cords, or ribbons to small or large cell neuroendocrine carcinoma (NEC G3, Ki-67 > 50%)
- MiNEN seems to be more common than pure NET or NEC, mostly in combination with HCC (Fig. 8.16)
- Tumor cells of NEN are positive for synaptophysin and chromogranin A (NEC may less frequently express chromogranin A compared to NET)

Differential Diagnosis

- Hepatic metastasis of NEN

Genetic Features

- Genetic features of hepatic NEN are poorly characterized
- Exome sequencing has shown recurrent mutations in epigenetic modifiers *SETB1*, *BPTF*, *MECP2* and *WDR5*), and genes involved in cell cycle (*TP53*, *ATM*, *MED12*, *DIDO1* and *ATAD5*) and neural development (*UBR4*, *MEN1*, *GLUL* and *GIGYF2*)

Prognosis and Predictive Factors

- Surgical resection of NET is the therapy of choice; recurrence may be observed even after several years
- Ki-67 labeling index is a strong prognostic factor in all NEN
- NEC and MiNEN are highly aggressive and rapidly lethal

Malignant Vascular Neoplasms

Definitions

- Epithelioid hemangioendothelioma (mEHE) is a malignant endothelial neoplasm composed of epithelioid cells in a myxohyaline or fibrous stroma, most commonly with a *WWTR-CAMTA1* (>90%) or *YAP1-TFE3* (~5%) gene fusion
- Kaposi sarcoma (KS) is an HHV8-associated vascular neoplasm characterized by disorganized endothelial cell growth
- Angiosarcoma (AS) is a malignant neoplasm showing endothelial differentiation and a variable degree of vessel formation

Clinical Features

- mEHE is often diagnosed incidentally and develops sporadically; multifocal liver involvement is frequently observed; extrahepatic disease (e.g., lungs, spleen) may be present
- Hepatic KS is caused by HHV8 and is usually clinically asymptomatic; it may develop as consequence of immunosuppressive therapy (e.g., following organ transplantation) or in the context of acquired immune deficiency syndrome (AIDS); multifocal hepatic disease is usually associated with AIDS
- Hepatic AS has its peak incidence in the seventh decade and may become clinically manifest as fulminant hepatic failure or with unspecific symptoms like abdominal pain or ascites; tumor rupture may lead to acute abdomen; etiological factors include exposure to carcinogens (e.g., vinyl chloride, arsenic, androgen steroids, and historically the contrast agent thorium dioxide) or local irradiation

Pathologic Features

- Macroscopically, mEHE shows a white and firm cut surface; histologically, mEHE reveals epithelioid, dendritic, or spindle-shape morphology and are embedded as cords or individual cells in a hyaline to myxohyaline stroma (Fig. 8.17)
  - Intracytoplasmic vacuoles (sometimes containing erythrocytes) are typically evident
  - Some tumors may show nuclear atypia and sometimes pleomorphic or multinucleated cells are seen
- *CAMTA1*-rearranged mEHE may be immunohistologically positive for CAMTA1, while TFE3 immunostaining may be seen independent of a *TFE3* gene rearrangement
- Hepatic KS manifests as uniform atypical spindle cells forming slit-like vascular spaces in portal or periportal location; eosinophilic hyaline globules and signs of recurrent hemorrhage may be seen
- The macroscopic appearance of AS is variable ranging from firm greyish to hemorrhagic tissue with cystic spaces; histologically, AS is a high-grade neoplasm with nuclear atypia of spindle or epithelioid cells (Fig. 8.18), prominent mitotic activity, and frequent necrosis
  - Typically, the tumor cells form buds and papillary-like projections and other complex structures like whorls, but very well-differentiated cases showing rather inconspicuous hobnail-like sinusoidal cells do occur
- All vascular tumors are positive for endothelial markers (e.g., CD31, CD34, ERG) (Figs. 8.17, 8.18, and 8.19) and some tumors with epithelioid appearance may show positivity for epithelial cell markers (e.g., K8, K18)

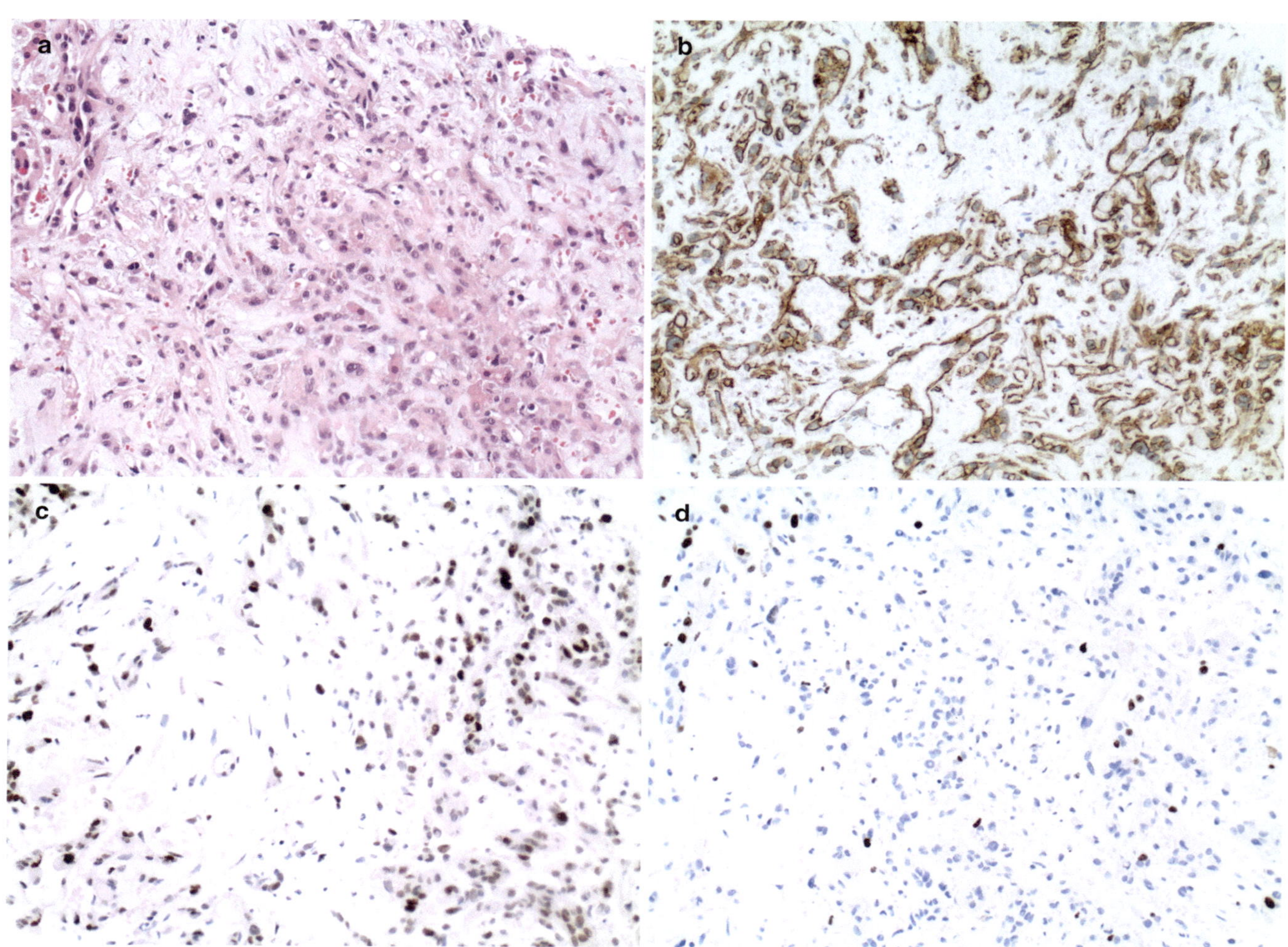

Fig. 8.17 Malignant epithelioid hemangioendothelioma. (**a**) Epithelioid neoplasm growing in cords and tubules within a hyaline stroma. (**b**) Tumor cells are positive for CD31. (**c**) p53 staining intensity of tumor cells is variable. (**d**) Proliferative activity of tumor cells is low (Ki-67 staining, original magnification 200-fold)

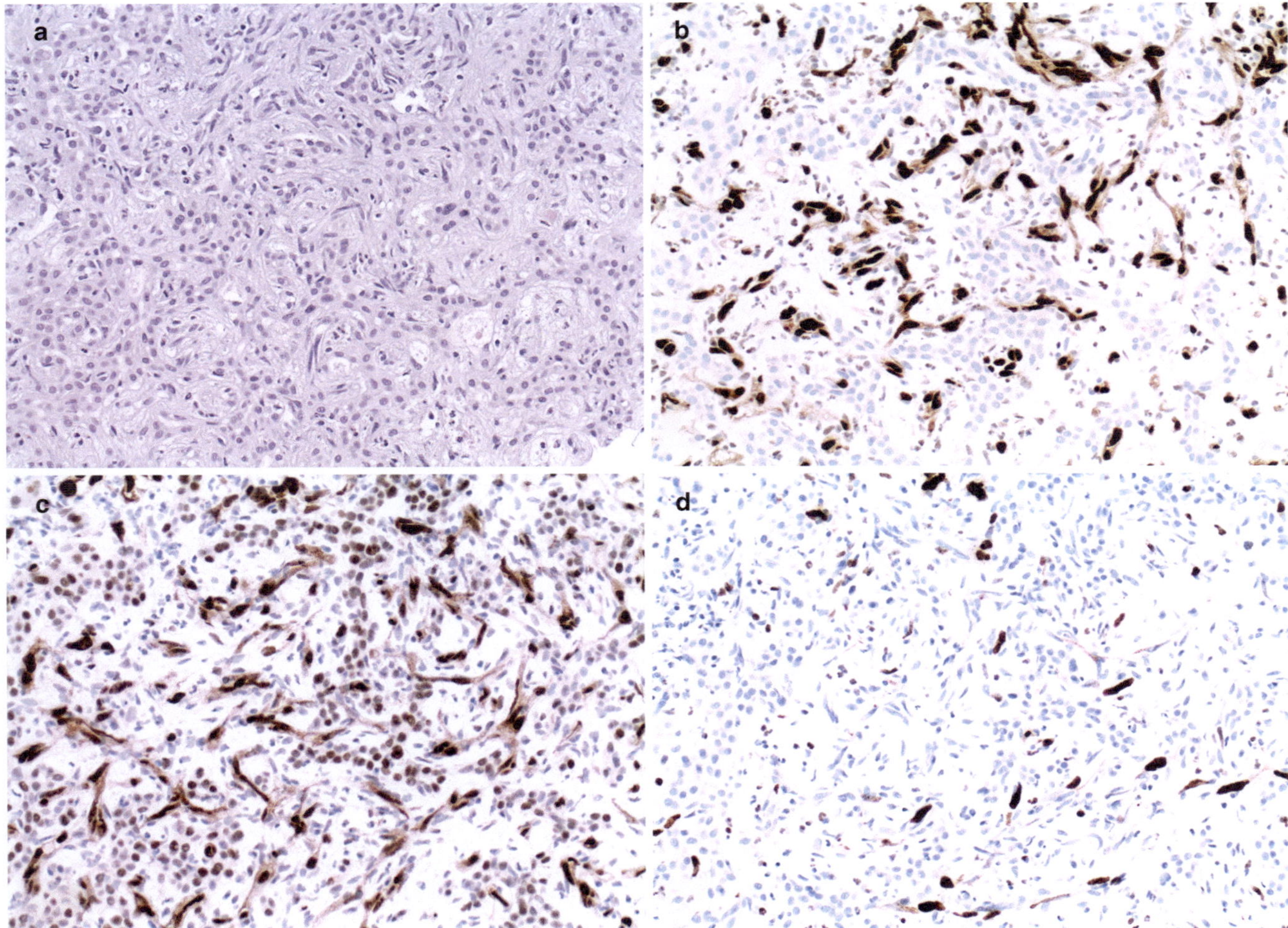

Fig. 8.18 Angiosarcoma. (**a**) Spindle cell tumor with intrasinusoidal growth and compressed liver cell plates. (**b**) Tumor cells are ERG-positive. (**c**) Constitutive strong p53 staining of tumor cells, while the p53 staining intensity of the surrounding hepatocytes is variable. (**d**) High proliferative activity of spindle-shaped tumor cells (Ki-67 staining, original magnification 200-fold)

Differential Diagnosis

- The main differential diagnoses of mEHE are sclerotic carcinoma (e.g., iCCA) and AS
- AS must be further distinguished from inflammatory or benign vascular disorders, Kaposi sarcoma, HCC, and liver metastases (epithelioid AS express keratins)
- If present a mutation type TP53 expression pattern (diffuse strong positivity or complete negativity) of liver endothelial cells may support the diagnosis of AS (Fig. 8.18)
- Hepatic small vessel neoplasm (anastomosing hemangioma) shows sinusoidal proliferation of spindle cells, but is morphologically bland, has a low Ki-67 labeling index, and aberrant expression of TP53 or MYC are not seen (Fig. 8.19)
  - Detection of GNAQ or GNA14 mutations may further support the diagnosis of hepatic small vessel neoplasm in difficult cases

Genetic Features

- Most mEHE reveal a *WWTR1-CAMTA1* gene fusion, few cases harbor a *YAP1-TFE3* gene fusion
- HHV8 can be detected in KS either by immunohistochemistry, PCR or in situ hybridization
- Hepatic AS are frequently ATRX-deficient and may show an associated alternative lengthening of telomers; vinyl chloride associated AS harbor increased *TP53* mutation frequencies; postradiation AS may show MYC overexpression due to *MYC* amplification

Prognosis and Predictive Factors

- The clinical course of mEHE is variable and cannot be predicted based on histology; distant metastases may occur in 20–30% of patients

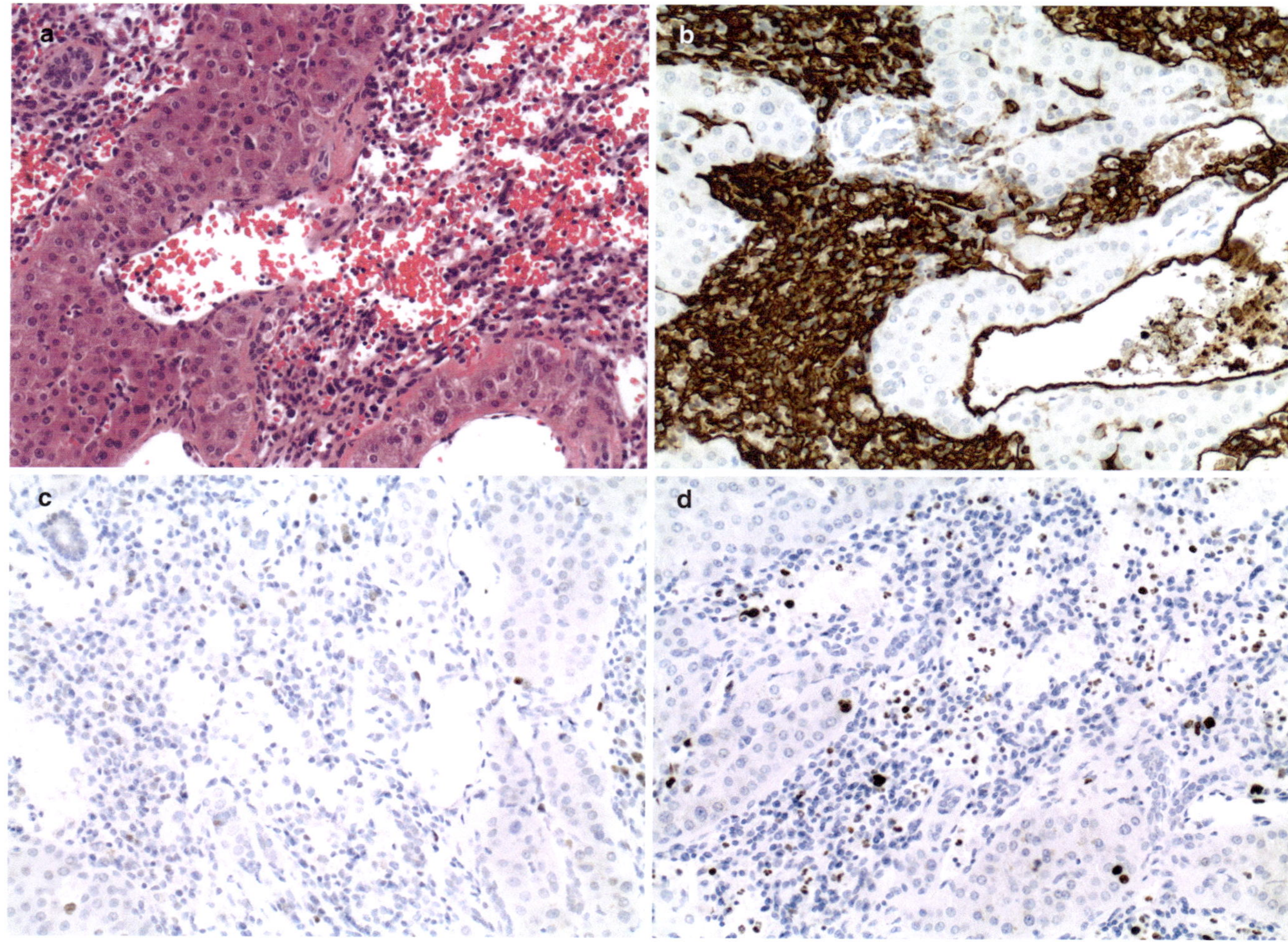

Fig. 8.19 Hepatic small vessel neoplasm. (**a**) The proliferation of mildly pleomorphic spindle cells in dilated sinusoids surrounded by mildly distorted and thickened hepatocellular trabeculae without atypia. (**b**) Spindle cells are CD31 positive. (**c**) p53 is variably expressed. (**d**) Proliferative activity is low (Ki-67, original magnification 200-fold)

- In rare cases, hepatic KS may undergo a rapid progression leading to liver and organ failure with high fatality rate; combined highly active antiretroviral therapy and systemic chemotherapy may improve morbidity and mortality of hepatic KS
- Hepatic AS is highly aggressive and survival >1 year is rare; older age, large tumor size, and high proliferative fraction are considered poor prognostic factors

Hepatic Angiomyolipoma

Definitions

- Hepatic angiomyolipoma (PEComa) is a mesenchymal neoplasm composed of distinctive, predominantly epithelioid cells showing variable expression of smooth muscle and neuroectodermal markers; angiomyolipoma additionally contains fat cells and thick-walled blood vessels

Clinical Features

- Hepatic PEComa shows a female predominance and mostly affects middle-aged patients
- Diagnosis is usually made incidentally; some patients present with unspecific abdominal symptoms; tumor rupture and intraabdominal bleeding are exceptionally rare
- Patients presenting with multifocal disease or manifestations in both liver and kidney typically suffer from tuberous sclerosis (5–10% of cases)

Pathologic Features

- Most hepatic angiomyolipomas contain a variable admixture of adipocytes, epithelioid cells, and thick-walled blood vessels, with epithelioid cells dominating in many cases (Fig. 8.20)
    - Extramedullary hematopoiesis may be detected
- In large tumors necrosis and hemorrhage may be seen; the proliferation activity is usually low
- PEComas express smooth muscle markers (e.g., smooth muscle actin) and neuroectodermal markers (e.g., HMB45, Melan-A, and MITF)

Differential Diagnosis

- Hepatic angiomyolipoma may be positive for KIT, which should not lead to confusion with gastrointestinal stroma tumors

- Hepatic angiomyolipoma may be positive for GS and is consistently negative for FABP1, which may require distinction from HCA (hepatocellular markers, e.g., Hepar1)

Genetic Features

- Mutations of *TSC2* (biallelic inactivation) are frequently detected and result in activated mTOR signaling; *TSC1* mutation is rare
- TFE3 immunostaining may be detected, but *TFE3* gene rearrangements have not been described in hepatic angiomyolipoma (in contrast to other PEComas) so far

Prognosis and Predictive Factors

- Nearly all primary hepatic angiomyolipomas are benign, but exceedingly rare malignant cases have been described

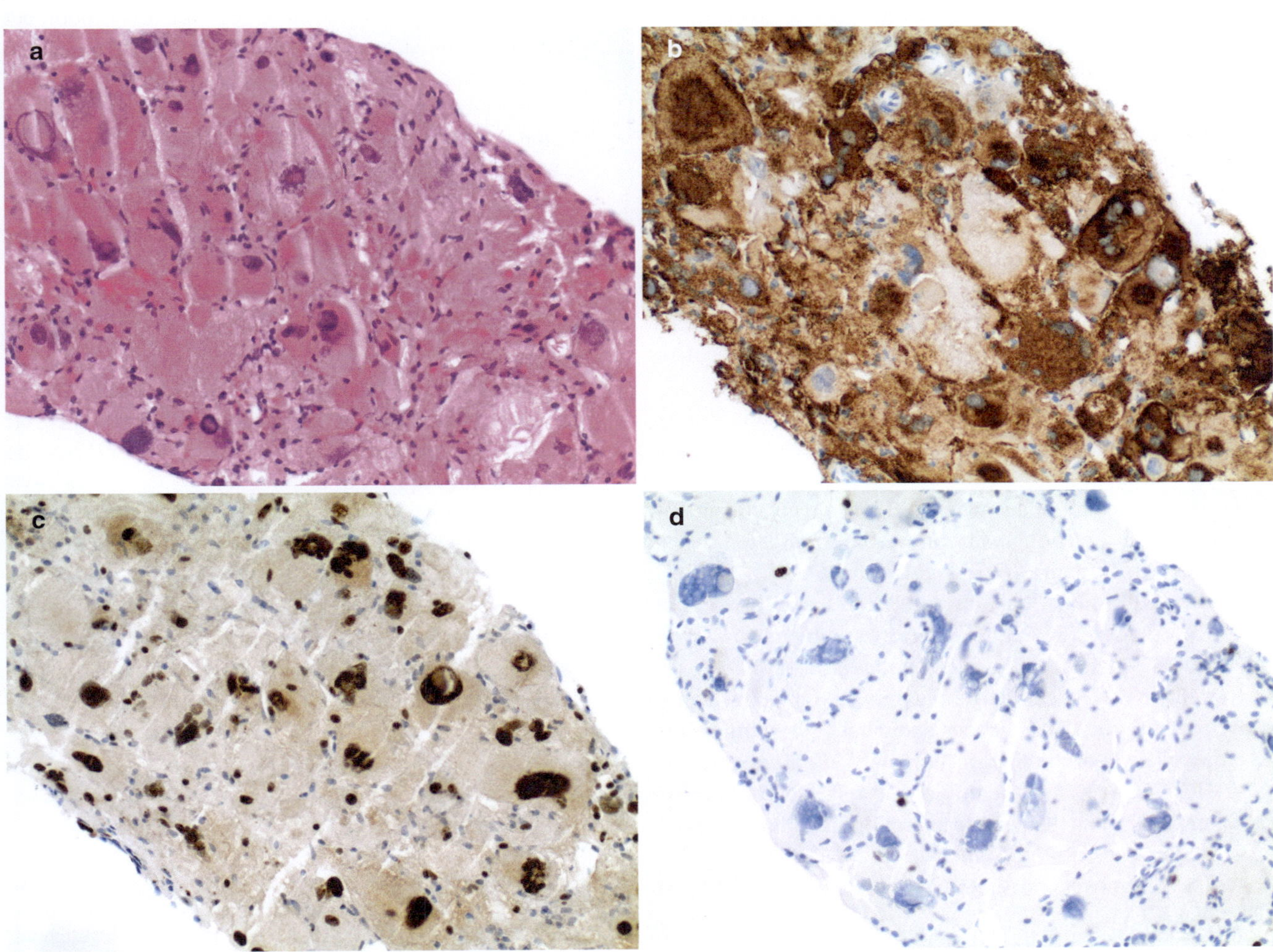

Fig. 8.20 Epithelioid angiomyolipoma consisting of (**a**) large epitheliod cells with abundant eosinophilic cytoplasm and large, pleomorphic nuclei. Tumor cells are positive for (**b**) HMB45 and (**c**) MITF. (**d**) The proliferative activity is very low (Ki-67, original magnification 200-fold)

- Marked nuclear atypia and mitotic activity (>1 mitosis/mm^2) may predict but not prove a malignant biological behavior and require exclusion of metastasis (from renal epithelioid PEComa)
- Treatment with mTOR inhibitors may induce response

Summary of Molecular Pathology of Liver Tumors

- Neoplasms of the liver are classified morphologically and immunohistochemically
- Molecular analyses support diagnosis in equivocal cases, aiming at the identification of genotypes-phenotype associations and allow for morpho-molecular tumor (sub) typing
- Hepatocellular carcinoma is the most common primary liver carcinoma
  - It usually develops in cirrhotic liver tissue via morphologically defined premalignant dysplastic nodules
  - The most prevalent somatic mutations include *TERT* promoter, *TP53*, and *CTNNB1*, while other mutations occur in low frequency but cluster in certain pathways (e.g., RAS/MAP, PI3K/AKT, JAK/STAT, TGFβ, NOTCH, IGF2)
  - Predictive biomarkers are scarce so far and have not entered clinical practice
- Intrahepatic cholangiocarcinoma represents the second most common liver cancer
  - The peripheral mass-forming type usually shows a small duct type phenotype and may develop via plasticity/transdifferentiation from hepatocytes; precursor lesions are not known
  - Precursor lesions of large duct iCCA include biliary intraepithelial neoplasia and intraductal papillary neoplasm
  - More than half of advanced iCCA show genetic alterations that are principally amenable to targeted therapies (including *IDH1*, *BAP1*, *BRCA1*, *BRCA2*, *BRAF*[V600E] mutations, microsatellite instability and in *RAS*-wildtype iCCA genetic rearrangements involving *FGFR2*, *NRG1*, *ALK*, *ROS1*, and *NTRK1–3*
- Combined hepatocellular-cholangiocarcinoma is a rare type of primary liver cancer showing both components either close to each other or intimately intermingled
  - HCC-CCA shares the etiological risk factors of HCC and iCCA and the genetic alterations recapitulate those reported in both HCC and iCCA
  - The prognosis of HCC-CCA is worse than HCC and like iCCA
  - HCC-CCA tends to behave like HCC with respect to vascular invasion
- Hepatoblastoma
  - HB is the most common pediatric liver tumor accounting for 1% of pediatric malignancies
  - HB is classified as epithelial (fetal, embryonal, SCUD, cholangioblastic, macrotrabecular, or mixed epithelial pattern) or mixed epithelial and mesenchymal
  - Alterations in the APC/β(beta)-catenin pathway are pivotal in the pathogenesis of hepatoblastoma
  - Prognosis is predicted by tumor stage using PRETEXT, posttreatment extent of tumor, age, serum AFP levels, and the histological subtype
- Hepatic neuroendocrine neoplasms
  - Are exceedingly rare and show morphological and immunohistological features of neuroendocrine differentiation ranging from well-differentiated neuroendocrine tumors to poorly differentiated neuroendocrine carcinoma
  - Typically present as a circumscribed mass-forming lesion; mixed neuroendocrine-neuroendocrine neoplasms with NET or NEC components and mostly in combination with HCC may occur
  - Molecular characterization of hepatic NEN is poor, but recurrent mutations have been described in epigenetic modifiers, cell cycle regulators, and genes involved in neural development
- Malignant vascular neoplasms
  - Are a rare group of tumors
  - mEHE is characterized by specific gene fusions, develops sporadically, and has a variable clinical course, which cannot be predicted based on histology
  - In immunosuppressed patients KS is induced by HHV8, which can be detected in tissue and confirms the diagnosis
  - Hepatic AS is a usually fatal disease developing in context of exposure to carcinogens or following radiation therapy; overexpression of TP53 or MYC may be helpful in establishing the diagnosis
- Hepatic angiomyolipoma (PEComa)
  - Shows expression of smooth muscle and neuroectodermal markers and may contain fat cells and thick-walled blood vessels
  - Multifocal disease or manifestations in both liver and kidney may develop in context of tuberous sclerosis (with *TSC2* mutation)

Further Reading

Agaimy A, Daum O, Markl B, Lichtmannegger I, Michal M, Hartmann A. SWI/SNF complex-deficient undifferentiated/rhabdoid carcinomas of the gastrointestinal tract: a series of 13 cases highlighting mutually exclusive loss of SMARCA4 and SMARCA2 and frequent co-inactivation of SMARCB1 and SMARCA2. Am J Surg Pathol. 2016;40:544–53.

Al Nassan A, Sughayer M, Matalka I, et al. INI1 (BAF 47) immunohistochemistry is an essential diagnostic tool for children with hepatic tumors and low alpha fetoprotein. J Pediatr Hematol Oncol. 2010;32:e79–81.

Alves VAF, Rimola J. Malignant vascular tumors of the liver in adults. Semin Liver Dis. 2019;39:1–12.

Antonescu CR, Dickson BC, Sung YS, et al. Recurrent YAP1 and MAML2 gene rearrangements in retiform and composite hemangioendothelioma. Am J Surg Pathol. 2020;44:1677–84.

Bayard Q, Caruso S, Couchy G, et al. Recurrent chromosomal rearrangements of ROS1, FRK and IL6 activating JAK/STAT pathway in inflammatory hepatocellular adenomas. Gut. 2020;69:1667–76.

Boyault S, Rickman DS, de Reynies A, et al. Transcriptome classification of HCC is related to gene alterations and to new therapeutic targets. Hepatology. 2007;45:42–52.

Calderaro J, Couchy G, Imbeaud S, et al. Histological subtypes of hepatocellular carcinoma are related to gene mutations and molecular tumour classification. J Hepatol. 2017;67:727–38.

Cancer Genome Atlas Research Network. Comprehensive and integrative genomic characterization of hepatocellular carcinoma. Cell. 2017;169:1327–41. e1323

Chaudhary P, Bhadana U, Singh RA, Ahuja A. Primary hepatic angiosarcoma. Eur J Surg Oncol. 2015;41:1137–43.

Di Tommaso L, Destro A, Fabbris V, et al. Diagnostic accuracy of clathrin heavy chain staining in a marker panel for the diagnosis of small hepatocellular carcinoma. Hepatology. 2011;53:1549–57.

Durnez A, Verslype C, Nevens F, et al. The clinicopathological and prognostic relevance of cytokeratin 7 and 19 expression in hepatocellular carcinoma. A possible progenitor cell origin. Histopathology. 2006;49:138–51.

Eichenmuller M, Trippel F, Kreuder M, et al. The genomic landscape of hepatoblastoma and their progenies with HCC-like features. J Hepatol. 2014;61:1312–20.

Errani C, Zhang L, Sung YS, et al. A novel WWTR1-CAMTA1 gene fusion is a consistent abnormality in epithelioid hemangioendothelioma of different anatomic sites. Genes Chromosomes Cancer. 2011;50:644–53.

Fan B, Malato Y, Calvisi DF, et al. Cholangiocarcinomas can originate from hepatocytes in mice. J Clin Invest. 2012;122:2911–5.

Farshidfar F, Zheng S, Gingras MC, et al. Integrative genomic analysis of Cholangiocarcinoma identifies distinct IDH-mutant molecular profiles. Cell Rep. 2017;18:2780–94.

Fazlollahi L, Hsiao SJ, Kochhar M, Mansukhani MM, Yamashiro DJ, Remotti HE. Malignant rhabdoid tumor, an aggressive tumor often misclassified as small cell variant of hepatoblastoma. Cancers (Basel). 2019;11:1992.

Finn RS, Qin S, Ikeda M, et al. Atezolizumab plus bevacizumab in unresectable hepatocellular carcinoma. N Engl J Med. 2020;382:1894–905.

Flucke U, Vogels RJ, de saint Aubain Somerhausen N, et al. Epithelioid Hemangioendothelioma: clinicopathologic, immunhistochemical, and molecular genetic analysis of 39 cases. Diagn Pathol. 2014;9:131.

Fukayama M, Miettinnen M, Lazar AJ. Mesenchymal tumours of the digestive system. In: Lokuhetty D, White VA, Watanabe R, Cree IA, editors. WHO classification of tumours—digestive system tumours, vol. 1. Lyon: WHO Press; 2019. p. 433–98.

Gill RM, Buelow B, Mather C, et al. Hepatic small vessel neoplasm, a rare infiltrative vascular neoplasm of uncertain malignant potential. Hum Pathol. 2016;54:143–51.

Goeppert B, Folseraas T, Roessler S, et al. Genomic characterization of cholangiocarcinoma in primary sclerosing cholangitis reveals therapeutic opportunities. Hepatology. 2020;72:1253–66.

Guichard C, Amaddeo G, Imbeaud S, et al. Integrated analysis of somatic mutations and focal copy-number changes identifies key genes and pathways in hepatocellular carcinoma. Nat Genet. 2012;44:694–8.

Hooks KB, Audoux J, Fazli H, et al. New insights into diagnosis and therapeutic options for proliferative hepatoblastoma. Hepatology. 2018;68:89–102.

Huang SC, Chuang HC, Chen TD, et al. Alterations of the mTOR pathway in hepatic angiomyolipoma with emphasis on the epithelioid variant and loss of heterogeneity of TSC1/TSC2. Histopathology. 2015;66:695–705.

Jimbo N, Nishigami T, Noguchi M, et al. Hepatic angiomyolipomas may overexpress TFE3, but have no relevant genetic alterations. Hum Pathol. 2017;61:41–8.

Joseph NM, Brunt EM, Marginean C, et al. Frequent GNAQ and GNA14 mutations in hepatic small vessel neoplasm. Am J Surg Pathol. 2018;42:1201–7.

Joseph NM, Tsokos CG, Umetsu SE, et al. Genomic profiling of combined hepatocellular-cholangiocarcinoma reveals similar genetics to hepatocellular carcinoma. J Pathol. 2019;248:164–78.

Kuo FY, Huang HY, Chen CL, Eng HL, Huang CC. TFE3-rearranged hepatic epithelioid hemangioendothelioma-a case report with immunohistochemical and molecular study. APMIS. 2017;125:849–53.

Kurebayashi Y, Ojima H, Tsujikawa H, et al. Landscape of immune microenvironment in hepatocellular carcinoma and its additional impact on histological and molecular classification. Hepatology. 2018;68:1025–41.

Longerich T, Mueller MM, Breuhahn K, Schirmacher P, Benner A, Heiss C. Oncogenetic tree modeling of human hepatocarcinogenesis. Int J Cancer. 2012;130:575–83.

Lotfalla MM, Folpe AL, Fritchie KJ, et al. Hepatic YAP1-TFE3 rearranged epithelioid hemangioendothelioma. Case Rep Gastrointest Med. 2019;2019:7530845.

Luchini C, Pelosi G, Scarpa A, et al. Neuroendocrine neoplasms of the biliary tree, liver and pancreas: a pathological approach. Pathologica. 2021;113:28–38.

Moeini A, Sia D, Zhang Z, et al. Mixed hepatocellular cholangiocarcinoma tumors: cholangiolocellular carcinoma is a distinct molecular entity. J Hepatol. 2017;66:952–61.

Mueller C, Waldburger N, Stampfl U, et al. Non-invasive diagnosis of hepatocellular carcinoma revisited. Gut. 2018;67:991–3.

Mullhaupt B, Durand F, Roskams T, Dutkowski P, Heim M. Is tumor biopsy necessary? Liver Transpl. 2011;17(Suppl 2):S14–25.

Nakamura H, Arai Y, Totoki Y, et al. Genomic spectra of biliary tract cancer. Nat Genet. 2015;47:1003–10.

Nault JC, Couchy G, Balabaud C, et al. Molecular classification of hepatocellular adenoma associates with risk factors, bleeding, and malignant transformation. Gastroenterology. 2017;152:880–94. e886

International Consensus Group for Hepatocellular Neoplasia. Pathologic diagnosis of early hepatocellular carcinoma: a report of the international consensus group for hepatocellular neoplasia. Hepatology. 2009;49:658–64.

Okamura R, Kurzrock R, Mallory RJ, et al. Comprehensive genomic landscape and precision therapeutic approach in biliary tract cancers. Int J Cancer. 2021;148:702–12.

Ong CK, Subimerb C, Pairojkul C, et al. Exome sequencing of liver fluke-associated cholangiocarcinoma. Nat Genet. 2012;44:690–3.

Paradis V, Fukayama M, Park YN, Schirmacher P. Tumours of the liver and intrahepatic bile ducts. In: Lokuhetty D, White VA, Watanabe R, Cree IA, editors. WHO classification of tumours—digestive system tumours, vol. 1. Lyon: WHO Press; 2019. p. 215–64.

Saha SK, Parachoniak CA, Ghanta KS, et al. Mutant IDH inhibits HNF-4alpha to block hepatocyte differentiation and promote biliary cancer. Nature. 2014;513:110–4.

Seehawer M, Heinzmann F, D'Artista L, et al. Necroptosis microenvironment directs lineage commitment in liver cancer. Nature. 2018;562:69–75.

Sekiguchi M, Seki M, Kawai T, et al. Integrated multiomics analysis of hepatoblastoma unravels its heterogeneity and provides novel druggable targets. NPJ Precis Oncol. 2020;4:20.

Sempoux C, Gouw ASH, Dunet V, Paradis V, Balabaud C, Bioulac-Sage P. Predictive patterns of glutamine synthetase immunohistochemical staining in CTNNB1-mutated hepatocellular adenomas. Am J Surg Pathol. 2021;45:477–87.

Sumazin P, Chen Y, Trevino LR, et al. Genomic analysis of hepatoblastoma identifies distinct molecular and prognostic subgroups. Hepatology. 2017;65:104–21.

Tate G, Suzuki T, Mitsuya T. Mutation of the PTEN gene in a human hepatic angiosarcoma. Cancer Genet Cytogenet. 2007;178:160–2.

Verlingue L, Malka D, Allorant A, et al. Precision medicine for patients with advanced biliary tract cancers: an effective strategy within the prospective MOSCATO-01 trial. Eur J Cancer. 2017;87:122–30.

Vokuhl C, Oyen F, Haberle B, von Schweinitz D, Schneppenheim R, Leuschner I. Small cell undifferentiated (SCUD) hepatoblastomas: all malignant rhabdoid tumors? Genes Chromosomes Cancer. 2016;55:925–31.

Xu AM, Zhang SH, Zheng JM, Zheng WQ, Wu MC. Pathological and molecular analysis of sporadic hepatic angiomyolipoma. Hum Pathol. 2006;37:735–41.

Xue R, Chen L, Zhang C, et al. Genomic and transcriptomic profiling of combined hepatocellular and intrahepatic cholangiocarcinoma reveals distinct molecular subtypes. Cancer Cell. 2019;35:932–47. e938

Yang P, Huang X, Lai C, et al. SET domain containing 1B gene is mutated in primary hepatic neuroendocrine tumors. Int J Cancer. 2019;145:2986–95.

Yasir S, Torbenson MS. Angiosarcoma of the liver: clinicopathologic features and morphologic patterns. Am J Surg Pathol. 2019;43:581–90.

Yong KJ, Chai L, Tenen DG. Oncofetal gene SALL4 in aggressive hepatocellular carcinoma. N Engl J Med. 2013;369:1171–2.

Molecular Pathology of Lung Tumors

Ying-Chun Lo and Neal I. Lindeman

Contents

Introduction

- The lung is one of the most affected sites of cancer in the body, with involvement by a variety of different types of neoplasms; primary lung carcinomas are mainly subdivided into nonsmall cell lung carcinoma (NSCLC) and small cell lung carcinoma (SCLC) categories
  - Common entities in the NSCLC category
    - Adenocarcinomas (>50%)
      - Minimally invasive adenocarcinoma: Small (less than 3 cm) adenocarcinoma with a predominantly lepidic pattern and less than 5 mm invasion
      - Invasive nonmucinous adenocarcinoma
        - The most common and "conventional" adenocarcinoma with morphological or immunohistochemical evidence of glandular differentiation
        - These commonly have mixed histologic features (Fig. 9.1), including lepidic, acinar, papillary, micropapillary, and solid patterns
      - Invasive mucinous adenocarcinoma: Primary lung adenocarcinoma with cytological features of goblet cell or columnar cell morphology and abundant intracytoplasmic mucin
    - Squamous cell carcinomas (~20%): NSCLC with morphological or immunohistochemical evidence of squamous cell differentiation
    - Adenosquamous cell carcinomas (~2–3%): NSCLC including components of both adenocarcinoma and

Y.-C. Lo
Department of Laboratory Medicine and Pathology, Mayo Clinic, Rochester, MN, USA

N. I. Lindeman (✉)
Department of Pathology and Laboratory Medicine, Weill Cornell Medicine, New York Presbyterian Hospital, New York, NY, USA
e-mail: nel4003@med.cornell.edu

L. Cheng et al. (eds.), *Molecular Surgical Pathology*, https://doi.org/10.1007/978-3-031-35118-1_9

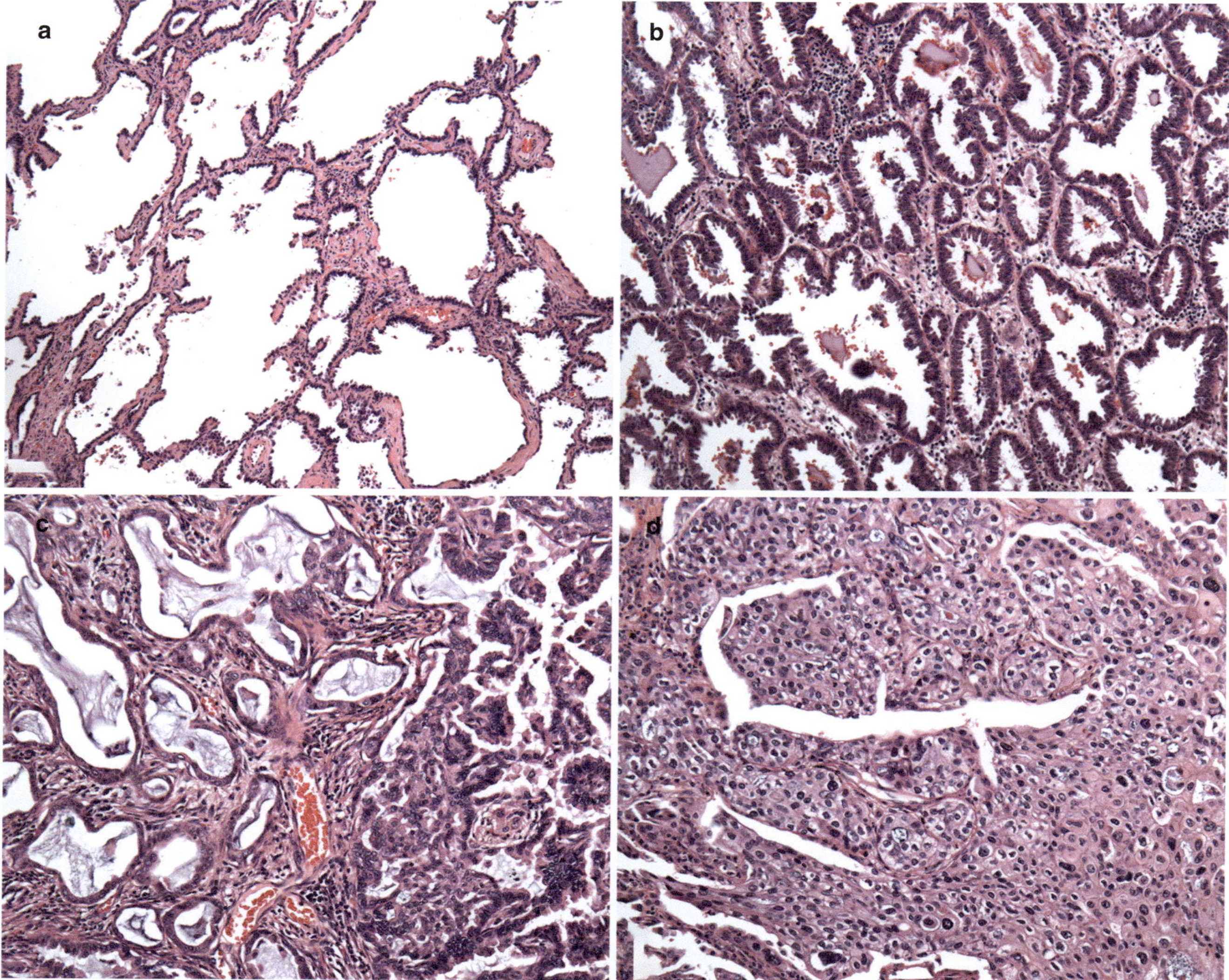

Fig. 9.1 Representative images of lung adenocarcinoma including (**a**) lepidic; (**b**) acinar; (**c**) combined acinar and papillary; and (**d**) solid patterns

squamous cell carcinoma, with each comprising at least 10% of the tumor

- Sarcomatoid carcinomas (~2–3%): NSCLC with spindle cell and/or giant cell morphology
- Large cell carcinomas (~1%): Undifferentiated NSCLC without architectural, cytological, or immunohistochemical features of adenocarcinoma, squamous cell carcinoma, or small cell carcinoma
- Neuroendocrine tumors (<5%)
  - Typical carcinoid tumor: Low-grade
  - Atypical carcinoid tumor: Intermediate-grade
  - Large cell neuroendocrine carcinoma: High-grade
- NSCLC, not otherwise specified, is a term used on biopsy to include tumors with limited representation that cannot be clearly defined based on morphologic and immunohistochemical grounds

- SCLCs (~15%)
  - Small cells characterized by scant cytoplasm, fine chromatin, nuclear molding, and lack of prominent nucleoli
  - High mitotic count, abundant necrosis, and neuroendocrine differentiation by immunohistochemistry are frequently observed
- Mesothelioma is a rare and aggressive cancer originated from mesothelial cells, particularly arising from the pleura, and can also arise from peritoneum and pericardium
  - Morphologically, mesothelioma can be classified into epithelioid, sarcomatoid, desmoplastic, and biphasic subtypes
  - Pleura is a common site for metastases
    - It is clinically important to differentiate epithelioid mesothelioma from metastatic carcinoma

♦ Immunohistochemistry and molecular studies are also frequently applied to differentiate sarcomatoid mesothelioma from sarcoma
  – Metastases to the lung are extremely common, as the predominant organ to receive venous and lymphatic drainage from the rest of the body
    ○ Solid cancers of all types can metastasize to the lung
    ○ Not uncommonly, cancers arising in other sites may present initially with lung metastases. Accordingly, evaluation of a new cancer in the lung should exclude the possibility of a metastasis from another organ
  – Other tumor types, including sarcomas and lymphomas, may arise in the lung, but will not be discussed further here

Clinical Features

- Primary lung cancer is the third most common cancer in the USA
  – Estimated 235,760 new cases in 2021
- Primary lung cancer is the most lethal cancer in the USA
  – Estimated 131,880 deaths in 2021
  – More deaths to lung cancer than the next two most lethal cancers (colorectal and pancreatic) combined
- Lung cancers are highly associated with tobacco exposure
  – Lung cancer was uncommon prior to the mass commercialization of tobacco products in the late nineteenth century
- Survival depends heavily on stage
  – Early stage (I–II) disease (ipsilateral lung involvement only, peribronchial, or intraparenchymal nodal metastases): survival is 53–92% at 5 years
  – Advanced stage (III and IV) disease (contralateral lung involvement, mediastinal, subcarinal, contralateral nodal, and/or distant metastases): survival is 13–36% and <10% at 5 years, respectively
- Death rates have declined significantly in the past two decades
  – Death rates declined from 58.9/100,000 person in 1992, to 51.7/100,000 person in 2006, further to 33.4/100,000 person in 2019
  – Death rates have declined due to decreased tobacco smoking and medical advances in diagnosis and treatment
- Clinical presentation can be nonspecific
  – Cough, pleuritic chest pain, dyspnea
  – Some patients present with symptoms from distant metastases

- Diagnosis is suspected by signs, symptoms, and radiology
  – Areas of consolidation on chest X-ray
  – Mass lesions or "ground glass opacities" by CT scan
- Diagnosis is confirmed by microscopic examination of a lesion
  – Patients often present in advanced stage with unresectable disease
  – Small biopsies and cytopathology specimens are common
- Treatment
  – Nonsmall cell carcinomas
    ○ Early stage: surgical resection
    ○ Advanced stage: chemotherapy, immunotherapy, molecular targeted therapy, and radiation therapy
  – Small cell carcinomas
    ○ Rare selective early stage: surgical resection and adjuvant chemotherapy
    ○ Most cases: chemotherapy, radiation therapy, and immunotherapy

Molecular Genetic Pathology

- Significant advances have been made in the past two decades concerning the molecular pathology and clinical applications thereof in adenocarcinoma of the lung (Fig. 9.2), which will be the primary focus of this section
- Molecular genetic testing related to immunotherapy for lung cancer will also be presented
- Data are emerging regarding the genetic alterations in squamous cell carcinomas, small cell carcinomas, and mesothelioma, which have not garnered widespread clinical treatment application as of January 2022

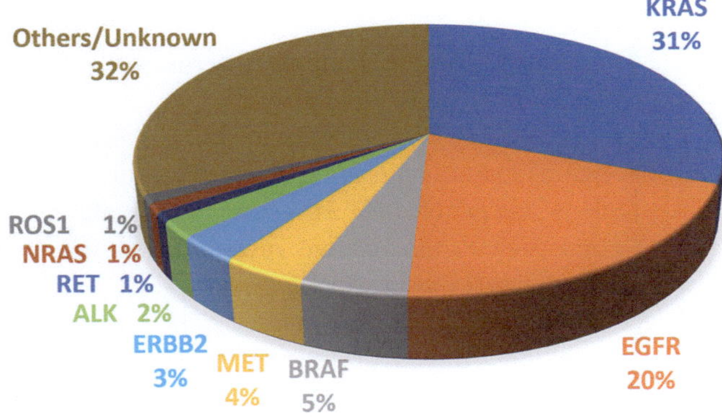

Fig. 9.2 Incidence rate of oncogenic driver gene alterations in 4744 nonsmall cell carcinomas, enriched for adenocarcinoma, sequenced by OncoPanel (DNA-based NGS panel) at Brigham and Women's Hospital, Boston, USA. These oncogenic driver alterations are usually mutually exclusive, except in rare occurrences

- These will be discussed subsequently. NUT carcinoma and thoracic SMARCA4-deficient undifferentiated tumor are two thoracic cancers with defined molecular features, which will also be discussed

Epidermal Growth Factor Receptor (*EGFR*)

- *EGFR* gene at 7p12
  - Also known as *HER1* or *ERBB1*
- Encoded protein epidermal growth factor receptor is a transmembrane receptor protein with cytoplasmic tyrosine kinase involved in transduction of growth factor signaling
  - Member of the ERBB family of receptor tyrosine kinases
- Variably worldwide, 15–40% of adenocarcinomas have to activate somatic mutations in *EGFR*, with the highest prevalence in East Asia
  - Mutations in the cytoplasmic tyrosine kinase domain of *EGFR* (predominantly exons 18–21) lead to constitutive activation of downstream RAS/MAPK signaling pathways in the absence of growth factor receptor binding
    - Exon 19 deletions (45–50%)
      - Variably-sized small in-frame deletions that include at least a four-residue LREA motif in codons 746–749
    - p.Leu858Arg (L858R) point mutation in exon 21 (40–45%)
    - Less common mutations (5–10%) include:
      - A variable point mutation involving glycine at codon 719 (exon 18)
      - p.L861Q point mutation (exon 21)
      - Missense point mutations in exons 18, 20, and 21
    - Exon 20 insertions (5%)
      - Small variable duplication/insertions involving codons 767–774
    - p.Thr790Met (T790M) point mutation in exon 20
- EGFR-mutant lung cancers are more common in women, nonsmokers, and patients of East Asian ancestry, although these clinical characteristics are insufficient to use for selecting patients for testing or for therapy
- Driver oncogenic alterations, including *EGFR* mutations, *KRAS* mutations, *ALK* rearrangements and others, are usually mutually exclusive, except in rare cases or in the posttreatment relapse setting
- The mutated *EGFR* allele is frequently amplified

Molecular Diagnostics

- Indication for testing: therapeutic selection
  - Tumors with the common *EGFR* mutations show better response rates and progression-free survival when treated with EGFR tyrosine kinase inhibitors (TKIs), i.e., gefitinib, erlotinib, and afatinib, than when treated with standard platinum-based chemotherapy
    - Objective response rate (tumor shrinkage by >25%) is ~70–90% in patients with *EGFR* mutations treated with an EGFR-TKI, but only 20–30% with chemotherapy
    - Median progression-free survival doubles for patients with *EGFR*-mutant tumors treated with an EGFR-TKI, as compared to chemotherapy
    - *EGFR* mutations do NOT predict response to therapeutic antibodies directed against the extracellular ligand-binding domain of the EGFR protein, such as cetuximab
  - Mutations in exon 20 are associated with resistance to first and second-generation of EGFR TKIs
    - Exon 20 duplications/insertions are associated with primary treatment resistance to conventional EGFR (TKIs) but sensitive to newly-designed EGFR-TKI (mobocertinib) and EGFR blocking antibody (amivantamab-vmjw)
    - T790M point mutation is found in approximately 50–60% of cases with acquired (secondary) resistance to conventional EGFR TKIs, i.e., to erlotinib, gefitinib, or afatinib, after initial response. NSCLC with T790M point mutation is sensitive to osimertinib, a third-generation EGFR-TKI
    - Rare germline Thr790Met mutations have been seen in families with inherited predisposition to lung cancer
  - Common resistance mechanism for EGFR-TKI treatment includes T790M point mutation, MET amplification, transformation to small cell carcinoma, and occasional alterations in other driver pathways
  - Data showed an overall survival benefit of using Osimertinib as front-line treatment for advanced NSCLC with an *EGFR* mutation, compared to conventional comparator EGFR-TKI. Osimertinib has become a standard first-line treatment for advanced NSCLC with an *EGFR* mutation
- Testing methods: PCR-based single gene assay, NGS panel (Fig. 9.3)

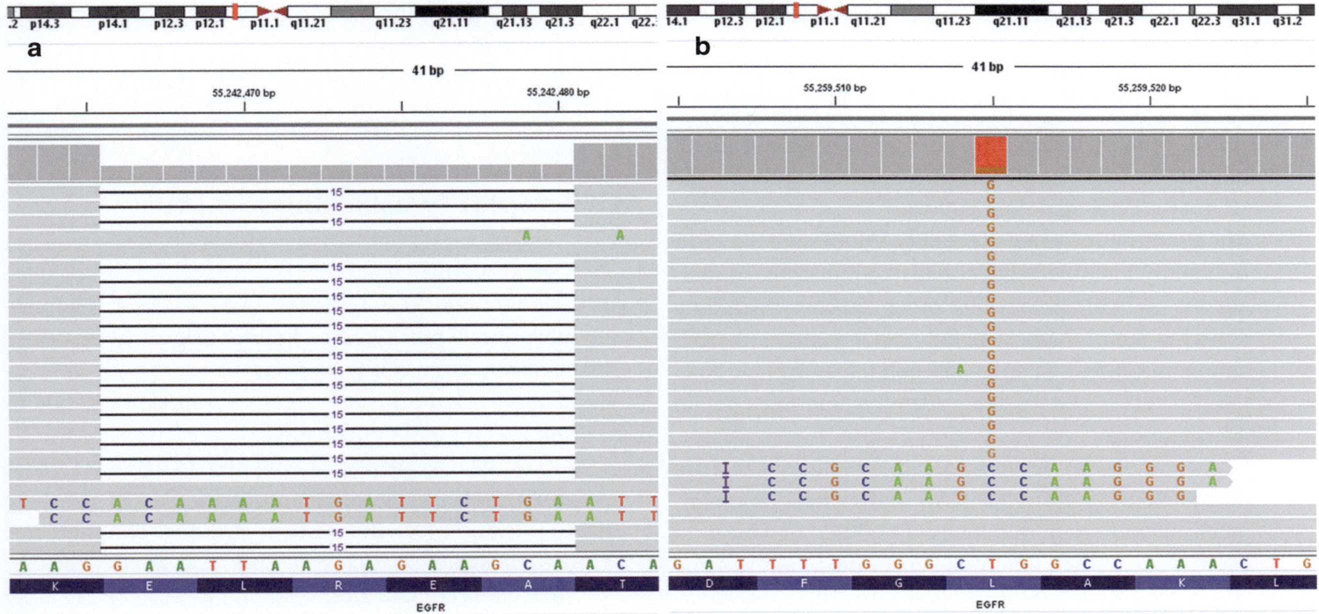

Fig. 9.3 NGS results demonstrated on Integrative Genomics Viewer, showing (**a**) EGFR exon 19 in-frame deletion, and (**b**) EGFR L858R mutation in exon 21

KRAS Proto-Oncogene, GTPase (*KRAS*)

- *KRAS* gene at 12p12
  - One of the earliest discovered oncogenes
- Encoded protein GTPase KRas is a membrane-associated G-protein
  - Also known as Kirsten rat sarcoma 2 viral oncogene homolog
- Approximately 25% of lung adenocarcinomas contain activating *KRAS* mutations, resulting in constitutive activation of the RAS/MAPK signaling pathway
  - Variable missense substitutions in codons 12 and 13 (>90%) and 61 (less common)
  - *KRAS* mutations are more common in patients with a history of tobacco smoking, enriched for G12C mutation
  - In nonsmoker patients with *KRAS* mutations, G12D is most common (>50%), followed by G12V and G12C
  - The most common molecular alterations in invasive mucinous adenocarcinoma and gastrointestinal adenocarcinoma are *KRAS* mutations, mainly G12D and G12V
- The mutated *KRAS* allele is often amplified

Molecular Diagnostics

- Test indications: treatment selection
  - KRAS G12C mutation is targetable by small molecule inhibitors binding specifically to the unique cysteine of G12C, sotorasib and adagrasib, preventing downstream signaling without affecting wild-type KRAS
  - Currently, other *KRAS* mutations are not associated with FDA-approved treatment
    - ○ Various targeted inhibitors are under development
  - Since *KRAS* mutations are usually mutually exclusive of *EGFR* mutations and *ALK* rearrangements, the presence of a *KRAS* mutation in a tumor can be used to exclude a patient from more expensive and time-consuming testing for *EGFR* mutations and *ALK* rearrangements
  - KRAS is downstream of EGFR. The presence of a *KRAS* mutation indicates the patient will be unlikely to respond to EGFR-targeted therapies
  - *KRAS* has very high homology to other *RAS* family genes, including *NRAS* and *HRAS*, of which the mutations in codons 12, 13, and 61 are also oncogenic
    - ○ This must be considered when designing a *KRAS* assay
- Testing methods: PCR-based single gene assay, NGS panel

B-Raf Proto-Oncogene, Serine/Threonine Kinase (*BRAF*)

- *BRAF* gene at 7q34
- Encoded protein serine/threonine-protein kinase B-raf belongs to the RAF family, playing a role in regulating the MAP kinase/ERK signaling pathway
  - Mutations in this gene, especially V600E, are the most frequently identified cancer-causing mutations in melanoma, as well as various other cancers, including non-Hodgkin lymphoma, colorectal cancer, thyroid carcinoma, NSCLC, and hairy cell leukemia
- Approximately 3–4% of lung adenocarcinomas contain *BRAF* mutations
  - Class I (45–50%): Kinase activating mutations that are independent of upstream RAS and signals as monomers, including p.Val600Glu (V600E) mutation in exon 15, other less common substitutions at the same codon like V600K/D/R/M, and rare small insertion-deletion mutations involving the V600 codon
    - Near half of *BRAF*-mutated NSCLC bears BRAF V600E mutation
    - Usually mutually exclusive with other oncogenic driver alterations
  - Class II (25–30%): Kinase activating mutations that are independent of upstream RAS and signals as constitutively active dimers, including mutations clustering at codons 601, 597, 469 and 464
  - Class III (20–25%): Mutations that demonstrate low to absent kinase activity and the activation of downstream signaling are dependent on RAS activity, including mutants scattered throughout hotspots in exons 11 (notably codons 466, 594, and 596) and 15
    - Those mutations appear to heterodimerize with CRAF to trigger ERK signaling and occasionally co-occur with oncogenic mutations in other genes in the RAS family.

Molecular Diagnostics

- Test indications: treatment selection
  - BRAF V600E mutation is targetable by combined administration of small molecule inhibitors of BRAF and MEK, i.e., combination of dabrafenib and trametinib
  - Currently, classes II and III BRAF mutations are not associated with FDA-approved treatment
    - Class II mutations may require dual inhibition of RAF and MEK signaling
    - Class III mutations may be responsive to MEK inhibitors
- Testing methods: IHC, PCR-based single gene assay, NGS panel

ALK Receptor Tyrosine Kinase (*ALK*)

- *ALK* gene at 2p23
- Encoded protein ALK tyrosine kinase receptor is a transmembrane receptor tyrosine kinase, originally described in anaplastic large cell lymphoma
  - Member of the insulin receptor superfamily, but its function is poorly understood
  - Also known as anaplastic lymphoma kinase or CD246
- Approximately 4–5% of adenocarcinomas of the lung have activation of the ALK kinase by chromosomal rearrangements
  - Most cases contain an inversion: inv(2)(p21p23)
    - Amino terminus of *EML4* fused to the entire cytoplasmic portion of *ALK*
  - Other *ALK* rearrangement partners have been reported, including *KIF5B, TFG, KLC1, TPR*
  - t(2;5), associated with *NPM1–ALK* fusion, the characteristic finding in anaplastic large cell lymphoma, has not been reported in lung cancer
  - *ALK* rearrangements are associated with younger age and are more common in males and never or light smokers; ethnic associations are less clear. However, clinical variables are inadequate for the selection of patients for testing or treatment
  - Resistance to ALK TKIs is ascribed to secondary mutations in the ALK tyrosine kinase domain, some of which are analogous to common resistance mutations in *EGFR* and *BCR–ABL*
    - L1196M confers high-level resistance and is analogous to T315I in BCR–ABL and T790M in EGFR
    - Other mutations that affect the crizotinib or ATP-binding sites include 1151Tins, L1152R, C1156Y, F1174L, L1198P, and D1203N
  - *ALK*-rearranged lung adenocarcinomas are associated with solid, micropapillary, and papillary-predominant growth patterns and signet ring or hepatoid cytomorphology

Molecular Diagnostics

- Test indications: treatment selection

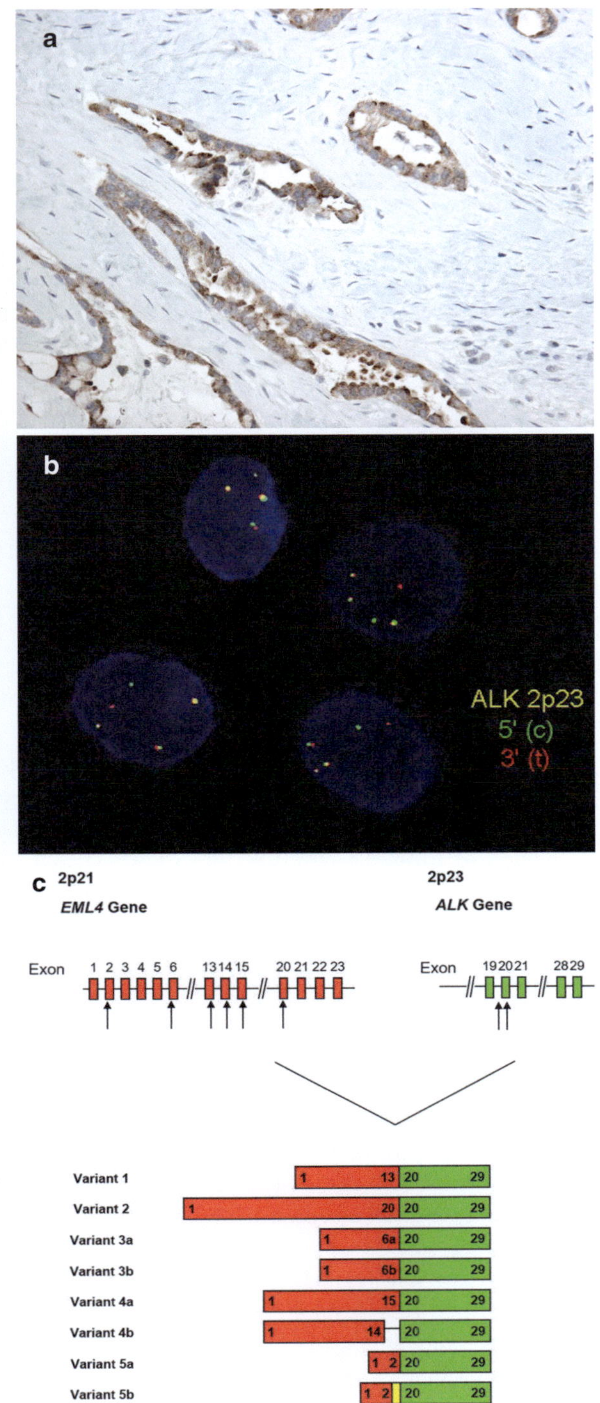

Fig. 9.4 (**a**) Immunohistochemistry for ALK using clone 5A4 in an *ALK*-translocated tumor. The ALK protein is expressed in the cytoplasmic compartment. (**b**) *ALK* FISH using *ALK* breakapart probes. The separation of the *green* and *red* signals indicates the presence of a balanced translocation involving the *ALK* locus and is typical of the *EML4-ALK* fusion. Additional copies of the fused (*yellow*) signal indicate the presence of polysomy at this locus, a common finding both in *ALK*-translocated and *ALK* wild-type tumors. (**c**) Schematic of *EML4-ALK* fusion products, variants 1–5. The breakpoints in *EML4* are highly variable. In most cases, the *ALK* breakpoint occurs in exon 20; however, the breakpoint in variant 5 occurs in intron 19. Variant 4b contains an 11 bp linker between *EML4* and *ALK*

- Patients with *ALK*-rearranged lung cancers respond better to treatment with a targeted inhibitor of the ALK tyrosine kinase, i.e., crizotinib and lorlatinib, than to conventional chemotherapy, with response rates of 50–65% and 82%, respectively, as compared with 20% with chemotherapy
  - Second-generation ALK inhibitors, i.e., ceritinib, alectinib, and brigatinib, have shown efficacy in treating crizotinib-resistant *ALK*-rearranged NSCLC
- Testing methods: IHC, FISH, RT-PCR-based single gene assay, NGS panel
  - IHC (Fig. 9.4a)
    - ○ Standard ALK antibodies used for anaplastic lymphoma are insufficiently sensitive to distinguish *ALK*-rearranged lung cancer from nonrearranged lung cancer without additional signal amplification strategies
  - Newer monoclonal antibodies (Cell Signaling clone D5F3, Novocastra clone 5A4) have been shown to correlate well with FISH
  - FISH (Fig. 9.4b)
    - ○ Split-apart probes to *ALK* enable the detection of *EML4–ALK* rearrangements as well as other less common *ALK* rearrangements
  - RT-PCR-based assay (Fig. 9.4c)
    - ○ At least 13 molecular variants of *EML4–ALK* have been reported involving the fusion of *EML4* exons 2, 6, 13, 14, 15, 17, 18, or 20 to *ALK* exon 20 or intron 19
    - ○ Due to the number of molecular and chromosomal fusion variants, 5′ and 3′, an RNA expression imbalance strategy can be used for detecting less common *ALK* rearrangements
  - NGS sequencing panel
    - ○ Higher sensitivity and easier panel design for RNA NGS panel compared to DNA NGS panel, due to the large number and size of ALK introns needed to capture
    - ○ Targeted sequencing of known *ALK* rearrangements
    - ○ Anchored multiplex PCR (AMP) targeted NGS can be used to detect less common *ALK* rearrangements without knowing the fusion partners

ROS Proto-Oncogene 1, Receptor Tyrosine Kinase (*ROS1*)

- *ROS1* gene at 6q22
- Encoded protein proto-oncogene tyrosine-protein kinase ROS is a type I integral membrane protein with tyrosine kinase activity, which functions as a growth or differentiation factor receptor

- Member of the sevenless subfamily of tyrosine kinase insulin receptor genes, with no known ligand and biological function in humans
- Approximately 3% of adenocarcinomas of the lung have activation of the ROS1 kinase by chromosomal rearrangements
  - The most commonly reported rearrangement partner is *CD74*
  - Various *ROS1* rearrangement partners have been reported, including *SLC34A2, SDC4, EZR, FIG, TPM3, LRIG3, GOPC, KDELR2, CCDC6,* and *MSN*
  - Clinicopathologically, *ROS1* rearrangements are similar to *ALK* rearrangements, which are associated with younger age and never or light smokers

Molecular Diagnostics

- Test indications: treatment selection
  - Patients with *ROS1*-rearranged lung cancers respond to treatment with a targeted inhibitor of the ROS1 tyrosine kinase, crizotinib, and entrectinib, with overall response rates of 66% and 78%, respectively
- Testing methods: IHC, FISH, RT-PCR-based single gene assay, NGS panel (RNA > DNA)
  - Similar to *ALK* rearrangement testing

Ret Proto-Oncogene (*RET*)

- *RET* gene at 10q11
- Encoded protein proto-oncogene tyrosine-protein kinase receptor Ret is a transmembrane receptor tyrosine kinase, binding ligands such as GDNF (glial cell-line derived neurotrophic factor) and other related proteins. Receptor dimerization activates downstream signaling pathways such as RAS/MAPK, PI3K/AKT, and JNK that play a role in cell differentiation, growth, migration, and survival
- Approximately 1–2% of adenocarcinomas of the lung have activation of the RET kinase by chromosomal rearrangements
  - Most common rearrangement partners are *KIF5B, CCDC6,* and *NCOA4*
  - Less common fusion partners include *TRIM33, ZNF477P, ERCC1, HTR4, CLIP1,* and others
  - *RET* rearrangements are also frequently seen in papillary thyroid carcinoma, seen in 10–20% of cases
  - Activating point mutations in *RET* can give rise to the hereditary cancer syndrome known as multiple endocrine neoplasia type 2 (MEN 2); however, not significantly associated with lung cancer

- Similar to *ALK* and *ROS1* rearrangements, *RET* rearrangements are associated with younger age and never or light smokers

Molecular Diagnostics

- Test indications: treatment selection
  - Patients with *RET*-rearranged lung cancers respond to treatment with a targeted inhibitor of the RET tyrosine kinase, i.e., selpercatinib and pralsetinib, with overall response rates of 64% and 57%, respectively
- Testing methods: FISH, RT-PCR-based single gene assay, NGS panel (RNA > DNA)
  - FISH is particularly challenging due to the small separation of probe signals associated with the most common *RET* rearrangements

MET Proto-Oncogene, Receptor Tyrosine Kinase (*MET*)

- *MET* gene at 7q31
- The encoded protein, hepatocyte growth factor receptor, is a receptor tyrosine kinase with ligand hepatocyte growth factor (HGF). Binding induces dimerization and activation of PI3K/AKT/MTOR, RAS/RAF/MEK/ERK, JAK/STAT, SRC, Wnt/β-catenin, and other signaling pathways, which plays a role in cellular survival, embryogenesis, and cellular migration and invasion
  - Also known as proto-oncogene c-MET or tyrosine-protein kinase MET
- Approximately 3–4% of adenocarcinomas of the lung have *MET* exon 14 skipping mutations, leading to loss of a regulatory domain, overactive Met-mediated signaling, and thus cell proliferation and tumor growth (Fig. 9.5)
  - Exon 14 encodes the 47-amino acid juxtamembrane domain of the Met receptor. Y1003, located in exon 14, is the target of E3 ubiquitin-protein ligase CBL and, therefore, a key regulatory region for ubiquitination and further lysosomal degradation, preventing Met from oversignaling
  - Mutations near intron 13/exon 14 junction (intron 13 splice acceptor site) or near exon 14/intron 14 junction (intron 14 splice donor site) disrupt proper splicing during transcription and cause exon 13 to be fused with exon 15 in the mature mRNA, thus so-called exon 14 skipping
  - Exon 14 skipping mutations are commonly single nucleic acid substitutions or small deletions/insertions
- *MET* amplification/polysomy
  - Seen in ~25% of patients with *EGFR* mutations who develop secondary resistance to EGFR TKIs (Fig. 9.6)

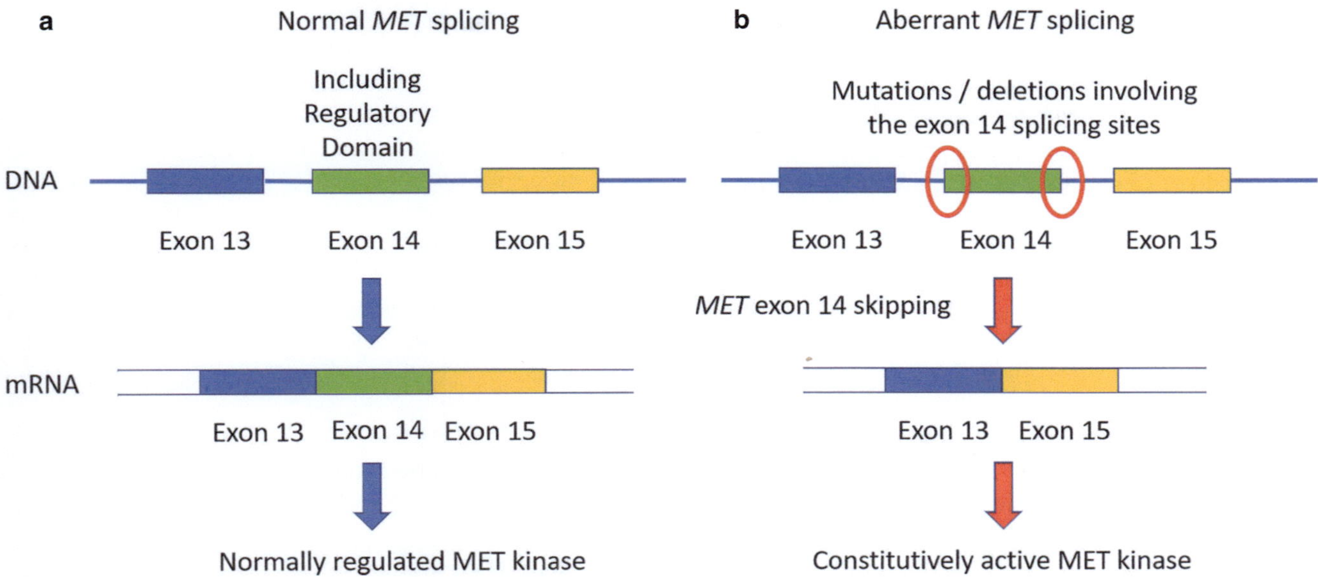

Fig. 9.5 (**a**) Exon 14 of *MET* encodes an E3-ubiquitin ligase c-cbl binding site, responsible for MET kinase regulation and lysosomal degradation. (**b**) Alterations involving exon 14 splicing sites disrupt proper exon 14 slicing, resulting in exon 14 skipping. MET kinase without sequence from exon 14 fails to be regulated, leading to constitutive activation

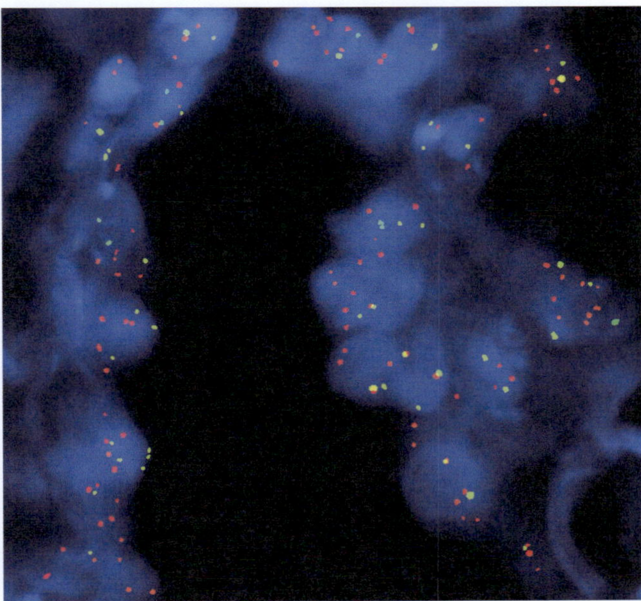

Fig. 9.6 *MET* FISH using *MET* (*red*) and Cep7 (centromere of chromosome 7) control (*green*) probes. This case shows polysomy at both the *MET* and Cep7 loci

Molecular Diagnostics

- Test indications: treatment selection
  - Lung cancer patients with *MET* exon 14 skipping mutation respond to targeted inhibitors of the MET tyrosine kinase, i.e., capmatinib and tepotinib, with overall response rates of 68% and 43%, respectively
  - Clinical trial data suggest that crizotinib, an ALK, ROS1, and MET inhibitor, is effective in treating patients with *MET*-amplified NSCLC
- Testing methods: FISH, RT-PCR-based single gene assay, NGS panel (DNA or RNA)

Neurotrophic Receptor Tyrosine Kinases 1/2/3 (*NTRK1/2/3*)

- *NTRK1* gene at 1q23, *NTRK2* gene at 9q21, *NTRK3* gene at 15q25
- Encoded proteins high-affinity nerve growth factor receptor, BDNF/NT-3 growth factor receptor, and NT-3 growth factor receptor are members of the neurotrophic tyrosine kinase receptor family. These kinases are membrane-bound receptors that activate the downstream MAPK pathway upon neurotrophin binding
  - Also known as neurotrophic tyrosine kinase receptors type 1, 2, and 3
  - Member of the sevenless subfamily of tyrosine kinase insulin receptor genes, with no known ligand and biologic function in humans
- Less than 1% of NSCLC have activation of the NTRK1/2/3 kinases by chromosomal rearrangements
  - A variety of *NTRK* rearrangement partners have been reported, including *ETV6*, *LMNA*, *TPM3*

Molecular Diagnostics

- Test indications: treatment selection
  - Patients with *NTRK*-rearranged solid tumors, regardless of tumor type, respond to treatment with a targeted inhibitor of the NTRK tyrosine kinases, i.e., larotrectinib and entrectinib, with overall response rates of 75% and 57%, respectively
- Testing methods: IHC, FISH, RT-PCR-based single gene assay, NGS panel (RNA > DNA)
  - Similar to *ALK* rearrangement testing

Other Alterations in Lung Adenocarcinoma

Less Common Oncogenic Mutations with Emerging Clinical Significance

- Several other genes have been identified as oncogenic drivers for lung adenocarcinoma. Although no FDA-approved targeted therapy of them for NSCLC yet as of January 2022, there are emerging drug candidates in the pipeline as potential future treatment options
- *ERBB2* (also known as *HER2*)
  - Seen in 2–4% of lung adenocarcinomas
  - Activating mutations (~2%) are commonly insertion mutations in exon 20 and occasionally SNVs in exons 19, 20, and 21
    - Second-generation dual EGFR/HER2 TKIs, e.g., afatinib, dacomitinib, and neratinib, have shown efficacy among patients with *ERBB2*-mutated NSCLC in preclinical studies and clinical reports
    - Trastuzumab was unable to show convincing treatment benefits for *ERBB2*-mutated NSCLC
  - Amplification-mediated *ERBB2* activation (~2%) typically seen in breast cancer is less common in lung cancer
    - Unlike breast and gastric cancers, *ERBB2* amplification in NSCLC does not predict response to current HER2-targeted therapy
  - FDA approved fam-trastuzumab deruxtecan-nxki for adult patients with unresectable or metastatic NSCLC whose tumors have activating HER2 (ERBB2) mutations in August 2022
- *BRCA1/2* loss of function alterations
  - Seen in ~1% of lung adenocarcinomas
  - PARP inhibitors have demonstrated good efficacy for breast, ovarian, and prostate cancers with homologous recombination deficiency (HRD)-positive status, including deleterious *BRCA* mutations
  - Indication of PARP inhibitors in *BRCA* loss NSCLC is uncertain. There are clinical trials ongoing
- *PIK3CA* mutations
  - Seen in ~2–4% of lung adenocarcinomas
  - Most common mutations are in exons 9 and 20, in particular, E545K in exon 9
  - Higher frequency in squamous cell carcinoma (up to 10%) compared to adenocarcinoma
  - More than half concurrent with alterations in other oncogenic driver genes
  - Targeted therapy for *PIK3CA*-mutated breast cancers has shown good efficacy. However, the indication for *PIK3CA*-mutated NSCLC is uncertain
- Other potential targets include *NRAS* mutations, *CDKN2A* loss, *NF1* loss, *TSC1/2* loss, *PTEN* loss, *MAP 2K1* mutations, etc.

Less Common Gene Fusions and Amplifications

- *NRG1*
  - *NRG1* rearrangement fusion has been found in lung-invasive mucinous adenocarcinoma (reported to be 30%), nonmucinous adenocarcinoma (less than 1%), squamous cell carcinoma, as well as breast cancer, pancreatic cancer, cholangiocarcinoma, and others
  - The most common rearrangement partners include *CD74*, *SLC3A2*, *SDC*
  - Preclinical studies are under investigation for targeted agents
- *FGFR1/2*
  - Amplification has been found in ~1% of lung adenocarcinoma
  - Preclinical studies are under investigation for targeted agents
  - Rarely, *FGFR* rearrangement has been reported

Other Tumor Suppressor Gene Mutations

- Other tumor suppressor gene mutations and loss of function alterations are commonly seen in genes like *TP53*, *LKB1/STK11*, and *RB1*
  - Commonly co-occur with or without oncogenic driver alterations
  - Often associated with poor prognosis
  - No effective treatments to date

Molecular Pathology Related to NSCLC Immunotherapy

- Immune checkpoint inhibitors, for example nivolumab (anti-PD-1), pembrolizumab (anti-PD-1), atezolizumab (anti-PD-L1), and ipilimumab (anti-CTLA4), are breakthrough immunotherapies in recent years that have significantly improved survival and quality of life of patients with NSCLC

- Mainstream predictive companion diagnostics are based on PD-L1 expression of tumor cells and/or tumor-infiltrating inflammatory cells by IHC
- The indication of many immune checkpoint inhibitors requires not only PD-L1 expression status but also tumor being negative for *EGFR* or *ALK* genomic aberrations
  - For NSCLC with oncogenic driver alterations, for example, *EGFR* mutation or *ALK* rearrangement, the use of immune checkpoint inhibitors after TKIs, or in combination with TKIs, has been disappointing
  - The efficacy data of frontline immunotherapy in such a setting remains conflicting
- Tumor mutational burden-high (TMB-H) is an indication for immune checkpoint inhibitor therapy (i.e., pembrolizumab) in all solid tumors, regardless of tumor type
  - Although variable between panels, TMB-H is defined as $\geq$10 mutations/megabase (mut/Mb)
    - However, different assays vary widely in their approach to measuring TMB, and standardization programs are underway to harmonize across different methods
  - Lung cancers have a high proportion of TMB-H tumors
    - Both squamous cell carcinoma and small cell carcinoma have more than 40% TMB-H tumors, whereas nonsquamous NSCLC have 35–40%
- Microsatellite instability-high (MSI-H) or mismatch repair deficiency (dMMR) is also indicative for immune checkpoint inhibitor therapy (i.e., pembrolizumab) in all solid tumors, regardless of tumor type
  - Tumors with microsatellite molecular testing that shows variation in microsatellite length in 30% or more of the tested microsatellites are designated as MSI-H
  - Less than 1% of lung cancer is MSI-H or dMMR
  - MSI is associated with hereditary cancer predisposition (Lynch syndrome), but may also be a sporadic finding in cancers
- *STK11/LKB1* alterations are associated with immunotherapy resistance
  - *LKB1/STK11* mutations are common, found in up to 35% of lung adenocarcinomas, but not in squamous cell carcinomas
  - *LKB1/STK11* is commonly inactivated by loss of function mutations, together with deletion/loss of heterozygosity
  - *STK11/LKB1* alterations are the most prevalent genomic driver of primary resistance to PD-1 axis inhibitors in *KRAS*-mutant lung adenocarcinoma
  - In vitro studies suggest that concurrent *LKB1* and *KRAS* mutations may predict response to MEK and mTOR inhibitors in lung adenocarcinomas

Considerations of Molecular Tests for NSCLC Management

- Selection of molecular tests depends on patients' clinical scenarios, availability and type of specimens, accessibility of different testing methods, turnaround time of assays, and financial considerations
- In general, it is preferred to use combined targeted panels that include most drug-targetable alterations rather than perform individual single-gene assays
- Tissue triage, prevention of tissue block exhaustion, and workflow optimization are critical issues for NSCLC management to save material for molecular testing and decrease overall turnaround time
- Cytology specimens (i.e., smears) are not inferior to FFPE specimens for molecular testing
- Cell-free DNA (cfDNA) NGS analysis is a powerful emerging tool for special clinical situations, especially when adequate tissue material is unable to be obtained, or for evaluation of early or occult recurrent disease

Important Molecular Alterations in Other Lung Carcinomas

Squamous Cell Carcinoma (SCC)

- Typically associated with tobacco exposure
- Approximately 30% have *PIK3CA* amplification
  - Response to targeted inhibitors has been investigated in preclinical studies
- Approximately 20% have *FGFR1* amplification
  - Response to targeted inhibitors has been shown in preclinical studies
- Approximately 3–20% have *MET* amplification
  - Response to targeted inhibitors has been shown in preclinical studies
- Approximately 2% have *PIK3CA* mutation
  - Missense substitutions in codons 542, 545, and 1047, as are seen in lung adenocarcinomas and in cancers of other organs
- SCC from never-smokers has a higher probability, compared to smokers, to contain oncogenic mutations like adenocarcinoma, particularly *EGFR* mutations

Small Cell Lung Carcinoma (SCLC)

- Almost exclusively associated with tobacco exposure
- Initially responds to radiation and chemotherapy but quickly relapses and is rapidly fatal

- Most common mutations are in tumor suppressor genes: *TP53*, *RB1*, and *PTEN*
- Studies of recurrent disease following TKI therapy have shown that *EGFR*-mutant adenocarcinomas can recur as small-cell carcinomas and retain the original *EGFR*-activating mutation
- Recent studies suggested small cell carcinoma can be subclassified into four subtypes based on the dominant expression of *ASCL1*, *NEUROD1*, *POU2F3*, or *YAP1/triple negative/inflamed*. SCLC-A and SCLC-N are associated with neuroendocrine characteristics, while SCLC-P and SCLC-Y often lack neuroendocrine markers staining on IHC studies

Genetically Defined Thoracic Tumor Entities

NUT Carcinoma

- Poorly differentiated carcinoma identified by the presence of nuclear protein in testis (*NUTM1*) gene rearrangement
- Also known as NUT midline carcinoma, t(15;19) carcinoma
- Primitive-looking cells with small round blue cell or squamous cell morphology
- Fusion partners: *BRD4* (>75%), *BRD3* (15%), *NSD3* (5%), and others (*ZNF532* and *ZNF592*)
- Bromodomain and extraterminal domain inhibitors (BETi), which interfere with BET proteins, are currently being evaluated in clinical trials

Molecular Diagnostics
- Testing methods: IHC, FISH, NGS panel (RNA > DNA)

Thoracic *SMARCA4*-Deficient Undifferentiated Tumor (*SMARCA4*-UT)

- High-grade malignant neoplasm with undifferentiated or rhabdoid phenotype and biallelic inactivation of *SMARCA4*
- Used to call: *SMARCA4*-deficient thoracic sarcoma, *SMARCA4*-deficient thoracic sarcomatoid tumor
- *SMARCA4* gene encodes BRG1 protein, which is part of the SWI/SNF chromatin-remodeling complex and binds BRCA1
- Transcriptional profiles of *SMARCA4*-UT are distinct from *SMARCA4*-deficient NSCLC but like those of malignant rhabdoid tumor
  - Reported that up to 10% of conventional NSCLCs have *SMARCA4* mutation

Molecular Diagnostics
- Testing methods: IHC, NGS panel (DNA)

Important Molecular Alterations in Mesothelioma

- Frequently mutated genes include *BAP1*, *NF2*, *TP53*, etc
- Biallelic loss of function of *BAP1* (often mutation and concurrent loss of heterozygosity) can be observed in up to 67% of malignant mesothelioma, higher in epithelioid mesothelioma than in sarcomatoid mesothelioma
- Homozygous deletion of *CDKN2A* (p16) can be observed in approximately 60% of malignant mesothelioma
- *MTAP* gene locus is adjacent to *CDKN2A/B*; thus, loss of expression of cytoplasmic MTAP has been used as a surrogate of CDKN2A/B loss
- Loss of *BAP1* and/or loss of *MTAP* on IHC, or homozygous deletion of *CDKN2A* on FISH have been useful clinical tools to differentiate malignant mesothelioma versus reactive mesothelial proliferation. Microarray works as well but is less frequently used
- Germline mutations in the *BAP1* gene have been associated with *BAP1*-tumor predisposition syndrome, which is associated with increased risks of uveal melanoma, cutaneous melanoma, malignant mesothelioma, and renal cell carcinoma

Acknowledgments
- The authors thank Dr. Lynette Sholl, Brigham and Women's Hospital and Harvard Medical School, for her contribution to the preparation of contents in the previous version of this chapter.

Further Reading

https://www.fda.gov/drugs
https://cancer.sanger.ac.uk/cosmic/
http://seer.cancer.gov/statfacts/html/lungb.html
Acquaviva J, Wong R, Charest A. The multifaceted roles of the receptor tyrosine kinase ROS in development and cancer. Biochim Biophys Acta. 2009;1795(1):37–52. https://doi.org/10.1016/j.bbcan.2008.07.006.
Amin M, Gress D, Meyer Vega L, Edge S, editors. AJCC cancer staging manual. 8th ed. New York: Springer; 2018.
Bergethon K, Shaw AT, Ignatius Ou SH, et al. ROS1 rearrangements define a unique molecular class of lung cancers. J Clin Oncol. 2012;30:863–70.
Arbour KC, Jordan E, Kim HR, Dienstag J, Yu HA, Sanchez-Vega F, et al. Effects of co-occurring genomic alterations on outcomes in patients with KRAS-mutant non-small cell lung cancer. Clin Cancer Res. 2018;24(2):334–40. https://doi.org/10.1158/1078-0432.CCR-17-1841.
Baine MK, Hsieh MS, Lai WV, Egger JV, Jungbluth AA, Daneshbod Y, et al. SCLC subtypes defined by ASCL1, NEUROD1, POU2F3, and YAP1: a comprehensive Immunohistochemical and Histopathologic

characterization. J Thorac Oncol. 2020;15(12):1823–35. https://doi.org/10.1016/j.jtho.2020.09.009.

Banno E, Togashi Y, Nakamura Y, Chiba M, Kobayashi Y, Hayashi H, et al. Sensitivities to various epidermal growth factor receptor-tyrosine kinase inhibitors of uncommon epidermal growth factor receptor mutations L861Q and S768I: what is the optimal epidermal growth factor receptor-tyrosine kinase inhibitor? Cancer Sci. 2016;107(8):1134–40. https://doi.org/10.1111/cas.12980.

Bergethon K, Shaw AT, Ou SH, Katayama R, Lovly CM, McDonald NT, et al. ROS1 rearrangements define a unique molecular class of lung cancers. J Clin Oncol. 2012;30(8):863–70. https://doi.org/10.1200/JCO.2011.35.6345.

Bose R, Kavuri SM, Searleman AC, Shen W, Shen D, Koboldt DC, et al. Activating HER2 mutations in HER2 gene amplification negative breast cancer. Cancer Discov. 2013;3(2):224–37. https://doi.org/10.1158/2159-8290.CD-12-0349.

Camidge DR, Otterson GA, Clark JW, Ignatius Ou SH, Weiss J, Ades S, et al. Crizotinib in patients with MET-amplified NSCLC. J Thorac Oncol. 2021;16(6):1017–29. https://doi.org/10.1016/j.jtho.2021.02.010.

Chuang JC, Stehr H, Liang Y, Das M, Huang J, Diehn M, et al. ERBB2-mutated metastatic non-small cell lung cancer: response and resistance to targeted therapies. J Thorac Oncol. 2017;12(5):833–42. https://doi.org/10.1016/j.jtho.2017.01.023.

Cocco E, Scaltriti M, Drilon A. NTRK fusion-positive cancers and TRK inhibitor therapy. Nat Rev. Clin Oncol. 2018;15(12):731–47. https://doi.org/10.1038/s41571-018-0113-0.

Costa DB, Jorge SE, Moran JP, Freed JA, Zerillo JA, Huberman MS, et al. Pulse Afatinib for ERBB2 exon 20 insertion-mutated lung adenocarcinomas. J Thorac Oncol. 2016;11(6):918–23. https://doi.org/10.1016/j.jtho.2016.02.016.

Dagogo-Jack I, Martinez P, Yeap BY, Ambrogio C, Ferris LA, Lydon C, et al. Impact of BRAF mutation class on disease characteristics and clinical outcomes in BRAF-mutant lung cancer. Clin Cancer Res. 2019;25(1):158–65. https://doi.org/10.1158/1078-0432.CCR-18-2062.

De Greve J, Teugels E, Geers C, Decoster L, Galdermans D, De Mey J, et al. Clinical activity of afatinib (BIBW 2992) in patients with lung adenocarcinoma with mutations in the kinase domain of HER2/neu. Lung Cancer. 2012;76(1):123–7. https://doi.org/10.1016/j.lungcan.2012.01.008.

Drilon A, Laetsch TW, Kummar S, DuBois SG, Lassen UN, Demetri GD, et al. Efficacy of larotrectinib in TRK fusion-positive cancers in adults and children. N Engl J Med. 2018;378(8):731–9. https://doi.org/10.1056/NEJMoa1714448.

Drilon A, Oxnard GR, Tan DSW, Loong HHF, Johnson M, Gainor J, et al. Efficacy of selpercatinib in RET fusion-positive non-small-cell lung cancer. N Engl J Med. 2020;383(9):813–24. https://doi.org/10.1056/NEJMoa2005653.

Drusbosky LM, Rodriguez E, Dawar R, Ikpeazu CV. Therapeutic strategies in RET gene rearranged non-small cell lung cancer. J Hematol Oncol. 2021;14(1):50. https://doi.org/10.1186/s13045-021-01063-9.

Dutt A, Ramos AH, Hammerman PS, et al. Inhibitor-sensitive FGFR1 amplification in human nonsmall cell lung cancer. PLoS One. 2011;6:e20351.

Eberhard DA, Johnson BE, Amler LC, Goddard AD, Heldens SL, Herbst RS, et al. Mutations in the epidermal growth factor receptor and in KRAS are predictive and prognostic indicators in patients with non-small-cell lung cancer treated with chemotherapy alone and in combination with erlotinib. J Clin Oncol. 2005;23(25):5900–9. https://doi.org/10.1200/JCO.2005.02.857.

Engelman JA, Zejnullahu K, Mitsudomi T, et al. MET amplification leads to gefitinib resistance in lung cancer by activating ERBB3 signaling. Science. 2007;316:1039–43.

Ettinger DS, Wood DE, Aisner DL, Akerley W, Bauman J, Chirieac LR, et al. Non-small cell lung cancer, version 5.2017, NCCN clinical practice guidelines in oncology. J Natl Compr Cancer Netw. 2017;15(4):504–35. https://doi.org/10.6004/jnccn.2017.0050.

Frampton GM, Ali SM, Rosenzweig M, Chmielecki J, Lu X, Bauer TM, et al. Activation of MET via diverse exon 14 splicing alterations occurs in multiple tumor types and confers clinical sensitivity to MET inhibitors. Cancer Discov. 2015;5(8):850–9. https://doi.org/10.1158/2159-8290.CD-15-0285.

Haratake N, Seto T. NTRK fusion-positive non-small-cell lung cancer: the diagnosis and targeted therapy. Clin Lung Cancer. 2021;22(1):1–5. https://doi.org/10.1016/j.cllc.2020.10.013.

Hong DS, Fakih MG, Strickler JH, Desai J, Durm GA, Shapiro GI, et al. KRAS(G12C) inhibition with sotorasib in advanced solid tumors. N Engl J Med. 2020;383(13):1207–17. https://doi.org/10.1056/NEJMoa1917239.

Huang Y, Wang R, Pan Y, Zhang Y, Li H, Cheng C, et al. Clinical and genetic features of lung squamous cell cancer in never-smokers. Oncotarget. 2016;7(24):35979–88. https://doi.org/10.18632/oncotarget.8745.

Jonna S, Feldman RA, Swensen J, Gatalica Z, Korn WM, Borghaei H, et al. Detection of NRG1 gene fusions in solid tumors. Clin Cancer Res. 2019;25(16):4966–72. https://doi.org/10.1158/1078-0432.CCR-19-0160.

Jordan EJ, Kim HR, Arcila ME, Barron D, Chakravarty D, Gao J, et al. Prospective comprehensive molecular characterization of lung adenocarcinomas for efficient patient matching to approved and emerging therapies. Cancer Discov. 2017;7(6):596–609. https://doi.org/10.1158/2159-8290.CD-16-1337.

Kwak EL, Bang YJ, Camidge DR, et al. Anaplastic lymphoma kinase inhibition in nonsmall-cell lung cancer. N Engl J Med. 2010;363:1693–703.

Lindeman NI, Cagle PT, Aisner DL, Arcila ME, Beasley MB, Bernicker EH, et al. Updated molecular testing guideline for the selection of lung cancer patients for treatment with targeted tyrosine kinase inhibitors: guideline from the College of American Pathologists, the International Association for the Study of Lung Cancer, and the Association for Molecular Pathology. J Thorac Oncol. 2018;13(3):323–58. https://doi.org/10.1016/j.jtho.2017.12.001.

Maemondo M, Inoue A, Kobayashi K, et al. Gefitinib or chemotherapy for nonsmall-cell lung cancer with mutated EGFR. N Engl J Med. 2010;362:2380–8.

McLeer-Florin A, Moro-Sibilot D, Melis A, et al. Dual IHC and FISH testing for ALK gene rearrangement in lung adenocarcinomas in a routine practice: a French study. J Thorac Oncol. 2012;7:348–54.

Mino-Kenudson M, Chirieac LR, Law K, et al. A novel, highly sensitive antibody allows for the routine detection of ALK-rearranged lung adenocarcinomas by standard immunohistochemistry. Clin Cancer Res. 2010;16:1561–71.

Nagano T, Tachihara M, Nishimura Y. Mechanism of resistance to epidermal growth factor receptor-tyrosine kinase inhibitors and a potential treatment strategy. Cell. 2018;7(11):212. https://doi.org/10.3390/cells7110212.

Nishino M, Klepeis VE, Yeap BY, Bergethon K, Morales-Oyarvide V, Dias-Santagata D, et al. Histologic and cytomorphologic features of ALK-rearranged lung adenocarcinomas. Mod Pathol. 2012;25(11):1462–72. https://doi.org/10.1038/modpathol.2012.109.

Ohashi K, Sequist LV, Arcila ME, Lovly CM, Chen X, Rudin CM, et al. Characteristics of lung cancers harboring NRAS mutations. Clin Cancer Res. 2013;19(9):2584–91. https://doi.org/10.1158/1078-0432.CCR-12-3173.

Paik PK, Drilon A, Fan PD, Yu H, Rekhtman N, Ginsberg MS, et al. Response to MET inhibitors in patients with stage IV lung adenocarcinomas harboring MET mutations causing exon 14 skipping.

Cancer Discov. 2015;5(8):842–9. https://doi.org/10.1158/2159-8290.CD-14-1467.

Perez-Moreno P, Brambilla E, Thomas R, Soria JC. Squamous cell carcinoma of the lung: molecular subtypes and therapeutic opportunities. Clin Cancer Res. 2012;18(9):2443–51. https://doi.org/10.1158/1078-0432.CCR-11-2370.

Planchard D, Besse B, Groen HJM, Souquet PJ, Quoix E, Baik CS, et al. Dabrafenib plus trametinib in patients with previously treated BRAF(V600E)-mutant metastatic non-small cell lung cancer: an open-label, multicentre phase 2 trial. Lancet Oncol. 2016;17(7):984–93. https://doi.org/10.1016/S1470-2045(16)30146-2.

Planchard D, Kim TM, Mazieres J, Quoix E, Riely G, Barlesi F, et al. Dabrafenib in patients with BRAF(V600E)-positive advanced non-small-cell lung cancer: a single-arm, multicentre, open-label, phase 2 trial. Lancet Oncol. 2016;17(5):642–50. https://doi.org/10.1016/S1470-2045(16)00077-2.

Qin A, Johnson A, Ross JS, Miller VA, Ali SM, Schrock AB, et al. Detection of known and novel FGFR fusions in non-small cell lung cancer by comprehensive genomic profiling. J Thorac Oncol. 2019;14(1):54–62. https://doi.org/10.1016/j.jtho.2018.09.014.

Ramalingam SS, Vansteenkiste J, Planchard D, Cho BC, Gray JE, Ohe Y, et al. Overall survival with osimertinib in untreated, EGFR-mutated advanced NSCLC. N Engl J Med. 2020;382(1):41–50. https://doi.org/10.1056/NEJMoa1913662.

Rekhtman N, Montecalvo J, Chang JC, Alex D, Ptashkin RN, Ai N, et al. SMARCA4-deficient thoracic Sarcomatoid tumors represent primarily smoking-related undifferentiated carcinomas rather than primary thoracic sarcomas. J Thorac Oncol. 2020;15(2):231–47. https://doi.org/10.1016/j.jtho.2019.10.023.

Rodig SJ, Mino-Kenudson M, Dacic S, et al. Unique clinicopathologic features characterize ALK-rearranged lung adenocarcinoma in the western population. Clin Cancer Res. 2009;15:5216–23.

Rudin CM, Poirier JT, Byers LA, Dive C, Dowlati A, George J, et al. Molecular subtypes of small cell lung cancer: a synthesis of human and mouse model data. Nat Rev Cancer. 2019;19(5):289–97. https://doi.org/10.1038/s41568-019-0133-9.

Sana ME, Quilliam LA, Spitaleri A, Pezzoli L, Marchetti D, Lodrini C, et al. A novel HRAS mutation independently contributes to left ventricular hypertrophy in a family with a known MYH7 mutation. PLoS One. 2016;11(12):e0168501. https://doi.org/10.1371/journal.pone.0168501.

Sequist LV, Waltman BA, Dias-Santagata D, et al. Genotypic and histological evolution of lung cancers acquiring resistance to EGFR inhibitors. Sci Transl Med. 2011;3:75ra26.

Schabath MB, Cote ML. Cancer Progress and priorities: lung cancer. Cancer Epidemiol Biomark Prev. 2019;28(10):1563–79. https://doi.org/10.1158/1055-9965.EPI-19-0221.

Scheffler M, Bos M, Gardizi M, Konig K, Michels S, Fassunke J, et al. PIK3CA mutations in non-small cell lung cancer (NSCLC): genetic heterogeneity, prognostic impact and incidence of prior malignancies. Oncotarget. 2015;6(2):1315–26. https://doi.org/10.18632/oncotarget.2834.

Schmid K, Oehl N, Wrba F, Pirker R, Pirker C, Filipits M. EGFR/KRAS/BRAF mutations in primary lung adenocarcinomas and corresponding locoregional lymph node metastases. Clin Cancer Res. 2009;15(14):4554–60. https://doi.org/10.1158/1078-0432.CCR-09-0089.

Schrock AB, Frampton GM, Suh J, Chalmers ZR, Rosenzweig M, Erlich RL, et al. Characterization of 298 patients with lung cancer harboring MET exon 14 skipping alterations. J Thorac Oncol. 2016;11(9):1493–502. https://doi.org/10.1016/j.jtho.2016.06.004.

Schuler M, Cho BC, Sayehli CM, Navarro A, Soo RA, Richly H, et al. Rogaratinib in patients with advanced cancers selected by FGFR mRNA expression: a phase 1 dose-escalation and dose-expansion study. Lancet Oncol. 2019;20(10):1454–66. https://doi.org/10.1016/S1470-2045(19)30412-7.

Sehgal K, Patell R, Rangachari D, Costa DB. Targeting ROS1 rearrangements in non-small cell lung cancer with crizotinib and other kinase inhibitors. Transl Cancer Res. 2018;7(Suppl 7):S779–86. https://doi.org/10.21037/tcr.2018.08.11.

Shaw AT, Kim DW, Nakagawa K, Seto T, Crino L, Ahn MJ, et al. Crizotinib versus chemotherapy in advanced ALK-positive lung cancer. N Engl J Med. 2013;368(25):2385–94. https://doi.org/10.1056/NEJMoa1214886.

Shaw AT, Yeap BY, Solomon BJ, et al. Effect of crizotinib on overall survival in patients with advanced nonsmall-cell lung cancer harbouring ALK gene rearrangement: a retrospective analysis. Lancet Oncol. 2011;12:1004–12.

Shigematsu H, Takahashi T, Nomura M, Majmudar K, Suzuki M, Lee H, et al. Somatic mutations of the HER2 kinase domain in lung adenocarcinomas. Cancer Res. 2005;65(5):1642–6. https://doi.org/10.1158/0008-5472.CAN-04-4235.

Sholl LM, Xiao Y, Joshi V, et al. EGFR mutation is a better predictor of response to tyrosine kinase inhibitors in nonsmall cell lung carcinoma than FISH, CISH, and immunohistochemistry. Am J Clin Pathol. 2010;133:922–34.

Sholl LM, Yeap BY, Iafrate AJ, et al. Lung adenocarcinoma with EGFR amplification has distinct clinicopathologic and molecular features in never-smokers. Cancer Res. 2009;69:8341–8.

Skoulidis F, Goldberg ME, Greenawalt DM, Hellmann MD, Awad MM, Gainor JF, et al. STK11/LKB1 mutations and PD-1 inhibitor resistance in KRAS-mutant lung adenocarcinoma. Cancer Discov. 2018;8(7):822–35. https://doi.org/10.1158/2159-8290.CD-18-0099.

Suh JH, Johnson A, Albacker L, Wang K, Chmielecki J, Frampton G, et al. Comprehensive genomic profiling facilitates implementation of the National Comprehensive Cancer Network Guidelines for lung cancer biomarker testing and identifies patients who may benefit from enrollment in mechanism-driven clinical trials. Oncologist. 2016;21(6):684–91. https://doi.org/10.1634/theoncologist.2016-0030.

Suzawa K, Toyooka S, Sakaguchi M, Morita M, Yamamoto H, Tomida S, et al. Antitumor effect of afatinib, as a human epidermal growth factor receptor 2-targeted therapy, in lung cancers harboring HER2 oncogene alterations. Cancer Sci. 2016;107(1):45–52. https://doi.org/10.1111/cas.12845.

Tsao AS, Tang XM, Sabloff B, et al. Clinicopathologic characteristics of the EGFR gene mutation in nonsmall cell lung cancer. J Thorac Oncol. 2006;1:231–9.

Turke AB, Zejnullahu K, Wu YL, et al. Preexistence and clonal selection of MET amplification in EGFR mutant NSCLC. Cancer Cell. 2010;17:77–88.

Vaishnavi A, Capelletti M, Le AT, Kako S, Butaney M, Ercan D, et al. Oncogenic and drug-sensitive NTRK1 rearrangements in lung cancer. Nat Med. 2013;19(11):1469–72. https://doi.org/10.1038/nm.3352.

WHO Classification of Tumours Editorial Board. World Health Organization classification of tumours pathology and genetics of tumours of the lung, pleura, thymus and heart. Lyon: IARC Press; 2021.

Yarchoan M, Albacker LA, Hopkins AC, Montesion M, Murugesan K, Vithayathil TT, et al. PD-L1 expression and tumor mutational burden are independent biomarkers in most cancers. JCI Insight. 2019;4(6):e126908. https://doi.org/10.1172/jci.insight.126908.

Yasuda H, Kobayashi S, Costa DB. EGFR exon 20 insertion mutations in non-small-cell lung cancer: preclinical data and clinical implications. Lancet Oncol. 2012;13(1):e23–31. https://doi.org/10.1016/S1470-2045(11)70129-2.

Yasuda H, Park E, Yun CH, Sng NJ, Lucena-Araujo AR, Yeo WL, et al. Structural, biochemical, and clinical characterization of epidermal growth factor receptor (EGFR) exon 20 insertion mutations in lung cancer. Sci Transl Med. 2013;5(216):216ra177. https://doi.org/10.1126/scitranslmed.3007205.

Yu HA, Sima CS, Shen R, Kass S, Gainor J, Shaw A, et al. Prognostic impact of KRAS mutation subtypes in 677 patients with metastatic lung adenocarcinomas. J Thorac Oncol. 2015;10(3):431–7. https://doi.org/10.1097/JTO.0000000000000432.

Molecular Pathology of Breast Tumors

10

Yesim Gökmen-Polar and Sunil S. Badve

Contents

Introduction

- Breast cancer is the most common cancer affecting women, affecting more than 250,000 women each year in the United States alone
- It is one of the first solid cancers where laboratory research has had a large impact on the routine clinical management of patients, ranging from detection to diagnosis and therapy
- Molecular approaches to pathology have had an enormous influence, especially in the areas of diagnosis and therapeutic decision-making
- The topic of molecular pathology in breast cancer is very large and evolving far too rapidly to cover completely in a single chapter
- This chapter will therefore primarily focus on reviewing aspects that are already in routine clinical use, some of the more promising applications on the horizon, and scientific questions that are currently at the forefront of translational research
- From an etiological point of view, the molecular pathology of breast cancer is the result of abnormalities occurring in important normal processes, including the gross, microscopic, and molecular anatomy of the breast, breast development, and adult physiology—which is where we begin

Normal Characteristics of the Female Human Breast

Gross, Microscopic, and Molecular Anatomy

- The size and the extent of the adult female breast varies enormously. On average, it is about 10–12 cm in diameter, 5–8 cm in thickness, and weighs about 700 g. Weight may almost double during pregnancy and lactation. Anatomy books describe the extent from just under the clavicle to the sixth rib. For simplicity, the breast is typically divided into four quadrants (Q): upper outer (UOQ), upper inner (UIQ), lower outer (LOQ), and lower inner (LIQ). Other important regions are the areola/nipple complex and the lymph nodes in the axillary tail extending from the UOQ. Lymphatic (and vascular) drainage is important as the main pathway for breast cancer cells to metastasize. Most regions of the breast, especially the

Y. Gökmen-Polar · S. S. Badve (✉)
Department of Pathology and Laboratory Medicine, Emory
University School of Medicine, Atlanta, GA, USA
e-mail: sbadve@emory.edu

UOQ and LOQ, drain to the axillary nodes, although the LIQ and UIQ also drain to a chain of internal mammary nodes of breast cancer, also beneath the sternum and extending upwards

- There are no internal landmarks or quadrants within the breast and the structure is often likened to an inverted tree with 10–20 main duct opening at the nipple. These ducts as they progress towards the chest wall branch in an irregular manner creating a very complicated network. Although there are no well-demarcated lobes, this term is sometimes used to define regions within the breast. Microscopically, the breast consists of thousands of lobules, which are small grape-like clusters of glands lined by epithelial cells specialized to produce milk. Small ducts join to form larger ducts that eventually exit through the nipple, transmitting milk. The small lobules and smallest ducts connected to them, together often referred to as the terminal duct lobular unit (TDLU)

- The entire normal ductal and lobular system is delineated from the mesenchymal stroma ("connective tissue") by a continuous basement membrane (BM). The lumens of the ducts and lobules are generally lined by two distinct layers of cells; an outer layer directly on top of the BM referred to as a myoepithelial cell (MECs), and an inner layer directly on top of the MECs referred to as a luminal epithelial cells (LECs)—although LECs also have many subtle points of attachment with the BM interspersed with the MECs. Nearly, all LECs typically express large amounts of keratin proteins, particularly CK8, CK18, and CK19. MECs express abundant CK5 and CK6 but are generally negative for keratins found in LECs, and they do not express ER or PR. MECs also typically express several other molecules distinct from LECs, including smooth muscle actin (SMA), calponin, S100, p63, and CD10, which appear to be important in certain specialized normal functions such as contraction of duct lumens to expel milk and to maintain normal cell polarity within ducts. The MECs are thought to actively suppress the invasion of cancer cells. The differential expression patterns of markers in LECs and MECs have been used in the so-called "intrinsic molecular subtypes". In this classification, cancers that express the LEC keratins are referred to as being of the "luminal" subtype while others that express keratins normally associated with MECs are referred to as "basal" subtype breast cancers. As discussed later these phenotypes are transient and the switch from Luminal to Basal phenotype is not uncommon. There is a common misconception that luminal and basal breast cancers evolve from genetically altered LECs and MECs, respectively, partly because of molecular similarities including keratins—which is probably not true. The "stem" cell origin of all breast cancers is far from clear and a topic of much debate and research. A proportion of LECs (10–30%) also express nuclear estrogen receptors (ER) and progesterone receptors (PR). ER and PR are important mediators of growth and differentiation stimulated by the hormones, estrogen and progesterone. Although only a minority of luminal cells express ER/PR, the vast majority (~75%) of cancer cells express these receptors and this constitutes an important pathway for promoting tumor growth

- Recent studies have shown that histologically normal appearing breast epithelial cells may not be normal at the molecular level. Morphologically silent biological abnormalities may predispose the cells to premalignant or malignant transformation. For example, chromosomal gains and losses have been observed in normal breast epithelium. Although the overall frequency of imbalances is quite low, it is significantly higher in normal cells adjacent to cancer cells than normal cells at a distance. Some of these genetic defects may be shared with the adjacent cancer, although the majority are not and appear to be random. Other studies have shown that breast tissue, especially in women at high risk for breast cancer, may contain patches of histologically normal appearing cells in which activity of the p16 tumor suppressor gene is suppressed. Compared to adjacent cells with normal p16 function, these cells show increased proliferation and elevated expression of cyclooxygenase 2 (COX2), and the latter appears to be associated with the development of many types of cancers. There are likely to be many other acquired and inherited molecular abnormalities in otherwise normal appearing cells (Figs. 10.1 and 10.2)

Breast Development

- The molecular mechanisms responsible for human breast development are poorly understood because it is extremely difficult to study directly. Most of what we know is inferred from animal studies, particularly involving genetically engineered mice, where the effect of altering specific genes on breast development can be directly observed. However, it must be noted that mice (unlike humans) do not have a normal TDLU structure and terminal differentiation does not occur till pregnancy. Furthermore, in most of these models, there is a linear progression of normal epithelium to invasive cancer with or without intervening stage of hyperplasia or in situ carcinoma; this obligate progression is not observed in humans. Despite this, there are probably many important parallels in breast development among all mammals, and

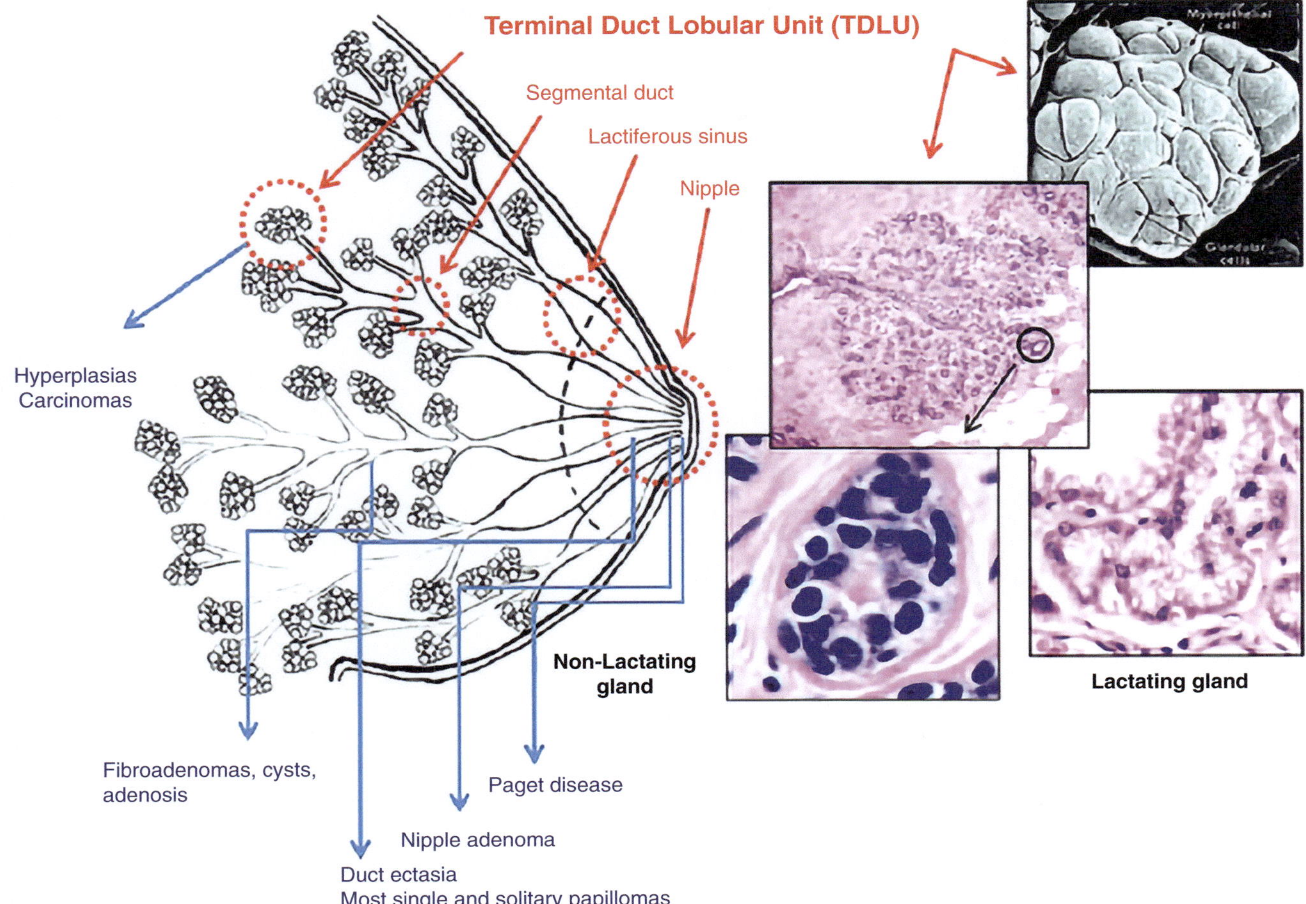

Fig. 10.1 Anatomy of the adult mature human breast. Correlation between compartments and different distinct pathologic processes arising in the breast

studies using mice and other models almost certainly reveal molecular mechanisms shared with humans. Undoubtedly, many normal developmental mechanisms play a central role in the development and progression of breast cancers. For example, cells in the earliest potential precursors of breast cancer, referred to as hyperplasias, demonstrate suppression of molecular pathways involved in adult differentiation, and reactivation of embryonic pathways, which is also true of later stages such as the progression of ductal carcinoma in situ (DCIS) to invasive breast cancer (IBC)

- Mammary glands are derived from ectodermal buds or ingrowths along mammary lines in the embryo. Between 14 and 18 weeks of gestation, distinct mesenchymal and ductal compartments start to develop. By 28 weeks, there are two clearly defined cell compartments (LECs and MECs). The ductal and lobular system continues to develop and mature throughout the second half of gesta-

tion, as well as the areola and nipple. Many genes are known to play critical roles in regulating development. For example, BCL2, which suppresses apoptosis, increases dramatically beginning at about 18 weeks and plays an important role in duct formation by inducing cells in the center of solid cords of primitive epithelial cells to die, forming patent lumens. Ductal budding and branching depend on prolactin which sensitizes cells to the growth-stimulating effects of insulin. Aldosterone promotes the differentiation of buds into ducts and lobules, forming primitive TDLUs. ER is expressed in LECs by the third trimester and PR, 2–3 months after birth. Genetic alterations of these regulatory molecules can play important roles in the development and progression of breast cancer, in general, by promoting "embryonic" growth in an inappropriate setting

- There are no structural or known molecular differences between male and female breasts during the postnatal

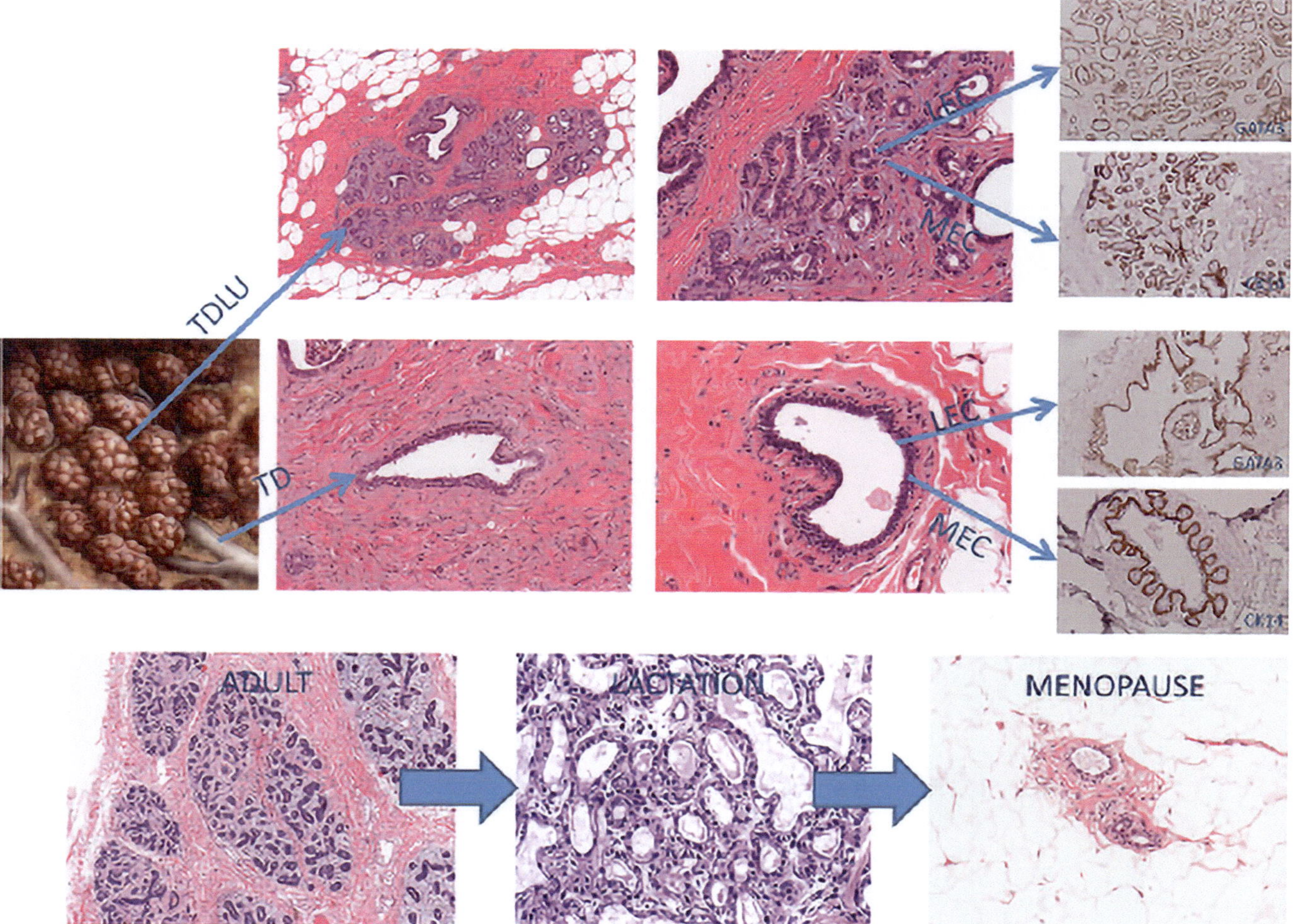

Fig. 10.2 Breast histology. Differences between luminal epithelial cells (LEC) and myoepithelial cell layer (MEC) compartments. GATA 3 is a representative marker of LEC in both TDLU and TD. CK14 is a distinctive marker of MEC. *Lower*: Histological changes associated with lactation and menopause. During lactation, the acini are closely packed, with a reduced amount of stroma; secretory material in the lumens is seen. With menopause, there is a marked reduction of acini and ducts, with replacement by fat

period. At birth, nipple ducts finally open onto the surface. Closely after birth, prolactin, estrogen, and progesterone decrease dramatically, resulting in the involution of newborn breast tissue. During this time, apocrine and cystic changes become prominent, which are also common in post- menopausal breasts. Between 2 years of age and puberty, the breasts are very small, and the main constituents are scattered small ducts/lobules embedded within a dense collagenous stroma. Pubertal changes are characterized by greatly increased growth of stroma, MECs, and LECs, which are prominently caused by increased levels of estrogen, although full differentiation requires other hormones and growth factors as well, including insulin, cortisol, thyroxin, prolactin, and growth hormone. ER is necessary for duct elongation, and ER knockout mice only develop rudimentary ducts without terminal end buds or alveolar buds. Interestingly, these glands are highly resistant to cancer development. PR is necessary for duct elongation and alveolar development, which are lacking in PR-knockout mice

- After menarche, prominent cyclical developmental changes occur with the menstrual cycle. Early on TDLUs develop more alveoli with each successive cycle. From menarche on, the mammary gland is fully anatomically and functionally developed to support pregnancy and lactation. Our group (Ramakrishnan et al.) was among the first to document the histological changes with the menstrual cycle and propose a schema for "dating" breast epithelium

- During pregnancy, the proliferation of essentially all types of cells, especially LECs, dramatically increases, mediated by increasing levels of estrogen, progesterone, ER, and PR. After delivery circulating ER and PR decrease to low levels, in preparation for lactation. Once lactation begins, cell proliferation ceases as the cells ter-

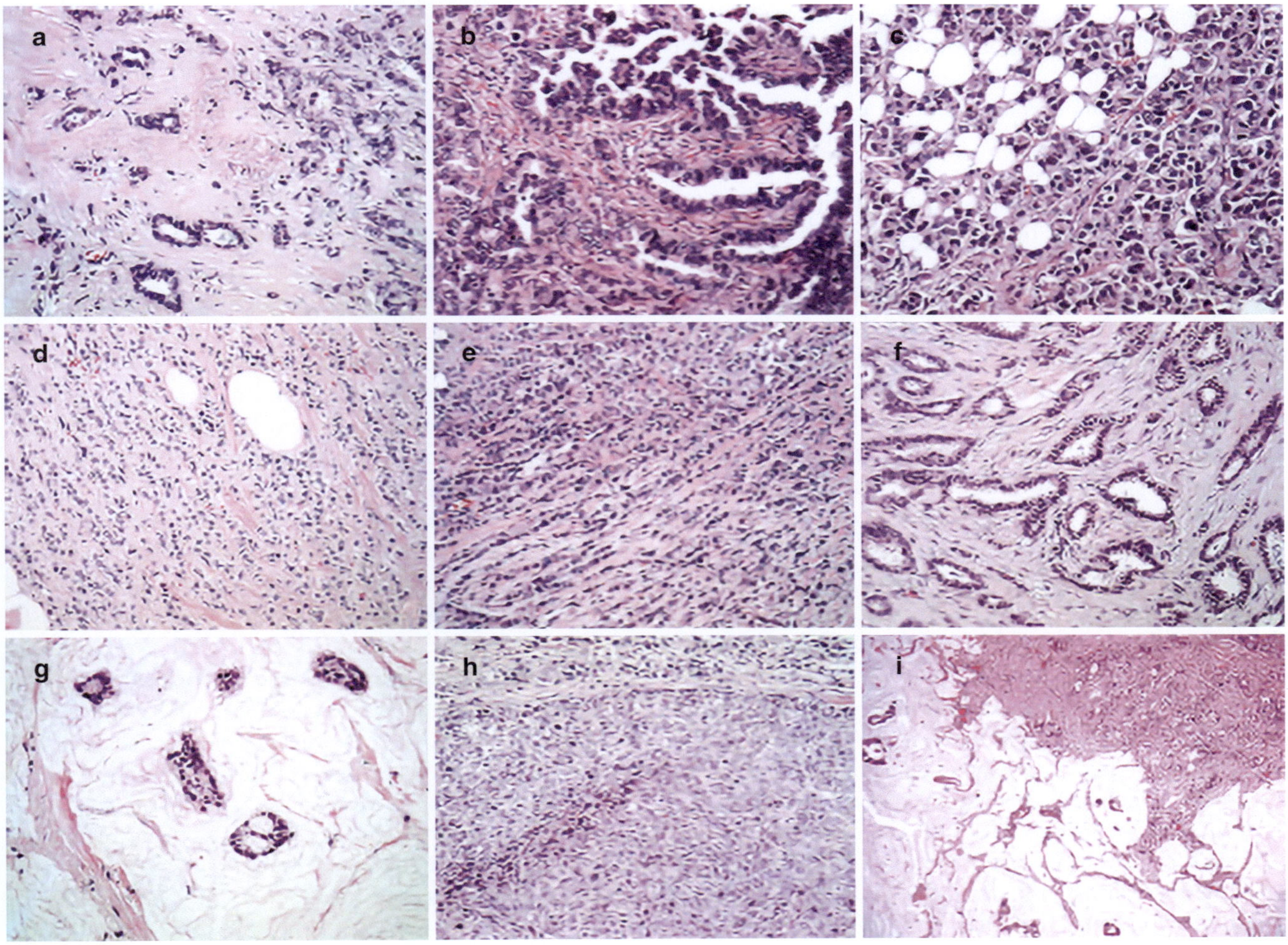

Fig. 10.3 Breast cancer pathology, common histologic subtypes. Grading invasive carcinomas depend on the degree of tubular formation, nuclear features, and mitotic index. Invasive ductal carcinoma of no special type. (**a**) Grade 1; (**b**) Grade 2; (**c**) Grade 3. (**d–i**) common histologic types. (**d, e**) Invasive lobular carcinoma; (**f**) invasive tubular carcinoma; (**g**) invasive mucinous carcinoma; (**h**) invasive medullary carcinoma; (**i**): invasive ductal carcinoma with mucinous features

minally differentiate to produce milk. When lactation ceases, secretory LECs undergo apoptosis, alveoli collapse, and the mammary gland involutes back to the non-pregnant condition, although the ductal system post-pregnancy retains a somewhat more complex ductal framework than prior to pregnancy

- It has been proposed that in the adult female breast, there is a relatively large reserve of normal stem cells which support the dynamic changes in growth and differentiation associated with menstrual cycling, pregnancy, and lactation. Presumably, various genetic alterations of normal stem cells may give rise to precancerous or cancer stem cells, which eventually grow uncontrollably. However, there are probably other sources of cancer stem cells, including the dedifferentiation of mature LECs due to specific mutations

- After menopause, both lobules and ducts are decreased in number. Intralobular stroma is replaced by collagen and the breast stroma undergoes replacement by fat (Fig. 10.3)

Traditional Pathological Classification and Biomarkers in Routine Clinical Practice

- *Histological classification*: Breast cancers are typically graded based on tubule formation, nuclear atypia, and mitotic activity. This grading system is prognostic and also to some extent predictive. Histologically, breast cancer is subclassified into almost 20 different subtypes the commonest being ductal and lobular subtypes. Molecular analysis of some of the uncommon subtypes has revealed specific alterations

- Adenoid cystic carcinomas of the breast show a characteristic translocation t(6;9), which creates a MYB–NFIB fusion transcript
- Secretory carcinomas also have an associated translocation, t(12;15) with the conformation of an ETV6– NTRK3 fusion transcript
- Micropapillary carcinomas have a distinct set of gene clusters on their own, including high-rate FGFR1 amplification
- Metaplastic breast cancers are a mixture of adenocarcinoma with metaplastic elements, homologous (squamous and spindle metaplasia) or heterologous (chondroid, osteoid, and skeletal muscle). They are typically associated with PI3K/AKT mutations and over 90% are HER2 and ER-negative. A mouse model with BRCA1 inactivation and wild-type allele of TP53 shows classical morphologic features of metaplastic carcinomas as well as activation of the WNT pathway

Estrogen Receptor and Progesterone Receptor

- The documentation of the importance of ovarian function in breast cancer dates to at least 1896 when Sir George Beatson reported regression of advanced breast cancer in women who underwent oophorectomy
- It was only later that the ovaries were identified as the source of estrogen and the receptors were identified in normal and diseased breast epithelia. ER controls essential developmental and physiological processes. It interacts with the receptor as estradiol, regulates growth and differentiation, and helps maintain homeostasis. Studies have shown that dysregulation of ER and PR during development are important in carcinogenesis
- The presence of hormone receptors in cancer has led to the measurement of ER and PR as a standard of practice in the evaluation of patients with primary breast cancer. The measurements can be performed accurately on formalin-fixed paraffin-embedded (FFPE) tissue by using immunohistochemistry (IHC) and the results have a good correlation with those of biochemical testing (ligand binding assays) and with quantitative reverse transcriptase polymerase chain reaction (RT-qPCR)
- The effects and actions of estradiol are mediated through interaction with two nuclear receptor proteins, ERalpha (ERa) and ERbeta (ERb), located in chromosomes 6q and 14q, respectively, which are encoded by two separate genes ESR1 and ESR2, respectively. Both ERa and ERb show substantial homology in the DNA binding domain. The role of ERb in breast cancer remains controversial even 20 years after its discovery. This is in part due to the

multiple splice variants which have distinct as well as overlapping expression patterns and functions. Hereafter, ERa will simply be referred to as ER
- The "classical or Genomic" function of ER involves the binding of 17bestradiol to ER located in the cell nucleus. This induces receptor dimerization, which binds to estrogen response elements (EREs) on many other genes, which are then indirectly regulated by estrogen and ERa. ERE-activated genes perform many important functions, including inhibition of apoptosis and stimulation of the cell cycle. There is cross-talk with other mitogenic pathways (RAS, RAF, and CYCLIN D1)
- In addition to the classical pathway there are other functions of ERa, often referred to as "non-genomic" functions. This pathway is responsible for the rapid/early impact of ER stimulation
- The downstream impact of estrogen is modulated by a number of protein regulators known as coactivators and corepressors. These coregulators are responsible for chromatin remodeling and thereby controlling access to EREs and the transcriptional start site (TSS). An additional class of proteins, termed Pioneer factors, also influence which of the 10,000+ targets of ERE will be activated for transcription. FOXA1 is a classic pioneer factor and its expression in breast cancer is of prognostic utility. Histone acetylation, through acetyltransferases, correlates with a more actively transcribed state of chromatin regulation, whereas methylation favors more tightly coiled chromatin, which is less accessible to transcription and less gene expression
- ER status is highly predictive of clinical benefit from endocrine therapy in both adjuvant and metastatic disease settings. ER+ tumors are more likely to respond to hormonal therapy, and have a better prognosis when compared to ER− tumors
  - Harvey et al. showed in a cohort of 1982 patients, using ligand binding assays (LBA) >3 fmol/mg and, retrospectively IHC (Allred Score > 2 or 1–10% weakly positive cells) (Fig. 10.4), showed IHC to be a stronger predictor of disease-free survival (DFS) in patients receiving endocrine therapy when compared to LBA
  - Elledge et al., in a cohort of 205 patients, showed a significant correlation between IHC ER and clinical response in patients with advanced metastatic disease (ER-negative 25%, intermediate 46%, and high 66%)
- Accurate measurements of ER are of considerable importance because it represents one of the strongest predictive factors of responsiveness to endocrine management. In some cases, endocrine therapy alone is an

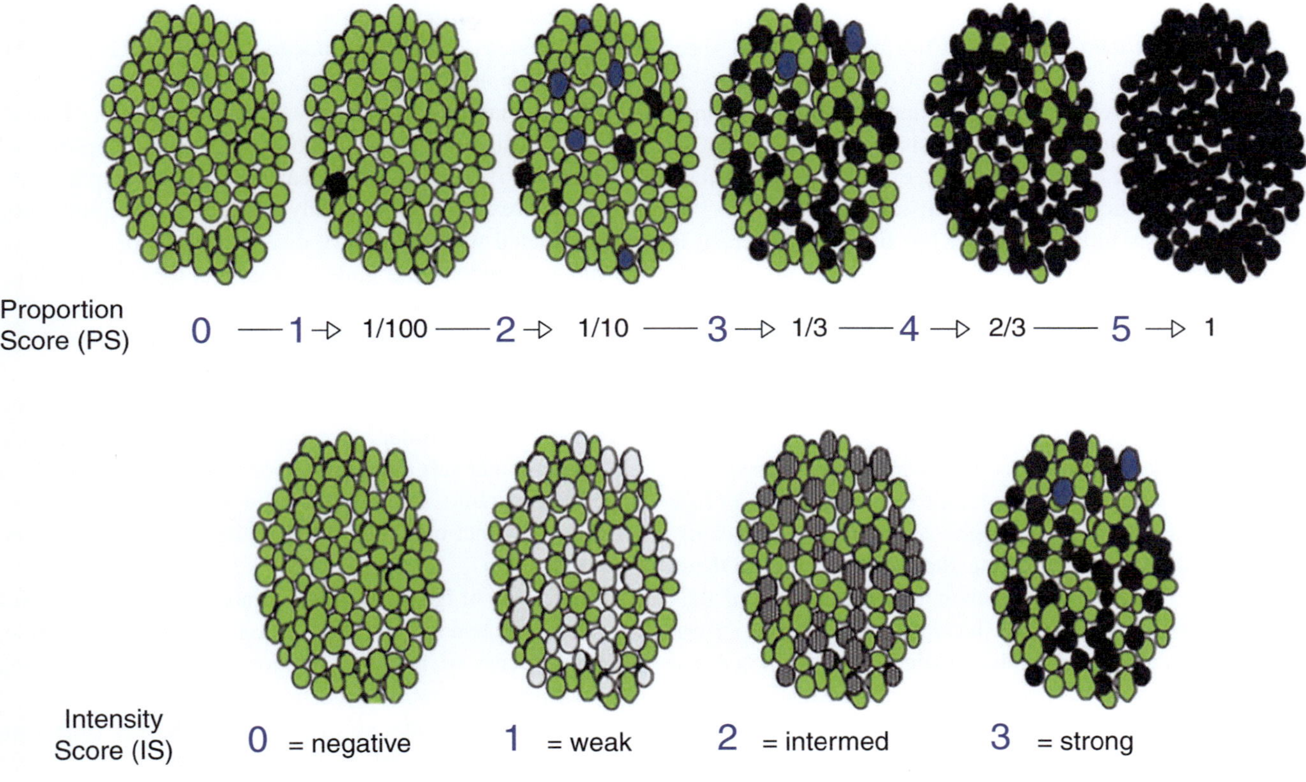

Fig. 10.4 Algorithm for scoring biomarkers (ER, PR) according to recent ASCO guidelines. Allred Score. A combination of a number of cells (proportion score) and intensity of staining (intensity score) is used. (Adapted from Allred et al.)

option, without additional cytotoxic therapy. About 70–80% of breast cancers are ER+ and 20–30% are ER−. Only 70% of ER+ tumors show clinical response to estrogen manipulation but measuring ER expression alone is insufficient to distinguish responders from non-responders. A significant fraction of patients with ER+ disease eventually develop resistance to endocrine therapy

- Clinical progression of ER+ breast cancer typically correlates with hormone resistance. Loss of response and decreased ER expression are associated with a more aggressive clinical course. Epigenetic alterations of the ER promoter, including methylation of the ESR1 gene, are thought to be important events in the development of ER− breast cancers

Progesterone Receptor

- Progesterone has an essential role in regulating breast maturation. A clear role in carcinogenesis has been shown in animal models, particularly with respect to the induction, maintenance, and progression of the neoplastic phe-

notype. Epidemiological studies have documented an increased risk of breast cancer in long-term users of progestin-only containing hormone-replacement therapy (HRT) regimens

- ER is important for regulating PR expression. Colocalization studies show that PR-expressing cells also express ER. In fact, PR expression is regarded as a marker of an intact ER axis. However, discrepancies exist: PR+/ER− phenotype exists in a small (<5%) of breast cancers by IHC, it is even lower by RT-qPCR (<1%). These patients are eligible for hormonal therapy. The relative risk of disease recurrence is higher in patients with ER+/PR− cancers, compared to ER+/PR+ tumors

- About 60% of breast cancers show nuclear expression of PR. Progesterone effects are mediated through the intracellular proteins PRA and PRB in addition to other splice variants. Both are coded from the same gene using two distinct translation initiation sites. The ratio of PRA/PRB appears important, however, most assays are not devised to detect the different isoforms

Ki67 and ER+ Breast Cancer

- ER+ cancers tend to show the full morphological spectrum of breast cancer ranging from bland tumors to extremely aggressive appearing lesions. Proliferation is an important measure of prognosis in these tumors. This can be assessed by mitotic counts, expression of proliferation markers such as Ki67 and Histone H3

- Ki67 analysis in breast cancer has been controversial in part due to the lack of negative control. There are multiple assays and scoring methods including global and hotspot counting methods. The International Ki67 Working Group has identified several key causes of variability and sought to develop methods for improving the consistency between pathologists (Fig. 10.5)

- Recently, the monarchE clinical trial randomized ER+ high-risk patients based on their expression of Ki67 (20% cutoff). Patients with high Ki67 had significantly improved survival following the addition of a CDK4/6 inhibitor (abemaciclib) to their treatment regimen than those with low levels. This has resulted in FDA approval of Ki67 as a companion diagnostic. It must be noted that all patients in the trial benefited from abemaciclib despite their level of Ki67; this has resulted in some researchers questioning the FDA approval

Human Epidermal Growth Factor Receptor 2 Gene

- The recognition of the importance of HER2 in breast cancer treatment has led to its routine analysis in breast cancer. Based on the expression of these three markers, it is common to divide breast cancers in ER + HER2−, HER2+, and triple-negative subtypes

- The human epidermal growth factor receptor 2 gene, more commonly referred to as HER2, is amplified in 15–25% of human breast cancers. HER2 amplification and overexpression are highly correlated, are significantly associated with aggressive disease (i.e., poor prognostic factors), and are the molecular targets for specific therapies, such as trastuzumab

- HER2 is a proto-oncogene located on chromosome 17. It encodes a tyrosine–kinase receptor residing in the surface membrane of breast epithelial cells. It forms complexes with similar proteins (ERBB1, ERBB3, and ERBB4) and acts as receptors for several ligands, such as EGF, heregulin, and amphiregulin. It regulates many normal cell functions, including proliferation, survival, and apoptosis

- Expression of HER2 is seen in almost every breast epithelial cell if the analysis is performed using a frozen section. The process of formalin fixation and tissue processing destroys a number of HER2 molecules on the cell surface; this results in gradation of staining with only cells with millions of copies of HER2 protein being regarded as overexpressing HER2. Alcoholic fixatives can result in larger numbers of molecules being retained and "false positivity" of the tumor and staining even in normal lobules. Similarly, decalcification can result in excessive destruction of the protein and false negativity. The FDA-

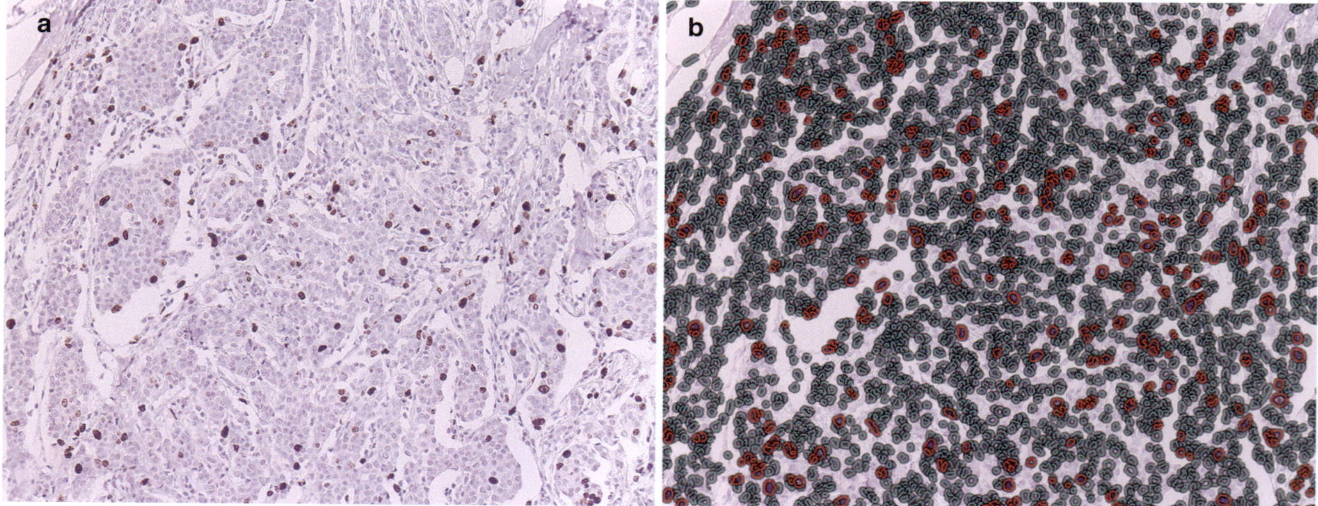

Fig. 10.5 Digital analysis of Ki67 expression using ImageJ-based algorithm. The algorithm counted 11.5% of the cells as positive. (Badve lab. unpublished data)

approved assays therefore require tissues to be fixed in non-alcoholic, neutral buffered formalin and are not approved for decalcified tissues

- The overall relationship between HER2 and clinical outcome is complex and varies with the clinical setting. A weak but significant association between poor outcome and a positive HER2 (overexpression or amplification) in patients receiving no additional therapy after initial surgery is seen. But this only represents a small fraction of patients today. Most patients typically receive some form of adjuvant treatment. Some studies have shown that HER2+ breast cancers are resistant to certain types of cytotoxic chemotherapy (e.g., the combination of cyclophosphamide, methotrexate, and 5-fluorouracil) but sensitive to others (e.g., anthracyclines and taxanes). In general, it is accepted that HER2+ cancers appear to be associated with relative, but not absolute, resistance to endocrine therapies in general. However, this issue remains very controversial. The most promising and useful findings are based on studies showing that HER2+ cancers respond favorably to antibody-based therapies, targeting specifically the HER2 protein, such as trastuzumab. The joint analysis of NCCTG – N9831 and NSABP-B31 clinical trials, which randomized patients with HER2+ cancer to adjuvant chemotherapy +/− trastuzumab, showed a 52% relative improvement (10–15% absolute) in disease-free survival with the monoclonal antibody when combined with chemotherapy
- A long and persistent controversy in the evaluation of the HER2 status by protein expression through IHC, or gene amplification by FISH exists. However, many studies have shown that, when properly performed, a very strong correlation between the two methods exists, and they are equivalent and complementary in the clinical practice
  - Owens et al. observed a similar frequency of HER2 amplified cases by IHC (20%) among 116,736 specimens and FISH (22%) among 6556 specimens
  - Most clinical trials using trastuzumab enroll patients with IHC positive, reflex FISH positive, or ISH alone
- The NSABP B-47 clinical trial conclusively documented that breast cancers having little or no protein overexpression, a normal gene copy number, and do not respond to trastuzumab
- The treatment of HER2-positive disease has dramatically evolved within the last decade with the availability of several new agents. These include additional monoclonal antibodies to different epitopes of HER2 as well as the development of antibody-drug combinations such as T-DM1, which uses trastuzumab as a carrier to target toxic chemotherapeutic agents to the cancer cells. Small molecular inhibitors such as lapatinib and more recently Neratinib and Tucatinib have been developed and their efficacy has been documented in clinical trials. These agents are now FDA-approved and are particularly useful in the treatment of brain metastasis as they can penetrate the blood-brain-barrier

HER2-low

- Novel antibody drug combinations such as trastuzumab deruxtecan (T-DXd) are promising to revolutionize the treatment. This agent combines trastuzumab with a Topo I isomerase inhibitor and has been documented to be extremely potent even in patients that are traditionally HER2–. Efficacy was observed in HER2-low (IHC 2+ or 1+) patients in the Destiny trials. Data from the Destiny-03 trial shows improved OS and DFS in patients; dramatic improvement is also seen in patients with brain metastases. This data was confirmed in the Destiny B04 clinical trial (Modi et al. NEJM 2022). It is possible that T-DXd may have efficacy in patients with even very focal staining (HER2-ultralow) patients (Table 10.1; Fig. 10.6); this is being evaluated in Destiny-06 clinical trial

TNBC

- The definition of TNBC has increasingly become controversial as it is unclear whether patients with very low levels of ER expression should be included in this category. Furthermore, the efficacy of novel agents in the HER2-low category will result in shrinkage of this subgroup
- Lack of Triple biomarkers: Triple-negative breast cancers (TNBCs) are usually high histological grade tumors with high proliferation, necrosis, pushing borders, and lymphocytic infiltrate. This subtype more commonly occurs in younger individuals, often of African-American or Hispanic ancestry. The tumors usually show high-initial response to cytotoxic chemotherapy, although the majority relapses and the overall prognosis is very poor. These features are similar to those seen in tumors of patients with BRCA1 mutation and the BRCA1 pathway is dysfunctional

Table 10.1 Interpretation of HER2 staining pattern and identification of IHC HER2-low category

Staining pattern	Score	Interpretation	Clinical implication for novel therapeutics
No membrane staining is observed	IHC 0	HER2-negative	Ineligible for therapy
Faint, partial staining of the membrane in 10% or less of cancer cells	IHC 0	HER2-negative	Ineligible for therapy
Faint, partial staining of the membrane in greater than 10% of the cancer cells	IHC1+	HER2-low	Eligible
Weak-to-moderate complete staining of the membrane in greater than 10% of the cancer cells	IHC 2+; reflex to HER2 ISH	HER2-low; if FISH negative	Eligible
Weak-to-moderate complete staining of the membrane in greater than 10% of the cancer cells	IHC 2+; reflex to HER2 ISH	HER2-positive; if FISH positive	Ineligible
Intense complete staining of the membrane in greater than 10% of the cancer cells	IHC 3+	HER2-positive	Ineligible

Modified from Modi et al. New Engl J Med 2022

- Lehmann et al. have subclassified TNBCs based on their gene expression profiles into six groups and these have been more recently condensed into four subtypes
  - Basal 1
  - Basal 2
  - Mesenchymal
  - Stem cell-like
- TNBCs have been further classified based on their levels of immune infiltration into inflamed, immune-activated, and immune desert subtypes. Patients with high levels of tumor-infiltrating lymphocytes (TILs) have a good prognosis
  - Loi et al. in the BIG 2–98 study documented the importance of TILs in TNBCs
  - Adams et al. in a large study based on two clinical trials confirmed the importance of stromal TILs in breast cancer
  - A meta-analysis of the data identified 30% as a good cutoff between high and low TILs
- The recognition of TNBC as immunological cancer has led to the application of immune checkpoint inhibitors (ICIs) in the treatment. PD-L1 is a commonly used marker for the selection of patients for ICIs. The expression is assessed using the combined positive score (CPS) or the immune score using IHC methods. The assessment of this marker is difficult and requires training to achieve consistency

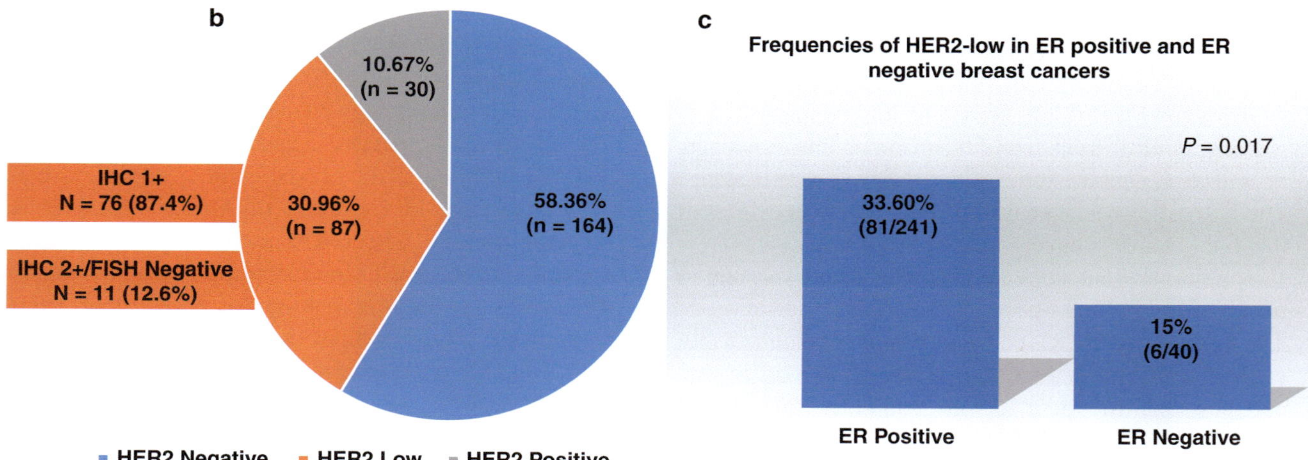

Fig. 10.6 Incidence of HER2-low breast cancers. (**a**) A table showing the case distribution in the study population; (**b**) A Pie chart showing the incidence of HER2-negative, HER2-low, and HER2-positive in the study population; (**c**) incidence of HER2-low breast cancers in estrogen receptor (ER)-positive and ER-negative breast cancers. (From Zhang H et al. Mod Pathol 2022 Feb 19. doi: 10.1038/s41379-022-01019-5)

- The KEYNOTE trials using Pembrolizumab in metastatic and neoadjuvant settings showed improved survival resulting in FDA approval for the agent. The role of PD-L1 analysis in a neoadjuvant context is uncertain as the response does not correlate with expression levels of the marker

Immunohistochemistry

- Immunohistochemistry on FFPE tissue replaced LBAs for testing in the late 1980s. Harvey et al. compared the predictive abilities of LBA and IHC using the 6F11 antibody in a large cohort of patients with newly diagnosed breast cancer. This cohort received a variety of types of adjuvant therapy that ranged from none to endocrine alone, chemotherapy alone, or a combination of the above. Receptor status was scored as the sum of the proportion and average intensity scores of positive staining tumor cells (Allred Score on a scale ranging from 0 to 8). Based on clinical outcomes in patients with adjuvant endocrine therapy, patients with Allred Score > 3 (corresponding to as few as 1–10% positive cells) had a substantially and statistically significant better prognosis than patients with scores less than 3 (<1% positive cells). The predictive ability of IHC was superior to LBA previously performed in the same tumors. Of note, a major limitation of this study was that all specimens analyzed were frozen samples collected for LBA and only later formalin-fixed and processed
- There is no gold standard available for IHC assays of ER and PR. A relevant standard would be any assay with specific preanalytic and analytic components conformed to assays whose results have been validated against clinical benefit from endocrine therapy (clinical validation). Several assays/reagents are available
- ER status can also be determined at the RNA level. Badve et al. compared the levels of the ER/PR protein by IHC and of mRNA by RT-PCR (by Onco*type* DX Assay) and showed a discordance rate of 9% and 12%, respectively when comparing individual genes. The concordance was vastly improved when the combination of ER/PR (i.e., hormone receptor) status was assessed
- A laboratory that performs ER testing should validate its proposed or existing assay against one of the clinically validated assays and demonstrate acceptable concordance. To be considered acceptable, the results of the assay must be initially 90% concordant with those of the clinically validated assay for the ER+ and PR+ categories, and 95% concordant for the ER– or PR– categories
- The cutoff for distinguishing "positive" from "negative" cases should be 1% ER+ positive tumor cells. Patients whose breast tumors show at least 1% ER+ cells are candidates for endocrine therapy and those with less are not.

The percentage of stained tumor cells provides valuable predictive and prognostic information to inform treatment strategies. The cutoff of 1% was chosen based on excellent concordance obtained at this cutoff between (local and central) IHC and RT-qPCR analysis in the E2197 cohort (Badve et al.)

- Several studies have described the relationship between hormone-receptor levels and patient outcomes. Overall survival, DFS, recurrence/relapse-free survival, 5-year survival, time to treatment failure, response to endocrine therapy, and time to recurrence were positively related to ER levels
- PR status provides additional predictive value independent of ER values, especially among premenopausal women. Its predictive value has been demonstrated in retrospective studies using 1% as the cutoff point. Among patients who received adjuvant endocrine therapy, the best cutoff for both DFS ($p = 0.0021$) and OS ($p = 0.0014$) was a total PR Allred Score > 2, which corresponds to greater than 1% of carcinoma cells exhibiting weakly positive staining. In patients with metastatic breast cancer who received first-line endocrine therapy on relapse, a correlation between PR status and response to endocrine therapy was found at a 1% staining threshold ($p = 0.044$) or response to tamoxifen at 10% ($p = 0.021$). Patients with carcinomas >1% PR staining had better survival after relapse ($p = 0.0008$)
- Reporting results for ER and PR: The percentage and proportion of tumor cells staining positively should be recorded and reported. All tumor areas of the tissue section on the slide should be evaluated. This can be achieved manually by counting cells or through image analysis
- The intensity of the staining should be recorded and reported as weak, moderate, or strong. This measurement should represent an estimate of the average staining of the intensity of the positively stained tumor cells on the entire section relative to the intensity of the positive controls run on the same batch. A cutoff of a minimum of 1% of the tumor cells positive for ER/PR for a specimen is POSITIVE
- The current ASCO-CAP guidelines recognize a new category termed ER-low, which contains cases showing between 1% and 10% expression of ER. It is thought that these tumors are more likely to behave like TNBCs. These patients might be eligible for many TNBC clinical trials
- Expression of ER in less than 1% of the tumor cells is NEGATIVE. Such patients do not receive meaningful benefits from endocrine therapy
- Any specimen lacking intrinsic elements (normal breast epithelium) that are negative on ER and/or PR assay should be repeated using another tumor block or another specimen and reported as not interpretable rather than as negative

- "Not interpretable" receptor results refer to samples that did not conform to preanalytic specifications of the guidelines, were processed using procedures that did not conform to guideline specifications of the lab operating procedures, or the assay used to analyze the specimen was not validated and controlled as specified in the guideline. Examples of circumstances leading to not interpretable results include testing of needle biopsies or cytology samples fixed in alcohol, use of fixatives other than 10% NBF, biopsies fixed for intervals shorter than 6 h or longer than 72 h, samples where fixation was delayed more than 1 h, samples with prior decalcification, and samples without internal or external controls

- Negative ER and PR interpretations in tumors that characteristically have an ER+ phenotype (e.g., lobular, tubular, and mucinous carcinomas) should be confirmed by retesting

- ER and PR should be documented in all newly diagnosed breast cancers. Recurrences should also always be tested to exclude prior false negatives and to document changes in biological behavior

- The NSABP B-24 clinical trial compared placebo versus tamoxifen after lumpectomy and radiation. There was a significant reduction (40–50%) in subsequent breast cancer (ipsilateral and contralateral) restricted to patients with DCIS ER+ at 10 years follow-up. This analysis has significant limitations as only a subset of samples was centrally tested. Analysis of the entire cohort with results from local and central labs did not substantiate the lack of benefit in ER– patients. Despite this, it has become routine practice to test DCIS lesions for ER and PR expression

Guidelines for Hormone Receptor and HER2 Testing in Breast Cancer

- To improve the quality of testing for ER/PR and HER2, the American Society of Oncologists (ASCO) and the College of American Pathologists (CAP) jointly developed and recently published guidelines for pathologists to follow. Compliance with the guidelines is now mandatory for laboratories in the US to receive CAP accreditation

ASCO-CAP Guidelines for HER2 Assessment

- ASCO and the CAP jointly developed and published guidelines to improve the quality of HER2 testing and these have been updated several times since the original publication

- A POSITIVE HER2 test is defined as a result of 3+ surface protein expression (defined as uniform intense membrane staining of >30% of invasive tumor cells) or FISH result of amplified HER2 gene copy number (average of >6 copies/nucleus for test systems without internal control probe) or HER2/CEP17 ratio of more than 2.2, where CEP17 is a centromeric probe for chromosome 17 on which the HER2 gene resides

- Originally, FISH testing results were reported as either positive or negative. However, the ASCO-CAP guidelines defined an intermediate range (referred to as an equivocal range) but its clinical significance remains unclear. Much of the confusion using this term comes from the need to define the need for trastuzumab treatment. There is also significant variation in the intermediate (equivocal) ranges for both the IHC and FISH assays. The EQUIVOCAL range for IHC consists of samples scored 2+, which includes up to 15% of samples. An equivocal result (2+) is complete membrane staining that is either nonuniform or weak in intensity but with obvious circumferential distribution in at least 10% of cells. Some, but not all these samples may have HER2 gene amplification and require additional testing to define the true HER2 status. The equivocal range for FISH assays is defined as HER2/CEP17 ratios from 1.8 to 2.2 or average gene copy numbers between 4.0 and 6.0 for systems without an internal control probe. Equivocal results of a single test require additional action, which should be specified in the report. Equivocal results by IHC should follow the confirmatory FISH analysis. Counting additional cells or repeating the test confirms equivocal FISH results. If the results remain equivocal, confirmatory IHC is recommended

- About 3% of patients have ratios of 2.0–2.2 and were previously included in treatment arms with trastuzumab. Retrospective analysis of these patients has shown that they respond to trastuzumab. This called for a revisiting of the older guidelines

- Polysomy 17 is a vague term, seen in up to 8% of tumors. If polysomy 17 is defined as three or more copies of CEP17, most are not associated with protein or mRNA overexpression

- The current ASCO-CAP guidelines recognized five categories of FISH results referred to as categories 1 to 5 based on the total number of copies and HER2/CEP17 ratios

- Discordant results (IHC3+/FISH– or IHC < 3+/FISH+) have been documented in approximately 4% of cases. The significance of this is unclear

- A NEGATIVE HER2 test is defined as either an IHC result of 0 or 1+ for cellular membrane protein expression (no staining or weak, incomplete membrane staining in any proportion of tumor cells), or a FISH result showing HER2/CEP17 ratio of less than 1.8 or an average of fewer than four copies of HER2 gene per nucleus for systems without an internal control probe

- The ASCO/CAP guidelines establish that to classify a test as positive or negative, the laboratory must have performed concordance testing with a validated FISH assay and confirmed that only 5% or less of samples classified as either + or − disagree with the validated assay on an ongoing basis. Equivocal cases are not expected to be 95% concordant, but rather subjected to a confirmatory test
- The ASCO-CAP guidelines have not yet addressed the definition of HER2-low

ASCO-CAP Guidelines for Hormone Receptor Assessment

- The ASCO-CAP guidelines for the assessment of hormone receptors have been published and they recommend optimal fixation in neutral buffered formalin (6–72 h) and processing
- The ASCO-CAP guidelines defined ER positivity as a nuclear expression in greater than 1% of the tumor cells (Fig. 10.4)
  - Badve et al., in a cohort of 776 patients treated in the ECOG randomized trial (E2197) showed that this cut-off correlated with RT-qPCR levels as well as was associated with good concordance between local and central pathology assessments. This is the basis of the current 1% cutoff
- Recently, the guidelines have been modified to identify cases with low levels of expression (1–10%) as ER-low. These patients tend to have more aggressive diseases and tend to have a poor response to endocrine therapy

Molecular Subtypes (Intrinsic Subtypes)

- Perou et al. in 2000 performed the first molecular classification of tumors, not considering the histology but a description of gene expression profiles of different breast tumors to identify molecular portraits (often referred to as intrinsic subtypes) of breast carcinomas. This work described four molecular subtypes: luminal, normal breast-like, HER2, and basal-like. Subsequently, luminals were further subdivided into Luminal A, Luminal B, and Luminal C; the latter is no longer used
- The definition of subtypes has changed over the years as well documented by Weigelt et al. The current definition is based on the PAM50 assay that used 50 genes for the classification. A version of this assay, Prosigna, is commercially available for prognostication but is not FDA-approved for tumor subtyping
- Luminal tumors are thought to be reminiscent of "normal luminal epithelial cells" that express CK8/18+. Lum A is ER+ and enriched with genes associated with active ER pathway, low levels of proliferation-related genes, low histological grade, and generally good prognosis. The Lum B tumors are typically higher grade, with high proliferation indexes, and worse outcomes, and a significant proportion are HER2+. Available data show that Lum A and Lum B tumors cannot be separated based on proliferation
- The normal breast-like subtype has gene expression profiles similar to fibroadenomas and normal breasts enriched in adipose tissue genes. They are relatively poorly characterized, and their prognostic significance is unclear. Whether the normal breast-like group is a true subtype or whether it may be an artifact caused by contamination of samples with normal tissue has been debated
- There is a relatively poor correlation between clinical HER2+ tumors and Her2-enriched subtype. A good number of HER2 amplified, ER+ cancers fall into the Lum B category
- The relationship of intrinsic molecular subtypes to special histological subtypes of breast cancer: Some studies, mainly using microarray-based technology, have shown that at the transcriptional level, tubular, mucinous, and lobular subtypes are more homogeneous than invasive ductal carcinomas or no special type (IDC/NST). Tubular, mucinous, and neuroendocrine carcinomas are typically included in the luminal phenotype. Adenoid cystic, secretory, and metaplastic are considered basal-like cancers
- The basal subtype expresses genes found in normal basal/MECs of the breast, such as CK5, CK14, p-cadherin, caveolins 1–2, CD44, and EGFR. A minority has EGFR amplification. However, unlike MECs, they also express certain proteins characteristic of LECs, such as CK8, CK18, and KIT
- In the neoadjuvant settings, pathologic complete response (pCR) has been used to determine response to chemotherapy. pCR is only seen in 20–30% of patients (with the use of standard anthracycline and taxane-based chemotherapy): Different rates have been shown across the different molecular subtypes with the highest rates observed in Basal-like and HER2 subtypes
- Three additional ER− molecular subtypes have been also described: One, referred to as "molecular apocrine," is similar to the HER2 subtype but shows activation of androgen receptor signaling; another, referred to as "Interferon subtype," is characterized as STAT1; and the third referred to as the "claudin-low" group, typically demonstrate a cancer stem cell-like phenotype
- A recent ECOG clinical trial tried to assess the relevance of basal subtype in TNBCs with reference to therapy. Basal-like tumors did not have an improved response to

cisplatin-based therapy and the trial was closed early due to poor outcomes in the treatment arm (as compared to the control)

- Recently, several studies have questioned whether intrinsic subtyping is reproducible or stable and whether it has any useful clinical significance. More importantly, in the AURORA study, 37% of the patients showed a transition of the intrinsic subtype between the primary and metastatic tumors
- IHC-based subtyping: The use of IHC has been used as a surrogate to microarray analysis to define the intrinsic molecular subtypes: Expression by IHC of ER, PR, and luminal CKs (CK8 and CK18), lack of HER2 overexpression, and low Ki67 are typical of Lum A. Expression of ER, PR, and luminal CKs, and HER2 overexpression are seen in Lum B. Absence of ER and PR, and HER2, and expression of basal CKs (CK5/6) define basal-like tumors. Similarly, the expression of higher levels of Ki67 has been used to categorize tumors as "Lum B"
- It is important to understand that molecular subtypes do not add much additional information of prognostic significance compared to the current standards of histologic subtypes and pathologic grading. Currently, there are no specific therapies for the molecular subtypes. The sub-classification of TNBCs into basal-like tumors is not supported by any clinical data. A clinical trial which randomized patients to receive cisplatinum-based therapy on molecular subtypes had to be stopped due to poor outcomes in the treatment arm

Multigene Prognostic Indices in ER+/ HER– Tumors

Oncotype DX

- Oncotype DX is a prognostic test measuring the RNA expression of 21 genes, which provides a recurrence score (RS; range 0–100) using FFPE tumor samples. The genes include proliferation markers (Ki67, survivin, cyclin D1), invasion-related (MMP11, cathepsin), HER2, ER, PR, and others (GSTM1, CD68, BCL2), as well as five housekeeping genes used to normalize expression overall. The RS quantifies the likelihood of disease recurrence based on studies in women with early-stage hormone estrogen receptor (ER) + only breast cancer and assesses the likely benefit from certain types of chemotherapy. Scores are reported as low (<18), intermediate (18–31), or high (>31) relative to the risk of recurrence
- The assay was developed using the NSABP B-14 and B-20 clinical trials. In these initial studies, most benefits from chemotherapy was observed in patients categorized as "high scores". Similarly, most benefits from endocrine therapy was observed in patients with low and intermedi-

ate scores. The lack of significant benefit of chemotherapy in intermediate group patients was a major concern and resulted in the design of the TAILORx trial (detailed below)

- Initial studies have demonstrated that treatment is modified in 31% of patients who are tested by Oncotype DX, including omission of presumed unnecessary chemotherapy in 22%. Based on these findings, it is estimated that the gene expression assay is likely to result in an overall cost saving, as well as reduced toxicity and quality of life improvements for patients. The test has also shown similar prognostic and predictive significance in women with receptor-positive node-positive received adjuvant treatment with the aromatase inhibitor anastrozole, and in cancer patients receiving neoadjuvant hormonal therapy and chemotherapy
- The TAILORx study is an important phase III clinical trial that was designed to help optimize the use of adjuvant endocrine and chemotherapy in patients with node-negative receptor-positive breast cancer. Based on their recurrence score, women will be assigned to three different treatment groups: women with a recurrence score higher than 25 will receive chemotherapy plus hormonal therapy (the standard of care); women with a recurrence score lower than 11 will receive hormonal therapy alone; and women with a recurrence score of 11–25 will be randomly assigned to receive adjuvant hormonal therapy, with or without chemotherapy. The study is primarily designed to evaluate the effect of chemotherapy on those with a recurrence score of 11–25. Data from the study shows that patients with low and intermediate recurrence scores (<25), which constitute approximately 70% of patients, could safely avoid chemotherapy. A subset analysis shows there may be some benefit in pre-menopausal patients with scores between 20–25
- The RxPonder trial was similarly designed to examine the benefit of adding chemotherapy in patients with node-positive ER+/HER2– breast cancers. Similar to TAILORx, little or no benefit was observed in postmenopausal patients with node-positive breast cancers with intermediate RS scores

The Mammaprint

- This 70-gene prognostic index was validated as clinically useful in studies of younger women with node-negative breast cancer by classifying them into low risk and risk for disease recurrence. Although initially requiring frozen tumor samples, it has been adapted to FFPE samples. Genes involved in the regulation of cell cycle, invasion, and angiogenesis heavily weigh it. Genes of interest do not include known prognostic markers such as ER, PR, and HER2. High-risk patients are most likely to benefit from cytotoxic chemotherapy. In contrast, the low-risk group typically responds very well to endocrine therapy without chemotherapy

- The Microarray in Node-Negative Disease May Avoid Chemotherapy (MINDACT) trial sought to validate the utility of the MammaPrint signature's prognostic value. This trial assessed all patients by the standard clinicopathologic prognostic factors included in the adjuvant setting and by the 70-gene signature assay. If both traditional and molecular assays predicted a high-risk status, the patient received adjuvant cytotoxic chemotherapy and hormonal therapy if ER+. If both assays indicated a low risk, no chemotherapy is given and ER+ patients are given adjuvant hormonal therapy only. When there was discordance between the traditional clinicopathologic prognostic factor prediction of risk and the 70-gene signature prediction of risk, the patients were randomized to receive treatment based on either the genomic or the clinical prediction results. The primary goal of the study was to confirm that breast cancer patients with a "low risk" molecular prognosis by MammaPrint and a "high risk" clinical prognosis can be safely spared chemotherapy without affecting distant metastases-free survival (DMFS). The trial confirmed the clinical utility of the assay
- It is possible that patients with ultra-low scores might require minimal or no adjuvant therapy

Prosigna (PAM50) Assay

- This assay was developed to efficiently determine intrinsic molecular subtypes (discussed later) based on evaluating 50 selected genes using next-generation sequencing and FFPE tissue samples. The test provides a risk of relapse score (ROR) initially based on studies of patients with node-negative breast cancer who did not receive adjuvant systemic therapy
- The ability of ROR to predict prognosis has been confirmed in an independent cohort of patients with ER+ breast cancer. In these studies, ROR was a better predictor than standard clinicopathologic variables, including Ki67, PR, and histological grade

Breast Cancer Index (BCI)

- The Breast Cancer Index (BCI) is an algorithmic gene expression-based signature comprised of two functional biomarker panels, the two-gene ratio, HOXB13/IL17BR (H/I), and the molecular grade index (MGI) that evaluate estrogen signaling and tumor proliferation, respectively. The MGI panel consists of five genes (*BUB1B*, *CENPA*, *NEK2*, *RACGAP1*, and *RRM2*). The BCI test reports both a prognostic as well as a predictive result. MGI and H/I are integrated to generate a prognostic BCI score quantifying both the risk of overall (0–10 years) and late

(5–10 years) distant recurrence. The BCI HOXB13/IL17BR ratio (BCI-H/I) has been shown to predict endocrine therapy in several studies. In the extended endocrine therapy setting, BCI predicted to benefit from an additional 5 years of letrozole after adjuvant tamoxifen in the MA.17 study

- In a recent analysis, the value of BCI-H/I was assessed in the NSABP-B-42 study for the prediction of extended letrozole therapy benefit. The benefit of recurrence-free interval could not be confirmed, although additional follow-up data may be necessary for a definitive analysis

Endopredict

- EndoPredict is an FFPE-based 12-gene assay developed to predict the likelihood of distant recurrence in ER+/HER2. The EndoPredict test also includes the size of cancer and whether cancer is in the lymph nodes when calculating the risk score. EndoPredict test results are given as an EPclin Risk Score, a number between 1.1 and 6.2 that maps to a percentage risk of recurrence as a continuous variable. In addition, scores can be dichotomized into low-risk or high-risk (higher than a 10% risk of recurrence) using a cutoff of 3.3287. EP has also been validated in multiple studies

Concordance between Assays

- The assays are not concordant, and patients identified as high risk by one assay are not necessarily high risk by other assays. This has resulted in some degree of confusion and misuse of terminology. Studies comparing the different assays have been performed and the results vary depending on the cohort of patients used. A definite comment about the superiority of one assay over the others is not possible

Familial and Hereditary Breast Cancer

- Familial and hereditary breast cancer (HBC) are similar in many respects, both are associated with increased incidence of cancer in a family. The major difference is that the root cause of HBC can be located in a specific genetic alteration. The cause of cancer in FBC is uncertain and could be multi-factorial including genetic

BRCA1 and BRCA2

- In 1990, Hall et al. described a linkage-specific site of breast cancer on chromosome 17q. The BRCA1 gene was later cloned. Subsequently, a second gene located in chromosome 13q was cloned, BRCA2. BRCA1 and BRCA2 are the major well-characterized genes contributing to HBC. Patients with the mutation are candidates for prophylactic bilateral mastectomy

- The BRCA Exchange dataset is composed of information from existing clinical databases—the Breast Cancer Information Core, ClinVar, and the Leiden Open Variation Database—as well as population databases and data from clinicians, clinical laboratories, and researchers worldwide. It currently includes more than 20,000 unique BRCA1 and BRCA2 variants. More than 6100 variants in the database have been classified by an expert panel, the Evidence-based Network for the Interpretation of Germline Mutant Alleles, and approximately 3700 of these variants are known to cause disease. In view of a large number of variants, testing should be performed in consultation with physicians and genetic counselors. Most commercial assays test for common alterations in BRCA genes, in the event of a strong family history additional testing of the entire gene can be performed
- Cancers in BRCA1 mutation carriers are characterized by a significantly earlier onset of breast cancer (average, 45, beginning at the age of 20), an excess of bilateralism, a greater frequency of multiple primary cancers (such as breast and ovary), and an autosomal dominant pattern of inheritance. The lifetime risk of breast cancer in BRCA1 mutation carriers is about 85%. Phenotypically, most BRCA1 mutated tumors are basal-like breast cancers: highly proliferative, poorly differentiated, and genomically unstable
- BRCA2 mutated cancers have more variable phenotypes than BRCA1, including a much higher proportion of low histological grade. In BRCA2 cancers, ER/PR expression appears to be similar to non-BRCA cancer—a single study has even shown higher levels. Mutations of the BRCA2 gene are also linked to other types of cancer, including pancreatic, prostate, and melanoma

Non-BRCA Hereditary Breast Cancer

- Non-BRCA HBC represents approximately 50% of cases in the general population. Overall, their clinical pathological features are statistically similar to sporadic breast cancer patients overall, including histological subtypes and grade, proliferation, p53 status, and intrinsic subtypes
- Recent studies have identified a number of other genes such as ATM, CHEK2, PALB2, PTEN, TP53, and among others have been associated with breast cancer. A number of extended panels are now commercially available for genetic testing, including:
  - Limited panels that include up to 15 genes: panels consisting of genes associated primarily with breast cancer (ATM, BARD1, BRCA1, BRCA2, BRIP1, CDH1, CHEK2, FANCC, MRE11A, NBN, NF1, PALB2, PTEN, STK11, and TP53); most of these have established clinical management guidelines
  - Panels containing up to 42 genes including all the above genes plus genes associated with other commonly assessed cancer types (i.e., breast, gynecologic, and gastrointestinal) (APC, AXIN2, BMPR1A, CDKN2A, DICER1, EPCAM, GREM1, KIT, MEN1, MLH1, MSH2, MSH6, MUTYH, PDGFRA, PMS2, POLD1, POLE, RAD51C, RAD51D, SDHA, SDHB, SDHC, SDHD, SMAD4, SMARCA4, TSC1, TSC2, and VHL)
  - Large, comprehensive panels of up to 79 genes including all the genes in groups A and B plus an expanded list of genes for other tumor types (i.e., prostate, sarcoma, and brain)
- Germline mutations of CDH1 (E-cadherin), which are very rare, confer a 40–70% lifetime risk of hereditary diffuse gastric carcinoma, and a 39–52% of ILC. E-cadherin is an adhesion protein, which is lost in sporadic ILC through somatic mutations
- Li–Fraumeni syndrome: Lynch et al. described extended kindred with a broad spectrum of cancers: sarcoma, breast cancer and brain tumors, lung and laryngeal cancers, leukemia, lymphoma, and adrenocortical carcinomas (SBLA syndrome). It is caused by a TP53 germline mutation. The penetrance is variable with two age-specific models: one in childhood and the second in adult life
- Cowden syndrome is a cancer-associated genodermatosis, also referred to as multiple hamartoma syndrome. It has an autosomal dominant pattern of inheritance and is associated with distinctive mucocutaneous lesions and cancer of the breast, thyroid, and female genitourinary tract
- Germline mutations of the PTEN gene (also seen in Bannayan–Riley–Ruvalcaba syndrome). Cutaneous manifestations include trichilemmomas, which are pathognomonic. Also, multiple facial papules, acral and palmoplantar keratosis, skin tags, and lipomas. Merkel cell carcinoma can occur

Familial Breast Cancer

- Familial breast cancer (FBC) is described as breast cancer within a family history of one or more first or second-degree relatives affected. A patient with one or more first-degree relatives with breast cancer in this category has a substantial excess lifetime risk of breast cancer when compared to patients in the general population. The relative risk increases with one, two, and three first degree relatives compared to women without affected pedigree. There may be a combination of genetic and nongenetic (i.e., environmental) factors that contributed to the development of cancers within a family. In such instances, where an alteration in a single major gene is not likely or is not identified, individuals may still face elevated risks of cancer

Next-Generation Sequencing Assays Molecular Breast Pathology to Advance Personalized Treatment of Cancer

- Whole genome sequencing (WGS): The use of rapidly evolving techniques that combines whole genome, deep generation sequencing, and next-generation sequencing have provided novel insights into the understanding of mutational analysis in breast cancer. The Cancer Genome Atlas (TCGA) has identified the mutational profile of more than 1000 breast cancer patients. Unfortunately, apart from p53 and PI3KCA, most mutations are uncommon (in the range of 5% or less)
- In a study of 560 patients, Nik-Zainal et al. identified an average of 6214 single base pair substitutions, 665 small insertions and deletions, and 140 structural variants per cancer
- One of the next major challenges in breast cancer research will be to determine which of the mutations are the "drivers" for developing breast cancer

p53 Mutations
- TP53 is mutated in up to 30% of sporadic breast cancers, and in the vast majority of basal/TNBCs. The gene is located on chr 17 and encodes a nuclear transcription factor normally involved in cellular pathways activated in response to stress by inhibiting the proliferation and inducing apoptosis, of cells damaged in a variety of ways. p53 acts as a transcriptional activator of genes involved in the inhibition of the cell cycle, blood vessel formation, stimulation of apoptosis, and promotion of DNA repair. TP53 mutations can affect any region of the gene and there are no "hotspots". The mutations can result in gene inactivation or uncommonly in increased function. About 75% are single nucleotide substitutions leading to the substitution of a single amino acid, and the remaining 25% are insertions, deletions, and nonsense mutations. Mutations in one allele are associated with inactivation of the other one by loss of heterozygosity (LOH) in most affected breast cancers. Mutation of the gene often results in stabilization of the protein and consequent detection by nuclear p53 expression by IHC. However, lack of expression by IHC does not rule out a mutation. Somatic mutations of TP53 occur in IBCs and DCIS. In both settings, they are associated with increased tumor size and grade, as well as axillary metastasis. p53 status is a strongly unfavorable prognostic factor for relapse-free survival and overall survival only in a triple-negative group in patients treated with adjuvant anthracycline-containing chemotherapy. A relationship between p53 status and response to radiation has been suggested. Currently, there are no specific therapies for p53 mutant tumors

PI3KCA Mutations
- These are common in ER+ breast cancer and most alterations are seen in limited areas of the genome ("hotspots"). The mutations have been targeted for therapeutics. After several attempts, a successful, relatively non-toxic, agent has been developed. The SOLAR-1 phase III randomized clinical trial documented that Alpelisib (PI3Kalpha inhibitor) prolonged progression-free survival among patients with PIK3CA-mutated, HR-positive, HER2-negative advanced breast cancer who had received endocrine therapy previously

ESR1 Mutations
- ERα is mutated in a low percentage (~5%) of primary breast cancers. Suzanne Fuqua's group initially described mutations in S47T, K531E, and Y537N of ESR1 and provided basic science data documenting their role in ER resistance. Mutations lead to constitutive activity and increased binding to co-activator proteins
- ESR1 mutations in circulating tumor DNA (ctDNA) also predicted the disease progression of patients treated with AIs better than the measurement of cancer antigens or total levels of cell-free DNA. A recent trial showed that switching therapy from AI to fulvestrant upon detection of ctDNA ESR1 mutations was associated with clinical benefit
- ESR1 Mutations can be detected in up to 25–30% of patients with metastatic breast cancer. Recent results from the BioItaLEE trial suggest that tracking circulating tumor DNA (ctDNA) is useful in the detection of the acquisition of mutations. Furthermore, switching therapy from AI to SERM or SERD might prove beneficial

Hormonally Directed Therapeutics

- Targeting the ER pathway has been successfully attempted in multiple different ways that include decreasing estrogen synthesis, anti-receptor agents as well as downstream targeting
- Anti-receptor agents have dominated the field for the past few decades and have been the standard of care
  - Tamoxifen is a partial agonist (both antagonistic and agonistic effects) of the ER receptor and induces dimerization and nuclear translocation and is designated as a selective ER modulator (SERM). One of the minor concerns of the drug is the increased risk of thromboembolism and endometrial cancer. In spite of these, Tamoxifen is the most commonly prescribed in pre-menopausal women. The ATLAS and ATOM trials have documented that 10 years of therapy might be superior to 5 years of endocrine therapy

- Fulvestrant directly binds to ER monomers, inhibits dimerization, and suppresses activation, thereby functioning as a pure antiestrogen (selective estrogen receptor downregulator; SERD). Although there were some concerns regarding its activity, the recent usage of a 500 mg dose has unequivocally established its role in the treatment in both metastatic and adjuvant settings. Recent phase 3 clinical trials have documented the clinical benefit of newer orally active SERDs such as Elacestrant (Fulvestrant needs to be injected). A number of other agents including PROTACs are in advanced phase 2 or phase 3 clinical agents
- Aromatase Inhibitors (AIs). In the last decade, prospective randomized clinical trials have shown the superiority of AIs over tamoxifen in postmenopausal receptor-positive women
  - Anastrozole, letrozole, and exemestane are aromatase inhibitors (AI) which block the conversion of adrenally produced precursor compounds to estrogenic molecules. Recent trials also showed the benefits of estrogen deprivation persist for many years even after completion of the initial hormonal therapy in reducing both unilateral and contralateral breast cancers
- Suppression of Ovarian function. Another approach to decreasing hormonal levels is to suppress ovarian function using pharmacological agents. The SOFT and TEXT trials have tested this approach using Triptorelin in addition to Tamoxifen or AIs (exemestane) and documented the utility of this approach. At a median follow-up of 12 or 13 years, there continues to be a reduction in recurrences and deaths in both pre-menopausal and postmenopausal women getting ovarian suppression (triptorelin)
- A number of downstream pathways have also been effectively targeted in ER+ tumors. These include the PI3 kinase pathway, AKT pathway, mTOR inhibitors as well as epigenetic targeting using histone deacetylating agents
- Recent studies with anti-HER2 agents have shown activity in patients with low levels of HER2 expression (2+ or 1+ by IHC; FISH not amplified). Trastuzumab deruxtecan (T-DXd; DS8201), an antibody drug conjugate has been to dramatically improve progression-free survival and overall survival in HER2-low patients. It appears that about 50% of ER+ patients may be candidates for this ADC. Further details regarding the utility of T-DXd in the adjuvant setting is being studied in Destiny 04 and 06 trials, which are currently enrolling patients

Recent Developments in the Breast Cancer Landscape that Revolutionize the Treatment Strategies

- The landscape of breast cancer has been primarily determined by the therapeutic targets and their agents that are available. In the early days, only ER was targetable, and the classification was binary ER+ and ER−. The recognition of HER2 as a target for therapeutics resulted in the current three classes (ER+, HER2+, and TNBCs). However, newer developments are promising to alter the status quo

CDK4/6 Inhibitors in Breast Cancer-Co-Targeting CDk4/6 with Endocrine Therapy

- Based on the knowledge that proliferation is essential for tumor growth, targeting the cell cycle has been attempted in many forms including cell cycle proteins, kinases, and mitotic spindle proteins. Targeting of cell cycle kinases, more specifically, cyclin-dependent kinases (CDKs) has been successful in breast cancer. CDKs are a family of at least 20 kinases and targeting CDK4/6 using inhibitors has led to meaningful improvement in progression-free survival in patients with metastatic breast cancer. At least three drugs (palbociclib, ribociclib, and abemaciclib) are approved for the therapy of metastatic breast cancer. Recent results have shown that Abemaciclib also has potent activity in stage breast cancer while palbociclib seems to have relatively weak activity. The reasons for the differences are not clear but could be related to the structural differences of the molecules as well as their potency to inhibit other CDKs

Landscape of HER2-low Metastatic Breast Cancer and Antibody Drug Conjugates (ADCs) for HER2+ Disease

- HER2 has long been the target for therapeutics in breast cancer. Trastuzumab-based therapies have required the expression of high levels of HER2 (3+) or FISH amplification for the therapy to be effective. However, a novel antibody-drug conjugate (Trastuzumab deruxtecan) is effective in patients with lower levels (1+ and 2+) of HER2 expression. HER2-low patients constitute approximately 50% of ER+ breast cancer and 25–30% of TNBCs. These successes promise to make HER2+ breast cancer the most common subtype of breast cancer
- HER3 has also been recently successfully targeted for therapeutics using an antibody-drug conjugate. The response to the agent was independent of the levels of HER3 expression

The Immune Landscape of Breast Cancer and Immune Checkpoint Inhibitors

- Pioneering work done by our group (Adams et al.), amongst others, has documented the important prognostic role of tumor-infiltrating lymphocytes in TNBCs. The recognition of the immune nature of breast cancer has resulted in the application of immune checkpoint inhibitors in breast cancer; this has been predominantly in TNBC but also in HER2+ and ER+ cancers

- Significant successes have been reported both in metastatic and neoadjuvant settings with the use of anti-PD-1/PD-L1 directed therapies
- Pembrolizumab is the major drug being used for immune therapy either alone or in combination with traditional chemotherapy; there seems to be only weak single-agent activity. The KEYNOTE 355 and 522 clinical trials established the utility of Pembrolizumab in metastatic and early stage triple-negative tumors. The role of testing for PD-L1 in breast cancer is unclear as the response in the neo-adjuvant setting seems to be not related to the expression of PD-L1. Testing if performed should be undertaken using the Agilent PharmDx kit, which contains the 22C3 antibody (Figs. 10.7 and 10.8)
- Atezolizumab had received accelerated approval based on the results of ImPassion130 trial, but the manufacturer has withdrawn the indication in breast cancer based on the results of ImPassion131 clinical trial

Clinical Utility of Trop-2 in Breast Cancer

- Trop-2 is expressed in all breast cancer subtypes, including TNBC, and has been linked with poor prognosis and decreased survival. TNBC has been shown to have high membrane expression of Trop-2, with up to 88% of primary and metastatic tumors having moderate to strong Trop-2 staining. Anti-Trop 2 antibody-drug conjugate, Sacituzumab govitecan, has demonstrated activity in TNBC resulting in FDA approval. Recent analyses seem to suggest that response to sacituzumab may not be relative to the levels of Trop-2 expression. Recent data from the TROPiCS-02 (ASCO 2022) seem to suggest that this agent has activity even in ER+ disease. Furthermore, another antibody-drug conjugate (datopotamab deruxtecan) directed against the same target has also been found to be effective in breast cancer

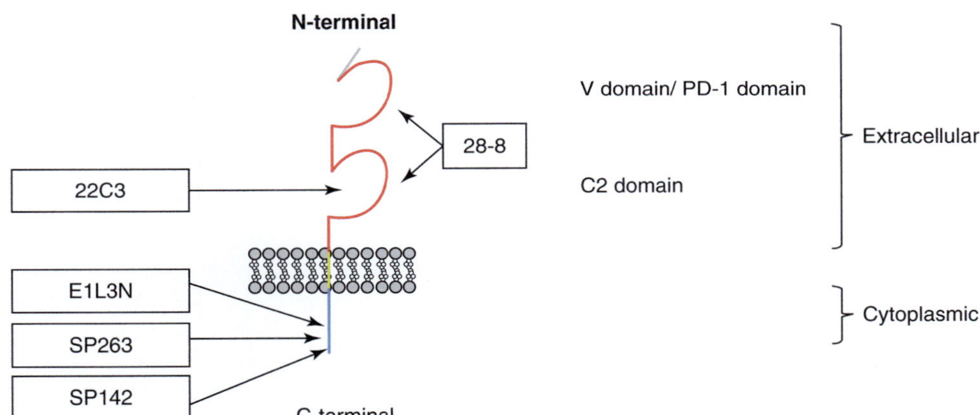

Fig. 10.7 SP263 and SP142 bind identical epitopes in the cytoplasmic domain of PD-L1. E1L3N also binds an epitope in the cytoplasmic domain; however, it is different from SP263 and SP142. 22C3 and 28-8 bind epitopes on distinct surfaces of the extracellular domain of PD-L1, which are different from one another. Mutated residues are in red (ECD) and green (CD). CD cytoplasmic domain, CLIPS Chemical Linkage of Peptides onto Scaffolds, ECD extracellular domain, HDX-MS hydrogen-deuterium exchange mass spectrometry, Ig immunoglobulin, PD-1 programmed cell death-1, PD-L1 programmed cell death ligand-1, SP signal peptide, TM transmembrane domain. (From Lawson et al. Mod Pathol 2020;33:518–530)

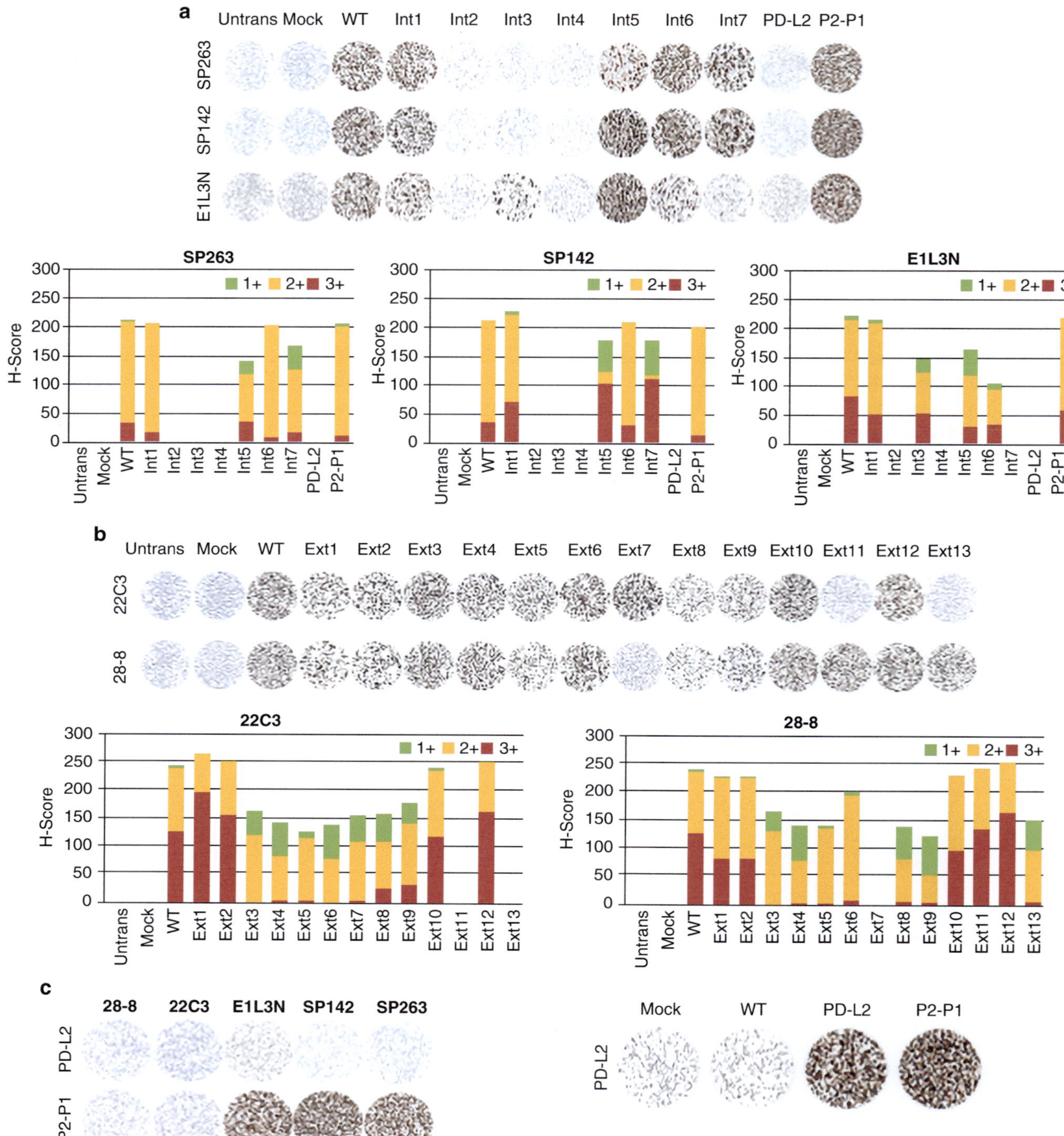

Fig. 10.8 Immunohistochemical staining using PD-L1 antibodies. *H*-scores and representative images of Expi293F cell lines transfected with (**a**) PD-L1 WT, Int1, Int2, Int3, Int4, Int5, Int6, Int7, WT PD-L2, and PD-L2-PD-L1 fusion constructs stained with SP263 and SP142 assays and E1L3N laboratory-developed test. (**b**) *H*-scores and representative images of Expi293F cell lines transfected with PD-L1 WT, Ext1, Ext2, Ext3, Ext4, Ext5, Ext6, Ext7, Ext8, Ext9, Ext10, Ext11, Ext12, and Ext13 constructs stained with 22C3 and 28-8 assays. The proportion of total *H*-score contributed by cells staining at each intensity is represented by red (3+), orange (2+), and green (1+). (**C**) Immunohistochemical staining of PD-L2 WT and P2-P1 transfected cell lines with SP263, SP142, E1L3N, 22C3, 28-8, and PD-L2 antibodies. PD-L1 programmed cell death ligand-1, PD-L2 programmed cell death ligand-2, P2-P1 construct containing the extracellular domain of PD-L2 fused to the last 27 residues of the PD-L1 cytoplasmic domain, untrans untransfected, WT wild-type. (From Lawson et al. Mod Pathol 2020;33:518–530)

Summary and Conclusions

- Our understanding of the molecular biology of breast cancer continues to evolve and this will undoubtedly affect the strategies employed to prevent and manage the disease
- The efficacy of "old" targeted therapies is being improved by the development of novel agents such as oral SERDs against ER+ cancers and antibody-drug conjugates (ADCs) for HER2+ disease
- The scope of these targets is also being expanded as documented by the efficacy of ADCs in HER2-low tumors. Novel targets such as Trop-2 are being identified and targeted
- The precision medicine approaches promise to identify specific mutations that can be used for targeting cancers as well as monitoring minimal residual disease
- The landscape of breast cancer is truly changing, and this is being reflected in the increasing lifespan of the patients
- A major area in urgent need of improvement is to identify causes of "late" recurrence in breast cancer and modalities to treat it

Further Reading

Abd El-Rehim DM, Pinder SE, Paish CE, et al. Expression of luminal and basal cytokeratins in human breast carcinoma. J Pathol. 2004;203:661–71.

Adams S, Gray RJ, Demaria S, et al. Prognostic value of tumor-infiltrating lymphocytes in triple-negative breast cancers from two phase III randomized adjuvant breast cancer trials: ECOG 2197 and ECOG 1199. J Clin Oncol. 2014;32:2959–66.

Aftimos P, Oliveira M, Irrthum A, et al. Genomic and Transcriptomic analyses of breast cancer primaries and matched metastases in AURORA, the breast international group (BIG) molecular screening initiative. Cancer Discov. 2021;11:2796–811.

Allison KH, Hammond MEH, Dowsett M, et al. Estrogen and progesterone receptor testing in breast cancer: ASCO/CAP guideline update. J Clin Oncol. 2020;38:1346–66.

Allred DC, Carlson RW, Berry DA, et al. NCCN task force report: estrogen receptor and progesterone receptor testing in breast cancer by immunohistochemistry. J Natl Compr Canc Netw. 2009;7(Suppl 6):S1–S21; quiz S22–S23.

Allred DC, Harvey JM, Berardo M, Clark GM. Prognostic and predictive factors in breast cancer by immunohistochemical analysis. Mod Pathol. 1998;11:155–68.

Amgad M, Stovgaard ES, Balslev E, et al. Report on computational assessment of tumor infiltrating lymphocytes from the International Immuno-Oncology Biomarker Working Group. NPJ Breast Cancer. 2020;6:16.

Anderson E. The role of oestrogen and progesterone receptors in human mammary development and tumorigenesis. Breast Cancer Res. 2002;4:197–201.

Andre F, Ismaila N, Allison KH, et al. Biomarkers for adjuvant endocrine and chemotherapy in early-stage breast cancer: ASCO guideline update. J Clin Oncol. 2022;40:1816–37.

Audeh W, Blumencranz L, Kling H, Trivedi H, Srkalovic G. Prospective validation of a genomic assay in breast cancer: the 70-gene MammaPrint assay and the MINDACT trial. Acta Med Acad. 2019;48:18–34.

Badve S, Turbin D, Thorat MA, et al. FOXA1 expression in breast cancer--correlation with luminal subtype a and survival. Clin Cancer Res. 2007;13:4415–21.

Badve SS, Baehner FL, Gray RP, et al. Estrogen- and progesterone-receptor status in ECOG 2197: comparison of immunohistochemistry by local and central laboratories and quantitative reverse transcription polymerase chain reaction by central laboratory. J Clin Oncol. 2008;26:2473–81.

Badve SS, Penault-Llorca F, Reis-Filho JS, et al. Determining PD-L1 status in patients with triple-negative breast cancer: lessons learned from IMpassion130. J Natl Cancer Inst. 2022;114:664–75.

Bardia A, Hurvitz SA, Tolaney SM, et al. Sacituzumab Govitecan in metastatic triple-negative breast cancer. N Engl J Med. 2021;384:1529–41.

Bardia A, Messersmith WA, Kio EA, et al. Sacituzumab govitecan, a Trop-2-directed antibody-drug conjugate, for patients with epithelial cancer: final safety and efficacy results from the phase I/II IMMU-132-01 basket trial. Ann Oncol. 2021;32:746–56.

Barnes DM, Harris WH, Smith P, Millis RR, Rubens RD. Immunohistochemical determination of oestrogen receptor: comparison of different methods of assessment of staining and correlation with clinical outcome of breast cancer patients. Br J Cancer. 1996;74:1445–51.

Beatson GT. On the treatment of inoperable cases of carcinoma of the mamma: suggestions for a new method of treatment, with illustrative cases. Trans Med Chir Soc Edinb. 1896;15:153.

Beca F, Lee SSK, Pareja F, et al. Whole-exome sequencing and RNA sequencing analyses of acinic cell carcinomas of the breast. Histopathology. 2019;75:931–7.

Bjornstrom L, Sjoberg M. Mechanisms of estrogen receptor signaling: convergence of genomic and nongenomic actions on target genes. Mol Endocrinol. 2005;19:833–42.

Burstein HJ, Curigliano G, Thurlimann B, et al. Customizing local and systemic therapies for women with early breast cancer: the St. Gallen international consensus guidelines for treatment of early breast cancer 2021. Ann Oncol. 2021;32:1216–35.

Buyse M, Loi S, van't Veer L, et al. Validation and clinical utility of a 70-gene prognostic signature for women with node-negative breast cancer. J Natl Cancer Inst. 2006;98:1183–92.

Cardoso F, Piccart-Gebhart M, Van't Veer L, Rutgers E, Consortium T. The MINDACT trial: the first prospective clinical validation of a genomic tool. Mol Oncol. 2007;1:246–51.

Cardoso F, Van't Veer L, Rutgers E, Loi S, Mook S, Piccart-Gebhart MJ. Clinical application of the 70-gene profile: the MINDACT trial. J Clin Oncol. 2008;26:729–35.

Carey LA, Loirat D, Punie K, et al. Sacituzumab govitecan as second-line treatment for metastatic triple-negative breast cancer-phase 3 ASCENT study subanalysis. NPJ Breast Cancer. 2022;8:72.

Cheang MC, Chia SK, Voduc D, et al. Ki67 index, HER2 status, and prognosis of patients with luminal B breast cancer. J Natl Cancer Inst. 2009;101:736–50.

Cheang MC, Voduc KD, Tu D, et al. Responsiveness of intrinsic subtypes to adjuvant anthracycline substitution in the NCIC.CTG MA.5 randomized trial. Clin Cancer Res. 2012;18:2402–12.

Chia SK, Bramwell VH, Tu D, et al. A 50-gene intrinsic subtype classifier for prognosis and prediction of benefit from adjuvant tamoxifen. Clin Cancer Res. 2012;18:4465–72.

Conneely OM, Lydon JP. Progesterone receptors in reproduction: functional impact of the a and B isoforms. Steroids. 2000;65:571–7.

Cortes J, Cescon DW, Rugo HS, et al. Pembrolizumab plus chemotherapy versus placebo plus chemotherapy for previously untreated locally recurrent inoperable or metastatic triple-negative breast cancer (KEYNOTE-355): a randomised, placebo-controlled, double-blind, phase 3 clinical trial. Lancet. 2020;396:1817–28.

Cui X, Schiff R, Arpino G, Osborne CK, Lee AV. Biology of progesterone receptor loss in breast cancer and its implications for endocrine therapy. J Clin Oncol. 2005;23:7721–35.

Cuzick J, Sestak I, Cawthorn S, et al. Tamoxifen for prevention of breast cancer: extended long-term follow-up of the IBIS-I breast cancer prevention trial. Lancet Oncol. 2015;16:67–75.

Delaloge S, Piccart M, Rutgers E, et al. Standard Anthracycline based versus Docetaxel-Capecitabine in early high clinical and/or genomic risk breast cancer in the EORTC 10041/BIG 3-04 MINDACT phase III trial. J Clin Oncol. 2020;38:1186–97.

Derakhshan F, Reis-Filho JS. Pathogenesis of triple-negative breast cancer. Annu Rev. Pathol. 2022;17:181–204.

Dieci MV, Radosevic-Robin N, Fineberg S, et al. Update on tumor-infiltrating lymphocytes (TILs) in breast cancer, including recommendations to assess TILs in residual disease after neoadjuvant therapy and in carcinoma in situ: a report of the international Immuno-oncology biomarker working group on breast cancer. Semin Cancer Biol. 2018;52:16–25.

Downs-Kelly E, Yoder BJ, Stoler M, et al. The influence of polysomy 17 on HER2 gene and protein expression in adenocarcinoma of the breast: a fluorescent in situ hybridization, immunohistochemical, and isotopic mRNA in situ hybridization study. Am J Surg Pathol. 2005;29:1221–7.

Dowsett M, Allred C, Knox J, et al. Relationship between quantitative estrogen and progesterone receptor expression and human epidermal growth factor receptor 2 (HER-2) status with recurrence in the Arimidex, Tamoxifen, alone or in combination trial. J Clin Oncol. 2008;26:1059–65.

Dowsett M, Cuzick J, Wale C, et al. Prediction of risk of distant recurrence using the 21-gene recurrence score in node-negative and node-positive postmenopausal patients with breast cancer treated with anastrozole or tamoxifen: a TransATAC study. J Clin Oncol. 2010;28:1829–34.

Dowsett M, Sestak I, Regan MM, et al. Integration of clinical variables for the prediction of late distant recurrence in patients with estrogen receptor-positive breast cancer treated with 5 years of endocrine therapy: CTS5. J Clin Oncol. 2018;36:1941–8.

Early Breast Cancer Trialists' Collaborative Group, Correa C, McGale P, et al. Overview of the randomized trials of radiotherapy in ductal carcinoma in situ of the breast. J Natl Cancer Inst Monogr. 2010;2010:162–77.

El Bairi K, Haynes HR, Blackley E, et al. The tale of TILs in breast cancer: a report from the International Immuno-Oncology Biomarker Working Group. NPJ Breast Cancer. 2021;7:150.

Elledge RM, Green S, Pugh R, et al. Estrogen receptor (ER) and progesterone receptor (PgR), by ligand-binding assay compared with ER, PgR and pS2, by immuno-histochemistry in predicting response to tamoxifen in metastatic breast cancer: a southwest oncology group study. Int J Cancer. 2000;89:111–7.

Fan C, Oh DS, Wessels L, et al. Concordance among gene-expression-based predictors for breast cancer. N Engl J Med. 2006;355:560–9.

Fisher B, Redmond C, Brown A, et al. Influence of tumor estrogen and progesterone receptor levels on the response to tamoxifen and chemotherapy in primary breast cancer. J Clin Oncol. 1983;1:227–41.

Fisher B, Redmond C, Brown A, et al. Treatment of primary breast cancer with chemotherapy and tamoxifen. N Engl J Med. 1981;305:1–6.

Francis PA, Pagani O, Fleming GF, et al. Tailoring adjuvant endocrine therapy for premenopausal breast cancer. N Engl J Med. 2018;379:122–37.

Geyer CE Jr, Tang G, Mamounas EP, et al. 21-gene assay as predictor of chemotherapy benefit in HER2-negative breast cancer. NPJ Breast Cancer. 2018;4:37.

Gilfillan S, Fiorito E, Hurtado A. Functional genomic methods to study estrogen receptor activity. J Mammary Gland Biol Neoplasia. 2012;17:147–53.

Gnant M, Dueck AC, Frantal S, et al. Adjuvant Palbociclib for early breast cancer: the PALLAS trial results (ABCSG-42/AFT-05/BIG-14-03). J Clin Oncol. 2022;40:282–93.

Goldstein LJ, Gray R, Badve S, et al. Prognostic utility of the 21-gene assay in hormone receptor-positive operable breast cancer compared with classical clinicopathologic features. J Clin Oncol. 2008;26:4063–71.

Gonzalez-Ericsson PI, Stovgaard ES, Sua LF, et al. The path to a better biomarker: application of a risk management framework for the implementation of PD-L1 and TILs as immuno-oncology biomarkers in breast cancer clinical trials and daily practice. J Pathol. 2020;250:667–84.

Guerini-Rocco E, Piscuoglio S, Ng CK, et al. Microglandular adenosis associated with triple-negative breast cancer is a neoplastic lesion of triple-negative phenotype harbouring TP53 somatic mutations. J Pathol. 2016;238:677–88.

Gusterson B. Do 'basal-like' breast cancers really exist? Nat Rev Cancer. 2009;9:128–34.

Haibe-Kains B, Desmedt C, Loi S, et al. A three-gene model to robustly identify breast cancer molecular subtypes. J Natl Cancer Inst. 2012;104:311–25.

Hammond ME, Hayes DF, Dowsett M, et al. American Society of Clinical Oncology/College of American Pathologists guideline recommendations for immunohistochemical testing of estrogen and progesterone receptors in breast cancer. J Clin Oncol. 2010;28:2784–95.

Harbeck N, Rastogi P, Martin M, et al. Adjuvant abemaciclib combined with endocrine therapy for high-risk early breast cancer: updated efficacy and Ki-67 analysis from the monarchE study. Ann Oncol. 2021;32:1571–81.

Harvey JM, Clark GM, Osborne CK, Allred DC. Estrogen receptor status by immunohistochemistry is superior to the ligand-binding assay for predicting response to adjuvant endocrine therapy in breast cancer. J Clin Oncol. 1999;17:1474–81.

Hewitt SC, Harrell JC, Korach KS. Lessons in estrogen biology from knockout and transgenic animals. Annu Rev. Physiol. 2005;67:285–308.

Hortobagyi GN, Stemmer SM, Burris HA, et al. Overall survival with Ribociclib plus Letrozole in advanced breast cancer. N Engl J Med. 2022;386:942–50.

Howell A, Cuzick J, Baum M, et al. Results of the ATAC (Arimidex, Tamoxifen, alone or in combination) trial after completion of 5 years' adjuvant treatment for breast cancer. Lancet. 2005;365:60–2.

Hudecek J, Voorwerk L, van Seijen M, et al. Application of a risk-management framework for integration of stromal tumor-infiltrating lymphocytes in clinical trials. NPJ Breast Cancer. 2020;6:15.

Jacob L, Witteveen A, Beumer I, et al. Controlling technical variation amongst 6693 patient microarrays of the randomized MINDACT trial. Commun Biol. 2020;3:397.

Jankowitz RC, Cooper K, Erlander MG, et al. Prognostic utility of the breast cancer index and comparison to adjuvant! Online in a clinical case series of early breast cancer. Breast Cancer Res. 2011;13:R98.

Jasani B, Douglas-Jones A, Rhodes A, et al. Measurement of estrogen receptor status by immunocytochemistry in paraffin wax sections. Methods Mol Med. 2006;120:127–46.

Jerevall PL, Ma XJ, Li H, et al. Prognostic utility of HOXB13:IL17BR and molecular grade index in early-stage breast cancer patients from the Stockholm trial. Br J Cancer. 2011;104:1762–9.

Jhaveri K, Drago JZ, Shah PD, et al. A phase I study of Alpelisib in combination with Trastuzumab and LJM716 in patients with PIK3CA-mutated HER2-positive metastatic breast cancer. Clin Cancer Res. 2021;27:3867–75.

Johnston SRD, Harbeck N, Hegg R, et al. Abemaciclib combined with endocrine therapy for the adjuvant treatment of HR+, HER2-, node-positive, high-risk, early breast cancer (monarchE). J Clin Oncol. 2020;38:3987–98.

Kalinsky K, BaR-low WE, Gralow JR, et al. 21-gene assay to inform chemotherapy benefit in node-positive breast cancer. N Engl J Med. 2021;385:2336–47.

Kos Z, Roblin E, Kim RS, et al. Pitfalls in assessing stromal tumor infiltrating lymphocytes (sTILs) in breast cancer. NPJ Breast Cancer. 2020;6:17.

Kuiper GG, Enmark E, Pelto-Huikko M, Nilsson S, Gustafsson JA. Cloning of a novel receptor expressed in rat prostate and ovary. Proc Natl Acad Sci U S A. 1996;93:5925–30.

Levin ER, Pietras RJ. Estrogen receptors outside the nucleus in breast cancer. Breast Cancer Res Treat. 2008;108:351–61.

Liedtke C, Hatzis C, Symmans WF, et al. Genomic grade index is associated with response to chemotherapy in patients with breast cancer. J Clin Oncol. 2009;27:3185–91.

Loi S, Drubay D, Adams S, et al. Tumor-infiltrating lymphocytes and prognosis: a pooled individual patient analysis of early-stage triple-negative breast cancers. J Clin Oncol. 2019;37:559–69.

Loi S, Michiels S, Salgado R, et al. Tumor infiltrating lymphocytes are prognostic in triple negative breast cancer and predictive for trastuzumab benefit in early breast cancer: results from the FinHER trial. Ann Oncol. 2014;25:1544–50.

Loi S, Salgado R, Adams S, et al. Tumor infiltrating lymphocyte stratification of prognostic staging of early-stage triple negative breast cancer. NPJ Breast Cancer. 2022;8:3.

Lopes Cardozo JMN, Byng D, Drukker CA, et al. Outcome without any adjuvant systemic treatment in stage I ER+/HER2- breast cancer patients included in the MINDACT trial. Ann Oncol. 2022;33:310–20.

Lopes Cardozo JMN, Drukker CA, Rutgers EJT, et al. Outcome of patients with an ultralow-risk 70-gene signature in the MINDACT trial. J Clin Oncol. 2022;40:1335–45.

Lu YS, Im SA, Colleoni M, et al. Updated overall survival of Ribociclib plus endocrine therapy versus endocrine therapy alone in pre- and Perimenopausal patients with HR+/HER2- advanced breast cancer in MONALEESA-7: a phase III randomized clinical trial. Clin Cancer Res. 2022;28:851–9.

Ma XJ, Salunga R, Dahiya S, et al. A five-gene molecular grade index and HOXB13:IL17BR are complementary prognostic factors in early stage breast cancer. Clin Cancer Res. 2008;14:2601–8.

Ma XJ, Wang Z, Ryan PD, et al. A two-gene expression ratio predicts clinical outcome in breast cancer patients treated with tamoxifen. Cancer Cell. 2004;5:607–16.

Mamounas EP, Tang G, Paik S, et al. 21-gene recurrence score for prognosis and prediction of taxane benefit after adjuvant chemotherapy plus endocrine therapy: results from NSABP B-28/NRG oncology. Breast Cancer Res Treat. 2018;168:69–77.

Marotti JD, Collins LC, Hu R, Tamimi RM. Estrogen receptor-beta expression in invasive breast cancer in relation to molecular phenotype: results from the nurses' health study. Mod Pathol. 2010;23:197–204.

Mayayo-Peralta I, Prekovic S, Zwart W. Estrogen receptor on the move: cistromic plasticity and its implications in breast cancer. Mol Asp Med. 2021;78:100939.

Mayer EL, Dueck AC, Martin M, et al. Palbociclib with adjuvant endocrine therapy in early breast cancer (PALLAS): interim analysis of a multicentre, open-label, randomised, phase 3 study. Lancet Oncol. 2021;22:212–22.

Mayer EL, Fesl C, Hlauschek D, et al. Treatment exposure and discontinuation in the PALbociclib CoLlaborative adjuvant study of Palbociclib with adjuvant endocrine therapy for hormone receptor-positive/human epidermal growth factor receptor 2-negative early breast cancer (PALLAS/AFT-05/ABCSG-42/BIG-14-03). J Clin Oncol. 2022;40:449–58.

Metzger-Filho O, Catteau A, Michiels S, et al. Genomic grade index (GGI): feasibility in routine practice and impact on treatment decisions in early breast cancer. PLoS One. 2013;8:e66848.

Miller K, Rhodes A, Jasani B. Variation in rates of oestrogen receptor positivity in breast cancer again. BMJ. 2002;324:298.

Modi S, Jacot W, Yamashita T, et al. Trastuzumab Deruxtecan in previously treated HER2-low advanced breast cancer. N Engl J Med. 2022;387(1):9–20.

Modi S, Saura C, Yamashita T, et al. Trastuzumab Deruxtecan in previously treated HER2-positive breast cancer. N Engl J Med. 2020;382:610–21.

Nakshatri H, Badve S. FOXA1 as a therapeutic target for breast cancer. Expert Opin Ther Targets. 2007;11:507–14.

Nakshatri H, Badve S. FOXA1 in breast cancer. Expert Rev. Mol Med. 2009;11:e8.

Nguyen B, Sanchez-Vega F, Fong CJ, et al. The genomic landscape of carcinomas with mucinous differentiation. Sci Rep. 2021;11:9478.

Nielsen TO, Parker JS, Leung S, et al. A comparison of PAM50 intrinsic subtyping with immunohistochemistry and clinical prognostic factors in tamoxifen-treated estrogen receptor-positive breast cancer. Clin Cancer Res. 2010;16:5222–32.

O'Shaughnessy J, Brufsky A, Rugo HS, et al. Analysis of patients without and with an initial triple-negative breast cancer diagnosis in the phase 3 randomized ASCENT study of sacituzumab govitecan in metastatic triple-negative breast cancer. Breast Cancer Res Treat. 2022;195(2):127–39.

Pagani O, Francis PA, Fleming GF, et al. Absolute improvements in freedom from distant recurrence to tailor adjuvant endocrine therapies for premenopausal women: results from TEXT and SOFT. J Clin Oncol. 2020;38:1293–303.

Paik S, Shak S, Tang G, et al. A multigene assay to predict recurrence of tamoxifen-treated, node-negative breast cancer. N Engl J Med. 2004;351:2817–26.

Pan H, Gray R, Braybrooke J, et al. 20-year risks of breast-cancer recurrence after stopping endocrine therapy at 5 years. N Engl J Med. 2017;377:1836–46.

Pareja F, Geyer FC, Marchio C, Burke KA, Weigelt B, Reis-Filho JS. Triple-negative breast cancer: the importance of molecular and histologic subtyping, and recognition of low-grade variants. NPJ Breast Cancer. 2016;2:16036.

Pareja F, Weigelt B, Reis-Filho JS. Problematic breast tumors reassessed in light of novel molecular data. Mod Pathol. 2021;34:38–47.

Parker JS, Mullins M, Cheang MC, et al. Supervised risk predictor of breast cancer based on intrinsic subtypes. J Clin Oncol. 2009;27:1160–7.

Penault-Llorca F, Filleron T, Asselain B, et al. The 21-gene recurrence score(R) assay predicts distant recurrence in lymph node-positive, hormone receptor-positive, breast cancer patients treated with adjuvant sequential epirubicin- and docetaxel-based or epirubicin-based chemotherapy (PACS-01 trial). BMC Cancer. 2018;18:526.

Perez EA, Baehner FL, Butler SM, et al. The relationship between quantitative human epidermal growth factor receptor 2 gene expression by the 21-gene reverse transcriptase polymerase chain reaction assay and adjuvant trastuzumab benefit in alliance N9831. Breast Cancer Res. 2015;17:133.

Perez EA, Roche PC, Jenkins RB, et al. HER2 testing in patients with breast cancer: poor correlation between weak positivity by immunohistochemistry and gene amplification by fluorescence in situ hybridization. Mayo Clin Proc. 2002;77:148–54.

Perou CM, Sorlie T, Eisen MB, et al. Molecular portraits of human breast tumours. Nature. 2000;406:747–52.

Persons DL, Tubbs RR, Cooley LD, et al. HER-2 fluorescence in situ hybridization: results from the survey program of the College of American Pathologists. Arch Pathol Lab Med. 2006;130:325–31.

Piccart M, van 't Veer LJ, Poncet C, et al. 70-gene signature as an aid for treatment decisions in early breast cancer: updated results of the phase 3 randomised MINDACT trial with an exploratory analysis by age. Lancet Oncol. 2021;22:476–88.

Prat A, Cheang MC, Martin M, et al. Prognostic significance of progesterone receptor-positive tumor cells within immunohistochemically defined luminal a breast cancer. J Clin Oncol. 2013;31:203–9.

Pruneri G, Gray KP, Vingiani A, et al. Tumor-infiltrating lymphocytes (TILs) are a powerful prognostic marker in patients with triple-negative breast cancer enrolled in the IBCSG phase III randomized clinical trial 22-00. Breast Cancer Res Treat. 2016;158:323–31.

Pusztai L, Mazouni C, Anderson K, Wu Y, Symmans WF. Molecular classification of breast cancer: limitations and potential. Oncologist. 2006;11:868–77.

Regan MM, Viale G, Mastropasqua MG, et al. Re-evaluating adjuvant breast cancer trials: assessing hormone receptor status by immunohistochemical versus extraction assays. J Natl Cancer Inst. 2006;98:1571–81.

Rhodes A, Jasani B, Balaton AJ, et al. Study of interlaboratory reliability and reproducibility of estrogen and progesterone receptor assays in Europe. Documentation of poor reliability and identification of insufficient microwave antigen retrieval time as a major contributory element of unreliable assays. Am J Clin Pathol. 2001;115:44–58.

Rhodes A, Jasani B, Balaton AJ, Miller KD. Immunohistochemical demonstration of oestrogen and progesterone receptors: correlation of standards achieved on in house tumours with that achieved on external quality assessment material in over 150 laboratories from 26 countries. J Clin Pathol. 2000;53:292–301.

Roche PC, Suman VJ, Jenkins RB, et al. Concordance between local and central laboratory HER2 testing in the breast intergroup trial N9831. J Natl Cancer Inst. 2002;94:855–7.

Rugo HS, O'Shaughnessy J, Boyle F, et al. Adjuvant abemaciclib combined with endocrine therapy for high-risk early breast cancer: safety and patient-reported outcomes from the monarchE study. Ann Oncol. 2022;33:616–27.

Salgado R, Denkert C, Demaria S, et al. The evaluation of tumor-infiltrating lymphocytes (TILs) in breast cancer: recommendations by an international TILs working group 2014. Ann Oncol. 2015;26:259–71.

Sano MV, Martorana F, Lavenia G, et al. Ribociclib efficacy in special populations and analysis of patient-reported outcomes in the MONALEESA trials. Expert Rev. Anticancer Ther. 2022;22:343–51.

Schmid P, Cortes J, Dent R, et al. Event-free survival with pembrolizumab in early triple-negative breast cancer. N Engl J Med. 2022;386:556–67.

Schwartz CJ, da Silva EM, Marra A, et al. Morphologic and genomic characteristics of breast cancers occurring in individuals with lynch syndrome. Clin Cancer Res. 2022;28:404–13.

Sgroi DC, Carney E, Zarrella E, et al. Prediction of late disease recurrence and extended adjuvant letrozole benefit by the HOXB13/IL17BR biomarker. J Natl Cancer Inst. 2013;105:1036–42.

Solin LJ, Gray R, Baehner FL, et al. A multigene expression assay to predict local recurrence risk for ductal carcinoma in situ of the breast. J Natl Cancer Inst. 2013;105:701–10.

Sorlie T, Perou CM, Tibshirani R, et al. Gene expression patterns of breast carcinomas distinguish tumor subclasses with clinical implications. Proc Natl Acad Sci U S A. 2001;98:10869–74.

Tanos T, Rojo L, Echeverria P, Brisken C. ER and PR signaling nodes during mammary gland development. Breast Cancer Res. 2012;14:210.

Tubbs RR, Pettay JD, Roche PC, Stoler MH, Jenkins RB, Grogan TM. Discrepancies in clinical laboratory testing of eligibility for trastuzumab therapy: apparent immunohistochemical false-positives do not get the message. J Clin Oncol. 2001;19:2714–21.

van 't Veer LJ, Dai H, van de Vijver MJ, et al. Gene expression profiling predicts clinical outcome of breast cancer. Nature. 2002;415:530–6.

van de Vijver MJ, He YD, van't Veer LJ, et al. A gene-expression signature as a predictor of survival in breast cancer. N Engl J Med. 2002;347:1999–2009.

Viale G, Regan MM, Maiorano E, et al. Prognostic and predictive value of centrally reviewed expression of estrogen and progesterone receptors in a randomized trial comparing letrozole and tamoxifen adjuvant therapy for postmenopausal early breast cancer: BIG 1-98. J Clin Oncol. 2007;25:3846–52.

Viale G, Regan MM, Maiorano E, et al. Chemoendocrine compared with endocrine adjuvant therapies for node-negative breast cancer: predictive value of centrally reviewed expression of estrogen and progesterone receptors--international breast cancer study group. J Clin Oncol. 2008;26:1404–10.

Winer EP, Lipatov O, Im SA, et al. Pembrolizumab versus investigator-choice chemotherapy for metastatic triple-negative breast cancer (KEYNOTE-119): a randomised, open-label, phase 3 trial. Lancet Oncol. 2021;22:499–511.

Wirapati P, Sotiriou C, Kunkel S, et al. Meta-analysis of gene expression profiles in breast cancer: toward a unified understanding of breast cancer subtyping and prognosis signatures. Breast Cancer Res. 2008;10:R65.

Wolff AC, Hammond ME, Hicks DG, et al. Recommendations for human epidermal growth factor receptor 2 testing in breast cancer: American Society of Clinical Oncology/College of American Pathologists clinical practice guideline update. J Clin Oncol. 2013;31:3997–4013.

Wolff AC, Hammond ME, Schwartz JN, et al. American Society of Clinical Oncology/College of American Pathologists guideline recommendations for human epidermal growth factor receptor 2 testing in breast cancer. J Clin Oncol. 2007;25:118–45.

Wolmark N, Mamounas EP, Baehner FL, et al. Prognostic impact of the combination of recurrence score and quantitative estrogen receptor expression (ESR1) on predicting late distant recurrence risk in estrogen receptor-positive breast cancer after 5 years of Tamoxifen: results from NRG Oncology/National Surgical Adjuvant Breast and Bowel Project B-28 and B-14. J Clin Oncol. 2016;34:2350–8.

Zaret KS, Carroll JS. Pioneer transcription factors: establishing competence for gene expression. Genes Dev. 2011;25:2227–41.

Molecular Pathology of Ovarian Tumors

11

Brian S. Finkelman, Kruti P. Maniar, and Ie-Ming Shih

Contents

Introduction

- Ovarian cancer is the fifth most common cause of cancer-related death in women in the US, and the leading cause of death among gynecologic malignancies
- The 5-year survival for ovarian cancer is 48% overall and is only 29% among those presenting with distant metastasis (59% of cases)
  - This dismal prognosis is the main driver of the high mortality burden of ovarian cancer, as its incidence is about one-third that of endometrial cancer, the most common gynecologic malignancy in the US

B. S. Finkelman
Department of Pathology and Laboratory Medicine, University of Rochester Medical Center, Rochester, NY, USA

K. P. Maniar (✉)
Department of Pathology and Laboratory Medicine, NorthShore , University of Chicago Pritzker School of Medicine, Chicago, IL, USA
e-mail: kmaniar@northshore.org

I.-M. Shih
Departments of Pathology, Oncology, and Gynecology and Obstetrics, Johns Hopkins University, Baltimore, MD, USA

- Understanding the pathophysiology of ovarian neoplasms, particularly the molecular basis of disease, is crucial in improving diagnostic and treatment modalities

Classification of Ovarian Epithelial Neoplasms

- The traditional idea of a progression from well to poorly differentiated carcinoma for all ovarian cancer subtypes has been replaced by a comprehensive, dualistic model of ovarian epithelial neoplasia based on morphologic and molecular data (Table 11.1)

Table 11.1 Characteristics of type I vs. type II ovarian tumors

	Type I	Type II
Histotypes	Low-grade serous carcinoma	High-grade serous carcinoma
	Mucinous carcinoma	Carcinosarcoma
	Endometrioid carcinoma	Undifferentiated carcinoma
	Clear cell carcinoma	
	Malignant Brenner tumor	
Precursors/origin	Benign adenomas → borderline (atypical proliferative) tumors → malignant neoplasms	Arise from the tubal epithelium
Clinical behavior	Slow growing	Aggressive
	Often confined to the ovary at the time of diagnosis	Rapid progression
		Early metastasis
Molecular abnormalities	MAPK signaling pathway (*KRAS/NRAS/BRAF*)	*TP53*
	Wnt/β(Beta)-catenin/Cyclin D1	*CDKN2*/p16
	PI3K/Akt2/PTEN pathway	*BRCA1* and *BRCA2*
	ARID1a	*Akt2* (*PI3K/Akt2/PTEN* pathway)
	Microsatellite instability	*Notch3*
	HNF1-β(beta)	*HBXAP* (*Rsf-1*)
	PPP2R1A	*NAC1*
	EGFR/HER2neu	HLA-G
		Cyclin E1
		EGFR/HER2

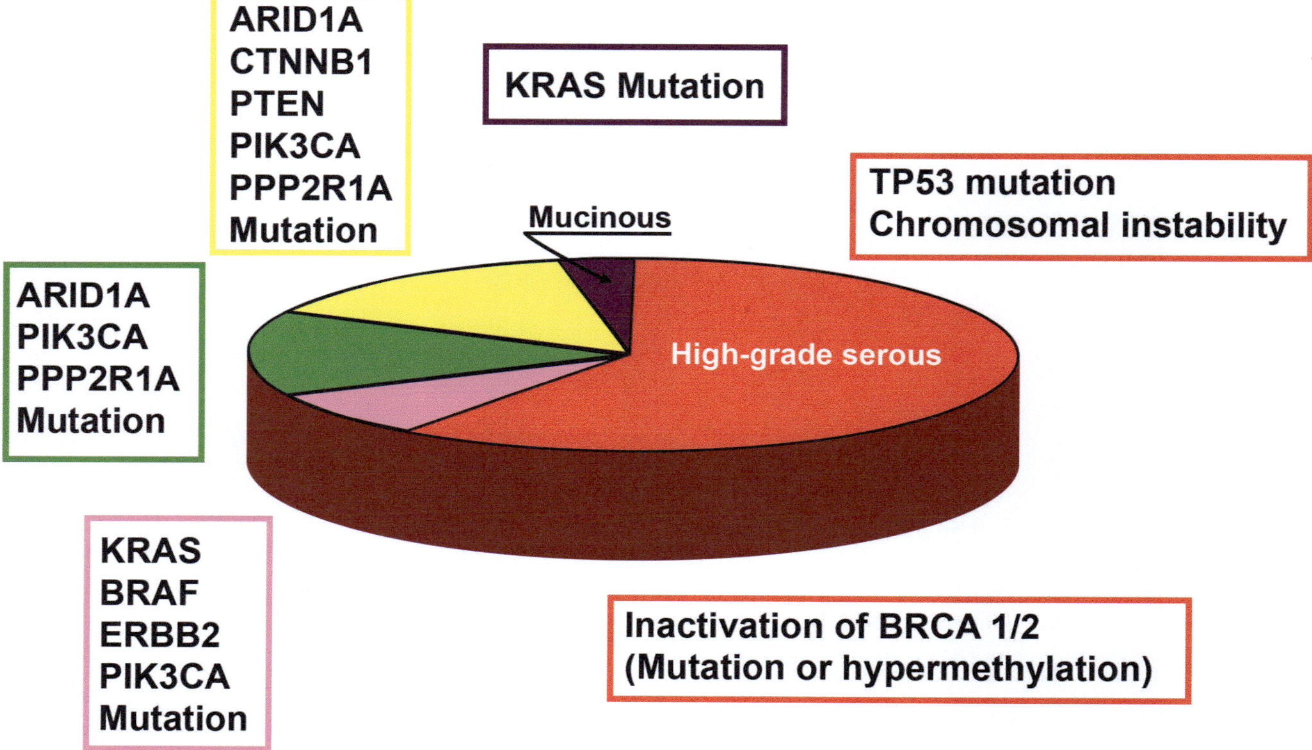

Fig. 11.1 Prevalence of epithelial ovarian cancer histotypes and their associated molecular abnormalities. (Reprinted from Human Pathology, Kurman and Shih 2011, with permission from Elsevier)

- This classification is described below, with a more detailed description of the molecular abnormalities included in subsequent sections (Fig. 11.1)

Type I Tumors

- Type I tumors are composed of several diverse histotypes including low-grade serous carcinoma (LGSC), mucinous carcinoma, endometrioid carcinoma, malignant Brenner tumor, and clear cell carcinoma
- Precursors/origin
  - These tumors are thought to develop in a stepwise fashion from benign to borderline to malignant tumors
    - Borderline tumors
      - The term "borderline tumor" refers to an entity intermediate in behavior between cystadenomas and carcinomas
      - Findings in recent years have led to the refinement of this category and the histologic and behavioral spectrum it encompasses
      - For serous tumors, two categories have been defined based on behavior: serous borderline (atypical proliferative) tumors (SBT), which is a typical borderline tumor with or without noninvasive implants; and the micropapillary/cribri-form variant of SBT, a term synonymous with noninvasive LGSC
      - Mucinous borderline tumors (of intestinal type) are relatively indolent even if they contain areas of intraepithelial carcinoma or foci of microinvasion (<5 mm), and therefore are best categorized as mucinous borderline (atypical proliferative) tumors (MBT), with qualification as necessary ("with intraepithelial carcinoma" or "with microinvasion")
      - Seromucinous borderline (atypical proliferative) tumor is associated with endometriosis and endometrioid tumors
  - Low-grade serous tumors are believed to progress from adenomas to borderline tumors to micropapillary/cribriform borderline tumors (noninvasive LGSC), and finally to invasive LGSC
  - Mucinous tumors may arise in some cases from the tubal–peritoneal junction, based on their association with Walthard nests and Brenner tumors, and the presence of transitional metaplasia at the tubal–peritoneal junction in some salpingectomy specimens
  - Many endometrioid and clear cell carcinomas have also been found to be associated with benign or borderline-like lesions, as well as with endometriosis, which is thought to be the benign precursor of these tumors

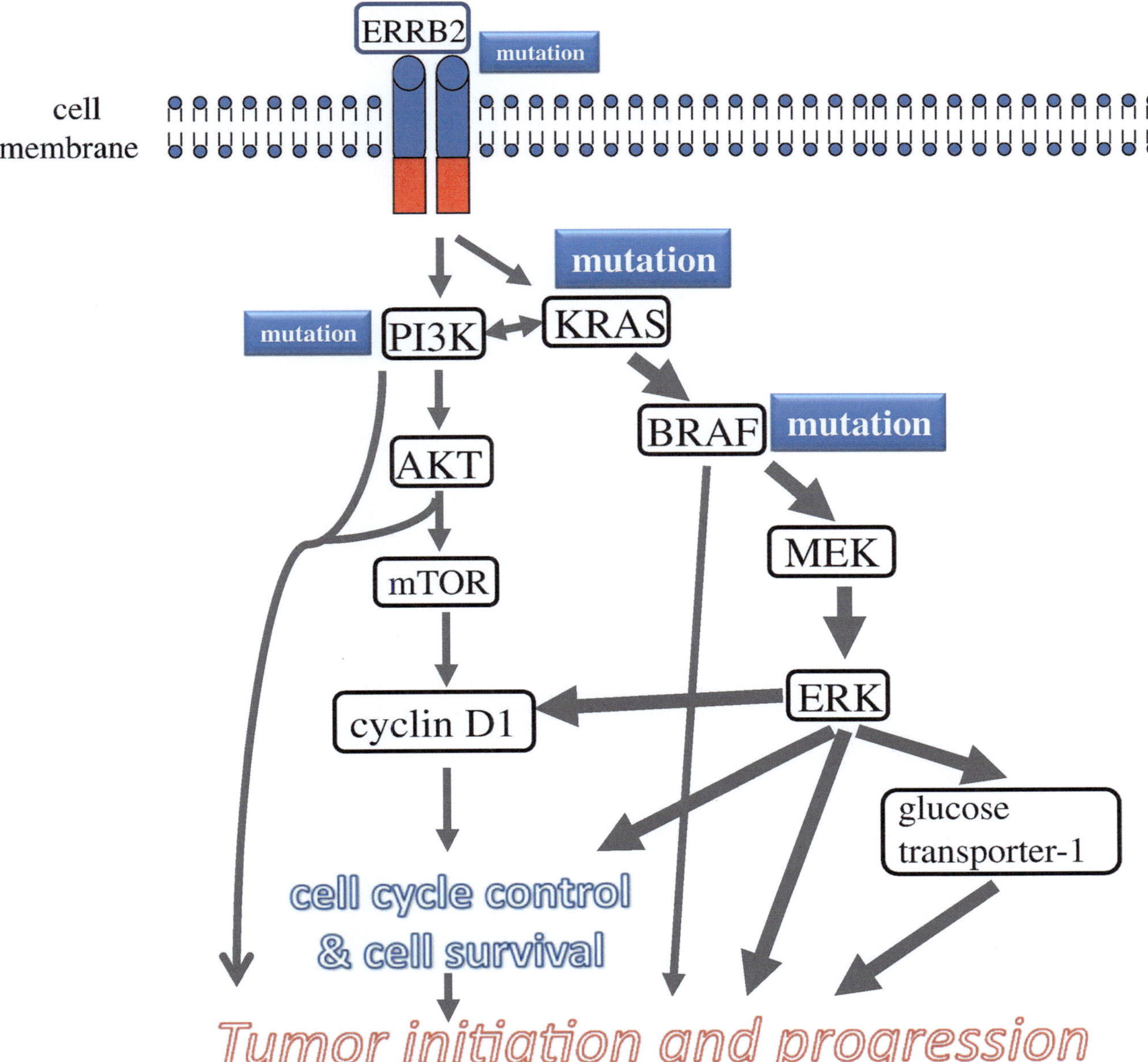

Fig. 11.2 Interaction of pathways involved in the pathogenesis of LGSC and other type I tumors. (Reprinted from Human Pathology, Kurman and Shih (2011), with permission from Elsevier)

- Clinical behavior
  - These are indolent tumors which grow to a large size while largely remaining confined to the ovary at diagnosis
  - Although LGSC demonstrates a pattern of spread similar to its high-grade counterpart, it behaves in a more indolent fashion and is associated with a better prognosis
- Summary of molecular findings (Fig. 11.2)

  - Endometrioid and clear cell carcinomas commonly have abnormalities in *ARID1A*, *CCND1*/β(beta)-catenin, and the *PI3K/Akt2/PTEN* pathway
  - LGSC commonly has abnormalities in the MAPK signaling pathway (*KRAS*, *NRAS*, and *BRAF*) and the *PI3K/Akt2/PTEN* pathway
  - Mucinous carcinoma commonly has mutations in *KRAS*

- Seromucinous carcinoma commonly has mutations in *KRAS*, *ARID1A*, and the *PI3K/Akt2/PTEN* pathway. Some authors suggest that virtually all cases of seromucinous carcinoma can be re-classified as either endometrioid carcinoma or LGSC based on morphologic, immunohistochemical, and molecular findings, and the current (5th Ed) World Health Organization classification of Female Genital Tumors categorizes these as a subtype of endometrioid carcinoma
- Type I tumors generally lack mutations in *TP53*, unlike type II tumors, and therefore this helps in differentiating the two categories

Type II Tumors

- This category includes high-grade serous carcinoma (HGSC), carcinosarcoma (malignant mixed Müllerian tumor/MMMT), and undifferentiated carcinoma
- Type II tumors comprise about two-thirds to three-quarters of ovarian cancer cases
- Histopathology
  - Tumors in this category are high-grade in appearance, with complex architecture and significant nuclear atypia
  - Necrosis and high mitotic activity are common
  - HGSC is exclusively epithelial in differentiation, while carcinosarcoma displays both epithelial and mesenchymal differentiation
- Precursors/origin
  - Most type II tumors appear to arise from tubal epithelium, either from intraepithelial carcinomas which shed cells that implant on the ovary, or from normal tubal epithelium that implants on the ovary to form inclusion cysts from which serous carcinoma can develop
  - Evidence that supports an origin from the distal fallopian tube includes:
    - Gene expression profiling has found a significant correlation between serous carcinomas and the normal fallopian tube
    - Higher rates of tubal hyperplasia, dysplasia, and occult carcinoma (particularly in the distal tube or fimbriae), as well as *TP53* mutations within dysplastic foci, have been found in prophylactic salpingo-oophorectomy specimens compared to resections for other causes
    - In patients with concurrent serous tubal intraepithelial carcinoma (STIC) and ovarian serous carcinoma, identical *TP53* mutations have been found in both components

- Prophylactic bilateral salpingectomy reduces the risk of ovarian cancer in the general population by about 50%
- Clinical behavior
  - These tumors behave aggressively, with rapid progression and early metastasis
- Summary of molecular findings
  - Type II tumors commonly have mutations in *TP53*, chromosomal instability, and inactivation of *BRCA1* and *BRCA2*

Molecular Pathways and Alterations by Tumor Type

Type I Ovarian Tumors

Low-Grade Serous Tumors

Introduction

- This category of tumors includes SBT and LGSC
- Low-grade serous tumors of all types commonly demonstrate a papillary architecture and psammoma bodies (Fig. 11.3)
- SBT demonstrates papillary epithelial proliferation with hierarchical branching
  - Foci of invasion less than 5 mm are permitted (SBT with microinvasion)
  - Extraovarian implants may be noninvasive or invasive
- The micropapillary/cribriform variant of SBT, or noninvasive LGSC, demonstrates an appearance similar to that of SBT but with a micropapillary or cribriform epithelial proliferation
- LGSC also demonstrates cells with a higher nuclear–cytoplasmic ratio and slightly more cytologic atypia than SBT
- The following discussion focuses largely on the molecular pathology of LGSC but also addresses findings in SBT and benign serous tumors where relevant

Genetic Pathways: Functions, Role in Pathogenesis, and Frequency of Abnormalities

- *MAPK* signaling pathway
  - MAPK (mitogen-activated protein kinase), also known as ERK (extracellular signal-regulated protein kinase), is a downstream target of RAS, RAF, and MAPK/ERK kinase (MEK), and is upstream of mTOR
  - *MAPK* responds to growth factors and other signals by promoting cell proliferation and opposing cell death

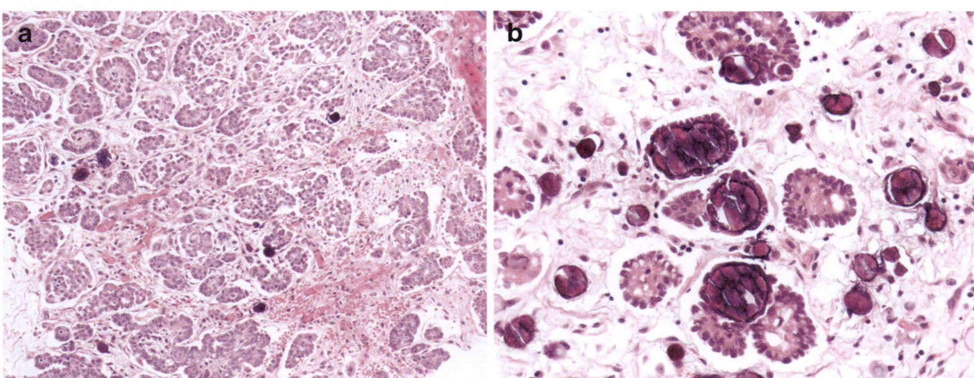

Fig. 11.3 Low-grade serous carcinoma. Low-power (**a**) demonstrates micropapillae surrounded by clear spaces, with interspersed psammoma bodies. High-power (**b**) demonstrates uniform cells with mild cytologic atypia

and is important in mediating drug-induced apoptosis in tumor cells

- *KRAS, NRAS,* and *BRAF* are oncogenes involved in the activation of the MAPK pathway
- *USP9X* and *EIF1AX* also commonly mutated in LGSC, and both genes have been linked to the regulation of mTOR
- Frequency of mutations
  - *KRAS* mutations have been found in 22–37% of SBT and up to 33% of LGSC
  - *NRAS* mutations have been found in 26% of LGSC and have not been identified in SBT
  - *BRAF* mutations have been found in up to 42% of SBT and up to 36% of LGSC
  - Overall, 60–88% of SBT express mutations in either *KRAS* or *BRAF*
    - With rare exceptions, these mutations are mutually exclusive
  - Mutations in *KRAS* and *BRAF* help distinguish low-grade serous tumors from HGSC, as these mutations are found in only up to 12% of HGSC
  - Of note, serous cystadenomas adjacent to *KRAS*- or *BRAF*-mutated SBTs were found to have identical mutations in 86% of cases, suggesting that mutation of these two genes precedes progression to SBT
    - Others have found these mutations in early SBT, supporting their early role in tumorigenesis
  - More recent data suggest that *KRAS* and *BRAF* mutations may be less common, and that *PIK3CA* mutations may be more common, in East Asian patients than in patients of European and African ancestry
- Immunohistochemical findings
  - By immunohistochemical staining, activated (phosphorylated) MAPK was found to be expressed in 71% of SBT, 80–81% of LGSC, and 41% of HGSC
  - In low-grade serous tumors, MAPK immunoexpression correlates with mutations in *KRAS* and *BRAF*.
- CDKN2/p16

- p16, encoded by *CDKN2* (*p16ink4*) on 9p21, is a tumor suppressor gene involved in the Rb pathway and is discussed in more detail below (see section "HGSC and Carcinosarcoma")
- Although found more frequently overexpressed in HGSC, p16 has been found by some authors to be expressed in a significant percentage of low-grade serous tumors
- Immunohistochemistry
  - Overexpression of p16 by immunohistochemical analysis has been found in 27.3% of LGSC in one study, and as many as 85% of SBT in another
    - A possible loss of expression as tumors progress from SBT to LGSC has been proposed
- *ERBB2* mutation prevalence is variable, seen in 0–5% of SBT and in 0–30% of LGSC
- Genomic copy number aberrations (CNA) are seen in 100% of LGSC and 61.4% of SBT
  - Most frequent copy number alterations include loss of 1p (20%), 9p (14%), and 19 (11%), as well as gain of 7/7q (18%), 8/8q (20%), and 12/12p (17%)
  - Loss of 9p, in particular, more frequent in LGSC (53%) than SBT (2%)
    - Includes locus of *CDKN2*
  - CNA frequency is not as high as in HGSC

Clinical Implications

- *MAPK* signaling pathway
  - The constitutive activation of the *MAPK* signaling pathway in type I tumors suggests a role for MAPK kinase (MEK) inhibitors in the treatment
    - In fact, treatment of *KRAS*- or *BRAF*-mutated ovarian cancer cell lines with a MEK inhibitor was found to cause significant apoptosis and growth inhibition
  - Treatment with cisplatin may induce activation of MAPK, with subsequent development of cisplatin resistance

- o Furthermore, treatment with a proteasome inhibitor sensitizes cisplatin-resistant ovarian cancer cells to cisplatin-induced cell death, indicating a potential role for proteasome inhibitors along with cisplatin in *MAPK*-activated tumors
  - Patients with both *MAPK* expression and paclitaxel sensitivity have significantly better 5-year survival than those without these two characteristics (74.9% vs. 31%)
  - Patients with *KRAS*- or *BRAF*-mutated tumors may have better overall survival than LGSC patients without these mutations
  - Clinical studies of MEK inhibitors to date have had mixed results, other larger trials are ongoing
  - Low prevalence of *KRAS* and *BRAF* mutations and high prevalence of *PIK3CA* mutations in East Asian populations has led some investigators to propose the use of therapies targeting the PIK3CA/AKT pathway in these populations
  - Recent preclinical data suggest that metformin either alone or in combination with MEK inhibitors may inhibit the growth of LGSC cancer cell lines, suggesting a possible role in LGSC management
  - Early clinical data with BRAF inhibitors have shown mixed results, with a minority of patients showing durable treatment responses
- CDKN2/p16
  - Following FDA approvals in hormone-receptor–positive breast cancer, phase II clinical trials are currently underway testing the effect of CDK4/6 inhibitors in LGSC

Summary

- The most common molecular abnormalities in low-grade serous tumors are in the *MAPK* signaling pathway
  - *MAPK* expression is common in SBT and LGSC and correlates with mutations in *KRAS* and *BRAF*
- Mutations in *KRAS* and *BRAF* are useful in distinguishing LGSC from HGSC, in which they occur much less fre-

quently, although prevalence may vary in different ethnic populations
- A possible role exists for MEK inhibitors, either alone or in combination with proteasome inhibitors or metformin, for the management of LGSC
- A possible role also exists for BRAF inhibitors, such as vemurafenib, dabrafenib, and encorafenib
- Aberrant expression of p16 is more common in HGSC but has also been found in a subset of LGSC, with some authors finding a correlation with lower-stage and lower-grade tumors
  - Clinical trials of CDK4/6 inhibitors in LGSC are currently underway

Endometrioid and Clear Cell Carcinomas

Introduction

- Endometrioid carcinomas of the ovary, most commonly found in women in their 50s, demonstrate a histologic appearance similar to that of their uterine counterparts (Fig. 11.4)
  - Well-differentiated endometrioid carcinomas are composed of branching and confluent glands lined by tall columnar stratified cells
  - Grading is based primarily on the extent of nonsquamous solid architecture, with grade 1 having <5% solid areas, grade 2 with 5–50%, and grade 3 with >50%
  - Nuclear atypia is variable and can be significant, and may also be used as a criterion to assign a tumor one grade higher than that indicated by architecture
  - Various types of metaplasia may be seen, including squamous, secretory, and mucinous
  - Endometrioid carcinomas tend to demonstrate expansile invasion but also may be infiltrative
  - About 85–90% of endometrioid carcinomas are believed to originate from endometriosis, which can show a similar mutation profile

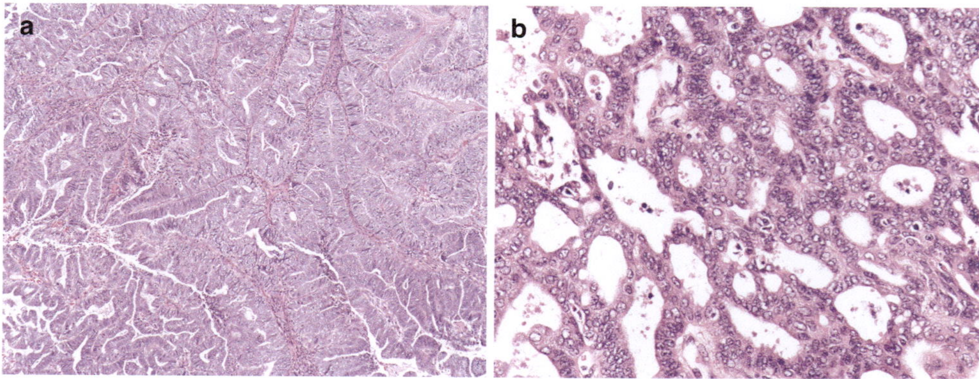

Fig. 11.4 Endometrioid carcinoma, FIGO grade 1. (**a**) At low-power, cribriform and complex papillary architecture are appreciated. (**b**) High power demonstrates mild to moderate tumor cell atypia and smooth luminal borders

Fig. 11.5 Clear cell carcinoma. (**a**) This example demonstrates the predominantly papillary architecture with hyalinized fibrovascular cores at low power. (**b**) Higher power demonstrates cells with clear cytoplasm, a range of nuclear atypia, and frequent "hobnailing" into luminal spaces

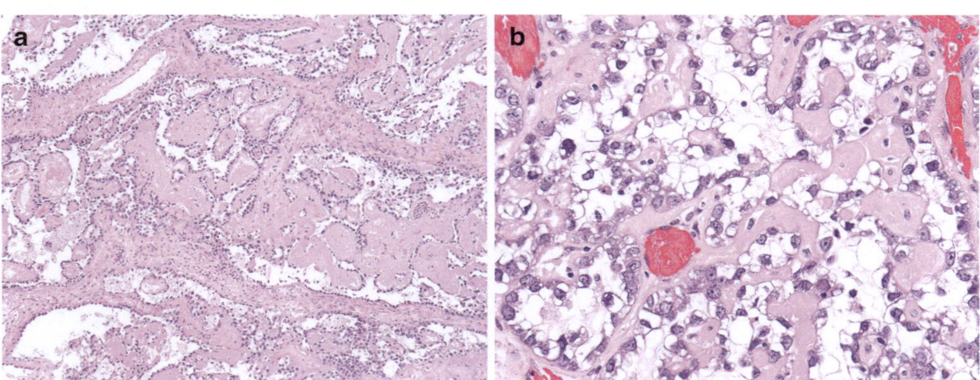

- Four molecular subtypes of ovarian endometrioid carcinoma have been proposed, analogous to the subtypes defined by The Cancer Genome Atlas (TCGA) in endometrial endometroid carcinoma
  - Hypermutated due to mismatch repair deficiency (8–14%)
  - Ultramutated due to *POLE* mutations (5–10%)
  - *TP53*-mutated (9–24%)
  - No specific molecular profile, defined as the absence of the above alterations (69–73%)
- Clear cell carcinomas, which affect a similar age group, display diverse but distinct histologic appearances (Fig. 11.5)
  - Common patterns include papillary, tubulocystic, and solid, and these often coexist in a single tumor
    - Papillary areas commonly have hyalinized stroma
    - Tubulocystic carcinomas demonstrate tubules and cysts of varying sizes lined by tumor cells
    - Solid areas demonstrate sheets of polygonal cells with clear cytoplasm
  - A variable number of cells may demonstrate eosinophilic cytoplasm rather than the classic clear cytoplasm
    - PAS-positive hyaline globules in the cytoplasm may also be seen
  - A spectrum of nuclear atypia is observed, often within the same tumor
  - Clear cell carcinomas are classified as high grade by definition
- Endometrioid and clear cell carcinomas are discussed together in this section due to the significant overlap of molecular pathways involved in their respective pathogeneses

Genetic Pathways: Functions, Role in Pathogenesis, and Frequency of Abnormalities

- Wnt/β(beta)-catenin pathway
  - The gene *CTNNB1* on 3p22.1 encodes β(beta)-catenin, a protein involved in the Wnt signaling pathway, which plays a role in the regulation of cell proliferation and differentiation
  - Missense mutations of *CTNNB1* frequently result in constitutive activation of the Wnt signaling pathway in endometrioid carcinomas
  - Mouse models suggest that defects in Wnt/β(beta)-catenin signaling often co-occur with defects in PI3K/Pten signaling, suggesting a joint role in carcinogenesis
  - Frequency of mutations
    - Mutations in *CTNNB1* have been found in 31–53% of ovarian endometrioid tumors, primarily in exon 3 of the gene
    - Although *CTNNB1* mutations have been linked to microsatellite instability in colon cancers, only rare cases of ovarian endometrioid carcinoma have been found to have both *CTNNB1* mutations and microsatellite instability
      - While *CTNNB1* mutations are always associated with nuclear staining for β(beta)-catenin, microsatellite instability is associated with a membranous β(beta)-catenin staining pattern
      - These abnormalities therefore likely represent two independent mechanisms of pathogenesis
  - Immunohistochemistry
    - Nuclear β(beta)-catenin immunohistochemical staining has been found in 38–85% of endometrioid carcinomas and up to 5.5% of clear cell carcinomas
      - In endometrioid carcinomas, immunopositivity has been strongly correlated with the presence of *CTNNB1* mutations
      - Nuclear expression of β(beta)-catenin in endometrioid carcinomas has been associated with squamous differentiation in these tumors
- Cyclin D1
  - The protein Cyclin D1, encoded by the oncogene *CCND1* on 11q13, is a target of the β(beta)-catenin pathway, with activation of the pathway resulting in increased expression of cyclin D1

- Cyclins are involved in the regulation of cyclin-dependent kinases (CDKs) protein, with cyclin D1 functioning specifically in allowing the cell to progress from G1 to the S phase
- Immunohistochemistry
  - Although some studies have found no association between cyclin D1 overexpression and histotype, others show a more frequent association with endometrioid carcinomas, with immunohistochemical positivity being found in 32% of ovarian endometrioid carcinomas and 6% of clear cell carcinomas
- *PI3K/Akt2/PTEN* pathway
  - The *PI3K/Akt2/PTEN* pathway is involved in the regulation of apoptosis, angiogenesis, cell proliferation and growth, and cell metabolism
  - Activation of the pathway can be the result of amplification of *PIK3CA* or *Akt2*, activating mutations in *PIK3CA*, or inactivating mutations of *PTEN*
    - *PIK3CA* is an oncogene located on chromosome 3q26.32, and encodes the *PI3K* catalytic subunit
    - *Akt2* is an oncogene located on chromosome 19q13.1–13.2 and encodes a protein–serine/threonine kinase
    - *PTEN* (phosphatase and tensin homolog deleted on chromosome 10) is a tumor suppressor gene located on chromosome 10q23.3
  - Frequency of mutations/amplifications
    - Mutation or amplification of *PIK3CA* has been found in about 30% of ovarian cancers overall, and up to 50% of endometrioid and clear cell carcinomas
    - *PTEN* is mutated in 14–31% of endometrioid carcinomas (mostly of low grade and low stage), up to 8.3% of clear cell carcinomas, and 20.6% of endometrial cysts
      - ◆ Most of the identified mutations have been frameshift mutations
      - ◆ LOH at the 10q23.3 locus has been found in 42% of endometrioid carcinomas, 27.3% of clear cell carcinomas, 56.5% of endometrial cysts, and 0% of normal endometrium
    - Mutations in *Akt2* are more often seen in HGSC (see below)
- *ARID1A*
  - The gene *ARID1A* encodes the protein BAF250a, which binds to AT-rich DNA sequences and functions as a component of a complex (SWI/SNF) involved in regulating the expression of cell proliferation genes
  - Because many cases of *ARID1A* mutations show both alleles to be affected, it is hypothesized that *ARID1A* is a tumor suppressor gene
  - Frequency of mutations

- Mutations in *ARID1A* have been found in 46–57% of ovarian clear cell carcinomas, as well as 71% of ovarian clear cell carcinoma cell lines
- Endometrioid carcinomas have also been found to harbor *ARID1A* mutations at a frequency of up to 30%
- Mutations have not been found in HGSC
- Immunohistochemical findings
  - Mutations in *ARID1A* in both endometrioid carcinomas and clear cell carcinomas have been correlated with a loss of immunopositivity for BAF250a, with 73% of *ARID1A*-mutated clear cell carcinomas and 50% of *ARID1A*-mutated endometrioid carcinomas showing this loss
    - ◆ Loss of expression is specific for these tumors vs. HGSC
  - *ARID1A* may be mutated earlier than *HNF1-β(beta)* in tumorigenesis and frequently coexists with *PIK3CA* mutations
- *HNF1-β(beta)*
  - HNF1-β(beta) (hepatocyte nuclear factor-1-β(beta), also known as vHNF-1 or LFB3), along with the related protein HNF1-α(alpha), has been shown to play a role in transcriptional activation during embryogenesis
  - *HNF1-β(beta)* mRNA expression levels have been found to be several times higher in clear cell carcinomas vs. other ovarian tumors
    - The mechanism of upregulation may be related to CpG island hypomethylation
  - Immunohistochemical findings
    - Almost all clear cell carcinomas have been found to be immunopositive for HNF1-β(beta), with most other tumors showing either no staining or weak/patchy positivity
      - ◆ Nuclear staining is also absent in endometriosis and in the normal ovarian surface epithelium
    - Molecular profiling of clear cell carcinomas, however, has failed to find mutations in the *HNF1A* gene
- Microsatellite instability
  - Microsatellite instability refers to the inactivation of DNA mismatch repair genes, with a resulting increase in mutation frequency of oncogenes and tumor suppressor genes, and a consequently increased risk of neoplastic transformation in various tissue types
  - This is the mechanism responsible for Lynch syndrome (hereditary nonpolyposis colorectal cancer, or HNPCC), a hereditary cancer syndrome in which loss of function of the mismatch repair genes *MLH1, MSH2,* and *MSH6* is frequently found (see also the section on "Familial/Hereditary Ovarian Cancer" below)

- In addition to colorectal cancers, patients with Lynch syndrome are at increased risk of developing cancers at numerous other sites, including the upper gastrointestinal tract, urinary system, and female genital tract, particularly the endometrium and ovary
- Mechanisms of microsatellite instability in ovarian cancers include frameshift mutations in the coding tracts of *BAX*, *IGFIIR*, and *MSH3*, as well as *MLH-1* promoter hypermethylation
  - Loss of *hMSH2* expression has also been demonstrated
- Frequency of mutations
  - The overall frequency of microsatellite instability in sporadic ovarian cancers was found to be 17%
  - Endometrioid tumors show a frequency of 12–50%, and clear cell carcinomas show a frequency of 6%
  - Microsatellite instability is uncommon in other ovarian tumor types
- Immunohistochemical findings
  - Loss of hMLH-1 nuclear staining has been reported to occur in 12–14% of endometrioid carcinomas and 6% of clear cell carcinomas, with most tumors being of high-grade but low stage
- Therapeutic implications
  - Use of the immune checkpoint inhibitor pembrolizumab has been demonstrated to be beneficial in the treatment of solid noncolorectal MMR-deficient tumors
  - Data specifically in ovarian cancer, including endometrioid and clear cell carcinoma, are more limited
- POLE
  - DNA polymerase epsilon (POLE) is an enzyme involved in recognizing and repairing errors during DNA synthesis
  - Mutations involving the exonuclease domain have been identified in several human cancers, including endometrial endometrioid carcinoma
  - Frequency of mutations
    - *POLE* mutations have been identified in about 5–10% of ovarian endometrioid carcinomas and are associated with favorable prognosis
- *MAPK* signaling pathway
  - Mutations in *KRAS* and *BRAF,* both oncogenes involved in the activation of the *MAPK* pathway, are more commonly associated with mucinous tumors and LGSC, but may also be mutated in endometrioid and clear cell carcinomas. (See sections "Mucinous Tumors" and "Low-Grade Serous Tumors" for more details on these two genes and the MAPK pathway.)
  - Frequency of mutations
    - *KRAS* mutations have been found in up to 33% of endometrioid carcinomas and up to 7% of clear cell carcinomas
    - *BRAF* mutations have been found in up to 9% of endometrioid carcinomas and up to 25% of clear cell carcinomas, though other studies have shown much lower frequencies around 1%
- *TP53*
  - Although found with the greatest frequency in serous tumors, *TP53* mutations have also been reported to occur in a subset of endometrioid and clear cell carcinomas, particularly those of advanced stage and high grade
  - Frequency of mutations
    - *TP53* mutations have been reported to occur in 7–42% of endometrioid carcinomas and 8–18% of clear cell carcinomas, with the highest frequency (75%) in grade 3 endometrioid carcinomas
    - In a mouse model, *TP53* mutations were found with lower frequency in tumors which had defects in the Wnt/β(beta)-catenin or *PI3K/Akt2/PTEN* signaling pathways, suggesting two separate pathways in the development of low vs. high-grade endometrioid carcinomas
- PPP2R1A
  - The serine–threonine protein phosphatase PP2A is a family of holoenzymes containing a heterodimer core with a catalytic subunit and a regulatory subunit (PPP2R1A or PPP2R1B)
    - PPP2R1A acts as a scaffold in this complex, which is involved in regulating cell growth and proliferation
  - The heterozygous and clustered nature of mutations in this gene suggest that PPP2R1A is an oncogene in tumorigenesis
  - Frequency of mutations
    - Mutations in the *PPP2R1A* gene have been found in 7–12% of endometrioid carcinomas, 4–9% of clear cell carcinomas, as well as in 3/7 ovarian clear cell carcinoma cell lines
- *ERBB2* (HER2)
  - Amplification of *ERBB2* (*HER2*), as determined by fluorescence in situ hybridization (FISH), has been seen in 9–20% of ovarian clear cell carcinomas and has not been identified in ovarian endometrioid carcinoma

Clinical Implications

- Wnt/B(beta)-catenin pathway
  - The presence of *CTNNB1* mutations or β(beta)-catenin nuclear immunopositivity in endometrioid carcinomas has been associated with better differentiation, low grade, early stage, and a favorable prognosis

- One group found cyclin D1 expression to be inversely correlated with tumor grade, suggesting a better prognosis for tumors with high cyclin D1 expression
- *HNF1-β(beta)*
  - Silencing of *HNF1-β(beta)* in ovarian cancer cell lines results in significantly more apoptosis compared to controls. It is therefore an important potential target for novel ovarian cancer therapies
  - Given its relative specificity for clear cell carcinomas compared to other ovarian cancer histotypes, HNF1-β(beta) is also a useful immunohistochemical marker for diagnostic purposes
- Microsatellite instability
  - PCR analysis of tumor tissue using microsatellite markers as well as detecting loss of expression of mismatch repair genes by immunohistochemistry are useful methods of assessing for microsatellite instability in patients deemed to be at risk
  - Patients with tumors with microsatellite instability, as assessed by immunohistochemistry for DNA mismatch repair proteins, appear to have excellent survival
  - Similarly, excellent survival is also seen in patients with *POLE*-mutated tumors
- Patients with *BRAF* mutations could potentially benefit from BRAF inhibitors, such as vemurafenib, dabrafenib, and encorafenib

Summary

- Endometrioid and clear cell carcinomas show significant similarities and overlap in their molecular pathology
- The Wnt/β(beta)-catenin pathway, including cyclin D1, is most commonly aberrant in endometrioid carcinomas
  - Mutations in *CTNNB1*, which correlate with immunohistochemical positivity for β(beta)-catenin, are commonly seen, and have been linked to squamous differentiation of the tumor
  - Cyclin D1 immunopositivity may also be seen
- The *PI3K/Akt2/PTEN* pathway is also affected in these two tumor types, with *PIK3CA* abnormalities common in both, and *PTEN* mutations more common in endometrioid carcinomas
  - Of note, endometrioid carcinomas with *PTEN* mutations have been associated with a lower grade and a better prognosis
- *ARID1A* is a putative tumor suppressor gene, with aberrations which can be seen in both tumor types, although they are more frequently associated with clear cell carcinomas
  - In *ARID1A*-mutated carcinomas, loss of the BAF250a protein is seen by immunohistochemistry

- Expression of *HNF1-β(beta)* is relatively specific for clear cell carcinomas, and immunohistochemical staining for the protein, therefore, serves as a useful diagnostic tool
- Microsatellite instability and *POLE* mutation can be seen in endometrioid carcinomas, and are associated with excellent prognosis
  - Microsatellite instability is seen with increased frequency in patients with Lynch syndrome
  - A small percentage of clear cell carcinomas also demonstrate microsatellite instability
- *MAPK* signaling pathway defects, namely mutations in *KRAS* and *BRAF*, may also be present in endometrioid and clear cell carcinomas
- Although *TP53* mutations are more frequently associated with type II tumors, they also occur in a significant number of high-grade endometrioid carcinomas and, less commonly, in clear cell carcinomas
- Clear cell carcinomas and endometrioid carcinomas also rarely demonstrate abnormalities in the gene *PPP2R1A*

Mucinous Tumors

Introduction

- The category of ovarian mucinous tumors includes MBT and mucinous carcinoma, both of which are often unilateral and may grow to a large size prior to resection
- Two types of MBT have been described: the gastrointestinal type and the endocervical-like or seromucinous type
- The more common subtype is the gastrointestinal type, a generally multicystic tumor with an intestinal-type mucinous lining (Fig. 11.6)
- Mucinous cystadenomas can arise in association with mature cystic teratoma or Brenner tumors
- Endocervical-like or seromucinous borderline tumor demonstrates both mucinous and serous-type lining cells, and may also demonstrate endometrioid or eosinophilic epithelium (Fig. 11.7)
  - These tumors are more often bilateral and small, with an architecture resembling that of SBT, and are more frequently associated with endometrioid and clear cell carcinomas
- MBT may display architectural complexity
- The presence of marked nuclear atypia without invasion warrants a diagnosis of mucinous intraepithelial carcinoma
  - The prognosis of these tumors is still favorable
- Mucinous carcinomas, the majority of which are of the gastrointestinal type, are uncommon compared to other types of ovarian tumors (Fig. 11.8)

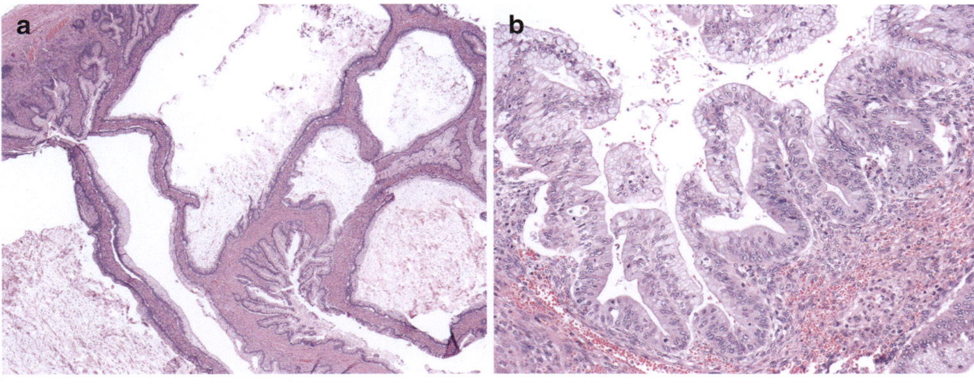

Fig. 11.6 Mucinous borderline tumor. (**a**) Low power demonstrates variably sized, simple to complex cystic spaces filled with mucin. (**b**) At higher power, lining cells demonstrate mild to moderate atypia and areas of stratification

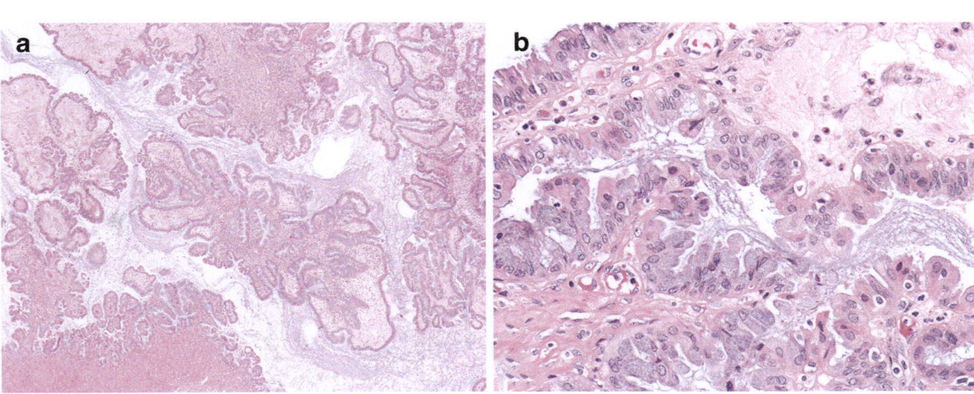

Fig. 11.7 Seromucinous borderline tumor. (**a**) Low power demonstrates complex papillary architecture with hierarchical branching. Some papillae demonstrate stromal edema. (**b**) The lining cells consist of varying proportions of endocervical-type mucinous cells, eosinophilic cells, and ciliated cells

- A well-differentiated architecture is typical, and grading is best determined based on nuclear features
- Invasion in these tumors may be destructive and infiltrative or expansile
- Mucinous carcinomas often coexist with adjacent MBT

Genetic Pathways: Functions, Role in Pathogenesis, and Frequency of Abnormalities

- MAPK signaling pathway
  - *KRAS* (vi-Ki-ras2 Kirsten rat sarcoma 2 viral oncogene homolog) and *BRAF* (v-raf murine sarcoma viral oncogene homolog B1) are both members of the RAS–RAF–MEK–ERK–MAP kinase pathway (see section "Low-Grade Serous Tumors" above), and are also downstream activators of the EGFR pathway
  - Both KRAS and BRAF function as oncogenes
  - Frequency of mutations
    - *KRAS* mutations have been found in 13–56% of mucinous adenomas, 33–79% of mucinous borderline tumors, and 10–75% of mucinous carcinomas
    - Codon 12 is the most common site of mutation in the *KRAS* gene
    - *BRAF* mutations are relatively uncommon in mucinous tumors, found in 9–23% of mucinous carcinomas and in 0–10% of mucinous borderline tumors
    - All *BRAF* mutations were found in exon 15, with most involving codon 600
    - *NRAS* mutations are less common, seen in about 9% of mucinous carcinomas
    - As in other tumor types, *KRAS* and *BRAF* mutations have been found to be mutually exclusive in most cases
- *CDKN2*/p16
  - Mutations in CDKN2A have been seen in about 15% of mucinous tumors overall
- *TP53*
  - Mutations in *TP53* have been seen in about half of mucinous carcinomas, and much less frequently in benign and borderline tumors
- *ERBB2* (HER2)
  - HER2 amplification, as assessed by immunohistochemistry, FISH, and copy number variant analysis, has been seen in about 18–27% of mucinous carcinomas
- Other mutated genes include *RNF43, ELF3, GNAS, ERBB3* and *KLF5*

Clinical Implications

- *KRAS* mutations have been demonstrated to be more common in lower-stage tumors, but no association with prognosis has been found

Fig. 11.8 Mucinous carcinoma. (**a**) Low power demonstrates crowded mucinous glands with complex intraglandular architecture. The invasive pattern is expansile rather than infiltrative. (**b**) At higher power, areas of moderate nuclear atypia and intraglandular cribriform are seen

Fig. 11.9 Benign Brenner tumor. (**a**) Nests of epithelial cells within a fibromatous stroma are seen at low power. (**b**) At high power, the nests are composed of uniform ovoid cells with pale eosinophilic cytoplasm, small nucleoli, and frequent nuclear grooves

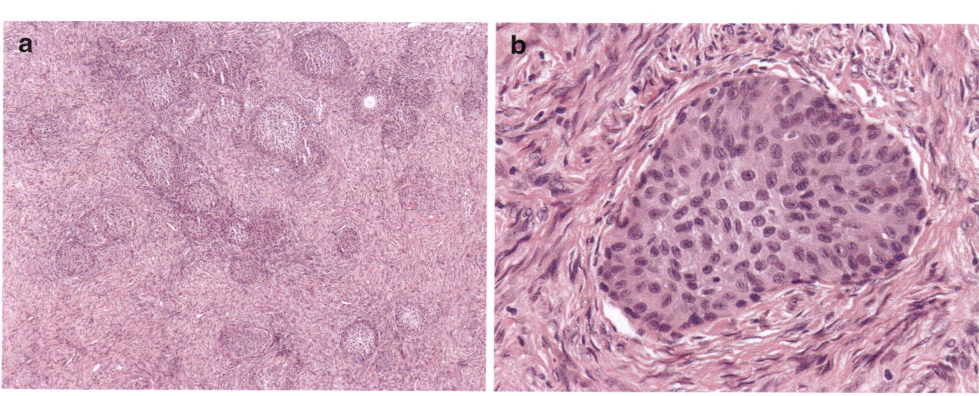

- Patients with *BRAF* mutations could potentially benefit from BRAF inhibitors, such as vemurafenib, dabrafenib, and encorafenib
- Patients with HER2 positivity could potentially be treated with anti-HER2 therapies, such as trastuzumab

Summary

- The most commonly mutated gene in mucinous tumors of the ovary is *KRAS,* with mutations found in adenomas, MBT, and carcinomas
- *BRAF* mutations are much less common, but when present have been found to be mutually exclusive with *KRAS* mutations
- Alterations in *CDKN2A*, *TP53*, and *ERBB2* (HER2) also commonly seen

Brenner Tumors

- Benign Brenner tumor
  - Benign Brenner tumors account for approximately 5% of benign ovarian epithelial tumors
  - Characterized by small oval to irregular nests of bland urothelial epithelium within a dense fibromatous stroma (Fig. 11.9)
  - The etiology of Brenner tumors is not well understood, but they may be derived from Walthard nests in the fallopian tube in some cases
  - Mutations in Brenner tumor are seen at low frequency in a variety of genes, and amplification of *MYC* has been reported
- Borderline Brenner tumor
  - Borderline Brenner tumors are rare and are presumed to arise from benign Brenner tumors
  - Typically, large and cystic, and histologically resemble low-grade papillary urothelial neoplasms of the urinary tract
  - Mutations are not well characterized, but mutations in *CDKN2A*, *KRAS*, and *PIK3CA* have been reported
  - Mutations in *TP53* have not been identified
- Malignant Brenner tumor
  - Malignant Brenner tumors arise from benign and borderline Brenner tumors, and account for <5% of all Brenner tumors
  - Tumors resemble invasive urothelial carcinoma, with highly pleomorphic nucleoli, variably prominent nucleoli, and increased mitotic activity
  - Mutations in *PIK3CA* and *MDM2* amplifications have been reported
  - Mutations in the *TERT* promoter and *TP53* have not been found, and lack of *TERT* promoter mutation

could help distinguish malignant Brenner tumor from metastatic urothelial carcinoma
- Key molecular alterations and exact pathogenesis remain to be fully elucidated

Type II Ovarian Tumors

High-Grade Serous Carcinoma and Carcinosarcoma

Introduction

- HGSC is the most common type of ovarian cancer and usually occurs in the sixth and seventh decades
  - Patients often present at an advanced stage, with abdominal and pelvic dissemination of tumor
  - Architecturally, these carcinomas are complex, with papillary, glandular, cribriform, and solid patterns; necrosis is common (Fig. 11.10)
  - High-grade cytology is seen, with marked nuclear atypia and high mitotic activity
  - Overall survival is generally poor
- Carcinosarcoma, or malignant mixed Mullerian tumor (MMMT), is characterized by both epithelial and stromal malignant components (Fig. 11.11)

  - The epithelial component may be comprised of any ovarian carcinoma type, most often HGSC or endometrioid carcinoma
  - The stromal component demonstrates a sarcomatous appearance and may contain heterologous elements
  - The frequent expression of epithelial markers in the sarcomatous component, as well as the demonstration of monoclonality in these tumors, supports the idea that these are carcinomas with sarcomatoid differentiation

- o Some have referred to them as "metaplastic carcinomas"
- o These tumors often recur as HGSC and show a metastatic pattern similar to HGSC
- HGSC and carcinosarcoma are discussed together in this section because of similar molecular aberrations, particularly in *TP53*, as well as their putative similar origin from STIC

Genetic Pathways: Functions, Role in Pathogenesis, and Frequency of Abnormalities

- *TP53*
  - *TP53* is a tumor suppressor gene located on chromosome 17p, encoding the transcription factor p53, which is involved in regulation of apoptosis

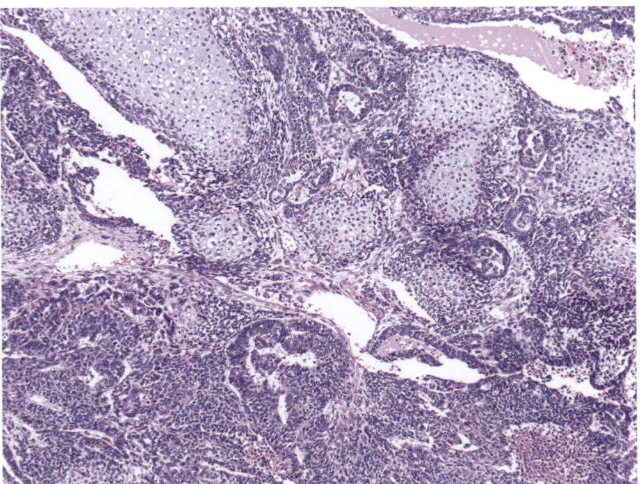

Fig. 11.11 Carcinosarcoma. The epithelial component (lower half of the image) consists of a high-grade adenocarcinoma with serous features. The mesenchymal component (upper half of the image) consists of predominantly heterologous chondroid elements in this example

Fig. 11.10 High-grade serous carcinoma. (**a**) At low power, the tumor demonstrates a typical "labyrinthine" pattern composed of fused complex papillae and intervening slit-like spaces. (**b**) High power demonstrates marked nuclear atypia and pleomorphism as well as frequent mitotic figures

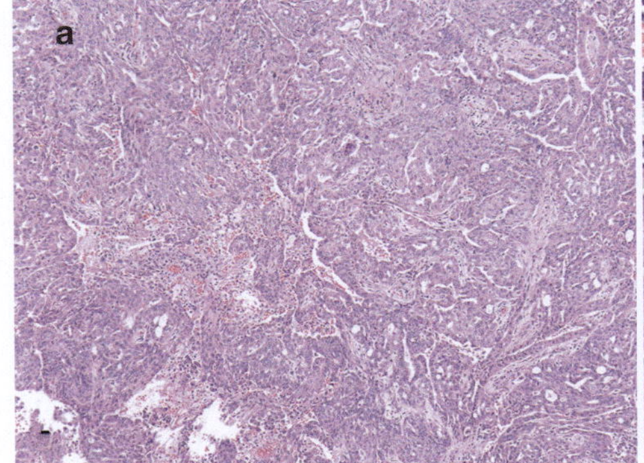
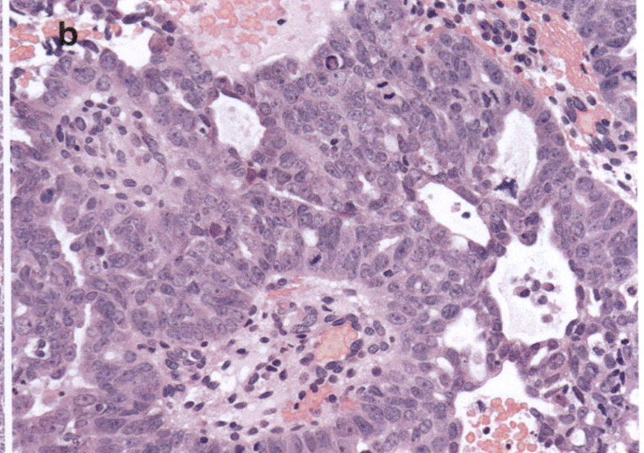

- o *TP53* has frequently been found to be inactivated in diverse tumor types
- − Mutations in *TP53* are the most common and significant molecular abnormality found in type II ovarian carcinomas
- − Frequency of mutations
  - o Mutations in *TP53* are found in 40–60% of all advanced ovarian cancer cases, and as many as 79% of all malignant ovarian or similar peritoneal epithelial tumors
  - o Among pelvic (ovarian, tubal, and peritoneal) HGSC, more than 96% were found to have *TP53* mutations, including tumors of low stage
    - ♦ Most were missense mutations in exons 4–8
    - ♦ Subsequent analysis of the cases lacking *TP53* mutations suggests that essentially all de novo HGSC contains *TP53* somatic mutations/deletions, except for very rare cases of HGSC that arise from a low-grade serous precursor lesion
  - o Carcinosarcoma has also been found to have *TP53* mutations, with identical mutations and LOH patterns in the carcinoma and sarcoma components
  - o In two cases of carcinosarcoma arising in serous carcinoma, the carcinosarcoma was found to have the same *TP53* mutation as the serous carcinoma, supporting the idea of the sarcomatous component arising from the carcinoma
- − Immunohistochemical findings
  - o Most HGSC and carcinosarcomas are p53-immunopositive, with a significant correlation between immunopositivity and *TP53* gene mutations
- • *CDKN2*/p16
  - − The gene *CDKN2* (also known as *p16ink4* or *MTS1*), located on 9p21, encodes the cyclin-dependent kinase inhibitor p16
  - − p16 is a tumor suppressor which binds CDK4 and CDK6, inhibiting the activity of the CDK4–6/cyclin D enzyme complex, which is required for the phosphorylation of Rb and resulting progression of the cell cycle
  - − Immunohistochemical findings
    - o Overexpression of p16 by immunohistochemical analysis has been found in the majority of HGSC, with 83.3% of these showing diffuse positivity, compared to less than 30% of LGSC
    - o Some groups have found differing results (see the section on "LGSC" above); however, the diffuse expression of p16 by immunohistochemistry is still generally most consistent with HGSC
- • Telomere length
  - − Telomeres are the noncoding ends of eukaryotic chromosomes, consisting of guanine-rich simple tandem repeats
- − Telomeres function to protect the chromosome from end-to-end fusions, exonuclease activity, and other damage
- − In normal cells, telomeres shorten with each replication cycle, eventually contributing to the onset of replicative senescence
- − In immortal cell lines, such as tumor cells, the enzyme telomerase is activated, functioning to maintain telomere length, and allowing the cell to continue dividing indefinitely
- − Frequency of abnormalities
  - o STIC has been found to have telomeres shorter than those of normal tubal epithelium in 82% of cases
    - ♦ This shortening of telomeres may be due to ovulation-induced oxidative stress, resulting in chromosomal instability and contributing to the development of STIC
    - ♦ Those STICs which acquire the ability to maintain telomere length may then progress to HGSC
  - o Most HGSC demonstrate shorter telomeres than the associated normal tubal epithelium
    - ♦ This is also true of metastatic malignant cells in ascites specimens compared to the accompanying benign cells
  - o Telomerase activity is seen more frequently and to a greater degree in invasive carcinomas compared to normal ovaries and benign and borderline serous tumors
  - o Telomerase activity also helps distinguish malignant from benign cells in ascites specimens
- • *BRCA1* and *BRCA2*
  - − *BRCA1* and *BRCA2* are primarily linked to hereditary ovarian cancers. See the separate section on "Hereditary/Familial Ovarian Cancer" below for information on familial cases
  - − Sporadic *BRCA1* and *BRCA2* mutation
    - o Both genes and their related pathways may play a role in sporadic cancers, which can be categorized into "*BRCA1*-like" and "*BRCA2*-like" based on gene expression profiles
    - o Epigenetic effects or changes in downstream effectors of *BRCA1* and *BRCA2* may be responsible
    - o Results from The Cancer Genome Atlas project have demonstrated somatic mutations in *BRCA1* or *BRCA2* in 3% of HGSC
    - o 11–31% of lost *BRCA1* expression has been found to be due to DNA hypermethylation rather than mutation
    - o Alterations have also been found in other homologous recombination genes, with approximately half of all HGSC found to have homologous recombination defects

- Tumors with *BRCA1* mutations often demonstrate solid, pseudoendometrioid, and transitional cell carcinoma-like morphology ("SET" features), necrosis, and tumor-infiltrating lymphocytes
- *PI3K/Akt2/PTEN* pathway
  - Abnormalities in this pathway are more commonly found in endometrioid and clear cell carcinomas (discussed previously)
  - However, the oncogenes *Akt2* and *PIK3CA* are more frequently amplified in HGSC than in other tumor types
  - Frequency of abnormalities
    - HGSC is *Akt2*-amplified in 18.2–29% of cases
      - Normal ovarian tissue, benign tumors, borderline tumors, and LGSC show no amplification of the gene
    - Although *PIK3CA* mutations are infrequently seen in HGSC, *PIK3CA* amplifications are seen in approximately 13% of these
    - Recurrent mutations in *PIK3CA* and *PTEN* and frequent amplification of chromosomal segments containing *PIK3CA* have also been identified in carcinosarcoma
- MAPK
  - *MAPK* (mitogen-activated protein kinase) is most frequently expressed in low-grade serous tumors (see above) but has also been found in HGSC
  - Immunohistochemical findings
    - By immunohistochemical staining, activated (phosphorylated) MAPK was found to be expressed in 41% of HGSC
    - In contrast to LGSC, HGSC demonstrating *MAPK* expression all had wildtype *KRAS* and *BRAF*
- *Notch3*
  - Notch receptors are membrane receptors which play a role in cell fate regulation, cell proliferation, and cell death during development
  - The *Notch3* gene, located at 19p13.2, encodes one such Notch receptor
  - Frequency of abnormalities
    - Overexpression of *Notch3* is more common in HGSC (amplification frequency of 19.5%, overexpression in 66%) vs. low-grade serous tumors and nonneoplastic epithelium
  - Immunohistochemical findings
    - Immunohistochemical staining for Notch3 (both nuclear and cytoplasmic) has been found in 55% of ovarian carcinomas but not in the normal ovarian surface epithelium
      - The intensity of staining is correlated with the DNA copy ratio
- *HBXAP (Rsf-1)*

- The gene *Rsf-1* (*HBXAP*, Hepatitis B virus x-associated protein), located at 11q13.5, encodes a protein which partners with hSNF2H to form the RSF complex; this complex is involved in chromatin remodeling
- Frequency of abnormalities
  - Amplification of the 11q13.5 locus has been found in 13.2–15.7% of HGSC, with *Rsf-1* found to have the most significantly amplified mRNA expression among genes at this locus
  - No amplification is seen in low-grade tumors and normal ovaries
- Immunohistochemical findings
  - Immunohistochemical staining for Rsf-1 correlates with the presence of gene amplification
  - A correlation has also been found between the intensity of Rsf-1 nuclear immunostaining and that for hSNF2H, with evidence suggesting that Rsf-1 may stabilize the hSNF2H protein
- *NAC1*
  - NAC1 (nucleus accumbens 1), encoded by the gene *NAC1* on 19p13, is a member of the BTB/POZ domain family and contains a domain which may play a role in chromatin organization and transcription
  - The role of NAC1 in ovarian cancer pathogenesis may also be partly mediated by its negative regulation of the growth inhibitor Gadd45GIP1 (DNA-damage-inducible 45-gamma interacting protein)
  - Immunohistochemical findings
    - Immunopositivity for NAC1 is stronger in serous carcinomas than in benign tumors or normal ovaries, with high immunointensity seen more frequently in HGSC compared to LGSC
    - Higher staining intensity and mRNA levels were also found in recurrent tumors compared to primary tumors
- Histone genes
  - High frequency of mutations in the genes encoding histone H2A and H2B has been identified in carcinosarcoma of the ovary and uterus
  - Frequent amplification of the segment of chromosome 6p containing the histone gene cluster with these genes is also seen
  - Mutation of histone proteins has been associated with increased expression of markers of epithelial-mesenchymal transition and may play a role in sarcomatous transformation
- HLA-G
  - HLA-G (human leukocyte antigen G) is a major histocompatibility (MHC) protein, the expression of which has been shown to facilitate evasion of immunosurveillance by tumor cells and has been linked to multiple nonovarian cancers

- Immunohistochemical findings
  - By immunohistochemical analysis, 61% of HGSC have been found to express HLA-G, with a discrete membranous staining pattern
  - Expression has not been found in low-grade serous tumors or normal ovarian surface epithelium
- By PCR analysis, the HLA-G isoforms 1 and 5 were found to predominate in HGSC
- Cyclin E1 (*CCNE1*)
  - Cyclin E, encoded by the gene *CCNE1* at 19q13, is involved in promoting the progression of the cell cycle from S1 to the G phase
  - Frequency of abnormalities
    - Amplification of the *CCNE1* locus has been found to be specific for HGSC vs. LGSC or normal ovarian tissue, with a frequency of 32.2–36.1%
  - Immunohistochemical findings
    - High cyclin E1 expression by immunohistochemical analysis has been correlated with the amplification of the *CCNE1* gene

Clinical Implications

- *TP53*
  - *TP53* gene mutations and overexpression have been linked to cisplatin resistance, resulting from the inability of the mutated protein to activate apoptosis
  - The evidence for the prognostic significance of *TP53* mutations is still somewhat contradictory, with some evidence pointing to these mutations as a negative prognostic factor, others finding no correlation, and one study even finding a short-term survival benefit
    - Additional data is needed to more definitively define the role of p53 in prognosis
  - Immunohistochemical staining for p53 is useful in differentiating type I and type II tumors
- Telomere length and telomerase activity
  - The increased telomerase activity seen in HGSC suggests a potential utility for telomerase inhibitors in treatment; several studies have explored this potential
  - The cytokine interferon-β(beta) (IFN-β(beta)), which inhibits tumor cell growth, was found to suppress telomerase activity in ovarian cancer cells
  - Inhibition of hTERT (human telomerase reverse transcriptase), the major site of transcriptional regulation of the enzyme, has demonstrated rapid inhibition of growth in ovarian cancer cell lines
- *CDKN2/p16*
  - The combination of block positivity for p16 and loss of staining for Rb by immunohistochemistry has been associated with significantly increased overall survival in ovarian HGSC

- These tumors may have increased sensitivity to cytotoxic chemotherapy
- *BRCA1* and *BRCA2*
  - In comparison to women with sporadic ovarian cancer, those with *BRCA1*- and *BRCA2*-mutated cancers have better outcomes
  - Patients with epigenetically silenced *BRCA1* have survival similar to those with wildtype *BRCA1*
  - *PARP1* (poly-ADP-ribose-polymerase) is a nuclear enzyme required for base excision repair of single-strand breaks
    - *PARP* inhibitors are effective in tumors with defects in DNA repair, including those with *BRCA1* and *BRCA2* mutations, by inducing a catastrophic level of DNA damage in cancer cells
    - Based on the results of large clinical trials, *PARP* inhibitors are now standard of care as maintenance therapy for patients with advanced ovarian cancer and either germline or somatic mutations in *BRCA1* or *BRCA2*
- *PI3K/Akt2/PTEN* pathway
  - Amplification of *Akt2* has been associated with undifferentiated histology and with age over 50 years; a trend towards higher mortality is also observed
- MAPK
  - A greater expression of *MAPK* is seen in high-grade tumors from younger patients
- *Notch3*
  - Tumors with *Notch3* overexpression may be amenable to targeted therapy, either by γ(gamma)-secretase inhibitors or by disruption of Notch3 and ligand binding
  - γ(gamma)-secretase inhibitors prevent activation of Notch3, and are found to inhibit proliferation and promote apoptosis in *Notch3*-expressing cancer cell lines
- *HBXAP (Rsf-1)*
  - *Rsf-1* may have prognostic significance, as patients with HGSC and *Rsf-1* amplification demonstrate shorter overall survival compared to those with non-amplified tumors
  - Implications for treatment also exist, with silencing of Rsf-1 in overexpressing cell lines resulting in significant inhibition of growth, and expression of *Rsf-1* in cell lines being associated with paclitaxel resistance
- NAC1
  - The intensity of NAC1 immunostaining was found to be predictive of recurrence within 1 year in patients with advanced-stage HGSC status post optimal debulking and standard chemotherapy
  - *NAC1* expression may also be associated with resistance to paclitaxel and resulting in shorter survival in paclitaxel-treated patients

○ This resistance may be mediated by Gadd45GIP1, which is also a potential target for treatment

- HLA-G
  - An HLA-G-specific ELISA test has been developed to measure sHLA-G, a product of the HLA-G5 isoform
    ○ Using this test, sHLA-G was found in almost all malignant ascites samples and at significantly higher levels than in benign samples, indicating potential use as a diagnostic tumor marker
  - HLA-G may also have prognostic implications, with an association between the presence of HLA-G-expressing tumor cells in effusions and better survival
- Cyclin E1/*CCNE1*
  - High cyclin E1 expression has been associated with poor clinical outcomes in multiple studies
  - High-level amplification of cyclin E1 may be mutually exclusive with germline BRCA1/2 mutation
  - Coamplification of CCNE1 and BRD4 may be particularly associated with platinum resistance and poor outcomes

Summary

- Mutations in *TP53* are the most common genetic abnormality in HGSC and in ovarian cancer overall
  - The presence of these mutations is helpful for diagnosis and is thought to confer a worse prognosis, although data on the latter are somewhat contradictory
  - In carcinosarcoma, identical *TP53* mutations have been found in both the epithelial and stromal components
- p16 is commonly overexpressed in HGSC
- The enzyme telomerase, expressed in a wide variety of tumors, is also found in ovarian carcinomas and has demonstrated potential as a target for therapy
- Expression of *MAPK*, although more common in type I tumors, may also be seen in HGSC
  - *MAPK*-expressing HGSC are wildtype for *KRAS* and *BRAF*
- Within the *PI3K/Akt2/PTEN* pathway, amplification of *Akt2* is most associated with HGSC and has been associated with undifferentiated histology and a worse prognosis
  - *PIK3CA* amplification is also seen in HGSC
- *BRCA1* and *BRCA2* are the genes responsible for most familial cases of breast and ovarian cancer
  - Patients with ovarian carcinomas in this setting have better outcomes than those with sporadic cancers
  - *PARP* inhibitors have become standard therapy in patients with somatic or germline *BRCA1* or *BRCA2* mutations

- These genes and their related pathways may also be involved in sporadic cancers
- Several other genetic abnormalities have also been described in HGSC, including overexpression of *Notch3, HBXAP, NAC1, HLA-G,* and *Cyclin E*
  - Many of these abnormalities have diagnostic and therapeutic relevance, as discussed above

Undifferentiated/Dedifferentiated Carcinoma

- Undifferentiated tumors lack evidence of specific differentiation, and dedifferentiated tumors contain both an undifferentiated and a differentiated component (Fig. 11.12)
- Tumors usually present at a high stage, and median survival in one study was 9 months
- Associated with low-grade endometrioid carcinoma in a subset of tumors, suggesting tumor progression
- Similar to pathogenesis to counterpart in the uterine corpus, which is more common
- Molecular pathways
  - Inactivating mutations in chromatin remodeling genes in SWI/SNF complex lead to loss of expression of *SMARCA2, SMARCA4 (BRG1), SMARCB1 (INI1),* and/or both *ARID1A* and *ARID1B* in undifferentiated component
  - Abnormalities in DNA mismatch repair proteins are seen in about half of the tumors
  - Somatic mutations were seen in *PIK3CA, CTNNB1, TP53, FBXW7,* and *PPP2R1A*, with the same mutations seen in undifferentiated and concurrent differentiated components

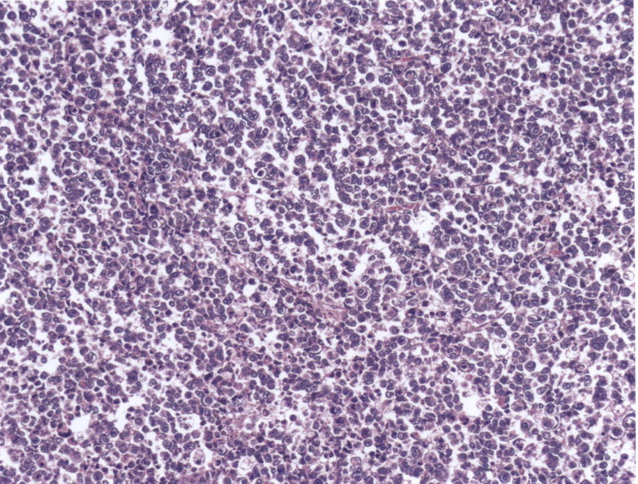

Fig. 11.12 Undifferentiated carcinoma. The tumor consists of sheets of discohesive tumor cells with high-grade nuclear features. No glandular or other architectural patterns are appreciated

Molecular Abnormalities Not Associated with Specific Histology

- EGFR family
  - The EGFR (epidermal growth factor receptor) family, also known as the ERBB or HER family, is a group of transmembrane receptors which include both EGFR and HER2 (*ERBB2*)
  - Activation of EGFR family receptors by ligand binding leads to the activation of multiple different signaling pathways, including *MAPK* and *PI3K/Akt2/PTEN*, with effects on cell survival, proliferation, and differentiation (Fig. 11.2)
  - Frequency of abnormalities
    - Amplification of *EGFR* has been found in up to 22% of ovarian cancers overall
      - Activating mutations may also be seen
    - Amplifications of *HER2* have been found in 23% of borderline tumors and 8–66% of ovarian carcinomas, both of various subtypes
      - Frequency in specific histotypes is discussed above
  - Immunohistochemical findings
    - EGFR expression by immunohistochemistry has been found in as many as 64.5% of invasive ovarian carcinomas overall, including mucinous, serous, and endometrioid carcinomas
    - HER2 overexpression by immunohistochemistry may overestimate gene amplification by FISH in ovarian HGSC
    - Recent evidence suggests that EGFR and HER2 expression may be mutually exclusive in HGSC
  - Clinical implications
    - There may be a role for EGFR inhibitors in patients with *EGFR* mutations, although the evidence remains unclear
      - The response rate has been relatively low, possibly because the mutations most often found in ovarian cancers are not the same as those found in non-small cell lung cancers
    - EGFR expression has been associated with poor outcomes in multiple studies, as well as with higher tumor grade, abnormal *TP53* expression, larger residual tumor size, and a higher proliferation index
    - Some studies suggest that *HER2* amplification/overexpression may be associated with higher-stage tumors and poor prognosis, although conflicting data have also been seen
    - Trastuzumab and pertuzumab are monoclonal anti-HER2 antibodies commonly used in the treatment of *HER2*-amplified breast cancer
      - Clinical utility is limited by the low frequency of HER2 overexpression and low response rates among patients with HER2 overexpression
      - In ovarian cancer, overall response rates for these two drugs individually have been relatively low, but some evidence suggests a potentially greater response rate using both in combination
    - The mutual exclusivity of EGFR and HER2 expression in HGSC may limit the usefulness of combination anti-HER2 and anti-EGFR therapy for this disease
- DNA methylation
  - DNA methylation is an epigenetic alteration which has been shown to occur aberrantly in a wide variety of human neoplasms
  - Global hypomethylation results in the activation of oncogenes, while tumor suppressor genes can be silenced via hypermethylation of CpG islands within their promoter regions
  - Hypermethylation of various genes has been described in ovarian cancers
    - For example, promoter hypermethylation of specific genes (*CDKN2, E-cadherin, RAR-β(beta), H-cadherin, APC, GSTP1, MGMT,* and *RASSF1A*) increases in frequency from benign cystadenomas to invasive carcinomas
  - Hypomethylation has also been shown to progressively increase from non-neoplastic ovarian tissue to carcinoma
  - Other examples of aberrant methylation status have been discussed earlier in relation to specific genetic loci
  - Clinical implications
    - The detection of DNA methylation status has promising potential as a screening tool, particularly with the development of sensitive assays to detect the methylation status of multiple genes, and the potential to detect biomarkers in fluids draining the tumor site
    - Strong hypomethylation in ovarian cancer tissue has been associated with advanced-stage and high grade
- PD-1/PD-L1 axis
  - PD-1/PD-L1 expression has been reported in ovarian carcinomas of various histology
    - Tumor cell expression of PD-L1 seen in about half of ovarian cancer overall
      - Higher expression levels seen in tumor-infiltrating lymphocytes
      - Lower expression than seen in many other solid malignancies

- o Approximately 45% of ovarian clear cell carcinomas showed increased expression of PD-1/PD-L1 in one study
- o PD-L1 expression has been reported in ~25% of high-grade serous carcinoma
- o Higher expression of PD-L1 in ovarian cancer may be associated with mutations in *TP53* and *BRCA1/2* and with worse survival in high-grade serous carcinoma
- – Tumor mutational burden (TMB), another biomarker of potential response to immune checkpoint inhibitors, is lower in ovarian cancer than many other solid malignancies
- – Initial clinical trials of immune checkpoint inhibitors in ovarian cancer have largely been disappointing
  - o JAVELIN Ovarian 100 phase 3 trial of avelumab (PD-L1 inhibitor) discontinued due to futility
  - o IMagyn050/GOG3015/ENGOT-OV39 phase 3 trial of atezolizumab (PD-L1 inhibitor) showed no significant difference in survival
- – Pembrolizumab was approved in 2017 for use in metastatic solid malignancies with microsatellite instability and/or deficiency in DNA mismatch repair (MMR) genes, including ovarian cancer, based on the KEYNOTE-158 trial
  - o Trial only included 15 patients with ovarian cancer, with an objective response rate seen in about one third
- – Many clinical trials of immunotherapy agents in ovarian cancer remain ongoing, with active research to determine biomarkers that may best predict which patient subsets may respond best, given low overall response rates

Molecular Abnormalities Associated with Rarer/Nonepithelial Ovarian Cancers

- Adult granulosa cell tumor
  - – Nearly all tumors have a recurrent somatic *FOXL2* mutation, p.C134W
  - – The effect of this mutation on tumor pathogenesis is unclear
  - – While tumors are typically positive for FOXL2 by immunohistochemistry, positive expression is frequently seen in other sex-cord stromal tumors and is not specific for adult granulosa cell tumor
  - – The *FOXL2* p.C134W mutation is associated with increased expression of aromatase, which may cause estrogenic symptoms
- Juvenile granulosa cell tumor
  - – Mutations in *AKT1* and *GNAS* have been reported in approximately 60% and 30% of juvenile granulosa cell tumors

- – Inhibition of AKT could be a potential avenue for targeted therapy
- Sertoli-Leydig cell tumor
  - – Hotspot mutations in the RNase IIIb domain of *DICER1* is seen in about half of cases
    - o These are seen as germline mutations in about two-thirds of cases tested
    - o May play a role in inducing sertoliform differentiation and causing androgenic symptoms
    - o *DICER1* mutation is associated with retiform differentiation, presence of heterologous elements, and moderately to poorly differentiated histology, with one study finding that 100% of moderately to poorly differentiated tumors harbored *DICER1* mutations
    - o See also section below: "Hereditary/Familial Ovarian Cancers"
  - – *FOXL2* p.C134W mutation is seen in 0–22% of tumors and is mutually exclusive with *DICER1*
- Germ cell tumors
  - – Chromosome 12 abnormalities are frequently seen in dysgerminoma (~80%), yolk sac tumor (~60%), and embryonal carcinoma (~80%)
  - – Typically seen as isochromosome 12p or 12p amplification
  - – Mutations in *KIT* (exon 17, codon 816) are seen in 30–50% of dysgerminoma and amplification of *KIT* is seen in 30%
- Small cell carcinoma of the ovary of hypercalcemic type
  - – Somatic or germline mutations in *SMARCA4* are detected in virtually all cases
- Sex cord tumor with annular tubules
  - – Commonly associated with Peutz-Jeghers syndrome
  - – Syndromic cases have germline *STK11* gene mutations, located on chromosome 19p13.3
    - o See also section below: "Hereditary/Familial Ovarian Cancers"

Hereditary/Familial Ovarian Cancers

- Familial ovarian cancers tend to occur at a younger age compared to sporadic cancers
- Approximately 10–17% of ovarian cancer occurs in patients with a known predisposing genetic mutation, primarily *BRCA1* and *BRCA2*
- *BRCA1* and *BRCA2*
  - – Both *BRCA1* and *BRCA2* are tumor suppressor genes that are involved in DNA double-strand break repair
  - – *BRCA1*, located on 17q21, and *BRCA2*, located on 13q12, have both been linked to a hereditary predisposition for breast cancer
  - – Mutations in *BRCA1* and *BRCA2* have also been linked to ovarian cancers

- It has been found that most breast-ovarian cancer families carry *BRCA1* mutations, and most of the remainder carry *BRCA2* mutations
- *BRCA1* mutations particularly predispose to serous carcinoma, which comprises 90% of ovarian cancers in *BRCA1* mutation carriers
- *BRCA1* and *BRCA2* can also be mutated in sporadic ovarian cancers (see above section: "HGSC and Carcinosarcoma" ➜ "*BRCA1* and *BRCA2*")
- PARP inhibitors are the standard treatment for patients with both germline and somatic *BRCA1* and *BRCA2* mutations (see above section: "HGSC and Carcinosarcoma" ➜ "Clinical Implications")
- Lynch syndrome
  - Lynch syndrome, also known as hereditary nonpolyposis colorectal cancer or HNPCC, results from germline mutations in one or more of the DNA mismatch repair (MMR) genes *MLH1, MSH2, MSH6,* and *PMS2*
  - Patients most commonly develop colorectal, skin, and endometrial cancers
  - The risk for ovarian cancer varies among each of the four genes. Cumulative incidence by age 75 is 11% for *MLH1*, 17.4% for *MSH2*, 10.8% for *MSH6*, and 3.0% for *PMS2*
  - The most common histologic types associated with Lynch syndrome are mixed (mucinous/endometrioid/clear cell; 33%), endometrioid (25%), serous (22%), clear cell (12%,) and mucinous (4%)
  - Most patients (65%) have FIGO stage I or II cancer
  - The average age at diagnosis is 45.3 years, with a median of 43 years and a range of 19–82 years
  - Use of the immune checkpoint inhibitor pembrolizumab has been demonstrated to be beneficial in the treatment of solid noncolorectal MMR-deficient tumors
  - See also the section "Endometrioid and Clear Cell Carcinomas" ➜ "Microsatellite Instability"
- *BRIP1, RAD51C,* and *RAD51D*
  - *BRIP1* (*BRCA1*-interacting protein C-terminal helicase 1), encodes the Fanconi anemia group J protein
  - *BRIP1, RAD51C* (*RAD51* homolog C) and *RAD51D* (*RAD51* paralog D) all code for proteins which interact with BRCA1/2 and which are involved in DNA repair
  - The odds ratios for ovarian cancers in patients with these mutations compared to controls are 4.94 for *BRIP1*, 5.59 for *RAD51C*, and 6.94 for *RAD51D*
  - Per one study, these three genes overall comprise a 10% contribution to hereditary ovarian cancer, which would make them the largest contribution to hereditary ovarian cancer after *BRCA1/BRCA2*

- Mutations in these genes are associated primarily with high-grade serous histology
- As with *BRCA1* and *BRCA2* mutations, ovarian cancer patients with these mutations could potentially benefit from treatment with PARP inhibitors
- *DICER1*
  - Germline mutations causing a loss of function in *DICER1* are associated with multiple tumors including thyroid neoplasms, pleuropulmonary blastoma, cystic nephroma, ovarian Sertoli-Leydig cell tumor, and gynandroblastoma
  - The syndrome demonstrates autosomal dominant inheritance with reduced penetrance
  - Approximately two-thirds of Sertoli-Leydig cell tumor patients with *DICER1* tumor mutations have been found to have a germline mutation
  - See also the section "Molecular Abnormalities Associated with Rarer/Nonepithelial Ovarian Cancers" ➜ "Sertoli-Leydig Cell Tumors"
- Peutz-Jeghers syndrome
  - Peutz-Jeghers syndrome (PJS) is an autosomal dominant disease caused by a germline mutation in the tumor suppressor gene *STK11/LKB1* on chromosome 19p13.3
  - Patients commonly develop multiple hamartomatous intestinal polyps and mucocutaneous melanin macules and can develop a variety of benign and malignant neoplasms
  - Approximately one-third of patients with ovarian sex cord tumors with annular tubules (SCTAT) have PJS
    - SCTATs associated with PJS, in comparison with sporadic SCTATs, are more likely to be multifocal, bilateral, small, and associated with benign behavior
    - See also the section "Molecular Abnormalities Associated with Rarer/Nonepithelial Ovarian Cancers"
  - Recently, an unusual adnexal neoplasm characterized by *STK11* mutations and an association with PJS has been described
    - While most of these tumors arose in a paratubal location, a small number appeared to arise from the ovary
    - The exact histogenesis of these tumors is not yet characterized, and they demonstrate variable morphology overlapping with Wolffian tumors, sex cord–stromal tumors, and adenocarcinomas
    - 50% of the reported cases were associated with PJS
- Several other genes, including *BARD1, PALB2, NBN, CHEK2,* and *ATM*, have also been proposed to increase the risk of ovarian cancer, although some studies have not found evidence of a significant association

Conclusions

- Studies of the molecular characteristics of ovarian cancer have led to numerous new insights into the pathophysiology of this disease
- Differences in the involved molecular pathways have supported the division of epithelial ovarian cancers into type I tumors, which are slow-growing tumors thought to develop in a stepwise manner from benign and borderline precursors, and type II tumors, which are more aggressive and arise in a de novo fashion
- Although there is much overlap in the molecular pathways involved in the various ovarian tumor types, each histologic type is associated with certain characteristic abnormalities
- These insights into the specific aberrations present in each ovarian cancer histotype have translated into insights on new diagnostic, therapeutic, and prognostic modalities
- In particular, the use of *PARP* inhibitors has become the standard of care in certain patients with high-grade serous carcinoma of the ovary and *BRCA1* or *BRCA2* mutations
- Several familial ovarian cancer syndromes have been described, with the majority secondary to *BRCA1* and *BRCA2* mutations
- Familial ovarian cancers tend to occur at a younger age compared to sporadic ovarian cancer
- Characteristic mutations in several rarer ovarian cancer subtypes have been identified in recent years
- Continued efforts to better understand the molecular characteristics of ovarian cancer promise to offer further insights into its pathophysiology and best clinical management, with the hopes of ultimately reducing the burden of this high-mortality disease

Further Reading

Abeln EC, Smit VT, Wessels JW, et al. Molecular genetic evidence for the conversion hypothesis of the origin of malignant mixed müllerian tumours. J Pathol. 1997;183:424–31.

Afify AM, Werness BA, Mark HF. HER2/neu oncogene amplification in stage I and stage III ovarian papillary serous carcinoma. Exp Mol Pathol. 1999;66:163–9.

Ahmed AA, Etemadmoghadam D, Temple J, et al. Driver mutations in TP53 are ubiquitous in high grade serous carcinoma of the ovary. J Pathol. 2010;221:49–56.

Al-Agha OM, Huwait HF, Chow C, et al. FOXL2 is a sensitive and specific marker for sex cord-stromal tumors of the ovary. Am J Surg Pathol. 2011;35:484–94.

Anglesio MS, Kommoss S, Tolcher MC, et al. Molecular characterization of mucinous ovarian tumours supports a stratified treatment approach with HER2 targeting in 19% of carcinomas. J Pathol. 2013;229:111–20.

Anglesio MS, Wang Y, Yang W, et al. Cancer-associated somatic DICER1 hotspot mutations cause defective miRNA processing and reverse-strand expression bias to predominantly mature 3p strands through loss of 5p strand cleavage. J Pathol. 2013;229:400–9.

Arend RC, Davis AM, Chimiczewski P, et al. EMR 20006-012: a phase II randomized double-blind placebo controlled trial comparing the combination of pimasertib (MEK inhibitor) with SAR245409 (PI3K inhibitor) to pimasertib alone in patients with previously treated unresectable borderline or low grade ovarian cancer. Gynecol Oncol. 2020;156:301–7.

A study of MEK162 vs. physician's choice chemotherapy in patients with low-grade serous ovarian, fallopian tube or peritoneal cancer. https://ClinicalTrials.gov/show/NCT01849874.

Auner V, Kriegshäuser G, Tong D, et al. KRAS mutation analysis in ovarian samples using a high sensitivity biochip assay. BMC Cancer. 2009;9:111.

Barton CA, Hacker NF, Clark SJ, et al. DNA methylation changes in ovarian cancer: implications for early diagnosis, prognosis and treatment. Gynecol Oncol. 2008;109:129–39.

Bellacosa A, de Feo D, Godwin AK, et al. Molecular alterations of the AKT2 oncogene in ovarian and breast carcinomas. Int J Cancer. 1995;64(4):280–5.

Bennett JA, Pesci A, Morales-Oyarvide V, et al. Incidence of mismatch repair protein deficiency and associated clinicopathologic features in a cohort of 104 ovarian endometrioid carcinomas. Am J Surg Pathol. 2019;43:235–43.

Bennett JA, Young RH, Howitt BE, et al. A distinctive adnexal (usually paratubal) neoplasm often associated with Peutz-Jeghers syndrome and characterized by STK11 alterations (STK11 adnexal tumor): a report of 22 cases. Am J Surg Pathol. 2021;45:1061–74.

Berchuck A, Kohler MF, Marks JR, et al. The p53 tumor suppressor gene frequently is altered in gynecologic cancers. Obstet Gynecol. 1994;170:246–52.

Bessiere L, Todeschini AL, Auguste A, et al. A hot-spot of in-frame duplications activates the oncoprotein AKT1 in juvenile granulosa cell tumors. EBioMedicine. 2015;2:421–31.

Bookman MA, Darcy KM, Clarke-Pearson D, et al. Evaluation of monoclonal humanized anti-HER2 antibody, trastuzumab, in patients with recurrent or refractory ovarian or primary peritoneal carcinoma with overexpression of HER2: a phase II trial of the gynecologic oncology group. J Clin Oncol. 2003;21:283–90.

Buza N, Wong S, Hui P. FOXL2 mutation analysis of ovarian sex cord-stromal tumors: genotype-phenotype correlation with diagnostic considerations. Int J Gynecol Pathol. 2018;37:305–15.

Campbell IG, Russell SE, Choong DY, et al. Mutation of the PIK3CA gene in ovarian and breast cancer. Cancer Res. 2004;64:7678–81.

Cancer Genome Atlas Research Network. Integrated genomic analyses of ovarian carcinoma. Nature. 2011;474:609–15.

Catasús L, Bussaglia E, Rodrguez I, et al. Molecular genetic alterations in endometrioid carcinomas of the ovary: similar frequency of beta-catenin abnormalities but lower rate of microsatellite instability and PTEN alterations than in uterine endometrioid carcinomas. Hum Pathol. 2004;35:1360–8.

Champer M, Miller D, Kuo DY. Response to trametinib in recurrent low-grade serous ovarian cancer with NRAS mutation: a case report. Gynecol Oncol Rep. 2019;28:26–8.

Chan AM, Enwere E, McIntyre JB, et al. Combined CCNE1 high-level amplification and overexpression is associated with unfavourable outcome in tubo-ovarian high-grade serous carcinoma. J Pathol Clin Res. 2020;6:252–62.

Chao WR, Lee MY, Lin WL, et al. Assessing the HER2 status in mucinous epithelial ovarian cancer on the basis of the 2013 ASCO/CAP guideline update. Am J Surg Pathol. 2014;38:1227–34.

Cheng L, Roth LM, Zhang S, et al. KIT gene mutation and amplification in dysgerminoma of the ovary. Cancer. 2011;117:2096–103.

Cheng L, Zhang S, Talerman A, et al. Morphologic, immunohistochemical, and fluorescence in situ hybridization study of ovarian embryonal carcinoma with comparison to solid variant of yolk sac tumor and immature teratoma. Hum Pathol. 2010;41:716–23.

Cherniack AD, Shen H, Walter V, et al. Integrated molecular characterization of uterine carcinosarcoma. Cancer Cell. 2017;31:411–23.

Chetrit A, Hirsh-Yechezkel G, Ben-David Y, et al. Effect of BRCA1/2 mutations on long-term survival of patients with invasive ovarian cancer: the National Israeli Study of Ovarian Cancer. J Clin Oncol. 2008;26:20–5.

Cho KR. Ovarian cancer update: lessons from morphology, molecules, and mice. Arch Pathol Lab Med. 2009;133:1775–81.

Choi JH, Sheu JJ, Guan B, et al. Functional analysis of 11q13.5 amplicon identifies rsf-1 (HBXAP) as a gene involved in paclitaxel resistance in ovarian cancer. Cancer Res. 2009;69:1407–15.

Coatham M, Li X, Karnezis AN, et al. Concurrent ARID1A and ARID1B inactivation in endometrial and ovarian dedifferentiated carcinomas. Mod Pathol. 2016;29:1586–93.

Counter CM, Hirte HW, Bacchetti S, et al. Telomerase activity in human ovarian carcinoma. Proc Natl Acad Sci U S A. 1994;91:2900–4.

Crasta J, Ravikumar G, Rajarajan S, et al. Expression of HER2 and EGFR proteins in advanced stage high-grade serous ovarian tumors show mutual exclusivity. Int J Gynecol Pathol. 2021;40:49–55.

Cristescu R, Mogg R, Ayers M, et al. Pan-tumor genomic biomarkers for PD-1 checkpoint blockade-based immunotherapy. Science. 2018;362:eaar3593.

Cuatrecasas M, Catasus L, Palacios J, et al. Transitional cell tumors of the ovary: a comparative clinicopathologic, immunohistochemical, and molecular genetic analysis of Brenner tumors and transitional cell carcinomas. Am J Surg Pathol. 2009;33:556–67.

Cuatrecasas M, Erill N, Musulen E, et al. K-ras mutations in nonmucinous ovarian epithelial tumors: a molecular analysis and clinicopathologic study of 144 patients. Cancer. 1998;82:1088–95.

Cuatrecasas M, Villanueva A, Matias-Guiu X, et al. K-ras mutations in mucinous ovarian tumors: a clinicopathologic and molecular study of 95 cases. Cancer. 1997;79:1581–6.

Datar RH, Naritoku WY, Li P, et al. Analysis of telomerase activity in ovarian cystadenomas, low-malignant-potential tumors, and invasive carcinomas. Gynecol Oncol. 1999;74:338–45.

Davidson B, Elstrand MB, McMaster MT, et al. HLA-G expression in effusions is a possible marker of tumor susceptibility to chemotherapy in ovarian carcinoma. Gynecol Oncol. 2005;96:42–7.

de Kock L, Terzic T, McCluggage WG, et al. DICER1 mutations are consistently present in moderately and poorly differentiated Sertoli-Leydig cell tumors. Am J Surg Pathol. 2017;41:1178–87.

Despierre E, Lambrechts D, Neven P, et al. The molecular genetic basis of ovarian cancer and its roadmap towards a better treatment. Gynecol Oncol. 2010;117:358–65.

Dimova I, Zaharieva B, Raitcheva S, et al. Tissue microarray analysis of EGFR and erbB2 copy number changes in ovarian tumors. Int J Gynecol Cancer. 2006;16:145–51.

Dominguez-Valentin M, Sampson JR, Seppala TT, et al. Cancer risks by gene, age, and gender in 6350 carriers of pathogenic mismatch repair variants: findings from the prospective lynch syndrome database. Genet Med. 2020;22:15–25.

Drew Y, Calvert H. The potential of PARP inhibitors in genetic breast and ovarian cancers. Ann N Y Acad Sci. 2008;1138:136–45.

Faratian D, Zweemer AJ, Nagumo Y, et al. Trastuzumab and pertuzumab produce changes in morphology and estrogen receptor signaling in ovarian cancer xenografts revealing new treatment strategies. Clin Cancer Res. 2011;17:4451–61.

Farley J, Brady WE, Vathipadiekal V, et al. Selumetinib in women with recurrent low-grade serous carcinoma of the ovary or peritoneum: an open-label, single-arm, phase 2 study. Lancet Oncol. 2013;14:134–40.

Farley J, Smith LM, Darcy KM, et al. Cyclin E expression is a significant predictor of survival in advanced, suboptimally debulked ovar-

ian epithelial cancers: a gynecologic oncology group study. Cancer Res. 2003;63:1235–41.

Fleming NI, Knower KC, Lazarus KA, et al. Aromatase is a direct target of FOXL2: C134W in granulosa cell tumors via a single highly conserved binding site in the ovarian specific promoter. PLoS One. 2010;5:e14389.

Friedlander ML, Russell K, Millis S, et al. Molecular profiling of clear cell ovarian cancers: identifying potential treatment targets for clinical trials. Int J Gynecol Cancer. 2016;26:648–54.

Fujita M, Enomoto T, Yoshino K, et al. Microsatellite instability and alterations in the hMSH2 gene in human ovarian cancer. Int J Cancer. 1995;64:361–6.

Fulvestrant plus abemaciclib in women with advanced low grade serous carcinoma. https://ClinicalTrials.gov/show/NCT03531645.

Gallardo A, Matias-Guiu X, Lagarda H, et al. Malignant müllerian mixed tumor arising from ovarian serous carcinoma: a clinicopathologic and molecular study of two cases. Int J Gynecol Pathol. 2002;21:268–72.

Gamallo C, Palacios J, Moreno G, et al. Beta-catenin expression pattern in stage I and II ovarian carcinomas: relationship with beta-catenin gene mutations, clinicopathological features, and clinical outcome. Am J Pathol. 1999;155:527–36.

Gatalica Z, Snyder C, Maney T, et al. Programmed cell death 1 (PD-1) and its ligand (PD-L1) in common cancers and their correlation with molecular cancer type. Cancer Epidemiol Biomark Prev. 2014;23:2965–70.

Gemignani ML, Schlaerth AC, Bogomolniy F, et al. Role of KRAS and BRAF gene mutations in mucinous ovarian carcinoma. Gynecol Oncol. 2003;90(2):378–81.

Gershenson DM, Sun CC, Wong KK. Impact of mutational status on survival in low-grade serous carcinoma of the ovary or peritoneum. Br J Cancer. 2015;113:1254–8.

Gordon AN, Finkler N, Edwards RP, et al. Efficacy and safety of erlotinib HCl, an epidermal growth factor receptor (HER1/EGFR) tyrosine kinase inhibitor, in patients with advanced ovarian carcinoma: results from a phase II multicenter study. Int J Gynecol Cancer. 2005;15:785–92.

Gordon MS, Matei D, Aghajanian C, et al. Clinical activity of pertuzumab (rhuMAb 2C4), a HER dimerization inhibitor, in advanced ovarian cancer: potential predictive relationship with tumor HER2 activation status. J Clin Oncol. 2006;24:4324–32.

Gorringe KL, Cheasley D, Wakefield MJ, et al. Therapeutic options for mucinous ovarian carcinoma. Gynecol Oncol. 2020;156:552–60.

Goulvent T, Ray-Coquard I, Borel S, et al. DICER1 and FOXL2 mutations in ovarian sex cord-stromal tumours: a GINECO group study. Histopathology. 2016;68:279–85.

Gras E, Catasús L, Arguelles R, et al. Microsatellite instability, MLH-1 promoter hypermethylation, and frameshift mutations at coding mononucleotide repeat microsatellites in ovarian tumors. Cancer. 2001;92:2829–36.

Hall J, Paul J, Brown R. Critical evaluation of p53 as a prognostic marker in ovarian cancer. Expert Rev. Mol Med. 2004;6:1–20.

Havrilesky L, Darcy KM, Hamdan H, et al. Prognostic significance of p53 mutation and p53 overexpression in advanced epithelial ovarian cancer: a gynecologic oncology group study. J Clin Oncol. 2003;21:3814–25.

Helder-Woolderink JM, Blok EA, Vasen HF, et al. Ovarian cancer in lynch syndrome; a systematic review. Eur J Cancer. 2016;55:65–73.

Hersmus R, Stoop H, van de Geijn GJ, et al. Prevalence of c-KIT mutations in gonadoblastoma and dysgerminomas of patients with disorders of sex development (DSD) and ovarian dysgerminomas. PLoS One. 2012;7:e43952.

Ho CL, Kurman RJ, Dehari R, et al. Mutations of BRAF and KRAS precede the development of ovarian serous borderline tumors. Cancer Res. 2004;64:6915–8.

Hoang LN, McConechy MK, Kobel M, et al. Polymerase epsilon exonuclease domain mutations in ovarian endometrioid carcinoma. Int J Gynecol Cancer. 2015;25:1187–93.

Hoei-Hansen CE, Kraggerud SM, Abeler VM, et al. Ovarian dysgerminomas are characterised by frequent KIT mutations and abundant expression of pluripotency markers. Mol Cancer. 2007;6:12.

Hogdall EV, Christensen L, Kjaer SK, et al. Distribution of HER-2 overexpression in ovarian carcinoma tissue and its prognostic value in patients with ovarian carcinoma: from the Danish MALOVA ovarian cancer study. Cancer. 2003;98:66–73.

Hortobagyi GN, Stemmer SM, Burris HA, et al. Ribociclib as first-line therapy for HR-positive, advanced breast cancer. N Engl J Med. 2016;375:1738–48.

Hsu CY, Bristow R, Cha MS, et al. Characterization of active mitogen-activated protein kinase in ovarian serous carcinomas. Clin Cancer Res. 2004;10:6432–6.

Hunter SM, Anglesio MS, Ryland GL, et al. Molecular profiling of low grade serous ovarian tumours identifies novel candidate driver genes. Oncotarget. 2015;6:37663–77.

Ishibashi T, Nakayama K, Razia S, et al. High frequency of PIK3CA mutations in low-grade serous ovarian carcinomas of Japanese patients. Diagnostics (Basel). 2019;10:13.

Ishibashi M, Nakayama K, Yeasmin S, et al. A BTB/POZ gene, NAC-1, a tumor recurrence-associated gene, as a potential target for taxol resistance in ovarian cancer. Clin Cancer Res. 2008;14:3149–55.

Jamieson S, Butzow R, Andersson N, et al. The FOXL2 C134W mutation is characteristic of adult granulosa cell tumors of the ovary. Mod Pathol. 2010;23:1477–85.

Jazaeri AA, Yee CJ, Sotiriou C, et al. Gene expression profiles of BRCA1-linked, BRCA2-linked, and sporadic ovarian cancers. J Natl Cancer Inst. 2002;94:990–1000.

Jelinic P, Mueller JJ, Olvera N, et al. Recurrent SMARCA4 mutations in small cell carcinoma of the ovary. Nat Genet. 2014;46:424–6.

Jinawath N, Vasoontara C, Yap KL, et al. NAC-1, a potential stem cell pluripotency factor, contributes to paclitaxel resistance in ovarian cancer through inactivating Gadd45 pathway. Oncogene. 2009;28:1941–8.

Jones S, Wang TL, Shih I, et al. Frequent mutations of chromatin remodeling gene ARID1A in ovarian clear cell carcinoma. Science. 2010;330:228–31.

Kalfa N, Ecochard A, Patte C, et al. Activating mutations of the stimulatory g protein in juvenile ovarian granulosa cell tumors: a new prognostic factor? J Clin Endocrinol Metab. 2006;91:1842–7.

Karnezis AN, Wang Y, Keul J, et al. DICER1 and FOXL2 mutation status correlates with clinicopathologic features in ovarian Sertoli-Leydig cell tumors. Am J Surg Pathol. 2019;43:628–38.

Kato N, Tamura G, Motoyama T. Hypomethylation of hepatocyte nuclear factor-1beta (HNF-1beta) CpG Island in clear cell carcinoma of the ovary. Virchows Arch. 2008;452:175–80.

Khani F, Diolombi ML, Khattar P, et al. Benign and malignant Brenner tumors show an absence of TERT promoter mutations that are commonly present in urothelial carcinoma. Am J Surg Pathol. 2016;40:1291–5.

Kindelberger DW, Lee Y, Miron A, et al. Intraepithelial carcinoma of the fimbria and pelvic serous carcinoma: evidence for a causal relationship. Am J Surg Pathol. 2007;31:161–9.

Kobel M, Hoang LN, Tessier-Cloutier B, et al. Undifferentiated endometrial carcinomas show frequent loss of core switch/sucrose non-fermentable complex proteins. Am J Surg Pathol. 2018;42:76–83.

Kounelis S, Jones MW, Papadaki H, et al. Carcinosarcomas (malignant mixed müllerian tumors) of the female genital tract: comparative molecular analysis of epithelial and mesenchymal components. Hum Pathol. 1998;29:82–7.

Kuhn E, Ayhan A, Bahadirli-Talbott A, et al. Molecular characterization of undifferentiated carcinoma associated with endometrioid carcinoma. Am J Surg Pathol. 2014;38:660–5.

Kuhn E, Ayhan A, Shih Ie M, et al. The pathogenesis of atypical proliferative Brenner tumor: an immunohistochemical and molecular genetic analysis. Mod Pathol. 2014;27:231–7.

Kuhn E, Meeker A, Wang TL, et al. Shortened telomeres in serous tubal intraepithelial carcinoma: an early event in ovarian high-grade serous carcinogenesis. Am J Surg Pathol. 2010;34:829–36.

Kuo KT, Mao TL, Jones S, et al. Frequent activating mutations of PIK3CA in ovarian clear cell carcinoma. Am J Pathol. 2009;174:1597–601.

Kupryjanczyk J, Thor AD, Beauchamp R, et al. p53 gene mutations and protein accumulation in human ovarian cancer. Proc Natl Acad Sci U S A. 1993;90:4961–5.

Kurman RJ, Shih I. The origin and pathogenesis of epithelial ovarian cancer: a proposed unifying theory. Am J Surg Pathol. 2010;34:433–43.

Kurman RJ, Shih I. Molecular pathogenesis and extraovarian origin of epithelial ovarian cancer–shifting the paradigm. Hum Pathol. 2011;42:918–31.

Lassus H, Sihto H, Leminen A, et al. Gene amplification, mutation, and protein expression of EGFR and mutations of ERBB2 in serous ovarian carcinoma. J Mol Med (Berl). 2006;84:671–81.

Ledermann J, Colombo N, Oza A, et al. Avelumab in combination with and/or following chemotherapy vs. chemotherapy alone in patients with previously untreated epithelial ovarian cancer: results from the phase 3 javelin ovarian 100 trial. Gynecol Oncol. 2020;159:13–4.

Lee JH, Lee SY, Lee JH, et al. p21 WAF1 is involved in interferon-beta-induced attenuation of telomerase activity and human telomerase reverse transcriptase (hTERT) expression in ovarian cancer. Mol Cells. 2010;30:327–33.

Luo Y, Yi Y, Yao Z. Growth arrest in ovarian cancer cells by hTERT inhibition short-hairpin RNA targeting human telomerase reverse transcriptase induces immediate growth inhibition but not necessarily induces apoptosis in ovarian cancer cells. Cancer Investig. 2009;27:960–70.

Marabelle A, Le DT, Ascierto PA, et al. Efficacy of pembrolizumab in patients with noncolorectal high microsatellite instability/mismatch repair-deficient cancer: results from the phase II KEYNOTE-158 study. J Clin Oncol. 2020;38:1–10.

Mert I, Chhina J, Allo G, et al. Synergistic effect of MEK inhibitor and metformin combination in low grade serous ovarian cancer. Gynecol Oncol. 2017;146:319–26.

Milea A, George SH, Matevski D, et al. Retinoblastoma pathway deregulatory mechanisms determine clinical outcome in high-grade serous ovarian carcinoma. Mod Pathol. 2014;27:991–1001.

Min A, Im SA, Yoon YK, et al. RAD51C-deficient cancer cells are highly sensitive to the PARP inhibitor olaparib. Mol Cancer Ther. 2013;12:865–77.

Moes-Sosnowska J, Szafron L, Nowakowska D, et al. Germline SMARCA4 mutations in patients with ovarian small cell carcinoma of hypercalcemic type. Orphanet J Rare Dis. 2015;10:32.

Moore K, Colombo N, Scambia G, et al. Maintenance olaparib in patients with newly diagnosed advanced ovarian cancer. N Engl J Med. 2018;379:2495–505.

Moore KN, Bookman M, Sehouli J, et al. LBA31 primary results from IMagyn050/GOG 3015/ENGOT-OV39, a double-blind placebo (pbo)-controlled randomised phase III trial of bevacizumab (bev)-containing therapy +/– atezolizumab (atezo) for newly diagnosed stage III/IV ovarian cancer (OC). Ann Oncol. 2020;31:S1161–2.

Moujaber T, Etemadmoghadam D, Kennedy CJ, et al. BRAF mutations in low-grade serous ovarian cancer and response to BRAF inhibition. JCO Precis Oncol. 2018;2:1–14.

Makarla PB, Saboorian MH, Ashfaq R, et al. Promoter hypermethylation profile of ovarian epithelial neoplasms. Clin Cancer Res. 2005;11:5365–9.

Marquez RT, Baggerly KA, Patterson AP, et al. Patterns of gene expression in different histotypes of epithelial ovarian cancer correlate with those in normal fallopian tube, endometrium, and colon. Clin Cancer Res. 2005;11:6116–26.

Matz M, Coleman MP, Sant M, et al. The histology of ovarian cancer: worldwide distribution and implications for international survival comparisons (CONCORD-2). Gynecol Oncol. 2017;144:405–13.

Mayr D, Hirschmann A, Löhrs U, Diebold J. KRAS and BRAF mutations in ovarian tumors: a comprehensive study of invasive carcinomas, borderline tumors and extraovarian implants. Gynecol Oncol. 2006;103:883–7.

McAlpine JN, Wiegand KC, Vang R, et al. HER2 overexpression and amplification is present in a subset of ovarian mucinous carcinomas and can be targeted with trastuzumab therapy. BMC Cancer. 2009;9:433.

McConechy MK, Anglesio MS, Kalloger SE, et al. Subtype-specific mutation of PPP2R1A in endometrial and ovarian carcinomas. J Pathol. 2011;223:567–73.

McConechy MK, Ding J, Senz J, et al. Ovarian and endometrial endometrioid carcinomas have distinct CTNNB1 and PTEN mutation profiles. Mod Pathol. 2014;27:128–34.

Medeiros F, Muto MG, Lee Y, et al. The tubal fimbria is a preferred site for early adenocarcinoma in women with familial ovarian cancer syndrome. Am J Surg Pathol. 2006;30:230–6.

Mok SC, Bell DA, Knapp RC, et al. Mutation of K-ras protooncogene in human ovarian epithelial tumors of borderline malignancy. Cancer Res. 1993;53:1489–92.

Morand S, Devanaboyina M, Staats H, et al. Ovarian cancer immunotherapy and personalized medicine. Int J Mol Sci. 2021;22:6532.

Moreno-Bueno G, Gamallo C, Perez-Gallego L, et al. Beta-catenin expression pattern, beta-catenin gene mutations, and microsatellite instability in endometrioid ovarian carcinomas and synchronous endometrial carcinomas. Diagn Mol Pathol. 2001;10:116–22.

Nakayama K, Nakayama N, Davidson B, et al. A BTB/POZ protein, NAC-1, is related to tumor recurrence and is essential for tumor growth and survival. Proc Natl Acad Sci U S A. 2006a;103:18739–44.

Nakayama K, Nakayama N, Kurman RJ, et al. Sequence mutations and amplification of PIK3CA and AKT2 genes in purified ovarian serous neoplasms. Cancer Biol Ther. 2006b;5:779–85.

Nakayama K, Nakayama N, Jinawath N, et al. Amplicon profiles in ovarian serous carcinomas. Int J Cancer. 2007a;120:2613–7.

Nakayama K, Nakayama N, Wang TL, et al. NAC-1 controls cell growth and survival by repressing transcription of Gadd45GIP1, a candidate tumor suppressor. Cancer Res. 2007b;67:8058–64.

Nazlioglu HO, Ercan I, Bilgin T, et al. Expression of p16 in serous ovarian neoplasms. Eur J Gynaecol Oncol. 2010;31:312–4.

Noe M, Ayhan A, Wang TL, et al. Independent development of endometrial epithelium and stroma within the same endometriosis. J Pathol. 2018;245:265–9.

Obata K, Morland SJ, Watson RH, et al. Frequent PTEN/MMAC mutations in endometrioid but not serous or mucinous epithelial ovarian tumors. Cancer Res. 1998;58:2095–7.

O'Neill CJ, McBride HA, Connolly LE, et al. High-grade ovarian serous carcinoma exhibits significantly higher p16 expression than low-grade serous carcinoma and serous borderline tumour. Histopathology. 2007;50:773–9.

Park JT, Li M, Nakayama K, et al. Notch3 gene amplification in ovarian cancer. Cancer Res. 2006;66:6312–8.

Parra-Herran C, Lerner-Ellis J, Xu B, et al. Molecular-based classification algorithm for endometrial carcinoma categorizes ovarian endometrioid carcinoma into prognostically significant groups. Mod Pathol. 2017;30:1748–59.

Patch AM, Christie EL, Etemadmoghadam D, et al. Whole-genome characterization of chemoresistant ovarian cancer. Nature. 2015;521:489–94.

Pavanello M, Chan IH, Ariff A, et al. Rare germline genetic variants and the risks of epithelial ovarian cancer. Cancers (Basel). 2020;12:3046.

Peiro G, Diebold J, Mayr D, et al. Prognostic relevance of hMLH1, hMSH2, and BAX protein expression in endometrial carcinoma. Mod Pathol. 2001;14:777–83.

Perego P, Giarola M, Righetti SC, et al. Association between cisplatin resistance and mutation of p53 gene and reduced bax expression in ovarian carcinoma cell systems. Cancer Res. 1996;56:556–62.

Petersen S, Wilson AJ, Hirst J, et al. CCNE1 and BRD4 co-amplification in high-grade serous ovarian cancer is associated with poor clinical outcomes. Gynecol Oncol. 2020;157:405–10.

Pfarr N, Darb-Esfahani S, Leichsenring J, et al. Mutational profiles of Brenner tumors show distinctive features uncoupling urothelial carcinomas and ovarian carcinoma with transitional cell histology. Genes Chromosomes Cancer. 2017;56:758–66.

Pfisterer J, Du Bois A, Bentz EK, et al. Prognostic value of human epidermal growth factor receptor 2 (her-2)/neu in patients with advanced ovarian cancer treated with platinum/paclitaxel as first-line chemotherapy: a retrospective evaluation of the AGO-OVAR 3 trial by the AGO OVAR Germany. Int J Gynecol Cancer. 2009;19:109–15.

Piek JM, van Diest PJ, Zweemer RP, et al. Dysplastic changes in prophylactically removed fallopian tubes of women predisposed to developing ovarian cancer. J Pathol. 2001;195:451–6.

Pohl G, Ho CL, Kurman RJ, et al. Inactivation of the mitogen-activated protein kinase pathway as a potential target-based therapy in ovarian serous tumors with KRAS or BRAF mutations. Cancer Res. 2005;65:1994–2000.

Rahman M, Nakayama K, Rahman MT, et al. PPP2R1A mutation is a rare event in ovarian carcinoma across histological subtypes. Anticancer Res. 2013;33:113–8.

Rahman N, Stratton MR. The genetics of breast cancer susceptibility. Annu. Rev. Genet. 1998;32:95–121.

Ramalingam P, Croce S, McCluggage WG. Loss of expression of SMARCA4 (BRG1), SMARCA2 (BRM) and SMARCB1 (INI1) in undifferentiated carcinoma of the endometrium is not uncommon and is not always associated with rhabdoid morphology. Histopathology. 2017;70:359–66.

Rambau PF, McIntyre JB, Taylor J, et al. Morphologic reproducibility, genotyping, and immunohistochemical profiling do not support a category of seromucinous carcinoma of the ovary. Am J Surg Pathol. 2017;41:685–95.

Ramos P, Karnezis AN, Craig DW, et al. Small cell carcinoma of the ovary, hypercalcemic type, displays frequent inactivating germline and somatic mutations in SMARCA4. Nat Genet. 2014;46:427–9.

Ramus SJ, Song H, Dicks E, et al. Germline mutations in the BRIP1, BARD1, PALB2, and NBN genes in women with ovarian cancer. J Natl Cancer Inst. 2015;107:djv214.

Ray-Coquard I, Pautier P, Pignata S, et al. Olaparib plus bevacizumab as first-line maintenance in ovarian cancer. N Engl J Med. 2019;381:2416–28.

Ribociclib and letrozole treatment in ovarian cancer. https://ClinicalTrials.gov/show/NCT03673124.

Ross JS, Yang F, Kallakury BV, et al. HER2/neu oncogene amplification by fluorescence in situ hybridization in epithelial tumors of the ovary. Am J Clin Pathol. 1999;111:311–6.

Rubin SC, Blackwood MA, Bandera C, et al. BRCA1, BRCA2, and hereditary nonpolyposis colorectal cancer gene mutations in an unselected ovarian cancer population: relationship to family history and implications for genetic testing. Am J Obstet Gynecol. 1998;178:670–7.

Ryland GL, Hunter SM, Doyle MA, et al. Mutational landscape of mucinous ovarian carcinoma and its neoplastic precursors. Genome Med. 2015;7:87.

Sato N, Tsunoda H, Nishida M, et al. Loss of heterozygosity on 10q23.3 and mutation of the tumor suppressor gene PTEN in benign endometrial cyst of the ovary: possible sequence progression from benign endometrial cyst to endometrioid carcinoma and clear cell carcinoma of the ovary. Cancer Res. 2000;60:7052–6.

Schilder RJ, Sill MW, Chen X, et al. Phase II study of gefitinib in patients with relapsed or persistent ovarian or primary peritoneal carcinoma and evaluation of epidermal growth factor receptor mutations and immunohistochemical expression: a gynecologic oncology group study. Clin Cancer Res. 2005;11:5539–48.

Schlosshauer PW, Deligdisch L, Penault-Llorca F, et al. Loss of p16INK4A expression in low-grade ovarian serous carcinomas. Int J Gynecol Pathol. 2011;30:22–9.

Schultz KAP, Stewart DR, Kamihara J, et al. DICER1 tumor predisposition. In: Adam MP, Ardinger HH, Pagon RA, et al., editors. GeneReviews((R)). Seattle, WA: University of Washington, Seattle; 1993.

Seidman JD, Yemelyanova A, Zaino RJ, et al. The fallopian tube-peritoneal junction: a potential site of carcinogenesis. Int J Gynecol Pathol. 2011;30:4–11.

Shah SP, Kobel M, Senz J, et al. Mutation of FOXL2 in granulosa-cell tumors of the ovary. N Engl J Med. 2009;360:2719–29.

Sheu JJ, Choi JH, Yildiz I, et al. The roles of human sucrose nonfermenting protein 2 homologue in the tumor-promoting functions of rsf-1. Cancer Res. 2008;68:4050–7.

Shih I, Davidson B. Pathogenesis of ovarian cancer: clues from selected overexpressed genes. Future Oncol. 2009;5:1641–57.

Shih I, Kurman RJ. Ovarian tumorigenesis: a proposed model based on morphological and molecular genetic analysis. Am J Pathol. 2004;164:1511–8.

Shih Ie M, Panuganti PK, Kuo KT, et al. Somatic mutations of PPP2R1A in ovarian and uterine carcinomas. Am J Pathol. 2011;178:1442–7.

Shih I, Sheu JJ, Santillan A, et al. Amplification of a chromatin remodeling gene, rsf-1/HBXAP, in ovarian carcinoma. Proc Natl Acad Sci U S A. 2005;102:14004–9.

Sieben NL, Macropoulos P, Roemen GM, et al. In ovarian neoplasms, BRAF, but not KRAS, mutations are restricted to low-grade serous tumours. J Pathol. 2004;202:336–40.

Siegel RL, Miller KD, Jemal A. Cancer statistics, 2020. CA Cancer J Clin. 2020;70:7–30.

Singer G, Kurman RJ, Chang HW, et al. Diverse tumorigenic pathways in ovarian serous carcinoma. Am J Pathol. 2002;160:1223–8.

Singer G, Oldt R III, Cohen Y, et al. Mutations in BRAF and KRAS characterize the development of low-grade ovarian serous carcinoma. J Natl Cancer Inst. 2003a;95:484–6.

Singer G, Rebmann V, Chen YC, et al. HLA-G is a potential tumor marker in malignant ascites. Clin Cancer Res. 2003b;9:4460–4.

Soslow RA, Han G, Park KJ, et al. Morphologic patterns associated with BRCA1 and BRCA2 genotype in ovarian carcinoma. Mod Pathol. 2012;25:625–36.

Southey MC, Goldgar DE, Winqvist R, et al. PALB2, CHEK2 and ATM rare variants and cancer risk: data from COGS. J Med Genet. 2016;53:800–11.

Stadlmann S, Gueth U, Reiser U, et al. Epithelial growth factor receptor status in primary and recurrent ovarian cancer. Mod Pathol. 2006;19:607–10.

Sui L, Tokuda M, Ohno M, et al. The concurrent expression of p27(kip1) and cyclin D1 in epithelial ovarian tumors. Gynecol Oncol. 1999;73:202–9.

Suszynska M, Ratajska M, Kozlowski P. BRIP1, RAD51C, and RAD51D mutations are associated with high susceptibility to ovarian cancer: mutation prevalence and precise risk estimates based on a pooled analysis of ~30,000 cases. J Ovarian Res. 2020;13:50.

Tafe LJ, Garg K, Chew I, et al. Endometrial and ovarian carcinomas with undifferentiated components: clinically aggressive and frequently underrecognized neoplasms. Mod Pathol. 2010;23:781–9.

Tafe LJ, Muller KE, Ananda G, et al. Molecular genetic analysis of ovarian Brenner tumors and associated mucinous epithelial neoplasms: high variant concordance and identification of mutually

exclusive RAS driver mutations and MYC amplification. Am J Pathol. 2016;186:671–7.

Tan H, Mei L, Huang Y, et al. Three novel mutations of STK11 gene in Chinese patients with Peutz-Jeghers syndrome. BMC Med Genet. 2016;17:77.

The WHO Classification of Tumours Editorial Board. WHO classification of tumours: female genital tumours. 5th ed. Lyon: International Agency for Research on Cancer (IARC); 2020.

Trametinib in treating patients with recurrent or progressive low-grade ovarian cancer or peritoneal cavity cancer. https://ClinicalTrials.gov/show/NCT02101788.

Tsuchiya A, Sakamoto M, Yasuda J, et al. Expression profiling in ovarian clear cell carcinoma: identification of hepatocyte nuclear factor-1 beta as a molecular marker and a possible molecular target for therapy of ovarian clear cell carcinoma. Am J Pathol. 2003;163:2503–12.

Vang R, Levine DA, Soslow RA, et al. Molecular alterations of TP53 are a defining feature of ovarian high-grade serous carcinoma: a rereview of cases lacking TP53 mutations in the cancer genome atlas ovarian study. Int J Gynecol Pathol. 2016;35:48–55.

Van Nieuwenhuysen E, Busschaert P, Neven P, et al. The genetic landscape of 87 ovarian germ cell tumors. Gynecol Oncol. 2018;151:61–8.

Vermeij J, Teugels E, Bourgain C, et al. Genomic activation of the EGFR and HER2/neu genes in a significant proportion of invasive epithelial ovarian cancers. BMC Cancer. 2008;8:3.

Verri E, Guglielmini P, Puntoni M, et al. HER2/neu oncoprotein overexpression in epithelial ovarian cancer: evaluation of its prevalence and prognostic significance. Clinical study. Oncology. 2005;68:154–61.

Wang C, Horiuchi A, Imai T, et al. Expression of BRCA1 protein in benign, borderline, and malignant epithelial ovarian neoplasms and its relationship to methylation and allelic loss of the BRCA1 gene. J Pathol. 2004;202:215–23.

Wang J, Zhou JY, Wu GS. Bim protein degradation contributes to cisplatin resistance. J Biol Chem. 2011;286:22384–92.

Wang Q, Lou W, Di W, et al. Prognostic value of tumor PD-L1 expression combined with CD8(+) tumor infiltrating lymphocytes in high grade serous ovarian cancer. Int Immunopharmacol. 2017;52:7–14.

Wang Y, Chen J, Yang W, et al. The oncogenic roles of DICER1 RNase IIIb domain mutations in ovarian Sertoli-Leydig cell tumors. Neoplasia. 2015;17:650–60.

Wen WH, Reles A, Runnebaum IB, et al. p53 mutations and expression in ovarian cancers: correlation with overall survival. Int J Gynecol Pathol. 1999;18:29–41.

Widschwendter M, Jiang G, Woods C, et al. DNA hypomethylation and ovarian cancer biology. Cancer Res. 2004;64:4472–80.

Wiegand KC, Shah SP, Al-Agha OM, et al. ARID1A mutations in endometriosis-associated ovarian carcinomas. N Engl J Med. 2010;363:1532–43.

Wieser V, Gaugg I, Fleischer M, et al. BRCA1/2 and TP53 mutation status associates with PD-1 and PD-L1 expression in ovarian cancer. Oncotarget. 2018;9:17501–11.

Willner J, Wurz K, Allison KH, et al. Alternate molecular genetic pathways in ovarian carcinomas of common histological types. Hum Pathol. 2007;38:607–13.

Witkowski L, Carrot-Zhang J, Albrecht S, et al. Germline and somatic SMARCA4 mutations characterize small cell carcinoma of the ovary, hypercalcemic type. Nat Genet. 2014;46:438–43.

Woo JS, Apple SK, Sullivan PS, et al. Systematic assessment of HER2/neu in gynecologic neoplasms, an institutional experience. Diagn Pathol. 2016;11:102.

Wu R, Zhai Y, Fearon ER, et al. Diverse mechanisms of beta-catenin deregulation in ovarian endometrioid adenocarcinomas. Cancer Res. 2001;61:8247–55.

Wu R, Hendrix-Lucas N, Kuick R, et al. Mouse model of human ovarian endometrioid adenocarcinoma based on somatic defects in the

Wnt/beta-catenin and PI3K/Pten signaling pathways. Cancer Cell. 2007;11:321–33.

Xu Y, Bi R, Xiao Y, et al. Low frequency of BRAF and KRAS mutations in Chinese patients with low-grade serous carcinoma of the ovary. Diagn Pathol. 2017;12:87.

Yamamoto S, Tsuda H, Takano M, et al. Loss of ARID1A protein expression occurs as an early event in ovarian clear-cell carcinoma development and frequently coexists with PIK3CA mutations. Mod Pathol. 2012;25:615–24.

Yoon SH, Kim SN, Shim SH, et al. Bilateral salpingectomy can reduce the risk of ovarian cancer in the general population: a meta-analysis. Eur J Cancer. 2016;55:38–46.

Young RH, Welch WR, Dickersin GR, et al. Ovarian sex cord tumor with annular tubules: review of 74 cases including 27 with Peutz-Jeghers syndrome and four with adenoma malignum of the cervix. Cancer. 1982;50:1384–402.

Zhao S, Bellone S, Lopez S, et al. Mutational landscape of uterine and ovarian carcinosarcomas implicates histone genes in epithelial-mesenchymal transition. Proc Natl Acad Sci U S A. 2016;113:12238–43.

Zhu J, Wen H, Bi R, et al. Prognostic value of programmed death-ligand 1 (PD-L1) expression in ovarian clear cell carcinoma. J Gynecol Oncol. 2017;28:e77.

Molecular Pathology of Endometrial Tumors

Sonia Gatius, Nuria Eritja, and Xavier Matias-Guiu

Contents

S. Gatius
Department of Pathology, Hospital Universitari Arnau de Vilanova, University of Lleida, Institute of Biomedical Research of Lleida (Irblleida), Lleida, Spain

N. Eritja
University of Lleida, Institute of Biomedical Research of Lleida (Irblleida), Lleida, Spain

X. Matias-Guiu (✉)
Department of Pathology, Hospital Universitari Arnau de Vilanova, University of Lleida, Institute of Biomedical Research of Lleida (Irblleida), Lleida, Spain

Department of Pathology, Hospital Universitari de Bellvitge, Bellvitge Institute of Biomedical Research (IDIBELL), University of Barcelona, Barcelona, Spain
e-mail: fjmatiasguiu.lleida.ics@gencat.cat

Introduction

- In Western countries, endometrial carcinoma is the most common cancer of the female genital tract accounting for 10–20 per 100,000 person-years
- Endometrial carcinoma occurs in peri- and postmenopausal women, although it may also develop in premenopausal women, particularly in the setting of hyperestrogenism and hereditary nonpolyposis colon cancer (HNPCC) syndrome
- Etiological factors include unopposed estrogenic stimulation (anovulatory cycles, estrogen administration), obesity, tamoxifen treatment, or insulin resistance
- Prognosis is favorable for patients with low-grade, and early-stage tumors but the outcome for patients with metastatic/recurrent tumors remains poor
- After surgery (which is the most usual treatment), patients with tumors with a high risk of recurrence receive adjuvant radiotherapy. Chemotherapy is restricted to metastatic/recurrent and high-grade endometrial carcinomas
- However, traditional chemotherapeutic regimes are less effective in comparison with cancers of other organs, which emphasize the importance of identifying new

Table 12.1 Clinicopathological features of types I and II endometrial carcinomas

	Type I	Type II
Age	Pre- and perimenopausal	Postmenopausal
Unopposed estrogen	Present	Absent
Hyperplasia precursor	Present	Absent
Grade	Low	High
Myometrial invasion	Minimal	Deep
Histologic type	Endometrioid	Nonendometrioid
Behavior	Stable	Progressive

molecular targets. Endometrial carcinoma patients presenting systemic disease and/or overt metastasis represent nowadays a therapeutic challenge

- From a clinical viewpoint, endometrial carcinoma falls into two different types (types I and II) (Table 12.1)
    - Type I tumors are low-grade and estrogen-related endometrioid–endometrial carcinomas (EEC) that usually develop in perimenopausal women and coexist with or are preceded by endometrial hyperplasia
    - Type II tumors are high-grade nonendometrioid endometrial carcinomas (mainly serous and clear cell carcinomas), unrelated to estrogen stimulation, which may arise in endometrial polyps or from precancerous lesions that develop atrophic endometrium and tend to occur in older women
- Although the type I versus type II classification is interesting for educational and epidemiological purposes, it is not useful for tumor stratification from the pathologic viewpoint, because there are significant overlapping features at the clinical, pathological, and molecular levels

WHO Classification of Endometrial Carcinoma (WHO 2020)

- The WHO Classification of Tumours distinguishes several histopathological types of endometrial carcinoma particularly based on their microscopic appearance (Fig. 12.1)
- These types have different histological and molecular features, precursor lesions, and natural history
- Histological typing has consistently been demonstrated as an important biologic predictor in endometrial carcinoma, and determines the extent of the initial surgical procedure, and subsequent use of adjuvant therapy
- EEC represents the paradigm of type I endometrial carcinoma and accounts for 80% of endometrial cancer cases
    - EEC develops because of the proliferation of endometrial glands, with a spectrum of changes that goes from endometrial hyperplasia (reactive proliferation secondary to estrogenic stimulation) to endometrioid intraepithelial neoplasia/atypical endometrial hyperplasia (true preneoplastic precursor)

- Although a proportion of EECs arise from endometrial hyperplasia, in many cases the nonneoplastic endometrium appears atrophic or weakly proliferative
- Low-grade (grades 1 and 2) endometrioid carcinomas (EEC 1–2), represent the most frequent tumors and are usually associated with a favorable outcome
- EEC 1–2 is usually composed of cells arranged in a branching, maze-like glandular or complex papillary pattern of growth (Fig. 12.2)
- EEC 1–2 may show some specific types of terminal differentiation such as squamous and mucinous differentiation or specific patterns of growth such as villoglandular, small nonvillous papillae, microglandular, sex cord-like formations, and hyalinization and sertoliform structures
- Mucinous carcinoma is now considered a variant of EEC 1–2, because it is at one extreme in the spectrum of low-grade endometrioid tumors with mucinous differentiation, very similar to conventional EEC 1–2 in terms of prognosis, molecular features, and risk factors
- High-grade endometrioid carcinoma (EEC 3) accounts for 10% of the cases
    - It is characterized by a solid growth pattern associated with mostly moderate nuclear atypia and increased mitosis (Fig. 12.2)
    - Application of the TCGA-molecular surrogate (see below) has shown that this is a heterogeneous group of tumors with different prognoses
- The other types of endometrial carcinomas, accounting for 20% of the tumors, are generically considered type II tumors, and are occasionally referred to as nonendometrioid carcinomas
- Serous carcinoma represents 10% of endometrial carcinomas
    - Serous carcinoma is characterized by marked nuclear pleomorphism and prominent nucleoli, and typically a papillary architecture
        - However, serous carcinoma can be glandular or even solid, raising problems in differential diagnosis with EEC
    - The tumor usually invades the myometrium deeply and there is extensive lymphovascular space invasion
    - Serous carcinomas are not preceded by endometrial hyperplasia
    - The noninvasive type (formerly called intraepithelial carcinoma) is part of the spectrum of serous carcinoma and can develop peritoneal metastasis
- Clear cell carcinoma is infrequent; tumor cells show cuboidal, polygonal, hobnail, or flat appearances, with clear or eosinophilic cytoplasm

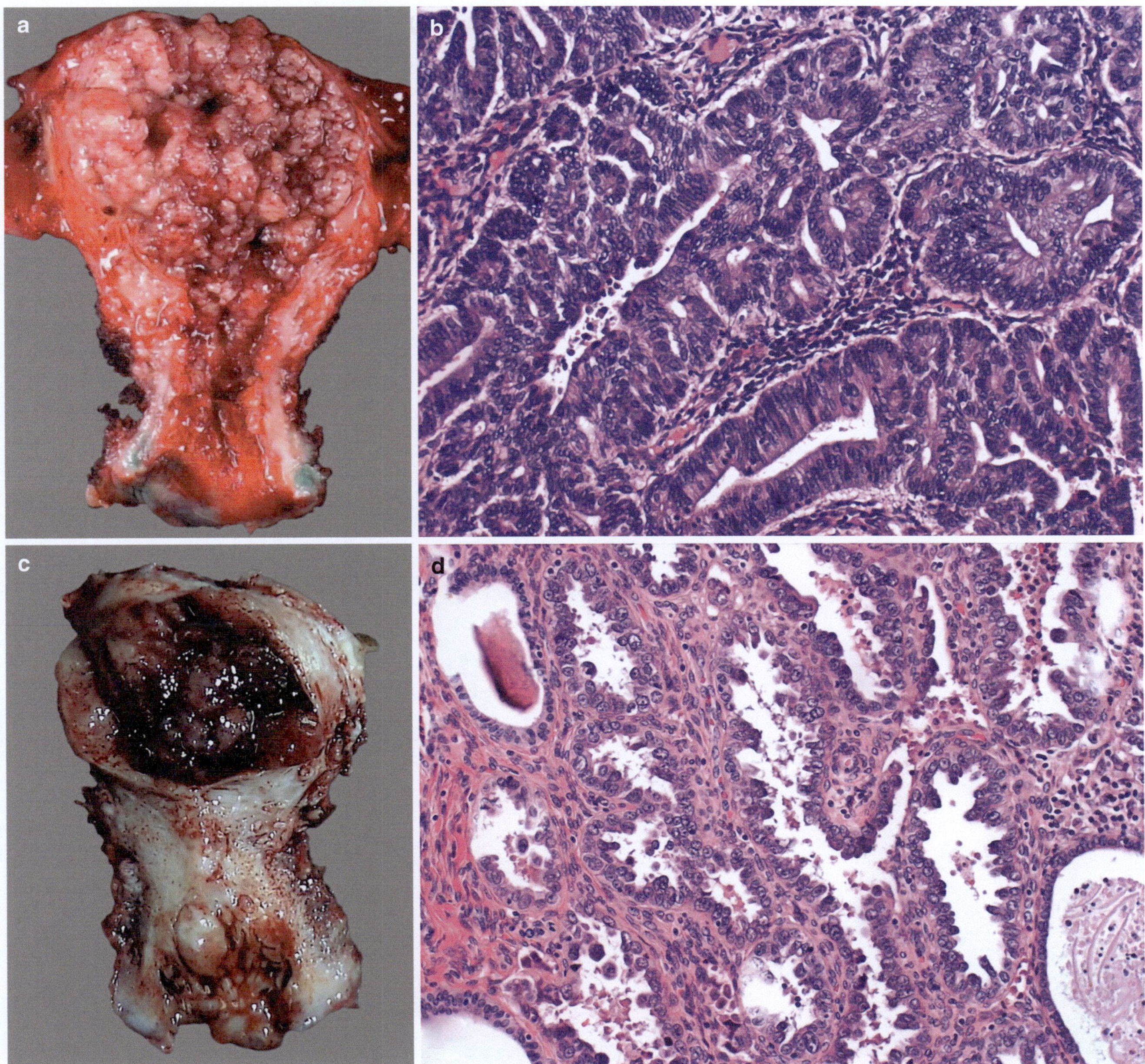

Fig. 12.1 (**a**) Endometrioid endometrial carcinoma. A large irregular mass fills the endometrial cavity; (**b**) Grade I endometrioid adenocarcinoma. The tumor is composed of well-defined glands with cribriform patterns and mild cytologic atypia; (**c**) Serous endometrial carcinoma. An irregular polypoid mass grows in atrophic endometrium; (**d**) Uterine serous carcinoma. The tumor shows pseudoglands and poorly formed papillae with marked cytologic atypia

- Strict adherence to diagnostic criteria, architectural, and cytological is necessary since clear cells are present in EEC and serous carcinoma
- Clear cell carcinoma is characterized by three major architectural patterns (tubulocystic, papillary, and solid) which are frequently admixed
- Undifferentiated carcinoma is infrequent; it is composed of small to intermediate-sized, discohesive cells of relatively uniform size growing in solid sheets

- Occasionally, undifferentiated carcinoma is associated with a second component of differentiated carcinoma (usually EEC 1–2) and the term dedifferentiated carcinoma is used
- Undifferentiated carcinoma is characterized by a monotonous proliferation of medium-sized, epithelial cells
- Undifferentiated carcinoma is associated with poor prognosis in most of the cases
- Over 50% of cases present with advanced-stage disease and 75% of patients die of tumors

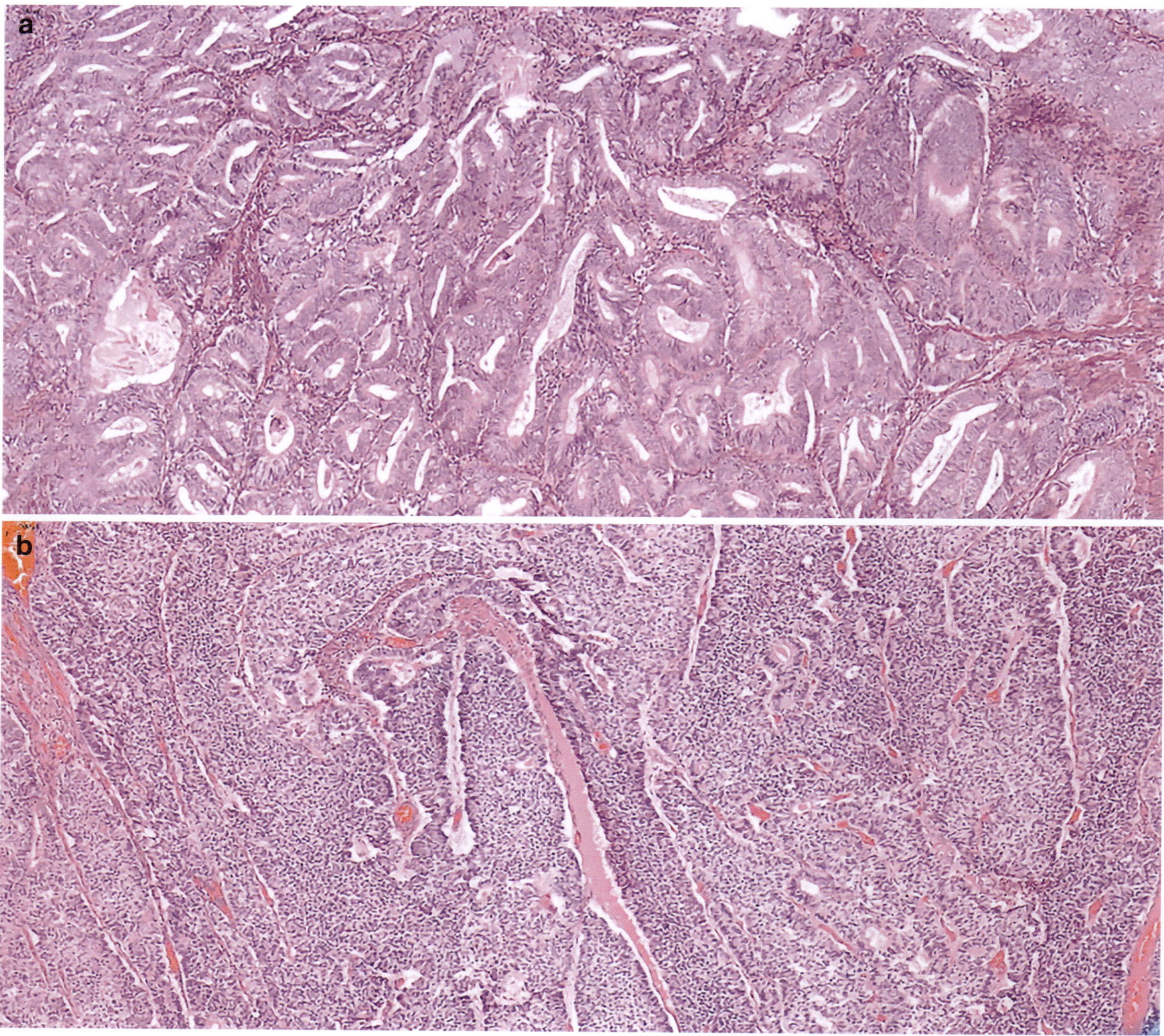

Fig. 12.2 (**a**) Low-grade endometrioid endometrial carcinoma. The tumor is composed of irregular and cribriform glands lined by cells with moderate nuclear pleomorphism. (**b**) High-grade endometrioid endometrial carcinoma. The tumor shows a solid growth pattern and moderate nuclear atypia

- Prognosis of dedifferentiated carcinoma is identical to that of undifferentiated carcinoma
- Mixed carcinomas are composed of two or more discrete histological types of endometrial carcinoma, of which at least one component is either serous or clear cell
- Carcinosarcoma, also called malignant mixed müllerian tumors (MMMT), formerly included in the group of mixed epithelial and stromal tumors, is now classified as a type of endometrial carcinoma
  - The tumor shows the typical biphasic pattern, and a microscopic appearance analogous to sarcomatoid or metaplastic carcinomas of other organs
  - Epithelial elements have usually the features of serous carcinoma, but may also have the appearance of EEC
  - Sarcomatous components may be either homologous or heterologous
  - Even if carcinosarcoma is currently classified as endometrial carcinoma, it should be regarded as a special type since it is associated with a worse prognosis than that of the ordinary endometrial carcinoma
  - The metastatic pattern of carcinosarcoma supports the theory of neoplasia driven by the epithelial component

- Myometrial infiltration, lymphovascular involvement, and metastases display more often the epithelial elements than the sarcomatous components of the tumor
- The new WHO classification includes novel tumor types, such as squamous cell carcinoma, mesonephric-like adenocarcinoma, and gastrointestinal-type mucinous carcinoma, which are very unusual
- Neuroendocrine carcinomas of the endometrium are usually high-grade, similar to the most frequent pulmonary counterparts

Histological Grade

- Evaluation of histopathological grade in endometrial carcinoma is very important in risk stratification and decision on the extent of surgical treatment and administration of adjuvant therapy
- Serous carcinoma, clear cell carcinoma, undifferentiated carcinoma, and carcinosarcoma are considered as high-grade by definition and thus do not need to be graded
- EEC is graded using FIGO grading criteria, based on architectural features: grade 1, 2, and 3 tumors, respectively, exhibiting ≤5%, 6–50%, and >50% solid nonglandular growth
  - In EEC with squamous differentiation, the grade of the tumor should be assessed in the nonsquamous areas
  - The presence of severe cytological atypia in most cells (> 50%) increases the grade by one level
  - However, a discordant high nuclear grade in association with a low architectural (endometrioid-like) pattern should raise the possibility that the tumor is not an EEC, but a serous carcinoma
- WHO 2020, recommends binary grading, whereby EEC 1–2 are classified as low-grade and EEC 3 as high grade

Reproducibility of Histological Typing

- There are problems in interobserver reproducibility regarding histopathological typing
- While diagnosis is highly reproducible in EEC 1–2, which account for 70% of the cases, and in typical serous carcinoma and clear cell carcinoma, there is a poor interobserver agreement in approximately 10% of tumors
- Although the molecular profile of EEC and serous carcinoma, is different, there is a subset of tumors with overlapping molecular features, in the grey zone between EEC and serous carcinoma, which correspond to the subset of tumors with poorly reproducible microscopic appearance (around 10% of endometrial carcinomas)

- EEC with mismatch repair deficiency (MMRD) or POLE mutation, show spatial heterogeneity and may contain areas that mimic serous carcinoma
- Distinction between EEC 3 and serous carcinoma is usually done by microscopic examination
  - Differential diagnosis may be difficult and subjected to interobserver variation in some cases
  - Immunohistochemistry can be of help
- The typical immunohistochemical profile of EEC includes positive immunoreaction for cytokeratins, vimentin, and estrogen and progesterone receptors
- The typical immunohistochemical profile for serous carcinoma includes strong immunoreaction for p53 and p16, as well as intact mismatch repair proteins (MMRP)
  - Whereas p53 is expressed in only 10–35% of EEC—usually in high-grade tumors—70–90% of serous carcinoma show p53 immunoreaction
  - Similarly, p16 shows patchy, weak to moderate, (mosaic-pattern) staining in EEC but diffuse and strong immunoreaction in 95% of serous carcinoma
  - MMRD is a feature of EEC, and it is very infrequent among serous carcinomas
- Other potentially useful markers include beta-catenin, PTEN, and ARID1A
  - Nuclear immunoreaction for beta-catenin and inactivation (lack of immunoreaction) of the PTEN and ARID1A tumor suppressor genes are seen in EEC
  - The squamous morules of complex atypical hyperplasia and well-differentiated adenocarcinoma typically show nuclear immunoreaction for beta-catenin
- In difficult cases, a wide panel of immunohistochemical markers, or molecular analysis (POLE mutation) may be necessary
- Targeted next-generation sequencing is a good tool to help in the differential diagnosis

Molecular Features of Endometrioid– Endometrial Carcinoma

- The molecular alterations involved in the development of EEC differ from those of nonendometrioid carcinoma
  - cDNA analysis has clearly shown that EEC and nonendometrioid carcinoma exhibit different gene expression profiles
- In single-gene analysis, EEC shows microsatellite instability (MSI), and mutations in the *PTEN, KRAS, PIK3CA,* and beta-catenin genes
- According to TCGA, the most frequently mutated genes in EEC were *PTEN* (77.7%), *PIK3CA* (53.1%), *PIK3R1* (37.1%), *CTNNB1* (36.6%), *ARID1A* (35.4%), *KRAS* (24.6%), and *CTCF* (20.6%)

Microsatellite Instability/Mismatch Repair Deficiency

- It is estimated that around 30% of endometrial carcinoma show MSI/MMRD
- MSI/MMRD is secondary to alterations in mismatch repair (MMR) genes, such as MLH-1, MSH-2, MSH-6, PMS-2
- MSI/MMRD leads to the accumulation of mutations in repetitive DNA sequences, ie, microsatellites, and high mutation load
- Mutations in some repetitive mononucleotide tracts located within the coding sequence of some genes involved in cell proliferation, cell differentiation, DNA repair, and apoptosis, such as BAX, IGFIIR, hMSH3, hMSH6, MBD4, CHK-1, CASP-5, ATR, ATM, BML, RAD-50, BCL-10, and APAF-1, are secondary events in endometrial carcinoma with MSI/MMRD
- Around 3% of patients with endometrial carcinoma, are associated with hereditary nonpolyposis colon cancer (HNPCC; Lynch syndrome), and around 27% have sporadic endometrial carcinoma with MSI/MMRD
- Endometrial carcinoma patients from HNPCC kindreds have an inherited germline mutation in *MLH-1, MSH-2, MSH-6,* or *PMS-2*; nevertheless, endometrial carcinoma only develops after the instauration of a deletion or mutation in the contralateral *MLH-1, MSH-2, MSH-6, or PMS-2* allele in endometrial cells
- In sporadic endometrial carcinoma, MSI/MMRD, is secondary to somatic MLH-1 promoter hypermethylation in 23%, double somatic mutations in 3.5%
- Abnormal methylation of MLH-1 may also be detected in atypical hyperplasia, suggesting that it may be an early event in the pathogenesis of EEC that precedes the development of MSI/MMRD
- MSI/MMRD occurs almost exclusively in EEC, but also in a minority of nonendometrioid carcinoma (a subset of undifferentiated carcinoma, and very few serous carcinomas, clear cell carcinoma, and carcinosarcoma)
- Characteristic microscopical features of EEC arising in this setting are
  - Poor differentiation
  - Crohn-like lymphoid reaction
  - Lymphangioinvasive growth
  - Tumor-infiltrating lymphocytes
- MSI/MMRD can be identified by PCR-based molecular techniques (NIH, Promega, Idylla) (Fig. 12.3), next-generation sequencing, or IHC
- MSI is the preferred term when molecular tests are used, while MMRD is used when IHC is performed
- It is generally accepted that IHC is the best test for MSI/MMRD

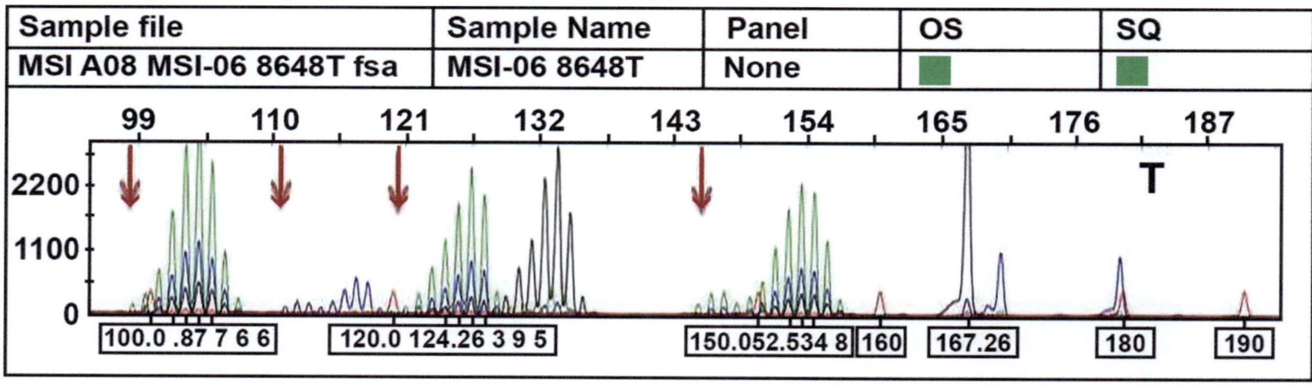

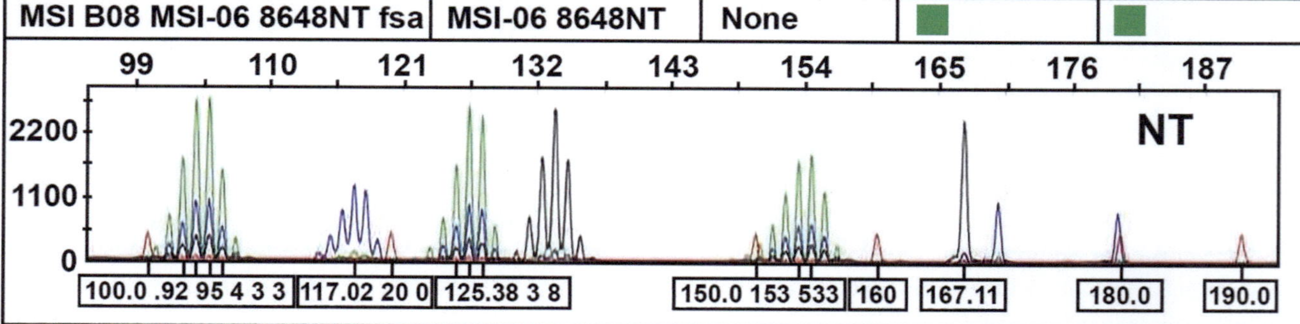

Fig. 12.3 Promega MSI Analysis System of endometrioid carcinoma. Results show microsatellite instability in the tumor sample

- IHC approach consists of assessment of the expression of four MMR proteins (MLH-1, PMS-2, MSH-6, and MSH-2)
  - A simplified cost-effective version includes only the minor dimeric partners PMS-2 and MSH6, with expanded analysis of MLH-1 when PMS-2 is lost, and of MSH-2 when MSH-6
- PCR-based tests are important when IHC is not contributory, or when IHC results do not correlate with familiar history or clinical setting
- Systematic clinical screening of personal and family history misses a significant proportion of HNPCC women since up to 75% of them do not fulfill the revised Bethesda Guidelines criteria, which suggests that MSI/MMRD tests should be performed in all endometrial carcinoma samples, irrespective of the age of the patient
- It is necessary to perform MLH-1 promoter methylation assays in all endometrial carcinoma with abnormal expression of MLH1
  - If MLH1 promoter methylation is present, the tumor is most likely to be sporadic
  - Whenever there is a lack of MLH-1 promoter methylation or abnormal expression of MSH2 or MSH6, there is still a possibility that there is a somatic mutation in any of these genes (Fig. 12.4)
  - Germline mutation analysis is necessary to confirm that the patient has Lynch syndrome
- Several international guidelines recommend universal testing for MSI/MMRD for four reasons
  - Diagnostic, since MSI/MMRD is considered a marker for EEC
  - It is part of the algorithm to identify patients with Lynch syndrome
  - It is part of the surrogate to stratify patients according to The Cancer Genome Atlas (TCGA) based molecular classification (see below)
  - Predictive for the potential utility of immune checkpoint inhibitor therapy

Phosphatase and Tensin Homolog (*PTEN*)

- *PTEN* is frequently abnormal in endometrial carcinoma
  - LOH at chromosome 10q23 occurs in 40% of endometrial carcinoma
  - Somatic *PTEN* mutations are common in endometrial carcinoma
  - They are almost exclusively restricted to EEC and occur in 37–61% of cases
- Concordance between MI status and *PTEN* mutations suggests that *PTEN* could be a likely candidate to be targeted for mutations in the MI-positive endometrial carcinoma

- Somatic *PTEN* mutations often coexist with other alterations in the PI3K pathway, including *PIK3CA*, *PIK3R1* and/or *KRAS* mutations
- In agreement with Knudson's two-hit proposal, LOH at 10q23 frequently coexists with somatic *PTEN* mutations
  - The coexistence of both alterations leads to the activation of the PI3K/AKT pathway, which plays a key role in the regulation of cellular homeostasis
- For some time, PTEN immunohistochemistry was poorly reproducible
  - After the publication of optimization protocols, PTEN immunostaining has been found to be a good surrogate marker for PTEN mutations, sometimes used in protocols for distinguishing between EEC and serous carcinoma
- *PTEN* mutations are occasionally detected in normal endometrial cells in premenopausal women by lack of PTEN expression by IHC
  - It is thought that, in the absence of estrogenic excess, these glands are shed during menses
  - When a patient is under estrogen stimulation as in the setting of obesity, PTEN-null glands may overgrow and give rise to endometrial hyperplasia, endometrioid intraepithelial neoplasia, and ultimately, EEC
- *PTEN* mutations have been detected in endometrial hyperplasias with and without atypia (19% and 21%, respectively), both regarded as precursors of EEC
- Although the prognostic significance of *PTEN* mutations in endometrial carcinoma is controversial, their association with favorable prognostic factors has been reported
  - In the TCGA endometrial carcinoma study, PTEN was found to be altered in the majority (>75%) of endometrial carcinoma associated with better prognosis (POLE-mutated, MSI/MMRD or nonspecific molecular profile subtype) while rarely altered (<10%) in the most aggressive group of tumors (copy number high)
- Patients with Cowden syndrome, with inherited germline *PTEN* mutations have increased cancer risk, including endometrial carcinoma (Fig. 12.5)

PIK3CA and *PIK3R1*

- Mutations in *PIK3CA* (p110α) (alpha) contribute to the alteration of the *PI3K/AKT* signaling pathway in endometrial carcinoma
  - They are predominantly located in the helical (exon 9) and kinase (exon 20) domains, but can also occur in exons 1–7
  - *PIK3CA* mutations are infrequent in endometrial hyperplasia
- *PIK3CA* mutations occur in 24–39% of the cases and frequently coexist with *PTEN* mutations

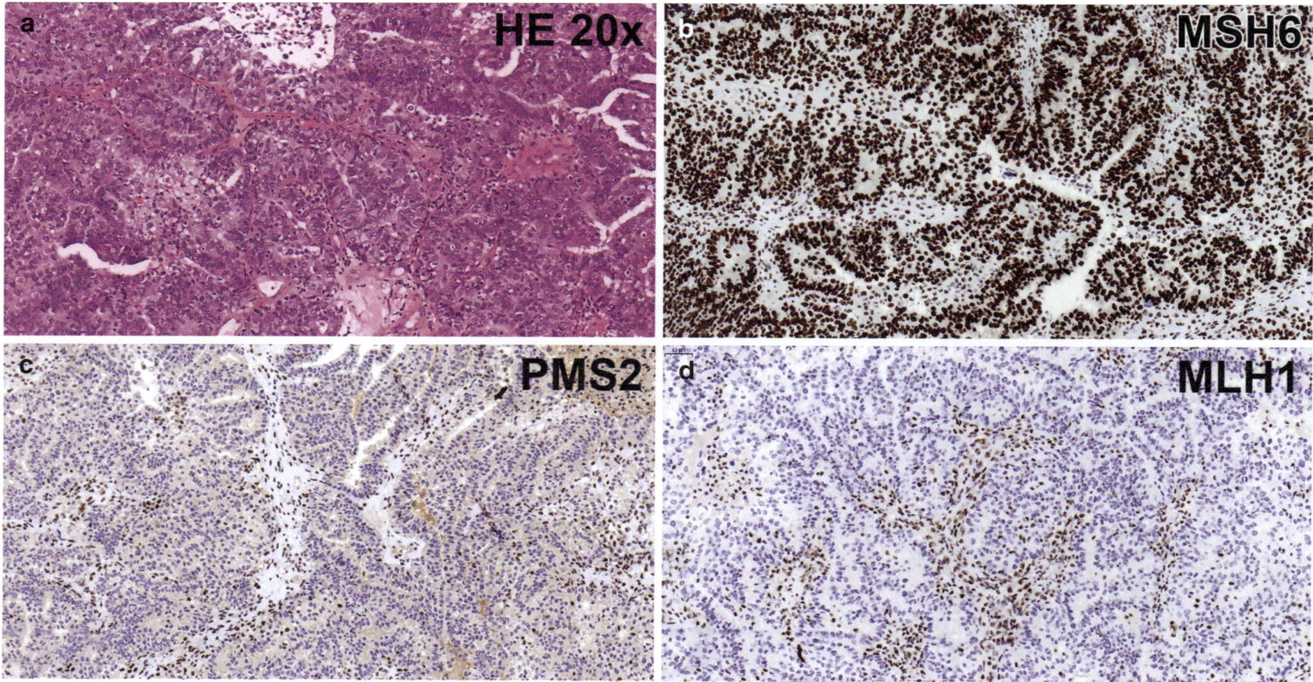

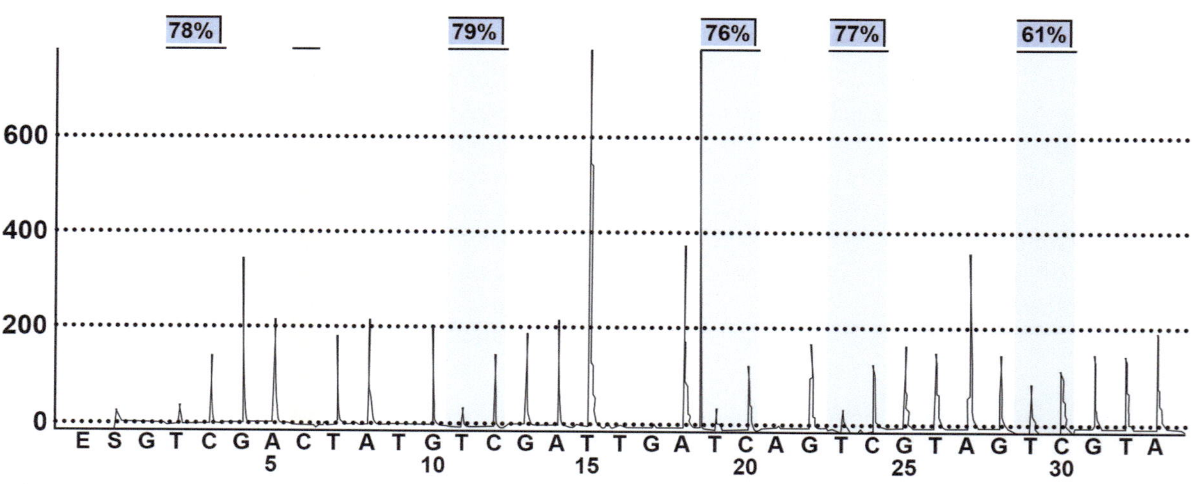

Sequence to analyze:

YGGATAGTGATTTTTAAYGYGTAAGYGTATA

Position	1	2	3	4	5
Quality	Passed	Passed	Passed	Passed	Passed
Meth (%)	78	79	76	77	61

No warnings

Fig. 12.4 Endometrial endometrioid adenocarcinoma (H&E, **a**) with MSH6 preserved expression (**b**) and loss of expression of PMS2 (**c**) and MLH1 (**d**). Pyrosequencing assay shows MLH-1 promoter hypermethylation (**e**)

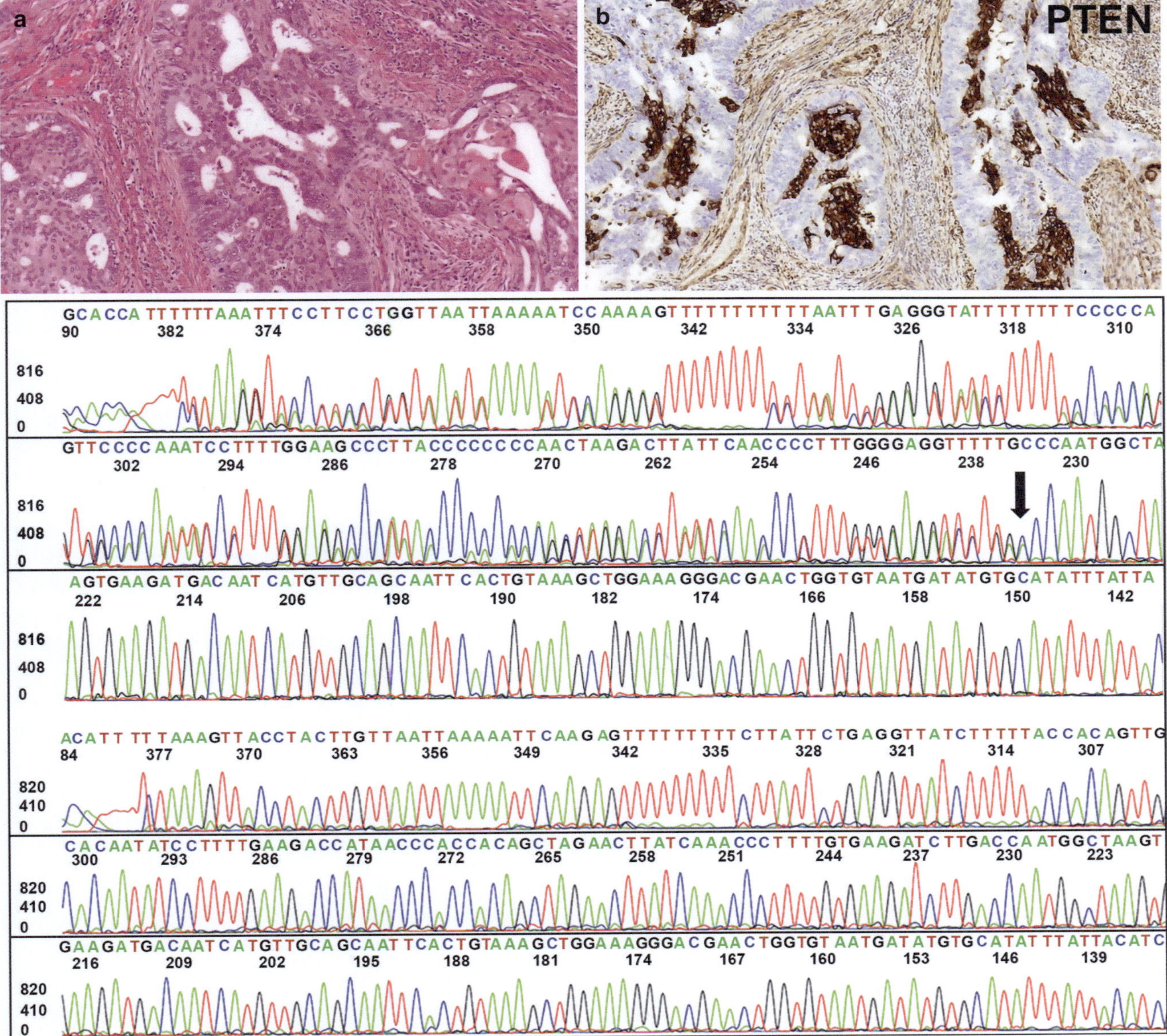

Fig. 12.5 Endometrioid carcinoma (H&E, **a**) in a Cowden syndrome patient, showing (**b**) loss of expression of PTEN and retained expression in tumor morules. Sequencing analysis of the normal tissue of the patient detected a mutation consisting of the insertion of a nucleotide in exon 5 of the PTEN gene (c.326_327insC), which produces a change in the (**c**) reading pattern of the protein, causing a stop codon (p.Q110fs * 5)

- Although initially described in EEC, *PI3KCA* mutations also occur in nonendometrioid carcinoma, and in mixed EEC–nonendometrioid carcinoma
- Mutations in *PIK3RI* (p85α) (alpha), the inhibitory subunit of *PI3K*, have been detected in 43% of EEC, and 12% of nonendometrioid carcinoma
  - Distribution of *PIK3R1* mutations is nonrandom; most mutations are localized to the p85α-nSH2 (alpha) and -iSH2 domains that mediate binding to p110α (alpha)
- In EEC, *PIK3R1* mutations frequently coexist with *PTEN* and *KRAS* mutations but tend to be mutually exclusive with PIK3CA mutations

RAS–MAPK Pathway

- The RAS–RAF–MEK–ERK signaling pathway plays an important role in endometrial tumorigenesis
- The frequency of *KRAS* mutations in endometrial carcinoma ranges between 10% and 30%
  - In some series, *KRAS* mutations are more frequent in EEC with microsatellite instability
- *RASSF1A* inactivation by promoter hypermethylation may contribute significantly to increasing the activity of the RAS–RAF–MEK–ERK signaling pathway

- Endometrial carcinoma frequently shows inactivation of *SPRY-2* by promoter methylation
  - *SPRY2* is involved in the negative regulation of the FGFR pathway
  - Reduced SPRY2 immunoexpression is seen in almost 20% of endometrial carcinoma, and is strongly associated with increased cell proliferation
- Somatic mutations in the receptor tyrosine kinase *FGFR2* have been found in 10–12% of endometrial carcinoma, particularly in EEC (16%)
  - *FGFR2* mutations and *KRAS* mutations are mutually exclusive events

Beta-Catenin

- The beta-catenin gene (*CTNNB1*) maps to 3p21
  - Appears to be important in the function of both APC and E-cadherin
  - A component of the E-cadherin–catenin unit, important for cell differentiation and maintenance of the normal tissue architecture
  - Important in signal transduction
  - Increased cytoplasmic and nuclear levels result in transcriptional activation through the *LEF/Tcf* pathway
- Mutations in exon 3 of *CTNNB1* occur in 14–44% of endometrial carcinoma and result in stabilization of the protein, cytoplasmic, and nuclear accumulation, and participation in signal transduction and transcriptional activation through the formation of complexes with DNA-binding proteins (Fig. 12.6). *CTNNB1* mutations appear to be independent of MI and mutational status of *PTEN* and *KRAS*
- Although there is a good correlation between *CTNNB1* mutations and beta-catenin nuclear immunoreaction, other genes of the *Wnt/beta-catenin/LEF-1* pathway may be responsible for the stabilization and putative transcription activator role of beta-catenin in the EEC
- Beta-catenin alterations have been described in endometrial hyperplasias and grade 1 EECs that contain squamous metaplasia (morules)
- Prognostic significance of beta-catenin mutations in endometrial carcinoma is controversial; however, tend to occur in tumors with favorable prognosis in general
  - In the TCGA endometrial carcinoma study, beta-catenin alterations were more common in endometrial carcinoma associated with better prognosis (POLE-mutated, MSI/MMRD or nonspecific molecular profile subtype) while rarely altered in the most aggressive group of tumors (copy number high)
  - However, beta-catenin mutations could be associated with worse prognosis in the group of nonspecific profile

Fig. 12.6 (**a, b**) Different immunoexpression patterns of CTNNB1: (**a**) Membranous immunoreaction; (**b**) Membranous and nuclear immunostaining in squamous morules. (**c**) Missense mutation in exon 3 of CTNNB1

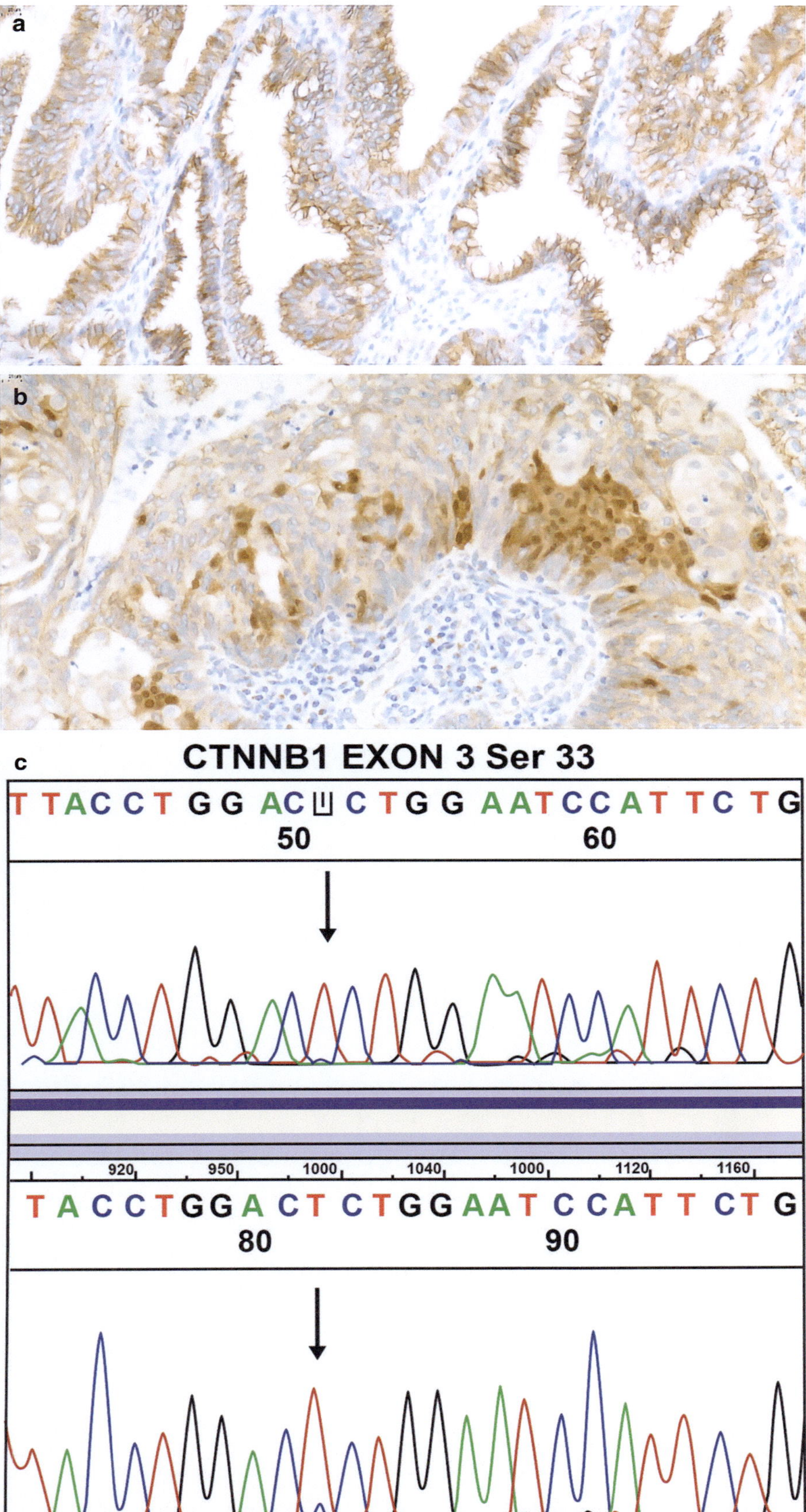

ARID1A

- The SWI/SNF chromatin remodeling complex modulates gene transcription by remodeling nucleosomes in an ATP-dependent manner and is involved in diverse cellular processes such as tissue differentiation, proliferation, and DNA repair

- The SWI/SNF is mutated in >20% of all human cancers, and the ARID1A (BAF250A) subunit is especially prone to somatic mutations in gynecologic cancers
- *ARID1A* has an important role in DNA damage response (DDR) maintaining genomic integrity
- *ARID1A* gene mutations are present in 40% of EEC 1–2 whereas ARID1A protein expression is lost in ~25% of these tumors (Fig. 12.7)

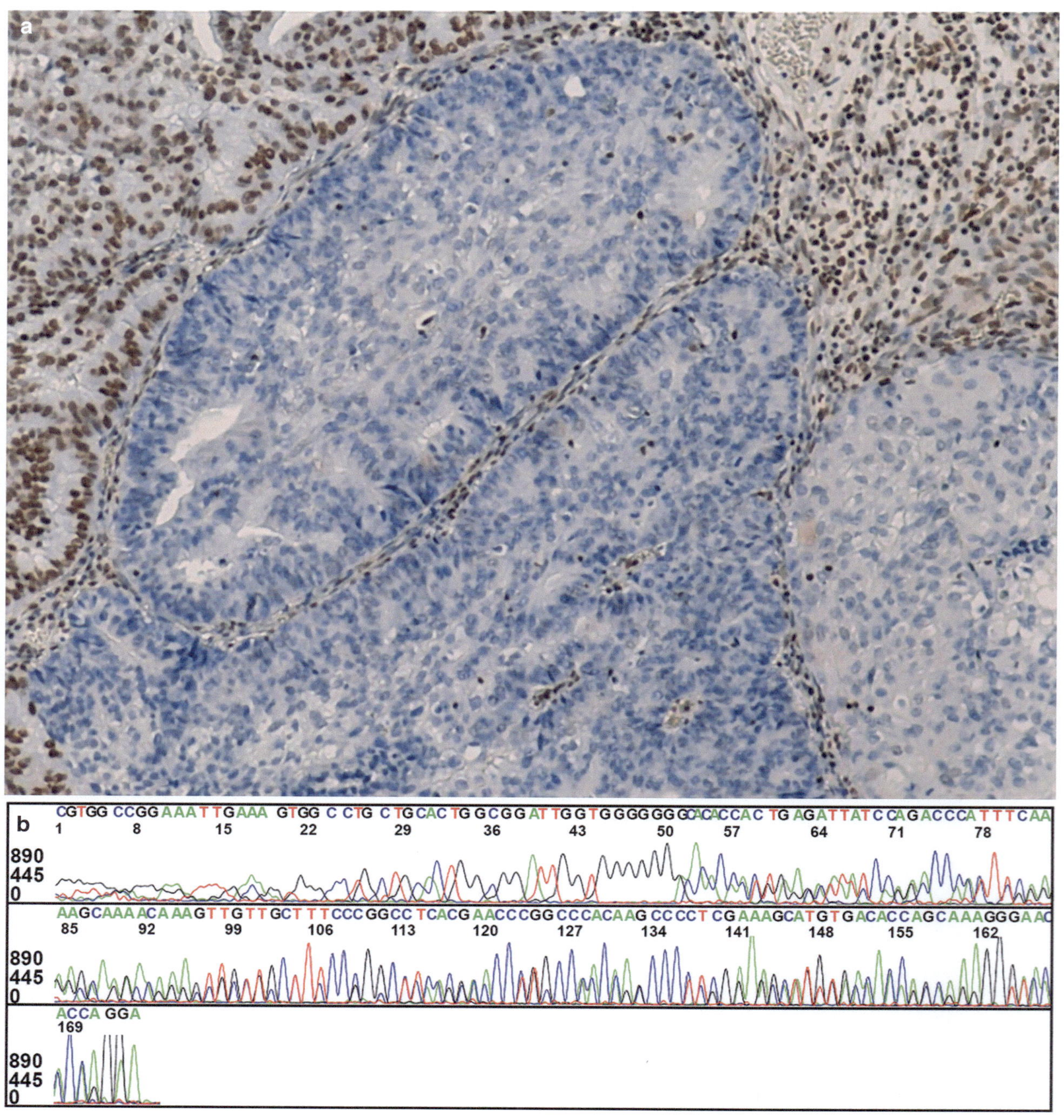

Fig. 12.7 (**a**) Endometrial endometrioid carcinoma with loss of expression of ARID1a. (**b**) Sequencing analysis of the tumor detected a mutation consisting of a nucleotide deletion in exon 20 of ARID1a gene (c.5548 del G) which produces a change in the reading pattern of the protein, causing a stop codon (p.D1850Tfs *33)

- *ARID1A* loss has been observed in focal areas of atypical endometrial hyperplasia, revealing clonal loss, in association with malignant transformation and concurrent endometrial carcinomas
- ARID1A expression in the endometrial biopsy or curettage of an endometrial carcinoma is associated with a significantly increased risk and higher FIGO stages in subsequent hysterectomy specimen

Molecular Features of Serous Carcinoma

- In single gene analysis, serous carcinomas have been found to exhibit alterations of *TP53*, loss of heterozygosity (LOH) on several chromosomes (chromosomal instability), as well as other molecular alterations (*STK15, p16*, E-cadherin, and *C-erbB2*)
- The genes most frequently mutated in serous carcinoma, according to TCGA are TP53 (90.7%), PIK3CA (41.9%), FBXW7 (30.2%), and PPP2R1A (36.6%)
  - Additional studies using exome-sequencing analysis have shown mutations in TAF-1 (30%)

TP53

- Whereas *TP53* mutations occur in over 90% of serous carcinoma, they are present in only 10–20% of EEC (3% EEC 1, 11% EEC2, 37% EEC3)
- The TP53 mutational spectrum is different in serous carcinoma than in EEC
  - TP53 hotspot mutations are significantly more frequent in serous carcinoma (46%), than in EEC (15%)
- The frequency of TP53 mutations varies according to the TCGA-molecular subtype of the tumor
  - They are more frequent in copy number high tumors (91%), and POLE mutated tumors (35%) in comparison with endometrial carcinoma with MSI/MMRD (8%) of low-copy number/nonspecific molecular profile (1%)
- P53 protein can induce apoptosis or prevent a cell from dividing if there is DNA damage
  - Mutation of the *TP53* gene diminishes the cell's ability to repair DNA damage before entry to S-phase, lead-

ing to a greater chance that mutations will be fixed in the genome and passed to successive generations of cells (Fig. 12.8)

Other Alterations

- Inactivation of the cell cycle regulator *p16* is also more frequent in serous carcinoma (40%) than in EEC (10%)
  - Although the underlying mechanism is unclear, it probably involves deletion and promoter hypermethylation
- Reduced expression of E-cadherin is frequent in serous carcinoma and may be caused by LOH or promoter hypermethylation
  - LOH at 16q22.1 is seen in almost 60% of serous carcinoma but only in 22% of EEC
- *C-erbB2* overexpression and amplification are seen more frequently in serous carcinoma (43%) than in EEC (29%)
  - This is interesting from the viewpoint of targeted therapy
  - *C-erbB2* alterations are occasionally present only in part of the tumor
- *FBXW7* mutations are frequent
- *CCNE1* amplification is common
- serous carcinoma shows chromosomal instability, widespread chromosomal gains and losses, and aneuploidy
- cDNA arrays have demonstrated that serous carcinomas usually show overexpression of genes (*STK-15, BUB1, CCNB2*) involved in the regulation of the mitotic spindle checkpoint
  - One of these genes, *STK-15*, essential for chromosome segregation and centrosome functions, is frequently amplified in serous carcinoma (60%)
- New biomarkers of serous carcinoma are EpCAM, claudin-3, and claudin-4 receptors, serum amyloid A, folate-binding protein, mesothelin, LRP-1, and IMP2
- Serous carcinoma, like the tumors that are associated with p53 alterations (copy number high) is occasionally associated with homologous recombination deficiency
  - They sometimes occur in the setting of germline BRCA 1 or 2 germline mutations

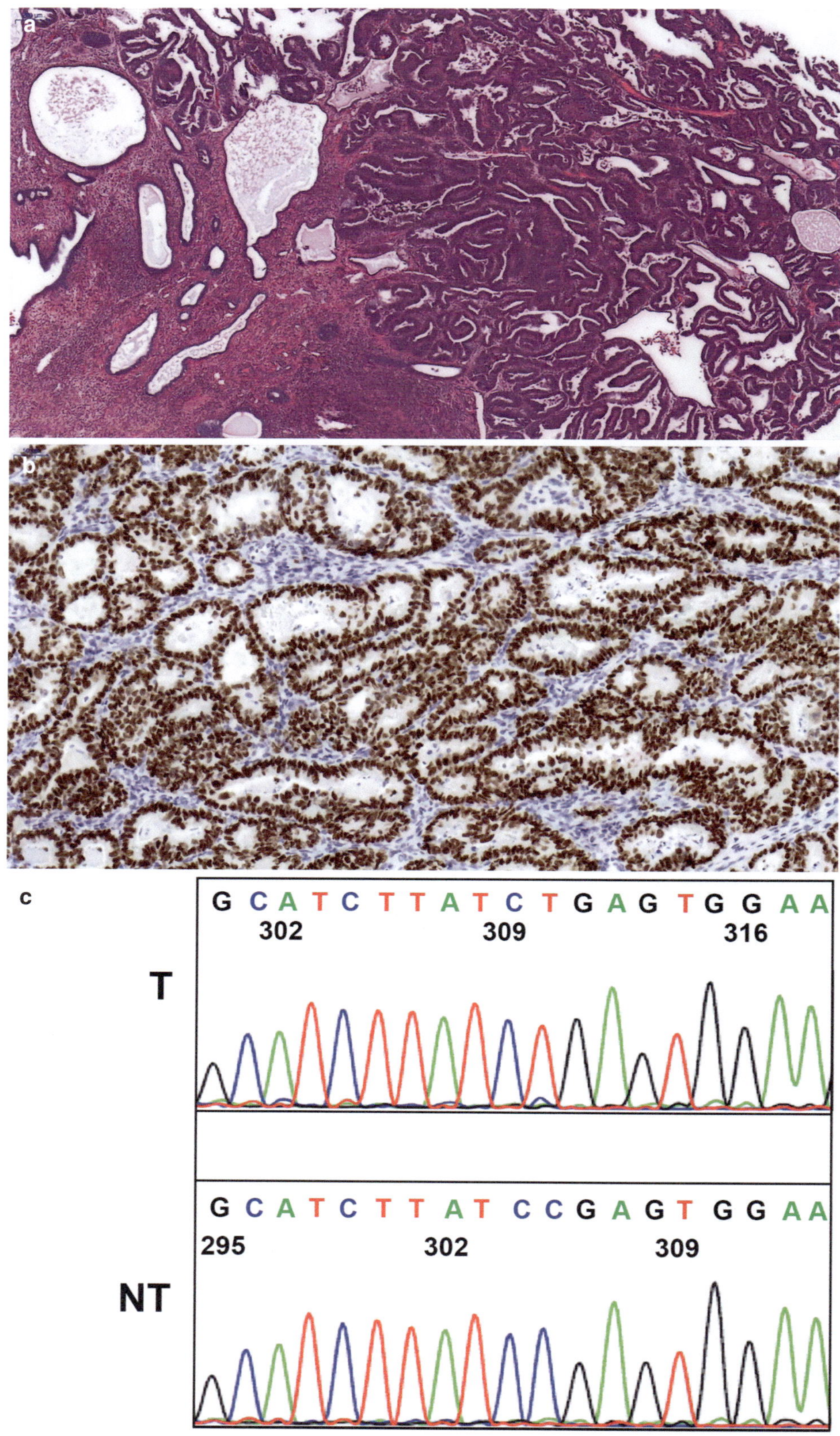

Fig. 12.8 (**a**) Uterine serous carcinoma developing in a preexisting endometrial polyp. (**b**) Serous carcinoma showing a p53 mutated pattern by immunohistochemical analysis. (**c**) Sequencing analysis showed a missense mutation in p53 gene that causes a stop codon (p.R196*)

Molecular Features of Clear Cell Carcinoma

- The genes most frequently mutated in clear cell carcinoma are TP53 (39.7%), PIK3CA (23.8%), PIK3R1 (15.9%), ARID1A (15.9%), PPP2R1A (15.9%), SPOP (14.3%), and TAF1 (9.5%),
- A small subset of tumors shows MSI/MMRD (11.3%), and POLE mutations, and may be associated with improved prognosis, although the evidence is small
- clear cell carcinoma show immunoreactivity for HNF1beta, Napsin A, and AMACR
- Mutation of the AT-rich interactive domain-containing protein 1A (*ARID1A*) gene and loss of the corresponding protein BAF250a has been described as a frequent event (almost 50% of cases) in clear cell carcinomas of the ovary, and endometrium (26%)
- Some clear cell carcinomas molecularly resemble serous carcinoma and others molecularly resemble EEC

Molecular Features of Mixed Endometrioid–Nonendometrioid Adenocarcinomas

- Endometrial carcinomas showing an admixture of EEC and nonendometrioid carcinoma (serous and/or clear cell carcinomas) are classified as mixed carcinomas and prognosis depends on the proportion of the most aggressive component
- It has been suggested that, in mixed carcinomas, the non-endometrioid carcinoma component develops because of tumor progression—through *TP53* and *PIK3CA* mutations—from a preexisting EEC, since these tumors frequently retain the molecular alterations of typical EEC
  - This hypothesis would explain not only the existence of mixed EEC–nonendometrioid carcinoma, but also the presence of MI, as well as alterations in *PTEN, KRAS*, or beta-catenin, in nonendometrioid carcinoma
- Mixed EEC–nonendometrioid carcinoma may exhibit overlapping features with EEC (mixed EEC–nonendometrioid carcinoma morphology, early age at presentation, evidence of estrogen stimulation or preexisting hyperplasia, coexistence of *TP53* mutations, and MI or *PTEN* mutations)

- Several studies with sequencing strategies in the microdissected tumor components, have confirmed a common clonal origin\ in a significant proportion of cases

Molecular Features of Endometrioid Carcinomas with Ambiguous Features

- Some endometrial carcinomas exhibit overlapping or intermediate features between EEC and serous carcinoma
  - In these tumors, it is not possible to delineate two different components, i.e., one of EEC, and another of serous carcinoma
  - The term "endometrial carcinoma with ambiguous features" has been proposed for such cases
  - This is a heterogeneous group of tumors both at the microscopic and molecular features

Molecular Features of Undifferentiated Carcinoma and Dedifferentiated Carcinoma

- WHO defines undifferentiated carcinoma of the endometrium as an epithelial tumor with no overt cell lineage differentiation
- In dedifferentiated carcinoma, the undifferentiated component is regarded because of tumor progression from the coexisting low-grade EEC
  - Sequencing strategies in the microdissected tumor components, have confirmed a common clonal origin in a significant proportion of cases
- Although MSI/MMRD is the predominant molecular feature of undifferentiated carcinoma, TP53 mutations may also be found (Fig. 12.9)
- POLE mutations have been detected in a few cases, associated with improved prognosis
- PI3K pathway mutations involving PTEN, PIK3CA or PIK3R1 are frequent (>50% of cases)
- Inactivating mutations in genes of the SWI/SNF complex proteins (SMARCA4-BRG1, SMARCB2-INI1, ARID1A, ARID1B) are frequent, commonly associated with de-differentiation

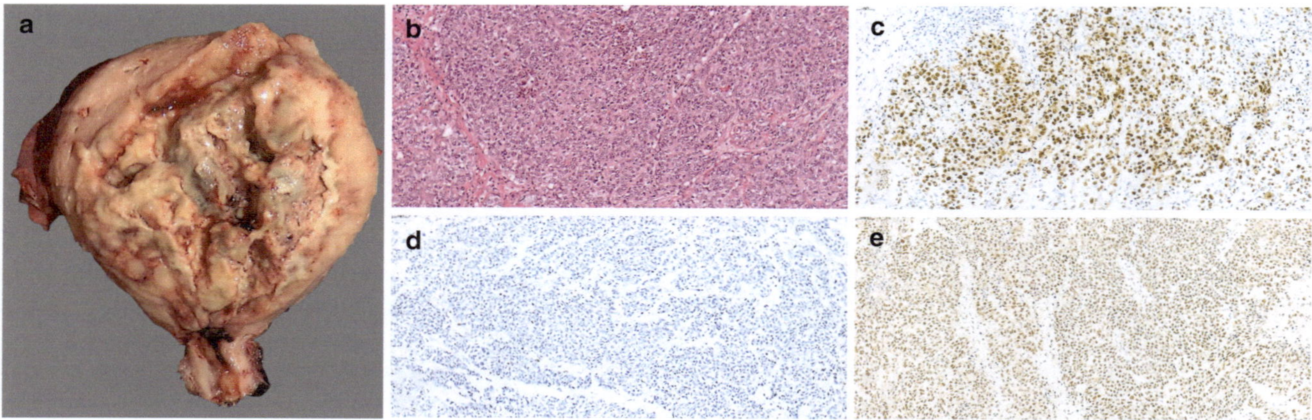

Fig. 12.9 (**a**) Undifferentiated Carcinoma (undifferentiated carcinoma) of the endometrium. The endometrial cavity is fulfilled with a large necrotic mass that extends to the low uterine segment. (**b**) The tumor is composed of epithelial cells with no overt cell lineage differentiation. (**c**) The tumor shows TP53 mutated pattern of expression and (**d**) loss of expression of MSH6. (**e**) PMS2 expression is preserved in this case

Molecular Features of Carcinosarcoma (Malignant Mixed Müllerian Tumors)

- Immunohistochemical and molecular genetics studies support the clonal nature of the two—epithelial and mesenchymal—components in carcinosarcoma, supporting the hypothesis that they represent in fact metaplastic (sarcomatoid) carcinomas
    - Expression of epithelial markers in the sarcomatous components occurs in a large proportion of cases (Fig. 12.10)
- Carcinosarcoma cell lines can differentiate into epithelial, mesenchymal, or both components
    - Chromosome X inactivation studies, LOH, and gene mutation analyses all have shown that the epithelial and mesenchymal elements share common genetic alterations
- Carcinosarcoma probably occurs through epithelial-to-mesenchymal transition (EMT) in endometrial carcinomas
    - EMT is a process of cellular transdifferentiation in which epithelial cells lose polarity and cell–cell contacts, reorganize their cytoskeleton, and acquire expression of mesenchymal phenotype
- Carcinosarcoma show expression of genes that repress epithelial markers (E-cadherin) and enhance expression of mesenchymal markers, including proteins involved in skeletal muscle development
    - Carcinosarcoma has revealed a microRNA signature typical of EMT
    - TCGA study identified an epithelial-mesenchymal phenotypic diversity
- According to TCGA, the genes more frequently mutated in carcinosarcoma are TP53 (91%), FBXW7 (39%), PIK3CA (35%), PPP2R1A (28%), PTEN (19%), ARID1A (12%), and KRAS (12%), similar to serous carcinoma and EEC
- A small subset of tumors shows MSI/MMRD (11.3%), and POLE mutations, and may be associated with improved prognosis, although the evidence is small

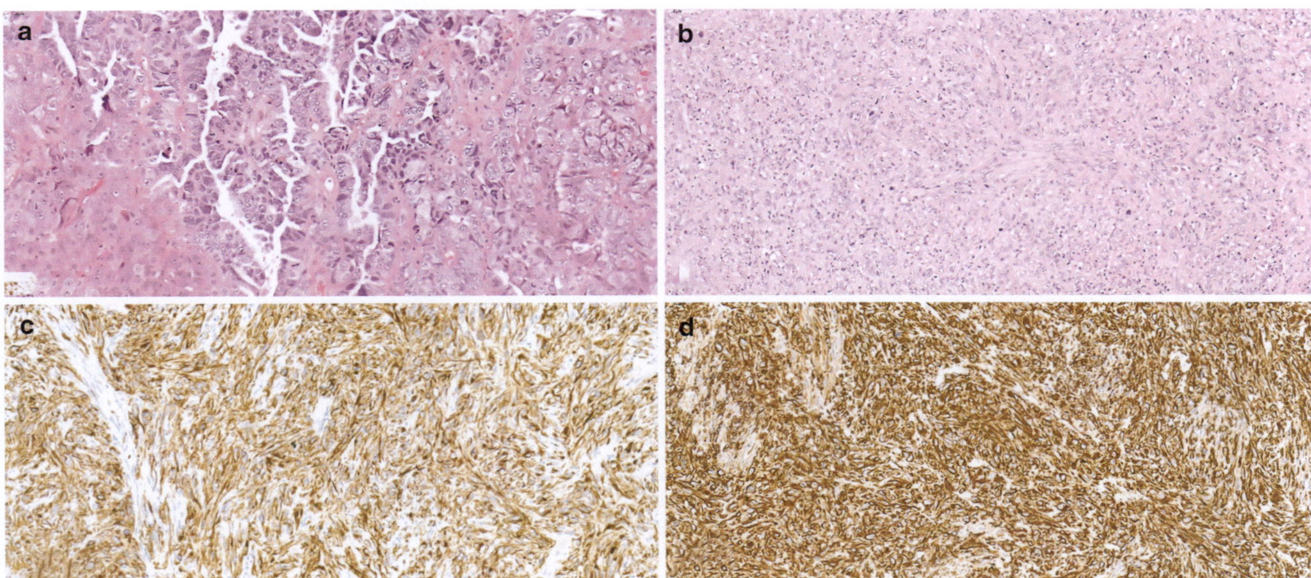

Fig. 12.10 Carcinosarcoma (Malignant mixed müllerian tumor). The tumor shows a biphasic pattern. (**a**) Epithelial component with serous carcinoma features. (**b**) The mesenchymal component shows sarcomatoid features. (**c**) CKAE1/AE3 expression in the sarcomatoid component. (**d**) Vimentin expression in sarcomatoid component

Molecular Features of Mesonephric-Like Carcinomas

- *KRAS* mutations are frequent

Molecular Aspects of Staging

- Tumor staging is very important in assessing prognosis, tumor risk stratification, and decision-making process
- Important pathologic aspects in tumor staging
  - Extent of myometrial invasion
  - Stromal cervical invasion
  - Adnexal involvement
  - Lymph node metastasis
  - Extension to adjacent or distant organs
- Deep myometrial invasion is an important prognostic factor of endometrial carcinoma
  - Usually correlates with high histological grade, vascular invasion, cervical involvement, and lymph node metastasis
  - Associated with a high risk of recurrence
  - EEC may exhibit various patterns of myometrial invasion, including diffuse infiltration or expansile-type invasion (Fig. 12.11)
  - A distinctive pattern of myometrial invasion designated microcystic, elongated, and fragmented (MELF) change shows glands lined by attenuated epithelium with luminal neutrophilic infiltrate resembling endothelium
    - MELF most likely represents (Fig. 12.11)
  - Epithelial to mesenchymal transition (EMT) is important in the myometrial invasion, particularly of MELF type
    - EMT can be induced by different signals and pathways, such as those mediated by TGFbeta, tyrosine kinase receptors, and/or Wnt, depending on the specific cellular context
  - Activation of one or more of these pathways frequently converges in a group of transcription factors such as Snail1, Slug, ZEB1, ZEB2, E47, E2–2, and Twist, most of them capable of repressing E-cadherin, a master regulator of cell adhesion and polarity
  - Comparison between endometrial carcinoma samples from the most superficial tumor and the myoinvasive front has shown increases in SNAIL, SLUG, HMGA2, and TWIST mRNA expression, and a decrease in E-cadherin expression, at the myoinvasive front
  - ETS transcription factors activate matrix-degrading proteases and are related to EMT
    - Upregulation of ERM/ETV5, an ETS transcription factor, is associated with early myometrial invasion and correlates with increased matrix metalloproteinase (MMP)-2
  - A higher expression of matrix metalloproteinases (MMP-2, MMP-9) in endometrial carcinoma is associ-

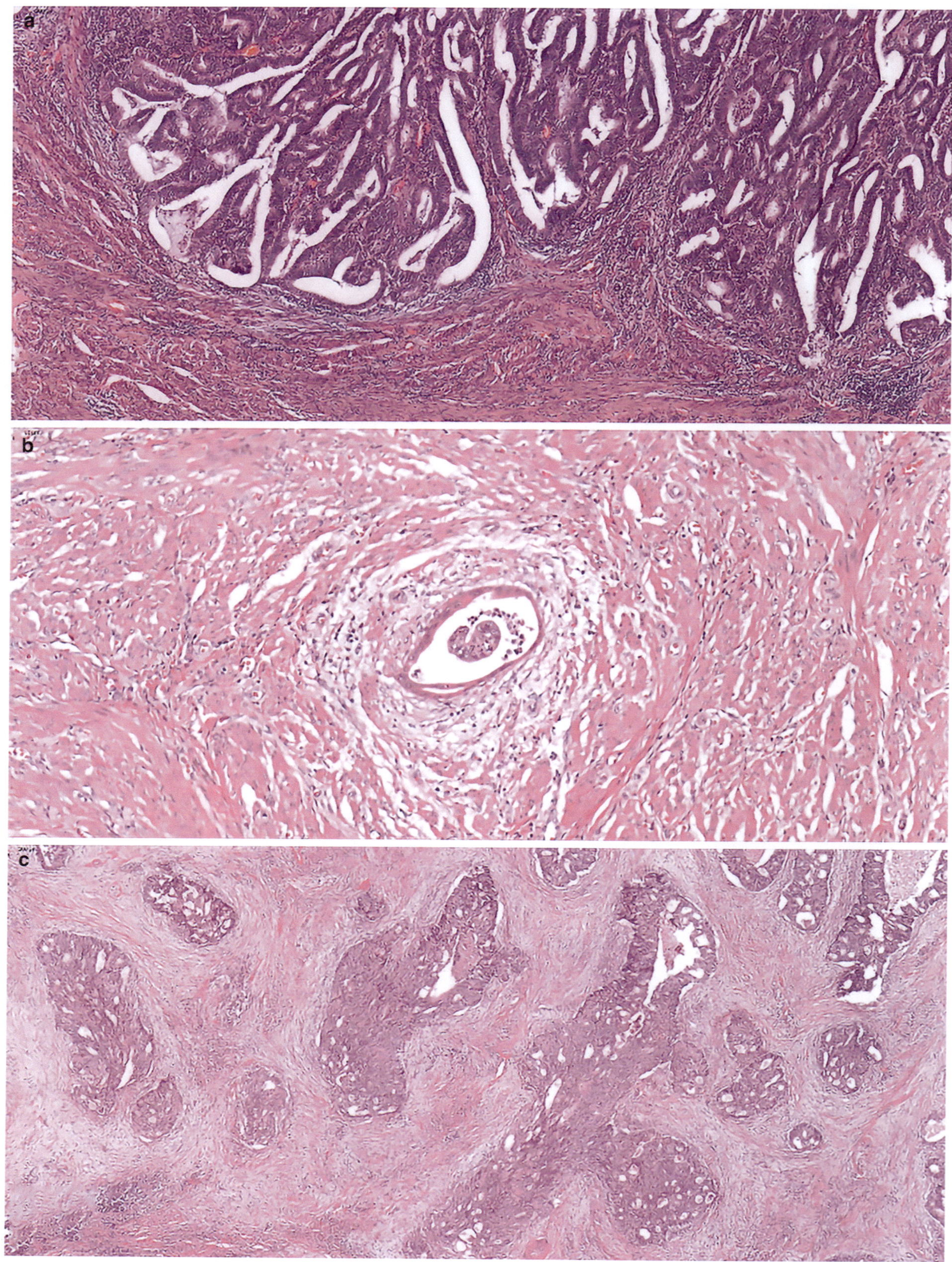

Fig. 12.11 Different invasion patterns of endometrioid endometrial carcinoma. (**a**) Expansile pattern composed of neoplastic glands with well-defined margins that appear to push into the underlying myometrium without a desmoplastic response. (**b**) Microcystic, elongated, and fragmented (MELF) pattern of invasion. (**c**) Invasive patterns with marked stromal desmoplasia

ated with invasive and aggressive behavior in nonendometrioid carcinoma

- o Increases of MMP-7 have been seen, because of beta-catenin nuclear accumulation, in endometrial carcinoma with CTNNB1 mutations
- – Transcription factor RUNX1/AML1 is upregulated in endometrial carcinoma during the invasion
- Adnexal involvement is a controversial issue; for many years, it has been important to distinguish between endometrial carcinoma with ovarian metastasis and synchronous primary tumors of the endometrium and the ovaries, because the prognosis was indolent in some cases, particularly for EEC 1–2
  - – Recent molecular studies have shown that the vast majority of ovarian tumors that occur in the setting of endometrial carcinoma, are indeed metastatic
    - o Although there is always the possibility of the coincidental independent primary in the endometrium and the ovaries, this situation seems to be exceedingly unusual
  - – This is an evolving field, and it is not clear at this time why a subset of metastatic tumors is associated with a good prognosis
    - o Potential explanations include:
      - ◆ That clonal ovarian metastasis occurs early in the process of endometrial tumor development, thereby allowing tumors in each site to acquire additional, sometimes distinct genetic abnormalities
      - ◆ Tumor cells follow retrograde uterine/transtubal spread, possibly with ovarian implantation, rather than destructive invasion
  - – From a practical point of view, WHO 2020 recommends low-risk tumors be managed conservatively (as if they were two independent primaries) when fulfilling the following criteria
    - o Low-grade endometrioid morphology
    - o No more than superficial myometrial invasion
    - o Absence of lymphovascular space invasion
    - o Absence of additional metastases
- Lymph node metastasis is also prognostically relevant
  - – According to TNM8, macrometastases are >2 mm, micrometastases are >0.2 to 2 mm and/or >200 cells, and isolated tumor cells are up to 0.2 mm and ≤200 cells
  - – Macrometastases are regarded as pN1 or pN2 depending on location (pelvic for pN1, para-aortic for pN2), micrometastases as pN1mi or pN2mi (depending again on the location of the involved lymph nodes) and isolated tumor cells are pN0(i+)
  - – Sentinel node biopsy has emerged as an alternative to lymphadenectomy, because allows the detection

of a high percentage of lymph node-positive cases by accurate analysis of one or a few lymph nodes by ultra-staging of the lymph nodes in combination with IHC

- – A molecular approach to detect and quantify transcripts of CK19 in lymph nodes, by OSNA (One Step Nuclear Amplification) has proven to be helpful, and in the process of implementation in Japan and Europe
- Liquid biopsy represents an opportunity to improve tumor staging
  - – There are ongoing studies on the technical validation and prognostic value of detecting circulating tumor cells (CTCs), circulating tumor DNA, or exosomes in peripheral blood or peritoneal washes from patients with endometrial carcinoma

Intratumor Heterogeneity

- Intratumoral heterogeneity is frequent in endometrial carcinoma, both at the morphologic and molecular level
- Discrepancies in diagnosis (histological type and grade) in the same tumor between the diagnostic biopsy and the surgically resected specimen are important and are explained by the fact that the material obtained in the biopsy is not always representative of the real components of the tumor in the hysterectomy specimen, because of tumor heterogeneity
- Assessing the molecular basis of tumor heterogeneity may help in understanding the mechanisms involved in tumor progression and metastasis in endometrial carcinoma
  - – Intratumor heterogeneity analysis allows for the identification of subclonal variants present at low frequencies and nonuniform distribution across different primary tumor regions but may become predominant in tumor metastasis
- Tumor heterogeneity may play a significant role in resistance to treatment; this is particularly relevant for tumors that contain different cell populations with different levels of sensitivity to radiation therapy or chemotherapy
  - – Interestingly, postradiation tumor recurrences have frequently different morphologic features in comparison with the corresponding primary tumor, suggesting that radiation may select some clones in some heterogeneous tumors
  - – C-erb B2 alterations, as biomarkers for targeted therapy in serous carcinoma, are occasionally heterogeneously distributed in different areas of the tumors

cDNA Array Results

- cDNA array studies have demonstrated that the expression profiling of EEC differs from that of nonendometrioid carcinoma
  - In one study, 191 genes exhibited greater than twofold differences between 10 EECs and 16 nonendometrioid carcinomas
    - One of the genes, *TFF3*, was significantly upregulated in EECs, while increased expression of *FOLR* was seen in nonendometrioid carcinomas
- In another study, a different expression profile involving 66 genes was seen in EEC and nonendometrioid carcinoma
  - Estrogen-regulated genes were upregulated in EEC
  - Nonendometrioid carcinoma showed increased expression of genes involved in the regulation of the mitotic spindle checkpoint
- Differential expression of 1055 genes between EECs and serous carcinomas was seen in another investigation
  - Genes upregulated in serous carcinomas were *IGF2, PTGS1*, and *p16*
  - Genes upregulated in EEC included *TFF3, FOXA2*, and *MSX2*
- Another analysis identified 315 genes that statistically distinguished EEC from nonendometrioid carcinoma
- Endometrial carcinomas with microsatellite instability and stable endometrial carcinomas also have different gene expression profiles
  - Two members of the secreted frizzled-related protein family (SFRP1 and SFRP4) are downregulated more frequently in endometrial carcinoma with microsatellite instability
- Ovarian and uterine tumors with beta-catenin alterations show similar gene expression profile
- The gene expression profiles of similar histological subtypes of ovarian and endometrial carcinomas show that clear cell carcinoma has a similar profile regardless of the organ of origin
  - Differences were seen when comparing EEC and serous carcinoma of ovarian and endometrial origin

TCGA-Based Molecular Classification

- The Cancer Genome Atlas Research Network (TCGA) performed an integrating genomic, transcriptomic, and proteomic characterization of endometrial carcinoma. Exome sequence analysis revealed four groups of tumors

- Group 1, with somatic inactivating mutations in POLE exonuclease and very high mutation rates (18 × 106 mutations/Mb) (ultramutated), usually high-grade EEC (7%), associated with good prognosis (Fig. 12.12). They show a mutation profile characterized by mutations in PTEN (94%), PIK3CA (71%), PIK3R1 (65%), FBXW7 (82%), ARID 1A (76%), KRAS (53%), and ARID5B (47%)
- Group 2 included EEC with microsatellite instability (MSI) (hypermutated), frequently with MLH-1 promoter hypermethylation and high mutation rates (28%), associated with an intermediate prognosis. They have a high mutation rate (18 × 106 mutations/Mb), and show mutations in PTEN (88%), RPL22 (33%), KRAS (35%), PIK3CA (54%), PIK3R1 (40%), and ARID 1A (37%)
- Group 3 tumors included EEC with low copy number alterations, also designated tumors with nonspecific molecular profile (NSMP) (39%), associated with an intermediate prognosis. They show a low mutation rate (2.9 × 106 mutations/Mb) and show mutations in PTEN (77%), CTNNB1 (52%), PIK3CA (53%), PIK3R1 (33%), and ARID 1A (42%)
- Group 4 (Serous-like or copy-number high) (26%) showed a low mutation rate, but frequent TP53 mutations, and a worse prognosis, and was predominantly composed of most (but not all) serous carcinoma, but also some EEC (many high-grade); they show low mutation rate (2.3 × 106 mutations/Mb), with frequent mutations in TP53 (92%), PPP2R1A (22%) and PIK3CA (47%)
  - They show chromosomal instability, with recurrent amplifications (MYC, ERBB2, CCNE1, FGFR3, and SOX17)
- A TCGA-based molecular classification surrogate, composed of three immunohistochemical markers (p53, MSH-6, and PMS-2) and one molecular test (mutation analysis of POLE), was proposed to bring TCGA molecular-based classification into clinical practice
  - The TCGA-surrogate approach was validated in several studies
  - According to this surrogate, tumors with pathogenic POLE mutations correspond to ultramutated tumors
    - MSH-6 or PMS-2 abnormal expression defines tumors in the hypermutated group
    - Abnormal expression of p53 (mutated pattern), characterizes high copy number group
    - Finally, NSMP is defined by the absence of POLE mutation, and normal expression pattern for MSH-6, PMS-2, and p53

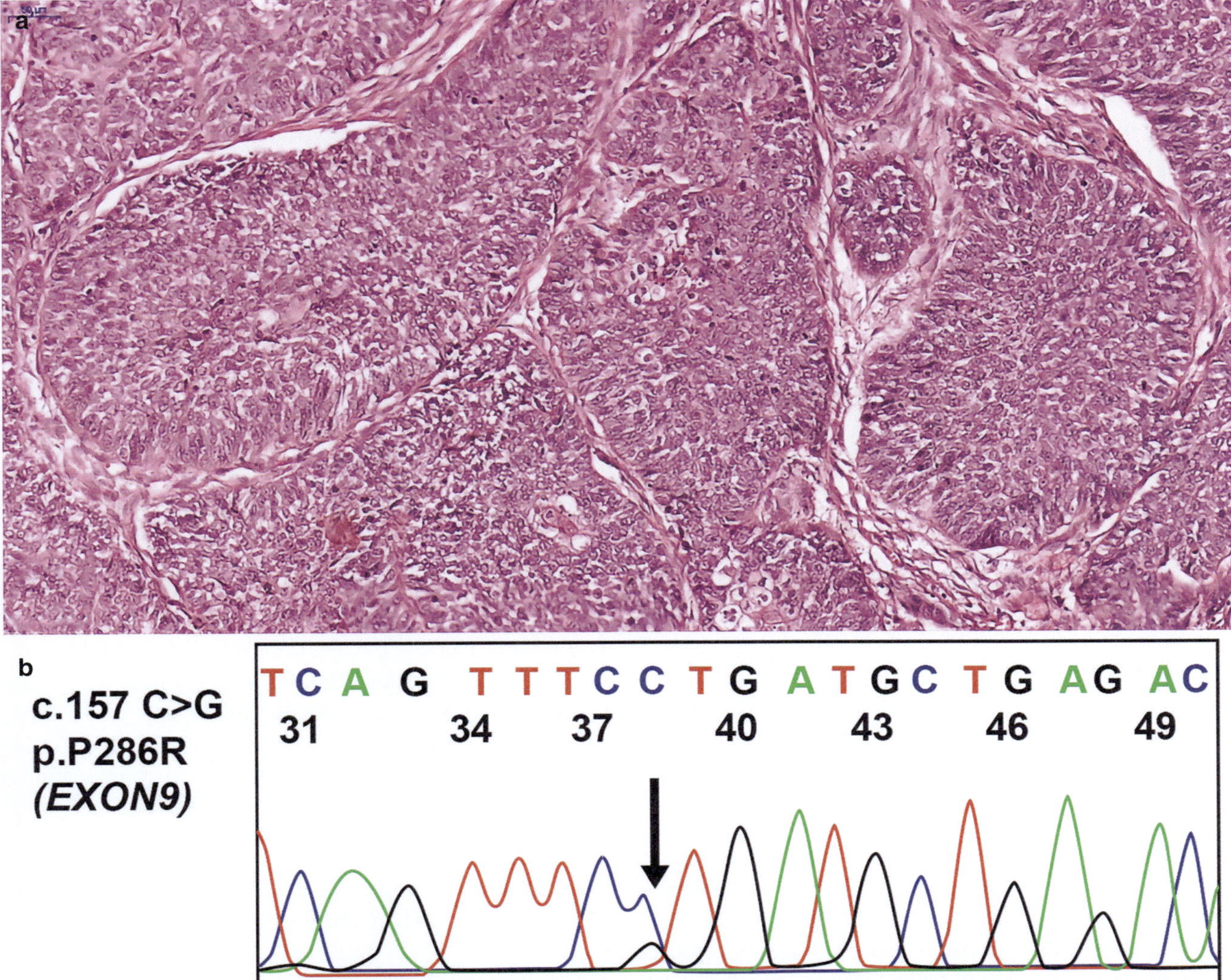

Fig. 12.12 (**a**) High-grade endometrioid endometrial carcinoma. (**b**) Sequencing analysis detected a POLE mutation in exon 9

- The TCGA surrogate approach has been shown to be particularly helpful in the group of high-grade EEC, including the cases in the grey zone between EEC and serous carcinoma
  - Application of the TCGA surrogate shows that there is a group of high-grade EEC with excellent prognosis (tumors with pathogenic POLE mutations) and a group with very bad prognosis (p53-abnormal tumors)
    - EEC 3 with MSI or NSMP, has an intermediate prognosis
  - The frequencies in EEC 3 were: POLE mutated 13%, microsatellite unstable 36%, LCN/NSMP endometrial carcinoma 30%, and serous-like/high copy number 21%
- Application of TCGA to the other types of endometrial carcinoma is also very interesting
  - Preliminary data show that could be applicable to clear cell carcinoma, undifferentiated carcinoma, and carcinosarcoma, at least to identify the small subset of POLE-mutation tumors, that show improved prognosis
- Vast majority of EEC 1–2 is LCN/NSMP or MSI, with POLE-mutated, or p53-abnormal tumors accounting for about 10%
  - This raises the question of whether the TCGA molecular-based surrogate is cost-effective in this group of tumors
- Vast majority (95%) of serous carcinoma are p53 abnormal/CNH
- There is still a need for standardization on methodology for assessing POLE mutations, and there is a need for a complete catalog of pathogenic POLE mutations

- The biological significance of "double classifiers" (tumors exhibiting more than one alteration (for example POLE and p53 abnormalities; or MSI/MMRD and p53 alterations) is still an evolving field
- There is still room for other prognostic biomarkers that may be potentially useful in the big group of EEC 1–2 that belong to the NSMP subgroup, such as L1CAM expression, Estrogen receptor immunoreactivity, or mutations in CTNNB1

Current Targeted Therapies

- Adjuvant chemotherapy and radiotherapy in high-risk endometrial carcinoma
  - Adjuvant chemotherapy and radiotherapy significantly improves recurrence-free survival compared to radiotherapy therapy alone in p53 abnormal/CNH group patients
  - Adjuvant chemotherapy and radiotherapy do not increase 5-year overall survival in high-risk endometrial carcinoma
- Dostarlimab
  - As been approved for the treatment of MSI (hypermutated group) recurrent or advanced endometrial carcinoma
  - Dostarlimab is an antiprogrammed death-1 (PD-1) antibody that binds with high affinity to the PD-1 receptor
  - In one study, with recurrent or advanced MSI tumors ($n = 104$) who had progressed on platinum-based doublet chemotherapy, the confirmed objective response rate was 42%; 13% of patients had a confirmed complete response, and 30% of patients had a confirmed partial response
  - In another more recent study analyzing the response of MSI recurrent or advanced endometrial carcinoma, the objective response rate detected was 44.7%
- Pembrolizumab and Lenvatinib
  - In July 2021, the FDA approved a combination of both compound sfor patients with advanced endometrial carcinomas that are not MSI/MMRD, who have disease progression following prior systemic therapy, and who are not candidates for curative surgery or radiation
  - Pembrolizumab is a monoclonal antibody targeting programmed death receptor-1 (PD-1)
  - Lenvatinib is a multikinase inhibitor of VEGFR 1–3, FGFR 1–4, PDGFRα, RET, and KIT
  - Study 309/KEYNOTE-775 enrolled 827 patients with advanced endometrial carcinoma previously treated with at least one prior platinum-based chemotherapy regimen

- The results obtained showed that lenvatinib plus pembrolizumab reduced the risk of disease progression or death by 44% and the risk of death by 38%, based on the median progression free survival and median overall survival; in the experimental versus chemotherapy (doxorubicin or paclitaxel) arms
- HER2 overexpression is very frequent in serous carcinomas and other p53 abnormal endometrial carcinomas; its overexpression and/or gene amplification appears to be a poor prognostic
  - Trastuzumab is a humanized antibody targeting HER2/neu has recently demonstrated its significant clinical activity in patients with HER2-amplified serous advanced/recurrent carcinomas, when combined with carboplatin/paclitaxel
  - DHES0815A is a novel antibody-drug which binds specifically to HER2 overexpressing tumors
    - In preclinical and in vitro studies, DHES0815A has demonstrated cellular growth-inhibition effects on HER2/ne, u positive endometrial cancer cells without collateral damage to surrounding cells
- Hormone therapy is the preferred front-line systemic therapy for patients with advanced/recurrent low-grade carcinomas without rapidly progressive disease
  - Progestogens are recommended
  - Alternative hormonal therapy options include aromatases inhibitors, tamoxifen, and fulvestrant

Potential-Targeted Therapies Approaches

- The high percentage of alterations observed in the PI3K/AKT/mTOR pathway in endometrial carcinoma encourages the investigation towards the development of specific targeted small inhibitors acting on the three main molecular steps of this pathway (PI3K/AKT/mTOR), that may be used as potential anticancer drugs
  - In particular, the inhibitors of the PI3K/AKT/mTOR pathway fall into four main categories: mTOR inhibitors, PI3K inhibitors, AKT inhibitors, and dual mTOR/PI3K inhibitors (Fig. 12.13) although the most promising results have been seen with mTOR inhibitors and Dual PI3K–mTOR inhibitors
  - mTOR inhibitors
    - mTOR is the downstream effector of AKT; upon activation, mTOR-Raptor activates S6K and inhibits 4EBP1 to accelerate mRNA translation
    - Several mTOR inhibitors are available for clinical trials: First-generation mTOR inhibitors, which suppress the mTORC1 but not the mTORC2: CCI-779 (temsirolimus), RAD001 (everolimus), and AP23573 (ridaforolimus)

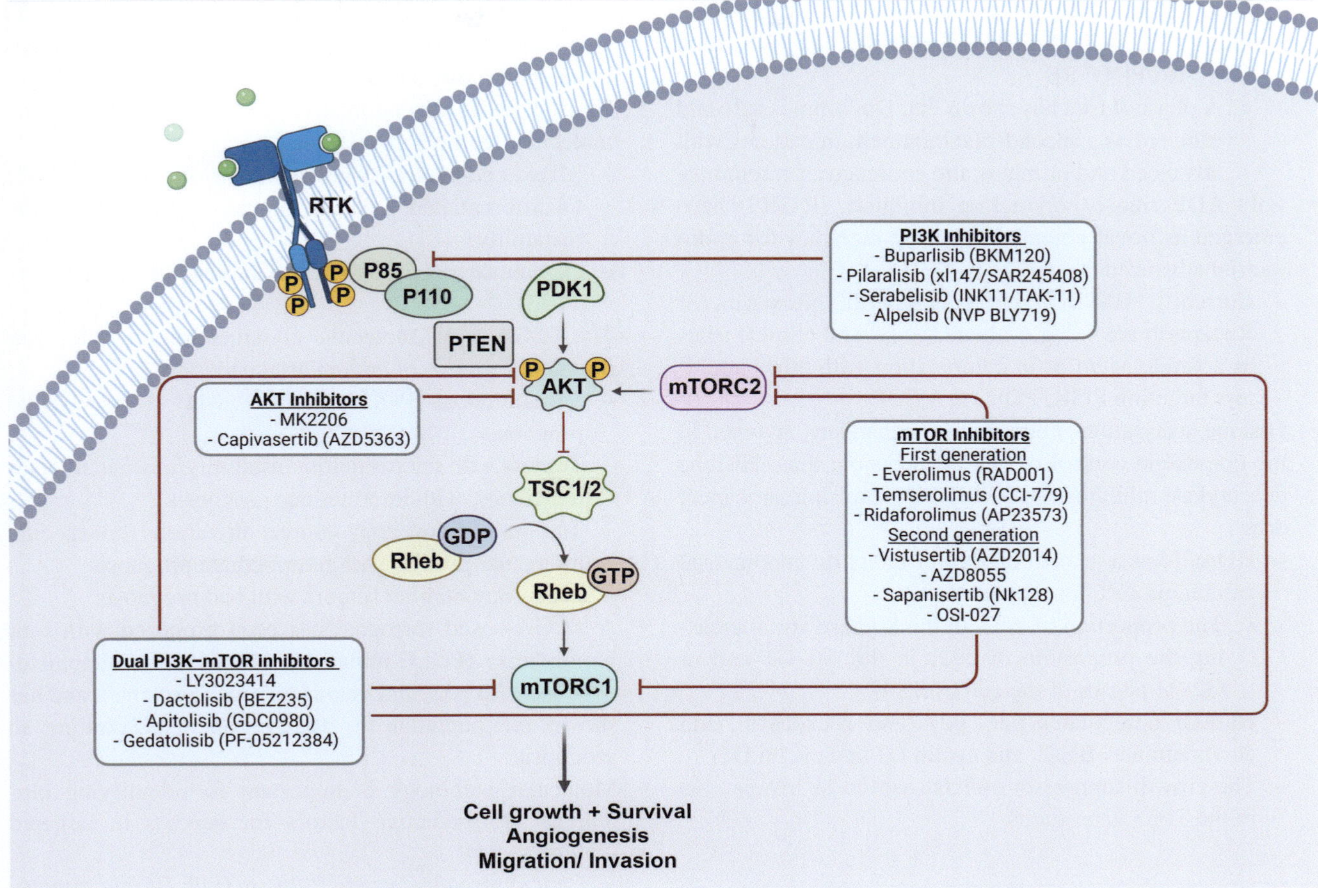

Fig. 12.13 PI3K/AKT/mTOR pathway in endometrial carcinoma with potential target inhibitors

- ♦ In the last years, dual mTOR inhibitors (known also as second-generation inhibitors) have been developed
- ♦ These inhibitors can suppress both the mTORC1 and mTORC2 and their main advantage is the considerable decrease of AKT phosphorylation and better inhibition of mTORC1
- ♦ These inhibitors include AZD2014 (vistusertib), INK 128 (Sapanisertib) and OSI-027
- – Dual PI3K–mTOR inhibitors
  - ○ The dual mTOR/PI3K inhibitors are designed to bind the ATP-binding site of both class I PI3Ks and mTORC1/2 and should lead to more complete suppression of the PI3K/AKT/mTOR pathway
  - ○ PF-05212384 (gedatolisib), has proved effective in two clinical trials in patients with recurrent endometrial carcinoma following platinum-containing chemotherapy
- • Receptor tyrosine kinases are frequently mutated in endometrial carcinoma (Table 12.2) and such dysregulation is closely associated with cancer development and progres-

Table 12.2 Mutation frequency of the most relevant receptor tyrosine kinases in endometrial carcinomas

RTK	Mutation frequency (%)
FGFR2	12.80
KIT	7.20
FLT1	6.90
MET	6.80
KDR	6.30
PDFGFRα	6.00
HER2	5.10
EGFR	4.90
FLT4	4.90
PGFRβ	4.70
ABL1	3.90
BRAF	3.80
BCR	3.60
SRC	1.40

sion; thus, a great effort is being made to develop novel molecules able to target receptor tyrosine kinases
- – The receptor tyrosine kinase inhibitor Dovitinib target kinases implicated in pathogenic angiogenesis, tumor

growth, and cancer progression, such as FLT3, c-Kit, FGFR1/FGFR3, VEGFR1/VEGFR2/VEGFR3, and PDGFRα/PDGFRβ

    o A phase II trial has shown that Dovitinib is safe and efficient as a second-line treatment in patients with advanced and/or metastatic endometrial carcinoma

- Poly ADP ribose polymerase inhibitors (PARPi) have emerged as novel potential targeting therapies for endometrial advanced cancers
  - Currently, PARPi such as Olaparib, Niraparib, or Rucaparib are being evaluated in several clinical trials as a single agent or in combination with other inhibitors targeting PI3K, PD-1, or VEGF
- Histone acetylation is one of the mechanisms involved in the epigenetic control of gene expression; thus, Histone deacetylase inhibitors (HDACi) are promising anticancer drugs
  - HDACI has a growth inhibitory effect on endometrial carcinoma cell lines, by decreasing
    - The proportion of cells in the S phase and increasing the proportion of cells in the G0–G1 and or G2–M phases of the cell cycle
  - HDACI upregulates p21, p27, and E-cadherin, and downregulates Bcl-2, and cyclin D1 and cyclin D2
  - The growth-suppressor effects seem to be irrespective of the TP53 gene status

Summary

- The two more frequent histologic types of endometrial carcinomas (endometrioid and serous) show specific molecular features and different gene expression profiles
- The main molecular features of endometrioid carcinomas
  - Microsatellite instability
  - Mutations of *PTEN, PIK3CA, KRAS*, and beta-catenin genes
- In HNPCC patients, microsatellite instability analysis and immunoreactivity of mismatch repair proteins are important to confirm the diagnosis of hereditary endometrial carcinoma
- The main features of serous carcinomas
  - *TP53* mutations
  - Inactivation of *p16* and E-cadherin, *c-erbB2* amplification
  - Alterations in genes involved in the regulation of the mitotic spindle checkpoint (*STK-15*)
  - LOH at multiple loci indicating chromosomal instability
- Some nonendometrioid carcinomas probably arise from preexisting endometrioid carcinomas

- This is the most likely reason why some tumors exhibit combined or mixed features at the clinical, pathological, and molecular levels
- Unusual types of endometrial carcinoma show specific molecular features
  - Mixed endometrioid–nonendometrioid tumors (TP53)
  - Dedifferentiated carcinomas (microsatellite instability)
  - Carcinosarcoma (epithelial to mesenchymal transition, TP53)
- The TCGA-based molecular classification identifies four prognostic groups of endometrial carcinomas
  - Ultramutated (POLE-mutated) with a very good prognosis
  - Tumors with microsatellite instability/mismatch repair deficiency, with intermediate prognosis
  - Tumors with low copy number alterations (nonspecific molecular profile) with intermediate prognosis
  - High copy number tumors with bad prognosis
- A TCGA-based surrogate has been proposed with four biomarkers (POLE-mutation analysis, immunohistochemistry for p53, and mismatch repair proteins), and has shown the potential to stratify patients according to prognosis
- Molecular pathology is important for identifying biomarkers as predictive factors for success in targeted therapies
  - Candidate pathways are PI3K, mTOR, EGFR, apoptosis, and histone acetylation
  - C-erb B2 in serous carcinoma, and immune checkpoint inhibitors for tumors with mismatch repair deficiency are being implemented in clinical practice

Further Reading

Buza N, Roque DM, Santin AD. HER2/neu in endometrial cancer: a promising therapeutic target with diagnostic challenges. Arch Pathol Lab Med. 2014;138(3):343–50.

Cho KR, Cooper K, Croce S, Djordevic B, Herrington S, Howitt B, et al. International Society of Gynecological Pathologists (ISGyP) endometrial cancer project: guidelines from the special techniques and ancillary studies group. Int J Gynecol Pathol. 2019;38(Suppl 1):S114–22.

Concin N, Creutzberg CL, Vergote I, Cibula D, Mirza MR, Marnitz S, et al. ESGO/ESTRO/ESP guidelines for the management of patients with endometrial carcinoma. Virchows Arch. 2021;478(2):153–90.

Costas L, Frias-Gomez J, Guardiola M, Benavente Y, Pineda M, Pavón M, et al. New perspectives on screening and early detection of endometrial cancer. Int J Cancer. 2019;145(12):3194–206.

de Boer SM, Powell ME, Mileshkin L, Katsaros D, Bessette P, Haie-Meder C, et al. Adjuvant chemoradiotherapy versus radiotherapy alone for women with high-risk endometrial cancer (PORTEC-3): final results of an international, open-label, multicentre, randomised, phase 3 trial. Lancet Oncol. 2018;19(3):295–309.

Kandoth C, Schultz N, Cherniack AD, Akbani R, Liu Y, Shen H, et al. Integrated genomic characterization of endometrial carcinoma. Nature. 2013;497(7447):67–73.

León-Castillo A, de Boer SM, Powell ME, Mileshkin LR, Mackay HJ, Leary A, et al. Molecular classification of the PORTEC-3 trial for high-risk endometrial cancer: impact on prognosis and benefit from adjuvant therapy. J Clin Oncol. 2020;38(29):3388–97.

Makker V, Taylor MH, Aghajanian C, Oaknin A, Mier J, Cohn AL, et al. Lenvatinib plus pembrolizumab in patients with advanced endometrial cancer. J Clin Oncol. 2020;38(26):2981–92.

Megino-Luque C, Moiola CP, Molins-Escuder C, López-Gil C, Gil-Moreno A, Matias-Guiu X, et al. Small-molecule inhibitors (SMIs) as an effective therapeutic strategy for endometrial cancer. Cancers (Basel). 2020;12(10):2751.

Murali R, Davidson B, Fadare O, Carlson JA, Crum CP, Gilks CB, et al. High-grade endometrial carcinomas: morphologic and Immunohistochemical features, diagnostic challenges and recommendations. Int J Gynecol Pathol. 2019;38(Suppl 1):S40–63.

Oaknin A, Tinker AV, Gilbert L, Samouëlian V, Mathews C, Brown J, et al. Clinical activity and safety of the anti-programmed death 1 monoclonal antibody dostarlimab for patients with recurrent or advanced mismatch repair-deficient endometrial cancer: a nonrandomized phase 1 clinical trial. JAMA Oncol. 2020;6(11):1766–72.

Piulats JM, Guerra E, Gil-Martín M, Roman-Canal B, Gatius S, Sanz-Pamplona R, et al. Molecular approaches for classifying endometrial carcinoma. Gynecol Oncol. 2017;145(1):200–7.

Soslow RA, Tornos C, Park KJ, Malpica A, Matias-Guiu X, Oliva E, et al. Endometrial carcinoma diagnosis: use of FIGO grading and genomic subcategories in clinical practice: recommendations of the International Society of Gynecological Pathologists. Int J Gynecol Pathol. 2019;38(Suppl 1):S64–74.

WHO. Female genital tumours. WHO classification of tumours, vol. 4. 5th ed. Lyon: Editorial World Health Organization; 2020.

Molecular Pathology of Kidney Tumors

Khaleel I. Al-Obaidy, Zainab I. Alruwaii,
Sambit K. Mohanty, Liang Cheng, and Sean R. Williamson

Contents

K. I. Al-Obaidy
Department of Pathology and Laboratory Medicine, Henry Ford
Health, Detroit, MI, USA

Z. I. Alruwaii
Department of Pathology, Regional Laboratory and Blood Bank,
Eastern Province, Dammam, Saudi Arabia

S. K. Mohanty
Department of Pathology and Laboratory Medicine, Advanced
Medical Research Institute, Bhubaneswar, Odisha, India

L. Cheng
The Legorreta Cancer Center at Brown University, Department of
Pathology and Laboratory Medicine, Warren Alpert Medical
School of Brown University, Lifespan Academic Medical Center,
Providence, RI, USA

S. R. Williamson (✉)
Robert J Tomsich Pathology and Laboratory Medicine Institute,
Cleveland Clinic, Cleveland, OH, USA

Introduction

- Renal cancer is the sixth leading site of new cancer diagnosis for men and eighth for women, with an estimated 76,080 patients newly diagnosed in 2021, according to American Cancer Society statistics
  - Carcinoma of renal tubular origin (92%) makes up most of these tumors, followed by carcinoma of the renal pelvis (7%) and nephroblastoma (Wilms tumor, 1%)

© The Author(s), under exclusive license to Springer Nature Switzerland AG 2023
L. Cheng et al. (eds.), *Molecular Surgical Pathology*, https://doi.org/10.1007/978-3-031-35118-1_13

- Renal neoplasms are a distinctive and heterogeneous group of entities
  - Clinical behavior varies significantly between lesions
    - Sometimes histologic appearance is deceptively bland
    - Cytologic/histologic similarity can be significant across both benign and malignant neoplasms

- Molecular and cytogenetic alterations have been of great historical importance in establishing the classification of renal neoplasms
  - Modern molecular methodologies are already playing a significant role in classification, such as:
    - Resolution of challenging differential diagnoses (Fig. 13.1)

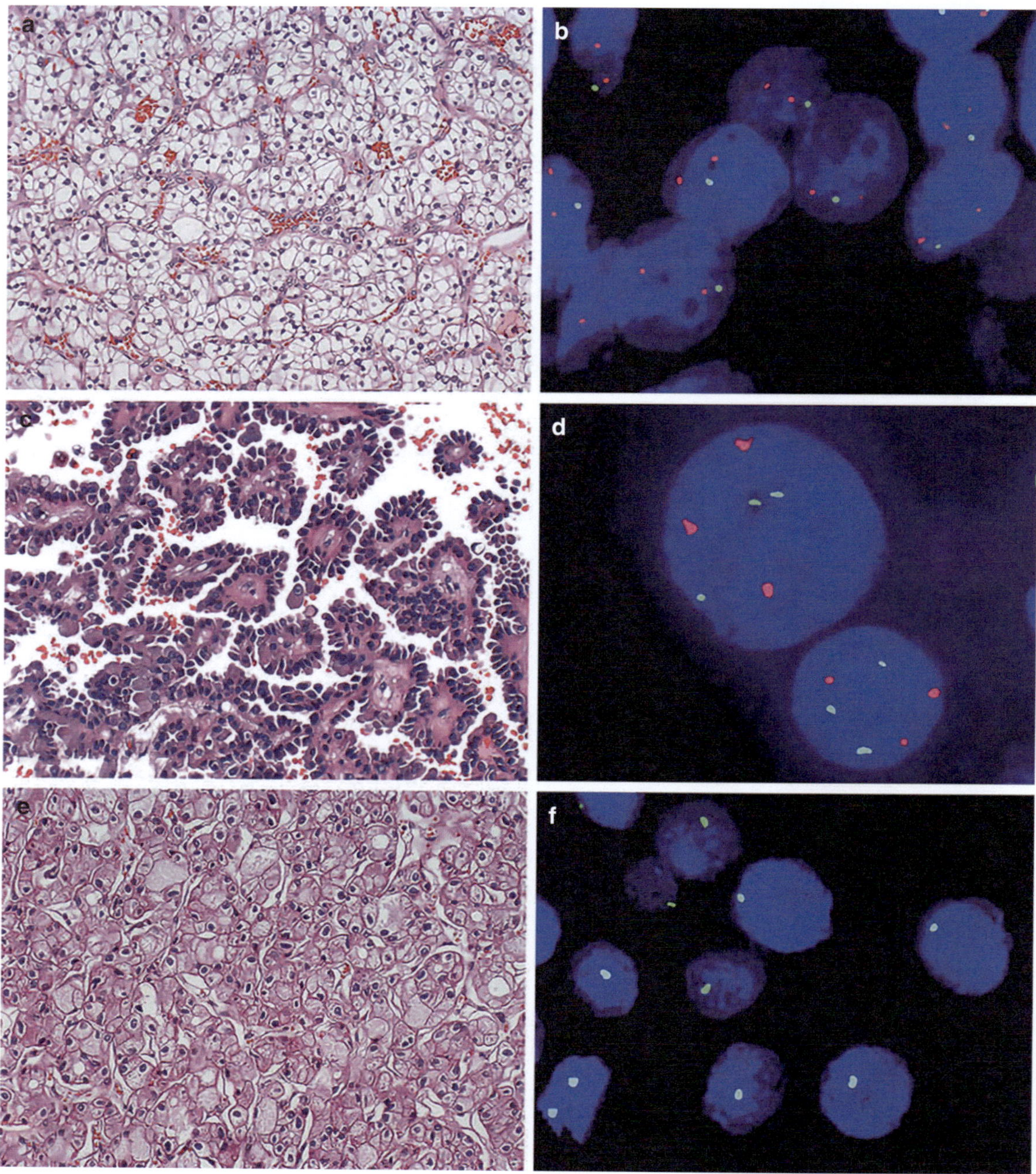

Fig. 13.1 Histopathologic and cytogenetic features of renal cell neoplasms by FISH. Clear cell renal cell carcinoma (**a**) frequently shows chromosome 3p deletion (**b**) indicated by the presence of a single 3p signal (green) in cells with two chromosome 3 centromere signals (red) per cell. Papillary renal cell carcinoma (**c**) commonly shows trisomy 7 (three green signals) and 17 (three red signals) (**d**). Most chromophobe renal cell carcinomas (**e**) show complex losses of chromosomes, with monosomy of 1, 2, 6, 10, and 17. Each chromophobe tumor cell possesses only a single chromosome 10 signal (**f**). In contrast, a predominantly oncocytic neoplasm with features largely resembling oncocytoma (**g**) does not exhibit multiple chromosomal deletions. Each of the two nuclei in (**h**) shows two fluorescent signals for chromosome 17 (disomic pattern) (from Cheng et al. 2009; with permission from Elsevier)

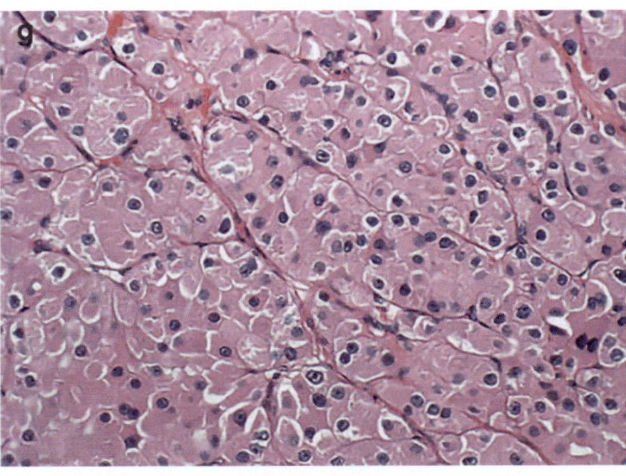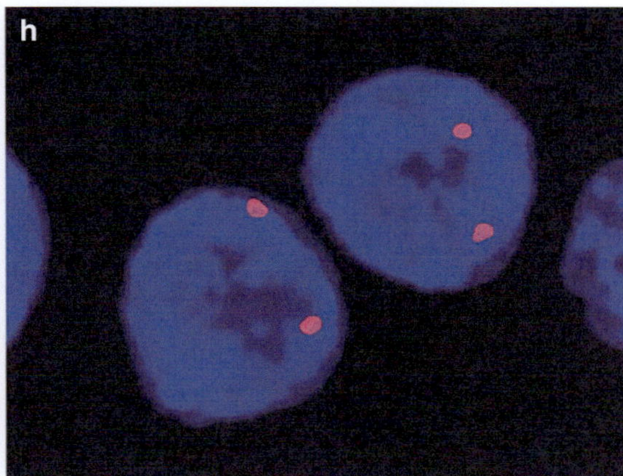

Fig. 13.1 (continued)

- o Subdivision of new diagnostic entities
- o Direction of targeted therapy
- Several genetic syndromes with a predisposition to the development of renal tumors have been carefully studied, resulting in an increased understanding of underlying cytogenetic and molecular events

Genetic Renal Neoplasia Syndromes

- Although most renal neoplasms are sporadic, approximately 3–5% are associated with inherited syndromes (Table 13.1)
- Investigation into their molecular and genetic mechanisms has led to an improved understanding of both the syndromes themselves and sporadic renal neoplasms (Fig. 13.2)

von Hippel–Lindau Disease

- von Hippel–Lindau (VHL) disease is a genetic syndrome with autosomal dominant inheritance, associated with germline inactivating mutation of the *VHL* tumor suppressor gene
  - The *VHL* gene is located at chromosome 3p25.3
  - Estimated incidence of 1 in 35,000 live births
  - Greater than 90% penetrance at 65 years of age
  - Patients with VHL disease generally have one mutated copy of the *VHL* gene
    - o Tumorigenesis is thought to occur with inactivation of the second copy, via loss of heterozygosity (LOH), promoter hypermethylation, or mutation, in a "two-hit" model (Fig. 13.3)
    - o Fewer total steps are required in the development of a renal tumor, compared to sporadic tumors

Molecular Pathogenesis

- The *VHL* gene codes for VHL protein, a 213-amino acid member of the ubiquitin ligase family
- *VHL* is key in cell cycle control and gene regulation, particularly in the regulation of hypoxia-inducible factor (*HIF*)
  - HIFα
    - o A transcription factor involved in the response to changes in oxygen supply in the cellular environment
    - o Important in physiologic response to ischemia and hypoxia, as well as tumor growth and angiogenesis
    - o Under normoxic status, 2 HIF1α are hydroxylated by propyl hydroxylase enzyme facilitating their binding to normal VHL ubiquitin–ligase complex with elongin B (ELOB), elongin C (ELOC), RBX1, NEDD8, and cullin 2 (CUL2), which then are targeted for ubiquitin-mediated proteasomal degradation
    - o In hypoxic or iron-deficient conditions, HIF1α is not hydroxylated, and therefore not degraded; instead, it forms a complex with HIF1β to bind to hypoxia-response element sequences in the nucleus, resulting in activation of several downstream genes, including *VEGFA*, *PDGFB*, *GLUT1*, and *TGFA* (Fig. 13.4)
    - o Loss of *VHL* in tumor cells, therefore, leads to accumulation of HIF1α and upregulation of these downstream genes, with key roles that promote tumor growth, support, and spread, such as the following:
      - ♦ Cell proliferation
      - ♦ Angiogenesis
      - ♦ Metastatic potential
      - ♦ Glucose transportation

Table 13.1 A summary of the clinical and genetic findings in hereditary cancer syndromes of the kidney

Syndrome	Mode of inheritance	Chromosomal location	Gene	Renal manifestations	Extrarenal manifestations
von Hippel–Lindau disease	Autosomal dominant	3p25.3	*VHL*	– Renal cysts (multiple and bilateral) – CCRCC (multiple and bilateral)	– Hemangioblastoma (retinal and CNS) – Endolymphatic sac tumors – Pancreatic cysts – Pancreatic NET – Epididymal cystadenomas in males – Broad ligament cystadenomas in females – Pheochromocytoma
Hereditary papillary renal carcinoma	Autosomal dominant	7q21–q31	*MET*	PRCC-1 (multiple and bilateral)	– None
Hereditary leiomyomatosis and renal cell carcinoma	Autosomal dominant	1q42.3–q43	*FH*	HLRCC-associated renal cell carcinoma	– Cutaneous and uterine leiomyomata
Tuberous sclerosis	Autosomal dominant	9q34 and 16p13.3	*TSC1* and *TSC2*	– Renal cystic disease – Angiomyolipoma – Oncocytoma – CCRCC – Chromophobe-like RCC – Renal cell carcinoma with fibromyomatous stroma – ESC-RCC – EVT	– Cutaneous lesions (facial angiofibromas, periungual fibromas, shagreen patches, and hypopigmented macules) – CNS lesions (cortical tubers, subependymal nodules, and subependymal giant cell astrocytomas) – Cardiac rhabdomyomas – Pulmonary lymphangioleiomyomatosis – Retinal astrocytic hamartomas
Birt–Hogg–Dubé syndrome	Autosomal dominant	17p11.2	*FLCN*	– "Hybrid" tumors – ChrRCC – Renal oncocytosis	– Cutaneous tumors (fibrofolliculomas; trichodiscomas, acrochordons) – Pulmonary cysts – Medullary thyroid carcinoma
Succinate dehydrogenase (SDH) germline mutations	Autosomal dominant	5p15, 1p36, 1q21, 11q23	*SDHA, SDHB, SDHC, SDHD*	– SDH-deficient RCC	– Pheochromocytoma/paraganglioma – SDHB-deficient GIST
Constitutional chromosome 3 translocations	Autosomal dominant	See Table 13.2	See Table 13.2	– CCRCC (multiple and bilateral)	– None
BAP1 mutations and familial kidney cancer	Autosomal dominant	3p21	*BAP1*	– No specific subtype	– Uveal melanoma – Cutaneous melanocytic tumors (melanoma and atypical Spitz nevi) – Mesothelioma
Cowden syndrome	Autosomal dominant	10q23.31	PTEN	– No specific subtype	– Multiple hamartomas – Breast carcinoma – Endometrial carcinoma – Thyroid carcinoma

From Al-Obaidy et al. 2022; with permission from Elsevier

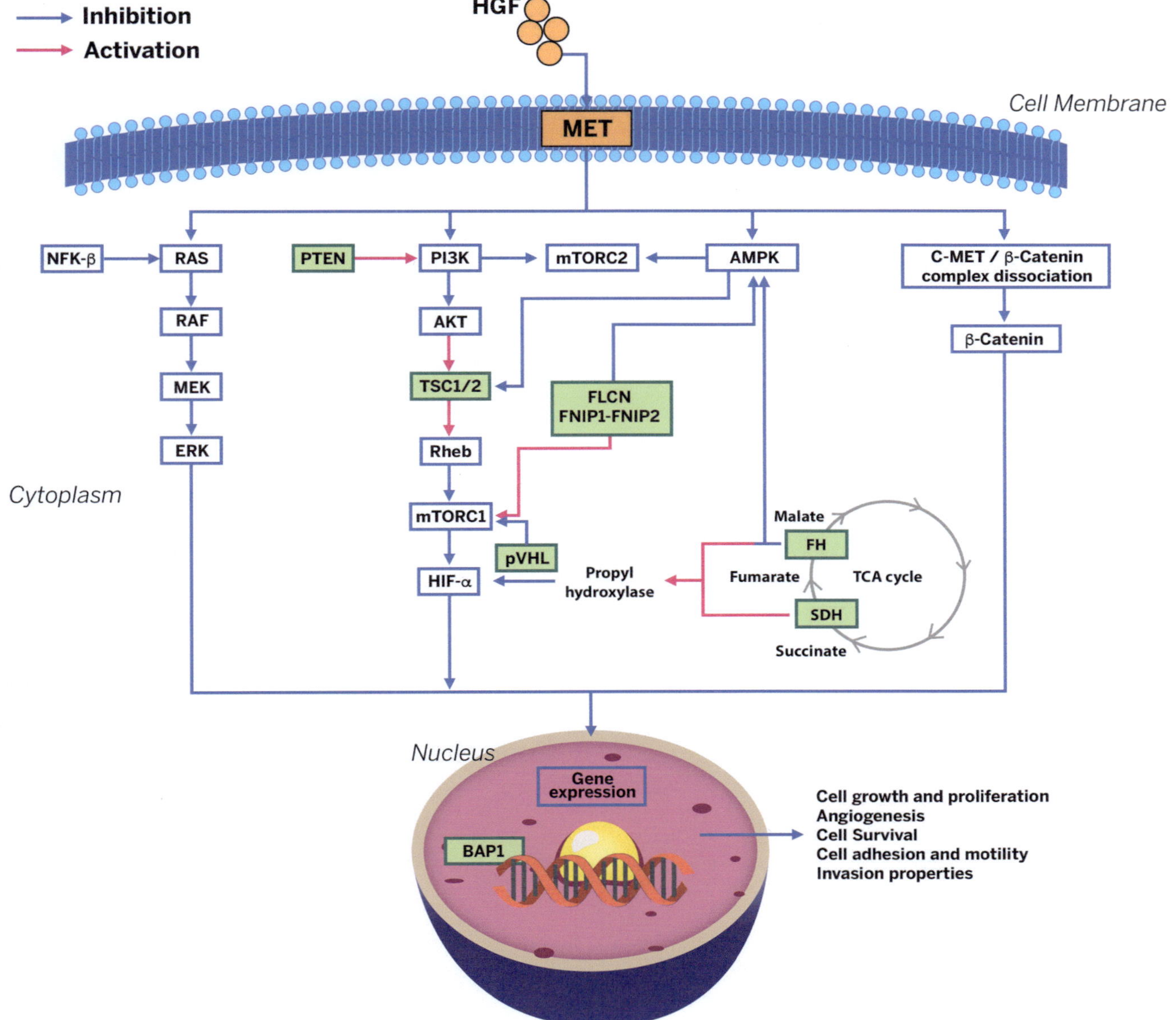

Fig. 13.2 A summary of signal transduction pathways associated with hereditary renal cell carcinoma syndromes (from Al-Obaidy et al. 2022; with permission from John Wiley & Sons)

– Carbonic anhydrase 9 (CA9)
  - A membrane protein whose expression is VHL–HIF1α-dependent
  - Expression is induced by hypoxic conditions
  - It is involved in intracellular and extracellular pH regulation
  - As such, membranous immunohistochemical positivity is present in clear cell renal cell carcinoma, usually even in high-grade and sarcomatoid forms
    ♦ In VHL patients, single renal tubular epithelial cells and cells within renal cysts have been found to show positivity

♦ Diffuse positivity may be helpful in supporting clear cell subtype
♦ Focal positivity can be observed with ischemia/necrosis
♦ However, positivity can be observed in non-renal cell tumors, so staining should be interpreted in the context of the clinical features, morphology, and other markers (e.g., PAX8)
– Mechanistic target of rapamycin kinase (mTOR)
  - Inactivation of *VHL* upregulates the mTOR pathway

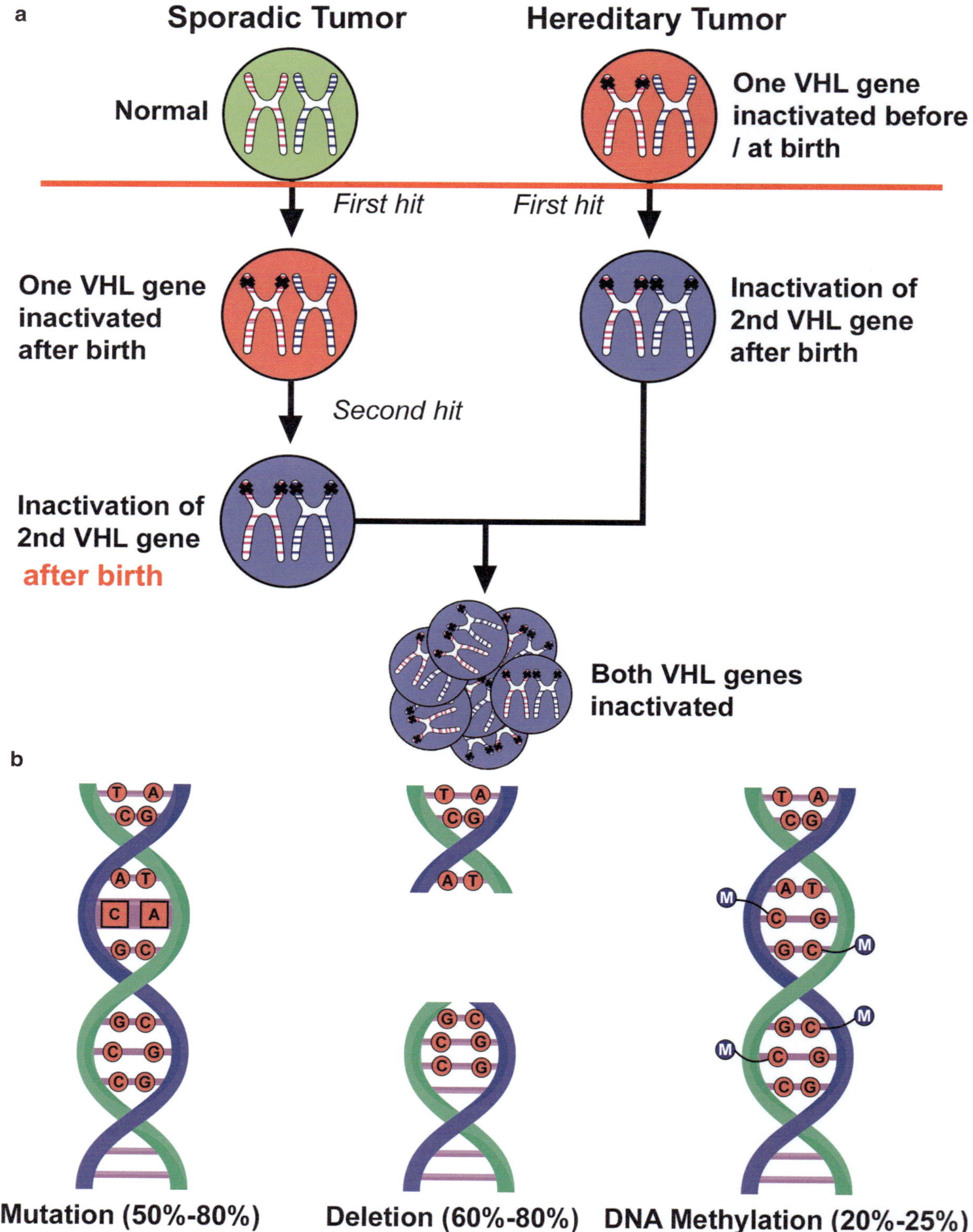

Fig. 13.3 Pathways of VHL -associated carcinogenesis. The VHL tumor suppressor gene plays a critical role in both sporadic and hereditary renal cancers (**a**). Both copies of the VHL gene must become inactivated to initiate tumor development. In familial clear cell renal carcinoma, one copy of the VHL gene has already been inactivated at birth; consequently, a single additional VHL gene inactivation initiates tumorigenesis. Sporadic tumors require two steps to inactivate both VHL alleles to initiate tumorigenesis. Further genetic alterations may also be required for initiating the development of a tumor. VHL gene inactivation may occur through one of several different pathways, including genomic mutation (50–80%), deletion (60–80%), or abnormal DNA methylation (20–25%) (**b**). Different mechanisms are frequently involved in carcinogenesis. Mutation of one allele and deletion of the remaining wild-type allele are often seen in clear cell renal carcinomas (from Cheng et al. 2009; with permission from Elsevier)

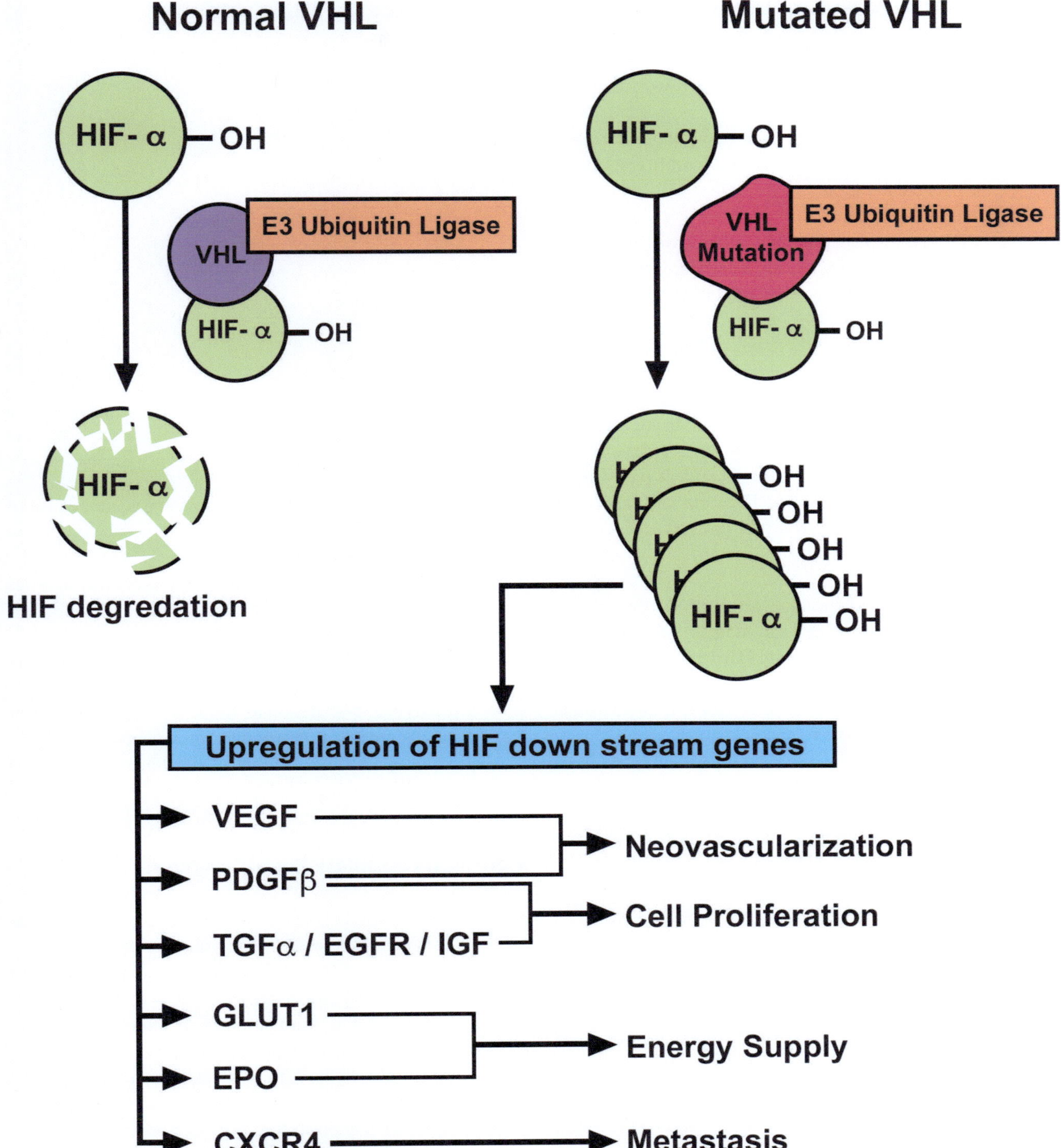

Fig. 13.4 Mechanisms of *VHL*-associated carcinogenesis. The normal *VHL* gene product forms a complex with E3 ubiquitin ligase, which induces the degradation of HIFα to regulate the activity of genes downstream from *HIF*. When both copies of the *VHL* gene are inactivated, HIF degradation is retarded and HIF accumulates to higher levels, triggering upregulation of HIF downstream gene activities. The major upregulated factors include *VEGF*, *PDGFβ*, *TGFα*, *EGFR*, *IGF*, *GLUT1*, *EPO*, and *CXCR4*, resulting in enhanced neovascularization, stimulation of cell growth, enhanced glucose transportation, and cancer progression. HIF downstream upregulation is associated with tumorigenesis in both familial and sporadic clear-cell renal cancers. *VHL* von Hippel–Lindau; *VEGF* vascular endothelial growth factor; *PDGF* platelet-derived growth factor; *TGF* transforming growth factor; *EGFR* epidermal growth factor receptor; *IGF* insulin-like growth factor; *GLUT1* glucose transporter protein 1; *EPO* erythropoietin; *CXCR4* chemokine (C-X-C motif) receptor 4 (from Cheng et al. 2009; with permission from Elsevier)

- o mTOR is a serine/threonine protein kinase involved in monitoring of cellular and environmental nutrition and energy status
- o It affects protein translation, cell growth, angiogenesis, and apoptosis
- o Downstream targets include phosphatidylinositol 3 kinase, involved in cell survival, proliferation, and neovascularization
- – Primary cilium
  - o VHL has also been found to play a role in the structure of the primary cilium, a cellular sensory mechanism involved in the inhibition of epithelial proliferation
  - o VHL functions in association with phosphatidylinositol 3 kinase and glycogen synthase kinase 3β (GSK3B) in the maintenance of the primary cilium
  - o Defects of the primary cilium have been implicated in the formation of renal cysts
  - o Cyst-dependent and cyst-independent pathways of development for clear cell carcinoma have been proposed, particularly in patients with VHL disease
- – CXCR4
  - o CXCR4 is a chemokine receptor involved in the metastasis of renal cell carcinomas and other organs
  - o Regulation occurs through HIF1α and mTOR
  - o Inhibition of this pathway represents a potential therapeutic target in the management of metastatic renal cell carcinoma
- • The VHL syndrome is classified into two subtypes based on the pheochromocytoma development:
  - – Type 1 (low risk)
  - – Type 2 (high risk); type 2 is divided into:
    - o 2A (low risk of RCC development)
    - o 2B (high risk of RCC development)
    - o 2C (pheochromocytoma only)

Clinical and Pathologic Manifestations

- • Renal manifestations
  - – Multiple, bilateral renal cysts
    - o 66% of patients
    - o Cysts have been hypothesized to represent a potential precursor for clear cell carcinoma, as some small carcinomas can be identified arising in a cyst; however, VHL patients also develop small, solid carcinomas without demonstrable association with a cyst (Fig. 13.5)
  - – Clear cell renal cell carcinoma
    - o Clear cell renal cell carcinoma is variably reported to develop in up to 70% of patients by the age of 60 years

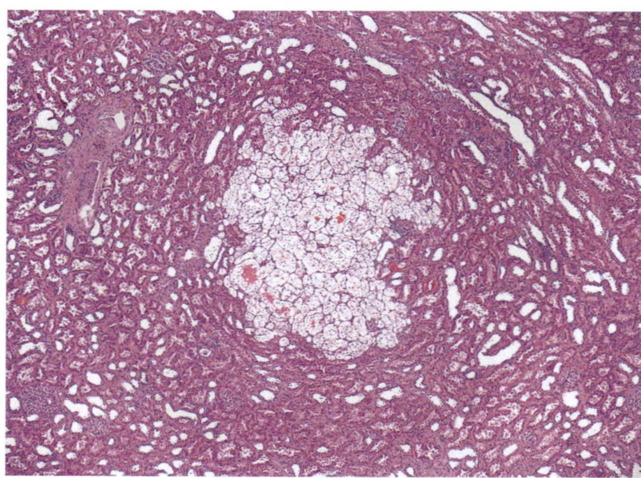

Fig. 13.5 A small, poorly circumscribed, and unencapsulated focus of clear cell carcinoma in a patient with VHL disease, arising without an identifiable cyst

- o Metastatic disease represents a significant cause of death in VHL patients
- • Nonrenal manifestations
  - – Capillary hemangioblastomas of the central nervous system and retina (60–84% of patients)
  - – Pheochromocytomas (18% of patients)
  - – Pancreatic cysts (70%) and neuroendocrine tumors (9% of patients)
  - – Epididymal cystadenomas (54% of male patients) or broad ligament cystadenomas (in women)
  - – Endolymphatic sac tumors of the inner ear (14% of patients)

Hereditary Papillary Renal Carcinoma

- • Hereditary papillary renal carcinoma is a genetic syndrome with autosomal dominant inheritance
- • High penetrance (90% likelihood of developing renal cell carcinoma by 80 years of age)
- • It is associated with the late onset of multiple, bilateral, papillary renal cell carcinomas (former type 1)
- • Generally, onset in the fifth to seventh decades
- • Numerous tumors often present within the same kidney
- • A form with early onset has also been described (second to third decades)

Molecular Pathogenesis

- • Activating mutation of *MET* protooncogene (MET protooncogene, receptor tyrosine kinase) at chromosome 7q31.3 (126 kb genomic region with 20 exons)

- It encodes MET protein (1390 amino acid)
- MET is a tyrosine kinase receptor, whose ligand is HGF (hepatocyte growth factor)
  - Increased *MET* is implicated in papillary tumors of other organs, including thyroid, ovary, and colon carcinomas
- Mutation of the tyrosine kinase domain results in constitutive activation without ligand binding
  - Results in cell proliferation, neovascularization, and cell motility
- Duplication of the mutated chromosome 7 is frequently seen (trisomy 7)
- Activating *MET* mutation seen in only a small subset of sporadic papillary renal cell carcinoma cases (13%)

Hereditary Leiomyomatosis and Renal Cell Carcinoma

- Hereditary leiomyomatosis and renal cell carcinoma (HLRCC) is a genetic syndrome with autosomal dominant inheritance and incomplete penetrance

Molecular Pathogenesis

- The syndrome is associated with a mutation of the *FH* (fumarate hydratase) gene at chromosome 1q42.3–q43
  - *FH* includes 10 exons and codes for a 500-amino acid peptide
  - The FH protein is involved in the conversion of fumarate to malate in the Krebs (TCA) cycle
  - This leads to intracellular accumulation of fumarate
    - Fumarate competitively inhibits the propyl hydroxylase enzyme necessary for HIF hydroxylation, creating a pseudohypoxia status, similar to that seen in VHL
  - Biallelic inactivation is found in most tumors, suggesting a role of *FH* as a tumor suppressor gene

Clinical and Pathologic Manifestations

- Renal manifestations
  - Renal tumors termed fumarate hydratase-deficient renal cell carcinoma
  - Develops in only a subset of patients (15–20%)
  - It is often unifocal/unilateral, with early onset
  - Tumors are aggressive, often presenting at an advanced stage, with many patients dying of metastatic disease
  - Morphologically, tumors have features of the former type 2 papillary carcinoma, including papillary, solid, alveolar, tubular, glandular, and sheet-like architecture
    - Unique morphologic features include a prominent eosinophilic nucleolus, surrounded by a clear halo (mimicking the inclusions seen in cytomegalovirus infection) (Fig. 13.6) and heterogeneous morphologic patterns within the same tumor
    - A renal cell carcinoma resembling tubulocystic carcinoma with an abrupt transition to high-grade infiltrative carcinoma is particularly suspicious for FH-deficient renal cell carcinoma
- Nonrenal manifestations
  - Leiomyomata of the skin and uterus

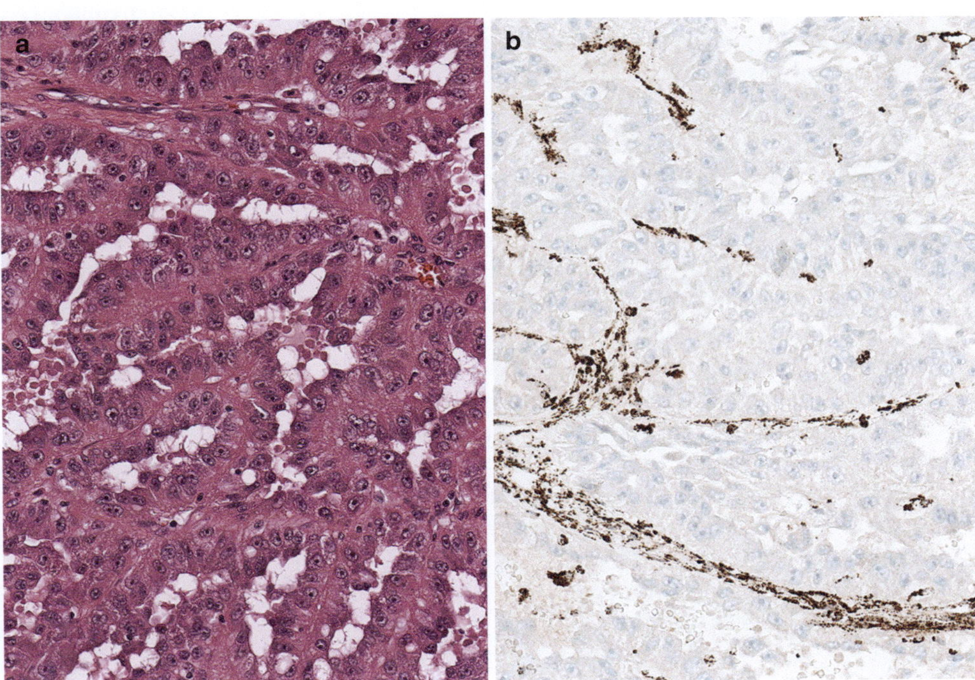

Fig. 13.6 Fumarate hydratase-deficient renal cell carcinoma shows tubular structures with eosinophilic cells and prominent nucleoli (**a**). There is an abnormal absence of fumarate hydratase staining using immunohistochemistry (**b**), a helpful surrogate for molecular alterations

- In contrast to renal carcinoma, most patients (85–98%) develop uterine and cutaneous leiomyomata
  - Development of uterine leiomyomata at a young age (under 30 years) may prompt consideration of the diagnosis of HLRCC
  - Morphologically, the leiomyomata may show similar nuclear features to the kidney tumors (prominent eosinophilic nucleolus with "halo")
- Leiomyosarcoma occurs in a small subset of cases

Tuberous Sclerosis

- Tuberous sclerosis (tuberous sclerosis complex or TSC) is an autosomal dominant with near-complete penetrance and variable expressivity
- Estimated to affect 1 in 6000 people
- Two-thirds (~70%) result from sporadic de novo mutations, whereas 30% are inherited
- Cases are thought to be more frequent than historically detected, due to improvements in imaging modalities and increased utilization of imaging
- Tuberous sclerosis is associated with *TSC1* or *TSC2* (TSC complex subunit 1 or 2) mutations

Molecular Pathogenesis

- *TSC1*
  - Located at chromosome 9q34; protein also known as hamartin (130 kD protein)
  - Involved in cell adhesion through ezrin–radixin–moesin family of actin-binding proteins and GTPase RHOA
  - Proposed tumor suppressor function, due to tumor formation with loss of both wild-type alleles
- *TSC2*
  - Located at 16p13.3
    - Protein also known as tuberin protein with GTPase-activating activity for RAS-related protein (*RAP1GAP*)
  - *TSC2–PKD1* contiguous gene syndrome
    - *PKD1* is adjacent to *TSC2* on chromosome 16p13 and mutations disrupting both genes are associated with severe, early-onset polycystic kidney disease
- Since TSC1 and TSC2 form a complex together, similar syndromic phenotypes are seen as a result of mutation of either gene
- Normally, TSC1and TSC2 form a heterodimer complex
- Most of the TSC functions are related to the GAP domain of TSC2
- When stimulated, TSC1/TSC2 complex regulates mTOR activity through the RHEB-GTP

- After growth factor stimulation, the TSC1/TSC2 complex is phosphorylated, leading to a decrease in its GTPase-activating activity
- Stimuli such as hypoxia and low energy (energy deficit) lead to its phosphorylation and increase in GTPase-activating activity
- The activated RHEB (Ras homolog, mTORC1 binding) stimulates mTOR which leads to phosphorylation cascades resulting in cell growth and proliferation
- Lack of TSC1/TSC2 complex function due to genetic alterations, leads to constant dysregulated activation of the RHEB and mTOR pathway
- Targeted therapy with inhibitors of the mTOR pathway (rapamycin/sirolimus, everolimus) has shown promise in control of angiomyolipoma growth, though tumors potentially resume growth after cessation of therapy
- TSC2 also regulates *VEGF* expression (a component of the VHL/HIF pathway)
- Most tumors demonstrate a benign clinical course; however, some cases show progressive growth or outright aggressive behavior, suggesting that additional genetic events may be involved

Clinical and Pathologic Manifestations

- Renal manifestations
  - Renal cystic disease
    - Variable cyst formation in approximately 45% of patients, ranging from microcystic disease, undetectable by imaging, to a polycystic kidney phenotype
  - Renal tumors (50–80% of patients)
    - Angiomyolipoma (multifocal and bilateral), present in 75–80% of affected children greater than 10 years of age
    - Clear cell renal cell carcinoma (multifocal)
      - Some previous reports may not represent newer emerging entities
    - Chromophobe-like renal cell carcinoma
    - Renal cell carcinoma with fibromyomatous (leiomyomatous) stroma
    - Eosinophilic solid and cystic renal cell carcinoma
- Nonrenal manifestations
  - Skin lesions
    - Facial angiofibromas ("adenoma sebaceum")
    - Periungual fibromas
    - Shagreen patches (peau chagrin)
    - Hypopigmented macules
  - Central nervous system lesions
    - Cortical tubers
    - Subependymal nodules
    - Subependymal giant cell astrocytomas

- Tumors of other organs
  - Cardiac rhabdomyomas (in up to 50% of patients)
  - Pulmonary lymphangioleiomyomatosis
  - Retinal astrocytic hamartomas ("phakomas")

Birt–Hogg–Dubé Syndrome

- Birt–Hogg–Dubé syndrome is an autosomal dominant tumor syndrome with incomplete penetrance

Molecular Pathogenesis

- Associated with mutation of the *FLCN* or *BHD* gene with proposed tumor suppressor function
- It is located at 17p11.2, encoding folliculin (64-kD protein)
- More than 50 different germline mutations have been described
- In most cases, the mutation results in truncation and a dysfunctional gene product
- Normally, FLCN protein forms complex with folliculin interacting proteins 1 and 2 (FNIP1, 2) which interacts with PRKAA2 (AMPK) related to the energy- and nutrient-sensing pathways
- In a low-energy state, AMPK is phosphorylated which leads to negative regulation of mTOR either directly or through TSC2 phosphorylation, inhibiting protein synthesis and cell growth
- Also, activated AMPK upregulates and directly phosphorylate *PGC-1α/PPARGC1A-TEAM* signaling axis, mitochondrial gene transcription factors
  - Alterations in this mitochondrial gene are only seen in BHD-associated tumors
- The loss of *FLCN* function promotes the mTOR overactivation leading to cellular proliferation and tumor development

Clinical and Pathological Manifestations

- Renal manifestations
  - Multicentric renal epithelial tumors with unusual features and varied histologic subtypes, even within the same kidney
    - "Hybrid" tumors with overlapping features between chromophobe carcinoma and oncocytoma, the most common type
      - A single tumor may include multiple subpopulations of cells, resembling both chromophobe carcinoma and oncocytoma
      - The chromophobe renal cell carcinoma component is identical to nonsyndromic tumors, whereas the oncocytoma component usually lacks the loose connective tissue stroma and central scar
      - Chromophobe renal cell carcinoma, the second most common type
    - Renal oncocytosis (Multiple microscopic foci of oncocytic cells in the renal cortex)
    - Clear cell and papillary renal cell carcinomas are less frequent
    - Renal cysts
- Nonrenal manifestations
  - Benign cutaneous tumors
    - Fibrofolliculomas in the forehead, scalp, face, neck
    - Trichodiscomas
    - Acrochordons
  - Pulmonary cysts leading to spontaneous pneumothorax
  - Medullary thyroid carcinoma

Succinate Dehydrogenase Germline Mutations

- Similar to fumarate hydratase in HLRCC, succinate dehydrogenase (SDH) is an enzyme involved in the Krebs (TCA) cycle, whose mutations have been implicated in the development of tumors, particularly pheochromocytoma and paraganglioma
- The SDH enzyme consists of four subunits, encoded by four mitochondrial complex II genes (*SDHA*, *SDHB*, *SDHC*, and *SDHD*), located on chromosomes 5p15, 1p36, 1q21, and 11q23, respectively, along with two assembly factors (*SDHAF1* and *SDHAF2*) located on chromosomes 19q13 and 11q12, respectively
- Germline mutations of the subunits of succinate dehydrogenase (mitochondrial complex II genes) are associated with pheochromocytoma/paraganglioma syndromes (PHEO/PGL)
  - *SDHB* gene (PGL4), *SDHC* (PGL3), and *SDHD* (PGL1); sometimes called "Carney–Stratakis" syndrome when combined with gastrointestinal stromal tumor (GIST)
  - Mutation of *SDHAF2* (formerly *SDH5*) may be the cause of PGL2 syndrome

Molecular Pathogenesis

- The pathogenic role in tumorigenesis is not fully understood
- However, the accumulation of intracellular succinate causes HIFα prolyl hydroxylase inhibition and genome-wide hypermethylation

Clinical and Pathologic Manifestations

- Renal manifestations
  - Renal tumors occur in a subset of some patients, particularly those with *SDHB* mutations (estimated 14%)
  - Distinctive morphologic features include the following:
    ○ Bubbly eosinophilic cytoplasm with intracytoplasmic vacuoles/inclusions and indistinct cell borders
    ○ Cystic areas with a lobulated/circumscribed tumor border and entrapment of normal tubules/glomeruli (Fig. 13.7)
    ○ Some tumors may be poorly differentiated or sarcomatoid
    ○ Prognosis appears to be good in nonsarcomatoid cases
- Nonrenal manifestations
  - Pheochromocytoma
  - Paraganglioma
  - *SDH*-deficient GIST (GIST, type 2)
- Immunohistochemical staining for *SDHB* can be performed and reveals a strong positive granular cytoplasmic reaction (mitochondrial) in normal tissues, with the abnormal absence of staining in the associated tumors

Constitutional Chromosome 3 Translocation

- Constitutional chromosome 3 translocation outside of the setting of VHL disease has also been found to be associated with an increased risk of developing bilateral, multifocal clear cell renal cell carcinoma (Table 13.2)
- A variety of breakpoints involved in translocations or insertions have been described, including the following:
  - t(1:3)(q32:q13.3)
    ○ Disrupts *NORE1A* and *LSAMP*
  - t(2:3): t(2:3)(q33:q21), t(2:3)(q35:q21), t(2;3)(q37.3;q13.2), t(2;17)(q21.1;q11.2)

Table 13.2 A summary of the reported constitutional chromosome rearrangements

Rearrangement	Affected gene
t(1;3)(q32;q13.3)	*NORE1A & LSAMP*
t(2;3)(q33;q21)	*DIRC1*
t(2;3)(q35;q21)	*DIRC2 & DIRC3*
t(2;3)(q37.3;q13.2)	
t(2;17)(q21.1;q11.2)	
t(3;4)(p13;p15)	*KCNIP4*
t(3;4)(p13;p16)	
t(3;4)(q21;q31)	*FBXW7*
t(3;6)(p13;q25.1)	*STXBP5*
t(3;6)(p14.2;p12)	
t(3;6)(q12;q15)	
t(3;6)(q11.2;q13)	
t(3;6)(q22;q16.2)	
t(3;8)(p14.2;q24.1)	*FHIT and TRC8*
t(3;12)(q13.2;q24.1)	
t(3;14)(q13.3;q23)	
t(3;15)(p11;q21)	
Ins(3;13)(p24.2;q32q21.2)	
Inv(3)(p14.2q12)	

From Al-Obaidy et al. 2022; with permission from John Wiley & Sons

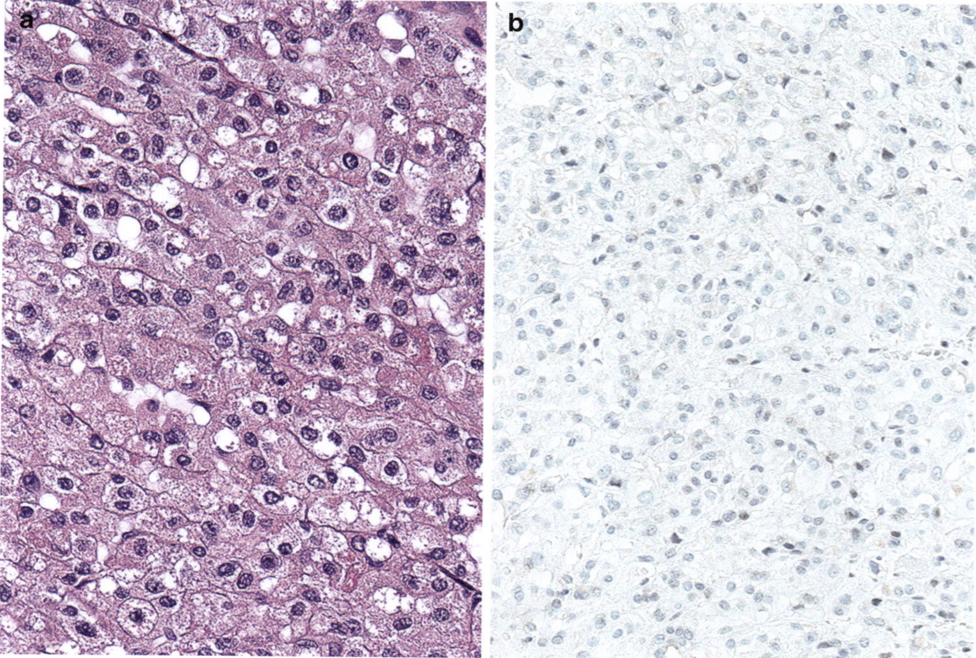

Fig. 13.7 Succinate dehydrogenase-deficient renal cell carcinoma shows sheets of cells with eosinophilic, vacuolated cytoplasm (**a**) and corresponding abnormal negative succinate dehydrogenase staining using immunohistochemistry (**b**)

- ○ t(2:3)(q33:q21) disrupts *DIRC1*
- ○ t(2:3)(q35:q21) disrupts *DIRC2* and *DIRC3*
- – t(3:4): t(3:4)(p13:p16), t(3:4)(q21;q31) and t(3;4)(p13;p15)
  - ○ t(3;4)(p13;p15) results in disruption of *KCNIP4*
  - ○ t(3;4)(q21;q31) results in disruption of *FBXW7*
- – t(3;6): t(3;6)(p13;q25.1), t(3;6)(q12:q15), t(3;6)(q11.2;q13), t(3;6)(p14.2;p12) and t(3;6)(q22;q16.2)
  - ○ t(3;6)(p13;q25.1) results in disruption of *STXBP5*
- – t(3;8): t(3;8)(p14;q24) and t(3;8)(p13;q24)
  - ○ t(3;8)(p14;q24) results in disruption of *FHIT* and *TRC8*
- – t(3;12)(q13.2;q24.1)
- – t(3;14)(q13.3;q23)
- – t(3;15)(p11;q21)
- – ins(3;13)(p24.2;q32q21.2)
  - ○ Also associated with the development of lung and prostate cancers
- – ins(3;13)(p24.2;q32q21.2)
- – inv(3)(p14.2q12)
- A "three-hit" model of carcinogenesis has been proposed, in contrast to the "two-hit" model of VHL disease

  - – First: Germline balanced translocation
  - – Second: Somatic loss of the chromosome 3 translocate
  - – Third: Somatic mutation of the remaining *VHL* gene allele
  - – Tumors often have later onset compared to VHL disease patients, perhaps due to a longer time course of progression through the three events

BAP1 Mutations and Familial Kidney Cancer

- *BAP1* (BRCA1 associated protein-1) is a tumor suppressor gene
- Located on the short arm of chromosome 3 (3p21)
- It encodes a nuclear deubiquitinase and is involved in the regulation of transcription, cell cycle and growth, and response to DNA damage
- Germline mutations are inherited as an autosomal dominant syndrome that predisposes patients to develop multiple tumors

Clinical and Pathologic Manifestations

- Renal manifestations
  - – Renal cell carcinoma
    - ○ The phenotype associated with *BAP1* mutation is not fully elucidated

- ○ Both clear cell and non-clear cell renal cell carcinomas have been described
- ○ Somatic *BAP1* mutations in clear cell renal cell carcinoma were found to be associated with high WHO/ISUP grade and poor prognosis
- Nonrenal manifestations
  - – Uveal melanoma
  - – Cutaneous melanocytic tumors (melanoma and atypical Spitz nevi)
  - – Mesothelioma

Cowden Syndrome (PTEN Hamartoma Tumor Syndrome)

- Cowden syndrome (CS) is an autosomal dominant syndrome caused by germline alterations of the *PTEN* tumor suppressor gene
- It is located at chromosome 10q23.31
- It is involved in the inhibition of cell cycle progression, induction of cell death, and modulation of arrest signal through antagonizing the function of PI3K and blocking the activation of the mTOR pathway
- It is characterized by the development of multiple hamartomas and associated with an increased risk of different malignancies, including breast, endometrium, and thyroid
- The risk of renal cell carcinoma development in patients with CS is probably underestimated but >30-fold increased risk of RCC development described
- No specific phenotypic subtype has yet been associated with *PTEN* mutation

Malignant Neoplasms

Clear Cell Renal Cell Carcinoma

- The most common subtype of adult renal neoplasm, making up approximately 60–75% of surgically removed tumors
  - – Believed to recapitulate cells of the proximal tubule

Light Microscopy

- Tumor cells with variably abundant clear cytoplasm (due to loss of cytoplasmic lipid/glycogen during histologic processing) forming nested, tubular, and alveolar/cystic architecture with a prominent fine vascular network
- Nuclei range from small and lymphocyte-like to large and hyperchromatic, with prominent large nucleoli and bizarre, irregular nuclear contours

- Classification is often readily accomplished based on the presence of typical clear cell histopathologic features, however, higher-grade tumors may have confounding morphology, such as:
  - Eosinophilic cytoplasm
  - Pseudopapillary architecture (loss of cohesive growth)
  - Bizarre giant tumor cells
  - Poorly-differentiated carcinoma-like morphology
- Application of molecular techniques may aid the differential diagnostic process in challenging cases

Molecular Characteristics

- Chromosome 3p
  - Deletion of chromosome 3p is believed to represent one of the key events in carcinogenesis
  - Seen in 70–90% of clear cell renal cell carcinoma tumors
    - Detection by fluorescence in situ hybridization (FISH) (Fig. 13.1a, b), LOH studies, or comparative genomic hybridization (CGH)
  - Less commonly found in other subtypes of renal neoplasms, but probably not entirely specific
    - Also reported in unclassified renal cell carcinoma, carcinoma with *TFEB*/6p21 amplification, and perhaps some papillary renal cell carcinoma
  - 3p25, 3p12–14, and 3p21 frequently involved
  - Introduction of normal chromosome 3p into tumor cell lines results in suppression of tumorigenesis
    - *VHL* gene
      - *VHL* gene inactivation (chromosome 3p25.3) is common in up to 75% of sporadic clear cell renal cell carcinoma cases (as seen in VHL disease, where germline inactivation of *VHL* is present)
        - Inactivation may occur through mutation, deletion, or abnormal DNA-methylation
      - In sporadic cases, one gene copy may be inactivated by mutation (as in patients with VHL disease) with the second copy lost through deletion in a similar "two-hit" model
        - Mouse xenografts of clear cell carcinoma with homozygous VHL loss of function mutation result in tumor formation
        - Introduction of wild-type *VHL* gene copies into the mice results in markedly decreased or absent tumor formation
        - However, additional steps beyond *VHL* mutation also appear to be required in models of tumor development
      - Prognostic significance of *VHL* inactivation is unclear

- Nuclear and cytoplasmic expression of pVHL by immunohistochemistry associated with lower nuclear grade and lower tumor stage
- Crucial in tumorigenesis via the HIF and mTOR pathways (see section "von Hippel–Lindau Disease")
- Deletion of regions flanking the *VHL* gene suggests a role for potential additional tumor suppressor genes in the pathogenesis of clear cell renal cell carcinoma
  - In addition to VHL, several other key genes on chromosome 3p are now recognized to be important in clear cell renal cell carcinoma
    - BAP1
      - Tumor suppressor gene associated with chromatin remodeling
      - Mutated in up to 15–20% of clear cell renal cell carcinoma
      - Thought to be associated with higher tumor grade and more aggressive behavior
      - Recently shown that BAP1-mutated tumors may have more eosinophilic cytoplasm, hyaline globules, papillary/pseudopapillary morphology, higher nuclear grade, and frequent keratin 7/AMACR positivity
    - PBRM1
      - Tumor suppressor gene required for stability of the SWI/SNF chromatin remodeling complex
      - Negative regulator of cell proliferation
      - Up to 40% of clear cell renal cell carcinomas mutated for *PBRM1*
      - Alterations thought to be associated with more favorable behavior
    - SETD2
      - Tumor suppressor gene, that functions as a histone methyltransferase
      - Up to 12% mutated in clear cell renal cell carcinoma
      - Thought to be associated with more aggressive behavior
- Other chromosomal regions affected in clear cell carcinoma
  - 5q
    - Second most common chromosomal region involved in clear cell renal cell carcinoma, after chromosome 3p
    - Allelic duplications at 5q31.1
    - Trisomy or partial trisomy of chromosome 5, including 5q22–qter
    - Chromosome 3 and 5 translocation, leading to loss of 3p13–pter and duplication of 5q22–qter

- Gain of 5q may be associated with better prognosis in clear cell renal cell carcinoma
- 9p and 14q
  - Loss is seen in a subset of cases and has been associated with poorer prognosis
  - Associated with higher histologic grade and tumor stage
  - 9p loss occurs primarily at 9p21 (the site of p16) and 9p22–23 (*PTCH* gene)
  - 8q gain and LOH at 8p may also participate in tumor progression
- Other chromosomal sites sometimes involved in clear cell carcinoma: 6q, 9q, 10q, 13p, 17p

Multifocality

- Up to 25% of patients undergoing nephrectomy for renal cell carcinoma have multifocal tumors
- Some studies of clear cell renal cell carcinoma have found a common clonal origin by LOH analysis, raising the possibility of "satellite" tumors representing intrarenal metastases
- However, multifocality is not necessarily associated with increased risk of progression and metastasis, suggesting independent origin, perhaps by "field-effect"
- Other studies have found discordant patterns of 3p deletion, X chromosome inactivation, and LOH analysis, supporting independent origin in a significant number of cases (46%)

Molecular Differential Diagnosis

- A spectrum of other renal neoplasms may have significant overlap in light microscopic appearance with clear cell renal cell carcinoma, leading to the potential utility for molecular diagnostic studies
  - Confirming clear cell renal cell carcinoma
    - Clear cell vs. non-clear cell renal cell carcinoma may have implications for treatment, especially in the metastatic setting
    - In these situations, the only tissue may represent a small biopsy from either the primary tumor or a metastatic site
    - Morphology may be nonclassic
    - Strong membranous CA9 staining in general favors clear cell subtype amongst known renal cell carcinomas
    - However, molecular testing, especially for *VHL* mutation, may assist in confirming clear cell type from these unusual morphologies (Fig. 13.8)
  - Clear cell papillary renal cell tumor
    - A renal neoplasm with overlapping morphologic and immunohistochemical features (*see* section "Clear Cell Papillary Renal Cell Tumor"), yet with highly favorable behavior
    - Clear cell papillary renal cell tumor has been found to lack chromosome 3p and *VHL* gene abnormalities, including promoter hypermethylation; however, the *HIF* pathway may be upregulated by other mechanisms

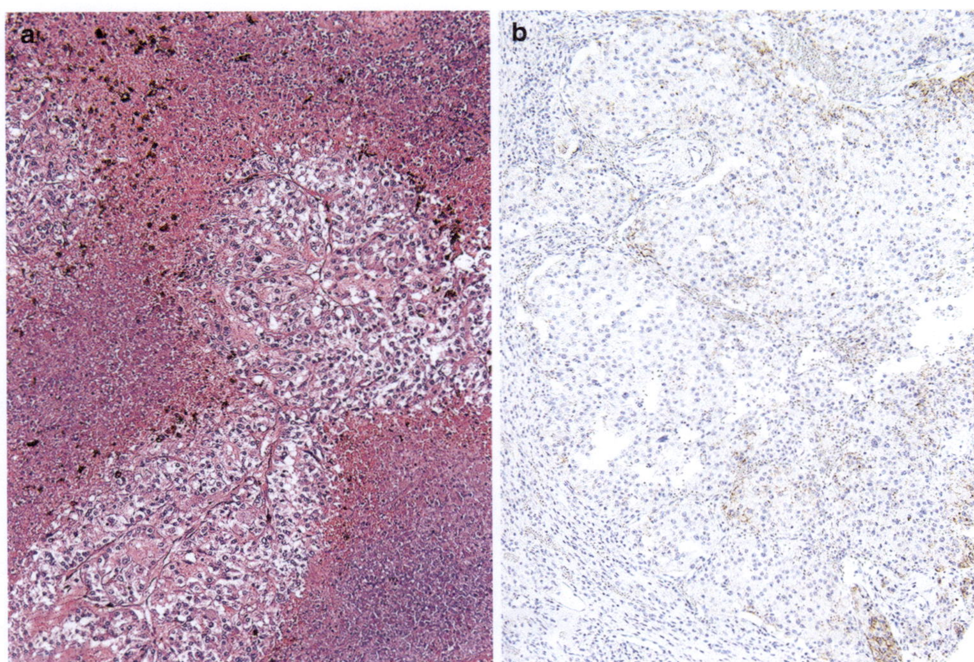

Fig. 13.8 This renal mass shows extensive necrosis and varied tumor morphology (**a**). CA9 is predominantly negative, with only focal staining (**b**); however, molecular testing in this tumor revealed *VHL* mutation, supporting clear cell type

- MITF (*TFE3* or *TFEB*) translocation carcinoma
  ○ Can include a prominent component of cells with clear cytoplasm, resembling clear cell carcinoma at the light microscopic level (sometimes intermingled with components including eosinophilic cytoplasm and papillary architecture; *see* section "MIT Translocation Carcinoma")
  ○ Tumors characteristically harbor fusions involving the *TFE3* gene at the Xp11.2 breakpoint, with a variety of fusion partners, including *ASPL*, *PRCC*, *SFPQ*, *NONO*, and *CLTC*
  ○ Using immunohistochemistry, epithelial markers may show limited positivity or negative results
  ○ Translocation carcinomas are typically positive for TFE3 immunohistochemistry
  ○ CA9 is usually only focally positive (6%)

Papillary Renal Cell Carcinoma

- Papillary renal cell carcinoma is the second most common surgically removed renal tumor (approximately 10–15%), following clear cell carcinoma
  - Believed to recapitulate the proximal convoluted tubule

Light Microscopy

- Papillary renal cell carcinoma is characterized by a predominant pattern of tubulopapillary architecture, often including foamy macrophages within papillae and/or psammoma bodies
- Immunohistochemical features include positivity for keratin 7 and alpha-methylacyl-CoA racemase (AMACR), with limited or negative CA9
  - Former type 1
    ○ Papillae are lined by a single layer of smaller cells with pale cytoplasm and round to ovoid nuclei, imparting a basophilic overall appearance
    ○ Seen in the hereditary papillary renal carcinoma syndrome (*see* section "Hereditary Papillary Renal Carcinoma")
    ○ Associated with better prognosis and survival when compared to the former category of type 2 tumors
  - Former type 2
    ○ Subclassification into type 1 vs. type 2 is no longer recommended
    ○ However, the previous type 2 category included larger, pseudostratified cells with more abundant eosinophilic cytoplasm
    ○ It is now thought that this morphologic pattern likely includes several diagnostic entities, such as

FH-deficient carcinoma (*see* section "Hereditary Leiomyomatosis and Renal Cell Carcinoma"), translocation carcinoma, and others

Molecular Characteristics

- Polysomy 7 and 17
  - Trisomy 7
    ○ Commonly present in papillary renal carcinoma, although a nonspecific finding
      ◆ Seen in several human neoplasms, including those of epithelial, mesenchymal, and neural origins
      ◆ Also identifiable in benign conditions, such as nodular prostatic hyperplasia and benign renal epithelial cells
      ◆ Approximately 75% of sporadic papillary carcinomas exhibit trisomy of chromosome 7 (Fig. 13.1c, d)
      ◆ Contains genes for both *MET* and ligand *HGF*
      ◆ However, activating *MET* mutation at chromosome 7q31 is only seen in a smaller proportion of sporadic papillary carcinoma tumors, in contrast to tumors of the hereditary papillary renal carcinoma syndrome
  - Trisomy 17 (Fig. 13.1c, d)
    ○ Common in papillary carcinoma (80–90% by FISH)
    ○ Less commonly found in other human tumors and other subtypes of renal cell carcinoma
    ○ Characterized by the following:
      ◆ Full trisomy of chromosome 17
      ◆ Isochromosome 17q
      ◆ Duplication of the 17q21–qter region (Fig. 13.9)
  - Chromosome 3
    ○ Although abnormalities of chromosome 3p are generally associated with clear cell carcinoma, deletions are also less commonly reported in papillary renal cell carcinoma
    ○ Abnormalities involving 3q may be seen in both clear cell and papillary carcinoma
    ○ Unbalanced translocation with reduplication of the normal chromosome 3 resulting in partial trisomy of 3q is seen more often in papillary carcinoma
  - Other chromosomal regions
    ○ Allelic loss at 7q31.1–31.2 is seen in some cases
    ○ Contains the aphidicolin-inducible fragile site FRA7G
    ○ Fragile sites are thought to be involved in tumorigenesis due to chromosomal breakage and resulting translocation, deletion, or amplification
    ○ Gains of 8, 12q, 16q, and 20q

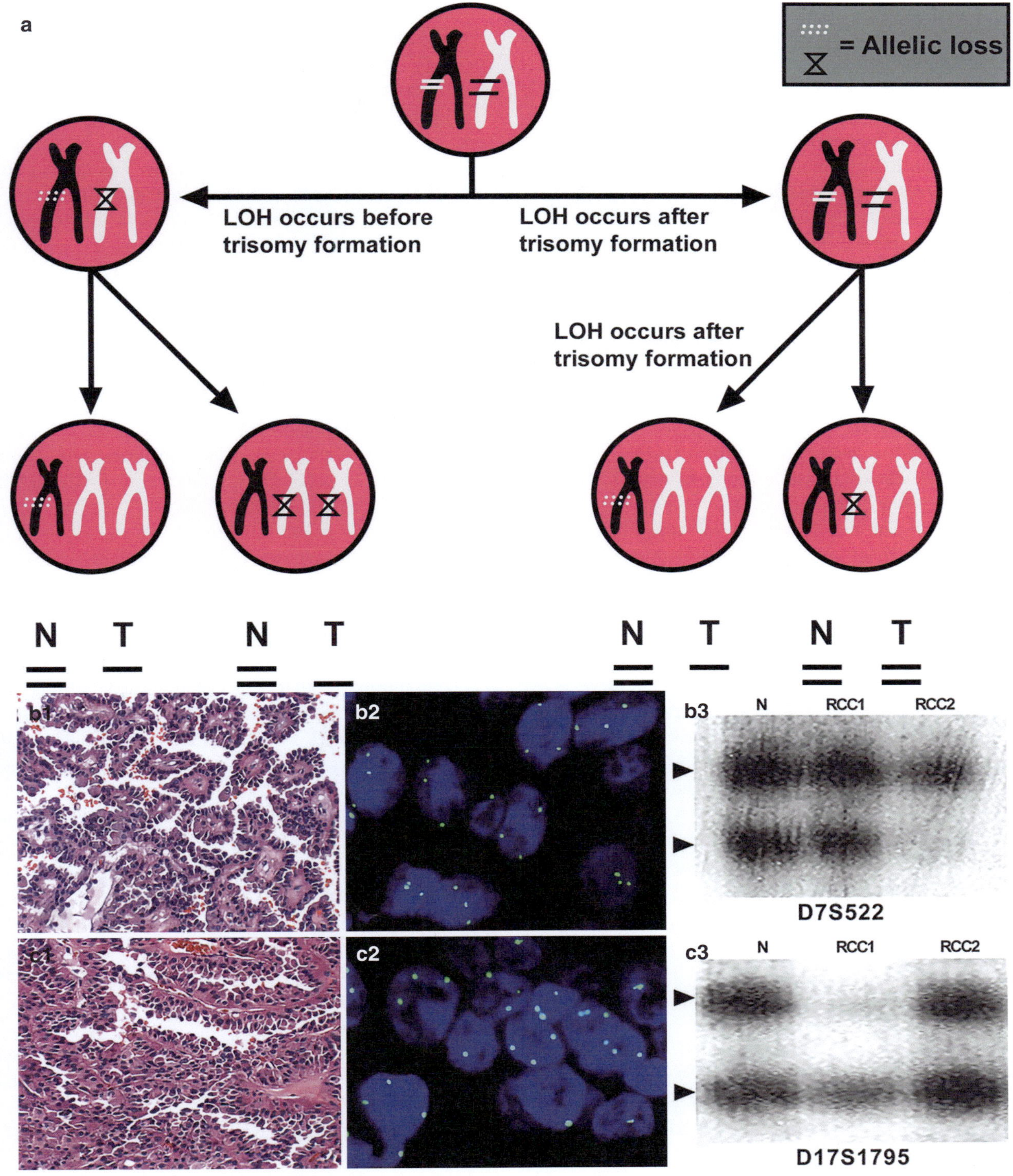

Fig. 13.9 Proposed mechanism resulting in trisomic chromosomes 7 and 17 (**a**). Acentromeric misdivision results in nondisjunction of the pair of chromosomes during cell mitosis and trisomic daughter cells. Loss of heterozygosity (LOH) could occur before trisomy formation (**a**, *left*) in which case allelic loss of either the upper or lower allele could occur (**a**, *bottom left*). LOH could occur after trisomy formation (**a**, *right*), in which case only allelic loss on the nonduplicated chromosome could be detected. Allelic loss on only one copy of the duplicated chromosome would not be detectable by LOH analysis (**a**, *bottom right*). Histopathologic, FISH, and LOH features of papillary renal cell carcinoma (**b**, **c**). Cases of papillary renal cell carcinoma (**b1–b3** and **c1–c3**). Papillary renal cell carcinoma is composed of branching papillae and covered with a single layer of cells with eosinophilic cytoplasm (**b1** and **c1**). FISH of the corresponding tumor with a centromeric probe of chromosomes 7 and 17 showing groups of tumor cells with trisomic chromosome 7 (**b2**) or 17 (**c2**; *green signals*). Microsatellite analysis on chromosomes 7 (D7S522) and 17 (D17S1795) revealed different LOH patterns between different tumor foci (**b3** and **c3**) (from Jones et al. 2005; with permission)

- o Duplication of 20q11.2 and 20q13.2 thought to contain genes involved in the development of papillary carcinoma
  - o Loss of 1p, 4q, 6q, 11p, 13q, 14q, 18, 21q, X and Y
  - – Other genes
    - o Leucine-rich repeat kinase 2 (*LRRK2*) has been found to be amplified and overexpressed in papillary renal cell carcinoma
    - o Downregulation of LRRK2 compromises MET activation and downstream signaling, suggesting a cooperative role with MET in tumor growth and survival of papillary carcinoma
- Pathway analysis
  - – Classic papillary renal cell carcinoma (former type 1): Significant enrichment in
    - o WNT signaling (Chr 11 and Chr 7)
    - o MET activation (Chr 7)
    - o NOTCH signaling (Chr 17)
    - o DNA damage bypass related pathways (Chr 16)
  - – Former type 2 tumors
    - o Significant enrichment in pathways implicated in either tumor metastasis or cell cycle, including Chr 8q
    - o Dysregulation of G2-M checkpoint genes
    - o Meiosis and meiotic recombination pathways were enriched in the CNV regions (Chr 5p)
    - o Metastasis augmenting pathways were detected among the aberrant CNV regions, as were gap junction degradation, and dopa decarboxylase-mediated attractive signaling (Chr 5p and 8q)
    - o miRNA enhanced pathways included collagen degradation and fibroblast growth factor receptor (FGFR), important in migration, invasion, and survival of cancer cells as well as activation of PI3K/AKT and mTOR pathway

Multifocality

- Multifocality is most common in papillary renal cell carcinoma, compared to other subtypes
- However, concordant patterns of allelic loss are usually not seen between multifocal tumors (5%)
- Multifocal and bilateral tumors may show cytogenetic heterogeneity between lesions, supporting independent origin, rather than intrarenal metastasis

Molecular Differential Diagnosis

- Clear cell papillary renal cell tumor
  - – Areas of clear cell change are sometimes present within papillary renal cell carcinoma, bringing clear cell papillary renal cell tumor into the differential diagnosis
  - – Copy number abnormalities of chromosomes 7 and 17 have been reported infrequently in clear cell papillary renal cell tumors, although less frequently than papillary carcinoma
  - – Co-expression of CA9 and keratin 7 is seen in clear cell papillary renal cell carcinoma, in contrast to papillary renal cell carcinoma, which lacks CA9 positivity expression but is usually positive for keratin 7 and AMACR
- Mucinous tubular and spindle cell carcinoma
  - – Characterized by extensive tubular architecture with morphologic similarity to papillary renal cell carcinoma
  - – Copy number abnormalities of chromosomes 7 and 17 and loss of chromosome Y are not found
  - – Multiple other genetic abnormalities are present including losses of chromosomes 1, 4, 6, 8, 9, 10, 12q, 13, 14, 15, 16q 17, 20q, and 22
- Metanephric adenoma
  - – Areas of solid growth in papillary renal cell carcinoma may show morphologic overlap with metanephric adenoma
  - – Metanephric adenoma usually shows a normal karyotype by cytogenetics, with infrequent abnormalities of chromosomes other than 7 and 17, making molecular studies useful in resolving the differential diagnosis
  - – Recurrent *BRAF* V600E mutation in metanephric adenoma, contrasting with the recurrent *MET* mutations in papillary renal cell carcinoma
- Clear cell renal cell carcinoma with pseudopapillary change
  - – Areas of noncohesive growth in clear cell renal cell carcinoma may impart a papillary appearance
  - – FISH analysis for abnormalities of chromosomes 7, 17, and 3p may be helpful in resolving the differential diagnosis, combined with other typical light microscopic and immunohistochemical features
- MITF translocation renal cell carcinoma
  - – Prominent papillary architecture is sometimes present in renal cell carcinoma associated with *TFE3* and *TFEB* translocation
  - – Nuclear labeling for TFE3 protein by immunohistochemistry in TFE3 translocation carcinoma; cathepsin K, melan A, and HMB45 in TFEB translocation carcinoma with confirmation by molecular testing supports the diagnosis of translocation carcinoma
  - – Chromosome 7, 17, and Y abnormalities absent
  - – Next-generation sequencing (NGS) studies may be more helpful than FISH to confirm the diagnosis in some cases, as chromosomal inversions may yield a subtle or false-negative FISH pattern

- Implications of molecular genetic alterations on therapy
  - *MET*: The use of tyrosine kinase inhibitors, targeting the tyrosine kinase domain of MET has been proposed as a potential therapy in both hereditary and sporadic papillary renal cell carcinoma

Hereditary Leiomyomatosis RCC Syndrome–Associated RCC

- See the section "Genetic Renal Neoplasia Syndrome"

Chromophobe Renal Cell Carcinoma

- Less common than clear cell carcinoma or papillary carcinoma, making up approximately 5% of renal tumors (similar in incidence to oncocytoma)
  - Believed to recapitulate intercalated cells of the collecting ducts

Light Microscopy

- Chromophobe carcinoma is composed of solid nests, sheets, and trabecular bands of generally large, polygonal tumor cells with flocculent to eosinophilic cytoplasm and prominent ("plant cell"-like) cell borders
- The eosinophilic variant of chromophobe carcinoma is composed predominantly of cells with eosinophilic cytoplasm, leading to considerable differential diagnostic overlap with oncocytoma and potential utility for molecular genetic studies
- Both classic and eosinophilic areas may be intermixed within the same tumor

Molecular Characteristics

- Chromosomal alterations
  - Multiple complex losses of chromosomes 1, 2, 6, 10, 13, 17, and 21 by cytogenetics, restriction fragment length polymorphism analysis, CGH, FISH (Fig. 13.1e, f), and microsatellite analysis
  - LOH has also been identified at 9p23 (43%), 18q22 (30%), 5q22 (28%), and 8p (28%)
- Genetic alterations
  - *TP53* gene mutation is present in a subset of tumors
  - The folliculin *FLCN* (or *BHD*) gene is mutated in the setting of the Birt–Hogg–Dubé syndrome; however, such abnormalities are not found in sporadic cases
    - Chromophobe renal cell carcinoma has been found to have distinct heteroplasmic mtDNA mutations

- Downregulation of ATP5A1, the alpha subunit of complex V of the respiratory chain, has been demonstrated by 2D gel electrophoresis of mitochondrial proteins
- The possibility that oncocytoma represents a precursor to chromophobe carcinoma has been entertained (with the eosinophilic variant constituting an intermediate form)
  - However, both classic and eosinophilic types show similar complex losses of multiple chromosomes, while oncocytoma more frequently has a normal karyotype of loss of chromosome 1
  - Quantitative reverse transcription PCR (RT-PCR) and or immunohistochemistry reveal
    - Differential expression of *AP1M2*, *MAL2*, *PROM2*, *PRSS8*, and *FLJ20171* genes, leading to effective separation of chromophobe carcinoma from oncocytoma
    - *CD82* and *S100A1* as markers of chromophobe carcinoma and *AQP6* as a marker of oncocytoma
    - MAL2 and CLDN8 staining highlights the distal nephron; CLDN8 staining is seen in both entities; however, MAL2 is limited to chromophobe carcinoma
  - *FOXI1*, *RHCG*, and *LINC01187* in classic, eosinophilic, and metastatic chromophobe RCCs identified by next-generation sequencing; however, these are not specific

Molecular Differential Diagnosis

- Oncocytoma
  - Losses of chromosomes 2, 6, 10, and 17 by FISH studies may be helpful in excluding the diagnosis of oncocytoma and supporting the diagnosis of chromophobe renal cell carcinoma
  - Virtual karyotyping and microRNA expression studies have been proposed as a diagnostically useful modality for differentiating chromophobe renal cell carcinoma from oncocytoma and eosinophilic variants of clear cell carcinoma

Collecting Duct Carcinoma

- Collecting duct carcinoma (or carcinoma of the collecting ducts of Bellini) is a rare but highly aggressive renal neoplasm
  - Believed to recapitulate the phenotype of the distal nephron/collecting duct epithelium
  - Diagnosis of exclusion only after mimics are thoroughly argued against

Light Microscopy

- Tumors are characterized by trabecular, papillary, solid, and glandular architecture, with extensively infiltrating growth

Molecular Characteristics

- Cytogenetic studies
  - Monosomy of chromosomes 1, 6, 14, 15, and 22
  - Frequent allelic loss on chromosomal arms 1q, 6p, 8p, 13q, and 21q
  - LOH on 1q is observed in two-thirds of tumors
  - LOH at 6p, 8p, 13q, and 21q
  - Deletion of 1q32.1–32.2 was identified in 57–69%, suggesting that this region may contain a tumor suppressor gene involved in carcinogenesis
  - Loss of chromosome arm 8p, which may be important in the aggressive behavior of collecting duct carcinoma
    - In clear cell renal cell carcinoma, it has been associated with high-stage and aggressive behavior
  - No losses of chromosome 3p, although *VHL* allelic loss has been reported in some cases
  - Allelic loss of 9p, seen frequently in urothelial carcinoma, shows variables ranging from absent to 50% of cases
- Genetic mutations
  - Mutations in *EGFR*, *RET*, *NF2*, and *TSC2* were reported
  - Homozygous deletion of *CDKN2A*
  - Some authors have found *SMARCB1* abnormalities; however, these are likely better classified as renal medullary carcinoma/renal cell carcinoma with medullary phenotype
  - Loss of tumor suppressor gene *RB* (retinoblastoma) has also been reported to be present in a significant number of cases
  - Expression of MET by immunohistochemistry represents an area of similarity between collecting duct carcinoma, urothelial carcinoma, and papillary carcinoma
- Molecular pathway analysis
  - Deregulation of genes involved in cancer and cell cycle pathways
    - Upregulation of *KRT17*
    - Downregulation of *cubilin* and solute carrier genes (*SLC3A1, SLC9A3, SLC26A7, and SLC47A2*)

Molecular Differential Diagnosis

- Urothelial carcinoma
  - Urothelial carcinoma may have considerable overlap with collecting duct carcinoma at the microscopic level
  - In general, gains of chromosomes 3, 7, and 17 and loss of 9p21 (as tested with the UroVysion probe set) are seen preferentially in urothelial carcinoma rather than collecting duct carcinoma; however, some overlap in the genetic characteristics of the two tumors is possible (such as variable allelic loss at 9p)
  - In our interpretation, GATA3 or p63 positivity would favor urothelial carcinoma
  - Although not specifically studied, *TERT* promoter mutation is common in urothelial carcinoma and may suggest urothelial carcinoma rather than collecting duct carcinoma, which is a diagnosis of exclusion
- Renal medullary carcinoma
  - Morphology and immunohistochemistry substantially overlap between medullary carcinoma and collecting duct carcinoma; however, we would consider *SMARCB1* (INI1) abnormality to be diagnostic of medullary carcinoma, as collecting duct carcinoma is a diagnosis of exclusion

Renal Medullary Carcinoma

- Renal medullary carcinoma is a rare, aggressive malignancy, prone to affect individuals with sickle cell trait, although patients without hemoglobinopathies have also been reported

Light Microscopy

- Tubular, cribriform, reticular, microcystic, and adenoid-cystic-like architecture with marked desmoplasia and an acute inflammatory reaction
- Cells are often eosinophilic, with prominent nucleoli
- Sarcomatoid features are sometimes present
- Overlap in appearance with collecting duct carcinoma raises the possibility that the two entities are related
- Careful examination of the vasculature may show sickle-shaped red blood cells

Molecular Characteristics

- Inactivation of *SMARCB1* with loss of expression of INI1 protein by immunohistochemistry is characteristic, representing a point of similarity to the rhabdoid tumor of the kidney and distinction from other renal tumors (such as urothelial carcinoma and other subtypes of renal cell carcinoma)

- This is thought to be linked to the downregulation of *P16INK4a* and upregulation of cyclin D1, promoting cycle progression
- INI1 expression is also retained in renal cell carcinomas with a rhabdoid morphology
- Other less common alterations include
  - Loss of chromosomes 9, 13, and 22 or partial loss of chromosomes 8p or 7q
  - Gains of chromosomes 7, 8, 10, and 11
  - t(9;22) and t(10;16) with *BCR::ABL* gene rearrangement by FISH

MiT Family Translocation Carcinoma

- MiT family translocation renal cell carcinoma is a variably aggressive type of renal cell carcinoma that constitutes around 5% of all renal cell carcinomas in adults and up to 50% in children
- Tumors are characterized by genetic rearrangements involving members of the MiT family of transcription factors, mainly the *TFE3* and *TFEB* genes, although rearrangement of *MITF* itself has been recently reported; the MiT/TFE fusion activates genes involved in mTORC1 signaling, antioxidant stress response, ROS sensing, and oxidative stress and xenobiotics responses
- Histologic features are variable but generally include clear and eosinophilic cells in various architectural formations

TFE3-Rearranged/Xp11.2 Translocation Renal Cell Carcinoma

- Renal carcinomas associated with Xp11.2 translocation include a spectrum of tumors with unique light microscopic features and specific translocations involving the *TFE3* transcription factor gene (transcription factor binding to IGHM enhancer 3), located at chromosome Xp11.2
- *TFE3* translocation carcinoma is thought to be the most common subtype of RCC in children
- Makes up a small percentage of renal tumors in adults (estimated 1% or less); however, gene fusion may be undetected in a subset of adult cases, which shows extensive morphologic similarity with clear cell renal cell carcinoma or papillary renal cell carcinoma; overall numbers of cases in adults may, therefore, be greater than those in children

Light Microscopy

- Shows variable admixtures of cells with clear cytoplasm and/or eosinophilic cytoplasm, as well as varied papillary and nested/solid architectures, sometimes with psammoma bodies (Fig. 13.10a, b)
- Tumors associated with *ASPSCR1::TFE3* gene fusion tend to exhibit a characteristic light microscopic appearance including voluminous clear cytoplasm with nested and papillary architecture
- In contrast, tumors with *PRCC::TFE3* fusion tend to demonstrate more compact architecture with less abundant clear cytoplasm
- In contrast to other types of renal cell carcinoma, markers of epithelial differentiation are less frequently positive using immunohistochemistry, such as epithelial membrane antigen (EMA) or keratins; similarly, vimentin may be negative, contrasting to other types of renal cell carcinoma
- A subset of tumors is positive for cathepsin K, although this varies depending on the gene fusion partner; this can be helpful to distinguish translocation renal cell carcinomas from other adult renal cell carcinomas using immunohistochemistry

Molecular Characteristics

- The *TFE3* gene is a member of the melanocyte-inducing transcription factor/transcription factor E (MITF–TFE) family, which also includes *TFEB*, *TFEC*, and *MITF* (involved in melanocyte development)
- Such genes have been implicated in various tumors, such as melanoma, clear cell sarcoma, alveolar soft part sarcoma, and some perivascular epithelioid cell neoplasms (PEComa); the former so-called "melanotic Xp11 translocation renal cancer" with *SFPQ::TFE3* fusion are now thought to better represent *TFE3*-rearranged PEComas (Fig. 13.10c, d) (*see also* section "angiomyolipoma")
- A variety of fusion partners have been implicated in the translocations with *TFE3*, including the following:
  - *ASPSCR1* gene—t(X;17)(p11.2;q25)
    - *ASPSCR1::TFE3* gene fusion (unbalanced) is seen also in alveolar soft part sarcoma
  - *ASPSCR1::TFE3* fusion protein appears to transactivate the *MET* promoter, increasing *MET* mRNA expression with associated high levels of MET protein by IHC and Western blot methods
  - *PRCC* gene—t(X;1)(p11.2;q21)
  - *SFPQ* gene—t(X;1)(p11.2;p34)
  - *NONO* gene—inv(X)(p11;q12)
  - CLTC gene—t(X;17)(p11.2;q23)
  - Many less frequently encountered gene partners have been reported, such as *RBM10*, *NEAT1*, *KATA6A*, *MED15*, *GRIPAP1*, *PARP14*, etc.
  - Other fusions with undetermined gene partners, including the following

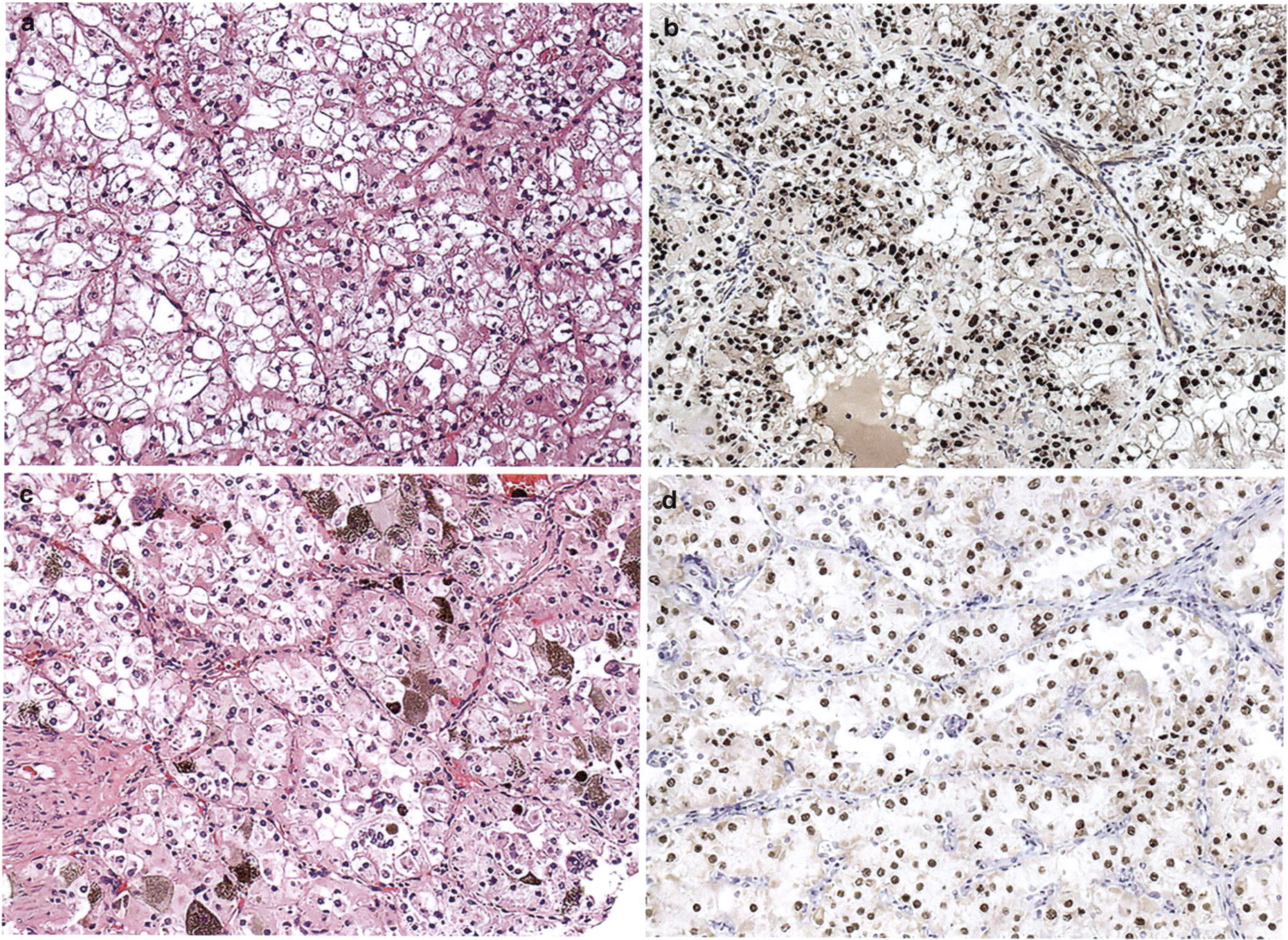

Fig. 13.10 *TFE3* translocation carcinoma, characterized by a mixed population of cells with clear and eosinophilic cytoplasm (**a**). Strong nuclear overexpression of TFE3 protein can be detected by immunohistochemistry (**b**). The entity formerly described as melanotic Xp11 translocation renal cancer is now thought to be better classified as a *TFE3*-rearranged PEComa (**c**). This tumor demonstrates scattered cells with prominent cytoplasmic pigmentation and (**d**) nuclear overexpression of TFE3 using immunohistochemistry

- - t(X;3)(p11.2;q23)
  - t(X;19)(p11.2;q13.1)
  - t(X;10)(p11.2;q23)
- Evidence of *TFE3* gene fusion can be demonstrated by immunohistochemical nuclear labeling for the TFE3 protein or by FISH break-apart probe
- Negative FISH has been reported in a subset of tumors with paracentric inversion with fusion partners including *RBM10, NONO, RBMX,* and *GRIPAP1*, which can be detected by RNA-ISH or RNAseq and reverse transcriptase-polymerase chain reaction

Molecular Differential Diagnosis

- Clear cell renal cell carcinoma
  - A significant component of clear cell cytology may raise the differential diagnosis between clear cell carcinoma and *TFE3* translocation carcinoma
  - Low positivity for typical markers of clear cell carcinoma by immunohistochemistry, such as keratin, EMA, and vimentin, coupled with nuclear expression of TFE3 protein may aid in resolving the differential diagnosis
  - FISH studies utilizing a break-apart probe for the *TFE3* gene are useful for verification
  - NGS fusion panels are also very helpful, as these may confirm not only rearrangement but also identify the fusion partner
    - Select fusion partners with chromosomal inversion yield subtle positive or false-negative FISH but can be detected by sequencing
- Papillary renal cell carcinoma
  - Papillary architecture, psammoma bodies, and variable amounts of eosinophilic cytoplasm may raise consideration for PRCC
  - Evaluation of *TFE3* by either FISH, sequencing, or immunohistochemistry (less preferred) can resolve this distinction

- Implications of molecular genetic alterations on therapy
  - Due to the transactivation of the *MET* promoter by the *ASPSCR1::TFE3* fusion protein, therapy directed against the MET tyrosine kinase has been proposed as a potential therapeutic strategy
  - Elevated expression of phosphorylated S6 in *TFE3* translocation renal cell carcinoma suggests the mTOR pathway as another potential therapeutic target

TFEB-Rearranged/*T*(6;11) Renal Cell Carcinoma

- Renal cell carcinomas with *TFEB* rearrangements, usually in the form of t(6;11)(p21.1;q12–13), are rare tumors, seen preferentially in young patients, although tumors in the sixth decade or later have also been reported
- Many tumors have shown generally indolent behavior, although metastases and aggressive behavior with the death of disease have been reported in a smaller subset of cases

Light Microscopy

- Morphologic features may show significant overlap with other subtypes of renal cell carcinoma, such as clear cell carcinoma or *TFE3*-rearranged carcinoma, including nested or solid growth with eosinophilic, granular, or clear cytoplasm
- The originally-described classic morphology, although not always present (Fig. 13.11a) includes clear cells and a second population of smaller cells with dense nuclear chromatin and less abundant cytoplasm, centered around collections of hyaline (basement membrane) material

- Cathepsin K has been reported to be a highly sensitive and specific marker in these tumors with consistent and diffuse staining pattern
- Coexpression of HMB45 and Melan A is also helpful
- Like *TFE3*-rearranged carcinoma, *TFEB*-rearranged carcinomas may be negative for epithelial markers but are positive for tubular markers such as PAX8

Molecular Characteristics

- Tumors most commonly show fusion of the 5′ aspect of the *MALAT1* gene (also known as Alpha) at 11q12, with the *TFEB* gene at 6p21
- Alternative partners include *COL21A1*, *CADM2*, and *KHDRBS2* genes

TFEB-Amplified Renal Cell Carcinoma

- Tumors with *TFEB* amplification have been described
- Patients with *TFEB*-amplified carcinomas are older than those with *TFEB*-rearranged carcinomas
  - Morphologically, they exhibit higher grade features and less frequent biphasic appearance than *TFEB*-rearranged tumors
  - The *VEGFA* gene is also located at 6p21 and is typically included in the amplicon
- Morphology can include oncocytic and tubulopapillary features with high-grade nuclei (Fig. 13.11b)
  - Behavior appears to be aggressive, with metastasis and death in 46% of tumors
  - A small percentage of *TFEB*-amplified carcinomas found to also harbor *TFEB* translocation

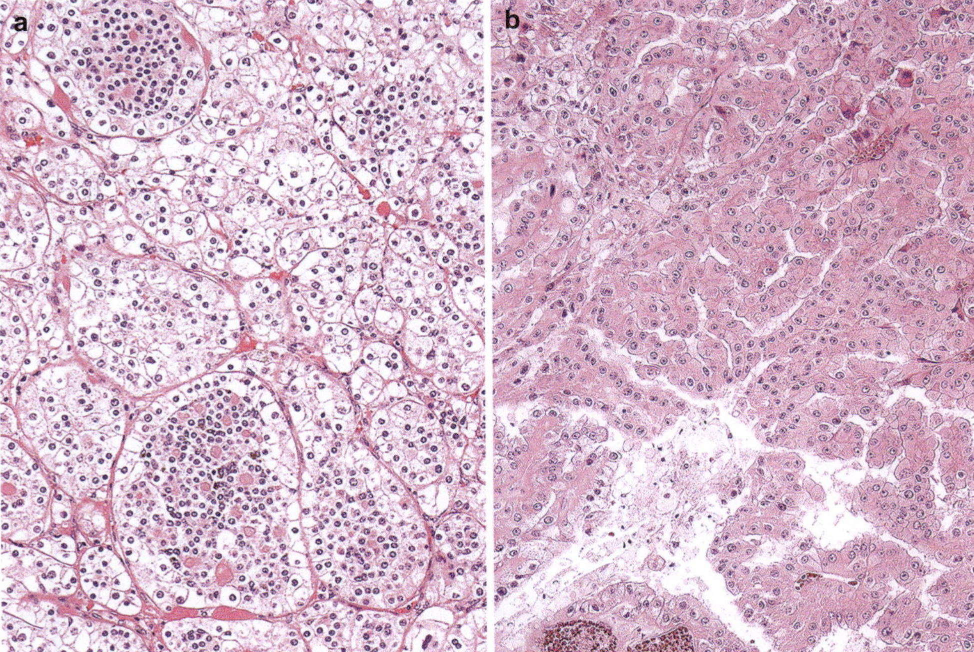

Fig. 13.11 *TFEB* rearranged renal cell carcinoma, characterized by nests of clear cells and a second population of smaller cells, centered around collections of hyaline (basement membrane) material (**a**). *TFEB* amplified renal cell carcinoma with oncocytic cells arranged in tubulopapillary architectures with high-grade nuclei (**b**)

- Loss of chromosome 3p has been noted in a subset, despite lack of *VHL* alteration or CA9 labeling, suggesting that 3p loss is not entirely specific for clear cell renal cell carcinoma
- Significant upregulation of TFEB protein levels results from promoter substitution, likely leading to changes in the expression of downstream genes
- A few tumors have also been found to have losses of chromosomes 1 and 22
- Similar to the TFE3 protein expression seen in Xp11 translocation carcinoma, nuclear immunohistochemical staining for TFEB protein may be helpful in establishing the diagnosis
- As the *MALAT1* gene lacks introns, both RT-PCR and DNA PCR have been utilized in confirmation of the gene fusion

Molecular Differential Diagnosis

- A number of entities may be considered in the differential diagnosis of carcinomas associated with *TFEB* rearrangements, such as clear cell renal cell carcinoma, angiomyolipoma (and epithelioid angiomyolipoma), and *TFE3* translocation carcinoma (particularly tumors associated with *ASPSCR1::TFE3* fusion)
- Molecular studies may be useful in resolving this differential diagnosis, combined with the unique light microscopic features, and immunohistochemical expression of melanocytic markers and TFEB protein

SDH-Deficient Renal Carcinoma

- See the section "Genetic Renal Neoplasia Syndrome"

Mucinous Tubular and Spindle Cell Carcinoma

- Mucinous tubular and spindle cell carcinoma is a unique neoplasm with distinctive morphologic features, including small cuboidal arranged in elongated tubules or sheets, merging with bland spindled cells in a mucinous background
- Lesions are generally low-grade and low-stage; however, rare sarcomatoid differentiation has been reported
- Metastasis is infrequent

Molecular Characteristics

- Significant morphologic overlap with classic papillary renal cell carcinoma (former type 1) has raised a challenging differential diagnosis and speculation that the two entities are related

- However, gains of chromosomes 7 and 17 and losses of chromosome Y are generally lacking in FISH studies
- CGH and cytogenetics have demonstrated losses of chromosomes 1, 4, 6, 8, 9, 13, 14, 15, and 22 and gain of chromosomes 12q, 16q, 17, and 20q in smaller numbers of cases, suggesting that complex karyotypic abnormalities are present in the tumor
- Characteristic recurrent chromosomal losses and somatic mutations in the Hippo signaling pathway genes leading to potential *YAP1* activation with increased YAP1 protein expression have been described
- *VSTM2A* and *IRX5* were found to be cancer-specific and lineage-specific biomarkers

Tubulocystic Renal Cell Carcinoma

- Tubulocystic carcinoma is an uncommon renal neoplasm, characterized microscopically by tubular and cystic architecture, lined by a single layer of cells with eosinophilic cytoplasm, prominent nucleoli, and hobnail features
- Generally, well-differentiated
- Tumors with poorly differentiated foci have been reported, but more recently it is thought that these represent FH-deficient renal cell carcinoma (HLRCC tumors)
- Low grade, with rare metastatic events

Molecular Characteristics

- Relatively little is known about the molecular genetics of these unusual lesions
  - Gene expression profile showed overexpression of *VIM*, *TP53*, and *AMACR*
- The variable presence of chromosomal gains of 7 and 17 with loss of Y has led some authors to consider that the lesion bears a close relationship with papillary renal cell carcinoma (*see* section "Papillary Renal Cell Carcinoma")
  - Along these lines, the gene expression profile has been found to be similar but not identical to papillary renal cell carcinoma by some investigators
- Alternative hypotheses have proposed that tubulocystic carcinoma represents a low-grade collecting duct carcinoma; however, gene expression profiles show significant differences between the two entities

Acquired Cystic Disease-Associated Renal Cell Carcinoma

- Patients with end-stage renal disease and acquired cystic kidney disease are prone to the development of various types of renal cell carcinoma, including clear cell carcinoma, papillary carcinoma, and clear cell papillary renal cell carcinoma

- However, acquired cystic disease-associated renal cell carcinoma appears to represent a unique tumor subtype in this setting with distinctive histologic and possibly molecular features

Light Microscopy

- Unique features in these tumors include abundant eosinophilic cytoplasm, variable solid, cribriform, tubulocystic, and papillary architecture
- Deposits of calcium oxalate crystals are a characteristic and distinctive feature (Fig. 13.12)

Molecular Characteristics

- Chromosomal alterations
  - Gains of chromosomes 1, 2, and 6, with or without gains of 10, or normal complements of these chromosomes
  - Mixed trisomy and monosomy for chromosomes 3 and 16, with additional monosomy of chromosome 9 in the setting of sarcomatoid change
  - Another case revealed gains of chromosomes 3, 7, 16, and X, and loss of Y by cytogenetics
  - These molecular genetic characteristics suggest the distinction of these unusual tumors from other well-known subtypes of renal cell carcinoma
- Recurrent mutations in the *KMT2C* gene and *TSC2* were reported by next generation sequencing

Thyroid-Like Follicular Renal Cell Carcinoma

- Thyroid-like follicular renal cell carcinoma, first described in 2004, is rare
- It was considered a provisional entity in the 2016 WHO classification of kidney tumors due to the small number of reported cases and limited data regarding its pathogenesis
- Occurs more frequently in females
- Most patients appear to be cured by surgical resection, although metastases have been reported in 6 patients
- Tumors range from 8 to 70 mm in diameter

Light Microscopy

- Tumors are mostly well-circumscribed and unencapsulated
- Composed of variably sized, small, and large follicle-like cysts, filled by eosinophilic secretions of variable density
- Clear spaces separating the lining epithelial cells and luminal secretions
- The cells have minimal eosinophilic to amphophilic cytoplasm with high nuclear to cytoplasmic ratios and predominantly oval to elongated overlapping nuclei (Fig. 13.13)
- Variable stromal quantity and composition, ranging from stroma poor to rich areas

Molecular Characteristics

- Comparative genomic hybridization analysis showed complex karyotyping in some cases, including gains of 7q36, 8q24, 12, 16, 17p11-q11, 17q24, 19q, 20q13, 21q22.3, and Xp; and losses of 1p36, 3, and 9q21–33
- Overexpression of cell cycle regulatory genes and mixed lineage leukemia/trithorax homolog was reported in three tumors
- Recurrent *EWSR1::PATZ1* fusions by next-generation sequencing have been recently reported
- No known pathogenic gene mutations or copy number alterations identified (two analyzed tumors)

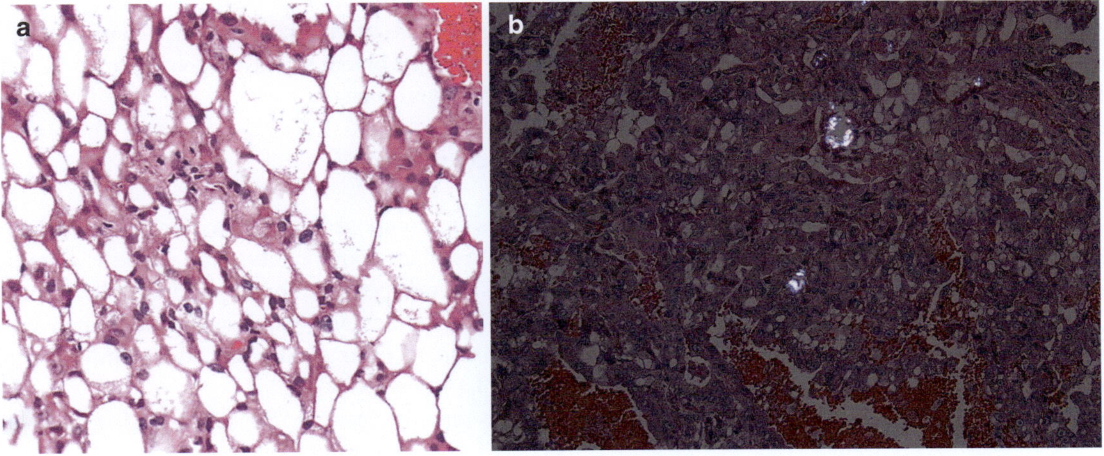

Fig. 13.12 Acquired cystic disease-associated renal cell carcinoma shows sieve-like architecture (**a**) with associated oxalate crystals under polarization (**b**)

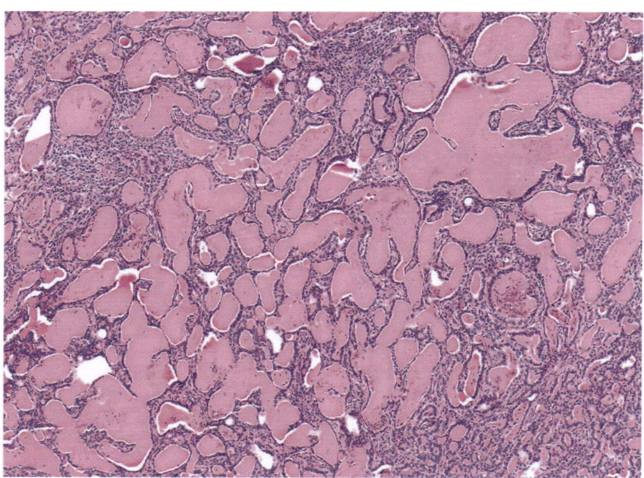

Fig. 13.13 Thyroid-like follicular renal cell carcinoma shows follicular-like cystic architecture, filled with eosinophilic, colloid-like secretions

Molecular Differential Diagnoses

- Metastasis from primary thyroid carcinoma
  - Both are positive for PAX8, however, metastasis from the thyroid almost invariably is positive for TTF1 and often thyroglobulin, which are negative in thyroid-like follicular renal cell carcinoma
  - Most of the follicular and papillary, follicular variant, thyroid carcinomas have *NRAS* and *HRAS* or *BRAF* mutations respectively, contrasting with thyroid-like follicular renal cell carcinomas which lack these mutations
- Atrophic kidney-like lesion
  - These are composed of variably sized cysts with dense eosinophilic secretions; however, the follicles are typically lined by flattened epithelium with occasional hobnail morphology and microcalcifications
  - Immunohistochemically, these tumors/lesions are consistently positive for WT1 in contrast to thyroid-like follicular renal cell carcinoma which is negative for WT1
- Papillary renal cell carcinoma
  - It may also show foci that resemble thyroid-like follicular renal cell carcinoma when the papillary cores become markedly edematous imparting a thyroid-like follicular pattern

ALK Rearrangement-Associated Renal Cell Carcinoma

- First reported in 2011
- Majority appear indolent; however, an aggressive disease with metastasis and death could occur
  - Treatment with ALK inhibitors has shown promise in individual cases with metastases
- Has been reported in pediatric African American patients with sickle cell trait, typically occurring in the renal medulla and exhibiting *VCL::ALK* and *TPM3::ALK* fusions

Light Microscopy

- Pediatric tumors have morphologic similarities to renal medullary and collecting duct carcinomas, with the neoplastic cells showing prominent vacuolization
- Adult tumors are usually cortical and show a heterogeneous morphology, including papillary, solid, tubular, tubulocystic, cribriform, trabecular, spindle cell, and signet-ring cells with eosinophilic cytoplasm of variable appearances (rhabdoid, vacuolated, pleomorphic giant cell, small cell (metanephric adenoma-like) and mucinous tubular and spindle RCC-like morphologies

Molecular Characteristics

- *ALK* gene rearrangement resulting in fusion with various partner genes leading to aberrant *ALK* activation and formation of oncogenic chimeric proteins
  - This can be detected by IHC, FISH, or sequencing
- Several *ALK* fusion partners have been identified, including *VCL, TPM3, EML4, STRN, HOOK1, PLEKHA7, CLIP1, KIF5B*, and *KIAA1217*

Renal Cell Carcinoma with Fibromyomatous Stroma

- Various names have been used to refer to this tumor, including mixed renal tumor with carcinomatous and fibroleiomyomatous components, RCC associated with prominent angioleiomyoma-like proliferation, and clear cell RCC with smooth muscle stroma, among others
- It has indolent behavior, although a few tumors with lymph node involvement have been reported in patients with TSC syndrome

Light Microscopy

- Tumors are composed of admixed epithelial and stromal (fibromuscular) components with variable extents; the epithelial component forms nodules of elongated, branching tubules and papillae that are lined by cells with abundant clear or slightly eosinophilic cytoplasm (Fig. 13.14)
- Diffuse positive reaction for keratin 7 is typical and is generally required for the diagnosis

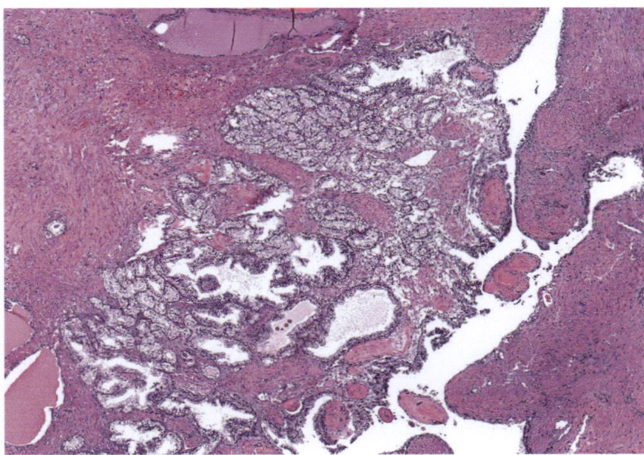

Fig. 13.14 Renal cell carcinoma with fibromyomatous stroma shows nests and glands of clear cells separated by fibromuscular stroma

- Also positive for high molecular weight keratin and CA9, like clear cell papillary renal cell tumor; however, CD10 is often positive, contrasting to clear cell papillary renal cell tumor

Molecular Characteristics

- Recurrent mutations involving the TSC/mTOR pathway
- A subset of tumors with similar morphology has shown mutations involving *ELOC* (elongin C [formerly known as TCEB1]), typically associated with monosomy 8
  - It is currently debated whether these should be grouped together despite different molecular pathways
  - Fewer *ELOC* mutated carcinomas have been reported so far
- Tumors lack alterations of 3p, contrasting with clear cell RCC

Eosinophilic Solid and Cystic Renal Cell Carcinoma

- Eosinophilic solid and cystic renal cell carcinoma is a recently described, mostly sporadic, tumor type that occurs preferentially in women
- Rare tumors with similar morphology were reported in patients with tuberous sclerosis complex
- The majority are indolent tumors, but rare metastases have been reported

Light Microscopy

- Morphologically, the tumors form macro- and microcystic spaces, with solid areas containing compact acinar or nested growth
  - The cells have abundant eosinophilic cytoplasm, with coarse cytoplasmic granules ('stippling'), round to oval nuclei, and hobnail-shaped cells (Fig. 13.15)
- Using immunohistochemistry, characteristic immunoreactivity for keratin 20 is present in most tumors, with negative or a lesser amount of keratin 7 reactivity
  - Cathepsin K has been found to be reactive in the majority of tumors as well

Molecular Characteristics

- Copy number gains included 16p13.3-16q23.1 (nearly whole chromosome), 7p21.2-7q36.2 (nearly whole chromosome), 13q14.2, and 19p12. Copy number losses included Xp11.21 and 22q11.23
- Recurrent and mutually exclusive, somatic biallelic mutations in the TSC genes, *TSC1* and *TSC2*, have been found by next-generation sequencing

Unclassified Renal Cell Carcinoma

- Unclassified renal cell carcinoma is a diagnostic category for tumors which fail to fit well into one of the known subtypes of renal cell carcinoma
- Features that may contribute to this inability to definitively classify a tumor include the following
  - A mixture of microscopic features of two or more distinct subtypes of renal neoplasm (mixed papillary and solid/tubular architecture or overlapping cytologic features of more than one subtype)
  - Oncocytic, granular, eosinophilic cytoplasm (which may be seen as a common endpoint in several renal cell carcinoma subtypes, as well as oncocytoma)
  - Spindle cell component (such as a sarcomatoid carcinoma without overtly identifiable features of a particular epithelial subtype)
  - Other unusual features in a tumor that appears to be of primary renal origin
- This morphologic heterogeneity is also reflected molecularly
  - A study of 62 high-grade unclassified renal cell carcinoma showed recurrent somatic mutations in 29 genes, including most commonly *NF2*, *SETD2*, *BAP1*, *KMT2C*, and *mTOR*
- Not surprisingly, molecular diagnostic studies may aid in resolving the differential diagnosis for such cases
  - Virtual karyotyping with single nucleotide polymorphism (SNP) microarrays has been proposed as a diagnostically practical method, often able to successfully categorize otherwise challenging cases

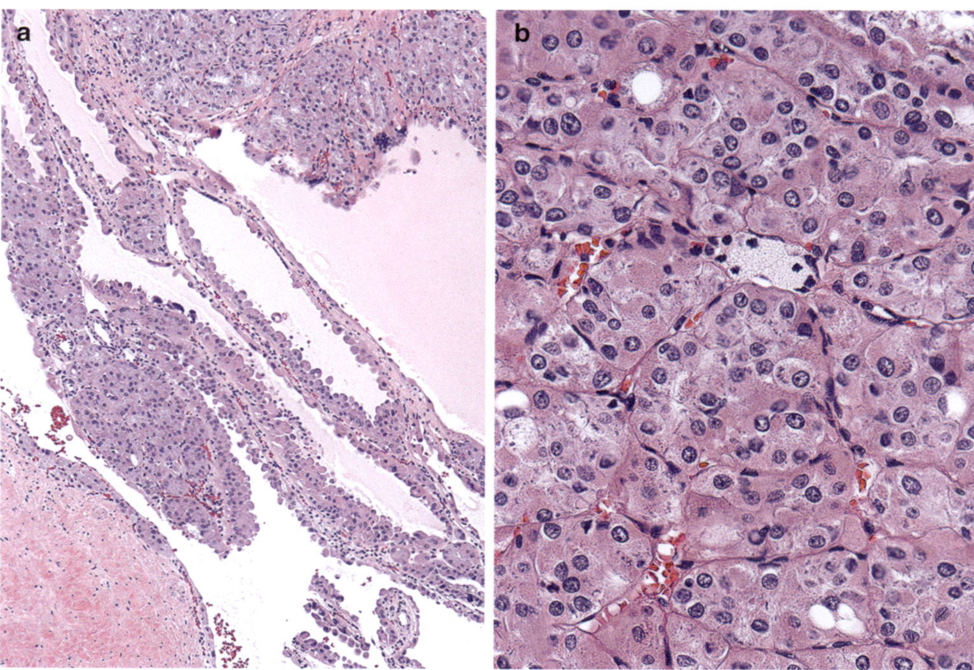

Fig. 13.15 Eosinophilic solid and cystic renal cell carcinoma is composed of eosinophilic cells lining cysts (**a**) and forming solid nests (**b**) with basophilic stippling of the cytoplasm

- Other methods, such as FISH, for the characteristic chromosomal abnormalities of common renal neoplasms may be helpful (such as chromosome 3p, 7, 17, Y, and others)

Sarcomatoid Renal Cell Carcinoma

- Sarcomatoid transformation in renal cell carcinoma represents an end stage of dedifferentiation and does not constitute a histologic subtype of its own accord
  - May arise from the clear cell, papillary, chromophobe, collecting duct carcinomas, unclassified renal cell carcinoma, etc
  - Estimated 5–10% of renal cell carcinomas, variable between histologic subtypes
- Characterized by highly aggressive/malignant behavior and poorer response to therapy

Molecular Characteristics

- Although some molecular alterations are preserved in the epithelial and sarcomatoid components (such as chromosome 3p or *VHL* abnormalities in clear cell carcinoma with a sarcomatoid component), other alterations appear to be unrelated to the original neoplasm
- Strong expression of clear cell carcinoma-specific markers using immunohistochemistry (such as HIF1A, CA9, and GLUT1) is often preserved in sarcomatoid tumors, compared to sarcomatoid carcinoma originating from other subtypes of renal cell carcinoma (non-clear cell),

which lack these markers (although VEGF expression may be present)
  - These markers may be of utility in establishing a clear cell carcinoma origin for challenging cases
- X chromosome inactivation studies generally reveal similar patterns of X chromosome inactivation in clear cell carcinoma and its sarcomatoid component, However, LOH patterns may have significant differences between the two components in the same patient, supporting the hypothesis that tumors arise from the same progenitor cell but undergo genetic divergence over their evolution
- Although typical chromophobe carcinoma (containing an epithelial component only) is characterized by multiple complex losses of Y, 1, 2, 6, 10, 13, 17, and 21, sarcomatoid tumors have been found to have frequent gains of several of these chromosomes, including 1, 2, 6, 10, and 17, by FISH
- Increased chromosomal imbalances with 1q and 8q gains, losses of 9q, 15q, 18p/q, and 22q compared to non-sarcomatoid RCC
- Other studies have found little or no similarity in the pattern of genetic alteration between the epithelial and sarcomatoid components
- E- to N-cadherin switching, dissociation of β-catenin from the cell membrane, and increased expression of Snail and secreted protein acidic and rich in cysteine (SPARC), were found supporting interpretation of sarcomatoid renal carcinoma as an example of epithelial–mesenchymal transition
- A study of 26 sarcomatoid RCCs using tumor microdissection from mixed parent histologies by targeted

sequencing showed that sarcomatoid RCC harbored frequent mutations in *TP53*, *VHL*, *CDKN2A*, and *NF2*

- Gain of mesenchymal characteristics is thought to be associated with an increased ability to migrate and metastasize
- Although a great deal of variability exists in the pattern of genetic abnormalities in sarcomatoid components, molecular studies may be helpful in a subset of cases for determining the histologic subtype of the original carcinoma

Other Tumors

Eosinophilic Vacuolated Tumor

- Eosinophilic vacuolated tumor was initially described as a "high-grade oncocytic tumor" (abbreviated as HOT) and "sporadic renal cell carcinoma with eosinophilic and vacuolated cytoplasm
- Thought initially to be sporadic; however, it was later also identified in rare patients with tuberous sclerosis complex
- It is typically detected incidentally in patients of a broad age range from 25 to 73 years and occurs more frequently in women
- All reported cases to date were found to have indolent behavior, without any evidence of local recurrence or metastatic disease

Light Microscopy

- Oncocytic tumor characterized by the finding of large intracytoplasmic vacuoles
- Typically has solid microscopic architecture, in some cases focally admixed with nested and tubulocystic areas
- Thick-walled vessels are always found at the periphery
- Entrapped tubules are common, particularly at the border with the normal renal parenchyma
- Cells often have large intracytoplasmic vacuoles with WHO/ISUP grade 3 nuclei
- Typically, is positive for KIT (CD117), CD10, PAX8, keratin AE1/AE3, antimitochondrial antigen-antibody, and cathepsin K in a great majority of tumors, albeit focally in some
- Keratin 7 expression is typically restricted only to scattered cells, usually not exceeding 5–10%
- Typically, negative for vimentin
- Fumarate hydratase and SDHB are retained/positive

Molecular Characteristics

- Loss of chromosome 1, 19p, and/or 19q, and loss of heterozygosity at 16p11 and 7q3133

- Complete losses or gains of other chromosomes have not been found
- TSC/mTOR (*MTOR*, *TSC2*, and *TSC1*) mutations (either germline or somatic leading to mTORC1 activation), seem to be the key molecular genetic findings in eosinophilic vacuolated tumors with low mutational rates
- Rare coexistent *RICTOR* missense mutation is also seen
- Molecular testing is not necessary for a great majority of tumors, as they can be distinguished from their mimickers primarily based on their characteristic morphologic and immunohistochemical features
- In cases with overlapping morphology or in cases where morphology is not convincing, analysis of mTOR pathway genes would be useful to establish the diagnosis

Epithelioid Angiomyolipoma and Other Renal PEComas

- In addition to angiomyolipoma, other members of the PEComa family of tumors affecting the kidney include epithelioid angiomyolipoma and several other variants, such as:
  - Microscopic angiomyolipoma and intraglomerular angiomyolipoma
  - Angiomyolipoma with epithelial cysts
  - Oncocytoma-like angiomyolipoma
  - Lymphangioleiomyomatosis of the renal sinus
- PEComas are considered to exhibit the phenotype of the perivascular epithelioid cell (PEC), a unique cell type without a known normal counterpart
  - Notable for coexpression of markers of myogenic and melanocytic differentiation, suggesting a neural crest origin or acquisition of melanocytic expression translocation or mutational event
- Epithelioid angiomyolipoma and other PEComas may be seen sporadically and in the context of tuberous sclerosis (see section "Tuberous Sclerosis")

Molecular Characteristics

- Loss of heterozygosity of the *TSC2* gene has been reported in some sporadic epithelioid angiomyolipomas, with variable frequency
- Sporadic renal angiomyolipomas and PEComa of other organs show activation of the PI3K/AKT/mTOR pathway, with the expression of phospho-S6 kinase and phospho-S6
- Other genetic abnormalities besides disruption of the *TSC* genes may be involved in the pathogenesis of PEComas
- A subset of pure epithelioid angiomyolipomas/PEComas harboring TFE3 translocation, previously referred to as "melanotic Xp11 translocation cancers", with *SFPQ*

being the most commonly reported partner gene, has been described

- These tumors are now regarded as a unique type of PEComas, known as "*TFE3*-rearranged/Xp11 translocation PEComas"
- They usually affect young patients who do not have TSC
  - Immunohistochemically, they have minimal to no immunoreactivity for muscle markers, and genetic analysis reveals lacking *TSC1* or *TSC2* gene alterations
  - Despite the morphologic overlap with translocation renal cell carcinoma, *TFE3*-rearranged PEComas are negative for PAX8

- Losses of chromosomes 1p, 17p, 18p, and 19 and gains of 2q, 3q, 5q, 12q, and X have been identified in renal angiomyolipoma with a similar distribution of abnormalities in PEComa

Adult Nephroblastoma (Wilms Tumor)

- Occurrence of nephroblastoma or Wilms tumor in adult patients is unusual (approximately 3% of cases), showing similar light microscopic features to those of pediatric patients
  - Recent studies using high-resolution genomic analysis revealed more pronounced genetic complexity than seen in pediatric cases, suggesting its distinct biological status compared to pediatric tumors
  - Uniparental disomies of most chromosomes, microdeletions of genes involved in tumor formation (*LRP1B*, *FHIT*, and *WWOX*), and organogenesis (*NEGR1* and *ZFPM2*)
  - In contrast, allelic loss patterns have revealed similar abnormalities by restriction fragment length polymorphism in both adult and pediatric patients

Neuroendocrine Tumors/Carcinomas

- Well-differentiated neuroendocrine tumors and neuroendocrine carcinomas (small cell carcinoma) of the kidney are uncommon neoplasms with similar morphologic features to neuroendocrine tumors seen in other organs
- Histogenesis of neuroendocrine tumors in the kidney is not completely clear; hypotheses have included the following:
  - Neuroendocrine differentiation of a primitive totipotential cell line
  - Metastasis from an occult primary tumor elsewhere
  - Misplaced progenitor cells or teratomatous cells
    - Well-differentiated neuroendocrine tumor has been reported in association with horseshoe kidney or renal teratoma

- Tumors have been found to generally lack reactivity for PAX2 and PAX8 by immunohistochemistry, in contrast to other tumors of renal origin
- In small cell carcinoma, frequent association with other urothelial carcinoma components supports a multipotent urothelial stem cell as a potential origin rather than intrinsic urinary tract neuroendocrine cells

Molecular Characteristics

- Relatively few studies have investigated renal neuroendocrine tumors at the molecular genetic level
  - The most frequently mutated genes are *CDH1* and *TET2*, followed by *AKT3, ROS1, PIK3R2, BCR* and *MYC*
  - FISHstudy demonstrated complex chromosomal abnormalities indicative of a high degree of chromosome instability with gain of multiple chromosomes, loss of the short arm of chromosome 3, numerical/structural aberrations of chromosome 13, loss of *TP53*, and amplification of *MYC* gene in renal small cell carcinoma
  - FISH for translocation involving the *EWSR1* gene has been suggested as a helpful marker in distinguishing renal carcinoid tumors from Ewing family of tumor/primitive neuroectodermal tumor
  - Studies have shown that well-differentiated neuroendocrine tumors of the kidney lack *ATRX* and *DAXX* (death-domain associated protein X) mutations unlike the gastroenteropancreatic and pulmonary neuroendocrine tumors

Primitive Neuroectodermal Tumor/Ewing Family of Tumors

- Primitive neuroectodermal tumors may sometimes arise primarily within the kidney (peripheral primitive neuroectodermal tumor or Ewing family tumors)
  - The majority show t(11;22)(q24;q12) with a fusion transcript between the *EWS* gene (22q12) and the *ETS*-related oncogene, *FLI1* (11q24) in 85–95% of cases, depending on the method
  - Variant translocations of *EWS* with other *ETS* -related oncogenes (*ERG* at 21q22), (*E1AF* at 7p22), (*FEV* at 2q33), and (17q12) have been found
  - Differential diagnosis can be challenging, including the following:
    - Intrarenal neuroblastoma
    - Carcinoid tumor
    - Desmoplastic small round cell tumor
    - Embryonal rhabdomyosarcoma
    - Non-*EWSR1* rearranged round cell sarcomas

- o Solitary fibrous tumor with pure round cell morphology
- o Neuroendocrine carcinoma/small cell carcinoma
- o Hematolymphoid neoplasms
- o Wilms tumor (particularly blastemal predominant tumors)
- Although immunohistochemical staining has largely been considered useful in resolving difficult cases (positive for CD99, FLI1, and negative for WT1), molecular methods are recommended in many cases, as considerable overlap may be present
- Reverse transcriptase polymerase chain reaction (RT-PCR) for the *EWSR1::FLI1* fusion transcript and/or break-apart FISH for *EWSR1* may be used to confirm the diagnosis

Urothelial Carcinoma

- Urothelial carcinoma in the upper urinary tract largely shows similar histologic features to tumors arising in the bladder
- In some cases, urothelial carcinoma may extensively infiltrate the kidney, raising a challenging clinicopathologic differential diagnosis with other malignancies, such as high-grade papillary renal cell carcinoma, renal medullary carcinoma, collecting duct carcinoma, and metastatic carcinoma to the kidney

Molecular Characteristics

- Urothelial carcinoma frequently exhibits gains of chromosomes 3, 7, and 17 and loss of 9p21
  - FISH probes for abnormalities of these chromosomes (UroVysion) may support a diagnosis of urothelial carcinoma
  - However, other tumors involving the urinary tract may also show positivity with the probe set, including:
    - o Primary bladder tumors: squamous cell carcinoma, adenocarcinoma, and urothelial carcinoma with squamous differentiation
    - o Primary renal tumors: clear cell, papillary, chromophobe, and sarcomatoid renal cell carcinomas
    - o Secondary tumors: adenocarcinoma of colonic, prostatic, and cervical origin
- Mutation of the *FGFR3* gene in superficial papillary neoplasms, associated with frequent recurrence and less frequent progression to invasion
- Mutation of *TP53* in high-grade, invasive urothelial carcinomas
- *TERT* promoter mutations are common in urothelial carcinoma in general

- Mutation of mismatch repair genes, including *MLH1, MSH2, MSH6,* and *PMS2*

  - In the setting of hereditary nonpolyposis colorectal cancer syndrome (Lynch syndrome), patients are predisposed to various tumors other than colorectal neoplasms—Other sites of involvement include endometrium, ovary, small bowel, stomach, hepatobiliary, skin, brain, and urinary tract (particularly the upper urinary tract)
  - Defective DNA mismatch repair leads to a more rapid accumulation of errors in microsatellite regions

Molecular Differential Diagnosis

- FH-deficient RCC
  - May exhibit destructive/infiltrative pattern, although usually admixed with more typical renal cell elements
  - Abnormal negative FH immunohistochemistry or *FH* mutation supports the diagnosis
- Metastasis from another cancer
  - Metastases to the kidney may be solitary/unilateral and extend into the renal pelvis, mimicking urothelial carcinoma
  - Immunohistochemical markers specific to organ sites (e.g., TTF1 for lung) may aid in confirming the site of origin
  - Molecular alterations of the non-renal cancers may be tested in particularly challenging tumors
- Medullary carcinoma
  - Loss of expression of INI1 (SMARCB1) by immunohistochemistry is seen in renal medullary carcinoma, a feature that may be of diagnostic utility in differentiating tumors from urothelial carcinoma
- Collecting duct carcinoma
  - Diagnosis of exclusion, once considerations such as urothelial carcinoma, metastasis from another cancer, medullary carcinoma, and FH-deficient carcinoma are excluded
- Papillary renal cell carcinoma
  - Uncommonly, papillary renal cell carcinoma may show a high-grade, infiltrative tubular growth pattern with areas that may mimic urothelial carcinoma, particularly in the setting of biopsy specimens
    - o FH-deficient cancer should be considered with this pattern
  - Presence of the characteristic numerical abnormalities of chromosomes 7, 17, or Y may be used to support a diagnosis of papillary carcinoma; however, these chromosome alterations are not entirely specific

Unknown/Low Malignant Potential Neoplasms

Multilocular Cystic Renal Neoplasm, Low Malignant Potential

- Multilocular cystic renal cell carcinoma is a unique variant of renal cell carcinoma with an excellent prognosis, characterized histologically by multiple cystic spaces lined by clear cells, generally with low nuclear grade and small aggregates of clear cells within the fibrous septa
  - Aggregates of clear cells do not expand the septa or otherwise form a significant solid component, differentiating the lesion from clear cell carcinoma with a cystic component
- Although molecular studies in this uncommon variant are limited in number, the presence of chromosome 3p deletion in a significant number of tumors (74%) supports a molecular relationship to clear cell RCC
  - *VHL* gene mutation has been identified in approximately 25% of tumors, somewhat lower than the rates found in clear cell carcinoma in general, perhaps due to difficulty in obtaining cellular areas for analysis
  - Similar expressions of PAX2, GSK3β, PTEN, and CA9 are identifiable by immunohistochemistry, supporting a similar pathogenesis to that of clear cell carcinoma
    - However, strong nuclear expression of p27 is preserved in multilocular cystic renal cell carcinoma, suggesting an area of distinction from clear cell carcinoma

Clear Cell Papillary Renal Cell Tumor

- Originally described as clear cell papillary renal cell carcinoma in 2006, the 2022 WHO Classification has proposed reclassification of this neoplasm as a "tumor," rather than carcinoma, due to favorable behavior
- Morphology is characterized by tubular/ductular, cystic, and branched papillary architectures, composed of cells with clear cytologic features and low nuclear grade
- Tumors were originally described in the setting of end-stage renal disease and acquired cystic kidney disease, but occur more often in kidneys unaffected by these abnormalities
- Using immunohistochemistry, tumors express keratin 7 and CA9 (with negative staining for AMACR and CD10), a phenotype that overlaps between clear cell renal cell carcinoma and papillary renal cell carcinoma

Molecular Characteristics

- Chromosomal alterations
  - Lack of abnormalities of 3p or the *VHL* gene (including promoter hypermethylation), supporting distinction from clear cell carcinoma
    - However, coexpression of CA9, HIF1A, and GLUT1 suggests upregulation of the HIF pathway by a non–VHL-dependent mechanism
  - Monosomy for chromosome 3 and rare 3p losses have been reported
  - Gain of chromosomes 3, 7, 10, 12, 17, 18, loss of chromosomes 1q, 18p, 9, 19p, or complex abnormalities have been reported; however, contrasting to papillary renal cell carcinoma, trisomy of 7 and 17 is infrequent rather than common
  - A minority of cases show low copy number gains of chromosome 7 and/or 17, in contrast to the more frequent gains in papillary carcinoma
  - The genomic profile of renal cell carcinoma in end-stage renal disease, including clear cell papillary tumor, by analysis of genomic copy number aberrations, reveal similar genomic profiles to papillary renal cell carcinoma
- Few reports of recurrent *MET* mutations by next-generation sequencing

Molecular Differential Diagnosis

- Clear cell renal cell carcinoma
  - Perhaps the most likely entity to be considered in the differential diagnosis, due to the prominent clear cell cytology and areas of compact tubular/ductular growth, resembling and sometimes nearly identical to the solid areas of clear cell carcinoma
  - Absence of 3p or *VHL* gene abnormalities by molecular methods may be helpful in supporting the diagnosis of clear cell papillary tumors for challenging cases, combined with the coexpression of keratin 7 and CA9 by immunohistochemistry
  - Other immunohistochemical markers include frequent positivity for GATA3 and high molecular weight keratin, suggesting a distal nephron phenotype
- Papillary renal cell carcinoma
  - Papillary renal cell carcinomas may have areas of clear cell change that may raise the possibility of clear cell papillary tumor, particularly in biopsy specimens
  - Although a specific genetic alteration to differentiate the two lesions is lacking, papillary renal cell carcinomas more frequently have gains of chromosomes 7 and

17, compared to clear cell papillary tumors, in which low copy number gains of these chromosomes are reported in a minority of cases

- Multilocular cystic renal cell neoplasm of low malignant potential
    - Clear cell papillary tumors may also have extensive cystic growth; however, a solid component with branched glands favors interpretation as clear cell papillary tumor
    - Similar to clear cell renal cell carcinoma, in general, 3p abnormalities by FISH can be detected in the majority of cases of multilocular cystic renal cell carcinoma, while they have not been demonstrated in clear cell papillary tumor

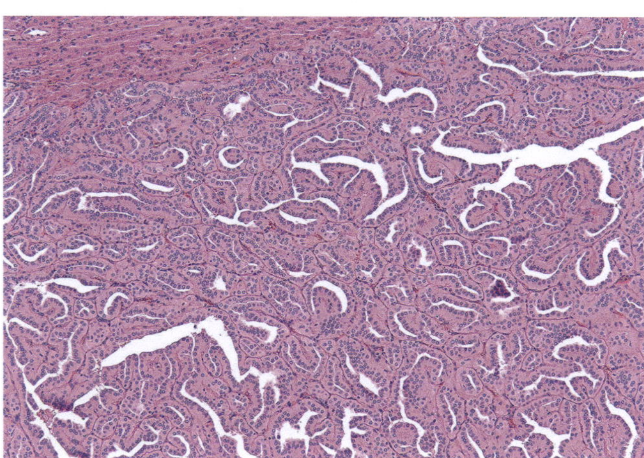

Fig. 13.16 Papillary renal neoplasm with reverse polarity shows papillary and tubular architecture with eosinophilic cytoplasm and apical localization of the nuclei

Papillary Renal Neoplasm with Reverse Polarity

- Various names have been used to refer to this tumor, including papillary renal cell carcinoma with oncocytic cells and nonoverlapping low-grade nuclei and papillary renal cell carcinoma, type 4/oncocytic low-grade
- Uniformly low-grade, indolent tumor
- Typically, small (mean, 1.6 cm) and may have a surrounding capsule

Light Microscopy

- Tumors are predominantly formed of thin arborizing papillary or tubulopapillary architectures
    - Some papillae may expand and become thicker and hyalinized
    - Tubular pattern may be predominant in some neoplasms (Fig. 13.16)
- The cells are cuboidal cells with eosinophilic finely granular cytoplasm, containing apically located nuclei opposite to the basement membrane
- Intracytoplasmic clear vacuoles, clear cell changes, and hobnail features can be seen
- No large prominent nucleoli (low WHO/ISUP grade), psammoma bodies, intracellular hemosiderin, necrosis, or mitotic figures are present
- Characteristically, these are positive for GATA3 and L1CAM and are negative for vimentin and, to a lesser extent, α-methylacyl-CoA-racemase (AMACR/p504s) by immunohistochemistry

Molecular Characteristics

- Characteristic recurrent *KRAS* mutations, identified in lesions as small as 0.6 mm
- Gene set enrichment analysis (GSEA) analysis found an enrichment of 'Singh *KRAS* dependency signature' when

compared with normal kidney tissue or the papillary renal cell carcinomas

- Unsupervised mRNA clustering showed papillary renal neoplasms with reverse polarity to have a different expression profile compared with papillary renal cell carcinomas
- Methylation studies showed a different clustering pattern from CCRCC and PRCC

Low-Grade Oncocytic Tumor

- Describes a subset of eosinophilic renal tumors with "oncocytic" features with indolent clinical behavior

Light Microscopy

- Tumors have predominantly solid with tuboloreticular and trabeculated architectures, that are lined by eosinophilic cells with bland low-grade, round to oval nuclei that often show delicate perinuclear halos (Fig. 13.17)
- Areas of sharply delineated edematous stroma, containing loosely connected strands of tumor cells or individual cells
- Characteristically, the tumor shows diffuse reaction for keratin 7 and negative KIT (CD117), which contrasts with renal oncocytoma and chromophobe renal cell carcinoma

Molecular Characteristics

- Microarray-based comparative genomic hybridization showed deletions of 19p13, 19q13, and 1p36
- Genetic alterations in mTOR/TSC pathway with possible activation of *mTOR* gene were identified

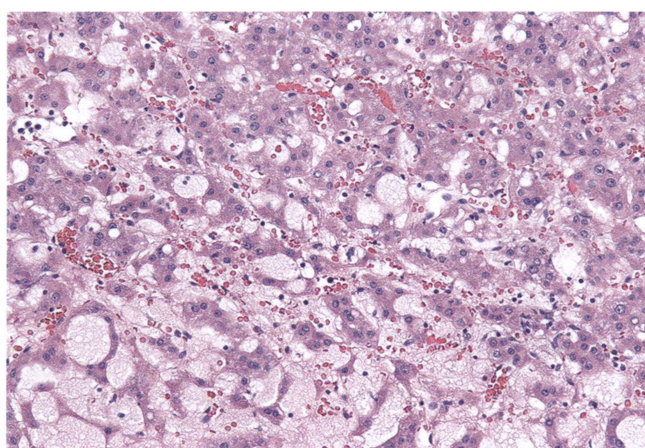

Fig. 13.17 The low-grade oncocytic tumor shows an oncocytoma-like tumor with eosinophilic cells loosely distributed in edema

- Low or null FOXI1 mRNA expression (in contrast to renal oncocytoma and ChRCC), distinct transcriptomic profiles and clustering attributes, frequent mutations involving the mTOR signaling pathway

Atrophic Kidney-Like Lesion

- The term "atrophic kidney-like tumor" was used in 2014, but was later changed to "atrophic kidney-like lesion"
- Additional tumors in the literature, likely representing examples of this lesion were reported as "thyroid-like follicular carcinoma"; however, it is now thought that there are differences between this entity and thyroid-like carcinoma
- All tumors have shown benign behavior, with a tendency to occur at younger age

Light Microscopy

- Presents as a circumscribed cortical nodule with a thick muscular capsule
- Formed of a variably sized follicular-like cysts, containing dense eosinophilic secretions and scattered psammomatous or coarse amorphic microcalcifications
- The lining cells are round to flattened, focally showing hobnail-shaped cells, and detached cells within the lumens
- Intraluminal glomerular-like tufts, suggesting that these represent cystically dilated glomeruli
- The tissue between the cysts contains atrophic tubules and small collapsed glomeruli (Fig. 13.18)
- These features suggest a localized/segmental form of nonneoplastic glomerulocystic and atrophic tubular change

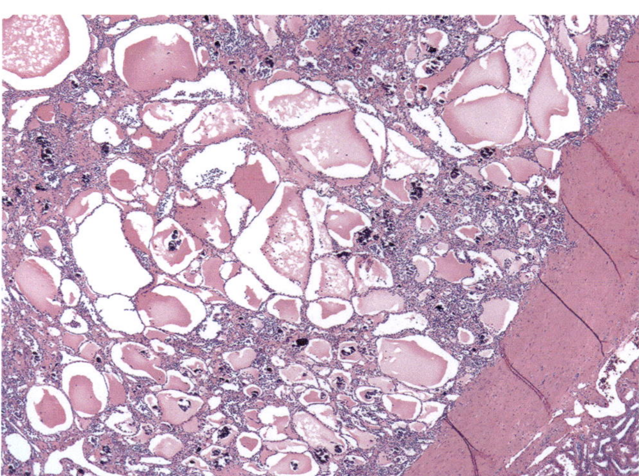

Fig. 13.18 The atrophic kidney-like lesion shows variably sized follicular-like cysts, containing dense eosinophilic secretions and scattered microcalcifications, surrounded by a thick fibromuscular capsule

Benign Neoplasms

Angiomyolipoma

- Relatively rare benign neoplasm composed of variable amounts of smooth muscle, adipose tissue, and blood vessels
- It belongs to the perivascular epithelioid cell tumors (PEComa) family of neoplasms, defined as "mesenchymal tumors composed of histologically, ultrastructurally, and immunohistochemically distinctive perivascular epithelioid cells"
- Represents a primary manifestation in 60–80% of patients with TSC who carry germline inactivation mutations involving *TSC1* or *TSC2* genes (*see also* section "Tuberous Sclerosis")
- Also found sporadically in approximately 1 in 300 individuals without TSC
  - Approximately 80% of patients with angiomyolipoma do not have TSC
  - Usually unifocal, compared to bilateral and multifocal involvement in TSC patients
- Similar to tuberous sclerosis syndrome, loss of *TSC2* has been identified in sporadic angiomyolipoma with variable frequency ranging from and thought to be more frequent than *TSC1* mutations
- Proposed to arise from renal mesenchymal precursor cells
  - RT-PCR of mRNA for gp100 (the antigenic target of HMB45) can be detected in low levels in proximal and distal tubules of the normal kidney
- Genetic alterations of tuberous sclerosis complex in *TSC1* or *TSC2* genes causes activation of mTOR pathway

- By immunohistochemical methods, tumors show activation of the mTOR pathway with an expression of phospho-S6 kinase and phospho-S6
- Tumors also characterized by positivity for both markers of smooth muscle and melanocytic origin
- mTOR activates cathepsin K expression, demonstrated by IHC in all lesions
- Constitutive activation of mTOR pathway leads to TFE3/TFEB overexpression with aberrant immunoreactivity for TFE3 in human kidney and angiomyolipoma cell lines with *TSC2* loss
- An overlapping phenotype between epithelioid angiomyolipoma and MiT family translocation carcinoma, TFEB in particular, exists
  - In contrast to MiT family translocation carcinoma, angiomyolipoma demonstrates negative PAX8 immunoreactivity and gene rearrangement studies
- A subset of pure epithelioid angiomyolipomas/PEComas harboring TFE3 translocation, with *SFPQ* being the most commonly reported partner gene, has been described
  - These tumors were initially referred to as "melanotic Xp11 translocation cancers" but are now regarded as a unique type of PEComas, known as "*TFE3*-rearranged/Xp11 translocation PEComas"
  - They usually affect young patients who do not have TSC
    - Immunohistochemically, they have minimal to no immunoreactivity for muscle markers, and genetic analysis reveals lacking *TSC1* or *TSC2* gene alterations
    - Despite the morphologic overlap with translocation renal cell carcinoma, *TFE3*-rearranged PEComas are negative for PAX8
- Losses of chromosomes 1p, 17p, 18p, and 19 and gains of 2q, 3q, 5q, 12q, and X have been identified in renal angiomyolipoma, similar to the abnormalities seen in PEComa
- Implications of molecular genetic alterations on therapy
  - mTOR inhibitors may be used as a treatment of TSC-associated angiomyolipomas and angiomyolipomas in patients who are not candidates for surgical treatment
  - However, concern for the resumption of tumor growth after cessation of therapy has been raised

Oncocytoma

- Renal oncocytoma is a benign neoplasm, characterized by a nested or trabecular architecture, composed of cells with abundant, granular eosinophilic cytoplasm and round, generally uniform nuclei

- Tremendous overlap may exist with chromophobe renal cell carcinoma, particularly the eosinophilic variant, so it is unclear exactly where the cutoff should be drawn between these two tumor types
- Oncocytoma has demonstrated genetic alterations involving mitochondrial genome-encoded enzymes or control regions, resulting in respiration defects
  - They are also characterized by the dramatic accumulation of mitochondria due to increased production and impaired autophagy
- Two subsets have been identified based on molecular alterations
  - Aneuploid oncocytomas characterized by loss of chromosomes 1, Y, 14 q, 21, and X,
    - Many tumors have normal complements of chromosomes (Fig. 13.1g, h); however, abnormalities of chromosome 1 are the most common nondisomic finding in sporadic and familial cases
    - Similar findings have been demonstrated utilizing cytogenetics, FISH, CGH, and SNP-based oligoarray methods
    - Loss of a tumor suppressor gene residing on chromosome 1p has been proposed as an early genetic event in the development of oncocytoma
  - Diploid oncocytomas with rearrangement/translocation of 11q12–13: t(5;11) that demonstrated a breakpoint flanked by the markers D11S443/D11S146 and the *CCND1/BCL1* locus
    - Translocations have included the following:
      - t(5;11)(q35;q13)
      - t(9;11)(p23;q12)
      - t(9;11)(p23;q13)
    - Other chromosomal partners, including 1, 6, 7, and 8
    - FISH reveals close proximity of *CCND1* (*PRAD1*, *BCL1*) to the 11q13.3 breakpoint
    - In combination with cyclin D1 overexpression by immunohistochemistry, these findings suggest a role for cyclin D1 in oncocytomas with the 11q translocation
    - An abundance of mitochondria is a key feature in oncocytoma, imparting the granular/oncocytic cytoplasmic characteristics
      - 11q13 includes several genes for mitochondrial proteins, including *UCP2*, *UCP3*, *NDUFC2*, and *SDHD*
      - Mitochondrial protein 2D electrophoresis in oncocytoma reveals downregulation of NDUFS3 from complex I of the respiratory chain and upregulation of COX5A, COX5B, and ATP5H from complex IV and V

Molecular Differential Diagnosis

- Due to the prominent morphologic overlap with the eosinophilic variant of chromophobe carcinoma, many studies have been directed at the identification of markers that are useful in distinguishing the two tumors
  - LOH was found in chromosomes 1, 2, 6, 10, 13, 17, and 21 at frequencies of 90%, 90%, 96%, 86%, 85%, 90%, and 72%, respectively, in chromophobe carcinoma; although the loss of one may be found in oncocytoma, the others are typically lacking

Papillary Adenoma

- Papillary adenoma is a common benign tubular proliferation, often found incidentally in kidneys resected for other lesions or at autopsy
- Microscopically, papillary adenoma is an unencapsulated tumor of less than 15 mm with papillary or tubular architecture with or without calcification
- Four morphologic types have been described
  - Type A characterized by papillae and tubules covered by cells with scant cytoplasm, often with psammoma bodies, morphologically similar to papillary renal cell carcinoma
  - Type B consists of broad papillae with lymphocytes in the cores; this overlaps with what has been recently referred to as "distal tubular hyperplasia"
  - Type C exhibits cysts that are lined by columnar cells, macrophages, and psammoma bodies
  - Type D shows papillae lined by large eosinophilic cells with apically located nuclei and lacking nuclear stratification; this is now identified as "papillary renal neoplasm with reverse polarity" due to the shared histomolecular features
- The molecular relationship between papillary adenomas and papillary renal cell carcinomas is incompletely understood
  - Similar gains of chromosomes 7 and 17, and loss of Y are often present
  - Progression to papillary carcinoma through gains of additional chromosomes, such as 12, 16, and 20 has been postulated
  - However, FISH studies have demonstrated gains of chromosomes 12, 16, and 20 in small papillary adenomas
  - Frequencies of gains of chromosomes 7, 17, 16, 12, and 20, and loss of the Y chromosomes were similar in both papillary adenomas and papillary renal cell carcinomas

- As such, chromosomal alterations do not appear to be reliable for differentiating papillary adenoma from papillary carcinomas
- Type B/distal tubular hyperplasia lacks trisomy 7 or 17 or loss of Y chromosome

Cystic Nephroma

- Cystic nephroma is an uncommon benign cystic neoplasm with similar histologic features but different and distinct molecular alterations in the adult and pediatric age group

Adult Cystic Nephroma

- Microscopically, the lesion is characterized by variably sized cystic spaces, lined by flattened, cuboidal, or hobnail cells with a spindle cell stroma, sometimes imparting an ovarian stroma-like appearance
- Overlap in characteristics with MEST has led some authors to suggest the term "renal epithelial and stromal tumor (REST)" as a unifying name for the two lesions
- Unlike pediatric cystic nephroma, adult cyst nephroma lacks *DICER1* mutations
- Molecular genetic characteristics of cystic nephroma are not completely understood
  - Gene expression profiling has found a similar profile in cystic nephroma and mixed epithelial and stromal tumor (MEST) of the kidney, distinct from other renal neoplasms, including urothelial carcinoma, chromophobe carcinoma, oncocytoma, clear cell carcinoma, papillary carcinoma, and normal kidney tissue

Pediatric Cystic Nephroma

- Pediatric cystic nephroma exhibits similar morphologic features as adult cystic nephroma with multiple cysts separated by fibrous septa that lack immature nephroblastic cells
  - The tumor mainly occurs in children under 4 years old, with higher prevalence in males
- 86% of pediatric cases demonstrated somatic or germline *DICER1* alterations
  - *DICER1* encodes an RNA endonuclease (located at chromosome 14q31)
- Familial cases are associated with germline *DICER1* mutations
- Patients with germline mutation have also been found to develop the following:

- Pleuropulmonary blastoma (pleuropulmonary blastoma familial tumor and dysplasia syndrome)
- Ovarian sex cord-stromal tumors (including Sertoli–Leydig tumor)
- Wilms tumor
- Intraocular medulloepithelioma
- Medulloblastoma/PNET
- Germ cell tumor
- Rhabdomyosarcoma
- Multinodular goiter

Mixed Epithelial and Stromal Tumor

- MEST of the kidney is a biphasic neoplasm, composed of a spindle-shaped cell stromal component and variable epithelial component
- Studies investigating molecular genetic features are limited
  - Clonality investigation found the same pattern of non-random X chromosome inactivation in the epithelial and stromal components for most tumors, supporting the theory that both components are neoplastic and arise from a common origin
  - Studies directed at differentiating MEST from congenital mesoblastic nephroma have found the two lesions to be distinct, with MEST lacking the typical genetic features of mesoblastic nephroma including (12;15)(p13;q25), resulting in *ETV6::NTRK3* gene fusion (cellular variant) and abnormalities of chromosomes 8, 11, and 17 by FISH
  - One study found t(1;19)(p22; p13.1) in a tumor from a male patient
  - Gene expression profiling studies showed:
    - A similar gene expression profile to cystic nephroma, distinct from urothelial carcinoma, chromophobe renal cell carcinoma, oncocytoma, clear cell renal cell carcinoma, papillary renal cell carcinoma, and normal kidney tissue
    - Insulin-like growth factor 2 (*IGF2*) showed the greatest degree of differential expression in MEST compared to normal kidney tissue and other renal neoplasms (32-fold higher)
    - Carbonic anhydrase 2 (*CA2*) showed 16-fold lower expression in MEST
  - A case of malignant MEST with rhabdoid features has been described, lacking the *SS18::SSX1* or *SS18::SSX2* fusion transcripts seen in synovial sarcoma
    - Molecular studies may be helpful in ruling out the diagnosis of synovial sarcoma, in which epithelial cysts may be embedded in the spindle cell stroma

Juxtaglomerular Cell Tumor

- Juxtaglomerular cell tumor is a rare renal neoplasm, thought to arise from the specialized smooth muscle of the juxtaglomerular apparatus
  - Tumors are associated with the production of renin and, therefore, uncontrolled hypertension and hypokalemia
  - Light microscopic features include sheets of polygonal or spindled cells, round nuclei, abundant eosinophilic/granular cytoplasm, and distinct cell borders; less commonly, epithelial channels and/or papillary architecture may be present
- In a study comparing the lesion with endocrine tumors of the pancreas, the juxtaglomerular cell tumors showed differential expression of several proteins by immunohistochemistry
  - Nuclear accumulation of cyclin D1, p21, and p27 was present, with the absence of cyclin D3, p53, p16(INK4a), and MDM2
  - BCL2 protein was strongly expressed, and RB was moderately expressed
- Multiple methodologies have revealed loss of chromosome 9 as a frequent event, as well as loss of chromosome 11 (or 11q)
- Other combinations of abnormalities have been identified as potential pathogenetic events, including the following:
  - Additional loss or monosomy of chromosomes X, 6, 15, and 21
  - Gain of chromosomes 3, 4, 10, 13, 17, 18
  - Upregulation of genes: *BM1, KIT, PIP4K2A, TLX1, TNIP3*

Metanephric Adenoma

- Metanephric adenoma is a rare neoplasm, with generally benign behavior
- Tumors are highly cellular, composed of tightly packed acini, and branching tubular structures, lined by cells with scant cytoplasm and small, uniform nuclei (Fig. 13.19)
  - Psammoma bodies and papillary structures may be present, leading to significant resemblance to papillary renal cell carcinoma (particularly the solid pattern)

Molecular Characteristics

- Analysis by a variety of molecular genetic methods has demonstrated normal karyotypes and diploid histograms by flow cytometric DNA content analysis in many tumors; however, other studies have demonstrated the following:

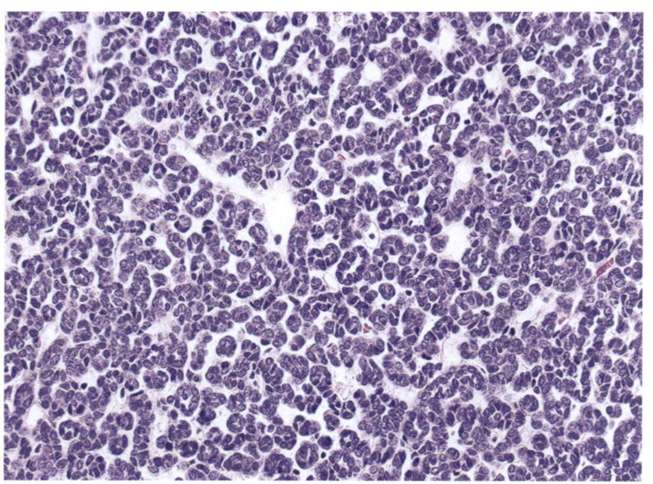

Fig. 13.19 A metanephric adenoma is composed of small tubules lined by cytologically bland cells with small nuclei and scant cytoplasm

- Frequent gains of chromosome 19 in a study of nine tumors (19p more frequently than 19q)
- Other cases in the study revealed a mixture of multiple chromosomal imbalances, normal karyotypes, or abnormalities of 11q
- Deletion of chromosome 2p, with alteration of 2p13, suggesting the site of a tumor suppressor gene
- Allelic imbalances were detected for chromosomes 2p, 7, 8p, 12q, 16q, and 20
- Balanced pericentric inversion involving the short and long arms of chromosome 9, inv(9)(p12q13)
- Balanced translocation t(9;15)(p24;q24) and balanced paracentric inversion inv(12)(q13q15)
- Dual t(1;22)(q22;q13) and t(15;16)(q21;p13) translocations
- Mutational studies identified *BRAF* V600E gene mutation in 80–90% of all metanephric adenomas
  - Detection of *BRAF* VE600E is aided by BRAF immunostaining with high sensitivity

Molecular Differential Diagnosis

- Papillary renal cell carcinoma
  - *BRAF* mutations and expression are absent
  - Chromosome panels specific for papillary renal cell carcinoma (7, 17, and Y) may be helpful in discrimination from metanephric adenoma
- ALK rearrangement-associated renal cell carcinoma (see section "ALK Rearrangement-associated Renal Cell Carcinoma")
- Wilms tumor
  - Common chromosomal abnormalities seen in Wilms tumor have been for the most part absent in metanephric adenoma (such as gain of 1q, 7q, and 12 and loss of 11p and 16q)

Renomedullary Interstitial Cell Tumor (Medullary Fibroma)

- Renomedullary interstitial cell tumor (formerly known as medullary fibroma) is a frequent incidental finding at autopsy or examination of the kidney for other reasons, generally comprising a small (1–5 mm) medullary nodule
- Light microscopic features include small stellate to polygonal cells in a background of loose stromal material, sometimes with deposits of amyloid
- Ultrastructural features more in keeping with medullary interstitial cells rather than fibroblasts led to the proposal of the designation of renomedullary interstitial cell tumor rather than medullary fibroma
- Little is known regarding the molecular genetic characteristics of these tumors

Pediatric Neoplasms

General Molecular Characteristics

- Pediatric renal tumors are diverse, affecting different age groups and with different biological origins and clinical behaviors
  - Widening use of molecular techniques, including cytogenetics and sequencing, has led to appropriate diagnosis, classification, and treatment for these tumors (Table 13.3)
- Pediatric cystic nephromas are associated with germline or somatic mutations in *DICER1*; they are distinct from cystic partially differentiated nephroblastoma and adult cystic nephromas
- Metanephric adenomas, adenofibromas, and stromal tumors are related and carry somatic BRAF mutations
- In addition to *NTRK3* fusions, congenital mesoblastic nephromas contain a variety of fusions or alterations in receptor tyrosine kinases or downstream effector molecules
- Alterations in clear cell sarcoma of the kidney include *YWHAE::NUTM2B* fusions, *BCOR* internal tandem duplications, and other rare alterations of BCOR, linking them with other *BCOR*-altered tumors
- Rhabdoid tumors of the kidney are characterized by alterations in *SMARCB1*, causing loss of INI1 expression and downstream effects on numerous critical cell proliferation pathways

Table 13.3 Summarizes the salient clinicopathologic features, key molecular abnormalities, and recommended molecular diagnostic assays in pediatric renal tumors

Tumors	Salient clinicopathologic features	Key molecular abnormalities	Useful diagnostic assay(s)
Pediatric cystic nephroma	– Molecularly distinct from cystic partially differentiated nephroblastoma and adult cystic nephroma – *DICER1* tumor predisposition syndrome (e.g., pleuropulmonary blastoma, ovarian Sertoli-Leydig cell tumors, thoracic or uterine cervical embryonal rhabdomyosarcoma, multinodular goiter) – Cystic nephroma is an indication for germline testing, often paired with somatic testing	– Somatic (usually missense) or germline (usually loss of function) *DICER1* mutations	– *DICER1* mutations are detected by next-generation sequencing (NGS), but exon-based sequencing may miss intronic mutations – Rare larger deletion events or loss of heterozygosity may require specific deletion/duplication technologies (microarray or multiplex ligation-dependent probe amplification) – Interpretation of germline status complicated by mosaicism
Metanephric tumors	– The established molecular link between metanephric adenoma, metanephric adenofibroma, and metanephric stromal tumor, which in most cases distinguishes them from nephroblastic lesions – Immunohistochemistry for BRAF V600E may be helpful in difficult cases and offers a rapid molecular correlation, but will be negative in cases with other BRAF mutations	– *BRAF* V600E mutations, and rarely other codon V600 mutations	– Testing for *BRAF* codon V600 mutations available by multiple modalities, including real-time polymerase chain reaction, NGS, and traditional sequencing technologies
Congenital mesoblastic nephroma	– Pan-TRK immunohistochemistry may allow rapid characterization of *NTRK* fusion–positive cases – Identification of *NTRK* or other targetable mutations may allow the use of inhibitor therapies in advanced-stage patients	– *ETV6::NTRK3* fusion is present in most cellular and a subset of mixed congenital mesoblastic nephromas – A subset of cellular and mixed cases has a growing set of alternative fusions or alterations in *NTRK* genes, *RET,* and *BRAF,* among others – Most classic and fusion-negative cellular and mixed cases have *EGFR* internal tandem duplications	– Fusions can be detected by karyotype, fluorescence in situ hybridization, reverse transcriptase polymerase chain reaction, or RNA-based NGS technologies – If negative, consider NGS covering *BRAF* and *EGFR*
Clear cell sarcoma of the kidney	– Common molecular alterations and morphology links clear cell sarcoma of the kidney with undifferentiated round cell sarcomas, primitive myxoid mesenchymal tumor of infancy, and other tumors	– Most tumors have an internal tandem duplication of the *BCOR* gene – Minority of tumors have a *YWHAE::NUTM2B* fusion; rare tumors have *BCOR::CCNB3* fusions	– Fusions can be detected by karyotype, fluorescence in situ hybridization, reverse transcriptase polymerase chain reaction, or RNA-based NGS – *BCOR* internal tandem duplications may be detected by polymerase chain reaction-based fragment analysis or by DNA or RNA-based NGS technologies but may require specific analysis pipeline considerations
Rhabdoid tumor of the kidney	– Morphologic and molecular overlap with rhabdoid tumors of other sites – Germline mutations are associated with rhabdoid tumor predisposition syndrome – Loss of SMARCB1/INI1 immunohistochemistry nuclear staining is characteristic of rhabdoid tumors – Potential role for *SMARCB1/INI1* in epigenetic regulation of other genes is a target for new therapeutic strategies	– Somatic or germline *SMARCB1* mutations – Overall, very low tumor mutational burden	– Tumor diagnosis is usually confirmed by immunohistochemistry, without the need for molecular testing – Identification of germline mutations may involve sequencing of tumor and germline specimens – Larger deletion events or loss of heterozygosity may require specific deletion/duplication technologies, like microarray or multiplex ligation-dependent probe amplification

(continued)

Table 13.3 (continued)

Tumors	Salient clinicopathologic features	Key molecular abnormalities	Useful diagnostic assay(s)
Translocation renal cell carcinoma	– Histology moderately correlates with fusion type, but extensive overlap with other translocation renal cell carcinomas and fusion-negative renal cell carcinomas	– *TFE3* fusions most common: Primary partners are *ASPSCR1* and *PRCC*; others described	– Most fusions can be detected by karyotype or fluorescence in situ hybridization, but uncommon *TFE3* fusion partners show cryptic inversions not detected by FISH (*NONO, RBM10, RBMX, GRIPAP1*)
	– Immunohistochemistry is available for TFE3 and TFEB but sensitivity and specificity are affected by tissue, antibody, and technical factors	– *TFEB* fusions with *MALAT1* are less common – *MITF* fusion is very rare, with only a few cases reported	– Consider NGS-based RNA testing of patients with morphologic features of translocation renal cell carcinomas, but with negative FISH

Nephroblastoma (Wilms Tumor)

- Nephroblastoma or Wilms tumor is the most common pediatric renal malignancy and a relatively common solid tumor of childhood, composed of a triphasic population of blastemal, tubular, and stromal components
- Believed to originate from the nephrogenic rests
- The majority of nephroblastoma cases are sporadic
  - 10–20% of patients with sporadic nephroblastoma have a heterozygous or homozygous mutation of *WT1*
    - Several inherited tumor syndromes confer an increased risk of development of nephroblastoma (although germline mutations are the source of only approximately 5–15% of cases), including:
      - Beckwith–Wiedemann syndrome
        - Associated with abnormality of 11p15.5 and abnormality of the *WT2* gene
        - Most children with the Beckwith–Wiedemann syndrome do not develop nephroblastoma; however, their risk is markedly increased compared to the general population
      - Wilms tumor, aniridia, genitourinary abnormalities, and mental retardation syndrome (WAGR)
        - Contiguous gene syndrome with larger deletions of 11p13 affecting adjacent genes, including *WT1* and *PAX6*
        - Risk for development of nephroblastoma is significant, though much lower in aniridia patients who do not have involvement of the *WT1* gene
      - Denys–Drash syndrome
        - Associated with nephropathy and gonadal dysgenesis
        - Approximately 90% risk for development of nephroblastoma, associated with a point mutation of *WT1* gene
      - Frasier syndrome
        - Development of nephroblastoma is uncommon

Molecular Characteristics

- Copy number alterations, including
  - Loss of genetic material of chromosome 11p, the location of the *WT1* and *WT2* genes
    - *WT1* on chromosome 11p13 encodes a transcription factor of the zinc finger family involved in the survival and differentiation of renal stem cells
  - Other abnormalities include gains of chromosomes 1q, 7q, and 12 and loss of 16q
    - Loss of heterozygosity for 1p and 16q has been associated with relapse
- Other genetic loci associated with the development of nephroblastoma have included the following:
  - *WT2* on chromosome 11p15.5
  - *WT3* on chromosome 16q
  - *WT4* (*FWT1*) on chromosome 17q12–q21
  - *WT5* on chromosome 7p15–p11.2
  - *FWT2* on chromosome 19q
  - *CTNNB1* (beta-catenin)
    - Located at 3p21, mutations are seen in approximately 15% of tumor
    - Involved in the Wnt signaling pathway
  - *WTX* (Wilms tumor on X, FAM123B)
    - Tumor suppressor gene located at Xq11.1
    - Involving the single X allele in male patients or the active X in female patients
  - TP53
    - Mutations of *TP53* (17p13.1) have been associated with the presence of unfavorable, (anaplastic) histology, metastasis, and relapse
  - Other genes, including *FBXW, BRCA2, HACE1*, and *GPC38*
- The diagnostic molecular assays recommended include immunohistochemistry, karyotyping, FISH, reverse transcriptase-polymerase chain reaction, or RNA-based NGS technologies

Clear Cell Sarcoma

- Clear cell sarcoma of the kidney is a rare pediatric renal malignancy, composed of epithelioid or spindle-shaped tumor cells, arranged in nests and cords
- A background of myxoid stromal material and fine, vesicular nuclear chromatin impart a clear appearance
- A propensity for bone metastases led to the original name "bone metastasizing renal tumor of childhood"
- Association with a tumor predisposition syndrome is not a characteristic feature

Molecular Characteristics

- Internal tandem duplication (ITD) of the *BCOR* gene in most tumors
- *YWHAE::NUTM2B* fusion in a minority (approximately 12%) of tumors (A recurrent translocation between 10q and 17p)
- *BCOR::CCNB3* fusions in rare tumors
  - Most fusion-negative cases were found to have a recurrent ITD in the *BCOR* gene
  - This ITD occurs at the end of the last exon of the gene and is always in the frame
  - This alteration seems to be specific for clear cell sarcoma of the kidney and is mutually exclusive with the *YWHAE::NUTM2B* fusion
  - The histologic and clinical characteristics of fusion-positive and ITD–positive cases are similar
- Other molecular genetic abnormalities, including the following:
  - t(10;17) with a breakpoint at the *TP53* locus on chromosome 17p13 However, most tumors have lacked abnormality of *TP53*, except for rare cases showing positivity by immunohistochemistry in the setting of anaplasia (similar to nephroblastoma), these translocations have included the following:
    - ○ t(10;17)(q11;p12)
    - ○ t(10;17)(q22;p13) recently reported to involve the *FAM22* and *YWHAE* genes
    - ○ t(10;17)(q22;p13), del(14) (q24.1q31.1)
  - A complex karyotype including deletion of 14q23, loss of chromosome 11p, t(2;22)(q21;q11), loss of imprinting for IGF2, gain of 1q, loss of 10q, loss of terminal 4p, loss of chromosome 19, and gain of 19p
  - A significant number of tumors has shown normal karyotypes or normal CGH profiles
- Common molecular alterations and morphology links clear cell sarcoma of the kidney with undifferentiated round cell sarcomas, primitive myxoid mesenchymal tumor of infancy, and other tumors

  - Gene expression profiling of the *BCOR* ITD cases shows marked overlap with *BCOR::CCNB3*–fused undifferentiated sarcomas
  - Rare *YWHAE::NUTM2B* fusions were also reported in undifferentiated round cell sarcomas, leading to additional work to clarify the relationship between undifferentiated round cell sarcomas and clear cell sarcoma of the kidney
  - Similar *BCOR* ITD mutations were found in patients with the clinicopathologically similar tumors of clear cell sarcoma of the kidney, undifferentiated round cell sarcoma, and primitive myxoid mesenchymal tumor of infancy
  - As may be expected based on these relationships, cases of clear cell sarcoma of the kidney with *BCOR::CCNB3* fusions have also been reported and may be seen in a slightly older patient population
  - Additional tumors that share these molecular features now include a subset of high-grade brain tumors and uterine sarcomas
- The diagnostic molecular assays recommended include karyotyping, fluorescent in situ hybridization, reverse transcriptase-polymerase chain reaction, or RNA-based NGS technologies
  - *BCOR* ITD may be detected by PCR-based fragment analysis or by DNA or RNA-based NGS technologies but may require specific analysis pipeline considerations
  - *BCOR* ITD is exonic and can be detected by both DNA-based and RNA-based NGS methods, although, because of its location at the end of the last exon, bioinformatics analysis may require additional steps
  - If sending a tumor for sequencing, the ability of the laboratory to detect these alterations should be verified
    - ○ RNA-based sequencing that can detect both the fusions and the *BCOR* ITD is the single most effective molecular test for the diagnosis
- The ITD can also be detected by techniques like capillary gel electrophoresis that analyze the size of the region

Rhabdoid Tumor

- Rhabdoid tumor of the kidney is a highly aggressive pediatric renal neoplasm, characterized by sheets of tumor cells that overrun the normal architecture of the kidney
- Vesicular chromatin, prominent nucleoli, and hyaline cytoplasmic inclusions are frequently present (Fig. 13.20)
- Similar tumors have been described in a variety of anatomic sites, designated "malignant extrarenal rhabdoid tumor,"

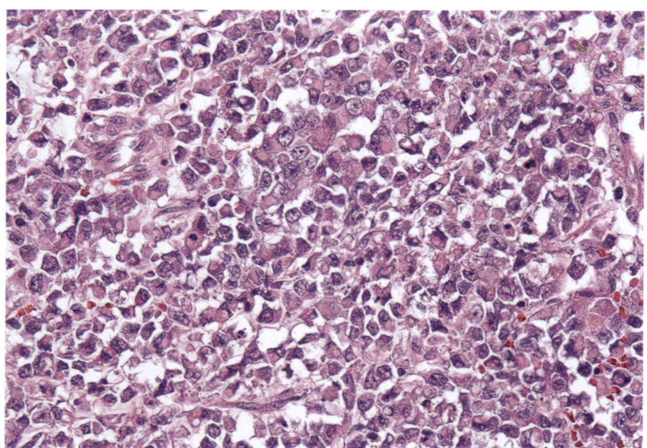

Fig. 13.20 Rhabdoid tumor is composed of malignant cells with prominent nucleoli and eccentric densely eosinophilic cytoplasm

"malignant rhabdoid tumor of soft tissue," or "atypical teratoid/rhabdoid tumor" (in the central nervous system)
- Tumors are associated with a very poor prognosis

Molecular Characteristics

- Deletions of the long arm of chromosome 22, the common overlapping regions were mapped to the *SMARCB1* tumor suppressor gene, previously known as *BAF47, hSNFS, SNF5,* and *INI1*
  - The SMARCB1/INI1 protein is part of the SWI/SNF chromatin remodeling complex, which has an active role in transcriptional activation and repression
  - Inactivation of SMARCB1/INI1 affects numerous critical cell pathways such as cyclin D1 regulated cell cycle progression, sonic hedgehog, WNT/beta-catenin, *RB*, and *MYC*
  - SMARCB1/INI1 and the SWI/SNF complex have a role in histone acetylation and, thus, transcriptional activation; the complex functions in opposition to the polycomb repressive complex 2, which includes EZH2 as its catalytic subunit and is involved in histone methylation and transcriptional repression
  - Inactivation of *SMARCB1* leads to overexpression of EZH2 and trimethylation of histone H3K27, causing transcriptional silencing of tumor suppressor genes
  - The participation of many pathways in the oncogenesis of rhabdoid tumors of the kidney and other sites has led to the investigation of a variety of targeted therapies, including EZH2 inhibitors and immune checkpoint inhibitors
- Loss-of-function mutations, resulting in biallelic loss or inactivation of the gene in both renal and extrarenal rhabdoid tumors

- Germline mutations were identified in a subset of cases, leading to the description of the rhabdoid tumor predisposition syndrome, type 1
  - The syndrome is associated with earlier presentation, bilateral primary renal tumors, and rhabdoid tumors at other sites
  - About one-third of patients have an underlying germline mutation, with a large deletion, monosomy of chromosome 22, or copy-neutral loss of heterozygosity as the second hit
  - The rhabdoid tumor predisposition syndrome, type 2 is associated with mutations in the *SMARCA4* gene, but no cases of rhabdoid tumor of the kidney with *SMARCA4* mutations have been reported to date
  - A subset of cases may have a reduction of INI1 protein expression due to epigenetic mechanisms
- Germline mutations of *SMARCB1* are present in a subset of patients, presenting at an earlier age of 6 months, associated with a worse prognosis
- Germline *SMARCB1* mutation has also been described to predispose to schwannomatosis and meningioma, though interestingly schwannoma and rhabdoid tumor rarely occur together, perhaps due to variable penetrance
  - Some patients within an affected family may also present with schwannomatosis, usually without rhabdoid tumors, depending on the specific familial mutation; regular tumor surveillance by MRI is recommended for affected patients
- Other kidney tumors may show rhabdoid features; however, loss of expression of INI1 protein by immunohistochemistry is a helpful diagnostic feature of rhabdoid tumor
  - Expression of INI1 is preserved in other tumors, such as nephroblastoma, mesoblastic nephroma, Ewing family of tumors, desmoplastic small round cell tumor, rhabdomyosarcoma, and renal cell carcinoma
- Rhabdoid tumors have one of the lowest tumor mutational burdens of all cancer types, suggesting that *SMARCB1* and its protein are the near-sole drivers of oncogenesis and tumor behavior in the following ways:
- Molecular genetic studies for the characteristic genetic abnormalities of other tumors may be helpful in challenging cases and include analysis of the tumor tissue for both mutations and deletions, followed by screening for the detected mutations in the blood, although, depending on the sequencing approach used, some mutations may not be detected
- Larger deletion events or loss of heterozygosity may require specific deletion/duplication technologies, such as microarray or multiplex ligation-dependent probe amplification

Congenital Mesoblastic Nephroma

- Congenital mesoblastic nephroma is the most common congenital renal neoplasm, generally occurring in the first year of life, though making up only 2–4% of pediatric renal tumors
- Two forms are recognized, with similar histopathologic and molecular genetic characteristics to infantile fibromatosis and infantile fibrosarcoma
  - Cellular congenital mesoblastic nephroma
    - Is remarkable for greater cellularity, decreased cytoplasmic volume, vesicular nuclear chromatin, and a pushing border
    - A high mitotic rate and areas of necrosis may be present
  - Classic congenital mesoblastic nephroma
    - Characterized by interlacing fascicles of fibroblastic cells with thin to fusiform nuclei, eosinophilic cytoplasm, collagen deposition, and low mitotic activity
    - Frequently, the tumor intermingles with the adjacent structures' renal parenchyma

Molecular Characteristics

- Cellular congenital mesoblastic nephroma have shown t(12;15) (p13;q25), associated with *ETV6::NTRK3* fusion
  - This fusion is also seen in infantile fibrosarcoma, supporting the morphologic similarity between the two lesions
  - This fusion is absent in classic congenital mesoblastic nephroma
- Aneuploidy of chromosomes 8, 11, and 17 is also frequently seen in cellular congenital mesoblastic nephroma
  - Classic congenital mesoblastic nephroma have diploid karyotyping
- A subset of cellular and mixed cases has a growing set of alternative fusions or alterations in *NTRK* genes, *RET* and *BRAF*, among others
  - Most classic, fusion-negative cellular and mixed cases have *EGFR* internal tandem duplications
  - Pan-TRK IHC may allow rapid characterization of *NTRK* fusion–positive cases
  - Identification of *NTRK* or other targetable mutations may allow the use of inhibitor therapies in advanced-stage patients
- The diagnostic molecular assays include karyotyping, FISH, RT-PCR, or RNA-based NGS technologies; if negative, sequencing covering *BRAF* and *EGFR* can be performed

Metanephric Tumors

- The family of metanephric tumors comprises metanephric adenoma, metanephric adenofibroma (previously known as nephrogenic adenofibroma), and metanephric stromal tumor
- These tumors are related by their overlapping histopathologic spectrum of features and by the presence of a *BRAF* mutation in most tumors, throughout the histologic spectrum
- Although metanephric adenomas are more common in adults, metanephric adenofibromas are seen in children and young adults and metanephric stromal tumors are primarily seen in younger children
- The light microscopic features include stellate/spindled tumor cells, with thin hyperchromatic nuclei
  - Tumors are believed to interact with various renal elements, resulting in "onion skin" concentric rings around blood vessels (angioplasia), entrapment of renal tubules, and sometimes juxtaglomerular cell hyperplasia or heterologous elements (glial/cartilaginous)

Molecular Characteristics

- Copy number alterations
  - Molecular studies of metanephric adenoma have shown chromosomal abnormalities, including complex gains of the long arm of chromosome 17 and amplification of the short arm of chromosome 19, although many cases showed no gains or losses
  - A recent case demonstrated partial triplication of the segment between bands 17q22 and 17q24.3 and duplication of the segment between bands 17q24.3 and 17q25.3 by cytogenetic analysis and FISH studies
- Genetic mutations
  - Molecular studies have demonstrated *BRAF* V600E mutations, with rare reports of other *BRAF* codon V600 mutations
    - Evaluation of the *BRAF* V600E mutation by immunohistochemistry is available, although *BRAF* codon V600 mutations other than the V600E, such as the V600D, are not detected by the antibody
  - Broader sequencing analysis of metanephric adenomas showed variable mutations in other neoplasia-associated genes, including NF1, NOTCH1, and SPEN, and rare fusions between BRAF and other genes
    - Some patients with neurofibromatosis also develop juxtaglomerular cell hyperplasia, renal artery aneurysms, and renovascular angiodysplasia, similar to the features seen in metanephric stromal tumors,

leading to the hypothesis that neurofibromatosis and possibly nephroblastoma are linked to metanephric stromal tumor

- The diagnostic molecular assays recommended include immunohistochemistry for BRAF V600E (helpful in difficult cases and offers a rapid molecular correlation, but will be negative in cases with other *BRAF* mutations); testing for *BRAF* codon V600 mutations available by multiple modalities, including real-time PCR, NGS, or traditional sequencing technologies

Neuroblastoma

- Rarely neuroblastoma may present as a true intrarenal mass, raising the differential diagnosis of other pediatric renal neoplasms
  - Invasion of the kidney by adjacent neuroblastoma, in contrast, is more common

Molecular Characteristics

- Amplification of *MYCN* in a subset of cases
- Other abnormalities include:
  - Amplification of *BIRC5, MDM2,* and *LIN28B*
  - *PHOX2B* Mutation, which can be tested by immunohistochemistry
  - *ALK* rearrangement
  - Alterations in PI3K/Akt/mTOR and RAS-MAPK pathways, as well as epigenetic regulators

Renal Cell Carcinoma Associated with Neuroblastoma

- Renal cell carcinoma arising in patients with long-term survival from neuroblastoma has been described
- Although therapy may play a role in the development of these tumors, occasional patients develop renal cell carcinoma without treatment or simultaneously with the neuroblastoma

Light Microscopy

- Tumors have shown a variety of morphologic features, including solid or papillary architecture, abundant eosinophilic cytoplasm, or alternatively features of clear cell renal cell carcinoma

Molecular Characteristics

- It has been recently recognized that a significant subset of these tumors likely represents MITF family translocation

renal cell carcinoma and can be diagnosed with similar techniques to other MITF renal cell carcinomas

Pediatric Cystic Nephroma

- Pediatric cystic nephromas are grossly multicystic, identical in appearance to cystic partially differentiated nephroblastomas, with thin septa and lacking solid areas
  - The cyst linings are smooth, and the cysts may contain clear fluid
  - The cysts are lined by flattened to the cuboidal or hobnailed epithelium, and the septa are composed of fibrous tissue and may have a layer of spindle cells
  - Mature tubular structures may be present in the septa, but no blastemal or primitive epithelial elements are seen

Molecular Characteristics

- Germline *DICER1* mutations
  - A high proportion of histologically defined pediatric cystic nephromas were shown to have *DICER1* mutations, whereas mutations were absent in patients with cystic partially differentiated nephroblastoma and adult cystic nephromas
  - 12% of patients with pleuropulmonary blastoma had or had a family member with cystic nephroma
  - Rare exceptions of *DICER1* mutated adult cystic nephromas most likely represent pediatric patients with cystic nephromas that were not recognized in childhood
  - *DICER1*-associated tumors have biallelic mutations in *DICER1*
    - They may carry an autosomal dominant germline plus a somatic mutation or may be entirely somatic
    - Most somatic mutations clustering in hot spots in the ribonuclease IIIB domain and germline mutations more commonly spread throughout the gene and result in loss of function
    - Patients with a *DICER1*-associated tumor in the absence of family history may benefit from testing of both the tumor and a germline specimen by sequencing and deletion/duplication testing
    - Pediatric cystic nephroma is considered a major indication for germline testing
- The diagnostic molecular assays recommended include the detection of *DICER1* mutations by NGS as exon-based sequencing may miss intronic mutations
- Rare larger deletion events or loss of heterozygosity may require specific deletion/duplication technologies (microarray or multiplex ligation-dependent probe amplification)

Further Reading

Abro B, Kaushal M, Chen L, et al. Tumor mutation burden, DNA mismatch repair status and checkpoint immunotherapy markers in primary and relapsed malignant rhabdoid tumors. Pathol Res Pract. 2019;215(6):152395.

Adeniran AJ, Shuch B, Humphrey PA. Hereditary renal cell carcinoma syndromes: clinical, pathologic, and genetic features. Am J Surg Pathol. 2015;39:e1–e18.

Alaghehbandan R, Michal M, Kuroda N, Hes O. Thyroid-like follicular carcinoma of the kidney: an emerging renal neoplasm with curiously misplaced histologic features. A case report. Int J Surg Pathol. 2017;25:379–80.

Alaghehbandan R, Trpkov K, Tretiakova M, et al. Comprehensive review of numerical chromosomal aberrations in chromophobe renal cell carcinoma including its variant morphologies. Adv Anat Pathol. 2021;28:8–20.

Albert CM, Davis JL, Federman N, et al. TRK fusion cancers in children: a clinical review and recommendations for screening. J Clin Oncol. 2019;37(6):513–24.

Al-Hussain TO, Cheng L, Zhang S, Epstein JI. Tubulocystic carcinoma of the kidney with poorly differentiated foci: a series of 3 cases with fluorescence in situ hybridization analysis. Hum Pathol. 2013;44:1406–11.

Ali EM, Elnashar AT. Adult Wilms' tumor: review of literature. J Oncol Pharm Pract. 2012;18(1):148–51.

Al-Obaidy KI, Bridge JA, Cheng L, et al. EWSR1-PATZ1 fusion renal cell carcinoma: a recurrent gene fusion characterizing thyroid-like follicular renal cell carcinoma. Mod Pathol. 2021;34:1921–34.

Al-Obaidy KI, Eble JN, Cheng L, et al. Papillary renal neoplasm with reverse polarity: a morphologic, immunohistochemical, and molecular study. Am J Surg Pathol. 2019;43:1099–111.

Al-Obaidy KI, Eble JN, Nassiri M, et al. Recurrent KRAS mutations in papillary renal neoplasm with reverse polarity. Mod Pathol. 2020;33:1157–64.

Al-Obaidy KI, Saleeb RM, Trpkov K, et al. Recurrent KRAS mutations are early events in the development of papillary renal neoplasm with reverse polarity. Mod Pathol. 2022;35(9):1279–86.

Amin MB, Gupta R, Ondrej H, et al. Primary thyroid-like follicular carcinoma of the kidney: report of 6 cases of a histologically distinctive adult renal epithelial neoplasm. Am J Surg Pathol. 2009;33:393–400.

Amin MB, MacLennan GT, Gupta R, et al. Tubulocystic carcinoma of the kidney: clinicopathologic analysis of 31 cases of a distinctive rare subtype of renal cell carcinoma. Am J Surg Pathol. 2009;33:384–92.

Anderson J, Gibson S, Sebire NJ. Expression of ETV6-NTRK in classical, cellular and mixed subtypes of congenital mesoblastic nephroma. Histopathology. 2006;48(6):748–53.

Antonescu CR, Dickson BC, Swanson D, et al. Spindle cell tumors with RET gene fusions exhibit a morphologic spectrum akin to tumors with 714 Treece NTRK gene fusions. Oncogene. 2020;39(6):1361–77.

Argani P, Antonescu CR, Couturier J, et al. PRCC-TFE3 renal carcinomas: morphologic, immunohistochemical, ultrastructural, and molecular analysis of an entity associated with the t(X;1)(p11.2;q21). Am J Surg Pathol. 2002;26:1553–66.

Argani P, Antonescu CR, Illei PB, et al. Primary renal neoplasms with the ASPL-TFE3 gene fusion of alveolar soft part sarcoma: a distinctive tumor entity previously included among renal cell carcinomas of children and adolescents. Am J Pathol. 2001;159:179–92.

Argani P, Aulmann S, Illei PB, et al. A distinctive subset of PEComas harbors TFE3 gene fusions. Am J Surg Pathol. 2010;34:1395–406.

Argani P, Aulmann S, Karanjawala Z, et al. Melanotic Xp11 translocation renal cancers: a distinctive neoplasm with overlapping features of PEComa, carcinoma, and melanoma. Am J Surg Pathol. 2009;33:609–19.

Argani P, Beckwith BJ. Metanephric stromal tumor: report of 31 cases of a distinctive pediatric renal neoplasm. Am J Surg Pathol. 2000;24(7):917–26.

Argani P, Fritsch M, Kadkol SS, et al. Detection of the ETV6-NTRK3 chimeric RNA of infantile fibrosarcoma/cellular congenital mesoblastic nephroma in paraffin-embedded tissue: application to challenging pediatric renal stromal tumors. Mod Pathol. 2000;13(1):29–36.

Argani P, Hicks J, De Marzo AM, et al. Xp11 translocation renal cell carcinoma (RCC): extended immunohistochemical profile emphasizing novel RCC markers. Am J Surg Pathol. 2010;34:1295–303.

Argani P, Kao YC, Zhang L, et al. Primary renal sarcomas with BCOR-CCNB3 gene fusion: a report of 2 cases showing histologic overlap with clear cell sarcoma of kidney, suggesting further link between BCOR-related sarcomas of the kidney and soft tissue. Am J Surg Pathol. 2017;41(12):1702–12.

Argani P, Lae M, Hutchinson B, et al. Renal carcinomas with the t(6;11)(p21;q12): clinicopathologic features and demonstration of the specific alphaTFEB gene fusion by immunohistochemistry, RTPCR, and DNA PCR. Am J Surg Pathol. 2005;29(2):230–40.

Argani P, Lee J, Netto GJ, et al. Frequent BRAF V600E mutations in metanephric stromal tumor. Am J Surg Pathol. 2016;40(5):719–22.

Argani P, Lui MY, Couturier J, et al. A novel CLTC-TFE3 gene fusion in pediatric renal adenocarcinoma with t(X;17)(p11.2;q23). Oncogene. 2003;22:5374–8.

Argani P, Olgac S, Tickoo SK, et al. Xp11 translocation renal cell carcinoma in adults: expanded clinical, pathologic, and genetic spectrum. Am J Surg Pathol. 2007;31:1149–60.

Argani P, Pawel B, Szabo S, et al. Diffuse strong BCOR immunoreactivity is a sensitive and specific marker for clear cell sarcoma of the kidney (CCSK) in pediatric renal neoplasia. Am J Surg Pathol. 2018;42(8):1128–31.

Argani P, Perlman EJ, Berslow NE, et al. Clear cell sarcoma of the kidney: a review of 351 cases from the National Wilms Tumor Study Group Pathology Center. Am J Surg Pathol. 2000;24(1):4–18.

Argani P, Reuter VE, Zhang L, et al. TFEB-amplified renal cell carcinomas: an aggressive molecular subset demonstrating variable melanocytic marker expression and morphologic heterogeneity. Am J Surg Pathol. 2016;40:1484–95.

Argani P, Zhang L, Reuter VE, Tickoo SK, Antonescu CR. RBM10-TFE3 renal cell carcinoma: a potential diagnostic pitfall due to cryptic Intrachromosomal Xp11.2 inversion resulting in false-negative TFE3 FISH. Am J Surg Pathol. 2017;41:655–62.

Arroyo MR, Green DM, Perlman EJ, et al. The spectrum of metanephric adenofibroma and related lesions. Am J Surg Pathol. 2001;25(4):433–44.

Arva NC, Bonadio J, Perlman E, et al. Diagnostic utility of PAX8, PASX2, and NGFR immunohistochemical expression in pediatric renal tumors. Appl Immunohistochem Mol Morphol. 2018;26(10):721–6.

Astolfi A, Fiore M, Melchionda F, et al. BCOR involvement in cancer. Epigenomics. 2019;11(7):835–55.

Astolfi A, Melchiona F, Perotti D, et al. Whole transcriptome sequencing identifies BCOR internal tandem duplication as a common feature of clear cell sarcoma of the kidney. Oncotarget. 2015;6(38):40934–9.

Bahubeshi A, Bal N, Frio TR. Germline DICER1 mutations and familial cystic nephroma. J Med Genet. 2010;47(12):863–6.

Bakouny Z, Sadagopan A, Ravi P, et al. Integrative clinical and molecular characterization of translocation renal cell carcinoma. Cell Rep. 2022;38:110190.

Barrisford GW, Singer EA, Rosner IL, Linehan WM, Bratslavsky G. Familial renal cancer: molecular genetics and surgical management. Int J Surg Oncol. 2011;2011:658767.

Biegel JA, Zhou JY, Rorke LB, et al. Germ-line and acquired mutations of INI1 in atypical teratoid and rhabdoid tumors. Cancer Res. 1999;59(1):74–9.

Biegel JA. Molecular genetics of atypical teratoid/rhabdoid tumors. Neurosurg Focus. 2006;20(1):E11.

Bolande RP. Congenital mesoblastic nephroma of infancy. Perspect Pediatr Pathol. 1973;1:227–50.

Boman F, Hill DA, Williams GM, et al. Familial association of pleuropulmonary blastoma with cystic nephroma and other renal tumors: a report from the international Pleuropulmonary Blastoma registry. J Pediatr. 2006;149:850–4.

Bratslavsky G, Gleicher S, Jacob JM, et al. Comprehensive genomic profiling of metastatic collecting duct carcinoma, renal medullary carcinoma, and clear cell renal cell carcinoma. Urol Oncol. 2021;39:367.e361–5.

Brunelli M, Eble JN, Zhang S, Martignoni G, Cheng L. Gains of chromosomes 7, 17, 12, 16, and 20 and loss of Y occur early in the evolution of papillary renal cell neoplasia: a fluorescent in situ hybridization study. Mod Pathol. 2003;16:1053–9.

Brunelli M, Gobbo S, Cossu-Rocca P, et al. Chromosomal gains in the sarcomatoid transformation of chromophobe renal cell carcinoma. Mod Pathol. 2007;20:303–9.

Bugert P, Von Knobloch R, Kovacs G. Duplication of two distinct regions on chromosome 5q in non-papillary renal-cell carcinomas. Int J Cancer. 1998;76:337–40.

Cairns P. Renal cell carcinoma. Cancer Biomark. 2010;9:461–73.

Cajaiba MM, Dyer L, Geller JI, et al. The classification of pediatric and young adult renal cell carcinomas registered on the children's oncology group (COG) protocol AREN03B2 after focused genetic testing. Cancer. 2018;124(16):3381–9.

Cajaiba MM, Khanna G, Smigh EA, et al. Pediatric cystic nephromas: distinctive features and frequent DICER1 mutations. Hum Pathol. 2016;48:81–7.

Cajaiba MM, Khanna G, Smith EA, et al. Pediatric cystic nephromas: distinctive features and frequent DICER1 mutations. Hum Pathol. 2016;48:81–7.

Calderaro J, Masliah-Planchon J, Richer W, et al. Balanced translocations disrupting SMARCB1 are Hallmark recurrent genetic alterations in renal medullary carcinomas. Eur Urol. 2016;69:1055–61.

Calderaro J, Moroch J, Pierron G, et al. SMARCB1/INI1 inactivation in renal medullary carcinoma. Histopathology. 2012;61:428–35.

Caliò A, Brunelli M, Segala D, et al. Comprehensive analysis of 34 MiT family translocation renal cell carcinomas and review of the literature: investigating prognostic markers and therapy targets. Pathology. 2020;52:297–309.

Calio A, Warfel KA, Eble JN. Papillary adenomas and other small epithelial tumors in the kidney: an autopsy study. Am J Surg Pathol. 2019;43:277–87.

Capovilla M, Couturier J, Molinié V, et al. Loss of chromosomes 9 and 11 may be recurrent chromosome imbalances in juxtaglomerular cell tumors. Hum Pathol. 2008;39:459–62.

Carlo MI, Mukherjee S, Mandelker D, et al. Prevalence of Germline mutations in cancer susceptibility genes in patients with advanced renal cell carcinoma. JAMA Oncol. 2018;4:1228–35.

Chami R, Yin M, Marrano P, et al. BRAF mutations in pediatric metanephric tumors. Hum Pathol. 2015;46(8):1153–61.

Chang HY, Hang JF, Wu CY, et al. Clinicopathological and molecular characterisation of papillary renal neoplasm with reverse polarity and its renal papillary adenoma analogue. Histopathology. 2021;78:1019–31.

Chen M, Ye Y, Yang H, et al. Genome-wide profiling of chromosomal alterations in renal cell carcinoma using high-density single nucleotide polymorphism arrays. Int J Cancer. 2009;125:2342–8.

Chen N, Nie L, Gong J, et al. Gains of chromosomes 7 and 17 in tubulocystic carcinoma of kidney: two cases with fluorescence in situ hybridisation analysis. J Clin Pathol. 2014;67:1006–9.

Chen X, Dou FX, Cheng XB, Guo AT, Shi HY. Clinicopathologic characteristics of thyroid-like follicular carcinoma of the kidney: an analysis of five cases and review of literature. Zhonghua Bing Li Xue Za Zhi. 2016;45:687–91.

Chen YB, Mirsadraei L, Jayakumaran G, et al. Somatic mutations of TSC2 or MTOR characterize a morphologically distinct subset of sporadic renal cell carcinoma with eosinophilic and vacuolated cytoplasm. Am J Surg Pathol. 2019;43:121–31.

Chen YB, Xu J, Skanderup AJ, et al. Molecular analysis of aggressive renal cell carcinoma with unclassified histology reveals distinct subsets. Nat Commun. 2016;7:13131.

Cheng L, MacLennan GT, Zhang S, et al. Evidence for polyclonal origin of multifocal clear cell renal cell carcinoma. Clin Cancer Res. 2008;14:8087–93.

Choueiri TK, Atkins MB, Rose TL, et al. A phase 1b trial of the CXCR4 inhibitor mavorixafor and nivolumab in advanced renal cell carcinoma patients with no prior response to nivolumab monotherapy. Investig New Drugs. 2021;39:1019–27.

Choueiri TK, Cheville J, Palescandolo E, et al. BRAF mutations in metanephric adenoma of the kidney. Eur Urol. 2012;62:917–22.

Classe M, Malouf GG, Su X, et al. Incidence, clinicopathological features and fusion transcript landscape of translocation renal cell carcinomas. Histopathology. 2017;70:1089–97.

Compérat E, Couturier J, Peyromaure M, Cornud F, Vieillefond A. Benign mixed epithelial and stromal tumor of the kidney (MEST) with cytogenetic alteration. Pathol Res Pract. 2005;200:865–7.

Cossu-Rocca P, Eble JN, Delahunt B, et al. Renal mucinous tubular and spindle carcinoma lacks the gains of chromosomes 7 and 17 and losses of chromosome Y that are prevalent in papillary renal cell carcinoma. Mod Pathol. 2006;19:488–93.

Crino PB, Nathanson KL, Henske EP. The tuberous sclerosis complex. N Engl J Med. 2006;355:1345–56.

Curatolo P, Bombardieri R, Jozwiak S. Tuberous sclerosis. Lancet. 2008;372:657–68.

D'Alterio C, Buoncervello M, Ierano C, et al. Targeting CXCR4 potentiates anti-PD-1 efficacy modifying the tumor microenvironment and inhibiting neoplastic PD-1. J Exp Clin Cancer Res. 2019;38:432.

Damayanti NP, Budka JA, Khella HWZ, et al. Therapeutic targeting of TFE3/IRS-1/PI3K/mTOR Axis in translocation renal cell carcinoma. Clin Cancer Res. 2018;24:5977–89.

Davis CF, Ricketts CJ, Wang M, et al. The somatic genomic landscape of chromophobe renal cell carcinoma. Cancer Cell. 2014;26:319–30.

Davis CJ, Barton JH, Sesterhenn IA, et al. Metanephric adenoma: clinicopathological study of fifty patients. Am J Surg Pathol. 1995;19(10):1101–14.

Davis JL, Vargas SO, Rudzinski ER, et al. Recurrent RET gene fusions in paediatric spindle mesenchymal neoplasms. Histopathology. 2020;76(7):1032–41.

De Luise M, Girolimetti G, Okere B, et al. Molecular and metabolic features of oncocytomas: seeking the blueprints of indolent cancers. Biochim Biophys Acta Bioenerg. 2017;1858:591–601.

De Waele L, Lagae L, Mekahli D. Tuberous sclerosis complex: the past and the future. Pediatr Nephrol. 2015;30:1771–80.

Debelenko LV, Raimondi SC, Daw N, et al. Renal cell carcinoma with novel VCL-ALK fusion: new representative of ALK-associated tumor spectrum. Mod Pathol. 2011;24:430–42.

Delahunt B, Thompson KJ, Ferguson AF, et al. Familial cystic nephroma and pleuropulmonary blastoma. Cancer. 1993;71:1338–42.

Demellawy DE, Cundiff CA, Nasr A, et al. Congenital mesoblastic nephroma: a study of 19 cases using immunohistochemistry and ETV6-NTRK3 fusion gene rearrangement. Pathology. 2016;48(1):47–50.

Deml KF, Schildhaus HU, Comperat E, et al. Clear cell papillary renal cell carcinoma and renal angiomyoadenomatous tumor: two variants of a morphologic, immunohistochemical, and genetic distinct entity of renal cell carcinoma. Am J Surg Pathol. 2015;39:889–901.

Dhillon J, Amin MB, Selbs E, et al. Mucinous tubular and spindle cell carcinoma of the kidney with sarcomatoid change. Am J Surg Pathol. 2009;33:44–9.

Dhillon J, Tannir NM, Matin SF, et al. Thyroid-like follicular carcinoma of the kidney with metastases to the lungs and retroperitoneal lymph nodes. Hum Pathol. 2011;42:146–50.

DiNatale RG, Gorelick AN, Makarov V, et al. Putative drivers of aggressiveness in TCEB1-mutant renal cell carcinoma: an emerging entity with variable clinical course. Eur Urol Focus. 2021;7:381–9.

Ding Y, Wang C, Li X, et al. Novel clinicopathological and molecular characterization of metanephric adenoma: a study of 28 cases. Diagn Pathol. 2018;13:54.

Dong L, Huang J, Huang L, et al. Thyroid-like follicular carcinoma of the kidney in a patient with skull and meningeal metastasis: a unique case report and review of the literature. Medicine (Baltimore). 2016;95:e3314.

Doros LA, Rossi CT, Yang J. DICER1 mutations in childhood cystic nephroma and its relationship to DICER1-renal sarcoma. Mod Pathol. 2014;27(9):1267–80.

Eaton KW, Tooke LS, Wainwright LM, et al. Spectrum of SMARCB1/INI1 mutations in familial and sporadic rhabdoid tumors. Pediatr Blood Cancer. 2011;56(1):7–15.

Eble JN, Delahunt B. Emerging entities in renal cell neoplasia: thyroid-like follicular renal cell carcinoma and multifocal oncocytoma-like tumours associated with oncocytosis. Pathology. 2018;50:24–36.

Elliott A, Bruner E. Renal medullary carcinoma. Arch Pathol Lab Med. 2019;143:1556–61.

Fan R. Primary renal neuroblastoma—a clinical pathologic study of 8 cases. Am J Surg Pathol. 2012;36(1):94–100.

Fanburg-Smith JC, Hengge M, Hengge UR, et al. Extra-renal rhabdoid tumors of soft tissue: a clinicopathologic and immunohistochemical study of 18 cases. Ann Diagn Pathol. 1998;2:351–62.

Fanelli GN, Fassan M, Dal Moro F, et al. Thyroid-like follicular carcinoma of the kidney: the mutational profiling reveals a BRAF wild type status. Pathol Res Pract. 2019;215:152532.

Farcaş M, Gatalica Z, Trpkov K, et al. Eosinophilic vacuolated tumor (EVT) of kidney demonstrates sporadic TSC/MTOR mutations: next-generation sequencing multi-institutional study of 19 cases. Mod Pathol. 2022;35(3):344–51.

Fernandes FG, Silveira HCS, Junior JNA, et al. Somatic copy number alterations and associated genes in clear-cell renal-cell carcinoma in Brazilian patients. Int J Mol Sci. 2021;22:2265.

Fischer J, Palmedo G, von Knobloch R, et al. Duplication and overexpression of the mutant allele of the MET proto-oncogene in multiple hereditary papillary renal cell tumours. Oncogene. 1998;17:733–9.

Fittschen A, Wendlik I, Oeztuerk S, et al. Prevalence of sporadic renal angiomyolipoma: a retrospective analysis of 61,389 in- and out-patients. Abdom Imaging. 2014;39:1009–13.

Folpe AL, Kwiatkowski DJ. Perivascular epithelioid cell neoplasms: pathology and pathogenesis. Hum Pathol. 2010;41(1):1–15.

Foulkes WD, Kamihara J, Evans DGR. Cancer surveillance in Gorlin syndrome and rhabdoid tumor predisposition syndrome. Clin Cancer Res. 2017;23(12):e62–7.

Foulkes WD, Priest JR, Duchaine TF. DICER1: mutations, microRNAs and mechanisms. Nat Rev Cancer. 2014;14:662–72.

Furtwaengler R, Reinhard H, Leuschner I, et al. Mesoblastic nephroma—a report from the Gesellschaft fur Padiatrische Onkologie und Hamatologie (GPOH). Cancer. 2006;106(10):2275–83.

Gasparre G, Romeo G, Rugolo M, Porcelli AM. Learning from oncocytic tumors: why choose inefficient mitochondria? Biochim Biophys Acta. 1807;2011:633–42.

Geller JI, Ehrlich PF, Cost NG, et al. Characterization of adolescent and pediatric renal cell carcinoma: a report from the Children's Oncology Group Study AREN03B2. Cancer. 2015;121(14):2457–64.

Geller JI, Roth JJ, Biegel JA. Biology and treatment of rhabdoid tumor. Crit Rev. Oncog. 2015;20(3–4):199–216.

Giannikou K, Malinowska IA, Pugh TJ, et al. Whole exome sequencing identifies TSC1/TSC2 biallelic loss as the primary and sufficient driver event for renal angiomyolipoma development. PLoS Genet. 2016;12:e1006242.

Gobbo S, Eble JN, Grignon DJ, et al. Clear cell papillary renal cell carcinoma: a distinct histopathologic and molecular genetic entity. Am J Surg Pathol. 2008;32:1239–45.

Gonçalves AF, Adlesic M, Brandt S, et al. Evidence of renal angiomyolipoma neoplastic stem cells arising from renal epithelial cells. Nat Commun. 2017;8:1466.

Gooskens SL, Gadd S, van den HeuvelEibrink MM, et al. BCOR internal tandem duplications in clear cell sarcoma of the kidney. Genes Chromosomes Cancer. 2016;55(6):549–50.

Gooskens SL, Houwing ME, Vujanic GM. Congenital mesoblastic nephroma 50 years after its recognition: a narrative review. Pediatr Blood Cancer. 2017;64(7):1–9.

Gooskens SL, Kenny C, Lazaro A, et al. The clinical phenotype of YWHAE-NUTM2B/E positive pediatric clear cell sarcoma of the kidney. Genes Chromosomes Cancer. 2016;55:143–7.

Gossage L, Eisen T. Alterations in VHL as potential biomarkers in renal-cell carcinoma. Nat Rev. Clin Oncol. 2010;7:277–88.

Grubb RL 3rd, Franks ME, Toro J, et al. Hereditary leiomyomatosis and renal cell cancer: a syndrome associated with an aggressive form of inherited renal cancer. J Urol. 2007;177:2074–9; discussion 2079–2080.

Gu L, Peng C, Zhang F, Fang C, Guo G. Sequential everolimus for angiomyolipoma associated with tuberous sclerosis complex: a prospective cohort study. Orphanet J Rare Dis. 2021;16:277.

Guo J, Tretiakova MS, Troxell ML, et al. Tuberous sclerosis-associated renal cell carcinoma: a clinicopathologic study of 57 separate carcinomas in 18 patients. Am J Surg Pathol. 2014;38:1457–67.

Gupta S, Argani P, Jungbluth AA, et al. TFEB expression profiling in renal cell carcinomas: clinicopathologic correlations. Am J Surg Pathol. 2019;43:1445–61.

Gupta S, Jimenez RE, Herrera-Hernandez L, et al. Renal neoplasia in tuberous sclerosis: a study of 41 patients. Mayo Clin Proc. 2021;96:1470–89.

Gupta S, Johnson SH, Vasmatzis G, et al. TFEB-VEGFA (6p21.1) co-amplified renal cell carcinoma: a distinct entity with potential implications for clinical management. Mod Pathol. 2017;30:998–1012.

Gupta S, Rowsey RA, Cheville JC, Jimenez RE. Morphologic overlap between low grade oncocytic tumor (LOT) & eosinophilic variant of chromophobe renal cell carcinoma. Hum Pathol. 2022;119:114–6.

Moch H, Humphrey PA, Ulbright TM, Reuter VE, editors. WHO classification of tumours of the urinary system and male genital organs. Lyon: International Agency for Research on Cancer; 2016.

Hakimi AA, Tickoo SK, Jacobsen A, et al. TCEB1-mutated renal cell carcinoma: a distinct genomic and morphological subtype. Mod Pathol. 2015;28:845–53.

Halat S, Eble JN, Grignon DJ, et al. Multilocular cystic renal cell carcinoma is a subtype of clear cell renal cell carcinoma. Mod Pathol. 2010;23:931–6.

Han H, Betrannd KC, Patel KR, et al. BCORCCNB3 fusion-positive clear cell sarcoma of the kidney. Pediatr Blood Cancer. 2019;26:e28151.

Han JM, Sahin M. TSC1/TSC2 signaling in the CNS. FEBS Lett. 2011;585:973–80.

Hang JF, Chung HJ, Pan CC. ALK-rearranged renal cell carcinoma with a novel PLEKHA7-ALK translocation and metanephric adenoma-like morphology. Virchows Arch. 2020;476:921–9.

He H, Trpkov K, Martinek P, et al. "High-grade oncocytic renal tumor": morphologic, immunohistochemical, and molecular genetic study of 14 cases. Virchow Arch. 2018;473:725–38.

Hennigar RA, Bekwith JB. Nephrogenic adenofibroma: a novel kidney tumor of young people. Am J Surg Pathol. 1992;16(4):325–34.

Herlitz L, Hes O, Michal M, et al. "Atrophic kidney"-like lesion: clinicopathologic series of 8 cases supporting a benign entity distinct from thyroid-like follicular carcinoma. Am J Surg Pathol. 2018;42:1585–95.

Hes O, de Souza TG, Pivovarcikova K, et al. Distinctive renal cell tumor simulating atrophic kidney with 2 types of microcalcifications. Report of 3 cases. Ann Diagn Pathol. 2014;18:82–8.

Hes O, Trpkov K. Do we need an updated classification of oncocytic renal tumors?: emergence of low-grade oncocytic tumor (LOT) and eosinophilic vacuolated tumor (EVT) as novel renal entities. Mod Pathol. 2022;35(9):1140–50.

Hill DA, Ivanovich J, Priest JR. DICER1 mutations in familial pleuropulmonary blastoma. Science. 2009;325(5943):965.

Hollmann TJ, Hornick JL. INI1-deficient tumors: diagnostic features and molecular genetics. Am J Surg Pathol. 2011;35:e47–63.

Hoot AC, Russo P, Judkins A, et al. Immunohistochemical analysis of hSNF5/INI1 distinguishes renal and extra-renal malignant rhabdoid tumors from other pediatric soft tissue tumors. Am J Surg Pathol. 2004;28(11):1485–91.

Hsieh JJ, Le VH, Oyama T, et al. Chromosome 3p loss-orchestrated VHL, HIF, and epigenetic deregulation in clear cell renal cell carcinoma. J Clin Oncol. 2018;36:JCO2018792549.

Huang W, Goldfischer M, Babayeva S, et al. Identification of a novel PARP14-TFE3 gene fusion from 10-year-old FFPE tissue by RNA-seq. Genes Chromosomes Cancer. 2015;54:500–5.

Humphrey PA, Moch H, Cubilla AL, Ulbright TM, Reuter VE. The 2016 WHO classification of tumours of the urinary system and male genital organs-part B: prostate and bladder tumours. Eur Urol. 2016;70:106–19.

Inamura K. Translocation renal cell carcinoma: an update on clinicopathological and molecular features. Cancers. 2017;9:111.

Inoue T, Matsuura K, Yoshimoto T, et al. Genomic profiling of renal cell carcinoma in patients with end-stage renal disease. Cancer Sci. 2012;103:569–76.

Isaacs JS, Jung YJ, Mole DR, et al. HIF overexpression correlates with biallelic loss of fumarate hydratase in renal cancer: novel role of fumarate in regulation of HIF stability. Cancer Cell. 2005;8:143–53.

Ishida U, Kato K, Kigasawa H, et al. Synchronous occurrence of pleuropulmonary blastoma and cystic nephroma: possible genetic link in cystic lesions of the lung and the kidney. Med Pediatr Oncol. 2000;35(1):85–7.

Ito T, Pei J, Dulaimi E, et al. Genomic copy number alterations in renal cell carcinoma with Sarcomatoid features. J Urol. 2016;195:852–8.

Jager S, Handschin C, St-Pierre J, Spiegelman BM. AMP-activated protein kinase (AMPK) action in skeletal muscle via direct phosphorylation of PGC-1alpha. Proc Natl Acad Sci U S A. 2007;104:12017–22.

Jehangir S, Kurian JJ, Selvarajah D. Recurrent and metastatic congenital mesoblastic nephroma: where does the evidence stand? Pediatr Surg Int. 2017;33:1183–8.

Jenkins TM, Rosenbaum J, Zhang PJ, et al. Thyroid-like follicular carcinoma of the kidney with extensive sarcomatoid differentiation: a case report and review of the literature. Int J Surg Pathol. 2019;27:678–83.

Jeung JA, Cao D, Selli BW, et al. Primary renal carcinoid tumors: clinicopathologic features of 9 cases with emphasis on novel immunohistochemical findings. Hum Pathol. 2011;42(10):1554–61.

Jones TD, Eble JN, Wang M, et al. Clonal divergence and genetic heterogeneity in clear cell renal cell carcinomas with sarcomatoid transformation. Cancer. 2005;104:1195–203.

Jones TD, Eble JN, Wang M, et al. Molecular genetic evidence for the independent origin of multifocal papillary tumors in patients with papillary renal cell carcinomas. Clin Cancer Res. 2005;11:7226–33.

Joshi S, Tolkunov D, Aviv H, et al. The genomic landscape of renal oncocytoma identifies a metabolic barrier to tumorigenesis. Cell Rep. 2015;13:1895–908.

Joshi VV, Beckwith JB. Multilocular cyst of the kidney (cystic nephroma) and cystic partially differentiated nephroblastoma. Cancer. 1989;64(2):466–79.

Jung SJ, Chung JI, Park SH, Ayala AG, Ro JY. Thyroid follicular carcinoma-like tumor of kidney: a case report with morphologic, immunohistochemical, and genetic analysis. Am J Surg Pathol. 2006;30:411–5.

Kalimuthu SN, Chetty R. Gene of the month: SMARCB1. J Clin Pathol. 2016;69:484–9.

Kapur P, Gao M, Zhong H, et al. Eosinophilic vacuolated tumor of the kidney: a review of evolving concepts in this novel subtype with additional insights from a case with MTOR mutation and concomitant chromosome 1 loss. Adv Anat Pathol. 2021;28:251–7.

Kapur P, Gao M, Zhong H, et al. Germline and sporadic mTOR pathway mutations in low-grade oncocytic tumor of the kidney. Mod Pathol. 2022;35(3):333–43.

Karlsson J, Valind A, Gisselsson D. BCOR internal tandem duplication and YWHAE-NUTM2B/E fusion are mutually exclusive events in clear cell sarcoma of the kidney. Genes Chromosomes Cancer. 2016;55(2):120–3.

Kenny C, Bausenwein S, Lazaro A, et al. Mutually exclusive BCOR internal tandem duplications and YWHAE-NUTM2 fusions in clear cell sarcoma of kidney: not the full story. J Pathol. 2016;238(5):617–20.

Klacz J, Wierzbicki PM, Wronska A, et al. Decreased expression of RASSF1A tumor suppressor gene is associated with worse prognosis in clear cell renal cell carcinoma. Int J Oncol. 2016;48:55–66.

Klomp JA, Petillo D, Niemi NM, et al. Birt-Hogg-Dube renal tumors are genetically distinct from other renal neoplasias and are associated with up-regulation of mitochondrial gene expression. BMC Med Genet. 2010;3:59.

Knezevich SR, Garnett MJ, Pysher TJ, et al. eTV6- NTRK3 gene fusions and trisomy 11 establish a histogenetic link between mesoblastic nephroma and congenital fibrosarcoma. Cancer Res. 1998;58(22):5046–8.

Kohn L, Svenson U, Ljungberg B, Roos G. Specific genomic aberrations predict survival, but low mutation rate in cancer hot spots, in clear cell renal cell carcinoma. Appl Immunohistochem Mol Morphol. 2015;23:334–42.

Komai Y, Fujiwara M, Fujii Y, et al. Adult Xp11 translocation renal cell carcinoma diagnosed by cytogenetics and immunohistochemistry. Clin Cancer Res. 2009;15:1170–6.

Kondo T, Sasa N, Yamada H, et al. Acquired cystic disease-associated renal cell carcinoma is the most common subtype in long-term dialyzed patients: central pathology results according to the 2016 WHO classification in a multi-institutional study. Pathol Int. 2018;68:543–9.

Kordes U, Gesk S, Fruhwald MC, et al. Clinical and molecular features in patients with atypical teratoid rhabdoid tumor or malignant rhabdoid tumor. Genes Chromosomes Cancer. 2010;49:176–81.

Kozman D, Kao CS, Nguyen JK, Smith SC, Kehr EL, Tretiakova M, Przybycin CG, Williamson SR, Argani P, Eng C, Campbell SC, McKenney JK, Alaghehbandan R. Renal Neoplasia Occurring in Patients With PTEN Hamartoma Tumor Syndrome: Clinicopathologic Study of 12 Renal Cell Carcinomas From 9 Patients and Association With Intrarenal "Lipomas". Am J Surg Pathol. 2023. Epub ahead of print.

Kunju LP, Wojno K, Wolf JS Jr, Cheng L, Shah RB. Papillary renal cell carcinoma with oncocytic cells and nonoverlapping low grade nuclei: expanding the morphologic spectrum with emphasis on clinicopathologic, immunohistochemical and molecular features. Hum Pathol. 2008;39:96–101.

Kuroda N, Maris S, Monzon FA, et al. Juxtaglomerular cell tumor: a morphological, immunohistochemical and genetic study of six cases. Hum Pathol. 2013;44:47–54.

Kuroda N, Ohe C, Mikami S, et al. Review of acquired cystic disease-associated renal cell carcinoma with focus on pathobiological aspects. Histol Histopathol. 2011;26:1215–8.

Kuroda N, Tamura M, Hamaguchi N, et al. Acquired cystic disease-associated renal cell carcinoma with sarcomatoid change and rhabdoid features. Ann Diagn Pathol. 2011;15:462–6.

Kuroda N, Trpkov K, Gao Y, et al. ALK rearranged renal cell carcinoma (ALK-RCC): a multi-institutional study of twelve cases with identification of novel partner genes CLIP1, KIF5B and KIAA1217. Mod Pathol. 2020;33:2564–79.

Kuroda N, Yamashita M, Kakehi Y, et al. Acquired cystic disease-associated renal cell carcinoma: an immunohistochemical and fluorescence in situ hybridization study. Med Mol Morphol. 2011;44:228–32.

La Rosa S, Bernasconi B, Micello D, et al. Primary small cell neuroendocrine carcinoma of the kidney: morphological, immunohistochemical, ultrastructural, and cytogenetic study of a case and review of the literature. Endocr Pathol. 2009;20(1):24–34.

Ladanyi M, Lui MY, Antonescu CR, et al. The der(17)t(X;17)(p11;q25) of human alveolar soft part sarcoma fuses the TFE3 transcription factor gene to ASPL, a novel gene at 17q25. Oncogene. 2001;20:48–57.

Lallier M, Bouchard S, DiLorenzo M, et al. Pleuropulmonary blastoma: a rare pathology with an even rarer presentation. J Pediatr Surg. 1999;34:1057–9.

Launonen V, Vierimaa O, Kiuru M, et al. Inherited susceptibility to uterine leiomyomas and renal cell cancer. Proc Natl Acad Sci U S A. 2001;98:3387–92.

Lawrie CH, Larrea E, Larrinaga G, et al. Targeted next-generation sequencing and non-coding RNA expression analysis of clear cell papillary renal cell carcinoma suggests distinct pathological mechanisms from other renal tumour subtypes. J Pathol. 2014;232:32–42.

Lee HY, Yoon CS, Sevenet N. Rhabdoid tumor the kidney is a component of the rhabdoid predisposition syndrome. Pediatr Dev Pathol. 2002;5(4):395–9.

Lee RS, Stewar C, Carter SL. A remarkably simple genome underlies highly malignant pediatric rhabdoid cancers. J Clin Invest. 2012;122(8):2983–8.

Lei L, Stohr BA, Berry S, et al. Recurrent EGFR alterations in NTRK3 fusion negative congenital mesoblastic nephroma. Pract Lab Med. 2020;21:e00164.

Lerma LA, Schade GR, Tretiakova MS. Co-existence of ESC-RCC, EVT, and LOT as synchronous and metachronous tumors in six patients with multifocal neoplasia but without clinical features of tuberous sclerosis complex. Hum Pathol. 2021;116:1–11.

Letouze E, Martinelli C, Loriot C, et al. SDH mutations establish a hypermethylator phenotype in paraganglioma. Cancer Cell. 2013;23:739–52.

Li Y, Pawel B, Hill DA, et al. Pediatric cystic nephroma are morphologically, immunohistochemically, and genetically distinct from adult cystic nephroma. Am J Surg Pathol. 2017;41(4):472–81.

Li Y, Pawel BR, Hill DA, Epstein JI, Argani P. Pediatric cystic nephroma is morphologically, immunohistochemically, and genetically distinct from adult cystic nephroma. Am J Surg Pathol. 2017;41:472–81.

Li Y, Reuter VE, Matoso A, et al. Re-evaluation of 33 'unclassified' eosinophilic renal cell carcinomas in young patients. Histopathology. 2018;72:588–600.

Linehan WM, Spellman PT, Ricketts CJ, et al. Comprehensive molecular characterization of papillary renal-cell carcinoma. N Engl J Med. 2016;374:135–45.

Looyenga BD, Furge KA, Dykema KJ, et al. Chromosomal amplification of leucine-rich repeat kinase-2 (LRRK2) is required for oncogenic MET signaling in papillary renal and thyroid carcinomas. Proc Natl Acad Sci U S A. 2011;108:1439–44.

Louie BH, Kurzrock R. BAP1: not just a BRCA1-associated protein. Cancer Treat Rev. 2020;90:102091.

Lynch ED, Ostermeyer EA, Lee MK, et al. Inherited mutations in PTEN that are associated with breast cancer, cowden disease, and juvenile polyposis. Am J Hum Genet. 1997;61:1254–60.

Malinowska I, Kwiatkowski DJ, Weiss S, et al. Perivascular epithelioid cell tumors (PEComas) harboring TFE3 gene rearrangements lack the TSC2 alterations characteristic of conventional PEComas: further evidence for a biological distinction. Am J Surg Pathol. 2012;36:783–4.

Malouf GG, Ali SM, Wang K, et al. Genomic characterization of renal cell carcinoma with sarcomatoid dedifferentiation pinpoints recurrent genomic alterations. Eur Urol. 2016;70:348–57.

Malouf GG, Su X, Yao H, et al. Next-generation sequencing of translocation renal cell carcinoma reveals novel RNA splicing partners and frequent mutations of chromatin-remodeling genes. Clin Cancer Res. 2014;20:4129–40.

Mangray S, Breese V, Jackson CL, et al. Application of BRAF V600E mutation analysis for the diagnosis of metanephric adenofibroma. Am J Surg Pathol. 2015;39(9):1301–4.

Manivel JC, Priest JR, Watterson J, et al. Pleuropulmonary blastoma. The so-called pulmonary blastoma of childhood. Cancer. 1988;62:1516–26.

Manley BJ, Hakimi AA. Molecular profiling of renal cell carcinoma: building a bridge toward clinical impact. Curr Opin Urol. 2016;26:383–7.

Marino-Enriquez A, Ou WB, Weldon CB, Fletcher JA, Perez-Atayde AR. ALK rearrangement in sickle cell trait-associated renal medullary carcinoma. Genes Chromosomes Cancer. 2011;50:146–53.

Marsden L, Jennings LJ, Gadd S, et al. BRAF exon 15 mutations in pediatric renal stromal tumors: prevalence in metanephric stromal tumors. Hum Pathol. 2017;60:32–6.

Marsh DJ, Coulon V, Lunetta KL, et al. Mutation spectrum and genotype-phenotype analyses in Cowden disease and Bannayan-Zonana syndrome, two hamartoma syndromes with germline PTEN mutation. Hum Mol Genet. 1998;7:507–15.

Martignoni G, Bonetti F, Chilosi M, et al. Cathepsin K expression in the spectrum of perivascular epithelioid cell (PEC) lesions of the kidney. Mod Pathol. 2012;25(1):100–11.

Martignoni G, Pea M, Gobbo S, et al. Cathepsin-K immunoreactivity distinguishes MiTF/TFE family renal translocation carcinomas from other renal carcinomas. Mod Pathol. 2009;22:1016–22.

McGarrah PW, Westin GFM, Hobday TJ, et al. Renal neuroendocrine neoplasms: a single-center experience. Clin Genitourin Cancer. 2020;18(4):e343–9.

Mehra R, Smith SC, Divatia M, Amin MB. Emerging entities in renal neoplasia. Surg Pathol Clin. 2015;8:623–56.

Mehra R, Vats P, Cao X, et al. Somatic bi-allelic loss of TSC genes in eosinophilic solid and cystic renal cell carcinoma. Eur Urol. 2018;74:483–6.

Mehra R, Vats P, Cieslik M, et al. Biallelic alteration and dysregulation of the hippo pathway in mucinous tubular and spindle cell carcinoma of the kidney. Cancer Discov. 2016;6:1258–66.

Mester JL, Zhou M, Prescott N, Eng C. Papillary renal cell carcinoma is associated with PTEN hamartoma tumor syndrome. Urology. 2012;79:1187.e1–7.

Mitchell TJ, Turajlic S, Rowan A, et al. Timing the landmark events in the evolution of clear cell renal cell cancer: TRACERx renal. Cell. 2018;173:611–623.e617.

Moch H, Cubilla AL, Humphrey PA, Reuter VE, Ulbright TM. The 2016 WHO classification of tumours of the urinary system and male genital organs-part a: renal, penile, and testicular tumours. Eur Urol. 2016;70:93–105.

Moch H, Humphrey PA, Ulbright TM, et al., editors. WHO classification of tumours of the urinary system and male genital organs. 4th ed. Lyon: International Agency for Research on Cancer; 2016.

Moch H, Ohashi R. Chromophobe renal cell carcinoma: current and controversial issues. Pathology. 2021;53:101–8.

Monzon FA, Alvarez K, Peterson L, et al. Chromosome 14q loss defines a molecular subtype of clear-cell renal cell carcinoma associated with poor prognosis. Mod Pathol. 2011;24:1470–9.

Morini A, Drossart T, Timsit MO, et al. Low-grade oncocytic renal tumor (LOT): mutations in mTOR pathway genes and low expression of FOXI1. Mod Pathol. 2022;35(3):352–60.

Muglia VF, Prando A. Renal cell carcinoma: histological classification and correlation with imaging findings. Radiol Bras. 2015;48:166–74.

Muir TE, Cheville JC, Lager DJ. Metanephric adenoma, nephrogenic rests, and Wilms' tumor: a histologic and immunophenotypic comparison. Am J Surg Pathol. 2001;25(10):1290–6.

Muscara MJ, Simper NB, Gandia E. Thyroid-like follicular carcinoma of the kidney. Int J Surg Pathol. 2017;25:73–7.

Napolioni V, Curatolo P. Genetics and molecular biology of tuberous sclerosis complex. Curr Genomics. 2008;9:475–87.

Nemes K, Frühwald MC. Emerging therapeutic targets for the treatment of malignant rhabdoid tumors. Expert Opin Ther Targets. 2018;22(4):365–79.

Netto GJ, Cheng L. Emerging critical role of molecular testing in diagnostic genitourinary pathology. Arch Pathol Lab Med. 2012;136(4):372–90.

Nickerson ML, Warren MB, Toro JR, et al. Mutations in a novel gene lead to kidney tumors, lung wall defects, and benign tumors of the hair follicle in patients with the Birt-Hogg-Dube syndrome. Cancer Cell. 2002;2:157–64.

Nikiforova MN, Lynch RA, Biddinger PW, et al. RAS point mutations and PAX8-PPAR gamma rearrangement in thyroid tumors: evidence for distinct molecular pathways in thyroid follicular carcinoma. J Clin Endocrinol Metab. 2003;88:2318–26.

O'Meara E, Stack D, Lee CH, et al. Characterization of the chromosomal translocation t(10;17)(q22;p13) in clear cell sarcoma of kidney. J Pathol. 2012;227:72–80.

Oshiro Y, Hida AI, Tamiya S, et al. Bilateral atrophic kidney-like tumors. Pathol Int. 2014;64:478–80.

Osunkoya AO, Young AN, Wang W, Netto GJ, Epstein JI. Comparison of gene expression profiles in tubulocystic carcinoma and collecting duct carcinoma of the kidney. Am J Surg Pathol. 2009;33:1103–6.

Palese MA, Ferrer F, Perlman E, et al. Metanephric stromal tumor: a rare benign pediatric mass. Urology. 2001;58(3):462.

Pan CC, Epstein JI. Detection of chromosome copy number alterations in metanephric adenomas by array comparative genomic hybridization. Mod Pathol. 2010;23(12):1634–40.

Parilla M, Alikhan M, Al-Kawaaz M, et al. Genetic underpinnings of renal cell carcinoma with Leiomyomatous Stroma. Am J Surg Pathol. 2019;43:1135–44.

Parwani AV, Husain AN, Epstein JI, Beckwith JB, Argani P. Low-grade myxoid renal epithelial neoplasms with distal nephron differentiation. Hum Pathol. 2001;32:506–12.

Pavlovich CP, Walther MM, Eyler RA, et al. Renal tumors in the Birt-Hogg-Dube syndrome. Am J Surg Pathol. 2002;26:1542–52.

Pawel BR. SMARCB1-deficient tumors of childhood: a practical guide. Pediatr Dev Pathol. 2017;21(1):6–28.

Perrino CM, Grignon DJ, Williamson SR, et al. Morphological spectrum of renal cell carcinoma, unclassified: an analysis of 136 cases. Histopathology. 2018;72:305–19.

Pettinato G, Manivel JC, Wick MR, et al. Classical and cellular (atypical) congenital mesoblastic nephroma: a clinicopathologic, ultrastructural, immunohistochemical, and flow cytometric study. Hum Pathol. 1989;20(7):682–90.

Pierson CR, Schober MS, Wallis T, et al. Mixed epithelial and stromal tumor of the kidney lacks the genetic alterations of cellular congenital mesoblastic nephroma. Hum Pathol. 2001;32:513–20.

Pillay N, Ramdial PK, Cooper K, Batuule D. Mucinous tubular and spindle cell carcinoma with aggressive histomorphology—a sarcomatoid variant. Hum Pathol. 2008;39:966–9.

Pinto A, Signoretti S, Hirsch MS, Barletta JA. Immunohistochemical staining for BRAF V600E supports the diagnosis of metanephric adenoma. Histopathology. 2015;66:901–4.

Pinto A, Signoretti S, Hirsch MS, et al. Immunohistochemical staining for BRAF V600E supports the diagnosis of metanephric adenoma. Histopathology. 2015;66(6):901–4.

Pivovarcikova K, Grossmann P, Hajkova V, et al. Renal cell carcinomas with tubulopapillary architecture and oncocytic cells: molecular analysis of 39 difficult tumors to classify. Ann Diagn Pathol. 2021;52:151734.

Plank TL, Yeung RS, Henske EP. Hamartin, the product of the tuberous sclerosis 1 (TSC1) gene, interacts with tuberin and appears to be localized to cytoplasmic vesicles. Cancer Res. 1998;58:4766–70.

Priest JR, Watterson J, Strong L, et al. Pleuropulmonary blastoma: a marker for familial disease. J Pediatr. 1996;128:220–4.

Przybycin CG, Harper HL, Reynolds JP, et al. Acquired cystic disease-associated renal cell carcinoma (ACD-RCC): a multiinstitutional study of 40 cases with clinical follow-up. Am J Surg Pathol. 2018;42:1156–65.

Punnett HH, Halligan GE, Zaeri N, et al. Translocation 1;17 in clear cell sarcoma of the kidney: a first report. Cancer Genet Cytogenet. 1989;41(1):123–8.

Rakheja D, Weinberg AG, Tomlinson GE, et al. Translocation (10;17)(q22;p13): a recurring translocation in clear cell sarcoma of kidney. Cancer Genet Cytogenet. 2004;154(2):175–9.

Rakowski SK, Winterkorn EB, Paul E, et al. Renal manifestations of tuberous sclerosis complex: incidence, prognosis, and predictive factors. Kidney Int. 2006;70:1777–82.

Rakozy C, Schmahl GE, Bogner S, Storkel S. Low-grade tubular-mucinous renal neoplasms: morphologic, immunohistochemical, and genetic features. Mod Pathol. 2002;15:1162–71.

Rao Q, Chen JY, Wang JD, et al. Renal cell carcinoma in children and young adults: clinicopathological, immunohistochemical, and VHL gene analysis of 46 cases with follow-up. Int J Surg Pathol. 2011;19:170–9.

Rao Q, Liu B, Cheng L, et al. Renal cell carcinomas with t(6;11)(p21;q12): a clinicopathologic study emphasizing unusual morphology, novel alphaTFEB gene fusion point, immunobiomarkers, and ultrastructural features, as well as detection of the gene fusion by fluorescence in situ hybridization. Am J Surg Pathol. 2012;36(9):1327–38.

Rao Q, Shen Q, Xia QY, et al. PSF/SFPQ is a very common gene fusion partner in TFE3 rearrangement-associated perivascular epithelioid cell tumors (PEComas) and melanotic Xp11 translocation renal cancers: clinicopathologic, immunohistochemical, and molecular characteristics suggesting classification as a distinct entity. Am J Surg Pathol. 2015;39:1181–96.

Reid-Nicholson MD, Motiwala N, Drury SC, et al. Chromosomal abnormalities in renal cell carcinoma variants detected by Urovysion fluorescence in situ hybridization on paraffin-embedded tissue. Ann Diagn Pathol. 2011;15(1):37–45.

Reid-Nicholson MD, Ramalingam P, Adeagbo B, Cheng N, Peiper SC, Terris MK. The use of Urovysion fluorescence in situ hybridization in the diagnosis and surveillance of non-urothelial carcinoma of the bladder. Mod Pathol. 2009;22(1):119–27.

Richard S, Graff J, Lindau J, Resche F. Von Hippel-Lindau disease. Lancet. 2004;363:1231–4.

Richards FM, Phipps ME, Latif F, et al. Mapping the Von Hippel-Lindau disease tumour suppressor gene: identification of germline deletions by pulsed field gel electrophoresis. Hum Mol Genet. 1993;2:879–82.

Rohan SM, Xiao Y, Liang Y, et al. Clear-cell papillary renal cell carcinoma: molecular and immunohistochemical analysis with emphasis on the von Hippel-Lindau gene and hypoxia-inducible factor pathway-related proteins. Mod Pathol. 2011;24:1207–20.

Roy A, Kumar V, Zorman B, et al. Recurrent internal tandem duplications of BCOR in clear cell sarcoma of the kidney. Nat Commun. 2015;6:8891.

Rubin BP, Chen CJ, Morgan TW, et al. Congenital mesoblastic nephroma t(12;15) is associated with ETV6-NTRK3 gene fusion: cytogenetic and molecular relationship to congenital (infantile) fibrosarcoma. Am J Pathol. 1998;153(5):1451–8.

Rudzinski ER, Lockwood CM, Stohr BA, et al. PanTRK immunohistochemistry identifies NTRK rearrangements in pediatric mesenchymal tumors. Am J Surg Pathol. 2018;42(7):927–35.

Saleeb RM, Brimo F, Farag M, et al. Toward biological subtyping of papillary renal cell carcinoma with clinical implications through histologic, Immunohistochemical, and molecular analysis. Am J Surg Pathol. 2017;41:1618–29.

Saleeb RM, Farag M, Ding Q, et al. Integrated molecular analysis of papillary renal cell carcinoma and precursor lesions unfolds evolutionary process from kidney progenitor-like cells. Am J Pathol. 2019;189:2046–60.

Saleeb RM, Srigley JR, Sweet J, et al. Melanotic MiT family translocation neoplasms: expanding the clinical and molecular spectrum of this unique entity of tumors. Pathol Res Pract. 2017;213:1412–8.

Salles DC, Asrani K, Woo J, et al. GPNMB expression identifies TSC1/2/mTOR-associated and MiT family translocation-driven renal neoplasms. J Pathol. 2022;257(2):158–71.

Sancak O, Nellist M, Goedbloed M, et al. Mutational analysis of the TSC1 and TSC2 genes in a diagnostic setting: genotype—phenotype correlations and comparison of diagnostic DNA techniques in tuberous sclerosis complex. Eur J Hum Genet. 2005;13:731–41.

Sanders ME, Mick R, Tomaszewski JE, Barr FG. Unique patterns of allelic imbalance distinguish type 1 from type 2 sporadic papillary renal cell carcinoma. Am J Pathol. 2002;161:997–1005.

Sansal I, Sellers WR. The biology and clinical relevance of the PTEN tumor suppressor pathway. J Clin Oncol. 2004;22:2954–63.

Satoh F, Tsutsumi Y, Yokoyama S, et al. Comparative immunohistochemical analysis of developing kidneys, nephroblastomas and related tumors: considerations on their histogenesis. Pathol Int. 2000;50(6):458–71.

Schmidt D, Harms D, Zieger G. Malignant rhabdoid tumor of the kidney. Histopathology, ultrastructure and comments on differential diagnosis. Virchows Arch A Pathol Anat Histopathol. 1982;398(1):101–8.

Schmidt L, Junker K, Weirich G, et al. Two north American families with hereditary papillary renal carcinoma and identical novel mutations in the MET proto-oncogene. Cancer Res. 1998;58:1719–22.

Schmidt LS, Nickerson ML, Angeloni D, et al. Early onset hereditary papillary renal carcinoma: germline missense mutations in the tyrosine kinase domain of the met proto-oncogene. J Urol. 2004;172:1256–61.

Schmidt LS, Nickerson ML, Warren MB, et al. Germline BHD-mutation spectrum and phenotype analysis of a large cohort of families with Birt-Hogg-Dube syndrome. Am J Hum Genet. 2005;76:1023–33.

Schraml P, Hergovich A, Hatz F, et al. Relevance of nuclear and cytoplasmic von Hippel Lindau protein expression for renal carcinoma progression. Am J Pathol. 2003;163:1013–20.

Schreiner A, Daneshmand S, Bayne A, et al. Distinctive morphology of renal cell carcinomas in tuberous sclerosis. Int J Surg Pathol. 2010;18:409–18.

Schultz KAP, Williams GM, Kamihara J. DICER1 and associated conditions: identification of at-risk individual and recommended surveillance strategies. Clin Cancer Res. 2018;24(10):2251–61.

Selak MA, Armour SM, MacKenzie ED, et al. Succinate links TCA cycle dysfunction to oncogenesis by inhibiting HIF-alpha prolyl hydroxylase. Cancer Cell. 2005;7:77–85.

Seo AN, Yoon G, Ro JY. Clinicopathologic and molecular pathology of collecting duct carcinoma and related renal cell carcinomas. Adv Anat Pathol. 2017;24:65–77.

Sevenet N, Lellouch-Tubiana A, Schofield D, et al. Spectrum of hSNF5/INI1 somatic mutations in human cancer and genotype-phenotype correlations. Hum Mol Genet. 1999;8:2359–68.

Shah A, Lal P, Toorens E, et al. Acquired cystic kidney disease-associated renal cell carcinoma (ACKD-RCC) harbor recurrent mutations in KMT2C and TSC2 genes. Am J Surg Pathol. 2020;44:1479–86.

Shah RB, Stohr BA, Tu ZJ, et al. "Renal cell carcinoma with Leiomyomatous Stroma" harbor somatic mutations of TSC1, TSC2, MTOR, and/or ELOC (TCEB1): clinicopathologic and molecular characterization of 18 sporadic tumors supports a distinct entity. Am J Surg Pathol. 2020;44:571–81.

Sharma G, Kakkar N, Singh SK, et al. Primary primitive neuroectodermal tumour of the kidney in adults: experience of managing 12 cases with systematic review and pooled analysis of literature. Int J Clin Pract. 2021;75(12):e14971.

Shehabeldin AN, Ro JY. Neuroendocrine tumors of genitourinary tract: recent advances. Ann Diagn Pathol. 2019;42:48–58.

Shen Q, Rao Q, Xia QY, et al. Perivascular epithelioid cell tumor (PEComa) with TFE3 gene rearrangement: clinicopathological, immunohistochemical, and molecular features. Virchows Arch. 2014;465:607–13.

Shimizu M, Yokota J, Mori N, et al. Introduction of normal chromosome 3p modulates the tumorigenicity of a human renal cell carcinoma cell line YCR. Oncogene. 1990;5:185–94.

Shuch B, Ricketts CJ, Vocke CD, et al. Germline PTEN mutation Cowden syndrome: an underappreciated form of hereditary kidney cancer. J Urol. 2013;190:1990–8.

Siadat F, Trpkov K. ESC, ALK, HOT and LOT: three letter acronyms of emerging renal entities knocking on the door of the WHO classification. Cancers. 2021;12:168.

Siegel RL, Miller KD, Jemal A. Cancer statistics, 2020. CA Cancer J Clin. 2020;70:7–30.

Sigauke E, Rakheja D, Maddox DL, et al. Absence of expression of SMARCB1/INI1 in malignant rhabdoid tumors of the central nervous system, kidneys and soft tissue: an immunohistochemical study with implications for diagnosis. Mod Pathol. 2006;19(5):717–25.

Simon RA, di Sant'agnese PA, Palapattu GS, et al. Mucinous tubular and spindle cell carcinoma of the kidney with sarcomatoid differentiation. Int J Clin Exp Pathol. 2008;1:180–4.

Siroky BJ, Yin H, Bissler JJ. Clinical and molecular insights into tuberous sclerosis complex renal disease. Pediatr Nephrol. 2011;26(6):839–52.

Skala SL, Wang X, Zhang Y, et al. Next-generation RNA sequencing-based biomarker characterization of chromophobe renal cell carcinoma and related oncocytic neoplasms. Eur Urol. 2020;78:63–74.

Smith NE, Deyrup AT, Marino-Enriquez A, et al. VCL-ALK renal cell carcinoma in children with sickle-cell trait: the eighth sickle-cell nephropathy? Am J Surg Pathol. 2014;38:858–63.

Smith PS, Whitworth J, West H, et al. Characterization of renal cell carcinoma-associated constitutional chromosome abnormalities by genome sequencing. Genes Chromosomes Cancer. 2020;59:333–47.

Sredni ST, Tomita T. Rhabdoid tumor predisposition syndrome. Pediatr Dev Pathol. 2015;18(1):49–58.

Sterlacci W, Verdorfer I, Gabriel M, Mikuz G. Thyroid follicular carcinoma-like renal tumor: a case report with morphologic, immunophenotypic, cytogenetic, and scintigraphic studies. Virchows Arch. 2008;452:91–5.

Sugimoto R, Uesugi N, Yamada N, et al. Sarcomatoid change associated with epithelial-mesenchymal transition in mucinous tubular and spindle cell carcinoma of the kidney: a case report. Int J Clin Exp Pathol. 2019;12:2767–71.

Sukov WR, Hodge JC, Lohse CM, et al. TFE3 rearrangements in adult renal cell carcinoma: clinical and pathologic features with outcome in a large series of consecutively treated patients. Am J Surg Pathol. 2012;36:663–70.

Szponar A, Yusenko MV, Kovacs G. High-resolution array CGH of metanephric adenomas: lack of DNA copy number changes. Histopathology. 2010;56(2):212–6.

Tajima S, Waki M, Doi W, et al. Acquired cystic disease-associated renal cell carcinoma with a focal sarcomatoid component: report of a case showing more pronounced polysomy of chromosomes 3 and 16 in the sarcomatoid component. Pathol Int. 2015;65:89–94.

Teegavarapu PS, Rao P, Matrana M, et al. Neuroendocrine tumors of the kidney: a single institution experience. Clin Genitourin Cancer. 2014;12(6):422–7.

Thomlinson GE, Breslow NE, Dome J, et al. Rhabdoid tumor of the kidney in the national Wilms' tumor study: age at diagnosis as a prognostic factor. J Clin Oncol. 2005;23:7641–5.

Tickoo SK, Alden D, Olgac S, et al. Immunohistochemical expression of hypoxia inducible factor-1alpha and its downstream molecules in sarcomatoid renal cell carcinoma. J Urol. 2007;177:1258–63.

Tickoo SK, de Peralta-Venturina MN, Harik LR, et al. Spectrum of epithelial neoplasms in end-stage renal disease: an experience from 66 tumor-bearing kidneys with emphasis on histologic patterns distinct from those in sporadic adult renal neoplasia. Am J Surg Pathol. 2006;30:141–53.

Tomlinson GE, Nisen PD, Timmons CG, et al. Cytogenetics of a renal cell carcinoma in a 17-month old child. Evidence of Xp11.2 as a recurring breakpoint. Cancer Genet Cytogenet. 1991;57(1):11–7.

Tomlinson IP, Alam NA, Rowan AJ, et al. Germline mutations in FH predispose to dominantly inherited uterine fibroids, skin leiomyomata and papillary renal cell cancer. Nat Genet. 2002;30:406–10.

Tong K, Hu Z. FOXI1 expression in chromophobe renal cell carcinoma and renal oncocytoma: a study of the cancer genome atlas transcriptome-based outlier mining and immunohistochemistry. Virchows Arch. 2021;478:647–58.

Toutain J, VuPhi Y, Doco-Fenzy M, et al. Identification of a complex 17q rearrangement in a metanephric stromal tumor. Cancer Genet. 2011;204(6):340–3.

Trpkov K, Abou-Ouf H, Hes O, et al. Eosinophilic solid and cystic renal cell carcinoma (ESC RCC): further morphologic and molecular characterization of ESC RCC as a distinct entity. Am J Surg Pathol. 2017;41:1299–308.

Trpkov K, Bonert M, Gao Y, et al. High-grade oncocytic tumour (HOT) of kidney in a patient with tuberous sclerosis complex. Histopathology. 2019;75(3):440–2.

Trpkov K, Hes O, Bonert M, et al. Eosinophilic, solid, and cystic renal cell carcinoma: clinicopathologic study of 16 unique, sporadic neoplasms occurring in women. Am J Surg Pathol. 2016;40:60–71.

Trpkov K, Hes O. New and emerging renal entities: a perspective post-WHO 2016 classification. Histopathology. 2019;74:31–59.

Trpkov K, Williamson SR, Gao Y, et al. Low-grade oncocytic tumour of kidney (CD117-negative, cytokeratin 7-positive): a distinct entity? Histopathology. 2019;75:174–84.

Trpkov K, Williamson SR, Gill AJ, et al. Novel, emerging and provisional renal entities: the genitourinary pathology society (GUPS) update on renal neoplasia. Mod Pathol. 2021;34:1167–84.

Trpkov K, Williamson SR, Gill AJ, et al. Novel, emerging and provisional renal entities: the genitourinary pathology society (GUPS) update on renal neoplasia. Mod Pathol. 2021;34:1392–424.

Tsuda M, Davis IJ, Argani P, et al. TFE3 fusions activate MET signaling by transcriptional up-regulation, defining another class of tumors as candidates for therapeutic MET inhibition. Cancer Res. 2007;67:919–29.

Turbiner J, Amin MB, Humphrey PA, et al. Cystic nephroma and mixed epithelial and stromal tumor of kidney: a detailed clinicopathologic analysis of 34 cases and proposal for renal epithelial and stromal tumor (REST) as a unifying term. Am J Surg Pathol. 2007;31:489–500.

Turner RM, Tomaszewski JJ, Fox JA, et al. Metanephric adenofibroma. Can J Urol. 2013;20(2):6737–8.

Udager AM, Pan J, Magers MJ, et al. Molecular and immunohistochemical characterization reveals novel BRAF mutations in metanephric adenoma. Am J Surg Pathol. 2015;39:549–57.

Ueno-Yokohata H, Okita H, Nakasato K, et al. Consistent in-frame internal tandem duplications of BCOR characterize clear cell sarcoma of the kidney. Nat Genet. 2015;47(8):861–3.

Ursani NA, Robertson AR, Schieman SM, Bainbridge T, Srigley JR. Mucinous tubular and spindle cell carcinoma of kidney without sarcomatoid change showing metastases to liver and retroperitoneal lymph node. Hum Pathol. 2011;42:444–8.

van den Heuvel-Eibrink MM, Grundy P, Graf N, et al. Characteristics and survival of 750 children diagnosed with a renal tumor in the first seven months of life: a collaborative study by the SIOP/GPOH/SFOP, NWTSG, and UKCCSG Wilms tumor study groups. Pediatr Blood Cancer. 2008;50(6):1130–4.

Varshney N, Kebede AA, Owusu-Dapaah H, et al. A review of Von Hippel-Lindau syndrome. J Kidney Cancer VHL. 2017;4:20–9.

Velickovic M, Delahunt B, Grebe SK. Loss of heterozygosity at 3p14.2 in clear cell renal cell carcinoma is an early event and is highly localized to the FHIT gene locus. Cancer Res. 1999;59:1323–6.

Versteege I, Sevenet N, Lange J. Truncating mutations of hSNF5/INI1 in aggressive paediatric cancer. Nature. 1998;394(6689):203–6.

Vokuhl C, Nourkami-Tutdibi N, Furtwangler R, et al. ETV6-NTRK3 in congenital mesoblastic nephroma: a report of the SIOP/GPOH nephroblastoma study. Pediatr Blood Cancer. 2018;65(4):1–6.

Vujanic GM, Kelsey A, Perlman EJ, et al. Anaplastic sarcoma of the kidney: a clinicopathologic study of 20 cases of a new entity with polyphenotypic features. Am J Surg Pathol. 2007;31:1459–68.

Wach S, Taubert H, Weigelt K, et al. RNA sequencing of collecting duct renal cell carcinoma suggests an interaction between miRNA and target genes and a predominance of deregulated solute carrier genes. Cancers (Basel). 2019;12:64.

Walpole S, Pritchard AL, Cebulla CM, et al. Comprehensive study of the clinical phenotype of germline BAP1 variant-carrying families worldwide. J Natl Cancer Inst. 2018;110:1328–41.

Wang J, Papanicolau-Sengos A, Chintala S, et al. Collecting duct carcinoma of the kidney is associated with CDKN2A deletion and SLC family gene up-regulation. Oncotarget. 2016;7:29901–15.

Wang L, Zhang Y, Chen YB, et al. VSTM2A overexpression is a sensitive and specific biomarker for mucinous tubular and spindle cell carcinoma (MTSCC) of the kidney. Am J Surg Pathol. 2018;42:1571–84.

Wang X-M, Zhang Y, Mannan R, et al. TRIM63 is a sensitive and specific biomarker for MiT family aberration-associated renal cell carcinoma. Mod Pathol. 2021;34:1596–607.

Warren M, Hiemenz MC, Schmidt R. Expanding the spectrum of DICER1-associated sarcomas. Mod Pathol. 2020;33:164–74.

Weeks DA, Beckwith JB, Mierau GW, et al. Rhabdoid tumor of kidney. A report of 111 cases from the National Wilms' tumor study pathology center. Am J Surg Pathol. 1989;13(6):439–58.

Wegert J, Vokuhl C, Collord G, et al. Recurrent intragenic rearrangements of EGFR and BRAF in soft tissue tumors of infancy. Nat Commun. 2018;9(1):2378.

Wei S, Al-Saleem T. The pathology and molecular genetics of sarcomatoid renal cell carcinoma: a mini-review. J Kidney Cancer VHL. 2017;4:19–23.

Williamson SR, Al-Obaidy KI, Cheng L, et al. Distal tubular hyperplasia: a proposal for a unique form of renal tubular proliferation distinct from papillary adenoma. Am J Surg Pathol. 2021;45:516–22.

Williamson SR, Grignon DJ, Cheng L, et al. Renal cell carcinoma with chromosome 6p amplification including the TFEB gene: a novel mechanism of tumor pathogenesis? Am J Surg Pathol. 2017;41:287–98.

Williamson SR. Clear cell papillary renal cell carcinoma: an update after 15 years. Pathology. 2021;53:109–19.

Wobker S, Matoso A, Pratilas C, et al. Metanephric adenoma-epithelial Wilms tumor overlap lesions: an analysis of BRAF status. Am J Surg Pathol. 2019;43(9):1157–69.

Wong MK, Ng CCY, Kuick CH. Clear cell sarcomas of the kidney are characterized by BCOR gene abnormalities, including exon 15 internal tandem duplications and BCOR-CCNB3 gene fusion. Histopathology. 2018;72(2):320–9.

Woodward ER, Skytte AB, Cruger DG, Maher ER. Population-based survey of cancer risks in chromosome 3 translocation carriers. Genes Chromosomes Cancer. 2010;49:52–8.

Wu MK, Cotter MB, Pears J. Tumor progression in DICER1-mutated cystic nephroma—witnessing the genesis of anaplastic sarcoma of the kidney. Hum Pathol. 2016;53:114–20.

Wu MK, Goudie C, Druker H. Evolution of renal cysts to anaplastic sarcoma of kidney in a child with DICER1 syndrome. Pediatr Blood Cancer. 2016;63(7):1272–5.

Wu MK, Vujanic GM, Fahiminiya S. Anaplastic sarcomas of the kidney are characterized by DICER1 mutations. Mod Pathol. 2018;31(1):169–78.

Xie Z, Yadav S, Lohse CM, et al. Collecting duct carcinoma: a single-institution retrospective study. Urol Oncol. 2022;40(1):13.e9–13.e18.

Yang XJ, Zhou M, Hes O, et al. Tubulocystic carcinoma of the kidney: clinicopathologic and molecular characterization. Am J Surg Pathol. 2008;32:177–87.

Yeh YA, Constantinescu M, Chaudoir C, et al. Renal cell carcinoma with leiomyomatous stroma: a review of an emerging entity distinct from clear cell conventional renal cell carcinoma. Am J Clin Exp Urol. 2019;7:321–6.

Yoshida Y, Nobusawa S, Nakata S, et al. CNS highgrade neuroepithelial tumor with BCOR internal tandem duplication: a comparison with its counterparts in the kidney and soft tissue. Brain Pathol. 2018;28(5):710–20.

Yousuf H, Kumar S, Al-Moundhri M. Rarest of the rare metastatic Tubulocystic carcinoma of kidney. Cureus. 2020;12:e12117.

Zafar A, Wang W, Liu G, et al. Molecular targeting therapies for neuroblastoma: Progress and challenges. Med Res Rev. 2021 Mar;41(2):961–1021.

Zhang Y, Xia M, Jin K, et al. Function of the c-met receptor tyrosine kinase in carcinogenesis and associated therapeutic opportunities. Mol Cancer. 2018;17:45.

Zhou M, Kort E, Hoekstra P, et al. Adult cystic nephroma and mixed epithelial and stromal tumor of the kidney are the same disease entity: molecular and histologic evidence. Am J Surg Pathol. 2009;33:72–80.

Zhou M, Yang XJ, Lopez JI, et al. Renal tubulocystic carcinoma is closely related to papillary renal cell carcinoma: implications for pathologic classification. Am J Surg Pathol. 2009;33:1840–9.

Jiayu Chen, William G. Nelson, Karen Sfanos,
Srinivasan Yegnasubramanian, and Angelo M. De Marzo

Contents

Introduction

Biology of Prostatic Epithelium

- The prostate is not fully developed until after puberty. The normal adult prostate epithelium consists of a single layer of columnar luminal cells situated above a single layer of cuboidal basal cells (Fig. 14.1). Rare neuroendocrine cells are interspersed throughout the epithelium.

J. Chen · W. G. Nelson · K. Sfanos · S. Yegnasubramanian
A. M. De Marzo (✉)
Departments of Pathology, Urology and Oncology, The Sidney Kimmel Comprehensive Cancer Center at Johns Hopkins, Johns Hopkins University School of Medicine, Baltimore, MD, USA
e-mail: ademarz@jhmi.edu

- Basal cells separate luminal cells from the basement membrane and often extend cytoplasmic projections that intercalate between the basolateral aspects of the luminal cells.
- Basal cells express nuclear p63, keratins 5 and 14, and many have low levels (albeit non-negative) of AR, NKX3.1, and HOXB13. They do not express PSA/KLK3.
- Basal cells are traditionally thought to be the stem/progenitor cells of prostate epithelium; in mice, especially when the tissue is damaged or inflamed, they can proliferate and differentiate into luminal cells.
  - Prostatic epithelial cell turnover is slow in normal conditions; proliferation is very infrequent in normal luminal cells, occurring more frequently in normal-basal cells.
  - Most proliferation in normal epithelium is found in the basal cell compartment (e.g., 70% of cells expressing Ki67).
  - Putative multipotent basal progenitor cells are enriched near the proximal ducts/urethra.
- Loss of basal cells is a hallmark feature of prostate adenocarcinoma.
- Mature luminal cells carry out the differentiated functions of the prostate, including secretion of PSA and other components into the acinar lumens to contribute to the ejaculate fluid.
- Luminal cells express high levels of "differentiation markers" including the androgen receptor (AR), prostate-specific antigen (PSA, encoded by KLK3) prostate-specific acid phosphatase, NKX3.1, HOXB13, keratins 8 and 18, and FOXA1.
- Maintenance of this differentiated status requires androgens; castration results in decreased expression of androgen-regulated genes (e.g., NKX3.1 and KLK3/PSA), apoptosis of many luminal cells, an atrophic cuboidal appearance of the remaining luminal cells, and a prominence of the basal layer.

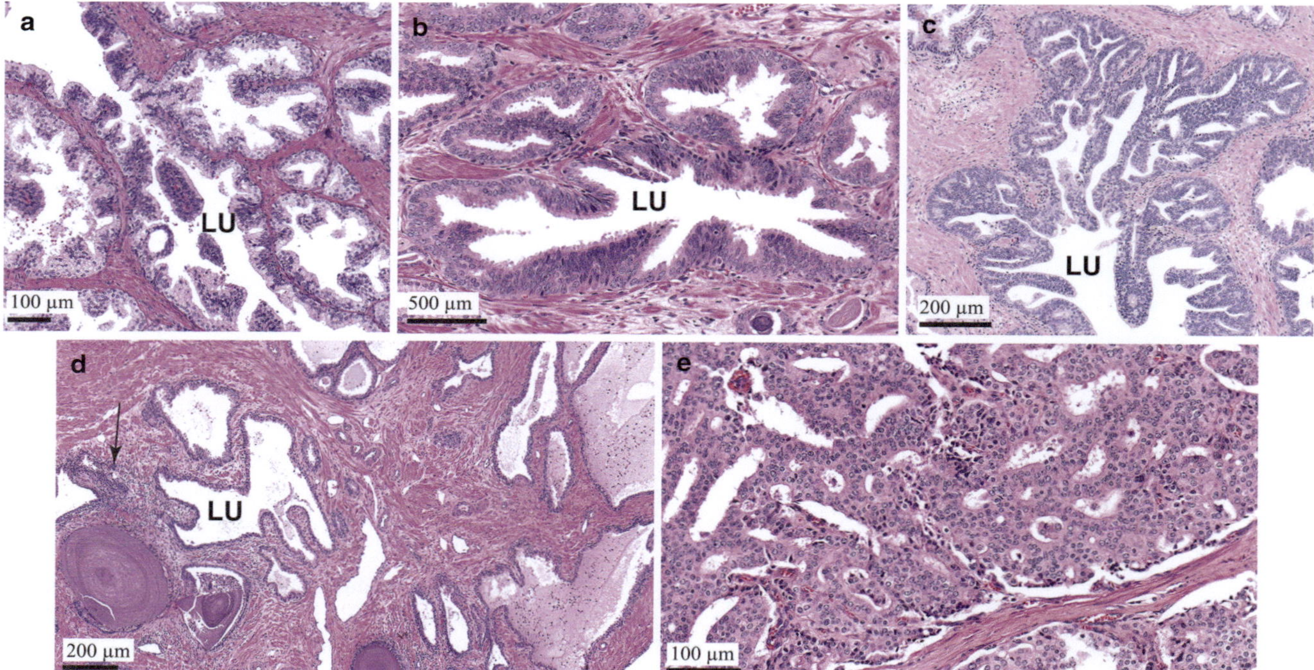

Fig. 14.1 Histopathology of normal, preneoplastic, intraductal, and invasive adenocarcinoma. (**a**) Normal appearing epithelium and stroma. (**b**) High-grade PIN. (**c**) Intraductal carcinoma. (**d**) Simple atrophy/PIA. (**e**) Invasive adenocarcinoma of the prostate, Gleason pattern 4, large cribriform

- Atrophy occurring in luminal cells after androgen withdrawal or blockade is considered "hormonal" or diffuse atrophy.
- The nature of the "true" long-lived stem cells in the prostate is somewhat controversial; most evidence suggests this activity is predominantly in the basal compartment with enrichment toward the urethra, but some studies indicate that both basal and luminal cells can each self-renew; and at times, give rise to both cell types.
- In regions of focal atrophy (that are not associated with androgen withdrawal/blockade and often accompanied by chronic inflammation; referred to as proliferative inflammatory atrophy or PIA), there are variable numbers of "intermediate" luminal cells that have reduced yet variable levels of AR and differentiation markers, and many express keratins typical of both basal and luminal cells (e.g., keratin 5); they show a relatively high proliferative fraction and these cells appear to be efficient progenitor cells in stem-like cell assays.
- Recent studies using single-cell RNA sequencing have shown cellular heterogeneity within the luminal populations:
    - One population is mainly secretory, and the other is secretory but contains more stem cell properties.
- Neuroendocrine cells are very rare, encompassing <1% of all epithelial cells.
- They express chromogranin A, synaptophysin, neuron-specific enolase, neural cell adhesion molecule, fork-head-box A2, and CXC chemokine receptor 2; they are negative for AR and PSA.
- Primary prostatic adenocarcinomas (by far the most common histological type) nearly always (except for very rare primary tumors characterized by p63 nuclear positivity) have phenotypic features of luminal cells, suggesting the cell of origin for prostate cancer is a luminal cell.
- The prostatic stroma consists of abundant smooth muscle cells along with nerves (controlling smooth muscle function during ejaculation), blood vessels, indistinct fibroblasts, and scattered immune cells.

Epidemiology and Etiology of Prostate Cancer

- Incidence and Mortality.
    - Prostate cancer (prostatic adenocarcinoma) is the most common noncutaneous malignancy in men.
    - While low-grade prostate cancers (Gleason score 6 or grade group 1) may remain clinically indolent for many years, higher grade lesions may progress to lethal metastatic disease and death.
    - Prostate cancer is second only to lung cancer in cancer-related deaths in males, with 34,130 men estimated to die of this disease in the United States in 2021.
    - In the United States, the lifetime risk of a prostate cancer diagnosis is roughly 1 in 9, yet the risk of dying is roughly 1 in 41;

- This indicates the need to determine which tumors are potentially aggressive and life threatening and which are not.
- Globally, there are more than 1.2 million new cases and deaths exceed 350,000 annually.
- The incidence of prostate cancer has been rising in a number of Asian countries.
- The major risk factors for the development and progression of prostate cancer include advancing age, family history/germline genetics, and race.
- The sharp increased risk with age results in particularly high levels of cases in regions with high life expectancy.
- The disease disproportionately impacts African American/Black men, with an approximately twofold higher incidence and mortality compared with non-Hispanic White men.
  - Recent results indicate that, despite the increased incidence and mortality, when controlling for grade and stage of disease, and access to high-quality care, the rate of progression to metastatic disease and death rate from prostate cancer in Black men is not different than in White men.
  - Recent studies also suggest improved outcomes after radiation therapy for Black men as compared with White men after treatment for localized disease.
- Environmental Factors
  - Environmental exposures are implicated in prostate cancer since men who emigrate from South East Asia

to North American or Australia develop a higher risk of prostate cancer within 1 generation;
- Dietary factors are believed to underlie these risks.
  ○ Diets rich in red meats and well done meats have been implicated.
- The most well-recognized precursor to invasive prostate cancer is high-grade prostatic intraepithelial neoplasia (PIN).
- Chronic inflammation may drive disease development through increased oxidative damage and sublethal and lethal injury to epithelial cells leading to regeneration and development of PIA (Fig. 14.2).
- PIA lesions contain intermediate/progenitor liminal cells that may lead to PIN, and/or at times directly to early invasive carcinomas.
- Recent evidence suggests that bacterial infection in association with chronic inflammation may at times drive the development of TMPRSS2:ERG gene fusions in PIA lesions.
- Obesity and weight gain are associated with increased disease recurrence after primary treatment.

Clinical Features

- The widespread use of prostate-specific antigen (PSA) serum testing starting in the 1990s in the US revolutionized the early detection of prostate cancer, resulting in a

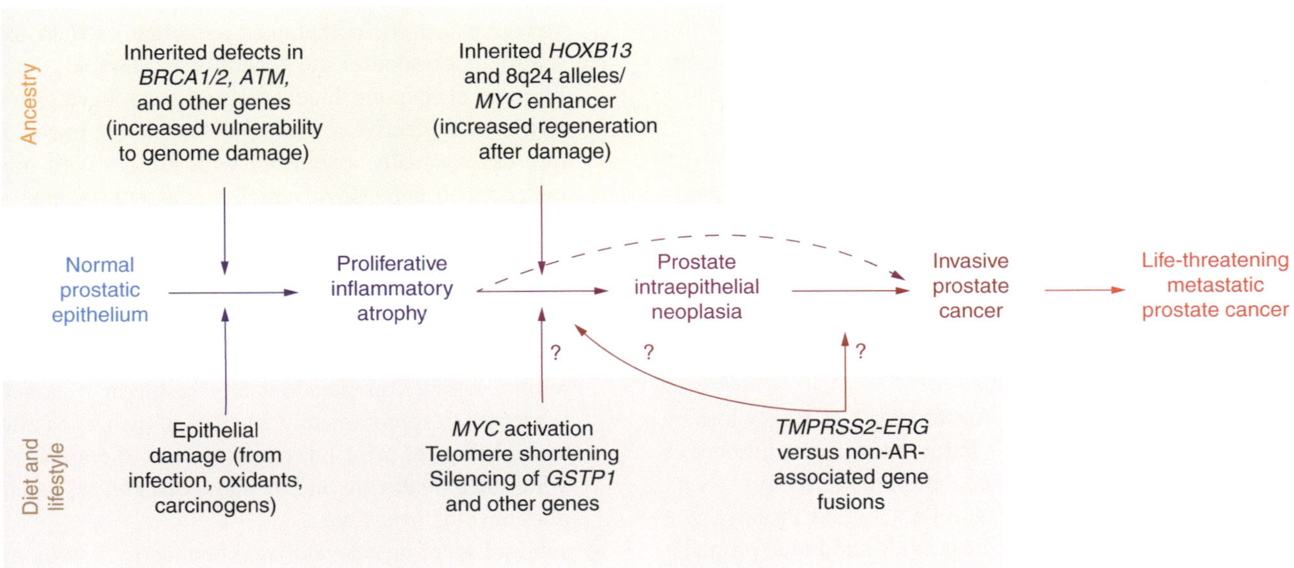

Fig. 14.2 Diet, lifestyle, and ancestry converge to produce proliferative inflammatory atrophy to drive the molecular pathogenesis of prostate cancer. Inherited vulnerability to cell and genome damage repair and response sensitizes prostate cells to infections, inflammation, and carcinogens, leading first to proliferative inflammatory atrophy and then to neoplastic transformation and malignant progression. Gene rearrangements could occur via AR-dependent mechanisms, like TMPRSS2-ERG, or non-AR-dependent mechanisms. (From Nelson 2022 J Clin Invest. with permission)

marked increase in detection, and a stage migration at diagnosis from predominantly metastatic to mostly clinically localized.

- The death rate for prostate cancer has been decreasing over the last few decades, and some of this decrease is likely attributable to PSA screening and early detection followed by radical treatment of primary cancers with surgery, and/or radiation therapy, the latter with or without combined androgen deprivation.

- Despite the success of PSA-driven early detection and treatment, PSA testing has also led to overtreatment of nonlife threatening disease (Grade Group 1 or GS6) in which men are subjected to potential serious side effects from radical treatment, yet do not stand to benefit because their disease would not progress to a symptomatic or life-threatening aggressive form in their lifetime.

- To decrease overtreatment, recommendations regarding screening changed (US Preventative Task Force 2012 grade D recommendation) such that there is now less PSA screening, especially in men over age 75, as well as an increasing use of active surveillance in men that are diagnosed with low grade (Grade Group 1) disease.
  - However, recent data indicate that while the incidence of localized prostate cancer decreased after recommendations to reduce PSA screening were implemented, there has been an increase in men presenting with metastatic disease, indicating a potentially clinically detrimental cost of such reduced screening.

- Advances in prostate imaging, especially multiparametric MRI, are improving diagnostic accuracy and increasing the safe use of active surveillance.

- Newer types of imaging are promising to improve this even further, including PET imaging for PSMA and using combined information from both mpMRI and PET-PSMA imaging.

- Clinically, the major known determinants of indolent versus aggressive disease and treatment decisions are largely based on the pathological grade from needle biopsies, as well as the serum PSA and clinical and pathological disease stage.

- For patients with intermediate risk disease, the clinical course is quite variable and not well predicted by Gleason grading and clinical staging.
  - A number of single biomarkers such as PTEN loss by IHC or FISH, and Ki67 Index, as well as a number of commercial RNA-based multiplex assays (e.g., Genomic Health Oncotype Dx, Myriad Prolaris, and GenomeDx Decipher) can provide additional prognostic information beyond typical clinico-pathological variables; however, none are used routinely or widely in standard clinical practice.

- For clinically localized prostate cancer, radical prostatectomy or radiation therapy (with or without combined androgen deprivation treatment) remains the mainstay of treatment.

- After primary treatment, combined histopathological and clinical features are often used in algorithms such as the Cancer of the Prostate Risk Assessment Post-Surgical score (CAPRA-S) as prognostic tools; studies are continuing to determine whether the addition of biomarkers, such as those indicated above, can add prognostic value beyond these features.

- Approximately 30% of men with intermediate or high-risk adenocarcinoma that are treated with curative intent experience disease recurrence, which generally starts out as biochemical recurrence (increased serum PSA).

- Many men with biochemical recurrence are treated with combined androgen deprivation therapy (> 99% of primary adenocarcinomas express high levels of the androgen receptor), and some of these recurrences become castration resistant even prior to metastatic disease development. Others, whether treated or not for biochemical recurrence, develop distant metastatic disease.

- The most frequent site of metastasis is bone, with lymph nodes and liver also being involved commonly. Other metastatic sites may include the lungs and adrenals.

- Patients with metastatic disease (or at times with local recurrence or biochemical recurrence) are treated with combined androgen deprivation therapy, which results in initial responses in nearly all men; however, nearly all men progress to castration resistant metastatic disease (many are subsequently treated with second and third line hormonal therapies that can provide benefit but are not curative).
  - Taxane-based chemotherapies are often used in this setting, but responses are generally not durable.
  - Immune checkpoint blockade treatments have generally been ineffective so far in prostate cancer, except in rare cases usually associated with tumors with mismatch repair defects and a high mutational burden.
  - Some men with homologous recombination repair defects, such as those caused by BRCA2 mutations, appear to benefit from PARP inhibitors.
  - An additional promising approach in the CRPC setting is the administration of intermittent high dose testosterone, which can paradoxically result in treatment responses in approximately 20–30% of men, even after several lines of prior hormonal or other therapies. As with other treatments in late-stage disease, resistance does develop over time.

- A small subset of men develop neuroendocrine carcinoma (with a spectrum including overt small cell neuroendocrine carcinoma (SCNC), very rare large cell neuroendocrine carcinoma, to very high grade poorly adenocarcinomas with prominent neuroendocrine features), or an otherwise androgen receptor reduced or negative disease; this can occur very rarely in a primary hormone naive state, or more commonly after a number of rounds of androgen deprivation therapy. Current estimates range from between 5 and 20% of late-stage cases. In almost all cases, these tumors appear to arise from lineage plasticity occurring in a preexisting clonally related concomitant adenocarcinoma. Mechanistic studies, along with molecular studies of clinical samples, suggests that combined complete inactivation of RB1 and TP53 mutations are key drivers of this transition. Since these tumors are highly resistant to all standard chemotherapies, the biological/molecular nature of these lesions is under intense study (see below).

Histopathology of Prostate Cancer

Precursor Lesions

- PIN (prostatic intraepithelial neoplasia) is defined as the presence of cells with morphological features of adenocarcinoma, often with cellular crowding and pseudostratification, present within preexisting ducts and acini (Fig. 14.1b). The diagnosis of high-grade PIN in almost all cases requires marked nucleolar enlargement in at least 10% of the cells.
- Low-grade PIN has similar features but lacks the pervasive nucleolar enlargement.
- Most early prostatic adenocarcinomas are likely derived from high-grade PIN (HGPIN), although some have been associated more directly with PIA without HGPIN.
- At times it may be difficult to distinguish high-grade PIN from intraductal spread of adenocarcinoma.

Intraductal Carcinoma (IDC-P)

- Intra-acinar/intraductal spread of prostatic adenocarcinoma (Fig. 14.1c) occurs frequently in cases from grade groups 3–5. In prostatectomies, it can be recognized by the expansion of preexisting ducts and acini by carcinoma cells, which often show a cribriform or solid pattern.
  - Intraductal carcinoma has strict diagnostic criteria when diagnosed on needle biopsy. However, this criteria likely results in an underestimate of actual intracinar/intraductal spread of preexisting adenocarcinoma into benign glands/acini, and efforts are underway to better distinguish HGPIN from intraductal carcinoma molecularly.
    - The presence of IDC-P is a prognostic marker that is often associated with higher grade cancer, higher

cancer-specific mortality, as well as distant metastasis at initial presentation.
  - Loss of PTEN is also a promising, albeit not highly sensitive, biomarker to distinguish intraductal carcinoma vs. high-grade PIN, since it is common to lose PTEN in intraductal carcinoma, but not in HGPIN.

Acinar Adenocarcinoma

- Most prostate cancers are acinar adenocarcinomas that arise from the peripheral zone of the prostate, less commonly from the transition zone (the site of most benign prostatic hyperplasia) and very infrequently from the central zone (Fig. 14.3).
- Most cases are multifocal that often have proven to be clonally distinct; often, there is a dominant nodule that is also the highest grade lesion.
  - The increased sophistication of molecular diagnostic techniques has allowed for the molecular distinction of separately arising lesions within the prostate.
  - These findings have confirmed the tumor heterogeneity in prostate cancer even within the same patient.
  - ERG IHC is a useful marker for a more rapid assessment of multifocality in prostate cancer.
  - A recent study using a combination of multiple proteins and DNA markers identified interfocal molecular heterogeneity in ~60% of primary prostate tumor samples as well as ~10% collision tumors as evidenced as discordant ERG/SPINK1 status.
- Diagnosis
  - Criteria for invasive carcinoma include several features that together aid in the final diagnosis.
  - A characteristic hallmark in almost all carcinomas is that many of the tumor cells contain enlarged prominent nucleoli.
    - Tumor cells also frequently show hyperchromasia, and in almost all cases, nuclear enlargement.
    - In well differentiated carcinomas, atypical glands are often smaller than benign/normal glands, have straight liminal borders and infiltrate into the stroma between benign glands.
  - Diagnostically, specific features for carcinoma include perineural invasion, glomeruloid formations, mucinous fibroplasia, or seminal vesicle invasion.
  - In difficult cases, one can employ basal cell-specific staining to demonstrate the absence of basal cells (keratins 5/14 or p63; or a combination of AMACR, p63, and basal cell keratins referred to as a PIN4 stain).

Prostate zones

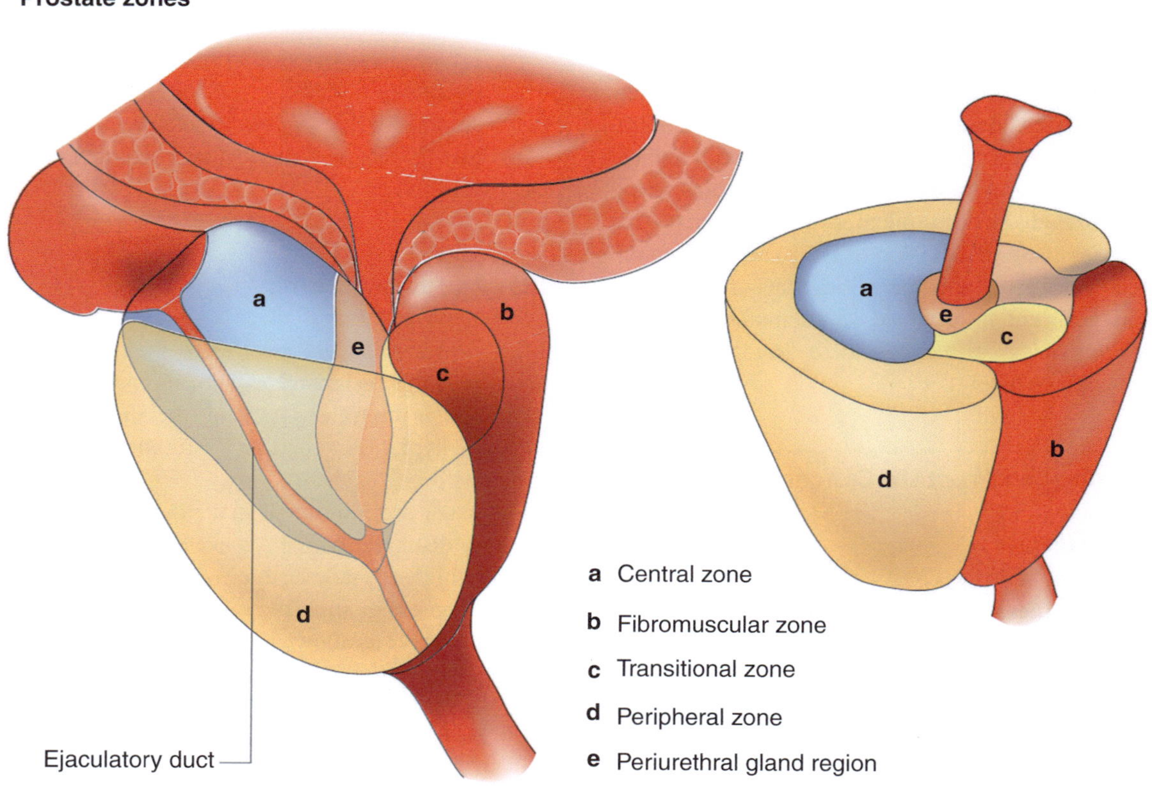

	Prostate zone		
	Peripheral	Transition	Central
Focal atrophy	High prevalence	Medium-high prevalence	None
Acute inflammation	None	None	None
Chronic inflammation	Medium-high prevalence	Medium-high prevalence	None
Benign prostatic hyperplasia	None	High prevalence	None
High-grade PIN	Medium-high prevalence	Low prevalence	None
Carcinoma	Medium-high prevalence	Low prevalence	None

Legend: ▮ High prevalence ▮ Medium-high prevalence □ Low prevalence □ None

Fig. 14.3 Zonal predisposition to prostate disease. Most cancer lesions occur in the peripheral zone of the gland, fewer occur in the transition zone and almost none arise in the central zone. Most benign prostate hyperplasia (BPH) lesions develop in the transition zone, which might enlarge considerably beyond what is shown. The inflammation found in the transition zone is associated with BPH nodules and atrophy, and the latter is often present in and around the BPH nodules. Acute inflammation can be prominent in both the peripheral and transition zones, but is quite variable. The inflammation in the peripheral zone occurs in association with atrophy in most cases. Although carcinoma might involve the central zone, small carcinoma lesions are virtually never found here in isolation, strongly suggesting that prostatic intraepithelial neoplasia (PIN) lesions do not readily progress to carcinoma in this zone. Both small and large carcinomas in the peripheral zone are often found in association with high grade PIN, whereas carcinoma in the transition zone tends to be of lower grade and is more often associated with atypical adenomatous hyperplasia or adenosis, and less often associated with high grade PIN. The various patterns of prostate atrophy, some of which frequently merge directly with PIN and at times with small carcinoma lesions, are also much more prevalent in the peripheral zone, with fewer occurring in the transition zone and very few occurring in the central zone. Upper drawings are adapted from an image on Understanding Prostate Cancer website. PIN, prostatic intraepithelial neoplasia. (From De Marzo 2007 Nat Rev. Cancer, with permission)

  – There are a number of histological features, that if present as the sole finding on needle biopsy, can make it difficult to render a clear diagnosis. These include atrophic carcinoma, foamy gland carcinoma, pseudo-hyperplastic carcinoma, and mucinous carcinoma.
  – Very rare tumors express nuclear p63 diffusely. These distinct lesions have bland nuclei and so far have not been shown to be aggressive.
- Other histological patterns of acinar adenocarcinoma include atrophic glands, pseudohyperplastic adenocarcinoma, microcystic, and foamy gland carcinoma. The primary significance of these patterns are that each can be misconstrued as benign on needle biopsies, as they may mimic benign glands.
- Adenocarcinomas frequently show mucinous differentiation and at times may show prominent extracellular mucin.
- Carcinomas may also contain signet ring-like cells, which often, albeit not always, appear as Gleason pattern 5. These cells accumulate lipid and not mucin.

Ductal Adenocarcinomas

- Tumor cells are columnar with hyperchromasia, basally located nuclei, and a pseudostratified appearance.
- The glands may be cribriform, or show prominent papillary infoldings with fibrovascular cores.
- Most are found mixed with acinar adenocarcinomas (usually grade group 3 or higher), but in rare cases these may be present as a lone component at the prostatic urethra as distinct papillary lesions seen on cystoscopy and found by transurethral resection.
- There is generally no known histological or molecular distinction between prostatic ducts and acini, unless one observes a long duct radiating from the urethra outward. Thus, other than convention and the fact that at times they appear to arise near the urethra, there is not a strong biological basis for referring to these as ductal versus adenocarcinomas.
- These tumors tend to present with relatively low PSA levels for their volume and behave somewhat aggressively, often with visceral metastases.

Rare Subtypes

Neuroendocrine Carcinoma (NEPC)

- These come in two major types, those that are very well differentiated and traditionally considered carcinoid tumors and those that are poorly differentiated, which include a spectrum from SCNC to large cell neuroendocrine carcinoma (LCNC, much more rare).
- Carcinoid tumors of the prostate are very rare and will not be considered further.
- Most primary SCNC are mixed with acinar or other subtypes. While they may express androgen receptor (AR), it is usually at low levels and most show low or absent PSA expression, although other prostate restricted markers such as NKX3.1 may still be expressed; again, often at low levels.
- An evolving panel of neuroendocrine markers is being employed to better classify these lesions; newer markers include loss of YAP1, loss of cyclin D1, loss of RB1, and strong expression of FOXA2 and INSM1. Traditional markers such as chromogranin, synaptophysin, and CD56 may be positive, but not in all cases.
- Molecular studies have shown that NEPC can be driven to arise from adenocarcinoma cells by transdifferentiation/lineage plasticity after concomitant inactivation of both alleles of RB1 and TP53 with upregulation of SOX2 and EZH2.
- It is still possible that a small subset of NEPC may arise directly from prostatic basal cells and/or from pre-existing neuroendocrine cells in the tumor.
- Another recent study has found extensive reprogramming of the FOXA1 transcriptome in a series of NEPC xenografts that was required for maintenance of the NEPC phenotype.
- SCNC and LCNC are extremely aggressive lesions and most patients succumb to metastatic disease within a few years of diagnosis.

Other Histological Variants/Patterns of Differentiation

- Many primary acinar adenocarcinomas contain neuroendocrine cells (staining positive for chromogen and/or synaptophysin that are present in numbers from a scattered few to relatively frequent), but these tumors do not behave like SCNC and LCNC.
- Some poorly differentiated tumors with prominent Gleason pattern 5 sheet-like differentiation appear to be hybrids with parts of the tumor showing evidence of neuroendocrine differentiation with low/negative AR staining and signaling (e.g., PSA and/or NKX3.1 expression) and others showing retained strong AR staining and signaling.
- Some poorly differentiated carcinomas, along with SCNC, have been referred to as "anaplastic" or more recently, "aggressive variant" carcinomas, although these terms also relate to clinical behavior as very aggressive; and, they tend to be at least somewhat responsive to platinum based therapies.

- Other rare histological subtypes of prostatic carcinoma include sarcomatoid carcinoma, PIN-like ductal carcinoma, pleomorphic giant cell adenocarcinoma, and squamous carcinoma.

Grading of Adenocarcinoma

- Grading of adenocarcinoma of the prostate has been based on the Gleason system for several decades. Since 2005, several modifications have been made by the International Society of Urological Pathology (ISUP 2014) and more recently the Genitourinary Pathology Society (GUPS).
- The system is based on the fact that there are consistent glandular architectural patterns of invasive prostatic adenocarcinoma and that more than one pattern is often present in a given tumor lesion.
- Each pattern is given a numeric value, from 1 to 5, based on increasing levels of architectural distortion starting from glands that appear nearly benign, to those consisting of sheets of cells lacking acinar formation.
- Traditionally, to arrive at a Gleason score, one takes the most common pattern and adds it to the second most common pattern (e.g., $3 + 4 = 7$).
- In needle biopsies, however, one now takes the most common and the highest grade.
- The adoption of grade groups (GGs) has occurred that start at GG1 (Gleason score of 6 in GUPS system) and end at GG5.
- In prostatectomies, the grade can include a tertiary pattern and if this is deemed greater than 5%, then it becomes incorporated as the secondary grade.
- More recent developments have added an estimation of the percentage of pattern 4 in Gleason 7 cancers and the presence of cribriform patterns (Fig. 14.1e), although precisely how to define this, and the ability to distinguish it from intraductal carcinoma, is still somewhat in flux.
- Currently, there are a few differences between the ISUP 2019 and GUPS systems. For example, the 2019 ISUP allows for some grade $3 + 4 = 7$ lesions to be included as GG1, whereas the GUPS system does not.
  - Therefore, for precise communication with clinicians, pathologists should designate which system they are using when reporting grade groups.
- Clinical progression is uncommon in low-grade (e.g., Gleason 6 = GG1 and low volume GG2) cases, and many men now elect not to undergo immediate definitive treatment but instead opt for "active surveillance".
- Despite this, within the middle of the grade groupings, there is a wide variation in disease progression and additional tools and treatment approaches are needed.

- Such tools are being developed and evaluated, including a number of molecular biomarkers.

Artificial Intelligence in Prostate Cancer Histopathology

- This field is moving rapidly and recent work indicates that AI-based systems can perform as well or better than expert genitourinary pathologists at diagnosing and grading prostate cancer on needle biopsies.
- Many additional studies are underway to determine precisely how AI-based technologies, using digitally scanned slides, will augment the ability of pathologists to diagnose, grade, and predict outcomes and response to treatments worldwide.

Molecular Features of Prostate Cancer

Germline Alterations

- While there is not a specific gene, such as *APC* for hereditary colorectal cancer, that when inherited in mutant form severely increases the risk of prostate cancer, family and twin studies implicate a strong hereditary contribution.
- Large-scale genome-wide association (GWAS) studies have implicated many loci and some genes and variants have consistent associations from multiple studies including:
  - *HOXB13* (17q21)
    - Encodes a homeobox transcription factor that is expressed in adult tissues in a prostate and distal GI tract-specific manner.
    - Germline mutations/variants associated with increased risk of prostate cancer are enriched in the conserved homeodomain that interacts with homeobox cofactor MEIS1.
    - The *HOXB13* G84E variant is higher in men of European ancestry among affected men and those diagnosed at a younger age or with a family history of prostate cancer.
      - May be associated with pseudo-hyperplastic features, less frequent ERG rearrangements, and more SPINK1 overexpression.
      - A recent study shows that wild-type HOXB13 binds to HDAC3, repressing lipogenic regulators, whereas HOXB13 G84E does not; this was reported to result in increased expression of key prostate cancer growth regulators including FASN (encoding fatty acid synthase).
    - *HOXB13* G132E is associated with increased risk in Japanese and Chinese men.

- o An African variant (X285K, a stop-loss mutation resulting in a longer protein) is associated with early onset and increased disease aggressiveness.
- *MYC*
  - *MYC* is located on chromosome 8q24, a region that undergoes somatic copy number increases in aggressive prostate cancer.
  - Several inherited variants located on chromosome 8q24 near *MYC* have been associated with an increased risk for prostate cancer (approximately 15 independent risk variants).
    - o The majority of these are more frequent in African American men than men with European ancestry.
      - One such rare variant (rs72725854 [A>G/T] (~6% frequency of the African ancestry specific "T" risk allele) is localized within a prostate cancer-specific enhancer region that can modulate expression of *MYC*, and several nearby long noncoding RNAs including *PCAT1* and *PVT1*, sensitizing then to androgen regulation.
- DNA repair genes: studies implicate germline mutations in DNA repair related genes that impart increased risk of overall and aggressive (higher grade) cancers.
  - Homologous recombination (HR) pathway for double-strand DNA repair: *BRCA2, BRCA1, ATM, CHEK2, PALB2* (and other repair genes).
    - o A recent study estimating the prevalence of germline *BRCA2* mutations in the United Kingdom resulted in an estimated 8.6-fold increased risk of prostate cancer by age 65, which corresponds to an absolute risk of 15% by age 65.
    - o Germline mutations in DNA repair genes including *BRCA1, BRCA2, ATM,* and *CHEK2* have been associated with more aggressive prostate cancer and worse outcomes.
      - In patients with biallelic inactivation of *BRCA2* in their cancers, approximately 50% inherited a mutated inactive allele.
        - *CHEK2*
          - *CHEK2* encodes a cell cycle checkpoint kinase that is activated by DNA damage and leads to either cell cycle arrest until the DNA is repaired, or apoptosis.
          - Germline mutations in *CHEK2* are associated with a higher risk of prostate cancer (found in 1–2% of cases).
            - One of the most common *CHEK2* mutations, c.1100delC, is enriched in lethal prostate cancer in European American patients compared to indo-

lent prostate cancer or patients from other origins.
- A recent study involving a small number of prostate cancer patients suggested frequent co-occurrence of germline *CHEK2* mutations and somatic *CDK12* mutations, indicative of a potential synergistic effect.
  - *PALB2*
    - Encodes a protein that links BRCA1 and BRCA2 during the HR process of DNA double-strand break repair.
    - Germline mutation prevalence is approximately 0.29% in a Polish population and was associated with more aggressive disease, lower 5-year survival, and a higher all-cause mortality rate.
- o Castration-resistant prostate cancer patients carrying germline homologous recombination defects showed better response to platinum treatment.
- o Prostate cancer patients carrying germline or somatic mutations in DNA repair genes, especially *BRCA2*, showed higher sensitivity to PARP (poly-ADP ribosylase) inhibitors (e.g., olaparib and rucaparib).
  - DNA mismatch repair genes (MMR).
    - o *MLH1, MSH2, MSH6,* and *PMS2* – canonical genes.
    - o Germline mutation frequency is approximately 1% in advanced prostate cancer and much less so in localized disease.
    - o Several studies have reported potentially favorable responses to checkpoint blockade immunotherapy in prostate cancer patients with MMR-deficiency/MSI-H, potentially through a higher presence of tumor-infiltrating T cells recognizing neoepitopes in those tumors with a high mutational burden.
  - *AR*
    - o Studies have shown that in populations with a higher incidence of prostate cancer (African Americans), *AR* may have shorter polymorphic polyglutamine repeats, which are associated with increased receptor activity.
    - o This is in contrast to populations with a low incidence of prostate cancer (Asians) who have been reported to have longer polymorphic polyglutamine repeats.
    - o This has led to speculation that the length of these repeats affects prostate cancer susceptibility—however, there is conflicting evidence for this.

- *TP53*
  - A recent study found germline *TP53* mutations occur in ~0.6% of prostate cancer patients; and for Li-Fraumeni syndrome (LFS) patients with a germline *TP53* mutation, the incidence of prostate cancer is 25-fold higher compared to the general population.
  - Tumors harboring *TP53* germline mutations often present with higher grade and stage, with 2/3 of them also possessing a somatic second allele inactivation.
  - Mutational hotspots in these prostate tumors are different from the classical LFS *TP53* mutations.
- Clinical relevance on hereditary cancer genetic testing
  - Due to the high prevalence of hereditary genetic mutations in high grade and mCRPC, germline genetic testing is rapidly developing and helps to determine optimal disease management options for screening, active surveillance, and precision therapies.
  - Testing has progressed from single-gene to multigene panels.
    - Common testing options include genes in the DNA damage repair pathways, *TP53* and *HOXB13*.
  - Multiple organizations, including the National Comprehensive Cancer Network, have provided guidelines for germline testing criteria for prostate cancer.
    - Guidelines are consistent in recommending genetic testing to men with prostate cancer with any of the following characteristics: metastatic disease, high or very high risk (based on stage and Gleason pattern), Ashkenazi Jewish ancestry, or intraductal or cribriform histology, or family history of mutations in known related cancer-risk genes.
  - Sample types mainly include saliva, blood, cheek, and/or buccal swabs and sometimes skin punch biopsies.
    - Some institutions also use other methods such as immunohistochemistry to detect germline mutations including MMR for high-grade prostate cancer (GS9-10).

Somatic Genomic Alterations in Prostate Cancer

- As in other cancers, there is a stepwise acquisition of molecular alterations during the development and progression of prostate cancer.
- Whole genome and whole exome sequencing efforts revealed gene fusions to ETS family members, with *TMPRSS2–ERG* rearrangements as the most common somatic genomic alterations. A number of additional driver genes undergo recurrent point mutations (e.g., *SPOP, FOXA1, TP53, PTEN; KMT2C, KDM6A, CHD1, ATM,* etc.) (Fig. 14.4).
- There are frequent copy number alterations and complex genome rearrangements.
- The importance of epigenetic mechanisms in tumorigenesis is well established.

ETS Gene Fusions

- The ETS (E26 transformation-specific) gene family encodes a group of transcription factors that all share a conserved ETS domain responsible for DNA-binding activity.

Fig. 14.4 Mutational significance in 1013 prostate cancers and enrichment of genomic alterations in metastatic tumors. (**a**) Recurrently mutated genes (*n* = 97). Genes are ordered by frequency, and mutations are stratified by mutation type and, for missense mutations, by recurrence. Recurrence is defined via http://cancerhotspots.org/, http://oncokb.org/, and COSMIC; truncating mutations are defined as frameshift, nonsense, splice, and nonstop. (**b**) Mutations in epigenetic regulators and chromatin remodelers are significantly enriched in ETS-fusion-negative tumors. P values are calculated using a two-tailed Fisher's exact test and shown for ETS fusions as compared to all epigenetic mutations (including those co-occurring with SPOP and CUL3) and for ETS fusions as compared to non-overlapping mutations in epigenetic modifiers only. (**c**) Cohort-wide view of mutations in epigenetic regulators and chromatin remodelers, which affect 20% of samples. Samples are shown from left to right (only the 202 tumors with alterations are shown, out of 1013), and gene alterations are color-coded by mutation type and, for missense mutations, by assumed driver status; mutations are assumed to be drivers if they have been previously reported and entered into COSMIC or annotated in OncoKB or variants of unknown significance (VUS). (**d**) Most genomic alterations are enriched in metastatic disease. Alteration percentages in metastatic samples (*n* = 333) are shown on the x axis, and those in primary samples (*n* = 680) are shown on the y axis. The significance of enrichment (two-sided Fisher's test q value or weighted permutation test) is shown by the size of the dots. Genes in bold have a significant enrichment of mutations using Fisher's test and weighted permutation test correcting for mutation burden. (From Armenia 2018 Nat. Genet. with permission)

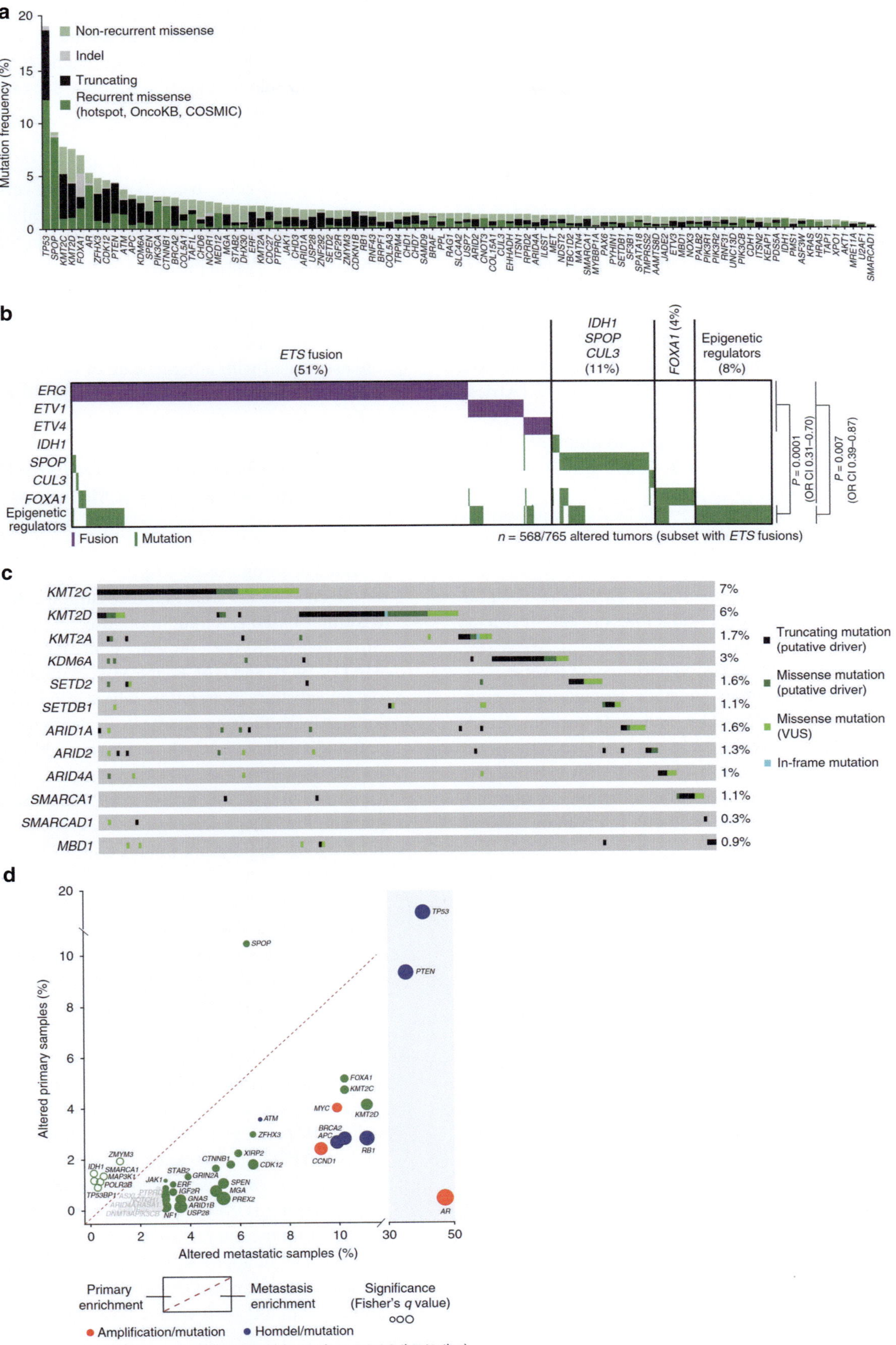

- These proteins play an important role in normal development with distinct spatial-temporal-specific expression patterns.
- Five ETS genes, *ERG, ETV1, ETV4, ETV5,* and *FLI1,* have been identified to be rearranged in prostate cancer, leading to overexpression of transcripts with truncations at the 5′ ends.
  - *ERG* (21q22) is the most commonly rearranged member, ranging from 20 to 50% of both localized and metastatic prostate cancer.
  - Significant variation of the prevalence among men in different racial/ancestral and ethnic backgrounds is seen.
    ○ The prevalence is approximately 50% in White men of European ancestry, as low as 17% in Asians, and approximately 25% in Black men (African American and African Caribbean).
  - *ETV1* (7p21) is the second most common rearranged gene, found in up to 8–10% of all prostatic adenocarcinomas.
  - *ETV4* (17q21), *ETV5* (3q27), and *FLI1* (11q24) are rearranged in 1–5% of cases.
- The most common 5′ fusion partner (~85% of all cases harboring ETS rearrangements) is *TMPRSS2* (21q22), which encodes an androgen-inducible, prostate-restricted transmembrane serine protease.
  - Other less frequent 5′ fusion partners are diverse.
- In addition, some primary prostatic carcinomas overexpress full-length *ETV1, ETV4,* and *FLI1* without detectable gene rearrangement.
- For *TMPRSS2-ERG*, gene fusions occur through two predominant mechanisms: interstitial deletion on chromosome 21 or translocations without intervening genetic loss.
- The fusion of the gene leads to the androgen-mediated overexpression of the particular ETS transcription factor via the *TMPRSS2* regulatory region, which leads to incomplete cellular differentiation and modified AR transcriptional output, enhanced NOTCH signaling, as well as increased cellular migration and invasion.
- Very recent work suggests ERG expression may facilitate prostate carcinogenesis by blocking oncogene-induced senescence in prostatic luminal cells.
- The ETS gene fusions occur as very early events in the development of prostatic adenocarcinomas, either at the stage of PIN or right at the onset of invasion, although rare examples have been reported in low-grade PIN and PIA lesions.
- Diagnostic implications
  - *TMPRSS2–ERG* fusions (and hence ERG protein overexpression) have a >95% specificity for prostate cancer, or high-grade PIN.

- Positive staining of ERG protein by immunohistochemistry can be useful as an aid to diagnosis in lesions suspicious for, but not diagnostic, of cancer by H&E alone (negative staining does not help in these cases).
- Prognostic implications.
  - While there is some evidence to indicate that the mechanism of gene fusion (i.e., deletion versus translocation) may relate to outcome, in general there is not an increased risk of aggressive disease in tumors that are ETS gene fusion positive.
  - Animal studies have shown synergy in disease progression in combination with PTEN loss, although in humans, tumors with PTEN loss that are ERG negative are associated with a higher rate of death due to prostate cancer than PTEN-negative and ERG-positive lesions.

Other Apparently Mutually Exclusive (with ETS Alterations) and Truncal Somatic Mutations

- *SPOP*
  - Located at chromosome 17q21
  - *SPOP* encodes the substrate-recognition component of a Cullin3-based E3-ubiquitin ligase.
  - Structurally, SPOP protein contains 3 domains, MATH, BTB, and BACK; the MATH domain is essential for substrate recognition, whereas the later two can interact with their counterparts in another SPOP protein and facilitate homodimerization, which is critical for the ubiquitin ligase function.
  - Upon ubiquitylation, many SPOP substrates are targeted to the 26S proteasome and degraded.
  - In terms of single point mutations, *SPOP* is the most commonly mutated gene in primary prostate cancer, occurring in ~10% of the cases, and less frequently mutated in metastatic disease (~5%).
    ○ This lower frequency in metastatic lesions may relate to the fact that SPOP-mutated tumors tend to be more responsive to hormonal therapies (see below).
  - SPOP mutations and ETS rearrangements are generally mutually exclusive in prostate cancer (Fig. 14.4), while gene deletions of *CHD1*, a chromatin remodeler, and SPOP paralogue SPOPL have been seen concurrently in SPOP mutant cancers.
    ○ Mutations in other genes in the ubiquitin-proteasome (USP) and ligase family also occur in both primary and metastatic prostate cancer, with a frequency of approximately 1–2%.
  - Point mutations in *SPOP* are always restricted to the substrate-binding cleft within the MATH domain. As a

result, heterodimers formed by wild-type SPOP and mutant SPOP can lead to unstable substrate recognition and thus less ubiquitination (removing the brake for degradation of oncogenic proteins via a dominant-negative effect). Another mutant (Q165P) impairs dimerization and substrate degradation.

- o Several oncogenic proteins in prostate tumorigenesis were found to be SPOP substrates, including AR and its co-activators TRIM24, SRC-3, and BET proteins, which leads to upregulation of AR signaling.
  - ♦ Since AR signaling is key in SPOP mutant prostatic carcinomas, patients carrying SPOP mutations tend to respond better to androgen deprivation therapies in various clinical settings, including neo-adjuvant treatments for primary tumors, CSPC and CRPC.
- o Wild-type SPOP is known to facilitate homologous recombination during double-stranded DNA break repair by promoting degradation of 53BP1 which induces NHEJ and inhibits HR. SPOP mutants, therefore, can induce HR defects and chromosomal instability as 53BP1 is no longer degraded. Since this is similar to loss of BRCA1, this may lead to increased sensitivity to PARP inhibition and radiation therapy.
- o PD-L1 has also been identified as a SPOP substrate, thus tumors with mutated SPOP showed elevated PD-L1, potentially making them more susceptible to PD-1/PD-L1 inhibitors.

- • *FOXA1*
  - – Located at chromosome 14q21.
  - – Generally mutually exclusive to ETS gene rearrangements.
  - – Encodes a transcriptional pioneer factor that induces an open chromatin conformation and subsequent recruitment of transcription factors such as AR.
  - – Under physiological conditions, FOXA1 induces a prostatic luminal cell phenotype.
  - – FOXA1 is overexpressed at the mRNA level in a step-wise manner going from normal epithelium to primary tumors and then to metastasis.
  - – FOXA1 mutations occur in the protein coding regions in 10–13% of prostate cancers across all stages.
    - o Many FOXA1 mutations, which are mainly missense and in-frame indels, occur in the forkhead (FKHD) DNA-binding domain and frequently reside within its wing 2 region that directly contacts DNA.
      - ♦ Wing 2-associated mutations in FOXA1 can lead to faster nuclear de-compacting activity and thus promote an oncogenic luminal AR transcription program.

- o The second most frequent mutations consist of frameshift truncations toward the C terminal regulatory domain; the resulting truncated protein is able to replace the wild-type protein and drive a WNT metastasis program and is enriched in mCPRC cases.
- – Other common FOXA1 alterations are structural variants that are mainly in the forms of tandem duplications and translocations without changing the protein coding sequence of FOXA1.
  - o Both types of structural variants can lead to FOXA1 overexpression.
  - o These are present in approximately 8% of primary cancers and enriched in up to 22% mCRPC cases.
  - o Therefore, the overall cumulative frequency of genomic alterations in mCRPC is ~35% for FOXA1.
- – One recent study reported overexpression of FOXA1 at mRNA level in NEPC, although to a less extent compared to prostatic adenocarcinomas, and its importance in maintaining neuroendocrine features through its binding to relevant regulatory elements in the genome.

- • *CDK12*
  - – A tumor suppressor gene located on chromosome 17q12.
  - – Encodes cyclin-dependent kinase 12, which heterodimerizes with cyclin K, functioning in DNA repair, splicing, and differentiation.
  - – Recurrent deleterious CDK12 mutations occur in 2%–4% primary prostate cancers and in 4.7%–11% of mCRPCs; they can be monoallelic or biallelic.
  - – Carcinomas with biallelic inactivation of CDK12 show a distinct form of genetic instability; while they are baseline diploid, there are numerous focal copy number gains representing tandem duplications, without high-level amplifications or widespread deletions (Fig. 14.5).
  - – Cases may contain high neoantigen burdens from gene fusions from focal tandem duplications, imparting increased immunogenicity.
    - o It is not clear yet whether these tumors consistently contain increased tumor-infiltrating lymphocytes, since results so far have been mixed.
  - – CDK12 alterations are associated with a high Gleason score at diagnosis and worse survival.
  - – A clinical trial conducted on mCRPC patients in a Chinese population reported a higher prevalence (15.4%) of CDK12 loss-of-function alterations than Western populations and an unfavorable response to abiraterone.

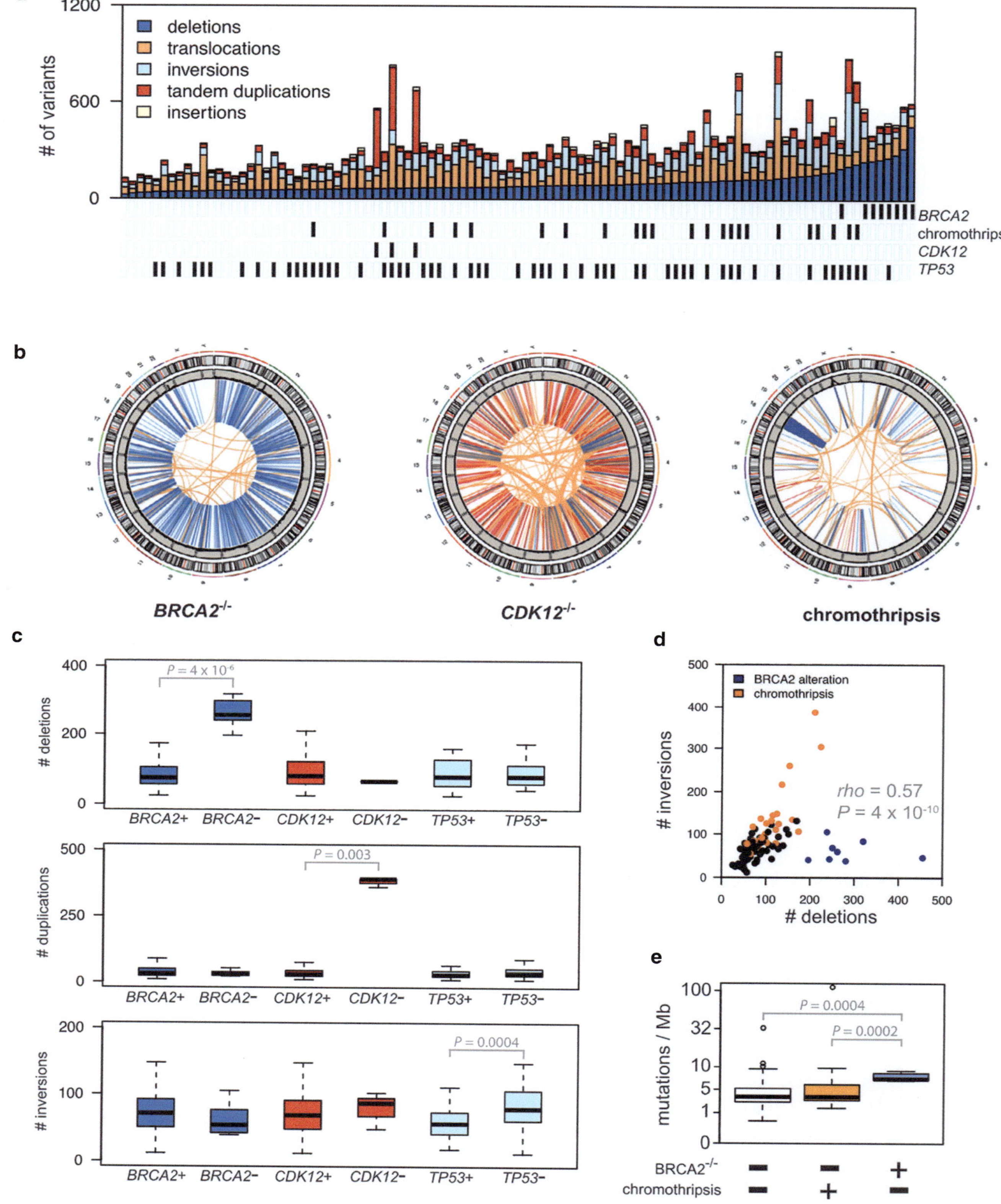

Fig. 14.5 DNA Repair Alterations Are Associated with Structural Variation Frequency (**a**) Top: structural variant frequency by sample, sorted by deletion frequency. Bottom: presence of chromothripsis or biallelic inactivating alterations in BRCA2, CDK12, or TP53. (**b**) Circos plots illustrating BRCA2 inactivation (left), CDK12 inactivation (center), and chromothripsis (right). Colors as in (**a**). (**c**) Box and whiskers plots showing association between biallelic inactivating alterations in BRCA2, CDK12, or TP53 and the frequencies of deletions, tandem duplications, and inverted rearrangements respectively. (**d**) Counts of inverted rearrangements and deletions per sample. Samples with biallelic BRCA2 loss drawn in blue, samples bearing chromothripsis drawn in orange. (**e**) Box and whisker plots showing mutation frequency in the presence of biallelic loss of BRCA2 and chromothripsis. (From Quigley 2018 Cell with permission)

- *IDH1*
  - Located on chromosome 2q34.
  - Encodes cytoplasmic isocitrate dehydrogenase 1 (IDH1) that catalyzes the decarboxylation of isocitrate to generate α-ketoglutarate (α-KG) while replenishing the NADPH pool using NADP(+).
  - Missense mutations of *IDH1* have been found in 1–2% of prostate cancers with almost all occurring in codon 132 (R132).
    - Such mutations confer a novel function of the IDH1 protein (neomorphic) that instead of generating α-ketoglutarate and NADPH, it converts isocitrate to d-2-hydroxyglutarate (2-HG) and consumes NADPH.
  - Prostate tumors harboring *IDH1* R132 mutations show high levels of genome-wide hypermethylation with numerous epigenetically silenced genes.
    - This occurs through 2-HG-mediated inhibition of α-KG-dependent DNA demethylases including the TET family proteins.
  - *IDH1* mutant prostate tumors usually do not possess other commonly observed oncogenic drivers such as ETS gene fusions and have been proposed as a distinct molecular subtype of prostate cancer; however, it is difficult to be sure of this since IDH1 mutations are so rare.

Other Genetic Alterations in Prostate Cancer

- Telomere shortening
  - Somatic telomere shortening occurs in most cases of high-grade PIN and adenocarcinoma.
  - Such shortening can lead to chromosome instability.
  - A prognostic biomarker has been proposed which consists of a combination of telomere shortening in stromal cells in the immediate tumor microenvironment and variability in telomere length in tumor cells.
- Tumor Suppressor Genes/Chromosomal deletions
  - Chromosome 8p
    - Deletions and loss of heterozygosity on the short arm of chromosome 8 (8p) are very common in prostate cancer.
    - The most well-studied gene in this area is *NKX3.1*.
      - *NKX3.1* codes for a prostate-restricted homeobox protein involved in developmental regulation and protection against oxidative damage from free radical effects.
      - Loss of *NKX3.1* generally involves one allele only (can be germline or somatic but is usually somatic).
      - Since its expression is maintained (albeit at somewhat reduced levels compared to normal luminal cells) in the vast majority of prostatic adenocarcinomas and is not seen in most other tumor types, NKX3.1 has proven useful as part of a panel of IHC stains, as a marker for prostate cancer, in cases in which very high-grade cancers are present at the bladder neck that are difficult distinguish between prostate and bladder cancer, as well as in metastatic lesions of unknown primary origin.
  - *PTEN*
    - A tumor suppressor gene located on 10q23.
    - PTEN is a lipid and protein phosphatase whose best-known function is to dephosphorylate PIP3, which counterbalances PI3 kinase—a protein involved in the PI3K–AKT-mTOR pathway important for cell growth, proliferation, and survival.
    - Inactivated biallelically in 20–50% of all prostate cancers, with higher rates in high grade and metastatic disease.
      - In the majority of cases, there are large deletions encompassing the *PTEN* locus, which are either homozygous or accompanied by a mutation in the other allele.
        - The significance of single copy loss is still unclear.
    - Loss of PTEN by FISH (chromosome 10q) or IHC is associated with higher Gleason score and advanced stage.
    - Loss of PTEN is associated with a poor prognosis, including an increased rate of biochemical recurrence, shorter time to metastasis, and decreased survival, the latter mostly occurring in ERG-negative cases.
    - If PTEN loss is detected in a lower grade tumor, the chances of there being a higher grade tumor present nearby is higher.
      - Some labs are employing IHC for PTEN (an excellent surrogate for genomic loss) in all GG1 cancers.
    - Loss of PTEN is often seen concurrently with ETS rearrangements.
    - PTEN loss has also been associated with an immunosuppressive tumor microenvironment in prostate cancer.
    - PTEN status as a predictive biomarker has also been examined:
      - Loss has been associated with less effective AR-targeted therapy, although its loss does not preclude such therapy.
      - In early-phase clinical trials, PTEN loss has been associated with response to AKT inhibitors.

- *CDKN1B*
  - *CDKN1B* encodes p27, a cyclin-dependent kinase inhibitor, which can show single copy genomic loss or at times mutations and/or biallelic inactivation; even without genetic alterations, there is commonly decreased p27 protein in PIN and adenocarcinoma lesions.
  - One mechanism by which p27 is also downregulated is by the PI3K–AKT signaling pathway.
  - Loss of p27 has been associated with a poor prognosis in prostate cancer in a number of studies.
- Other tumor suppressor genes
  - Deletions and/or mutations of tumor suppressor genes common to other cancers are also seen in prostate cancer.
    - *TP53* (mutations are present in approximately 5% of primary tumors but in upward of 50% of mCRPC).
      - Associated with elevated genomic inversion events in mCRPC.
      - Missense mutations in *TP53* often associated with p53 protein overexpression have demonstrated prognostic value including an association with biochemical recurrence and prostate-specific death in localized primary tumors.
    - Single copy loss of RB1 is common in primary tumors, but biallelic inactivation of *RB1* is more common in advanced disease; although loss of both RB1 alleles is infrequent except in SCNC, where it is present in >80% of cases.
  - DNA damage response pathways
    - Multiple studies have uncovered that mutations in DNA damage response (DDR) pathways are commonly observed in prostate cancer, both germline (discussed above) and somatically.
    - Overall mutations of DDR genes are present in 10–19% of primary localized prostate cancers and are further enriched (23–27%) in mCRPC.
    - When DNA damage occurs, the cell has a cascade of pathways to sense the damage, transduce the signal, and resolve the damage depending on the type of lesion:
      - When the DNA lesion is limited to one strand, including single-strand breaks (SSBs), intra-strand cross-links and base mismatches, the cell responds with base excision repair (BER), nucleotide excision repair (NER), and mismatch repair (MMR) pathways, respectively.
      - Double-strand breaks (DSBs) can be repaired mainly by two pathways, homologous recombination repair (HR) and nonhomologous end-joining (NHEJ).
        - Error-free HR uses sister chromatids as a template to repair the break, yet NHEJ is error-prone and repairs the DSBs by ligating the DNA ends.
      - *BRCA2, CDK12, ATM, FANCA, PALB2, RAD50,* and *BRCA1* are the most frequently mutated HR genes and *MSH2, MLH1, MSH6, PMS2* are the most frequently mutated MMR genes in prostate cancer.
    - *BRCA1/2*
      - *BRCA1* is located on chromosome 17q and *BRCA2* on 13q.
      - The encoded proteins BRCA1 and BRCA2 play an important role in DNA repair, specifically in HR for double-strand breaks.
      - Alterations in *BRCA1/2* are usually homozygous deletions or loss-of-function mutations:
        - *BRCA1* is altered in ~1% of prostate cancer in both localized and metastatic stages.
        - *BRCA2* gene alterations are found in 3% of primary prostate cancer and 5.3–13% of mCRPC, one of the most frequently altered DDR genes.
          - mCRPC tumors harboring biallelic *BRCA2* mutations show significantly higher genomic deletion events as well as tumor mutational burden.
      - Several clinical trials, including ongoing ones, have shown promising responses to PARP inhibitors (PARPi) in metastatic prostate cancer patients with *BRCA1/2* mutations (germline or somatic).
    - *ATM*
      - Located on chromosome 11q.
      - Encodes a kinase that senses DSBs and initiates DDR by phosphorylating various proteins in relevant pathways.
      - Inactivating mutations of ATM represent the second most frequently mutated DDR gene in both localized prostate cancer (4%) and mCRPC.
    - *PALB2*
      - Also involved in the Fanconi anemia pathway, if germline homozygously inactivated, leads to FA phenotype.
      - Somatically mutated or biallelically inactivated in ~2% of patients across all stages of prostate cancer.
    - Mismatch repair (MMR) pathway genes
      - The MMR system is responsible for repairing base–base mispairs and small insertions/deletions of DNA mainly occurring during DNA replication.
      - There are 8 genes encoding protein components of the MMR system, among which *MLH1, MSH2, MSH6,* and *PMS2* are most frequently mutated in prostate cancer.

- ♦ Overall mutation prevalence of MMR genes is less than 5% and is often associated with higher Gleason score and advanced disease at diagnosis.
  - Homozygous deletion and hypermutation are two common types of alterations of MMR genes in prostate cancer.
- ♦ A large portion of prostate tumors harboring MMR gene mutations demonstrate MMR protein(s) loss and/or microsatellite instability high (MSI-H) and a high tumor mutation burden.
- ♦ Several studies have reported potentially favorable responses to checkpoint blockade immunotherapy, such as pembrolizumab, in prostate cancer patients with MMR-deficiency/MSI-H, potentially through a higher presence of tumor-infiltrating T cells.
- o Genes in the Fanconi anemia (FA) pathway
  - ♦ The Fanconi anemia DNA repair pathway is responsible for recognizing and resolving interstrand cross-links (ICL) of DNA during replication.
  - ♦ The FA pathway thus includes many genes in the HR pathway, such as *FANCD1/BRCA2*, *FANCN/PALB2*, and *FANCS/BRCA1* whose significance in prostate cancer is discussed elsewhere in this chapter.
  - ♦ *FANCA* is an FA/HR gene that is recurrently mutated in prostate cancer (in 3–8% of all cases), mainly in the form of missense mutations and homozygous deletions.
    - FANCA protein mainly interacts with BRCA1 during the HR process.
    - Recent studies reported that prostate cancer patients possessing biallelic *FANCA* loss showed response to PARPi.
- o CDK12 is discussed elsewhere in this chapter.
- Hormonal Pathway Genes
  - – *AR*
    - o The androgen receptor (encoded by *AR*) is a ligand regulated (physiologically by testosterone and dihydrotestosterone) prostate master transcription factor.
    - o Upon ligand binding, AR is translocated to the nucleus and binds to thousands of sites throughout the genome (these sites together constitute the AR "cistrome").
    - o AR is highly expressed in normal prostatic luminal cells and is associated with prostatic epithelial and stromal morphogenesis and epithelial cellular differentiation. Binding to its main ligand, DHT, is required for luminal cell survival (for many luminal cells) and for proper differentiated function.
    - o The protein product is expressed in most prostatic adenocarcinomas and its inhibition by castration, medical castration, or by AR antagonists is a key well-known treatment for locally advanced and metastatic prostate cancer.
    - o AR "constitutive" activation is found in the majority of mCRPC—those cancers that no longer respond to castrate levels of testosterone and DHT in the circulation.
      - ♦ *AR* gene amplification, activating point mutations, and AR enhancer amplification are only seen to any degree in mCRPC. These alterations are accompanied by high levels of *AR* mRNA and protein expression (much higher than in primary tumors).
      - ♦ These alterations are thought to increase the sensitivity to very low androgen levels, which are derived from the adrenals and at times have been shown to be produced endogenously by the tumor.
      - ♦ These findings regarding AR support the concept of oncogene addiction in prostate cancer.
        - In this case the need for androgen signally for proliferation and prevention of cell death is inherent to prostatic cancer cells.
      - ♦ AR amplification, which is commonly seen in mCRPC, increases the sensitivity to lower levels of AR ligands in circulation.
        - Detection of amplified *AR* in circulating tumor cells (CTCs) or circulating tumor DNA (ctDNA) in clinical trials has been associated with treatment resistance to enzalutamide and abiraterone.
      - ♦ AR variants (AR-Vs)
        - AR-Vs are truncated AR proteins lacking the AR ligand-binding domain, potentially resulting from rearrangements in the *AR* gene and/or alternative splicing of the *AR* mRNA.
        - Without the ligand-binding domain, AR-Vs can be constitutively activated and drive the AR-dependent transcriptional programs even in the absence of ligands.
        - AR-V7 is one of the most well-studied AR-Vs that is rare in primary prostate cancer but is commonly seen in patients treated with hormonal deprivation therapies, especially in those with mCRPC.
        - Detection of AR variants, especially AR-V7, in mCRPC tumor samples or CTC from mCRPC patients has been associated with

treatment resistance to enzalutamide and abiraterone as well as favorable response to taxane-based therapies.

- *HSD3B1*
  - encodes 3β-hydroxysteroid dehydrogenase-1, an enzyme that catalyzes the initial rate-limiting step in converting dehydroepiandrosterone (DHEA) to testosterone (T) and dihydrotestosterone (DHT),
  - A germline variant or somatic mutation of this gene at nucleotide position 1245 from A to C can lead to resistance to protein degradation by the ubiquitin proteasome system.
    - HSD3B1 (1245C) is thus called an "adrenal permissive" allele as it increases potent AR ligand (T/DHT) production using adrenal androgen precursors.
      - The allele frequency of HSD3B1 (1245C) ranges from 8 to 34% depending on the ancestry, in that it is higher in men from Europe, and lower in men from Asia.
      - The 1245C allele is also selected for in patients undergoing androgen deprivation therapy, either through acquiring somatic mutations or loss of heterozygosity.
    - A few clinical studies have shown potential prognostic value for the 1245C allele after ADT as patients with low-volume prostate cancer carrying 1245C showed worse outcome and shorter survival.
    - The presence of the 1245C allele in mCRPC patients is also associated with poor outcome after they were treated with enzalutamide and abiraterone, suggesting potential predictive value.
- Oncogenes
  - *MYC*
    - *MYC* is located at 8q24 and in primary untreated tumors, low-level amplification is associated with high Gleason score, advanced stage, and disease progression.
    - Overexpression of MYC mRNA and protein, decoupled from 8q24 gain, arises as an early event in prostate cancer, including almost all PIN lesions.
    - Upwards of 80–90% of all prostate cancers overexpress MYC mRNA and protein.
      - MYC protein is also highly expressed in late-stage castration-resistant disease.
    - MYC overexpression results in a profound transcriptional reprogramming of prostate luminal epithelial cells characterized by the induction of genes related to nucleolar function, ribosome biogenesis, and cell proliferation.
      - MYC expression also reprograms the AR cistrome, blunting AR-induced gene expression at loci associated with classic AR signaling, which appears to occur by preventing pause release at AR target genes, but not by reducing AR binding to such regions.
        - Further, MYC overexpression increases AR binding at loci associated with FOXA1 occupancy.
      - The structural organization of the 8q24 locus has come into sharp focus recently. Epigenetic regulation of MYC mRNA overexpression has recently been tied to long- range interactions between distant enhancer regions and the MYC promoter. The region contains a number of noncoding RNAs that may be coexpressed with MYC including *PCAT1* and *PVT1*.
      - Upon androgen deprivation, MYC upregulation is often seen, which may contribute to development of castration resistance; conversely, supraphysiological levels of testosterone can lead to MYC downregulation and tumor regression in some patients.
        - One mechanism of such suppression appears to be that androgen disrupts the interaction between a super enhancer and the MYC promoter by redistributing and/or sequestering transcriptional coactivators between these two regions.
      - FISH for chromosome 8q24 amplification, encompassing the MYC locus, has shown prognostic value.
        - When present in combination with PTEN loss, MYC copy number gain is associated with higher Gleason score as well as prostate cancer-specific death.
- *EZH2*
  - EZH2 is a histone lysine methyltransferase involved in chromatin remodeling as part of the PRC2 polycomb repressive complex.
  - It is overexpressed in all phases of prostate cancer including the precursor lesion, high grade PIN.
  - EZH2 promotes proliferation, invasion, and tumorigenicity of prostate cancer cells.
  - Upregulation of EZH2 in prostate cancer can result from:
    - Gene amplification.
    - By deletion of its negative regulator mir-101.
    - Transcriptional regulation by *ETS* gene family members.
    - Transcriptional regulation directly by MYC.
    - Downregulation of other negative regulators mir-26a and mir-26b, which are themselves negatively regulated by MYC.
    - Gain of function mutations.

- o Noncanonical functions of EZH2 have been identified recently including transcriptionally activating AR and posttranslationally methylating FOXA1 protein to improve protein stability and promote oncogenic phenotypes.
  - *SPINK1*
    - o SPINK1 is a protein with a high homology to the epidermal growth factor receptor (EGFR) and has been found to be overexpressed in some prostate cancers, particularly in *ETS*-fusion negative cases.
    - o Prostate cancers harboring SPINK1 overexpression have been associated with faster progression to biochemical recurrence and castration resistance.
    - o Androgen deprivation therapy can induce SPINK1 upregulation.
- Genome/chromosome alterations
  - In general, the tumor mutational burden in primary prostatic adenocarcinomas tends to be quite low except in rare cases of mismatch repair deficiency.
    - o mCRPC, however, does possess a higher tumor mutational burden compared to mCSPC and primary tumors (Fig. 14.4).
  - Copy number alterations, including gains and losses, are common in prostate cancer, although there appears to be a high fraction of grade group 1 cancers that are relatively "quiet" in this regard with few copy number changes.
    - o The fraction of the genome altered can be prognostic.
  - Complex chromosome alterations
    - o Chromoplexy
      - ◆ defined as complex genomic structure rearranged chromosome segments formed in a chain in an interdependent manner.
      - ◆ More frequently observed in ETS-rearragement positive tumors.
      - ◆ May account for loss of tumor suppressor genes and upregulation of known oncogenes.
      - ◆ Mechanistically may result from AR-induced double-stranded break and TOP2B-mediated chromatin reorganization.
    - o Chromothripsis
      - ◆ Complex genomic structures formed by up to thousands of shattered chromosomal segments in a single catastrophic event; usually involve only one chromosome or one arm of a chromosome.
      - ◆ A recent report suggested ~50% prevalence of chromothripsis events in prostate cancer, which can contribute to oncogene amplification and loss of tumor-suppressor genes, similar to chromoplexy.

- ◆ TP53 inactivation and polyploidy are two potential predisposing factors for chromothripsis.

Epigenetic Alterations in Prostate Cancer

- Three major epigenetic marks are found to be commonly altered in prostate cancer including histone acetylation, histone methylation, and DNA methylation.
- Each mark has its corresponding regulation machinery consisting of epigenetic writers, erasers, readers, preservers, and remodelers.
- Collectively as a group, epigenetic machinery genes are the most frequently mutated in prostate cancer, found in ~15–20% of all cases and mostly are potentially inactivating.
- Mutations in these epigenetic machinery genes are significantly associated with higher Gleason score at diagnosis, and are significantly enriched in tumors without ETS gene fusions or other known drivers.
- *SChLAP1*
  - Long-noncoding RNA *SChLAP1* has been found to be overexpressed in ~25% of prostate cancer and has shown an antagonistic effect on the function of chromatin-remodeling complex SWI/SNF by interfering with its genomic binding ability.
  - Overexpression of SChLAP1 in prostate cancer has been associated with higher Gleason score and pT stage, intraductal/cribriform histology, increased biochemical recurrence, metastasis, and prostate cancer-specific lethality.
- DNA methylation
  - DNA hypermethylation is one of the most consistent epigenetic alterations in prostate cancer.
  - Many DNA hypermethylation alterations are associated with higher grade and/or stage, disease recurrence, as well as lethal prostate cancer and NEPC.
  - CpG hypermethylation
    - o Involves the methylation of deoxycytidine residues within CpG dinucleotides, usually in the upstream regulatory regions of specific genes and often leads to gene repression.
    - o The most well understood gene affected in prostate cancer is *GSTP1*.
      - ◆ *GSTP1* encodes a protein that is part of a family of enzymes that counteract damage from reactive chemical species via a glutathione-mediated conjugation mechanism.
      - ◆ CpG hypermethylation results in silencing of the gene and increased sensitivity to genetic damage from oxidative stress.
      - ◆ This somatic genome alteration has been found in approximately 90–95% of all prostate cancers

and can be detected in blood, urine, and prostatic fluid.

- ◆ CpG hypermethylation of *GSTP1* is present in ~70% of PIN and between 4 and 6% of prostate atrophy lesions but is not present in normal appearing prostatic epithelium, even in the microscopic vicinity of carcinoma.
- ◆ Prostate cancers that retain GSTP1 expression are substantially enriched in African American patients, especially those with positive ERG expression.
- ○ Other genes known to be affected recurrently by CpG hypermethylation in prostate cancer include:
  - ◆ *APC*
  - ◆ *RASSF1a*
  - ◆ *ENDRB*
  - ◆ *PTGS2*
  - ◆ *MDR1*
- ○ There is a global reduction of 5-hydroxymethylcytosine in prostate cancer
  - ◆ 5-hydroxymethylcytosine (5hmC) is one of the major oxidized products of 5-methylcytosine, the most common DNA methylation mark.
  - ◆ This oxidative reaction is catalyzed by a family of TET proteins (10–11 translocation), TET1-3, that are mutated at times in prostate cancer.
  - ◆ The resulting 5hmC could be detected, excised, and repaired with nonmethylated cytosine through the base excision pathway, leading to DNA demethylation.
  - ◆ This reduced DNA demethylation may then contribute to the DNA hypermethylation commonly seen in prostate cancer, as well as in many normal stem cell compartments.
- – Global DNA hypomethylation in repetitive elements is also observed in prostate cancer, usually at later stages of disease; the clinical significance of this type of epigenetic change is under active research.
- Histone Modifications
  - – Histones, the "DNA packaging protein," can be subjected to a variety of post-translational modifications including methylation and acetylation.
  - – Histone acetylation generally is associated with transcriptional activation while de-acetylation is correlated with transcriptional repression.
  - – Histone methylation can be associated with either activation or repression.
  - – H3K27me3 global reduction:
    - ○ is highly correlated with the 5hmC amount in prostate cancer and is associated with the stem cell/progenitor cell phenotype.
    - ○ occurs as early as in PIN and is continuously present in more advanced stages of prostate cancer; mechanistically linked to MYC overexpression.

Tumor Microenvironment in Prostate Cancer

- Growing evidence has suggested a pivotal role of the tumor microenvironment including the immune cell populations for prostate cancer initiation and progression.
- Prostate cancer is generally considered "immune cold" with low levels of inflammatory infiltrates in the tumors and often show only very limited responses to immune checkpoint blockade.
- Recent studies have offered some insights into the potential mechanisms leading to such an "immune desert model" for prostate cancer.
  - – Prostate tumors usually possess a low tumor mutational burden, especially in primary lesions, unless they harbor mutations in MMR genes (discussed above), resulting in low mutation-related tumor-specific neoantigens that can be recognized as foreign by the immune system.
  - – Prostate cancer cells generally express little-to-no PD-L1, one of the critical immune co-inhibitory checkpoint molecules, on their cell surface, suggesting that the lack of immune recognition is not from PD-L1 upregulation of tumor cells.
  - – Alternatively, some studies indicated a repressed adaptive immune microenvironment for prostate cancer.
    - ○ Cytotoxic CD8+ T cells in prostate tissues from prostate cancer patients often concurrently express PD-1, suggestive of an "exhaustion phenotype" of T cells.
    - ○ The number of FOXP3+ regulatory T cells (Treg) is found to be increased somewhat in prostate cancer samples.
    - ○ Innate immune cells, including mast cells in benign tissues and protumorigenic M2 macrophages, may contribute to prostate cancer progression.
    - ○ MHC molecules (major histocompatibility complex) that facilitate immune system recognition by presenting foreign molecules including neoantigens on the cell surface are found to be downregulated in prostate cancer cells.
- The detailed immune landscape in the tumor microenvironment of prostate cancer across disease stages is still an active area of research with advances in multiplex phenotyping techniques, single cell transcriptomics, and spatial transcriptomics promising to markedly augment our knowledge in this area in the near future.

Prognostic Utility of Somatic Tissue-based Genetic Testing

- DNA Based Testing
  - – Many patients with high grade and metastatic cancers are having tumor tissues tested for somatic DNA alter-

ations including mutations, gene fusions, and copy number alterations using panel-based testing.

- Commercial examples of such tests include those from Foundation Medicine, Tempus, and Caris.
- While none of these are employed as standard of care, increased use is occurring to determine whether patients may be candidates for PARP inhibitors (e.g., with mutations in genes involved in HR repair defects) or checkpoint in inhibitor therapies (e.g., MMR defects).

- RNA Based Testing
  - An emerging understanding of prostate cancer biology through microarray studies and RNA sequencing efforts has led to develop multiple tissue-based testing for prognosis and risk stratification.
  - These include Prolaris (Myriad Genetics), OncotypeDx Genomic Prostate Score, and Decipher (GenomeDx).
  - These tests use RT-PCR or microarrays to measure expression of a panel of genes in various pathways, mainly cell proliferation/cell cycle (Prolaris) but also androgen signaling, stromal response and cellular organization (GenomeDx).
    - OncotypeDx GPS was designed for needle biopsies while Prolaris Decipher has been used for both biopsies and radical prostatectomy samples.
  - These tests provide a score that shows prognostic values in terms of biochemical recurrence, metastasis, as well as prostate cancer-specific lethality. While none are used routinely in clinical practice, they show potential for molecular profiling to augment our ability to tailor patients for adjuvant therapies and for selection for specific treatments in clinical trials.

Suggested Readings

Abida W, Armenia J, Gopalan A, Brennan R, Walsh M, Barron D, et al. Prospective genomic profiling of prostate cancer across disease states reveals germline and somatic alterations that may affect clinical decision making. JCO Precis Oncol. 2017;2017 https://doi.org/10.1200/PO.17.00029.

Abida W, Cheng ML, Armenia J, Middha S, Autio KA, Vargas HA, et al. Analysis of the prevalence of microsatellite instability in prostate cancer and response to immune checkpoint blockade. JAMA Oncol. 2019;5:471–8.

Adams EJ, Karthaus WR, Hoover E, Liu D, Gruet A, Zhang Z, et al. FOXA1 mutations alter pioneering activity, differentiation and prostate cancer phenotypes. Nature. 2019;571:408–12.

Antonarakis ES, Lu C, Wang H, Luber B, Nakazawa M, Roeser JC, et al. AR-V7 and resistance to enzalutamide and abiraterone in prostate cancer. N Engl J Med. 2014;371:1028–38.

Antonarakis ES, Shaukat F, Isaacsson Velho P, Kaur H, Shenderov E, Pardoll DM, et al. Clinical features and therapeutic outcomes in men with advanced prostate cancer and DNA mismatch repair gene mutations. Eur Urol. 2019;75:378–82.

Aparicio A, Xiao L, Tapia ELN, Hoang A, Ramesh N, Wu W, et al. The aggressive variant prostate carcinoma (AVPC) molecular signature

(-MS) and platinum-sensitivity in castration resistant prostate cancer (CRPC). J Clin Oncol. 2017;35:5013.

Armenia J, Wankowicz SAM, Liu D, Gao J, Kundra R, Reznik E, et al. The long tail of oncogenic drivers in prostate cancer. Nat Genet. 2018;50:645–51.

Arora K, Barbieri CE. Molecular subtypes of prostate cancer. Curr Oncol Rep. 2018;20:58.

Arriaga JM, Panja S, Alshalalfa M, Zhao J, Zou M, Giacobbe A, et al. A MYC and RAS co-activation signature in localized prostate cancer drives bone metastasis and castration resistance. Nat Cancer. 2020;1:1082–96.

Asrani K, Torres AFC, Woo J, Vidotto T, Tsai HK, Luo J, et al. Reciprocal YAP1 loss and INSM1 expression in neuroendocrine prostate cancer. J Pathol. 2021;255(4):425–37.

Baca SC, Prandi D, Lawrence MS, Mosquera JM, Romanel A, Drier Y, et al. Punctuated evolution of prostate cancer genomes. Cell. 2013;153:666–77.

Baca SC, Takeda DY, Seo J-H, Hwang J, Ku SY, Arafeh R, et al. Reprogramming of the FOXA1 cistrome in treatment-emergent neuroendocrine prostate cancer. Nat Commun. 2021;12:1979.

Bernasocchi T, Theurillat J-PP. SPOP-mutant prostate cancer: translating fundamental biology into patient care. Cancer Lett. 2021;529:11–8.

Butler W, Huang J. Neuroendocrine cells of the prostate: histology, biological functions, and molecular mechanisms. Precis Clin Med. 2021;4:25–34.

Cancer Genome Atlas Research Network. The molecular taxonomy of primary prostate cancer. Cell. 2015;163:1011–25.

Cortés-Ciriano I, Lee JJ-K, Xi R, Jain D, Jung YL, Yang L, et al. Comprehensive analysis of chromothripsis in 2,658 human cancers using whole-genome sequencing. Nat Genet. 2020;52:331–41.

Deshmukh D, Xu J, Yang X, Shimelis H, Fang S, Qiu Y. Regulation of p27 (Kip1) by ubiquitin E3 ligase RNF6. Pharmaceutics. 2022;14:802.

Dong B, Fan L, Yang B, Chen W, Li Y, Wu K, et al. Use of circulating tumor DNA for the clinical management of metastatic castration-resistant prostate cancer: a multicenter, real-world study. J Natl Compr Canc Netw. 2021;19:905–14.

Faisal FA, Murali S, Kaur H, Vidotto T, Guedes LB, Salles DC, et al. CDKN1B deletions are associated with metastasis in African American men with clinically localized, surgically treated prostate cancer. Clin Cancer Res. 2020;26:2595–602.

Fang L, Li D, Yin J, Pan H, Ye H, Bowman J, et al. TMPRSS2-ERG promotes the initiation of prostate cancer by suppressing oncogene-induced senescence. Cancer Gene Ther. 2022;29(10):1463–76.

Flavin R, Pettersson A, Hendrickson WK, Fiorentino M, Finn S, Kunz L, et al. SPINK1 protein expression and prostate cancer progression. Clin Cancer Res. 2014;20:4904–11.

Fontugne J, Cai PY, Alnajar H, Bhinder B, Park K, Ye H, et al. Collision tumors revealed by prospectively assessing subtype-defining molecular alterations in 904 individual prostate cancer foci. JCI Insight. 2022;7:e155309.

George RS, Htoo A, Cheng M, Masterson TM, Huang K, Adra N, et al. Artificial intelligence in prostate cancer: definitions, current research, and future directions. Urol Oncol. 2022;40(6):262–70.

Ghiam AF, Cairns RA, Thoms J, Dal Pra A, Ahmed O, Meng A, et al. IDH mutation status in prostate cancer. Oncogene. 2012;31:3826.

Giri VN, Morgan TM, Morris DS, Berchuck JE, Hyatt C, Taplin M-E. Genetic testing in prostate cancer management: considerations informing primary care. CA Cancer J Clin. 2022;72(4):360–71.

Guedes LB, Almutairi F, Haffner MC, Rajoria G, Liu Z, Klimek S, et al. Analytic, preanalytic, and clinical validation of p53 IHC for detection of TP53 missense mutation in prostate cancer. Clin Cancer Res. 2017;23:4693–703.

Guo H, Wu Y, Nouri M, Spisak S, Russo JW, Sowalsky AG, et al. Androgen receptor and MYC equilibration centralizes on developmental super-enhancer. Nat Commun. 2021;12:1–18.

Gurel B, Iwata T, Koh CM, Jenkins RB, Lan F, Van Dang C, et al. Nuclear MYC protein overexpression is an early alteration in human prostate carcinogenesis. Mod Pathol. 2008;21:1156–67.

Ha Chung B, Horie S, Chiong E. The incidence, mortality, and risk factors of prostate cancer in Asian men. Prostate Int. 2019;7:1–8.

Haffner MC, Chaux A, Meeker AK, Esopi DM, Gerber J, Pellakuru LG, et al. Global 5-hydroxymethylcytosine content is significantly reduced in tissue stem/progenitor cell compartments and in human cancers. Oncotarget. 2011;2:627–37.

Haffner MC, Pellakuru LG, Ghosh S, Lotan TL, Nelson WG, De Marzo AM, et al. Tight correlation of 5-hydroxymethylcytosine and Polycomb marks in health and disease. Cell Cycle. 2013;12:1835–41.

Haffner MC, Weier C, Xu MM, Vaghasia A, Gürel B, Gümüşkaya B, et al. Molecular evidence that invasive adenocarcinoma can mimic prostatic intraepithelial neoplasia (PIN) and intraductal carcinoma through retrograde glandular colonization. J Pathol. 2016;238:31–41.

Heaphy CM, Joshu CE, Barber JR, Davis C, Lu J, Zarinshenas R, et al. The prostate tissue-based telomere biomarker as a prognostic tool for metastasis and death from prostate cancer after prostatectomy. bioRxiv. 2021.

Heaphy CM, Yoon GS, Peskoe SB, Joshu CE, Lee TK, Giovannucci E, et al. Prostate cancer cell telomere length variability and stromal cell telomere length as prognostic markers for metastasis and death. Cancer Discov. 2013;3:1130–41.

Hernández-Llodrà S, Segalés L, Safont A, Juanpere N, Lorenzo M, Fumadó L, et al. SPOP and FOXA1 mutations are associated with PSA recurrence in ERG wt tumors, and SPOP downregulation with ERG-rearranged prostate cancer. Prostate. 2019;79:1156–65.

Hieronymus H, Murali R, Tin A, Yadav K, Abida W, Moller H, et al. Tumor copy number alteration burden is a pan-cancer prognostic factor associated with recurrence and death. Elife. 2018;7:e37294.

Hinsch A, Brolund M, Hube-Magg C, Kluth M, Simon R, Möller-Koop C, et al. Immunohistochemically detected IDH1R132H mutation is rare and mostly heterogeneous in prostate cancer. World J Urol. 2018;36:877–82.

Horak P, Weischenfeldt J, von Amsberg G, Beyer B, Schütte A, Uhrig S, et al. Response to olaparib in a PALB2 germline mutated prostate cancer and genetic events associated with resistance. Cold Spring Harb Mol Case Stud. 2019;5:a003657.

Jamaspishvili T, Berman DM, Ross AE, Scher HI, De Marzo AM, Squire JA, et al. Clinical implications of PTEN loss in prostate cancer. Nat Rev Urol. 2018;15:222–34.

Jernberg E, Bergh A, Wikström P. Clinical relevance of androgen receptor alterations in prostate cancer. Endocr Connect. 2017;6:R146–61.

Kidd SG, Carm KT, Bogaard M, Olsen LG, Bakken AC, Løvf M, et al. High expression of SCHLAP1 in primary prostate cancer is an independent predictor of biochemical recurrence, despite substantial heterogeneity. Neoplasia. 2021;23:634–41.

Kron KJ, Murison A, Zhou S, Huang V, Yamaguchi TN, Shiah Y-J, et al. TMPRSS2-ERG fusion co-opts master transcription factors and activates NOTCH signaling in primary prostate cancer. Nat Genet. 2017;49:1336–45.

Lancho O, Herranz D. The MYC enhancer-ome: long-range transcriptional regulation of MYC in cancer. Trends Cancer Res. 2018;4:810–22.

Liu W, Xie CC, Thomas CY, Kim S-T, Lindberg J, Egevad L, et al. Genetic markers associated with early cancer-specific mortality following prostatectomy. Cancer. 2013;119:2405–12.

Lozano R, Castro E, Aragón IM, Cendón Y, Cattrini C, López-Casas PP, et al. Genetic aberrations in DNA repair pathways: a cornerstone of precision oncology in prostate cancer. Br J Cancer. 2020;124:552–63.

Lu C, Brown LC, Antonarakis ES, Armstrong AJ, Luo J. Androgen receptor variant-driven prostate cancer II: advances in laboratory investigations. Prostate Cancer Prostatic Dis. 2020;23:381–97.

Luo J, Attard G, Balk SP, Bevan C, Burnstein K, Cato L, et al. Role of androgen receptor variants in prostate cancer: report from the 2017 mission androgen receptor variants meeting. Eur Urol. 2018;73:715–23.

Ma TM, Romero T, Nickols NG, Rettig MB, Garraway IP, Roach M 3rd, et al. Comparison of response to definitive radiotherapy for localized prostate cancer in Black and White men: a meta-analysis. JAMA Netw Open. 2021;4:e2139769.

Maxwell KN, Cheng HH, Powers J, Gulati R, Ledet EM, Morrison C, et al. Inherited TP53 variants and risk of prostate cancer. Eur Urol. 2022;81(3):243–50.

McKay RR, Sarkar RR, Kumar A, Einck JP, Garraway IP, Lynch JA, et al. Outcomes of Black men with prostate cancer treated with radiation therapy in the Veterans Health Administration. Cancer. 2021;127:403–11.

Michl J, Zimmer J, Tarsounas M. Interplay between Fanconi anemia and homologous recombination pathways in genome integrity. EMBO J. 2016;35:909–23.

Nakayama M, Bennett CJ, Hicks JL, Epstein JI, Platz EA, Nelson WG, et al. Hypermethylation of the human glutathione S-transferase-pi gene (GSTP1) CpG island is present in a subset of proliferative inflammatory atrophy lesions but not in normal or hyperplastic epithelium of the prostate: a detailed study using laser-capture microdissection. Am J Pathol. 2003;163:923–33.

Nelson WG, Brawley OW, Isaacs WB, Platz EA, Yegnasubramanian S, Sfanos KS, et al. Health inequity drives disease biology to create disparities in prostate cancer outcomes. J Clin Invest. 2022;132:e155031.

Nicholas TR, Strittmatter BG, Hollenhorst PC. Oncogenic ETS factors in prostate cancer. In: Dehm SM, Tindall DJ, editors. Prostate cancer: cellular and genetic mechanisms of disease development and progression. Cham: Springer; 2019. p. 409–36.

Nizialek E, Lotan TL, Isaacs WB, Yegnasubramanian S, Paller CJ, Antonarakis ES. The somatic mutation landscape of germline CHEK2-altered prostate cancer. J Clin Orthod. 2021;39:5084.

Ozbek B, Ertunc O, Erikson A, Vidal ID, Alexandre CG, Guner G, et al. Multiplex immunohistochemical phenotyping of t cells in primary prostate cancer. medRxiv. 2021.

Park SH, Fong K-W, Kim J, Wang F, Lu X, Lee Y, et al. Posttranslational regulation of FOXA1 by Polycomb and BUB3/USP7 deubiquitin complex in prostate cancer. Sci Adv. 2021;7:eabe2261.

Parolia A, Cieslik M, Chu S-C, Xiao L, Ouchi T, Zhang Y, et al. Distinct structural classes of activating FOXA1 alterations in advanced prostate cancer. Nature. 2019;571:413–8.

Pećina-Šlaus N, Kafka A, Salamon I, Bukovac A. Mismatch Repair Pathway. Genome Stability and Cancer. Front Mol Biosci. 2020;7:122.

Pellakuru LG, Iwata T, Gurel B, Schultz D, Hicks J, Bethel C, et al. Global levels of H3K27me3 track with differentiation in vivo and are deregulated by MYC in prostate cancer. Am J Pathol. 2012;181:560–9.

Pham M-TN. Topoisomerase 2 beta facilitates chromatin reorganization during androgen receptor induced transcription and contributes to chromoplexy in prostate cancer. Johns Hopkins University. 2021. Available: https://jscholarship.library.jhu.edu/handle/1774.2/66729

Qiu X, Boufaied N, Hallal T, Feit A, de Polo A, Luoma AM, et al. MYC drives aggressive prostate cancer by disrupting transcriptional pause release at androgen receptor targets. Nat Commun. 2022;13:1–17.

Quigley DA, Dang HX, Zhao SG, Lloyd P, Aggarwal R, Alumkal JJ, et al. Genomic hallmarks and structural variation in metastatic prostate cancer. Cell. 2018;174:758–769.e9.

Rebello RJ, Oing C, Knudsen KE, Loeb S, Johnson DC, Reiter RE, et al. Prostate cancer. Nat Rev Dis Primers. 2021;7:9.

Rescigno P, Gurel B, Pereira R, Crespo M, Rekowski J, Rediti M, et al. Characterizing CDK12-mutated prostate cancers. Clin Cancer Res. 2021;27:566–74.

Russo J, Giri VN. Germline testing and genetic counselling in prostate cancer. Nat Rev Urol. 2022;19(6):331–43.

Sabharwal N, Sharifi N. HSD3B1 genotypes conferring adrenal-restrictive and adrenal-permissive phenotypes in prostate cancer and beyond. Endocrinology. 2019;160:2180–8.

Saunders EJ, Kote-Jarai Z, Eeles RA. Identification of germline genetic variants that increase prostate cancer risk and influence development of aggressive disease. Cancers. 2021;13:760.

Schiewer MJ, Knudsen KE. DNA damage response in prostate cancer. Cold Spring Harb Perspect Med. 2019;9:a030486.

Schlomm T, Iwers L, Kirstein P, Jessen B, Köllermann J, Minner S, et al. Clinical significance of p53 alterations in surgically treated prostate cancers. Mod Pathol. 2008;21:1371–8.

Sedhom R, Antonarakis ES. Clinical implications of mismatch repair deficiency in prostate cancer. Future Oncol. 2019;15:2395–411.

Sena LA, Kumar R, Sanin DE, Thompson EA, Rosen DM, Dalrymple SL, et al. Prostate cancer androgen receptor activity dictates efficacy of Bipolar Androgen Therapy. bioRxiv. 2022.

Shrestha E, Coulter JB, Guzman W, Ozbek B, Hess MM, Mummert L, et al. Oncogenic gene fusions in nonneoplastic precursors as evidence that bacterial infection can initiate prostate cancer. Proc Natl Acad Sci U S A. 2021;118:e2018976118.

Teng M, Zhou S, Cai C, Lupien M, He HH. Pioneer of prostate cancer: past, present and the future of FOXA1. Protein Cell. 2021;12:29–38.

Thomas L, Sharifi N. Germline HSD3B1 genetics and prostate cancer outcomes. Urology. 2020;145:13–21.

Thomas DJ, Robinson M, King P, Hasan T, Charlton R, Martin J, et al. p53 expression and clinical outcome in prostate cancer. Br J Urol. 1993;72:778–81.

Tiwari R, Manzar N, Bhatia V, Yadav A, Nengroo MA, Datta D, et al. Androgen deprivation upregulates SPINK1 expression and potentiates cellular plasticity in prostate cancer. Nat Commun. 2020;11:384.

Trabzonlu L, Kulac I, Zheng Q, Hicks JL, Haffner MC, Nelson WG, et al. Molecular pathology of high-grade prostatic intraepithelial neoplasia: challenges and opportunities. Cold Spring Harb Perspect Med. 2018;9:a030403.

Vidal I, Zheng Q, Hicks JL, Chen J, Platz EA, Trock BJ, et al. GSTP1 positive prostatic adenocarcinomas are more common in Black than White men in the United States. PLoS One. 2021;16:e0241934.

Walavalkar K, Saravanan B, Singh AK, Jayani RS, Nair A, Farooq U, et al. A rare variant of African ancestry activates 8q24 lncRNA hub by modulating cancer associated enhancer. Nat Commun. 2020;11:3598.

Wang D, Ma J, Botuyan MV, Cui G, Yan Y, Ding D, et al. ATM-phosphorylated SPOP contributes to 53BP1 exclusion from chromatin during DNA replication. Sci Adv. 2021;7:eabd9208.

WHO Classification of Tumors: Urinary and Male Genital Tumors. International Agency for Research on Cancer; 2022; 2022.

Williams JL, Greer PA, Squire JA. Recurrent copy number alterations in prostate cancer: an in silico meta-analysis of publicly available genomic data. Cancer Genet. 2014;207:474–88.

Wokołorczyk D, Kluźniak W, Stempa K, Rusak B, Huzarski T, Gronwald J, et al. PALB2 mutations and prostate cancer risk and survival. Br J Cancer. 2021;125:569–75.

Wu Y, Yu H, Zheng SL, Na R, Mamawala M, Landis T, et al. A comprehensive evaluation of CHEK2 germline mutations in men with prostate cancer. Prostate. 2018;78:607–15.

Yegnasubramanian S, De Marzo AM, Nelson WG. Prostate cancer epigenetics: from basic mechanisms to clinical implications. Cold Spring Harb Perspect Med. 2019;9:a030445.

Molecular Pathology of Urinary Bladder Tumors

15

George J. Netto and Liang Cheng

Contents

G. J. Netto (✉)
Department of Pathology and Laboratory Medicine, Perelman
School of Medicine at the University of Pennsylvania,
Philadelphia, PA, USA
e-mail: george.netto@pennmedicine.upenn.edu

L. Cheng
The Legorreta Cancer Center at Brown University, Department of
Pathology and Laboratory Medicine, Warren Alpert Medical
School of Brown University, Lifespan Academic Medical Center,
Providence, RI, USA
e-mail: liang_cheng@brown.edu

Introduction

- 82,790 new cases of bladder cancers (BC) were diagnosed in the United States in 2023 with 16,710 deaths
- BC is the fourth most common cancer in American males; ranked eighth in mortality
  - Due to high rate of recurrence and need for cystoscopy and surveillance, BC incurs the highest cost per patient; $3 billion annual cost to US health care system
- BC presents unique opportunities for early detection by novel molecular detection methods (liquid biopsy)
- Urothelial carcinoma (UC) accounts for more than 90% of all bladder cancers
- Urothelial carcinoma (UC) manifests as two distinct phenotypes
  - Superficial, non–muscle-invasive bladder cancer (NMIBC; 75% of new cases)
    - 50% of NMIBC recur
      Approximately 10–20% will progress to MIBC
    - NMIBC is treated by is transurethral resection biopsy (TURB), with or without intravesical chemotherapy, and immune therapy with bacillus Calmette-Guérin (BCG)
  - The second phenotype is muscle invasive bladder cancer (MIBC).
    - Represents 20–30% of all BC; only 15% with prior history of superficial disease (progression)
    - MIBC patients have a poor (50–60%) overall survival rate, despite aggressive combined treatment modalities that include cystectomy, chemotherapy, and immunotherapy

Oncogenic Pathways in Urothelial Carcinoma of the Urinary Bladder

- Two distinct pathogenic pathways
  - NMIBC originate from benign urothelium through a process of urothelial hyperplasia
    - ○ Three primary genetic alterations are associated with the pathogenesis pathway of NMIBC
    - ○ Tyrosine kinase receptor FGFR3
      - ♦ HRAS
      - ♦ PIK3CA
      - ♦ RAS-MAPK and PI3K-AKT pathway alterations are responsible for promoting cell growth in urothelial neoplasia
- ○ Activating mutations in RAS lead to activation of MAPK and PI3K pathways
- ○ Activating mutations in upstream tyrosine kinase receptor FGFR3 seems to be mutually exclusive with RAS mutations given that both signal through a common downstream pathway in urothelial oncogenesis
- ○ PIK3CA and FGFR3 mutations generally cooccur, suggesting a potential synergistic additive oncogenic effect of PIK3CA mutations
- ○ Telomerase reverse transcriptase (*TERT*) promoter mutation is the most common mutation that occurs in approximately 60–80% of bladder cancers across different grades and stages (Figs. 15.1 and 15.2)

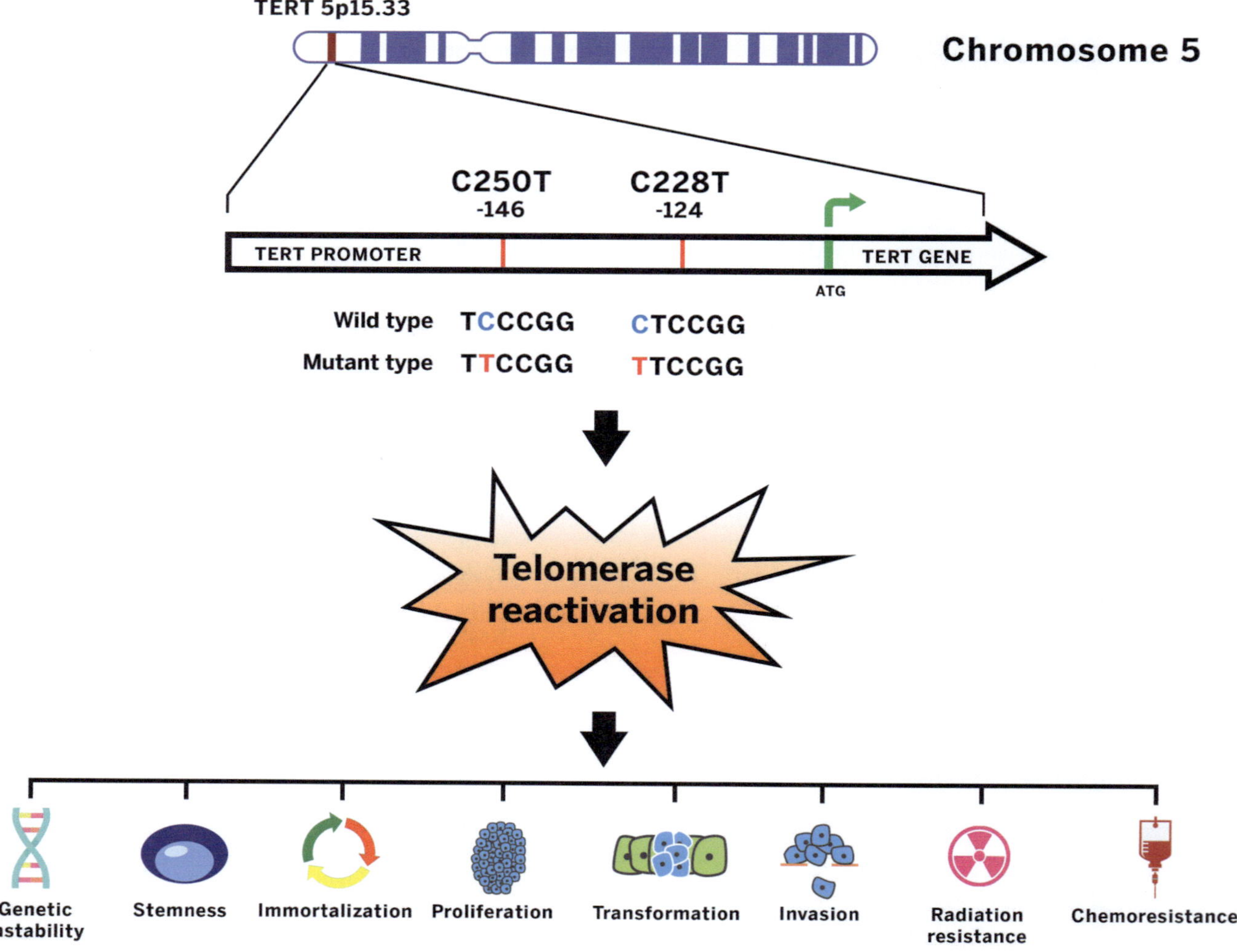

Fig. 15.1 *TERT* promoter mutation and its clinical impacts. The two major *TERT* promoter mutations, C228T and C250T, are located at −124 and −146 base pairs upstream of ATG start codon. *TERT* promoter mutations create a TTCCGG motif, a consensus E-26 transcription factor/ternary complex factor binding site, which upregulates the TERT expression leading to reactivation of telomerase. *TERT* promoter mutation has been implicated in cell genetic instability, stemness, immortalization, cell proliferation, transformation, invasion, and chemo-/radiotherapy resistance (from Cheng L et al. *Hum Pathol*. 2023; with permission)

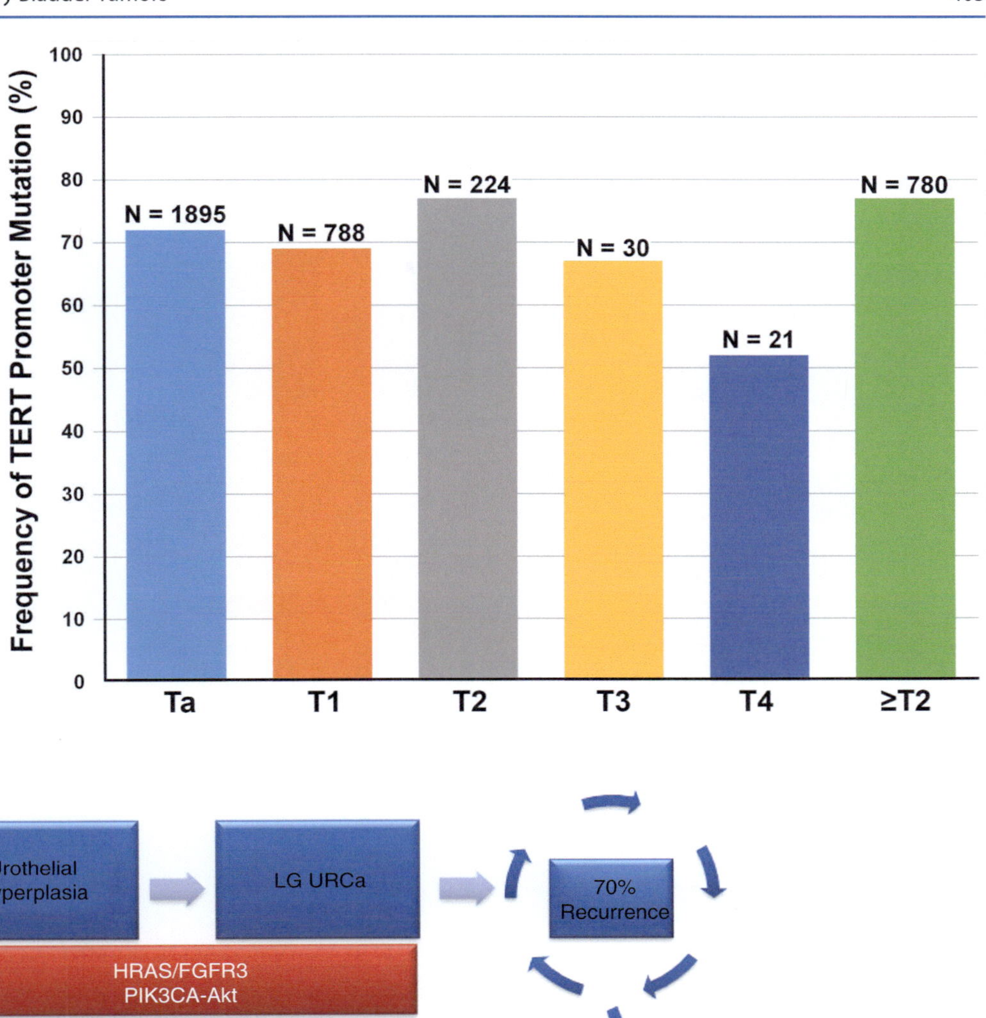

Fig. 15.2 Prevalence of *TERT* promoter mutation across different pathologic stages of bladder cancer (from Cheng L et al. *Hum Pathol.* 2023; with permission)

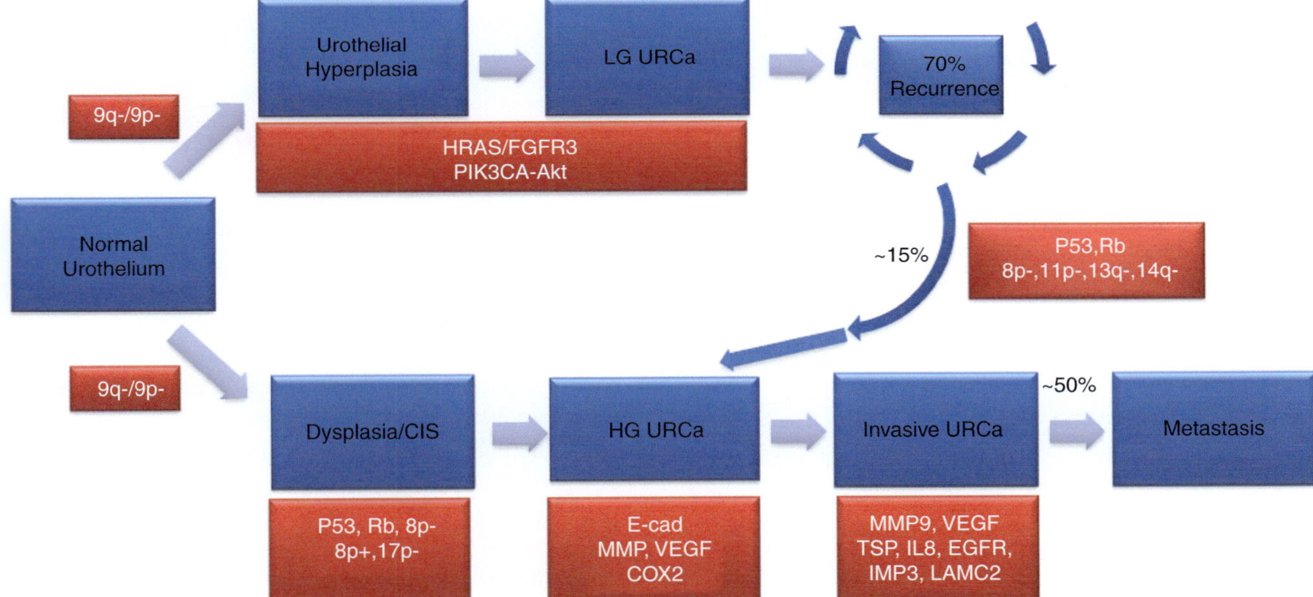

Fig. 15.3 Divergent molecular pathways of oncogenesis in NMIBC and MIBC urothelial carcinoma of urinary bladder; genetic alterations are depicted in key stages of disease progression. URCa, urothelial carcinoma of urinary bladder; LG, noninvasive low grade; HG-URCa, noninvasive high grade (from Netto GJ et al. *Arch Pathol Lab Med.* 2012;36:372; with permission)

- Majority of MIBC originate through progression from dysplasia to flat carcinoma in situ (CIS) and high-grade noninvasive lesions
    - Genetic instability facilitates the accumulation of genetic alterations
    - Pathogenic pathway primarily involves alterations in tumor suppressor genes involved in cell cycle control, including TP53, p16, and RB (Figs. 15.3 and 15.4)
- Progression of the subset of NMIBC into MIBC is similarly based on alterations in *TP53* and *RB* tumor suppressor genes (Fig. 15.3)

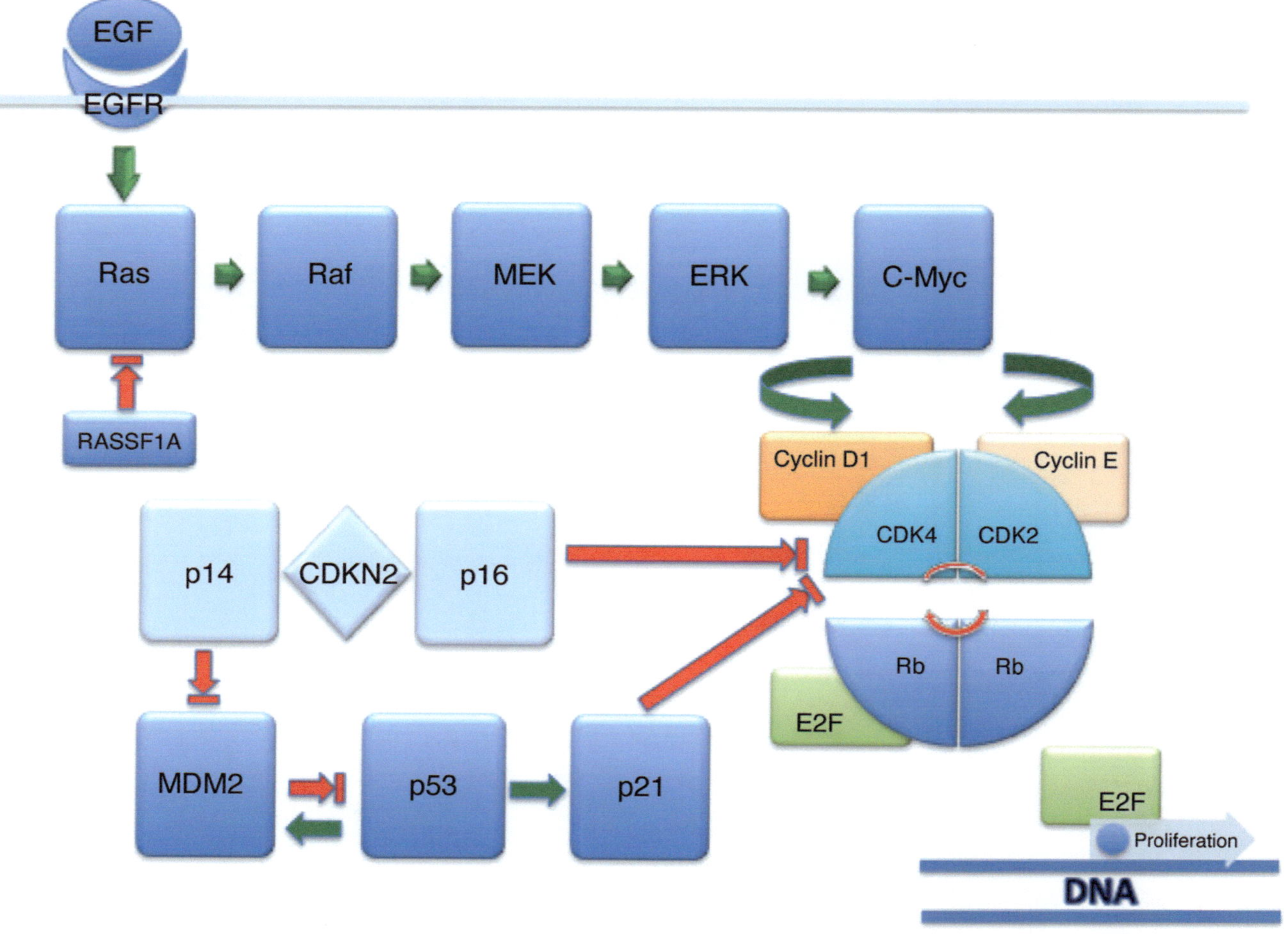

Fig. 15.4 Receptor tyrosin kinase (EGFR/RAS/MEK/ERK) and cell-cycle regulator (p14, p16, p53, p21, Cyclin D1, Cyclin E, and Rb) pathways in urothelial carcinoma; *green* and *red arrows* represent stimulation and inhibition, respectively (from Netto GJ et al. *Arch Pathol Lab Med.* 2012;36:372; with permission)

Prognostic Biomarkers in Bladder Cancer

General Background

- Established clinicopathologic prognostic parameters for NMIBC
  - pT stage
  - WHO/ISUP grade
  - Tumor size
  - Tumor multifocality
  - Presence of CIS
  - Frequency and rate of prior recurrences
- Prognostic parameters to accurately predict progression in NMIBC are actively sought to identify patients in need of vigilant surveillance and aggressive treatment

- The translational field of molecular prognostication, theranostics, and targeted therapy in BC has sharply gained momentum with our understanding of oncogenetic pathways and intrinsic molecular subtypes of MIBC
- A rigorous validation process ought to precede the incorporation of molecular biomarkers in clinical management (Table 15.1)
  - Initial retrospective discovery studies need to be validated in large independent cohorts
  - Robustness of the proposed biomarker tested in well-controlled, multi-institutional randomized prospective study

Table 15.1 Established clinicopathologic and potential molecular prognostic parameters in nonmuscle invasive urothelial carcinoma and muscle invasive urothelial carcinoma

Clinicopathologic prognostic parameters	
Nonmuscle invasive urothelial carcinoma	**Muscle invasive urothelial carcinoma**
WHO/ISUP grade	pTNM
pT stage	Lymphovascular invasion (LVI)
Presence of associated CIS/dysplasia	Resistance to neoadjuvant chemotherapy
Disease duration	
Time to and frequency of recurrences	*Divergent histology*
Multifocality	Micropapillary
Tumor size (>3 cm)	Plasmacytoid
Failure of prior BCG Rx	Nested
Presence of LVI	Lymphoepithelioma-like
Depth of lamina propria invasion	Sarcomatoid
Emerging molecular prognostic markers	
Nonmuscle invasive urothelial carcinoma	**Muscle invasive urothelial carcinoma**
Proliferation index (Ki-67, MIB1, S phase)	p53 inactivation/accumulation
FGFR3 mutation/overexpression (protective)	Alterations of Rb expression
mG (FGFR#/MIB1)	Loss of p21 expression
p53 inactivation/accumulation	Alteration of p16 expression
DNA ploidy status	Loss of E-cadherin
Multitarget FISH	
HRAS	*RTK*
ERBB3, ERBB4 overexpression (protective)	EGFR overexpression
Loss of E-cadherin	ERBB2/HER2 overexpression/amplification
Cell cycle control	*mTOR-Akt pathway*
Downregulation of Rb expression	mTOR
Downregulation of p21 expression	Phos S6 expression (protective)
Downregulation of p27 expression	
Cyclin D3 overexpression	
Cyclin D1 overexpression	
Multi-biomarker immunoexpression analysis (p53, p27, Ki-67, Rb, p21)	
Angiogenesis markers	*Angiogenesis markers*
VEGF overexpression	VEGF overexpression
HIF1A overexpression	HIF1A overexpression
TSP1 overexpression	TSP1 overexpression
Genomic and gene expression array panels	
Nonmuscle invasive urothelial carcinoma	**Muscle invasive urothelial carcinoma**
Epigenetic alterations	*Epigenetic alterations*
RASSF1 promoter hypermethylation	RASSF1 promoter hypermethylation
DAPK promoter hypermethylation	CDH1 (E-cadherin) promoter hypermethylation
APC promoter hypermethylation	EDNRB promoter hypermethylation
CDH1 (E-cadherin) promoter hypermethylation	

Chromosomal Numerical Alterations–Early Culprit of Genetic Instability in Bladder Cancer

- Chromosome 9 alterations are the earliest genetic alterations in both arms of divergent pathways of BC development
- Additional structural/numerical somatic chromosomal alterations are also common, such as gains of chromosomes 3q, 7p, and 17q and deletion of 9p21 (p16 locus)
  - A multitarget interphase fluorescence in situ hybridization (FISH)-based urine cytogenetic assay was developed based on the above numerical chromosomal alterations and is commercially available (Fig. 15.5); see section, Biomarkers of Early Detection
  - Have potential diagnostic and prognostic value
  - Initially, FDA approved for surveillance, subsequently gained approval for screening in high-risk (smoking exposure) patients with hematuria
  - Sensitivity range of 69–87% and specificity range of 89–96% have been reported
  - May enhance the sensitivity of routine urine cytology and can be used in combination as a reflex testing in cases with atypical cytology
- Some studies have pointed to potential prognostic role for multitarget FISH analysis
  - Low-risk, FISH-positive patients, defined as 9p21 loss/Ch3 abnormalities, have a higher rate of recurrence compared to FISH-negative patients
  - The recurrence rate is even greater in patients with a high-risk positive FISH (Ch7/Ch17 abnormality)
  - Using bladder washings and formalin-fixed, paraffin-embedded transurethral biopsy samples, loss of 9p21 predicts recurrence but not progression in NMIBC
  - Urine cytology and FISH in post-BCG bladder washings predict failure to BCG therapy in patients with NMIBC disease

Receptor Tyrosine Kinase Alterations

- Numerous studies have pointed to the potential prognostic value of evaluating the expression of receptor tyrosine kinases (RTK), such as FGFR3, EGFR, and other ERB family members (HER2 and ERBB3) in NMIBC and MIBC
- *FGFR3* mutations commonly occur in NMIBC; theoretically, they can be used alone or combined with *RAS* and *PIK3CA* oncogenes as markers of early recurrence during surveillance
  - Sensitive PCR assays can detect *FGFR3* mutations in voided urine

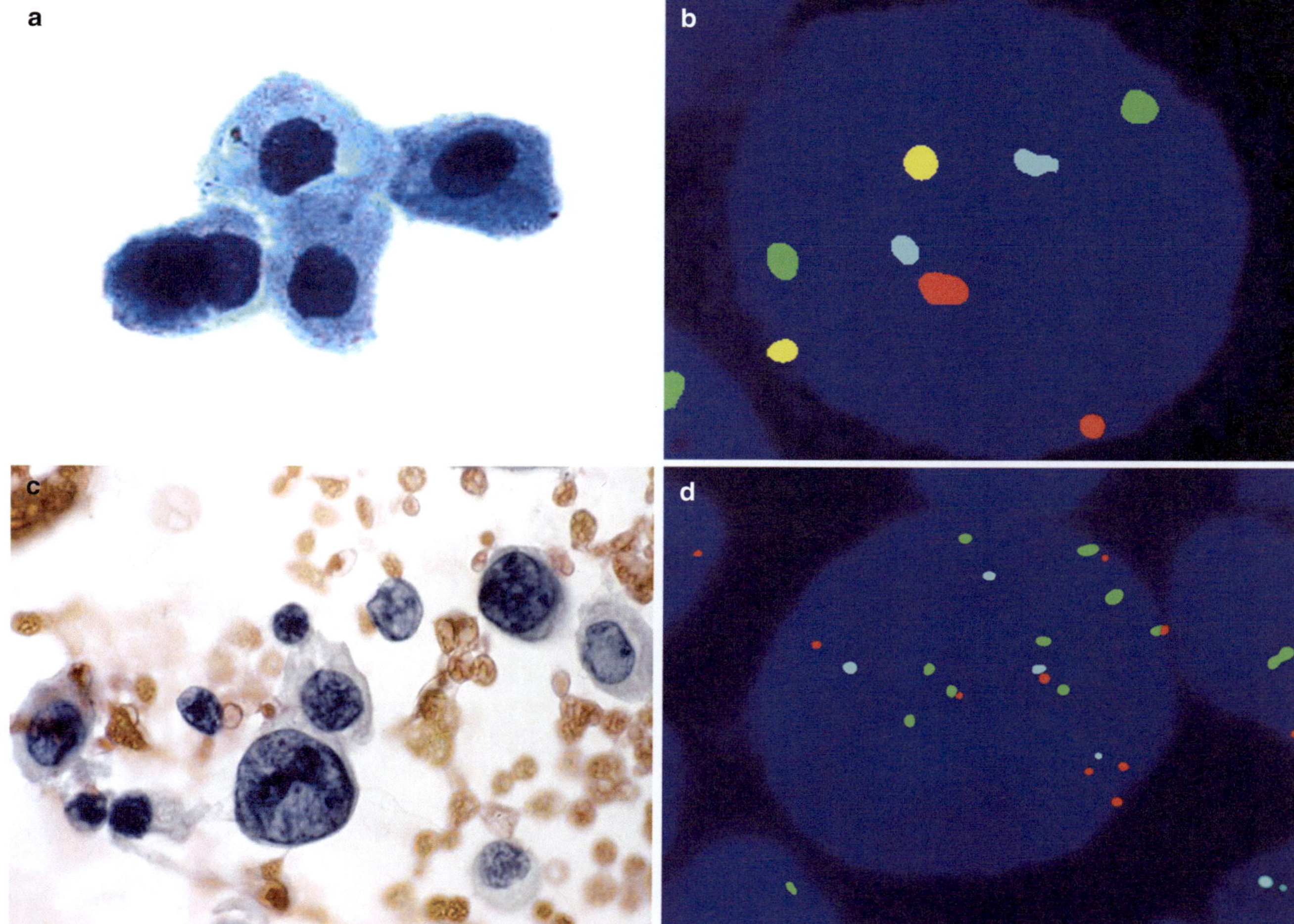

Fig. 15.5 Detection of urothelial carcinoma by the UroVysion FISH analysis. Normal urothelial cells (**a**) showed two signals from each probe for CEP3 (*red*), CEP7 (*green*), CEP17 (*aqua*), and 9p21 (*gold*) (**b**). Malignant urothelial cells (**c**) demonstrated gaining of chromo-somes as indicated by 8 *red* (CEP3), 9 *green* (CEP7), 4 *aqua* (CEP17), and loss of 9p21 as indicated by the absence of *yellow* signals (9p21) (**d**) (from Cheng L et al. *Hum Pathol* 2011;42:455; with permission)

- Positive urine sample associated with concomitant or future recurrence in 81% of NMIBC cases
- A predictive value of 90% was achieved in patients with consecutive *FGFR3*+ urine samples
- Superior to cytology (78% vs 0%) in detecting post-TURB recurrence in NMIBC harboring *FGFR3* mutations in primary tumors

- A multiplex PCR assay has been developed for mutational analysis detecting the most frequent mutation hot spots of *HRAS, KRAS, NRAS, FGFR3, TERT* promoter, and *PIK3CA* in formalin-fixed, paraffin-embedded TURB samples
  - Evidence of at least one mutation in up to 88% of low-grade NMIBC samples was demonstrated
  - Revealed *FGFR3* mutations to be more common among low malignant potential neoplasms (77%) and TaG1/TaG2 tumors (61%/58%) than among TaG3 (34%) and T1G3 tumors (17%) NMIBC

- A molecular grade parameter (mG) based on a combination of *FGFR3* gene mutation status and MIB1 index was proposed by van Rhijn et al. as an alternative to pathologic grade in NMIBC
  - mGsubsequently was compared to the European Organization for Research and Treatment of Cancer (EORTC) NMIBC risk calculator (weighted score of six variables including WHO 1973 grade, stage, presence of CIS, multiplicity, size, and prior recurrence rate)
    - mG was more reproducible than the pathologic grade (89% vs 41–74%)
    - NMIBC *FGFR3* mutations significantly correlated with favorable disease parameters, whereas

increased MIB1 was frequently seen with pT1, high grade, and high EORTC risk scores

- o The addition of mG for progression increased the predictive accuracy of EORTC score (see below section on molecular grading)

p53, Cell Cycle Regulators, and Proliferation Activity Index

- p53 alterations are a strong independent predictor of disease progression in BC (NMIBC, MIBC, as well as CIS)
- p53 is predictive of increased sensitivity to chemotherapeutic agents that lead to DNA damage
- Among other G1-S phase cell cycle regulators, cyclin D3, cyclin D1, p16, p21, and p27 have also been evaluated as prognosticators in NMIBC
- A synergistic prognostic role for combining p53 evaluation with other cell cycle control elements, such as pRB, cyclin E1, p21, and p27, is emerging in both NMIBC and MIBC
  - In NMIBC, synchronous immunohistochemical alterations in all four tested markers (p53, p21, pRB, and p27) have significantly lower disease-free survival (DFS) compared to patients with only three markers
  - The negative predictive effect was decreased with decreasing number of altered markers (3 vs 2 vs 1)
  - Combining p53, p27, and Ki-67 assessment in pT1 radical cystectomy specimens improved the prediction of DFS and disease-specific survival (DSS)
- A similar synergistic prognostic role for p53, pRB, and p21 expression has been demonstrated in patients undergoing cystectomy for MIBC
- Tumor proliferation index measured immunohistochemically by either Ki-67 or MIB1 was consistently shown to be a prognosticator
  - MIB1 in NMIBC plays a prognostic role as an element of the above described mG.
  - Ki-67 index in NMIBC TURB biopsy is predictive of DFS and DSS
- Similar role for proliferation index as prognosticator is established in MIBC
  - Building on initial findings of significance in an organ-confined subset of MIBC, a bladder consortium multi-institutional trial confirmed the role of proliferation index, measured in cystectomy specimens
  - Ki-67 improved prediction of both DFS and DSS when added to standard prediction models, supporting a role for stratifying patients for perioperative systemic chemotherapy
- Despite the above evidence, cell cycle marker assessment never became a part of standard of care in BC

Epigenetic Alterations

- Epigenetic analysis is also gaining momentum in BC as a noninvasive diagnostic and prognostic tool for screening and surveillance (Table 15.1)
- Hypermethylation analysis at 11 CpG promoter islands, performed by Catto et al., using methylation-specific PCR (MSP)
  - Promoter methylation was found in 86% of all tumors, and the incidence was relatively higher in upper tract tumors compared to BC
  - Methylation was associated with advanced tumor stage and higher tumor progression and mortality rates
  - Methylation at the *RASSF1A* and *DAPK* gene promoters was associated with disease progression independent of tumor stage and grade on multivariate analysis
- Five loci associated with progression (RASSF1a, *CDH1* [E-cadherin], TNFSR25, EDNRB, and *APC*) were found using quantitative MSP at 17 candidate gene promoters
  - Multivariate analysis revealed that the overall degree of methylation was more significantly associated with subsequent progression and death than tumor stage
  - An epigenetic predictive model developed using artificial intelligence techniques identified likelihood and timing of progression with 97% specificity and 75% sensitivity
- The diagnostic role of promoter hypermethylation using MSP assay in four genes (*ECDH1, p16, p14,* and *RASSF1A*) in primary tumor DNA and urine sediment DNA
  - MSP detected hypermethylation in the urine of 80% of tested patients
  - Hypermethylation analysis *of CDH1, p14,* or *RASSF1A* in urine sediment DNA detected 85% of superficial and low-grade BC and 79% of high-grade and 75% of invasive BC
  - The study highlighted the great potential of such test in detecting NMIBC
- Potential diagnostic role for methylation-specific multiplex ligation-dependent probe amplification assay (MS-MLAP) analysis of 25 tumor suppressor genes has been shownby Cabello et al.
  - The tumor suppressor genes included *PTEN, CD44, WT1, GSTP1, BRCA2, RB1, TP53, BRCA1, TP73, RARB, VHL, ESR1, PAX5A, CDKN2A,* and *PAX6*
  - *BRCA1, WT1,* and *RARB* were found to be the most frequently methylated tumor suppressor genes with significant diagnostic accuracies in two additional validation sets
- Assessment of promoter hypermethylation is giving additional insights on BC oncogenesis

- Promoter hypermethylation of CpG Islands and "shores" controlling miRNA expression is one such example

Intrinsic Molecular Genomic Subtypes of Urothelial Carcinoma

- Genomic studies have validated previously deciphered genetic pathways of BC development and unmasked additional crucial driver genetic alterations
- Earlier array-based gene expression studies highlighted differentially expressed genetic signatures that can predict recurrence and progression
- Subsequent integrated multiplatform genomic, transcriptomic, proteomic, and epigenomic studies defined clinically relevant molecular subtypes of BC (Fig. 15.6)
  - Two main genomic molecular circuitries were reported by Lindgren et al.
    - The first is characterized by *FGFR3* alterations, overexpression of CCND1, and deletions in 9q and CDKN2A
    - The second is characterized by E3F3 amplifications, RB1 and PTEN deletions, gains of 5p, and overexpression of CDKN2A

- Advanced tumors in both groups demonstrated *TP53/MDM2* gene alterations
- The first study to point to a significantly worse prognosis associated with the gene expression signature of a keratinized/squamous phenotype (CK6+)
- Choi et al. also showed the aggressive behavior of this molecularly defined subtype, termed "basal-like," and two additional intrinsic molecular subtypes
  - The "basal-like" subtype is characterized by p63 activation, squamous differentiation, positive CK5/6, EGFR, and CD44 expression and lack of CK20 (Fig. 15.7) and appeared sensitive to neoadjuvant chemotherapy (NAC)
  - The "luminal" subtype is typically enriched for activating *FGFR3* mutations, active estrogen receptor pathway, and ERBB2 and PPARγ expression profile (all potential targets of therapy)
  - The third subtype, characterized by wildtype *TP53* gene expression signature (p53-like), is resistant to NAC
  - Tumors from the luminal and "basal-like" subtypes also displayed the *TP53* wildtype expression signature upon resistance to chemotherapy

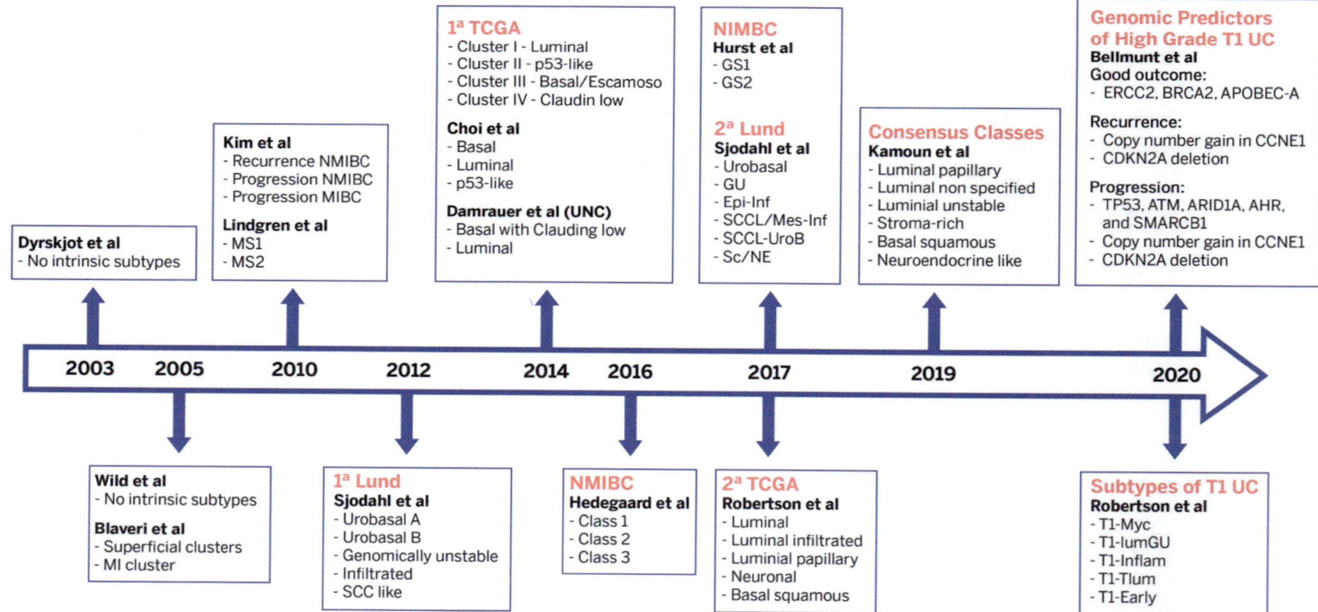

Fig. 15.6 Evolving concepts in molecular classification of bladder cancer (from Lopez-Beltran A et al. *Hum Pathol* 2021;113:67; with permission)

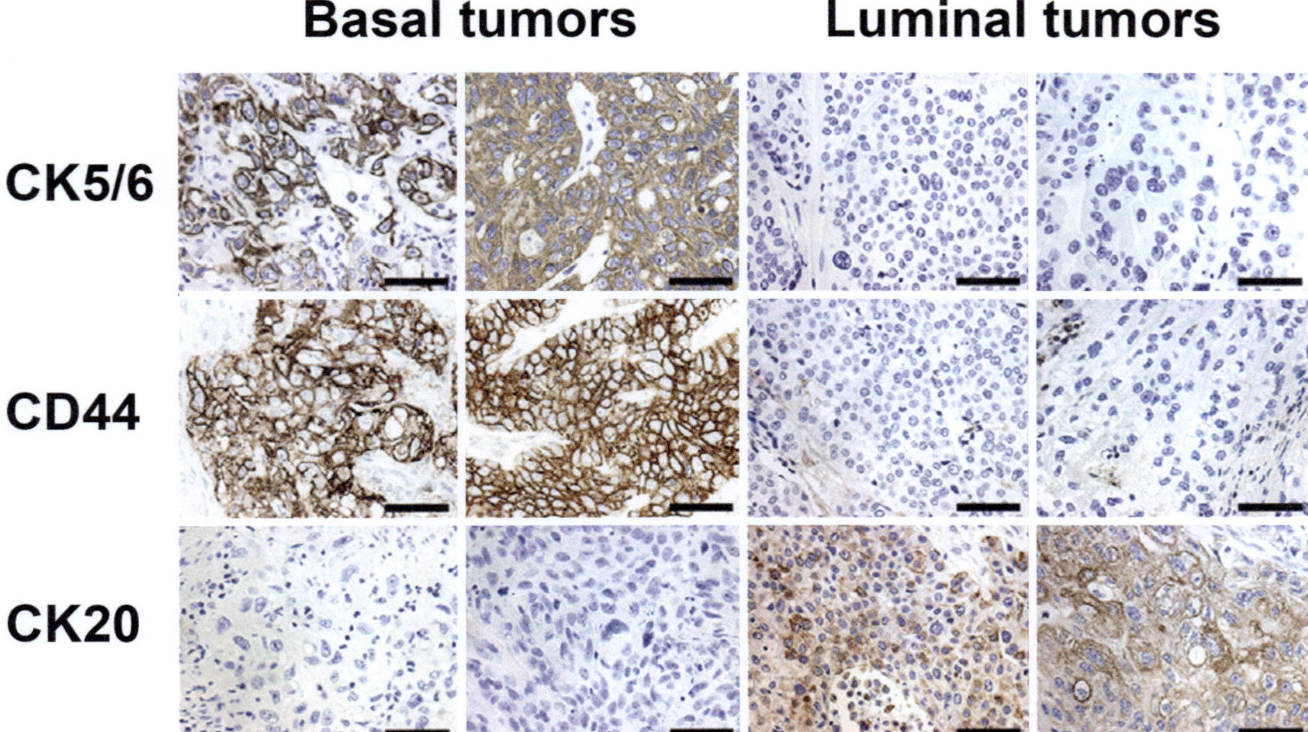

Fig. 15.7 Immunohistochemical analysis of basal and luminal markers expression in bladder cancer. The basal and luminal immunoprofiles (CK5/6+, CD44+, CK20− and CK5/6−, CD44−, CK20+, respectively) are shown in basal *(left)* and luminal *(right)* tumors that were categorized as such based on gene expression profiling is shown. (Modified with permission from Choi W et al. *Cancer Cell.* 2014;25[:152])

Bladder Cancer TCGA 2014

- 131 MIBC
- 302 mutations, 204 segmental copy number alterations, and 22 rearrangements on average per tumor
- Recurrent driver mutations in 32 genes involved in cell-cycle regulation, chromatin regulation, kinase signaling pathways, and nine additional genes (e.g., *MLL2*, *ERCC2*, *ELF3*, *KLF5*, *RXRA*, and *CDKN1A*)
- Integration of mRNA and miRNA and protein expression analysis revealed four major expression clusters
  - "Papillary-like" cluster (cluster I), enriched for *FGFR3* gene alterations, papillary morphology and together with cluster II share expression of luminal urothelial differentiations markers (activated expression of ER, GATA3, uroplakin, and ERBB2); cluster III "basal/squamous-like" was characterized by CK5/6 and EGFR expression (Fig. 15.8)

Bladder Cancer TCGA 2017

- 412 MIBC
- Recurrent driver mutations in 58 genes
- Overall mutational load in BC is primarily ascribed to APOBEC-induced mutagenesis signature
- Clustering by mutational signatures identified a high-mutation subset of BC with 75% 5-year survival
- mRNA expression clustering refined and expanded the initial TCGA clusters classification (Fig. 15.9)
- Three subgroups within the luminal subtype along with a basal-squamous subtype as well a distinct neuronal cluster for a total of five intrinsic subtypes
- Luminal-papillary subtype (35%)
  - Characterized by *FGFR3* mutations, fusions with TACC3, and/or amplification; and papillary histology
  - They have lower risk for progression and are candidates for FGFR3 inhibitors considering their low response to cisplatin-based NAC
- Luminal-infiltrated subtype (19%)
  - Characterized by the lowest purity and high expression of EMT and myofibroblast markers
  - Demonstrates medium level of CD274 (PD-L1) and CTLA4 immune marker expressions that may account for their response to immune checkpoint therapy
- Luminal subtype (6%)
  - Shows high expression of luminal markers (e.g., uroplakin, GATA3, and FOXA1) and KRT20

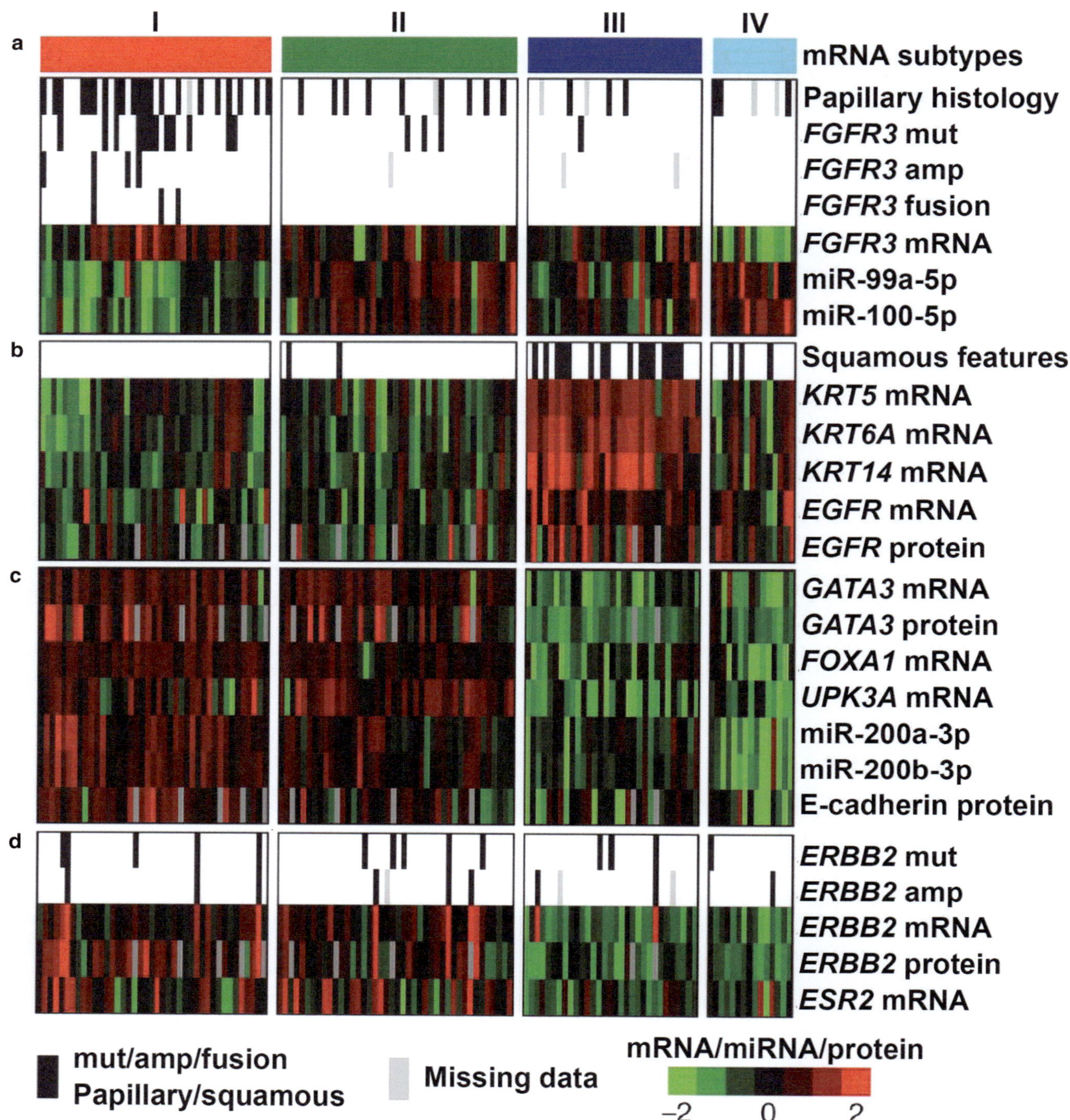

Fig. 15.8 Integrated analysis of mRNA, miRNA, and protein data pointed to four clusters of urothelial carcinoma based on expression profiles. (**a**) Papillary histology, FGFR3 alterations, FGFR3 expression, and reduced FGFR3-related miRNA expression were encountered in cluster I. (**b**) High expression of epithelial markers and stem/progenitor cytokeratins is demonstrated in cluster III tumors that occasionally display a squamous histology. (**c**) Luminal breast and urothelial differentiation markers are enriched in clusters I and II. (**d**) ERBB2 mutation and estrogen receptor beta (ESR2) expression are enriched in both clusters I and II (adapted with permission from Cancer Genome Atlas Research Network. *Nature*.2014;507:315)

- Basal-squamous subtype (35%)
    - Characterized by higher incidence in women
    - Presence of squamous differentiation (includes tumors without definitive squamous differentiation)
    - High expression of basal and stem-like markers (CD44, KRT5, KRT6A, and KRT14)
    - High expression of CD274 (PD-L1) and CTLA4 and immune infiltration

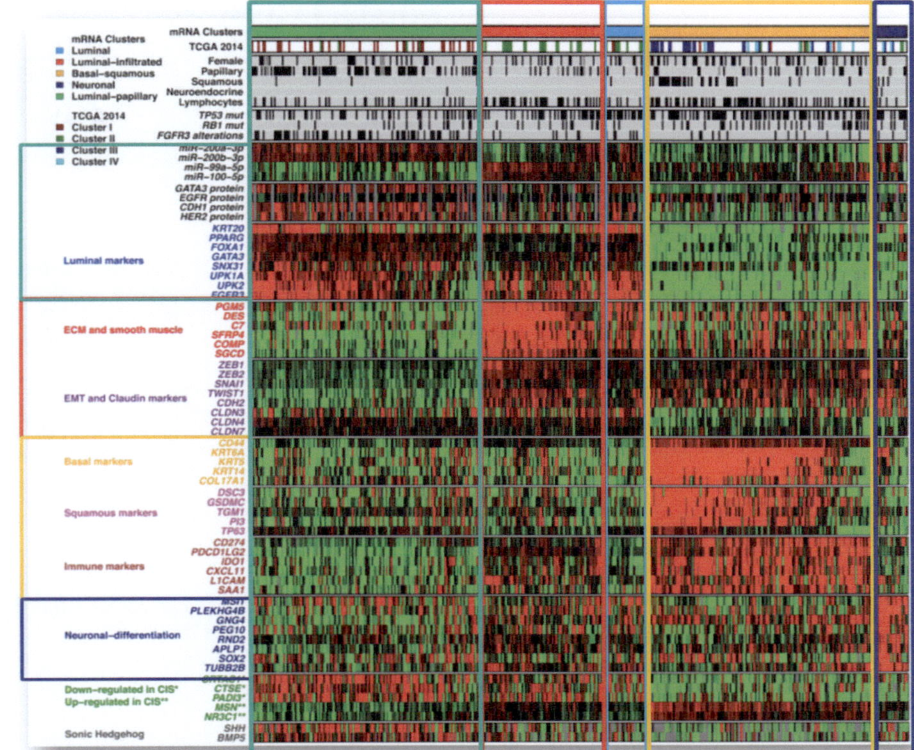

Fig. 15.9 Five mRNA expression subtypes: luminal-papillary, luminal-infiltrated, luminal, basal-squamous, and neuronal. Covariates: 4 previously reported TCGA subtypes; selected clinical covariates and key genetic alterations; normalized expression for miRNAs and proteins; log2 (fold change against the median expression across samples) for selected genes, for labeled gene sets. Samples within the three luminal subtypes, the basal-squamous subtype, and the neuronal subtype are ordered by luminal, basal, and neuroendocrine signature scores, respectively (modified with permission from Robertson G et al. *Cancer Cell.* 2017;171:540)

- This subtype has higher likelihood of response to cisplatin-based NAC and immune checkpoint therapy
- Neuronal subtype (5%)
  - Associated with the worst clinical outcome
  - Majority with p53/cell-cycle pathway alterations
    - Only a subset associated with mutations in both *TP53* and *RB1* (the hallmark alteration in small cell/ neuroendocrine)
  - Assigning a tumor to this subtype is based on expression of neuroendocrine markers by mRNA-seq or immunohistochemistry, given the lack of typical small cell neuroendocrine morphology in the majority of tumors
  - Like neuroendocrine-type tumors of other sites, etoposide-cisplatin based therapy may be effective (Fig. 15.10)
- To date, a total of six MIBC molecular classifications have been proposed
  - Derived from largely nonoverlapping datasets
  - Share some subtype-specific molecular features and some overlap of their subtypes
- 2020 consensus reconciled the published classification schemes in a "Consensus Molecular Classification of Muscle Invasive Bladder Cancer" proposal

- The consensus includes six molecular classes
  - Luminal papillary (24% of MIBC)
  - Luminal nonspecified (8%)
  - Luminal unstable (15%)
  - Stroma-rich (15%)
  - Basal/squamous (35%)
  - Neuroendocrine-like (3%)
- These classes differ in underlying oncogenic mechanisms, in their infiltration by associated tumor immune microenvironment (immune and stromal cells), and histological and clinical characteristics features and outcomes (Figs. 15.11 and 15.12)
- The hope is that adopting this classification will help address unanswered questions stability of these molecular subtypes within a given tumor (intratumoral heterogeneity within primary vs metastasis) and after treatment (Table 15.2)
- Future studies should determine whether molecular subtyping can be incorporated into clinical decision making

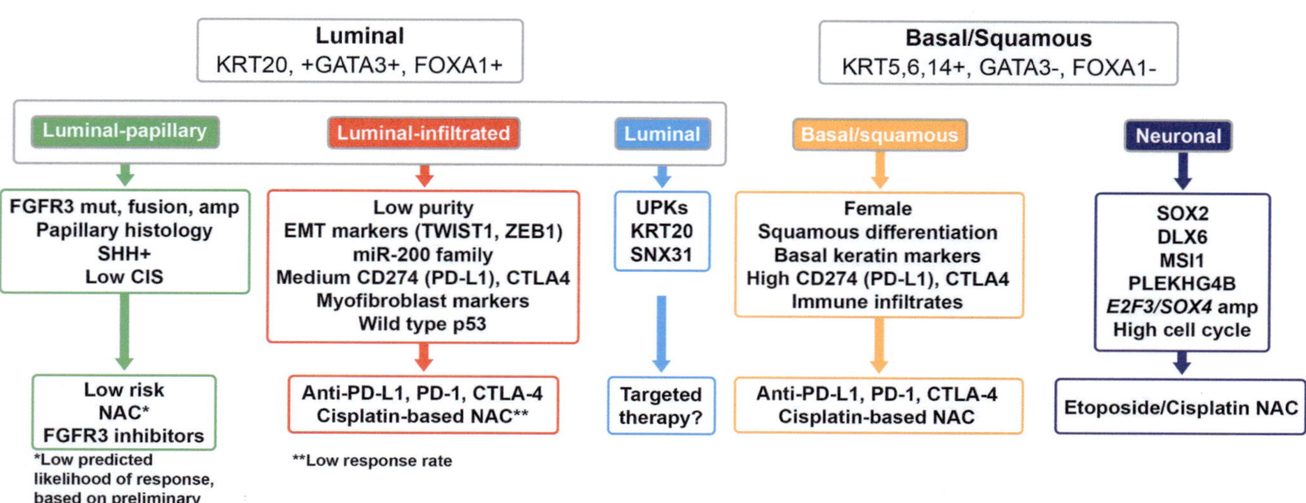

Fig. 15.10 Proposed schema of expression-based, subtype-stratified therapeutic approach as a framework for prospective hypothesis testing in clinical trials (modified with permission from Robertson G et al. *Cancer Cell.* 2017;171:540)

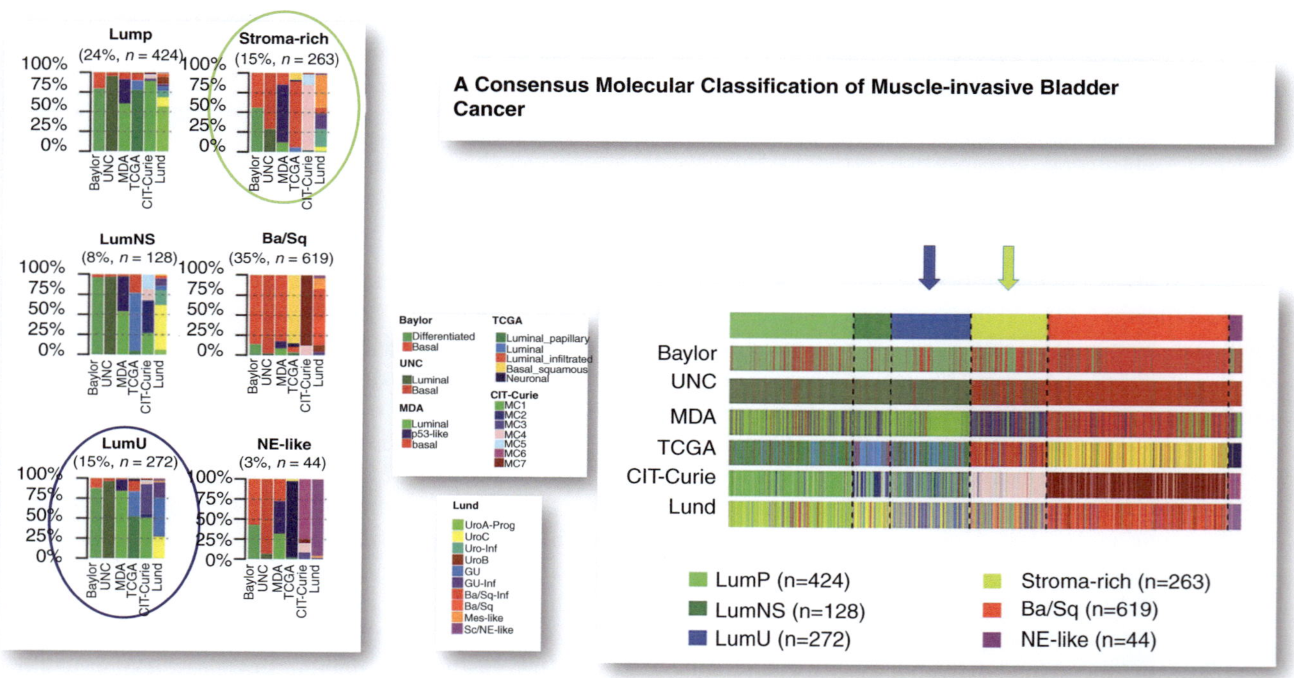

Fig. 15.11 Summary of six consensus classes and their relationships to input molecular subtypes. The samples are grouped by their predicted consensus class labels: LumP, LumNS, LumU, stroma-rich, Ba/Sq, and neuroendocrine (NE)-like. For each consensus class, a bar plot shows the proportion of samples assigned in each input subtype of each input classification system. Relationship between subtyping results from the six input classification schemes. Samples are ordered by predicted consensus classes. Ba/Sq, basal/squamous; LumNS, luminal nonspecified; LumP, luminal papillary; LumU, luminal unstable; MCL, Markov cluster algorithm; MDA, MD Anderson Cancer Center; MIBC, muscle-invasive bladder cancer; TCGA, the Cancer Genome Atlas; UNC, University of North Carolina(modified with permission from Kamoun A et al. *Eur Urol.* 2020;77:420).

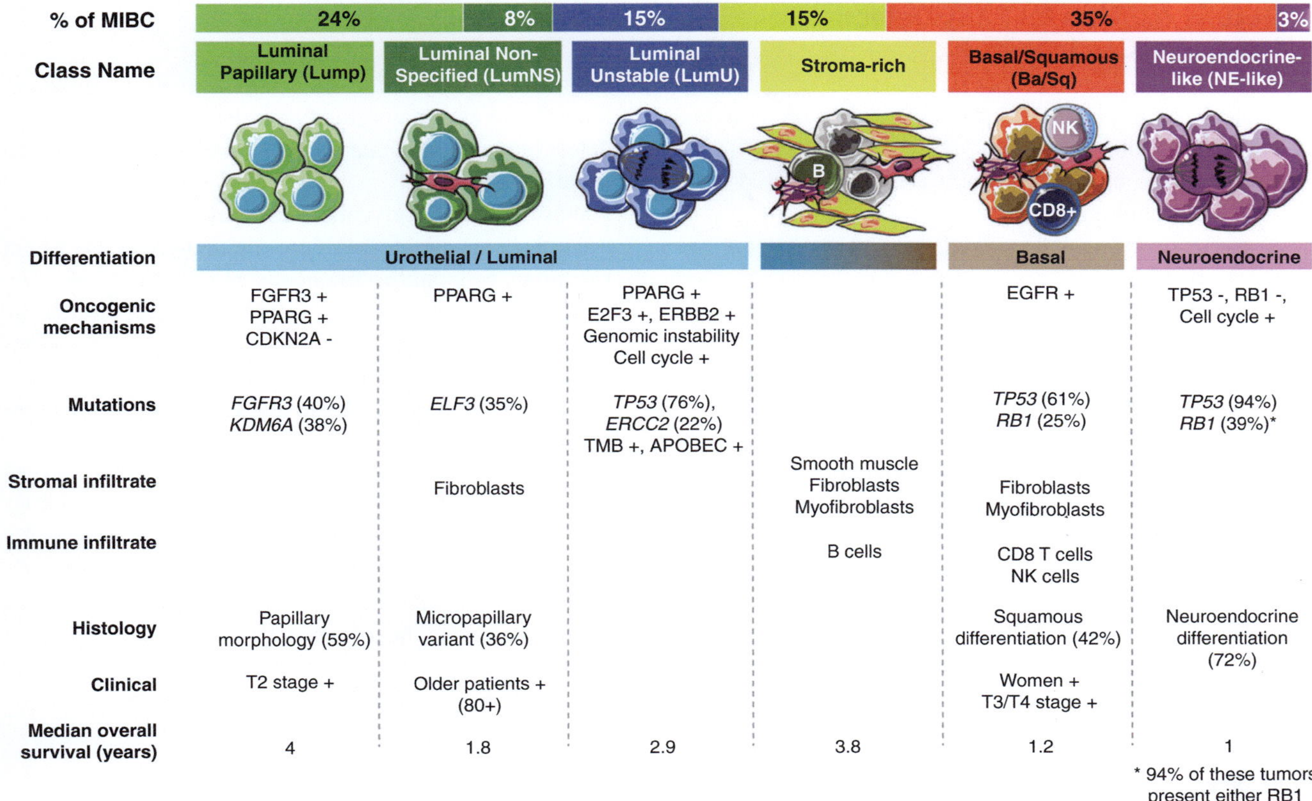

Fig. 15.12 Summary of the main characteristics of the consensus classes. From top to bottom, the following characteristics are presented: proportion of consensus classes in the 1750 tumor samples; consensus class names; schematic graphical representation of tumor cells, and their microenvironments (immune cells, fibroblasts, and smooth muscle cells); differentiation-based color scale showing features associated with consensus classes, including a luminal-to-basal gradient and neuroendocrine differentiation; and a table displaying the dominant characteristics such as oncogenic mechanisms, mutations, stromal infiltrate, immune infiltrate, histology, clinical characteristics, and median overall survival. Abbreviations: Ba/Sq = basal/squamous; EGFR = estimated glomerular filtration rate; LumNS = luminal nonspecified; LumP = luminal papillary; LumU = luminal unstable; MIBC = muscle-invasive bladder cancer; NE = neuroendocrine; NK = natural killer (modified with permission from Kamoun A et al. *Eur Urol*.2020;77:420)

Table 15.2 Reported characteristics of the molecular subtypes of MIBC cancer and their and therapeutic implications

	Luminal-papillary (24%)	Luminal nonspecified (8%)	Luminal unstable (15%)	Stroma rich (15%)	Basal-squamous (35%)	Neuroendocrine-like (3%)
Immunohistochemistry panel (positive)	GATA3, FOXA1 FGFR3, CK20, Uroplakin 2	GATA3, FOXA1 FGFR3, CK20, Uroplakin 2	GATA3, FOXA1 FGFR3, CK20, Uroplakin 2	Vimentin, desmin, SMA	CK5/6, CK14, Desmoglein 3, STAT3	INSM1, synaptophysin, chromogranin, CD56
Potential targeting therapy	Low risk, FGFR3 inhibitors, low sensitivity to NAC	Response to ICI, low sensitivity to NAC	Response to ICI Sensitivity to NAC	Not specified	Response to ICI EGFR inhibitors, cisplatin NAC	Response to ICI, combined chemotherapy
Additional terminology[a]	Lund Uro A MDA luminal TCGA luminal-papillary UNC luminal	Lund Uro C MDA p53-like and luminal TCGA luminal UNC luminal	Lund genomically unstable MDA luminal TCGA luminal UNC luminal	Lund mesenchymal-like MDA p53-like TCGA basal-squamous UNC basal	Lund basal-squamous and basal-squamous infiltrated/Uro B MDA basal TCGA basal-squamous UNC basal	Lund small cell/ NE-like TCGA neuronal, UNC basal

Activated mutations (m), amplifications (a), fusions (f), deletions (d); *ICI* immune checkpoint inhibitors, *MDA* MD Anderson Cancer Center, *MIBC* muscle invasive bladder cancer, *NAC* neoadjuvant chemotherapy, *NE* neuroendocrine, *TCGA* The Cancer Genome Atlas, *TMB* tumor mutation burden, *UNC* University of North Carolina

[a] Some of the proposed categories in other classifications present overlapping features with limited correspondence between them. (Modified from Lopez-Beltran A et al. *Hum Pathol* 2021;113:67; with permission

Molecular Markers for Early Detection of Urothelial Carcinoma

- Urine cytology is currently the most widely used method for BC screening
    - Diagnostic cytologic criteria are largely based upon cell morphology
        - Accuracy is hampered by the element of subjectivity
    - Cytology is highly effective in detecting high-grade cancers
        - Sensitivity and specificity in detecting low-grade UC are poorer
    - Accumulated knowledge in the molecular processes involved in carcinogenesis has resulted in the development of many new markers for diagnosis, surveillance
    - UroVysion, BTA-Stat/BTA-TRAK, NMP22, and ImmunoCyt/uCyt are currently available testing methods
        - Relatively widely used
        - Possessing either Food and Drug Administration (FDA) clearance or approval

UroVysion

- A multicolor, multitargeted FISH assay
- Uses chromosome enumeration probes for chromosomes 3, 7, and 17 and a locus-specific indicator probe for 9p21
- Polysomy of one or more of these three chromosomes or deletion of the 9p21 locus was chosen for their ability to detect common abnormalities in urothelial neoplasia (Fig. 15.5)
- May be helpful for follow-up of known BC, further evaluation of suspicious urine cytology findings, post-BCG follow-up, or as a general adjunct to urinary cytology
- Overall sensitivity varies between 69% and 87%, but significantly lowers for low-grade and low-stage tumors
- May be potentially useful as a grading tool
- UroVysion patterns may predict the risk of recurrence and DFS of such patients
- UroVysion is particularly attractive because it is an objective rather than a subjective assessment of urothelial cell abnormalities
- Diagnosis of urothelial CIS may be especially challenging, as lesions are not always cystoscopically identifiable
    - More extensive investigation of CIS may reveal clinical circumstances for which this method is particularly useful
    - This technique is being proposed as an aid for resolution of histologically challenging biopsies as well as cytologic samples

BTA-Stat

- BTA-Stat (bladder tumor antigen) is a point-of-care (qualitative) immunoassay using two monoclonal antibodies to detect human complement factor H-related protein in the urine
- Factor H, a soluble glycoprotein regulator of complement activation, appears to have an immuno-protective effect for tumor cells and is frequently released into urine by urothelial neoplasms
- The counterpart to BTA-Stat is BTA-TRAK, a quantitative standard ELISA assay
- Both tests improve in sensitivity for detection of high-grade lesions
- Sensitivity results have varied in different studies
    - Sensitivity of the test improved with grade from 50% (grade 1) to 72% (grade 2), and 91% (grade 3)
- As a screening modality, BTA-Stat is associated with false-positive results that are usually caused by inflammatory conditions in the urinary tract

NMP22

- A nuclear matrix protein usually present in very low quantities in the urine of a normal individual, but with greatly elevated levels in the urine of patients with BC
- NMP22 is a quantitative sandwich ELISA test using two antibodies recognizing two epitopes
- The sensitivity and specificity of NMP22 are 56% and 85%, respectively, compared to 16% and 99% for cytology
- A nomogram has been developed to better predict the probability of UC recurrence and progression, using the NMP22 test
    - Reliability is uncertain due to the reported variability in the diagnostic performance of the test between different institutions

ImmunoCyt

- This test relies upon the visualization of tumor-associated antigens in urothelial carcinoma cells using a panel of fluorescent-labeled monoclonal antibodies including two mucin-like proteins and carcinoembryonic antigen
- ImmunoCyt showed a specificity of 79% for grade 1, 84% for grade 2, and 92% for grade 3 tumors
- The combination of cystoscopy and ImmunoCyt testing provided 100% sensitivity in UC detection
    - Combining cystoscopy and cytology marginally improved upon the sensitivity of cystoscopy alone

- The major advantage of ImmunoCyt over other tests is its sensitivity in detecting both low-grade and high-grade tumors
- Novel early detection assays have harnessed the power of next-generation sequencing (NGS) technology applied to urine cell-free DNA (cfDNA) or cellular DNA; these assays detect variable sets of genetic alterations of BC

UroSEEK

- When combined with cytology, it achieved a sensitivity of 95% and a specificity of 93% for early detection of BC
- The assay detects *TERT* promoter mutations (an alteration that we found to occur in up to 80% of BC) together with 10 additional genes, which include *FGFR3, PIK3CA, HRAS, KRAS, TP53, CDKN2A,* and *ERBB2*

Circulating Tumor Cells

- Background
  - Originate from the primary tumor
  - Migrate to distant body sites through the blood stream
  - At times, establish new colonies at these sites, forming detectable metastases
- Presence of circulating tumor cells (CTC) in whole blood before and during radical cystectomy is a parameter for determining the need for adjuvant or even perioperative chemotherapy
  - Results and conclusions gained from studies have been contradictory and inconclusive
  - It is notable that mobilization of tumor cells from the primary site is necessary but not sufficient to produce distant metastases
- Attempts at molecular detection of CTC in the blood have been hampered by a paucity of molecules specific to UC
  - Immunodetection, a technique that relies upon the use of antibodies specifically chosen to attach to BC cells, such as certain cytokeratins, uroplakin, or MUC7
  - Such detection methods may lack sufficient sensitivity and specificity for BC tumor cells
- CTC numbers may be useful indicators of the risk of metastasis
  - Detection of CTC correlates with an increased risk of metastasis
- Use of PCR-based technologies to detect CTC in the blood is a field of intense investigation
  - Several BC cell markers, such as UPII, CK20, EGFR, and MUC7, have been analyzed for use as candidate detection molecules

- The sensitivity of these techniques is well documented, but their specificity for diagnostic purposes remains debatable
- Peripheral blood of patients was studied by nested RT-PCR assay for uroplakins Ia, Ib, II, and III and EGFR
- The combination of uroplakin Ia/II detected 75% of the CTC, with a specificity of 50%
- In one study, MUC7 positivity was detected in 38% peripheral blood samples from patients with Ta or T1 BC, and in 78% of patients with advanced-stage BC (≥T2)
- These findings suggest the potential utility of these novel approaches

Other Markers

- Novel markers are also being evaluated for their utility in detecting cancer cells in urine sediment, including epigenetic markers, telomerase, and survivin
- Microsatellite analysis of urine samples has been used for the surveillance of patients after treatment for UC
  - Data suggest that microsatellite alteration is a strong predictor for tumor recurrences
- Single-nucleotide polymorphism (SNP) analysis has shown that 100% of urine DNA samples from patients with bladder tumors had 24 or more SNP DNA alterations
  - This suggests that the HuSNP chip is a valuable tool for the detection of BC
- Efforts to develop DNA methylation markers in urine sediments for detection of BC are also underway
  - Studies of the methylation status of Wnt-antagonist genes have shown significantly higher methylation levels of Wnt antagonists in bladder tumors than in normal bladder mucosa
  - The overall sensitivity was 77% and specificity was 67%

Molecular Grading and Staging of Bladder Cancer

Molecular Grading

- The morphological heterogeneity is influenced by molecular events involved in carcinogenesis
  - The morphological criteria used for UC grading have been continuously updated
  - Current classification of urothelial neoplasms is based on an attempt to reconcile molecular genetic and pathologic findings

- Most of the WHO 2004 categories have been successfully validated by expression and genome profiling and by identification of distinctive genetic alterations
- Mutations in the *FGFR3* and *TP53* genes define two independent and distinct pathways in superficial papillary and invasive/flat UC (Figs. 15.7, 15.8, and 15.9)
  - Tumors characterized by these two pathways present as heterogeneous groups with distinct phenotypes and genotypes, and with markedly different biological behaviors and clinical outcomes
  - *FGFR3* mutations are usually present in low-grade papillary carcinomas with limited genetic instability, whereas high-grade UC are characterized by *TP53* mutation
  - *TP53* mutations have been found to be almost always mutually exclusive of *FGFR3* mutations
    - A fact that could potentially be exploited as a tool for the molecular grading of UC
- Data suggest that pathologic grading alone may not accurately predict tumor behavior
- Molecular grading may potentially discriminate high-risk cases from low-risk cases in patients with similar pathological grades
  - Combining morphological and molecular grading markers may allow better risk stratification for patients with UC
  - In one study of *FGFR3* mutation status and three molecular markers (MIB1 or Ki-67, p53, and P27kip1), three molecular grades (mG) could be identified
    - mG1 has *FGFR3* mutation/normal MIB1 expression and portends a favorable prognosis
    - mG2 has either mutated *FGFR3* or elevated MIB1 and is associated with an outcome intermediate between mG1 and mG3 tumors
    - mG3 tumors with wildtype *FGFR3* and high expression of MIB1 were associated with a poor clinical outcome
  - The molecular variables are more reproducible than the pathologic grade
  - Molecular grading provides a new, simple, and highly reproducible tool for clinical decision-making in UC patients
- Differential gene expression profiles can also be used to stratify tumor grade
  - Microarray data show clear distinctions between low- vs high-grade tumors in bladder washing samples of patients with low-grade and high-grade UC
- Recent investigations suggest a regulatory role for miRNA in UC
  - miRNA alterations occur in a tumor phenotype-specific manner

- High-grade UC tumors are characterized by miRNA upregulation, including mRNA-21, which suppresses p53 function
- Low-grade UC tumors are characterized by downregulation of miRNAs-99a/100, leading to upregulation of *FGFR3* even before its mutation
- A retrospective study was performed to evaluate differences in chromosomal aberrations in recurrent UC
  - The number of chromosomal aberrations differed significantly between tumor grades, regardless of whether grading was done using WHO 1973 or WHO 2004 grading parameters
  - The most frequent gains of chromosomal material were found on 19p, 7q, 16, 19q, 89, 12q, and 20, and the most frequent losses of chromosomal material were detected on 9, 13q, 5q, 8p, 11p, and 18q
  - Chromosomal aberrations correlated well with both grading systems
  - High-grade tumors showed aberrations usually associated with higher grade chromosomal alteration panels (1p+, 16p+, −2, and −5q) and poor clinical outcome
  - Polysomy of chromosome 17 by FISH was not seen in G1 tumors, but was seen in G2 tumors (28%) and G3 tumors (100%)
- Distinct molecular pathways of development for superficial papillary UC and those of flat invasive UC are well established
  - Molecular markers efficiently distinguish low-grade bladder tumors from high-grade tumors
  - Low-grade (G1–2) tumors possess few molecular alterations apart from deletions involving chromosome 9 and activating mutations of the *FGFR3*
  - Loss of 11p and inactivation of *TP53* are more commonly seen in tumors of higher grade

Molecular Staging

- Tumor staging is critical in predicting the disease course of an affected individual
  - American Joint Committee on Cancer TNM staging system is the most common tool to predict outcomes of BC patients
    - Offers general outcome estimates based on classic pathologic criteria
    - Predictive accuracy of TNM staging alone is limited
  - Nomograms combining molecular markers and classic pathologic criteria have been developed to improve prediction of clinical outcomes after cystectomy

- *FGFR3* and *TP53* mutations are frequently found in superficial papillary and invasive disease, respectively
  - *FGFR3* mutations are associated with low-stage tumors, whereas *TP53* mutations are associated with high-stage tumors
  - *FGFR3*mut/*TP53*wt is the most prevalent genotype in pTa tumors
    - *FGFR3*wt/*TP53*wt is the second most prevalent
  - *FGFR3*wt/*TP53*wt is the most frequent genotype in pT1 tumors, followed by *FGFR3*mut/*TP53*wt and *FGFR3*wt/*TP53*mut
  - *FGFR3*wt/*TP53*wt genotype accounted for 53% of cases of pT2–pT4 tumors
    - *FGFR33*wt/*TP53*mut was observed in 42% tumors
  - There is significant overlap with the TNM stages, and molecular staging subcategorized patients more accurately
  - Molecular staging reflects tumor behavior more closely and may be more useful than traditional staging methods
- Chromosomal instability could potentially be used as a staging parameter
  - Identification of genomic instability independently enhances the accuracy of molecular staging of UC
- Expression microarray analysis of divergent sets of bladder tumors could further classify tumors into more homogeneous and clinically relevant subgroups
  - Unsupervised hierarchical clustering successfully classified the samples into two subgroups containing superficial (pTa and pT1) vs MIBC (pT2–pT4) tumors
  - Supervised classification had a 91% success rate, separating superficial from MIBC tumors based on expression of a gene panel
  - Tumors could also be classified into transitional vs squamous subtypes (89% success rate) and good vs bad prognosis (78% success rate)
- Polysomy 17 is related to the stage of UC
  - microRNAs (miRNAs) are thought to play roles in cancer development, differentiation, and progression
  - Specific groups of miRNAs are differentially expressed in various cancers and may influence tumor phenotype and behavior
  - The specific role of miRNAs in the metastatic process is largely unknown
- Expression and genome profiling of BC allow reasonably good correlations between molecular findings and pathologic stage
  - Stage pTa UC are characterized by genetic stability, since they commonly lack *TP53* mutations
  - Chromosomal changes are predominantly limited to chromosome 9, whereas the majority of pT1 UC show increased genetic instability with chromosomal changes in 17p, 13q, and 8p

Molecular Detection of Lymph Node Metastasis

- Approximately 25% of patients undergoing radical cystectomy with pelvic lymph node dissection are found to have lymph node metastases
  - Lymph node metastasis is predictive of poor clinical outcome
  - Molecular markers enable the detection of micrometastatic disease with high sensitivity and specificity and potentially guide therapeutic decision-making
  - When molecular findings are taken into consideration, 25% of microscopically negative lymph nodes appear to harbor metastases

Targeted Therapy and Predictive Markers in Bladder Cancer

- The TCGA and genomic profiling studies uncovered a wide range of molecular therapeutic targets in >70% of MIBC (Fig. 15.13)
- Molecular targets include *PI3KCA/AKT/mTOR* pathway, *RTK/MAPK* pathways, such as *EGFR*, *FGFR3*, and *ERBB2*; *ER* pathway; immune response checkpoint modulators; and chromatin regulation and remodeling targets among others (Fig. 15.14)
  - Small molecule pan-FGFR inhibitors (e.g., BGJ398 and Erdafitinib) have demonstrated encouraging results in BC tumors harboring activating *FGFR* mutations or translocations (luminal-papillary subtypes)
  - *EGFR* inhibitors may be effective in chemotherapy naïve tumors with EGFR or ERBB2 overexpression
  - *mTOR* pathway inhibitors in combination with *MEK* inhibitors and inhibitors of cell cycle regulators (aurora kinase, *PLK1*, and cyclin-dependent kinase 4) are under investigation
- Antibody–drug conjugates are currently being pursued as new therapeutic modalities in BC
  - In 2019, a novel nectin-4-targeting antibody–drug conjugate (enfortumab vedotin) was approved by FDA for the treatment of platinum-refractory and immune

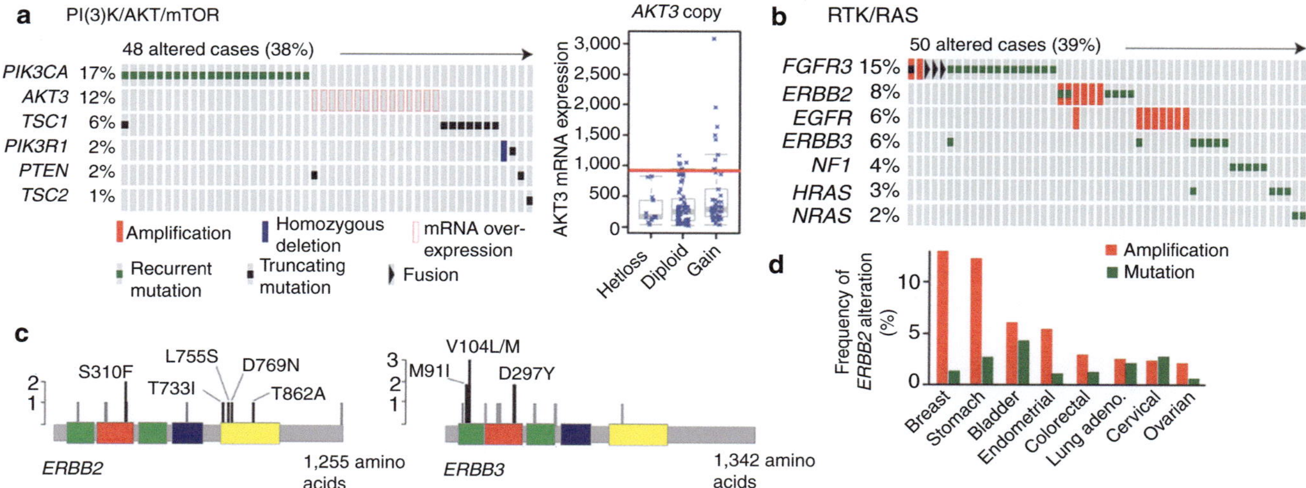

Fig. 15.13 Targets of therapy suggested by TCGA study in bladder cancer. (**a**) Mutually exclusive alterations in the PI3K/AKT/mTOR pathway are shown. AKT3 overexpression is found in 10% of samples, independent of copy number (*right panel*). (**b**) Alterations in receptor tyrosine kinases (amplification, mutation, or fusion) are found in 45% of tumors. (**c**) Recurrent mutations in ERBB2 and ERBB3. The mutations shown in black are either recurrent in the TCGA data set or reported in COSMIC. (**d**) ERBB2 amplification and recurrent mutations in other cancers profiled by TCGA. Hetloss, Heterozygous loss (*adapted with permission from Cancer Genome Atlas Research Network. Nature.2014;507:315*)

checkpoint blockade-refractory advanced urothelial carcinoma

- In addition to nectin-4, antibody–drug conjugates targeting Trop-2, HER2, and EpCAM are under investigation
  - Encouraging preclinical results have been achieved with trastuzumab conjugated with a cytotoxic agent DM1 (derivative of maytansine 1; T-DM1) in HER2+ tumors
  - Based on prior HER2-targeted trials in BC, evidence of tumor HER2 positivity by either immunohistochemistry or FISH could be used to guide therapy
  - In contrast to breast cancer, the majority of HER2 overexpression in BC are not associated with *HER2* gene amplification

- As an alternative to targeting intrinsic tumor growth pathways, targeted therapy can modulate the tumor vasculature to improve uptake of chemotherapy
  - In a recent study ("RANGE" trial, Petrylak et al.), advanced BC patients resistant to platinum-based chemotherapy were treated with docetaxel with or without Ramucirumaban (anti-VEGFR2) antibody
    - Progression-free survival was prolonged significantly in patients who received the combination; the first study to show superior
    - Progression-free survival for a combined cytotoxic and targeted therapy combination over chemotherapy alone

- Molecular biomarkers predictive of response to neoadjuvant chemotherapy
  - Certain intrinsic molecular subtypes (e.g., basal-squamous) have shown a better likelihood of response
  - A clinical trial to compare the clinical efficacy of the two frontline chemotherapy regimens (gemcitabine plus cisplatin vs MVAC) and the ability of novel gene expression profiling-based algorithm (CoXEN [Co-eXpression ExtrapolatioN]) to predict complete pathologic response was recently completed by SWAG with promising results (Flaig et al.)
  - Assessment of genetic alterations in ERCC2 and other DNA damage repair genes have also emerged as predictors of response to cisplatin-based chemotherapy

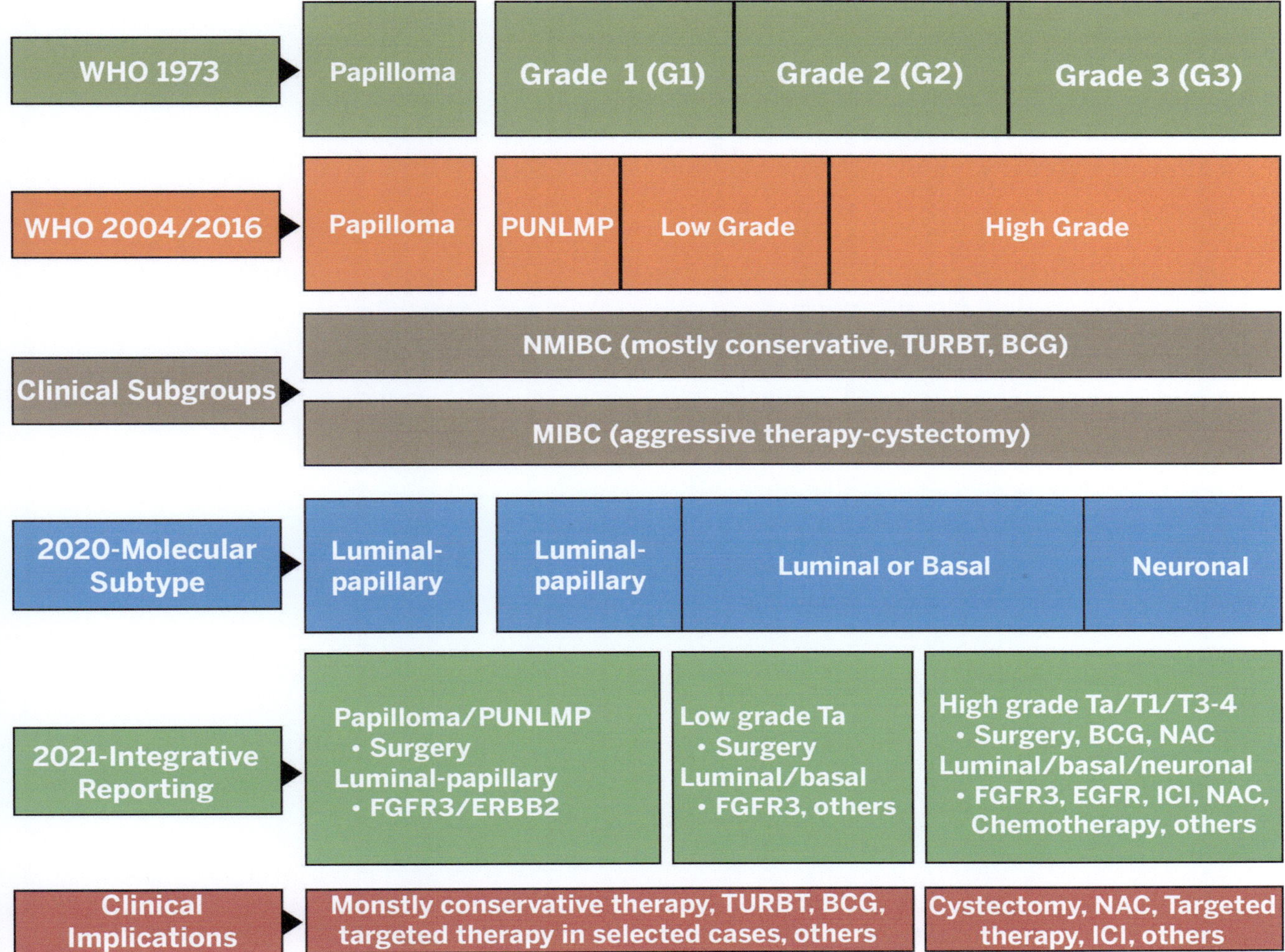

Fig. 15.14 Diagrammatic representation of combined reporting of bladder urothelial tumors. This is an evolving process from WHO 1973 to WHO 2004/2016 to molecular subtyping using immunohistochemistry. Advantages of combined reports include providing the clinician with more information sooner, leading ultimately to a more personalized approach to current therapies (from Lopez-Beltran A et al. *Hum Pathol* 2021;113:67; with permission)

Immunotherapy in Bladder Cancer

- Advanced BC has emerged as a strong candidate for treatment with numerous immune check inhibitor (ICI) agents
- Anti-PD-1 antibodies, nivolumab and pembrolizumab, were the first two anti-PD-1 monoclonal antibodies to receive FDA approval (Table 15.3)
  - Both are IgG4 humanized monoclonal antibodies that function predominantly by steric interference of the PD-1–PD-L1 interaction and have been shown to have very similar molecular and preclinical characteristics
- In 2014, the FDA granted MPDL3280A, a PDL1 monoclonal antibody inhibitor, a breakthrough therapy designation
  - This was based on the results of a phase I trial of patients with metastatic urothelial BC, which showed up to a 43% response rate in patients with 2+/3+ PDL1 positivity by immunohistochemistry
- Phase II and phase III studies (IMvigor 210 and IMvigor 211, respectively) evaluating anti-PDL1 agent atezolizumab in adjuvant therapy in patients with PD-L1+ high-risk MIBC (NCT02450331) followed
  - While the primary analysis of the phase II trial showed significant objective response that was highest in tumors with 2+/3+ PDL1 expression in immune cell, in phase III trial, Atezolizumab was not associated with longer overall survival compared to chemotherapy in patients with platinum-refractory metastatic disease

Table 15.3 PD-L1 assays in bladder cancer

Companion diagnostic assay	IO-Drug Trials	Trials	Target	PD- protein expression evaluation	Cut-off	Staining characteristics and pitfalls	Prescription
SP142 (Ventana)	Atezolizumab [Tecentriq]	IMvigor 210 trial IMvigor 211 ABACUS trial	PD-L1	IC score (%) $= \dfrac{positive\,IC\,area}{total\,tumor\,area}$	ICs ≥5%	• Plasma cells have to be excluded from scoring. • Dot−/ant-like staining pattern.	Currently necessary for first-line therapy stratification for Atezolizumab
22C3 (Dako)	Pembrolizumab [Keytruda]	KEYNOTE-052 KEYNOTE-045 PURE-01 trial KEYNOTE-057	PD-1	CPS $= \dfrac{positive\,TC + IC}{total\,TC} \times 100$	CPS ≥ 10	• Plasma cells have to be excluded from scoring. • Neutrophil granulocytes not included. • Mostly weak staining intensity.	Currently necessary for first-line therapy stratification for Pembrolizumab
SP263 (Ventana)	Durvalumab [Imfinzi]	NCT01693562	PD-L1	TC (%) $= \dfrac{positive\,TC}{total\,TC}$ IC (%) $= \dfrac{positive\,IC}{total\,IC}$	IC ≥ 25% or/and TC ≥ 25%	• Plasma cells have to be excluded from scoring. • Mostly strong staining intensity.	Currently not prescribed and only explored in ongoing clinical trials (potentially required in the future)
28-8 (Dako)	Nivolumab [Opdivo]	CheckMate 032 NCT01928394 Checkmate 275	PD-1	TPS (%) $= \dfrac{positive\,TC}{total\,TC}$	TC ≥ 1%	• Homogenous tumor cell staining. • Moderate-strong staining intensity.	Currently not prescribed and only explored in ongoing clinical trials (potentially required in the future)
73-10 (Dako)	Avelumab [Bavencio]	JAVELIN NCT01772004	PD-L1	TC (%) $= \dfrac{positive\,TC}{total\,TC}$	TC ≥ 5%	Not available	No FDA approved assay

IC: immune cells; *TC*: tumor cells; *CPS*: combined positive score; *TPS*: tumor proportion score (modified from Lopez-Beltran A et al. *Hum Pathol* 2021;113:67; with permission).

- The above suggest that a better selection of patients according to the biological characteristics of the disease is required
  - Although durable responses are invaluable, most patients do not respond to ICI, therefore predictive biomarkers are direly needed
- Assessing PD-L1 immunoexpression has not been found to consistently predict response; even among trials involving the same agent (e.g., pembrolizumab KEYNOTE-012 and KEYNOTE-052 vs KEYNOTE-045)
- Companion diagnostic biomarkers, PD-1 or PD-L1, immunohistochemistry evaluations have presented logistical challenges (Table 15.3)
  - Different antibody clones
  - Variable thresholds
  - Different cell types and counting methods (tumor cells, immune cells, or a combination) have contributed to the logistical challenge
- In addition to PD-L1 expression, tumor mutational burden, computational predictions of tumor neoantigens/MHC class I interaction, TGFB expression, and presence of microsatellite instability/mismatch repair deficiency are predictors of response to ICI therapy in BC
- Different molecular subtypes have shown differential response to ICI with the luminal infiltrated and basal-squamous subtypes having a favorable response
- Finally, novel prognostic and predictive biomarkers to include long noncoding RNA, miRNA, proteomic signatures, cell-free DNA, and a variety of tumor immune microenvironment markers are currently under investigation in tumor tissue as well as in liquid biopsy

Summary

- Current approaches to the diagnosis and management of BC will continue to evolve based on our sharper understanding of the complex molecular mechanisms involved in BC development
- The current paradigm of clinicopathology-based prognostic approach to predict progression in NMIBC should be supplemented by a molecular guided approach based on some of the markers listed in Table 15.1 and the molecular grading and molecular staging approaches summarized above
- Recently proposed molecular classifications provide important insights into the biology of MIBC; the consensus molecular classification scheme should be tested for its applicability as a feasible tool in routine clinical use (Fig. 15.14)
- Several new targeted therapy agents and immunotherapy strategies are under investigation in randomized clinical

trials in combination with standard chemotherapy agents either as first-line treatment or on a maintenance basis to prolong response in patients with advanced BC

Suggested Reading

Al-Ahmadie H, Netto GJ. Updates on the genomics of bladder cancer and novel molecular taxonomy. Adv Anat Pathol. 2020;27:36–43.

Al-Ahmadie H, Netto GJ. Molecular pathology of urothelial carcinoma. Surg Pathol Clin. 2021;14:403–14.

Al-Obaidy KI, Cheng L. Fibroblast growth factor receptor (FGFR) gene: pathogenesis and treatment implications in urothelial carcinoma of the bladder. J Clin Pathol. 2021;74:491–5.

Cancer Genome Atlas Research Network. Comprehensive molecular characterization of urothelial bladder carcinoma. Nature. 2014;507:315–22.

Catto JW, Azzouzi AR, Rehman I, et al. Promoter hypermethylation is associated with tumor location, stage, and subsequent progression in transitional cell carcinoma. J Clin Oncol. 2005;23:2903–10.

Cheng L, Davidson DD, Maclennan GT, et al. The origins of urothelial carcinoma. Expert Rev Anticancer Ther. 2010;10:865–80.

Cheng L, Zhang S, Alexander R, et al. Sarcomatoid carcinoma of the urinary bladder: the final common pathway of urothelial carcinoma dedifferentiation. Am J Surg Pathol. 2011a;35:e34–46.

Cheng L, Zhang S, Maclennan GT, et al. Bladder cancer: translating molecular genetic insights into clinical practice. Hum Pathol. 2011b;42:455–81.

Cheng L, Lopez-Beltran A, Bostwick DG. Bladder pathology. Wiley-Blackwell: Hoboken, NJ; 2012.

Cheng L, Davison DD, Adams J, et al. Biomarkers in badder cancer: translational and clinical implications. Crit Rev Oncol Hematol. 2014;89:73–111.

Cheng L, Zhang S, Wang L, et al. Fluorescence in situ hybridization in surgical pathology: principles and applications. J Pathol Clin Res. 2017;3:73–99.

Cheng L, Zhang S, Wang M, et al. Biological and clinical perspectives of TERT promoter mutation detection on bladder cancer diagnosis and management. Hum Pathol. 2023;133:56–75.

Choi W, Porten S, Kim S, et al. Identification of distinct basal and luminal subtypes of muscle-invasive bladder cancer with different sensitivities to frontline chemotherapy. Cancer Cell. 2014;25:152–65.

Damrauer JS, Beckabir W, Klomp J, et al. Collaborative study from the Bladder Cancer Advocacy Network for the genomic analysis of metastatic urothelial cancer. Nat Commun. 2022;13:6658.

Flaig TW, Tangen CM, Daneshmand S, et al. A Randomized Phase II Study of Coexpression Extrapolation (COXEN) with Neoadjuvant Chemotherapy for Bladder Cancer (SWOG S1314; NCT02177695). Clin Cancer Res. 2021;27:2435–41.

Flaig TW, Spiess PE, Abern M, et al. NCCN Guidelines® Insights: Bladder Cancer, Version 2.2022. J Natl Compr Canc Netw. 2022;20:866–78.

Hussain MH, MacVicar GR, Petrylak DP, et al. Trastuzumab, paclitaxel, carboplatin, and gemcitabine in advanced human epidermal growth factor receptor-2/neu-positive urothelial carcinoma: results of a multicenter phase II national cancer institute trial. J Clin Oncol. 2007;25:2218–24.

Kamoun A, de Reynies A, Allory Y, et al. A consensus molecular classification of muscle-invasive bladder cancer. Eur Urol. 2020;77:420–33.

Knowles MA, Hurst CD. Molecular biology of bladder cancer: new insights into pathogenesis and clinical diversity. Nat Rev Cancer. 2015;15:25–41.

Kouba EJ, Cheng L. Understanding the genetic landscape of small cell carcinoma of the urinary bladder and implications for diagnosis, prognosis and treatment. JAMA Oncol. 2017;3:1570–8.

Lindgren D, Frigyesi A, Gudjonsson S, et al. Combined gene expression and genomic profiling define two intrinsic molecular subtypes of urothelial carcinoma and gene signatures for molecular grading and outcome. Cancer Res. 2010;70:3463–72.

Lindskrog SV, Prip F, Lamy P, et al. An integrated multi-omics analysis identifies prognostic molecular subtypes of non-muscle-invasive bladder cancer. Nat Commun. 2021;12:2301.

Lopez-Beltran A, Cimadamore A, Montironi R, Cheng L. Molecular pathology of urothelial carcinoma. Hum Pathol. 2021;113:67–83.

Lotan Y, de Jong JJ, Liu VYT, et al. Patients with muscle-invasive bladder cancer with nonluminal subtype derive greatest benefit from platinum based neoadjuvant chemotherapy. J Urol. 2022;207:541–50.

Magers MJ, Cheng L. Practical molecular testing in a clinical genitourinary service. Arch Pathol Lab Med. 2020;144:277–89.

Netto GJ. Molecular diagnostics in urologic malignancies: a work in progress. Arch Pathol Lab Med. 2011;135:610–21.

Netto GJ, Cheng L. Emerging critical role of molecular testing in diagnostic genitourinary pathology. Arch Pathol Lab Med. 2012;36:372–90.

Netto GJ, Epstein JI. Theranostic and prognostic biomarkers: genomic applications in urological malignancies. Pathology. 2010;42:384–94.

Netto GJ, Amin MB, Berney DM, et al. The 2022 World Health Organization Classification of Tumors of the Urinary System and Male Genital Organs-Part B: Prostate and Urinary Tract Tumors. Eur Urol. 2022;82:469–82.

Petrylak DP, de Wit R, Chi KN, et al. Ramucirumab plus docetaxel versus placebo plus docetaxel in patients with locally advanced or metastatic urothelial carcinoma after platinum-based therapy (RANGE): overall survival and updated results of a randomised, double-blind, phase 3 trial. Lancet Oncol. 2020;21:105–20.

Robertson AG, Kim J, Al-Ahmadie H, et al. Comprehensive molecular characterization of muscle-invasive bladder cancer. Cell. 2017;171:540–56.

Siefker-Radtke AO, Necchi A, Park SH, et al. Efficacy and safety of erdafitinib in patients with locally advanced or metastatic urothelial carcinoma: long-term follow-up of a phase 2 study. Lancet Oncol. 2022;23:248–5.

Sjödahl G, Abrahamsson J, Holmsten K, et al. Different responses to neoadjuvant chemotherapy in urothelial carcinoma molecular subtypes. Eur Urol. 2022;81:523–32.

Sonpavde GP, Mouw KW, Mossanen M. Therapy for muscle-invasive urothelial carcinoma: controversies and dilemmas. J Clin Oncol. 2022;40:1275–80.

Thomas J, Sonpavde G. Molecularly targeted therapy towards genetic alterations in advanced bladder cancer. Cancers (Basel). 2022;14:1795.

Tran L, Xiao JF, Agarwal N, et al. Advances in bladder cancer biology and therapy. Nat Rev Cancer. 2021;21:104–21.

van Rhijn BW, Vis AN, van der Kwast TH, et al. Molecular grading of urothelial cell carcinoma with fibroblast growth factor receptor 3 and MIB-1 is superior to pathologic grade for the prediction of clinical outcome. J Clin Oncol. 2003;21:1912–21.

van Rhijn BW, Zuiverloon TC, Vis AN, et al. Molecular grade (FGFR3/MIB-1) and EORTC risk scores are predictive in primary non-muscle-invasive bladder cancer. Eur Urol. 2010;58:433–41.

Warrick JI, Knowles MA, Yves A, et al. Report from the International Society of Urological Pathology (ISUP) Consultation Conference on molecular pathology of urogenital cancers. II. Molecular pathology of bladder cancer: progress and challenges. Am J Surg Pathol. 2020;44:e30–46.

Yang Y, Miller CR, Lopez-Beltran A, et al. Liquid biopsies in the management of bladder cancer: next-generation biomarkers for diagnosis, surveillance and treatment response prediction. Crit Rev Oncog. 2017;22:389–401.

Yang Y, Kaimakliotis HZ, Williamson SR, et al. Micropapillary urothelial carcinoma of urinary bladder displays immunophenotypic features of luminal and p53-like subtypes and is not a variant of adenocarcinoma. Urol Oncol. 2020;38:449–58.

Yu EY, Petrylak DP, O'Donnell PH, et al. Enfortumab vedotin after PD-1 or PD-L1 inhibitors in cisplatin-ineligible patients with advanced urothelial carcinoma (EV 201): a multicentre, single-arm, phase 2 trial. Lancet Oncol. 2021;22:872–82.

Molecular Pathology of Testicular Cancer

Katharina Biermann, Liang Cheng,
and Leendert H. J. Looijenga

Contents

Introduction

- The testis is a complex organ with multiple functions, including generation of [mature] germ cells (spermatogenesis) and hormone production (i.e., testosterone) (Fig.16.1). The first is dependent on the second, although visa versa is not the case.
- In other words, testis might be completely functional regarding formation of hormones, in spite of a complete lack of germ cell formation and production, resulting in infertility
- To allow these processes to occur at the proper time and place, various cell types and structures are required. Most of them are initiated during early embryogenesis, while they further develop at different time points, even up to adult life
- The most obvious structures within the testis are the seminiferous tubules and the interstitial space. These compartments contain specific types of cells, dependent on age in various stages of maturation
- Testicular functions, i.e., germ cell formation and hormone production, are regulated by a highly sophisticated network of specific cell types in distinct compartments (Fig. 16.2)
- In the interstitial compartment, i.e., the stromal space in between the seminiferous tubules, different cell types and (microscopic) structures are present, from which only the Leydig cells are testis specific
- The Leydig cells produce androgens (testosterone) when stimulated by luteinizing hormone (LH) produced by the pituitary gland
- In addition, INSL3 is formed, required for the first phase of testicular descent. Other cells and microscopic structures of the interstitial compartment include vascular structures, fibroblasts, macrophages, and lymphocytes
- The intratubular compartment is separated from the interstitial space by a highly organized barrier composed of both cells (e.g., peritubular cells) and extracellular matrix components (basal lamina)
- Within the seminiferous tubule in principle two types of cells are present under physiological conditions. These are the Sertoli cells and the germ cells (Fig. 16.2)
- The Sertoli cells are nursing the germ cells, from the initial embryonic phase to the mature spermatozoa. In the adult (postpubertal) testis, the germ cells present, i.e., spermatogonia, undergo a process of both mitosis and meiosis, including defined steps of further maturation
- The spermatogonia are situated at the inner side of the basal lamina, under the tight junctions formed by the Sertoli cells
- The developmental stages that follow are spermatocytes (undergoing meiosis) and spermatids, and finally spermatozoa. This process of mitosis and meiosis, followed by

K. Biermann
Department of Pathology, Erasmus MC-University Medical Center Rotterdam, Rotterdam, The Netherlands

L. Cheng (✉)
The Legorreta Cancer Center at Brown University, Department of Pathology and Laboratory Medicine, Warren Alpert Medical School of Brown University, Lifespan Academic Medical Center, Providence, RI, USA

L. H. J. Looijenga
Princess Maxima Center for Pediatric Oncology, Utrecht, The Netherlands

L. Cheng et al. (eds.), *Molecular Surgical Pathology*, https://doi.org/10.1007/978-3-031-35118-1_16

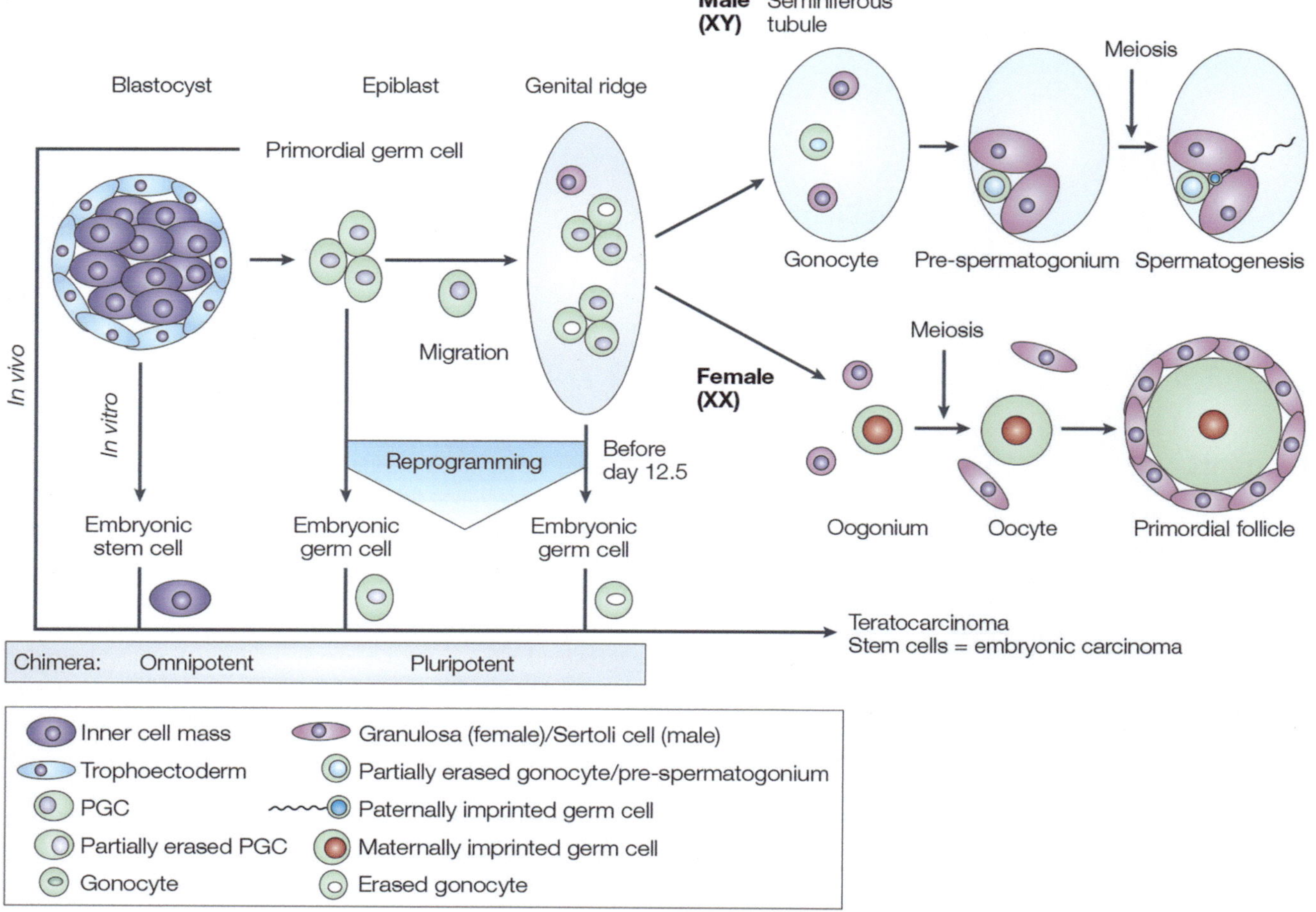

Fig. 16.1 Schematic representation of normal embryonic development and origin of the germ-cell lineage. The primordial germ cells (PGCs) originate in the epiblast, which can be identified by the alkaline phosphatase reactivity and by staining for octamer binding transcription factor 3/4. These cells migrate to the genital ridge, after which they are referred to as gonocytes. They differentiate either to pre-spermatogonia or oocytes. Embryonic stem cells (ESCs) are derived from the inner cell mass, whereas embryonic germ cells (EGCs) can be isolated from PGCs until day 12 of development. The ESCs show a biparental pattern of genomic imprinting, whereas in EGCs this is erased. ESCs and EGCs can give rise to pluripotent teratomas, of which the embryonal carcinoma cells are the stem cells. Teratomas can also be formed directly from PGCs in vivo. During spermatogenesis, the paternal pattern of genomic imprinting is established, whereas the maternal pattern is formed during oogenesis. The timing of meiotic I arrest is different between male and female germ cells. (Modified from Oosterhuis JW, Looijenga LH Nat Rev. Cancer 2005;5;210. Springer Nature)

spermiogenesis, is highly dependent on production of androgens (Fig. 16.1)

- Because of the epidemiological characteristics, this chapter will focus on germ cell tumors (GCTs), although at the end some characteristics of sex cord—stromal tumors will be discussed
- The majority of GCTs do not arise from adult germ cells, as found after puberty, but from early (embryonic) germ cells blocked in their normal maturation during fetal development (Fig. 16.3)
- The only exception is the rare type III GCT (spermatocytic seminoma) (see below). The prerequisite for development of a type I or type II GCTs is thus the escape from the strictly regulated maturation process from an embryonic germ cell to a (pre-) spermatogonium

  - Primordial germ cell (PGCs) arise from the proximal epiblast
  - PGCs retain an intrinsic, although suppressed capacity to pluripotency
  - PGCs move along the hindgut to the developing genital ridges (to develop into testes in an XY chromosomal constitution)
  - Migration of PGCs is regulated by the stem cell factor (SCF/KITLG)–c-KIT pathway
  - PGCs lose their biparental pattern of genomic imprinting (erasement)

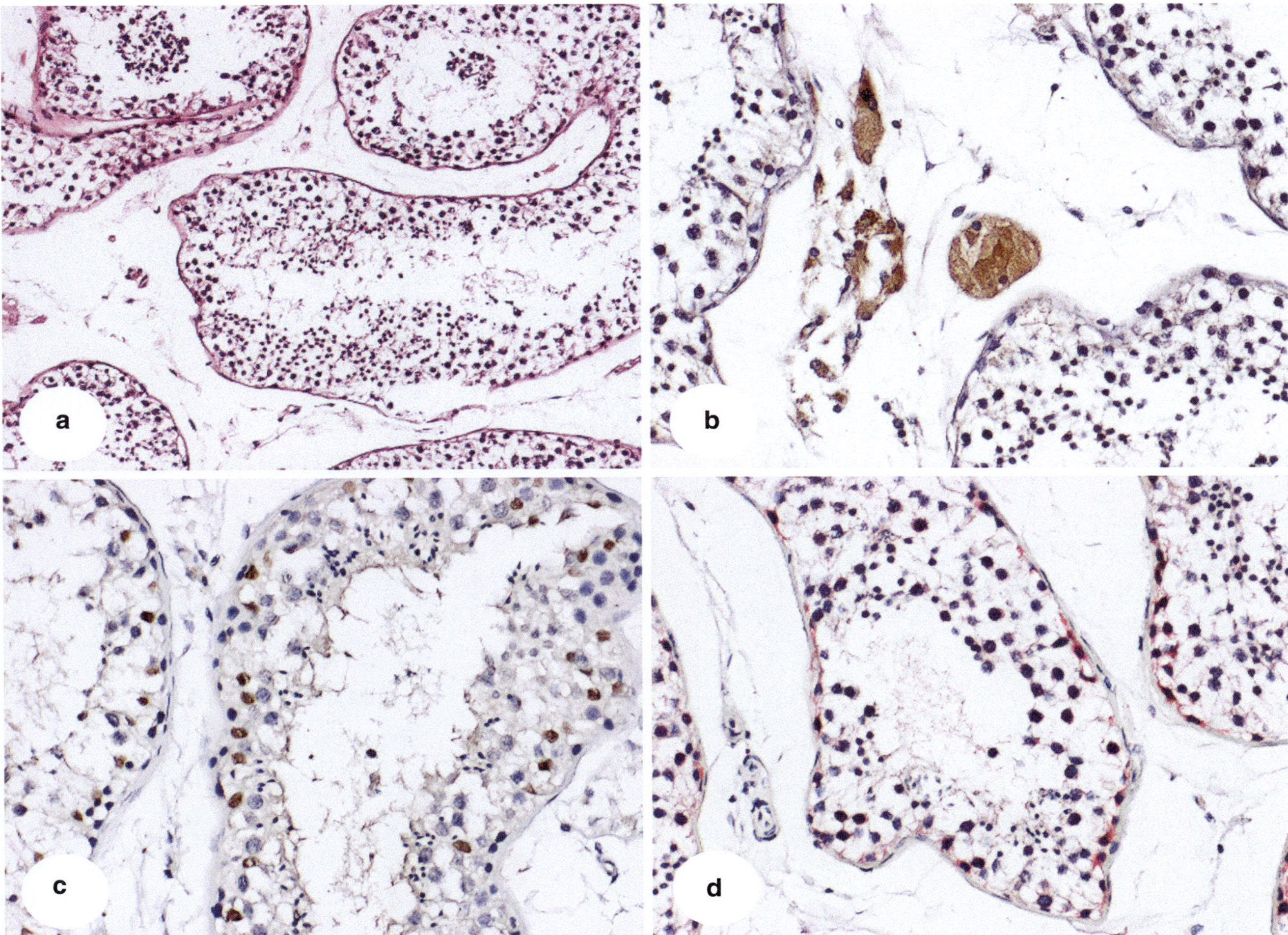

Fig. 16.2 Representative examples of normal adult testis histology, including (**a**) hematoxylin and eosin (H&E), 100×; (**b**) LH-R, (**c**) SOX9, (**d**) TSPY, all 200×. The immunohistochemical markers used are informative to identify the germ cells in the stage of spermatogonia, Sertoli cells, and Leydig cells, respectively

- Germ cells entering the genital ridge are called gonocytes
- During the second and third trimester of pregnancy, gonocytes mature into prespermatogonia
- Prespermatogonia lose expression of embryonic germ cell markers
- Proper maturation is required for spermatogenesis
- Spermatogenesis is activated by androgens at puberty
- Gonocytes can be delayed or blocked in the maturation process in a suboptimal microenvironment (i.e., cryptorchidism)
- Presence of germ cells with embryonic (PGC/gonocyte-like) characteristics after the first year of life indicates a maturation defect
- Delayed/blocked gonocytes can survive in postnatal testis
- Blocked gonocytes can transform and progress to neoplasm(s)

Germ Cell Tumors

General Classification

- In principle, all cells of the testis can give rise to neoplasms. This results in the fact that in the testis an enormous variety of histological variants of tumors can be observed, significantly influenced by age, amongst others
- Overall, it is of relevance to distinguish two main subgroups: GCTs and non-GCTs. Again within these categories, a large numbers of histological subgroups can be distinguished
- Although officially incorrect, the GCTs of the testis are often referred to as testicular cancer, mainly based on epidemiological criteria; they are the most frequent type of neoplasm of this organ

Fig. 16.3 Schematic representation of normal germ cell development (in black) and malignant germ cell development (in red). Embryogenesis starts with fertilization and generation of pluripotent stem cells, referred to as embryonic stem cells (ES), which are responsible for formation of the various differentiation lineages (both somatic and extra-embryonic). All these cells have a biparental pattern of genomic imprinting (GI) (represented in orange), resulting from a pure male- (blue) and female- (red) mature germ cell. The primordial germ cell (PGC) erases this biparental pattern, which in the male differentiation lineage in fully becoming paternal, via the stages pre-spermatogonia (A,B), primary and secondary spermatocyte and spermatid leading to fully matured sperm. The types of germ cell tumors originate from different stages of stem cell/ germ cell development, representing their developmental potential. Type I are the teratomas and yolk sac tumors found in neonates and infants, Type II are the seminomas and nonseminomas diagnosed in adolescents and young adults, and the Type III are the spermatocytic seminomas, predominantly occurring in elderly males. The timing of birth, start of meiosis, and puberty are indicated

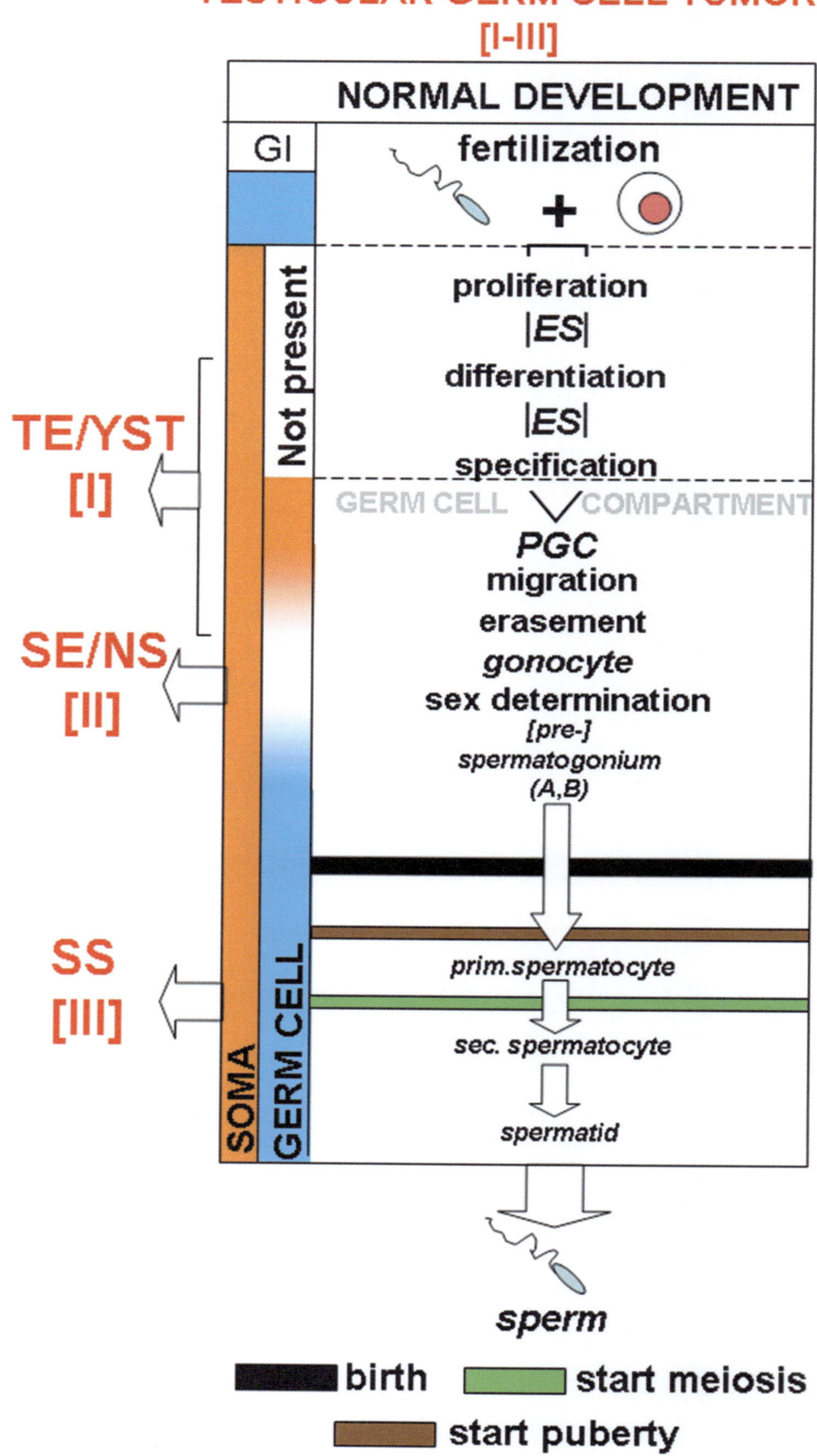

- Understanding the existence of the various types of neoplasms of the testis and to get insight into their pathogenesis it requires knowledge about the normal anatomy and physiology of both the developing and mature testis
- This resulted in novel information about the origin of especially the various types of GCTs and identification of informative diagnostic markers
- GCTs represent the most frequent neoplasm, followed by the sex cord–stromal tumors (Leydig cell and Sertoli cell tumors), and others (lymphoma, etc.)
- Based on morphological criteria, GCTs are subdivided into seminoma, , embryonal carcinoma, yolk sac tumor, teratoma, choriocarcinoma, and spermatocytic tumor. These can be either pure or (inter)mixed in composition

- In contrast to the histological description of GCTs, on which all pathological classification systems are based, an alternative is presented. This has been appreciated by the World Health Organization as well as specialized pathologists in the field
- According to developmental potential, cell of origin, age at clinical presentation, pattern of genomic imprinting, and molecular characteristics, testicular GCTs can be classified into three entities (type I–III), each with their own pathogenesis and pattern of (identified) risk factors (Fig. 16.4)
- There are a number of changes in recent 2022 WHO classification of testicuilar tumors (Table 16.1)

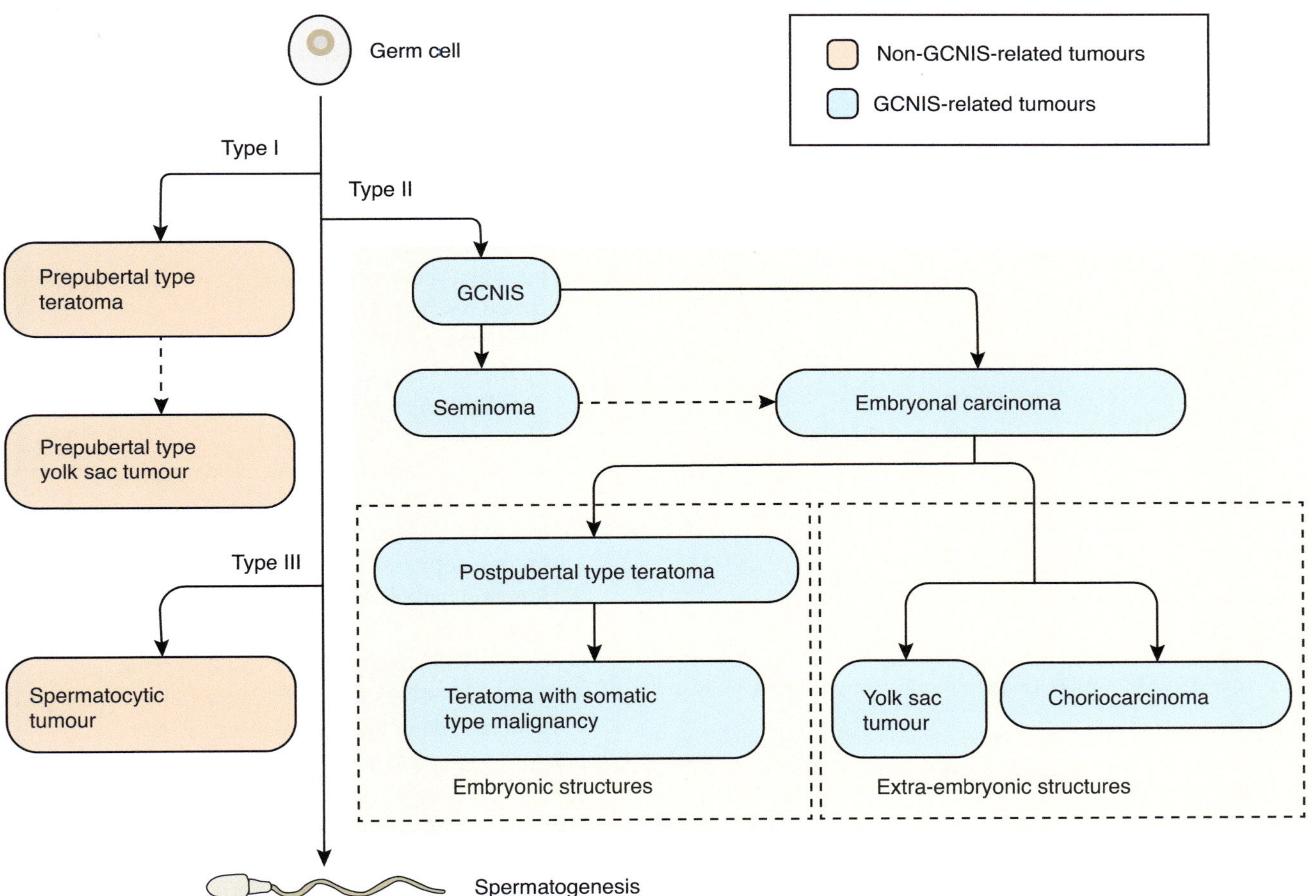

Fig. 16.4 Schematic representation of the types of testicular germ cell tumors. Nongerm cell neoplasia in situ (GCNIS)-related germ cell tumors include prepubertal type teratomas and yolk sac tumors (also known as type I testicular germ cell tumors (TGCTs); yolk sac tumors can originate from teratomas), as well as spermatocytic tumors (also known as type III TGCTs). The type I TGCTs originate from a pre-erased or partially erased embryonic germ cell or embryonic stem cell, and the type III TGCTs arise from a fully erased and subsequently paternally imprinted germ cell. GCNIS-related TGCTs (also known as type II TGCTs) originate from a fully erased embryonic germ cell and can be composed of seminoma (representing the embryonic germ cell lineage) as well as nonseminoma. Nonseminoma are pluripotent; elements found during intrauterine development, including both embry-onic structures (in postpubertal type teratoma, possibly with somatic type malignancy) and extra-embryonic structures (in yolk sac tumors and in choriocarcinoma), can be histologically detected. As Type II TGCTs can develop all these histological elements, they are true omnipotent or totopotent tumors. Type I TGCTs predominantly present before puberty, whereas type II TGCTs (mostly) and type III TGCTs (always) clinically manifest after puberty. All TGCTs represent the aberrant development of the physiological germ cell (at different phases of maturation) toward full spermatogenesis, which explains their histological diversity and their specific pattern of diagnostic biomarkers. Dashed arrows indicate a possible but not well established pathway of cancer progression (adapted from Cheng L et al. Nat Rev. Dis Primers. 2018;4:29. Springer Nature)

Table 16.1 WHO Classification of the testicular tumors, fifth edition, 2022

Germ cell tumors derived from germ cell neoplasia in situ
Noninvasive germ cell neoplasia
 Germ cell neoplasia in situ
 Specific forms of intratubular germ cell neoplasia
 Gonadoblastoma
Germinoma family of tumors
 Seminoma
Nonseminomatous germ cell tumors
 Embryonal carcinoma
 Yolk sac tumor, postpubertal-type
 Choriocarcinoma
 Placental site trophoblastic tumor
 Epithelioid trophoblastic tumor
 Cystic trophoblastic tumor
 Teratoma, postpubertal-type
 Teratoma with somatic-type malignancy
Mixed germ cell tumors of the testis
 Mixed germ cell tumors
Germ cell tumors of unknown type
 Regressed germ cell tumors
Germ cell tumors unrelated to germ cell neoplasia in situ
 Spermatocytic tumor
 Teratoma, prepubertal-type
 Yolk sac tumor, prepubertal-type
 Testicular neuroendocrine tumor, prepubertal-type
 Mixed teratoma and yolk sac tumor, prepubertal-type
Sex cord stromal tumors of the testis
Leydig cell tumor
 Leydig cell tumor
Sertoli cell tumors
 Sertoli cell tumor
 Large cell calcifying sertoli cell tumor
Granulosa cell tumors
 Adult granulosa cell tumor
 Juvenile granulosa cell tumor
Fibroma thecoma family of tumors
 Tumors in the fibroma the coma group
Mixed and other sex cord stromal tumors
 Mixed sex cord stromal tumor
 Signet ring stromal tumor
 Myoid gonadal stromal tumor
 Sex cord stromal tumor NOS

* Key changes in 2022 WHO Classification of Urinary and Male Genital Tumors:
• Change in nomenclature from "primitive neuroectodermal tumor" to "embryonic-type neuroectodermal tumor" and "testicular carcinoid" to "testicular neuroendocrine tumor"
• Seminoma is included in the "germinoma" family of tumors
• Gonadoblastoma is included under the noninvasive lesions derived from the nongerm cell carcinoma in situ (GCNIS)
• New entities like signet ring stromal tumor and myoid gonadal stromal tumor are described in the sex cord stromal tumors (SCST) of the testis
• Mixed/undifferentiated SCST have been separated into individual entities as mixed SCST and "SCST, NOS"
NOS Not otherwise specified
Adapted from Mohanty SK, Lobo A, Cheng L. The 2022 revision of the World Health Organization classification of tumors of the urinary system and male genital organs: advances and challenges. Hum Pathol (2023 in press, PMID: 36084769)

Type I (Pediatric) Germ Cell Tumor

• Clinical features
  – Tumor of predominantly neonates and infants (incidence of testicular GCT is 1–2 cases per million person-years)
  – First peak of incidence in first 2 years (mean age, 20 months for teratoma), second peak at age of 10
  – Can be found in sacrococcygeal region, retroperitoneum, intracranial, and mediastinum
  – Mostly benign (in contrast to type II teratomas, see below)
  – Malignant progression can occur, leading to yolk sac tumor (malignant)
  – No risk factors have been identified so far and no increase in incidence in the general population have been reported
  – No familial predisposition seems to be significant
• Gross and microscopic features
  – Teratoma
    ○ Most common GCT in pediatric population
    ○ Heterogeneous, mixed cystic, and solid mass with gray or brown cut surface
    ○ Mixture of differentiated (mature) somatic tissue, possibly containing (immature) neuroepithelial structures (Fig. 16.5)
    ○ Derivates of all three germinal (somatic) layers might be present (pluripotent)
    ○ Cartilage, and fetal mesenchymal tissue is often observed
    ○ Immature neural tissue might be intermixed in various quantities and include nests glands and tubules lined by immature embryonal-like cells with high mitotic activity
    ○ Somatic malignant transformation is rare
  – Yolk sac tumor
    ○ Can be primary malignancy in the testis
    ○ Can be histologically indistinguishable from type II yolk sac tumor (see below) (Fig. 16.5)
• Precursor lesions and cell of origin
  – Type I GCT, both teratoma and yolk sac tumor, arises from early immature embryonic stem or germ cells. This is supported by
    ○ Pattern of genomic imprinting—similarly to embryonic stem cells or early PGCs (biparental or partially erased)
    ○ Mouse teratoma models histologically represent type I GCTs: PTEN-knockout mouse, Ap2gamma-knockout mouse, Fhit-knockout mouse, p53-knockout mouse, Kit-ligand-deficient mouse in the 129/Sv background

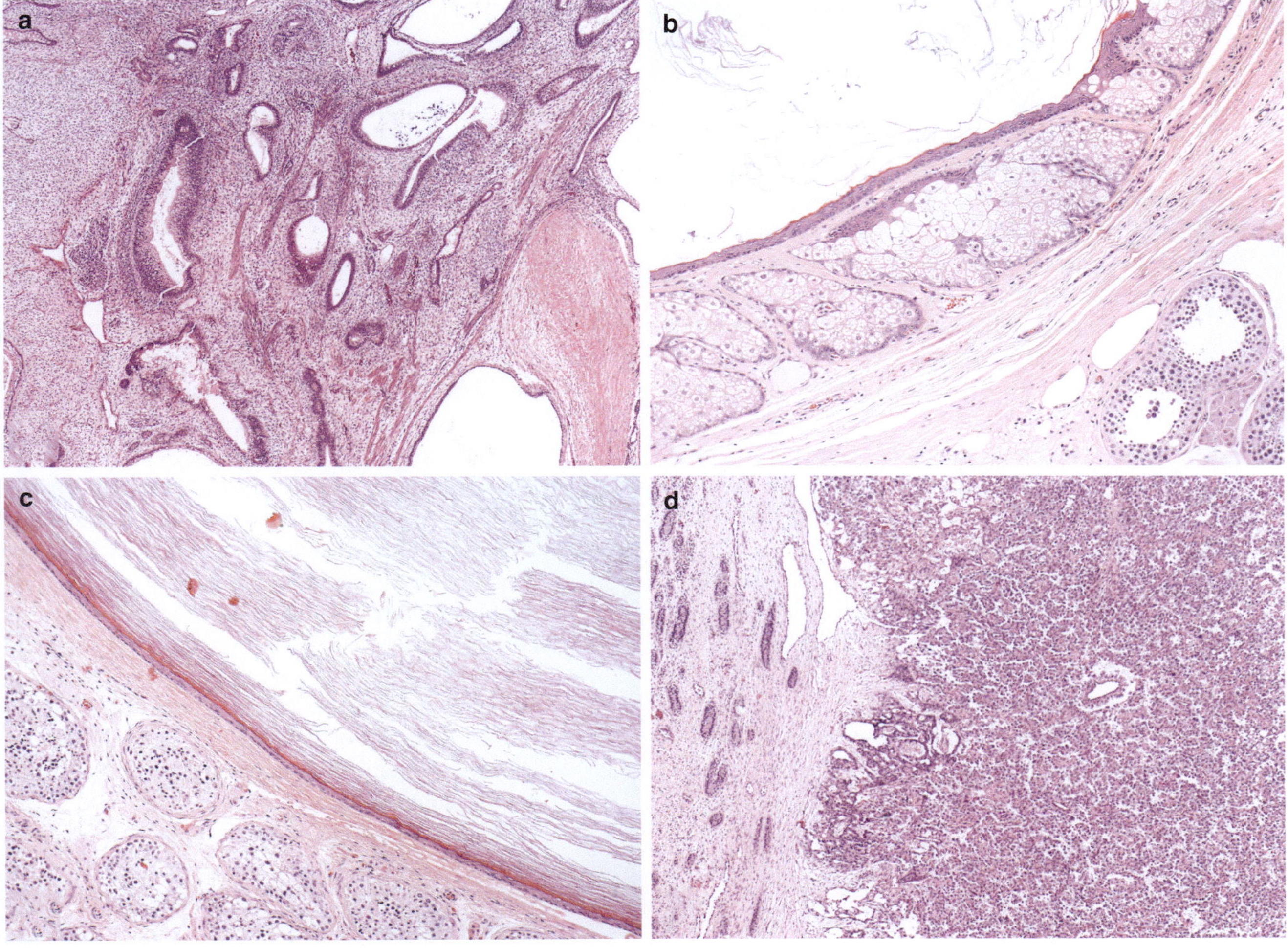

Fig. 16.5 Prepubertal-type germ cell neoplasms. (**a**) Prepubertal-type teratoma. (**b**) Dermoid cyst. (**c**) Epidermoid cyst. (**d**) Prepubertal-type yolk sac tumor. (adapted from Cheng L et al. Hum Pathol. 2017;59:10. Elsevier)

- – No precursor lesions of type I GCT are histologically identified yet
- Molecular features
- Type I teratomas show normal chromosomal content (diploidy). In contrast, type I yolk sac tumors are aneuploid, with recurrent chromosomal changes including:
  - – Loss of part(s) of 1p, 4, 6q
  - – Gain of part(s) of 1q, 12p13, 20q, and 22
  - – No genetic mutations have been found so far (teratoma and yolk sac tumor)

Type II (Adult) Germ Cell Tumors

- Clinical features
  - – Account for 60% of all malignancies diagnosed in Caucasian males between 20 and 40 years of age
  - – Incidence of 6–11 per 100,000, although dependent of ethnic background
  - – Asian and Blacks significant lower incidence, not influenced by migration
  - – Plateau in rise related to World War II
  - – Highest incidence in the northern European countries (Denmark, Germany, Norway, and Sweden)
  - – Significant rise in incidence during last decades (3–6%)
  - – Family predisposition involved
  - – No high penetrance cancer susceptibility gene identified
  - – Nonseminomas develop earlier than seminomas (median age, 25 vs. 35 years)
  - – In immunocompromised patients (HIV), seminoma present clinically at the age of nonseminoma
  - – Bilateral tumors occur in up to 5% of the patients (synchronous or metachronous)
  - – Can also be found in retroperitoneal region, intracranial site, and mediastinum

- Risk factors
  - Are associated with the aberrant germ cell maturation during the fetal development
  - Clinical predisposition is related to the testicular dysgenesis syndrome (TDS)
  - TDS includes a spectrum of disorders of the male reproductive system including cryptorchid testis, hypospadias, microlithiasis, sub- or infertility; widely accepted view on its pathogenesis is that environmental endocrine-disrupting chemicals act on Leydig cells and/or testicular Sertoli cells, resulting in abnormal development of the testis
  - Specific types of disorder of sex development (DSD)
    - DSD (previously intersex) is defined as a congenital condition in which development of a chromosomal, gonadal, or anatomical sex is atypical.
    - The risk for type II GCTs is specifically related to hypovirilization and gonadal dysgenesis, related to presence of part of the Y chromosome (likely TSPY as candidate)
  - Genome-wide association studies have implicated single nucleotide polymorphism (SNPs) related to SCF (KITLG) DMRT1, SPRY4, HTERT/CLPM1L, ATF7IP, and BAK1 genes as risk modifiers
  - Low and high birth weight suggested to be associated with increased risk
  - Most likely a combined action between genetic and environmental factors is the most important determinant in risk determination, referred to as GENVIRONMENT (Figs. 16.6 and 16.7)
- Prognosis
  - Type II GCTs are malignant neoplasms, with a high tendency to metastasize to retroperitoneal lymph nodes and different other organs, influenced by histological composition, patholologic stage (Tables 16.2 and 16.3). In spite of this, they overall show a good prognosis.
  - In patients with metastasized disease, three prognostic groups are identified: good, intermediate, and poor (Table 16.4)
  - Overall 10-year survival rate over 90%
  - Radiotherapy is effective for metastatic seminoma
  - Worse prognosis in patients with disseminated choriocarcinoma
  - Late relapses (after 2 years) can occur in nonseminomas, often with worse prognosis
  - Most patients with metastasized disease cured using cisplatin-based chemotherapy
    - Sensitivity to DNA-damaging agents (including irradiation and chemotherapy) supposed to be mul-

tifactorial related to the embryonic germ cell origin.
  - Embryonal stem cells are sensitive due to lack of DNA repair mechanisms combined with no G_1 arrest checkpoint. In addition, low level for apoptosis induction of germ cells is involved in preventing transmission of mutated DNA to the next generation
- Retroperitoneal lymph node dissection might be indicated
- Long-term side effects of the chemo(radio)therapeutic treatment are common, including subfertility, fatigue, cardiovascular complications, metabolic syndrome, and, less frequently, secondary cancer
- Serum markers
  - Three principal tumor serum markers for type II GCTs are available and used according to the guidelines for primary diagnosis, staging, monitoring of therapeutic response, and follow-up
  - Alpha-fetoprotein (AFP) (half-life of 4.5 days) is elevated in up to 70% of patients. Predominantly generated by the yolk sac component
  - Beta subunit of the human choriogonadotropin (HCG) (half-life of 24–36 h) elevated in 50% of patients. Predominantly generated by the choriocarcinoma component
  - Marijuana usage can result in false-positive hCG finding
  - Lactate dehydrogenase (LDH1), being less specific. Elevated in 40–60% of patients
  - If these tumor markers do not decline expected based on half-life after treatment, residual disease is likely
  - Normal level of the markers does not prove absence of disease (only 40–50 and 30% of relapses in patients under active surveillance for clinical stage I disease and after systemic chemotherapy are associated with marker increases)
- Circulating microRNAs
  - Levels of miR-371a-3p correlate with primary tumor mass, clinical stage, and International Germ Cell Cancer Collaborative Group risk groups
  - Serial measurements of circulating microRNAs mirror treatment efficacy in all clinical stages
  - Circulating miRNA levels, particularly of miR-371a-3p, have potential for incorporation in clinical practice and may aid in clinical decision-making
- Gross, microscopic, and immunohistochemical features
  - Histologically, different variants of type II GCT can be identified. These are subdivided into seminoma and nonseminomatous GCTs
  - Nonseminoma defines a cancer with the following histological types: embryonal carcinoma, teratoma, yolk

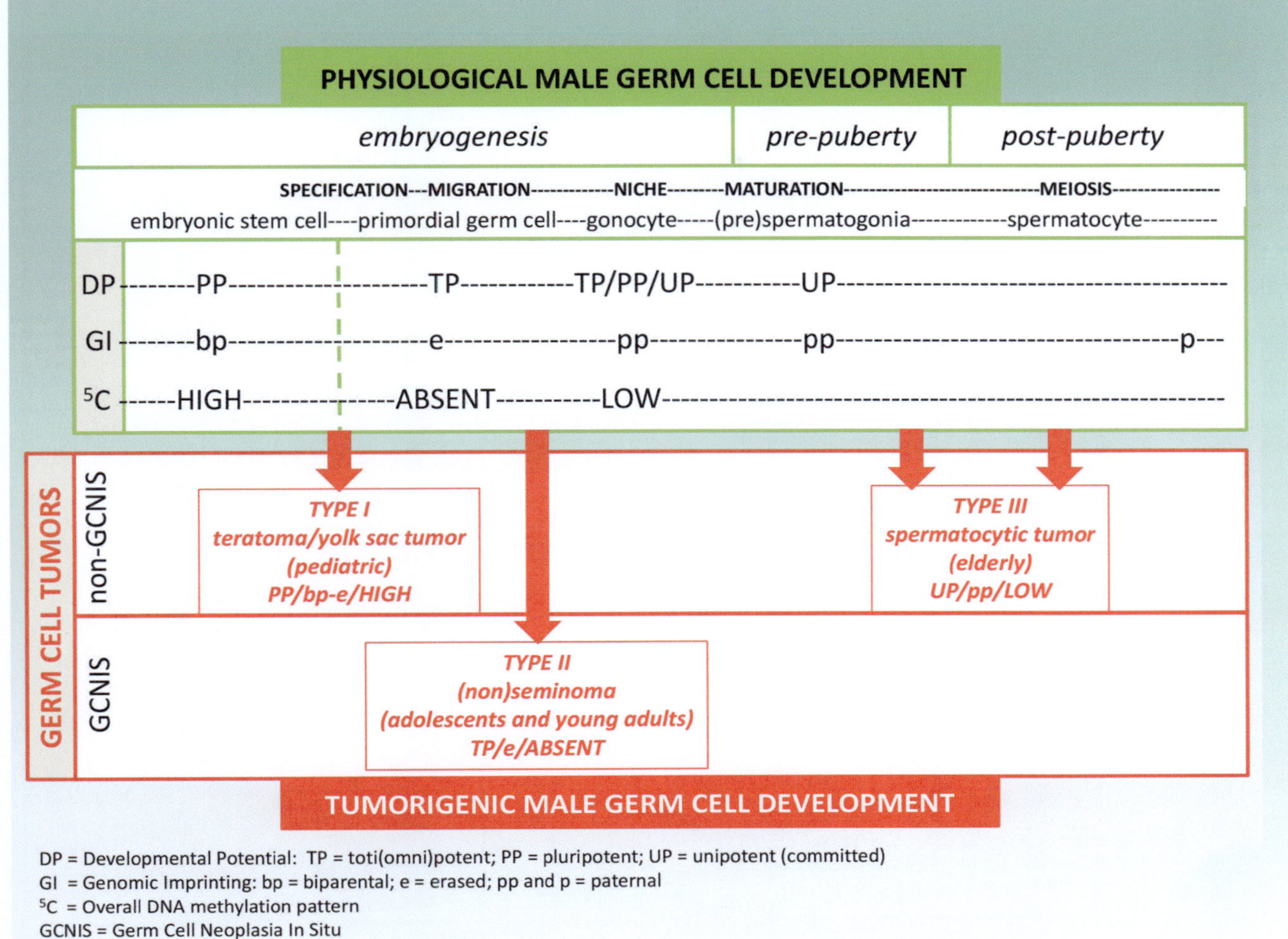

Fig. 16.6 Schematic representation of the physiological process of male germ cell development, during embryogenesis, pre- and postpuberty. Their developmental potential (DP), pattern of genomic imprinting (GI), and overall DNA methylation (^{5}C) are indicated. These are represented as being totipotent (TP), pluripotent (PP), and unipotent (UP); biparental (bp), erased (e), and paternal (pp and p); high, absent, and low, respectively. These characteristics are retained in the derived GCTs, including the Type I (teratoma/yolk sac tumors, paediatric age), Type II (seminomas and nonseminomas, adolescents and young adults), and Type III (spermatocytic tumor, elderly age). Note that two different pathways can be followed to generate the spermatocytic tumors. The ages indicated are those at clinical presentation, although variations exist. Types I and III GCTs are classified as nongerm cell neoplasia in situ related and the Type II as germ cell neoplasia in situ related

sac tumor, and choriocarcinoma (Figs. 16.8, 16.9, and 16.10)
- Overall, about 40% of type II GCTs are seminomas and 60% nonseminomas. Nonseminomatous GCTs show mostly a mixture of different histological components
- Immunohistochemistry is important in diagnosis and differential diagnosis (Table 16.5)
- The different histological components are described below in more detail
- Seminoma
  - Solid tumors with gray, white, or pink surface grossly
- Microscopically, the tumor consists of sheets or lobules separated by fibrous septa with lymphoid infiltrate (Fig. 16.11)
- Round to polygonal tumor cells with clear or eosinophilic cytoplasm
- Nuclei are central and contain prominent nucleoli
- Epithelioid cells, Langhans giant cells, or sarcoid-like granulomas can be present
- Up to 25% contain syncytiotrophoblastic giant cells
- Atypical seminoma shows a greater degree of polymorphism and a higher mitotic rate; however, the clinical significance of this subtype is doubtful

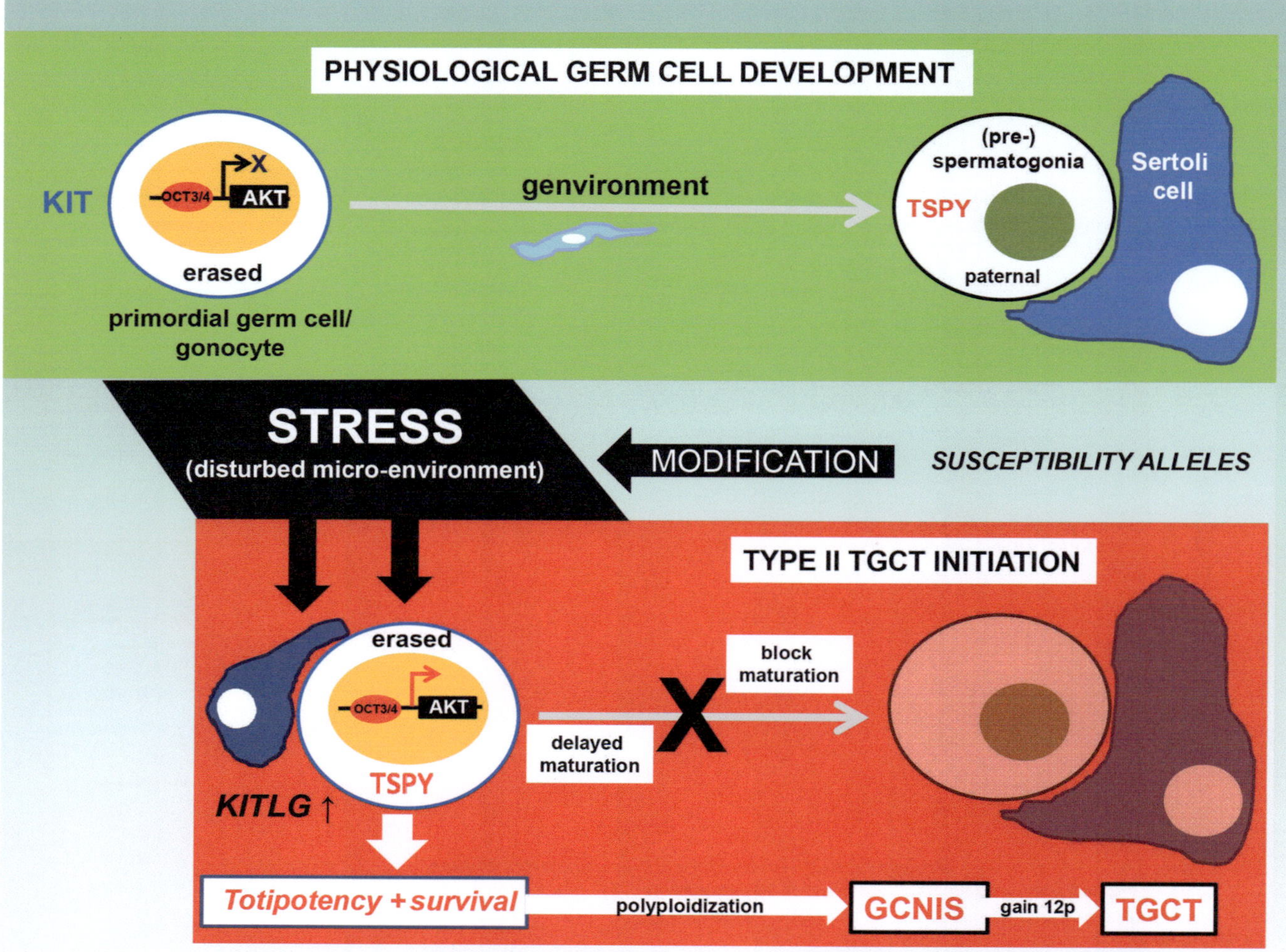

Fig. 16.7 (**a**) Molecular characteristics during pathogenesis of testicular germ cell tumors. Abbreviations: 1–4 N—cell ploidy (1–4 sets of chromosomes), KIT mut—KIT mutated, KIT wt—KIT wild type, GCNIS—germ cell neoplasia in situ, i(12p)—isochromosome 12p, miR—micro RNA (Modified from Shen H et al. Cell Rep 2018;23:3392. Open Access). (**b**) Schematic representation of the proposed pathogenetic model for the formation of the precursor lesion of Type II GCTs of the testis (i.e., germ cell neoplasia in situ [GCNIS]). In green (upper panel), the physiological maturation process of the primordial germ cell/gonocyte to (pre)spermatogonia is represented, in which the combinatory role of the epigenetic genomic constitution of the germ cell and the microenvironment is of importance. This results in downregulation of OCT3/4 (and other embryonic germ cell markers) and the stem cell factor receptor (c-KIT), associated with retention of the inactive state of the AKT promoter during the first year of postnatal life. In addition, expression of testis-specific protein on the Y chromosome (TSPY) is induced as well as the uniparental (i.e., paternal) pattern of genomic imprinting is established. As such, no coexpression of OCT3/4-KIT and TSPY is observed during normal development. In case of exposure to stress during a sensitive window of embryonic development, either due to aberrant genetic constitution or environmental factors (likely in combination), the microenvironment of the embryonic germ cell is changed. This might result in upregulation of the KITLG (stem cell factor), either in supportive cells or in the germ cells itself (represented in the red area). This will result in retention of OCT3/4 expression and activation of the AKT promoter, combined with expression of TSPY in an erased germ cell. The sensitivity of an individual for these particular changes is likely related to the genomic susceptibility alleles (risk SNPs). In this process, delayed maturation of embryonic germ cells can be observed, while a block is required for the formation of pre-GCNIS as intermediate between gonocyte and GCNIS. Polyploidization is an early step in the formation of GCNIS, while gain of 12p is recurrent in the progression to invasiveness. It remains to be proven whether all GCNIS will progress to an invasive TGCT (Type II testicular GCT, either seminoma or nonseminomas). The disturbed formation of prespermatogonia is also in line with the observation that sub−/infertility is a risk factor for this type of GCT

Table 16.2   2017 American Joint Committee on Cancer (AJCC) TNM classification system for testicular cancer

T classification

pTX: Primary tumor cannot be assessed

pT0: No evidence of primary tumors

pTis: Germ cell neoplasia in situ

pT1: Tumor limited to testis (including rete testis invasion) without lymphovascular invasion

pT1a: Tumor smaller than 3 cm in size[a]

pT1b: Tumor 3 cm or larger in size[a]

pT2: Tumor limited to testis (including rete testis invasion) with lymphovascular invasion OR tumor invading hilar soft tissue or epididymis or penetrating visceral mesothelial layer covering the external surface of tunica albuginea with or without lymphovascular invasion

pT3: Tumor invades spermatic cord with or without lymphovascular invasion

pT4: Tumor invades scrotum with or without lymphovascular invasion

N classification

pNX: Regional lymph nodes cannot be assessed

pN0: No regional lymph node metastasis

pN1: Metastasis with a lymph node mass 2 cm or smaller in the greatest dimension and less than or equal to five nodes positive, none larger than 2 cm in greatest dimension

pN2: Metastasis with a lymph node mass larger than 2 cm but no larger than 5 cm in greatest dimension; or more than five nodes positive, none larger than 5 cm; or evidence of extranodal extension of tumor

pN3: Metastasis with a lymph node mass larger than 5 cm in greatest dimension

M classification

M0: No distant metastases

M1a: Nonregional nodal or lung metastases

M1b: Distant metastasis other than nonregional nodal or lung

TNM descriptors: "p" indicates pathologic classification. For identification of special cases of TNM or pTNM classifications, the "m" suffix and "y" and "r" prefixes are used. The "m" suffix indicates the presence of multiple primary tumors in a single site and is recorded in parentheses: pT(m)NM. The "y" prefix indicates those cases in which classification is performed during or following initial multimodality therapy. The "r" prefix indicates a recurrent tumor when staged after a documented disease-free interval and is identified by the "r" prefix: rTNM

[a] Subclassifications of pT1 apply ONLY to pure seminoma

Table 16.3   2017 American Joint Committee on Cancer (AJCC) prognostic stage grouping system for testicular germ cell tumors

Stage	Tumor (T)	Node (N)	Metastasis (M)	Serum tumor markers (S)[a]
Stage 0	pTis	N0	M0	S0
Stage I				
Stage I	PT1-4	N0	M0	SX
Stage IA	pT1	N0	M0	S0
Stage IB	pT2	N0	M0	S0
	pT3	N0	M0	S0
	pT4	N0	M0	S0
Stage IS	Any pT/Tx	N0	M0	S1-3
Stage II				
Stage II	Any pT/Tx	N1–3	M0	SX
Stage IIA	Any pT/Tx	N1	M0	S0
	Any pT/Tx	N1	M0	S1
Stage IIB	Any pT/Tx	N2	M0	S0
	Any pT/Tx	N2	M0	S1
Stage IIC	Any pT/Tx	N3	M0	S0
	Any pT/Tx	N3	M0	S1
Stage III				
Stage III	Any pT/Tx	Any N	M1	SX
Stage IIIA	Any pT/Tx	Any N	M1a	S0
	Any pT/Tx	Any N	M1a	S1
Stage IIIB	Any pT/Tx	N1–3	M0	S2
	Any pT/Tx	Any N	M1a	S2
Stage IIIC	Any pT/Tx	N1–3	M0	S3
	Any pT/Tx	Any N	M1a	S3
	Any pT/Tx	Any N	M1b	Any S

(continued)

Table 16.3 (continued)

Stage	Tumor (T)	Node (N)	Metastasis (M)	Serum tumor markers (S)[a]
Definition of Serum Markers (S)				
SX	Marker studies not available or not performed			
S0	Marker study levels within normal limits			
	LDH[b]	hCG (mIU/ml)	AFP (ng/ml)	
S1	$<1.5 \times N^b$	<5000	<1000	
S2	$1.5{-}10 \times N^b$ or	5000–50,000 or	1000–10,000	
S3	$>10 \times N^b$ or	>50,000 or	>10,000	

[a] The serum tumor markers, lactate dehydrogenase, alpha fetoprotein, and beta-human chorionic gonadotropin are assessed; elevation of these markers corresponds to a higher risk grouping
[b] N indicates the upper limit of normal for the LDH assay

Table 16.4 The International Germ Cell Collaborative Group prognostic grouping

Prognosis grouping (risk status)		Nonpulmonary visceral metastases or mediastinal primary metastases	Serum markers[a]			5-year PFS[b] (%)	5-Year OS[b] (%)
			AFP (ng/ml)	hCG (IU/l)	LDH		
Good	NSGCT	No	<1000	<5000	$<1.5 \times N^c$	89 (90%)	92 (95%)
	Seminoma	No	Normal	Any	Any	82 (87%)	86 (93%)
Intermediate	NSGCT	No	1000–10,000	5000–50,000	$1.5{-}10 \times N^c$	75 (76%)	80 (85%)
	Seminoma	Yes	Normal	Any	Any	67 (***)	72 (***)
Poor	NGSCT	Yes	>10,000	>50,000	$>10 \times N^c$	41 (55%)	48 (64%)
	Seminoma[d]	N/A	N/A	N/A	N/A	N/A	N/A

AFP alpha fetoprotein, *hCG* human chorionic gonadotropin, *LDH* lactate dehydrogenase, *PFS* progression-free survival, *OS* overall survival, *NSGCT* non-seminomatous germ cell tumor, *N/A* not applicable
[a] Markers used for risk classification post-orchiectomy
[b] Please refer to Cheng L et al. Testicular Cancer. Nat Rev Dis Primers. 2018;4:29
[c] N indicates the upper limit of normal for the LDH assay
[d] No seminoma cases classified as poor prognosis
*** Based on very few patients

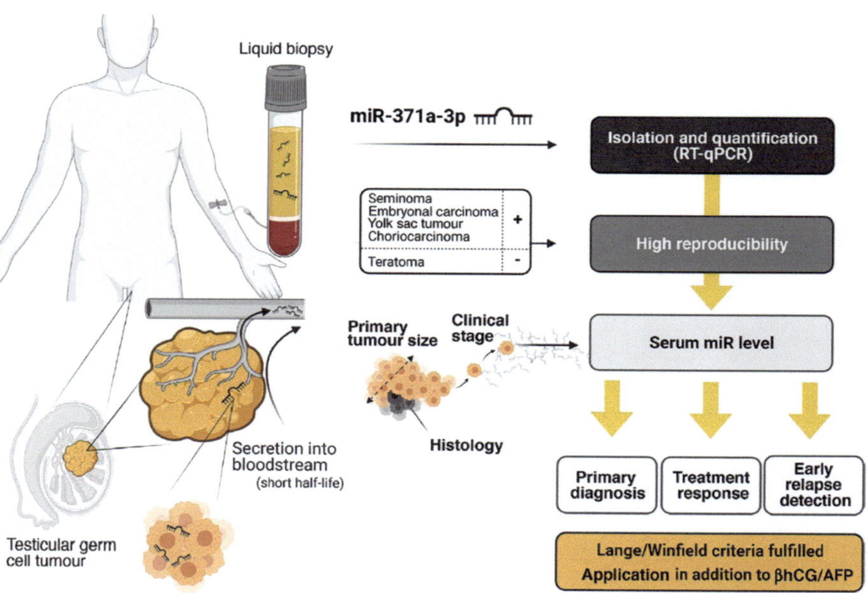

Fig. 16.8 Serum miR-371a-3p in testicular germ cell tumors. Biology, detection, and clinical application as a blood-based biomarker in compliance with the Lange-Winfield criteria for biomarkers. (1) The candidate substance is produced only by the malignancy itself. (2) It is secreted into body fluids. (3) It can be measured in reproducible fashion. (4) Levels in body fluids correlate with the amount of tumor present. (5) The substance can be detected even in early disease. (6) Measured levels correlate with response to treatment. (7) The half-life of the substance is short. *CSI*: Clinical stage I; *CSIIA*: Clinical stage IIA; *RPLND*: Retroperitoneal pelvic lymph-node dissection (Modified from Leao R et al. Eur Urol 2021;80:456. Elsevier)

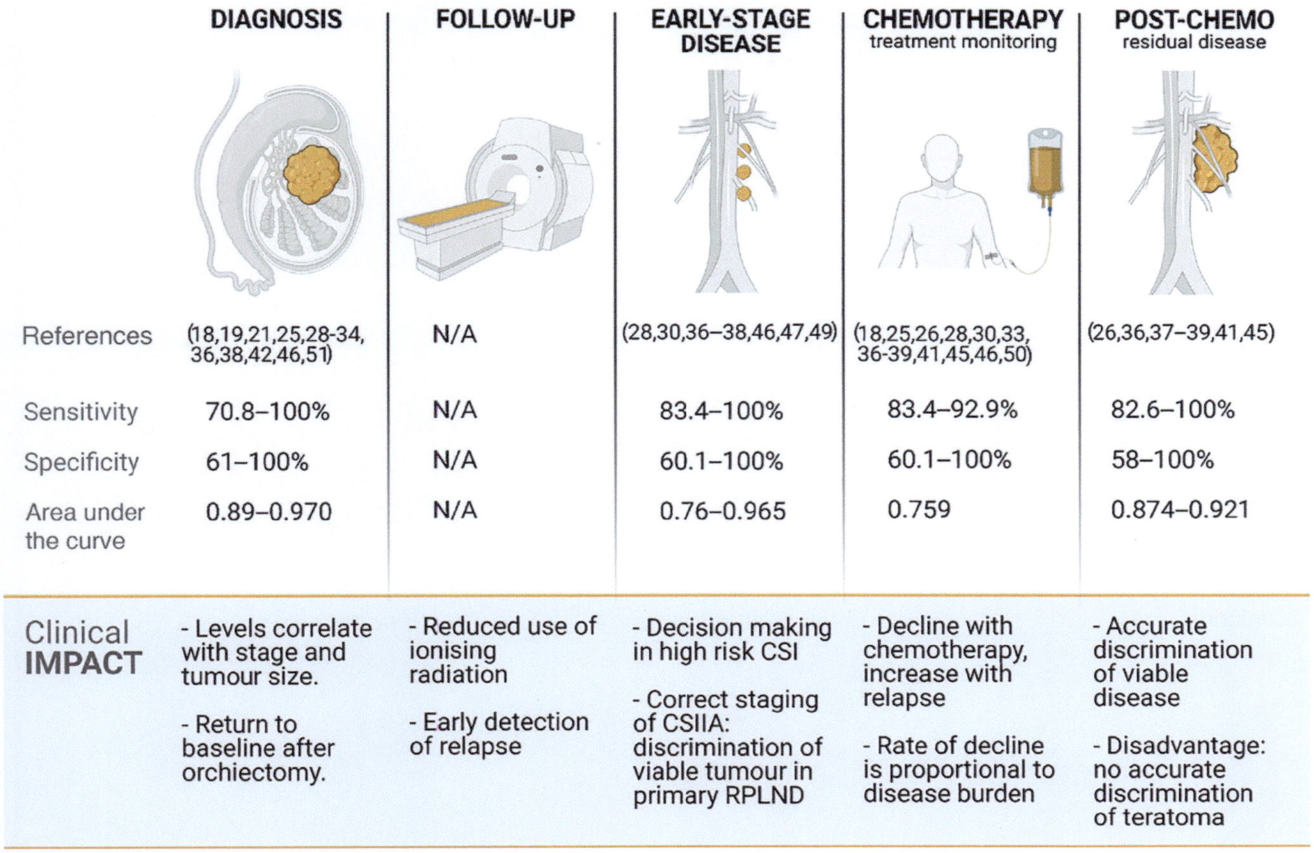

	DIAGNOSIS	FOLLOW-UP	EARLY-STAGE DISEASE	CHEMOTHERAPY treatment monitoring	POST-CHEMO residual disease
References	(18,19,21,25,28-34, 36,38,42,46,51)	N/A	(28,30,36–38,46,47,49)	(18,25,26,28,30,33, 36-39,41,45,46,50)	(26,36,37–39,41,45)
Sensitivity	70.8–100%	N/A	83.4–100%	83.4–92.9%	82.6–100%
Specificity	61–100%	N/A	60.1–100%	60.1–100%	58–100%
Area under the curve	0.89–0.970	N/A	0.76–0.965	0.759	0.874–0.921
Clinical IMPACT	- Levels correlate with stage and tumour size. - Return to baseline after orchiectomy.	- Reduced use of ionising radiation - Early detection of relapse	- Decision making in high risk CSI - Correct staging of CSIIA: discrimination of viable tumour in primary RPLND	- Decline with chemotherapy, increase with relapse - Rate of decline is proportional to disease burden	- Accurate discrimination of viable disease - Disadvantage: no accurate discrimination of teratoma

Fig. 16.9 Serum miR-371a-3p in clinically relevant scenarios. Summary of the literature on the use 09_of circulating miR-371a-3p in diagnosis, follow-up, early-stage disease, treatment monitoring, and post chemotherapy detection of residual disease, and the potential clinical impact of its use in each of these scenarios. *RT-qPCR*: real-time qualitative polymerase chain reaction; *beta-hCGL:* beta-human chorionicgonadotropin; *AFP*: alpha-fetoprotein (Modified from Leao R et al. Eur Urol 2021;80:456. Elsevier)

– There is nuclear staining for markers of undifferentiated germ cells, including OCT3/4, SALL4, NANOG, and AP2gamma (>90% of tumor cells are positive)

– OCT3/4 is nowadays the most sensitive and specific marker for seminoma (as well as for CIS and embryonal carcinoma), always nuclear in localization

– Variable membranous staining for markers of germ cell differentiation, including CD117 (c-KIT), D2-40, as well as for placental alkaline phosphatase (PLAP)

– In contrast to embryonal carcinoma (see below), no expression of EMA in seminoma

– Immunohistochemistry for cytokeratins can be positive without clinical impact

– SOX17 is positive in seminoma and can differentiate seminoma from embryonal carcinoma, which is SOX17 negative. However, normal PGCs and gonocytes as well as spermatogonia are positive as well

• Embryonal carcinoma
  – Solid tumor with gray to pink appearance and foci of hemorrhage and necrosis
  – Growth pattern varies from solid to papillary and syncytial (Fig. 16.12)
  – Typical epithelial-like cells with large irregular nuclei
  – Mitotic figures are frequent
  – Syncytiotrophoblastic giant cells might be scattered
  – There is a nuclear (and cytoplasmic) staining for markers of undifferentiated germ cells, including OCT3/4, SALL4, and NANOG (>90% of tumor cells are positive)
  – OCT3/4 also shows both a nuclear and cytoplasmic localization
  – Membranous staining for CD30, EMA, and PLAP
  – Markers differentially expressed in embryonal carcinoma vs. seminoma are SOX2, CD30 (exclusively

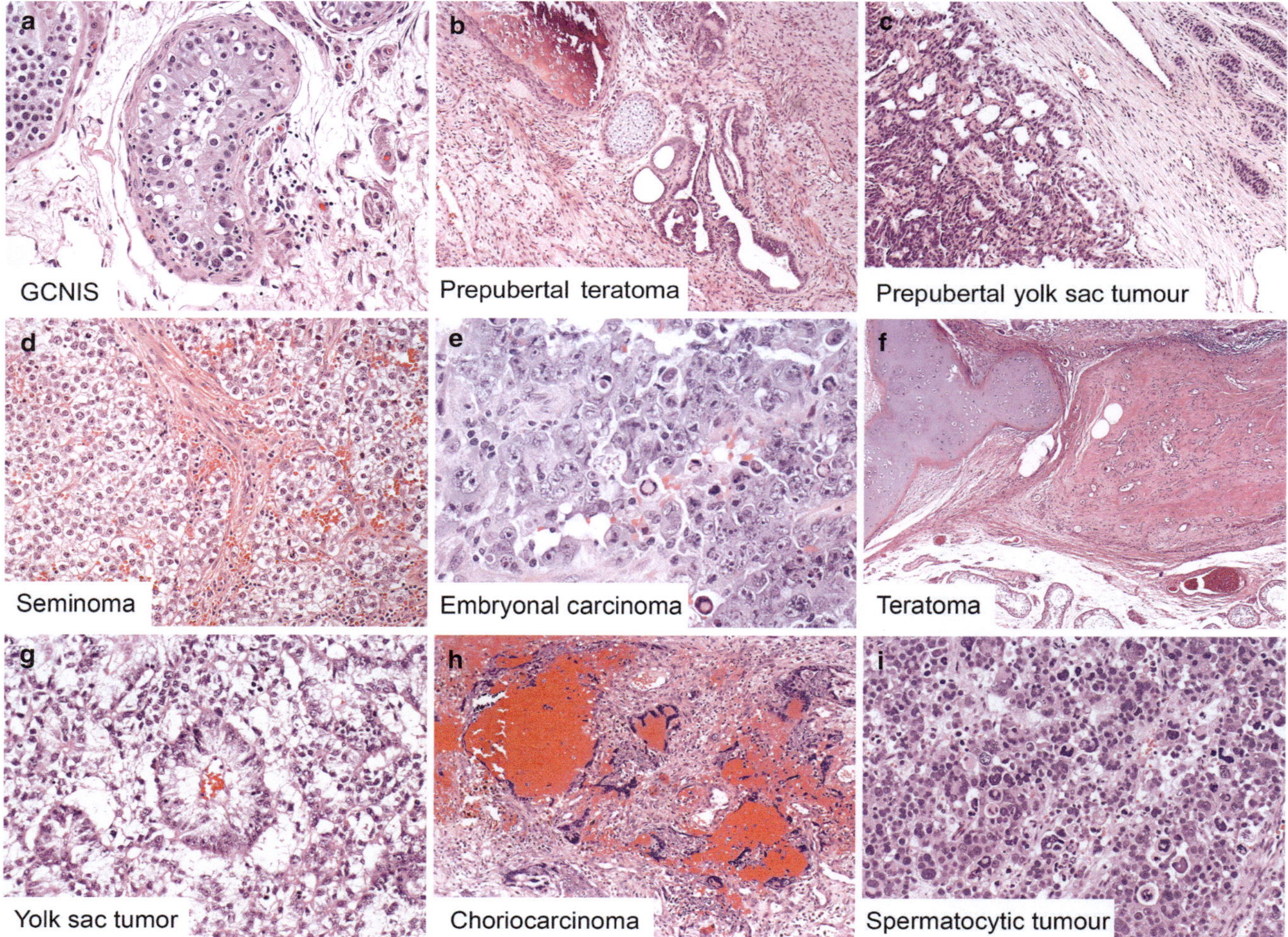

Fig. 16.10 Testicular germ cell tumors. The histological composition of different testicular germ cell tumors (TGCTs), as assessed by haematoxylin and eosin–staining under light microscopy, is shown. Germ cell neoplasia in situ (GCNIS) is the precursor lesion to most testicular cancers (**a**), although type I GCTs do not derive from GCNIS and are typically paediatric. Histologically, type I GCTs are prepubertal teratomas (**b**) or prepubertal yolk sac tumors (**c**). Type II GCTs do derive from GCNIS and, histologically, these are seminomas (**d**) or nonseminomas such as embryonal carcinoma (**e**), teratoma (**f**), yolk sac tumors (**g**), and choriocarcinoma (**h**). Type III GCTs do not derive from GCNIS and are spermatocytic tumors (**i**) (adapted from Cheng L et al. Nat Rev. Dis Primers. 2018;4:29. Springer Nature)

positive in embryonal carcinoma), and EMA of SOX17 (positive in seminoma) (Fig. 16.13)
- Vascular invasion is often the result of embryonal carcinoma
• Yolk sac tumor
    - Solid soft tumors, gray-white to yellow surface
    - Necrosis may be present
    - There are numerous patterns of differentiation: microcystic, macrocystic, endodermal sinus, papillary, glandular, solid, polyvesicular, vitelline, hepatoid, myxoid, and parietal pattern (Figs. 16.10 and 16.13)
    - Various patterns are usually admixed in one tumor
    - Foci of yolk sac tumor are frequently seen in nonseminomas
    - Pure yolk sac tumors are rare

- AFP and glypican 3 are variably expressed in yolk sac tumors and can be informative to discriminate from other components and cancers
- Combination of these markers can increase sensitivity for the detection
- Low-molecular-weight cytokeratins are positive in yolk sac tumors
• Choriocarcinoma
    - Tumor represents as nodules with hemorrhage
    - Composed of trophoblast-like cells and syncytial large cells
    - Frequently present in nonseminomatous
    - Rare in a pure form (<0.1% of all TGCT)
    - Pure choriocarcinoma is likely to present as a highly aggressive disease with hematogenous metastases

Table 16.5   Useful immunohistochemical stains in the differential diagnosis of testicular germ cell tumors

	EMA	AE1/3	CK7	SALL4	PLAP	OCT4	GATA3	D2–40	CD117	CD30	GPC3	AFP	SOX2	SOX17	HCG	Inihibin	Calretinin	SF-1	FOXL2	Melan-A
GCNIS	−	−	−/+	+	+	+	−	+	+	−	−	−	−	+	−	−	−	U	U	U
Seminoma	−	V	−/+	+	+	+	−	+	+	−	−	−	−	+	−	−	−	−	−	−
Embryonal carcinoma	−	+	−	+	+	+	−	−/+	−	+	−	−	+	−	−	−	−	−	−	−
Yolk sac tumor	−	+	−	+	+	−	+	−	−/+	−	+	+	−	+	−	−	−	−	−	−
Choriocarcinoma	+	+	+	+	+	−	+	V	−	−	+	−	−	−	+	−	−	−	−	−
Teratoma	+	+	+	+	V	−/+	−/+	−	−	−/+	−	−/+	V	V	−	−	−	−	−	−
Spermatocytic tumor	−	−	U	+	−	−	−	−	+[a]	−	U	−	U	U	−	−	−	U	U	−
Sertoli cell tumor	−/+	V	−	−	−	−	−	−	−	−	−	−	−	U	−	+	V	+	+	−
Leydig cell tumor	−	−	−	−	−	−	−	−	−	−	−	−	−	U	−	+	+	+	−/+	+
JGCT	−/+	−	−	−	−	−/+	U	−/+	−	−	−	−	−	U	−	+	+	+	+	−
AGCT	−	−	−	−	−	−/+	U	−/+	−	−	−	−	−	U	−	+	+	+	+	−/+

GCNIS germ cell neoplasia in situ, *JGCT* juvenile granulosa cell tumor, *AGCT* adult granulosa cell tumor, *EMA* epithelial membrane antigen, *PLAP* placental-like alkaline phosphatase, *D2-40* podoplanin, *GPC3* glypican 3, *HCG* human chorionic gonadotrophin, *SF-1* steroidogenic factor 1, *FOXL2*, forkhead box L2, −/+ usually negative, *V* variable, *U* unknown
[a]Approximately 50% of spermatocytic tumors are positive for CD117 (c-kit)
(adapted from Cheng L et al. Hum Pathol. 2017;59:10)

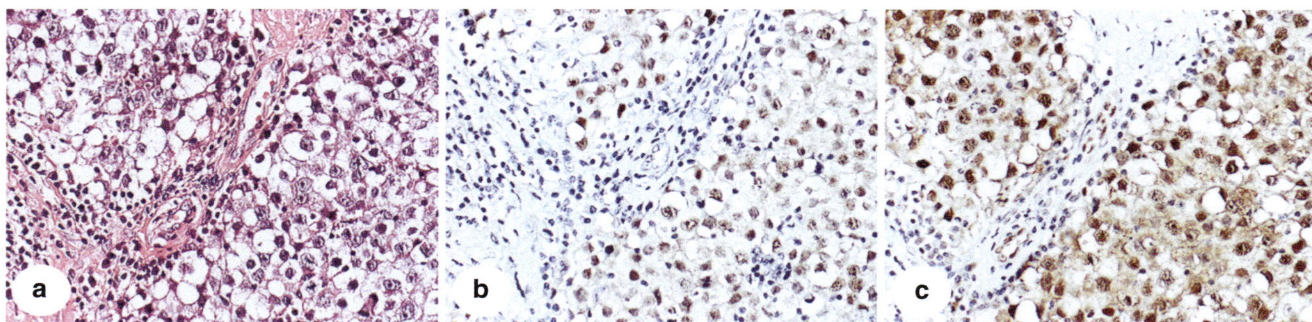

Fig. 16.11 Representative examples of seminoma, stained with (**a**) H&E, and the immunohistochemical markers (**b**) OCT3/4, and (**c**) SOX17

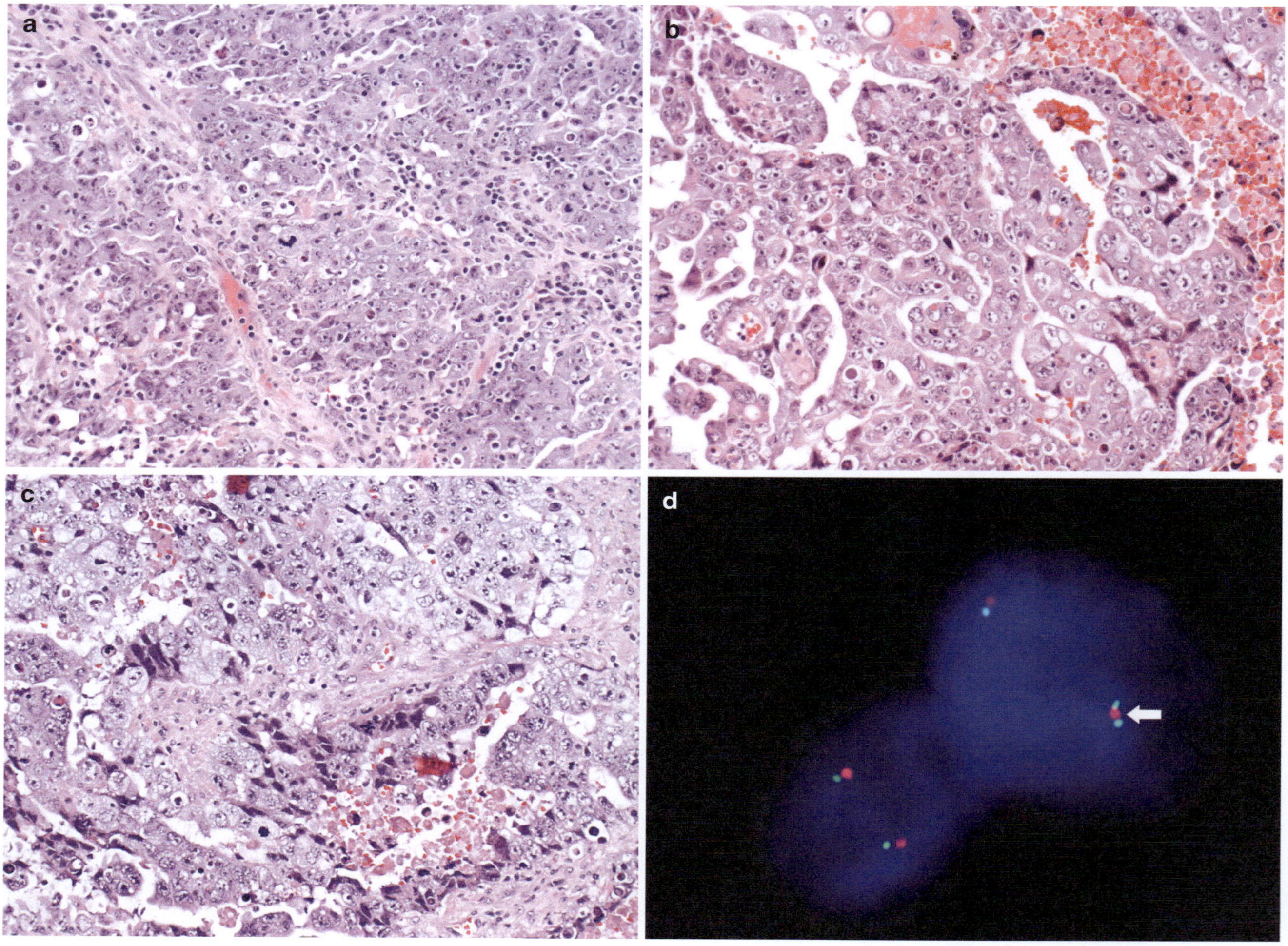

Fig. 16.12 Embryonal carcinoma. (**a**) Tumor cells are highly pleomorphic with syncytial growth and overlapping nuclei (200×). (**b**) Papillary formations are evident (original magnification 200×). (**c**) Syncytiotrophoblastic cells are present (original magnification 200×). (**d**) Dual color fluorescence in situ hybridization (FISH) shows isochromosome 12p (arrow), as evidenced by 2 12p signals (green) in close proximity to 1 centromeric 12 signal (red). The other copies of chromosome 12 (right) show a wild-type pattern, with 1 green 12p signal and 1 red centromeric signal (adapted from Cheng L et al. Hum Pathol. 2017;59:10. Elsevier)

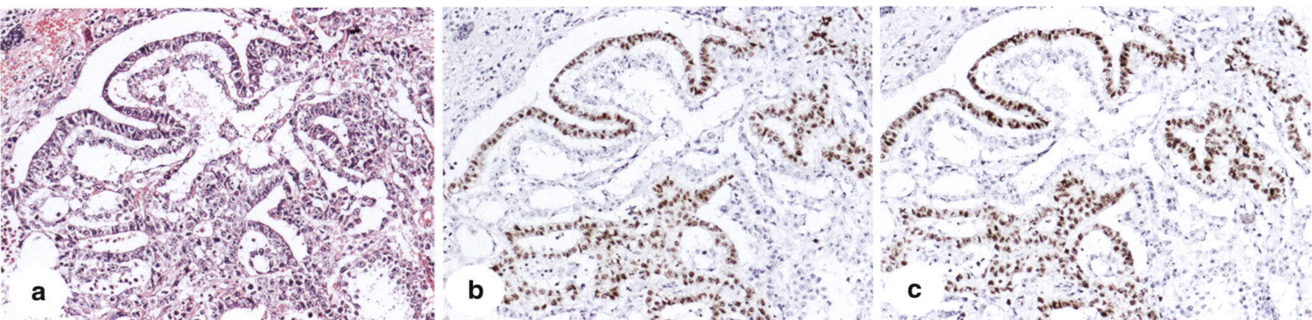

Fig. 16.13 Representative example of a mixed nonseminoma, containing an embryonal carcinoma and yolk sac tumor component, stained with (**a**) H&E, and the immunohistochemical markers (**b**) OCT3/4, and (**c**) SOX2

- Syncytiotrophoblasts are positive for beta-HCG, inhibin (alpha subunit), and EMA
- Cytokeratin is expressed in trophoblasts and syncytiotrophoblasts
- GATA3 is positive in 78% of choriocarcinoma
- Combination of these markers can increase sensitivity for the detection
- Low-molecular-weight cytokeratins are positive in yolk sac tumors
- Teratoma
  - Malignant tumor showing somatic differentiation with endoderm, ectoderm, and endoderm derivates
  - Teratomatous component is often present in nonseminomas
  - Various somatic malignancies might arise in the background of teratoma including sarcoma, PNET, nephroblastoma, carcinoma, and adenocarcinoma
  - No specific immunoprofile, the differentiated areas of teratoma show immunophenotype which is in accordance with the underlying cell type
  - AFP can be expressed in intestinal or hepatoid areas
  - Intratesticular epidermoid cyst represents a rare benign teratoma in the adult and should not be mixed up with the type II malignant teratoma
    - This benign teratoma shows in contrast to the malignant type II teratoma no CIS (see below) in the adjacent testis
    - Careful examination of the seminiferous tubules should be done under usage of the immunohistochemistry for OCT3/4 (or another marker for CIS)
    - Detection of CIS or isochromosome 12p (i12p) supports a malignant type II teratoma
- Precursor lesions and cell of origin
  - The precursor of all type II GCTs of the testis is the so-called carcinoma in situ (CIS) of the testis, also referred to as intratubular germ cell neoplasia unclassified (IGCNU), testicular intraepithelial neoplasia (TIN), or germ cell neoplasia in situ (GCNIS) (Fig. 16.14)
  - It is expected that all patients with CIS will eventually develop an invasive cancer (in the prospective study, 70% of the patients with CIS developed an invasive GCT within 7 years)
  - CIS cells are located at the inner side of the basal lamina of the seminiferous tubule, most frequently in a single row in close connection with Sertoli cells, under their interconnecting tight junctions
  - CIS is often present in the adjacent parenchyma of invasive type II GCTs, especially nonseminomas
  - Activated immune system as found in seminomas can also eradicate CIS
  - PGCs and CIS cells share the same pattern of genomic imprinting (erased), telomerase activity, and gene and protein expression profile
  - CIS shows homogeneous expression of markers of PGCs/gonocytes, including c-KIT, BLIMP1, AP2gamma, OCT3/4, NANOG, LIN28 (and many more)
  - OCT3/4 is the most specific and sensitive marker for CIS, strongly staining the nucleus of all CIS cells, but not normal spermatogonia
  - The CIS counterpart in dysgenetic gonads with a low level of virilization (i.e., no or limited testicular differentiation) is known as gonadoblastoma (Fig. 16.15)
    - Gonadoblastoma is composed on CIS-like cells intermixed with stromal cells expressing FOXL2 (granulosa differentiation), whereas Sertoli cells (SOX9 positive) are associated with CIS
    - Gonadoblastoma can mimic CIS, to be differentiated by SOX9 and FOXL2
    - Presence of gonadoblastoma may indicate the presence of DSD
  - Overdiagnosis of CIS is possible due to germ cell maturation delay. This can be avoided using immunohistochemical detection of SCF (KITLG) (Fig. 16.16)

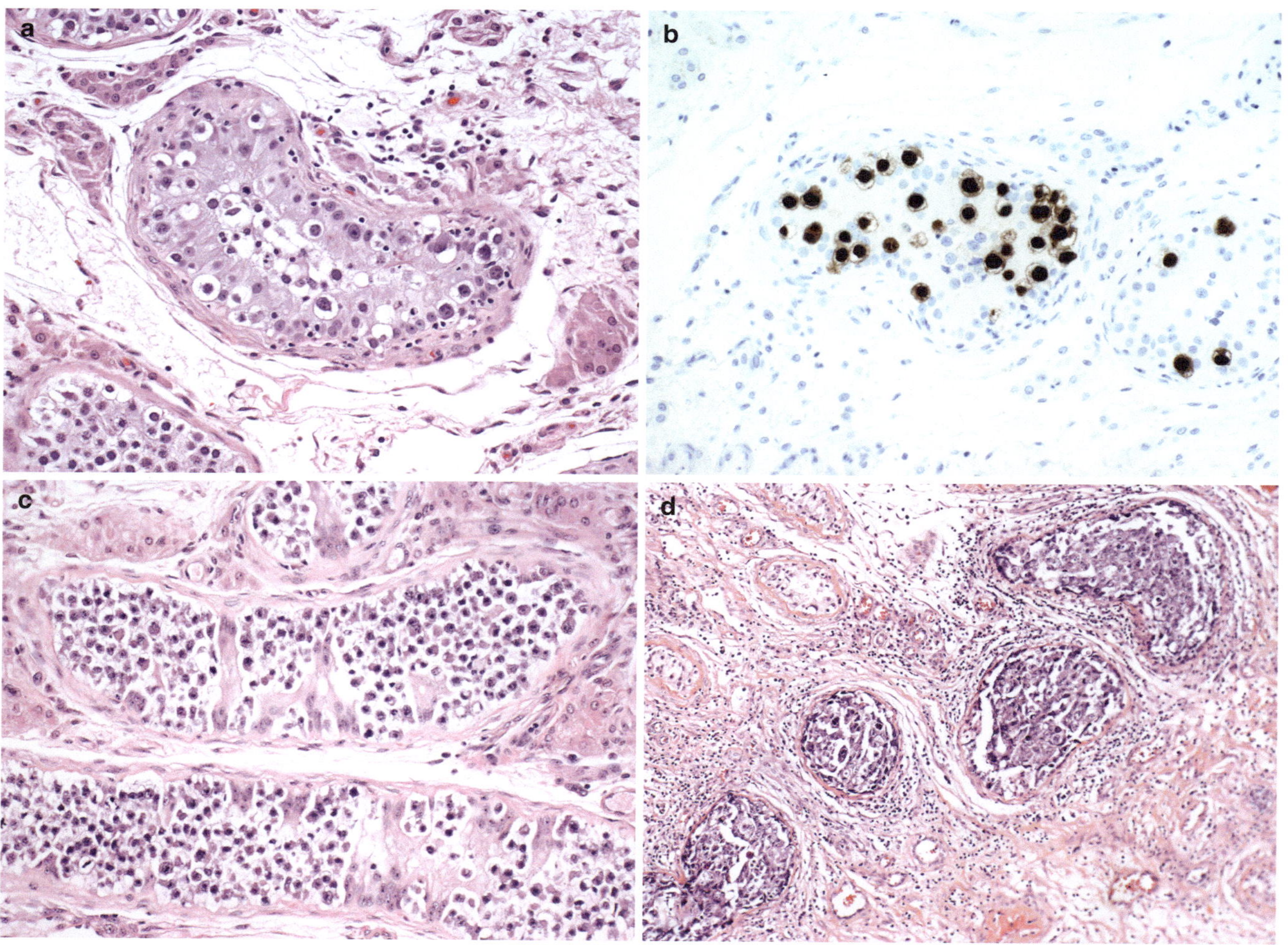

Fig. 16.14 Noninvasive germ cell neoplasia of the testis. (**a**) The cells of germ cell neoplasia in situ (GCNIS) have enlarged hyperchromatic nuclei with cytoplasmic clearing (original magnification 400×). (**b**) OCT4 immunostain highlights nuclei of neoplastic cells in GCNIS(original magnification 400×). (**c**) GCNIS with scattered syncytiotrophoblastic cells (original magnification 200×). (**d**) Intratubular embryonal carcinoma has a glandular pattern with prominent clefts (adapted from Cheng L et al. Hum Pathol. 2017;59:10. Elsevier)

- o In a testicular biopsy taken during the first year of life in an individual with possible germ cell maturation delay (e.g., cryptorchidism), overdiagnosis is possible.
- o Distinguishing morphological criteria are not strictly informative; immunohistochemistry with OCT3/4 is also discriminatory. It can be solved using immunohistochemistry for SCF, specifically present in the premalignant cells

- – No informative animal model has been identified yet
- • Molecular features (Tables 16.6 and 16.7)
  - – Chromosomal constitution
    - o Seminomas and CIS are hypertriploid and the nonseminomas hypotriploid type II GCT show various losses and gains of (parts of) chromosomes: loss of chromosomes 4, 5, 11, 13, 18, and Y and gain of chromosomes 7, 8, X, and 12

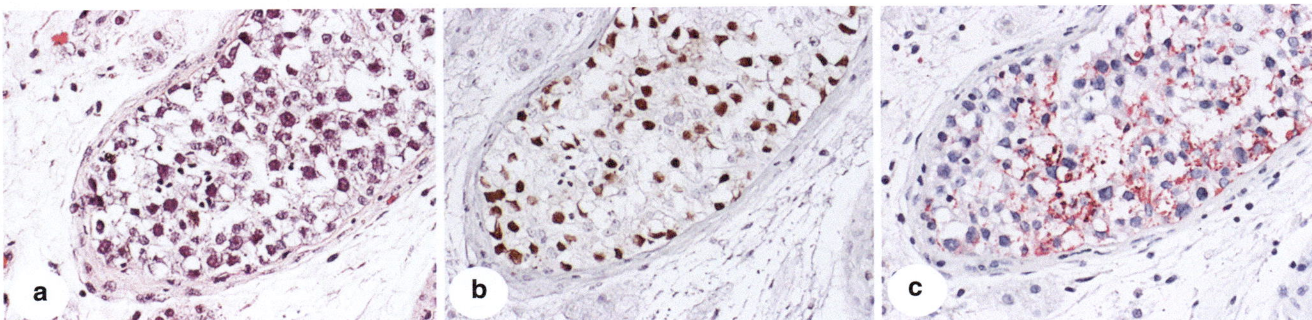

Fig. 16.15 Classical gonadoblastoma and variants. (**a**) Sectioned surface of the gonadal neoplasm shows yellow-tan classical gonadoblastoma with minute white calcifications in the lower part of the field and homogenous gray-white germinoma in the upper field. (**b**) An island of classical gonadoblastoma containing prominent dystrophic calcification to the right occupies the lower two-thirds of the field, and germinoma is seen in the upper field. (**c**) Islands of classical gonadoblastoma show strong nuclear expression of OCT4 in many but not all germ cells. (**d**) The sex cord elements of variably sized islands of gonadoblastoma show cytoplasmic expression of α-inhibin (adapted from Roth LM, Cheng L. Hum Pathol. 2020;100:47. Elsevier)

Fig. 16.16 Representative examples of a seminiferous tubules containing carcinoma in situ (CIS) cells and intratubular seminoma cells, stained with (**a**) H&E, and the immunohistochemical markers (**b**) OCT3/4, and (**c**) stem cell factor (KITLG)

Table 16.6 ISUP recommendations for reporting and molecular testing related to testicular GCTs

Topics	Recommendations	Remarks
Application of IHC for final diagnosis GCT and presence of GCNIS in biopsy and orchiectomy specimen	1. Presence of GCNIS must be documented, both in case of biopsy and orchiectomy	1. Reporting extent of spermatogenesis is highly recommended. Absence of testicular parenchyma as well as complete lack of spermatogenesis (i.e., germ cells) must be specified
	2. IHC is informative to identify GCNIS (i.e., using OCT3/4)	2. IHC for OCT3/4 is suboptimal in case of Stieve and Bouin fixatives
	3. Seminoma and embryonal carcinoma can be further distinguished based on supplemental IHC biomarkers in difficult cases (i.e., poor fixation)	3. IHC for OCT3/4 and SOX17 (for seminoma) and OCT3/4, CD30, and SOX2 (embryonal carcinoma). Lymphovascular invasion by embryonal carcinoma is of relevance for risk stratification of stage 1 nonseminomas
	4. Overdiagnosis of GCNIS in prepubertal testis (first few years of life) may be avoided using OCT3/4 staining	4. Morphologic characterization related to specific localization and clonal expansion can be informative, as well as IHC for TSPY and KITLG
	5. Presence of GB must be checked for, especially in case of nonscrotal testis, and seemingly presence of intratubular seminoma (mimicking GB)	5. GB is diagnostic for DSD, implicating GCT risk of contralateral gonad (testis?). IHC for FOXL2 (granulosa cells) is informative (compared with SOX9 of sertoli cells)
	6. Spermatocytic tumor is almost always recognizable via morphology alone; difficult cases can be distinguished using absence of staining for IHC markers typically positive in mimics	6. Spermatocytic tumor is in principle benign with exception of sarcomatous transformation; bilateral cases do occur
	7. IHC for AFP/GPC3 and β-hCG is helpful in confirming presence of YST and choriocarcinoma, respectively	7. No differential diagnostic IHC biomarkers are known for non-GCNIS and GCNIS-related YST. Nonchoriocarcinomatous trophoblastic tumors must be considered in DD
	8. SALL4 can be used as surrogate marker for GCT, especially in the setting of unknown history or unusual morphology in metastasis	8. SALL4 is nonspecific and should be interpreted with caution as other nongerm cell tumors may be positive
Performance of molecular testing	1. Molecular testing is informative to distinguish non-GCNIS vs. GCNIS-related GCTs (primary and metastatic)	1. Precaution has to be taken to include sufficient amount of tumor cells in the final analyses performed
	2. Non-GCNIS-related (benign) teratomas (type I) are diploid without specific genetic anomalies. This can be detected using (F)ISH, qPCR, as well as other more genome wide approaches ((array) CGH, SNP array, methylation arrays (450 K or EPIC)	2. Enucleation can be considered in case of absence of GCNIS and a diploid DNA content. Genome-wide approaches (see under 3) are most informative; intraoperative consultation may be helpful
	3. Gain of (the entire) 12p, on top of an overall aneuploid DNA content, is characteristic for most GCNIS-related GCTs, both seminomas and nonseminomas, although not for GCNIS itself (although being aneuploid). This can be detected using the techniques mentioned under point 3	3. Genome-wide approaches are more informative than a single target-based method
	4. P53 mutations or genomic MDM2 amplification seem to be the most prominent changes related to treatment resistance of GCNIS-related GCT so far. MSI seems to be of limited value. Possibly tumor heterogeneity must be kept in mind	4. Interference with this pathway might be an interesting approach to follow. Again, tumor organoids might be informative
	5. Gain of the entire chromosome 9 is characteristic for spermatocytic tumor. This can be detected using the techniques mentioned under point 3	5. Genome-wide CNV distinguished seminoma from spermatocytic tumor
	6. cKIT, KRAS, HRAS, and PI3CA are the most frequent mutations, especially found in seminoma (mainly without 12p)	6. Interference with this pathway might be an interesting approach to follow
	7. miR-371a-3p (and related family members) is informative to detect malignant GCT components, including YST of the non-GCNIS variant (type I). It can be detected both in tissue and body fluids (ie, liquid biopsies)	7. Efforts must be performed to prove the value of this molecular test compared with the golden standard AFP and hCG to promote clinical implementation. Liquid biopsy will likely find its usefulness in recent future

Bold indicates the most relevant in clinical practice at this moment

CGH comparative genomic hybridization, *DD* differential diagnosis, *DSD* disorder of sex development, *GB* gonadoblastoma, *hCG* human chorionic gonadotropin, *MSI* microsatellite instability, *qPCR* quantitative polymerase chain reaction. (Modified from Looijenga LHJ et al. Am J Surg Pathol. 2020;44:e66-e79)

Table 16.7 Molecular approaches for possible future implications

Molecular analysis in a research context technique and possible implications	Remark(s)
1. Defined SNP profiles are associated with development of GCNIS related (type II) GCTs (supporting an embryonic origin), partly overlapping with the non–GCNIS-related teratomas and YSTs (type I)	1. The power of the SNP profiling indicates that it might only be of clinical value in high risk populations (i.e., disorder of sex development and cryptorchidism). It is unknown at which step in the pathogenesis they act on
2. GCNIS and seminoma are completely demethylated, while the invasive nonseminomas show a lineage-specific pattern. DPP3A remained demethylated. Treatment-resistant seminomas can be hypermethylated. The XIST promotor is demethylated in GCNIS-related GCT of all histologies	2. The demethylated DNA pattern might be related to their genomic instability and might offer an alternative target for treatment as well as molecular target for monitoring (both in tissue and liquid biopsy)
3. Mutational load is very low in all GCTs, independent of being GCNIS or non-GCNIS related, while the mutations occurring are related to progression and can be heterogenous (including metastases)	3. This is likely based on the evolutionary mechanism preventing transmission of mutations to the next generation. Therefore, genome-wide approaches to identify hits for targeted treatment are expected to be unsuccessful on primary tumor analyses
4. HRAS and FGFR3 mutations are detectable in spermatocytic tumor, especially at a later age	4. On the basis of the overall benign behavior, this will not have a clinical impact
5. WNT activation, predominantly related to methylation-based and CNV-based anomaly, is identified in YSTs, both GCNIS and non-GCNIS related (types I and II)	5. Preclinical (in vivo) studies need to confirm these findings, in which the use of primary tumor cultures (organoids) might be informative in addition to the limited number of cell lines available

XIST indicates X-inactive-specific transcripts
Modified from Looijenga LHJ et al. Am J Surg Pathol. 2020;44:e66–e79)

- o Yolk sac tumors show recurrent chromosomal imbalances, including loss of 1p, 4, and 6q, and gain of 1q and 20q
- o All invasive tumors show gain of 12p, mostly due to formation of isochromosomes (i12p). Regional high-level amplification can also be observed. No obvious candidate gene(s) has been identified so far, although various have been suggested (CCND2, NANOG KRAS2, etc.) (Fig. 16.17a)
- o Studies indicate that gain of 12p is progression related (occurs when CIS cells become independent of interaction with Sertoli cells).
- o Cyclin D2 (CNND2) is expressed in all type II GCTs, as well as in CIS, while NANOG is expressed in CIS as well as seminoma and embryonal carcinoma
- o It is most likely that multiple genes located on 12p are relevant in the pathogenesis. It matches with the observation that gain of 12p can be found in extended in vitro cultures of human embryonic stem cells
- o X chromosome is gained in the majority of TGCT
  - ◆ Familial predisposition of type II GCT has been linked to the X chromosome
  - ◆ Additional X chromosome is also relevant in the context of Klinefelter syndrome patients, although these patients only develop mediastinal type II GCTs (not testicular)
  - ◆ A role of the X chromosome might also be suggested from data of patients with specific forms of DSD
  - ◆ Supernumerical X chromosomes are inactivated in nonseminomas by methylation. This, in parallel to normal embryogenesis, result from function of the non-(protein)-coding *XIST* gene
  - ◆ This phenomenon is correlated with hypomethylation of the promoter region, reported to be useful as molecular target for this type of cancer
- • Epigenetic modifications
  - – CIS and seminomas show a hypomethylated DNA status, in contrast to the various histological types of nonseminomas, this parallels normal embryogenesis
  - – Histone modification proteins BLIMP1 and PRMT5
    - o The complex of the transcription factor BLIMP1 and protein arginine methyltransferase-5 PRMT5 protein is expressed in CIS and seminoma
    - o Proposed function of BLIMP1/PRMT5 complex is suppression of premature differentiation and maintenance of pluripotency in PGCs/gonocytes by dimethylation of H2A/H4 at arginine 3 and repression of gene expression (see above)
    - o It suggests that histone H2A and H4 arginine 3 dimethylation suppress differentiation of CIS and seminoma, while loss of these histone modifications might induce reprogramming and differentiation to embryonal carcinomas and the various subtypes of differentiated nonseminomas
- • Embryonal stem cell genes

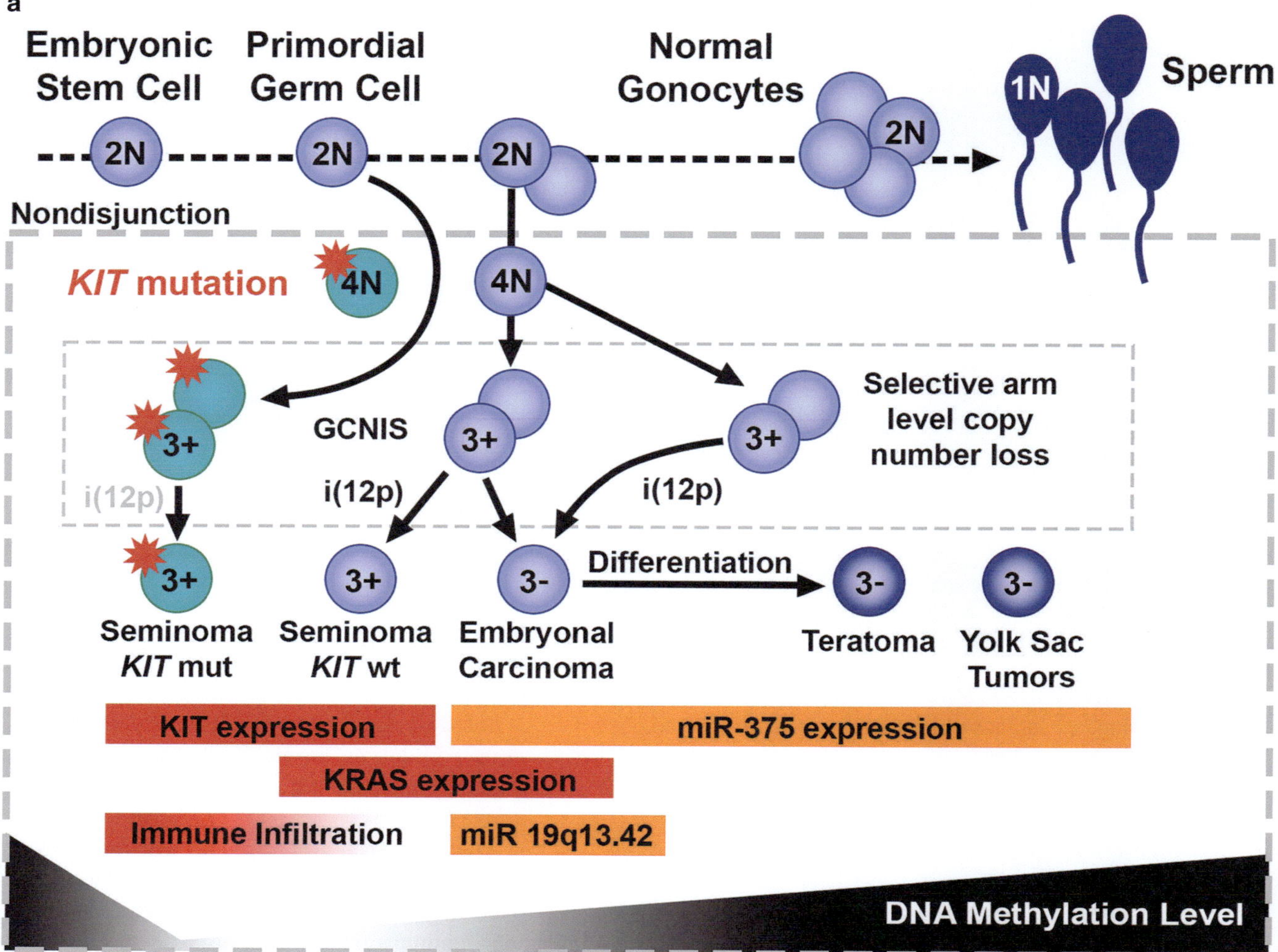

Fig. 16.17 Proposed pathogenetic model for the formation of germ cell neoplasia in situ. (**a**) During the physiological maturation of primordial germ cells or gonocytes to pre-spermatogonia or spermatogonia, OCT3/4, and c-KIT are downregulated. This downregulation results from the epigenetic genomic constitution of the germ cell and the influence of the microenvironment and ensures that the AKT promoter is kept in an inactive state during the first year of postnatal life. This is expected to result in aberrant survival of embryonic germ cells, with pluripotent characteristics, in the testis beyond the age of 6 months to a year and, possibly the formation of GCNIS. In addition, expression of testis specific protein on the Y chromosome (TSPY) is induced, and the paternal pattern of genomic imprinting is established. As such, OCT3/4-KIT and TSPY are not co-expressed during normal development. (**b**) The microenvironment of the embryonic germ cell is altered upon exposure to stress during embryonic development, facilitated by owing to an aberrant genetic constitution and/or environmental factors. This might result in the upregulation of the stem cell factor KITLG, either in supportive cells (that is, Sertoli cells) or in the germ cells themselves. KITLG expression results in the retention of OCT3/4 expression, activation of the AKT promoter, and the expression of TSPY, which is characteristic of an erased germ cell. The vulnerability sensitivity of an individual to undergo these embryonic germ cell niche-related changes is likely to depend on whether they express susceptibility single nucleotide polymorphisms (SNPs). In this germ cell niche affect ting process, delayed maturation of embryonic germ cells can be observed, whereas a maturation block toward pre-spermatogonia, being the physiological default pathway, is required for the formation of pre-GCNIS as an intermediate between gonocyte and GCNIS. Polyploidization is an early step in the formation of GCNIS, whereas gain of 12p is recurrent in the progression to invasiveness. It its unclear whether all GCNIS progress to invasive-type II TGCT. The disturbed formation of prespermatogonia associated with type II TGCT is also in line with the observation that subfertility and infertility is a risk factor for this type of TGCT (adapted from Cheng L et al. Nat Rev. Dis Primers. 2018;4:29. Springer Nature)

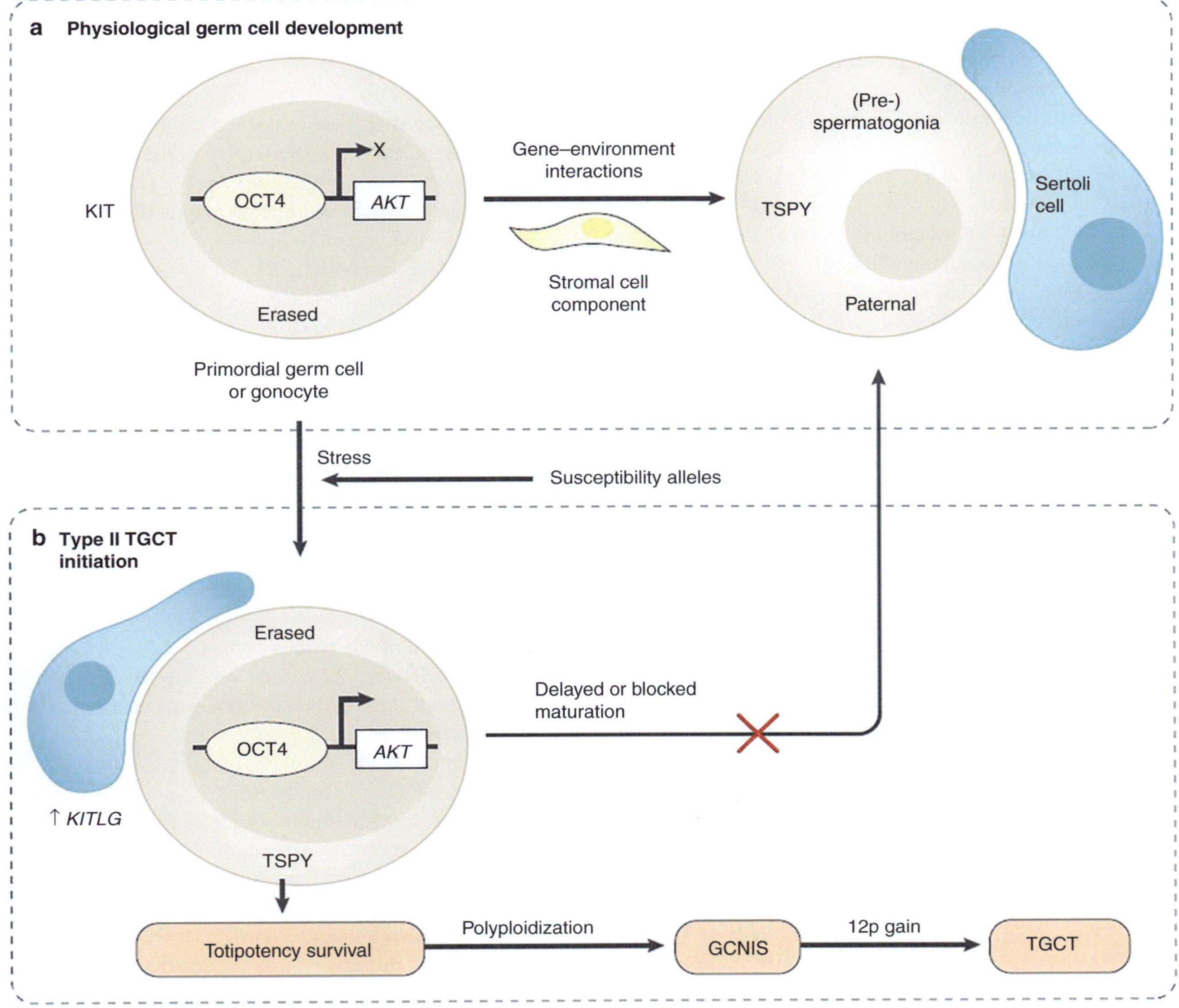

Fig. 16.17 (continued)

- Expression pattern of mRNA in seminoma and embryonal carcinoma shows high levels of mRNA of genes related to the pluripotency (OCT3/4, NANOG, and LIN28), in close similarity to the PGCs/gonocytes and embryonal stem cells
- In embryonal stem cells, pluripotency is regulated by interaction of POU5F1 (OCT3/4) with a member of the SOX family
- Studies in cell lines derived from seminoma and embryonal carcinoma provide evidence that the functional partner of POU5F1 in seminomas is SOX17 and in embryonal carcinomas SOX2

- The role of POU5F1/SOX17 complex is likely not in regulation of pluripotency but in prevention from apoptosis
- Two specific variants of the protein encoding OCT3/4 are recognized, of which the A (or I) type is a nuclear protein and is related to pluripotenc; the B (or II) variant is localized in the cytoplasm and is not related to regulation of pluripotency. Detection of OCT3/4 mRNA is hampered by existence of two variants as well as the presence of pseudogenes. This may result in false-positive RT-PCR observations in variety of non-GCTs
- Expression pattern of NANOG is similar to OCT3/4

- Involvement c-KIT (CD117)
  - Receptor tyrosine kinase c-KIT is expressed mainly in CIS and also, but less, in seminoma, but not in embryonal carcinoma
  - c-KIT is downregulated during the progression of CIS to seminomas
  - Amplification of c-KIT is found in a selected number of seminomas
  - Activating gene mutations in exon 17 are detected mostly in (bilateral) seminomas
  - It is postulated that activation of c-KIT in early germ cells during fetal phase of germ cell differentiation can potentially lead to survival of immature germ cells in the niche of spermatogonia (Fig. 16.17b)
  - It has also been proposed that c-KIT plays a role in the initiation of the germ cell malignancy but might not be relevant for the further steps of progression
- Mutational status
  - GCTs show overall an exceptional low mutation rate of genomic DNA
  - This is supported by mutation analysis of individual genes as well as by high-throughput investigation on the mutation status of the kinome
  - The uniqueness of this low mutation rate is likely to be (again) related to the embryonic cell of origin
  - Embryonic stem cells keep one of the two DNA strands protected against any form of mutations, the so-called immortal DNA strand. This reduces the probability of transmitting of the DNA anomalies to the next generation
  - BRAF and microsatellite instability
    - Overall, type II GCTs, with the exception of teratomas, show an exceptional sensitivity to DNA damaging agents. However, not all nonteratomatous elements are sensitive to cisplatin-based chemotherapy
    - Various putative mechanisms for resistance are proposed based on limited studies and number of cases. Intriguing findings are the role of microsatellite instability (MSI), BRAF mutations, disturbed apoptotic, etc
    - MSI is found in about 30% of the refractory cancers, in a significant number related to hypermethylation of the promoter region of hMLH1. Interestingly, this seems to be partially overlapping with activating mutations within the BRAF oncogene (V600E) in all patient groups

Type III Germ Cell Tumors (Spermatocytic Tumor)

- Clinical features
  - Formerly named "spermatocytic seminoma"
  - Rare tumor, up to 4% of all GCT of the testis
  - Incidence 0.4 per 1,000,000
  - Occur in older male compared to type II GCT, average age 52 years
  - Most tumors are unilateral
  - Bilateral tumors can occur
  - Serum markers (AFP, HCG, and LDH) are negative
  - Metastases from a pure spermatocytic seminoma are very rare
  - Sarcomatoid dedifferentiation rarely occurs within a spermatocytic seminoma and is associated with a progressive disease
  - No association with cryptorchidism
  - Excellent prognosis with surgery alone
- Gross, microscopic, and immunohistochemical features
  - Soft tumor with mucoid grayish-white cut surface, and friable texture
  - Background of edematous stroma
  - Tumor cells are of various size: large eosinophilic cells, small dark cells, and mono- or multinucleated giant cells (Fig. 16.18)
  - High mitotic activity
  - CIS/IGCNU/TIN is not present
  - Specific precursor lesions might be present (so-called intratubular spermatocytic seminoma in situ)
  - Immunohistochemical markers of classic seminoma (OCT3/4, PLAP, and c-KIT) are negative
  - Spermatogonial markers DMRT1, OCT2, SSX2–4, and SAGE1 are positive in spermatocytic seminoma (the last suggesting a heterogeneous origin) (Fig. 16.19)
  - In the histogenetic model, spermatocytic seminoma arise from mature spermatogonia or spermatocytes
- Molecular features
  - Lack isochromosome 12p (i12p)
  - Gain of chromosome 9 is consistent
  - DMRT1 is an interesting 9p gene
  - Paternal pattern of genomic imprinting
  - Many testis cancer antigens and genes related to spermatogenesis are expressed
  - HRAS and FGF3 genes are frequently mutated, especially in the cases in elderly men
  - The canine seminomas are the animal model for type III GCTs

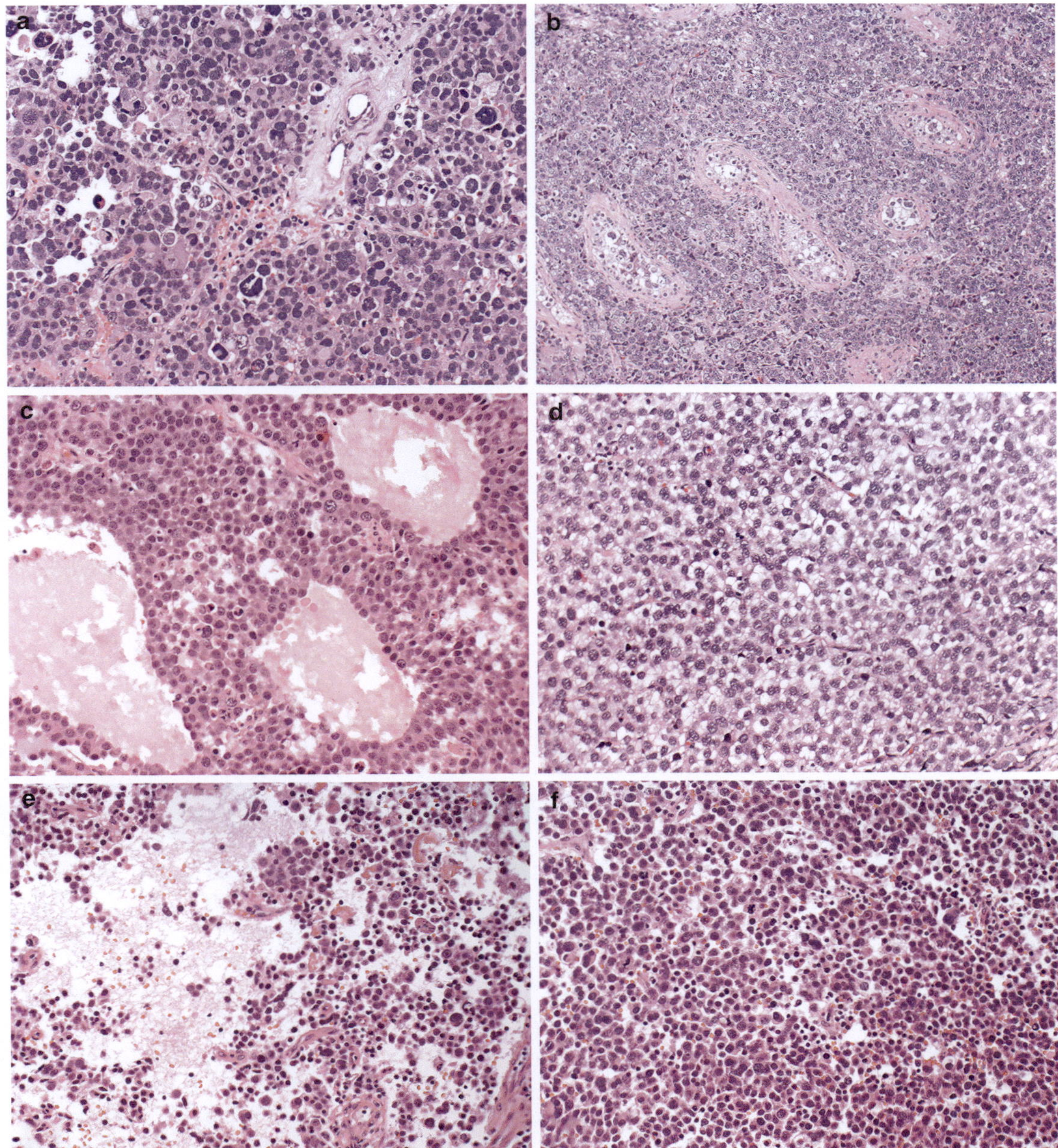

Fig. 16.18 Spermatocytic tumor. (**a**) The tumor is composed of a polymorphic tumor population. Note the presence of three cell types: small cells, intermediate-sized cells, and large giant cells. (**b**) Tumor infiltrates between seminiferous tubules. (**c**) Note the presence of pink eosinophilic materials in large cystic spaces. (**d**) This spermatocytic tumor shows solid growth with a uniform appearance of neoplastic germ cells with intermediate-sized nuclei. The lack of three distinct cell populations and the presence of clear cytoplasm in tumor cells may suggest the diagnosis of seminoma. (**e**) Clusters, nests, and single tumor cells are present in a tumor with extensive intercellular edema. Tumor cells have a plasmacytoid appearance. (**f**) Another case of spermatocytic tumor with discohesive growth. Cells have eosinophilic cytoplasm and occasional prominent nucleoli (adapted from Cheng L et al. Hum Pathol. 2017;59:10. Elsevier)

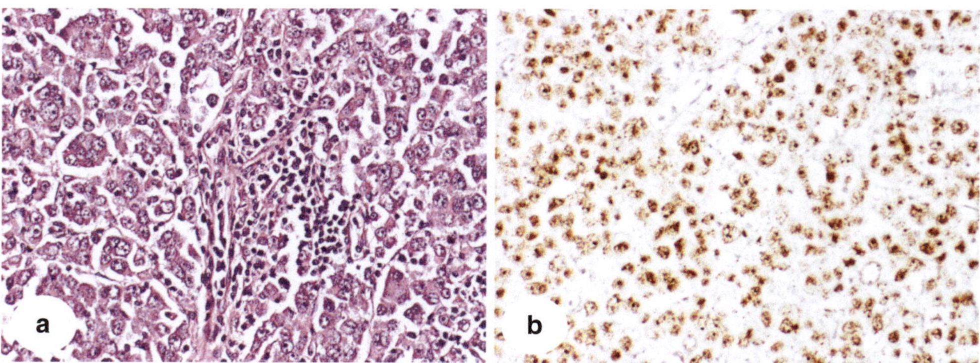

Fig. 16.19 Representative example of a spermatocytic tumor, stained with (**a**) H&E and (**b**) DMRT1

Selected Sex Cord-Stromal Tumors

- Sex cord tumors are rare; these tumors constitute about 4% of all testicular neoplasms
- Metastases can occur, mainly in adult patients, but no histological or molecular prognostic markers had been found so far to reliably predict the clinical behavior

Leydig Cell Tumors

- Clinical features
  - Two age peaks, in children mostly between 5 and 10, and in adults
  - Hormone-producing tumors: testosterone, androstenedione
  - Serum estrogen level might be elevated
  - In children, clinical presentation might be pubertas praecox or/and gynecomastia
  - Adults frequently present with testicular mass, 30% develop also gynecomastia
  - Bilaterality is rare
  - 10% of Leydig cell tumors are malignant
  - Malignant tumors do not respond to chemotherapy or radiation
- Gross, microscopic, and immunohistochemical features
  - Solid, yellowish tumor
  - Well circumscribed, sometimes nodular
  - Average tumor size 2–5 cm
  - Necrosis and extratesticular extension might occur
  - Typical histologic features include sheets of eosinophilic cells with distinct cell borders (Fig. 16.20)
  - Reinke crystals are pathognomonic and seen in up to 40% of the cases
  - Pseudoglandular and microcystic pattern can be seen
  - Lipomatous changes might occur
  - Malignant behavior correlates with the size (>5 cm), higher mitotic rate (>3 mitotic figures per 10 high power fields), necrosis, vascular invasion, invasion in the neighboring structures, and high proliferative activity
  - Tumors are positive for vimentin, inhibin, and LH-R

Sertoli Cell Tumors

- Clinical features
  - Very rare tumors (<1% of all testicular neoplasms)
  - Mean age at the time of the diagnosis 45 years
  - Some tumors are associated with genetic syndromes
  - Large cell calcifying Sertoli cell tumor (LCCST) can be sporadic or associated with Carney and Peutz–Jeghers syndromes
  - Estrogen production might lead to gynecomastia
  - Most Sertoli cell tumors are benign
- Gross, microscopic, and immunohistochemical features
  - Solid tumor, gray-whitish
  - Nests, tubules, sheets, mostly bland cytology (Fig. 16.21)
  - Frequently, stromal hyalinization is present
  - LCCST present with nests of large eosinophilic cells embedded with hyaline stroma (with calcifications in some cases)
  - In Peutz–Jeghers syndrome, multifocal bilateral Sertoli cell proliferation occur mostly in young patients
  - Immunohistochemically, Sertoli cell tumors are positive for inhibin, cytokeratin, vimentin, and SOX9

Granulosa Cell Tumors

- Clinical Features
  - In similarity to the ovary, juvenile and adult types are distinguished
  - Adult type is rare, metastases occur in up to 20% of the patients

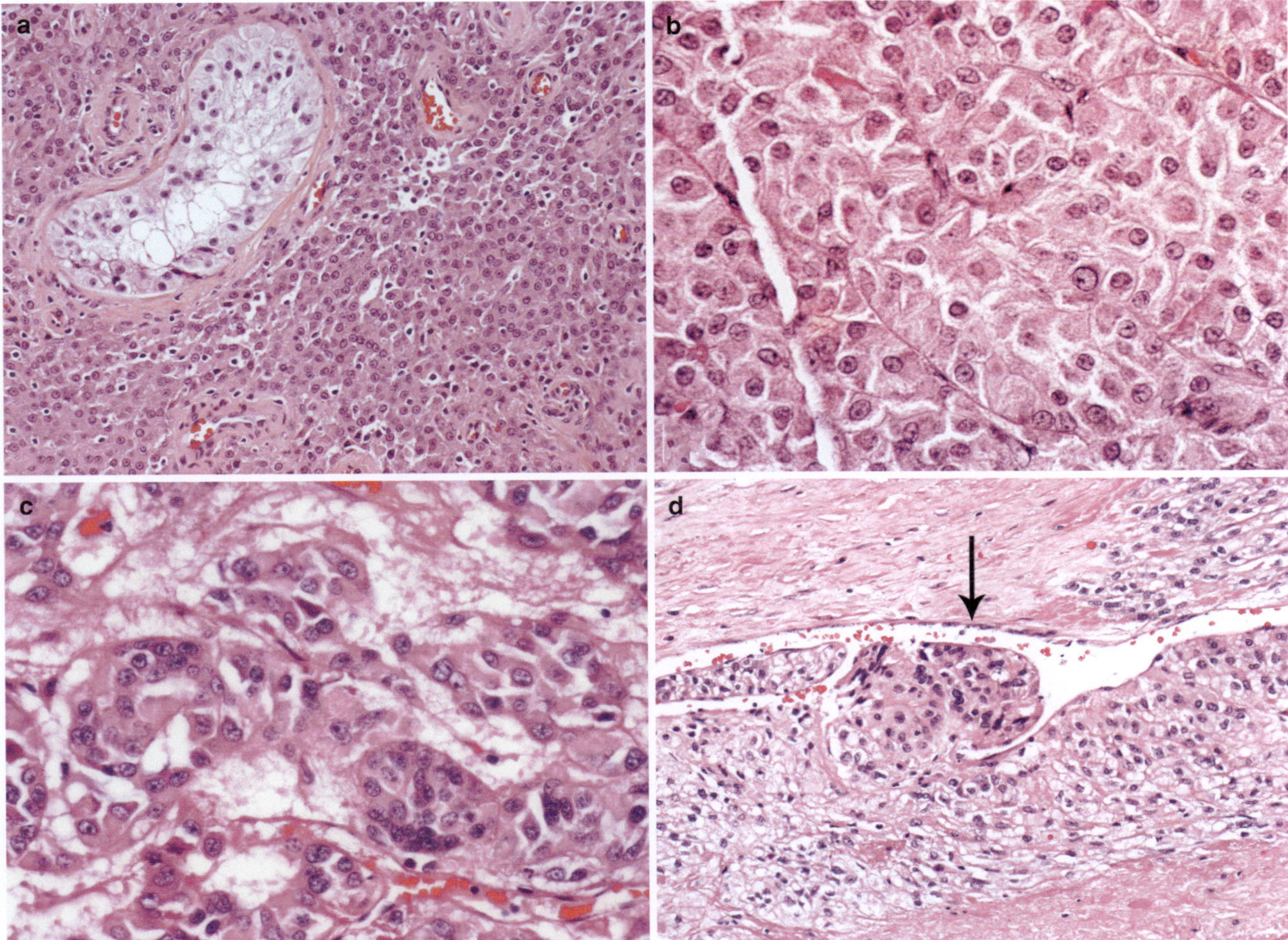

Fig. 16.20 Leydig cell tumor. (**a**) Histologically benign neoplasm is composed of polygonal cells with abundant eosinophilic cytoplasm surrounding an entrapped seminiferous tubule in the left upper part of the field (H&E, 200×). (**b**) The neoplastic cells have uniform bland nuclei with a small nucleolus and abundant eosinophilic cytoplasm (H&E, 600×). (**c**) Malignant Leydig cell tumor is composed of irregular clusters of neoplastic cells with enlarged pleomorphic nuclei and eosinophilic cytoplasm supported by a delicate vasculature (H&E, 400×). (**d**) Neoplasm extends into a thin-walled vascular space containing erythrocytes (arrow) (adapted from Roth L, Cheng L. Hum Pathol. 2017;65:1. Elsevier)

- Juvenile tumors occur mostly in maldescended testis, in young children
- Gross, microscopic, and immunohistochemical features
  - Adult type granulosa cell tumors
    - Solid and well circumscribed, with varying size
    - Several histological pattern can occur, including solid, trabecular, insular, and microfollicular
    - Typically, microfollicles surround eosinophilic material (Call–Exner bodies)
    - Tumor cells show grooved nuclei
    - Tumor cells are reactive for vimentin, inhibin, smooth muscle actin
  - Juvenile granulosa cell tumors
    - cell malignancy but might not be relevant for the further steps of progression
    - Often cystic
    - Solid and follicle-like zones are admixed (Fig. 16.22)
    - Follicles lined by stratified epithelium
    - Tumor cells are reactive for vimentin, inhibin, smooth muscle actin, and focally to anti-Müllerian hormone
- Molecular features of sex cord/gonadal stromal tumors
  - Activating mutations in LH-receptor had been detected in pediatric Leydig cell tumors
  - CTNNB1 genetic alterations are common in both testicular Leydig cell tumor and Sertoli cell tumor
  - Most ovarian adult-type granulosa cell tumors harbor a somatic missense mutation in the FOXL2 gene, but no information is available yet for the testicular granulosa cell tumors

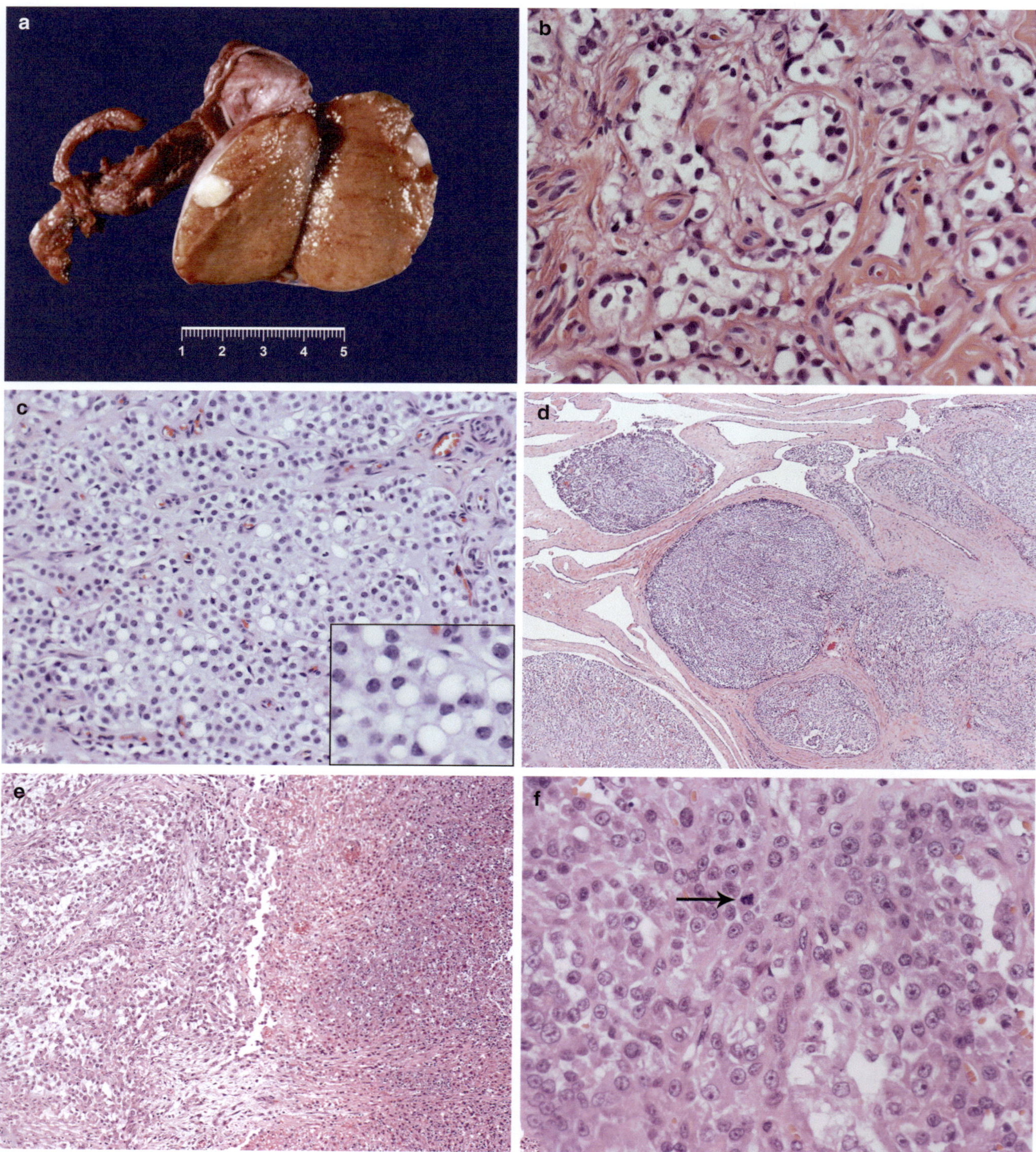

Fig. 16.21 Sertoli cell tumor, not otherwise specified (NOS). (**a**) The sectioned surface of the testis contains a small well-circumscribed ovoid white nodule near the tunica albuginea. (**b**) The neoplasm consists of irregular solid tubules within a fibrous stroma. The tumor cells have small, centrally located nuclei and finely vacuolated, lipid-rich cytoplasm (hematoxylin and eosin [H&E], original magnification, 400×). (**c**) Tumor cells have pale cytoplasm with cytoplasmic vacuolization due to lipid and occasional signet ring cell formation (H&E, 200×). Inset, Signet ring cell in the central portion of the field contains a large cytoplasmic vacuole that compresses a bland eccentrically located, crescent-shaped nucleus resembling a signet ring. (**d**) Malignant Sertoli cell tumor. Nodules of tumor infiltrate the rete testis (H&E, 40×). (**e**) The malignant Sertoli cell tumor has a diffuse growth pattern to the left with extensive tumor necrosis to the right (H&E, 100×). (**f**) The malignant Sertoli cell tumor shows moderate nuclear pleomorphism with enlarged nuclei and a distinct nucleolus. Note an abnormal mitotic figure in the upper central portion of the field (arrow) (adapted from Roth L, Cheng L. Hum Pathol. 2017;65:1. Elsevier)

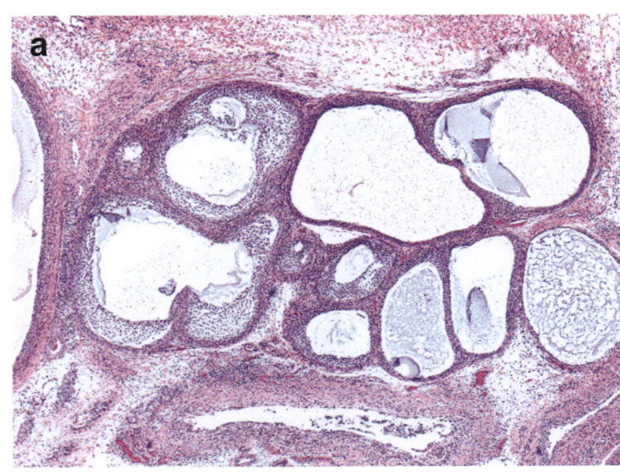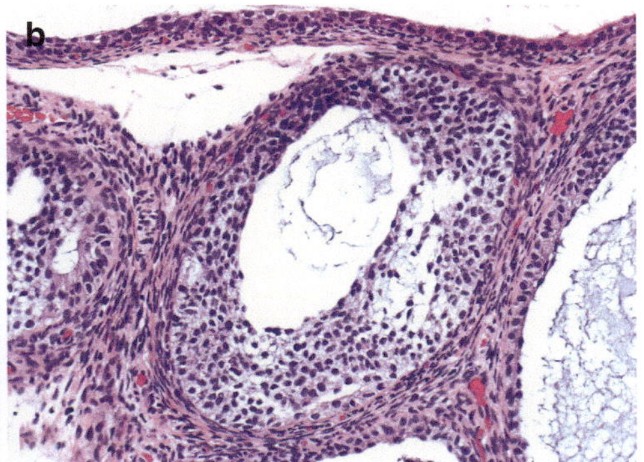

Fig. 16.22 Juvenile-type granulosa cell tumor. (**a**) Neoplasm forms confluent cystic follicles containing basophilic mucin (H&E, 100×). (**b**) The follicular structures are lined by layers of granulosa cells with small closely packed nuclei surrounded by neoplastic theca cells that have oval to elongated nuclei and indistinct cytoplasm. The granulosa cell nuclei are uniform and dense, lacking nuclear grooves (adapted from Roth L, Cheng L. Hum Pathol. 2017;65:1. Elsevier)

Table 16.8 Unmet needs in the management of germ cell tumors

Management area	Unmet need and; open questions	On-going and planned projects aimed at addressing unmet need
Risk factors	• Understanding the aetiology of the rising incidence of GCT, particularly among Hispanic men. • Understanding the genetics of familial GCT	• NCI registry (NCT00034424). • MSKCC registry (NCT02099734).
Mechanisms/ pathophysiology	• Improved understanding of the genetics and epigenetics underlying GCT development and cisplatin-sensitivity and -resistance	• The cancer genome atlas (TCGA) • Selective exon capture, whole exome & whole genome sequencing of GCTs as well as epigenetic studies
Stage I disease	• Identification of better risk factors for relapse following orchiectomy in nonseminoma and particularly in seminoma • Development of disease biomarkers that are more sensitive and specific than HCG, AFP, LDH, and CT scan • Avoiding overtreatment arising from use of pre-orchiectomy markers for decision-making and/or the false assumption that small retroperitoneal lymph nodes represent metastases	• National cohort studies in several European countries • Development of microRNAs as novel tumor markers for GCT • Dissemination of results of studies demonstrating superior outcomes at large volume centres and with adherence to guidelines
Advanced disease	• Improved risk stratification beyond the IGCCCG • Improved first-line treatment regimen for intermediate- and poor-risk patients beyond BEPx4 • Determination of the optimal second-line approach between conventional-dose and high-dose chemotherapy • Improved adherence to guidelines and referral to expert centres for difficult cases or specialized surgery (for example, RPLND) • Development of novel agents for relapsed or refractory disease	• IGCCCG-2 project • Ongoing studies of TIPx4 versus. BEPx4 (NCT01873326) and accelerated BEPx4 versus BEPx4 (NCT02582697) • TIGER study (A031102, NCT02375204) • Centralization of GCT care projects in some countries (for example, Germany, Norway, and Sweden) • Studies with immunotherapy combinations, the antibody-drug conjugate brentuximab, and the second generation taxane cabazitaxel
Survivorship	• Risk prediction models and identification of individual genomic risks for specific toxicities to allow the potential for tailoring treatment and for early intervention after treatment	• Platinum study • On-going survivorship research in many countries in North America and Europe

AFP alpha fetoprotein, *BEPx4* four cycles of bleomycin, etoposide and cisplatin, *GCT* germ cell tumor, *HCG* human chorionic gonadotropin, *IGCCG* International Germ Cell Collaborative Group, *MSKCC* Memorial Sloan Kettering Cancer Center, *NCI* National Cancer Institute, *RPLND* nerve-sparing retroperitoneal lymph node dissection, *TIPx4* four cycles of paclitaxel, ifosfamide, and cisplatin (adapted from Cheng L et al. Nat Rev. Dis Primers. 2018;4:29)

Summary

- Despite recent advances, there are still few number of unmet challenges in the managemenet of testicular cancer (Table 16.8)
- Neoplasms of the testis comprise a heterogeneous group of tumors, of which the majority is of the germ cell origin. GCTs of the testis can be classified by the cell of origin, as well as the molecular findings into three main groups
- Type I GCT is a common neoplasm in the neonates and children and arises from very early germ cells, which are blocked in their differentiation. The most common type I GCT is teratoma, which is a benign tumor. Yolk sac tumor

is a malignant type I GCT, developing from transformed early germ cells or, rarely, arises in a type I teratoma

- Type II GCT are malignant GCTs with complex morphology and different histological subtypes, all arising from a common precursor lesions, the carcinoma in situ
- Seminoma and embryonal carcinoma as well as carcinoma in situ share (partly) the same gene expression pattern and pattern of genomic imprinting as early fetal germ cells
- Transition from the precursor lesion to an invasive cancer is associated with gain of the short arm of chromosome 12, in which multiple genes might be involved
- Knowledge of cell of origin of type II GCT has led to the successful employment of OCT3/4 as a highly specific marker of CIS, seminoma, and embryonal carcinoma
- Type III GCTs represent a benign lesion composed of germ cells in the maturation stage of spermatogonia/spermatocyte. They consistently show gain of chromosome 9 and are positive for a number of immunohistochemical markers, including DMRT1

Suggested Reading

Almstrup K, Lobo J, Mørup N, et al. Application of miRNAs in the diagnosis and monitoring of testicular germ cell tumours. Nat Rev Urol. 2020;17:201–13.

Al-Obaidy KI, Chovanec M, Cheng L. Molecular characteristics of testicular germ cell tumors: pathogenesis and mechanisms of therapy resistance. Expert Rev Anticancer Ther. 2020;20:75–9.

Berney DM, Cree I, Rao V, et al. An introduction to the WHO 5th edition 2022 classification of testicular tumours. Histopathology. 2022;81:459–66.

Cheng L, Sung MT, Cossu-Rocca P, et al. LHJ: OCT4: biological functions and clinical applications as a marker of germ cell neoplasia. J Pathol. 2007;211:1–9.

Cheng L, Lyu B, Roth LM. Perspectives on testicular germ cell neoplasms. Hum Pathol. 2017;59:10–25.

Cheng L, Albers P, Berney DM, et al. Testicular Cancer. Nat Rev Dis Primers. 2018;4:29.

Cheng L, MacLennan GT, Bostwick DG, editors. Urologic surgical pathology. 4th ed. Philadelphia, PA: Elsevier; 2020.

Cheng L, Davidson DD, Montironi R, et al. Fluorescence in situ hybridization (FISH) detection of chromosomal 12p anomalies in testicular germ cell yumors. Methods Mol Biol. 2021a;2195:49–63.

Cheng L, Mann SA, Lopez-Beltran A, et al. Molecular characterization of testicular germ cell tumors using tissue microdissection. Methods Mol Biol. 2021b;2195:31–47.

Chovanec M, Cheng L. Molecular characterization of testicular germ cell tumors: chasing the underlying pathways. Future Oncol. 2019;15:227–9.

Chovanec M, Cheng L. Advances in diagnosis and treatment of testicular cancer. BMJ. 2022;379:e070499.

Emerson RE, Cheng L. Premalignancy of the testis and paratestis. Pathology. 2013;45:264–72.

Heijnsdijk EAM, Supit SJ, Looijenga LHJ, et al. Screening for cancers with a good prognosis: the case of testicular germ cell cancer. Cancer Med. 2021;10:2897–903.

King J, Adra N, Einhorn LH. Testicular cancer: biology to bedside. Cancer Res. 2021;81:5369–76.

Kozakova K, Mego M, Cheng L, et al. Promising novel therapies for relapsed and refractory testicular germ cell tumors. Expert Rev Anticancer Ther. 2021;21:53–69.

Leão R, Albersen M, Looijenga LHJ, et al. Circulating MicroRNAs, the next-generation serum biomarkers in testicular germ cell tumours: a systematic review. Eur Urol. 2021;80:456–66.

Lobo J, Gillis AJM, van den Berg A, et al. Identification and validation model for informative liquid biopsy-based microRNA biomarkers: insights from germ cell tumor in vitro, in vivo and patient-derived data. Cells. 2019;8:1637.

Lobo J, van Zogchel LMJ, Nuru MG, et al. Combining hypermethylated RASSF1A detection using ddPCR with miR-371a-3p testing: an improved panel of liquid biopsy biomarkers for testicular germ cell tumor patients. Cancers (Basel). 2021;13:5228.

Looijenga LH. Human testicular (non)seminomatous germ cell tumours: the clinical implications of recent pathobiological insights. J Pathol. 2009;218:146–62.

Looijenga LHJ, van der Kwast TH, Grignon D, et al. Report From the International Society of Urological Pathology (ISUP) Consultation Conference on Molecular Pathology of Urogenital Cancers: IV: Current and Future Utilization of Molecular-Genetic Tests for Testicular Germ Cell Tumors. Am J Surg Pathol. 2020;44:e66–79.

Mohanty SK, Lobo A, Cheng L. The 2022 revision of the World Health Organization classification of tumors of the urinary system and male genital organs: advances and challenges. Hum Pathol. 2023;136:123–43.

Oosterhuis JW, Looijenga LH. Testicular germ-cell tumours in a broader perspective. Nat Rev Cancer. 2005;5:210–22.

Oosterhuis JW, Looijenga LHJ. Human germ cell tumours from a developmental perspective. Nat Rev Cancer. 2019;19:522–37.

Roth LM, Cheng L. Mixed germ cell-sex cord stromal tumor of the testis with an intratubular component: a problem in differential diagnosis. Hum Pathol. 2016;51:51–6.

Roth LM, Cheng L. On the histogenesis of mixed germ cell-sex cord stromal tumour of the gonads. J Clin Pathol. 2017;70:22–227.

Roth LM, Cheng L. Gonadoblastoma: origin and outcome. Hum Pathol. 2020;100:47–53.

Roth LM, Lyu B, Cheng L. Perspectives on testicular sex cord-stromal tumors and those composed of both germ cells and sex cord-stromal derivatives with a comparison to corresponding ovarian neoplasms. Hum Pathol. 2017;65:1–14.

Shen H, Shih J, Hollern DP, et al. Integrated molecular characterization of testicular germ cell tumors. Cell Rep. 2018;23:3392–406.

Timmerman DM, Remmers TL, Hillenius S, Looijenga LHJ. Mechanisms of TP53 pathway inactivation in embryonic and somatic cells-relevance for understanding (Germ Cell) tumorigenesis. Int J Mol Sci. 2021a;22:5377.

Timmerman DM, Gillis AJM, Mego M, Looijenga LHJ. Comparative analyses of liquid-biopsy microRNA371a-3p isolation protocols for serum and plasma. Cancers (Basel). 2021b;13:4260.

Timmerman DM, Eleveld TF, Sriram S, et al. Chromosome 3p25.3 gain is associated with cisplatin resistance and is an independent predictor of poor outcome in male malignant germ cell tumors. J Clin Oncol. 2022;40:3077–87.

Molecular Pathology of Melanoma and Nonmelanoma Skin Tumors

Carlo De la Sancha, Amar Mirza, and Boris Bastian

Contents

C. De la Sancha · A. Mirza
Department of Pathology, University of California at San Francisco, San Francisco, CA, USA
e-mail: Carlo.DeLaSanchaVerduzco@ucsf.edu

B. Bastian (✉)
Helen Diller Family Comprehensive Cancer Center, Departments of Pathology and Laboratory Medicine, University of California at San Francisco, San Francisco, CA, USA
e-mail: Boris.Bastian@ucsf.edu

Introduction

- Approximately 9500 people in the United States are diagnosed with skin cancer every day
  - These estimates likely to be exceedingly conservative
  - Incidence rate dramatically rising
- Analogous trends have been observed worldwide, posing a considerable socioeconomic burden especially in Western countries
  - These numbers are more striking as incidence rates for most other cancers are either steady or decreasing
- Nonmelanoma skin cancer (NMSC), including basal cell carcinoma (BCC) and squamous cell carcinoma (SCC), accounts for up to 90% of all skin cancer cases
- Melanoma incidence is constantly on the rise
- Substantial scientific advances have been made in pathology, genetics, and basic research of skin cancer, dramatically altering the approach in diagnosing and treating both melanoma and NMSC
- Novel insights have been gained regarding inherited determinants of predisposition to skin neoplasms, mechanisms of ultraviolet light radiation (UVR)-induced oncogenesis, and molecular pathways regulating cutaneous cancer onset and progression; sizeable progress has been

L. Cheng et al. (eds.), *Molecular Surgical Pathology*, https://doi.org/10.1007/978-3-031-35118-1_17

made also in understanding the pathogenesis of less common cutaneous malignancies, including adnexal neoplasms, Merkel cell carcinoma (MCC), and dermatofibrosarcoma protuberans (DFSP)

- The once prevailing concept that cancer simply stems from the unregulated proliferation of a monoclonal and homogenous cell population is being replaced by a more complex notion, viewing tumors as heterogeneous collections of multiple subpopulations with different genetic background, molecular features, and responsiveness to external stimuli
- Established diagnostic classifications and prognostic schemes are being revisited considering recent molecular discoveries, new therapeutic agents are being developed targeting key drivers in skin cancer pathogenesis, and molecular diagnostics are becoming integral part of patients' management
- The introduction of next-generation genomics techniques and their rapid diffusion are destined to further revolutionize this field

Ultraviolet Radiation and Skin Cancer

- UVR exerts strong genotoxic effects; overwhelming epidemiological and molecular evidence point to solar UVR as a major pathogenic factor for both melanoma and non-melanoma skin cancer
- Based on radiation wavelength (λ), the UV spectrum is conventionally divided into UVA (λ320–400 nm), UVB (λ290–320 nm) and UVC (λ100–290 nm)
  - 95% of UVA and 5% of UBV wavelengths can reach the earth's surface
  - Depth of penetration of UVR into the skin is proportional to its wavelength: UVB does not penetrate past the basal layer of the epidermis, while UVA reaches into the papillary dermis and the superficial layers of the reticular dermis
  - With increasing wavelengths from UVB to UVA, the ability to excite and mutate the DNA molecule decreases exponentially. However, UVA is at least 50- to 100-fold more abundant in the natural sunlight and makes it a relevant mutagen for the skin
- Both UVB and UVA induce pyrimidine dimer DNA photoproducts that lead to UV-signature mutations
  - The two main photoproducts are cis-syn cyclobutane pyrimidine dimers and pyrimidine (6–4) pyrimidine photoproducts (6–4 PPs), which can be formed between any combination of thymine and cytosine: T:T, T:C, C:T, or C:T

- Pyrimidine dimer photoproducts are usually removed by DNA nucleotide excision repair (NER) mechanisms
  - Inherited genetic defects in the NER pathway such as xeroderma pigmentosum and Cockayne syndrome leads to UVR hypersensitivity and dramatically increased incidence of skin cancers
- Apoptosis in response to UV-induced damage also protects against photocarcinogenesis as it removes heavily damaged cells that may undergo malignant transformation
  - Melanocytes are relatively resistant to apoptosis, in contrast to keratinocytes
- If not removed prior to entering the S-phase of the cell cycle for DNA replication, cytosine bases in a pyrimidine dimer may undergo deamination to uracil or thymine. Translesional polymerase may then pair an adenine opposite to the deaminated C (U or T). Consequently, when NER eventually repairs the pyrimidine dimer, the introduced mutation will form the canonical UV light-induced C→T and, less commonly, C:C→T:T tandem mutations
- This distinctive spectrum of mutations, when occurring at a relatively increase rate compared to other base pair changes, are collectively called "UV signature" (>60% of C→T, or >5% of C:C→T:T transitions comprising the total mutation burden)

- UVA also damages DNA through indirect mechanisms. Excitation of endogenous chromophores such as melanin results in production of ROS and subsequent formation of oxidized DNA bases such as 8-oxo-7, 8-dihydro-2'-deoxyguanosine (8-oxo-dG); 8-oxo-dG may lead to G–T transversions and, to a lesser degree, G–A transitions
- UVR further promotes carcinogenesis through modulation in the immune system
  - Longstanding UVR exposure can result in a chronic proinflammatory state mediated by proinflammatory cytokines (i.e., IL-1 and TNF α, ROS and RNS intermediates, PAF, and eicosanoids (such as PGE2)
  - Immunosuppression can lead to tolerance to antigens exposed to the UVR-exposed skin, including UVR-induced melanoma antigens
    - Reduction of the number of Langerhans cells and their antigen-presenting ability
    - Th1 to Th2 shift in the adaptive immunity mediated by release of IL-4 and IL10, as well as downregulation of IL2
    - Production of cis-uronic acid, PAF, and PGE2
- UVR and NMSC

- Incidence of BCC, SCC precursors, and SCC strongly correlates with exposure to natural (solar) and artificial (indoor tanning) UVR
  - UVR is associated with ~90% of NMSC
  - NMSC show frequent mutations of TP53, which harbor UV signatures
- UV radiation may induce NMSC acting both as initiator and promoter
  - Keratinocytes can be initiated by UVR-induced TP53 mutations, which interfere with the induction of DNA repair, cell cycle arrest, and elimination of keratinocytes with irreparable DNA-damage by apoptosis. This promotes cancer progression by providing a survival advantage and by increasing the acquisition of subsequent mutations in response to UVR exposure
  - Critical oncogenic mutations induced by UV radiation involve NOTCH family members and HRAS in SCC, as well as PTCH1 in BCC
- UVR and Melanoma
  - More than 75% of all mutations in melanoma are UV-signature mutations. They are found in many mutated genes involved in melanoma development such as CDKN2A, PTEN, TP53, and TERT mutations
  - Epidemiologic and molecular data link UVR to different pathways of melanoma arising in sun-exposed skin: pathway I, II and III (described later)
- Melanin and photoprotection
  - Production of melanin by skin melanocytes represents the predominant mechanism of photoprotection in the skin
    - Levels of skin pigmentation correlate with melanoma and NMSC risk
  - Tyrosinase-mediated conversion of tyrosin to DOPA-quinone leads to two types of melanin:
    - Eumelanin (brown/black) shows high stability and exerts strong photoprotective effects by transforming UVR into heat through internal conversion, quenching ROS
    - Pheomelanin (yellow/red) is a weak photoprotective, has less stability and can be phototoxic by amplifying the UVR-induce production of ROS
    - Oxidation of guanine by ROS produces 8-oxoG, which in turn can pair with an adenine and eventually lead to a G→T mutation upon eventual base excision repair of the oxidized base
  - Given the overwhelming role of UVR in both melanoma and NMSC, photoprotection can potentially prevent a large fraction of these cancers
    - Important measures include use of sunscreen with a SPF of 15 or higher (against both UVA, and UVB), avoidance of excessive outdoor UVR (avoiding sun seeking, using protective clothing, and shade seek-ing), and avoidance of indoor UVR (tanning devices)
    - Regular self-examination, together with education regarding sun protection and whole body skin examination by a clinician in the presence of any change or emergence of new lesions
    - Other skin cancer prevention interventions include mass media campaigns, as well as environmental and legislative interventions

Melanoma

Overview

- Melanoma is a malignant neoplasm originating from melanocytes of the skin, mucosa, uvea, or very rarely, other tissues
- Melanoma is the fifth most common cancer type in the United States
  - Estimated new cases of invasive cutaneous melanoma in 2022: 99,780
  - Estimated deaths of invasive cutaneous melanoma in 2022: 7650
  - Approximately 2.3% of men and women will be diagnosed with cutaneous melanoma at some point during their lifetime
  - Cutaneous melanoma is most frequently diagnosed among people aged 65–74, with a median age at diagnosis of 65 years
- The age adjusted incidence rates for new cutaneous melanoma cases have been rising on average 1.4% each year from 2009 to 2018, while the age-adjusted death rates have been falling on average 3.2% each year over 2010–2019
  - Possible reasons of this rising trend include sun exposure behavior, increases in the number of biopsies and changes in diagnostic criteria
- Melanoma is more common in men than women and among individuals of fair complexion and correlates to exposure to UV radiation from sunlight or artificial sources
  - There are more new cases among whites, than in any other racial group
- The overall 5-year relative survival depends on the stage:
  - 99.3% for localized, 68% for regional, and 29.8% for metastatic disease

Pathogenesis

- Melanomas are comprised of multiple biologically distinct subtypes that vary in clinical and histologic presentation, epidemiology, age of onset, causative role of ultraviolet radiation (UVR), mutational processes that shape their cancer genomes and likely the cell of origin from which neoplasms arise

- Melanocytic neoplasms typically are initiated by gain-of-function alterations generated by point mutations, gene fusions, or gene amplifications that lead to constitutive activation of growth-promoting signaling pathways. This leads to a usually transient proliferation of the partially transformed melanocyte, which is kept in check by tumor suppressor mechanisms such as oncogene-induced senescence. These mechanisms can be overridden by subsequent acquisition of additional mutations that can promote their progression to intermediate or overtly malignant neoplasms. These alterations include loss of suppressor function through inactivating mutations, deletions, or epigenetic silencing, followed by activation of additional growth and survival-related genes
- The pathogenesis of melanoma varies according to the occurrence of specific genetic changes that drive each distinct neoplastic process. The mutational process of melanoma includes
  - Point mutations
    - Cutaneous melanoma has a high point mutation burden (~15 mutations per Mbp)
    - 70–80% are canonical UV-signature mutations (C→T transitions with a preceding pyrimidine base)
  - Structural alterations
    - Copy number alterations (CNA) accumulate during the progression of melanoma. They are rare in benign melanocytic neoplasms. Common alterations in melanoma include
      - Gains of chromosomes 1q, 5p, 6p, 7, 8q, 11q, 17q, and 20
      - Losses of chromosomes 6q, 8p, 9p, 9q, 10q, and 11q
      - A cut-off of ≥3 CNVs has demonstrated 85% sensitivity and 84% specificity to differentiate intermediate lesions from melanoma
  - Chromosomal rearrangements leading to constitutively active fusion proteins, or chromothripsis
- Important signaling pathways involved in the pathogenesis of melanoma (Fig. 17.1)

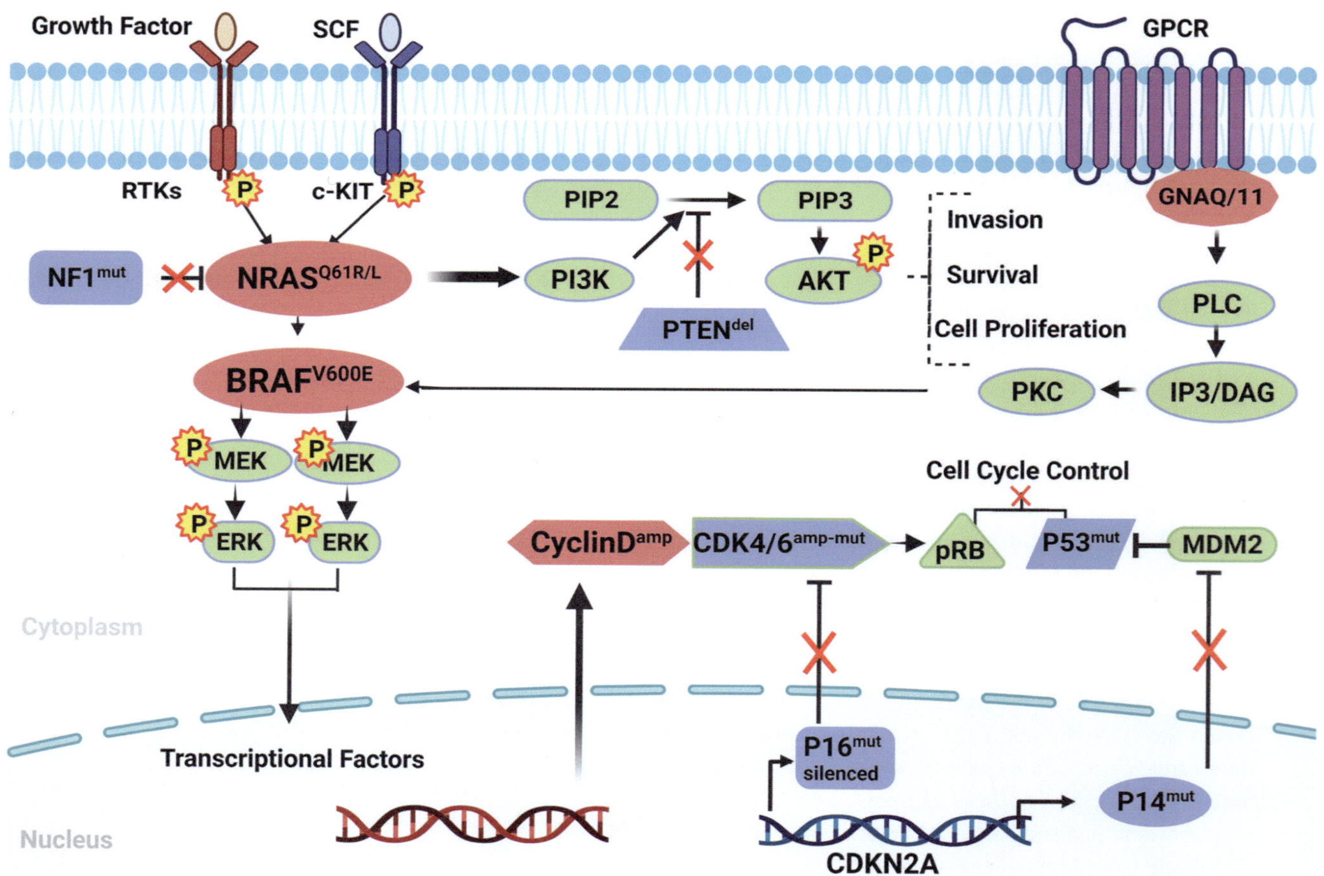

Fig. 17.1 Important signaling pathways of melanoma genesis and progression. Created with BioRender.com

- MAP/ERK pathway
  - Required for melanocyte proliferation and survival
  - Upon binding of growth factors to their receptor tyrosine kinases, activated GTP-bound RAS recruits RAFs to the plasma membrane, where they initiate a phosphorylation cascade that activates MEKs, which in turn activates ERK1/2. ERK1/2 translocates to the nucleus and controls transcriptional events involved in G1-S cell cycle progression
  - Includes BRAF, NRAS, and the tumor suppressor NF1
    - The mutations result in constitutive activation of the MAP/ERK pathway independent of physiologic receptor stimulation
    - BRAF V600E accounts for the majority of mutations, followed by BRAF V600K
    - Cutaneous melanomas show BRAF V600 hotspot mutations in 50%, NRAS mutations in 25%, and NF1 loss in ~10% of cases
    - Low-CSD melanomas are associated with BRAF V600E mutations, while NRAS, NF1, and nonV600E BRAF mutations are associated with high-CSD melanomas
    - RAS mutations result in constitutive activation of the MAP/ERK pathway as well as the PI3K/AKT pathway
    - NRAS (Q61, G12, G13) are the most common RAS mutations, followed by KRAS and HRAS (<5% of RAS-mutated melanomas)
    - RAS mutations impair GTP hydrolysis and result in an active, GTP-bound, protein that is constitutively activate
    - NF1 encodes a GTPase activating protein (GAP) that acts as a negative regulator of RAS by increasing its GTPase activity
- PI3K/AKT Pathway
  - Promotes cell proliferation, metabolism, motility, and cell survival
  - PI3K promotes the phosphorylation of PIP2 to PIP3, which recruits AKT to the plasma membrane and facilitates its phosphorylation and activation by PDK1. Once activated, AKT regulates proteins involved in cell growth, proliferation, motility, and adhesion
    - PTEN is a tumor suppressor that dephosphorylates PIP3 to PIP2, preventing the activation of AKT
    - Loss of PTEN has been strongly linked to progression of melanoma, and is seen in ~10–30% of cutaneous melanomas

- CDK4/RB pathway (CDKN2A)
  - Includes the tumor suppressor genes p16 (INK4A) and p14 (ARF), encoded by the CDKN2A locus
  - P16 protein inhibits CDK4 and CDK6, which normally phosphorylate the RB protein to promote the progression from G1 to S phase in the cell cycle
  - P14 prevents MDM2 from degrading p53, a key regulator of apoptosis, and cell cycle control
  - CDKN2A inactivation is linked to progression of melanoma and is present in up to 90% of all melanomas
  - CDKN2A is the most common altered gene in familial melanoma, present in ~1/3 of cases
  - CDK4 mutations or amplifications can be detected in some melanomas
- Gαq Pathway
  - GNAQ and GNA11 encode $G\alpha_q$ and $G\alpha_{11}$, alpha subunits of heterotrimeric G-proteins, acting downstream of G-protein-coupled receptors (GPCRs). Following receptor stimulation, they activate downstream effector such as phospholipase C β (PLCβ), which in turn cleaves phosphatidylinositol 4, 5-bisphosphate (PIP2) into diacylglycerol (DAG) and inositol 1, 4, 5-trisphosphate (IP3). These second messengers ultimately activate members of the protein kinase C (PKC) family and the MAPK/ERK pathway
  - Mutations in the Gαq pathway affect GNAQ and GNA11, and less commonly CYSLTR2 (a GPCR) or PLCB4 (encoding a homolog of phospholipase C β); they are mutually exclusive leading to constitutive activation
    - They are seen in virtually all uveal melanomas and melanomas arising in blue nevi
- TERT is a catalytic subunit of the enzyme telomerase, which lengthens telomeres
  - TERT Promotor mutations occur in approximately 70% of cutaneous, 30–41% of acral and 8% of mucosal melanomas, and are associated with poor prognosis
  - TERT amplifications are found in up to 50% of acral melanoma and 22% of mucosal melanomas
- KIT encodes the c-KIT receptor tyrosin kinase, which is involved in signaling pathways including the MAPK/ERK pathway, PI3K/AKT pathway, and MITF upregulation in melanocytes
  - KIT mutations or amplifications are seen in 39% of mucosal, 36% of acral, and 28% of melanomas on chronically sun-damaged skin

- BAP1 is a deubiquitinase that acts on histone H2A and is involved in chromatin modification and transcriptional regulation. It is a tumor suppressor gene and inactivation of BAP1 can occur because of deletions or loss of function mutations
  - o BAP1 mutations occur as secondary driver mutations in uveal melanomas (up to 40%) and melanomas arising in blue nevi (up to 17%)
  - o The autosomal dominant BAP1 tumor predisposition syndrome (BAP1-TPDS) is caused by germline mutations in BAP1 and it increases the risk of developing various tumors including BAP1 inactivated melanocytomas, uveal and cutaneous melanomas, among others (described in more detail below)
- Chromatin remodeling complexes: SWI/SNF and PRC
  - o The SWI/SNF or BAF complex is an ATP-dependent chromatin remodeling complex that regulates transcriptional activity by directly binding to the DNA and nucleosomes
    - ◆ Three SWI/SNF complexes have been identified: BAF, pBAF, and ncBAF. They all contain a catalytic ATPase subunit (SMARCA2 or SMARC4), exclusive subunits (BRM, ARID1A, and ARID1B in BAF; PBRM1, ARID2, and BRD7 in pBAF; GLTSCR1 and BRD9 in ncBAF), and share several associated proteins
    - ◆ The three complexes interact with various enhancers and promoters in a cell-type specific manner, which allows for a higher degree of specificity and regulation
    - ◆ Loss of function mutations in genes encoding SWI/SNF subunits are encountered in 20–25% of human cancers, and in over 30% of melanomas. ARID2 and SMARCA4 mutations have been found in 12% and in 5–10% of melanomas, respectively
    - ◆ Other mutations have been found in ARID1A, ARID1B, SMARCA2, and SMARCB1
  - o PRC (polycomb repressive complex) is a chromatin remodeling complex that mediates repressive histone modifications; PRC1 complex mediates histone ubiquitination and PRC2 mediates histone methylation
    - ◆ BMI1 is a component of PRC1, and a regulator of P16 and P14; it has found to be highly expressed in benign nevi, and lost during melanoma development
    - ◆ EZH2 is a catalytic subunit of PRC2 that can alter gene expression by trimethylation of Lys-27 in histone 3 (H3K27me3). Its overexpression is associated with disruption of tumor suppressive pathways (loss of P16 and increased

cyclin D1), melanoma progression, invasion, and metastatic disease
    - ◆ EZH2 activation by mutations, gene amplification and increased transcription have been found in 20% of melanoma patients
    - ◆ KDM5B is an H3K4 demethylase that promotes tumor immune evasion
      - ▪ In mouse melanoma models, depletion of KDM5B induces robust adaptive immune responses and enhances responses to immune checkpoint blockade
- RAC1
  - o RAC1 is a small Rho GTPase, whose downstream effectors include PAKs (PAK1-3), PI3K-β, AKT and WAVE2
  - o RAC1P29S mutations have been found in 4–9% of sun-exposed skin melanoma; it is an UVR-induced C>T mutant that activates the biological activity of RAC1 by fast-cycling of the enzyme between inactive GDP and active GTP states
  - o RAC1P29S promotes melanomagenesis driven by mutant BRAF or NF1 loss; it activates PAK, AKT, and the WAVE→ARP2/3→ SRF/MRTF cascade, which induces a melanocytic to mesenchymal transition to promote resistance of apoptosis and tumorigenesis
  - o RAC1P29S melanoma cells evade immune response by enhancing PDL1 expression and induce resistance to BRAF inhibitors through SRF/MRTF
- Genetic predisposition to melanoma involves several recognized high-penetrance, as well as low to moderate-penetrance susceptibility genes
  - High-penetrance genes include CDKN2A (accounts for 20–40% of high-risk families), CDK4, BAP1, and several telomerase genes (TERT, POT1, ACD, and TERF21P)
  - Low- to moderate-penetrance genes confer a smaller risk, but their combination with other genetic factors contribute to the risk of developing melanoma:
    - o MC1R is a G protein coupled receptor that regulates constitutional skin pigmentation and UV response
      - ◆ It increases eumelanin synthesis, and the ratio of eumelanin to pheomelanin, as well as enhances nucleotide excision repair
      - ◆ It is a highly polymorphic gene, with loss of function variants leading to red hair and light skin tone, poor tanning response, associated with melanoma risk
    - o MITF is a transcription factor that controls melanocyte development and pigmentation. It acts downstream of MC1R and increases expression of

multiple factors involved in melanin synthesis, including tyrosinase

- ◆ Germline mutation of MITF (E318K) have been reported in small numbers of melanoma families
- – Other less penetrant genes are HERC2/OCA2, TYR, TYRP1, SLC45A2, and ASIP

Diagnosis

- The gold standard for melanoma diagnosis is histopathology
  - – Although microscopic features vary according to the melanoma type, criteria for diagnosis include a combination of asymmetry, poor circumscription, ulceration, lack of maturation of neoplastic melanocytes, large confluent sheets of melanocytes, increased dermal mitotic activity, and cytologic atypia
- Most melanomas arise through stages of tumor progression, starting as a patch or plaque-like lesion involving the epidermis and/or superficial dermis (radial growth phase), followed by a deeper invasion of the dermis with the formation of a tumor nodule (vertical growth phase)
  - – Melanomas in vertical growth phase without an adjacent radial growth phase component are called nodular melanoma
- Ancillary tests can assist in the diagnosis of histopathologically ambiguous tumors and as prognostic tools in melanoma
  - – Immunohistochemistry with antibodies against S100, SOX10, Melan-A, HMB45, tyrosinase and MITF can be used to assess for melanocytic differentiation
    - ○ SOX10 and S100 have high sensitivity but lack specificity
    - ○ Melan-A and HMB45 are specific, but frequently negative for poorly differentiated melanocytic tumors
    - ○ HMB45 can be used to assess for dermal maturation of melanocytes
    - ○ Proliferation markers such as Ki-67 and H3 phospho-histone are increased in melanomas compared to nevi
    - ○ Loss of tumor suppressor genes such as p16 can provide insight into a heterozygous or homozygous loss of CDKN2A
    - ○ PRAME (preferentially expressed antigen in melanoma) is a tumor-associated antigen upregulated in melanomas, as well as other nonmelanocytic malignant neoplasms, including leukemia, synovial sarcoma, myxoid liposarcoma, NSCLC, breast, renal and ovarian carcinomas; normal tissues are not known to express PRAME, except for testis, ovary, placenta, adrenal, and endometrium
      - ◆ Diffuse nuclear PRAME expression (>75% of tumor cells) has been found in 87% of metastatic and 83.2% of primary melanomas
      - ◆ Sensitivity ranges from high in superficial spreading, lentigo maligna, acral, and nodular types, and low for desmoplastic melanomas (35% positivity)
      - ◆ ~15% of melanocytic nevi show focal positivity for PRAME
    - ○ Loss of expression of nuclear BAP1 can detect most forms of bi-allelic BAP1 inactivation
    - ○ Fusion oncogenes that can be detected by IHC include ALK, ROS, and NTRK
  - – Melanomas tend to have multiple DNA copy number change, whereas nevi do not show such aberrations
    - ○ Copy number alterations can be detected by genomewide approaches such as comparative genomic hybridization (CGH), SNP arrays and hybrid-capture based methods of next-generation sequencing (NGS)
    - ○ Commonly recurring copy number changes such as losses of chromosomes 9p21, 6q, and gains of 6p, 11q13, and 8q can be tested by fluorescence in situ hybridization (FISH) with probe sets demonstrating 94% sensitivity and 98% specificity for distinguishing nevi from melanoma
  - – NGS can provide information on activated oncogenes, inactivated tumor suppressor genes, copy number alterations and chromosomal rearrangements
    - ○ Detection of actionable mutations aids in guiding treatment, while information on mutational burden provides some estimate of the likelihood of response to immune checkpoint blockade therapy
  - – RNA-based gene expression analysis assesses mRNA levels in selected genes
    - ○ Commercial tests are available that aid in distinguishing unequivocal nevi from melanomas, but independent studies are needed to assess their clinical utility

Classification

- The first classification of melanoma proposed by Clark et al in 1973 used clinical and histomorphologic features to distinguish 4 major classes of cutaneous melanoma: superficial spreading melanoma (SSM), lentigo maligna melanoma (LMM), nodular melanoma (NM), and acro-lentiginous melanoma (ALM)
- The Cancer Genome Atlas network established in 2015 a genomic classification of cutaneous melanoma based on

significantly mutated genes: BRAF, RAS, NF1, and Triple Wild-type subtypes
- BRAF subtype: Largest genomic subtype defined by the presence of BRAF hot-spot mutations (52%). Most mutations involved V600E, followed by V600K, and V600R
  ○ The second most common was BRAF K601
- RAS subtype: Second most common subtype defined by the presence of RAS hot-spot mutations (28%)
  ○ NRAS were the most common mutations (Q61R, followed by Q61K, Q61L, and Q61H) Less frequent mutations were encountered in HRAS and KRAS
- NF1 subtype: Third most frequent subtype seen in 14%
- Triple wild-type subtype: Heterogeneous subgroup characterized by lack of hot-spot BRAF, N/H/K-RAS or NF1 mutations
  ○ Important genes in this category are GNAQ, GNA11, KIT, CTNNB1 and EZH2
- The WHO Classification of Skin tumors, 4th edition in 2018 is based on a proposal by Bastian in 2014, who integrated epidemiologic, clinical, histopathologic, and molecular features into multiple "pathways" which group different melanoma subtypes with their respective precursor lesions. The WHO currently distinguishes two groups and nine different pathways as described below

Melanomas Arising in Sun-Exposed Skin

Pathway I: Low-Cumulative Solar Damage (Low-CSD)/SSM

Epidemiology
- Most common form of melanoma in western countries; in people with fair skin, and poor tanning ability
- Associated with intermittent sun exposure and histologically with a low to moderate degree of elastosis, a marker of CSD
- Risk factors include UVR exposure from the sun and tanning beds
- Additional risk factors include total number of nevi, large size of nevi, and atypical/dysplastic nevi
- Peak incidence: 3rd to 6th decade

Clinical Features
- Occurs in intermittently sun exposed of the skin including trunk and extremities (except glabrous sites and nail apparatus), and bulbar conjunctiva, and is more common in patients with acquired nevi

- Commonest location in men is the back while in women is the back of the legs or calf region
- Presents as irregularly pigmented, asymmetrical patches or nodules, with irregular borders; can have multiple shades of brown, red, blue, black, gray, and white
- Often associated with a precursor nevus
- Can initially present as lymph node metastasis (melanoma of unknown primary)

Histopathology
- Absence of marked solar elastosis (grade I or II CSD)
- Enlarged, hyperpigmented tumor cells of round morphology arranged predominantly in nests, rather than single cells, along the dermal-epidermal junction
- Frequent pagetoid spread
- Frequent dusty pigmentation of neoplastic melanocytes
- Areas of partial or complete tumor regression within the dermis may be present

Precursor Lesions
- Benign melanocytic nevi
  - Melanocytic nevi are benign neoplasms defined by having only one pathogenic mutation (usually BRAF). Most are stable and will never progress
    ○ However, they may potentially acquire secondary pathogenic mutations and act as precursors of some melanomas
  - Characterized by a well circumscribed, symmetrical lesion with melanocytes arranged in regular clusters
    ○ If a dermal component is present, melanocytes mature toward the base of the nevus
    ○ There is lack of significant atypia and mitotic figures are generally absent
- Melanocytomas
  - BAP1 inactivated melanocytoma
    ○ Intermediate tumor characterized by the presence of epithelioid melanocytes with varying size and pleomorphism, and amphophilic cytoplasm, resembling those of Spitz tumors (ST), usually within a conventional nevus background
    ○ Epidermal thinning, often with attenuated rete ridge pattern (as opposed to ST)
    ○ Initial driving mutations: BRAF or NRAS + BAP1 biallelic inactivation (somatic and/or germline)
    ○ Rarely evolves to melanoma (melanoma in BIN)
  - Deep penetrating nevi
    ○ Intermediate tumor characterized by a wedge-shaped lesion extending to deep dermis and sometimes subcutaneous adipose tissue without maturation of neoplastic melanocytes
    ○ Melanocytes can have epithelioid to round to fusiform nuclei

- o Initial driving mutations: BRAF, MAP2K1, or NRAS mutations + mutations activating the WNT pathway: CTNNB1 (gain of function) or APC (loss of function)
  - o High-grade lesions are referred to as deep penetrating melanocytoma/MELTUMP
  - o A small fraction evolves to melanoma through the acquisition of additional mutations, including TERT promoter mutations
  - Pigmented epithelioid melanocytoma (PEM)
    - o Intermediate tumor characterized by a heavily pigmented, dome-shaped nodule composed of medium sized epithelioid cells, large epithelioid cells, and spindled cells
    - o Initial driving mutations: BRAF V600E + biallelic inactivation of PRKAR1A tumor suppressor gene or PRKCA gene rearrangement
    - o Rarely evolves to melanoma (melanoma in PEM)
- Melanoma in situ
  - Intermediate/high-grade dysplasia melanocytic lesion defined by the acquisition of additional pathogenic mutations (see below)
  - Asymmetric, poorly circumscribed lesion with lateral expansion of large nests of atypical melanocytes arranged irregularly along the DEJ junction; typically, abundant pagetoid spread of atypical melanocytes

Molecular Features

- Initiating driving mutations include BRAF V600E in 70%, and NRAS in 15%
- Secondary oncogenic mutations include TERT promoter mutations, additional activating mutations of the MAP-kinase pathway, and inactivation of CDKN2A, PTEN and TP53
  - TERT promotor mutations emerge early and can already be seen in melanoma in situ and some dysplastic nevi
  - Copy number increase of mutant BRAF, additional activating mutations in MEK genes and other MAP-kinase pathway mutations arise in primary melanomas
  - CDKN2A biallelic inactivation usually marks the transition to invasive melanoma
  - PTEN and TP53 are seen in advanced primary melanomas
- Copy-number alterations are seen in intermediate and in situ lesions and continue to accumulate during the progression to invasive and metastatic melanomas
  - Copy number gain of chromosome 1q, 6p, 7, 8q, 17q, and 20q
  - Loss of 9p, 10, and 21q

- Tumor heterogeneity is observed in primary melanomas reflecting branching evolution of competing subclones

Pathway II: High-CSD Melanoma/LMM

Epidemiology

- Second most common type of melanomas in western countries
- Associated with high cumulative exposure of UV radiation
  - Incidence increased in heavily sun-exposed population including outdoor workers
- Not related to increased number of acquired nevi
- Tend to affect older individuals (7th decade and older)
- More common in patients with non-melanoma skin cancers

Clinical Features

- Irregularly pigmented, asymmetrical patch or nodule in chronically sun exposed skin of the head, neck, lower arms, and lower legs
- Pigmentation is usually less than SSM, with some lesions being amelanotic, which can delay diagnosis
- Satellite or in-transit, lymph node, and distant metastases are seen in equal proportions

Histopathology

- Marked solar elastosis (grade III CSD)
- Lentiginous proliferation of single cells along the dermal-epidermal junction, effaced rete ridges and thinned epidermis; prominent nesting is not usually seen
- Subtle melanoma in situ often extends beyond the visible clinical border
  - Associated with local recurrences

Precursor Lesions

- Melanoma in situ, lentigo maligna type is an intermediate/high grade dysplasia melanocytic lesion characterized by an ill-defined lentiginous proliferation of atypical melanocytes along the DEJ junction replacing basal keratinocytes
  - There is epidermal atrophy, cytologic atypia, and usually a limited pagetoid upward scatter of melanocytes
- No benign precursor (nevus) identified

Molecular Features

- Rarely harbors canonical BRAF V600E mutations
- Initiating driving mutations include NF1, BRAF mutations other than V600E (e.g., BRAF V600K), NRAS, and KIT

- Secondary oncogenic mutations include CCND1, TERT, CDKN2A, RAC1, ARID2, PTEN, and TP53
- Copy number gain of chromosome 1q, 6p, 11q13, 17q, and 20q
- Copy number loss of 6q, 8p, 9p, 13, and 21q
- Very high mutation burden with predominant UV signatures (~20 mutations per Mbps of DNA), which may correlate with better responsiveness to checkpoint inhibitor therapy

Pathway III: Desmoplastic Melanoma (DM)

Epidemiology

- Accounts for 4% of primary cutaneous melanomas
- Most often occurring on high CSD skin of elderly adults with fair skin (7th decade or older)

Clinical Features

- Occurs most commonly on the chronically sun exposed skin of the head and neck, followed by extremities and trunk
- Presents as amelanotic nodules or plaques, or ill-defined scar-like lesion
  - May lead to delayed diagnosis, with a more advanced local stage than other melanoma subtypes (median tumor thickness of >2.0 mm)
- May arise de novo or in association with other melanoma subtypes, most often high CSD/lentigo maligna subtype
- Local recurrences are common, particularly after incomplete or narrow excision, or when neurotropism is present
- Distant metastases most commonly affect lungs, liver, and bone. Lymph node involvement is uncommon; particularly, in pure DMs (see below)

Histopathology

- Intradermal proliferation of spindled cells dispersed in a fibrotic stroma
- Lentiginous in situ component in 80% of cases; neurotropism in 30–40% of cases
  - If an in situ component is absent, additional diagnostic clues include the presence of hyperchromatic nuclei, and patchy lymphocytic infiltrates throughout the lesion
- Diffusely positive for S100 and SOX10 and usually negative for other melanocytic markers, including HMB45, MelanA/MART-1, and tyrosinase
- Two histologic subtypes are recognized
  - Pure DM: Spindled neoplastic cells represent >90% of the neoplastic cell population
    - Paucicellular, with prominent desmoplasia throughout the entire tumor with individual cells separated by delicate collagen fibers
  - Mixed DM: Presence of a distinct population of epithelioid melanocytes resembling high or low CSD melanoma, representing >10% of total neoplastic cells
    - Higher cellular density, higher mitotic index, and Ki-67 proliferation rate
    - The epithelioid component may be positive for HMB45, MelanA/MART-1
    - 3.5-fold greater risk for death or metastases than those with pure DM
- At diagnosis DMs typically have a greater tumor thickness than other melanomas but have a comparatively better overall survival
- Differential diagnosis includes low-grade spindle cell tumors, desmoplastic nevi, atypical leiomyomatous tumor/leiomyosarcoma, dermatofibroma, dermatofibrosarcoma protuberans, neurofibroma, mature and hyperplastic scars, and reactive fibrosing conditions

Precursor Lesions

- Melanoma in situ
- No benign precursor (nevus) identified

Molecular Features

- Lacks canonical BRAF and NRAS mutations (BRAF V600E and NRAS $^{Q61K/R}$)
  - Likely represent a melanoma that evolves by the slow accumulation of weakly oncogenic mutations
- Initiating driving mutations activate the MAPK and PI3K signaling cascades
  - Inactivation of NF1is the most common mutation
  - Other alterations in the pathway include amplifications of receptor tyrosin kinase genes: (EGFR, MET, and ERBB2) and mutations of MAP2K1, MAP3K1, BRAF (nonV600E), NRAS (noncanonical), RAC1, and SOS2
- Very high mutation burden (62 mutations/Mb average – among the most highly mutated cancers) with a strong UV signature, which makes it a good candidate for immune checkpoint blockade therapy
- Secondary pathogenic mutations include inactivation of CDKN2A and TP53

Melanomas Arising at Sun-Shielded Sites or Without Known Etiological Associations with UVR Exposure

Pathway IV: Malignant Spitz Tumor (Spitz Melanoma; SM)

Epidemiology

- Malignant counterpart of Spitz nevi, they can occur at any age, but more common >40 years

- Not associated with UV exposure, but rather with gene rearrangements

Clinical Features

- Usually larger than Spitz nevus, it presents as a changing or enlarging amelanotic or pigmented/variegated plaque, papule, or nodule, with or without ulceration
- Occurs most commonly on head and neck, and lower extremities, but can arise anywhere on the skin, including glabrous skin
- Frequent involvement of lymph nodes as first manifestation

Histopathology

- Asymmetrical, poorly circumscribed, often ulcerated lesion measuring often >1 cm
- Compound or intradermal proliferation of large atypical epithelioid or spindled melanocytes arranged in cohesive sheets, with high-grade cytologic atypia, lack of maturation, and increased mitotic figures (>6 mitosis/mm² in a child and >2 mitoses/mm² in an adult)
- In situ component with pagetoid spread may be extensive
- Can present with prominent lymphoid infiltrates and necrosis
- Typically, do not display distinguishing characteristics of Spitz nevi, such as Kamino bodies, and epidermal hyperplasia
- Lesions that fall short of the above-mentioned attributes may be classified as atypical Spitz tumor (AST) or melanocytic tumor of uncertain malignant potential
- By IHC, they may exhibit irregular staining for HMB45, loss of staining with MelanA/MART1 and p16, and a high Ki-67 proliferation index >20%

Precursor Lesions

- Spitz nevus
  - Benign melanocytic neoplasm with a specific set of initiating mutations, which include fusions of tyrosine and threonine kinases, RAS exchange factors, and mutations of HRAS
  - Symmetric, well circumscribed melanocytic lesion composed of varying proportions of spindled and epithelioid melanocytes
    - If a dermal component is present, there is maturation of melanocytes without significant atypia or increased mitotic activity
    - Artifactual clefting of papillary dermal nests from the overlying epidermis and "Kamino bodies" are usually present
- Atypical Spitz nevus (melanocytoma)
  - Overlapping features between typical Spitz nevus and Spitz melanoma

- Some atypical histologic features include increased proliferation rate, high cellularity, asymmetry, and morphologically distinct tumor cell populations
- Distinction with Spitz Melanoma can be often difficult

Molecular Features

- 20% of Spitz nevi harbor HRAS mutations (most commonly HRASQ$^{61K/R}$ in exon 3), typically accompanied by copy number increases of chromosome 11p
  - Leads to disruption of the GTPase activity of HRAS, with consequent active HRAS signaling through the MAP kinase and PI3 kinase pathways
  - Associated with a desmoplastic histologic phenotype (desmoplastic Spitz nevus)
- Initiating driving mutations in 50% of Spitz tumors include mutually exclusive translocations involving the receptor tyrosine kinases ALK, ROS, NTRK1, RET, and MET, MAP3K8, the serine threonine kinase BRAF, or the RASGRF1 activator of RAS, which lead independent constitutive activation of oncogenic-signaling pathways including the MAP-kinase, PI3, kinase, and STAT pathways
  - ALK fusion
    - Common fusion partners include TPM3 and DCTN1
    - Histopathologically, it shows characteristic plexiform, intersecting fascicles of fusiform melanocytes with smooth contours, and slightly vesicular chromatin; some tumors may display only epithelioid cells with pleomorphic nuclei, and focal plexiform pattern only
  - ROS fusion
    - Multiple fusion partners including PWWP2A, PPFIBP1, ERC1, MYO5A, CLIP1, HLA-A, KIAA1598, ZCCHC8, FIP1L1, and CAPRIN1, among others
  - NTRK1 and NTRK3 fusion
    - Most common fusion is LMNA-NTRK1
  - RET fusion
    - Common fusion partners include KIF5B and GOLGA5
  - MET fusion
    - Rare fusion, with fusion partners including TRIM4, and ZKSCAN1
  - BRAF fusion and amplification
    - Loss of the auto-inhibitory, N-terminal RAS binding domain
      - ◆ IHC is not specific, and fusions should be detected by FISH or NGS
      - ◆ BRAF amplification present in a small number of cases
  - MAP3K8 fusion

- o Multiple fusion partners including SVIL, DIP2C and UBL3
- o Fusion of the 5′ part of MAP3K8 comprising exons 1–8 fuses to one of the partner genes at 3′ end, leading to the replacement of autoinhibitory exon 9
  - RASGRF1 fusions
    - o Described fusions partners include CD63, EHBP1, and ABCC2
- Secondary oncogenic mutations include TERT promoter mutation and loss of CDKN2A

Pathway V: Acral Melanoma

- Two subtypes of melanoma originating from acral skin
  - BRAF V600E mutant acral melanomas
    - o Similar to melanomas occurring in low-CSD skin (pathway I)
    - o Often arise from nevi that harbor BRAFV600E, occurring in younger individuals of light complexion
    - o Characterized by fewer DNA copy number changes
  - Acral melanomas without BRAFV600E mutation
    - o Comprise the majority of acral melanomas (82.8%)
    - o Corresponds to the acral lentiginous melanoma variant first described by Coleman in 1980
    - o Characterized by low point mutation burden and high burden of structural rearrangements and copy number changes
- The following epidemiologic, clinical, histologic, and molecular features correspond to the latter subtype: acral melanomas without BRAF V600E mutation

Epidemiology

- Most common melanoma subtype in Asian, Hispanic, and African populations, mainly due to a reduced incidence of low and high CSD melanomas in these populations
  - Overall incidence is similar among all races
  - 2–3% of melanoma cases in the USA
- Usually affects people on their 6th decade or older but can also occur in younger people
- Tends to undergo a more aggressive clinical course, in part by often delayed diagnosis and muted response to immune checkpoint blockade and targeted therapy
- UV exposure is not a major causative factor
  - Significantly lower somatic mutation burden compared to melanomas arising in sun-exposed skin
- Possible association with mechanical or physical stress

Clinical Features

- Occurs in the glabrous skin of the volar aspects of palms and soles, fingers, toes, and nail apparatus (subungual melanoma)

- Brownish to black patch (radial growth phase) that enlarges over months to years to a thickened plaque with eventual nodular growth (vertical growth phase)
  - Can become eroded or ulcerated
- Subungual melanoma often starts as longitudinal melanonychia, followed by a pigmented patch that spreads over the nail plate, and onto the lateral or proximal nailfold
- Can rarely presents as a hypomelanotic or amelanotic lesion
- Satellite and in-transit and lymph node metastases may be the first manifestation

Histopathology

- Asymmetrical, poorly circumscribed lentiginous proliferation of atypical melanocytes with enlarged hyperchromatic nuclei and prominent dendrites without prominent pagetoid spread
- Moderate acanthosis and rete ridge elongation
- Lymphocytic infiltrates, pagetoid spread, atypia with mitosis, and dermal invasion may become evident over time
- Field effect (morphologically normal or slightly enlarged melanocytes with hyperchromatic nuclei beyond the periphery of the recognizable lesion, which share genomic abnormalities such as gene amplifications with the adjacent in situ and invasive melanoma) can be present likely contributing to recurrence
- Differential diagnosis includes acral nevi and Spitz nevi

Precursor Lesions

- Acral melanoma in situ
  - Intraepidermal atypical melanocytic proliferation of uncertain significance (IAMPUS) is a high-grade intraepidermal melanocytic dysplasia insufficient for a diagnosis of melanoma in situ
  - Can consist of a subtle proliferation of slightly atypical or normal appearing melanocytes along the dermo-epidermal junction, which have amplifications of oncogenes such as CCND1 (field cells)
- No definitive benign precursor (nevus) identified

Molecular Features

- Low burden of point mutations and high incidence of copy number variations and amplifications
- Somatic alterations in KIT, BRAF, NRAS, NF1, SPRED1, MAP2K1, KRAS, and HRAS
  - BRAF and RAS mutations are less frequent than in low-CSD melanomas
- Multiple gene amplifications including CCND1 (cyclin D1), CDK4, MITF, TERT, KIT/ PDGFRA, PAK1, GAB2, MDM2, RICTOR, YAP1, EP300, NOTCH2

- Fusion kinases including BRAF, NTRK3, ALK, and PRKCA and fusions of RASGRP1
- TERT translocations, amplifications, and promoter mutations
- Secondary driving mutations include inactivation of CDKN2A, PTEN, and TP53, which are associated with invasive melanoma

Pathway VI: Mucosal Melanoma (MM)

Epidemiology

- Accounts for 1–2% of all melanoma cases
- Affects with equal frequency all races, and therefore constitutes a considerable fraction of melanomas in non-white populations
- No association with UV radiation exposure
- Poor prognosis, in part because of delayed detection

Clinical Features

- Flat, pigmented patches or plaques that develop nodules which ulcerate at later progression stages
- Occurs on any mucosal membrane primarily the nasopharynx and paranasal sinuses, tarsal conjunctiva, anorectal region, and vulvovaginal region. Rarely in other sites such as cervix, penis, urinary tract, and gastrointestinal mucosa
- Lesions occurring in nasal sinuses or visceral organs are often recognized very late and present as polyps or tumors invading adjacent tissues
  - Symptoms may include pain, pruritus, bleeding, and, depending on the primary site, stuffiness, proptosis, and diplopia
- Differential diagnosis includes melanotic macules, atypical mucosal nevi, amalgam tattoo (oral cavity), lichen sclerosus with hemorrhage (vulva), primary acquired melanosis (conjunctiva), and hemorrhoids (anus)
- Frequent local recurrence and metastatic dissemination to regional lymph nodes and distant metastases

Histopathology

- Absence of solar elastosis
- Lentiginous growth of atypical melanocytes with limited pagetoid spread
- Vertical growth phase component shows atypical melanocytes of varying morphology. Ulceration and vascular invasion are common in advanced lesions
- Desmoplastic pattern sometimes seen

Precursor Lesions

- Mucosal melanoma in situ

- Atypical melanosis/IAMPUS is a high-grade intraepidermal melanocytic dysplasia insufficient for a diagnosis of melanoma in situ
- No definitive benign precursor (nevus) identified

Molecular Features

- Low mutation burden and high burden of copy number changes, including focused amplifications and deletions and structural rearrangements
- Amplification frequently involve CDK4, TERT, MDM2, MITF, CCND1, and NOTCH2
- Deletions of NF1, PTEN, SPRED1, CDKN2A, ATM, and ARID1B
- Recurrent somatic mutations involving SPRED1, SF3B1, NF1, KIT, NRAS, or BRAF
  - SPRED1, which interacts with NF1, also acting as a negative regulator of the MAP pathway, are found in 37% of MM, mostly of anorectal and vulvovaginal origin
  - Hotspot mutations in SF3B1, a factor involved in RNA splicing, lead mis-splicing of transcript and disruption of tumor suppressor genes in the SWI/SNF pathway, are a unique feature of mucosal melanoma and uveal melanoma
  - KIT mutations frequently co-occur with either NF1 or SPRED1 inactivation in approximately 30%, with NF1 and SPRED inactivated in a mutually exclusive pattern
  - NRASG12 and NRASG13 more prevalent than NRASQ61 mutation
  - NonV600E BRAF more prevalent than V600E BRAF mutation
- Additional mutations include CTNNB, MAP2K1, KRAS, ATRX, CHD8, TP53, PTEN, and CDKN2A
- Fusion kinases of relevance include BRAF (ZNF767-BRAF) and ALK (EML4-ALK)

Pathway VII: Melanoma Arising in a Congenital Nevus

Epidemiology

- Congenital nevi are divided into giant (>60cm), intermediate, and small and occur in ~ 1% of newborns. They can involve in extracutaneous sites, including the CNS (neurocutaneous melanosis)
- Melanomas occur in a small fraction of congenital nevi, typically in giant congenital nevi
  - Scalp and back are common primary sites
- Median age for melanoma development is 3 years, with 70% of them developing by puberty
- Prognosis is poor and may reflect delayed detection due to lack of distinction from background congenital nevus

Clinical Features

- Rapidly growing plaques or nodules, often with ulceration and color changes that may be difficult to discern from the background nevus
- Synchronous lymph node metastasis is a frequent finding

Histopathology

- Melanoma may develop in superficial or deep portions of the nevus
- The background congenital nevus is present in contiguity with the melanoma
- Melanocytes can display epithelioid, spindled, or small round blue cell morphology with increased mitotic activity and pagetoid spread
  - There is pleomorphism, occasional necrosis, and sharp circumscription from the adjacent nevus with an expansive pattern
- Tumor may express sarcomatous differentiation resembling malignant schwannoma, rhabdomyosarcoma, and liposarcoma
- Differential diagnosis primarily includes so called proliferative nodules, characterized by nodular proliferations of epithelioid and spindled melanocytes that are larger than the adjacent nevus, but lack high-grade cellular atypia and necrosis
  - In contrast to melanomas the neoplastic melanocytes of proliferative nodules tend to blend with the adjacent cell population of the nevus

Precursor Lesions

- Congenital nevus
  - Well-circumscribed, symmetric, melanocytic neoplasm usually extending into the lower reticular dermis and sometimes subcutaneous tissue
    - ○ Melanocytes tend to track along arrector pili muscles, neurovascular bundles, and eccrine ducts
    - ○ Maturation, lack of significant atypia or increased mitotic activity
  - Defined by single pathogenic mutations (see below)
    - ○ Most are stable and will never progress
    - ○ May potentially acquire secondary pathogenic mutations and act as precursors of melanoma
- Proliferative nodule in congenital nevus
  - Intermediate lesion without overt features of melanoma
- Melanoma in situ in congenital nevus

Molecular Features

- NRAS Q61 is the most common oncogenic mutation
- BRAF V600E mutations (mostly in small lesions) and BRAF fusions are found less frequently
  - In rare cases fusions of RAF1 or ALK

- Melanomas typically show partial copy-number gains and losses of arms or segments of chromosomes, in contrast to proliferative nodules, which tend to harbor gains or losses of entire chromosomes
- TERT promoter mutations can be found in some melanomas

Pathway VIII: Melanoma Arising in a Blue Nevus

Epidemiology

- Blue nevi are dermal proliferations of melanocytes (melanocytosis) with deep melanin pigmentation that gives rise to a blue macule or papule
  - Subtypes include cellular, epithelioid, sclerotic, and plaque-type
  - May occur sporadically or with increased incidence in patients with Carney complex (CNC)
- Melanomas arising in blue nevi are rare and can occur at any age, with the mean age at diagnosis at 45 years

Clinical Features

- Presents most commonly on the scalp, followed by face, trunk, and buttocks
- Present usually as large, fast-growing nodules, which may show evidence of a residual blue nevus
- Usually diagnosed late as tumors can be masked and be recognized only after an increase in size of the long-standing preexisting lesion
- Often aggressive, with frequent metastasis to lymph nodes, liver, lungs, and bone

Histopathology

- Cellular, dermal-based nodular proliferation of large melanocytes with marked cytologic atypia, frequent mitosis, and common necrosis and ulceration
  - Often adjacent blue nevus of the cellular variant
- Invasive nodule often destroys surrounding adnexal structures
  - The preexisting nevus can be found at the periphery of the tumor, when not totally destroyed
- Ki-67 is often >20% in the melanoma and <5% in the nevus
- When BAP1 is lost, it usually affects only large contiguous populations of malignant melanocytes, while retained in the nevus remnant
- Lesions that fall short of the above-mentioned attributes may be classified as atypical cellular blue nevus
- Differential diagnosis includes other hyperpigmented dermal lesions including atypical deep penetrating nevus, plexiform melanoma, and pigmented epithelioid melanocytoma

Precursor Lesions

- Blue nevus
  - Conventional blue nevus is a benign melanocytic neoplasm composed by dendritic-shaped melanocytes among dermal collagen in mid and upper dermis and a variable number of melanophages
  - Defined by common initiating mutations in GNAQ and GNA11 (see below)
  - Cellular blue nevus, in addition to dendritic melanocytes, have nests and fascicles of oval to spindled-shaped melanocytes often with intervening dendritic melanocytes and melanophages as seen in conventional blue nevus
- Atypical cellular blue nevus
  - Intermediate form of progression from cellular blue nevus to melanoma

Molecular Features

- The genetic features of melanoma in blue nevi are very similar to those of uveal melanoma
- Initiating driving mutations include GNAQ and GNA11 (p.Q209 and p.R183) and less frequently PLCB4 or CYSLTR2 mutations
  - Mutations on GNAQ, GNA11, PLCB4, and CYSLTR2 lead to the activation of the Gαq signaling pathway with downstream activation of the PKC and MAP/ERK pathway
  - The mutations are mutually exclusive of each other
- Biallelic inactivation of BAP1, point mutations of SF3B1, and EIF1AX mutations are associated with malignant transformation
- As in uveal melanoma, loss of BAP1 on chromosome 3p21, or gains/amplifications of c-Myc on chromosome 8q24 indicate a worse prognosis
- Gains of chromosomes 1q, 4p, 6p, and 8q and losses of 1p and 4q have been described
- Do not harbor BRAFV600E mutations

Pathway IX: Uveal Melanoma

Epidemiology

- Most common intraocular cancer accounting for ~5% of all melanomas in the United States
- Affects most commonly white population with an average age of diagnosis at 60 years
- Most cases arise de novo, but may sometimes arise from uveal nevi
- 90% of cases occur in the choroid. 10% occur in the ciliary body and iris
- Risk factors include uveal nevi, congenital ocular melanocytosis, and BAP1 tumor predisposition syndromes
- Ten-year mortality is 50%, remarkably higher than most other melanoma subtypes

Clinical Features

- Initial manifestations include blurred vision and visual field defects
- Choroidal melanomas can cause retinal pigment epithelium disruption and serous retinal detachment
  - Necrotic tumors may cause uveitis and glaucoma
- Optic nerve invasion is seen in ~5% of cases
- More than 50% of patients develop metastases, usually involving liver and bone
- Melanomas vary in pigmentation and occur as dome-shaped or mushroom-shaped tumors

Histopathology

- Solid nodules of spindled or epithelioid melanocytes with mitotic figures and occasional foci of necrosis
  - Lymphocytic infiltrates and pigmented-laden macrophages are common
  - Epithelioid morphology, lymphocytic infiltrates, and high mitotic account have been associated with more aggressive disease and higher mortality
- Immunohistochemical loss of BAP1 is associated with poor prognosis

Precursor Lesions

- Choroidal nevus
  - Mutations in GNAQ or GNA11, or other Gαq pathway mutations

Molecular Features

- Lacks BRAF, NRAS, and NF1 mutations
- Initiating driving mutations (usually mutually exclusive) include GNAQ or GNA11 mutations in over 90% of cases and less frequently CYSLTR2 and PLCB4
- Secondary somatic alterations include BAP1, SF3B1 and EIF1AX
  - Tumors with BAP 1 mutations are associated with high risk of metastasis and overall worse prognosis than those with SF3B1 or EIF1AX
- Recurrent copy number alterations include
  - Gains of 1q, 6p, and 8q and losses of 1p, 3, 8p, and 16q
    - BAP1 and MYC genes drive the copy number changes on chromosome 3 and 8q, respectively
    - Survival probability is worse when monosomy 3 and chromosome 8q gain coexist
    - Chromosome 6p gain is associated with a better prognosis
- Tertiary driver mutations include loss of function of CDKN2A, PBRM1, PIK3R2, and PTEN, gain of function of EZH2, PIK3CA, and MED12, and loss of heterozygosity in the GNAQ gain of function mutation
- Metastatic lesions acquire additional copies of 1q and 8q

- Uveal melanomas can be categorized by gene expression profiling into two molecular classes associated with metastatic risk
  - Class 1 tumors have a low metastatic risk
    - PRAME expression can identify patients within class 1 with increased risk of metastasis
  - Class 2 tumors have a high metastatic risk
    - Strongly associated with mutations in BAP1 on chromosome 3p21, usually accompanied by loss of the other copy of chromosome 3

Nodular Melanoma

- Invasive melanoma lacking a recognizable radial growth phase
- Can occur in any of the low and high CSD, and acral pathways and therefore their genomic features and epidemiology are heterogeneous
- Present as elevated, firm, usually symmetric, rapidly growing nodules, that range in pigmentation from very dark variants to hypo- and amelanotic variants
- Prognosis depends on depth of invasion, ulceration, and mitotic index as for other melanomas in the respective pathways
- Histologically, a radial growth phase may be lacking, but if present, the melanocytes in the epidermis do not extend more than 2 rete ridges beyond the dermal component
- The dermal component is often expansile with nests and sheets of atypical melanocytes with increased mitotic activity
- There is often an epidermal collarette adjacent to the dermal nodule
- Differential diagnosis includes BCC, SCC, vascular and adnexal tumors, sarcomas, lymphomas (in the amelanotic variants), and metastatic melanomas to the skin

Pediatric Melanoma

- Pediatric melanoma accounts for 1–3% of all pediatric malignancies. It can occur congenitally, during childhood (prepubescent; ≤10 years), and adolescence (postpubescent; 11–19 years), with most cases occurring after age 14 years
- Predisposition conditions include familial melanoma, giant congenital nevi, atypical nevus phenotype, xeroderma pigmentosum, and neurocutaneous melanosis
- Four different subtypes of pediatric melanoma are recognized
  - Congenital melanoma is rare and presents at birth or early in life; it can occur by transplacental metastatic dissemination from a melanoma of the mother or de novo, as a primary congenital melanoma
  - Melanoma arising in congenital nevi, usually in a large or giant congenital nevi (previously discussed)
  - Spitz Melanoma (previously discussed)
  - Conventional adult subtype melanoma shares the histological and clinical features of melanoma arising in low CSD in adults

Immunotherapy and Targeted Therapy

- Surgery is the main treatment option for most melanomas and is highly curative when performed at an early stage
- Immunotherapy and targeted therapy have dramatically improved the prognosis for patients with unresectable or metastatic melanoma
- The response to immunotherapy is associated with the mutational tumor burden
  - Neoplasms with high number of somatic mutations have more neoantigens capable of elicit an immune response
- Immunotherapy can be divided into four main groups:
  - Biological medications: Include cytokines such as IL2, interferons, and GM-CSF
    - High dose interleukin-2 (IL-2) is a T-cell growth factor that stimulates T-cell proliferation and cytotoxic activity
  - Vaccination: Aim to elicit immune responses against antigens expressed by melanoma cells, such as tumor-associated antigens, or neoantigens. Include vaccines targeting melanoma cells directly, dendritic cells-based vaccines, peptide- and RNA-based vaccines, and vector-based (viral) vaccines
    - Talimogene laherparepvec (T-VEC) is a genetically modified oncolytic virus (HSV1) with the integrated GM-CSF gene that targets tumor cells and harnesses the immune response
  - Adoptive cell therapy: Involves the collection of immune cells from the tumor or peripheral blood (peripheral blood lymphocytes or tumor-infiltrating lymphocytes), with subsequent selection, expansion, and activation in vitro. The lymphocytes are then reinfused into the patient to induce an anticancer immune response
  - Immune checkpoint inhibitors: Aim at blocking the inhibitory checkpoints that suppress the antitumor immune response; the main targets are CTLA-4, and PD-1
    - CTLA-4 is a member of the immunoglobulin superfamily and encodes a protein which transmits an inhibitory signal to T cells. The anti CTLA-4 anti-

body Ipilimumab was the first FDA-approved immune checkpoint inhibitor for metastatic melanoma and is often used in combination with anti PD-1 drugs
- ○ PD-1 is a cell surface molecule expressed by T cells with excessive exposure to antigens and is a negative regulator of T-cell activity. Its primary ligand is PD-L1, which is expressed in cancer cells and TIL
- ○ Anti PD-1 immunoglobulins (nivolumab and pembrolizumab) have been one the most effective drugs in melanoma, with treatment responses being durable even after ending immunotherapy
    - ♦ A PD-L1 IHC staining >1% is associated with improved outcomes. Still, given that up to a third of patients with PD-L1 <1% respond to nivolumab, or pembrolizumab, the value of PD-L1 staining alone is not used to exclude the use of these treatments
- Immunotherapy may be associated with immune-related adverse events (irAEs) in the dermatologic, gastrointestinal, pulmonary, endocrine, renal, ophthalmologic, rheumatic, cardiovascular, and hematologic system
- Targeted therapy: The characterization of the BRAF V600E mutation led to the development of highly specific kinase inhibitors that improved the survival of patients with advanced or metastatic melanoma from a median of 6 months obtained with chemotherapy to a median of 26–34 months
  - – BRAF inhibitors
    - ○ Vemurafenib was the first BRAF inhibitor approved in 2011, which provided a dramatic improvement of treatment outcome, followed by dabrafenib in 2013, and encorafenib in 2020
    - ○ BRAF inhibitors can induce keratoacanthoma and SCC in 15–20% due to the acquisition of RAS mutation and paradoxical activation of CRAF in keratinocytes. This effect can be counteracted by combination therapy with MEK inhibitors
  - – MEK inhibitors
    - ○ MEK is a kinase immediately downstream of RAF in the MAPK pathway
    - ○ Trametinib was first approved in 2013, followed by cobimetinib in 2015, and binimetinib in 2018
  - – Resistance of MAPK inhibition-targeted therapy is common and appears within 6–12 months by a variety of resistance mechanisms which include additional mutations in the MAP-kinase pathway, feedback reactivation of receptor tyrosine kinases and RAS, BRAF amplification and BRAFV600E splice variants
    - ○ Resistance to single-agent therapy prompted the development of combinatory regimens

- ○ Dabrafenib plus trametinib was first approved in 2014, followed by vemurafenib plus cobimetinib in 2015, and encorafenib plus binimetinib in 2018
- ○ Compared to previous single BRAF-targeted agents, combination therapy improved the clinical response rate from 50 to 60–70% and diminished side effects due to paradoxical activation of RAF
- – KIT inhibitors
- Currently, the first-line therapeutic approach for advanced-stage melanoma consists of immunotherapy with anti-PD1 antibodies or targeted therapy with BRAF and MEK inhibitors

Nonmelanoma Skin Cancer

- NMSC represent a heterogeneous group of cutaneous malignancies, including the most prevalent BCC, and SCC, as well as other rare carcinomas, such as MCC, and cutaneous adnexal carcinomas
- NMSC affects more than 3 million Americans per year
- UVR is estimated to cause 95% of NMSC
- The diagnosis and treatment of NMSC in the USA increased by 77% between 1994 and 2014

Basal Cell Carcinoma

Clinical Features
- BCC accounts for 80% of NMSC
- UV exposure is the most important risk factor for BCC, as evidenced by a high mutational burden with a UV-associated mutational signature
- Risk factors in addition to sun exposure include usage of tanning beds, phototherapy, photosensitizing agents, chronic arsenic exposure, ionizing radiation, and light complexion and poor tanning ability
- Immunosuppression is another important risk factor but less so than SCC
- Lesions present as skin-colored pearly papules, nodules, or plaques, with telangiectases, rolled borders, which can become ulcerated. Some BCCs are pigmented
- May arise in nevus sebaceous
- Surgical excision is curative. Incompletely excised lesions can regrow. Metastatic dissemination is very rare and associated with exceptionally large BCCs

Genetics and Pathogenesis
- Hedgehog signaling
  - – BCCs are characterized by mutations activating the hedgehog (Hh) signaling pathway

- Relevant components of the hedgehog pathway in mammals include the receptor for Hh, its negative regulator, PTCH1; its positive regulator, SMO; the downstream transcription factors GLI1 and GLI2; and the cytoplasmic protein which stabilizes inactive GLI transcription factors, SUFU (Fig. 17.2)
- The most frequent mutations found in BCCs are biallelic inactivation of PTCH1 (over 70%) resulting in Hh pathway activation, with gain of function mutations in SMO or SUFU in the remainder of cases
- Another commonly mutated gene is TP53, found in approximately 50% of cases
- Cell of origin
  - Controversy exists regarding the keratinocyte stem cell of origin for BCC
  - In irradiated PTCH1$^{+/-}$ mice BCCs primarily arise from the K15$^+$ stem cells in the bulge region of the hair follicle, with interfollicular epidermal stem cells contributing less frequently
  - Mice with constitutively active SMO implicate progenitor cells in the interfollicular epidermis and the upper infundibulum as additional cells of origin
- Genetic ablation of PTCH in different cell populations implicate various hair follicle stem cell populations. While most stem cell populations within the interfollicular epidermis do not efficiently produce tumors, the GLI1 positive stem cell of the touch dome are particularly susceptible to transformation
- Nevoid basal cell carcinoma syndrome (NBCCS), also known as basal cell nevus syndrome or Gorlin syndrome, is a rare autosomal dominant cancer predisposition syndrome caused by germline mutations that inactivate one of PTCH1 genes
  - Patients develop multiple BCCs at a young age in addition to odontogenic keratocysts and medulloblastoma through loss of the remaining allele
- Additional syndromic predispositions include Xeroderma pigmentosum, Rombo syndrome, Bazex syndrome, and oculocutaneous albinism
- Basosquamous transformation: While focal squamatization is a common finding in BCC, mixed tumors of BCCs acquiring MAPK activating mutations which induce basosquamous transformation have been reported

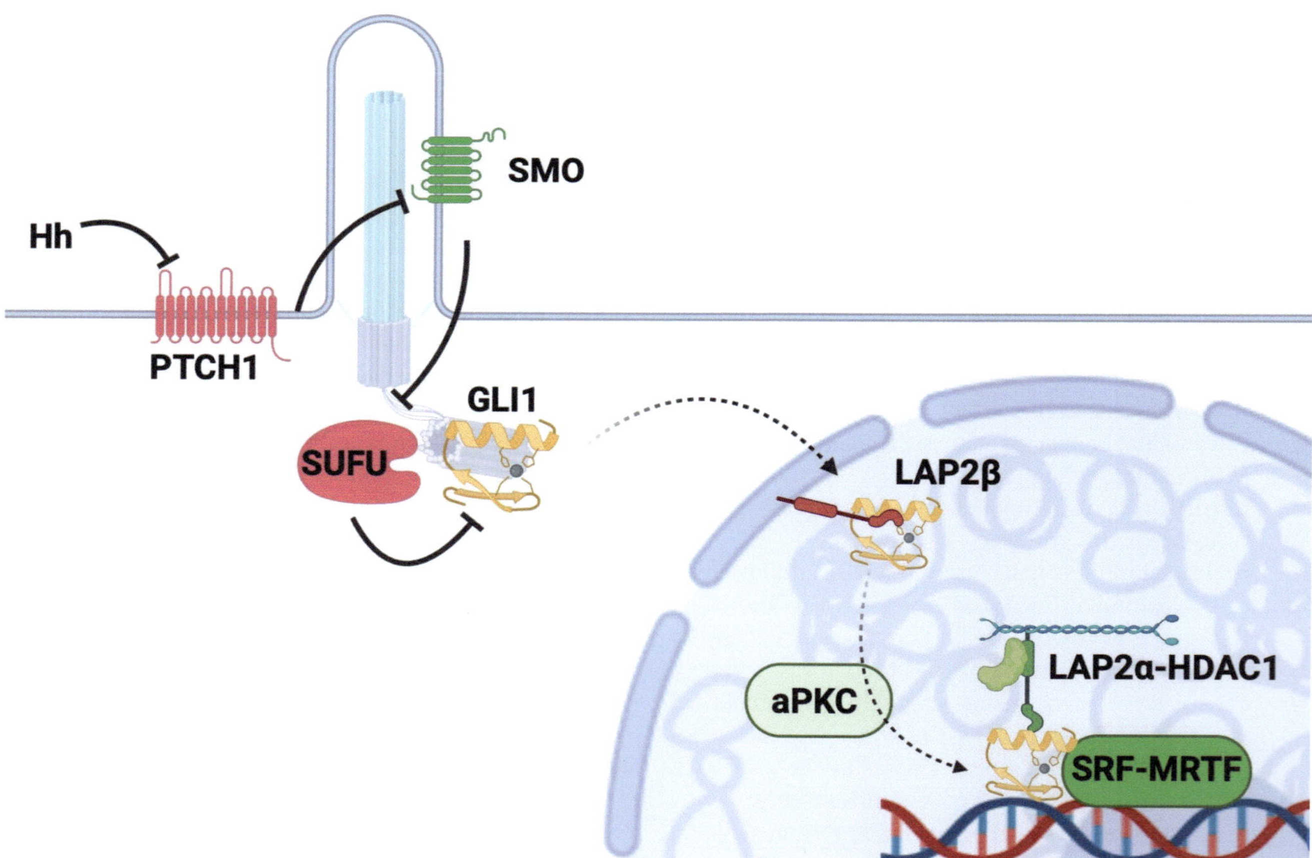

Fig. 17.2 Key members of the hedgehog signaling cascade in basal cell carcinoma. Created with BioRender.com

- Single cell RNA sequencing and murine studies demonstrate that c-FOS remodels chromatin to induce basosquamous transformation

Histopathology

- Classic morphology: BCCs are characterized by a densely cellular aggregates of basaloid keratinocytes reminiscent of the stratum basale of the epidermis with peripheral palisading of deeply basophilic atypical nuclei and surrounding clefts
  - Centrally scattered apoptotic cells, mitoses, regions of necrosis, lakes of mucin, or squamatization may be seen. The stroma is characteristically myxoid and cellular
- Histologic subtypes
  - Nodular: Most common morphology (80%); tumors arranged in dermal nodules which may connect to the overlying epidermis
  - Superficial: Second most common morphology (15%) typically found on the trunk or extremities; small aggregates of basaloid keratinocytes are connected to overlying epidermis and extend into the superficial dermis, "skip lesions" are common with intervening uninvolved regions
  - Infiltrative: Subtype associated with more frequent local recurrence; characterized by angulated aggregates of tumor in infiltrative arrays in the dermis
  - Micronodular: Subtype associated with more frequent local recurrence; poorly circumscribed tumors composed of small intradermal nests
  - Morpheic: Subtype associated with more frequent local recurrence; Narrow strands of basaloid keratinocytes embedded in a desmoplastic stroma
  - Pigmented: Various degrees of melanin pigmentation of tumor aggregates most seen in nodular and superficial types
    - May clinically be mistaken for a melanocytic neoplasm
  - Less common subtypes include: Infundibulocystic (chords of anastomosing cells with embedded infundibular-type cysts, typically facial) fibroepithelioma of Pinkus variant (abundant stroma with thin chords of anastomosing tumor creating a fenestrated or lace-like appearance)
- Ancillary testing
  - Immunohistochemistry for BerEp4 can assist in distinguishing BCC (positive) from SCC (negative)
- Mimics
  - Sebaceous carcinoma with minimal sebaceous differentiation can histopathologically mimic BCC
  - Androgen receptor (AR) can distinguish sebaceous carcinoma (positive) from BCC (negative, sparse staining)
  - Merkel cell carcinoma can mimic BCC and immunohistochemistry for CK20 can assist in the differential diagnosis
  - Hair follicle neoplasms including trichoepithelioma and trichoblastoma can also mimic BCC, particularly in cases where papillary mesenchymal bodies are not prominent
    - Identifying passenger Merkel cells (CK20+) in benign hair follicle tumors can help distinguish them from BCC

Molecular Therapies and Drug Resistance Mechanisms

- Targeted therapy using inhibitors of Smoothened (SMOi) is clinically effective in patients with inoperable and/or metastatic tumors and for patients with multiple BCCs in the context of NBCCS; drug resistance to SMOi, is a common problem
- SMOi-resistant recurrent BCCs occur by the selection of mutations to SMO which disrupt SMOi binding or mutations of other components of the Hh signaling pathways thereby re-establishing Hh signaling
- SMOi-sensitive recurrent BCCs following treatment have been described and appear to rely on a persistent population of slow cycling Lgr5+ follicular stem cells which rely on Wnt activation
- Noncanonical Hedgehog signaling:
  - Hyperactivation of protein kinase C ι/λ, an atypical PKC (aPKC) homolog classically associated with designating apical-basal polarity, modulates the acetylation dynamics of GLI1 via recruitment of the deacetylase HDAC1 in resistant BCCs
    - The acetylation of GLI1 segregates the transcription factor to the nuclear lamina in an inactivated nuclear-bound complex with LAP2β. LAP2α promotes GLI1 deacetylation via HDAC1 recruitment, releasing it from the nuclear lamina to associate with euchromatin (Fig. 17.2)
- TGF-β and AP1 establish an altered chromatin environment, which facilitates the recruitment of the SRF-MRTF coregulators to GLI1 target genes, enabling GLI1-driven noncanonical Hh signaling downstream of the primary cilia (Fig. 17.2)
- Next-generation molecular therapies, including topical HDAC inhibitors, are in clinical development with promising results

Squamous Cell Carcinoma

Overview

- SCC of the skin is a common malignant neoplasm originating from epidermal keratinocytes

- SCC is the second most common skin cancer in the USA, following BCC, and accounts for ~20% of NMSC (~1.8 million cases are diagnosed each year, and ~8000 deaths occur per year)
- The incidence of SCC increases 50- to 300-fold from age 45 to age 75
- Incidence is highest in non-Hispanic white men and women (360 and 150 cases per 100,000 individuals, respectively)
  - Incidence in dark-skinned individuals is 3 per 100,000 and tends to occur in the context of chronic inflammation, nonhealing wounds or scars independent of sun exposure
- Surgical excision is curative for most SCC but 1.5% to 5.2% are metastatic, with higher risk for men, increasing age, and immunocompromised patients
- Daily use of an SPF15 or higher sunscreen reduces the risk of developing SCC by about 40%

Pathogenesis

- Environmental and genetic factors and immunosuppression contribute to the development of SCC
- Environmental factors:
  - UVR exposure from the sun, tanning beds or PUVA therapy is the main environmental risk factor implicated in the development of SCC
    - Patients with high occupational exposure to UVR have a ~2 fold increased risk for SCC
    - Fair skin, light colored eyes and red hair is associated with an increased risk
  - 1% of SCCs arise in chronically inflamed skin (burns, chronic ulcers, nonhealing wounds sinus tracts, inflammatory dermatosis such as lichen sclerosus) or large scars,
  - Smoking may also be a risk factor
  - Ionizing radiation (environmental, therapeutic, or diagnostic radiation)
  - Arsenic (contaminated water or occupational exposure) and radon exposure
- Genetic risk factors:
  - Family history of skin cancer is associated with SCC risk, independent of known environmental and innate risk factors
- Inherited disorders
  - Xeroderma pigmentosum (XP) is an autosomal-recessive syndrome, caused by a genetic defect in the NER pathway that leads to UVR hypersensitivity and 1000-fold increased risk of skin cancer with tumors arising in multiplicity early in life
  - Epidermolysis bullosa is mainly an autosomal-dominant disease with mutations in KRT5 or KRT14, which encode keratins expressed in basal keratinocytes

  - The syndrome is associated with chronic blister formation and an increased risk for SCC
  - Oculocutaneous albinism is a group of autosomal-recessive disorders of melanin biosynthesis with an increased risk of early onset skin cancer, including SCC
  - Epidermodysplasia verruciformis is an autosomal-recessive disease characterized by increased susceptibility to beta type HPVs and an increased risk of early onset SCC, occurring in over 60% of patients
  - Additional genetic syndromes associated with an increased risk of early-onset SCC include Ferguson–Smith syndrome (keratoacanthomas), and hereditary syndromes associated with genomic instability including dyskeratosis congenita, Fanconi anemia, Rothmund–Tomson syndrome, Bloom syndrome, and Werner syndrome
- Chronic immunosuppression secondary to organ transplantation, HIV infection, or long-term glucocorticoid use increase the risk of developing SCC
  - Immunocompromised patients have a 65- to 250-fold increased risk of developing SCC
- Drugs: azathioprine, BRAF inhibitors, thiazide diuretics, and voriconazole are associated with an increased risk of developing SCC
  - Azathioprine is used as an immunosuppressive and mutagenic agent to prevent graft rejection in organ transplant patients; it is associated with UVA photosensitivity and carcinogenesis of the skin and results in a specific mutational signature in SCC

Squamous Cell Carcinoma Precursors

- Actinic keratosis is a precursor to SCC occurring on chronically sun-damaged skin (synonyms: solar keratosis, senile keratosis, actinic cheilitis when it occurs on the lip borders)
  - Often multiple lesions – "field effect"
  - Presents as an erythematous, thin keratotic patch or plaque, that is occasionally pigmented
  - Histopathologically presents with pleomorphic keratinocytes affecting primarily the basal layer with overlying parakeratosis, sparing the adnexal ostia
  - There is an accompanying superficial perivascular or lichenoid lymphohistiocytic infiltrate (rich in plasma cells in actinic cheilitis) and marked solar elastosis
  - Can be subdivided in different variants: hypertrophic, atrophic, bowenoid, acantholytic, lichenoid, and pigmented
  - Malignant transformation to SCC occurs in 1 in 1,000 to 10,000 lesions

- TP53 mutations are the most common alterations. Loss of heterozygosity in several chromosome arms has been reported
- Arsenic keratosis is a hyperkeratotic premalignant lesion that occurs in patients chronically exposed to arsenic
  - Characterized by melanosis, followed by yellowish small verrucous papules affecting the thenar and lateral borders of the palms, fingers, soles, heels, and toes
    - Average latency for skin lesions is more than 20 years
  - Histopathologically can present with hyperkeratosis, acanthosis, and papillomatosis with or without keratinocyte atypia
  - Arsenic exposure leads to chromosomal abnormalities, gene amplification, and impaired DNA repair
  - Can progress to invasive SCC, with many arsenical skin cancers harboring TP53 mutations
- PUVA Therapy
  - Results in PUVA keratoses in a dose-dependent fashion, with a 15–30% risk of developing SCC after 5–8 years of exposure
  - Frequently used as immunosuppression in patients with psoriasis
  - Characterized by keratotic pink patches or plaques
  - Histopathologically similar to actinic keratosis

Squamous Cell Carcinoma In Situ (Bowen disease)

- An intraepidermal carcinoma with specific histopathological and clinical features
- Histopathologically presents with pleomorphic keratinocytes, dyskeratotic cells, and atypical mitoses throughout the entire thickness of the epidermis with overlying parakeratosis
  - The basal layer is crowded ("eyeliner sign")
- Variants
  - Pagetoid Bowen disease
  - Clear cell Bowen disease
- Can be distinguished from Paget disease by negative IHC staining for CAM5.2 and CK7
- Risk factors include the same as those for SCC. In addition, high-risk HPV infection is the major cause of SCC in situ, particularly on the skin of the fingers and genitals
- Commonly affects chronically sun damaged skin but can involve sun-protected areas such as the genitalia
  - Bowen disease occurring on the glans penis of vulva is known as erythroplasia of Queyrat
- Presents as a scaly, erythematous, circumscribed plaque that expands slowly over many years
- 3–5% of all SCC in situ progress to invasive SCC

Invasive Squamous Cell Carcinoma

- Can develop on any cutaneous surface including head, neck, trunk, extremities, oral mucosa, periungual skin, and anogenital areas
  - Tumors arising on the ear, temporal skin, or at mucocutaneous interfaces tend to be more aggressive, with a ~ 10–30% rate of metastasis
- Well-differentiated tumors tend to appear as indurated, hyperkeratotic plaques, papules, or nodules, with or without ulceration
- Poorly differentiated lesions tend to be fleshy, granulomatous papules or nodules without the hyperkeratosis that characterizes well-differentiated lesions
  - They may present with ulceration, hemorrhage, or necrosis
- Histopathology: Irregular aggregates of neoplastic keratinocytes invading into the underlying dermis
- Variable degree of differentiation, ranging from well-differentiated tumors with keratinization and preservation of intercellular bridges to poorly differentiated tumors with complete absence of keratinization
- By immunohistochemistry, SCC is positive for p63, p40, EMA, Ck5/6, MNF116, and 34βE12. BerEp4 is negative
- SCC variants include keratoacanthoma, verrucous SCC, acantholytic SCC, spindle cell SCC, adenosquamous carcinoma, infiltrative and clear cell SCC (Table 17.1)
  - Additional uncommon variants are SCC with sarcomatoid differentiation, lymphoepithelioma-like carcinoma, pseudovascular SCC, and SCC with osteoclast-like giant cells
- Differential diagnosis of well-differentiated SCC includes pseudoepitheliomatous hyperplasia, syringometaplasia, perineural hyperplasia, endophytic keratoses, warts, cells traumatically introduced into scar, and squamoid adnexal tumors
- Differential diagnosis of poorly differentiated SCC include melanoma, atypical fibroxanthoma, sarcoma, and lymphoma
- High-risk prognostic features are tumor thickness >2 mm, full dermal or subcutaneous or perineural invasion, and an infiltrative growth pattern with narrow strands of tumor cells at the base

Molecular Features

- A recent meta-analysis in 2021, covering 105 SCC from 10 different studies calculated the tumor mutational burden and identified 5 distinct subtypes of cutaneous SCC based on established mutational signatures
  - Sporadic SCC (patients with no known comorbidities)
    - High mutation burden with most mutations attributable to UVR signature mutation – Signature 7

Table 17.1 Squamous cell carcinoma variants

SCC variant	Clinical features	Histopathology	Additional notes
Keratoacanthoma (KA)	• Well-differentiated variant of SCC. • Rapidly growing, dome-shaped tumor with a central keratin plug presenting on sun-damaged skin. • Marked tendency for spontaneous, complete involution within 3–6 months. • Rare causes reported to cause visceral metastases.	• Symmetrical, exoendophytic, cup-shaped proliferation of well-differentiated, ground glass-like keratinocytes with sharply demarcated peripheral border. • Central core with keratinous horn plug, frequent intraepithelial microabscesses, low mitotic rate. • Regressing KA: Increased number of apoptotic cells, loss of central keratin plug, flattened epithelium, underlying inflammatory foreign body reaction.	• Multiple KA (at times with sebaceous differentiation) are a characteristic feature of MTS and the AD Ferguson-smith syndrome, caused by inactivating germline mutation of TGFBR1. • Can be seen after immunosuppression, BRAF inhibitors, and tattoos. Polyomavirus 6 has been seen in KA. Polyomavirus 9 and 48 have been seen in KA from HIV patients.
Verrucous (VC) SCC	• Well-differentiated but locally aggressive variant of SCC. • "Variants" include: – Giant condyloma of Buschke and Löwenstein - genital skin. – Ackerman tumor— oral mucosa. – Carcinoma cuniculatum—plantar surface. • Can present as exophytic, white-gray tumors with broad and pushing base.	• Exoendophytic lesion with hyperkeratosis, acanthotic papillae, slender fibrovascular cores, prominent keratin craters between papillae, and pushing, club-shaped base tongues of intradermal growth. • Viral cytopathic effect is uncommon. • High grade areas and/or infiltrative borders are not a features of VC and raise the possibility of a mixed VC.	• No clear relation with HPV (PCR or ISH usually negative).
Adenosquamous carcinoma	• Aggressive variant of SCC. • Indurated keratotic plaques, usually indistinguishable from typical SCC.	• Mixed squamous and glandular differentiation. • Squamous component can be seen attached to the epidermis with occasional keratinizing cysts and desmoplastic stroma. • Variable glandular differentiation comprising 5–80% of the tumor (positive for CEA and CK7). • High mitotic rate, infiltrative pattern with frequent nerve, subcutis, muscle, and bone involvement.	• High rate of recurrence and metastasis.
Acantholytic SCC	• Rare, variant of SCC (<5% of SCCs) presenting mostly in chronically sun-exposed areas. • Same clinical features of classic SCC.	• Impaired cellular adhesion leads to acantholysis with resulting formation of pseudoglandular spaces (negative for mucin CK7, and CEA).	• Defective desmoglein 3 and e-cadherin expression.
Spindle cell SCC	• Rare, poorly differentiated variant of SCC. • Presents as an exophytic plaque or nodule indistinguishable from conventional SCC.	• Pleomorphic spindle cell proliferation arising in a background of marked solar elastosis. • High mitotic activity. • Keratinization is uncommon. • Can extend into the subcutis. • IHC positive for squamous markers usually required for diagnosis.	• Besides sun-exposed skin, it can also arise on mucocutaneous areas of the head and neck, urogenital track and distal penis. • Differential diagnosis includes atypical fibroxanthoma, desmoplastic melanoma, undifferentiated pleomorphic sarcoma, among others.
Clear cell SCC	• Clinical features indistinguishable from other types of SCC.	• Squamous cells with unilocular PAS+ intracytoplasmic vacuoles (glycogen) and focal keratinization. • Clear cell component is usually >25%.	• Prognosis comparable to other SCC types.

- SCC from immunosuppressed patients without usage of azathioprine
  - Low mutation burden with mutations attributable to UVR signature mutation – Signature 7
- SCC from immunosuppressed patients with usage of azathioprine
  - High mutation burden with high proportions of azathioprine associated mutational signature – Signature 32
- SCC from patients with recessive dystrophic epidermolysis bullosa
  - Low mutation burden, with mutations attributable to APOBEC-mediated mutagenesis – Signatures 2 and 13
- SCC from patients with XP
  - High mutation burden with high frequency of C➔T transitions. but no signature 7

- The same meta-analysis nominated 30 driving gene mutations affecting six important pathways (Fig. 17.3)
  - NOTCH Pathway, mutation occurring in 79.5% of tumors
    - The Notch pathway regulates cell proliferation, differentiation, and death; upon binding to its ligand, the transmembrane receptor NOTCH undergoes proteolytic cleavage that leads to the release of Nid, which enters the nucleus and interacts with CLS. CLS is a DNA-binding protein that interacts with proteins such as the adaptor protein MAML to promote transcription
    - Mutations involved NOTCH1 (55.4%), NOTCH2 (36.1%), EP300 (21.7%), and CREBBP (15.7%)
  - P53 Pathway, inactivating mutations occurring in 71% of tumors

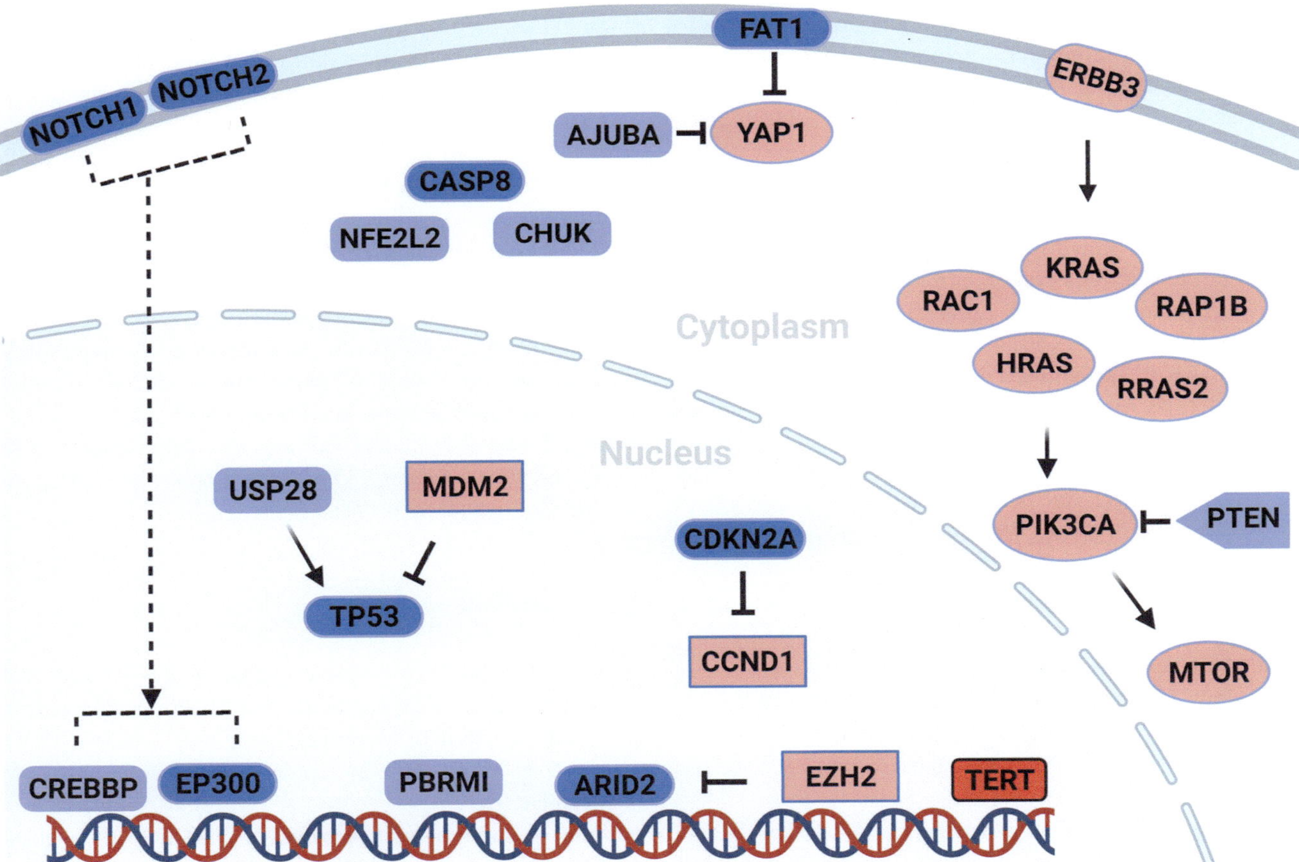

Fig. 17.3 Common driving gene mutation in the development of cutaneous squamous cell carcinoma. Genes in dark and light blue are tumor suppressor genes mutated in >20%, and < 20% of SCC cases, respectively. Genes in dark and light red are oncogenes mutated in >20%, and <20% of SCC cases, respectively. Created with BioRender. com

- o TP53 is a tumor suppressor gene located on chromosome 17p, encoding the transcription factor p53, which is involved in cell cycle-arrest and regulation of apoptosis
  - o Mutations: TP53 (66.3%), USP28 (12%) and MDM2 (2.4%)
- Rb pathway, inactivating mutations occurring in 38.6% of tumors (pathway described in melanoma chapter)
  - o Mutations: CDKN2A (34.9%) and CCND1 amplification (6%)
- Chromatin remodel pathway, mutations occurring in 38.6% of tumors
  - o Chromatin remodeling comprises the modification of the chromatin to regulate access of DNA to transcription factors and control gene expression
  - o Mutations: ARID2 (27.7%), PBRM1 (12%), and EZH2 (2.4%)
- Hippo pathway, activating mutations occurring in 37.3% of tumors
  - o Hippo pathway controls organ size by regulating cell proliferation, apoptosis, and stem cell self-renewal
  - o Mutations: FAT1 (30.1%), AJUBA (7.2%), and YAP1 (1.2%)
- Cellular stress pathway, inactivating mutations occurring in 32.5% of tumors
  - o Several proteins involved in apoptosis and other different cellular responses against stress stimuli
  - o Mutations: CASP8 (22.9%), CHUK (10.8%), and NFE2L2 (1.2%)
- MAPK/PI3K pathway, activating mutations occurring in 31.3% of tumors (pathway described in melanoma chapter)
  - o Mutations: HRAS (9.6%), PTEN, PIK3CA (6% each) RAP1B, RAC1 (2.4% each), and KRAS, RRAS2, ERBB3, MTOR (1.2% each)
- Immortalization
  - TERT promoter mutations are found in approximately 50% of SCC
- DNA copy number changes
  - Copy losses or LOH affect chromosomes 2q, 3p, 4p, 5q, 8p, 9p, 10p, 13q, 17q, and 17q
  - Copy gains affect chromosomes 3q, 5p, 7p, 11q, 18q, 19p, and 20q
- SNP associated with an increased risk of SCC:
  - 7 loci related to pigmentation: MC1R, ASIP, TYR, SLC45A2, OCA2, IRF4, and BNC2
  - 4 loci involved in immune evasion, apoptosis inhibition or other oncogenic pathways: 11q23.3 (CAOM1), 2p22.3, 7p21.1 (AHR), and 9q34.3 (SEC16A)

- Gene expression profile analysis revealed differentially expressed genes associated with SCC: CXCL8 (IL-8), MMP1, HIF1A, ITGA6, and ITGA2

Immunotherapy for SCC

- Until recently, cytotoxic chemotherapy was usually the standard of treatment for patients with unresectable or metastatic disease with limited efficacy
- The anti-PD-1 antibody cemiplimab was approved in 2018 for the treatment of locally advanced or metastatic SCC with an overall response rate (ORR) of 50%, and 17% complete response
- Pembrolizumab was approved in 2020 with an ORR of ~34% and 4% complete response

Merkel Cell Carcinoma

Overview

- MCC is an uncommon skin tumor with neuroendocrine differentiation with a high risk of metastatic dissemination
- 95% increased incidence from year 2000 to 2013 (2488 cases/year), with a further projected increased incidence to 3284 cases/year in 2025
- Most common in fair-skinned elderly adults with high cumulative sun-exposed skin
- Increased incidence, early onset, and more aggressive behavior are seen in immunosuppressed patients (HIV-infection, HSC or solid organ transplantation, chronic lymphocytic leukemia, and lymphomas)
- The 5-year overall survival rates are approximately 50% for localized disease, 35% for regional metastasis, and 14% for distant metastasis

Pathogenesis

- Cell of origin: Previously presumed to be derived from Merkel cells, emerging studies have proposed a dermal fibroblast origin of Merkel cell polyoma virus (MCPyV)-positive MCCs and an epidermal origin of MCPyV-negative MCCs
- Two main etiologies involved in MCC pathogenesis: Merkel cell polyoma virus infection (causing 60–80% of all MCC) and DNA damage from UVR exposure
  - Merkel cell polyomavirus induces MCC by the expression of two viral oncoproteins (ST and LT) that disrupt key tumor suppressor pathways (Fig. 17.4)

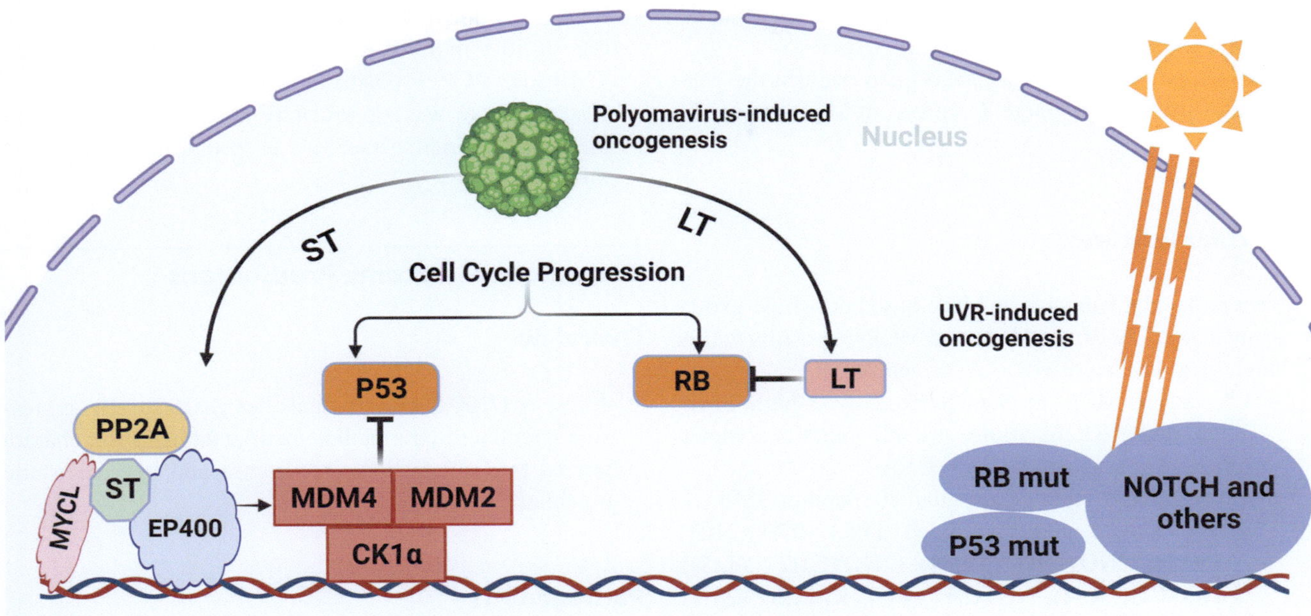

Fig. 17.4 Polyomavirus-induced and UVR-induced oncogenesis of Merkel cell carcinoma. Created with BioRender.com

- ○ ST forms a complex with PP2A, MYCL, and EP400. This complex promotes the transcription of genes inhibiting p53, including MDM2, MDM4, and CK1α
- ○ LT binds and inactivates the retinoblastoma tumor suppressor protein RB1
- – A high mutation burden caused by UVR exposure is found in the remainder of cases with common inactivating mutations of P53 and RB1. These tumors show no evidence of infection by MCPyV
- MCPyV-positive MCCs are associated with significantly longer overall survival, recurrent-free survival, and disease-specific survival

Table 17.2 The AEIOU acronym for clinical diagnosis of Merkel cell carcinoma

A – Asymptomatic/lack of tenderness
E – Expanding rapidly (doubling in <3 months)
I – Immunosuppression
O – Older than 50 years
U – UV-exposed sites

Clinical Features

- Predilection for chronically sun-damaged skin; most common affected sites are head, neck, extremities, and trunk
- Typically presenting as a painless, solitary, pink-purple, rapidly growing nodule
  - – The clinical features are summarized by the acronym AEIOU (Table 17.2)
- Frequently metastasizes to lymph nodes and distant organs such as liver, bone, pancreas, lung, and brain
- MCPyV-positive MCCs are associated with a larger size (>2cm) and greater tumor depth, while MCPyV-negative MCCs display greater frequency of immune compromise, ulceration, and coexistent SCC in situ

Histopathology

- Typically, a highly cellular intradermal and/or subcutaneous tumor that only rarely involves the epidermis of densely packed basophilic cells with a blastoid appearance
- Three histopathologic subtypes are identified:
- The intermediate variant is the most common: monotonously uniform cells with dense round nuclei and typical nuclear chromatin ("salt and pepper") pattern; sparse cytoplasm with paranuclear plaques consisting of intermediate filaments; diffuse and/or nested pattern of growth; high mitotic activity along with frequent single apoptotic cells and/or necrotic areas
  - – Additional variants include the small cell variant and the trabecular variant
  - – Rarely, expression of squamous, adnexal, or melanocytic differentiation and/or association with SCC, BCC, or melanoma
- Immunohistochemistry demonstrates positivity for common epithelial and neuroendocrine markers

- Tumor cells are positive for CK20 (expressed in a characteristic paranuclear dot-like pattern)
- The monoclonal antibody CM2B4 can confirm the presence of the MCPyV large T antigen in MCPyV-positive cases

Molecular Features

- Two molecular subtypes of MCC based on whole-exome sequencing: UV-driven MCC and MCPyV-positive/tumor mutational burden (TMB)-low driven MCC
  - UV driven MCC is characterized by a high TMB, TP53, and RB1 mutations, a UVR mutational signature, and absence of MCPyV DNA
    - High TMB often greater than 20 mutations/Mb
    - Top altered genes included TP53 (97%), RB1 (80%), NOTCH1 (45%), NOTCH2 (14%), NOTCH3 (6%), NOTCH4 (4%), KMT2D (26%), FAT1 (26%), LRP1B (23%), PIK3CA (21%), TERT (15%), and KMT2C (13%), PTEN (~10%)
    - Copy number alterations present in 24% of tumors
      - Gains of MYCL (6%) and MYC (4%)
      - Loss of RB1 (4%), PTEN (2%), TP53, LRP1B, CDKN2A, and NF1 (1% each)
  - MCPyV-positive/TMB-low MCC is defined by a low TMB, and MCPyV DNA integration
    - Low TMB, often less than 6 mutations/Mb
    - Top altered genes included TP53 (13%), RB1 (9%), and PTEN (7%)
- PI3K activation in both subtypes, suggesting a common requirement for alterations in this pathway
- Cluster analysis based on recurrent gene alterations showed that the molecular subgroups of MCC cluster with different tumor types:
  - The TMB-high subgroups clusters primarily with other neuroendocrine tumors (prostate and bladder neuroendocrine carcinoma and SCLC) and not with UV-associated skin cancers such as SCC, BCC, and melanoma
  - The TMB-low subgroup clusters primarily with carcinoid tumors, as well as other viral-driven tumors including HPV-positive head and neck SCC, cervical and anal SCC, and Kaposi sarcoma

Immunotherapy

- Avelumab, pembrolizumab, nivolumab, and ipilimumab are anti-PD-1 checkpoint inhibitors used for the treatment of locally advanced or metastatic MCC

- Response rate with checkpoint inhibitors have a response rate of 50% in TMB-high/UV-driven MCC and 41% of TMB-low/MCPyV-positive tumors
- Response rate was significantly correlated with line of therapy: 75% in first-line, 39% in second-line, and 18% in third line

Dermatofibrosarcoma Protuberans

Overview

- A rare (<1/100,000 individuals per year), locally aggressive sarcoma of intermediate malignant behavior, characterized by frequent local recurrence and low metastatic potential

Pathogenesis

- Cell of origin is considered to be fibroblastic in type
- The canonical gene fusion COL1A1-PDGFB is responsible for most cases, which brings the PDGFB gene under the control of the COL1A1 promoter. The rearrangement results in overexpression of PDGFB, which is proteolytically released from the COL1A1 sequence of the chimeric protein, with activation of the receptor PDGFRB, which activates the MAP-kinase pathway (Fig. 17.5)

Clinical Features

- Most common in young to middle-aged adults, but can occur at any age; congenital cases have been described
- Predilection for the shoulder and pelvic girdles, trunk, and proximal extremities
- Onset typically as a skin-colored or red-brownish firm plaque or nodule that can mimic a vascular neoplasm
- Slow, progressive, poorly defined growth, often reaching several centimeters in size; multinodular morphology in advanced stages
- 20–50% of local recurrences after surgical excision, but only rare risk for nodal or visceral metastases (<5%)
  - Local recurrence has been related to a fibrosarcomatous variant, and head and neck location

Histopathology

- Poorly circumscribed proliferation of small wavy to spindled cells in the deep dermis, displaying a monotonous storiform pattern

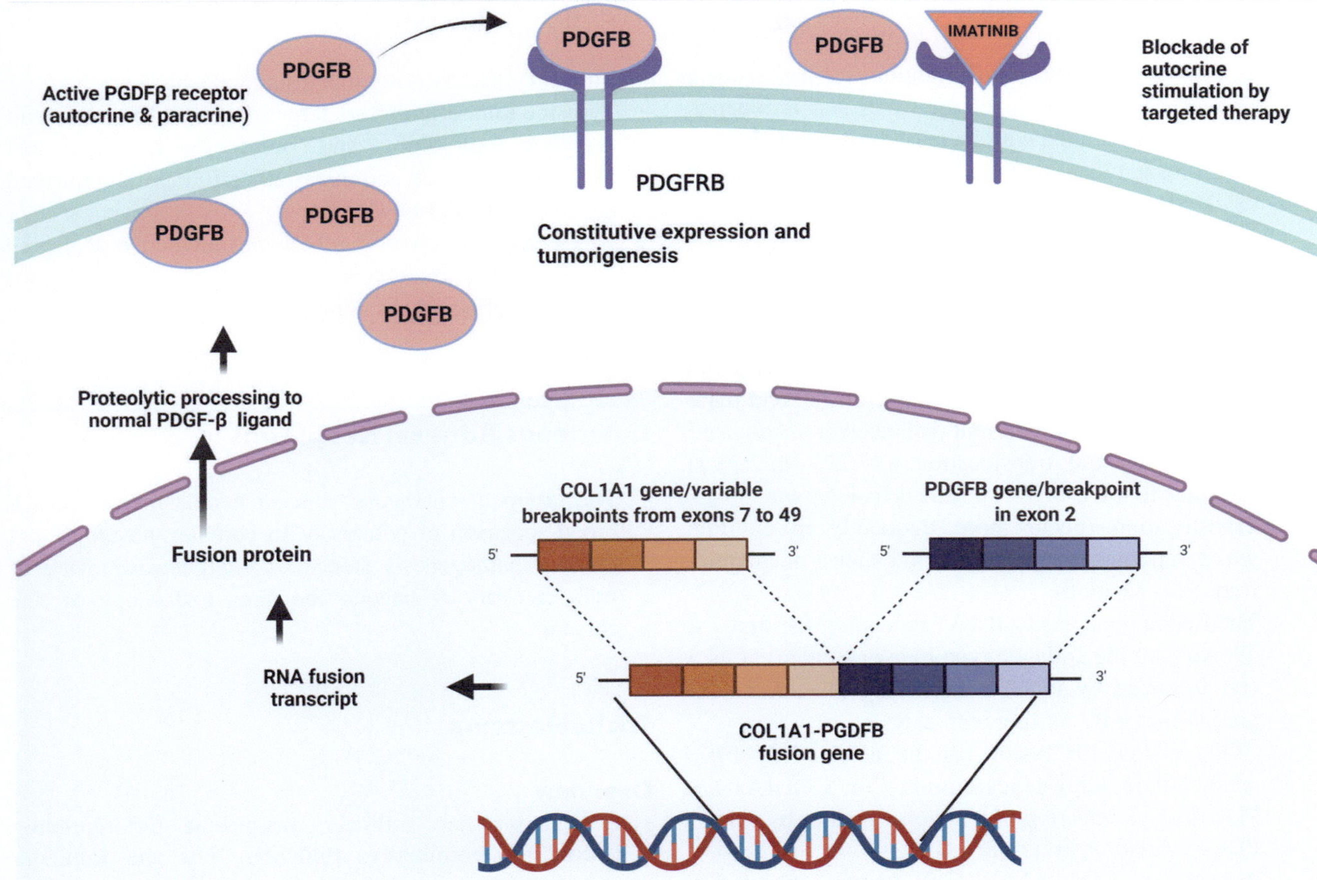

Fig. 17.5 Role of the COL1A1-PDGFB fusion gene in the pathogenesis of dermatofibrosarcoma protuberans. Created with BioRender.com

- Neoplastic cells exhibit uniform cytological features, with scant cytoplasm, mild degree of pleomorphism, and only few mitoses
- Striking tendency to invade the subcutaneous adipose tissue, initially along the adipose septa, later in a diffuse multilayered (honeycombing) pattern
- IHC positivity for CD34 (80–90% of cases) and apoD; negative staining for factor XIIIa, S100, CD31, and SMA
- Main differential diagnosis is dermatofibroma, which his associated with epidermal hyperplasia, collagen entrapment at the periphery of the lesion, and a less prominent storiform architecture with lack of honeycombing fat entrapment
- By IHC, dermatofibroma is positive for FXIIIA, CD68, CD163, and CD10, while negative or focally positive for CD34
- Several histologic variants exist
  - Giant cell fibroblastoma is most seen in childhood and shows in addition numerous multinucleated giant cells with wreath-like arrangement of nuclei (floret cells) and pseudovascular spaces in a loose, myxoid matrix
  - Myxoid DFSP is defined by the presence of >50% myxoid change, frequent branching small thin-walled vessels, and scattered mast cells
  - Myoid DFSP is characterized by pale nodules of spindle cells with eosinophilic cytoplasm that are positive for SMA and negative for CD34 and desmin; it has well-defined margins and is often centered on a blood vessel and associated with stromal hyalinization
  - Pigmented DFSP (Bednar tumor) contains numerous dendritic cells containing melanin pigment, which can be positive for S100
  - Rare variants include granular cell, sclerosing and atrophic DFSP
    - Multicentric atrophic DFSP is seen in association with deaminase-deficient severe combined immunodeficiency
  - Fibrosarcomatous transformation of DFSP can occur in ~10% of cases
    - Loss of the storiform growth pattern with abrupt transition to a fascicular or herringbone pattern, hypercellularity, cytologic atypia, increased mitotic

figures, and common loss or diminished expression of CD34

- o Fibrosarcomatous transformation is associated with metastases in ~13% of the cases and is independently associated with increased recurrences
- o Fibrosarcomatous transformation is seen in all DFSPs developing metastases

Molecular Features

- More than 95% of DFSPs feature supernumerary ring chromosomes containing centromere from 22 and material from 22 and 17, or recurrent (balanced or unbalanced) linear chromosomal translocation t(17;22) (q22;13.1), which leads to the COL1A1-PDGFB fusion gene
  - Translocations occur more frequently in children, while supernumerary ring chromosomes occur more frequently in adults
  - The fusion gene has COL1A1 (encoding the first 7 to 49 exons of the collagen type I alpha 1 chain) at its 5′ end followed by PDGFB (encoding platelet-derived growth factor B) starting with exon 2
  - COL1A1-PDGFB fusion can be detected by FISH using differentially labeled probes for COL1A1 and PDGFB and detection of a fusion signal or by using a PDGFB break-apart probe
  - A subset of COL1A1- PDGFB fusion genes (~5%) evades FISH testing, which can lead to misdiagnosis
    - o COL1A1 break-apart probe, next-generation of DNA or RNA may support the diagnosis of DFSP on these cases
    - o Cryptic COL1A1-PDGFB DFSPs are not clinicopathologically different from noncryptic cases, but they usually exhibit higher 5′-COL1A1 copy numbers (gains of 17q), which can be detected by comparative genomic hybridization
- Alternative, less common rearrangements include COL6A3-PDGFD and COL6A3-EMILIN2
  - Thought to activate a similar oncogenic autocrine loop involving PDGFRB signaling
  - COL6A3-PDGFD-positive DFSPs have been shown to predominate in females, and torso, especially the breast
  - EMILIN2-PDGFD-positive DFSPs are male predominant, preferentially subcutaneous and most of them exhibit fibrosarcomatous transformation
    - o A subset of EMILIN2-PDGFD DFSPs has been associated with homozygous deletion of CDKN2A
- Other rare translocations have been reported: TNC-PDGFD and MAP3K7CL-ERG fusion

Therapy

- First-line therapy is complete surgical excision
- Tyrosine kinase inhibitors targeting PDGFRB—imatinib mesylate, sunitinib, and sorafenib—are indicated in locally advanced and metastatic DFSP with confirmed COL1A1-PDGF fusion gene
  - Presence of chimeric gene is not predictive of degree of response to targeted therapy, but its absence is predictive of treatment failure

Cutaneous Adnexal Neoplasms

- The group of cutaneous adnexal neoplasms includes a broad spectrum of entities, with complex nosology and different pathogenesis. Hence, a comprehensive review of such category is beyond the aims and scope of this chapter

Trichoblastoma

Overview
- Benign biphasic follicular neoplasms differentiating toward the germinative follicular cells and follicular mesenchyme
- Sporadic tumors occur in the fifth and sixth decade of life, while syndromic tumors usually develop during puberty
- Most prevalent in women
- Most frequent neoplasm arising in nevus sebaceous (followed by syringocystadenoma papilliferum, trichilemmoma, and sebaceous adenoma)

Clinical features
- Most lesions occur in the head and neck area (frequently on the nose), followed by the trunk and extremities
- Present as solitary or multiple slow growing, asymptomatic, smooth, skin-colored papules, or nodules, sometimes associated with telangiectases
  - The presence of multiple lesions should raise the suspicion of Brooke–Spiegler syndrome (BRSS) or multiple familial trichoepithelioma

Histopathology
- Well-circumscribed dermal proliferation of basaloid cells with peripheral palisading but no cytologic atypia
- Surrounding stroma is abundant and composed by delicate fibrillary collagen bundles and spindled fibroblasts

- Resembles follicular papillae ("papillary mesenchymal bodies") and the perifollicular sheath
- A multinodular arrangement with the formation of distinct fibroepithelial units and clefts can be seen
  - Clefts are formed between the tumoral stroma and the peritumoral stroma, in contrast to BCC (clefts between tumor and stroma)
- Histologic variants include cribriform (conventional trichoepithelioma), small nodular, large nodular, retiform, racemiform, adamantanoid (cutaneous lymphadenoma), and columnar (desmoplastic trichoepithelioma)
  - Most neoplasms show a predominant variant accompanied by one or more minor variants (36.9%) or are composed by many variants without a clear predominant type; pure variants occur only in 29.5% of cases
  - Rare tumors show unusual growth patterns: Continental pattern and trabecular/garland-like pattern (3.4%)
  - Adamantanoid trichoblastoma has been deemed a distinct benign lymphoepithelial entity by some authors, given its distinct cytomorphology and cell markers
    - Epithelial nests of palisading basaloid cells, surrounding large polygonal cells with clear cytoplasm, and a dense lymphocytic infiltrate with occasional Reed Sternberg-like (RS-L) cells; epithelial cells positive for AR, and RS-L cells positive for Notch 1. Common EGFR mutations
- Main differential diagnosis is basal cell carcinoma (Table 17.3)
- Trichoblastic carcinoma is rare and is composed of a basaloid component that shows moderate to severe cytologic atypia, and a bland stromal component identical to that of trichoblastoma
- Trichoblastic carcinosarcoma is also rare, and both epithelial and mesenchymal components are malignant, with pleomorphic cells and abnormal mitotic figures

Molecular Features

- Loss of heterozygosity of the PTCH gene (pq22.3) has been reported in 48% of trichoblastoma
- Rare sporadic cases have been shown to carry HRAS and exon 3 CTNNB1 mutations
- EGFR, PIK3CA, and FGFR3 gain of function mutations have been detected in 56% of adamantanoid trichoblastoma (lymphadenomas)
- CTNNB1, SUFU, TP53, and CDKN2A mutations have been identified in trichoblastic carcinosarcomas
- The pathogenesis of several cases of multiple familial trichoepithelioma (MFT)-associated trichoepithelioma has been linked to germline mutations of CYLD-BRSS (described in the chapter on genetic/hereditary syndromes)

Table 17.3 Histological and immunohistochemical differences between trichoblastoma and basal cell carcinoma

Trichoblastoma	Basal cell carcinoma
Absent/focal mitotic figures	Mitotic figures
Absent/focal necrosis	Variable necrosis
No inflammatory infiltrate (except for adamantanoid variant)	Lymphocytic infiltrate
No connection with epidermis	Connection with epidermis
Follicular stroma +\− Mesenchymal bodies	Variable mucinous stroma
IHC: PHLDA1 (follicular stem cell) CK20 (Merkel cells) CD10 (stroma) CD34 (stroma) BCL2 (peripheral) CK5 CK6 CK15 BerEp4 P53	IHC: AR CD10 (epithelium) BCL2 (diffuse) CK5 CK6 CK15 BerEp4 P53

Cylindroma and Spiradenoma

Overview

- Cylindromas and spiradenomas are closely related benign cutaneous adnexal neoplasms with sweat gland differentiation
- They show similar histological features and may represent part of a morphological spectrum, further evidenced by rare cylindrospiradenoma hybrid tumors
- Malignant transformation is rare, but can result in the development of metastatic tumors with poor survival

Clinical Features

- Most tumors are sporadic and present as slow growing, skin-colored to erythematous dermal papules or nodules, with predilection for the head and neck (cylindromas), and the extremities (spiradenomas)
- Spiradenomas may be distinguished by bluish color and paroxysms of localized pain; glabrous surfaces of the skin are always spared
- Multiple cylindromas and spiradenomas together with trichoepitheliomas may be observed in BRSS and familial cylindromatosis

Histopathology

- Cylindromas are characterized by a dermal proliferation of closely set epithelial lobules arranged in a "jigsaw pattern"
  - In each lobule, peripheral basaloid cells surround larger pale cells; ductal differentiation is possible
  - Lobules are surrounded by prominent eosinophilic PAS-positive basement membrane material and exhibit eosinophilic PAS-positive globules within them

- Spiradenomas present as well-circumscribed epithelial nodules in the dermis. The two cell populations typical of cylindroma are intermixed in nodules of spiradenoma, together with numerous scattered lymphocytes
  - Occasional ductal and/or cystic structures
- Hybrid tumors with intermediate features between cylindroma and spiradenoma may often be seen ("cylindrospiradenoma")

Molecular Features

- Cylindroma and spiradenoma are related to oncogenic activation of the NF-kB pathway, triggered by mutations of either CYLD or ALPK1
  - CYLD is a tumor suppressor deubiquitinase that negatively regulates NF-kB activation
  - Most cases of cylindroma carry germline or somatic CYLD mutations compared to approximately one-third of cases of spiradenoma
  - CYLD loss of function mutations (germline or somatic) can be seen in ~30% of spiradenomas and 92% of cylindromas
    - Germline mutations of CYLD are seen in BRSS and its phenotypic variant familial cylindromatosis
  - ALPK1 is a member of the α-kinase family, which can activate the NF-kB pathway, and has previously been suggested to function as an oncogene
  - ALPK1 gain of function mutation (p.V1092A) is detected in 43% of spiradenomas and 28% of spiradenocarcinomas
  - TERT and TP53 mutations have been seen in a subset of spiradenocarcinomas
- MYB gene overexpression is a common feature of cylindromas and spiradenomas
  - c-MYB is a transcription factor with a DNA-binding domain that recognizes the consensus sequence C/TAACNG, frequently observed in the enhancers of genes associated with cell cycle progression, regulation of cell survival, and lineage specification
  - MYB overexpression may result from activated NF-kB, which can bind the MYB promoter. Another tempting explanation is that CYLD could alter chromatin dynamics at the MYB locus, since CYLD can negatively control histone deacetylases HDAC6 and HDAAC7
  - Overexpression of MYB leads to upregulation of anti-apoptotic genes BCL2 and BRC3, suggesting that MYB precipitates tumorigenesis through inhibition of apoptosis
  - MYB-NFIB fusions are seen in adenoid cystic carcinomas of the salivary gland and breast, which share morphologic and molecular features with cylindromas, supporting a pathogenetic link between these neoplasms
  - Conflicting data have been reported: One study reported MYB-NFIB fusions with MYB overexpression in a majority of sporadic cylindromas, while another study of the genomic analysis of 75 tumors (cylindromas, spiradenomas, and hybrid cylindrospiradenomas) revealed only overexpression of MYB, without an underlying fusion event
- Low somatic point mutation burden (0.04–2.88 mutations/Mb) and no significant presence of copy number alterations, except for high-grade spiradenocarcinomas, which harbor increased copy-number gains/losses
- Signature 1 (age-associated signature) is recurrently present across all tumor types (cylindroma, spiradenomas, and cylindrospiradenoma)
- Signature 7 (UV-associated signature) is present in cylindromas
- Signature 26 (mismatch repair-associated signature) is present in low-grade spiradenocarcinomas
- Cylindromas have also increased mutations in DNMT3A, which plays a role in the regulation of DNA methylation

Pilomatricoma

Overview

- A benign adnexal neoplasm characterized by follicular differentiation toward the matrix of the hair bulb
- Synonyms: pilomatrixoma, calcifying epithelioma of Malherbe

Clinical Features

- Occurs most commonly in the first two decades of life (mean age at 16 years), but may occur at any age
- Firm, solitary, pink–bluish, dermal nodule, or cyst on nonglabrous skin, especially head and neck region
- Usually asymptomatic, but inflammation and erythema may occur
- Commonly misdiagnosed (accuracy of clinical examination reported to be 16%)
- Pilomatricoma is associated with genetic diseases including Gardner syndrome, MYH-associated polyposis, Rubinstein–Taybi syndrome, Sotos syndrome, myotonic dystrophy, gliomatosis cerebri, and Turner syndrome, among others

Histopathology

- Well-circumscribed, nodulocystic tumor in the dermis
- Two cell populations with matrical differentiation
  - Basaloid cells often with mitoses at the periphery

- Eosinophilic cells with residual outlines of nuclei (shadow cells) in the center
- Transition between the two cell populations is often abrupt; at times, transitional cells with intermediate features and trichohyalin granules may be observed
- Varying degree of calcification, and even ossification, granulomatous inflammatory reaction, and fibrosis
- Diverse variants have been reported: Bullous, aggressive, superficial, perforating, proliferating, ossifying, cystic, pseudocystic, pigmented, acantholytic, and malignant
- Basaloid cells show prominent nuclear expression of β-catenin, but stain negatively for CK15 and SOX9

Molecular Features

- Pilomatricomas are characterized by increased β-catenin transcriptional activity and canonical Wnt signaling activation
  - Activating mutations in exon 3 of the CTNNB1 gene, encoding β-catenin, have been identified in 25–100% of sporadic pilomatricomas
  - These CTNNB1 mutations disrupt β-catenin phosphorylation by the axin/APC/GSK3β/CKIα complex required for its destruction and promote its stabilization and nuclear localization, leading to constitutive Wnt signaling
  - Gardner syndrome, which frequently features pilomatricomas and/or epidermal cysts with matrical changes, is a FAP variant; FAP-associated APC mutations also lead to β-catenin stabilization
- CTNNB1 activity is associated with matrical differentiation
  - Sporadic BCCs with foci of matrical differentiation harbored CTNNB1 mutations
    - Activating CTNNB1 mutations have been found in craniopharyngiomas, ameloblastomas, and calcifying odontogenic cysts, neoplasms which all may feature the abrupt transition between peripheral basaloid cells and central eosinophilic cells with shadow cells differentiation

Pilomatrical Carcinoma

- Recurrence of a pilomatricoma has been related to increased risk of malignant transformation
- Pilomatrical carcinoma is a rare, low-grade carcinoma distinguished by matrical differentiation and high propensity for recurrence
- Has CTNNB1 mutations similar to pilomatricoma
- Pilomatrical carcinomas show a significant predilection for adults and chronically sun-damaged skin and are rare in children
- Pathological features favoring a diagnosis of pilomatrical carcinoma include presence of bona fide necrosis, marked nuclear pleomorphism, or infiltrative growth

Sebaceous Neoplasms

Overview

- Sebaceous adenoma is a benign neoplasm composed predominantly by mature sebocytes, admixed with occasional immature basaloid seboblasts (<50% of tumor cells)
- Sebaceoma (synonym: sebaceous epithelioma) is a benign neoplasm composed predominantly by immature basaloid seboblasts (>50% of tumor cells) with admixed occasional mature sebocytes
- Sebaceous carcinoma (SC) is a malignant neoplasm demonstrating sebocytic differentiation

Clinical Features

- Sebaceous adenoma and sebaceoma are symptomatic, usually solitary, dome-shaped papules or nodules. Skin-colored to yellowish. SC presents as a frequently ulcerated rapidly growing erythematous nodule. Initial misdiagnosis of ocular SC as chalazion may hinder timely treatment
- All sebaceous neoplasms have a predilection for adults and for the head and neck region
  - SC can be periocular (has a predilection for the upper eyelid) or extraocular (favors the head and neck area, but may include trunk, extremities, and genitalia)
- Multiple lesions, at time with a cystic appearance, particularly when presenting outside the head and neck area may indicate Muir Torre syndrome (MTS)
  - Sebaceous adenoma exhibits the strongest association with MTS, followed by sebaceoma; SC has the weakest association
- Sebaceous adenoma and sebaceoma are benign. SC has a 30–40% risk of local recurrence, 20–25% risk of distant metastasis, and a 10–30% of tumor-related mortality

Histopathology

- Sebaceous adenoma and sebaceoma are well-circumscribed dermal tumors, with lobulated configuration typically emanating from the undersurface of the epidermis and comprised of two cell populations in variable proportion, mature sebocytes (usually prevailing toward the center), and immature basaloid seboblasts (more abundant at the periphery)
  - Mitotic figures may be observed in seboblasts, but no cytologic atypia or necrosis are found
  - Sebaceous adenoma is characterized by a more superficial location, smaller size, significant predominance of mature sebocytes (>50%), and multiple peripheral layers of immature seboblasts
  - Sebaceoma features deeper, larger proliferation of immature seboblasts (>50%), with foci of mature sebocytic differentiation

- Lesions with intermediate features may be observed indicating that sebaceous adenoma and sebaceoma represent a spectrum
- SC presents with infiltrative growth, seboblasts with cytologic atypia and frequent necrosis, and irregular areas of sebocytic differentiation
  - Intraepidermal, pagetoid spread of pale malignant cells is a frequent occurrence in periorbital SC
- Helpful IHC markers in the differential diagnosis include:
  - Factor XIIIa (AC-1A1), adipophilin, GATA3, AR, CK7, and EMA are consistently positive in sebaceous neoplasms
    - EMA is positive only in well-differentiated SC
    - CK7 is positive in ~50% of SC
  - The combination of cytoplasmic EMA positivity and lack of BerEP4 expression may aid in the distinction between sebaceoma and BCC
  - High nuclear Ki-67 and p53 staining may help in establishing the malignant nature of any sebaceous neoplasm
  - Complete loss of nuclear immunoreactivity for MLH1, MSH2, or MSH6 in lesional tissue may be employed as a screening assessment of mismatch repair (MMR) defects and microsatellite instability (MSI) (see Muir Torre Syndrome section)

Molecular Features

- The prevalence of MMR deficiency in sebaceous neoplasms occurs in 15–60%; however, most of these tumors do not occur on a background of MMR germline mutation (MTS)
  - Patients with MMR-deficient neoplasms without MTS can result from gene hypermethylation and somatic mutations
- The incidence of MTS in patients with sebaceous neoplasms ranges between 14 and 50%
  - The association is highest for multiple sebaceous neoplasms and those located outside the head and neck area with onset before age 60 years
  - Sebaceous adenoma appears to exhibit the strongest association with MTS, followed by sebaceoma. SC has the weakest association and relates mainly to extraocular tumors
- One study using whole-exome sequencing identified three distinct molecular/clinical subtypes in SC: SC with UV-damage associated mutational signatures, SC with microsatellite instability, and SC showing a paucity of somatic mutations
  - Overall, recurrent driver genes discovered across the 32 tumors included TP53, NOTCH1, NOTCH2, ZNF750, RREB1, KMT2D, HRAS, KRAS, and FAT3
  - SC with UV-damage-associated mutational signatures

- Develop on heavily sun-damaged skin
- Tumors are histologically poorly differentiated with an infiltrative growth pattern, sometimes with squamous differentiation
  - Transcriptomes of UV-SC are related to those in SCC and BCC
  - A candidate cell of origin of UV-damage-SC is a subpopulation of keratinocytes that acquire sebocytic differentiation secondary to somatic or epigenetic alterations
- High somatic mutation burden (>50 mutations per Mb): Predicts immunotherapy response
- SC with MSI
  - Usually occur on the trunk, a site almost completely spared in the other two subtypes
  - Germline or somatic inactivating mutations of MMR genes
  - Histopathologically, it shows greater differentiation with well-circumscribed borders
  - Lower mutational burden than UV-SC and higher than pauci-mutational SC
  - MMR-derived indels with MSI signatures
  - Recurrent mutation of RREB1 gene, which functions in adipocytic differentiation
- SC showing a paucity of somatic mutations
  - Mutational burden (1.2–5.5 mutations per Mb)
  - Tumors occur on the face, as ocular or extraocular
  - Ocular SC show moderately differentiation with scant chronic sun-induced skin damage
  - Harbor recurrent truncating mutations in the ZNF750 (especially ocular SC)
  - NOTCH 1 mutations in extraocular cases
- Another study using NGS of a targeted panel of cancer-associated genes identified two distinct subgroups of ocular SC
  - Type I ocular SC
    - Older age at presentation (mean age, 70 years)
    - High frequency of mutations in TP53 and/or RB1, which occasionally coexist with NOTCH mutations
    - More locally aggressive clinical phenotype including higher grade nuclear features and common local recurrence
  - Type II ocular SC
    - Younger age at presentation (mean age, 58 years)
    - Lack of TP53, RB, or NOTCH mutations
    - high-risk HPV infection in half of the cases
    - Uncommon local recurrence
- Mutations in TERT have been described in a subset of SC
  - Additional mutated genes include EGFR, ATRX, PDGFRA, CDKN2A, PTEN, and ACVR1
  - Mutations on EGFR, PDGFRA, and PTEN predict activation of the PI3K signaling cascade

Genetic/Hereditary Syndromes

Familial Melanoma

- Melanoma occurring in two or more first-degree relatives
- Accounts for ~10% of all malignant melanomas
- Caused by the following high-penetrance genes: CDKN2A, CDK4, POT1, ACD, TERF2IP, TERT, and BAP1 (Table 17.4)
  - Autosomal dominant germline mutations of CDKN2A account for 20–40% of cases in familial melanoma
    - Overall, it has a 30% and 67% penetrance by age 50 and 80 years, respectively. Although, penetrance can reach near 100% in patients with high-intensity UVR exposure and sunburns
    - Melanomas arise ~10 years earlier than in the general population
    - Increased risk also for pancreatic cancer (melanotic-pancreatic syndrome), astrocytomas (melanoma-astrocytoma syndrome), and multiple melanocytic nevi (familial atypical multiple mole melanoma – FAMMM), among others
  - Rest of high-penetrance genes including CDK4 mutation occur in <1% of cases in familial melanoma
- Intermediate-penetrance genes include MITF; involved in pigmentation
  - The most common variant of MITF is p.E418K, which enhances the binding of MITF to HIF1A promoter and increases its transcriptional activity
- Low-penetrance genes involve MC1R and ~20 loci associated with pigmentation, number of nevi, telomere biology, DNA repair, and cell cycle regulation
  - MC1R polymorphisms are associated with the "red hair color" phenotype, which is associated with impaired DNA repair and increased susceptibility to UVR damage and mutagenesis

Table 17.4 High-, intermediate-, and low-penetrance genes involved in familial melanoma

	Genes	Prevalence in familial melanoma
High-penetrance genes	CDKN2A	20–40%
	CDK4	<1%
	POT1	<1%
	ACD	<1%
	TERF2IP	<1%
	TERT	<1%
	BAP1	<1%
Intermediate-penetrance genes	MITF	1–5%
Low-penetrance genes	MC1R	70–90%

- Individuals carrying any MC1R variant have a higher risk of developing melanoma and nonmelanoma skin cancer, with even higher risk as the number of variants increases in the carrier

Xeroderma Pigmentosum

- Rare autosomal-recessive disorder with 100% penetrance characterized by genetic defects in the NER mechanism
- The incidence of XP is dependent on geographic distribution: 1 per 1 million people in Europe and the USA and 1 per 22 000 people in Japan
- Characterized by extreme sensitivity to UVR with subsequent damage in the skin, eyes, increased risk of melanoma and NMSC, and neurologic alterations
- Clinical features are heterogeneous and depend on the underlying genetic alteration:
  - Skin: lentigines, hyper/hypopigmentation, actinic keratosis, melanoma, and NMSC
    - Melanoma and NMSC are 2000-fold and 10,000-fold more frequent in XP patients compared to the general population, respectively, with a median age of onset of 10 years
  - Ocular: photophobia, keratitis, pterygia, pingueculae, corneal scarring, NMSC
  - Nervous system – mild to severe: peripheral neuropathy, sensorineural hearing loss, mental retardation, seizures, other
- XP is associated with mutations in 8 genes: 7 NER genes (XPA-XPG) and one XP variant with functional NER genes but a mutated DNA polymerase (Pol η) (Table 17.5)

Carney Complex

- Rare autosomal dominant multiple neoplasia syndrome, caused in most patients by loss of function mutations in the PRKAR1A gene
- Clinical features include skin pigmentations abnormalities, myxomas, endocrine tumors/ overactivity, gonadal and thyroid neoplasia, and psammomatous melanotic schwannomas
  - Skin manifestations include lentigines involving the lips, conjunctiva, canthi, and genital areas, and pigmented epithelioid melanocytoma
- Germline mutations in PRKARA1 gene on chromosome 17q24.2 are found in 37% of patients with sporadic CNC and more than 70% of patients with familial CNC
  - PKA is a cAMP-dependent serine-threonine kinase that is involved in the regulation of many cellular pro-

Table 17.5 Genes and characteristics associated with XP types

XP type	Gene involved	Normal protein function	Frequency (%)	Clinical features
XP-A	XPA	Helps repair damaged DNA via binding interactions with TFIIH, XPF, and XPG	30	Photosensitivity, poikiloderma, lentigines, skin cancer, neurodegeneration
XP-B	XPB/ERCC3	A 3′ to 5′ helicase - part of the TFIIH transcription factor complex	0.5	Photosensitivity, poikiloderma, lentigines, skin cancer, neurodegeneration
XP-C	XPC	Global nucleotide excision repair - binds to damaged DNA	27	Photosensitivity, poikiloderma, lentigines, skin cancer,
XP-D	XPD/ERCC2	A 5′ to 3′ helicase - part of the TFIIH transcription factor complex	15	Photosensitivity, poikiloderma, lentigines, skin cancer, neurodegeneration, brain tumors
XP-E	XPE/DDB2	Global nucleotide excision repair - binds to damaged DNA	1	Photosensitivity, poikiloderma, lentigines, skin cancer, neurodegeneration
XP-F	XPF/ERCC4	A 5′ endonuclease-forms a heterodimer with ERCC1	2	Photosensitivity, poikiloderma, lentigines, skin cancer, neurodegeneration, brain tumors
XP-G	XPG	A 3′ endonuclease, stabilizes open complex	1	Photosensitivity, poikiloderma, lentigines, skin cancer, neurodegeneration
XP-V	DNApol η	A translesion DNA polymerase	23.5	Milder photosensitivity and poikiloderma

cesses including transcription, metabolism, cell cycle progression, and apoptosis

- Inactivation of the regulatory subunit PRKAR1A leads to constitutive activation of the cAMP-PKA pathway with effects on pigmentation and cell proliferation in melanocytes
- More than 140 germline PRKARA1 alterations, single base substitutions, small and large deletions, insertions, and rearrangements have been identified

- The remaining cases are linked to a second affected locus on chromosome 2p16 (CNC2 locus); the responsible gene has not been identified
- The overall penetrance of CNC with pathogenic PRKARA1 variants is >95% by the age of 50, with only two variants known to result in incomplete penetrance (splice variant c.709(-7-2) del6 and substitution p.M1Vp). These mutations lead to a relatively mild CNC phenotype restricted mostly to psammomatous melanotic schwannomas
- In pigmented epithelioid melanocytomas, there is inactivation of the remaining wild-type allele of PRKRA1 in addition to a MAP-kinase pathway activating mutation such as in BRAF or NRAS that typically leads to loss of immunoreactivity for PRKRA1 protein

BAP-1 Tumor Predisposition Syndrome (BAP1-TPDS)

- Rare autosomal dominant tumor predisposition syndrome caused by germline mutation in the BAP 1 gene
- Clinical features include multiple BAP-1-inactivated melanocytomas of the skin and increased risk of developing uveal and cutaneous melanomas, peritoneal and pleural mesotheliomas, clear cell renal cell carcinomas, and basal cell carcinoma. Other cancer types include lung adenocarcinoma, meningioma, neuroendocrine carcinoma and paraganglioma, thyroid cancer, and cholangiocarcinoma, among others

- BAP-1 inactivated melanocytoma beginning in the second decade of life, patients usually develop multiple flesh-colored, well-circumscribed papules with characteristic histopathology (described in melanoma chapter). The number of lesions varies, typically ranging from a few to >50
  - Initiating mutations are BRAF or NRAS
  - Lesions are classified as intermediate lesions of uncertain malignant potential
  - Melanoma arising from BIN is infrequent
- BAP1 inactivating mutations occur in up to 40% of uveal melanomas, and up to 17% in melanomas arising in blue nevi

- BAP1 (BRCA1-associated protein I) is located on chromosome 3 (3p21.1); it is a nuclear tumor suppressor deubiquitinase that acts on histone H2A and is involved in chromatin modification and transcriptional regulation
  - Loss or inactivation of the wild-type BAP1 allele often occurs via deletions spanning its locus on chromosome 3p21.1 or monosomy 3
- Penetrance of BAP1-TPDS is incomplete; the risk of developing associated tumors may depend on additional environmental factors such as UVR exposure (melanoma) and asbestos exposure (mesothelioma)

Muir Torre Syndrome

- Rare autosomal dominant condition characterized by sebaceous neoplasms and/or keratoacanthomas and visceral malignancies

- High penetrance with variable expressivity
- Sebaceous neoplasms include sebaceous adenomas, sebaceoma, sebaceous carcinomas, keratoacanthomas, and basal cell carcinomas with sebaceous differentiation
  - Sebaceous neoplasms frequently involve the head and neck region
- Most frequent visceral malignancy is colorectal cancer, followed by the urogenital system (endometrium, ovary, bladder, kidney, and ureter). Other rare malignancies include breast, parotid, prostate, upper GI, larynx, and hematologic
  - Visceral malignancies from MTS are usually less aggressive than their sporadic counterparts
- Recognized as a distinct variant of hereditary nonpolyposis colorectal cancer
- Muir-Torre type 1: Caused by germline inactivating mutation of the mismatch repair enzymes: MLH1, MSH2, MSH6, and PMS2 that result in microsatellite instability (MSI)
  - Most mutated gene is MSH2 (90%)
  - Second most common mutated gene is MLH1 s(~10%)
    - Rare cases are associated with MSH6 or PMS2 mutations
    - In contrast with hereditary nonpolyposis colorectal cancer, in which germline variants are roughly equally distributed across all MMR genes
- Muir-Torre type 2: A low- to moderate-penetrance autosomal recessive syndrome, caused by biallelic inactivation of MUTYH, a base excision repair gene. Tumors do not have pathogenic variants in mismatch repair genes and have no MSI
- In patients with suspected MTS type 1, tumor testing for lack of MMR protein with IHC or MSI analysis can be used as a screening test
  - The positive predictive value for MTS using IHC has a wide range of variation within studies but is usually poor, ranging from 22% to 37%; although the NPV can be as high as 95%
  - DNA-based-MSI analysis is performed with five markers recommended by the National Cancer Institute: BAT25, BAT26, D2S123, D5S346, and D17S250 (Bethesda markers)
    - MSI-H (high) is defined when at least two of the five markers show instability
    - MSI-L (low) is defined when one of the markers show instability
    - MSI-S (stable) is defined when no marker shows instability
- In patients with abnormal results by IHC and/or MSI testing, especially if supported by a family history or clinical signs for MTS, additional genetic testing for MMR germline defects and genetic counseling should be considered
- Testing for variants in MUTYH can be considered for patients in whom a pathogenic variant in MMR genes is not detected

Brooke–Spiegler Syndrome

- Also known as CYLD cutaneous syndrome, is an autosomal dominant syndrome characterized by the progressive development of multiple cylindromas (favoring the scalp), spiradenomas, and/or trichoepitheliomas (cribriform trichoblastomas), appearing during childhood and early adolescence
- Penetrance increases with age, but some carriers remain free of clinically detectable lesions
- Marked phenotypical variability regarding both clinical and histologic findings
- Tumors are generally benign, but malignant transformation can occur in 5–10% of affected individuals
- 80–85% of BRSS are caused by germline mutations on CYLD, a tumor suppressor gene located on chromosome 16q12-13. The remaining wild-type CYLD allele is inactivated in neoplasms by somatic deletion, mutation, or LOH
  - CYLD gene encodes a deubiquitinase, which removes Lys-63-linked ubiquitin chains from proteins including TNF receptor-associated factors (TRAF2, TRAF6, and TRAF7), NF-kB essential modulator (NEMO), and BCL-3; this leads to downregulation of the nuclear factor kappa B (NF-kB) signaling
    - Loss of CYLD leads to increased NF-kB activity, promoting uncontrolled cell proliferation and development of tumors associated with BRSS
- Inactivation of the CYLD gene occurs by frameshift, nonsense, missense, splice-site mutations, and deletion.
  - There is no hotspot, but most mutations detected occur in exons 9–20, whereas exons 4–8 are spared
- Multiple familial trichoepithelioma (MFT) is a phenotypic variant characterized by the development of numerous trichoepitheliomas (cribriform trichoblastoma) only (confined to the face)
  - Germline CYLD mutations are seen in 40–50% of cases
- Recurrent pathogenic variants in DNMT3A and BCOR have been reported in a subset of tumors, suggesting that epigenetic dysregulation may also play a role in the pathogenesis of BRSS

Nevoid Basal Cell Carcinoma Syndrome (Gorlin Syndrome)

- Syndrome caused by mutations in hedgehog signaling components (PTCH1, PTCH2, and SUFU) which cause aberrant activation and development of hedgehog-pathway associated malignancies, affecting approximately 1:60,000 individuals
- Patients characteristically have innumerable basal cell carcinomas with microscopic lesions that may be clinically occult
- Patients may develop additional hedgehog-driven malignancies including medulloblastoma (1–5%), odontogenic keratocysts, cardiac and ovarian fibromas, meningioma, and rhabdomyosarcoma
- Additional clinical manifestations include central nervous system calcification (particularly the falx cerebri), palmar-plantar pits, skin tags, and facial milia
- Developmental abnormalities may include frontal bossing, macrocephaly, course facial features, cleft lip/palate, hypertelorism, rib malformation, scoliosis, polydactyly, syndactyly, and spina bifida
- De novo germline mutations account for 30–50% of affected patients
- Systemic smoothened inhibitor (vismodegib, sonidegib) has been successful in treating BCCs in the setting of NBCCS

Suggested Reading

Ultraviolet Light and Skin Cancer

Abdel-Malek ZA, Kadekaro AL, Swope VB. Stepping up melanocytes to the challenge of UV exposure. Pigment Cell Melanoma Res. 2010;23:171–86.

Autier P, Dore JF, Eggermont AM, et al. Epidemiological evidence that UVA radiation is involved in the genesis of cutaneous melanoma. Curr Opin Oncol. 2011;23:189–96.

Gerstenblith MR, Shi J, Landi MT. Genome-wide association studies of pigmentation and skin cancer: a review and meta-analysis. Pigment Cell Melanoma Res. 2010;23:587–606.

Kappes UP, Luo D, Potter M, et al. Short- and long-wave UV light (UVB and UVA) induce similar mutations in human skin cells. J Invest Dermatol. 2006;126:667–75.

Narayanan DL, Saladi RN, Fox JL. Ultraviolet radiation and skin cancer. Int J Dermatol. 2010;49:978–86.

Runger TM. How different wavelengths of the ultraviolet spectrum contribute to skin carcinogenesis: the role of cellular damage responses. J Invest Dermatol. 2007;127:2103–5.

Runger TM. C–>T transition mutations are not solely UVB-signature mutations, because they are also generated by UVA. J Invest Dermatol. 2008;128:2138–40.

Runger TM. Is UV-induced mutation formation in melanocytes different from other skin cells? Pigment Cell Melanoma Res. 2011;24:10–2.

Runger TM, Kappes UP. Mechanisms of mutation formation with long-wave ultraviolet light (UVA). Photodermatol Photoimmunol Photomed. 2008;24:2–10.

Von Thaler AK, Kamenisch Y, Berneburg M. The role of ultraviolet radiation in melanomagenesis. Exp Dermatol. 2010;19:81–8.

WHO Classification of Tumors Editorial Board. WHO classification of tumours: digestive system. Lyon, France: International Agency for Research on Cancer; 2019.

Wu J, Jiao Y, Dal Molin M, et al. Whole-exome sequencing of neoplastic cysts of the pancreas reveals recurrent mutations in components of ubiquitin-dependent pathways. Proc Natl Acad Sci U S A. 2011;108:1188–93.

Wu J, Matthaei H, Maitra A, et al. Recurrent GNAS mutations define an unexpected pathway for pancreatic cyst development. Sci Transl Med. 2011;3:92ra66.

Melanoma

Bahrami A, Barnhill RL. Pathology and genomics of pediatric melanoma: a critical reexamination and new insights. Pediatric Blood & Cancer. 2017;65(2):e26792.

Bastian BC, Kashani-Sabet M, Hamm H, et al. Gene amplifications characterize acral melanoma and permit the detection of occult tumor cells in the surrounding skin. Cancer Res. 2000a;60:1968–73.

Bastian BC, LeBoit PE, Pinkel D. Mutations and copy number increase of HRAS in Spitz nevi with distinctive histopathological features. Am J Pathol. 2000b;157:967–72.

Bastian BC, Olshen AB, LeBoit PE, et al. Classifying melanocytic tumors based on DNA copy number changes. Am J Pathol. 2003;163:1765–70.

Bastian BC, Xiong J, Frieden IJ, et al. Genetic changes in neoplasms arising in congenital melanocytic nevi: differences between nodular proliferations and melanomas. Am J Pathol. 2002;161:1163–9.

Bastian BC. The molecular pathology of melanoma: an integrated taxonomy of melanocytic neoplasia. Annu Rev Pathol. 2014;9(1):239–71.

Cancer Genome Atlas Network. Genomic classification of cutaneous melanoma. Cell. 2015;161(7):1681–96.

Chang D, Shain AH. The landscape of driver mutations in cutaneous squamous cell carcinoma. NPJ Genomic Med. 2021;6(1):1–10.

Chen LL, Jaimes N, Barker CA, Busam KJ, Marghoob AA. Desmoplastic melanoma: a review. J Am Acad Dermatol. 2013;68(5):825–33.

Cherepakhin OS, Argenyi ZB, Moshiri AS. Genomic and transcriptomic underpinnings of melanoma genesis, progression, and metastasis. Cancers. 2021;14(1):123.

Comito F, Pagani R, Grilli G, Sperandi F, Ardizzoni A, Melotti B. Emerging novel therapeutic approaches for treatment of advanced cutaneous melanoma. Cancers. 2022;14(2):271.

Costa S, Byrne M, Pissaloux D, Haddad V, Paindavoine S, Thomas L, Aubin F, et al. Melanomas associated with blue nevi or mimicking cellular blue nevi. Am J Surg Pathol. 2016;40(3):368–77.

Donati M, Kastnerova L, Martinek P, Grossmann P, Sticová E, Hadravský L, Torday T, Kyclova J, Michal M, Kazakov DV. Spitz tumors with ROS1 fusions: a clinicopathological study of 6 cases, including FISH for chromosomal copy number alterations and mutation analysis using next-generation sequencing. Am J Dermatopathol. 2020;42(2):92–102.

Elder DE, Massi D, Scolyer RA, Willemze R, editors. WHO classification of skin tumours. Lyon, France: International Agency for Research on Cancer; 2018.

Elder DE. Melanoma progression. Pathology. 2016;48(2):147–54.

Elder DE, Bastian BC, Cree IA, Massi D, Scolyer RA. The 2018 World Health Organization Classification of Cutaneous, Mucosal, and Uveal Melanoma: Detailed Analysis of 9 Distinct Subtypes Defined by Their Evolutionary Pathway. Arch Pathol Lab Med. 2020;144(4):500–22.

Fisher DE, Bastian, B. C. (Eds.). Melanoma. New York: Springer; 2019.

Gerami P, Li G, Pouryazdanparast P, Blondin B, Beilfuss B, Slenk C, Du J, Guitart J, Jewell S, Pestova K. A highly specific and discriminatory FISH assay for distinguishing between benign and malignant melanocytic neoplasms. Am J Surg Pathol. 2012;36(6):808–17.

Guo W, Wang H, Li C. Signal pathways of melanoma and targeted therapy. Signal Transduct Targeted Ther. 2021;6(1):424.

Kong Y, Chi Z, Si L, Sheng X, Cui C, Dai J, Ma M, Wu X, Tang H, Yu J, Yan J, Yu H, Xu T, Guo J. Whole genome and RNA sequencing reveal the distinct genomic landscape of acral melanoma. J Clin Oncol. 2017;35(15_suppl):9589.

Liang WS, Hendricks W, Kiefer J, Schmidt J, Sekar S, Carpten J, Craig DW, Adkins J, Cuyugan L, Manojlovic Z, Halperin RF, Helland A, Nasser S, Legendre C, Hurley LH, Sivaprakasam K, Johnson DB, Crandall H, Busam KJ, Zismann V. Integrated genomic analyses reveal frequent TERT aberrations in acral melanoma. Genome Res. 2017;27(4):524–32.

Mitchell TC, Feld E. Immunotherapy in melanoma. Immunotherapy. 2018;10(11):987–98.

Nassar KW, Tan AC. The mutational landscape of mucosal melanoma. Semin Cancer Biol. 2020;61(165):139–48.

Newell F, Kong Y, Wilmott JS, Johansson PA, Ferguson PM, Cui C, Li Z, Kazakoff SH, Burke H, Dodds TJ, Patch A-M, Nones K, Tembe V, Shang P, van der Weyden L, Wong K, Holmes O, Lo S, Leonard C, Wood S. Whole-genome landscape of mucosal melanoma reveals diverse drivers and therapeutic targets. Nat Commun. 2019;10(1):3163.

Ralli M, Botticelli A, Visconti IC, Angeletti D, Fiore M, Marchetti P, Lambiase A, de Vincentiis M, Greco A. Immunotherapy in the treatment of metastatic melanoma: current knowledge and future directions. J Immunol Res. 2020;2020(214):1–12.

Shain AH, Bagger MM, Yu R, Chang D, Liu S, Vemula S, Weier JF, Wadt K, Heegaard S, Bastian BC, Kiilgaard JF. The genetic evolution of metastatic uveal melanoma. Nat Genet. 2019;51(7):1123–30.

Shain AH, Garrido M, Botton T, Talevich E, Yeh I, Sanborn JZ, Chung J, Wang NJ, Kakavand H, Mann GJ, Thompson JF, Wiesner T, Roy R, Olshen AB, Gagnon A, Gray JW, Huh N, Hur JS, Busam KJ, Scolyer RA. Exome sequencing of desmoplastic melanoma identifies recurrent NFKBIE promoter mutations and diverse activating mutations in the MAPK pathway. Nat Genet. 2015;47(10):1194–9.

Shain AH, Joseph NM, Yu R, Benhamida J, Liu S, Prow T, Ruben B, North J, Pincus L, Yeh I, Judson R, Bastian BC. Genomic and transcriptomic analysis reveals incremental disruption of key signaling pathways during melanoma evolution. Cancer Cell. 2018;34(1):45–55.e4.

Shain AH, Yeh I, Kovalyshyn I, Sriharan A, Talevich E, Gagnon A, Dummer R, North J, Pincus L, Ruben B, Rickaby W, D'Arrigo C, Robson A, Bastian BC. The genetic evolution of melanoma from precursor lesions. N Engl J Med. 2015;373(20):1926–36.

Stark MS, Tell-Martí G, Martins da Silva V, Martinez-Barrios E, Calbet-Llopart N, Vicente A, Sturm RA, Soyer HP, Puig S, Malvehy J, Carrera C, Puig-Butillé JA. The distinctive genomic landscape of giant congenital melanocytic nevi. J Investig Dermatol. 2021;141(3):692–695.e2.

Wiesner T, He J, Yelensky R, Esteve-Puig R, Botton T, Yeh I, Lipson D, Otto G, Brennan K, Murali R, Garrido M, Miller VA, Ross JS, Berger MF, Sparatta A, Palmedo G, Cerroni L, Busam KJ, Kutzner H, Cronin MT. Kinase fusions are frequent in Spitz tumours and spitzoid melanomas. Nat Commun. 2014;5(1):2332–79.

Wiesner T, Kutzner H, Cerroni L, Mihm MJ, Busam KJ, Murali R. Genomic aberrations in spitzoid tumours and their implications for diagnosis, prognosis and therapy. Pathology. 2016;48(2):113–31.

Wolf Horrell EM, Boulanger MC, D'Orazio JA. Melanocortin 1 receptor: structure, function, and regulation. Front Genet. 2016;7(324):423–57.

Zhang T, Dutton-Regester K, Brown KM, Hayward NK. The genomic landscape of cutaneous melanoma. Pigment Cell Melanoma Res. 2016;29(3):266–83.

Nonmelanoma Skin Cancer

Alexandrov LB, Nik-Zainal S, Wedge DC, Aparicio SA, Behjati S, Biankin AV, et al. Signatures of mutational processes in human cancer. Nature. 2013;500(7463):415–21.

Atwood SX, Li M, Lee A, Tang JY, Oro AE. GLI activation by atypical protein kinase C ı/λ regulates the growth of basal cell carcinomas. Nature. 2013;494(7438):484–8.

Atwood SX, Sarin KY, Whitson RJ, Li JR, Kim G, Rezaee M, Ally MS, Kim J, Yao C, Chang ALS, Oro AE, Tang JY. Smoothened variants explain the majority of drug resistance in basal cell carcinoma. Cancer Cell. 2015;27(3):342–53.

Biehs B, Dijkgraaf GJP, Piskol R, Alicke B, Boumahdi S, Peale F, Gould SE, de Sauvage FJ. A cell identity switch allows residual BCC to survive Hedgehog pathway inhibition. Nature. 2018;562(7727):429–33.

Chahal HS, Lin Y, Ransohoff KJ, Hinds DA, Wu W, Dai H-J, Qureshi AA, Li W-Q, Kraft P, Tang JY, Han J, Sarin KY. Genome-wide association study identifies novel susceptibility loci for cutaneous squamous cell carcinoma. Nat Commun. 2016;7(1):3413–501.

Chang D, Shain AH. The landscape of driver mutations in cutaneous squamous cell carcinoma. NPJ Genomic Med. 2021;6(1):456–532.

Epstein EH. Basal cell carcinomas: attack of the hedgehog. Nat Rev Cancer. 2008;8(10):743–54.

Inman GJ, Wang J, Nagano A, Alexandrov LB, Purdie KJ, Taylor RG, Sherwood V, Thomson J, Hogan S, Spender LC, South AP, Stratton M, Chelala C, Harwood CA, Proby CM, Leigh IM. The genomic landscape of cutaneous SCC reveals drivers and a novel azathioprine associated mutational signature. Nature Commun. 2018;9(1):570–620.

Kilgour JM, Shah A, Urman NM, Eichstadt S, Do HN, Bailey I, Mirza A, Li S, Oro AE, Aasi SZ, Sarin KY. Phase II open-label, single-arm trial to investigate the efficacy and safety of topical remetinostat gel in patients with basal cell carcinoma. Clin Cancer Res. 2021;27(17):4717–25.

Kuonen F, Li NY, Haensel D, Patel T, Gaddam S, Yerly L, Rieger K, Aasi S, Oro AE. c-FOS drives reversible basal to squamous cell carcinoma transition. Cell Rep. 2021;37(1):109774.

Martínez-Jiménez F, Muiños F, Sentís I, Deu-Pons J, Reyes-Salazar I, Arnedo-Pac C, Mularoni L, Pich O, Bonet J, Kranas H, Gonzalez-Perez A, Lopez-Bigas N. A compendium of mutational cancer driver genes. Nature Reviews Cancer. 2020;20(10):555–72.

Mirza AN, Fry MA, Urman NM, Atwood SX, Roffey J, Ott GR, Chen B, Lee A, Brown AS, Aasi SZ, Hollmig T, Ator MA, Dorsey BD, Ruggeri BR, Zificsak CA, Sirota M, Tang JY, Butte A, Epstein E, Sarin KY. Combined inhibition of atypical PKC and histone deacetylase 1 is cooperative in basal cell carcinoma treatment. JCI Insight. 2017;2(21):206–45.

Mirza AN, McKellar SA, Urman NM, Brown AS, Hollmig T, Aasi SZ, Oro AE. LAP2 proteins chaperone GLI1 movement between the lamina and chromatin to regulate transcription. Cell. 2019;176(1-2):198–212.e15.

Peterson SC, Eberl M, Vagnozzi AN, Belkadi A, Veniaminova NA, Verhaegen ME, Bichakjian CK, Ward NL, Dlugosz AA, Wong SY. Basal cell carcinoma preferentially arises from stem cells within hair follicle and mechanosensory niches. Cell Stem Cell. 2015;16(4):400–12.

Pickering CR, Zhou JH, Lee JJ, Drummond JA, Peng SA, Saade RE, et al. Mutational landscape of aggressive cutaneous squamous cell carcinoma. Clin Cancer Res. 2014;20(24):6582–92.

Sánchez-Danés A, Larsimont J-C, Liagre M, Muñoz-Couselo E, Lapouge G, Brisebarre A, Dubois C, Suppa M, Sukumaran V, del Marmol V, Tabernero J, Blanpain C. A slow-cycling LGR5 tumour population mediates basal cell carcinoma relapse after therapy. Nature. 2018;562(7727):434–8.

Schmitt J, Haufe E, Trautmann F, Schulze H-J, Elsner P, Drexler H, Bauer A, Letzel S, John SM, Fartasch M, Brüning T, Seidler A, Dugas-Breit S, Gina M, Weistenhöfer W, Bachmann K, Bruhn I, Lang BM, Bonness S, Allam JP. Is ultraviolet exposure acquired at work the most important risk factor for cutaneous squamous cell carcinoma? Results of the population-based case-control study FB-181. Br J Dermatol. 2018;178(2):e161.

Shen L, Liu L, Yang Z, Jiang N. Identification of genes and signaling pathways associated with squamous cell carcinoma by bioinformatics analysis. Oncol Lett. 2015;11(2):1382–90.

Tokez S, Wakkee M, Kan W, Venables ZC, Mooyaart AL, Louwman M, et al. Cumulative incidence and disease-specific survival of metastatic cutaneous squamous cell carcinoma: a nationwide cancer registry study. J Am Acad Dermatol. 2022;86(2):331–8.

Wang GY, Wang J, Mancianti M-L, Epstein EH. Basal cell carcinomas arise from hair follicle stem cells in Ptch1+/− Mice. Cancer Cell. 2011;19(1):114–24.

Wang NJ, Sanborn Z, Arnett KL, Bayston LJ, Liao W, Proby CM, et al. Loss-of-function mutations in Notch receptors in cutaneous and lung squamous cell carcinoma. Proc Natl Acad Sci. 2011;108(43):17761–6.

Whitson RJ, Lee A, Urman NM, Mirza A, Yao CY, Brown AS, et al. Noncanonical hedgehog pathway activation through SRF–MKL1 promotes drug resistance in basal cell carcinomas. Nat Med. 2018;24(3):271–81.

Yilmaz AS, Ozer HG, Gillespie JL, Allain DC, Bernhardt MN, Furlan KC, et al. Differential mutation frequencies in metastatic cutaneous squamous cell carcinomas versus primary tumors. Cancer. 2017;123(7):1184–93.

Youssef KK, Van Keymeulen A, Lapouge G, Beck B, Michaux C, Achouri Y, Sotiropoulou PA, Blanpain C. Identification of the cell lineage at the origin of basal cell carcinoma. Nat Cell Biol. 2010;12(3):299–305.

Merkel Cell Carcinoma

DeCaprio JA. Molecular pathogenesis of merkel cell carcinoma. Annu Rev Pathol. 2021;16:69–91.

Harms KL, Zhao L, Johnson B, Wang X, Carskadon S, Palanisamy N, Palanisamy N, Rhodes DR, Mannan R, Vo JN, Choi JE, Chan MP, Fullen DR, Patel RM, Siddiqui J, Ma VT, Hrycaj S, McLean SA, Hughes TM, Bichakjian CK, Tomlins SA, Harms PW. Virus-positive Merkel cell carcinoma is an independent prognostic group with distinct predictive biomarkers. Clin Cancer Res. 2021;27(9):2494–504.

Harms PW, Vats P, Verhaegen ME, Robinson DR, Wu Y-M, Dhanasekaran SM, Palanisamy N, Siddiqui J, Cao X, Su F, Wang R, Xiao H, Kunju LP, Mehra R, Tomlins SA, Fullen DR, Bichakjian CK, Johnson TM, Dlugosz AA, Chinnaiyan AM. The distinctive mutational spectra of polyomavirus-negative merkel cell carcinoma. Cancer Res. 2015;75(18):3720–7.

Houben R, Adam C, Baeurle A, Hesbacher S, Grimm J, Angermeyer S, Henzel K, Hauser S, Elling R, Bröcker E-B, Gaubatz S, Becker JC, Schrama D. An intact retinoblastoma protein-binding site in Merkel cell polyomavirus large T antigen is required for promoting growth of Merkel cell carcinoma cells. Int J Cancer. 2011;130(4):847–56.

Houben R, Grimm J, Willmes C, Weinkam R, Becker JC, Schrama D. Merkel cell carcinoma and Merkel cell polyomavirus: evidence for hit-and-run oncogenesis. J Investig Dermatol. 2012;132(1):254–6.

Knepper TC, Montesion M, Russell JS, Sokol ES, Frampton GM, Miller VA, Albacker LA, McLeod HL, Eroglu Z, Khushalani NI, Sondak VK, Messina JL, Schell MJ, DeCaprio JA, Tsai KY, Brohl AS. The genomic landscape of merkel cell carcinoma and clinicogenomic biomarkers of response to immune checkpoint inhibitor therapy. Clin Cancer Res. 2019;25(19):5961–71.

Rollison DE, Giuliano AR, Becker JC. New virus associated with merkel cell carcinoma development. J Natl Compr Cancer Netw. 2010;8(8):874–80.

Schrama D, Peitsch WK, Zapatka M, Kneitz H, Houben R, Eib S, Haferkamp S, Moore PS, Shuda M, Thompson JF, Trefzer U, Pföhler C, Scolyer RA, Becker JC. Merkel cell polyomavirus status is not associated with clinical course of Merkel cell carcinoma. J Investig Dermatol. 2011;131(8):1631–8.

Wong HH, Wang J. Merkel cell carcinoma. Arch Pathol Lab Med. 2010;134(11):1711–6.

Wong SQ, Waldeck K, Vergara IA, Schröder J, Madore J, Wilmott JS, Colebatch AJ, De Paoli-Iseppi R, Li J, Lupat R, Semple T, Arnau GM, Fellowes A, Leonard JH, Hruby G, Mann GJ, Thompson JF, Cullinane C, Johnston M, Shackleton M. UV-associated mutations underlie the etiology of MCV-negative Merkel cell carcinomas. Cancer Res. 2015;75(24):5228–34.

Dermatofibrosarcoma Protuberans

Dadone-Montaudié B, Alberti L, Duc A, Delespaul L, Lesluyes T, Pérot G, Lançon A, Paindavoine S, Di Mauro I, Blay J-Y, de la Fouchardière A, Chibon F, Karanian M, MacGrogan G, Kubiniek V, Keslair F, Cardot-Leccia N, Michot A, Perrin V, Zekri Y. Alternative PDGFD rearrangements in dermatofibrosarcomas protuberans without PDGFB fusions. Mod Pathol. 2018;31(11):1683–93.

Greco A, Fusetti L, Villa R, Sozzi G, Minoletti F, Mauri P, Pierotti MA. Transforming activity of the chimeric sequence formed by the fusion of collagen gene COL1A1 and the platelet derived growth factor b-chain gene in dermatofibrosarcoma protuberans. Oncogene. 1998;17(10):1313–9.

Köster J, Arbajian E, Viklund B, Isaksson A, Hofvander J, Haglund F, et al. Genomic and transcriptomic features of dermatofibrosarcoma protuberans: unusual chromosomal origin of the COL1A1-PDGFB fusion gene and synergistic effects of amplified regions in tumor development. Cancer Genet. 2020;241:34–41.

Lee PH, Huang SC, Wu PS, Tai HC, Lee CH, Lee JC, Kao YC, Tsai JW, Hsieh TH, Li CF, Li WS, Liu TT, Su YL, Yu SC, Huang HY. Molecular characterization of dermatofibrosarcoma protuberans: the clinicopathologic significance of uncommon fusion gene rearrangements and their diagnostic importance in the exclusively subcutaneous and circumscribed lesions. Am J Surg Pathol. 2022;46(7):942–55.

Salgado R, Llombart B, Pujol MR, Fernández-Serra A, Sanmartín O, Toll A, Rubio L, Segura S, Barranco C, Serra-Guillén C, Yébenes M, Salido M, Traves V, Monteagudo C, Sáez E, Hernández T, de Álava E, Llombart-Bosch A, Solé F, Guillén C, Espinet B, López-Guerrero JA. Molecular diagnosis of dermatofibrosarcoma protuberans: a comparison between reverse transcriptase-polymerase chain reaction and fluorescence in situ hybridization methodologies. Genes Chromosomes Cancer. 2011;50(7):510–7.

Segura S, Salgado R, Toll A, Martín-Ezquerra G, Yébenes M, Sáez A, Solé F, Barranco C, Umbert P, Espinet B, Pujol RM. Identification of t(17;22)(q22;q13) (COL1A1/PDGFB) in dermatofibrosarcoma protuberans by fluorescence in situ hybridization in paraffin-embedded tissue microarrays. Human Pathol. 2011;42(2):176–84.

Simon M-P, Pedeutour F, Sirvent N, Grosgeorge J, Minoletti F, Coindre J-M, Terrier-Lacombe M-J, Mandahl N, Craver R, Blin N, Sozzi G, Turc-Carel C, O'Brien KP, Kedra D, Fransson I, Guilbaud C, Dumanski JP. Deregulation of the platelet-derived growth factor β-chain gene via fusion with collagen gene COL1A1 in dermatofibrosarcoma protuberans and giant-cell fibroblastoma. Nat Genet. 1997;15(1):95–8.

Cutaneous Adnexal Neoplasms

Corda G, Sala A. Cutaneous cylindroma: it's all about MYB. J Pathol. 2016;239(4):391–3.

Fehr A, Kovács A, Löning T, Frierson H, van den Oord J, Stenman G. The MYB-NFIB gene fusion-a novel genetic link between adenoid cystic carcinoma and dermal cylindroma. J Pathol. 2011;224(3):322–7.

Ferreira I, Wiedemeyer K, Demetter P, Adams DJ, Arends MJ, Brenn T. Update on the pathology, genetics and somatic landscape of sebaceous tumours. Histopathology. 2020;76(5):640–9.

Giang J, Biswas A, Mooyaart AL, Groenendijk FH, Dikrama P, Damman J. Trichoblastic carcinosarcoma with panfollicular differentiation (panfollicular carcinosarcoma) and CTNNB1 (beta-catenin) mutation. J Cutaneous Pathol. 2020;48(2):309–13.

Jones CD, Ho W, Robertson BF, Gunn E, Morley S. Pilomatrixoma: a comprehensive review of the literature. Am J Dermatopathol. 2018;40(9):631–41.

Kazakov DV, Sima R, Vanecek T, Kutzner H, Palmedo G, Kacerovska D, Grossmann P, Michal M. Mutations in Exon 3 of the CTNNB1 gene (β-Catenin Gene) in cutaneous adnexal tumors. Am J Dermatopathol. 2009;31(3):248–55.

Kolm I, Kastnerova L, Konstantinova AM, Michal M, Kazakov DV. Trichoblastoma: A consecutive series of 349 sporadic cases analyzed by Ackerman subtypes. Am J Dermatopathol. 2021;43(12):887–97.

Kraft S, Granter SR. Molecular pathology of skin neoplasms of the head and neck. Arch Pathol Lab Med. 2014;138(6):759–87.

Matt D, Xin H, Vortmeyer AO, Zhuang Z, Burg G, Böni R. Sporadic trichoepithelioma demonstrates deletions at 9q22.3. Arch Dermatol. 2000;136(5):657–60.

Monteagudo C, Fúnez R, Sánchez-Sendra B, González-Muñoz JF, Nieto G, Alfaro-Cervelló C, Murgui A, Barr RJ. Cutaneous lymphadenoma is a distinct trichoblastoma-like lymphoepithelial tumor with diffuse androgen receptor immunoreactivity, Notch1 ligand in reed-sternberg–like cells, and common EGFR somatic mutations. Am J Surg Pathol. 2021;45(10):1382–90.

Muñoz-Jiménez M-T, Blanco L, Ruano Y, Carrillo R, Santos-Briz Á, Riveiro-Falkenbach E, Requena L, Kutzner H, Garrido MC, Rodríguez-Peralto J-L. TERT promoter mutation in sebaceous neoplasms. Virchows Arch. 2021;479(3):551–8.

North JP, Golovato J, Vaske CJ, Sanborn JZ, Nguyen A, Wu W, Goode B, Stevers M, McMullen K, Perez White BE, Collisson EA, Bloomer M, Solomon DA, Benz SC, Cho RJ. Cell of origin and mutation pattern define three clinically distinct classes of sebaceous carcinoma. Nat Commun. 2018;9(1):1894.

Rajan N, Andersson MK, Sinclair N, Fehr A, Hodgson K, Lord CJ, Kazakov DV, Vanecek T, Ashworth A, Stenman G. Overexpression of MYB drives proliferation of CYLD-defective cylindroma cells. J Pathol. 2016;239(2):197–205.

Rashid M, van der Horst M, Mentzel T, Butera F, Ferreira I, Pance A, Rütten A, Luzar B, Marusic Z, de Saint Aubain N, Ko JS, Billings SD, Chen S, Abi Daoud M, Hewinson J, Louzada S, Harms PW, Cerretelli G, Robles-Espinoza CD, Patel RM. ALPK1 hotspot mutation as a driver of human spiradenoma and spiradenocarcinoma. Nat Commun. 2019;10(1):2213.

Shen A-S, Peterhof E, Kind P, Rütten A, Zelger B, Landthaler M, Berneburg M, Hafner C, Groesser L. Activating mutations in the RAS/mitogen-activated protein kinase signaling pathway in sporadic trichoblastoma and syringocystadenoma papilliferum. Human Pathol. 2015;46(2):272–6.

Tetzlaff MT, Curry JL, Ning J, Sagiv O, Kandl TL, Peng B, Bell D, Routbort M, Hudgens CW, Ivan D, Kim T-B, Chen K, Eterovic AK, Shaw K, Prieto VG, Yemelyanova A, Esmaeli B. Distinct biological types of ocular adnexal sebaceous carcinoma: HPV-driven and virus-negative tumors arise through nonoverlapping molecular-genetic alterations. Clin Cancer Res. 2018;25(4):1280–90.

Tetzlaff MT, Singh RR, Seviour EG, Curry JL, Hudgens CW, Bell D, Wimmer DA, Ning J, Czerniak BA, Zhang L, Davies MA, Prieto VG, Broaddus RR, Ram P, Luthra R, Esmaeli B. Next-generation sequencing identifies high frequency of mutations in potentially clinically actionable genes in sebaceous carcinoma. J Pathol. 2016;240(1):84–95.

Genetic/Hereditary Syndromes

Abbas O, Mahalingam M. Cutaneous sebaceous neoplasms as markers of Muir-Torre syndrome: a diagnostic algorithm. J Cutaneous Pathol. 2009;36(6):613–9.

Bertherat J, Horvath A, Groussin L, Grabar S, Boikos S, Cazabat L, Libe R, René-Corail F, Stergiopoulos S, Bourdeau I, Bei T, Clauser E, Calender A, Kirschner LS, Bertagna X, Carney JA, Stratakis CA. Mutations in regulatory subunit type 1A of cyclic adenosine 5′-monophosphate-dependent protein kinase (PRKAR1A): phenotype analysis in 353 patients and 80 different genotypes. J Clin Endocrinol Metab. 2009;94(6):2085–91.

Black JO. Xeroderma pigmentosum. Head Neck Pathol. 2016;10(2):139–44.

Burger B, Itin P. Muir-Torre syndrome. Dermatology. 2008;217(1):56–7.

Chhibber V, Dresser K, Mahalingam M. MSH-6: extending the reliability of immunohistochemistry as a screening tool in Muir–Torre syndrome. Mod Pathol. 2007;21(2):159–64.

Davies HR, Hodgson K, Schwalbe E, Coxhead J, Sinclair N, Zou X, et al. Epigenetic modifiers DNMT3A and BCOR are recurrently mutated in CYLD cutaneous syndrome. Nat Commun. 2019;10(1):1–9.

DiGiovanna JJ, Kraemer KH. Shining a light on xeroderma pigmentosum. J Investig Dermatol. 2012;132(3):785–96.

Fusaro RM. Familial atypical multiple mole melanoma syndrome (FAMMM). Arch Dermatol. 1983;119(1):2–3.

Helgadottir H, Höiom V, Jönsson G, Tuominen R, Ingvar C, Borg Å, Olsson H, Hansson J. High risk of tobacco-related cancers inCDKN2Amutation-positive melanoma families. J Med Genet. 2014;51(8):545–52.

Horvath A, Bertherat J, Groussin L, Guillaud-Bataille M, Tsang K, Cazabat L, Libé R, Remmers E, René-Corail F, Faucz FR, Clauser E, Calender A, Bertagna X, Carney JA, Stratakis CA. Mutations and polymorphisms in the gene encoding regulatory subunit type 1-alpha of protein kinase A (PRKAR1A): an update. Human Mutation. 2010;31(4):369–79.

John AM, Schwartz RA. Muir-Torre syndrome (MTS): an update and approach to diagnosis and management. J Am Acad Dermatol. 2016;74(3):558–66.

Kazakov DV. Brooke-Spiegler syndrome and phenotypic variants: an update. Head Neck Pathol. 2016;10(2):125–30.

Kirschner LS, Carney JA, Pack SD, Taymans SE, Giatzakis C, Cho YS, Cho-Chung YS, Stratakis CA. Mutations of the gene encoding the protein kinase A type I-α regulatory subunit in patients with the Carney complex. Nat Genet. 2000;26(1):89–92.

Lehmann AR, Fassihi H. Molecular analysis directs the prognosis, management and treatment of patients with xeroderma pigmentosum. DNA Repair. 2020;93:102907.

Mahalingam M. MSH6, Past and Present and Muir–Torre Syndrome—Connecting the Dots. Am J Dermatopathol. 2017;39(4):239–49.

Murali R, Wiesner T, Scolyer RA. Tumours associated with BAP1 mutations. Pathology. 2013;45(2):116–26.

Piccione M, Belloni Fortina A, Ferri G, Andolina G, Beretta L, Cividini A, De Marni E, Caroppo F, Citernesi U, Di Liddo R. Xeroderma pigmentosum: general aspects and management. J Personalized Med. 2021;11(11):1146.

Roberts ME, Riegert-Johnson DL, Thomas BC, Rumilla KM, Thomas CS, Heckman MG, et al. A clinical scoring system to identify patients with sebaceous neoplasms at risk for the Muir–Torre variant of Lynch syndrome. Genet Med. 2014;16(9):711–6.

Singh AD. Sebaceous adenoma of the eyelid in Muir-Torre syndrome. Arch Ophthalmol. 2005;123(4):562.

Soura E, Eliades PJ, Shannon K, Stratigos AJ, Tsao H. Hereditary melanoma: update on syndromes and management: genetics of familial atypical multiple mole melanoma syndrome. J Am Acad Dermatol. 2016;74(3):395–407.

Stratakis CA. Carney complex: a familial lentiginosis predisposing to a variety of tumors. Rev Endocr Metabolic Disord. 2016;17(3):367–71.

You J-F, Buhard O, Ligtenberg MJL, Kets CM, Niessen RC, Hofstra RMW, Wagner A, Dinjens WNM, Colas C, Lascols O, Collura A, Flejou J-F, Duval A, Hamelin R. Tumours with loss of MSH6 expression are MSI-H when screened with a pentaplex of five mononucleotide repeats. Br J Cancer. 2010;103(12):1840–5.

Zocchi L, Lontano A, Merli M, Dika E, Nagore E, Quaglino P, Puig S, Ribero S. Familial melanoma and susceptibility genes: a review of the most common clinical and dermoscopic phenotypic aspect, associated malignancies and practical tips for management. J Clin Med. 2021;10(16):3760.

Molecular Pathology of Head and Neck Tumors

18

Adam S. Fisch, Maie A. St. John, and Dipti P. Sajed

Contents

A. S. Fisch
Department of Pathology, Massachusetts General Hospital,
Harvard Medical School, Boston, MA, USA

M. A. St. John
Department of Head & Neck Surgery, David Geffen School of
Medicine at UCLA, Los Angeles, CA, USA
e-mail: MStJohn@mednet.ucla.edu

D. P. Sajed (✉)
Department of Pathology and Laboratory Medicine, David Geffen
School of Medicine at UCLA, Los Angeles, CA, USA
e-mail: dsajed@mednet.ucla.edu

Introduction

- Head and neck cancer is the sixth most common cancer worldwide and squamous cell carcinoma (SCC) is by far the most common malignancy
- Conventional SCC arises in multiple sites of the head and neck, especially the oral cavity and larynx
- There is significant overlap in the molecular features across these sites, an unsurprising finding when considering the common etiology of smoking- and alcohol-derived chemical exposures
- As incidence rises, survival has largely been stagnant despite multimodal treatment regimens, driving the search for targeted therapy
- Accordingly, the wave of immuno-oncology has reached head and neck SCC with some success, but the terrain is open for further discovery
- Molecular studies have also confirmed that the clinical distinction between conventional SCC and human papillomavirus (HPV)-associated oropharyngeal SCC is paralleled by the biological underpinnings detected on molecular testing, highlighting the need for specific biomarkers
- While other tumors of the head and neck are less common, by the very nature of the anatomic sites involved, treatment often leads to high morbidity, underscoring the significance of accurate diagnoses

- Particularly noteworthy are newly discovered, molecularly defined entities of the sinonasal tract, where characterization by molecular testing, or immunohistochemical surrogates, is necessary to identify specific tumors that are otherwise indistinguishable from morphological mimics
- Despite many advances in the molecular pathology of head and neck tumors, the pursuit of molecular markers in existing and emerging entities as they relate to diagnostics, risk stratification, and targetable therapeutic intervention is ongoing
- The last decade has seen an explosion of data that has refined our understanding of the molecular landscape of head and neck cancer, and the integration of these findings with our histomorphologic knowledge will enable continued evolution towards precision medicine

Conventional Squamous Cell Carcinoma

- Malignant neoplasm of the surface epithelium with squamous differentiation
- Most common malignancy of the head and neck (excluding nonmelanoma skin cancers); accounts for ~90% of malignancies at this site
- Usually de novo, but a small proportion of precursor dysplastic lesions evolve to invasive carcinoma
- Tobacco-derived carcinogens: primary risk factor that is synergistic with alcohol consumption
- Majority have conventional keratinizing histology with squamous pearls and intercellular bridges which are well to moderately differentiated (Fig. 18.1a and b)
- Variants include verrucous, papillary, spindle cell, basaloid, and carcinoma cuniculatum
- The Cancer Genome Atlas (TCGA) Network investigations into head and neck SCC (multiple sites) show nearly universal findings of inactivation of *TP53, CDKN2A*, and copy gains in chromosomes 3q (*TP63, SOX2, PIK3CA*) and 11q (*CCND1, FADD, CTTN*) in non-HPV-related SCC of the head and neck
- DNA signatures derived from variant analyses show that there is a higher rate of the smoking-related signature, COSMIC signature 4, in laryngeal SCC (82%) than in SCC of the oral cavity and oropharynx (44%)
  - The age-related COSMIC signature 5 of unclear etiology is also enriched in laryngeal SCC, is associated with increased age at this site in patients with no smoking history, and is present in decreasing proportions of tumors from patients with longer intervals since smoking
- Amplifications of receptor tyrosine kinases including *EGFR, ERBB2,* and *FGFR1* are frequently detected; the presence of *EGFR* amplification has been associated with

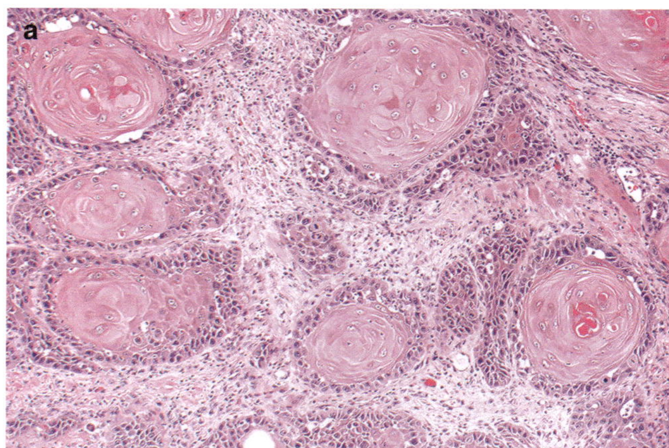

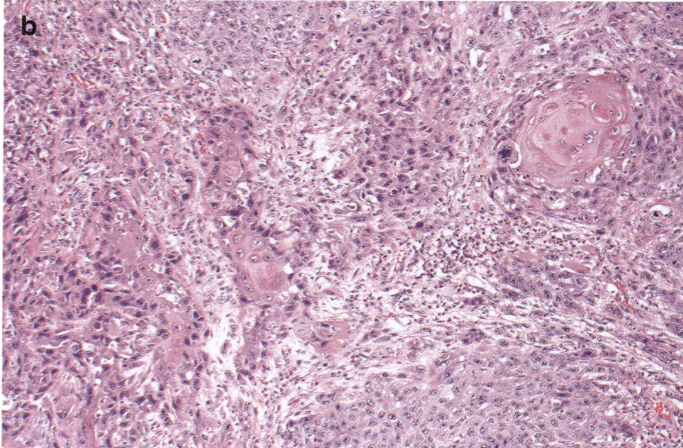

Fig. 18.1 (**a**) Invasive nests of conventional keratinizing squamous cell carcinoma showing squamous pearl and intercellular bridge formation. (**b**) A moderately differentiated tumor with squamous differentiation and infiltrating irregular nests composed of pleomorphic cells

worse overall survival in laryngeal, oral, and oropharyngeal sites and is an indication for targeted inhibition using cetuximab
- Focal deletions of tumor suppressor genes including primarily *NSD1, FAT1, NOTCH1,* and *SMAD4* have been identified
- Tumor clustering following transcriptome analyses has led to the identification of four main groups based on overexpression of unique gene sets
  - Classical—smoking-related genes, including *GPX2* and *NFE2L2*
  - Atypical—HPV-related with *CDKN2A, LIG1,* and *RPA2*
  - Mesenchymal—epithelial-to-mesenchymal transition genes including *VIM, DES, TWIST1,* and *HGF*
  - Basal—*COL17A1, TP63,* and *TGFA*
- Head and neck SCC is often found to possess activating and inactivating alterations in *TP63* and *TP73*, respectively, two genes from the *TP53* family that regulate cell survival and death

- Treatment approach is multimodal
  - Surgical resection with adjuvant radiation/chemoradiation
  - Immune checkpoint inhibitors (ICI), specifically pembrolizumab, have become the standard for treatment in recurrent or metastatic SCC (see separate section below)
  - Relatively recent trials of targeted inhibition (e.g., tipifarnib) in the treatment of advanced tumors with hotspot variants in *HRAS* have shown some promise
  - CDK4/6 inhibitors, such as palbociclib, have been studied with limited success in tumors with cell cycle pathway-altered (e.g., *CDKN2A*) tumors
  - EGFR-targeted therapies, including cetuximab and afatinib, have shown promise in the treatment of recurrent/metastatic *EGFR*-amplified tumors
  - TRK and FGFR inhibitors have been effective in tumors with *NTRK* and *FGFR3* fusions, respectively, which are both rare genomic events in SCC

Squamous Cell Carcinoma of the Oral Cavity

- Activating variants including *HRAS, PIK3CA,* and *BRAF* in the mitogenic signaling pathway with associated targeted therapies are found in ~60% of cases, and >90% of tumors have altered genes in the cell cycle pathway, predominantly *CDKN2A* inactivation and *CCND1* amplification
- A subset of tumors with lower levels of copy number alterations has *HRAS* activating variants, *CASP8* inactivating variants, and an improved outcome over those with more copy number variants
- Another small subset of tumors harbors inactivating alterations in *TGFBR2*
- Following the genomic events leading to development of hyperplasia and dysplasia (see oral epithelial dysplasia below), transformation to carcinoma in-situ includes

amplification of *CCND1*, which is associated with a poor clinical prognosis, and chromosomal losses in 11q13, 13q21, and 14q32; subsequent development of invasive carcinoma includes chromosomal losses of 4q27, 6p, chromosome 8, and 10q23

- From an epigenetic perspective, tumors of the oral cavity are often globally hypomethylated, with multifocal promoter hypermethylation events leading to inactivation of tumor suppressor genes such as *CDKN2A, CDH1,* and *MGMT*
- The main heritable syndromes conferring increased genetic susceptibility are Li Fraumeni, with inactivating alterations in *TP53* frequently leading to tumors that occur primarily in the oral cavity, and Fanconi anemia, with inactivating alterations in *FANCA* or one of numerous other genes leading to a 500–700-fold increased risk in developing SCC, especially following bone marrow transplant (Table 18.1)

Squamous Cell Carcinoma of the Hypopharynx, Larynx, Trachea, and Parapharyngeal Space

- In addition to smoking and alcohol, gastro-esophageal reflux/nutrition contributes to risk, albeit to a lesser extent
- Loss-of-heterozygosity (LOH) has been demonstrated in a large subset of tumors, primarily in regions of chromosomes 1p, 3p, 9p, 11q, and 17p; tumor suppressor inactivation secondary to genetic alterations, methylation, or LOH are frequently seen
- LOH in 11q, the chromosomal region containing *CCND1*, is thus far the only alteration associated with tumor grade
- *RB1* inactivation has been reported at varying rates
- Expression of *MMP13* is detected in more than half of tumors (no expression in normal mucosa) and correlates with more differentiated tumors and advanced local invasion

Table 18.1 Tumors with germline predisposition association

Affected tumor	Syndrome	Gene	Chromosome arm	Protein function
Squamous cell carcinoma (oral cavity)	Li Fraumeni	TP53	17p	Loss
	Fanconi anemia	FANCA	16q	Loss
		FANCC	9q	Loss
		FANCG	9p	Loss
Nasal chondromesenchymal hamartoma	DICER1 syndrome	DICER1	14q	Loss
Odontogenic keratocyst	Nevoid basal cell carcinoma (Gorlin) syndrome	PTCH1	9q	Loss
Central giant cell lesion	Cherubism	SH3BP2	4p	Gain
	Noonan	PTPN11	12q	Gain
		SOS1	2p	Gain
		RAF1	3p	Gain
	Neurofibromatosis	NF1	17q	Loss
Paraganglioma (see also endocrine chapter)	Carney–Stratakis syndrome/carney triad	SDHB	1p	Loss
		SDHC	1q	Loss
		SDHD	11q	Loss

Precursors to Conventional Squamous Cell Carcinoma

- Epithelial dysplasias manifest as a constellation of architectural and cytologic features without strict criteria; grading scheme is recommended by anatomic site

Oral Cavity Dysplasia

- Does not always follow tiered epithelial levels (i.e., 1/3, 2/3, full thickness) of severity as marked atypia may be present only at the basal layer and progress to invasive carcinoma without further surface involvement
- Mild, moderate, severe grades have poor inter-rater reproducibility; may improve with binary system
- A rare subset with conventional dysplastic features in addition to epithelial hyperplasia, marked karyorrhexis and apoptosis is found to be associated with transcriptionally active high-risk human papillomavirus (HR-HPV); prognosis, unlike in the oropharynx, is unclear
- Often have initial LOH in chromosome 9p21 or alternative inactivation of *CDKN2A*, followed by loss of the 3p21 and 17p13 (*TP53*) loci
- Recurrent inactivating alterations (by sequencing) in and copy number loss of *TP53* (by fluorescence in situ hybridization (FISH) of the 17p13.1 locus) are enriched with increasing histologic grade
- Transformation to carcinoma is associated with inactivating alterations in *TP53*, aneuploidy, and copy number alterations, particularly in *EGFR* and when multiple copy number changes are present

Laryngeal Dysplasia

- Assessed with a binary grading system; a greater percentage of high-grade dysplasias are capable of malignant progression
- There is significant molecular overlap with oral dysplasia
- Dysplasia and carcinoma show more *TERC* copy number gains (detected by FISH and immunohistochemistry) than normal-appearing squamous epithelium

Laryngeal Squamous Papilloma

- Benign epithelial neoplasm characterized by exophytic projections of squamous epithelium with fibrovascular cores
- May exhibit koilocytic features with wrinkled, hyperchromatic nuclei, and perinuclear halos

- Predominant molecular etiology is infection with low-risk human papillomavirus (LR-HPV) 6 and 11, detectable by PCR or in situ hybridization (ISH)
- Immunohistochemistry (IHC) for p16 may be positive and some of these lesions have detectable HR-HPV
- Recurrent respiratory papillomatosis is characterized by multiple lesions following transmission of LR-HPV, with a tendency for multiple recurrences and the rare possibility of malignant transformation

Human Papillomavirus-Associated Squamous Cell Carcinoma of the Oropharynx

- Malignant neoplasm arising from palatine and lingual tonsillar crypt epithelium
- Most common tumor at this site with a steeply rising incidence in the last few decades
- Mostly seen in developed countries, with a male predominance, often in the 5th decade with no significant tobacco history and a strong association with oral sex
- Predominantly nonkeratinizing morphology with a basaloid appearance growing in nests in lymphoid stroma of tonsil (Fig. 18.2a); occasionally other morphologies
- There is no precursor lesion such as dysplasia or carcinoma in situ as the tumors arise in the crypts and are all considered invasive with metastatic potential; rarely may involve the surface
- HR-HPV infection (majority type 16) and viral integration have a major role in oncogenic transformation
- HPV-encoded E6 and E7 oncoproteins inactivate the p53 (by degradation) and RB (by complex formation) tumor suppressors
- IHC for p16 (Fig. 18.2b) is an excellent surrogate marker for HPV integration (Fig. 18.2c): E7 promotes oncogene-induced senescence-related H3K27 demethylation of *CDKN2A*/p16^{INK4A} by KDM6B in HPV-infected cells, leading to increased expression of *CDKN2A*/p16^{INK4A}
- Chromogenic ISH targeting RNA and DNA from HR-HPV serotypes is used for confirmation of HPV infection-driven oncogenesis in specific contexts, at the discretion of the pathologist
- HPV-driven tumors have more frequent activating variants in *PIK3CA*, inactivation of *TRAF3,* and amplification of *E2F1*
- Better survival than conventional HNSCC despite presenting more frequently as lymph node metastases
- With HPV-positive tumors at this site, prognosis is independent of morphologies except for high-grade neuroendocrine carcinoma
- Given the younger age interval of the affected population, there are efforts to de-escalate treatment strategies, but

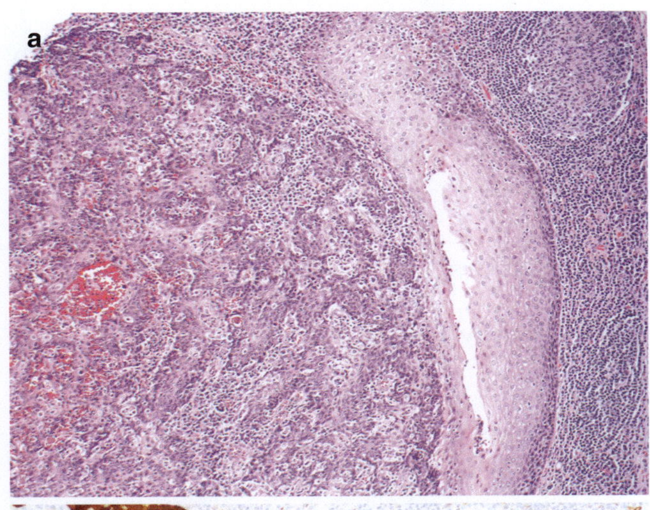

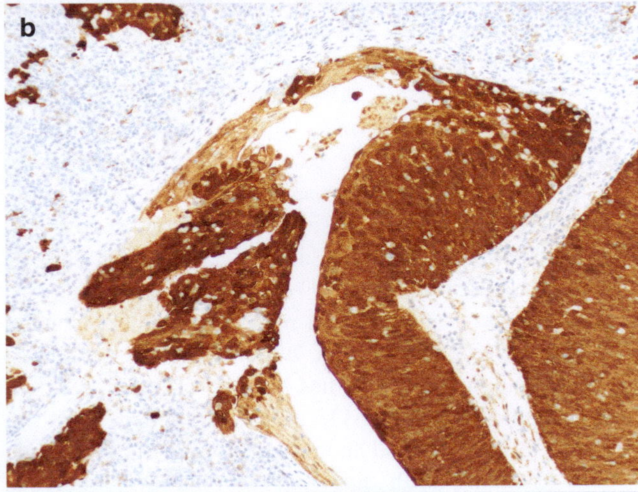

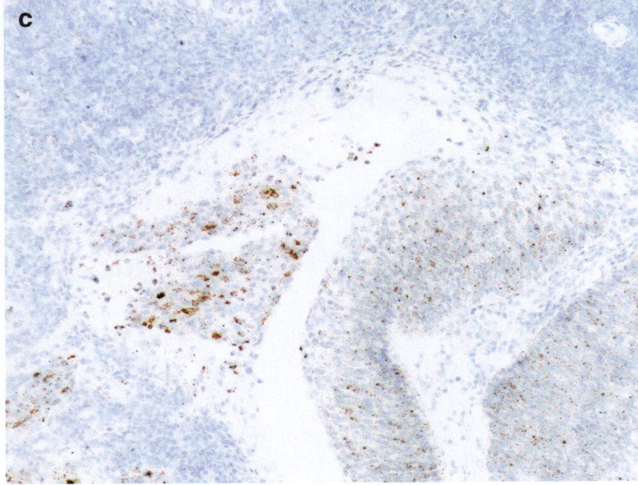

Fig. 18.2 (**a**) Human papillomavirus (HPV)-associated oropharyngeal squamous cell carcinoma characterized by basaloid neoplastic cells arising at the base of the normal-appearing tonsillar crypt epithelium, growing in nests within lymphoid stroma. (**b**) Positivity for p16 (defined by >70% cytoplasmic and nuclear staining) is a surrogate marker for (**c**) high-risk HPV (in situ hybridization) at this location

early results have not been exceptionally promising, and this approach remains investigational

- HPV-negative SCC of the oropharynx shares histological and molecular features with conventional SCC of oral and laryngeal sites
- FDA approval for the HPV vaccine in 2020 covering multiple oncogenic types may eventually begin to curtail rise in incidence

Immunotherapy for Head and Neck Cancer

- Carcinogen-related SCC without much without much improvement in survival rate as well as the unprecedented rise in HPV-driven SCC have driven a search for novel therapies
- The field of head and neck oncology has had some success in adopting the recent trend to exploit the immune system against tumor growth
- Inhibition of the PD-1/L1 checkpoint pathway with monoclonal antibodies (mAb) is the primary immuno-oncological treatment modality
- Three recent, large randomized phase 3 clinical trials have shown an overall survival benefit of ICIs in a subset of tumors over standard of care
  - CheckMate 141: Nivolumab (anti PD-1 mAb) versus single-agent systemic therapy (methotrexate, docetaxel, or cetuximab) in patients with platinum chemotherapy-refractory recurrent disease
  - KEYNOTE-040: Pembrolizumab (anti PD-1 mAb) versus systemic therapy (methotrexate, docetaxel, or cetuximab) in patients with platinum chemotherapy-refractory recurrent or metastatic disease
  - KEYNOTE-048: Pembrolizumab alone, pembrolizumab with a platinum and 5-fluorouracil (5-FU), or cetuximab with a platinum and 5-FU in patients with untreated, unresectable recurrent or metastatic disease
- There is also evidence that treatment with ICIs, specifically nivolumab with or without ipilimumab, may have a role in the neoadjuvant setting
- Based on these results, the FDA has approved:
  - Pembrolizumab and nivolumab for the treatment of platinum-refractory recurrent or metastatic HNSCC
  - Pembrolizumab as a first-line therapy in patients with unresectable or metastatic HNSCC with a combined positive score (CPS) ≥ 1 for PD-L1
  - Pembrolizumab with chemotherapy as a first-line therapy in patients with recurrent or metastatic HNSCC, regardless of PD-L1 CPS

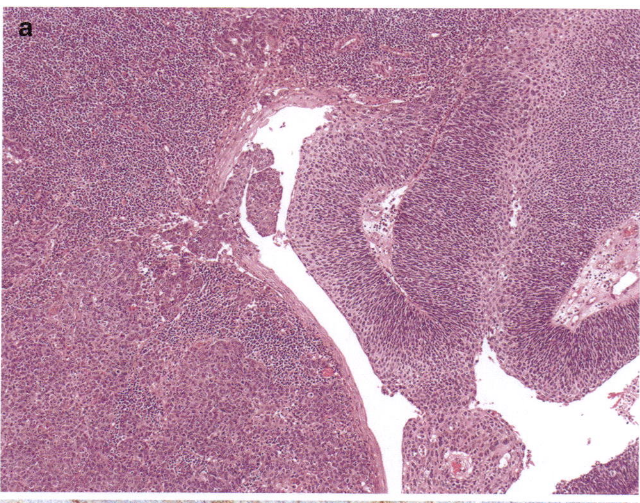

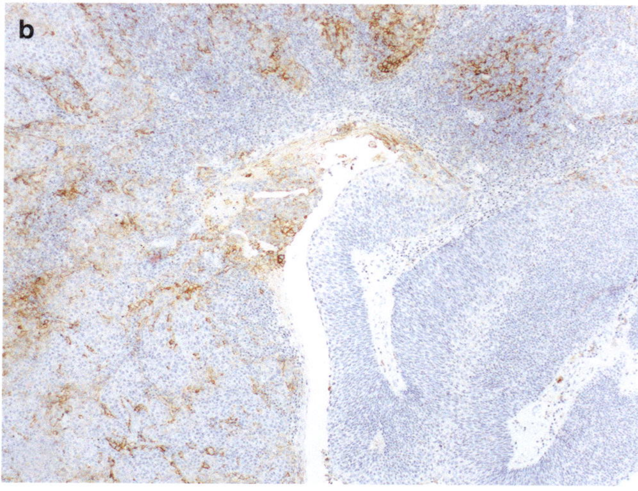

Fig. 18.3 (**a**) An HPV-associated squamous cell carcinoma arising in the tonsillar crypt epithelium, invading in nests. (**b**) Immunohistochemistry for PD-L1 showing differential staining within the tumor

- PD-L1 testing and reporting of CPS for HNSCC are incorporated into head and neck cancer program clinical workflows (Fig. 18.3a and b)
- Treatment algorithms may be refined as subgroups with differential responses emerge from continued analyses
- Prospective trials are actively being pursued
- Multiple biomarkers are being used to predict immunotherapy response
  - PD-L1 expression correlates with improved efficacy of ICIs; higher predictive value with CPS
  - Tumor mutational burden in HNSCC shown by some studies to be higher in responders, while others have shown no correlation
  - Tumor microenvironment: specific immune cells found to have CD3+, CD8+, and Foxp3+ tumor-infiltrating lymphocytes have been associated with a better prognosis

- HPV status is currently not recommended in decision making for ICI therapy, but research into HPV infection/transformation under immune control in the tonsillar crypt is ongoing
- Host microbiome: there is ongoing research that may define its role in stratification

Additional Oral Cavity Tumors

Ectomesenchymal Chondromyxoid Tumor

- Extremely rare, benign mesenchymal neoplasm of unknown origin
- Strong predilection for anterior dorsal tongue
- Lobules separated by fibrous bands, composed of polygonal/spindled/stellate cells arranged in cords, sheets, and reticulated patterns, in a background of chondromyxoid stroma
- An early study using FISH showed genetic alterations when probing for *EWSR1*, with a large proportion of tumors showing *EWSR1* gain, possibly attributable to chromosome 22 gain, and a smaller subset of tumors having *EWSR1* translocations; despite histomorphological overlap with pleomorphic adenoma, no cases showed *PLAG1* translocations
- A subsequent study showed recurrent fusions involving *RREB1::MKL2* in 90% of tumors by next-generation sequencing, a fusion that has been described in the context of extraglossal ECMTs as well as biphenotypic sarcomas of the sinonasal and oropharyngeal tracts; a single case of ECMT with divergent morphology harbored a fusion in *EWSR1::CREM*

GLI1-Altered Mesenchymal Neoplasms

- Rare, emerging entity with an uncertain histogenesis
- Predilection for the head and neck in young adults; tongue >> other HN sites
- Majority thus far have been shown to be indolent, but subset have metastasized
- Clinical behavior similar to a low-grade sarcoma
- Histologic spectrum is broad but tends to include multinodular growth of monomorphic, epithelioid cells within a capillary network (Fig. 18.4a and b)
- Protrusion into vascular spaces may be a helpful clue
- IHC not consistent but tumors can show S100 positivity (although *GLI1* amplified tumors even more variable; CDK4 and MDM2 positivity are helpful in these cases)
- Defined by pathognomonic alterations involving *GLI1*, including either
  - Gene fusions with *ACTB*, *PTCH1*, or *MALAT1* detected by FISH or NGS
  - Co-amplifications of *GLI1*, *CDK4*, and *MDM2* detected by FISH

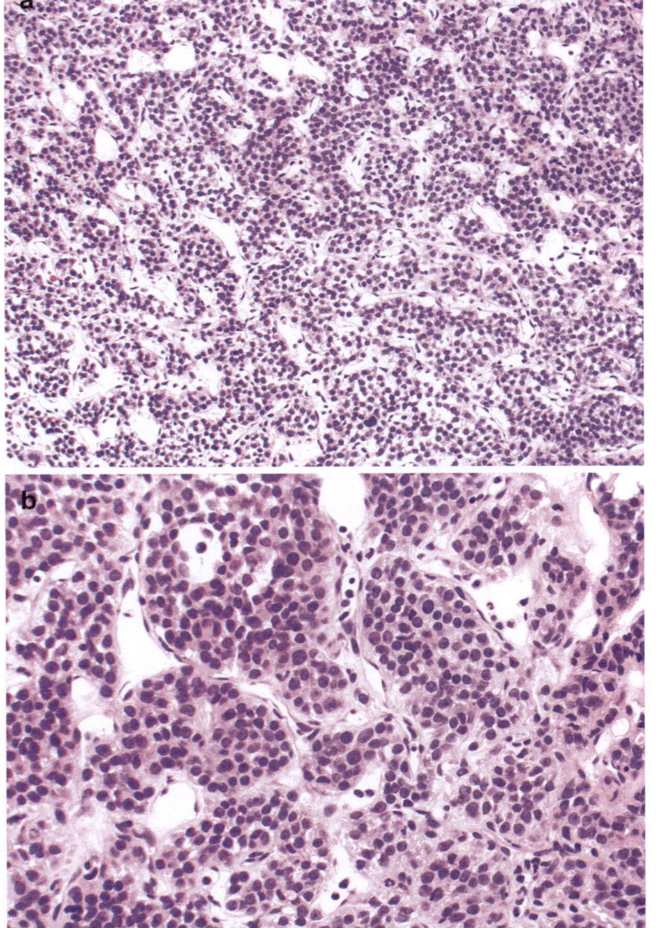

Fig. 18.4 (**a**) This tumor with an *ACTB::GLI1* fusion presented as an initial tongue mass and a thigh metastasis 9 years later. Histology at both sites demonstrated an anastomosing corded pattern with a prominent capillary network. (**b**) The cells seen here in prominent nests are monotonous with round occasionally larger nuclei with a single nucleolus and delicate pink cytoplasm

Nasal Cavity, Paranasal Sinus, and Skull Base Tumors

Sinonasal Papilloma

- Benign surface epithelial neoplasm of the sinonasal tract characterized by epithelial thickening and intraepithelial neutrophils and microabscesses

Inverted Type
- Most common sinonasal papilloma
- Nasal cavity (lateral)/maxillary sinus > other paranasal sinuses

- Characteristic inverted growth pattern of a multilayered squamous/respiratory epithelium with scattered muco-cytes (Fig. 18.5a)
- Malignant transformation is possible (5–15%), with cases having either:
  - *EGFR* alterations (almost all in-frame insertion-deletions) are found in a significant proportion of cases (~88%) and are associated with increased progression-free survival than those tumors not driven by *EGFR*
  - Small subset of tumors (Fig. 18.5b) driven by transcriptionally active LR-HPV (Fig. 18.5b inset) are mutually exclusive from those with *EGFR* alterations, showing condylomatous morphology, and possibly higher association with progression to malignancy

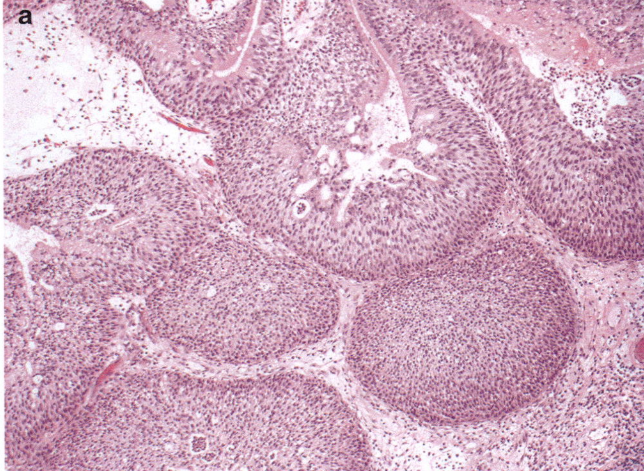

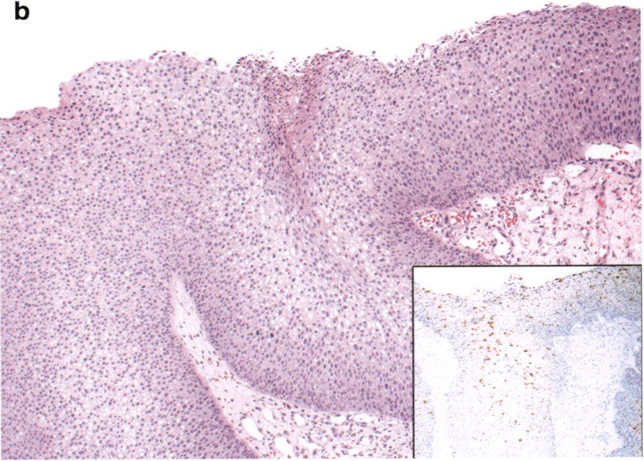

Fig. 18.5 (**a**) Thickened sinonasal epithelium growing in bulbous, inverted pattern with neutrophilic abscesses, characteristic of sinonasal inverted papilloma. (**b**) A small subset of tumors show thickened epithelium with inverted growth pattern and viral cytopathic change and are (inset) positive for low-risk human papillomavirus in-situ hybridization

– HPV DNA, including both low- and high risk, is detected in 20–25% overall, but in over 50% of cases with severe dysplasia or associated invasive SCC
– Approximately 10–25% show malignant transformation to SCC, with recurrent alterations in *TP53* and/or *CDKN2A*, including *CDKN2A* copy-number loss found exclusively in the malignant component, as well as copy-number gains in *TERT*, *SOX2*, *CCND1*, and/or *MYC* in a subset of cases

Oncocytic Type

- Has exophytic/endophytic growth of oncocytic columnar epithelium (Fig. 18.6a and b)
- Arises in the lateral nasal wall/maxillary and ethmoid sinuses

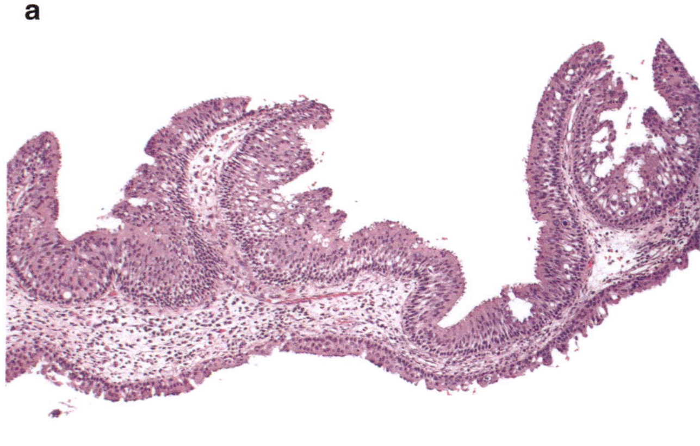

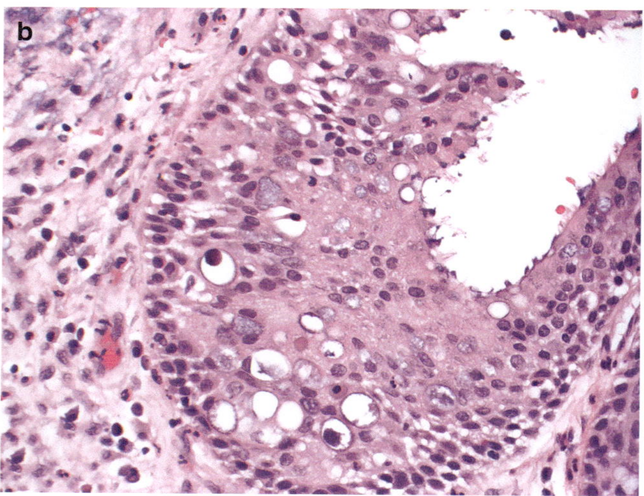

Fig. 18.6 (**a**) Oncocytic sinonasal papilloma showing both an exophytic and endophytic growth pattern. (**b**) A closer look shows prominent eosinophilic/granular cytoplasm, cilia, and numerous mucocytes, some with debris

- *KRAS* p.G12 and p.Q61 hotspot variants have been observed, the latter also appearing in cases with an associated SCC arising in the background of a sinonasal papilloma
- A small cohort of tumors with malignant progression to SCC showed hotspot alterations in *KRAS* which persisted in the malignant component, highlighting the clonal connection, as well as frequent sequence alterations in *TP53* and/or *CDKN2A* that were not found in the benign papilloma precursor
- Molecular studies have also shown recurrent copy-number gains in *TERT* in a subset of tumors during malignant progression

Exophytic Type

- Least common sinonasal papilloma showing broad papillary growth with fibrovascular cores
- Occurs mainly in the lower anterior nasal septum
- There is a significant association with LR-HPV 6 and 11

Types of Carcinoma

Squamous Cell Carcinoma

- May arise from malignant transformation of sinonasal papillomas, particularly the inverted type (Fig. 18.7a–c), with a large proportion acquiring alterations in *TP53* and/or *CDKN2A* in the process (see section on sinonasal papillomas for details)
- Investigations are ongoing into the efficacy of targeted irreversible inhibitors of *EGFR* for tumors arising from papilloma precursors driven by an *EGFR* alteration; reversible inhibitors often used for lung cancer have not shown success
- Two main types of SCC

Keratinizing Squamous Cell Carcinoma

- Malignant neoplasm with overt squamous differentiation arising from epithelial lining of nasal cavity and paranasal sinuses
- The sinonasal tract is the least common head and neck site: maxillary sinus > nasal cavity > ethmoid sinus
- Histologically similar to keratinizing SCC of other head and neck sites
- Very low rate of HPV infection overall (4%)

Nonkeratinizing Squamous Cell Carcinoma

- Malignant neoplasm with undifferentiated/immature features
- Found mainly in the maxillary sinus or nasal cavity

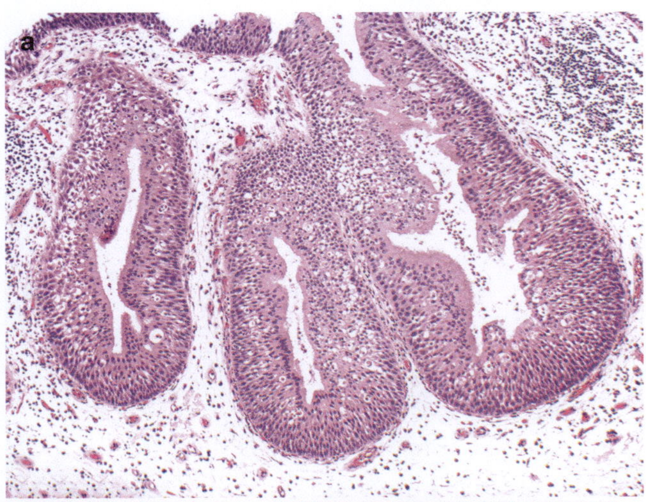

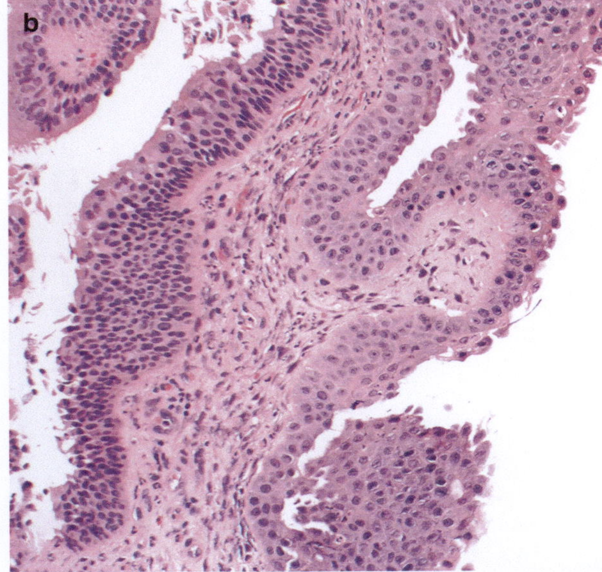

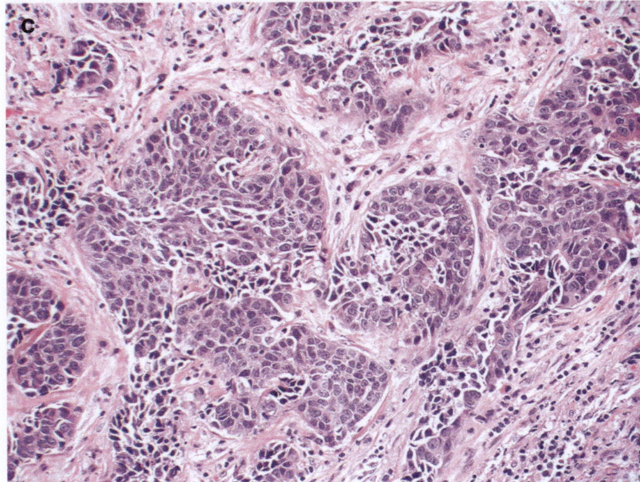

Fig. 18.7 (**a**) A sinonasal tumor demonstrating histologic features of a typical sinonasal inverted papilloma. (**b**) The same tumor showing a clear transition from benign cytology to high-grade dysplasia. (**c**) In other areas, the tumor demonstrates malignant transformation to an invasive carcinoma

- Expansive, ribbon-like growth of undifferentiated cells with a pushing border
- Up to ~40% have HPV, with a subset shown to harbor transcriptionally active HR-HPV; considered a second hotspot for HPV (first being oropharynx)
- Improved prognosis for patients with HPV-driven tumors

Intestinal-Type Adenocarcinoma

- Malignant epithelial neoplasm with glandular differentiation morphologically similar to intestinal adenocarcinoma
- The majority are related to occupational exposure to wood/leather dust
- Typically found in the ethmoid sinus >> nasal cavity > maxillary sinus
- Varied morphologies paralleling intestinal counterpart: colonic, papillary, solid, mucinous, signet ring; well to poorly differentiated
- Small proportion of tumors show *KRAS* variants, typically at hotspots leading to p.G12 and p.G13 alterations, with those patients having less aggressive tumors than patients with *KRAS*-wildtype tumors
- Approximately 40% of tumors have alterations in *TP53*
- Alterations in *CDKN2A* leading to inactivation are observed in approximately 60% of tumors
- Alterations in *EGFR*, including sequence variants and copy-number gains, are not a major factor in pathogenesis of these tumors

Nonintestinal-Type Adenocarcinoma

- Malignant glandular neoplasm classified in this category based on lack of histological features resembling salivary gland and intestinal-type adenocarcinoma
- Poorly characterized, morphologically heterogeneous group of tumors
- Prime territory for the discovery of novel/emerging entities as morphologically similar groups are revealed
  - *ETV6*-rearranged low-grade sinonasal adenocarcinoma
  - Sinonasal renal cell-like carcinoma
    - Histomorphologically defined entity
    - Molecular studies have yet to be performed

Neuroendocrine Carcinoma

- Rare tumor at this site, with morphological features like those seen in other primary sites; small cell type with finely granular chromatin, scant cytoplasm, and prominent crush artifact and large cell type with coarse chromatin; abundant mitotic figures and necrosis in both

- Next-generation sequencing of a patient cohort with neuroendocrine carcinoma of the head and neck, mostly of the paranasal sinuses and oropharynx, detected variants in *TP53*, *RB1*, and PI3K-PTEN pathway genes
- Fusion analysis in the same cohort yielded two cases with fusions: *FGFR3::TACC3* and *SEC11C::MYC*

Lymphoepithelial Carcinoma

- Uncommon tumor more frequently arising in the nasal cavity than the paranasal sinuses with a ~50% survival rate after 5 years
- Variable architecture composed of large tumor cells with vesicular nuclei, prominent nucleoli, and abundant amphophilic cytoplasm with an associated chronic inflammatory infiltrate
- Greater than 90% of cases are driven by Epstein-Barr virus (EBV) infection

NUT Carcinoma

- Highly lethal malignancy, rare but critical to identify
- Undifferentiated features with uniform, high-grade cells; abrupt, frank squamous differentiation may be a clue (Fig. 18.8a)
- Given morphologic overlap with other tumors, diagnostic gene fusions involving *NUTM1* gene, most often fused to *BRD4*, can be detected using molecular testing and an immunohistochemical marker for NUT (Fig. 18.8b)

SWI/SNF-Deficient Malignancies

- Have shown characteristic molecular alterations, including a definitive lack of *IDH1* and *IDH2* alterations on molecular testing and immunohistochemistry

SMARCB1 (INI-1)-Deficient Sinonasal Carcinoma

- Often in paranasal sinuses, predominantly ethmoid
- Undifferentiated basaloid or occasionally plasmacytoid/rhabdoid cells with high N/C ratio and uniform cytomorphology growing in sheets and nests (Fig. 18.9a)
- Defined by biallelic inactivation of *SMARCB1*, which codes for the INI-1 protein that is lost on IHC (Fig. 18.9b)
- *SMARCB1* gene deletion is the common molecular mechanism for inactivation, leading to loss of neighboring genes on chromosome 22q; a subset of tumors in one study showed a variety of co-occurring alterations, including sequence variants in *CTNNB1*, *TP53*, and/or *CDKN2A,* as well as copy-number gain of chromosome 7
- Some tumors exhibit glandular morphology, with a rare subset showing yolk sac tumor morphology and immunohistochemical staining patterns (glypican-3, SALL4, and

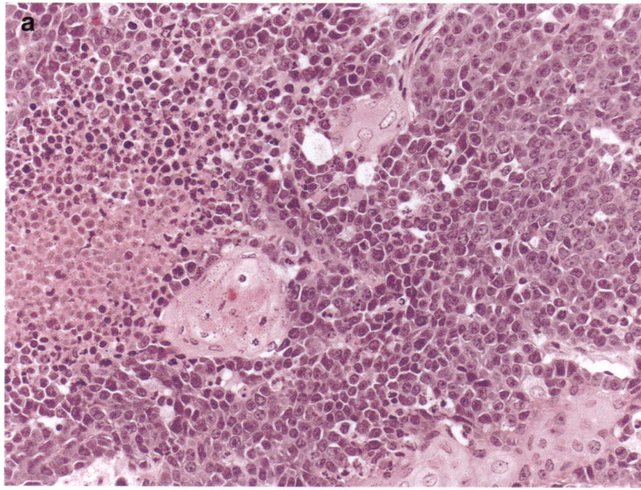

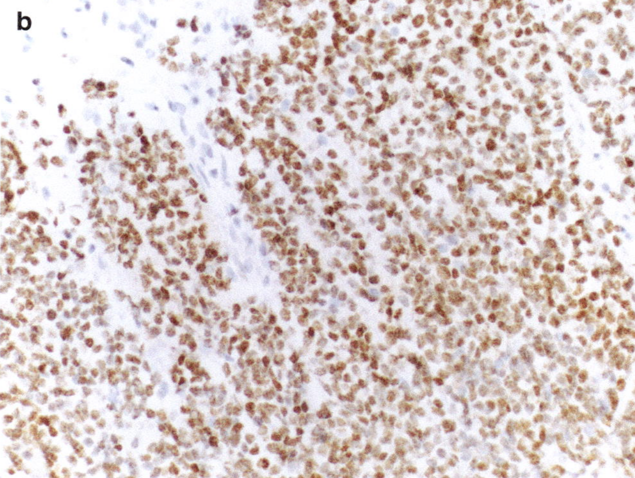

Fig. 18.8 (**a**) NUT carcinoma demonstrating undifferentiated cytomorphology with prominent nuclei and necrosis and abrupt keratinization which, if present, can be a clue. (**b**) Identified by immunohistochemistry for the NUT protein showing strong nuclear staining

alpha-fetoprotein); one study compared this type of tumor with a series of primary gonadal germ cell tumors with yolk sac differentiation and found preserved immunohistochemical staining of INI-1 in all 11 cases

SMARCA4 (BRG1)-Deficient Sinonasal Carcinoma

- Far less common than its *SMARCB1*-deficient counterpart
- Often found in the nasal cavity > paranasal sinuses
- Large cell morphology with frequent necrosis and mitotic activity which may be mistaken for a neuroendocrine tumor (Fig. 18.10a)
- Defined by inactivation of *SMARCA4*, which can be detected by molecular methods and using an immunohis-

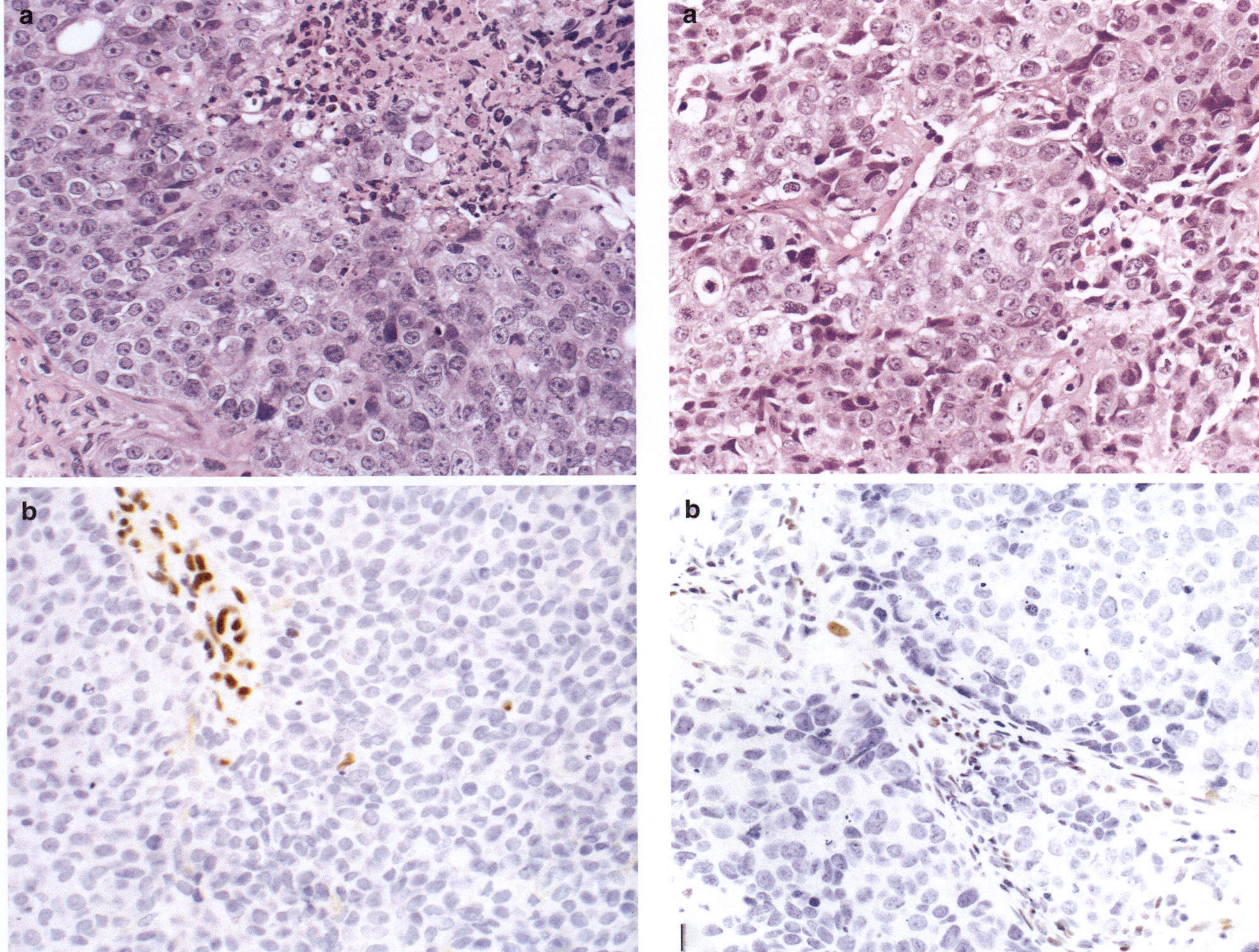

Fig. 18.9 (**a**) *SMARCB1* (INI-1)-deficient sinonasal carcinoma characterized by sheets of uniform cells with prominent nucleoli and adjacent necrosis. (**b**) Defined by complete loss of INI-1 in all neoplastic cells with a vessel and inflammatory cells retaining the protein

Fig. 18.10 (**a**) *SMARCA4* (BRG-1)-deficient sinonasal carcinoma showing sheets of large cells with coarse chromatin and frequent mitoses. (**b**) Defined by complete loss of BRG-1 in all neoplastic cells with adjacent stromal cells retaining the protein

tochemical antibody targeting BRG-1 (Fig. 18.10b), its coded protein
- Teratocarcinosarcomas, a rare and aggressive malignant tumor with carcinomatous, sarcomatous, and teratoma-like elements, have been shown to have recurrent *SMARCA4* inactivation on the genetic and protein (BRG-1) levels

IDH-Mutant Sinonasal Carcinoma
- Recurrent found hotspot variants, particularly *IDH2* p.R172 alterations, have been demonstrated in subsets of sinonasal undifferentiated carcinoma (Fig. 18.11)
- Recent studies have suggested the presence of *IDH2* alterations, detected by molecular methods as well as a p. R172 variant-specific immunohistochemical antibody, in

a more histologically diverse set of high-grade sinonasal malignancies including large cell neuroendocrine carcinoma, olfactory neuroblastoma, and non-intestinal adenocarcinoma
- Preliminary studies show that these tumors may behave better than wild-type counterparts
- The presence of *IDH1* or *IDH2* variants suggests the potential for therapy using targeted inhibitors or hypomethylating agents, which have been approved for use in other *IDH*-driven malignancies

Human Papillomavirus-Associated Multiphenotypic Sinonasal Carcinoma
- Malignant neoplasm exhibiting myoepithelial, ductal, and squamous differentiation
- Exclusive thus far to the sinonasal tract

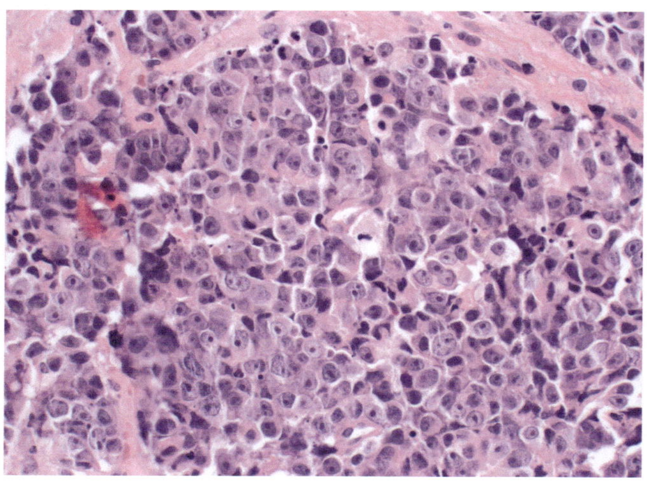

Fig. 18.11 Sheets of large, undifferentiated, pleomorphic cells with prominent nucleoli and frequent mitoses are observed in this *IDH* mutant sinonasal carcinoma (image courtesy of Dr. Jeffrey Krane)

- Basaloid cells growing in large, solid nests with cribriform/tubular foci (Fig. 18.12a) and frequent surface involvement of high-grade dysplasia (Fig. 18.12b), which can be a clue
- Harbors HR-HPV; type 33 most common (Fig. 18.12c)
- Relatively indolent despite high-grade histologic appearance: distant metastases are rare

Mesenchymal Tumors

Nasal Chondromesenchymal Hamartoma
- Rare benign mesenchymal neoplasm occurring in children
- Characterized by variably cellular stroma admixed with hyaline cartilage and bone
- Part of the spectrum of disease in pleuropulmonary blastoma (PPB)-associated *DICER1* tumor predisposition disorder; diagnosis of this lesion should prompt close clinical follow-up for manifestation of tumors associated with this syndrome (Table 18.1)

Sinonasal Glomangiopericytoma
- Low-grade soft tissue neoplasm exhibiting perivascular myoid differentiation
- Nasal cavity, occasionally extending to paranasal sinuses
- Syncytial growth of uniform, oval to spindled cells growing in many patterns including short fascicles, storiform, whorled, and solid (Fig. 18.13a); prominent

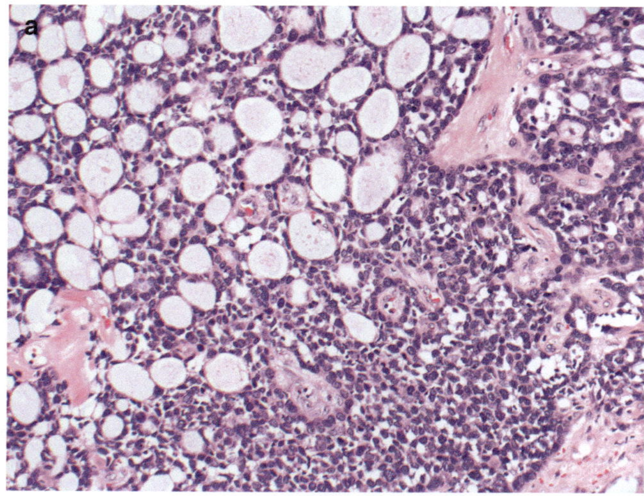

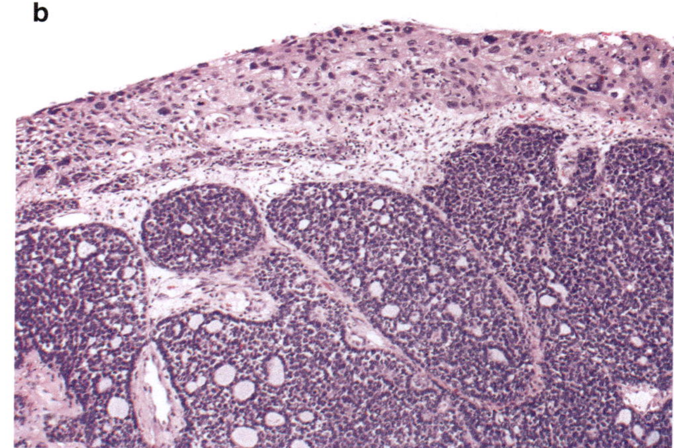

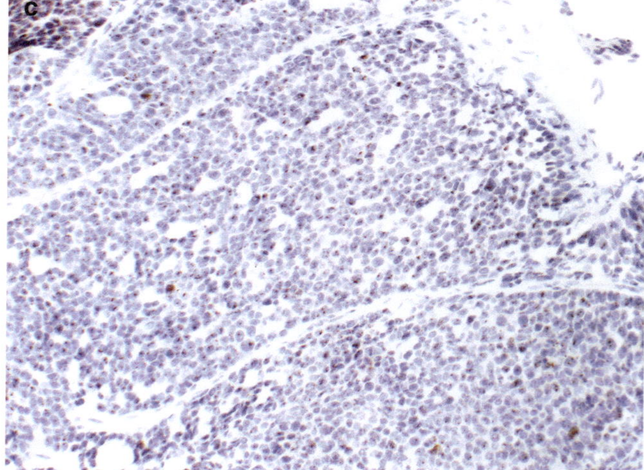

Fig. 18.12 (**a**) Neoplastic cells showing both solid growth and areas of punched out spaces reminiscent of adenoid cystic carcinoma in human papillomavirus (HPV)-associated multiphenotypic sinonasal carcinoma. (**b**) Frequent severe atypia of the overlying surface epithelium is observed. (**c**) Harbors transcriptionally active HPV with HPV 33 seen by ISH in a majority of tumors

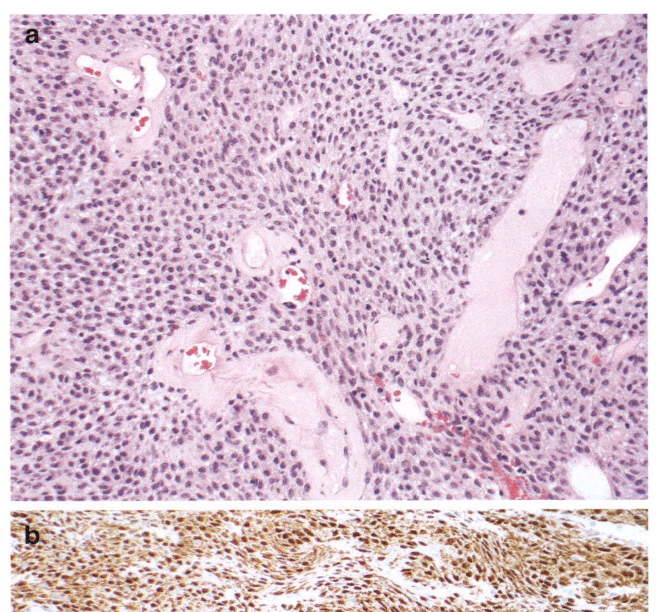

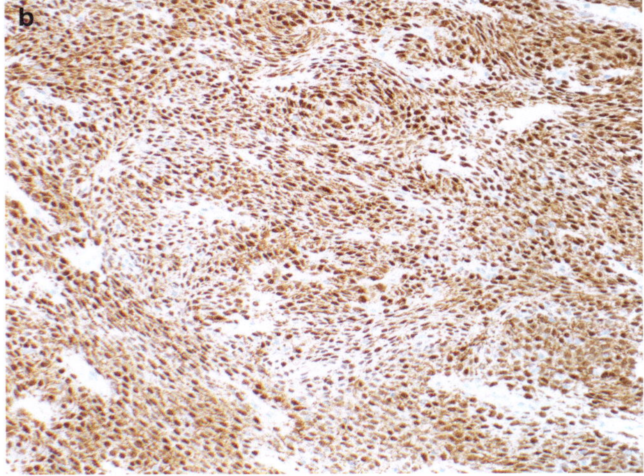

Fig. 18.13 (**a**) Cellular sheets of bland, uniform round/oval/spindled cells with prominent perivascular hyalinization seen in sinonasal glomangiopericytoma. (**b**) Diffuse nuclear β-catenin staining reflects a *CTNNB1* mutation

perivascular hyalinization; uninvolved zone of subepithelial space
- Very rarely malignant
- Nuclear β-catenin by immunohistochemistry is helpful (Fig. 18.13b), but not pathognomonic
- Often harbor hotspot sequence alterations in *CTNNB1* exon 3, a region that codes for a portion of the N-terminal region of the β-catenin protein; alterations in this region cause constitutive activation of the Wnt signaling pathway and nuclear translocation of the protein, leading to the typical immunohistochemical staining pattern seen
- Molecular testing, particularly for fusions, can be helpful in distinguishing from other key differential diagnoses, including solitary fibrous tumor (*NAB2::STAT6*), glomus tumor (*MIR143::NOTCH1/NOTCH2/NOTCH3*), or pericytoma (*ACTB::GLI1*)

Biphenotypic Sinonasal Sarcoma
- Low-grade malignant neoplasm with neural and myogenic differentiation (Fig. 18.14)
- Site-specific: upper nasal cavity, ethmoid > other paranasal sinuses
- Cellular proliferation of bland spindle cells with fascicular architecture, and distinctive invaginations of hyperplastic respiratory mucosa
- Locally aggressive but distant metastases have not been reported
- Harbor characteristic fusions involving *PAX3* exon 7 and *MAML3* exon 2, most often fused together as *PAX3::MAML3*, and less frequently one of these genes fused to alternative partners, e.g., *PAX3::FOXO1*, *PAX3::NCOA1*, *PAX3::NCOA2*, and *PAX3::WWTR1*

Spindle Cell/Sclerosing Rhabdomyosarcoma
- Malignant mesenchymal neoplasm demonstrating skeletal muscle differentiation
- Current division into 3 subgroups based on molecular and prognostic data (2 are included below given predilection for the head and neck)
  - Rhabdomyosarcoma with *TFCP2* Fusions
    ○ Newly recognized, rare entity, with current data showing strong predilection for a craniofacial, intraosseous location in young adults
    ○ Typical morphology shows loose clusters or fascicles of relatively uniform spindled and epithelioid cells with abundant cytoplasm and no obvious rhabdomyoblastic features

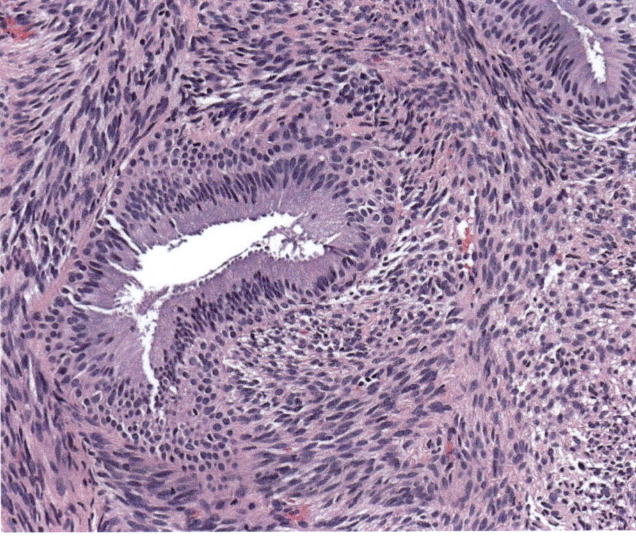

Fig. 18.14 Biphenotypic sinonasal sarcoma is composed of a cellular proliferation of bland spindle cells arranged in fascicles with a characteristic invagination of surface epithelium

- o Tumors show immunohistochemical evidence of rhabdomyoblastic differentiation (myogenin, MyoD1, desmin) with variable patterns and are often keratin and ALK positive
- o Characterized by fusions involving *TFCP2*, most often partnering with *EWSR1* or *FUS*, and occasionally show partial deletion of *ALK*
- o Thus far these tumors demonstrate a highly aggressive clinical course
- − *MYOD1*-mutant spindle cell/sclerosing rhabdomyosarcoma
  - o Head and neck region is the most common site in both children and adults with a wide age range
  - o Tumors can be purely spindled or sclerosing, arranged in long fascicles, or mixed with areas of primitive round cell morphology
  - o Tumors show minimal histologic evidence of rhabdomyoblastic differentiation but are diffusely positive for desmin and MyoD1; myogenin is more variable
  - o Most tumors harbor a *MYOD1* homozygous mutation in exon 1 (p.L122R) with the remainder being heterozygous
  - o A third of these tumors show coexisting *PIK3CA* mutations, and a few tumors show concurrent mutations in *HRAS*, *NRAS*, and *FGF4*; significance of these is not yet known
  - o Tumors show aggressive clinical course in both adults and children

Emerging Molecularly-Defined Entities

DEK::AFF2 Fusion-Associated Carcinoma

- Recently described; initial case discovered as an exceptional responder to immunotherapy
- Group of malignant epithelial neoplasms with squamous differentiation, focal to no keratinization, mixed exophytic/endophytic growth patterns, monotonous cytomorphology, and neutrophilic infiltrates
- Sinonasal/middle ear/skull base
- Aggressive clinical course
- Whether it is a distinct entity or a variant of SCC is yet to be determined
- All tumors in this group are defined by the pathognomonic fusion *DEK::AFF2*, with a recurrent exon 7 breakpoint in *DEK* and variable breakpoints in *AFF2*
- These tumors all lack HPV-driven oncogenesis as well as the driver alterations (e.g., in *EGFR* and *KRAS*) that are typical of other sinonasal neoplasms (see above)

Adamantinoma-Like Ewing Sarcoma

- Rare, enigmatic malignant neoplasm currently classified as a variant of Ewing sarcoma (ES)
- Initially described as "adamantinoma-like" given some overlapping features with tibial adamantinoma
- Harbors the *EWSR1::FLI1* fusion historically considered pathognomonic for ES and demonstrates corresponding immunohistochemical positivity for CD99 and NKX2.2
- Additionally demonstrates histopathological and immunohistochemical evidence of epithelial differentiation in the form of cohesion/squamous pearls and pankeratin/p40 positivity
- Presents predominantly in the head and neck: sinonasal cavity, salivary gland, and thyroid
- Infiltrative nests and sheets of uniform, basaloid cells with frequent mitoses
- Provisional clinicopathologic entity: whether ES family or carcinoma is to be determined

Hematolymphoid Tumors

Extranodal NK/T-Cell Lymphoma, Nasal Type

- Aggressive lymphoma of NK (predominant) or T-cell lineage
- Most common lymphoma in the sinonasal tract, with a predilection for the nasal cavity
- More prevalent in Asia and indigenous populations of Latin America
- Neoplastic cells show angiocentric and angiodestructive patterns with associated mixed inflammatory infiltrate and coagulative necrosis (Fig. 18.15a)
- Cytologic features may be insidiously nondescript (Fig. 18.15b); recognition of architecture is key
- Strong association with EBV infection (Fig. 18.15b inset) and occurs more frequently in the setting of immunosuppression
- Most cases show activation of the JAK-STAT pathway, with activating variants in *JAK3* (p.A572V and p.A573V) and *STAT3*, and inactivation of *PTPRK*, a negative regulator in the signaling cascade
- Inactivation of tumor suppressors is often detected, including *TP53* and *DDX3X*, often via complex copy number alterations that include loss of chromosome 6p21 where *PRDM1*, *PTPRK*, and *FOXO3* are located
- Epigenetic dysregulation has also been supported as a mechanism of oncogenesis following detection of recurrent alterations in *BCOR* and *KMT2D*
- Other lymphomas that are not typically enriched for the head and neck can involve this anatomic site

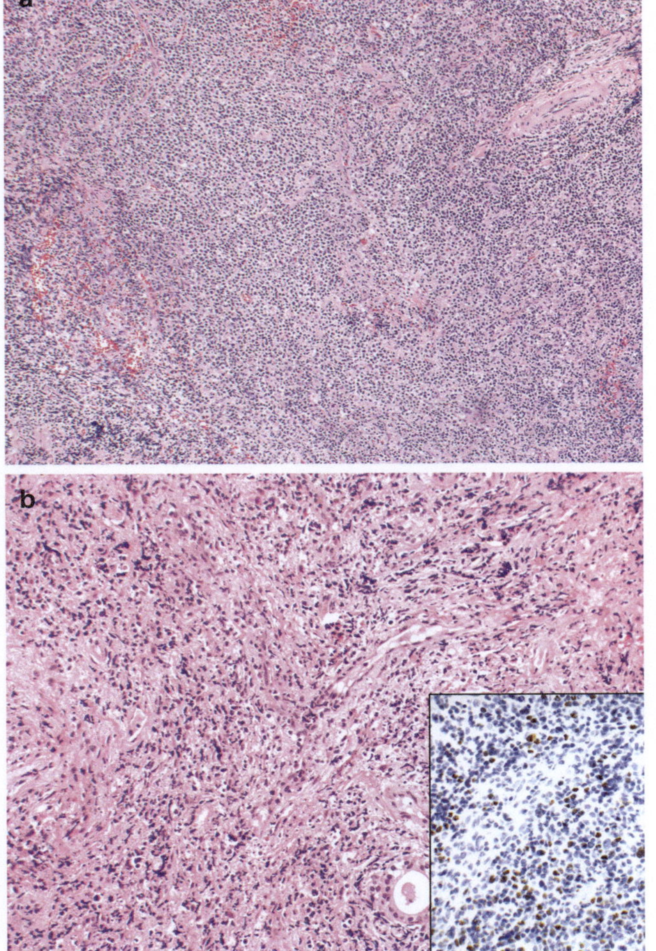

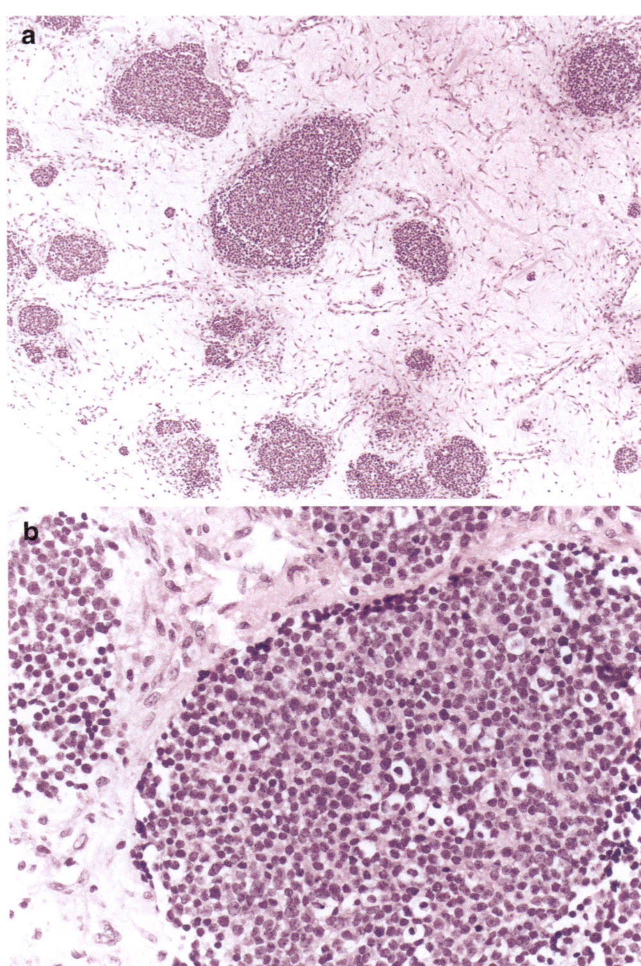

Fig. 18.15 (**a**) Extranodal NK/T-Cell Lymphoma, Nasal Type characterized by sheets of neoplastic cells showing angiodestructive growth patterns. (**b**) Can be easily mistaken for a mixed inflammatory background but a low threshold for EBV staining (inset) will reveal underlying neoplastic process

Fig. 18.16 (**a**) Lobular arrangement of an olfactory neuroblastoma in an edematous, vascular background. (**b**) Higher power reveals syncytial cells with focal areas of neurofibrillary matrix, nuclei with uniformly distributed chromatin and scattered larger cells with prominent nucleoli

Neuroectodermal Tumors

Olfactory Neuroblastoma

- Malignant neuroectodermal neoplasm arising from olfactory neuroepithelium found in the cribriform plate, superior turbinate, and the upper half of the nasal septum
- Histological features are assessed as part of Hyams grading system
  - Lower grade tumors show lobular architecture and uniform, round, blue cells with typical neuroendocrine "salt and pepper" chromatin in a neurofibrillary background (Fig. 18.16a and b)
  - Higher grade tumors grow in sheets, with nuclear pleomorphism, mitoses, and necrosis
  - Pseudorosettes and true rosettes may be seen but are neither necessary nor sufficient for diagnosis

- Tumors are locally aggressive, a subset recur and metastasize, and higher grade tumors have a worse prognosis
- *CCND1* amplification has been detected in a small subset of tumors, as has amplification of *FGFR3*, a potentially targetable alteration

Nasopharyngeal Tumors

Nasopharyngeal Angiofibroma

- Rare benign mesenchymal neoplasm characterized by numerous variably sized/shaped vessels in an equally variable cellular/fibrous stroma composed of spindled or stellate fibroblasts (Fig. 18.17)
- Exclusively in young males; associated with puberty

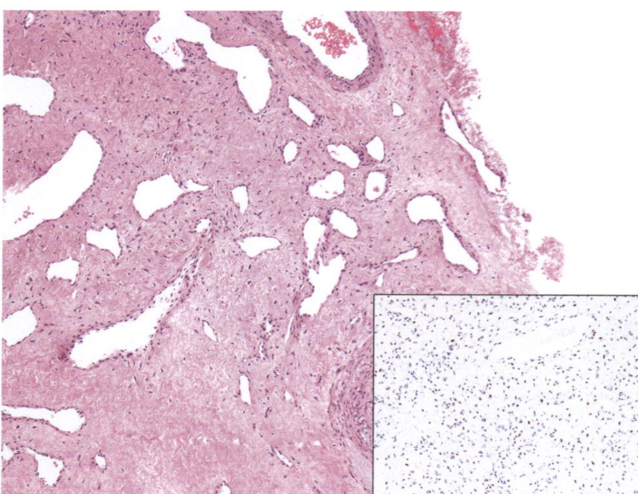

Fig. 18.17 Numerous vessels with variable thickness within a stroma composed of spindled/stellate cells which are positive for androgen receptor by immunohistochemistry (inset)

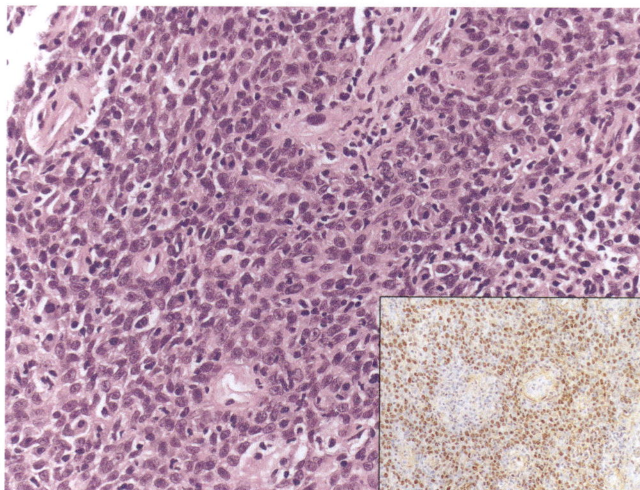

Fig. 18.18 Large cells with nonkeratinizing morphology growing in sheets is a typical appearance for the nasopharyngeal carcinoma and can be identified by ISH for EBV (inset)

- Arises in the nasopharynx and can be locally aggressive, sometimes extending into neighboring regions
- Diffuse nuclear β-catenin is observed in nearly all these tumors, the majority of which are secondary to *CTNNB1* somatic variants in the exon 3 hotspot region

Nasopharyngeal Carcinoma

- Rare malignant epithelial neoplasm with evidence of squamous differentiation
- Unbalanced geographic distribution with 70% of new cases in East/Southeast Asia
  - Presence of either of two nonsynonymous variants in *BALF2* from the EBV genome are associated with increased risk of nasopharyngeal carcinoma in southern China
  - Data show an association between the commonly found HLA-A-A*0207 allele among east Asian populations and susceptibility to nasopharyngeal carcinoma
- *MST1R* variation has been reported to confer increased susceptibility to nasopharyngeal carcinoma, particularly at c.G917A (p.R306H)
- Activation of the cyclin D1-RB pathway via *CDKN2A* inactivation or *CCND1* amplification have both been frequently detected in these tumors
- *ARID1A* is frequently inactivated in nasopharyngeal carcinoma, with *BAP1* and/or *TP53* inactivation and ERBB-PI3K pathway activation also occurring concurrently
- Global DNA methylation secondary to EBV infection also leads to broad silencing of other tumor suppressor genes

Nonkeratinizing Squamous Cell Carcinoma

- The most common form of nasopharyngeal carcinoma
- Large cells in syncytial sheets and trabeculae (undifferentiated, Fig. 18.18)
- Differentiated types may show intercellular bridges
- Strong association with EBV infection, which can be reliably detected by ISH (Fig. 18.18 inset)
- High propensity for lymph node metastasis
- A subset of these tumors is EBV-negative and associated with HR-HPV, particularly in areas where EBV is nonendemic

Keratinizing Squamous Cell Carcinoma

- Shares histological overlap with keratinizing SCC of other anatomic sites in the head and neck
- At least part of this tumor's etiology has been linked to secondary development following radiation treatment for nonkeratinizing nasopharyngeal carcinoma
- This tumor has not been studied extensively on the molecular level

Odontogenic Tumors

Ameloblastoma

- Benign epithelial odontogenic neoplasm with components resembling the dental organ (Fig. 18.19a and b)
- Mandible >>> maxilla

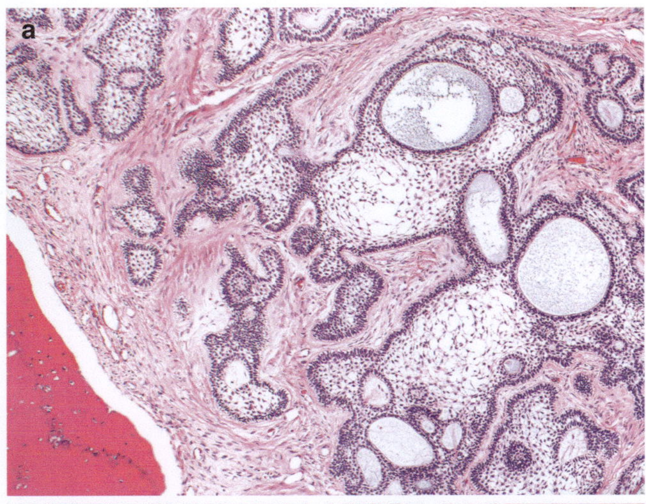

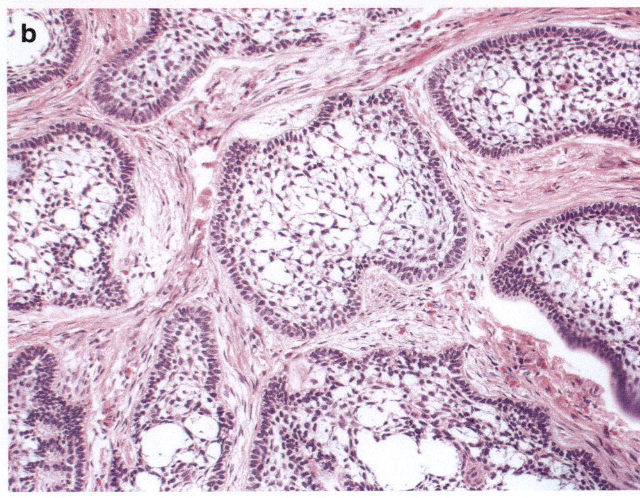

Fig. 18.19 (**a**) An intraosseous ameloblastoma growing in irregular nests and showing occasional cystic change. (**b**) Higher power of the follicular variant of ameloblastoma demonstrating characteristic basal palisading and central, loose stellate reticulum

- Intraosseous, unicystic (younger patients), peripheral
- Morphological types include follicular plexiform, acanthomatous, granular, basaloid, desmoplastic; tumors can show mixed morphology, and tumor type is not related to prognosis
- Typically harbor alterations in the MAPK pathway, most notably *BRAF* p.V600E in approximately 60% of cases; other mutually exclusive driver variants include those in *KRAS, HRAS,* or *NRAS* comprising roughly 20% of cases, or *FGFR2* alterations in about 10% of cases
- Presence of *BRAF* p.V600E appears to correlate with younger age of onset, likelihood of mandibular origin, and possibly longer recurrence-free survival, though there are also conflicting data indicating lower disease-free survival
- Treatment is primarily surgical, although the most affected genes in these tumors have established targeted inhibitors, including vemurafenib and dabrafenib against *BRAF*, trametinib against *MEK*, and regorafenib and ponatinib against *FGFR2*
- Targeted inhibition has been suggested to have a role in the context of tumor recurrence and metastasizing ameloblastoma

Ameloblastic Carcinoma

- Malignant morphologic counterpart of ameloblastoma with cytologic features of malignancy
- Extremely rare with a relatively low number of cases currently reported; most arise de novo
- Show MAPK pathway alterations and, like the benign counterpart ameloblastoma, can be driven by *BRAF* variants, specifically p.V600E
- There are reports of positive response to combined dabrafenib and trametinib treatment

Odontogenic and Non-Odontogenic Developmental Cysts

Odontogenic Keratocyst

- Odontogenic cyst arising from cells of the dental lamina characterized by a stratified squamous epithelium with a prominent, palisading basal layer (Fig. 18.20)
- Mandible >> maxilla
- Approximately 85% of patients with germline alterations in *PTCH1* leading to nevoid basal cell carcinoma syndrome (NBCCS), also known as Gorlin syndrome, have odontogenic keratocyst (Table 18.1); approximately 30% of sporadic OKC have *PTCH1* alterations
- Mechanisms of *PTCH1* inactivation include both sequence variants and loss of heterozygosity in chromosome 9q22-31
- FDA-approved small molecular inhibitors targeting the Sonic Hedgehog (SHH) pathway, including vismodegib and sonidegib, may have a role in future treatments for more aggressive tumors
- Patients with multiple cysts and cysts in the posterior mandible are more likely to have NBCCS

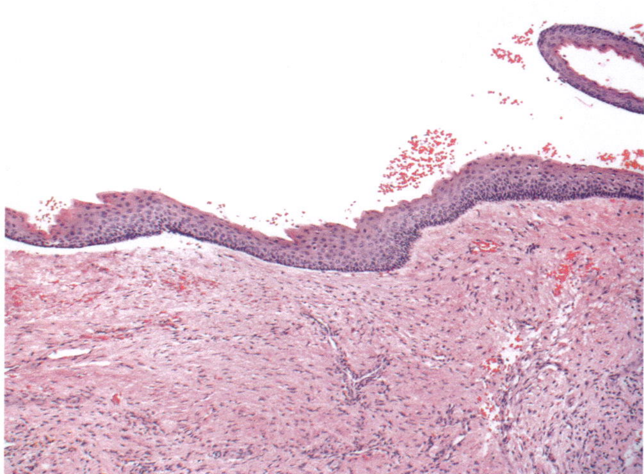

Fig. 18.20 Stratified squamous epithelium with a corrugated and keratinized surface layer, a palisading basal layer, and an artifactual clefting of the epithelium, all characteristic features of an odontogenic keratocyst

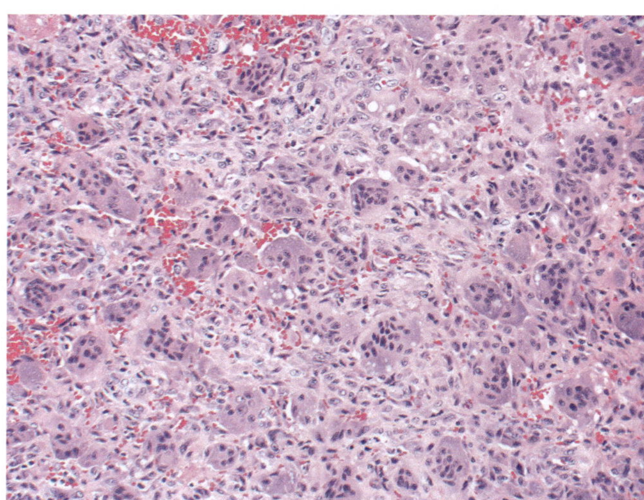

Fig. 18.21 Numerous osteoclast-like giant cells distributed haphazardly in a background of neoplastic stromal cells and scattered extravasated red blood cells is a histologic pattern common to central giant cell lesions, brown tumors of hyperparathyroidism, and syndrome-associated lesions

Calcifying Odontogenic Cyst

- Belongs to a group of heterogeneous jaw lesions with a ghost cell component that has been inconsistently classified as either a neoplasm or a developmental cyst; recent molecular data suggest the former
- Predominantly intraosseous and can occur with an associated odontoma
- Characterized by an epithelial lining of variable thickness that resembles the enamel organ with associated ghost cells that may amass and calcify
- Treated with enucleation with a good prognosis
- Two series have shown that 90% of tumors possess a sequence alteration in the *CTNNB1* gene, most occurring at either of two regulatory phosphorylation sites, p.S33 or p.S37, involved in the inactivation of Wnt/β-catenin signaling; one case had a simultaneous alteration in *APC*

Giant Cell Lesions

Central Giant Cell Lesion

- Benign osteolytic lesion of the jaw
- Young patients, typically under 20 years of age; most frequently in the mandible and in females
- Characterized by osteoclast-like giant cells in a background of variably cellular mononuclear stromal cells (Fig. 18.21)
- May be aggressive and can cause high morbidity based on age and location
- Alterations have been reported in oncogenes, including *BRAF, GNAS, HRAS, KRAS,* and *PDGFRB,* with some variants being clinically actionable

- These lesions harbor driver alterations distinct from the recurrent *H3F3A* variants found in giant cell tumor of bone which arises predominantly in long bones
- Possibility of targeted therapy: neoplastic stromal cells often express high levels of receptor activator of nuclear factor-kappa B ligand (RANKL), inducing giant cell formation; those with RANKL overexpression are often responsive to targeted treatment with denosumab, a RANKL monoclonal antibody
- Presence of multiple lesions should invoke consideration for germline testing as these may be the presenting features for syndromes involving the RAS/MAPK pathway (Table 18.1), including cherubism (*SH3BP2*), Noonan syndrome with and without lentigines (*PTPN11, SOS1,* and *RAF1*), and neurofibormatosis type I (*NF1*)

Peripheral Giant Cell Lesions

- Thought to form due to a reactive process
- Secondary to local irritation of the gingival/alveolar ridge mucosa

Ear

External Auditory Canal Squamous Cell Carcinoma

- Histological overlap with keratinizing SCC seen in other anatomic sites of the head and neck
- Based on whole exome sequencing of a small series, the most frequently altered gene is *TP53* in more than

half of cases, with other recurrent alterations in *CDKN2A, NOTCH1, NOTCH2, FAT1*, and *FAT3* identified as well
- HR-HPV 16 and 18 have been identified in some cases of SCC, though the oncogenic pathway has not been elucidated
- Considered an aggressive tumor, often showing metastases to lymph nodes and local tumor recurrence

Middle and Inner Ear Tumors

Endolymphatic Sac Tumor
- Epithelial neoplasm arising in the endolymphatic sac
- Proliferation of columnar epithelium in papillary architecture
- May be destructive of the temporal bone
- Approximately 4% of patients with von Hippel-Lindau disease develop endolymphatic sac tumor, and alterations in *VHL* are present in approximately 40% of sporadic tumors

Paraganglioma
- Neuroectodermal neoplasm derived from extra-adrenal paraganglia along parasympathetic nerves
- Predilection for the head and neck: carotid body>>jugulo tympanic>vagal>larynx
- Nested architecture composed of chief cells (Fig. 18.22a) and peripheral sustentacular cells (Fig. 18.22b)
- At least 30% are associated with a germline variant of known increased risk
- The most frequently altered genes in this anatomic region are *SDHD* and *SDHC*, associated primarily with carotid body and head and neck paragangliomas, respectively, followed by *SDHAF2* and *SDHB*, all playing a role in the succinate dehydrogenase complex
- Immunohistochemistry for SDHB has been demonstrated to show loss of staining in tumors (Fig. 18.21c) with germline alterations in any of the succinate dehydrogenase complex genes, and not in sporadic or cases with other germline alterations
- Patients with germline alterations in *SDHB* or with a history of neurofibromatosis type I, caused by *NF1* germline alterations, have an elevated risk for metastases with 23% and 12% occurrence, respectively
- Carney–Stratakis syndrome and Carney triad both have paragangliomas and gastrointestinal stromal tumors, and patients with the latter also have pulmonary chondromas (Table 18.1)
- Sporadic tumors more often harbor variants in *NF1* and rarely show variants in the succinate dehydrogenase complex
- See "Molecular Pathology of Endocrine Cancer" chapter for a broader discussion on paragangliomas

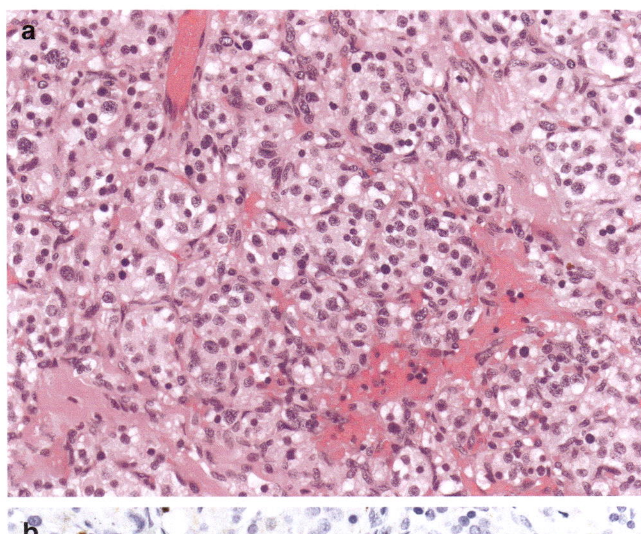

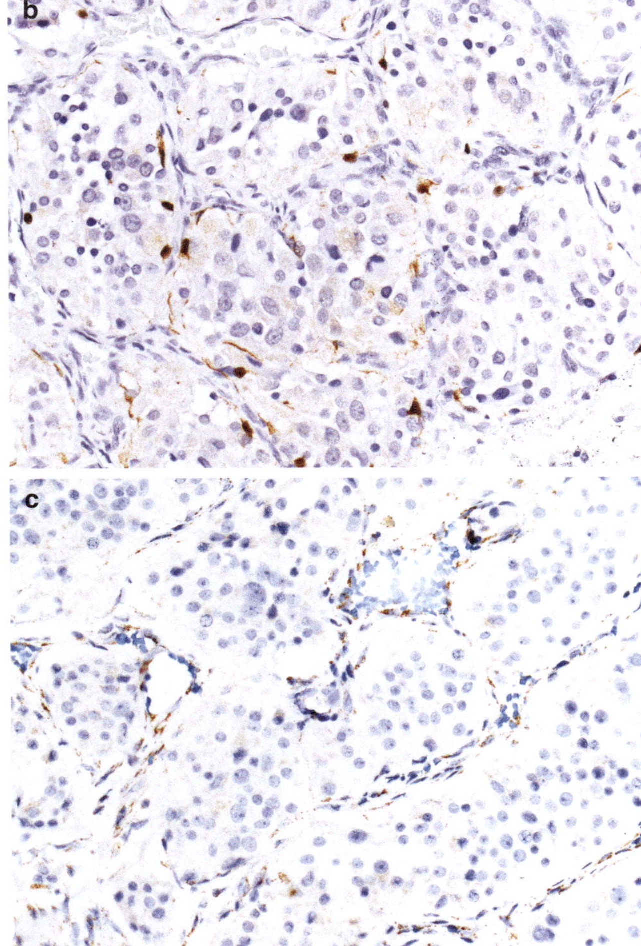

Fig. 18.22 (**a**) A carotid body tumor showing the typical nested appearance of a paraganglioma surrounded by a prominent fibrovascular network. (**b**) An S100 stain highlights scattered sustentacular cells within the fibrous septae. (**c**) A loss of SDHB staining in tumor cells (surrounding nonneoplastic cells are positive serving as an internal control) should prompt genetic testing for possible syndromic/inherited tumors

Suggested Reading

Abraham SC, Montgomery EA, Giardiello FM, Wu TT. Frequent beta-catenin mutations in juvenile nasopharyngeal angiofibromas. Am J Pathol. 2001;158(3):1073–8. https://doi.org/10.1016/s0002-9440(10)64054-0.

Agaimy A, Bishop JA. SWI/SNF-deficient head and neck neoplasms: An overview. Semin Diagn Pathol. 2021;38(3):175–82. https://doi.org/10.1053/j.semdp.2021.02.002.

Agaimy A, Hartmann A, Antonescu CR, et al. SMARCB1 (INI-1)-deficient Sinonasal Carcinoma: a series of 39 cases expanding the morphologic and clinicopathologic spectrum of a recently described entity. Am J Surg Pathol. 2017;41(4):458–71. https://doi.org/10.1097/PAS.0000000000000797.

Agaimy A, Jain D, Uddin N, Rooper LM, Bishop JA. SMARCA4-deficient sinonasal carcinoma. Am J Surg Pathol. 2020;44(5):8.

Agaram NP, LaQuaglia MP, Alaggio R, et al. MYOD1-mutant spindle cell and sclerosing rhabdomyosarcoma: an aggressive subtype irrespective of age. A reappraisal for molecular classification and risk stratification. Mod Pathol. 2019a;32(1):27–36. https://doi.org/10.1038/s41379-018-0120-9.

Agaram NP, Zhang L, Sung YS, et al. GLI1-amplifications expand the spectrum of soft tissue neoplasms defined by GLI1 gene fusions. Mod Pathol. 2019b;32(11):1617–26. https://doi.org/10.1038/s41379-019-0293-x.

Akbari M, Chen H, Guo G, Legan Z, Ghali G. Basal cell nevus syndrome (Gorlin syndrome): genetic insights, diagnostic challenges, and unmet milestones. Pathophysiology. 2018;25(2):77–82. https://doi.org/10.1016/j.pathophys.2017.12.004.

Alexiev BA, Tumer Y, Bishop JA. Sinonasal adamantinoma-like Ewing sarcoma: a case report. Pathol Res Pract. 2017;213(4):422–6. https://doi.org/10.1016/j.prp.2016.11.018.

Allanson BM, Low TH, Clark JR, Gupta R. Squamous cell carcinoma of the external auditory canal and temporal bone: an update. Head Neck Pathol. 2018;12(3):407–18. https://doi.org/10.1007/s12105-018-0908-4.

Andreasen S, Skálová A, Agaimy A, et al. ETV6 gene rearrangements characterize a morphologically distinct subset of sinonasal low-grade non–intestinal-type adenocarcinoma: a novel translocation-associated carcinoma restricted to the sinonasal tract. Am J Surg Pathol. 2017;41(11):1552–60. https://doi.org/10.1097/PAS.0000000000000912.

Andreasen S, Bishop JA, Hellquist H, et al. Biphenotypic sinonasal sarcoma: demographics, clinicopathological characteristics, molecular features, and prognosis of a recently described entity. Virchows Arch. 2018a;473(5):615–26. https://doi.org/10.1007/s00428-018-2426-x.

Andreasen S, Kiss K, Melchior LC, Laco J. The ETV6-RET Gene Fusion Is Found in ETV6-rearranged Low-grade Sinonasal Adenocarcinoma Without NTRK3 Involvement. Am J Surg Pathol. 2018b;42(7):985–8. https://doi.org/10.1097/PAS.0000000000001069.

Anzai T, Saito T, Tsuyama S, Toh M, Ikeda K, Ito S. A case of glomangiopericytoma at the nasal septum. Head Neck Pathol. 2018;12(4):572–5. https://doi.org/10.1007/s12105-017-0870-6.

Argyris PP, Bilodeau EA, Yancoskie AE, et al. A subset of ectomesenchymal chondromyxoid tumours of the tongue show EWSR1 rearrangements and are genetically linked to soft tissue myoepithelial neoplasms: a study of 11 cases. Histopathology. 2016;69(4):607–13. https://doi.org/10.1111/his.12973.

Bague-Liebert ME. [Prevention ... I am positive. Interview with M.E. Bague-Liebert). Interview by Nadine Wehvlin]. Rev Infirm 1989;39(19):10-11.

Bausch B, Wellner U, Peyre M, et al. Characterization of endolymphatic sac tumors and von Hippel-Lindau disease in the International Endolymphatic Sac Tumor Registry. Head Neck. 2016;38(Suppl 1):E673–9. https://doi.org/10.1002/hed.24067.

Benn DE, Robinson BG, Clifton-Bligh RJ. 15 YEARS OF PARAGANGLIOMA: Clinical manifestations of paraganglioma syndromes types 1-5. Endocr Relat Cancer. 2015;22(4):T91–103. https://doi.org/10.1530/ERC-15-0268.

Bezak B, Lehrke H, Elvin J, Gay L, Schembri-Wismayer D, Viozzi C. Comprehensive genomic profiling of central giant cell lesions identifies clinically relevant genomic alterations. J Oral Maxillofac Surg. 2017;75(5):955–61. https://doi.org/10.1016/j.joms.2016.10.027.

Bishop JA. OSPs and ESPs and ISPs, Oh My! An Update on Sinonasal (Schneiderian) Papillomas. Head Neck Pathol. 2017;11(3):269–77. https://doi.org/10.1007/s12105-017-0799-9.

Bishop JA, Guo TW, Smith DF, et al. Human papillomavirus-related carcinomas of the sinonasal tract. Am J Surg Pathol. 2013;37(2):185–92. https://doi.org/10.1097/PAS.0b013e3182698673.

Bishop JA, Andreasen S, Hang JF, et al. HPV-related multiphenotypic sinonasal carcinoma: an expanded series of 49 cases of the tumor formerly known as HPV-related carcinoma with adenoid cystic carcinoma-like features. Am J Surg Pathol. 2017;41(12):1690–701. https://doi.org/10.1097/PAS.0000000000000944.

Bishop JA, Gagan J, Paterson C, McLellan D, Sandison A. Nonkeratinizing squamous cell carcinoma of the sinonasal tract with DEK-AFF2: further solidifying an emerging entity. Am J Surg Pathol. 2021;45(5):718–20. https://doi.org/10.1097/PAS.0000000000001596.

Bodner L, Manor E, Friger MD, van der Waal I. Oral squamous cell carcinoma in patients twenty years of age or younger--review and analysis of 186 reported cases. Oral Oncol. 2014;50(2):84–9. https://doi.org/10.1016/j.oraloncology.2013.11.001.

Boikos SA, Xekouki P, Fumagalli E, et al. Carney triad can be (rarely) associated with germline succinate dehydrogenase defects. Eur J Hum Genet. 2016;24(4):569–73. https://doi.org/10.1038/ejhg.2015.142.

Bredell M, Rordorf T, Kroiss S, Rücker M, Zweifel DF, Rostetter C. Denosumab as a treatment alternative for central giant cell granuloma: a long-term retrospective cohort study. J Oral Maxillofac Surg. 2018;76(4):775–84. https://doi.org/10.1016/j.joms.2017.09.013.

Brown NA, Betz BL. Ameloblastoma: a review of recent molecular pathogenetic discoveries. Biomark Cancer. 2015;7(Suppl 2):19–24. https://doi.org/10.4137/BIC.S29329.

Brown NA, Rolland D, McHugh JB, et al. Activating FGFR2-RAS-BRAF mutations in ameloblastoma. Clin Cancer Res. 2014;20(21):5517–26. https://doi.org/10.1158/1078-0432.CCR-14-1069.

Brown NA, Plouffe KR, Yilmaz O, et al. TP53 mutations and CDKN2A mutations/deletions are highly recurrent molecular alterations in the malignant progression of sinonasal papillomas. Mod Pathol. 2021;34(6):1133–42. https://doi.org/10.1038/s41379-020-00716-3.

Brunner P, Bihl M, Jundt G, Baumhoer D, Hoeller S. BRAF p.V600E mutations are not unique to ameloblastoma and are shared by other odontogenic tumors with ameloblastic morphology. Oral Oncol. 2015;51(10):e77–8. https://doi.org/10.1016/j.oraloncology.2015.07.010.

Buchwald C, Franzmann MB, Jacobsen GK, Lindeberg H. Human papillomavirus (HPV) in sinonasal papillomas: a study of 78 cases using in situ hybridization and polymerase chain reaction. Laryngoscope. 1995;105(1):66–71. https://doi.org/10.1288/00005537-199501000-00015.

Burtness B, Harrington KJ, Greil R, et al. Pembrolizumab alone or with chemotherapy versus cetuximab with chemotherapy for recurrent or metastatic squamous cell carcinoma of the head and neck (KEYNOTE-048): a randomised, open-label, phase 3 study. Lancet. 2019;394(10212):1915–28. https://doi.org/10.1016/S0140-6736(19)32591-7.

Cancer Genome Atlas Network. Comprehensive genomic characterization of head and neck squamous cell carcinomas. Nature. 2015;517(7536):576–82. https://doi.org/10.1038/nature14129.

Carney JA, Stratakis CA. Familial paraganglioma and gastric stromal sarcoma: a new syndrome distinct from the Carney triad. Am J Med Genet. 2002;108(2):132–9. https://doi.org/10.1002/ajmg.10235.

Cazorla M, Hernández L, Nadal A, et al. Collagenase-3 expression is associated with advanced local invasion in human squamous cell carcinomas of the larynx. J Pathol. 1998;186(2):144–50. https://doi.org/10.1002/(SICI)1096-9896(1998100)186:2<144::AID-PATH147>3.0.CO;2-#.

Chai AWY, Lim KP, Cheong SC. Translational genomics and recent advances in oral squamous cell carcinoma. Semin Cancer Biol. 2020;61:71–83. https://doi.org/10.1016/j.semcancer.2019.09.011.

Chen CL, Hsu MM. Second primary epithelial malignancy of nasopharynx and nasal cavity after successful curative radiation therapy of nasopharyngeal carcinoma. Hum Pathol. 2000;31(2):227–32. https://doi.org/10.1016/s0046-8177(00)80224-5.

Chen YW, Guo T, Shen L, et al. Receptor-type tyrosine-protein phosphatase κ directly targets STAT3 activation for tumor suppression in nasal NK/T-cell lymphoma. Blood. 2015;125(10):1589–600. https://doi.org/10.1182/blood-2014-07-588970.

Chen YP, Chan ATC, Le QT, Blanchard P, Sun Y, Ma J. Nasopharyngeal carcinoma. Lancet. 2019;394(10192):64–80. https://doi.org/10.1016/S0140-6736(19)30956-0.

Chrisinger JSA, Wehrli B, Dickson BC, et al. Epithelioid and spindle cell rhabdomyosarcoma with FUS-TFCP2 or EWSR1-TFCP2 fusion: report of two cases. Virchows Arch. 2020;477(5):725–32. https://doi.org/10.1007/s00428-020-02870-0.

Chua MLK, Wee JTS, Hui EP, Chan ATC. Nasopharyngeal carcinoma. Lancet. 2016;387(10022):1012–24. https://doi.org/10.1016/S0140-6736(15)00055-0.

Cohen EEW, Soulières D, Le Tourneau C, et al. Pembrolizumab versus methotrexate, docetaxel, or cetuximab for recurrent or metastatic head-and-neck squamous cell carcinoma (KEYNOTE-040): a randomised, open-label, phase 3 study. Lancet. 2019;393(10167):156–67. https://doi.org/10.1016/S0140-6736(18)31999-8.

Contrera KJ, Woody NM, Rahman M, Sindwani R, Burkey BB. Clinical management of emerging sinonasal malignancies. Head Neck. 2020;42(8):2202–12. https://doi.org/10.1002/hed.26150.

Coppo P, Gouilleux-Gruart V, Huang Y, et al. STAT3 transcription factor is constitutively activated and is oncogenic in nasal-type NK/T-cell lymphoma. Leukemia. 2009;23(9):1667–78. https://doi.org/10.1038/leu.2009.91.

Cramer JD, Burtness B, Ferris RL. Immunotherapy for head and neck cancer: Recent advances and future directions. Oral Oncol. 2019;99:104460. https://doi.org/10.1016/j.oraloncology.2019.104460.

Dahia PLM. Pheochromocytoma and paraganglioma pathogenesis: learning from genetic heterogeneity. Nat Rev Cancer. 2014;14(2):108–19. https://doi.org/10.1038/nrc3648.

Dahlén A, Mertens F, Mandahl N, Panagopoulos I. Molecular genetic characterization of the genomic ACTB-GLI fusion in pericytoma with t(7;12). Biochem Biophys Res Commun. 2004;325(4):1318–23. https://doi.org/10.1016/j.bbrc.2004.10.172.

Dai W, Zheng H, Cheung AKL, et al. Whole-exome sequencing identifies MST1R as a genetic susceptibility gene in nasopharyngeal carcinoma. Proc Natl Acad Sci U S A. 2016;113(12):3317–22. https://doi.org/10.1073/pnas.1523436113.

de Mel S, Hue SSS, Jeyasekharan AD, Chng WJ, Ng SB. Molecular pathogenic pathways in extranodal NK/T cell lymphoma. J Hematol Oncol. 2019;12(1):33. https://doi.org/10.1186/s13045-019-0716-7.

Dickson BC, Antonescu CR, Argyris PP, et al. Ectomesenchymal chondromyxoid tumor: a neoplasm characterized by recurrent RREB1-MKL2 fusions. Am J Surg Pathol. 2018;42(10):1297–305. https://doi.org/10.1097/PAS.0000000000001096.

Dobashi A, Tsuyama N, Asaka R, et al. Frequent BCOR aberrations in extranodal NK/T-Cell lymphoma, nasal type. Genes Chromosom Cancer. 2016;55(5):460–71. https://doi.org/10.1002/gcc.22348.

Dogan S, Chute DJ, Xu B, et al. Frequent IDH2 R172 mutations in undifferentiated and poorly-differentiated sinonasal carcinomas. J Pathol. 2017;242(4):400–8. https://doi.org/10.1002/path.4915.

Dogan S, Cotzia P, Ptashkin RN, et al. Genetic basis of SMARCB1 protein loss in 22 sinonasal carcinomas. Hum Pathol. 2020;104:105–16. https://doi.org/10.1016/j.humpath.2020.08.004.

El-Mofty SK, Lu DW. Prevalence of high-risk human papillomavirus DNA in nonkeratinizing (cylindrical cell) carcinoma of the sinonasal tract: a distinct clinicopathologic and molecular disease entity. Am J Surg Pathol. 2005;29(10):1367–72. https://doi.org/10.1097/01.pas.0000173240.63073.fe.

El-Naggar A, Chan J, Grandis J, Takata T, Slootweg P. WHO classification of head and neck tumours. 4th ed. Lyon: IARC Press; 2017.

Ferris RL, Blumenschein G, Fayette J, et al. Nivolumab for recurrent squamous-cell carcinoma of the head and neck. N Engl J Med. 2016;375(19):1856–67. https://doi.org/10.1056/NEJMoa1602252.

Ferris RL, Blumenschein G, Fayette J, et al. Nivolumab vs investigator's choice in recurrent or metastatic squamous cell carcinoma of the head and neck: 2-year long-term survival update of CheckMate 141 with analyses by tumor PD-L1 expression. Oral Oncol. 2018;81:45–51. https://doi.org/10.1016/j.oraloncology.2018.04.008.

Fischer M, Uxa S, Stanko C, Magin TM, Engeland K. Human papilloma virus E7 oncoprotein abrogates the p53-p21-DREAM pathway. Sci Rep. 2017;7(1):2603. https://doi.org/10.1038/s41598-017-02831-9.

Flanagan AM, Speight PM. Giant cell lesions of the craniofacial bones. Head Neck Pathol. 2014;8(4):445–53. https://doi.org/10.1007/s12105-014-0589-6.

Foy JP, Pickering CR, Papadimitrakopoulou VA, et al. New DNA methylation markers and global DNA hypomethylation are associated with oral cancer development. Cancer Prev Res (Phila). 2015;8(11):1027–35. https://doi.org/10.1158/1940-6207.CAPR-14-0179.

Franchi A, Innocenti DRD, Palomba A, et al. Low prevalence of K-RAS, EGF-R and BRAF mutations in sinonasal adenocarcinomas. Implications for anti-EGFR treatments. Pathol Oncol Res. 2014;20(3):571–9. https://doi.org/10.1007/s12253-013-9730-1.

Fregnani ER, da Cruz Perez DE, Paes de Almeida O, et al. BRAF-V600E expression correlates with ameloblastoma aggressiveness. Histopathology. 2017;70(3):473–84. https://doi.org/10.1111/his.13095.

French CA, Miyoshi I, Kubonishi I, Grier HE, Perez-Atayde AR, Fletcher JA. BRD4-NUT fusion oncogene: a novel mechanism in aggressive carcinoma. Cancer Res. 2003;63(2):304–7.

Gale N, Poljak M, Kambic V, Ferluga D, Fischinger J. Laryngeal papillomatosis: molecular, histopathological, and clinical evaluation. Virchows Arch. 1994;425(3):291–5. https://doi.org/10.1007/BF00196152.

Gallo O, Sardi I, Pepe G, et al. Multiple primary tumors of the upper aerodigestive tract: is there a role for constitutional mutations in the p53 gene? Int J Cancer. 1999;82(2):180–6. https://doi.org/10.1002/(sici)1097-0215(19990719)82:2<180::aid-ijc5>3.0.co;2-p.

Gavrielatou N, Doumas S, Economopoulou P, Foukas PG, Psyrri A. Biomarkers for immunotherapy response in head and neck cancer. Cancer Treat Rev. 2020;84:101977. https://doi.org/10.1016/j.ctrv.2020.101977.

Glavac D, Volavsek M, Potocnik U, Ravnik-Glavac M, Gale N. Low microsatellite instability and high loss of heterozygosity rates indicate dominant role of the suppressor pathway in squamous cell carcinoma of head and neck and loss of heterozygosity of 11q14.3 correlates with tumor grade. Cancer Genet Cytogenet. 2003;146(1):27–32. https://doi.org/10.1016/s0165-4608(03)00109-2.

Glöss S, Jurmeister P, Thieme A, et al. IDH2 R172 mutations across poorly differentiated sinonasal tract malignancies: forty molecularly homogenous and histologically variable cases with favorable outcome. Am J Surg Pathol. 2021; https://doi.org/10.1097/PAS.0000000000001697. Publish Ahead of Print

Gomes CC, Diniz MG, Gomez RS. Review of the molecular pathogenesis of the odontogenic keratocyst. Oral Oncol. 2009;45(12):1011–4. https://doi.org/10.1016/j.oraloncology.2009.08.003.

Grein Cavalcanti L, Lyko KF, Araújo RLF, Amenábar JM, Bonfim C, Torres-Pereira CC. Oral leukoplakia in patients with Fanconi anaemia without hematopoietic stem cell transplantation. Pediatr Blood Cancer. 2015;62(6):1024–6. https://doi.org/10.1002/pbc.25417.

Guimarães LM, Diniz MG, Rogatto SR, Gomez RS, Gomes CC. The genetic basis of oral leukoplakia and its key role in understanding oral carcinogenesis. J Oral Pathol Med. 2021;50(7):632–8. https://doi.org/10.1111/jop.13140.

Ha PK, Califano JA. Promoter methylation and inactivation of tumour-suppressor genes in oral squamous-cell carcinoma. Lancet Oncol. 2006;7(1):77–82. https://doi.org/10.1016/S1470-2045(05)70540-4.

Haack H, Johnson LA, Fry CJ, et al. Diagnosis of NUT midline carcinoma using a NUT-specific monoclonal antibody. Am J Surg Pathol. 2009;33(7):984–91. https://doi.org/10.1097/PAS.0b013e318198d666.

Handley TPB, McCaul JA, Ogden GR. Dyskeratosis congenita. Oral Oncol. 2006;42(4):331–6. https://doi.org/10.1016/j.oraloncology.2005.06.007.

Hildesheim A, Apple RJ, Chen CJ, et al. Association of HLA class I and II alleles and extended haplotypes with nasopharyngeal carcinoma in Taiwan. J Natl Cancer Inst. 2002;94(23):1780–9. https://doi.org/10.1093/jnci/94.23.1780.

Hussain I, Husain Q, Baredes S, Eloy JA, Jyung RW, Liu JK. Molecular genetics of paragangliomas of the skull base and head and neck region: implications for medical and surgical management. J Neurosurg. 2014;120(2):321–30. https://doi.org/10.3171/2013.10.JNS13659.

Jares P, Fernández PL, Nadal A, et al. p16MTS1/CDK4I mutations and concomitant loss of heterozygosity at 9p21-23 are frequent events in squamous cell carcinoma of the larynx. Oncogene. 1997;15(12):1445–53. https://doi.org/10.1038/sj.onc.1201309.

Jiang L, Gu ZH, Yan ZX, et al. Exome sequencing identifies somatic mutations of DDX3X in natural killer/T-cell lymphoma. Nat Genet. 2015;47(9):1061–6. https://doi.org/10.1038/ng.3358.

Jo VY, Chau NG, Hornick JL, Krane JF, Sholl LM. Recurrent IDH2 R172X mutations in sinonasal undifferentiated carcinoma. Mod Pathol. 2017;30(5):650–9. https://doi.org/10.1038/modpathol.2016.239.

Johnson DE, Burtness B, Leemans CR, Lui VWY, Bauman JE, Grandis JR. Head and neck squamous cell carcinoma. Nat Rev Dis Primers. 2020;6(1):92. https://doi.org/10.1038/s41572-020-00224-3.

Kennedy WR, Werning JW, Kaye FJ, Mendenhall WM. Treatment of ameloblastoma and ameloblastic carcinoma with radiotherapy. Eur Arch Otorhinolaryngol. 2016;273(10):3293–7. https://doi.org/10.1007/s00405-016-3899-3.

Kılıç S, Kılıç SS, Kim ES, et al. Significance of human papillomavirus positivity in sinonasal squamous cell carcinoma. Int Forum Allergy Rhinol. 2017;7(10):980–9. https://doi.org/10.1002/alr.21996.

Kono M, Bandoh N, Matsuoka R, et al. Glomangiopericytoma of the nasal cavity with CTNNB1 p.S37C mutation: a case report and literature review. Head Neck Pathol. 2019;13(3):298–303. https://doi.org/10.1007/s12105-018-0961-z.

Koo GC, Tan SY, Tang T, et al. Janus kinase 3-activating mutations identified in natural killer/T-cell lymphoma. Cancer Discov. 2012;2(7):591–7. https://doi.org/10.1158/2159-8290.CD-12-0028.

Kreppel M, Zöller J. Ameloblastoma-clinical, radiological, and therapeutic findings. Oral Dis. 2018;24(1-2):63–6. https://doi.org/10.1111/odi.12702.

Kresty LA, Mallery SR, Knobloch TJ, et al. Frequent alterations of p16INK4a and p14ARF in oral proliferative verrucous leukoplakia. Cancer Epidemiol Biomark Prev. 2008;17(11):3179–87. https://doi.org/10.1158/1055-9965.EPI-08-0574.

Kubik M, Barasch N, Choby G, Seethala R, Snyderman C. Sinonasal renal cell-like carcinoma: case report and review of the literature. Head Neck Pathol. 2017;11(3):333–7. https://doi.org/10.1007/s12105-016-0774-x.

Kuo YJ, Lewis JS, Zhai C, et al. DEK-AFF2 fusion-associated papillary squamous cell carcinoma of the sinonasal tract: clinicopathologic characterization of seven cases with deceptively bland morphology. Mod Pathol. 2021;34(10):1820–30. https://doi.org/10.1038/s41379-021-00846-2.

Larque AB, Hakim S, Ordi J, et al. High-risk human papillomavirus is transcriptionally active in a subset of sinonasal squamous cell carcinomas. Mod Pathol. 2014;27(3):343–51. https://doi.org/10.1038/modpathol.2013.155.

Lasota J, Felisiak-Golabek A, Aly FZ, Wang ZF, Thompson LDR, Miettinen M. Nuclear expression and gain-of-function β-catenin mutation in glomangiopericytoma (sinonasal-type hemangiopericytoma): insight into pathogenesis and a diagnostic marker. Mod Pathol. 2015;28(5):715–20. https://doi.org/10.1038/modpathol.2014.161.

Lawson W, Schlecht NF, Brandwein-Gensler M. The role of the human papillomavirus in the pathogenesis of Schneiderian inverted papillomas: an analytic overview of the evidence. Head Neck Pathol. 2008;2(2):49–59. https://doi.org/10.1007/s12105-008-0048-3.

Lazo de la Vega L, McHugh JB, Cani AK, et al. Comprehensive molecular profiling of olfactory neuroblastoma identifies potentially targetable FGFR3 amplifications. Mol Cancer Res. 2017;15(11):1551–7. https://doi.org/10.1158/1541-7786.MCR-17-0135.

Le Loarer F, Laffont S, Lesluyes T, et al. Clinicopathologic and molecular features of a series of 41 biphenotypic sinonasal sarcomas expanding their molecular spectrum. Am J Surg Pathol. 2019;43(6):747–54. https://doi.org/10.1097/PAS.0000000000001238.

Lee S, Park HY, Kang SY, et al. Genetic alterations of JAK/STAT cascade and histone modification in extranodal NK/T-cell lymphoma nasal type. Oncotarget. 2015;6(19):17764–76. https://doi.org/10.18632/oncotarget.3776.

Leemans CR, Snijders PJF, Brakenhoff RH. The molecular landscape of head and neck cancer. Nat Rev Cancer. 2018;18(5):269–82. https://doi.org/10.1038/nrc.2018.11.

Lewis JS. Sinonasal squamous cell carcinoma: a review with emphasis on emerging histologic subtypes and the role of human papillomavirus. Head Neck Pathol. 2016;10(1):60–7. https://doi.org/10.1007/s12105-016-0692-y.

Lin DC, Meng X, Hazawa M, et al. The genomic landscape of nasopharyngeal carcinoma. Nat Genet. 2014;46(8):866–71. https://doi.org/10.1038/ng.3006.

Liu Y, Li DX, Tian C, Liu H, gang. Human telomerase RNA component (hTERC) gene amplification detected by FISH in precancerous lesions and carcinoma of the larynx. Diagn Pathol. 2012;7:34. https://doi.org/10.1186/1746-1596-7-34.

López F, García Inclán C, Pérez-Escuredo J, et al. KRAS and BRAF mutations in sinonasal cancer. Oral Oncol. 2012;48(8):692–7. https://doi.org/10.1016/j.oraloncology.2012.02.018.

Luzar B, Poljak M, Gale N. Telomerase catalytic subunit in laryngeal carcinogenesis – an immunohistochemical study. Mod Pathol. 2005;18(3):406–11. https://doi.org/10.1038/modpathol.3800275.

Mäkitie AA, Monni O. Molecular profiling of laryngeal cancer. Expert Rev Anticancer Ther. 2009;9(9):1251–60. https://doi.org/10.1586/era.09.102.

Marret G, Bièche I, Dupain C, et al. Genomic alterations in head and neck squamous cell carcinoma: level of evidence according to ESMO scale for clinical actionability of molecular targets (ESCAT). JCO Precis Oncol. 2021;5:215–26. https://doi.org/10.1200/PO.20.00280.

Marur S, Forastiere AA. Head and neck squamous cell carcinoma: update on epidemiology, diagnosis, and treatment. Mayo Clin Proc. 2016;91(3):386–96. https://doi.org/10.1016/j.mayocp.2015.12.017.

McLaughlin-Drubin ME, Park D, Munger K. Tumor suppressor p16INK4A is necessary for survival of cervical carcinoma cell lines. Proc Natl Acad Sci U S A. 2013;110(40):16175–80. https://doi.org/10.1073/pnas.1310432110.

Mechtersheimer G, Andrulis M, Delank KW, et al. RREB1-MKL2 fusion in a spindle cell sinonasal sarcoma: biphenotypic sinonasal sarcoma or ectomesenchymal chondromyxoid tumor in an unusual site? Genes Chromosom Cancer. 2021;60(8):565–70. https://doi.org/10.1002/gcc.22948.

Mehrad M, Stelow EB, Bishop JA, et al. Transcriptionally active HPV and targetable EGFR mutations in sinonasal inverted papilloma: an association between low-risk HPV, condylomatous morphology, and cancer risk? Am J Surg Pathol. 2020;44(3):340–6. https://doi.org/10.1097/PAS.0000000000001411.

Mehta P, Ebens, C. Fanconi Anemia. 2002 Feb 14 [Updated 2021 Jun 3]. In: Adam MP, Ardinger HH, Pagon RA, et al., editors. GeneReviews® [Internet]. Seattle (WA): University of Washington, Seattle; 1993-2021. Available from: https://www.ncbi.nlm.nih.gov/books/NBK1401/.

Mendez-Pena JE, Sadow PM, Nose V, Hoang MP. RNA chromogenic in situ hybridization assay with clinical automated platform is a sensitive method in detecting high-risk human papillomavirus in squamous cell carcinoma. Hum Pathol. 2017;63:184–9. https://doi.org/10.1016/j.humpath.2017.02.021.

Mito JK, Bishop JA, Sadow PM, et al. Immunohistochemical detection and molecular characterization of IDH-mutant sinonasal undifferentiated carcinomas. Am J Surg Pathol. 2018;42(8):1067–75. https://doi.org/10.1097/PAS.0000000000001064.

Montes-Mojarro IA, Chen BJ, Ramirez-Ibarguen AF, et al. Mutational profile and EBV strains of extranodal NK/T-cell lymphoma, nasal type in Latin America. Mod Pathol. 2020;33(5):781–91. https://doi.org/10.1038/s41379-019-0415-5.

Neumann TE, Allanson J, Kavamura I, et al. Multiple giant cell lesions in patients with Noonan syndrome and cardio-facio-cutaneous syndrome. Eur J Hum Genet. 2009;17(4):420–5. https://doi.org/10.1038/ejhg.2008.188.

Ohmoto A, Sato Y, Asaka R, et al. Clinicopathological and genomic features in patients with head and neck neuroendocrine carcinoma. Mod Pathol. 2021;34(11):1979–89. https://doi.org/10.1038/s41379-021-00869-9.

Orita Y, Gion Y, Tachibana T, et al. Laryngeal squamous cell papilloma is highly associated with human papillomavirus. Jpn J Clin Oncol. 2018;48(4):350–5. https://doi.org/10.1093/jjco/hyy009.

Pan S, Dong Q, Sun LS, Li TJ. Mechanisms of inactivation of PTCH1 gene in nevoid basal cell carcinoma syndrome: modification of the two-hit hypothesis. Clin Cancer Res. 2010;16(2):442–50. https://doi.org/10.1158/1078-0432.CCR-09-2574.

Park ES, Kim J, Jun SY. Characteristics and prognosis of glomangiopericytomas: a systematic review. Head Neck. 2017;39(9):1897–909. https://doi.org/10.1002/hed.24818.

Perrone F, Oggionni M, Birindelli S, et al. TP53, p14ARF, p16INK4a and H-ras gene molecular analysis in intestinal-type adenocarcinoma of the nasal cavity and paranasal sinuses. Int J Cancer. 2003;105(2):196–203. https://doi.org/10.1002/ijc.11062.

Pickering CR, Zhang J, Yoo SY, et al. Integrative genomic characterization of oral squamous cell carcinoma identifies frequent somatic drivers. Cancer Discov. 2013;3(7):770–81. https://doi.org/10.1158/2159-8290.CD-12-0537.

Priest JR, Williams GM, Mize WA, Dehner LP, McDermott MB. Nasal chondromesenchymal hamartoma in children with pleuropulmonary blastoma - A report from the International Pleuropulmonary Blastoma Registry registry. Int J Pediatr Otorhinolaryngol. 2010;74(11):1240–4. https://doi.org/10.1016/j.ijporl.2010.07.022.

Prime SS, Thakker NS, Pring M, Guest PG, Paterson IC. A review of inherited cancer syndromes and their relevance to oral squamous cell carcinoma. Oral Oncol. 2001;37(1):1–16. https://doi.org/10.1016/s1368-8375(00)00055-5.

Prime SS, Cirillo N, Cheong SC, Prime MS, Parkinson EK. Targeting the genetic landscape of oral potentially malignant disorders has the potential as a preventative strategy in oral cancer. Cancer Lett. 2021;518:102–14. https://doi.org/10.1016/j.canlet.2021.05.025.

Projetti F, Durand K, Chaunavel A, et al. Epidermal growth factor receptor expression and KRAS and BRAF mutations: study of 39 sinonasal intestinal-type adenocarcinomas. Hum Pathol. 2013;44(10):2116–25. https://doi.org/10.1016/j.humpath.2013.03.019.

Quintanilla-Martinez L, Kremer M, Keller G, et al. p53 Mutations in nasal natural killer/T-cell lymphoma from Mexico: association with large cell morphology and advanced disease. Am J Pathol. 2001;159(6):2095–105. https://doi.org/10.1016/S0002-9440(10)63061-1.

Rocco JW, Ellisen LW. p63 and p73: life and death in squamous cell carcinoma. Cell Cycle. 2006;5(9):936–40. https://doi.org/10.4161/cc.5.9.2716.

Rooper LM, Bishop JA. Soft tissue special issue: adamantinoma-like Ewing sarcoma of the head and neck: a practical review of a challenging emerging entity. Head Neck Pathol. 2020;14(1):59–69. https://doi.org/10.1007/s12105-019-01098-y.

Rooper LM, Jo VY, Antonescu CR, et al. Adamantinoma-like Ewing sarcoma of the salivary glands: a newly recognized mimicker of basaloid salivary carcinomas. Am J Surg Pathol. 2019;43(2):187–94. https://doi.org/10.1097/PAS.0000000000001171.

Rooper LM, Uddin N, Gagan J, et al. Recurrent Loss of SMARCA4 in Sinonasal Teratocarcinosarcoma. Am J Surg Pathol. 2020;44(10):1331–9. https://doi.org/10.1097/PAS.0000000000001508.

Rooper LM, Agaimy A, Dickson BC, et al. DEK-AFF2 carcinoma of the sinonasal region and skull base: detailed clinicopathologic characterization of a distinctive entity. Am J Surg Pathol. 2021; https://doi.org/10.1097/PAS.0000000000001741. Publish Ahead of Print

Rytkönen AE, Hirvikoski PP, Salo TA. Lymphoepithelial carcinoma: two case reports and a systematic review of oral and sinonasal cases. Head Neck Pathol. 2011;5(4):327–34. https://doi.org/10.1007/s12105-011-0278-7.

Sanderson RJ, Ironside J. a. D. Squamous cell carcinomas of the head and neck. BMJ. 2002;325(7368):822–7. https://doi.org/10.1136/bmj.325.7368.822.

Sato K, Komune N, Hongo T, et al. Genetic landscape of external auditory canal squamous cell carcinoma. Cancer Sci. 2020;111(8):3010–9. https://doi.org/10.1111/cas.14515.

Savage S. Dyskeratosis Congenita. 2009 Nov 12 [Updated 2019 Nov 21]. In: Adam MP, Ardinger HH, Pagon RA, et al., editors. GeneReviews® [Internet]. Seattle (WA): University of Washington, Seattle; 1993-2021. Available from: https://www.ncbi.nlm.nih.gov/books/NBK22301/.

Sawada K, Momose S, Kawano R, et al. Immunohistochemical staining patterns of p53 predict the mutational status of TP53 in oral epithelial dysplasia. Mod Pathol. 2021;17 https://doi.org/10.1038/s41379-021-00893-9.

Scholnick SB, Sun PC, Shaw ME, Haughey BH. el-Mofty SK. Frequent loss of heterozygosity for Rb, TP53, and chromosome arm 3p, but not NME1 in squamous cell carcinomas of the supraglottic larynx. Cancer. 1994;73(10):2472–80. https://doi.org/10.1002/1097-0142(19940515)73:10<2472::aid-cncr2820731005>3.0.co;2-b.

Sekine S, Sato S, Takata T, et al. Beta-catenin mutations are frequent in calcifying odontogenic cysts, but rare in ameloblastomas. Am J Pathol. 2003;163(5):1707–12. https://doi.org/10.1016/s0002-9440(10)63528-6.

Shah AA, Jain D, Ababneh E, et al. SMARCB1 (INI-1)-deficient adenocarcinoma of the sinonasal tract: a potentially under-recognized form of sinonasal adenocarcinoma with occasional yolk sac tumor-like features. Head Neck Pathol. 2020;14(2):465–72. https://doi.org/10.1007/s12105-019-01065-7.

Siegfried A, Romary C, Escudié F, et al. RREB1-MKL2 fusion in biphenotypic "oropharyngeal" sarcoma: new entity or part of the spectrum of biphenotypic sinonasal sarcomas? Genes Chromosom Cancer. 2018;57(4):203–10. https://doi.org/10.1002/gcc.22521.

South AP, den Breems NY, Richa T, et al. Mutation signature analysis identifies increased mutation caused by tobacco smoke associated DNA adducts in larynx squamous cell carcinoma compared with oral cavity and oropharynx. Sci Rep. 2019;9(1):19256. https://doi.org/10.1038/s41598-019-55352-y.

Stagner AM, Sajed DP, Nielsen GP, et al. Giant Cell lesions of the maxillofacial skeleton express RANKL by RNA in situ hybridization regardless of histologic pattern. Am J Surg Pathol. 2019;43(6):819–26. https://doi.org/10.1097/PAS.0000000000001257.

Stewart DR, Messinger Y, Williams GM, et al. Nasal chondromesenchymal hamartomas arise secondary to germline and somatic mutations of DICER1 in the pleuropulmonary blastoma tumor predisposition disorder. Hum Genet. 2014;133(11):1443–50. https://doi.org/10.1007/s00439-014-1474-9.

Stransky N, Egloff AM, Tward AD, et al. The mutational landscape of head and neck squamous cell carcinoma. Science. 2011;333(6046):1157–60. https://doi.org/10.1126/science.1208130.

Tan S, Pollack JR, Kaplan MJ, Colevas AD, West RB. BRAF inhibitor treatment of primary BRAF-mutant ameloblastoma with pathologic assessment of response. Oral Surg Oral Med Oral Pathol Oral Radiol. 2016;122(1):e5–7. https://doi.org/10.1016/j.oooo.2015.12.016.

Taoudi Benchekroun M, Saintigny P, Thomas SM, et al. Epidermal growth factor receptor expression and gene copy number in the risk of oral cancer. Cancer Prev Res (Phila). 2010;3(7):800–9. https://doi.org/10.1158/1940-6207.CAPR-09-0163.

Thompson LDR. Laryngeal dysplasia, squamous cell carcinoma, and variants. Surg Pathol Clin. 2017;10(1):15–33. https://doi.org/10.1016/j.path.2016.10.003.

Tsai ST, Li C, Jin YT, Chao WY, Su IJ. High prevalence of human papillomavirus types 16 and 18 in middle-ear carcinomas. Int J Cancer. 1997;71(2):208–12. https://doi.org/10.1002/(sici)1097-0215(19970410)71:2<208::aid-ijc14>3.0.co;2-d.

Tsang CM, Deng W, Yip YL, Zeng MS, Lo KW, Tsao SW. Epstein-Barr virus infection and persistence in nasopharyngeal epithelial cells. Chin J Cancer. 2014;33(11):549–55. https://doi.org/10.5732/cjc.014.10169.

Udager AM, Rolland DCM, McHugh JB, et al. High-frequency targetable EGFR mutations in sinonasal squamous cell carcinomas arising from inverted sinonasal papilloma. Cancer Res. 2015;75(13):2600–6. https://doi.org/10.1158/0008-5472.CAN-15-0340.

Udager AM, McHugh JB, Betz BL, et al. Activating *KRAS* mutations are characteristic of oncocytic sinonasal papilloma and associated sinonasal squamous cell carcinoma: *KRAS*-mutated oncocytic sinonasal tumours. J Pathol. 2016;239(4):394–8. https://doi.org/10.1002/path.4750.

Udager AM, McHugh JB, Goudsmit CM, et al. Human papillomavirus (HPV) and somatic EGFR mutations are essential, mutually exclusive oncogenic mechanisms for inverted sinonasal papillomas and associated sinonasal squamous cell carcinomas. Ann Oncol. 2018;29(2):466–71. https://doi.org/10.1093/annonc/mdx736.

van Nederveen FH, Gaal J, Favier J, et al. An immunohistochemical procedure to detect patients with paraganglioma and phaeochromocytoma with germline SDHB, SDHC, or SDHD gene mutations: a retrospective and prospective analysis. Lancet Oncol. 2009;10(8):764–71. https://doi.org/10.1016/S1470-2045(09)70164-0.

Veeramachaneni R, Walker T, Revil T, et al. Analysis of head and neck carcinoma progression reveals novel and relevant stage-specific changes associated with immortalisation and malignancy. Sci Rep. 2019;9(1):11992. https://doi.org/10.1038/s41598-019-48229-7.

Velleuer E, Dietrich R. Fanconi anemia: young patients at high risk for squamous cell carcinoma. Mol Cell Pediatr. 2014;1(1):9. https://doi.org/10.1186/s40348-014-0009-8.

Vos JL, Elbers JBW, Krijgsman O, et al. Neoadjuvant immunotherapy with nivolumab and ipilimumab induces major pathological responses in patients with head and neck squamous cell carcinoma. Nat Commun. 2021;12(1):7348. https://doi.org/10.1038/s41467-021-26472-9.

Walter V, Yin X, Wilkerson MD, et al. Molecular subtypes in head and neck cancer exhibit distinct patterns of chromosomal gain and loss of canonical cancer genes. PLoS One. 2013;8(2):e56823. https://doi.org/10.1371/journal.pone.0056823.

Weindorf SC, Brown NA, McHugh JB, Udager AM. Sinonasal papillomas and carcinomas: a contemporary update with review of an emerging molecular classification. Arch Pathol Lab Med. 2019;143(11):1304–16. https://doi.org/10.5858/arpa.2019-0372-RA.

Wenig BM. Lymphoepithelial-like carcinomas of the head and neck. Semin Diagn Pathol. 2015;32(1):74–86. https://doi.org/10.1053/j.semdp.2014.12.004.

Wiatrak BJ, Wiatrak DW, Broker TR, Lewis L. Recurrent respiratory papillomatosis: a longitudinal study comparing severity associated with human papilloma viral types 6 and 11 and other risk factors in a large pediatric population. Laryngoscope. 2004;114(11 Pt 2 Suppl 104):1–23. https://doi.org/10.1097/01.mlg.000148224.83491.0f.

Wong KCW, Hui EP, Lo KW, et al. Nasopharyngeal carcinoma: an evolving paradigm. Nat Rev Clin Oncol. 2021;18(11):679–95. https://doi.org/10.1038/s41571-021-00524-x.

Woo SB, Cashman EC, Lerman MA. Human papillomavirus-associated oral intraepithelial neoplasia. Mod Pathol. 2013;26(10):1288–97. https://doi.org/10.1038/modpathol.2013.70.

Wright JM, Vered M. Update from the 4th edition of the World Health Organization classification of head and neck tumours: odontogenic and maxillofacial bone tumors. Head Neck Pathol. 2017;11(1):68–77. https://doi.org/10.1007/s12105-017-0794-1.

Xu B, Chang K, Folpe AL, et al. Head and neck mesenchymal neoplasms with GLI1 gene alterations: a pathologic entity with distinct histologic features and potential for distant metastasis. Am J Surg Pathol. 2020;44(6):729–37. https://doi.org/10.1097/PAS.0000000000001439.

Xu B, Suurmeijer AJH, Agaram NP, Zhang L, Antonescu CR. Head and neck rhabdomyosarcoma with TFCP2 fusions and ALK overexpression: a clinicopathological and molecular analysis of 11 cases. Histopathology. 2021;79(3):347–57. https://doi.org/10.1111/his.14323.

Yang W, Lee KW, Srivastava RM, et al. Immunogenic neoantigens derived from gene fusions stimulate T cell responses. Nat Med. 2019;25(5):767–75. https://doi.org/10.1038/s41591-019-0434-2.

Zaini ZM, McParland H, Møller H, Husband K, Odell EW. Predicting malignant progression in clinically high-risk lesions by DNA ploidy analysis and dysplasia grading. Sci Rep. 2018;8(1):15874. https://doi.org/10.1038/s41598-018-34165-5.

Zamecnik M, Rychnovsky J, Syrovatka J. Sinonasal SMARCB1 (INI1) deficient carcinoma with yolk sac tumor differentiation: report of a case and comparison with INI1 expression in gonadal germ cell tumors. Int J Surg Pathol. 2018;26(3):245–9. https://doi.org/10.1177/1066896917741549.

Zhu X, Zhang F, Zhang W, He J, Zhao Y, Chen X. Prognostic role of epidermal growth factor receptor in head and neck cancer: a meta-analysis. J Surg Oncol. 2013;108(6):387–97. https://doi.org/10.1002/jso.23406.

Veronica K. Y. Cheung and Ruta Gupta

Contents

Introduction

- The functional units of the major and the minor salivary glands include multiple cellular components including acinar, ductal, and myoepithelial elements
  - All of these can give rise to benign or malignant salivary gland neoplasms, either individually or in combination with other cellular components
  - There is significant histologic and immunohistochemical overlap between benign and malignant salivary gland neoplasms raising significant diagnostic challenges
- The recent advances in the next-generation sequencing (NGS) technologies have identified recurrent genetic changes in multiple salivary gland neoplasms (Table 19.1)
- In addition to NGS, in situ hybridization (ISH) probes or immunohistochemical (IHC) stains are commercially available for most of these recurrent genetic alterations and can be utilized to increase the diagnostic accuracy in morphologically challenging cases (Table 19.1) or for targeted therapies

V. K. Y. Cheung · R. Gupta (✉)
Tissue Pathology and Diagnostic Oncology, Royal Prince Alfred Hospital, NSW Health Pathology, University of Sydney Central Clinical School, Sydney, Australia
e-mail: veronica.cheung@health.nsw.gov.au;
ruta.gupta@health.nsw.gov.au

Table 19.1 Salivary gland neoplasms with recurrent genetic alterations: characteristic histologic and immunohistochemical features, methods of detection of genetic alterations and selective differential diagnoses

Tumor	Morphology	Immunohistochemistry	Main molecular alterations with diagnostic/clinical Implications	Chromosomal rearrangement	Method of detection	Sensitivity (Sn)[a]/ specificity (Sp)[a]	Differential diagnoses
Acinic cell carcinoma (AciCC)	Serous acinar differentiation with intracytoplasmic zymogen granules	Pancytokeratin+, DOG1+, SOX10+, NR4A3+ Myoepithelial markers –	*SCPP-NR4A3* fusion	t(4;9)(q13;q31)	NR4A3 IHC. *NR4A3* FISH BA.	IHC Sp 97–100%, Sn 81–92% FISH Sp 100%, Sn 36–84%	MEC, SC
Adenoid cystic carcinoma (AdCC)	Biphasic tumor, cribriform and tubular architecture with luminal basophilic material	Pancytokeratin+, S100+, CKIT+, myoepithelial+, MYB+	*MYB-NFIB* fusion *MYBL1-NFIB* fusion	t(6;9) (q21-24;p13-23)	*MYB* FISH BA. MYB RNA ISH. MYB IHC.	FISH Sp100%, Sn 40–66% RNA ISH Sp 89%; Sn 92% IHC Sp 54%, Sn 94%	BCA, basaloid SCC, EMC, MyC, PA, PAC.
Basal cell adenoma (BCA)	Circumscribed, biphasic tumor with tubulo-trabecular, cribriform, or solid growth and reduplicated basement membrane matrix.	S100+ Epithelial cytokeratin+, CK7+ Myoepithelial+	*CTNNB1*-activating mutation *CYLD1*-activating mutation		Beta-catenin IHC. Only nuclear staining should be considered positive	Limited data due to rarity of the tumor	AdCC, MyC, PA
Carcinoma ex pleomorphic adenoma (CXPA)	Infiltrative growth of a malignant neoplasm on a background of a pleomorphic adenoma	Pancytokeratin+, myoepithelial+, PLAG1+,	*PLAG1-CTNNB1* *PLAG1-LIFR* *HMGA2* alteration	t(3;8)(p21;q12) t(5; 8) (p11;q12)	*PLAG1* FISH BA. PLAG1 IHC.	FISH Sp >90%, Sn 50–72% IHC Sp up to 60%, Sn 60–70%	Metastatic carcinoma, MyC, SDC, other salivary gland malignancies with HG transformation.
Epithelial myoepithelial carcinoma (EMC)	Biphasic pattern in tubules and cords (but various patterns seen) with basement membrane production	Pancytokeratin+, myoepithelial+	*HRAS* exon 3 hotspot mutation (Q61R, Q61K, G13R)		RAS Q61R IHC	IHC Sp 97%, Sn 70%	AdCC, HCCC, PA, myoepithelial neoplasms, basal cell adenoma/carcinoma
Hyalinising clear cell carcinoma (HCCC)	Infiltrative epithelial cells with clear to eosinophilic cytoplasm within sclerotic/hyalinized stroma PAS positive, dPAS sensitive	p40+, CK5/6+, p63+ Myoepithelial–,S100–SOX10–	*EWSR1-ATF1* fusion	t(12;22) (q13;q12)	*EWSR1* FISH BA	FISH Sp100%, Sn 82–87%	Clear cell variants of: EMC, MEC, MyC, oncocytic neoplasm, SCC
Intraductal carcinoma (IDC)	Cystic duct like structures with epithelial proliferation with surrounding myoepithelial cells (similar to breast ductal carcinoma in situ),	Intercalated: S100+, AR– Apocrine: AR+, GCDFP15+	*NCOA4-RET* fusion (intercalated duct type) *TRIM27-RET* fusion (apocrine type) Other alterations also described in both variants.		*RET* FISH BA	FISH Sn 47–60%	SC, SDC, MEC

Tumor	Histology	Markers	Genetic alteration	Translocation	Test	Sp/Sn[a]	Differential diagnosis
Mucoepidermoid carcinoma (MEC)	squamoid, intermediate, mucous cells in cystic and solid growth dPAS, mucicarmine positive	Pancytokeratin+, p63+, CK5/6+, p40+ S100–, SOX10–	*CRTC1-MAML2* fusion *CRTC3-MAML2* fusion	t(11;19)(q21;p13) t(11;15)(q21;q26)	*MAML2* FISH BA	FISH Sp 100%, Sn 82%	Mucocele, sialometaplasia, Warthin tumor, HCCC, SC, SDC, Oncocytic neoplasms
Microsecretory adenocarcinoma (MSA)	Infiltrative, microcystic, intercalated duct-like, intraluminal basophilic secretions	S100+, SOX10+, p63+ p40–,calponin–,mammaglobin–	*MEF2C-SS18* fusion	t(5;18)(q14;q11)	*SS18* FISH BA. RNA Seq fusion panels.	FISH Sn 92%	MEC, PAC, SC, SMA
Myoepithelial carcinoma (MyC)	Single population of myoepithelial cells showing variable cytology (epithelioid, plasmacytoid, spindle, and clear)	S100+, SOX10+, P63+, P40+, SMA+, other myoepithelial+	*PLAG1* rearrangement *EWSR1* rearrangement		*EWSR1* FISH BA *PLAG1* FISH BA	*EWSR1* FISH Sn 39% (clear cell)	AdCC, BCAc, EMC, PA, PAC
Pleomorphic adenoma (PA)	Biphasic epithelial component with mesenchymal component (myxoid, chondromyxoid, hyaline stroma)	Pancytokeratin+, myoepithelial+, PLAG1+	*PLAG1-CTNNB1* *PLAG1-LIFR* *HMGA2* alteration	t(3;8)(p21;q12) t(5; 8) (p11;q12)	*PLAG1* FISH BA. PLAG1 IHC.	FISH Sp 100%, Sn 50% IHC Sp 17–59%, Sn 70–80%	AdCC, BCA, CXPA, EMC, myoepithelioma
Polymorphous adenocarcinoma (PAC)/cribriform adenocarcinoma of salivary gland (CASG)	PAC: Architectural diversity but monotonous cytology CASG: Cribriform, solid or glomeruloid, nuclear monotony with clear chromatin, occasional grooves	CK7+, S100+, SOX-10+, p63+ p40–, myoepithelial–	PCA: *PRKD1 E710D* hotspot mutation CASG: *PRKD1/PRKD2/PRKD3* gene fusions with *ARID1A* or *DDX3X*		Hotspot mutation: NGS or PCR. *PRKD* FISH or RNA fusion panel.	Hotspot mutation: Sp 100%, Sn 75–90% FISH: Sn 80%(CAMSG), Sn ~10% (PAC)	AdCC, BCA/carcinoma, EMC, PA
Salivary duct carcinoma (SDC)	High-grade malignancy resembles high-grade breast carcinoma, large ducts and nests with cells arranged as Roman bridges and arches and central comedonecrosis. Apocrine cells with surface decapitation	CK7+, GATA3+, AR+, GCDFP15+ p40–,S100–,SOX-10–,CK5/6–	Androgen receptor (AR) gene splice site variation *HER2* (*ERBB2*) amplification		AR IHC. HER2 IHC, ISH	AR IHC Sp 100%, Sn 90% HER2 IHC/ISH Sn 30%	HG AdCC, HG MEC, IDC, PAC, oncocytic neoplasm, metastatic SCC
Sclerozing microcystic adenocarcinoma (SMA)	Infiltrative ducts and tubules with biphasic growth and bland cytology in sclerotic/hyalinized stroma	Epithelial: Cytokeratin+, CK7+ Myoepithelial: S100+, P63+, p40+, SMA+	*CDK11B* mutation		NGS or PCR	Limited data available	HCCC, PAC
Secretory carcinoma (SC)	Various growth pattern, cells with eosinophilic/vacuolated cytoplasm, eosinophilic bubbly secretions resemble colloid	CK7+, GATA3+, S100+, SOX-10+, mammaglobin+, MUC4+, pan-TRK+ Myoepithelial markers–	*ETV6-NTRK3* fusion	t(12;15)(p13;q25)	*ETV6* FISH BA. Pan-TRK IHC.	*ETV6* FISH Sp 100%, Sn >85% Pan-TRK IHC Sp 46–70%, Sn 85% (nuclear staining Sp 100%, Sn 69%)	AciCC, IDC, MEC, MSA, PAC

BA break apart probe, *FISH* fluorescent in situ hybridization, *HG* high grade, *IHC* immunohistochemistry, *ISH* in situ hybridization, *SCC* squamous cell carcinoma

Myoepithelial markers: P63, calponin, smooth muscle markers (smooth muscle actin, smooth muscle myosin heavy chain), GFAP

[a] Specificity and sensitivity of the test in diagnosing the particular salivary gland tumor from the differential diagnoses of relevant salivary gland neoplasms

Practical Tips for Utilizing Molecular Techniques in Salivary Gland Tumors

Fine-Needle Aspiration

- Fine needle aspiration (FNA) is often the first diagnostic procedure performed for most salivary gland neoplasms
- There can be significant morphologic overlap between common benign and malignant entities such as:
  - Pleomorphic adenoma (PA) and adenoid cystic carcinoma (AdCC)
  - Benign cystic lesions and mucoepidermoid carcinoma (MEC)
- It is advisable to collect sufficient material from the solid areas of the lesion for cell block preparation in all salivary gland lesions
- The cell blocks should ideally be prepared using the plasma thrombin clot method. Once the clot is formed, it is added to neutral buffered formalin and processed as per paraffin-embedded tissue sections
  - Reagents containing acetic acid (Zenker fixative) and picric acid (Bouin fluid) should be avoided
- Appropriately prepared cell blocks can be utilized for IHC, ISH, or fluorescent in situ hybridization (FISH) similar to formalin-fixed, paraffin-embedded histopathology sections (Table 19.1)

Tissue Preparation of Biopsies and Resection Specimens

- Tissue should ideally be fixed in neutral buffered formalin for at least 8–24 h prior to routine processing as per established protocols for histologic examination
  - Underfixation leads to suboptimal results with most molecular techniques, particularly ISH
  - Similarly, overfixation, use of nonbuffered formalin, or high temperatures leads to increased DNA fragmentation and suboptimal results with NGS
- Tissues from the oral cavity, particularly the palate and bony biopsies may require decalcification:
  - Ideally, some tumor tissue should be submitted for processing without decalcification at the time of gross examination
    These blocks should be clearly identified in the block key and utilized for molecular techniques
- If decalcification is inevitable, it should be performed with formic acid-based solution, or with a chelator such as EDTA

Tissue Selection for Testing

- Areas rich in tumor cells should be selected for molecular techniques
  - This is particularly important in tumors with large cystic areas such as MEC
  - Only the lesional cyst lining cells should be examined for ISH or IHC staining
    Lymphocytes, endothelial cells, and fibroblasts should be excluded based on the size and shape of the nuclei
- Areas with necrosis and inflammation should be avoided
- Correlation and comparison with concurrent hematoxylin and eosin stained section is critical while reporting IHC and ISH results

Malignant Neoplasms

Acinic Cell Carcinoma

- Acinic cell carcinoma (AciCC) accounts for approximately 6.5% of primary salivary gland neoplasms
- Majority arise in the parotid gland
  - Involvement of minor salivary gland tissues is rare
- Occur over a wide age range
  - 2nd most common primary salivary gland carcinoma in the pediatric population
- Generally present as slow growing, painless masses

Light Microscopy

- AciCC shows a variety of architectural patterns including solid, macro- or microcystic, follicular, or papillary
- Serous acinar differentiation with cells showing abundant intracytoplasmic basophilic zymogen granules is the hallmark of AciCC (Fig. 19.1a)
- The granules show intense staining with periodic acid Schiff (PAS) reagent and are resistant to digestion with diastase (DiPAS) (Fig. 19.1b)
- Immunohistochemically, in addition to pancytokeratins, AciCC show membranous (apical or complete membranous staining) with discovered on GIST 1 (DOG1) (Fig. 19.1c)
- In some instances, the tumor cells may show sparse or focal zymogen granules (zymogen granule poor AciCC)
  - May be predominantly composed of intercalated duct like cells with eosinophilic or amphophilic cells with centrally placed nucleus

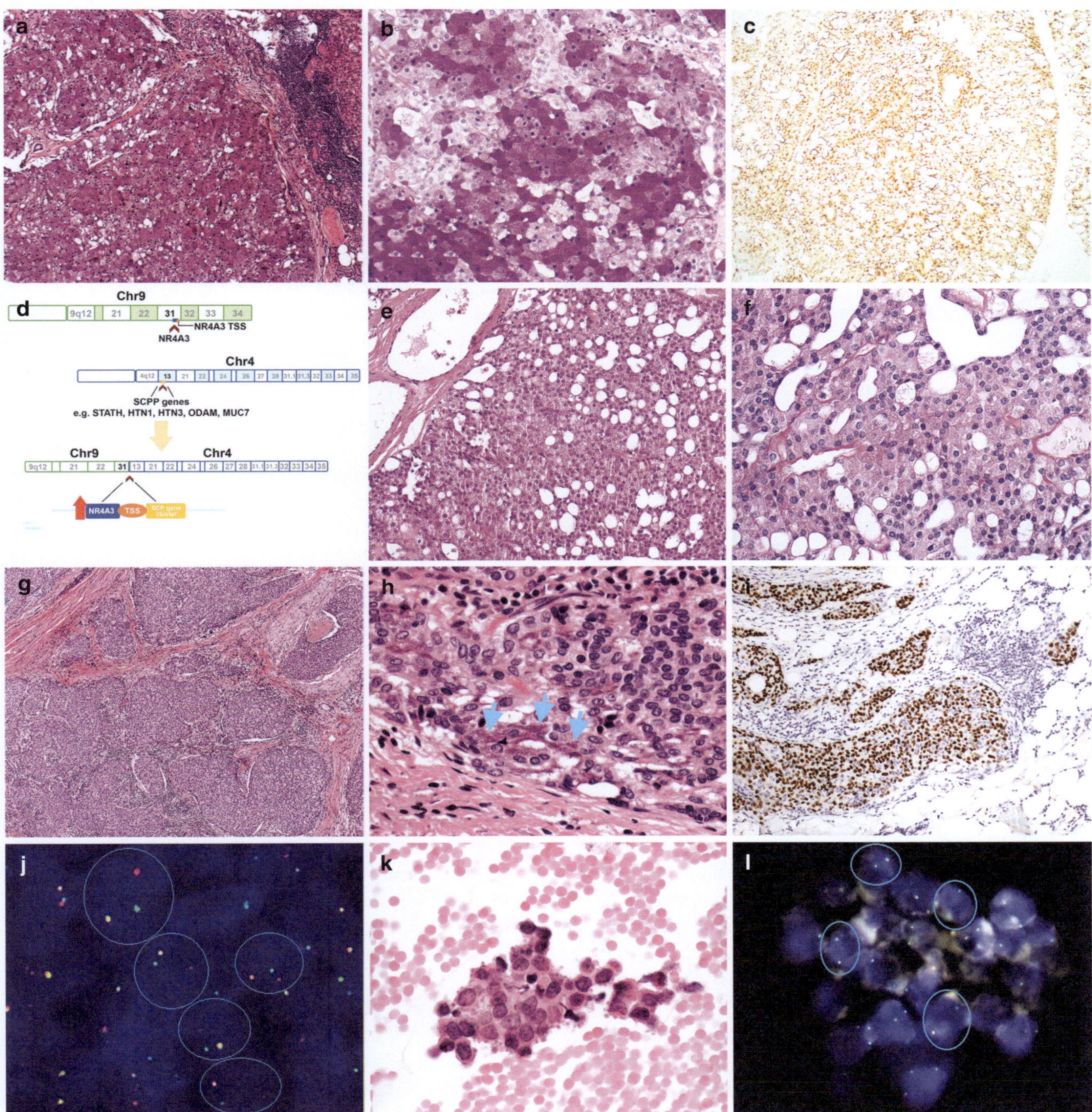

Fig. 19.1 Acinic cell carcinoma. (**a**) Acinic cell carcinoma characterized by polygonal cells showing abundant cytoplasmic basophilic zymogen granules and round nuclei. (**b**) The zymogenic granules are highlighted by PAS with diastase digestion. (**c**) Immunohistochemistry for DOG1 demonstrates with membranous and cytoplasmic staining of the tumor cells. (**d**) Enhancer hijacking in acinic cell carcinoma: t(4;9)(q13;q31) rearrangement bring enhancer regions from the structurally related genes encoding secretory Ca-binding phosphoproteins (SCPP) gene cluster in proximity to the transcription start site (TSS) of nuclear receptor subfamily 4 group A member 3 (NR4A3). The expression of SCPP gene cluster (STATH, HTN1, HTN3, ODAM, and MUC7) is limited to the salivary glands and here these are the most expressed genes. The break point of SCPP correlates with active enhancer H3K27ac with high chromatin marks. Placing the active enhancers of SCPP gene cluster in close association with the NR4A3 TSS leads to increased expression of NR4A3. (**e**) Zymogen granule poor acinic cell carcinoma. (**f**) PAS with diastase digestion shows extremely scanty granules. (**g**) Acinic cell carcinoma with high-grade transformation with areas of necrosis. (**h**) Extremely sparse zymogenic granules in acinic cell carcinoma with high-grade transformation. (**i**) IHC for NR4A3 demonstrating strong nuclear staining. (**j**) FISH for NR4A3 using break apart probe showing separation of the 3′ and 5′ signals (blue circles). (**k**) Cell block showing a high-grade epithelioid malignant neoplasm in a patient with history of AciCC. (**l**) FISH for NR4A3 using break apart probe performed on the cell block

- May show a large proportion of vacuolated cells with eccentrically displaced nucleus
- May predominantly be composed of clear cells reminiscent of clear cell carcinoma of the kidney
  - Most commonly thought to be due to a fixation artefact
- A proportion of AciCC also undergo high-grade transformation
  - AciCC with high-grade transformation show morphologic features of an undifferentiated carcinoma, and the areas resembling AciCC with zymogenic granules may be extremely sparse

NR4A3 Rearrangement

- Approximately 95% of AciCC demonstrate recurrent genetic translocation [t(4;9)(q13;q31)] leading to enhancer hijacking of *NR4A3* (Fig. 19.1d)
- This translocation places active enhancer regions of the secretory calcium-binding phosphoprotein (*SCPP*) at 4q13 in proximity of the nuclear receptor subfamily 4 group a member 3 (*NR4A3*) at 9q31
  - *SCPP* encodes phosphoproteins that are present in the saliva and are expressed at high levels in the salivary glands
  - Translocation of the active enhancer regions of *SCPP* near the regulatory regions of *NR4A3* leads to overexpression of NR4A3 through a phenomenon described as enhancer hijacking
- IHC detects the nuclear overexpression of NR4A3 in nearly 95–97% of AciCC
  - IHC for NR4A3 shows 3+ nuclear staining in 80–90% AciCC and 2+ or 1+ nuclear staining is rare (5–10%)
- FISH break apart probes for *NR4A3* rearrangements are also available but show lower sensitivity and specificity for detecting this rearrangement as compared with immunohistochemistry
  - FISH studies rely on identifying translocation through break apart of signals based on signal separation by a distance equal to at least 2–3 times the signal diameter
  - Such separation may not be seen following enhancer hijacking where the enhancer region of *SCPP* translocates to variable genetic regions in the regulators of *NR4A3*

Utility of *NR4A3* Rearrangement in the Diagnosis of AciCC

- Histologic diagnosis of AciCC with zymogen granules in a salivary gland is generally not difficult
  - Zymogen granule poor AciCC (Fig. 19.1e and f) can mimic other primary salivary gland neoplasms including MEC or secretory carcinoma (SC)

AciCC are rare in nonparotid locations and other salivary gland neoplasms such as SC or MEC should be excluded at these sites

Unlike AciCC, MEC demonstrates mucin, immunostaining with p40, CK5/6, and MAML2 rearrangement by FISH (Section 19.6)

Unlike AciCC, SC show staining with GATA3, S100, MUC4, mammaglobin, NTRK3, and lack NR4A3 expression

- AciCC with high-grade transformation can also pose a diagnostic challenge (Fig. 19.1g and h)
  - In these instances, IHC for detecting nuclear expression of NR4A3 can be a useful and cost-effective method of confirming a diagnosis of AciCC (Fig. 19.1i)
  - FISH with *NR4A3* break apart probe can also be useful (Fig. 19.1j)
- AciCC with high-grade transformation may present with metastases and can pose a diagnostic challenge on FNA
  - In these instances, information of previous AciCC is critical. A cell block (Fig. 19.1k) can be used for *NR4A3* FISH (Fig. 19.1l) or IHC
- Focal, weak nuclear staining with NR4A3 may rarely be seen in polymorphous adenocarcinoma (PAC) or PA
  - Both tumors show staining with p40 or p63, negative in AciCC

Other Genetic Changes in AciCC

- Approximately 2% of AciCC may show *NR4A2* rearrangement
- A subset of AciCC, particularly those with serous differentiation, can show *MSANTD3* aberrations including *MSANTD3* fusion with *HTN3* and *MSANTD3* copy number variations (CNV)

Adenoid Cystic Carcinoma

- Adenoid cystic carcinoma (AdCC) is one of the most common primary gland malignancies accounting for nearly 25% of primary salivary gland cancers
- AdCC predominantly involves the parotid gland followed by the minor salivary glands of the oral cavity
  - It may also arise in the sinonasal tract, lacrimal glands, ceruminal glands, tracheobronchial tree, and the breasts
- Oncologically adequate resection of AdCC can be difficult and the tumors tend to show a persistent and prolonged clinical course with local recurrences and distant metastases to the bones, lungs, or liver at 5–10 years
- A proportion of AdCC shows high-grade transformation and a rapidly progressive clinical course with regional lymph node and distant metastases

Light Microscopy

- AdCC typically show cribriform architecture with rounded nests showing punched out spaces containing basophilic material (Fig. 19.2a)
- Reduplicated basement membrane may also be present

- Tubular and solid patterns are also seen
- Biphasic carcinomas showing:
  - Dominant modified myoepithelial cells with scanty pale cytoplasm and enlarged angulated hyperchromatic nuclei

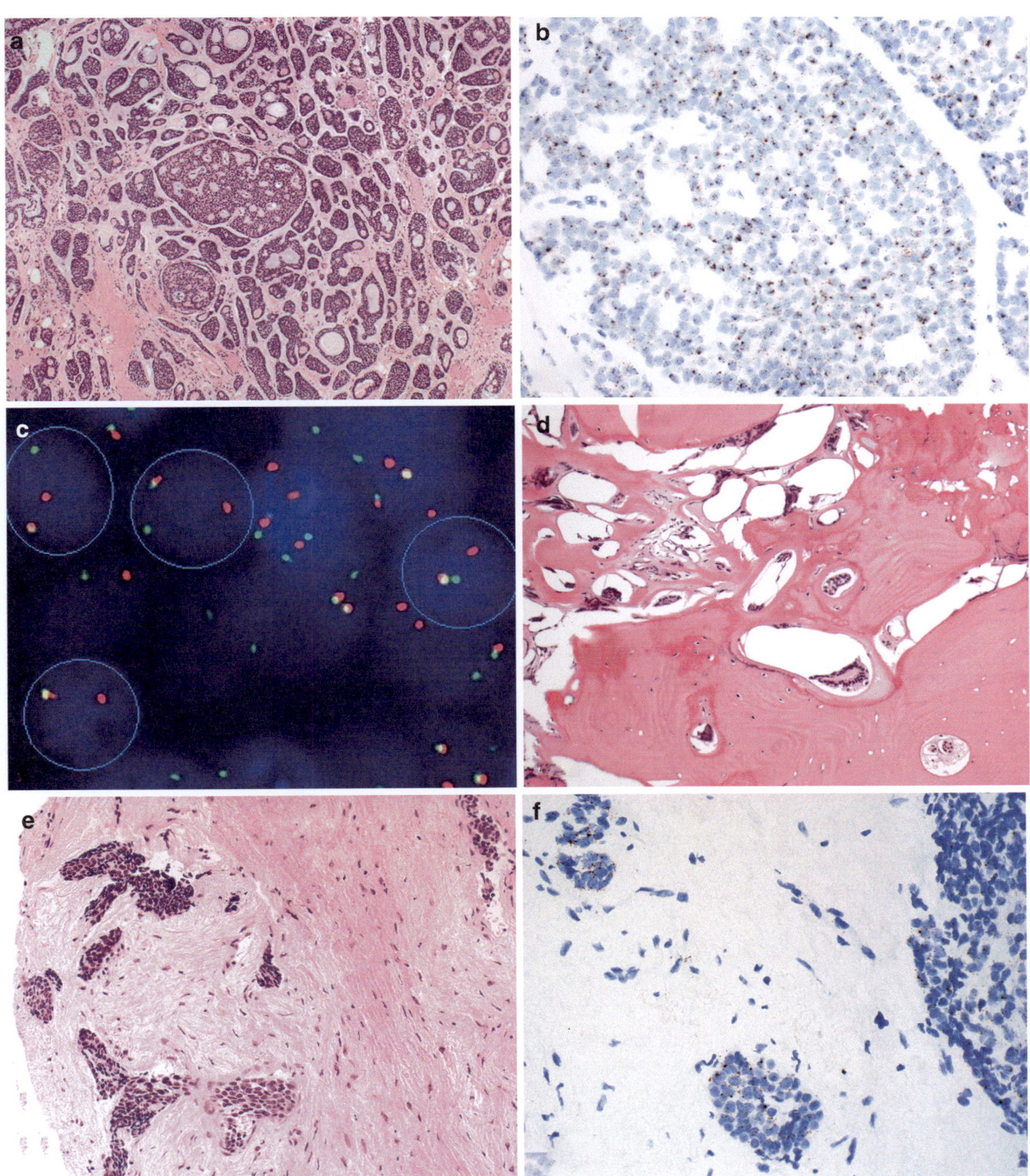

Fig. 19.2 Adenoid cystic carcinoma. (**a**) Characteristic cribriform architecture of adenoid cystic carcinoma. (**b**) MYB mRNA expression in AdCC. In situ hybridization demonstrates punctate nuclear and cytoplasmic signals, including clumped signals in AdCC. (**c**) FISH with MYB break apart demonstrating separation of the 3′ and 5′ red and green signals. (**d**) Scanty nests of metastatic AdCC in bone. (**e**) Metastatic AdCC lacking typical cribriform architecture. (**f**) MYB mRNA expression demonstrating punctate nuclear and cytoplasmic signals in the malignant cells. (**g**) AdCC with high-grade transformation with loss of typical cribriform architecture and areas of necrosis. (**h**) MYB mRNA expression demonstrating punctate nuclear and cytoplasmic signals in the malignant cells

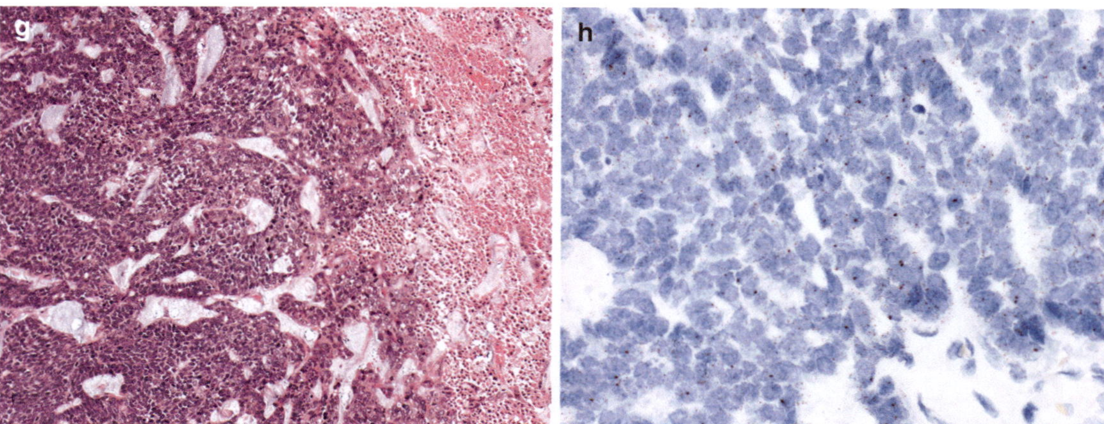

Fig. 19.2 (continued)

- ○ Demonstrate immunostaining with pancytokeratin, S100, p40, and myoepithelial markers such as p63, calponin, and smooth muscle actin
  - Scanty epithelial cells with eosinophilic cytoplasm and round nuclei
    - ○ Demonstrate immunostaining with pancytokeratin, S100, and CKIT 117
- Perineural and/or intraneural invasion is extremely common and may be seen beyond the main tumor mass

MYB Alterations

- AdCC are characterized by *MYB* (z-myb avian myeloblastosis viral oncogene homolog) or *MYBL1* activation due to:
  - Genetic translocation of *MYB* resulting in *MYB-NFIB* (nuclear factor I/B) fusion [(t (6;9) (q22–23; p23–24)] or
  - Enhancer hijacking of *MYB* or
  - Copy number variations of *MYB*
- *MYB* activation and overexpression can be identified using ISH techniques for MYB RNA in formalin fixed paraffin embedded sections in nearly 92–97% of AdCC (Fig. 19.2b)
  - Increased MYB RNA is seen irrespective of the underlying genetic mechanism of translocation, enhancer hijacking or copy number variation, thus imparting high sensitivity
  - The RNA expression is seen regardless of histologic pattern and is also seen in cases with high-grade transformation
- Rooper et al. suggest that presence of 3 or more punctate nuclear or cytoplasmic signals or at least a single clumped aggregate of signals in more than 30% of tumor cells should be used as a threshold for considering MYB RNA ISH test as positive

- FISH break apart probe for *MYB* rearrangement (Fig. 19.2c) is available but has lower sensitivity
  - FISH can only detect approximately 40–66% of cases in which the underlying mechanism is genetic translocation of *MYB*
- IHC for MYB is also available but can be difficult to optimize, demonstrates nonspecific staining, and is positive in a large variety of both salivary and nonsalivary gland tumors

Utility of *MYB* Alteration in the Diagnosis of AdCC

- Morphologically, typical AdCC rarely pose diagnostic challenges. However, heterogenous architectural patterns are common
  - AdCC can often mimic benign entities such as PA
  - AdCC can mimic malignant tumors including PAC, epithelial myoepithelial carcinoma (EMC), and basal cell adenocarcinoma
- ISH testing to detect MYB RNA can be performed on cell blocks in difficult cytology cases to distinguish AdCC from its mimics, particularly PA
- ISH testing to detect MYB RNA can be utilized in small biopsies from metastatic sites (Fig. 19.2d–f)
- AdCC with high-grade transformation demonstrates MYB RNA expression (Fig. 19.2g and h)
- MYB RNA expression has been reported to have a specificity of 89% for AdCC
  - MYB RNA expression has been described in a small number of other salivary gland tumors reported as basal cell adenocarcinoma and PAC
  - MYB RNA expression has been described in other head and neck malignancies that may show a cribriform pattern such as sinonasal undifferentiated carcinoma, basaloid squamous cell carcinoma, and HPV-related multiphenotypic sinonasal carcinomas

- MYB RNA ISH has been described as negative in other morphologically similar salivary gland tumors including PA, EMC, basal cell tumors, and myoepithelial carcinomas

Other Genetic Alterations Described in AdCC

- Many genetic alterations have been described in AdCC
  - 14q deletions in low-grade tumors and 1p, 6q, and 15q deletions in high-grade tumors
  - Single nucleotide variations (SNV) in *TP53, NOTCH, FGFR, PIK3CA, CDKN2A, ATM, TSC1, SMARCA2, CREBP*
  - Recurrent alterations in FGF/IGF/PI3K pathways
- Currently, detection of these alterations does not assist with diagnosis or prognosis
  - May carry therapeutic implications in the future

Epithelial Myoepithelial Carcinoma

- EMC is a rare, low-grade primary salivary gland neoplasm accounting for approximately 1% of all salivary gland tumors
- It predominantly occurs in the parotid gland followed by the other major and minor salivary glands

Light Microscopy

- EMC typically shows a lobulated or multinodular growth pattern (Fig. 19.3a)
- The typical examples of EMC are composed of tubules and cords
  - Cribriform, solid, fascicular, cystic, and papillary architecture can also occur
- Characteristic biphasic pattern with a variable proportion of epithelial cells surrounded by myoepithelial cells (Fig. 19.3b–d)
  - Luminal/central epithelial cells showing moderate amounts of eosinophilic cytoplasm
    - The luminal epithelial cells show staining with pancytokeratin and S100
  - The abluminal/peripheral myoepithelial cells showing clear cytoplasm
    - The abluminal myoepithelial cells show variable and weak staining with pancytokeratin and strong immunostaining with p63, p40, calponin, smooth muscle actin, and smooth muscle myosin heavy chain
- Basement membrane production is seen
- A variety of morphologic appearances are seen with both epithelial and myoepithelial cells showing clear cyto-plasm, oncocytic change, apocrine, or spindle cell appearance
- EMC can undergo high-grade transformation

HRAS Mutations

- *HRAS* exon 3 hot spot mutations have been identified in up to 85% of de novo EMC
- These predominantly include *HRAS Q61R* followed by
  - HRAS Q61K
  - HRAS G13R

Utility of Detecting *HRAS* Mutations in the Diagnosis of EMC

- NGS is the most reliable method for detecting these hotspot mutations
- *HRAS* mutations are seen in EMC that arise de novo
  - May not be present in EMC arising on a background of PA
- *HRAS* mutations are highly specific for EMC and are not seen in the various histologic mimics of EMC including PA, AdCC, and basal cell adenoma (BCA)/adenocarcinoma
  - A proportion of salivary duct carcinoma (SDC) can show *HRAS* mutations but show a different morphologic and immunohistochemical profile from EMC as SDC lacks a dual cell population
- A recent paper described immunostaining for RAS Q61R (SP174) in approximately 70% of EMC with staining restricted to the cytoplasm of myoepithelial cells (Fig. 19.3e)

Other Genetic Changes Observed in EMC

- EMC arising on a background of pleomorphic adenoma can show *PLAG1* or *HMGA2* alterations
- *PIK3CA, AKT1,* and *ARID1A* SNV have been described in EMC
- EMC with high-grade transformation can show *TP53* or *FBX7* SNV or *SMARCB1* deletion

Hyalinizing Clear Cell Carcinoma

- Hyalinizing clear cell carcinoma (HCCC) is a rare salivary gland tumor
- Arises from the minor salivary glands
  - Most commonly in the palate, buccal mucosa, tongue, and oropharynx
  - Extremely rare in the major salivary glands
- Most HCCC are associated with good outcomes following complete excision; however, perineural invasion, local

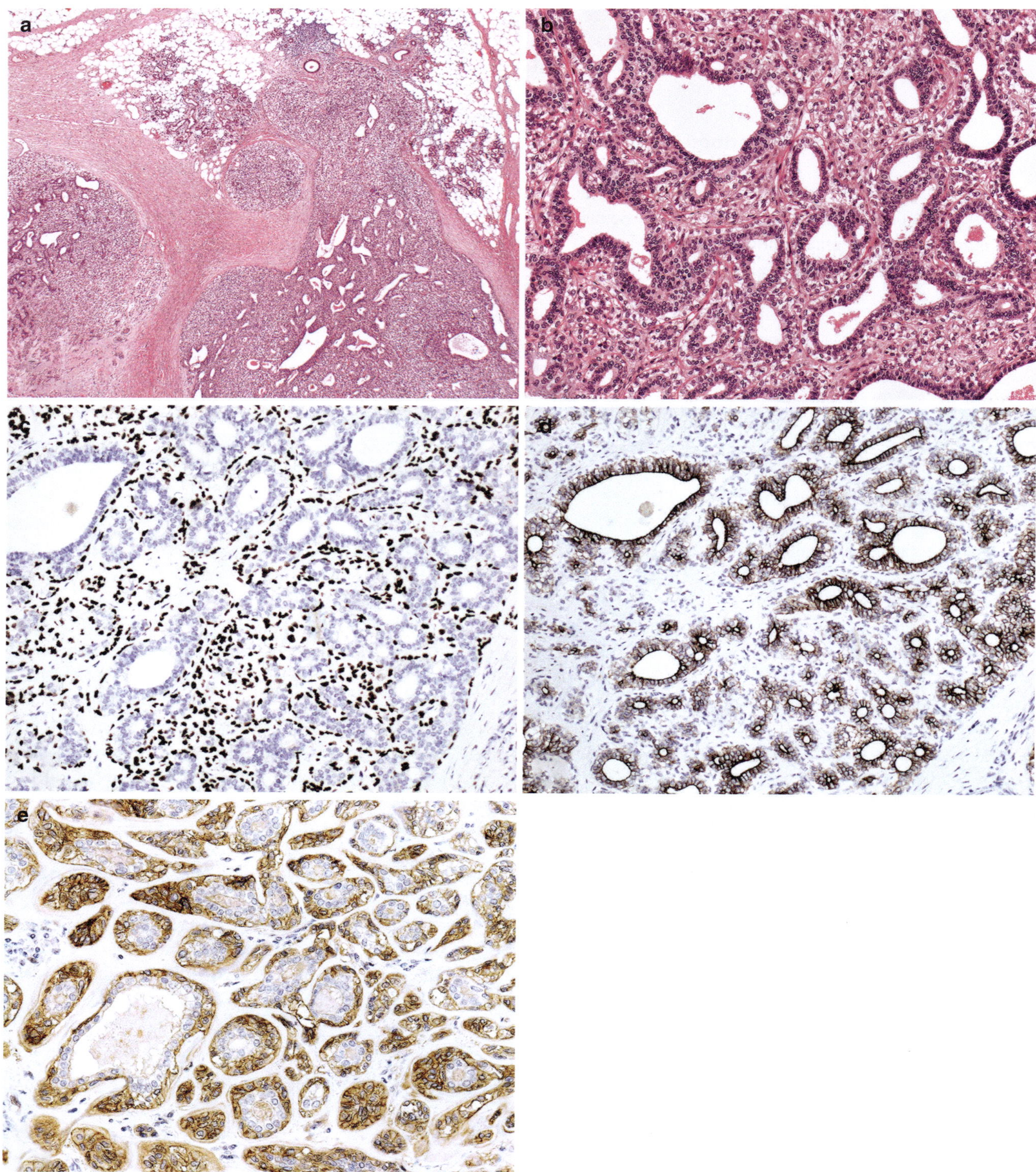

Fig. 19.3 Epithelial myoepithelial carcinoma. (**a**) Epithelial myoepithelial carcinoma demonstrating an expansile nodular growth pattern. (**b**) Epithelial myoepithelial carcinoma showing a biphasic pattern with abluminal myoepithelial cells with clear cytoplasm and ductal structures lined by epithelial cells with scanty eosinophilic cytoplasm and hyperchromatic nuclei. (**c**) IHC with p63 highlighting the myoepithelial cells. (**d**) IHC with CK7 highlighting the epithelial cells. (**e**) IHC with RAS Q61R antibody (SP174) shows cytoplasmic staining of myoepithelial cells (Image credit: Toshitaka Nagao, MD, PhD Professor and Chairman. Department of Anatomic Pathology, Tokyo Medical University. 6–7-1 Nishishinjuku, Shinjuku-ku, Tokyo, 160–0023, Japan. Telephone: +81-3-3342-6111, Fax: +81-3-3342-2062, E-mail: nagao-t@tokyo-med.ac.jp)

recurrence, and lymph node metastases have been described

- A histologically, immunophenotypically, and genetically similar entity, clear cell odontogenic carcinoma, also occurs in the gnathic bones

Light Microscopy

- HCCC (Figs 19.4a–d) characteristically shows:
  - Infiltrative nests, trabeculae, or cords of epithelial cells in a sclerotic or hyalinized stroma
  - The cells show clear to eosinophilic cytoplasm
  - The cells show strong intracytoplasmic PAS that is lost following diastase digestion
- There is strong and uniform immunostaining with squamous markers such as p40 (Fig. 19.4e), CK5/6, and p63 and lack of staining with myoepithelial markers

EWSR1 Rearrangement

- Nearly 87% of HCCC show recurrent chromosomal translocation, t(12;22)(q13;q12), resulting in fusion of the Ewing sarcoma RNA-binding protein 1 (*EWSR1*) gene on 22q12 with the activating transcription factor 1 (*ATF1*) gene on 12q13
  - *EWSR1* is a member of the TET family of transcription factors that encodes proteins involved in several cellular processes
  - *ATF1* is a part of the cAMP response element-binding protein (CREB)
    - ○ ATF family of transcription factors enhance cell growth and survival when phosphorylated by kinases including serine/threonine and cAMP-dependent protein kinase A
  - *EWSR1-ATF* gene fusion product acts as a constitutive transcriptional activator
- *EWSR1-CREM* fusion has also been described in a few HCCC

Utility of Detecting *EWSR1* Gene Rearrangement

- *EWSR1* gene rearrangement can be reliably detected using FISH (Fig. 19.4f)
- Histologic diagnosis of HCCC can be challenging due to its rarity and morphologic variations
  - HCCC can lack clear cells or hyalinized stroma
  - HCCC can also show squamous differentiation, or intracytoplasmic mucin
- HCCC can frequently mimic other neoplasms with prominent clear cell component like clear cell oncocytoma, squamous cell carcinoma with clear cell change, clear cell

variant of MEC, EMC, or myoepithelial carcinoma and rarely metastatic clear cell renal cell carcinoma

- In this morphologic context, demonstration of *EWSR1* rearrangement in addition to strong staining with CK5/6, p40 and lack of staining with myoepithelial markers can be useful in confirming a diagnosis of HCCC
  - None of the mimics of HCCC demonstrate *EWSR1* rearrangement
- Of note: *EWSR1* rearrangement is seen in myoepithelial carcinoma as well as in a variety of soft tissue tumors such as angiomatoid fibrous histiocytoma and clear cell sarcoma of soft parts
  - Thus, detection of *EWSR1* rearrangement in the absence of appropriate morphologic and immunophenotypic correlation is of limited value

Other Genetic Changes in HCCC

- Sequencing of HCCC has demonstrated *ATF1*-enriched gene signature indicating the *EWSR1-ATF1* fusion interferes with ATF1-regulated genomic responses
- Abnormalities in the insulin-like growth factor IGF/IGF1R pathway and PI3K/mTOR/FOXO pathway were also identified and could be potential targets for therapy

Intraductal Carcinoma

- Intraductal carcinoma (IDC) is a rare salivary gland tumor characterized by intraductal growth of tumor cells that resemble ductal carcinoma in situ (DCIS) of the breast
- They occur mostly in the parotid and affect mostly older adults

Light Microscopy

- The tumor is circumscribed but unencapsulated (Fig. 19.5a)
- The tumor shows cystic duct-like structures with epithelial proliferation in the form of solid nests, papillae, micropapillae, and cribriform structures similar to DCIS
- Diagnosis of IDC requires the presence of myoepithelial cells surrounding these tumor nests
  - Similar to the breast DCIS, the entire lesion may need to be examined and myoepithelial immunostains may be required to exclude an invasive component
- Four types of IDC have been described:
  - An intercalated ductal cell type with smaller cuboidal cells
    - ○ Positive for S100 and negative for AR and GCDFP-15

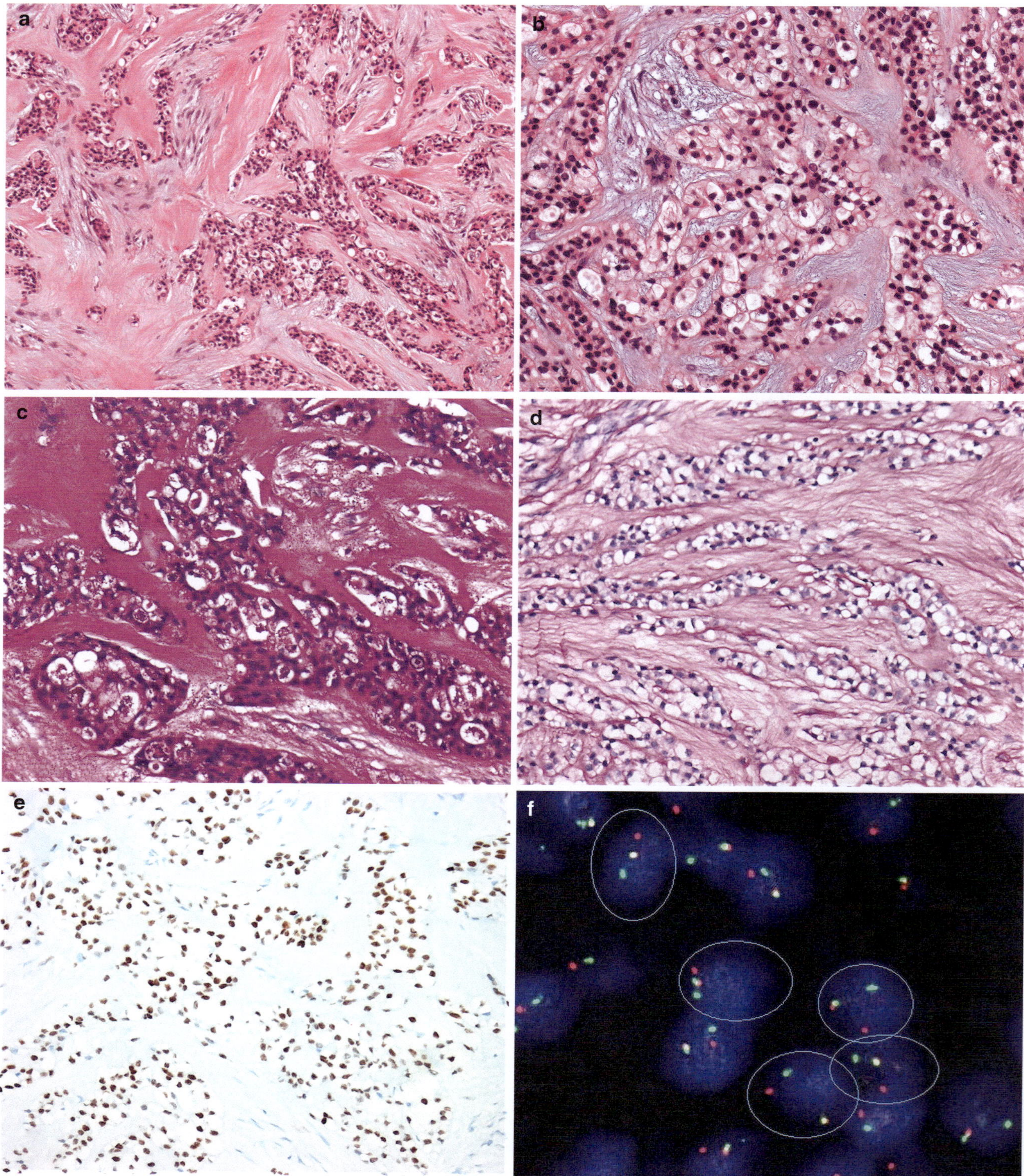

Fig. 19.4 Hyalinizing clear cell carcinoma. (**a**) HCCC involving the hard palate. (**b**) Nests and cords of polygonal cells with abundant clear cytoplasm. The cells are embedded in hyalinized stroma. (**c**, **d**) PAS staining (**c**) demonstrates strong, flocculent intracytoplasmic granular staining, that is lost following diastase digestion (**d**). (**e**) HCCC showing strong nuclear staining with p40. (**f**) FISH with EWSR1 break apart probe demonstrating clear separation of the red and green probes flanking the 3′ and 5′ ends of EWSR1

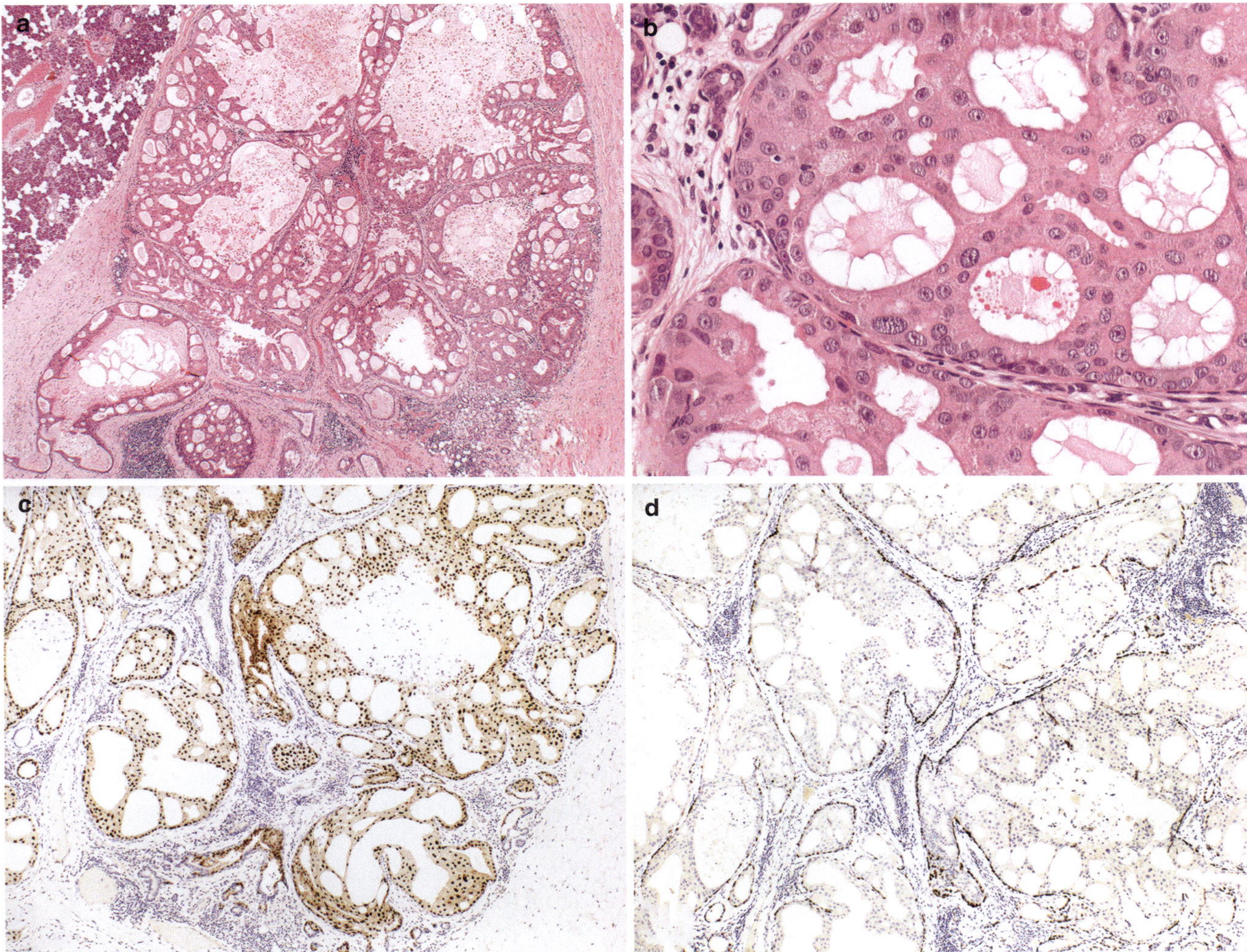

Fig. 19.5 Intraductal carcinoma. (**a**) A well demarcated but unencapsulated intraduct carcinoma. Cystically dilated ducts with cribriform structures reminiscent of Roman bridges and arches and papillae. (**b**) The cells demonstrate apocrine morphology with abundant granular cytoplasm, round nuclei with distinctive nucleoli. Flattened, spindle-shaped myoepithelial cells are present at the periphery of the nests. (**c**) Immunohistochemistry for AR showing strong nuclear staining. (**d**) Immunohistochemistry with p63 highlights the peripheral myoepithelial layer

- An apocrine variant with cells showing abundant eosinophilic cytoplasm, decapitation secretions, and round nuclei with distinctive nucleoli (Fig. 19.5b)
  - Positive for AR (Fig. 19.5c) and GCDFP-15 and negative for S100
- An oncocytic type composed of cuboidal cells with granular eosinophilic cytoplasm and round nucleus
  - Positive for S100 and SOX10 and negative for AR
- A mixed type with a combination of the above morphologies
- The peripheral myoepithelial cells show immunostaining with p63 (Fig. 19.5d), p40, CK5/6, calponin, and other myoepithelial stains in both types
- The cytologic atypia graded into three tiers: low, intermediate, and high grade

RET Rearrangement

- A variety of genetic changes have been described in these rare neoplasms,
  - The most common involving *RET* (rearranged during transfection) proto-oncogene
  - *RET* gene encodes a transmembrane receptor and is a member of tyrosine kinase family
  - *RET* is involved in activating signaling pathways that play a role in cell differentiation, growth, and survival
- The most common fusion partner in the intercalated duct type of IDC is *NCOA4*
  - Other fusion partners include *KIAA217, STRN-ALK, TUT1-ETV5* in intercalated duct type of IDC

- Apocrine type of IDC has been described to show *TRIM27-RET* fusion
  - Apocrine type of IDC may also show mutations in *TP53, PIK3CA,* and *HRAS*
- The oncocytic type IDC has been described to show *TRIM33-RET* fusion
  - *BRAF* (V600E) mutations have also been described

Utility of *RET* Rearrangement in the Diagnosis of Intraductal Carcinoma

- FISH for RET rearrangements is now widely available considering the therapeutic implications for RET-targeted therapies
- Targeted therapies are unlikely to be required for pure IDC considering their indolent behavior; however, detection of *RET* rearrangement can assist in diagnosis
  - Of note, *NCOA4-RET* fusion represents an intrachromosomal rearrangement with inversion.
    The separation between the two signals of a break apart probe may not be sufficient (2 signal diameter) for accurate detection of these rearrangements yielding false-negative results
- A significant number of cases may not show genetic fusions

Mucoepidermoid carcinoma

- MEC is the most common primary salivary gland malignancy
- MEC predominantly occurs in the parotid gland, followed by the other major and minor salivary glands
- MEC can rarely arise in the gnathic bones (central MEC) or in the tracheobronchial tree
- MEC occurs over a wide age range

- MEC is the most common salivary gland malignancy in children

Light Microscopy

- MEC is composed of variable amounts of squamoid, intermediate, mucous cells in a cystic, and solid growth pattern (Fig. 19.6a)
  - The mucous cells show mucicarmine (Fig. 19.6b) and diPAS positivity
  - All cell types show strong diffuse immunostaining with pancytokeratins, p63 (Fig. 19.6c), CK5/6 (Fig. 19.6d), and p40
- Immunostaining with membrane-bound mucins including MUC1, MUC4 is also seen
- Variant morphologies including Warthin-like MEC, oncocytic, and clear cell MEC can occur

MAML2 Rearrangement

- Nearly (90%) of MEC show *MAML2-CRTC1* fusion [t(11;19)(q21:p13)] or rarely *MAML2-CRTC3* fusion [t(11;15)(q21:q26)]
- *MAML2-CRTC1/3* fusions result in the N-terminal Notch binding domain of the coactivator *MAML2* (Mastermind-like 2) being replaced by the cyclic AMP-responsive element-binding protein (CREB) binding domain of *CRTC1* (CREB-regulated transcription coactivator)
- The resulting fusion oncogene has been linked to deregulation and activation of downstream signals in the cAMP/ CREB pathways
- FISH break apart probes targeting *MAML2* rearrangement are commercially available and can be used to detect *MAML2* translocation
- NGS techniques using fusion panels and PCR can also be used to detect *MAML2* rearrangement

Fig. 19.6 Mucoepidermoid carcinoma. (**a**) Cystic lesion with cysts lined by an admixture of mucous cells, squamoid, and intermediate cells. (**b**) Mucicarmine stain highlighting the intracytoplasmic mucin in the mucous cells. (**c**) Strong nuclear staining with p63 in all the malignant cells in MEC. (**d**) Strong membranous staining with CK5/6 in all the malignant cells in MEC. (**e, f**) Warthin-like MEC with cystic structures separated by a prominent lymphoid stroma. The higher magnification (F) shows nuclear enlargement, hyperchromasia of the cyst lining cells. Intracytoplasmic mucin is also seen. (**g**) FISH using MAML2 break apart probe demonstrates clear separation of the green and the red signals flanking the 3′ and 5′ end of MAML2. Evaluation of FISH in cystic lesion requires greater correlation with corresponding hematoxylin and eosin stained section to ensure that only the cyst lining lesional cells are evaluated and the lymphocytes and stromal cells are excluded. (**h, i**) MEC composed of cells with abundant granular cytoplasm and round nuclei with distinctive nucleoli. Confirmatory FISH for MAML2 demonstrating rearrangement. (**j, k**) MEC composed of cells with abundant clear cytoplasm. Intracytoplasmic mucin is rare. Confirmatory FISH for MAML2 demonstrating rearrangement. (**l, m**) Central MEC with a flattened cyst wall lined by squamoid cells with rare mucous cells. Confirmatory FISH for MAML2 demonstrating rearrangement

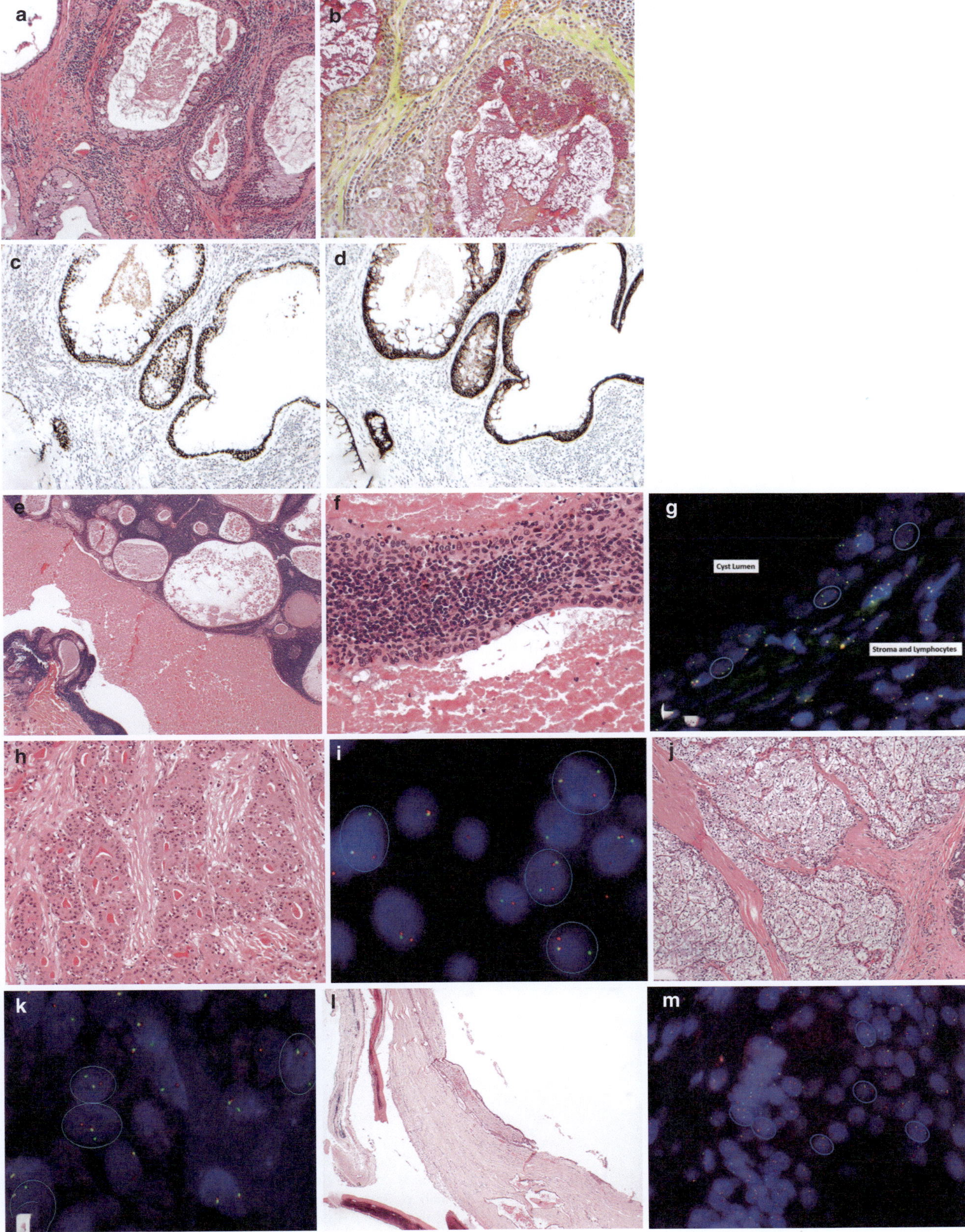
Cyst Lumen
Stroma and Lymphocytes

Utility of *MAML2* Rearrangement in Diagnosis of MEC

- The cytologic diagnosis of MEC can be extremely challenging, particularly in large cystic lesions
  - In these instances, targeting the cyst walls for the lining epithelium under radiologic guidance is useful
  - A cell block preparation can be used for evaluation of mucicarmine, dPAS, immunostains for p40, CK5/6, as well as FISH for *MAML2* rearrangement
- MECs can demonstrate a broad morphologic spectrum and mimic a range of benign and malignant entities
  - MECs with large cysts surrounded by areas of fibrosis can mimic obstructed ducts with metaplastic change of the lining epithelium, mucoceles, or necrotizing sialometaplasia
  - MECs with oncocytic change and prominent lymphoid stroma can mimic Warthin tumors (Fig. 19.6e–g) with squamous or goblet cell metaplasia following fine-needle aspiration and inflammation
  - MECs can show prominent oncocytic (Fig. 19.6h and i) or clear cell morphologies (Fig. 19.6j and k)
  - Cystic central MECs (Fig. 19.6l and m) can be difficult to distinguish from glandular odontogenic cysts or dentigerous cysts on small curetting that only partially sample the lesions
  - A range of salivary gland malignancies including HCCC, SC, mucinous variant of SDC, and mucinous adenocarcinoma can also mimic MEC
- FISH for *MAML2* rearrangement has high sensitivity and specificity for MEC
  - Care should be taken to evaluate the break apart probe signals only in the nuclei of the cyst lining lesional cells (Fig. 19.6g)

- The signals can be evaluated in the squamoid, intermediate, or mucinous cells
- Mimics of MEC have consistently demonstrated lack of *MAML2* rearrangement by FISH
- *MAML2* rearrangement is also seen in tracheobronchial MECs

Microsecretory Adenocarcinoma

- Microsecretory adenocarcinoma (MSA) is a recently described low-grade salivary gland neoplasm
- Currently, it has predominantly been described in minor salivary gland tissues with an occasional case in the parotid gland

Light Microscopy

- MSA demonstrates an infiltrative microcystic growth pattern (Fig. 19.7a)
  - Trabecular, cribriform and cord like architecture may also be present
- The cysts (Fig. 19.7b) are lined by cuboidal or flattened epithelial cells with scanty eosinophilic or clear cytoplasm in keeping with intercalated duct-like phenotype
- The cells show uniform round hyperchromatic nuclei
- The microcysts show intraluminal bubbly, basophilic secretions
- Mitoses are rare and necrosis is absent
- The tumor cells show strong diffuse immunostaining with S100, SOX10, and p63 and lack immunostaining with p40, mammaglobin and calponin

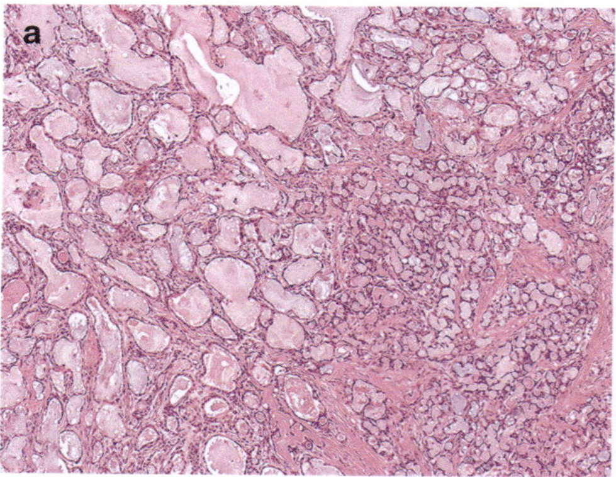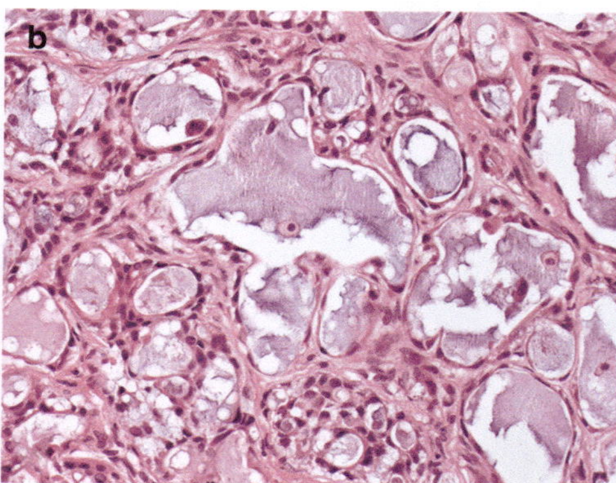

Fig. 19.7 Microsecretory adenocarcinoma. (**a**, **b**) MSA shows a microcystic and cribriform growth pattern with prominent intraluminal bubbly basophilic secretions. The cysts and glands are lined by cuboidal to flattened epithelial cells with scant cytoplasm. (Slide credit: Justin A. Bishop, MD, University of Texas Southwestern Medical Center, 6201 Harry Hines Blvd., Dallas, TX 75390-9073)

SS18 Rearrangement

- MSA has been recognized based on the detection of recurrent *MEF2C-SS18* [t(5;18)(q14;q11)] fusion in tumors with similar demographic, morphologic, and immunophenotypic characteristics
  - *SS18* is a member of SWI/SNF chromatin remodeling complex
    - The role of *SS18-SSX* fusion in synovial sarcoma is well established
  - *MEF2C* belongs to the MADS box transcription enhancer factor 2 family
- FISH using SS18 break apart probe, used in the diagnosis of synovial sarcoma, has a high sensitivity for detecting *SS18* rearrangement in MSA
- RNA Seq fusion panels can be used to detect *MEF2C-SS18* fusion

Utility of *SS18* in the Diagnosis of MSA

- MSA is a recently described entity and several differential diagnoses need to be considered as our understanding of the morphologic spectrum of this novel salivary gland neoplasm evolves:
  - SC (S100, SOX10, NTRK, mammaglobin +; p63 −; *ETV6* rearrangement) *(Section 19.12)*
  - PAC (S100 +, p63 and p40 demonstrate mutually exclusive staining pattern) *(Section 19.9)*
  - MEC (p63, p40, CK5/6 +; S100, SOX10 −; *MAML2* rearrangement) *(section 19.6)*
  - Sclerosing microcystic carcinoma (S100, p63 +) (Section 19.11) and
  - Rarely AdCC (MYB RNA detection) (Section "Practical Tips for Utilizing Molecular Techniques in Salivary Gland Tumors")
- *SS18* rearrangement is not seen in any other primary salivary gland neoplasm and has high specificity for MSA in the appropriate morphologic context

Other Genetic Changes in MSA
- *SS18* may show other fusion partners such as *ZBTB7A*

Myoepithelial Carcinoma

- Myoepithelial carcinoma (MyC) is a rare primary salivary gland malignancy composed almost entirely of myoepithelial cells
  - May arise de novo (50%)
  - May arise on a background of PA (2nd most common type of carcinoma ex-pleomorphic adenoma [CXPA] after SDC)
- Predominantly involves the parotid gland

Light Microscopy

- Multinodular, lobulated, expansile invasive front with solid, nested, fascicular architecture
- Composed entirely of myoepithelial cells
  - May have epithelioid, plasmacytoid, spindle shaped cells, clear or vacuolated cells
- Zonation with hypocellular, hyalinized, sclerotic or necrotic center, and hypercellular periphery
- The cells demonstrate immunostaining with S100, SOX-10, p63, p40, SMA, and other myoepithelial markers

PLAG1 and *HMGA2* Genetic Changes

- Most common genetic change identified in both de novo MyC as well as those arising as CXPA
  - *PLAG1* can show a variety of fusion partners including *TGFBR3* and *FGFR1*

EWSR1 Rearrangement

- *EWSR1* rearrangement detectable by FISH is seen in de novo MyC, particularly those with clear cells

Utility of Detecting *PLAG1* or *EWSR1* Rearrangement

- *PLAG1* or *EWSR1* rearrangement can be detected using FISH break apart probes
- Useful in distinguishing MyC from PAC, basal cell adenocarcinoma, or AdCC
  - *PLAG1* rearrangement is not useful in distinguishing myoepithelial carcinoma from PA or CXPA
  - *EWSR1* rearrangement is also seen in HCCC and a range of soft tissue neoplasms

Polymorphous Adenocarcinoma

- PAC, previously known as polymorphous low-grade adenocarcinoma, predominantly involves the minor salivary glands
  - It is the second most common primary salivary gland carcinoma involving the oral cavity
- Currently, some authorities recognize two subtypes of PAC
  - PAC, conventional type, and PAC, cribriform type
  - Both subtypes show recurrent genetic alterations in *PRKD* gene

Light Microscopy

- PAC histologically demonstrates architectural diversity (Fig. 19.8a and b) and cytologic monotony (Fig. 19.8c)
- The tumor has a subtly infiltrative growth pattern best appreciated on low magnifications
- The tumor shows a combination of solid, trabecular, cribriform, papillary, and single file architectural patterns
- The cells uniformly show scanty cytoplasm and round to ovoid nuclei with open chromatin (Fig. 19.8c)
  - Nuclear clearing may be seen
- Mitoses are rare
- Occasionally, cells with intracytoplasmic mucin, clear cell change, or oncocytic appearance (Fig. 19.8b) may be seen
- The stroma has a fibromyxoid quality
- Targetoid perineural invasion is extremely common toward the periphery of the tumor
- PAC, cribriform type (Fig. 19.9d) generally occurs in the base of the tongue
  - Shows a lobulated growth pattern
  - Predominance of cribriform, solid and glomeruloid architecture (Fig. 19.8d)
- PAC shows strong diffuse immunostaining with S100 (Fig. 19.9a), SOX10, CK7
  - Immunostaining for p63 and p40 are reported to be mutually exclusive with p63 + (Fig. 19.9b) and p40 − p63 immunostaining can be variable and patchy

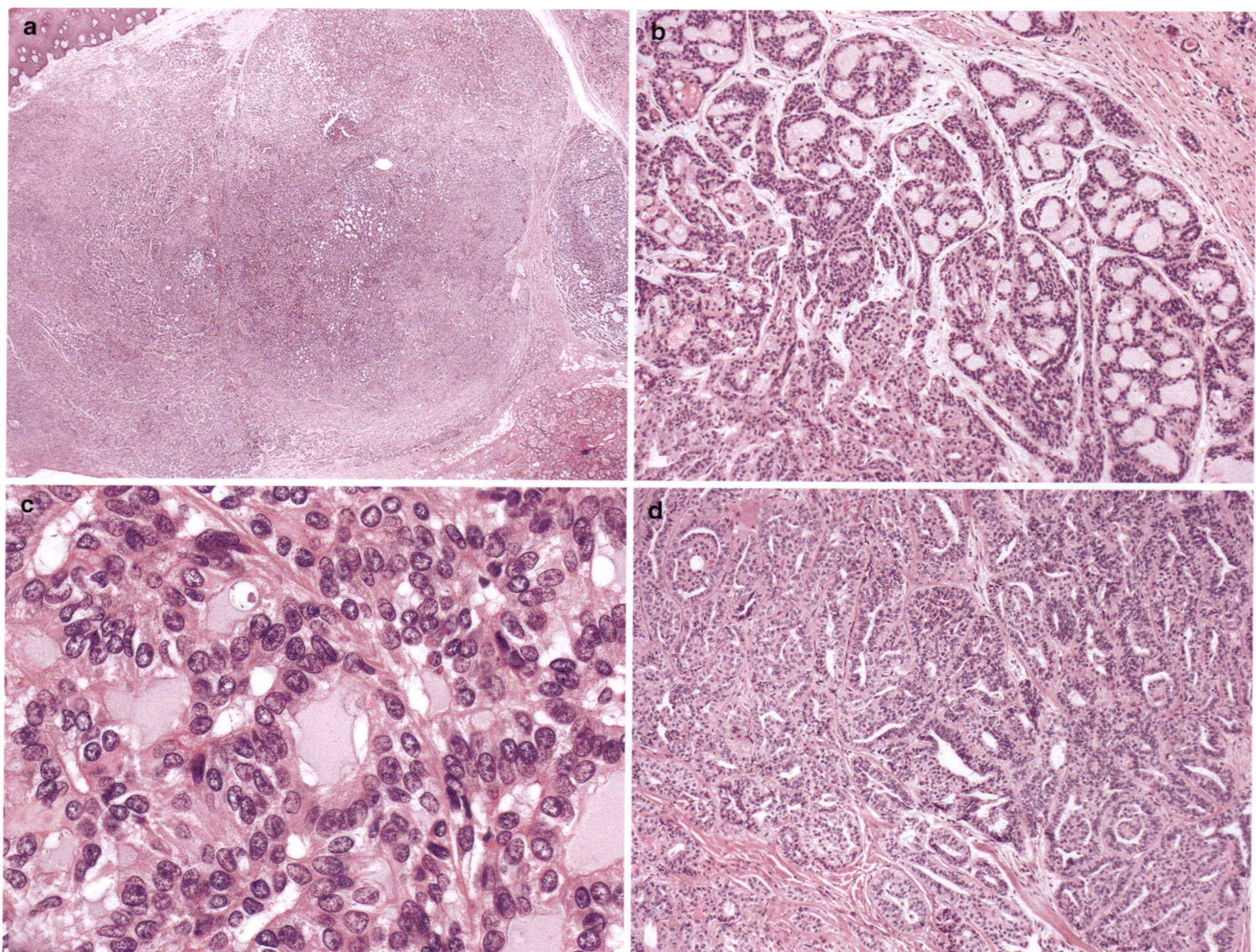

Fig. 19.8 Polymorphous adenocarcinoma. (**a**) Unencapsulated minor salivary gland tumor, deceptively well demarcated appearance on low magnification. (**b**) Cribriform structures embedded in grey chondroid stroma. Foci of oncocytic change are also present. (**c**) Cytologic monotony with uniform round to ovoid vesicular nuclei. (**d**) Polymorphous adenocarcinoma, cribriform type with glomeruloid architectures

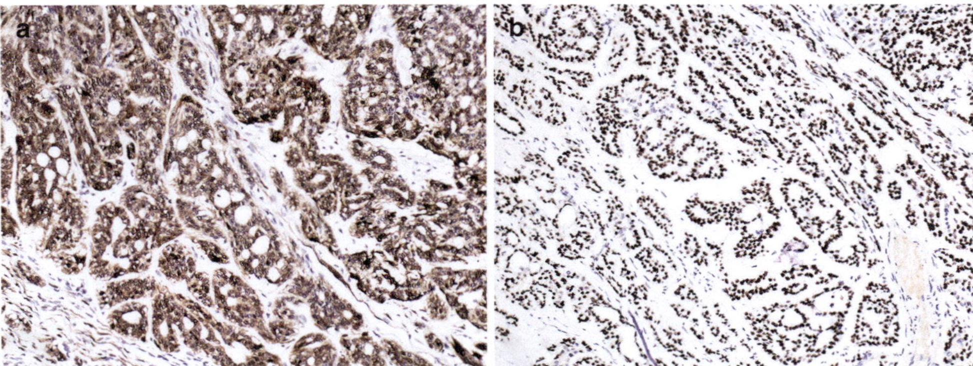

Fig. 19.9 Polymorphous adenocarcinoma. (**a**) Strong diffuse immunostaining with S100. (**b**) Nuclear immunostaining with p63

PRKD Gene Alterations

- Weinreb et al. have recently described *PRKD1 E710D* hotspot mutation in PAC, conventional type
  - *PR KD1 E710D* hotspot mutation can be detected by NGS or PCR in nearly 75–90% of PAC, conventional type
  - A small proportion (5–10%) demonstrate gene translocation/fusion of *PRKD1*, *PRKD2*, or *PRKD3*
- PAC, cribriform type, demonstrates *PRKD1* or *PRKD2* or *PRKD3* gene fusion with *ARID1A* or *DDX3X*
  - *PRKD1/2/3* fusions are detectable by FISH or NGS RNA fusion panels
  - *PR KD1 E710D* hot spot mutation detectable by NGS is also seen

Utility of Detecting *PRKD* Gene Alterations in PAC

- PAC, irrespective of the subtype, can be diagnostically challenging particularly on a small biopsy
- The most common differential diagnoses include AdCC followed by EMC or PA
  - The architectural diversity, monotony of the tumor cells, strong diffuse immunostaining with S100, and patchy immunostaining with myoepithelial markers should help identify PAC
  - NGS for *PRKD1 E710D* hot spot mutation or RNA fusion panel for *PRKD* gene fusions can be performed in difficult cases
- Currently, it is thought that tumors with *PRKD1* or *PRKD2* or *PRKD3* gene fusion (PAC, cribriform type) are more likely to develop lymph node metastases as compared with tumors with *PRKD1 E710D* hot spot mutation (PAC, conventional type)
  - It is unclear whether this reflects the true biologic behavior of the tumor or occurs due to the lymphovascular microanatomy of the base of tongue/oropharynx where PAC, cribriform type, arise. It is well recognized that tumors arising in the base of tongue/oropharynx have a higher propensity to metastasize

Salivary Duct Carcinoma

- Salivary duct carcinoma (SDC) is a high-grade, malignant carcinoma of the salivary gland that morphologically resembles high-grade breast carcinoma
- SDC represents around 2% of salivary gland neoplasms and 10% of salivary gland malignancies
- The tumor may arise de novo or as a component of CXPA
- SDC commonly arises in the parotid and presents as a rapidly growing mass in the older adult population and has a predilection for men
- Prognosis is poor with high rates of local recurrence and nodal/distant metastases

Light Microscopy

- The tumor is an infiltrative mass with large ducts and nests with central comedonecrosis, cribriform, and papillary patterns (Fig. 19.10a)
- The tumor cells typically have an apocrine appearance, sometimes with decapitation secretions (Fig. 19.10b)
- Mitoses are frequent
- Lymphovascular and perineural invasion is common
- Variants of SDC can show sarcomatoid, oncocytic, micropapillary, and mucin-rich morphology
- Immunohistochemically, SDC demonstrates immunostaining with CK7, GATA3, androgen receptor (AR), and lacks immunostaining with p40, S100, SOX10, or CK5/6
  - SDC is positive for AR in 70–90% cases (Fig. 19.10c)

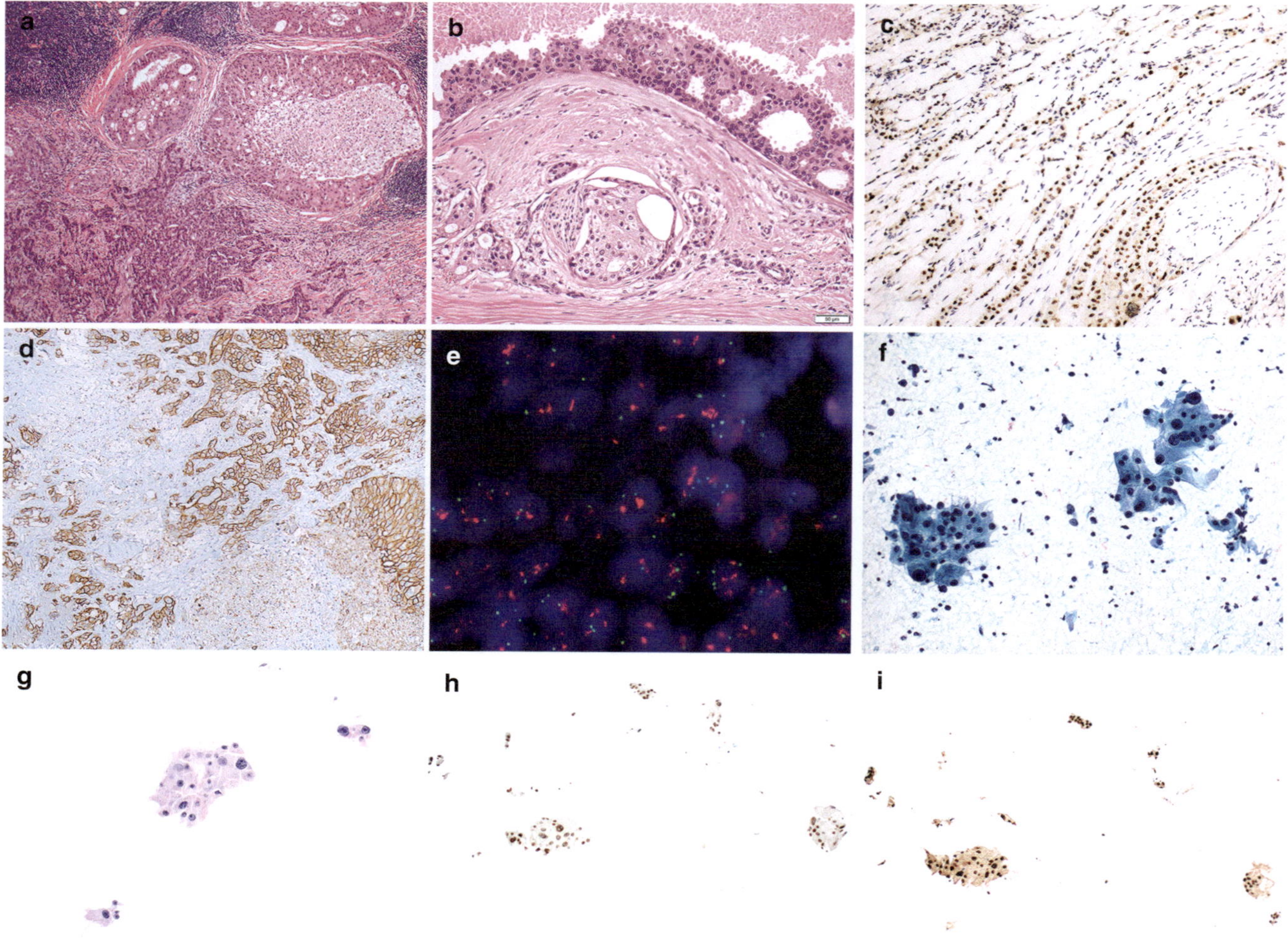

Fig. 19.10 Salivary duct carcinoma. (**a**) Salivary duct carcinoma showing dilated ducts with cribriform appearance and central comedonecrosis as well as infiltrative ductal pattern. (**b**) The cells lining the dilated ductal structures as well as the infiltrative nests show apocrine appearance with abundant granular cytoplasm and uniform round nuclei. Decapitation secretions are present. (**c**) IHC with AR showing strong nuclear staining in the malignant cells. (**d**) IHC with HER2 showing strong, complete membranous staining in the malignant cells. (**e**) FISH with HER2 (red signal) and centromere 17 (green signal) showing increase in HER2 copy numbers with approximately 10–15 copies per nucleus. (**f**) FNA of parotid gland: papanicolaou stain showing cohesive clusters of cells with moderate amounts of cytoplasm and enlarged round nuclei with distinctive nucleoli. (**g**) Cell block material with epithelioid cells with apocrine appearance. (**h**) Immunostaining with GATA3 performed on the cell block showing nuclear staining in most cells. (**i**) Immunostaining with AR performed on the cell block showing nuclear staining

Androgen Receptor

- Nuclear receptor that directly regulates gene transcription for homeostasis of reproductive organs
- Encoded by gene present on X chromosome
- Alternative splicing of AR is the most common alteration described in SDC
- Extra copies of chromosome X leading AR CNV have also been detected in approximately 40% of SDC
- Focal AR CNV or SNVs in AR are rare/not described in SDC
- IHC for AR is positive in up to 90% of SDC
  - The specificity of AR IHC for SDC is 100% among salivary gland neoplasms

Human Epidermal Growth Factor Receptor 2 (*HER2*)

- *HER2* is coded by an oncogene located on chromosome 17
- *HER2* amplification has been identified in up to 30% of SDC
  - IHC can show strong 3+ membranous staining of the malignant cells (Fig. 19.10d)
  - ISH can be used to identify amplification of *HER2* (Fig. 19.10e)
  - There are no established guidelines for interpreting IHC or ISH assays in SDC
    - o The ASCO guidelines for interpretation of *HER2* in breast carcinoma are most used

Utility of Detecting AR and HER2 in SDC

- SDC may mimic several high-grade neoplasms both within the salivary glands as well as at metastatic sites
  - In these instances, IHC for epithelial markers, GATA3 and AR can be helpful and can be performed on initial cytology cell block (Fig. 19.10f–i) material as well
- SDC have a poor prognosis with both regional and distant metastases with 2–3 years of diagnosis
- IHC for AR has nearly 90% sensitivity and 100% specificity for SDC
  - Most useful immunohistochemical stain for the diagnosis of SDC
    - May be used to guide androgen deprivation therapy
- Nearly 30% of SDC show HER2 amplification by ISH
  - Low diagnostic sensitivity but high specificity
  - May be used to guide treatment with trastuzumab

Other Genetic Alterations in SDC

- SDC are genomically unstable and harbor several genetic alterations
- Other genes frequently implicated in SDC include *TP53, PIK3CA, PTEN, HRAS,* and *NF1*
- Rearrangements involving *PLAG1* or *HMGA2* are seen in CXPA with SDC as the malignant component

Sclerosing Microcystic Adenocarcinoma

- Sclerosing microcystic adenocarcinoma (SMA) is a rare primary salivary gland neoplasm characterised by highly infiltrative tubules and ducts lined by epithelial and myoepithelial cells embedded in dense sclerotic stroma
  - Morphologically similar to microcystic adnexal carcinoma of the skin
- Predominantly involves the intraoral minor salivary glands
- Female predominance

Light Microscopy

- Highly infiltrative growth pattern (Fig. 19.11a)
- Ducts and tubules lined by cuboidal epithelial cells and peripheral myoepithelial cells (Fig. 19.11b)
- Cells show uniformly bland cytologic features (Fig. 19.11b)
  - Nuclear pleomorphism and mitoses are rare
- Sclerotic/hyalinized stroma
- Perineural invasion is common
- Immunohistochemistry highlights the biphasic nature of the cells lining the tubules and ducts

- Cytokeratins, CK7 positive in the epithelial cells (Fig. 19.11c)
- S100, p63 (Fig. 19.11d), p40, SMA, and other myoepithelial markers positive in the myoepithelial cells

CDK11B Mutations

- Loss of function mutations in *CDK11B* detected by NGS

Utility of Detecting *CDK11B* Mutations

- Diagnosis of sclerosing microcystic adenocarcinoma can be made based on morphologic and immunohistochemical characteristics alone
- Differential diagnoses such as PAC (p63 +/p40 –) and HCCC (*EWSR1* rearrangement) have distinctive immunohistochemical and molecular profiles
  - NGS for detection of *CDK11B* is generally not essential

Secretory Carcinoma

- Secretory carcinoma (SC) was previously known as mammary analogue secretory carcinoma
- Predominantly occurs in the parotid gland followed by minor salivary gland tissues of the oral cavity
  - SC can also arise in the breasts, sinonasal tract, and skin adnexal structures

Light Microscopy

- SC can be circumscribed or infiltrative and exhibit various growth patterns including cystic, tubular, papillary, and solid (Fig. 19.12a)
- The cells have ample eosinophilic to vacuolated cytoplasm and relatively uniform round to ovoid vesicular nuclei with fine chromatin and small nucleoli (Fig. 19.12b)
- Characteristic eosinophilic, bubbly secretions, resembling colloid are present (Fig. 19.12c)
- SC shows strong diffuse immunostaining with CK7, GATA3, S100 (Fig. 19.12d), SOX10, mammaglobin, and MUC4 (Fig. 19.12e)
  - The tumor cells lack staining with squamous or myoepithelial markers
- The tumor can demonstrate lymphovascular and/or perineural invasion as well as high-grade transformation with marked nuclear pleomorphism and frequent mitoses

ETV6 Rearrangement

- SC are defined by a characteristic *ETV6* translocation

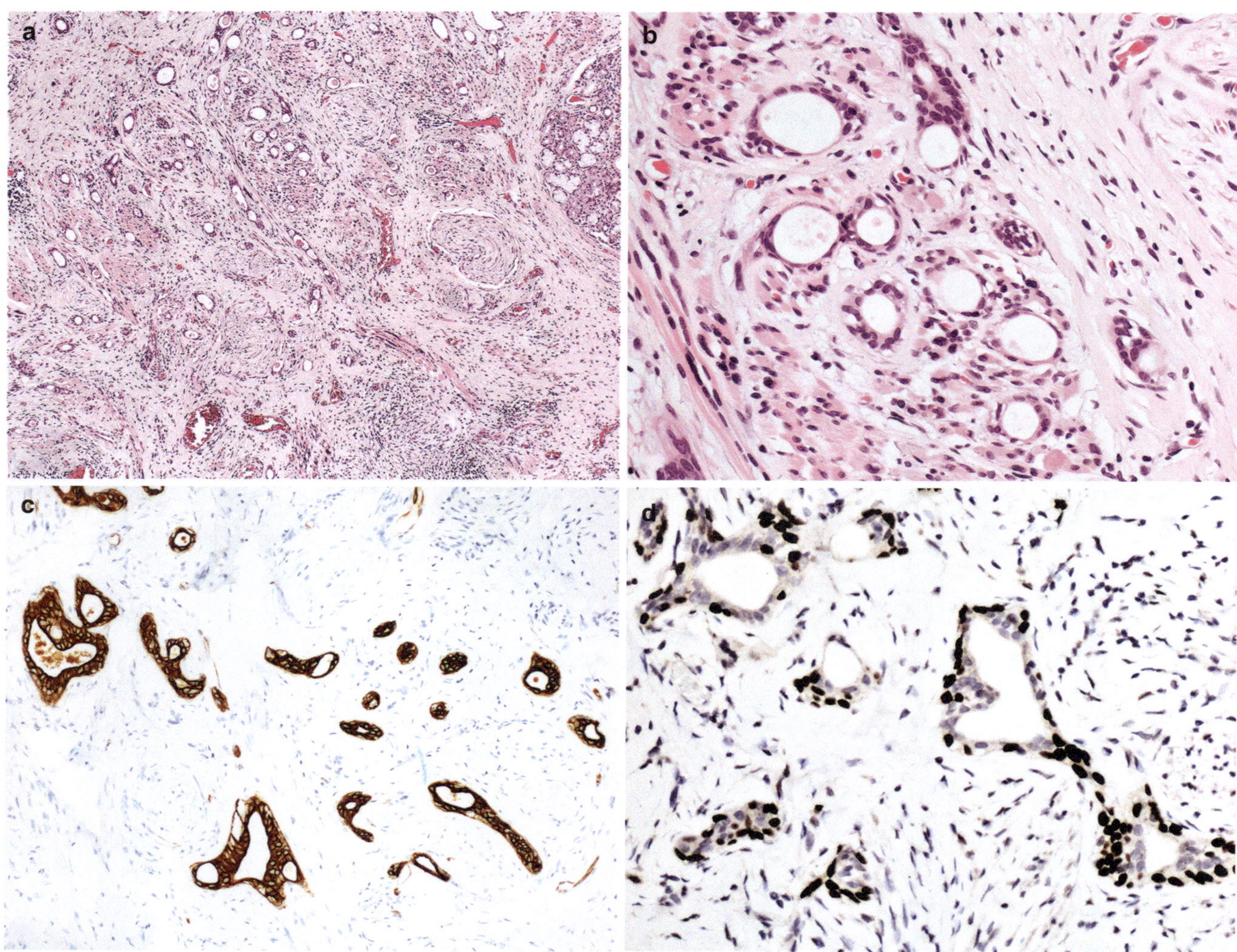

Fig. 19.11 Sclerosing microcystic adenocarcinoma. (**a**) Sclerosing microcystic adenocarcinoma arising from minor salivary gland showing small tubules and ducts with a diffusely infiltrative pattern. (**b**) Ducts lined by both epithelial and myoepithelial cells. Both cell types show bland cytologic features. (**c**) IHC with CK7 showing strong membranous staining in the lesional cells. (**d**) IHC with p63 showing nuclear staining in the myoepithelial cells

Fig. 19.12 Secretory carcinoma. (**a**) Secretory carcinoma showing a tubular and microcystic pattern. (**b**) Cuboidal cells of secretory carcinoma with eosinophilic cytoplasm and round nuclei. Occasional cells show vacuolated cytoplasm. (**c**) The intraluminal secretions have a colloid like appearance. (**d**) IHC with S100 showing strong diffuse staining in the malignant cells. (**e**) IHC with MUC4 showing strong diffuse staining in the malignant cells. (**f**) FISH break apart probe showing separation of the red and green signals flanking the 3′ and 5′ ends of the ETV6 locus demonstrating ETV6 rearrangement. (**g**) FISH break apart probe showing separation of the red and green signals flanking the 3′ and 5′ ends of the NTRK3 locus demonstrating NTRK3 rearrangement. (**h**) Strong nuclear staining with Pan NTRK IHC in secretory carcinoma

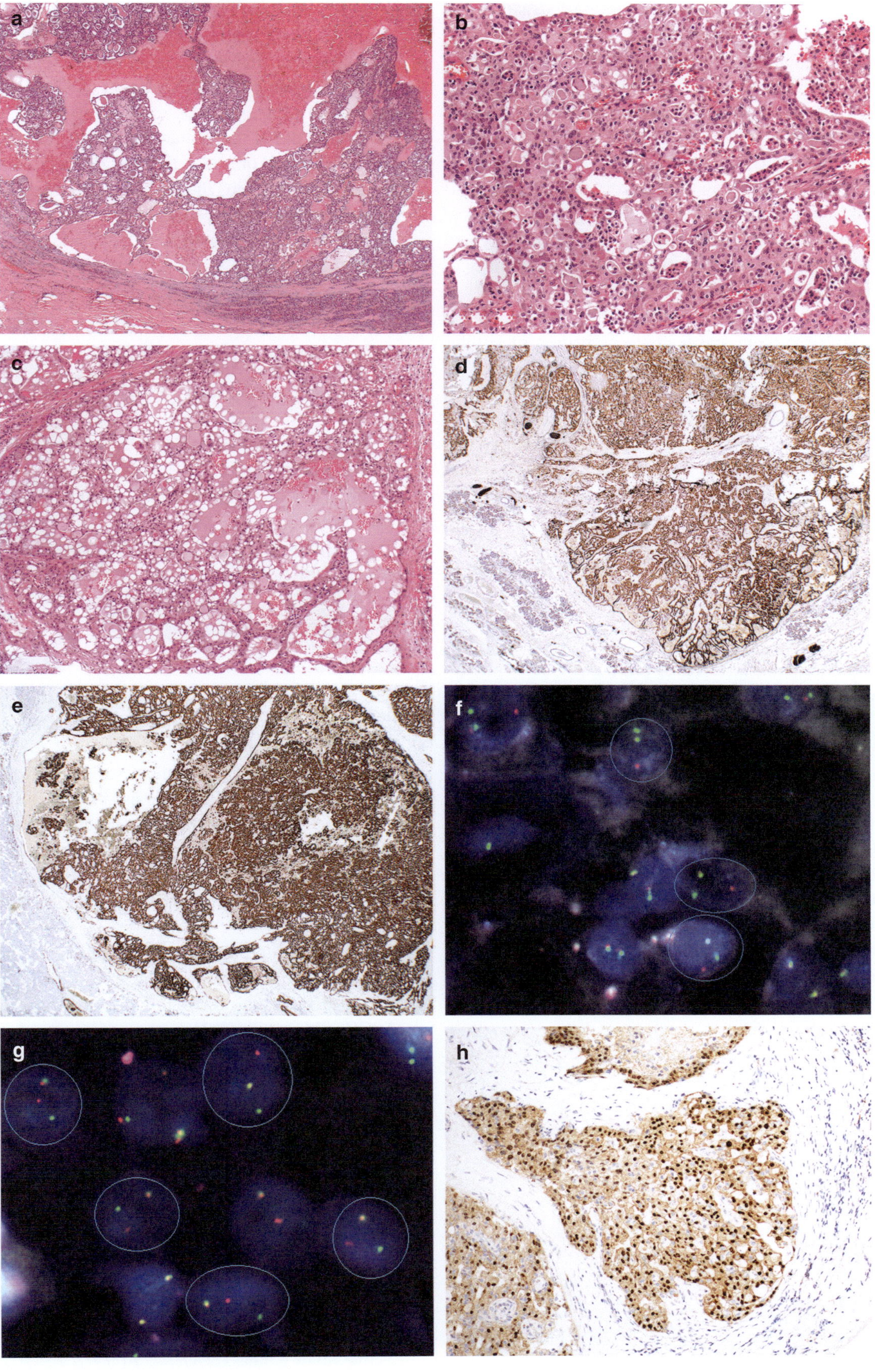

- The most frequent translocation is t(12;15)(p13;q25) leading to fusion of *ETV6-NTRK3*
  - *ETV6* is a member of a family of transcription factors involved in hematopoiesis and angiogenesis
  - *NTRK3* is involved in growth and development of the central nervous system by coding a surface receptor for neurotrophin-3
  - The fusion results in an oncoprotein that acts as a chimeric protein tyrosine kinase, leading to constitutive activation of Ras-MAP kinase mitogenic pathway and phosphatidyl inositol-3-kinase (PI3K)-Akt pathway
- Other fusion partners of *ETV6* include *MET* and *RET*
  - *MET* and *RET* both encode for surface receptor tyrosine kinase protein and leads to activation of similar pathways as *ETV6-NTRK3* fusion
- Copy number variations in *ETV6* have also been described
- Commercial break apart FISH probes are available for *ETV6* (Fig. 19.12f) as well as *NTRK3* (Fig. 19.12g)
- Immunohistochemistry for Pan NTRK is used to detect *NTRK1, NTRK2,* and *NTRK3* gene fusion in various cancers including SC

Utility of Detecting *ETV6* or *NTRK* Rearrangement in SC

- SC can generally be recognized by its distinctive morphologic and immunohistochemical profile
- SC may mimic zymogen granule poor AciCC, MSA, or MEC
- Immunohistochemistry for Pan-NTRK has 85% sensitivity for SC (Fig. 19.12h)
  - Positive nuclear staining can aid in distinguishing SC with *NTRK3* fusion from histologic mimics
  - Cytoplasmic/membranous staining may be seen in other non-NTRK rearranged salivary gland tumors including myoepithelial rich PA, PAC, and MEC
  - Identification of *NTRK* may also be important for therapeutic purposes
- Commercial break apart FISH probes for *ETV6* rearrangement are also available and have >85% sensitivity and specificity for SC
  - All other salivary gland neoplasms are consistently negative for *ETV6* rearrangement
- *ETV6* may have a non-NTRK fusion partner
  - These tumors may show an aggressive clinical course

Other Genetic Changes in SC

- The molecular landscape of SC continues to expand with additional novel fusions identified in case reports including:
  - *VIM-RET* fusion
  - *MYB-SMR3B*
  - *CTNNA1-ALK*

- *EGFR-SEPT14* fusion
- *NFIX-PKN1* translocation in a case of cutaneous SC
- Pathogenic missense mutation in *PRSS1, MLH1, MUTYH,* and *STK11* in a subset of SC with aggressive clinical course
- These additional molecular findings suggest that a negative *ETV6* FISH or *NTRK* FISH or IHC does not necessarily exclude SC in the appropriate morphologic and immunohistochemical context
  - Additional investigation with NGS fusion panels may be warranted
    - o Benign neoplasms with their malignant counterparts

Pleomorphic Adenoma

- PA is the most common salivary gland tumor in both adults and children
- It can occur in all glands but is predominantly found in the parotid
- Typical presentation is of a slow growing mass over a long period with locoregional symptoms relating to mass effect

Light Microscopy

- The tumor is characterized by biphasic growth with epithelial and myoepithelial cells, and a stromal component (Fig. 19.13a and b)
- The cellular element can show various cytomorphology (spindle, clear, plasmacytoid, and basaloid) and architecture (solid, tubules, trabecular, and cystic), and the stromal component can vary from myxoid to chondromyxoid, lipomatous, and chondro-osseous
- Squamous, oncocytic, and mucinous metaplasia are also common
- The variable morphological appearance can cause diagnostic difficulty in distinguishing PA from other salivary gland neoplasms including myoepithelioma, BCA, AdCC, EMC, and CXPA

PLAG1 or *HMGA2* Rearrangement

- Characteristic genetic rearrangement in PA involves 8q12 region where *PLAG1* is located and 12q14-15 that targets gene *HMGA2*
- *PLAG1* (pleomorphic adenoma gene 1) is a transcription factor that encodes a DNA binding zinc finger protein which is part of a family of cell cycle progression related proteins
  - Overexpression of *PLAG1* can result from translocation or intrachromosomal rearrangements with one of five other genes: *CTNNB1, FGFR1, LIFR, CHCHD7,*

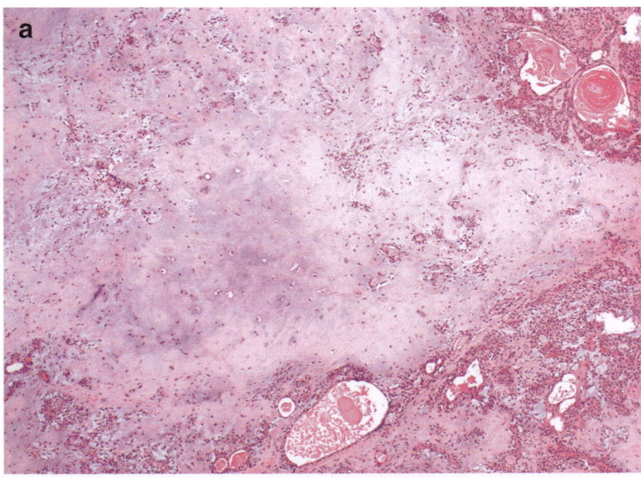

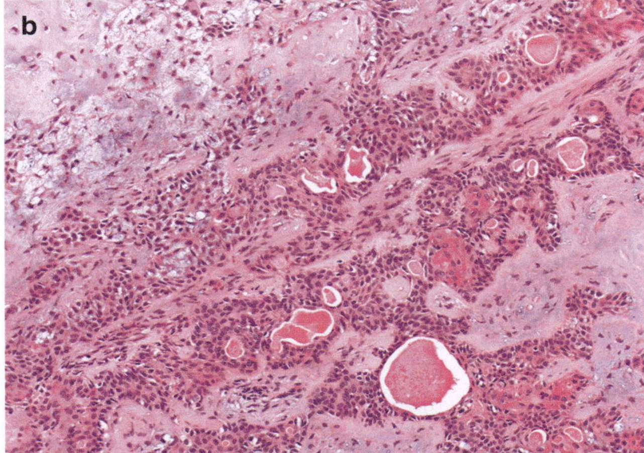

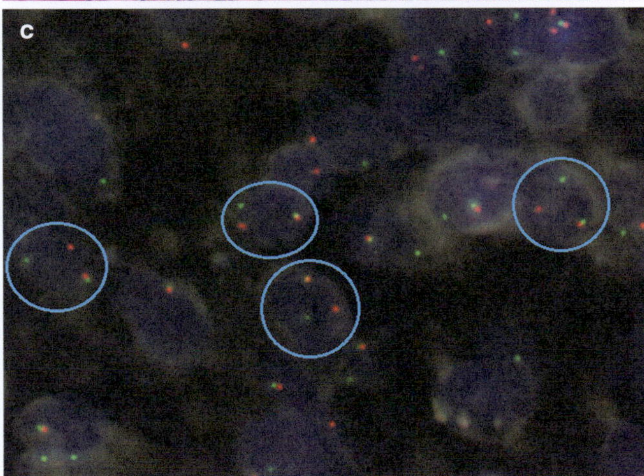

Fig. 19.13 Pleomorphic adenoma. (**a**, **b**) Epithelial and myoepithelial cells embedded in chondromyxoid stroma. The cells show bland cytologic features. (**c**) FISH break apart probe showing separation of the red and green signals flanking the 3′ and 5′ ends of the PLAG1 locus demonstrating PLAG1 rearrangement

TCEA1 that subsequently causes activation of the IGF-II signaling pathway
  – Enhancer hijacking of *PLAG1* is also described

  – *PLAG1* alterations are specific for PA as it has not been detected in other salivary gland tumors
- *HMGA2* is part of the high mobility group (HMG) protein family that affects gene transcription, recombination, and chromatin structure by binding to the minor groove of AT-rich DNA
  – *HMGA2* rearrangement leads to removal of the 3′ untranslated region and promotes gene expression, with subsequent activation of cell cycle regulators including *CCNA1* and *CCNB2*

Utility of Detecting *PLAG1* or *HMGA2* Rearrangement in PA

- Diagnosis of PA can usually be made on morphology alone when both epithelial and stromal components including chondromyxoid areas are readily identified
- PLAG1 immunohistochemistry is variably expressed by tumors with *PLAG1* and *HMGA2* rearrangements
  – A negative stain is useful to exclude a diagnosis of PA
    ○ Positive staining can be variable and weak and does not always reflect the presence of *PLAG1* rearrangement
  – Specificity is low as staining has also been reported in other salivary gland tumors including basal cell adenocarcinoma, epithelial myoepithelial carcinoma, myoepithelial carcinoma, and MEC
- *PLAG1* (Fig. 19.13c) or *HMGA2* gene rearrangement studies by FISH have been reported in 50% of PA cases
  – Other salivary gland neoplasms are negative for these rearrangements
  – FISH testing may be useful in distinguishing PA and CXPA from their histologic mimics
    ○ It should be noted that certain intrachromosomal *PLAG1* rearrangements such as those with *TCEA1* and *CHCHD1* cannot be detected by FISH

Carcinoma Ex-Pleomorphic Adenoma

- CXPA accounts for 12% of all salivary gland malignancies
- It typically presents as a rapidly growing mass in the setting of a preexisting or recurrent PA (Fig. 19.14a)
- Occurs in older adults with peak incidence in 6th to 7th decade

Light Microscopy

- Infiltrative growth and malignant cytological features together with the presence of a PA component (Fig. 19.14b) are the clue to diagnosing CXPA

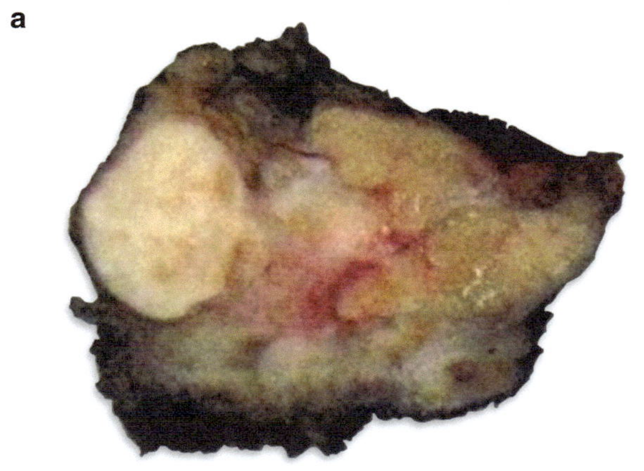

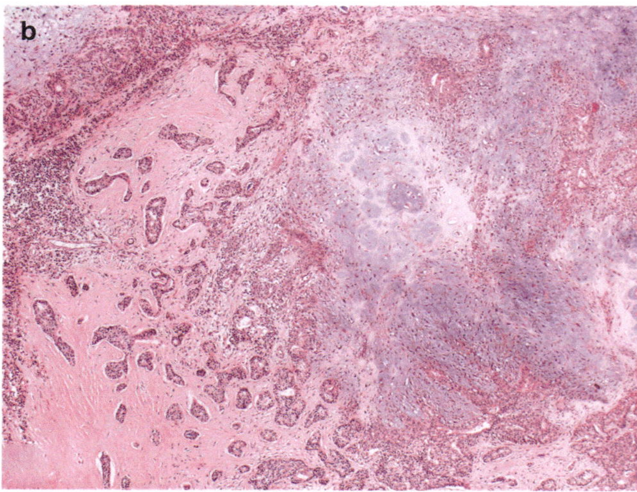

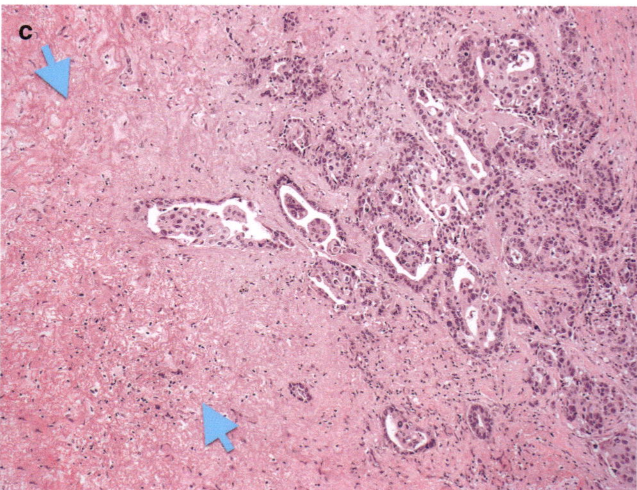

Fig. 19.14 Carcinoma ex pleomorphic adenoma. (**a**) Macroscopic appearance of CXPA showing a diffusely infiltrative neoplasm arising on a background of a well-demarcated nodule with glistening chondroid/myxoid appearance. (**b**) The PA component in CXPA may be readily recognizable. (**c**) The residual PA component in CXPA may be completely hyalinized/sclerotic (arrow)

- The PA component may be completely hyalinized or sclerotic (Fig. 19.14c)
- The carcinoma component can be of any type of salivary gland malignancy:
  - Salivary duct carcinoma (SDC) is the most common phenotype (Section 19.10)
  - The other phenotypes that may be seen include EMC (Section "Malignant Neoplasms"), myoepithelial carcinoma (Section 19.8) or carcinoma not otherwise specified

PLAG1 or *HMGA2* Rearrangement

- Specific details of pleomorphic adenoma gene 1 (*PLAG1*) on 8q12 and of high mobility group (*HMG*) are described in the PA section above

Utility of Detecting *PLAG1* or *HMGA2* Rearrangement in CXPA

- The diagnosis of CXPA is dependent on identifying an infiltrative neoplasm with malignant cytologic features arising on a background of a clinical history of rapid change in a long standing mass or histologic identification of a PA
- Detection of *PLAG1* or *HMGA2* by IHC or FISH does not assist in distinguishing CXPA from PA, cellular PA, or atypical PA

Other Genetic Changes in CXPA

- The SDC component of CXPA can show androgen receptor (AR) expression and *HER2* amplification
- CXPA developing from PA follows a multistep process where there is progressive loss of heterozygosity (LOH)
- CXPA demonstrates higher LOH of 12q and 7q, with alterations in 17p in late events of carcinogenesis
  - The amplified genes involved in these regions include *MDM2, MYC, PLAG1,* and *ERBB2*
  - Mutations of p53, cyclin D1, and p16 are also implicated in the transformation to CXPA

Basal Cell Adenoma

- BCA is a rare benign salivary gland neoplasm accounting for less than 5% of all salivary gland neoplasms
- Majority occurs in the parotid gland
- May occur in the context of familial cylindromatosis syndrome

Light Microscopy

- Well-circumscribed/encapsulated neoplasms demonstrating tubulotrabacular, membranous, cribriform, or solid architecture (Fig. 19.15a)
- Peripheral palisading of dark myoepithelial cells with central paler cells with eosinophilic cytoplasm
- Duct- like structures can be present
- The cells show vesicular nuclei
- Reduplicated basement membrane matrix is present
- IHC highlights the biphasic population with staining of the peripheral cells with myoepithelial markers (Fig. 19.15b) and staining of the luminal cells with CK7 (Fig. 19.15c)
  - Both cell types show strong staining with S100

CTNNB1 and *CYLD1* Mutations

- Activating mutations in *CTNNB1* have been described in approximately 50% of BCA, particularly BCA with tubulotrabecular architecture
  - *CTNNB1* codes for beta-catenin
  - *CTNNB1* mutations lead to nuclear accumulation of beta-catenin

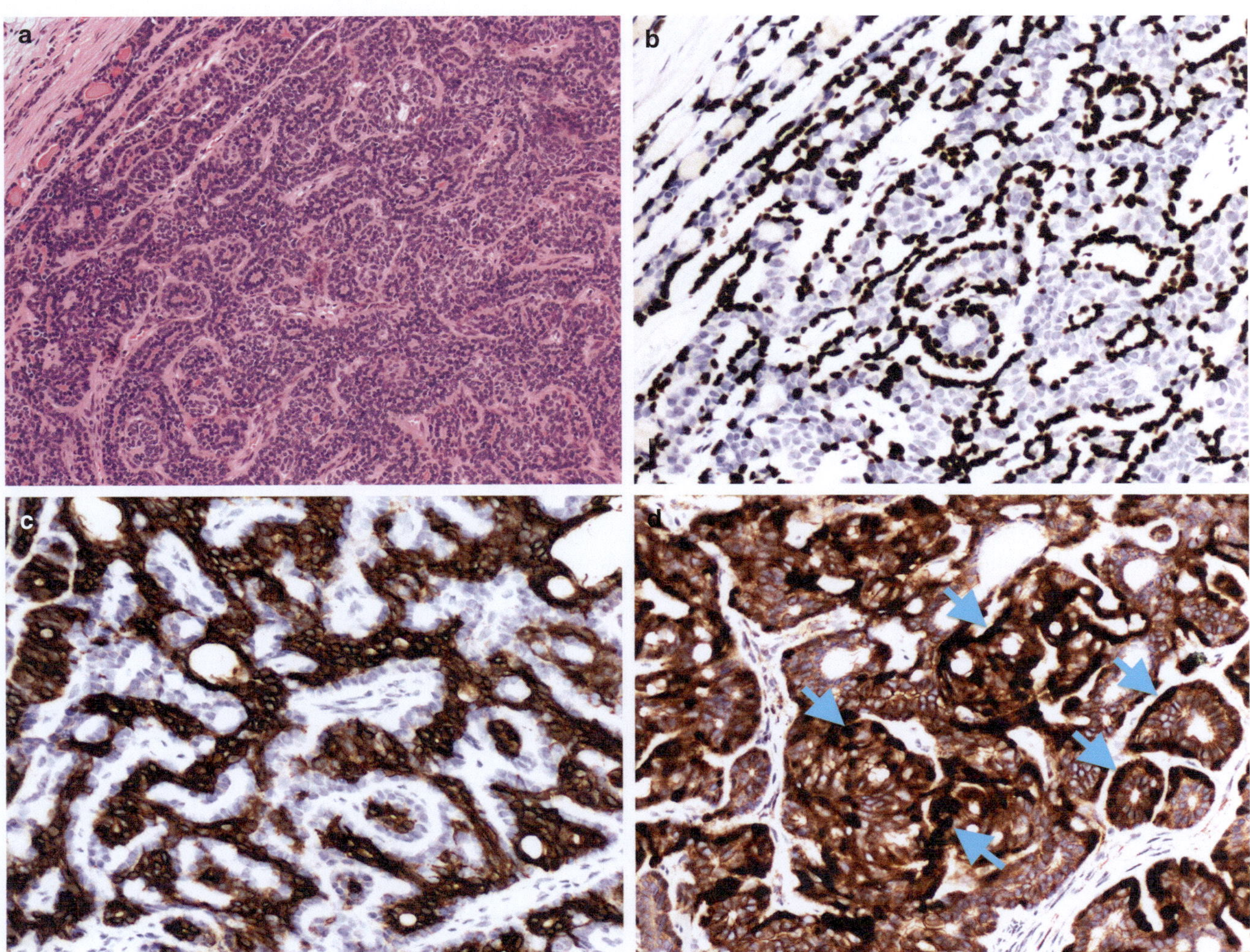

Fig. 19.15 Basal cell adenoma. (**a**) Well-circumscribed tumor showing trabecular architecture and reduplicated basement membrane. (**b**) IHC with p63 highlighting the peripheral myoepithelial cells. (**c**) IHC with CK7 highlighting the luminal epithelial cells. (**d**) IHC with beta-catenin. Only nuclear staining should be evaluated (Blue Arrows). Nuclear staining may be limited to the peripheral myoepithelial cells only

- Activating mutations in *CYLD1* have also been described in BCA, particularly membranous type BCA
  - *CYLD1* codes for a deubiquinating enzyme required for optimal functioning of Wnt pathway
  - Deregulation of Wnt pathway can lead to increased nuclear expression of beta-catenin

Detection of *CTNNB1* and *CYLD1* Mutations

- Immunohistochemistry for beta-catenin can be useful in detecting deregulation of CTNNB1 or Wnt pathway
  - Nuclear expression of beta-catenin may be limited to myoepithelial cells only (Fig. 19.15d)
- Sensitivity of beta-catenin for BCA is not very high as nuclear expression may not be seen in all cases
- Specificity of beta-catenin for BCA is not very high as expression, especially cytoplasmic staining, may be seen in a wide variety of tumors including PA, AdCC, and EMC
  - A small proportion of AdCC can show *CTNNB1* mutations
  - *CTNNB1* is the most common fusion partner of PLAG1 in PA

Basal Cell Adenocarcinoma

- A rare primary salivary gland malignancy with morphologic, immunohistochemical, and some molecular similarities with BCA
- Several studies indicate that basal cell adenocarcinoma arise on a background of BCA
- The membranous type of BCA is more likely to undergo malignant transformation
- *CYLD1* mutations have been described in basal cell adenocarcinoma
- *CTNNB1* mutations are thought to occur at a lower frequency in basal cell adenocarcinoma as compared with BCA
- Basal cell adenocarcinoma may show *PIK3Ca* mutations

Suggested Reading

Introduction

Freiberger SN, Brada M, Fritz C, Höller S, Vogetseder A, Horcic M, et al. SalvGlandDx - a comprehensive salivary gland neoplasm specific next generation sequencing panel to facilitate diagnosis and identify therapeutic targets. Neoplasia. 2021;23(5):473–87.

Goswami RS, Luthra R, Singh RR, Patel KP, Routbort MJ, Aldape KD, et al. Identification of factors affecting the success of Next Generation Sequencing testing in solid tumours. Am J Clin Pathol. 2016;145:222–37.

Luk PP, Selinger CI, Cooper WA, Mahar A, Palme CE, O'Toole SA, et al. Clinical utility of in situ hybridization assays in head and neck neoplasms. Head Neck Pathol. 2019;13(3):397–414.

Pantanowitz L, Thompson LDR, Rossi ED. Diagnostic approach to fine needle aspirations of cystic lesions of the salivary gland. Head Neck Pathol. 2018;12(4):548–61.

Skálová A, Stenman G, Simpson RHW, Hellquist H, Slouka D, Svoboda T, et al. The role of molecular testing in the differential diagnosis of salivary gland carcinomas. Am J Surg Pathol. 2018;42(2):e11–27.

Acinic Cell Carcinoma

Andreasen A, Varma A, Barasch N, Thompson LDR, Miettinen M, Rooper L, et al. The HTN3-MSANTD3 fusion gene defined a subset of acinic cell carcinoma of the salivary gland. Am J Surg Pathol. 2019;43:489–96.

Barasch N, Gong X, Kwei KA, Varma S, Biscocho J, Qu K, et al. Recurrent rearrangements of the Myb/SNAT-like DNA binding domain containing 3 gene (MSANTD3) in salivary gland acinic cell carcinoma. PLoS One. 2017;12(2):e0171265.

Chenevert J, Duvvuri U, Chiosea S, Dacic S, Cieply K, Kim J, et al. DOG1: a novel marker of salivary acinar and intercalated duct differentiation. Mod Pathol. 2012;25:919–29.

Chiosea SI, Griffith C, Asaad A, Seethala RR. The profile of acinic cell carcinoma after recognition of mammary analogue secretory carcinoma. Am J Surg Pathol. 2012;36:343–50.

Ellis GL, Corio RL. Acinic cell adenocarcinoma, a clinicopathologic analysis of 294 cases. Cancer. 1983;52:542–9.

Haller F, Bieg M, Will R, Korner C, Weichenhan D, Bott A, et al. Enhancer hijacking activates oncogenic transcription factor NR4A3 in acinic cell carcinoms of the salivary gland. Nat Commun. 2019a;10:368.

Haller F, Skálová A, Ihrler S, Markl B, Bieg M, Moskalev EA, et al. Nuclear NR4A3 immunostaining is a specific and sensitive novel marker for acinic cell carcinoma of the salivary glands. Am J Surg Pathol. 2019b;43:1264–72.

Haller F, Moskalev EA, Kuck S, Bieg M, Winkelmann C, Müller SK. etal. Nuclear NR4A2 (Nurr1) immunostaining is a novel marker for acinic cell carcinoma of the salivary glands lacking the classic NR4A3 (NOR-1) upregulation. Am J Surg Pathol. 2020;44(9):1290–2.

Owosho A, Tyler D, Adesina O, Odujoko O, Summersgill K. NR4A3 (NOR-1) immunostaining shows better performance than DOG1 immunostaining in acinic cell carcinoma of salivary gland: a preliminary study. J Oral Maxillofac Res. 2021;12(1):e4.

Skaugen JM, Seethala RR, Chiosea SI, Landau MS. Evaluation of NR4A3 immunohistochemistry and fluorescence in situ hybridization and comparison with DOG1 IHC for FNA diagnosis of acinic cell carcinoma. Cancer Cytopathol. 2021;129:104–13.

Viswanathan K, Beg S, He B, Zhang T, Cantley R, Lubin DJ, et al. NR4A3 immunostain is a highly sensitive and specific marker for acinic cell carcinoma in cytologic and surgical specimens. Am J Clin Pathol. 2021; https://doi.org/10.1093/ajcp/aqab099.

Wong KS, Marino-Enriques A, Hornick JL, Jo VY. NR4A3 immunohistochemistry reliably discriminates acinic cell carcinoma from mimics. Head and Neck Pathol. 2021;15:425–32.

Adenoid Cystic Carcinoma

Andersson MK, Mangiapane G, Nevado PT, Tsakaneli A, Carlsson T, Corda G, et al. ATR is a MYB regulated gene and potential therapeutic target in adenoid cystic carcinoma. Oncogenesis. 2020;9:5.

Brill LB, Kanner WA, Fehr A, Andren Y, Moskaluk CA, Loning T, et al. Analysis of MYB expression and MYB-NFIB gene fusion in adenoid cystic carcinoma and other salivary gland neoplasms. Mod Pathol. 2011;24:1169–76.

Frerich CA, Sedam HN, Kang H, Mitani Y, El-Naggar AK, Ness SA. N-Terminal truncated MYB with new transcriptional activity produced through use of an alternative MYB promoter in salivary gland adenoid cystic carcinoma. Cancers (Basel). 2019;12(1):45.

Ho AS, Kannan K, Roy DM, Morris LG, Ganly I, Katabi N, et al. The mutational landscape of adenoid cystic carcinoma. Nat Genet. 2013;45(7):791–8.

Mikse OR, Tchaicha JH, Akbay EA, Chen L, Bronson RT, Hammerman PS, et al. The impact of the MYB-NFIB fusion proto-oncogene in vivo. Oncotarget. 2016;7(22):31681–8.

Mitani Y, Liu B, Rao PH, Borra VJ, Zafereo M, Weber RS, et al. Novel MYBL1 gene rearrangements with recurrent MYBL1-NFIB fusions in salivary adenoid cystic carcinomas lacking t(6;9) translocations. Clin Cancer Res. 2016;22(3):725–33.

Nordkvist A, Mark J, Gustafsson H, Bang G, Stenman G. Non-random chromosome rearrangements in adenoid cystic carcinoma of the salivary glands. Genes Chromosomes Cancer. 1994;10:115–21.

Persson M, Andrein Y, Mark J, Horlings HM, Persson F, Stenman G. Recurrent fusion of MYB and NFIB transcription factor genes in carcinomas of the breast and head and neck. Proc Natl Acad Sci U S A. 2009;106(44):18740–4.

Persson M, Andren Y, Moskaluk CA, Frierson HF Jr, Cooke SL, Futreal PA, et al. Clinically significant copy number alterations and complex rearrangements of MYB and NFIB in head and neck adenoid cystic carcinoma. Gene Chromosomes Cancer. 2012;51(8):805–17.

Rettig EM, Talbot CC Jr, Sausen M, Jones S, Bishop JA, Wood LD, et al. Whole-genome sequencing of salivary gland adenoid cystic carcinoma. Cancer Prev Res (Phila). 2016;9(4):265–74.

Rooper LM, Lombardo KA, Oliai BR, Ha PK, Bishop JA. MYB RNA in situ hybridization facilitates sensitive and specific diagnosis of adenoid cystic carcinoma regardless of translocation status. Am J Surg Pathol. 2021;45(4):488–97.

Stenman G, Andersson MK, Andren Y. New tricks from an old oncogene. Cell Cycle. 2010;9(15):2986–95.

Stenman G, Licitra L, Said-Al-Naief N, van Zante A, Yarbrough WG. Chapter 7: Adenoid cystic carcinoma. In: El-Naggar AK, Chan JKC, Grandis JR, Takata T, Slootweg PJ, editors. World Health Organization Classification of head and neck tumours. 4th ed. Lyon: IARC; 2017. p. 164–5.

Stephens PJ, Davies HR, Mitani Y, Van Loo P, Shlien A, Tarpey PS, et al. Whole exome sequencing of adenoid cystic carcinoma. J Clin Invest. 2013;123:2965–8.

Togashi Y, Dobashi A, Sakata S, Sato Y, Baba S, Seto A, et al. MYB and MYBL1 in adenoid cystic carcinoma: diversity in the mode of genomic rearrangement and transcripts. Mod Pathol. 2018;31(6):934–46.

West RB, Kong C, Clarke N, Gilks T, Lipsick J, Cao H, et al. MYB expression and translocation in adenoid cystic carcinomas and other salivary gland tumours with clinicopathologic correlation. Am J Surg Pathol. 2011;35(1):92–9.

Epithelial Myoepithelial

De Cecio R, Cantile M, Fulciniti F, Botti G, Foschini MP, Losito NS. Salivary epithelial-myoepithelial carcinoma: clinical, morphological and molecular features. Pathologica. 2017;109:1–8.

Felisiak-Golabek A, Inaguma S, Kowalik A, Wasag B, Wang ZF, Zieba S, et al. SP174 antibody lacks specificity for NRAS Q61R and cross-reacts with HRAS and KRAS Q61R mutant proteins in malignant melanoma. Appl Immunohistochem Mol Morphol. 2018;26(1):40–5.

Hallani SE, Udager AM, Bell D, Fonseca I, Thompson LDR, Assaad A, et al. Epithelial-Myoepithelial carcinoma: frequent morphologic and molecular evidence of pre-existing pleomorphic adenoma, common HRAS mutations in PLAG1-intact and HMGA2-intact cases, and occasional TP53, FBXW7, and SMARCB1 alterations in high grade cases. Am J Surg Pathol. 2018;42(1):18–27.

Nakaguro M, Tanigawa M, Hirai H, Yamamoto Y, Urano M, Takahashi RH, et al. The diagnostic utility of RAS Q61R mutation-specific immunohistochemistry in epithelial-myoepithelial carcinoma. Am J Surg Pathol. 2021;45:885–94.

Urano M, Nakaguro M, Yamamoto Y, Hirai H, Tanigawa M, Saigusa N, et al. Diagnostic significance of HRAS mutations in epithelial-myoepithelial carcinomas exhibiting a broad histopathologic spectrum. Am J Surg Pathol. 2019;43(7):984–94.

Hyalinizing Clear Cell Carcinoma

Antonescu CR, Nafa K, Segal NH, Dal Cin P, Ladanyi M. EWS-CREB1: A recurrent variant fusion in clear cell sarcoma – Association with gastrointestinal location and absence of melanocytic differentiation. Clin Cancer Res. 2006;12:5356–62.

Antonescu CR, Dal Cin P, Nafa K, Teot LA, Surti U, Fletcher CD, Ladanyi M. EWSR1-CREB1 is the predominant gene fusion in angiomatoid fibrous histiocytoma. Genes Chromosomes Cancer. 2007;46:1051–60.

Antonescu CR, Katabi N, Zhang L, Sung YS, Seethala RR, Jordan RC, et al. EWSR1-ATF1 fusion is a novel and consistent finding in hyalinizing clear cell carcinoma of salivary gland. Genes Chromosomes Cancer. 2011;50(7):559–70.

Chapman A, Skalova A, Ptakova N, Martinek P, Goytain A, Tucker T, et al. Molecular profiling of hyalinizing clear cell carcinoma revealed a subset of tumors harboring a novel EWSR1-CREM fusion. Report of 3 cases. Am J Surg Pathol. 2018;42:1182–9.

Heft Neal ME, Gensterblum-Miller E, Bhangale AD, Kulkarni A, Zhai J, Smith J, et al. Integrative sequencing discovers an ATF1-motif enriched molecular signature that differentiates hyalinising clear cell carcinoma from mucoepidermoid carcinoma. Oral Oncol. 2021;117:105270.

Hernandez-Prera JC, Kwan R, Tripodi J, Chiosea S, Cordon-Cardo C, Najfeld V, et al. Reappraising hyalinising clear cell carcinoma: a population based study with molecular confirmation. Head Neck. 2017;39(3):503–11.

Milchgrub A, Gnepp DR, Vuitch F, Delgado R. Albores-Saavedra. Hyalinizing clear cell carcinoma of salivary gland. Am J Surg Pathol. 1993;18(1):74–82.

Shah AA, LeGallo RD, van Zante A, Firerson HF, Mills SE, Berean KW, et al. EWSR1 genetic rearrangement in salivary gland tumors. A specific and very common features of hyalinizing clear cell carcinoma. Am J Surg Pathol. 2013;37:571–8.

Thway K, Fisher C. Tumors with EWSR1-CREB1 and EWSR1-ATF1 fusions: the current status. Am J Surg Pathol. 2012;36:e1–e11.

Intraductal Carcinoma

Bishop JA, Gagan J, Krane JF, Jo VY. Low grade apocrine intraductal carcinoma: expanding the morphologic and molecular spectrum of an enigmatic salivary gland tumor. Head and Neck Pathol. 2020;14:869–75.

Bishop JA, Nakaguro M, Whaley RD, Ogura K, Imai H, Laklouk I, et al. Oncocytic intraductal carcinoma of salivary glands: a distinct variant with TRIM33-RET fusions and BRAF V600E mutations. Histopathology. 2021;79(3):338–46.

Loening T, Leivo I, Simpson RHW, Weinreb I. Chapter 7: Intraductal carcinoma. In: El-Naggar AK, Chan JKC, Grandis JR, Takata T, Slootweg PJ, editors. World Health Organization Classification of head and neck tumours. 4th ed. Lyon: IARC; 2017. p. 170–1.

Lu H, Graham RP, Seethala R, Chute D. Intraductal carcinoma of salivary glands harboring TRIM27-RET fusion with mixed low grade and apocrine types. Head Neck Pathol. 2020;14(1):239–45.

Palicelli A. Intraductal carcinomas of the salivary glands: systematic review and classification of 93 published cases. APMIS. 2020;128:191–200.

Skálová A, Vanecek T, Uro-Coste E, Bishop JA, Weinreb I, Thompson LDR, et al. Molecular profiling of salivary gland intraductal carcinoma revealed a subset of tumors harboring NCOA4-RET and novel TRIM27-RET fusions: a report of 17 cases. Am J Surg Pathol. 2018;42(11):1445–55.

Skálová A, Ptakova N, Santana T, Agaimy A, Ihrler S, Uro-Coste E, et al. *NCOA4-RET* and *TRIM27-RET* are characteristic gene fusions in sliavary intraductal carcinoma, including invasive and metastatic tumors. Is "intraductal" correct? Am J Surg Pathol. 2019;43:1303–13.

Todorovic E, Weinreb I. Intraductal carcinomas of the salivary gland. Surg Pathol Clin. 2021;14(1):1–15.

Weinreb I, Bishop JA, Chiosea AI, Seethala RR, Perez-Ordonez B, Zhang L, et al. Recurrent *RET* gene rearrangements in intraductal carcinomas of salivary gland. Am J Surg Pathol. 2018;42:442–52.

Mucoepidermoid Carcinoma

Bieńkowski M, Kunc M, Iliszko M, Kuźniacka A, Studniarek M, Biernat W. MAML2 rearrangement as a useful diagnostic marker discriminating between Warthin tumour and Warthin-like mucoepidermoid carcinoma. Virchows Arch. 2020;477(3):393–400.

Bishop JA, Cowan ML, Shum CH, Westra WH. MAML2 rearrangements in variant forms of mucoepidermoid carcinoma: ancillary diagnostic testing for the ciliated and Warthin-like variants. Am J Surg Pathol. 2018;42(1):130–6.

Coxon A, Rozenblum E, Park YS, Joshi N, Tsurutani J, Dennis PA, et al. MECT1-MAML2 fusion oncogene linked to the aberrant activation of Cyclic AMP/CREB regulated genes. Cancer Res. 2005;65:7137–44.

Garcia JJ, Hunt JL, Weinreb I, McHugh JB, Leon Barnes E, et al. Fluorescence in situ hybridization for detection of MAML2 rearrangements in oncocytic mucoepidermoid carcinomas: utility as a diagnostic test. Hum Pathol. 2011;42(12):2001–9.

Luk PP, Wykes J, Selinger CI, Ekmejian R, Tay J, Gao K, et al. Diagnostic and prognostic utility of Mastermind-like 2 (MAML2) gene rearrangement detection by fluorescent in situ hybridization (FISH) in mucoepidermoid carcinoma of the salivary glands. Oral Surg Oral Med Oral Pathol Oral Radiol. 2016;121:530–41.

Okumura Y, Nakano S, Murase T, Ueda K, Kawakita D, Nagao T, et al. Prognostic impact of CRTC1/3-MAML2 fusions in salivary gland mucoepidermoid carcinoma: a multi-institutional retrospective study. Cancer Sc. 2020;111(11):4195–204.

Seethala RR, Dacic S, Cieply K, Kelly LM, Nikiforova MN. A reappraisal of the MECT1/MAML2 translocation in salivary mucoepidermoid carcinomas. Am J Surg Pathol. 2010;34:1106–21.

Skálová A, Agaimy A, Stanowska A, Baneckova M, Ptakova N, et al. Molecular profiling of salivary oncocytic mucoepidermoid carcinomas helps to resolve differential diagnostic dilemma with low grade oncocytic lesions. AM J Surg Pathol. 2020;44:1612–22.

Tonon G, Modi S, Wu L, Kubo A, Coxon AB, Komiya T et al. t(11;19) (q21;p13) translocation in mucoepidermoid carcinoma creates a novel fusion product that disrupts a Notch signalling pathway (published correction appears in Nat Genet 2003;33(3):430) (letter). Nat Genet. 2003;33:208–13.

Microsecretory Adenocarcinoma

Bishop A, Weinreb I, Swanson D, Westra WH, Qureshi HS, Sciubba J, et al. Microsecretory adenocarcinoma: a novel salivary gland tumor characterized by a recurrent MEF2C-SS18 fusion. Am J Surg Pathol. 2019;43:1023–32.

Bishop JA, Koduru P, Veremis BM, Oliai BR, Weinreb I, Rooper LM, et al. SS18 break-apart fluorescence in situ hybridization is a practical and effective method for diagnosing microsecretory adenocarcinoma of salivary glands. Head Neck Pathol. 2021a;15(3):723–6.

Bishop JA, Sajed DP, Weinred I, Dickson BC, Bilodeau EA, Agaimy A, et al. Microsecretory adenocarcinoma of salivary glands: an expanded series of 24 cases. Head and Neck Pathol. 2021b; https://doi.org/10.1007/s12105-021-01331-7.

Kawakami F, Nagao T, Honda Y, Sakata J, Yoshida R, Nakayama H, et al. Microsecretory adenocarcinoma of the hard palate: a case report of a recently described entity. Pathol Int. 2020;70:781–5.

Rooper LM. Emerging Entities in Salivary Pathology: a practical review of sclerosing microcystic adenocarcinoma, microsecretory adenocarcinoma, and secretory myoepithelial carcinoma. Surg Pathol Clin. 2021;14(1):137–50.

Myoepithelial Carcinoma

Skálová A, WEinreb I, Hyrcza M, Simpson RHW, Laco J, Agaimy A, et al. Clear cell myoepithelial carcinoma of salivary glands showing EWSR1 rearrangement. Molecular analysis of 94 salivary gland carcinomas with prominent clear cell component. Am J Surg Pathol. 2015;39:338–48.

Skálová A, Agaimy A, Vanecek T, Baneckova M, Laco J, Prakova N, et al. Molecular profiling of clear cell myoepithelial carcinoma of salivary glands with EWSR1 rearrangement identifies frequent PLAG1 gene fusions but no EWSR1 fusion transcripts. Am J Surg Pathol. 2021;45:1–13.

Polymorphous Adenocarcinoma

Batsakis JG, Pinkston GR, Luna MA, Byers RM, Sciubba JJ, Tillery GW. Adenocarcinoms of the oral cavity: a clinicopathologic study of terminal duct carcinomas. J Layngol Otol. 1983;97(9):825–35.

Fonseca I, Assaad A, Katabi N, Seethala R, Weinreb I, Wenig BM. Chapter 7: Polymorphous adenocarcinoma. In: El-Naggar AK, Chan JKC, Grandis JR, Takata T, Slootweg PJ, editors. World Health Organization Classification of head and neck tumours. 4th ed. Lyon: IARC; 2017. p. 163–87.

Sebastiao APM, Xu B, Lozada JR, Pareja F, Geyer FC, Da Cruz PA, et al. Histologic spectrum of polymorphous adenocarcinoma of the salivary gland harbor genetic alterations affecting PRKD gene. Mod Pathol. 2020;33:65–73.

Weinreb I, Piscuoglio S, Martelotto LG, Waggott D, Ng CKY, Perez-Ordonez B, et al. Hotspot activating PRKD1 somatic mutations in polymorphous low grade adenocarcinomas of the salivary gland. Nat Genet. 2013;46:1166–9.

Weinreb I, Zhang L, Tirunagari LMS, Sung YS, Chen CL, Perez-Ordonez B, et al. Novel PRKD gene rearrangements and variant fusions in cribriform adenocarcinoma of salivary gland origin. Genes Chromosomes Cancer. 2014;53(10):845–56. https://doi.org/10.1002/gcc.22195.

Wysocki PT, Westra WH, Sidransky D, Brait M. Advancing toward a molecular characterisation of polymorphous low grade adenocarcinoma. Oral Oncol. 2017;74:192–3.

Xu B, Aneja A, Ghossein R, Katabi N. Predictors of outcome in the phenotypic spectrum of polymorphous low grade adenocarcinoma (PLGA) and cribriform adenocarcinoma of salivary gland (CASG). A retrospective study of 69 patients. Am J Surg Pathol. 2016;40:1526–37.

Salivary Duct Carcinoma

Chiosea SI, Williams L, Griffith CC, Thompson LDR, Weinreb I, Bauman JE, et al. Molecular characterisation of apocrine salivary duct carcinoma. Am J Surg Pathol. 2015;39(6):744–52.

Egebjerg K, Harwood CD, Woller NC, Kristensen CA, Mau-Sorensen M. HER2 positivity in histological subtypes of salivary gland carcinoma: a systematic review and meta-analysis. Front Oncol. 2021;11:693394.

Johnson CJ, Barry MB, Vasef MA, DeYoung BR. HER-2/neu expression in salivary duct carcinoma: an immunohistochemical and chromogenic in situ hybridization study. Appl Immunohistochem Mol Morphol. 2008;16:54–8.

Lewis JE, McKinney BC, Weiland LH, Ferreiro JA, Olsen KD. Salivary duct carcinoma. Clinicopathologic and immunohistochemical review of 26 cases. Cancer. 1996;77(2):223–30.

Liang L, Williams MD, Bell D. Expression of PTEN, androgen receptor, HER2/neu, cytokeratin 5/6, estrogen receptor-beta, HMGA2, and PLAG1 in salivary duct carcinoma. Head Neck Pathol. 2019;13:529–34.

Mitani Y, Rao PH, Maity SN, Lee YC, Ferrarotto R, Post JC, et al. Alterations associated with androgen receptor gene activation in salivary duct carcinoma of both sexes: potential therapeutic ramifications. Clin Cancer Res. 2014;20(24):6570–81.

Mueller SA, Gauthier MA, Blackburn J, Grady JP, Kraitsek S, Hajdu E, et al. Molecular patterns in salivary duct carcinoma identify prognostic subgroups. Mod Pathol. 2020;33(10):1896–909.

Simpson RHW. Salivary duct carcinoma: new developments – morphological variants including pure in situ high grade lesions; proposed molecular classifications. Head Neck Pathol. 2013;7:S48–58.

Skalova A, Starek I, Vanecek T, Kucerova V, Plank L, Szepe P, et al. Expression of HER-2/neu gene and protein in salivary duct carcinomas of parotid gland as revealed by fluorescence in-situ hybridization and immunohistochemistry. Histopathology. 2003;42:348–56.

Williams MD, Roberts D, Blumenschein GR, Temam S, Merrill SK, Rosenthal DI, et al. Differential expression of hormonal and growth factor receptors in salivary duct carcinomas. Am J Surg Pathol. 2007;31:1645–52.

Williams L, Thompson LDR, Seethala RR, Weinreb I, Assaad AM, Tuluc M, et al. Salivary duct carcinoma. The predominance of apocrine morphology, prevalence of histologic variants, and androgen receptor expression. Am J Surg Pathol. 2015;39:705–13.

Wolff AC, Hammond EH, Allison KH, Harvey BE, Mangu PB, et al. Human epidermal growth factor receptor 2 testing in breast cancer: American Society of Clinical Oncology/College of American Pathologists clinical practice guideline focused update. J Clin Oncol. 2018;36:2105–22.

Sclerosing Microcystic Adenocarcinoma

Jiang R, Marquez J, Tower JI, Jacobs D, Chen W, Mehra S, et al. Sequencing of sclerosing microcystic adenocarcinoma identified mutational burden and somatic variants associated with tumorigenesis. Anticancer Res. 2020;40(11):6375–9.

Mills AM, Policarpio-Nicholas MLC, Agaimy A, Wick MR, Mills SE. Sclerosing microcystic adenocarcinoma of the head and neck mucosa: a neoplasm closely resembling microcystic adnexal carcinoma. Head Neck Pathol. 2016;10(4):501–8.

Wood A, Conn BI. Sclerosing microcystic adenocarcinoma of the tongue: a report of 2 further cases and review of literature. Oral Maxillofac Pathol. 2018;125(4):e94–102.

Secretory Carcinoma

Black M, Liu CZ, Onozato M, Iafrate AJ, Darvishian F, Jour G, Cotzia P. Concurrent identification of novel *EGFR-SEPT14* fusion and *ETV6-RET* fusion in secretory carcinoma of the salivary gland. Head Neck Pathol. 2020;14:817–21.

Cipriani NA, Blair EA, Finkle J, Kraninger JL, Straus CM, Villaflor VM, et al. Salivary gland secretory carcinoma with high grade transformation, CDKN2A/B loss, distant metastasis and lack of sustained response to Crizotinib. Int J Surg Pathol. 2017;25(7):613–8.

Hung YP, Jo VY, Hornick J. Immunohistochemistry with a pan-TRK antibody distinguishes secretory carcinoma of the salivary gland from acinic cell carcinoma. Histopathology. 2019;75:54–62.

Ito Y, Ishibashi K, Masaki A, Fujii K, Fujiyoshi Y, Hattori H, et al. Mammary analogue secretory carcinoma of salivary glands: a clinicopathological and molecular study including 2 cases harboring ETV6-X fusion. Am J Surg Pathol. 2015;39:602–10.

Kastnerova L, Luzar B, Goto K, Grishakov V, Gatalica Z, Kamarachev J, et al. Secretory carcinoma of the skin; report of 6 cases including a case with novel NFIX-PKN1 translocation. Am J Surg Pathol. 2019;43:1092–8.

Knezevich SR, Garnett MJ, Pysher TJ, Beckwith B, Grundy PE, Sorensen PHB. ETV6-NTRK3 gene fusions and trisomy 11 establish a histogenetic link between mesoblastic nephroma and congenital fibrosarcoma. Cancer Res. 1998;58:5046–8.

Na K, Hernandez-Prera JC, Lim JY, Woo HY, Yoon SO. Characterisation of novel genetic alterations in salivary gland secretory carcinoma. Mod Pathol. 2020;33:541–50.

Rooper LM, Karantanos T, Ning Y, Bishop JA, Gordon SW, Kang H. Salivary secretory carcinoma with novel ETV6-MET fusion: expanding the molecular spectrum of a recently described entity. Am J Surg Pathol. 2018;42:1121–6.

Sasaki E, Masago K, Fujita S, Suzuki H, Hanai N, Hosoda W. Salivary secretory carcinoma harboring a novel ALK fusion: expanding the molecular characterisation of carcinomas beyond the ETV6 gene. Am J Surg Pathol. 2020;44:962–9.

Skálová A. Mammary analogue secretory carcinoma of salivary gland origin: an update and expanded morphologic and immunohistochemical spectrum of recently described entity. Head Neck Pathol. 2013;7:S30–6.

Skálová A, Venecek T, Sima R, Laco J, Weinreb I, Perez-Ordonez B, et al. Mammary analogue secretory carcinoma of salivary glands, containing the ETV6-NTRK3 fusion gene: a hitherto undescribed salivary gland tumour entity. Am J Surg Pathol. 2010;34:599–608.

Skálová A, Vanecek T, Majewska H, Laco J, Grossmann P, Simpson RHW, et al. Mammary analogue secretory carcinoma of salivary gland with high-grade transformation. Report of 3 cases with ETV6-NTRK3 gene fusion and analysis of TP53, B-catenin, EGFR and CCND1 genes. Am J Surg Pathol. 2014;38:23–33.

Skálová A, Vanecek T, Simpson RHW, Laco J, Majewska H, Baneckova M, et al. Molecular analysis of 25 ETV6 gene rearranged tumours with lack of detection of classical ETV6-NTRK3 fusion transcript by standard RT-PCR: Report of 4 cases harboring ETV6-X gene fusion. Am J Surg Pathol. 2016;40:3–13. https://doi.org/10.1097/PAS.0000000000000537.

Skálová A, Vanecek T, Martinek P, Wenreb I, Stevens TM, Simpson RHW, et al. Molecular profiling of mammary analogue secretory carcinoma revealed a subset of tumors harboring a novel ETV6-RET translocation, report of 10 cases. Am J Surg Pathol. 2018;42:234–46.

Skálová A, Baneckova M, Thompson LDR, Ptakova N, Stevens TM, Brcic L, et al. Expanding the molecular spectrum of secretory carcinoma of salivary glands with novel VIM-RET fusion. Am J Surg Pathol. 2020;44:1295–307.

Tognon C, Garnett M, Kenward E, Kay R, Morisson K, Sorensen PHB. The chimeric protein tyrosine kinase ETV6-NTRK3 requires both Ras-Erk1/2 and PI3-kinase-Akt signalling for fibroblast transformation. Cancer Res. 2001;61(24):8909–16.

Tognon C, Knezevich SR, Huntsman D, Roskelley CD, Melnyk N, Mathers JA, et al. Expression of the ETV6-NTRK3 gene fusion as a primary events in human secretory breast carcinoma. Cancer Cell. 2002;2(5):367–76.

Wai DH, Knezevich SR, Lucas T, Jansen B, Kay RJ, Sorensen PHB. The ETV6-NTRK3 gene fusion encodes a chimeric protein tyrosine kinase and transformed NIH3T3 cells. Oncogene. 2000;19:906–15.

Xu B, Al Rasheed MRH, Antonescu CR, Alex D, Frosina D, Ghossein R, et al. Pan-TRK immunohistochemistry is a sensitive and specific ancillary tool in diagnosing secretory carcinoma of salivary gland and detecting ETV6-NTRK3 fusion. Histopathology. 2020;76(3):375–82.

Pleomorphic Adenoma and Carcinoma Ex Pleomorphic Adenoma

Afshari MK, Fehr A, Nevado PT, Andersson MK, Stenman G. Activation of PLAG1 and HMGA2 by gene fusions involving the transcriptional regulator gene NFIB. Genes Chromosomes Cancer. 2020;59(11):652–60.

Antony J, Gopalan V, Smith RA, Lam A. Carcinoma ex pleomorphic adenoma: a comprehensive review of clinical, pathological and molecular data. Head Neck Pathol. 2012;6:1–9.

Bahrami A, Dalton JD, Shivakumar B, Krane J. PLAG1 alteration in carcinoma ex pleomorphic adenoma: immunohistochemical and fluorescence in situ hybridization studies of 22 cases. Head Neck Pathol. 2012;6:328–35.

Bell D, Bullerdiek J, Gnepp DR, Schwartz MR, Stenman G, Triantafyllou A. Chapter 7: Pleomorphic adenoma. In: El-Naggar AK, Chan JKC, Grandis JR, Takata T, Slootweg PJ, editors. World Health Organization Classification of head and neck tumours. 4th ed. Lyon: IARC; 2017. p. 186–7.

El-Naggar AK, Callender D, Coombes MM, Hurr K, Luna MA, Batsakis JG. Molecular genetic alterations in carcinoma ex-pleomorphic adenoma: a putative progression model? Genes Chromosomes Cancer. 2000;27:162–8.

Hernandez-Prera JC, Skálová A, Franchi A, Rinaldo A, Vander Poorten V, Zbären P, Ferlito A, Wenig BM. Pleomorphic adenoma: the great mimicker of malignancy. Histopathology. 2021 Sep;79(3):279–90.

Katabi N, Ghossein R, Ho A, Dogan S, Zhang L, Sung YS, et al. Consistent *PLAG1* and *HMGA2* abnormalities distinguish carcinoma ex-pleomorphic adenoma from its de novo counterparts. Hum Pathol. 2015;46:26–33.

Katabi N, Xu B, Jungbluth AA, Zhang L, Shao SY, Lane J, et al. PLAG1 immunohistochemistry is a sensitive marker for pleomorphic adenoma: a comparative study with PLAG1 genetic abnormalities. Histopathology. 2018;72(2):285–93.

Matsuyama A, Hisaoka M, Nagao Y, Hashimoto H. Aberrant PLAG1 expression in pleomorphic adenomas of the salivary gland: a molecular genetic and immunohistochemical study. Virchows Arch. 2011;458(5):583–92.

Persson F, Andren Y, Winnes M, Wedell B, Nordkvist A, Gunhildur G, et al. High resolution genomic profiling of adenomas and carcinoma of the salivary glands reveals amplification, rearrangement and fusion of HMGA2. Genes Chromosomes Cancer. 2009;48(1):69–82.

Basal Cell Adenoma and Adenocarcinoma

Fonseca I, Gnepp DR, Seethala R, Simpson RHW, Vielh P, Williams MD. Chapter 7: Basal cell adenocarcinoma. In: El-Naggar AK, Chan JKC, Grandis JR, Takata T, Slootweg PJ, editors. World Health Organization classification of head and neck tumours. 4th ed. Lyon: IARC; 2017. p. 169–70.

Jo VY, Sholl LM, Krane JF. Distinctive patterns of CTNNB1 (Beta-catenin) alterations insalivary gland basal cell adenoma and basal cell adenocarcinoma. Am J Surg Pathol. 2016;40:1143–50.

Li J, Fonseca I. Chapter 7: Basal cell adenoma. In: El-Naggar AK, Chan JKC, Grandis JR, Takata T, Slootweg PJ, editors. World Health Organization Classification of head and neck tumours. 4th ed. Lyon: IARC; 2017. p. 187–8.

Wilson TC, Ma D, Tilak A, Tesdahl B, Robinson RA. Next generation sequencing in salivary gland basal cell adenocarcinoma and basal cell adenoma. Head and Neck Pathol. 2016;10:494–500.

Yin P. (Rex) Hung and James R. Stone

Contents

Selected Primary Tumors of the Heart and Cardiovascular System

Introduction

- Most of the cardiac tumors are metastases from elsewhere or contiguous extensions of tumors originating from neighboring anatomic structures
- Primary cardiac tumors are uncommon, with a prevalence of <0.04% based on autopsy studies
- Of the primary cardiac tumors, ~90% is benign; primary malignant cardiac tumors are rare

Y. P. (Rex) Hung · J. R. Stone (✉)
Department of Pathology, Massachusetts General Hospital and Harvard Medical School, Boston, MA, USA
e-mail: yphung@mgh.harvard.edu; jrstone@mgh.harvard.edu

Benign Tumors

Myxoma

- Most common primary cardiac tumor
- Wide age distribution
- Predilection for women, with F:M ratio of 2:1
- Arises most commonly in the left atrium, followed by right atrium, seldom right/left ventricles
  - Typical location includes near the fossa ovalis of the atrial septum in the left atrium
  - Putative origin from multipotent cardiac stem/progenitor cells
- Signs and symptoms secondary to tumor embolization observed in up to 50% of patients
- Nonsyndromic in ~90% of patients
- Syndromic in ~10% of patients
  - Carney complex syndrome accounts for most of these syndromic patients
    - X-linked autosomal dominant
    - Characterized by multiple myxomas (cardiac, soft tissue, cutaneous), spotty skin pigmentations, endocrine tumors, and malignant melanotic nerve sheath tumor (psammomatous melanotic schwannoma)
    - Germline inactivating mutation involving *PRKAR1A*
  - Multifocal cardiac myxomas should raise concern for Carney complex syndrome
- Light microscopy (Fig. 20.1a)
  - Aggregates and nests of epithelioid to spindled or fusiform cells with eosinophilic cytoplasm, ovoid nuclei, and inconspicuous nuclei
  - Characteristic perivascular distribution of tumor cells surrounding small vessels
  - Abundant loose myxoid stroma, with variable hemorrhage and hemosiderin deposition
  - Foci of extramedullary hematopoiesis may be identified

L. Cheng et al. (eds.), *Molecular Surgical Pathology*, https://doi.org/10.1007/978-3-031-35118-1_20

- Glandular or squamous differentiation, or even heterologous cartilaginous or osseous elements in a subset of tumors
- Immunohistochemistry
  - Calretinin immunoreactivity in at least 70% of cardiac myxomas
  - Cytokeratin immunoreactivity in a subset of tumors, particularly those with glandular differentiation
  - Complete loss of PRKAR1A expression in ~30% of sporadic cases and >90% of syndromic cases
- *PRKAR1A*
  - Encodes protein kinase type I-alpha regulatory subunit, a critical component of the cAMP-dependent protein kinase A signaling pathway
  - Somatic mutations in *PRKAR1A* identified in <10% of nonsyndromic cases
  - Germline mutations in *PRKAR1A* identified in >50% of syndrome cases

Papillary Fibroelastoma

- Second most common primary cardiac tumor
- Wide age distribution
- Predilection for women
- Size usually smaller than 5 cm
- Originates on the cardiac valves
  - Most frequently involves aortic valve, followed by mitral valve and tricuspid/pulmonic valves
  - Resembles Lambl excrescences, which are small and typically arise at the closing edge near the Ariantus nodule of the aortic valve
- Light microscopy (Fig. 20.1b)
  - Central avascular stalk with delicate papillary frondlike projections
  - Core of the stalk composed predominantly of elastic tissues, which can be highlighted by elastic stain
  - Lined by inconspicuous flattened endothelial cells
- Immunohistochemistry
  - Surface endothelial cells express typical endothelial markers, including ERG, CD31, and CD34
- *Genetics*
  - Canonical activating *KRAS* mutations involving codons G12, G13, or Q61 in 30–80% of cases

Rhabdomyoma

- Most common primary cardiac tumor in infancy-childhood
- Associated with Tuberous Sclerosis Complex
- Spontaneous regression in some patients, including in some cases treated by mTOR inhibitor
- Arises most commonly in the myocardium of the ventricles

- Frequently multifocal
- Typically pedunculated, projecting into the cardiac chamber
- Light microscopy
  - Circumscribed borders
  - Spider cells: epithelioid cells with characteristic vacuolated cytoplasm and radiating strands of myofibrils
    Vacuolation due to accumulation of glycogen, which can be highlighted by periodic acid-Schiff (PAS) stain, but not after diastase treatment
  - Enlarged hyperchromatic nuclei may be present, simulating malignancy
- *Genetics*
  - Germline mutations involving *TSC1* or *TSC2*, including with low-level mosaicism in some patients

Fibroma

- Presents typically in children or young adults, often associated with arrhythmias or ventricular tachycardia
- Usually solitary and intramural
- Most frequently involves the left ventricle or interventricular septum
- Nonsyndromic in the majority of patients
- Syndromic in a small subset of patients
  - Gorlin syndrome: germline mutation in *PTCH1*
- Light microscopy
  - Bland fibroblastic-to-myofibroblastic spindled cells
  - Stromal background with variably dense collagen
  - Spindled cells infiltrate adjacent cardiac muscle, with entrapped cardiomyocytes often displaying reactive degenerative atypia
  - Calcifications and chronic inflammation often present
- Immunohistochemistry
  - Immunoreactivity for smooth muscle actin (SMA) in some cases
- *Genetics*
  - Somatic inactivating mutation of *PTCH1* reported in some cases
  - No mutations in *CTNNB1* detected

Lipomatous Hypertrophy of the Atrial Septum

- Clinical presentation often asymptomatic, as incidental findings noted on imaging or at autopsy
  - Sometimes, patients present with atrial arrhythmias
- Typically involves the atrial septum
  - Grossly appears as bilobed fatty thickening and may protrude into the atria
  - In contrast, cardiac lipoma is round, circumscribed, and encapsulated

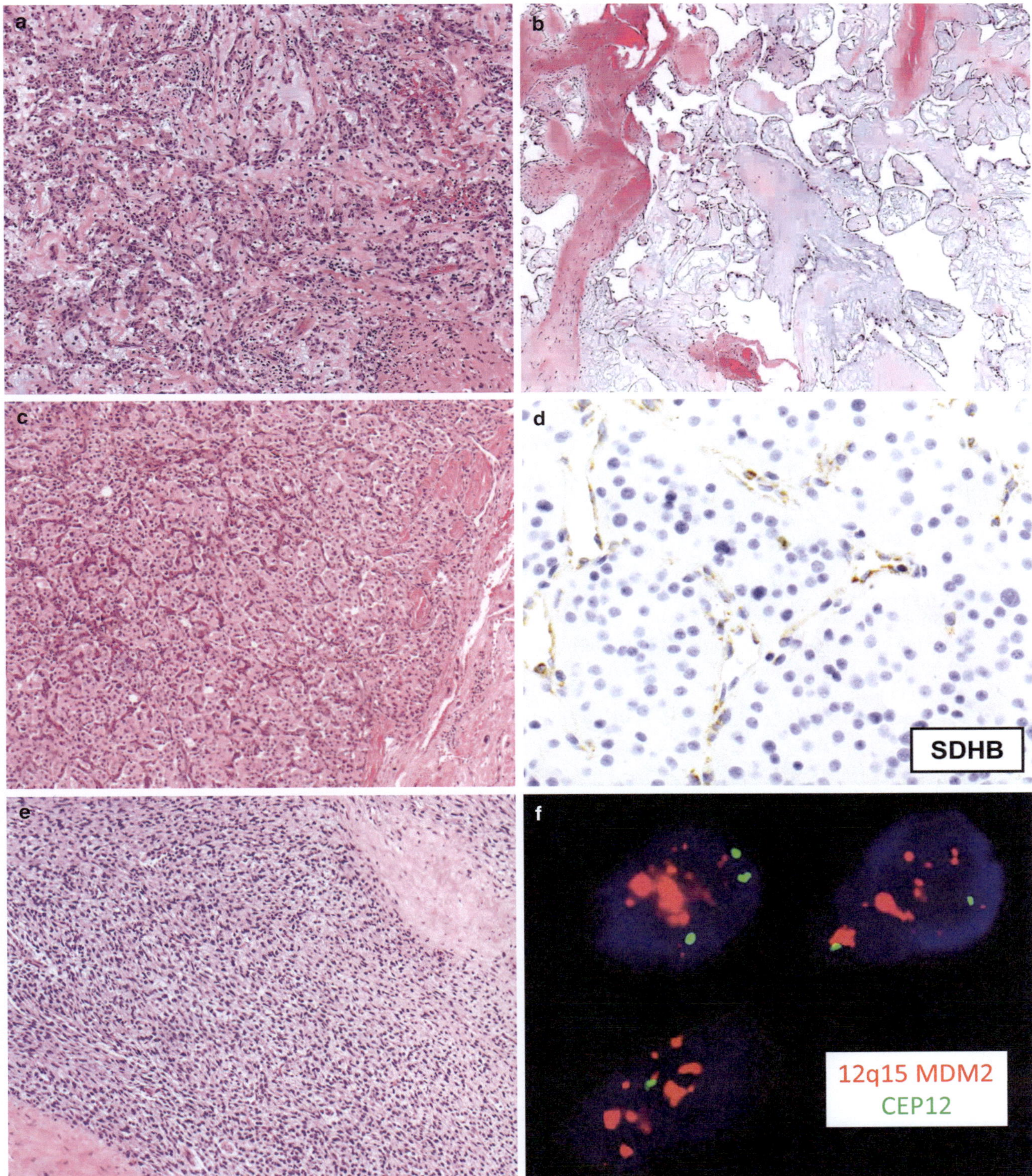

Fig. 20.1 Histopathologic and select immunohistochemical and cytogenetic features of cardiac tumors. (**a**) Microscopic appearance of a myxoma. (**b**) Microscopic appearance of a papillary fibroelastoma. (**c**) Microscopic appearance of a cardiac paraganglioma, with characteristic "zellballen" pattern, showing (**d**) complete loss of SDHB immunohistochemical expression; this tumor was confirmed by subsequent sequencing to harbor a pathogenic mutation involving SDHB. (**e**) Cardiac undifferentiated pleomorphic sarcoma, a subset of which harbors MDM2 gene amplification, as illustrated in this example (**f**) with MDM2 FISH showing amplification of MDM2 (red probes) as compared to the control chromosome 12 centromeric region (green probes)

- Light microscopy
  - Mature adipose tissue with entrapped hypertrophied atrial myocytes, which can show reactive/degenerative nuclear atypia
  - Unencapsulated
    - Unlike cardiac lipoma, which shows round circumscribed border with a thin capsule
- *Genetics*
  - *HMGA2* rearrangement detected in rare cases of both lipomatous hypertrophy of the atrial septum and cardiac lipoma, suggesting genetic overlap
  - No evidence of *MDM2* amplification

Paraganglioma

- No known age or sex predilection
- Functional tumors in >80% of patients, with norepinephrine secretion leading to palpitations, tachyarrhythmias, severe hypertension, or other signs/symptoms
- Involves frequently near the right atrium
  - Considered to arise from cardiac ganglia near the sinoatrial or atrioventricular node
- Light microscopy (Fig. 20.1c)
  - Classic "zellballen" pattern, with sheets of small uniform nests of epithelioid to spindly cells
  - Neuroendocrine cytomorphology, with round-to-ovoid nuclei, fine-speckled chromatin, and generally inconspicuous nucleoli
  - Each tumor cell nest encircled by delicate sustentacular cells
  - Histologically identical to paragangliomas of other sites
- Immunohistochemistry
  - Neuroendocrine cells typically express neuroendocrine markers, including synaptophysin, chromogranin, CD56, and neuron-specific enolase (NSE)
  - Neuroendocrine cells commonly show strong nuclear expression of GATA3
  - Sustentacular cells are positive for the S100 protein
  - Complete loss of succinate dehydrogenase B (SDHB) expression in some tumors, which suggests presence of molecular alterations involving genes encoding one of the components in the succinate dehydrogenase (SDH) complex (Fig. 20.1d)
- *Genetics*
  - Germline heterozygous mutations in one of the succinate dehydrogenase genes (*SDHB*, *SDHC*, and *SDHD*) in ~75% of cases
    - Mosaicism in SDHx mutation described in rare patients

Hemangioma

- Presentation: typically asymptomatic, incidentally identified on imaging or at autopsy
- No known age or sex predilection
- Involves most commonly right atrium or left ventricle, followed by other locations
- Light microscopy
  - Characterized by the presence of prominent vascular spaces, which can be of cavernous type, capillary type, or of both
  - No significant cytologic atypia
  - No endothelial multilayering
  - The vascular channels may infiltrate among adjacent cardiomyocytes
- Immunohistochemistry
  - Positive for vascular markers, including ERG, CD31, CD34, and factor VIII
- Genetics unknown

Cystic Tumor of the Atrioventricular Node

- Presentation: associated with sudden death in some cases, given its proximity to the conduction pathway
  - The most common primary cardiac tumor associated with sudden death
  - Predilection for women
- Located near the atrioventricular node
- Light microscopy
  - Microcystic appearance, characterized by nests and tubules of epithelioid-to-squamoid cells
  - Clear-to-eosinophilic cytoplasm, round-to-ovoid nuclei, and generally inconspicuous nucleoli
- Immunohistochemistry
  - Immunoreactivities for cytokeratin CK7, epithelial membrane antigen (EMA), and CEA in some tumors
- Genetics unknown

Malignant Tumors

Angiosarcoma

- Most common primary malignant cardiac tumor
- Originates most commonly in the right heart, particularly near the right atrioventricular sulcus
- Widespread metastases at the time of presentation common
- Dismal prognosis, with overall survival of <1 year

- Light microscopy
  - Infiltrating sheets of spindled to epithelioid cells
  - Presence of vasoformative features, with irregular anastomosing slit-like vascular spaces
  - Endothelial multilayering may be prominent
  - Cytologically overtly malignant, with nuclear hyperchromasia
  - Conspicuous mitoses and prominent necrosis with hemorrhage
- Immunohistochemistry
  - Endothelial markers CD31 and ERG are most sensitive markers
  - Immunoreactivity for D2-40 and CD34 seen in some cases
  - Immunoreactivity for cytokeratins and epithelial membrane antigen (EMA) seen in some cases, which may lead to diagnostic confusion with carcinomas
- Genetics
  - Molecular alterations of cardiac angiosarcomas appear similar to those of other sites
  - Frequent inactivating mutations in tumor suppressors including *TP53*, *CDKN2A*
  - Activating mutations in genes in the vascular endothelial growth factor (VEGF)/angiogenesis pathway, including *KDR*, *POT1*, *PLCG1*

Undifferentiated Pleomorphic Sarcoma

- Arises most commonly in the left ventricle, followed by other cardiac chambers
- Grossly often polypoid, protruding to the cavity
- Light microscopy
  - Wide histologic spectrum, with nondistinctive appearance
    - Variably cellular fascicles of spindled to pleomorphic cells
    - Some cells can appear myofibroblastic or show prominent herringbone pattern
  - Tumor cells are associated with variably myxoid to collagenous background
  - Variable cytologic atypia, often with hyperchromasia, vesicular nuclei, and prominent nucleoli
  - Necrosis and mitoses generally conspicuous

- A subset of tumors harbors heterologous elements, such as osteosarcomatous, chondrosarcomatous, and rarely rhabdomyosarcomatous components
- Immunohistochemistry
  - MDM2 immunoreactivity in a minority (~30%) of tumors
- Genetics (Fig. 20.1e, f)
  - *MDM2* amplification in ~30% of cardiac undifferentiated pleomorphic sarcomas, suggesting genetic overlap with intimal sarcoma
    - Frequent co-amplification of *CDK4* and *PDGFRA*
    - Occasional deletion of *CDKN2A*
    - Amplifications of *MDM4* or *CDK6* reported in MDM2-negative cases

Intimal Sarcoma

- Malignant mesenchymal tumor that arises in the intima of large arteries
  - Involves the pulmonary arteries in ~70% of cases
  - Involves the aorta in ~30% of cases
- Presentations often include signs/symptoms that suggest chronic pulmonary thromboembolic hypertension
  - Pulmonary artery intimal sarcoma noted in up to 4% of patients who undergo pulmonary thromboendarterectomy for chronic pulmonary hypertension
- Intraluminal involvement
- Light microscopy
  - Wide histologic spectrum, with nondistinctive appearance similar to those of cardiac undifferentiated pleomorphic sarcomas
  - Heterologous elements, such as osteosarcomatous, chondrosarcomatous, and rarely rhabdomyosarcomatous components, in a subset of tumors
- Immunohistochemistry
  - MDM2 immunoreactivity in >70% of tumors
- Genetics
  - Frequent amplification involving *MDM2*, rarely *MDM4* and *CDK6*
    MDM2, *MDM4*, and *CDK6* alterations appear mutually exclusive of one another
  - Frequent co-amplification of *CDK4* and *PDGFRA*
  - Occasional deletion of *CDKN2A*
  - Amplification of *EGFR* in a subset of tumors

Selected Hereditary/Genetic Syndromes

Introduction

- Hereditary cardiomyopathies can present as ventricular hypertrophy, heart failure, or sudden death
- Hereditary deposition and storage diseases often result in either myocyte vacuolization or accumulation of extracellular material
- Hereditary aortic diseases can cause aneurysms or dissections

Cardiomyopathies

Hypertrophic Cardiomyopathy

- One of the more common causes of sudden cardiac death in adolescents and young adults
- Afflicts 1 in 500 people
- Characterized by hypertrophy or thickening of the ventricular walls
- Thickening of the interventricular septum can cause sub-aortic stenosis
- Causes diastolic dysfunction, heart failure with preserved ejection fraction (HFpEF)
- Light microscopy
  - Enlarged myocytes with disarray
  - Interstitial and replacement fibrosis
  - Thickened walls of small intramyocardial arteries
- Genetics
  - Autosomal dominant inheritance
  - Typically results from mutations in genes encoding sarcomeric proteins: *ACTC1*, *MYBPC3*, *MYH7*, *MYL2*, *MYL3*, *TNNI3*, *TNNT2*, and *TPM1*

Arrhythmogenic Cardiomyopathy

- One of the more common causes of sudden cardiac death in adolescents and young adults in Italy
- Previously known as arrhythmogenic right ventricular cardiomyopathy
- Afflicts between 1 in 1,000 to 1 in 5,000 individuals
- Characterized by thinning or fatty replacement of the ventricular walls
- Predilection for involving the right ventricle, but may involve either or both ventricles
- Light microscopy
  - Fibrofatty replacement of the myocardium
- Immunofluorescence
  - Decreased staining for plakoglobin at desmosomes
- Genetics
  - Most often autosomal dominant inheritance

- Typically results from mutations in genes encoding desmosome proteins: *JUP*, *DSP*, *PKP2*, *DSG2*, and *DSC2*
- Homozygous mutation of *JUP* (Naxos disease) or *DSP* result in both heart and skin abnormalities
- Also reported with mutation of genes encoding non-desmosomal proteins: *TMEM43*, *LMNA*, *DES*, *CTNNA3*, *PLN*, *TGFB3*, *TTN*, *SCN5A*, and *CDH2*

Duchenne Muscular Dystrophy (DMD) and Becker Muscular Dystrophy (BMD)

- X-linked dilated cardiomyopathy associated with a systemic muscular dystrophy
- Prevalence 2 per 100,000 (BMD) and 6 per 100,000 (DMD)
- May affect female carriers
- Light microscopy
  - Patchy fibrofatty replacement of the myocardium
- Immunohistochemistry (Fig. 20.2)

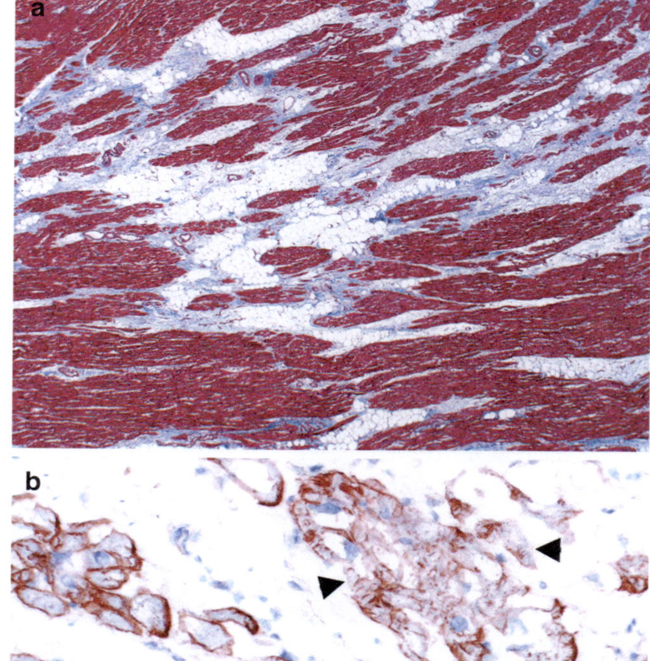

Fig. 20.2 Pathology of Becker cardiomyopathy. (**a**) Trichrome stain showing fibrofatty replacement of the myocardium. (**b**) Immunohistochemical stain for dystrophin showing patchy loss of dystrophin (arrowheads)

- Decreased staining for dystrophin (BMD) or complete loss of dystrophin (DMD)
- Genetics
  - X-linked
  - Mutations in the *DMD* gene cause a reduced amount or shortened dystrophin protein (BMD) or complete loss of dystrophin (DMD)

Desmin-Related Cardiomyopathy

- Variable clinical phenotype ranging from isolated skeletal myopathy to pure cardiac phenotypes including dilated cardiomyopathy, restrictive cardiomyopathy, and cardiac conduction disorders
- Rare disorder
- Light microscopy
  - Eosinophilic deposits within cardiac myocytes
- Immunohistochemistry
  - Staining of deposits within myocytes for desmin
- Electron microscopy
  - Electron-dense granulofilamentous aggregates within cardiac myocytes
- Genetics
  - Usually autosomal dominant
  - Mutations in the *DES* gene or in genes encoding desmin-associated proteins: *CRYAB*, *BAG3*, *FHL1*, *FLNC*, *MYOT*, *PLEC*, *TTN*, and *LDB3*

Nemaline Myopathy

- Congenital skeletal myopathy
- 1 in 50,000 births
- Cardiac involvement is rare and tends to occur in adulthood
- Immunohistochemistry
  - Staining of deposits within myocytes for alpha-actinin
- Electron microscopy
  - Rod-like aggregates of Z-band material comprised of polymers of Z-bands
- Genetics
  - Autosomal recessive mutations in *NEB* in 50% of cases
  - Autosomal dominant mutations in *ACTN1* in 15–25% of cases
  - May be caused by mutations in other genes including: *TPM2*, *TPM3*, *TNNT1*, *CFL2*, *KBTBD13*, *KLHL40*, *KLHL41*, *LMOD3*, and *MYO18B*

Phospholamban Cardiomyopathy

- Cardiomyopathy in patients of Dutch descent
- Responsible for 15% of idiopathic dilated cardiomyopathy in the Netherlands
- Light microscopy
  - Subepicardial fibrofatty replacement of the myocardium
- Immunohistochemistry
  - Staining of deposits within myocytes for phospholamban
- Genetics
  - Autosomal dominant
  - p.Arg14del mutation in *PLN* gene

Deposition and Storage Disorders

Pompe Disease (Glycogen Storage Disease Type IIa)

- Caused by deficiency of alpha-glucosidase
- Severity and age of onset related to degree of enzyme deficiency
- Light microscopy
  - Extensive myocyte vacuolization with eosinophilic material within vacuoles
- Special stains
  - Extensive PAS-positive deposits in myocytes, which are sensitive to diastase digestion
- Electron microscopy
  - Membrane-bound intracellular granular glycogen deposits
- Genetics
  - Autosomal recessive
  - Mutations in the *GAA* gene

Danon Disease (Glycogen Storage Disease Type IIb)

- Present in 1–6% of patients with unexplained left ventricular hypertrophy
- Light microscopy
  - Extensive myocyte vacuolization with eosinophilic material within vacuoles
- Special stains
  - Extensive PAS-positive deposits in myocytes, most of which are sensitive to diastase digestion
- Immunohistochemistry
  - Absence of LAMP2 staining
- Electron microscopy
  - Increased glycogen and lamellar bodies
- Genetics
  - X-linked
  - Caused by mutations in the *LAMP2* gene

Anderson-Fabry Disease

- Deficiency of lysosomal alpha-galactosidase A
- Incidence ~1 in 100,000 male live births
- Can cause cardiac left ventricular hypertrophy, usually noticeable between 20 and 40 years of age
- Light microscopy
  - Extensive myocyte vacuolization
  - Most of the inclusions are extracted during processing into paraffin
- Special stains
  - On frozen section myocyte inclusions stain for Sudan black and PAS and are diastase resistant
- Electron microscopy
  - Myelin bodies, inclusions with concentric or parallel lamellae spaced 4–5.5 nm apart
- Genetics
  - X-linked
  - Caused by mutations in the *GLA* gene

Hereditary Transthyretin Amyloidosis

- Systemic amyloidosis, which often most severely involves the heart
- Mutations render the transthyretin protein more amyloidogenic by destabilizing the tetrameric quaternary structure
- Age of onset and initial presentation (cardiomyopathy versus neuropathy) depend on specific mutation present
- Light microscopy
  - Amyloid deposition, amorphous eosinophilic extracellular material
- Special stains
  - Salmon pink staining of extracellular material on Congo red stain with regular light, and apple-green birefringence upon application of plane-polarized light
- Amyloid typing
  - Positive immunofluorescence for transthyretin
  - Transthyretin peptides by mass spectrometry, which may contain mutant sequences
- Genetics
  - Autosomal dominant
  - Caused by mutations in the *TTR* gene
  - Over 80 known amyloid-causing mutations
  - Val-122-Ile mutation is most common, present in 3.4% of African Americans

Hereditary Aortic Diseases

Marfan Syndrome

- Aortic disease characterized by ascending aortic aneurysms and dissections
- Associated with pectus excavatum, tall stature, arachnodactyly, lens ectopia, and mitral valve prolapse
- Light microscopy (Fig. 20.3)
  - Extensive translamellar mucoid extracellular matrix accumulation
- Special stains
  - Extensive loss of elastic fibers on elastic stain
- Genetics
  - Autosomal dominant
  - Caused by mutations in the *FBN1* gene
  - Almost 2,000 unique mutations identified

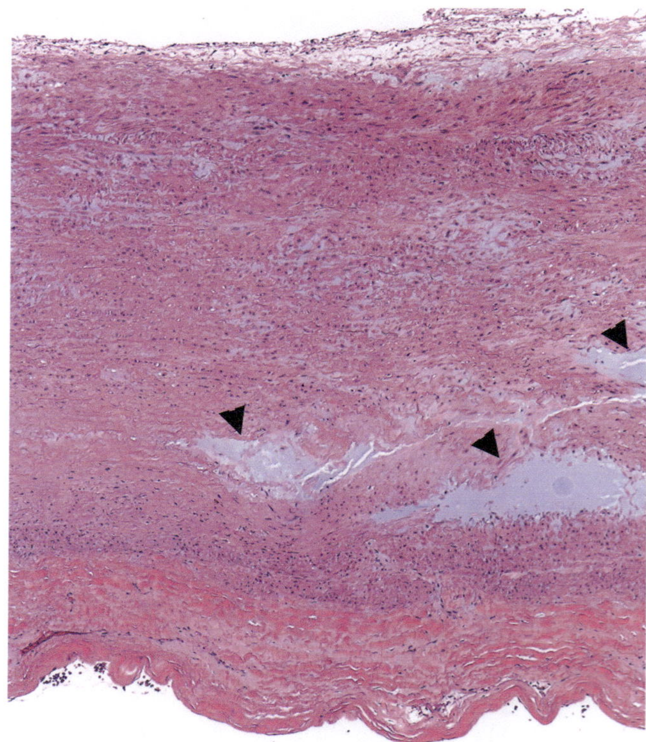

Fig. 20.3 Pathology of Marfan syndrome. Within the aortic wall, there is extensive translamellar mucoid extracellular matrix accumulation (arrowheads)

Vascular Ehlers-Danlos Syndrome

- Develop aneurysms and dissections of large and medium-sized arteries including the aorta
- Associated with thin skin with visible veins, easy bruising, thin lips, thin pinched nose, prominent ears, and visceral rupture
- Light microscopy
  - Mild translamellar mucoid extracellular matrix accumulation
- Genetics
  - Autosomal dominant
  - Caused by mutations in the *COL3A1* gene

Loeys–Dietz Syndrome

- Develop aneurysms and dissections of large and medium-sized arteries including the aorta
- Associated with wide/bifid uvula, hypertelorism, cleft palate, craniosynostosis, easy bruising, and visceral rupture
- Light microscopy
  - Mostly intralamellar mucoid extracellular matrix accumulation
- Special stains
  - Extensive loss of elastic fibers on elastic stain
- Genetics
  - Autosomal dominant
  - Caused by mutations in the *TGFBR1*, *TGFBR2*, *TGFB2*, and *SMAD3*

Suggested Reading

Ainsworth CD, Salehian O, Nair V, Whitlock RP. A bloody mass: rare cardiac tumor as a cause of symptomatic ventricular arrhythmias. Circulation. 2012;126(15):1923–31.

Bode-Lesniewska B, Zhao J, Speel EJ, Biraima AM, Turina M, Komminoth P, et al. Gains of 12q13-14 and overexpression of mdm2 are frequent findings in intimal sarcomas of the pulmonary artery. Virchows Arch. 2001;438(1):57–65.

Bois MC, Bois JP, Anavekar NS, Oliveira AM, Maleszewski JJ. Benign lipomatous masses of the heart: a comprehensive series of 47 cases with cytogenetic evaluation. Human pathology. 2014;45(9):1859–65.

Bois MC, Milosevic D, Kipp BR, Maleszewski JJ. KRAS mutations in papillary fibroelastomas: a study of 50 cases with etiologic and diagnostic implications. Am J Surg Pathol. 2020;44(5):626–32.

Burke AP, Rosado-de-Christenson M, Templeton PA, Virmani R. Cardiac fibroma: clinicopathologic correlates and surgical treatment. J Thoracic Cardiovasc Surg. 1994;108(5):862–70.

Calvete O, Garcia-Pavia P, Dominguez F, Mosteiro L, Perez-Cabornero L, Cantalapiedra D, et al. POT1 and damage response malfunction trigger acquisition of somatic activating mutations in the VEGF pathway in cardiac angiosarcomas. J Am Heart Assoc. 2019;8(18):e012875.

Cardot-Bauters C, Carnaille B, Aubert S, Crepin M, Boury S, Burnichon N, et al. A full phenotype of paraganglioma linked to a germline SDHB mosaic mutation. J Clin Endocrinol Metab. 2019;104(8):3362–6.

Chang JS, Chiou PY, Yao SH, Chou IC, Lin CY. Regression of neonatal cardiac rhabdomyoma in two months through low-dose everolimus therapy: a report of three cases. Pediatr Cardiol. 2017;38(7):1478–84.

Corrado D, Basso C, Judge DP. Arrhythmogenic cardiomyopathy. Circ Res. 2017;121(7):784–802.

Halushka MK, Angelini A, Bartoloni G, Basso C, Batoroeva L, Bruneval P, et al. Consensus statement on surgical pathology of the aorta from the Society for Cardiovascular Pathology and the Association for European Cardiovascular Pathology: II. Noninflammatory degenerative diseases - nomenclature and diagnostic criteria. Cardiovasc Pathol. 2016;25(3):247–57.

Jimbo N, Komatsu M, Itoh T, Hirose T. MDM2 dual-color in situ hybridization (DISH) aids the diagnosis of intimal sarcomas. Cardiovasc Pathol. 2019;43:107142.

Kirschner LS, Carney JA, Pack SD, Taymans SE, Giatzakis C, Cho YS, et al. Mutations of the gene encoding the protein kinase A type I-alpha regulatory subunit in patients with the Carney complex. Nat Genet. 2000;26(1):89–92.

Koelsche C, Benhamida JK, Kommoss FKF, Stichel D, Jones DTW, Pfister SM, et al. Intimal sarcomas and undifferentiated cardiac sarcomas carry mutually exclusive MDM2, MDM4, and CDK6 amplifications and share a common DNA methylation signature. Mod Pathol. 2021;34(12):2122–9.

Law KB, Feng T, Nair V, Cusimano RJ, Butany J. Cystic tumor of the atrioventricular node: rare antemortem diagnosis. Cardiovasc Pathol. 2012;21(2):120–7.

Leduc C, Jenkins SM, Sukov WR, Rustin JG, Maleszewski JJ. Cardiac angiosarcoma: histopathologic, immunohistochemical, and cytogenetic analysis of 10 cases. Human Pathol. 2017;60:199–207.

Li W, Teng P, Xu H, Ma L, Ni Y. Cardiac hemangioma: a comprehensive analysis of 200 cases. Ann Thorac Surg. 2015;99(6):2246–52.

Maleszewski JJ, Tavora F, Burke AP. Do "intimal" sarcomas of the heart exist? Am J Surg Pathol. 2014;38(8):1158–9.

Manzanilla-Romero HH, Weis D, Schnaiter S, Rudnik-Schoneborn S. Low-level mosaicism in tuberous sclerosis complex in four unrelated patients: comparison of clinical characteristics and diagnostic pathways. Am J Med Genet A. 2021;185(12):3851–8.

Martucci VL, Emaminia A, del Rivero J, Lechan RM, Magoon BT, Galia A, et al. Succinate dehydrogenase gene mutations in cardiac paragangliomas. Am J Cardiol. 2015;115(12):1753–9.

Mavrogeni S, Markousis-Mavrogenis G, Papavasiliou A, Kolovou G. Cardiac involvement in Duchenne and Becker muscular dystrophy. World J Cardiol. 2015;7(7):410–4.

Miller DV. Cardiac tumors. Surg Pathol Clin. 2012;5(2):453–83.

Miller DV, Wang H, Wang H, Fealey ME, Tazelaar HD. Beta-catenin mutations do not contribute to cardiac fibroma pathogenesis. Pediatr Dev Pathol. 2008;11(4):291–4.

Neuville A, Collin F, Bruneval P, Parrens M, Thivolet F, Gomez-Brouchet A, et al. Intimal sarcoma is the most frequent primary cardiac sarcoma: clinicopathologic and molecular retrospective analysis of 100 primary cardiac sarcomas. Am J Surg Pathol. 2014;38(4):461–9.

Pinard A, Jones GT, Milewicz DM. Genetics of thoracic and abdominal aortic diseases. Circ Res. 2019;124(4):588–606.

Ruberg FL, Grogan M, Hanna M, Kelly JW, Maurer MS. Transthyretin amyloid cardiomyopathy: JACC State-of-the-Art Review. J Am Coll Cardiol. 2019;73(22):2872–91.

Scalise M, Torella M, Marino F, Ravo M, Giurato G, Vicinanza C, et al. Atrial myxomas arise from multipotent cardiac stem cells. Eur Heart J. 2020;41(45):4332–45.

Scanlan D, Radio SJ, Nelson M, Zhou M, Streblow R, Prasad V, et al. Loss of the PTCH1 gene locus in cardiac fibroma. Cardiovasc Pathol. 2008;17(2):93–7.

Smith M, Sperling D. Novel 23-base-pair duplication mutation in TSC1 exon 15 in an infant presenting with cardiac rhabdomyomas. Am J Med Genet. 1999;84(4):346–9.

Stafford F, Thomson K, Butters A, Ingles J. Hypertrophic cardiomyopathy: genetic testing and risk stratification. Curr Cardiol Rep. 2021;23(2):9.

Tamborini E, Casieri P, Miselli F, Orsenigo M, Negri T, Piacenza C, et al. Analysis of potential receptor tyrosine kinase targets in intimal and mural sarcomas. J Pathol. 2007;212(2):227–35.

Teng F, Yang S, Chen D, Fang W, Shang J, Dong S, et al. Cardiac fibroma: a clinicopathologic study of a series of 12 cases. Cardiovasc Pathol. 2021;56:107381.

Tiberio D, Franz DN, Phillips JR. Regression of a cardiac rhabdomyoma in a patient receiving everolimus. Pediatrics. 2011;127(5):e1335–7.

Veinot JP, Nair V. Lysosomal storage disorders affecting the heart: a review. Cardiovasc Pathol. 2020;48:107217.

Wittersheim M, Heydt C, Hoffmann F, Buttner R. KRAS mutation in papillary fibroelastoma: a true cardiac neoplasm? J Pathol Clin Res. 2017;3(2):100–4.

Wold LE, Lie JT. Cardiac myxomas: a clinicopathologic profile. Am J Pathol. 1980;101(1):219–40.

Zhang PJ, Brooks JS, Goldblum JR, Yoder B, Seethala R, Pawel B, et al. Primary cardiac sarcomas: a clinicopathologic analysis of a series with follow-up information in 17 patients and emphasis on long-term survival. Human Pathol. 2008;39(9):1385–95.

Zhang Q, Wang T, Wang D, Liu J, Yu W, Liu X, et al. Somatic copy number losses on chromosome 9q21.33q22.33 encompassing the PTCH1 loci associated with cardiac fibroma. Cancer Genet. 2015;208(12):615–20.

Molecular Pathology of Endocrine Tumors

Lori A. Erickson

Contents

Thyroid Tumors

Papillary Thyroid Carcinoma

Definition

- Malignant tumor of follicular-cell derivation characterized by distinct cytologic features, including irregular nuclear shaped, prominent nuclear grooves, nuclear clearing, and cytoplasmic invagination into the nuclei

Clinical Features

- Most common endocrine malignancy in children and adults
- Incidence rate increasing worldwide with screening and diagnostic practices
- Children and young adults up to fourth and fifth decades are more commonly affected
- Females affected more commonly than males (3:1)
- Previous exposure to radiation increases incidence
- Often presents as a painless nodule with or without cervical lymphadenopathy
- More extensive tumors may be associated with tracheal and esophageal compression with dysphagia or stridor
- Rare cases present with distant metastasis
- Radiographically usually appears as a cold, solid, hypoechoic nodule, often with calcifications and may be cystic with a solid nodule

Pathologic Features

- Variable size, often 2–3 cm, but papillary microcarcinomas and very large tumors can occur
- Grossly white in color and firm, often appears infiltrative, may resemble a scar or may be cystic
- Conventional subtype has prominent papillae with fibrovascular cores (Fig. 21.1)
- Variable papillary architecture with neoplastic papillae containing thin fibrovascular cores covered by tumor cells with elongated nuclei, nuclear clearing, nuclear grooves, and may have cytoplasmic invagination into nuclei (Fig. 21.2)
- Immunohistochemistry shows positivity for TTF1, PAX8, thyroglobulin, and keratins and negativity for chromogranin-A, synaptophysin, calcitonin, and parathyroid hormone

L. A. Erickson (✉)
Department of Laboratory Medicine and Pathology, Mayo Clinic,
Rochester, MN, USA
e-mail: erickson.lori@mayo.edu

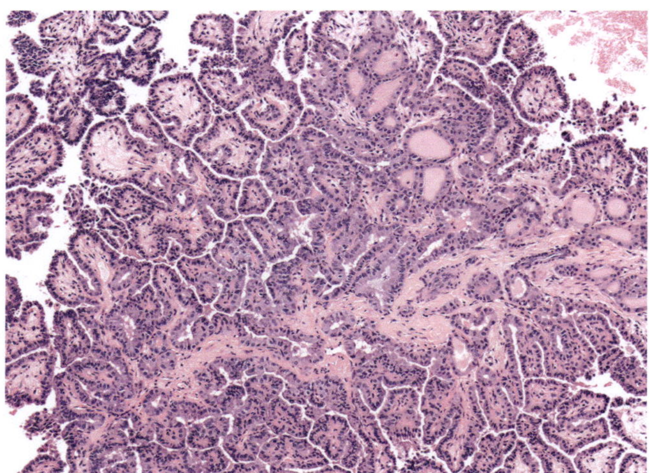

Fig. 21.1 Papillary thyroid carcinoma with a classic papillary architecture consisting of papillae lined by fibrovascular cores

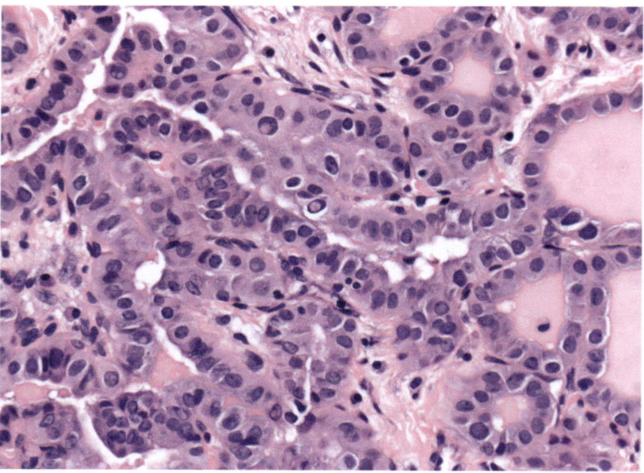

Fig. 21.2 Papillary thyroid carcinoma cells with elongated nuclei, nuclear clearing, nuclear grooves, and intranuclear holes

- HBME 1, galectin 3, and keratin 19 immunostains may help differentiate papillary thyroid carcinomas (PTC) from benign lesions but are not completely specific
- Numerous histologic subtypes (Table 21.1; Figs. 21.3, 21.4, 21.5, 21.6, 21.7, 21.8, 21.9, and 21.10)

Genetic Features

- Low somatic mutation burden
- Most genetic alterations are somatic mutations in *BRAF* and *RAS* and *RET* fusions
- *BRAF* mutations occur in 30–60% of cases, with *BRAF* V600E accounting for 98% of *BRAF* mutations and is usually seen in PTC with a papillary architecture (Table 21.2)

Table 21.1 Histologic subtypes of papillary thyroid carcinoma (PTC)

Subtypes of PTC	Features
Classic/ conventional	Often infiltrative, may have cystic features; classic papillary architecture with papillae containing thin fibrovascular cores lined by tumor cells with elongated nuclei, nuclear clearing, nuclear grooves, and may have cytoplasmic invagination into nuclei
Encapsulated	Classic type of PTC with a thick fibrous capsule which may or may not be infiltrated by tumor cells
Infiltrative follicular variant	Rare type of PTC with an exclusive or almost exclusive follicular growth pattern and lacks a capsule, behaves like a classic PTC, may have *BRAF* V600E mutation or RET translocations, *NTREK* or *ALK* fusions
Encapsulated follicular variant	Shows an exclusive or almost exclusive follicular growth pattern; encapsulated, often with capsular or vascular invasion, most common type of follicular variant of PTC, *RAS*-like tumor
Warthin-like	Composed of papillae lined by oncocytic cells with cytologic features of PTC, the papillary cores have prominent plasma cells and lymphocytes; overall pattern appears similar to tumor of salivary gland
Diffuse sclerosing	Nests and papillary formations, often with squamous metaplasia diffusely permeating the thyroid, with extensive lymphatic extension, sclerosis, and prominent psammoma bodies generally in association with chronic lymphocytic thyroiditis, association with radiation fall-out
Oncocytic	Oncocytic cells with nuclear features of PTC forming well-developed papillae and rarely follicles, rare in pure form, but focal oncocytic changes can be seen in various PTCs
Solid/trabecular	Solid, trabecular, or nested growth in >50% of the tumor, cytologic features of PTC, must differentiate from poorly differentiated thyroid carcinoma which may have tumor necrosis or high mitotic activity, features not seen in solid/trabecular PTC, uncommon in adults but common in children particularly following radiation exposure
Tall cell	Criteria vary over time, but they are generally composed of cells three times as tall as they are wide, and the tumor is composed of at least 30% tall cells, abundant eosinophilic cytoplasm with prominent cell membranes and intranuclear inclusions, occur in older age and are often large with extrathyroidal extension, angiolymphatic invasion, and may have regional and distant metastases
Hobnail	Cells with enlarged nuclei that bulge from the apical surface with a complex papillary and micropapillary architecture and nuclear features of PTC, often have high-grade features (mitotic count >5/10 high-power fields and/or tumor necrosis). Hobnail PTC may arise with other aggressive types of thyroid carcinoma including poorly differentiated and anaplastic. Hobnail PTC is a rare aggressive type of PTC often associated with rapid progression and death in 5 years, usually affecting older individuals, with female predominance. Tumors without mitotic activity or extrathyroidal extension often occurring in younger patients referred to as hobnail-like PTC and have a better prognosis

Table 21.1 (continued)

Subtypes of PTC	Features
Columnar	Columnar cells with prominent nuclear pseudostratification, pale eosinophilic to clear cytoplasm, may have subnuclear vacuoles similar in appearance to secretory endometrium, generally lack nuclear pseudoinclusions, prognosis is related to whether the tumor is encapsulated or infiltrative, may be positive for CDX2
Clear cell	Clear cell change can be seen in many types of thyroid carcinoma and can be mistaken for tumor secondarily involving the thyroid
PTC with fibromatosis/fasciitis-like stroma	Rare PTC with abundant cellular stroma resembling fibromatosis or nodular fasciitis
Spindle cell	PTC with predominance (>50%) of spindle cells with nuclear features of PTC

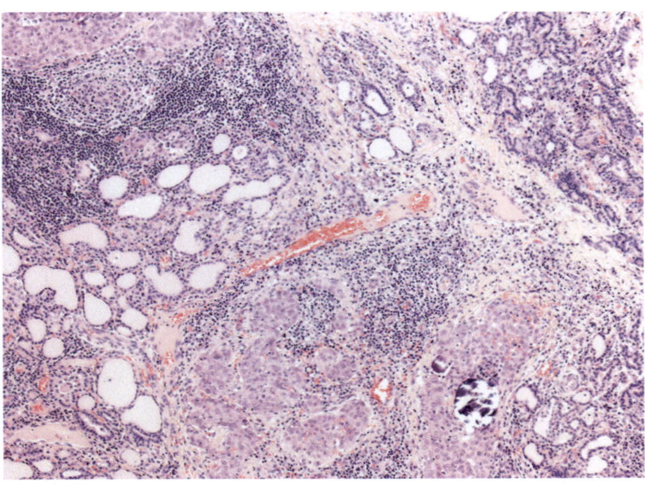

Fig. 21.5 Diffuse sclerosing papillary thyroid carcinoma with nests of cells, squamous metaplasia, and lymphatic extension, in a background of chronic lymphocytic thyroiditis

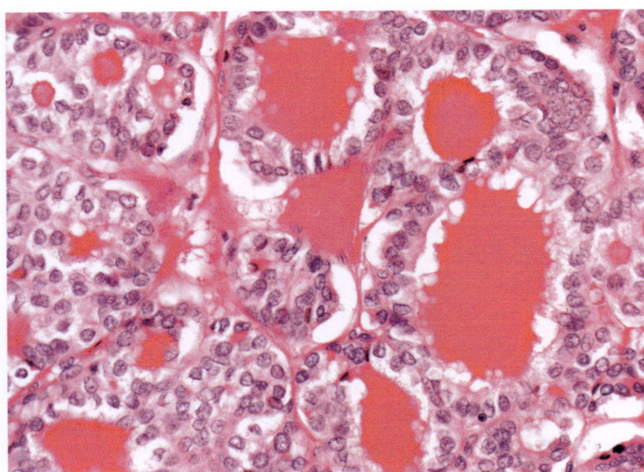

Fig. 21.3 Encapsulated follicular variant of papillary thyroid carcinoma shows an exclusive or almost exclusive follicular growth pattern with capsular and or vascular invasion

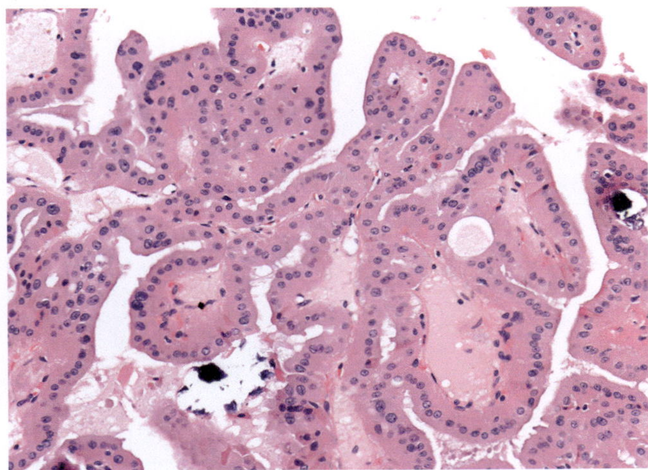

Fig. 21.6 Oncocytic variant of papillary thyroid carcinoma with oncocytic cells with papillary thyroid carcinoma nuclear features forming papillae

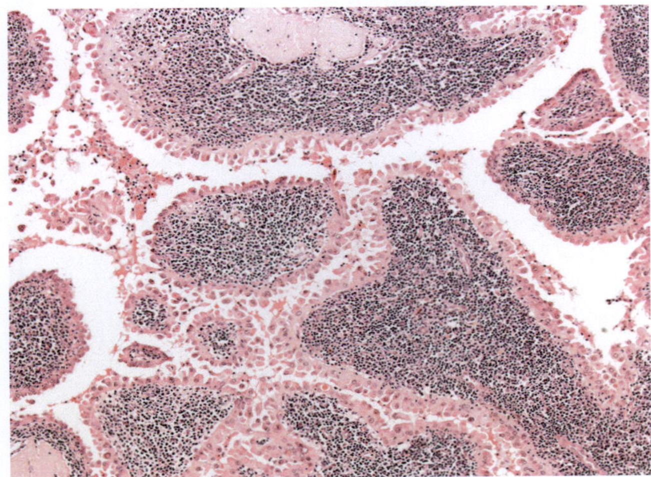

Fig. 21.4 Warthin-like variant of papillary thyroid carcinoma with oncocytic cells with papillary thyroid carcinoma cytologic features lining course with prominent plasma cells and lymphocytes

- *BRAF* mutation: *BRAF* maps to chromosome 7q34 and is a member of the RAF family
- *BRAF* V600E mutations activate the MAPK pathway
  - *BRAF* mutation is probably an early event in the pathogenesis
  - The frequency of *BRAF* mutations varies
  - Frequent in tall cell carcinomas, less frequent in conventional and columnar cell carcinoma, and is usually seen in the encapsulated follicular variant
  - Anaplastic carcinomas also have *BRAF* mutations that appear to be preserved when dedifferentiating from a *BRAF*-mutated PTC
  - V600E is the most common *BRAF* mutation (present in 95%)

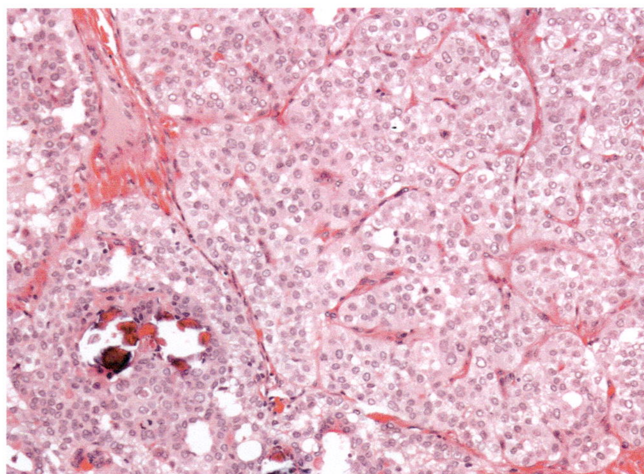

Fig. 21.7 Solid/trabecular subtype of papillary thyroid carcinoma with cytologic features of papillary thyroid carcinoma and solid or nested growth in greater than 50% of the tumor

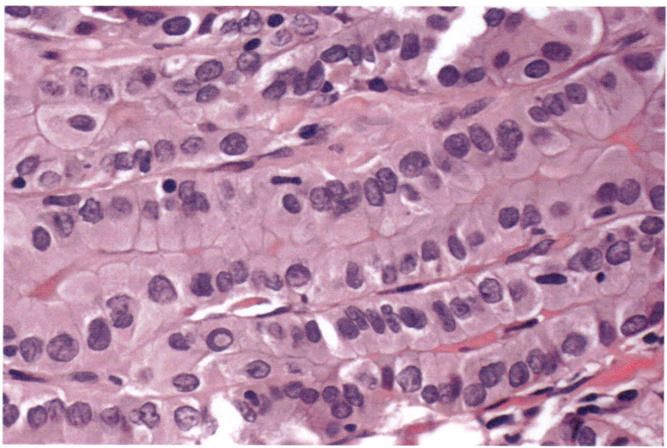

Fig. 21.8 Tall cell variant of papillary thyroid carcinoma with cells three times as tall as they are wide

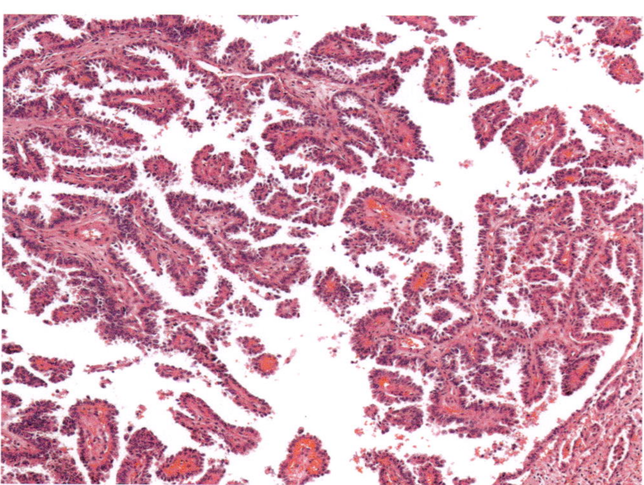

Fig. 21.9 Hobnail variant of papillary thyroid carcinoma with complex papillary and micropapillary architecture with cells with enlarged nuclei bulging from the apical surface of the papillae

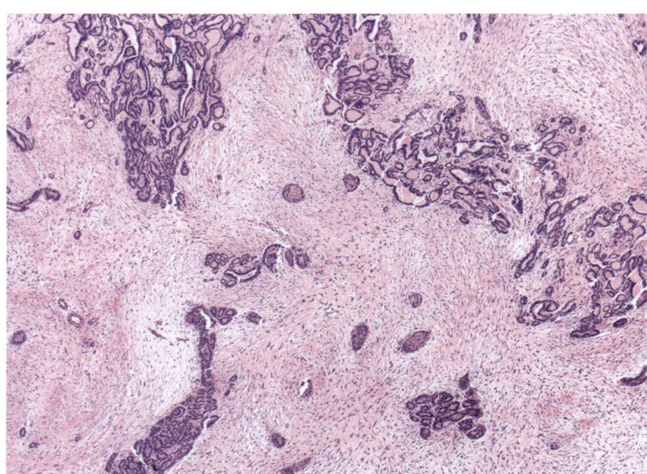

Fig. 21.10 Papillary thyroid carcinoma with fibromatosis/fasciitis-like stroma with abundant cellular stroma resembling fibromatosis or fasciitis

Table 21.2 BRAF mutations in papillary thyroid carcinomas (PTC)

Mutation	Distribution	Histologic variants of PTC
BRAF V600E	30–60%	Conventional PTC, tall cell variant
BRAF K601E	~3%	Encapsulated follicular variant of PTC
AKAP9/BRAF rearrangement	Rare	Conventional PTC

- Other *BRAF* mutations rarely found in thyroid carcinoma include K601E point mutation, small in-frame insertions or deletions surrounding codon 600, and *AKAP9–BRAF* rearrangement
- *RET/PTC* rearrangement is common in PTC (Table 21.3)
- *RET* protooncogene on chromosome 10q11.2
  - *RET* encodes a cell membrane receptor tyrosine kinase and consists of an extracellular domain, a transmembrane domain, and an intracellular tyrosine kinase domain
  - Wild-type *RET* is expressed at high levels in the calcitonin-producing C-cells but not in thyroid follicular cells
  - Chromosomal rearrangement leads to the activation of *RET/PTC*
  - The most common rearrangements include:
    - *RET/PTC1* formed by fusion with the H4 gene
    - *RET/PTC3* is formed by the fusion with the *NCOA4* (*ELE1*) gene
  - *RET/PTC1* and *RET/PTC3* are intrachromosomal paracentric inversions
  - In *RET/PTC2*, the genes are on chromosomes 10 and 17, resulting in *RET* fusion with the regulatory R1 alpha of the cAMP-dependent protein kinase 4
  - *RET/PTC* shows geographic variation: 11–43% in the United States; 40% in Canada; 29–35% in Italy; 3% in Saudi Arabia; and 85% in Australia

Table 21.3 RET/PTC rearrangements in papillary thyroid carcinoma

RET/PTC variant	Fused genes	Rearrangement	Prevalence (%)
RET/PTC1	H4	inv(10) (q11.2;q21)	10–43
RET/PTC2	R1alpha	t(10;17) (q11.2;q23)	5–10
RET/PTC3 and *RET/PTC4*	*NCOA4* (RF6,ELE1)	inv(10)(q11.2)	20–30

- – *RET/PTC* rearrangements in 10–43% of PTC
  - – *RET/PTC1* accounts for 60–70% of positive cases, *RET/PTC3* for 20–30% of cases, and *RET/PTC2* for <10% of cases
- *TRK* rearrangements
  - – *NTRK1* gene located on chromosome 1q22
  - – *TRK* rearrangement involves another tyrosine kinase gene *NTRK1* in PTC
  - – Encodes one of the nerve growth factor receptor genes expressed in neurons
  - – Involved in cell growth, differentiation, and survival
  - – In the thyroid, the gene is activated through chromosomal rearrangement with juxtaposition of the intracellular tyrosine kinase domain of *NTRK1* to the 5′ terminal sequence of different genes
  - – Two of the fusion partner genes include tropomyosin (TPM) and the translocated promoter region gene, both on the q arm of chromosome 1
  - – A third fusion partner, the *TF6* gene, is on chromosome 3, so there is a t(1;3) translocation
  - – *TRK* rearrangement promotes neoplastic transformation of thyroid cells
  - – *TRK* rearrangements are present in 10–15% of PTC
- *RAS* mutations
  - – *RAS* genes and their chromosomal locations include *NRAS* (1p13), *KRAS* (12p12), and *HRAS* (11p15)
  - – *RAS* genes all encode distinct 21 kDa proteins
  - – Somatic mutations in codons 12/13 and 61 of the three *RAS* genes are found in 40–50% of follicular carcinomas
  - – *RAS* point mutations are present in 10–20% of PTC
  - – *RAS* mutation in PTC is usually associated with the follicular variant pattern
  - – PTC with a follicular pattern often shows *RAS* mutation or rare *BRAF* K601E mutation
- *TERT* promoter mutations in 10%
- Infiltrative follicular variant of PTC may have *BRAF* V600E mutation or *RET* translocations or *NTREK* or *ALK* fusions
- Oncocytic PTC may have *BRAF* mutation, and somatic germline *GRIM19* mutations and *RET* rearrangements have been reported

Invasive Encapsulated Follicular Variant Papillary Carcinoma

Definition

- Encapsulated follicular variant of papillary thyroid carcinoma (EFVPTC) is a malignant, encapsulated, well-differentiated follicular-cell derived neoplasm that has an exclusive or almost exclusive follicular architecture, nuclear features of papillary thyroid carcinoma, and invasive growth

Clinical Features

- Occurs predominantly in adults generally in the fourth to fifth decades with a female predominance
- Accounts for approximately 15% of PTC in children
- Usually presents with a painless mass
- Like other RAS-like follicular pattern lesions (and unlike classic PTC), lymph node involvement is usually absent
- Majority occur sporadically, but can occur in PTEN hamartoma tumor syndrome, Carney complex, DICER syndrome
- Minimally invasive EFVPTC are quite indolent tumors when metastases are absent but are more aggressive tumors with vascular invasion (angioinvasive), and even more aggressive with extensive vascular invasion (>4 foci)
- Metastatic pattern is similar to follicular thyroid carcinoma (does not spread to lymph nodes)

Pathologic Features

- Encapsulated, solid tumors
- May be minimally invasive with capsular invasion only, encapsulated angioinvasive (with or without capsular invasion), or widely invasive (rare)
- Rare widely invasive EFVPTC show gross invasion through the thyroid parenchyma
- Follicular architecture (<1% true papillae) and capsular and/or vascular invasion, but nuclear features of PTC
- Nuclear features may be more subtle than in classic PTC, often foci around the periphery of the tumor with more characteristic nuclear features of PTC ("sprinkling sign")
- May have foci of solid or trabecular growth, but in these cases, poorly differentiated thyroid carcinoma must be excluded
- Immunopositive for keratin, TTF1, thyroglobulin, and PAX8

Genetic Features

- *RAS* mutations are the most common alteration (up to 60%)
- *PAX8-PPARG* rearrangements may occur
- *BRAF* K601E mutation may occur, but *BRAF* V600E mutation is extremely rare
- Other uncommon abnormalities involve *TERT* promoter, *EIF1AX*, and *TSHR*

Follicular Thyroid Carcinoma

Definition

- A malignant tumor derived from the thyroid follicle cells that shows invasion and has a follicular architecture and does not show the nuclear feature of PTC or high-grade features

Clinical Features

- Accounts for 5–15% of thyroid carcinomas
- Occurs predominantly in adults, most commonly in the fifth decade
- Rare in children
- Presents with a painless thyroid mass or detected on imaging
- Larger tumors may be associated with clinical symptoms of hoarseness, dysphagia, or stridor
- Radiologically often presents as a cold nodule
- Not associated with regional lymph node involvement
- Uncommonly (5–10%) may have distant metastases at diagnosis
- Majority are sporadic, but up to 10% may be hereditary

Pathologic Features

- Usually grossly encapsulated, solid tumors
- Follicular cells with a follicular architecture: microfollicular, macrofollicular, or normofollicular
- May have areas of solid or trabecular growth, but these tumors should be differentiated from poorly differentiated thyroid carcinomas which have a mitotic count of ≥3 per 10 high-power fields (HPF) and/or necrosis
- Lacks nuclear features of PTC
- Occasional rare features can be seen: clear cells, signet ring cells, spindle cells, or glomeruloid patterns
- Diagnosis requires invasive growth: capsular and/or vascular

- May be minimally invasive with capsular invasion only, may be encapsulated with vascular invasion (angioinvasive) with or without capsular invasion, or may be widely invasive (with gross invasion through the thyroid parenchyma, capsule may be only focally present, may also show vascular invasion) (Figs. 21.11 and 21.12)
- Tumors with limited vascular invasion (<4 foci) have a better prognosis than those with extensive vascular invasion (≥4 foci of invasion)
- Widely invasive follicular carcinomas show gross invasion through the thyroid parenchyma often in a multinodular pattern
- Follicular thyroid carcinomas with a mitotic count of ≥5 per 10 high-power fields (HPF) or necrosis are characterized as follicular thyroid carcinomas with high-grade features

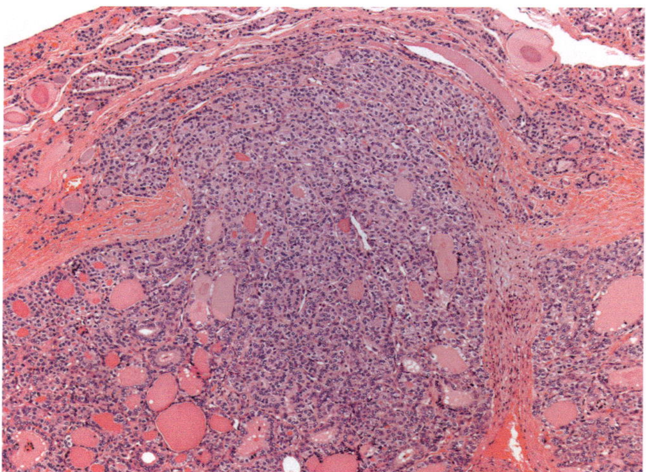

Fig. 21.11 Follicular thyroid carcinoma with capsular invasion

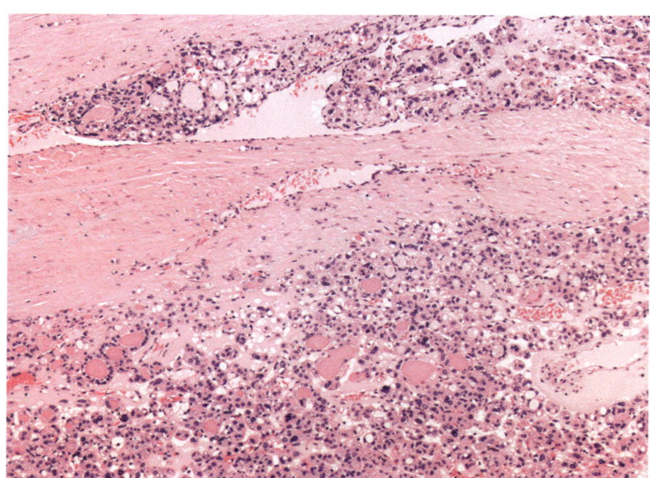

Fig. 21.12 Follicular thyroid carcinoma with vascular invasion

- Immunopositive for keratin, PAX8, thyroglobulin, and TTF1
- Prognosis depends on the extent of invasion and stage, with minimally invasive tumors having a very good prognosis, while widely invasive tumors may have a 50% mortality

Genetic Features

- *RAS* mutations in 40–50% of conventional follicular carcinomas (also in 20–40% of follicular adenomas)
  - Of *RAS* mutations, *NRAS* on 1p13 (often codon 61) is most common followed by *HRAS* on 11p15 (often codon 61) and *KRAS* on 12p12 (often codon 12/13)
  - *RAS* genes all encode distinct 21 kDa proteins
  - Somatic mutations in codons 12/13 and 61 of the three *RAS* genes are found in 40–50% of follicular carcinomas (and *RAS* point mutations are present in 10–20% of PTC usually with a follicular growth patterns)
- PAX8–PPARG
  - *PAX8* (2q13) encodes a paired domain transcription factor with an important role in thyroid development and follicular-cell differentiation
  - The protein binds to the promoters of thyroglobulin, thyroperoxidase, and sodium iodide symporter genes and regulates their thyroid-specific expression
  - There are several *PAX8* splice variants
  - Peroxidase proliferative-activated receptor (PPAR) is part of the rearrangement
  - PPARs are nuclear hormone receptors and control some of the genes involved in lipid metabolism
  - *PAX8/PPARG* fusion protein contains the partial homeobox domains of *PAX8* fused with the DNA binding ligand of *PPARG*
  - *PAX8/PPARG* is present in 10–35% of follicular carcinomas (2–10% of follicular adenomas) and has a lower prevalence in oncocytic cell carcinoma
  - Radiation-associated follicular carcinomas have a higher prevalence of *PAX8/PPARG*
  - *RAS* mutations and *PAX8-PPARG* rearrangements are generally mutually exclusive
- *TERT* promoter mutations in 15% (may occur with *RAS* mutations), and may be associated with distant metastases
- *EIF1AX* mutations in 5% (may occur with *RAS* mutations)
- *PTEN* mutations in 10%
- *PIK3CA* abnormalities in up to 30%
- Other abnormalities: somatic *DICER1* mutation, *CREB3L2-PPARG* gene fusion, somatic *NF1* mutation
- Follicular carcinomas usually lack *TP53* mutation (unlike oncocytic carcinomas)

- Can occur in *PTEN* hamartoma tumor syndrome (including Cowden syndrome), Werner syndrome, *DICER1* syndrome, and Carney complex
  - Approximately 25% of Cowden syndrome is associated with adenomatous nodules and follicular adenomas, as well as follicular thyroid carcinoma (which may occur in childhood)
  - Werner syndrome (*WRN*) is an autosomal recessive disease that can be associated with follicular thyroid carcinoma
  - *DICER1* syndrome is an autosomal-dominant disorder with a marked increased risk of thyroid carcinoma which can present childhood
  - Carney complex is an autosomal-dominant disorder (*PRKAR1A*) that includes thyroid follicular adenomas and carcinomas

Oncocytic Thyroid Carcinoma

Definition

- Oncocytic (Hürthle-cell) carcinoma is an invasive malignant thyroid follicular-cell neoplasm composed of at least 75% oncocytic cells that lacks nuclear features of PTC and lacks high-grade features

Clinical Features

- Approximately 3–4% of thyroid carcinomas
- Usually affects adults in their fifth to sixth decade, more common in women than men (2:1)
- Solitary, painless thyroid nodule, can be found incidentally
- Can be symptomatic, particularly with widely invasive tumors
- Unlike follicular carcinomas, which spread exclusively through the vascular system, oncocytic carcinomas can metastasize hematogenously and can metastasize to lymph nodes
- Usually sporadic, but can occur in *PTEN* mutation syndrome and Carney complex

Pathologic Features

- Solitary, usually encapsulated nodules, often a thick irregular capsule, often 3–4 cm or larger
- 75% of the tumor composed of oncocytic cells
- Lacks nuclear features of PTC
- Diagnosis requires invasive growth: capsular and/or vascular
- May be minimally invasive with capsular invasion only, may be encapsulated with vascular invasion (angioinva-

sive) with or without capsular invasion, or may be widely invasive (with gross invasion through the thyroid parenchyma, capsule may be only focally present, may also show vascular invasion) (Figs. 21.13 and 21.14)

- Tumors with limited vascular invasion (<4 foci) have a better prognosis than those with extensive vascular invasion (≥4 foci of invasion)
- Widely invasive oncocytic carcinomas show gross invasion through the thyroid parenchyma often in a multinodular pattern
- Oncocytic thyroid carcinomas with a mitotic count of ≥5 per 10 HPF or necrosis are characterized as oncocytic thyroid carcinomas with high-grade features
- May have areas of solid or trabecular growth; however, these tumors should be differentiated from poorly differentiated thyroid carcinomas which have a mitotic count of ≥3 per 10 HPF and/or necrosis
- Immunopositive for keratin, PAX8, thyroglobulin, and TTF1

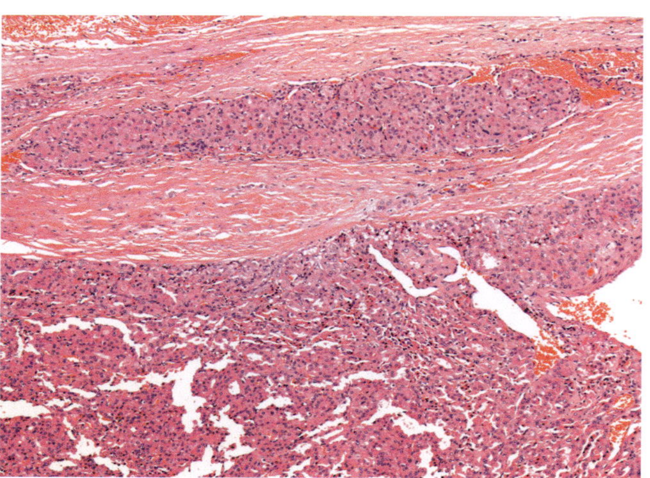

Fig. 21.13 Oncocytic thyroid carcinoma with vascular invasion

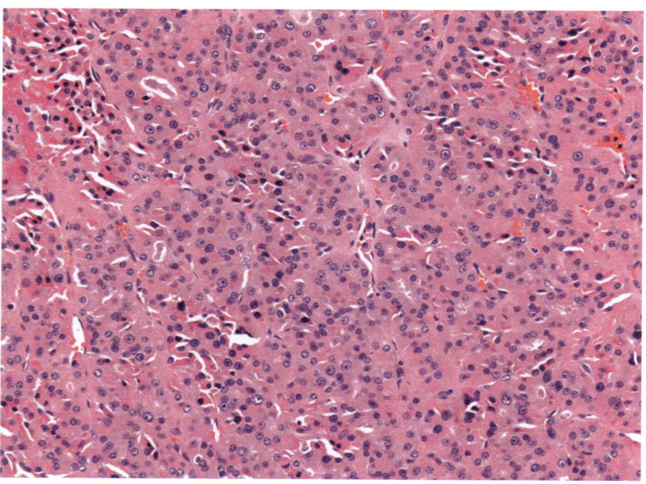

Fig. 21.14 Oncocytic thyroid carcinoma composed of oncocytic cells with deeply eosinophilic cytoplasm

Genetic Features

- Most occur sporadically but can occur in *PTEN* mutation syndrome and in Carney complex
- Widespread chromosome losses
- mtDNA mutation in 60%
- Often lack mutations seen in conventional thyroid carcinoma such as *PAX8-PPARG, BRAF, RAS, MEN1, PTEN,* and *EIF1AX*

Follicular-Derived Carcinomas, High-Grade

Definition

- A carcinoma of thyroid follicular cells with high-grade features defined by mitotic count and/or necrosis and lack anaplastic histology. These tumors may be poorly differentiated or may retain the distinctive morphology of well-differentiated carcinomas of follicular-cell derivation
- Two groups of follicular-cell-derived thyroid carcinomas with high-grade features have been delineated:
- Poorly differentiated thyroid carcinomas as defined by the Turin criteria (solid/trabecular/insular pattern of growth in a tumor diagnosed as malignant based on invasive properties; absence of conventional nuclear features of papillary carcinoma; and presence of at least one of the following: convoluted nuclei, mitotic count ≥3 per 10 high-power fields, tumor necrosis)
- Differentiated high-grade thyroid carcinomas retain their differentiation architecturally and cytologically (follicular or papillary) but have high-grade features (≥5 mitotic figures per 10 HPF and/or necrosis) and lack anaplastic features

Clinical Features

- Uncommon, comprises 1–6% of thyroid carcinomas
- Most common in adults in their fifth and sixth decades, rare in children
- Usually present as a large mass, often rapidly growing
- May arise in a background of multinodular goiter
- Solid, cold nodules, lack radioactive iodine uptake
- Distant metastasis at diagnosis in up to 25%
- May arise in association with a well-differentiated thyroid carcinoma or may develop subsequently
- Prognosis is intermediate between well-differentiated thyroid carcinomas and anaplastic thyroid carcinomas
- 5-year survival is 50–70%

Pathology Features

- Tumor is usually large and solid and may arise in a background of a multinodular goiter
- May appear encapsulated, but often widely invasive
- Conventional poorly differentiated thyroid carcinomas are diagnosed by Turin consensus criteria (solid/trabecular/insular pattern of growth in a tumor diagnosed as malignant based on invasive properties; absence of conventional nuclear features of papillary carcinoma; and presence of at least one of the following: convoluted nuclei, mitotic count ≥3 per 10 high-power fields, tumor necrosis) (Figs. 21.15, 21.16, 21.17, and 21.18)
- Oncocytic thyroid carcinomas may be poorly differentiated
- Differentiated high-grade thyroid carcinomas have increased mitotic activity (≥5 per 10 high-power fields)

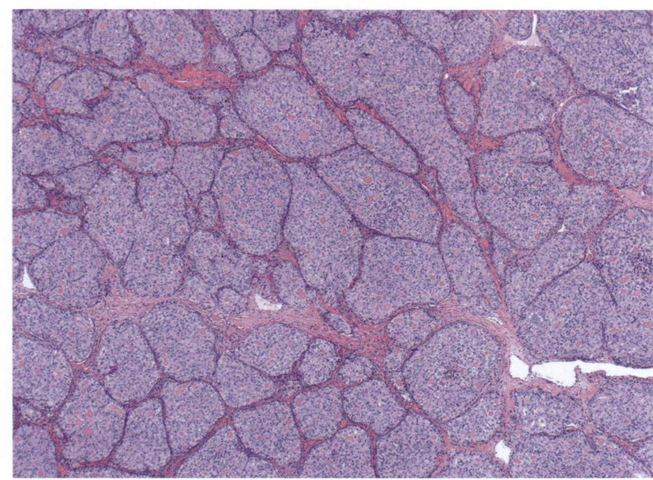

Fig. 21.17 Poorly differentiated thyroid carcinoma with an insular growth pattern

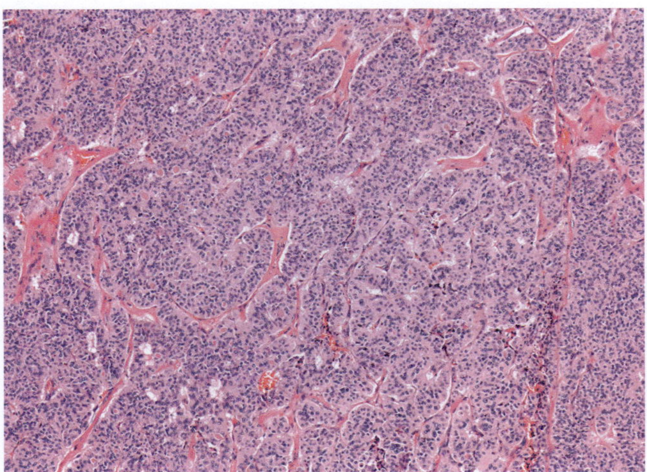

Fig. 21.15 Poorly differentiated thyroid carcinoma with areas of solid growth

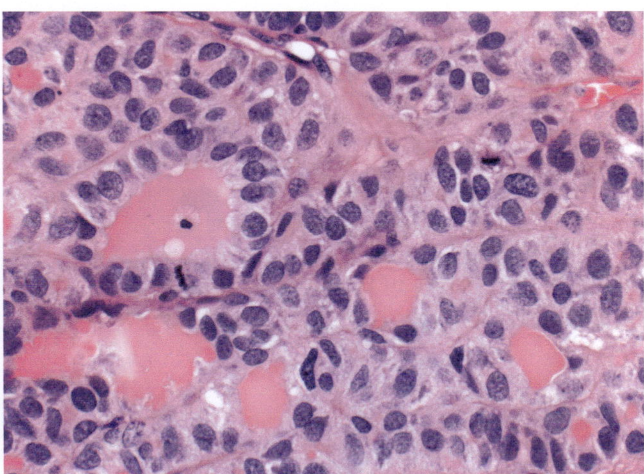

Fig. 21.18 Well-differentiated follicular thyroid carcinoma with high-grade features (mitotic rate of ≥5/10 high-power fields or necrosis)

or necrosis but maintain their well-differentiated status (are not poorly differentiated)
- Differentiated high-grade thyroid carcinomas may be high-grade PTC or high-grade follicular thyroid carcinoma
- Immunopositive for keratin, thyroglobulin, TTF1, and PAX8 is seen, while thyroglobulin may show decreased expression in poorly differentiated areas

Genetic Features

- *RAS* mutations in common in poorly differentiated carcinomas
- Tumors arising in association with a well-differentiated thyroid carcinoma retain the driver mutations (*RAS* or *BRAFV600E*) of the well-differentiated tumor
- *TP53* and *TERT* promoter mutations may occur as late genetic events

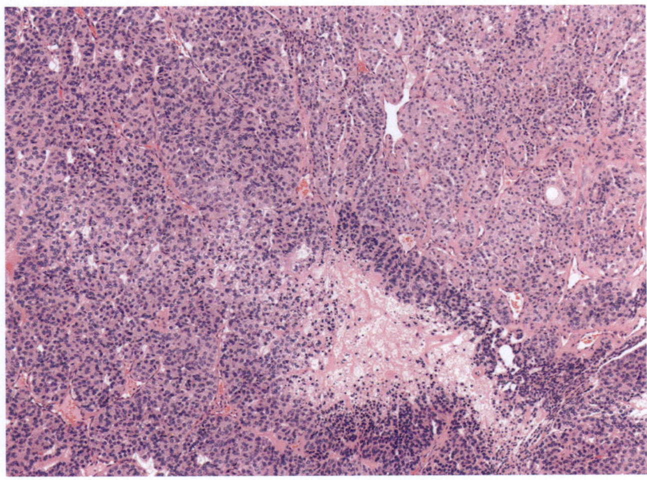

Fig. 21.16 Poorly differentiated thyroid carcinoma with necrosis

Anaplastic Thyroid Carcinoma

Definition

- Highly aggressive undifferentiated thyroid carcinoma composed of undifferentiated follicular thyroid cells
- Many undifferentiated carcinomas are associated with and appear to be derived from the dedifferentiation of papillary or follicular thyroid carcinomas

Clinical Features

- <1% of thyroid carcinomas
- Occurs in older patients: seventh and eighth decades of life
- Rapidly growing mass, widely infiltrative, with compression of adjacent neck structures
- May be associated with pain, hoarseness, breathing difficulty, or dysphagia
- >50% have lymph node metastasis at presentation, and up to 40% have distant metastasis (often to lung, bone, and brain)
- Mortality rate >90% at 3–6 months historically (almost always lethal), but with treatment advances, survival may be increased but remains <2 years
- Tumors found incidentally in a specimen removed for differentiated thyroid carcinoma may have a better prognosis

Pathologic Features

- Large, bulky, infiltrative tumors with hemorrhage and necrosis
- May replace the entire thyroid lobe and infiltrate the soft tissues of the neck
- Widely invasive neoplasms, often with necrosis and extensive vascular invasion marked cytologic atypia, and mitotic activity (Figs. 21.19 and 21.20)
- Usually, an admixture of epithelioid cells, spindle cells, and giant cells (Figs. 21.21, 21.22, and 21.23)
- Various patterns including spindle cell, epithelioid, giant cell, squamoid, angiomatoid, rhabdoid, and lymphoepithelioma-like and paucicellular
- Can have focal squamous features or be completely squamous (anaplastic thyroid carcinomas with squamous cell carcinoma phenotype) (Fig. 21.24)
- Usually have a predominating spindle or epithelioid cells, often with associated giant cells
- Heterologous elements (chondrosarcomatous, osteosarcomatous, and rhabdomyosarcomatous) may be seen

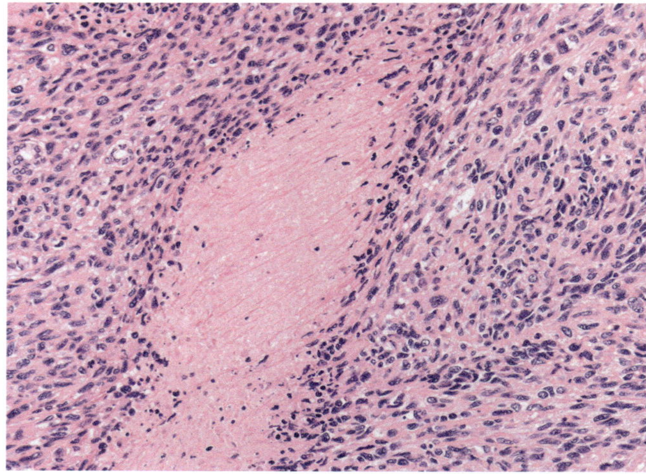

Fig. 21.19 Anaplastic thyroid carcinoma with necrosis

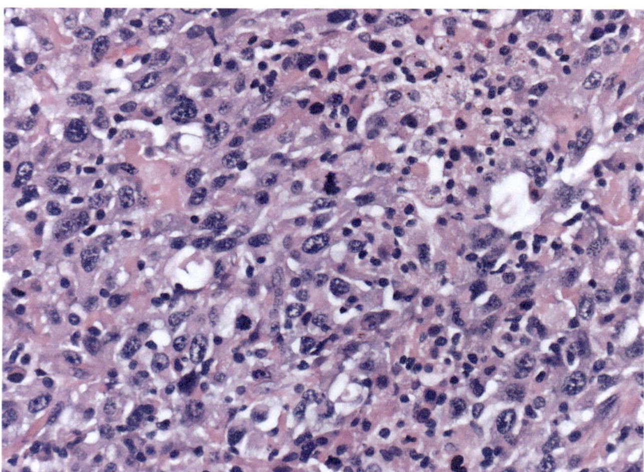

Fig. 21.20 Anaplastic thyroid carcinoma with marked cytologic atypia

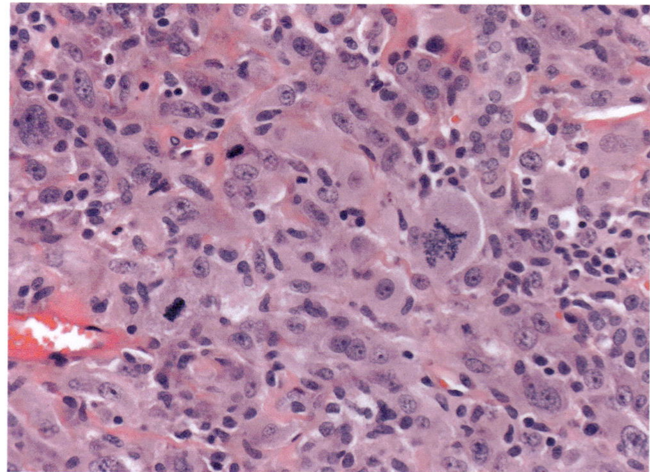

Fig. 21.21 Anaplastic thyroid carcinoma with prominent mitotic activity

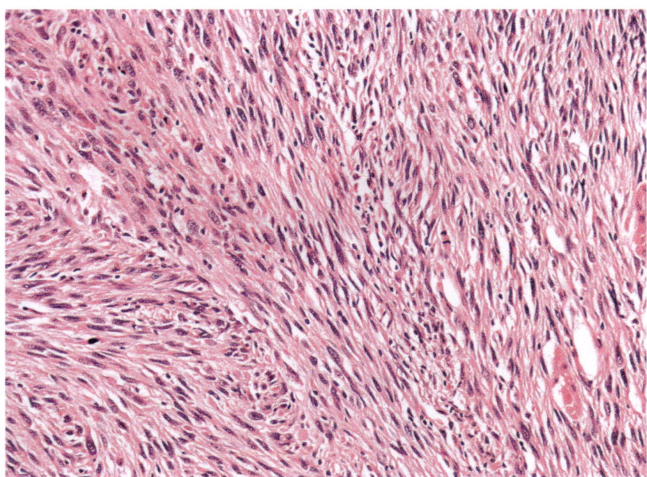

Fig. 21.22 Anaplastic thyroid carcinoma with marked spindle cells (sarcomatoid)

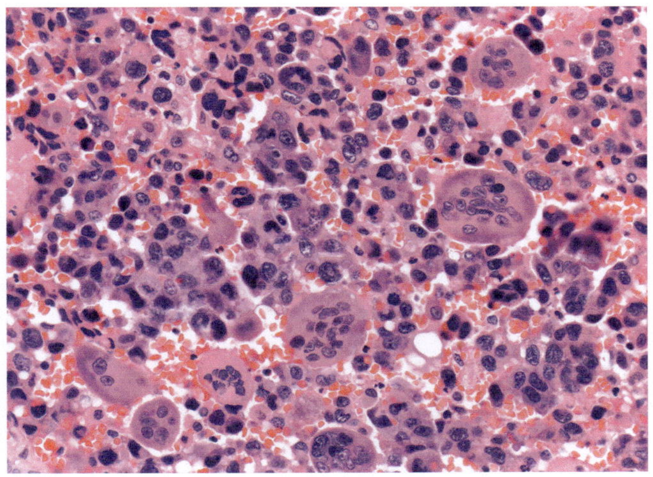

Fig. 21.23 Anaplastic thyroid carcinoma with prominent giant cells

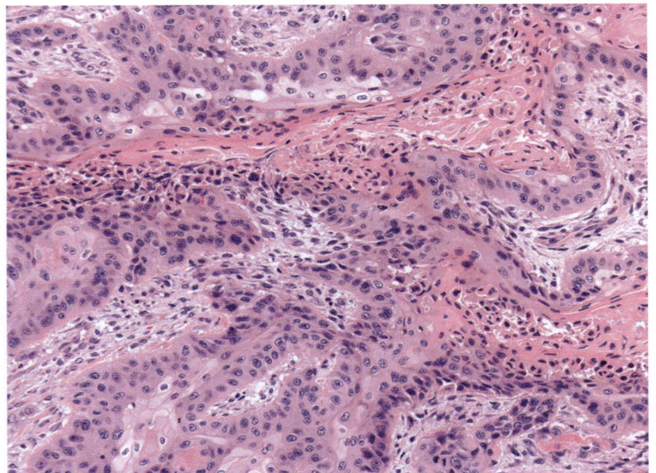

Fig. 21.24 Anaplastic thyroid carcinoma with squamous cell carcinoma phenotype

- In cases arising with well-differentiated thyroid carcinoma, remnants of the well-differentiated thyroid carcinoma (papillary, follicular, or oncocytic) may be present (Figs. 21.25 and 21.26)
- May also arise with associated poorly differentiated thyroid carcinoma
- Compared to conventional anaplastic thyroid carcinoma, anaplastic thyroid carcinoma with a squamous cell carcinoma phenotype is more frequently associated with previous or concurrent differentiated thyroid carcinoma

Genetic Features

- Somatic mutations are frequent, often complex, and numerous chromosomal aberrations

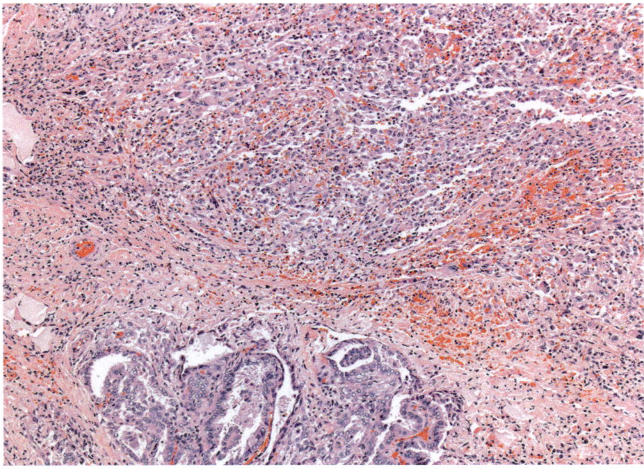

Fig. 21.25 Anaplastic thyroid carcinoma arising in association with a papillary thyroid carcinoma

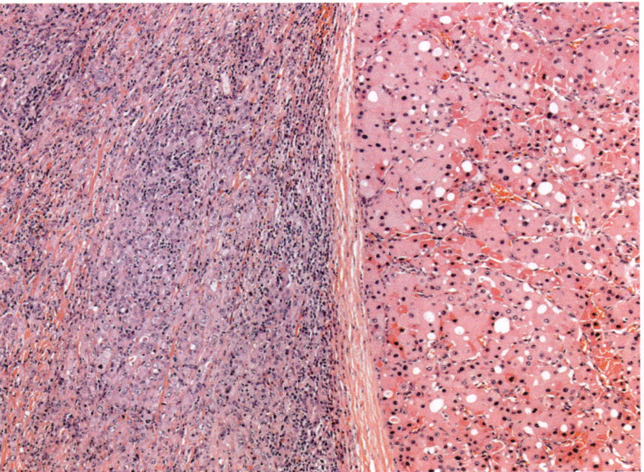

Fig. 21.26 Anaplastic thyroid carcinoma arising in association with an oncocytic (Hürthle) thyroid carcinoma

- More complex molecular alterations than poorly differentiated or well-differentiated thyroid carcinomas
- Anaplastic thyroid carcinoma arising with well-differentiated thyroid carcinoma will usually have preserved mutation pattern including early driver mutations
- A multistep process for anaplastic thyroid carcinoma development includes early driver mutations, such as *RAS* and *BRAF* V600E, and late events, including *TP53* and *TERT* promoter mutations, and abnormalities involving the PIK3CA-PTEN-AKT-mTOR pathway
- *TP53* (17q13.1) mutations are the most common (40–80% of anaplastic carcinomas)
  - p53 is a transcriptional transactivator
  - Regulating functions associated with the cell cycle, cell differentiation, angiogenesis, apoptosis, and DNA repair
  - p53 induces overexpression of p21
  - Immunohistochemical expression of p53 is not necessarily associated with *TP53* mutations
- *BRAF* mutations may occur in about 30–50% and are present in both the differentiated and undifferentiated components when anaplastic carcinoma is identified with a PTC
- *RAS* mutations in 10–50% and present in both the differentiated and undifferentiated components when an anaplastic carcinoma is identified with a differentiated thyroid carcinoma
- *TERT* mutations in 30–75%
- *CTNNB1* (3p22-3p21.3; β-catenin) mutations in 50–60%
  - B-catenin has a role in E-cadherin-mediated cell–cell adhesion
  - Point mutations in the phosphorylation sites of β-catenin in exon 3 stabilize the protein and leads to nuclear accumulation of β-catenin
  - Exon 3 mutation of *CTNNB1* in 25% of poorly differentiated carcinomas and 50–60% of anaplastic carcinomas
- *PIK3CA* mutations in 5–25%
  - Mutations in *PIK3CA* and *PTEN* involving the PI3K/AKT signaling portions
  - 5–25% of anaplastic carcinomas have *PIK3CA* mutations and 5–15% have *PTEN* mutations
  - *PIK3CA* mutations are most common in exon 20, which codes for the kinase domain and exon 9, which codes for the helical domain resulting in the AKT pathway activation
  - Increase in *PIK3CA* gene copy numbers in 30–40%
- *PTEN* (10q23.3) mutations in 10–25%
  - PTEN gene products have protein tyrosine phosphatase and three phosphoinositol phosphatase activities and are important in regulating cell migration, invasion, and proliferation

- Most *PTEN* mutations in anaplastic carcinomas are in the kinase domain in exon 20 and the helical domain in exons 4 or 9
- *EIF1AX* in 5–15% of anaplastic thyroid carcinoma

Hyalinizing Trabecular Tumor

Definition

- Hyalinizing trabecular tumor is a follicular-cell derived neoplasm composed of trabeculae of elongated/polygonal cells with hyaline cytoplasm admixed with intratrabecular hyaline material and prominent nuclear grooves, nuclear vacuoles, and membrane irregularities

Clinical Features

- <1% of thyroid neoplasms
- Most commonly (>80%) occurs in females with a mean age of 50 years
- May be found incidentally by ultrasonography in asymptomatic patients
- Hypoechoic and solid and lack microcalcifications on ultrasound

Pathologic Features

- Round or oval, circumscribed solid neoplasm, often appears lobulated, 0.5–7.5 cm
- Composed of trabeculae and nests of medium to large cells, elongated, perpendicularly oriented to the trabeculae with eosinophilic cytoplasm and finally granular hyaline-like material (Figs. 21.27 and 21.28)

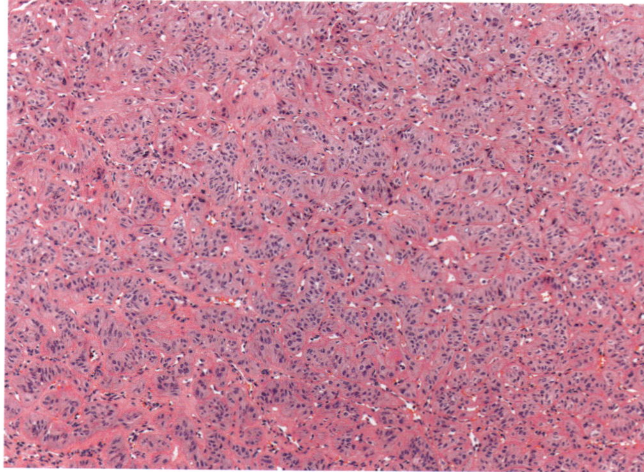

Fig. 21.27 Hyalinizing trabecular tumor composed of trabeculae of follicular cells admixed with intratrabecular hyaline material

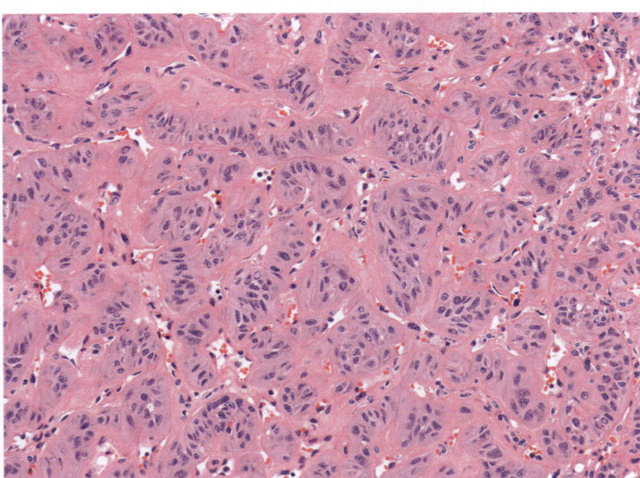

Fig. 21.28 Hyalinizing trabecular tumor composed of trabeculae and nests of elongated, medium to large cells perpendicularly oriented to the trabecula

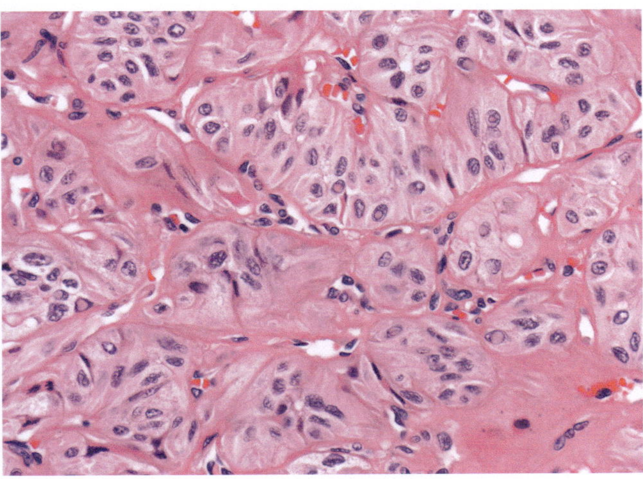

Fig. 21.29 Hyalinizing trabecular tumor with elongated cells with eosinophilic cytoplasm, medium to large nuclei with nuclear grooves and prominent intranuclear vacuoles

- May have yellow cytoplasmic bodies
- Medium to large nuclei with prominent grooves and vacuoles (Fig. 21.29)
- Variable intertrabecular stroma
- Very rare tumors with invasive features may be associated with aggressive behavior
- Positive for thyroglobulin and TTF1 and may show abnormal membranous staining with MIB1 (temperature variable)
- GLIS3 expression is associated with *PAX8-GLIS3* fusion

Genetic Features

- *GLIS* rearrangements in virtually all cases: *PAX8-GLIS3* and less often *PAX8-GLIS1*
- Absence of *RAS* or *BRAF* mutations

Cribriform Morular Thyroid Carcinoma

Definition

- Cribriform morular thyroid carcinoma (previously regarded as a variant of PTC but now considered a tumor of uncertain histogenesis) is a malignant thyroid tumor with a peculiar growth pattern secondary to constitutive activation of the WNT/β-catenin pathway that can occur in familial adenomatous polyposis or sporadically

Clinical Features

- Almost exclusively in young women (most <40 years of age)
- Approximately half are associated with familial adenomatous polyposis
- Tumors in the setting of familial adenomatous polyposis or more often multifocal and/or bilateral, where sporadic cases are usually solitary
- May present as a painless mass or incidentally discovered by ultrasound or may be symptomatic

Pathologic Features

- Well-circumscribed or encapsulated solid or cystic nodule, single and unilateral in sporadic disease but may be multifocal and bilateral in familial adenomatous polyposis
- Unusual admixture of growth patterns: cribriform, follicular, papillary, trabecular, and solid with morular/squamoid areas (Figs. 21.30 and 21.31)
- Anastomosing arches and bars form cribriform structures
- Trabecular areas with spindle cells
- Papillae and pseudopapillae covered by tall, cuboidal, or pseudostratified cells

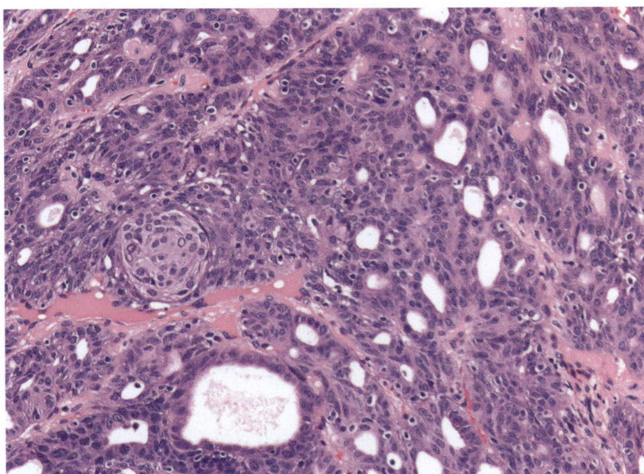

Fig. 21.30 Cribriform morular thyroid carcinoma with an admixture of growth patterns including cribriform, follicular, and solid

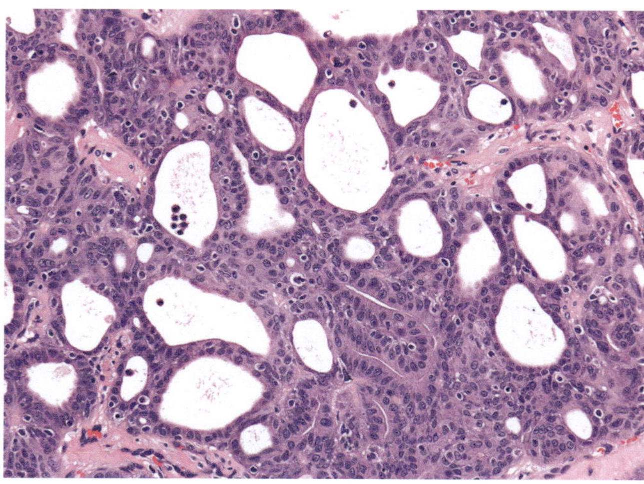

Fig. 21.31 Cribriform morular thyroid carcinoma with a follicular in tubular growth pattern

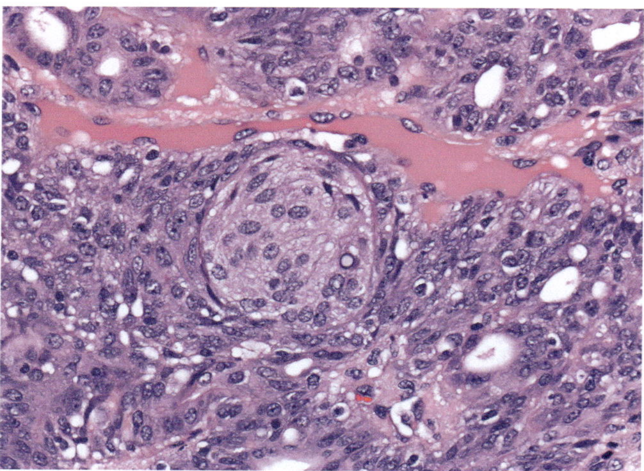

Fig. 21.32 Cribriform morular thyroid carcinoma with whorls of spindle cells forming squamoid morules

- Tubular follicles that lack colloid
- Whorls of spindle cells with nuclear clearing forming squamoid morules (Fig. 21.32)
- Unique aberrant strong staining for β-catenin
- Cribriform areas immunopositive for TTF1, keratin, estrogen, and progesterone receptors, weak to no staining with PAX8 (MRQ 50), negative for calcitonin, WT1, keratin 20 with negative or focal staining for thyroglobulin and PTEN
- Morules immunopositive for CDX2, CD5, CD10, and keratin 5 and are negative for TTF1, PAX8, thyroglobulin, estrogen, and progesterone receptors

Genetic Features

- Germline *APC* mutations (often in exon 15, codons 463, 1387, 1061) in cases associated with familial adenomatous polyposis
- Sporadic tumors may have somatic mutations involving the WNT/β-catenin pathway (*APC*, *CTNNB1*, and *AXIN1*)

Medullary Thyroid Carcinoma

Definition

- Malignant tumor of the thyroid gland composed of cells with C-cell differentiation that secrete calcitonin and other peptides and may develop sporadically or in a familial pattern

Clinical Features

- 2% of thyroid carcinomas and up to 8% of deaths from thyroid carcinoma
- About 1400 new cases per year in the United States
- Most occur sporadically, but 25% occur in familial setting
- Familial syndromes include multiple endocrine neoplasia (MEN) 2A, MEN2B, and familial medullary thyroid carcinoma (FMTC)
- Individuals with familial disease may develop carcinomas at a younger age
- May occur in early adults with MEN2A, often in very young including infants and children may be affected with MEN2B, whereas FMTC (familial isolated medullary thyroid carcinoma) individuals are usually older than those with MEN2A or MEN2B
- FMTC is considered a subtype of MEN2A
- Classical MEN2A is associated with medullary thyroid carcinoma (MTC), pheochromocytoma, and hyperparathyroidism
- MEN2A can also be associated with cutaneous lichen amyloidosis and Hirschsprung disease
- Individuals with familial isolated medullary thyroid carcinoma have only MTC
- Individuals with MEN2B have MTC, pheochromocytoma, mucosal neuromas, marfanoid body habitus, ganglioneuromatosis of the gastrointestinal tract, and may have ocular involvement
- Sporadic MTC generally presents as a solid mass

- Syndrome-associated MTC are often bilateral and/or multifocal tumors that may be preceded by C-cell hyperplasia
- Serum calcitonin and carcinoembryonic antigen are often elevated
- Elevated serum calcitonin may result in flushing, weight loss, and diarrhea
- Better prognosis: familial tumors diagnosed biochemically or by molecular, miR-224
- Worse prognosis: older age, male sex, present clinically (50–80% lymph node metastases, 10–15% distant metastasis at diagnosis), higher serum calcitonin, higher TNM stage, sporadic (vs hereditary), extrathyroidal extension, less extensive surgery, decreased Rb protein expression, somatic CDKN2C loss, possibly miR-183 & miR-375
- Meta-analysis including 23 studies and 964 sporadic MTC, *RET* mutation was associated with increased risk of lymph node and distant metastases, high stage, increased risk recurrence, higher mortality, while *RAS* mutation had no significant prognostic value
- Sporadic tumors have lymph node metastases in 50% and distant metastases in 15% at diagnosis
- Somatic copy number loss of *CDKN2C* in 20% of tumors and has been found to be associated with distant metastasis
- Medullary thyroid microcarcinomas can be associated with metastatic disease
- SEER study of 310 medullary microcarcinomas from 1988 to 2007, 37% (65 of 176) with lymph nodes removed had metastases

Pathologic Features

- MTC are usually single well-circumscribed and unencapsulated nodules measuring 2–3 cm in diameter but can vary greatly in size
- Most tumors occur in the upper and middle third of the thyroid gland
- Medullary microcarcinomas measure ≤1 cm
- Varying amounts of spindle and epithelioid cells (Figs. 21.33, 21.34, and 21.35)
- Rare cases have a true papillary pattern with fibrovascular stalks, but a pseudopapillary pattern is more common (Fig. 21.36)
- Can have a follicular (tubular/glandular) pattern
- Other variants including oncocytic (Fig. 21.37), clear cell, giant cell, spindle cell, paraganglioma-like, angiosarcoma-like, as well as very rare melanotic in squamous variants
- Small cell variant may be more aggressive than conventional
- Amyloid in the stroma in about 75% of case, and amyloid is calcitonin

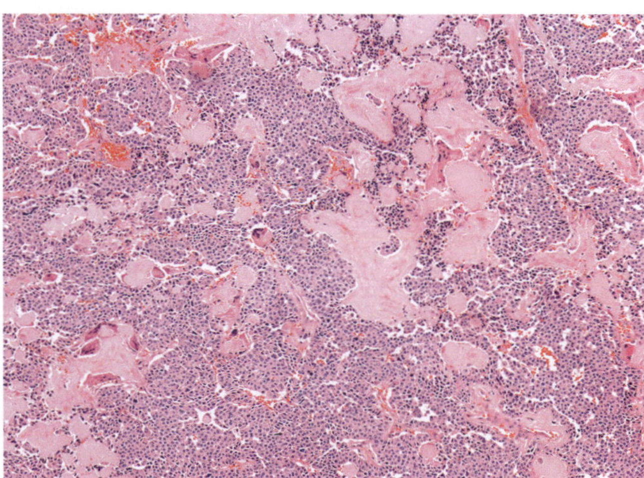

Fig. 21.33 Medullary thyroid carcinoma with predominantly nested and solid growth

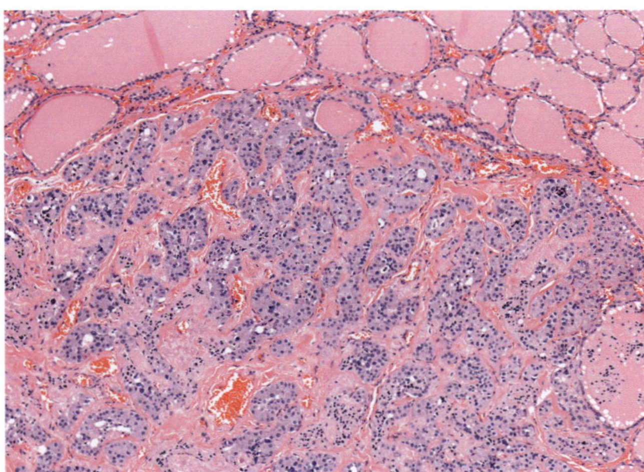

Fig. 21.34 Medullary thyroid carcinoma from a prophylactic thyroidectomy from a patient with familial medullary thyroid carcinoma

- Immunohistochemical positivity for chromogranin-A, synaptophysin, calcitonin, CEA, keratin, and sometimes somatostatin and negative for PAX8 (monoclonal, directed against C-terminus)
- Ultrastructural features include dense core secretory granules 100–600 μm in diameter

Genetic Features

- *RET* (10q11.1) protooncogene mutations
- *RET* encodes transmembrane receptor tyrosine kinase, which is involved in cell signals in pathways
- *RET* mutations occur in sporadic tumors and in tumors associated with MEN2A or MEN2B or in FMTC

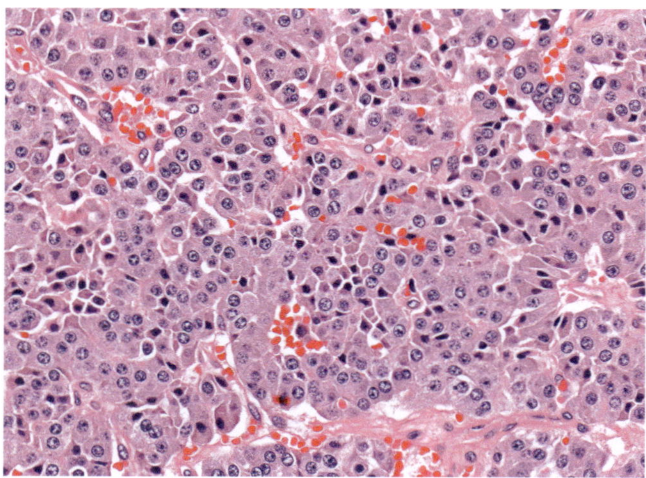

Fig. 21.35 Medullary thyroid carcinoma composed of epithelioid cells with somewhat stippled chromatin but predominantly small irregular nucleoli and eosinophilic to amphophilic cytoplasm

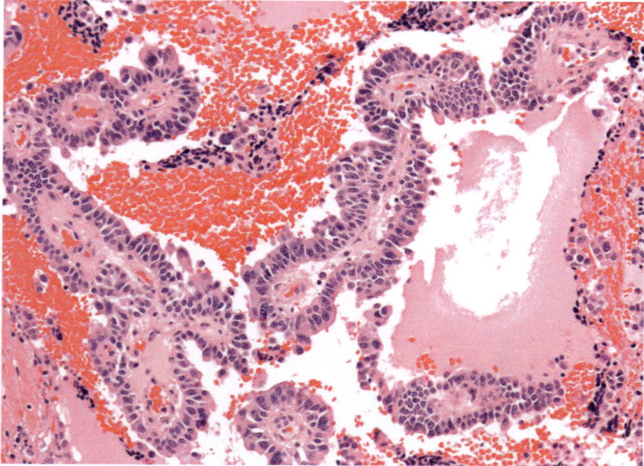

Fig. 21.36 Medullary thyroid carcinoma with a papillary growth pattern. Medullary thyroid carcinomas can have intranuclear holes similar to papillary thyroid carcinoma and can be confused with papillary thyroid carcinoma particularly in cases with a prominent papillary architectural pattern

- Germline *RET* mutations in certain functional regions are associated with MEN2A, MEN2B, and FMTC
- In MEN2A and FMTC, the mutations are present in exons 10 and 11
- Somatic *RET* mutations in 25–70% of sporadic tumors and involve codon 918
- Sporadic MTC without *RET* mutation may have sporadic *RAS* mutation in 10–80% of cases (*HRAS* 50–60%), *KRAS* (10–15%), and rarely *NRAS*
- *RAS* mutations are mutually exclusive with germline *RET* mutations
- Somatic *RAS* mutation essentially excludes germline *RET* mutation

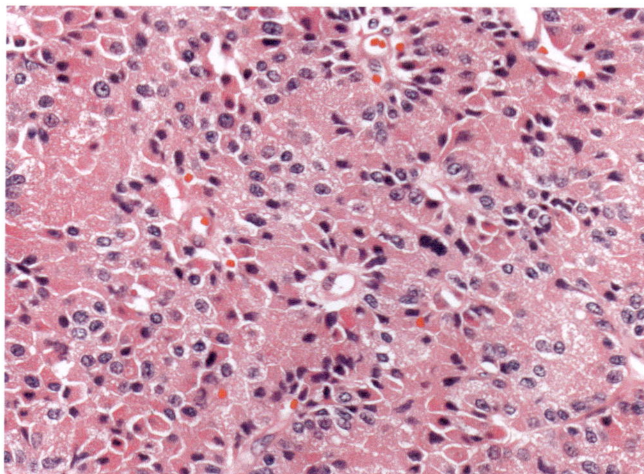

Fig. 21.37 Medullary thyroid carcinoma, oncocytic variant. The oncocytic variant of medullary thyroid carcinoma can be readily mistaken for an oncocytic thyroid or parathyroid tumor

Table 21.4 American Thyroid Association risk categories

2015	2009	*RET* codon	Thyroidectomy
"Highest risk" ATA-HST	D	M918T	Immediately, children first months of life, pheochromocytoma screening starting at 11 years
"High risk" ATA-H	C	C634, A883F*	<5 years old, pheochromocytoma screening starting at 11 years
"Moderate risk" ATA-MOD	A & B	Mutations other than M918T, C634, and A883F	<5 years or later if screen calcitonin every 6 months, pheochromocytoma screening starting at 16 years

- As 25% of MTCs are familial, germline *RET* mutation analysis is recommended for individuals diagnosed with MTC
- Virtually all FMTC cases are associated with *RET* germline mutation
- MEN2A: Most *RET* mutations (95%) in codon 634 (exon 11), but can also involve codons 609, 611, 618, and 620
- MEN2B: Most *RET* mutations (95%) involve codon 918 (exon 16)
- 75% of individuals with MEN2B have de novo *RET* mutations
- <5% of *RET* mutations in MEN2B involve codon 883 (A883F, exon 15), and this is generally associated with less aggressive disease
- Atypical MEN2B usually involves double *RET* germline mutation involving V804M and either Y806C, S904C, E805K or Q781R
- Risk categories defined by American Thyroid Association (Table 21.4)
- *RET* mutation: increased risk of lymph node and distant metastases, high stage, recurrence, and mortality, while *RAS* mutation no significant prognostic value

- Somatic copy number loss of *CDKN2C* in 20% and associated with distant metastasis
- *TERT* promoter methylation and *TERT* copy number gains have been reported, but *TERT* promoter mutations have not
- mRNAs 183 and 375 are more common in sporadic than hereditary disease, and microRNAs are associated with lateral lymph node metastasis, residual disease, distant metastasis, and mortality

Parathyroid Tumors

Parathyroid Adenoma

Definition

- Benign parathyroid neoplasm composed of chief, transitional, oncocytic, or clear cells or a mixture of cell types causing hypercalcemia and hyperparathyroidism

Clinical Features

- Cause >85% of primary hyperparathyroidism
- Females affected two to three times more often than males in sporadic cases
- Historically associated with symptoms but with automated serum calcium measurements, the incidence has increased, and patients are often asymptomatic
- Serum PTH and calcium elevated, but less than that seen in parathyroid carcinoma
- Inferior parathyroid glands may be affected slightly more often than the superior glands
- Parathyroid adenomas can also occur in ectopic locations
- Up to 15% of parathyroid adenomas may involve the right and left superior parathyroid glands (fourth pouch disease)
- Majority occur sporadically, but up to 10% may occur due to underlying genetic susceptibility; either syndromic (*CDC73*-related disease, such as hyperparathyroidism jaw tumor syndrome, as well as in MEN1, MEN2, MEN4, and MEN5) or nonsyndromic disease (such as familial isolated hyperparathyroidism)
- Although often referred to as "hyperplasia", the multiglandular involvement in hereditary hyperparathyroidism, such as in multiple endocrine neoplasia, is now recognized as due to multiple adenomas
- Preoperative localization studies and intraoperative PTH monitoring are critical
- Evaluation of parathyroid disease requires clinical, radiographic, surgical, and pathology input

Pathologic Features

- Enlarged gland usually >40–60 mg
- Chief cells are most common, and oncocytic adenomas comprise 3–6% of parathyroid adenomas, while lipoadenomas and water-clear cell parathyroid adenomas are uncommon (Figs. 21.38, 21.39, 21.40, 21.41, and 21.42)
- Oncocytic parathyroid adenomas tend to be larger than nononcocytic adenomas
- Rim of normal appearing or suppressed parathyroid may be present adjacent to the adenoma, but not specific as similar findings can be seen in 10% of cases of multiglandular disease
- Rim of normal-appearing parathyroid tissue is more often identified in smaller adenomas

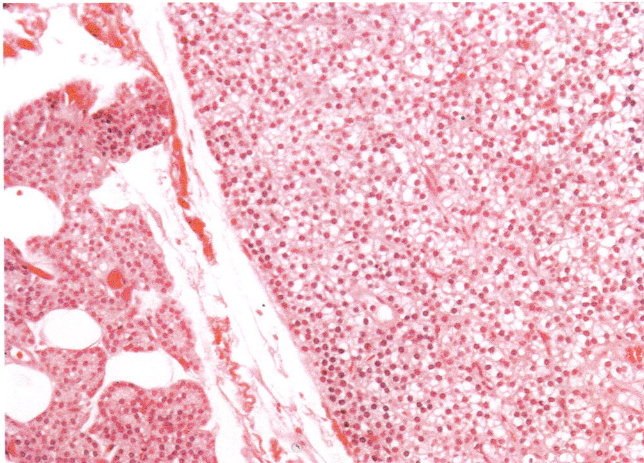

Fig. 21.38 Parathyroid adenoma with an adjacent rim of suppressed normal appearing parathyroid tissue. This tumor is composed predominantly of chief cells, which is the most common type of parathyroid adenoma

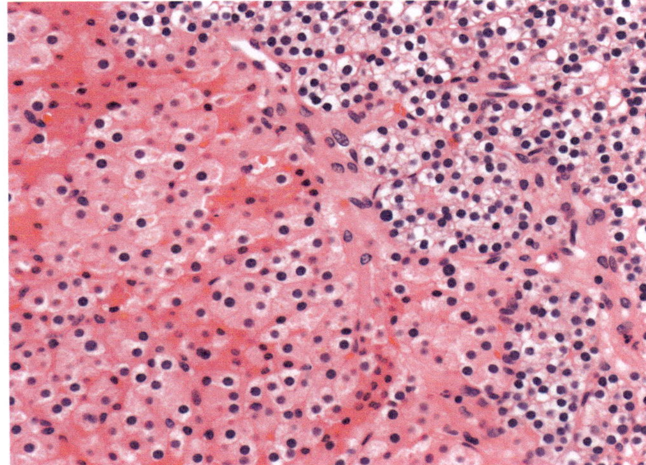

Fig. 21.39 Parathyroid adenoma with chief cells and oncocytic cells

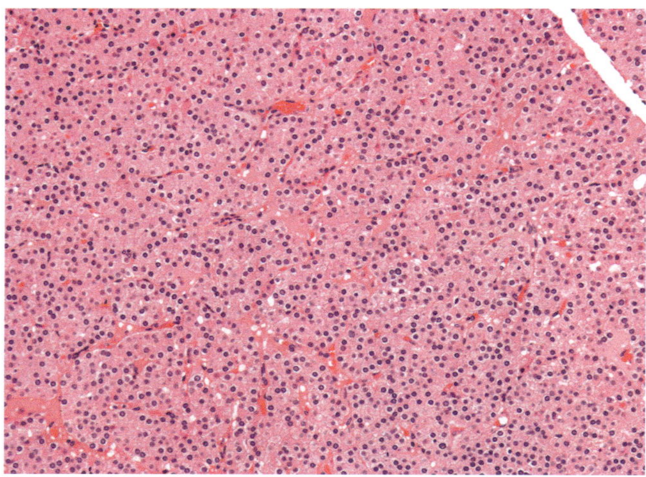

Fig. 21.40 Oncocytic parathyroid adenomas are less common than chief cell parathyroid adenomas, but they are functional tumors

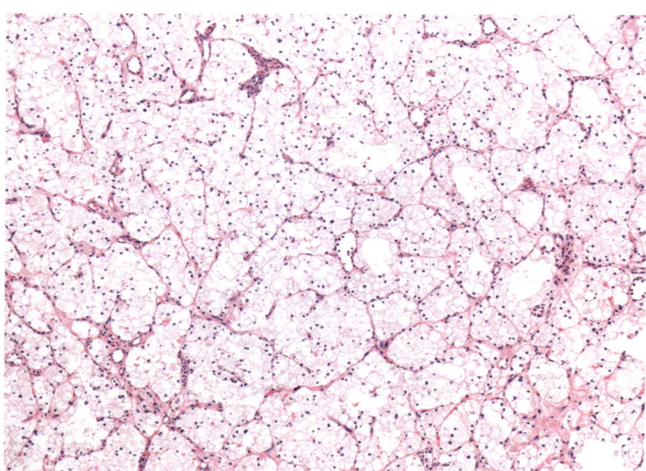

Fig. 21.41 Clear cell parathyroid adenoma is rare

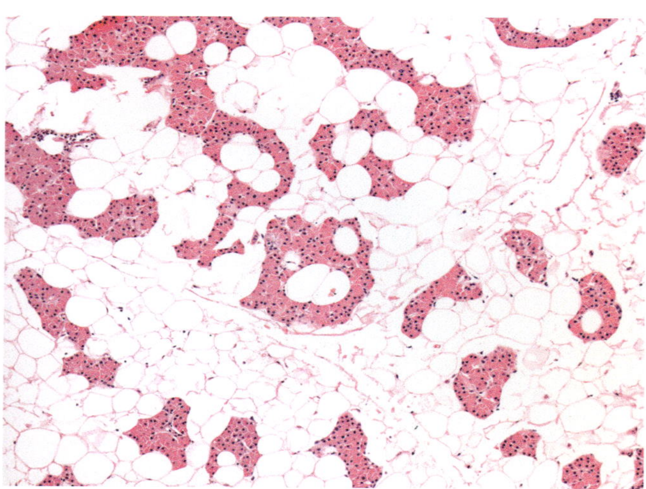

Fig. 21.42 Parathyroid lipoadenoma with abundant adipose tissue in this case but other mesenchymal tissue can also be seen with the associated parathyroid parenchymal tissue. This is an example of an oncocytic parathyroid lipoadenoma

- Parathyroid adenoma cannot be differentiated from parathyroid hyperplasia based on the histologic features of a single gland
- Cystic changes may be present and may be more common in the setting of hyperparathyroidism jaw tumor syndrome
- Cystic lesions or those associated with trauma, such as fine needle aspiration, may be associated with secondary fibrosis and hemosiderin deposition, and the gland may become adherent to adjacent structures
- Occasional mitotic figures may be present, but atypical mitoses are generally not seen
- Parathyroid adenomas are usually composed of chief cells but can be a mixture of cell types, including chief, oncocytic, transitional/oncocytic, clear cells and water-clear cells
- Oncocytic parathyroid adenomas are composed of >75% oncocytic cells
- Parathyroid lipoadenoma shows an unequivocal increase in parathyroid parenchymal tissue in association with abundant adipose and mesenchymal tissue
- May show fibrosis, occasional mitotic figures, and other atypical features, but they do not show unequivocal invasion as would be diagnostic of parathyroid carcinoma
- Immunopositive for parathyroid hormone and chromogranin-A and negative for TTF1, thyroglobulin, and PAX8 (C-terminus monoclonal antibody)
- Parafibromin is generally retained unless the adenoma is in the setting of germline disease
- Parathyroid adenomas in the setting of CDC73 disease (parafibromin deficient parathyroid tumors) often have microcystic change, sheet-like growth, nuclear enlargement, and eosinophilic cytoplasm with perinuclear clearing

Genetic Features

- Most parathyroid adenomas occur sporadically
- Sporadic parathyroid adenomas are most often associated 11q23 (loss of heterozygosity [LOH] of 11q13 or inactivation of *MENIN*) and rearrangement of *CCND1* with *PTH*
- Up to 10% may be hereditary and seen in *CDC73* (1q25, parafibromin)-related disorders such as hyperparathyroidism jaw tumor syndrome and in MEN1 (*MENIN*, 11q23, where the "hyperplasia" is more accurately classified as multiglandular adenomas), MEN2 (*RET*, 10q11.2), MEN4 (*CDKN1B*, 12p13 (p27)), and MEN5 (*MAX*) and nonsyndromic hereditary disease such as isolated familial hyperparathyroidism

Atypical Parathyroid Tumor

Definition

- Atypical parathyroid tumor is a parathyroid neoplasm with atypical cytological and architectural features but lacks unequivocal capsular, vascular, or perineural invasion or invasion into adjacent structures or metastases

Clinical Features

- 0.5–4.4% of primary hyperparathyroidism
- May be asymptomatic or have vague symptoms (fatigue, weakness, polyuria, polydipsia, or nausea)
- Palpable neck mass is generally absent with a parathyroid adenoma but can occur in 3% of atypical parathyroid tumors and in 13% of parathyroid carcinomas
- Serum calcium levels are elevated, often intermediate between that of adenoma and carcinoma
- Most occur sporadically, but they can occur with an underlying germline abnormality, such as *CDC73*-related disorder (hyperparathyroidism jaw tumor syndrome, multiple endocrine neoplasia type 1, and familial-isolated hyperparathyroidism)
- Cases with an underlying germline abnormality often occur two decades earlier than sporadic cases
- Females are more often affected than males in sporadic disease, while males and females are affected similarly in cases with an underlying germline abnormality
- Most have benign behavior, but behavior is difficult to predict with certainty (tumor of uncertain malignant potential), and long-term follow-up is recommended

Pathologic Features

- Usually solid, but it may be cystic
- May be adherent to adjacent structures, but no invasion of adjacent structures
- Cellular neoplasms composed predominantly of chief cells but may be composed of variable cell types
- May have atypical features, such as monotonous growth, fibrosis, fibrous bands, and mitotic activity, but lacks unequivocal capsular, perineural, or angiolymphatic invasion or invasion into adjacent structures or metastases (Figs. 21.43, 21.44, and 21.45)
- Term atypical parathyroid tumor is used for tumors with features worrisome for parathyroid carcinoma but lacking definitive invasion and metastases
- Immunopositive for parathyroid hormone, chromogranin-A, and usually retained parafibromin

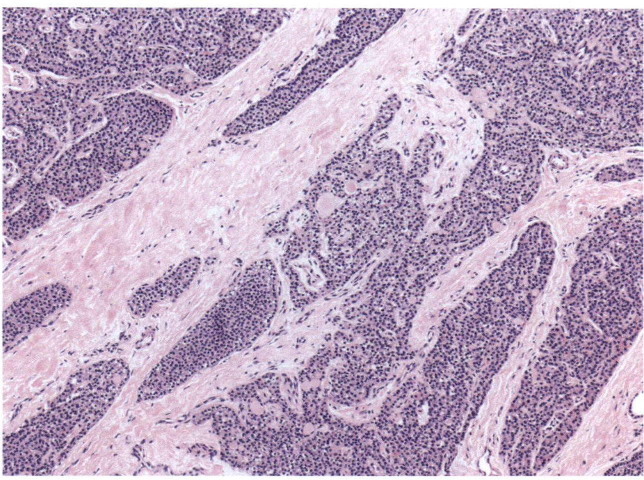

Fig. 21.43 Atypical parathyroid tumor with irregular growth in prominent fibrous bands but lacking unequivocal invasion as would be required for the diagnosis of parathyroid carcinoma

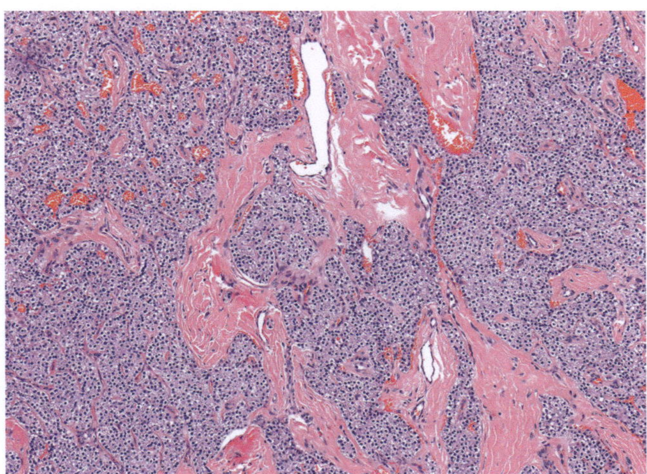

Fig. 21.44 Atypical parathyroid adenoma with fibrosis and marked cellularity and irregular growth

- Many markers suggested in literature to evaluate for malignancy (Ki67, p27, RB1, MDM2, p53, E-cadherin, PGP9.5, cyclin D1, parafibromin, etc.), but none are definitive
- Loss of nuclear parafibromin (parafibromin deficiency) is a concerning feature and may be helpful in predicting behavior of atypical parathyroid tumors

Genetic Features

- LOH *CDKN1A*, LOH *CDC73*, LOH *PTEN*, LOH *RB1*
- Germline *CDC73* mutation testing recommended in cases with loss of parafibromin (parafibromin deficient tumors)

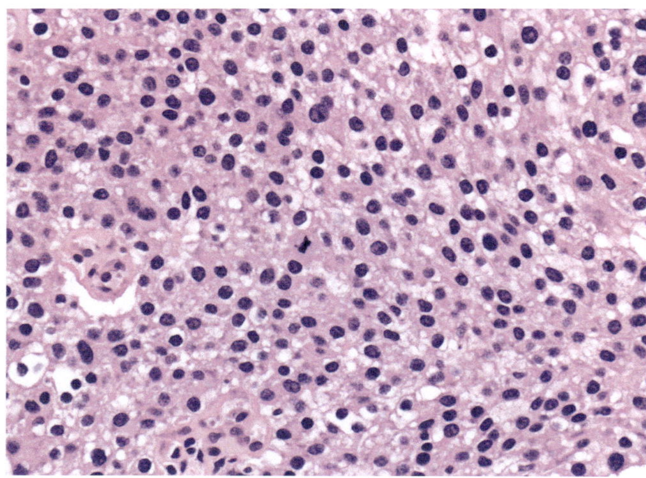

Fig. 21.45 Atypical parathyroid adenoma with mitotic activity. Mitotic features are a concerning feature in a parathyroid neoplasm, but they are not in the cells diagnostic of malignancy. Mitotic figures can be seen in benign and malignant parathyroid tissues and tumors

Parathyroid Carcinoma

Definition

- Malignant parathyroid neoplasm originating from parathyroid parenchymal cells

Clinical Features

- <1% of primary hyperparathyroidism
- Severe hyperparathyroidism, often symptomatic hyperparathyroidism
- Peak age usually in the fifth and sixth decades
- Equal sex distribution
- May have a palpable neck mass (highly unusual for adenoma and seen only in small subset of atypical parathyroid tumors)
- Serum calcium usually significantly elevated (>12–13.5 mg/dL) as well as PTH
- Most occur sporadically
- Significant increase in risk for parathyroid carcinoma in those with hyperparathyroidism jaw tumor syndrome (15% may have parathyroid carcinoma)
- Imaging may show invasive growth
- Recur locally or metastases to regional lymph nodes, lungs, and liver

Pathologic Features

- 2–10 g, but variable

- May see necrosis, vascular invasion, extension to adjacent soft tissues, or thyroid invasion
- Diagnosis requires invasive growth (capsular, angiolymphatic, perineural, invasion into adjacent structures) or metastases (Fig. 21.46)
- Composed mainly of chief cells, but oncocytic/oxyphilic carcinomas can occur
- Monotonous growth, but may have follicular growth mimicking a follicular thyroid carcinoma
- Often high nuclear to cytoplasmic ratios, macronuclei, mitotic figures, and atypical mitotic figures (Figs. 21.47, 21.48, and 21.49)
- Immunopositive for parathyroid hormone, chromogranin-A, often loss of nuclear parafibromin, and negative for TTF1 and thyroglobulin

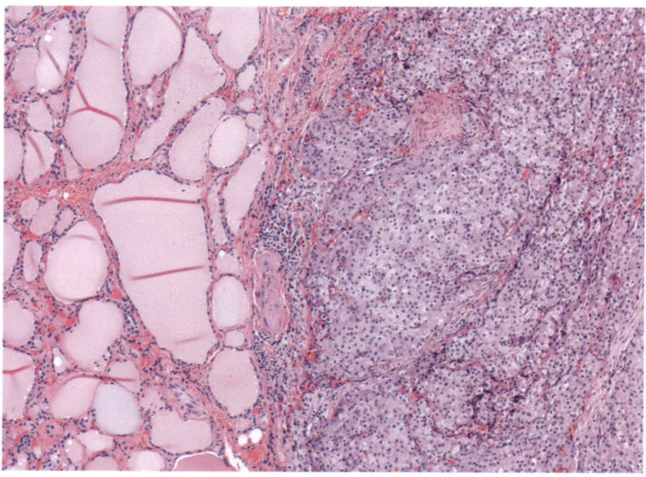

Fig. 21.46 Parathyroid carcinoma infiltrating the thyroid gland

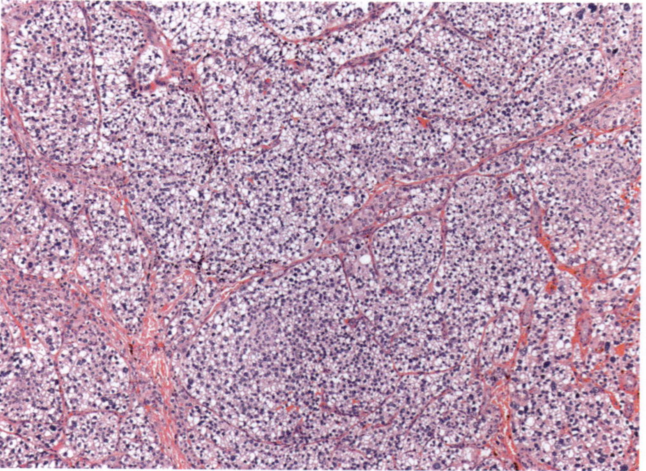

Fig. 21.47 Parathyroid carcinoma with marked cellularity forming irregular nests

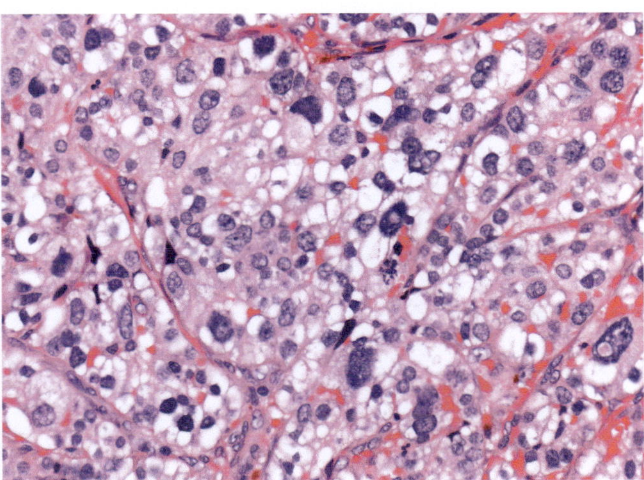

Fig. 21.48 Parathyroid carcinoma with marked nuclear atypia and mitotic activity. Parathyroid carcinoma often shows a monotonous growth pattern and cellular monotony. The cellular pleomorphism should not be mistaken for "endocrine atypia" which is seen in benign lesions, nor should cellular monotony be confused with a benign parathyroid lesion as parathyroid carcinoma often have a monotonous pattern of growth and cytomorphology

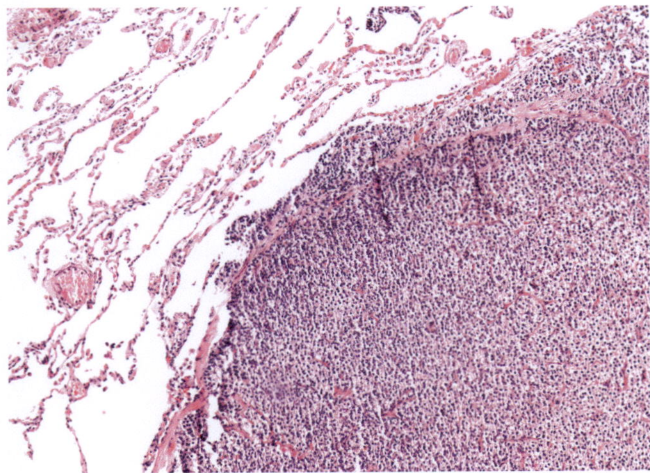

Fig. 21.49 Parathyroid carcinoma metastatic to the lung

Genetic Features

- Most occur sporadically but can occur in the setting of hyperparathyroidism jaw tumor syndrome and rarely in familial-isolated hyperparathyroidism
- Hyperparathyroidism jaw tumor syndrome has a significant risk of parathyroid carcinoma (10–15%)
- Germline *CDC73* (1q25) mutations or deletions cause hyperparathyroidism jaw tumor syndrome
- *CDC73* encodes parafibromin with frequent loss of nuclear parafibromin immunohistochemically in *CDC73*-associated parathyroid disease

- 40–80% of sporadic parathyroid carcinomas have somatic *CDC73* mutation or LOH
- Other genetic alterations involve *TP53*, *RB1*, *CCND1*, *NF1*, *KDR*, *PIK3CA*, *PTEN*, *APC*, *CDKN2*, and rarely *TERT* promoter or *MEN1*
- *DICER1* mutations associated with advanced disease

Adrenal Cortical Tumors

Definition

- Steroid-producing neoplasms which arise from adrenal cortical cells and may produce glucocorticoids or mineralocorticoids

Adrenal Cortical Nodular Disease

Definition

- A group of distinct pathological entities characterized by benign adrenal cortical nodular proliferations
- Subtypes include sporadic nodular adrenal cortical disease and bilateral micronodular adrenal cortical disease (isolated micronodular adrenal cortical disease), primary pigmented nodular adrenal cortical disease, and bilateral macronodular adrenocortical disease

Clinical Features

- The term "diffuse hyperplasia" remains appropriate for adrenal glands affected by pituitary ACTH production
- "Hyperplasia" for adrenal cortical disease such as "adrenal cortical hyperplasia" or "cortical nodular hyperplasia" is no longer an appropriate term for primary bilateral nodular adrenal cortical disease as the nodules are independent clonal proliferations
- Sporadic nodular adrenal cortical disease is more common than bilateral micronodular or macronodular adrenal cortical disease, may be unilateral or bilateral, and is composed of nonfunctional adrenal cortical nodule or nodules generally <1 cm
- Bilateral micronodular or macronodular adrenal cortical disease is usually associated with Cushing syndrome, involves both adrenal glands, and is due to an underlying genetic susceptibility in virtually all cases of bilateral micronodular adrenal cortical disease and a significant portion of cases of bilateral macronodular adrenal cortical disease
- Bilateral micronodular adrenal cortical disease includes isolated micronodular adrenal cortical disease and primary pigmented nodular adrenal cortical disease

- Bilateral micronodular adrenal cortical nodular disease occurs in children in young adults
- Primary pigmented nodular adrenal cortical disease occurs in over half of individuals with Carney complex but can occur in the absence of Carney complex (isolated PPNAD)
- Bilateral macronodular adrenal cortical disease usually occurs in adults

Pathologic Features

- Sporadic adrenal cortical nodular disease consists of small (<1 cm) cortical nodules (Fig. 21.50)
- Bilateral micronodular adrenal cortical disease has small (<1 cm) cortical nodules (which can be pigmented in the setting of PPNAD) composed of lipid-poor adrenal cortical cells
- Adrenal gland may be enlarged in bilateral macronodular adrenal cortical disease with an increased number of nodules identified in cases driven by *ARMC5* mutation
- Nodules in bilateral macronodular adrenal cortical disease are >1 cm and rich in clear cells

Genetic Features

- Germline disease activating the protein kinase A pathway in bilateral micronodular adrenal cortical disease
- Germline disease involving *ARMC5*, *MEN1*, *FH* (hereditary leiomyoma ptosis and renal cell cancer), *APC* (familial adenomatous polyposis), *GNAS* (McCune–Albright syndrome), *PDE11A*, *PDE8B*, and *EDNRA* in bilateral macronodular adrenal cortical disease

Adrenal Cortical Adenoma

Definition

- Benign epithelial neoplasm of adrenal cortical cells
- May be composed of adrenal cortical cells, which can range from cells of the zona glomerulosa, fasciculata to reticularis

Clinical Features

- Common and may be increasing in incidence, but this may be due to increasing imaging
- More common with increasing age
- May be associated with excess production of mineralocorticoids, glucocorticoids, or sex steroids
- Aldosterone-producing adrenal cortical disease causes of primary aldosteronism (Conn syndrome), which shows bilateral involvement in 60–70% of cases
- Cortisol-producing adenomas independent of pituitary disease
- Sex hormone-producing adenomas are rare
- Can also be nonfunctional or have subclinical hormone production

Pathologic Features

- Well-circumscribed solitary mass, usually <50 g, <5 cm, although size and weight can vary
- Central hemorrhage and degeneration may result in a larger tumor that must be distinguished from necrosis seen in adrenal cortical carcinoma (Fig. 21.51)

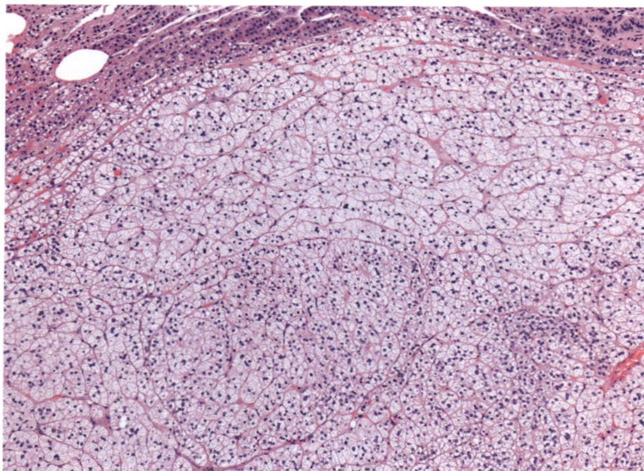

Fig. 21.50 Adrenal cortical nodular disease. The term "hyperplasia" is used for adrenal glands affected by pituitary ACTH production. This term is not used for primary sporadic nodular adrenal cortical disease or for micronodular or macronodular adrenal cortical disease

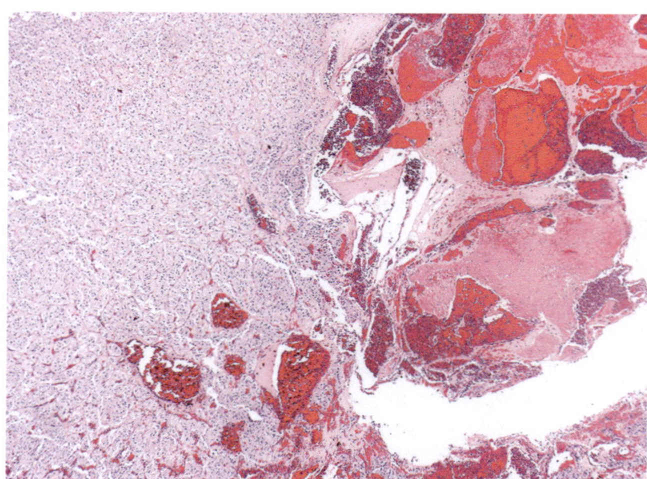

Fig. 21.51 Adrenal cortical adenoma showing cystic hemorrhagic degeneration. Central cystic hemorrhagic degeneration is not uncommon and can be mistaken for necrosis which would be an ominous finding in an adrenal cortical tumor

- Adrenal cortical adenomas can also show areas of myelo-lipomatous change which is of no known pathologic significance (Fig. 21.52)
- Usually composed of zona fasciculata cells, but may be zona glomerulosa or zona reticularis cells
- Cells have abundant lipid-laden cytoplasm and small nuclei, and may have a second component of more compact cells with lipid-poor cytoplasm (Figs. 21.53 and 21.54)
- In the absence of exogenous cortisol intake, nontumoral cortical atrophy is characteristic of autonomous cortisol secretion
- Aldosterone-producing adrenal cortical disease can be unifocal or multifocal and unilateral or bilateral, with micro tumors difficult to identify (Fig. 21.55)

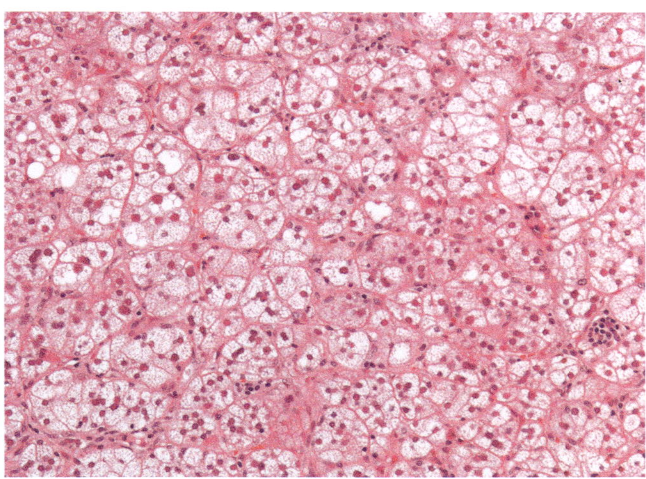

Fig. 21.54 Cortisol-producing adrenal cortical adenoma composed of cells with abundant lipid-laden cytoplasm

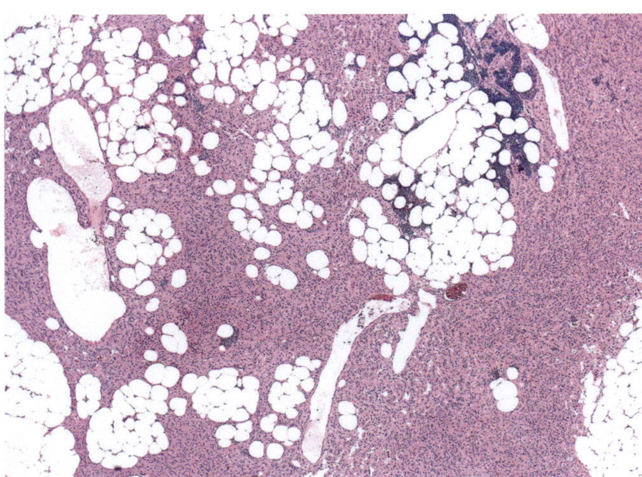

Fig. 21.52 Adrenal cortical adenoma showing myelolipomatous change, a relatively common finding without known pathologic significance

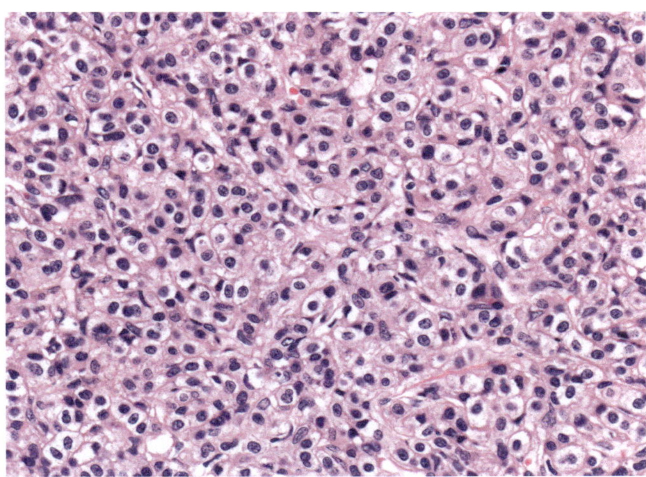

Fig. 21.55 Adrenal cortical adenoma with aldosterone production. Aldosterone-producing adrenal cortical disease can be unilateral and unifocal, but it is often multifocal and can be bilateral

- CYP11B2 immunohistochemistry may be needed to identify functional sites of aldosterone production in primary hyperaldosteronism
- Adrenal cortical adenomas show immunopositivity for steroidogenic factor 1, synaptophysin, Mart1/MelanA, and inhibin-alpha (particularly in functional tumors) and are negative for chromogranin-A

Genetic Features

- Most occur sporadically
- Can occur in hereditary syndromes: Li-Fraumeni (*TP53*), MEN1 (*MEN1*), Beckwith–Wiedemann syndrome (*H19* and *IGF2*), familial adenomatous polyposis (*APC*),

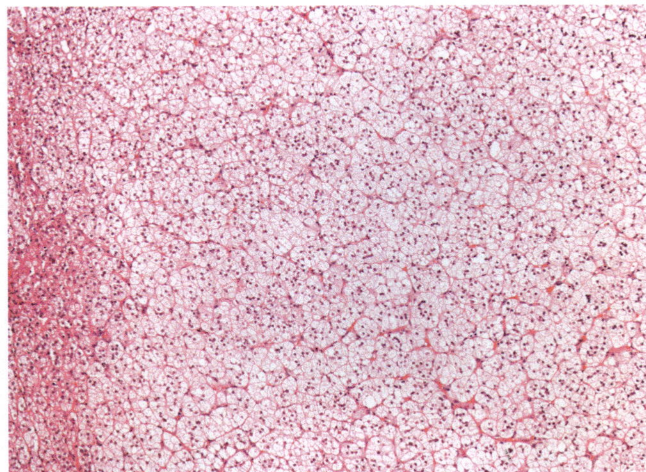

Fig. 21.53 Adrenal cortical adenoma, cortisol producing, forming a well-circumscribed solitary mass

Carney complex (*PRKAR1A*), familial hyperaldosteronism type 1 (*CYP11B1* and *CYP11TP53*), type II (*CLCN2*), type III (*KCNJ5*), type IV (*CACNAIH*), McCune–Albright, von Hippel–Lindau, and primary aldosteronism with seizure and neurological abnormalities (*CACNA1D*)

- *CYP11B2* encodes CYP11B2-aldosterone synthase
- Recently, genotypic phenotypic was reported in functional adrenal disease including the HISTALDO (histopathology of primary aldosteronism) classification system and CYP11B2 immunohistochemistry in aldosterone-producing adrenal cortical disease
- Genes involved in primary hyperaldosteronism: *KCNJ5* (11q24) and *CACNA1D* (3p14.3), *ATP1A1* (1p21), and *ATP2B3* (Xq28)
- Genes involved in adenomas associated with hypercortisolism: *PRKACA* (10p13.1), *PRKAR1A* (17q24.2), *ARMC5*, *GNAS*, and CTNNB1
- Adenomas associated with *CTNNB1* (Wnt/β-catenin) are often nonfunctional
- *TP53* mutations occur in 0–6%
- LOH may occur in one-third

Adrenal Cortical Carcinoma

Definition

- Malignant epithelial tumor of adrenal cortical cells

Clinical Features

- Rare, 1/1,000,000 population
- Bimodal, adults (40–60 years and children), females more often affected than males
- One-half with symptoms of functional tumor
- Virilization/feminization due to an adrenal mass in an adult is highly worrisome for malignancy
- Adrenal Cushing syndrome or a combination of Cushing syndrome and virilization/feminization are the most common functional types
- Aldosterone-producing adrenal cortical carcinomas are very rare
- Majority are sporadic, particularly in adults, but can occur in germline susceptibility syndromes
- Aggressive tumors with 5-year survival around 25%

Pathologic Features

- Large, often 10 cm or more, but can range in size, often with solid growth and necrosis and may have a trabecular growth pattern (Figs. 21.56 and 21.57)

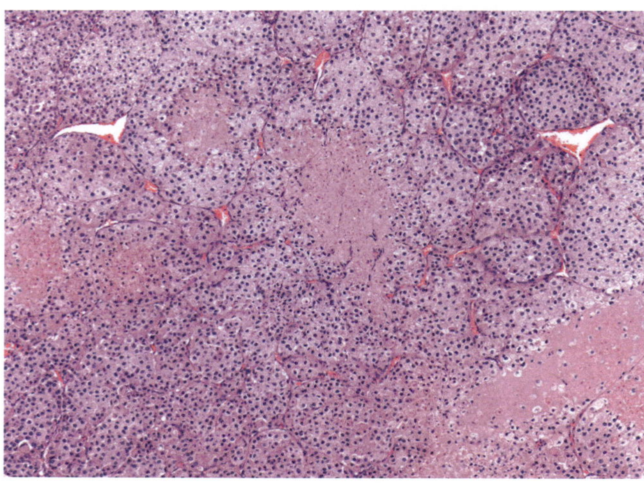

Fig. 21.56 Adrenal cortical carcinoma with solid growth and extensive necrosis

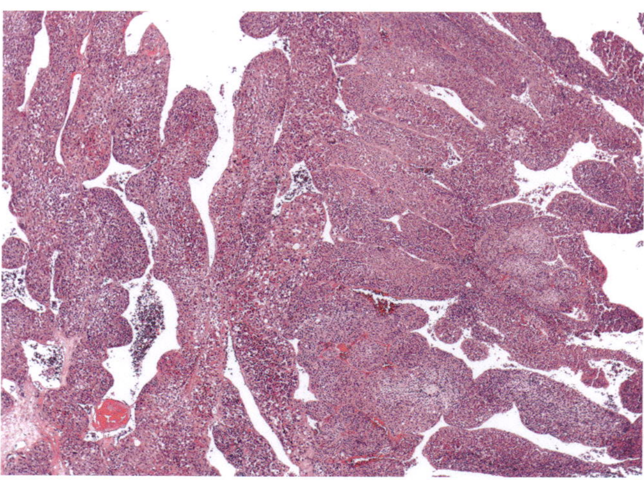

Fig. 21.57 Adrenal cortical carcinoma with an insular and trabecular growth pattern

- Increased mitotic activity including atypical mitotic figures, nuclear atypia, necrosis, and vascular invasion often present (Figs. 21.58, 21.59, and 21.60)
- Mitotic rate feature in most diagnostic algorithms and is the most important criterion for malignancy
- Mitotic rate and Ki-67 proliferative index used in grading and prognosis
- Mitotic rate in grading: low grade ≤20 mitotic figures/50 HPF; high grade >20 mitotic figures/50 HPF
- Many classification systems proposed: Weiss, VanSlooten, Aubert, Hough, Reticulin, Helsinki, etc
- Weiss system of 1984 and 1989 is most often used
  - Requires 3 of 9 features for diagnosis of malignancy: nuclear grade (Fuhrman grade III or IV), mitotic rate

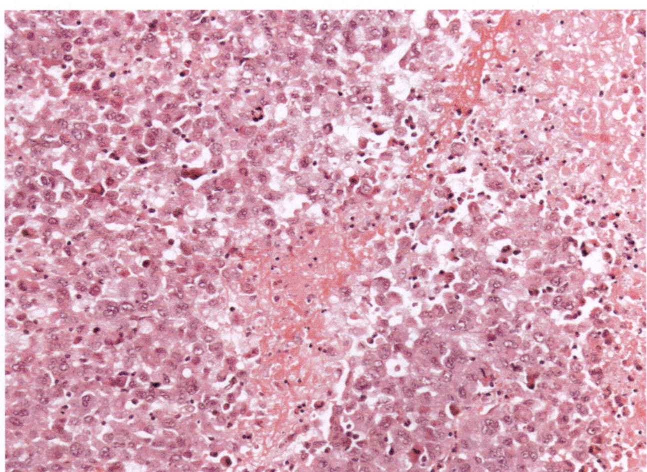

Fig. 21.58 Adrenal cortical carcinoma with necrosis

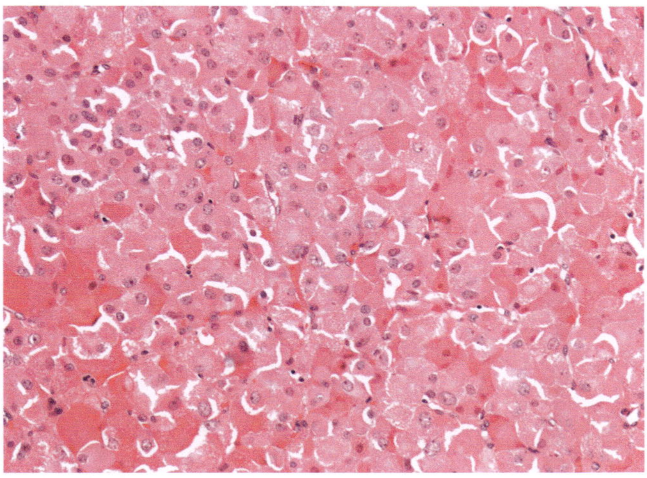

Fig. 21.61 Oncocytic adrenal cortical carcinoma required utilization of Lin Weiss Bisceglia system for classification

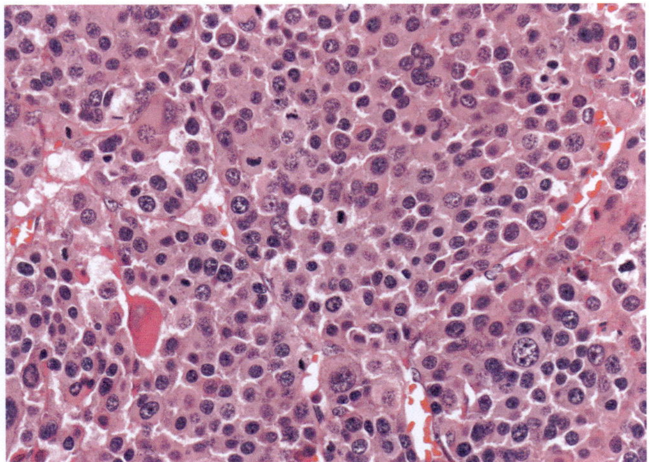

Fig. 21.59 Adrenal cortical carcinoma with marked cytologic atypia and mitotic activity

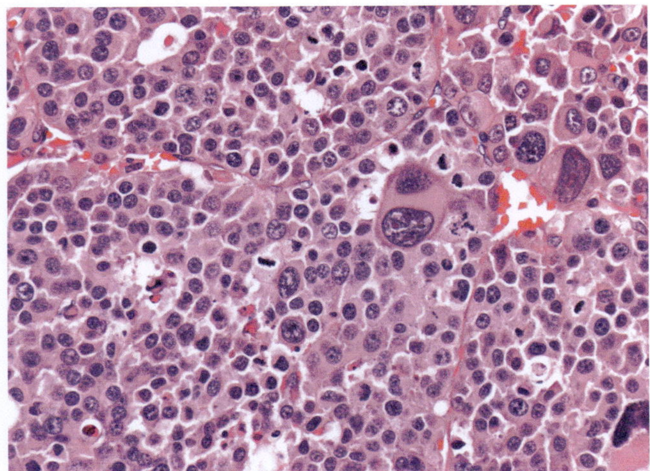

Fig. 21.60 Adrenal cortical carcinoma with mitotic activity and scattered multinucleated cells

>5/50 HPF, atypical mitotic figures, clear cells comprising ≤25% of the tumor, diffuse architecture in >1/3 of the tumor, necrosis, venous invasion, sinusoid invasion, capsular invasion

- Oncocytic adrenal cortical neoplasms (requires >90% oncocytic) classified by Lin, Weiss, Bisceglia system (Fig. 21.61)
- Three major criteria (high mitotic rate, atypical mitoses, and vascular invasion) and four minor criteria (large size >10cm or weight 200g, necrosis, capsular invasion, and sinusoidal invasion)
  - Malignant if one major criteria, uncertain malignant potential if one to four minor criteria, and benign if no major or minor criteria
- Reticulin algorithm and Helsinki score may be useful in oncocytic adrenal cortical tumors
- Myxoid adrenal cortical neoplasms may behave aggressively even with a Weiss score of 1 (Fig. 21.62)
- Weinke classification used for pediatric adrenal cortical tumors
- Immunopositive for synaptophysin, Mart1/MelanA steroidogenic factor 1, and alpha-inhibin (particularly in functional tumors) and negative for chromogranin-A

Genetic Features

- Numerous somatically mutated genes and can be a component of syndromes with germline mutations
- Most common somatically mutated genes: *TP53, CTNNB1, NF1, ATRX, MEN1, APC, EGFR, NOTCH1, GNAS, NRAS, RB1,* and *GNAS*
- *ZNRF3* shows biallelic inactivation in approximately 20% of cases

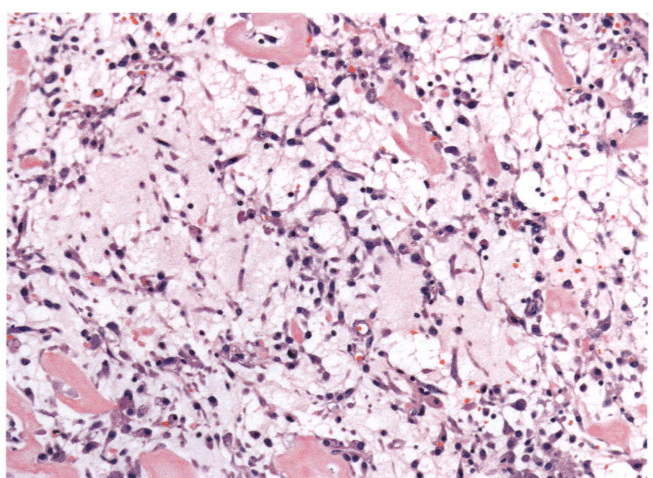

Fig. 21.62 Myxoid adrenal cortical tumors can be very difficult to classify as tumors with a Weiss score as low as 1 and can behave in an aggressive manner

- Hereditary syndromes: Li-Fraumeni (*TP53*), Lynch (*MSH2, MSH6, PMS2, MLH1, MEN1*, and familial adenomatous polyposis (*APC*)), Carney complex (*PRKAR1a*), Beckwith–Wiedeman (*H19* and *IGF2*), neurofibromatosis type 1 (*NF1*)
- Variety of other germline pathogenic events reported: *CHEK2, MUTYH, EGFR, ATM, FH, RET*, and *MSH*
- TP53
  - Most common mutations are inactivating *TP53* mutations
  - Adrenal cortical carcinoma is a component of Li-Fraumeni syndrome
  - Majority of children with adrenal cortical carcinoma have *TP53* mutation, while 3–5% of adults with adrenal cortical carcinoma may have *TP53* germline mutation
  - Somatic *TP53* mutation in 20–30% of sporadic adult adrenal cortical carcinomas
  - Allelic loss of 17p13 can occur in sporadic adult adrenal cortical carcinoma
  - *TP53* mutation is common in pediatric adrenal cortical carcinoma and is not predictive of outcome
  - *TP53* mutation in adult adrenal cortical carcinoma is associated with aggressive disease, large tumor size, high stage, and poor outcome
  - Although *TP53* mutation is less common in adult adrenal carcinoma, genetic counseling, and germline *TP53* testing may be helpful for all, as a study from the University of Michigan showed 7.5% of unselected patients with adrenal cortical carcinoma had germline *TP53* mutation (3 of 4 were over age 18)
- IGF2 overexpression and activation of Wnt/β-catenin pathway are common in adrenal cortical carcinoma
  - IGF1 and IGF2 overexpression in up to 80%
  - IGF2 (11p15) is associated with Beckwith–Wiedemann syndrome, which has alterations in 11p15 and is associated with adrenal cortical carcinoma
  - IGF2 has higher expression in adrenal carcinomas than adenomas
  - Wnt/β-catenin pathway is activated in both adrenal carcinomas and adenomas
  - Activation is usually due to *CTNNB1* (β-catenin gene) mutations
  - Activation is associated with decreased survival
  - *CTNNB1* mutations and abnormal β-catenin immunostaining are seen in adrenal cortical adenomas and carcinomas
  - *CTNNB1* mutations are common in large and nonsecreting adrenal cortical carcinomas suggesting Wnt/β-catenin activation may be associated with less differentiated carcinomas
- The Cancer Genome Atlas
  - Expanded adrenal cortical carcinoma driver genes: PRKAR1A, RPL22, TERF2, CCNB1, an NF1
  - Genes altered by somatic mutations, copy number, epigenetic silencing: TP53 (21%), ZNRF3 (19%), CDKN2A (15%), CTNNB1 (16%), TERT (14%), and PRKAR1A (11%)
  - Massive DNA loss followed by whole genome doubling frequency and associated with aggressive clinical course
  - Found three adrenal cortical carcinoma subtypes with distinct outcome and molecular alterations
- Low-risk adrenal cortical carcinomas: Lack mutations involving *TP53* or Wnt pathway, lack whole genome doubling, and have low levels of methylation
- High-risk adrenal cortical carcinomas: Have frequent mutations involving *TP53* or Wnt pathway, numerous nonrecurrent genetic alterations, and have a recurrent CpG island methylation

Pheochromocytomas and Paragangliomas

Definition

- Tumors associated with adrenal medulla and paraganglionic tissues, including pheochromocytomas and paragangliomas

Pheochromocytoma

Definition

- A tumor of chromaffin cells that arises in the adrenal medulla (intraadrenal paraganglioma, sympathetic paraganglioma)

Clinical Features

- May occur sporadically or in association with syndromes
- Any age, but most occur in the fourth to fifth decades
- Can be found incidentally or present with paroxysmal or sustained hypertension
- Hypertension, palpitation, tachycardia, tremors, and headaches may be present
- Elevated serum and urine catecholamines and metabolites of catecholamines, such as vanilmandelic acid
- Rate of metastatic disease varies with specific genetic mutations

Pathologic Features

- Circumscribed, unencapsulated tumors rising in the adrenal medulla usually 3–5 cm but size can vary
- Composed of nests of large polygonal cells with basophilic cytoplasm, referred to as the Zellballen pattern, although other growth patterns can occur (Figs. 21.63, 21.64, 21.65, and 21.66)
- Immunohistochemistry is positive for chromogranin-A and for synaptophysin and negative for cytokeratin
- Electron microscopy shows cytoplasmic dense core secretory granules
- S100 protein-positive sustentacular cells

Genetic Features

- Sporadic pheochromocytomas and paragangliomas can show somatic mutations in a variety of genes: *NF1, ATRX, VHL, HRAS, CDKN2A, HIF2A (EPAS1), RET, TP53, MET, MAX,* and *BRAF*
- Numerous hereditary susceptibility gene and genetic and/ or epigenetic alterations of at least 23 mutually exclusive

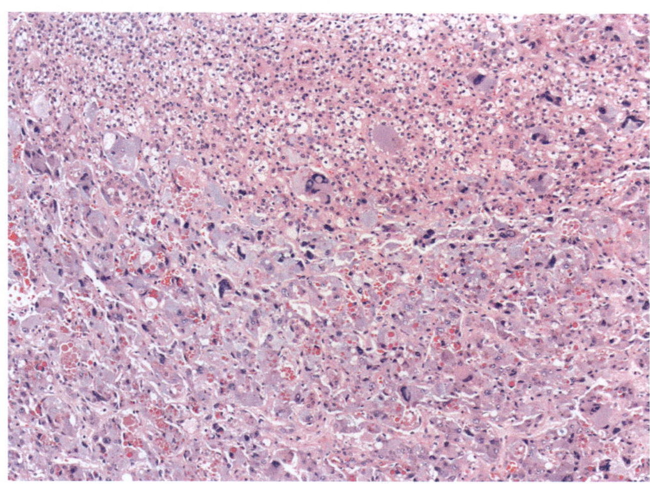

Fig. 21.64 Pheochromocytoma with nests of large polygonal cells with basophilic cytoplasm

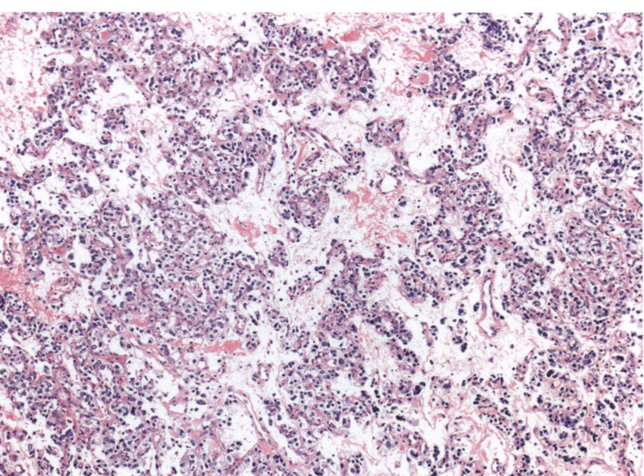

Fig. 21.65 Pheochromocytoma in the setting of Von Hippel–Lindau disease where the tumors have myxoid and edematous features

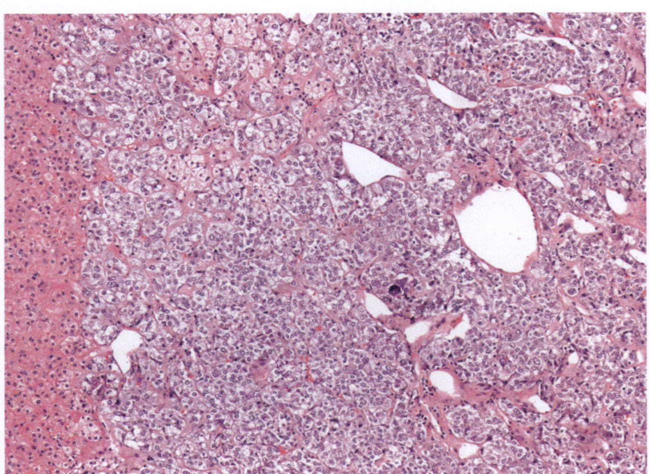

Fig. 21.63 Pheochromocytoma forming a circumscribed tumor in the adrenal medulla composed of nests of large polygonal cells with a Zellballen growth pattern

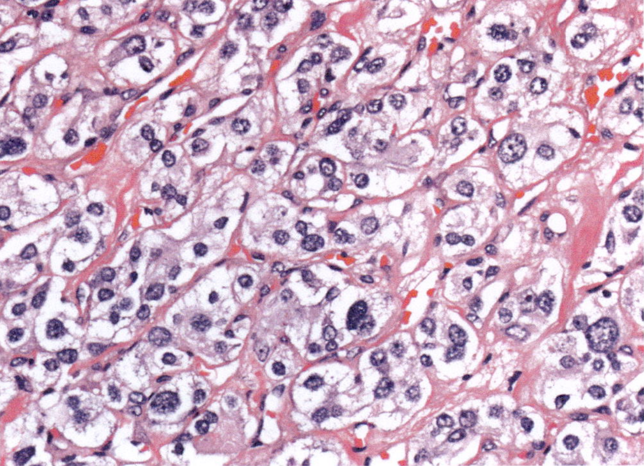

Fig. 21.66 Pheochromocytoma in the setting in von Hippel–Lindau disease where the tumors can have cytoplasmic clearing

genes are associated with paragangliomas and pheochromocytomas

- *MEN2A* and *MEN2B* mutations common in familial pheochromocytomas
- *VHL* mutations in 4% of pheochromocytomas
- *NF1* mutations present in about 4% of pheochromocytomas
- *PGL1* or *SDHD* rare in pheochromocytomas (approximately 4%)
- *PGL3* or *SDHC* mutations rare in pheochromocytomas
- *PGL4* or *SDHB* mutations in some pheochromocytomas (approximately 3%)
- *MEN2A* and *MEN2B* mutations are more common in familial tumors
- *NF1* mutations are uncommon in pheochromocytomas with metastatic disease
- *PGL1* or *SDHD* mutations present in a small percentage of pheochromocytomas with metastatic disease
- *PGL3* (*SDHC*) and *PGL4* (*SDHB*) mutations are rare in pheochromocytomas with metastatic disease

Paragangliomas

Definition

- Nonepithelial neuroendocrine tumors arising from the extraadrenal sympathetic and parasympathetic paraganglia

Clinical Features

- Tumors are distributed from the head and neck to the urinary bladder and organs of Zuckerkandl
- Head and neck paragangliomas arise from the extraadrenal paraganglia along parasympathetic nerves and are usually nonfunctional
- Sympathetic paragangliomas arise from extraadrenal paraganglia along the sympathetic chains of the prevertebral and paravertebral and sympathetic nerve fibers innervating organs and tissues and occur predominantly in the abdomen, retroperitoneum, pelvis, and thorax and occasionally near the cervical sympathetic ganglia
- Parasympathetic paraganglioma is derived from paraganglion cells of the autonomic nervous system and often referred to as head and neck paragangliomas; often associated with the vagus and glossopharyngeal nerves, although sympathetic paraganglioma can occur in association with the cervical sympathetic ganglia
- Most parasympathetic paragangliomas involve the carotid body or middle ear but can involve the vagal nerve trunk, sympathetic chain, or other areas in the head neck
- Multifocal tumors in familial cases

- Tumors may produce norepinephrine and less commonly epinephrine
- Rate of metastatic disease varies with location and specific genetic mutations

Pathologic Features

- Circumscribed, unencapsulated tumors
- Composed of nests of large polygonal cells with basophilic cytoplasm, referred to as Zellballen pattern, although other growth patterns can occur (Figs. 21.67 and 21.68)
- Growth patterns may have some genetic phenotypic correlation as the rosette pattern raises possibility of *SDHB* mutation

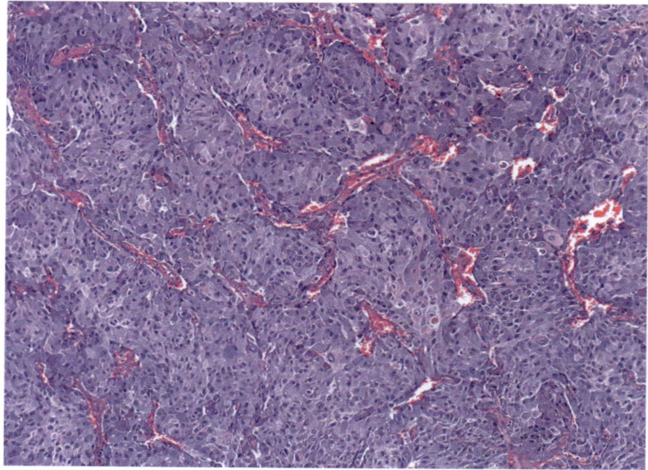

Fig. 21.67 Paraganglioma forming a circumscribed tumor

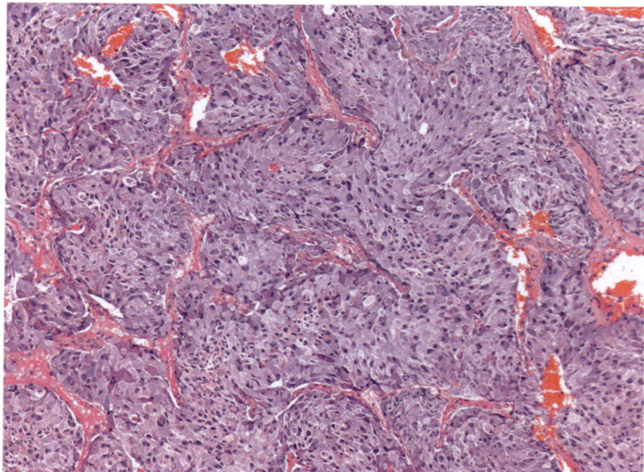

Fig. 21.68 Paraganglioma composed of nests of cells with amphophilic cytoplasm

Table 21.5 Familial pheochromocytomas (PHEO) and paragangliomas (PGL) and driver genes

Syndrome	Gene	Chromosome	Adrenal PHEO	PGL	Head and Neck PGL
Von Hippel–Lindau type 2	*VHL*	3p25–26	>50%	10–25%	Rare
MEN2	*RET*	10q11.2	50%	Rare	Rare
Neurofibromatosis type 1	*NF1*	17q11.1	5%	<5%	Rare
Familial PGL, MAX-related[a]	*MAX*	14q23.3	>50%	Rare <10%	Rare
Familial PGL, TMEM127-related[a]	*TMEM127*	2q11.2	>50%	Rare	<10%
Hereditary leiomyomatosis and renal cell cancer[a]	*FH*	1q42.1	<10%	<10%	<10%
Polycythemia pheochromocytoma paraganglioma	*EGLN1 (PHD2)*	1q42.1	<10%	<10%	–
PGL-somatostatinoma-polycythemia syndrome (Pacek-Zhuang)	*EPAS1*	2p21–p16	Rare	Rare	–
Familial paraganglioma type 1 (PGL1)	*SDHD*	11q23	<25%	<25%	>50% (80–90%)
Familial paraganglioma type 2 (PGL2)	*SDHAF2 (SDH5)*	11q12.2	<10% or none?	–	>50%
Familial paraganglioma type 3 (PGL3)	*SDHC*	1q21–23	<10%	Rare	>50%
Familial paraganglioma type 4 (PGL4)[a]	*SDHB*	1p36.1–p35	25–50% or 25%??	25–50% or 50–84%??	25–50% or 25%??
Familial paraganglioma type 5 (PGL5)	*SDHA*	5p15.33	25–50%	25–50%	25–50%

aIncreased frequency of metastasis

- Immunopositive for chromogranin-A, synaptophysin, dopamine beta-hydroxylase, and tyrosine hydroxylase and negative for cytokeratin
- Electron microscopy shows cytoplasmic dense core secretory granules
- S100 protein–positive sustentacular cells

Genetic Features

- Sporadic pheochromocytomas and paragangliomas can show somatic mutations a variety of genes, some of the most common are: *NF1, ATRX, VHL, HRAS, CDKN2A, HIF2A (EPAS1), RET, TP53, MET, MAX, BRAF* (Table 21.5)
- Numerous hereditary susceptibility gene and genetic and/or epigenetic alterations of at least 23 mutually exclusive genes are associated with paragangliomas and pheochromocytomas
- *PGL1* or *SDHD* mutations associated with head and neck paragangliomas
- *PGL2* or *SDH5* (*SDAF2*) mutations in head and neck paragangliomas
- *PGL3* or *SDHC* mutations in head and neck paragangliomas
- *PGL4* or *SDHB* mutations are commonly associated with metastatic disease in retroperitoneal paragangliomas
- Several inherited tumor syndromes associated with adrenal medullary tumors and paragangliomas

Hereditary Pheochromocytomas and Paragangliomas

- Up to 40% of paragangliomas and pheochromocytomas are associated with a germline pathogenic variant and even more with paragangliomas in children

- Males and females are equally affected, while women are more commonly affected by sporadic disease
- Individuals with multiple paragangliomas usually have a hereditary disease
- The Cancer Genome Atlas classified paragangliomas and pheochromocytomas into four groups:
  - Kinase signaling subtype (predominantly pheochromocytomas): Highest expression of PNMT; somatic and germline *NF1, RET, TMEM127*, and *HRAS* mutations and alterations in *BRAF*, NGFR, FGFR, and PKA subunits
  - Pseudohypoxia subtype (pheochromocytomas and paragangliomas): Germline *SDHB, SDHD*, and *VHL* mutations and somatic *VHL* and *EPAS1*
  - Wnt-altered subtype (pheochromocytomas): Driven by somatic *MAML3* gene fusions and *CSDE1* mutations
  - Cortical admixture subtype: *CYP11B2, CYP21A2, STAR*, and *MAX*
- Most common hereditary pheochromocytomas and paragangliomas
  - MEN2A (Sipple syndrome)
    - Autosomal dominant, *RET* (10q11.2)
    - Most *RET* mutations (95%) in codon 634 (exon 11), but can also involve codons 609, 611, 618, and 620
    - Up to 95% develop MTC, 50% develop pheochromocytoma, 15–30% develop hyperparathyroidism
    - Association with cutaneous lichen amyloidosis and Hirschprung disease in subset
  - MEN2B (Wagenmann-Froboese syndrome, MEN type 3, mucosal neuroma syndrome)
    - Autosomal dominant, *RET* (10q11.2)
    - Most *RET* mutations (95%) involve codon 918 (exon 16)
    - 75% of individuals with MEN2B have de novo *RET* mutations

- Less than 5% of *RET* mutations in MEN2B involve codon 883 (A883F, exon 15), and this is generally associated with less aggressive disease
  - Up to 95% develop MTC (often at very early age), 50% develop pheochromocytoma, association with mucosal neuromas and ganglioneuromatosis, thickened corneal nerves, and marfanoid habitus
- Von Hippel–Lindau
  - Autosomal dominant, *VHL* tumor suppressor gene (3p25.3)
  - *VHL* gene encodes E3 ubiquitin ligase protein involved in hypoxic response with HIF (hypoxia-inducible factor)
  - Central nervous system (cerebellar and spinal) hemangioblastoma, retinal capillary angiomas, retinal hemangioblastoma, pheochromocytomas and paragangliomas, renal cell carcinomas, renal cysts, pancreatic neuroendocrine tumors, pancreatic serous cystadenomas, endolymphatic sac rumors, café au lait spots
  - Type 1 VHL disease is associated with a low risk for pheochromocytomas (but a high risk for renal cell carcinoma and hemangioblastoma of central nervous system)
  - Type 2 VHL disease is associated with a high risk of pheochromocytoma
    - Type 2A: high-risk CNS hemangioblastoma, low-risk renal cell carcinoma
    - Type 2B: low-risk CNS hemangioblastoma, high-risk renal cell carcinoma
    - Type 2C: familial pheochromocytoma
  - Paragangliomas in Von Hippel–Lindau are associated with norepinephrine rather than epinephrine secretion
- Neurofibromatosis type 1 (von Recklinghausen disease)
  - Autosomal-dominant, *NF1* gene (17q11.2)
  - Multiple neurofibromas, café au lait patches, malignant peripheral nerve sheath tumors, optic nerve gliomas, lisch nodules, pheochromocytomas (<5%), rarely paraganglioma, duodenal carcinoid, skeletal dysplasias, learning disabilities
  - *NF1* gene encodes for the neurofibromin protein that downregulates p21
- Familial paraganglioma pheochromocytoma syndromes
  - Genes encoding subunits of succinate dehydrogenase (SDH) mitochondrial complex 2, a tumor suppressor
  - PGL type 1 on chromosome 11q23 associated with *SDHD* mutations: common in head and neck paraganglioma is in can be in abdominal paraganglioma send others

- PGL type 2 on chromosome 11q13.1 associated with *SDHAF2* (*SDH5*) mutations
- PGL type 3 on chromosome 1q21–23 associated with *SDHC* mutations: often in carotid body tumors
- PGL type 4 on chromosome 1q23–25 associated with *SDHB* mutations: abdominal paragangliomas
- PGL type 5 on chromosome 5p15 associated with *SDHA* mutations: abdominal paragangliomas, approximately 30% of SDH deficient gastrointestinal stromal tumors have *SDHA* mutation
- Tumors with *SDHB* mutations are more likely to be malignant

Suggested Readings

Thyroid Tumors

Abe I, Lam AK. Anaplastic thyroid carcinoma: current issues in genomics and therapeutics. Curr Oncol Rep. 2021;23(3):31.

Agrawal N, Jiao Y, Sausen M, et al. Exomic sequencing of medullary thyroid cancer reveals dominant and mutually exclusive oncogenic mutations in RET and RAS. J Clin Endocrinol Metab. 2013;98(2):E364–9.

Amacher AM, Goyal B, Lewis JS, et al. Prevalence of a hobnail pattern in papillary, poorly differentiated, and anaplastic thyroid carcinoma: a possible manifestation of high-grade transformation. Am J Surg Pathol. 2015;39(2):260–5.

Ambrosi F, Righi A, Ricci C, et al. Hobnail variant of papillary thyroid carcinoma: a literature review. Endocr Pathol. 2017;28(4):293–301.

Apel RL, Asa SL, LiVolsi VA. Papillary Hürthle cell carcinoma with lymphocytic stroma. "Warthin-like tumor" of the thyroid. Am J Surg Pathol. 1995;19(7):810–4.

Armstrong MJ, Yang H, Yip L, et al. PAX8/PPARγ rearrangement in thyroid nodules predicts follicular-pattern carcinoma, in particular the encapsulated follicular variant of papillary carcinoma. Thyroid. 2014;24(9):1369–74.

Asioli S, Erickson LA, Righi A, et al. Poorly differentiated carcinoma of the thyroid: validation of the Turin proposal and analysis of IMP3 expression. Mod Pathol. 2010a;23(9):1269–78.

Asioli S, Erickson LA, Sebo TJ, et al. Papillary thyroid carcinoma with prominent hobnail features: a new aggressive variant of moderately differentiated papillary carcinoma. A clinicopathologic, immunohistochemical, and molecular study of eight cases. Am J Surg Pathol. 2010b;34(1):44–52.

Asioli S, Erickson LA, Righi A, et al. Papillary thyroid carcinoma with hobnail features: histopathologic criteria to predict aggressive behavior. Hum Pathol. 2013;44(3):320–8.

Bai S, Baloch ZW, Samulski TD, et al. Poorly differentiated oncocytic (hürthle cell) follicular carcinoma: an institutional experience. Endocr Pathol. 2015;26(2):164–9.

Baloch Z, Mete O, Asa SL. Immunohistochemical biomarkers in thyroid pathology. Endocr Pathol. 2018;29(2):91–112.

Barletta JA, Nosé V, Sadow PM. Genomics and epigenomics of medullary thyroid carcinoma: from sporadic disease to familial manifestations. Endocr Pathol. 2021;32(1):35–43.

Basolo F, Pisaturo F, Pollina LE, et al. N-ras mutation in poorly differentiated thyroid carcinomas: correlation with bone metastases and inverse correlation to thyroglobulin expression. Thyroid. 2000;10:19–23.

Berho M, Suster S. The oncocytic variant of papillary carcinoma of the thyroid: a clinicopathologic study of 15 cases. Hum Pathol. 1997;28(1):47–53.

Boichard A, Croux L, Al Ghuzlan A, et al. Somatic RAS mutations occur in a large proportion of sporadic RET-negative medullary thyroid carcinomas and extend to a previously unidentified exon. J Clin Endocrinol Metab. 2012;97(10):E2031–5.

Boos LA, Dettmer M, Schmitt A, et al. Diagnostic and prognostic implications of the PAX8-PPARγ translocation in thyroid carcinomas-a TMA-based study of 226 cases. Histopathology. 2013;63(2):234–41.

Borrello MG, Smith DP, Pasini B, et al. RET activation by germline MEN2A and MEN2B mutations. Oncogene. 1995;11:2419–27.

Boyraz B, Sadow PM, Asa SL, et al. Cribriform-morular thyroid carcinoma is a distinct thyroid malignancy of uncertain cytogenesis. Endocr Pathol. 2021;32(3):327–35.

Cameselle-Teijeiro JM, Peteiro-González D, Caneiro-Gómez J, et al. Cribriform-morular variant of thyroid carcinoma: a neoplasm with distinctive phenotype associated with the activation of the WNT/β-catenin pathway. Mod Pathol. 2018;31(8):1168–79.

Cancer Genome Atlas Research Network. Integrated genomic characterization of papillary thyroid carcinoma. Cell. 2014;159(3):676–90.

Capezzone M, Robenshtok E, Cantara S, et al. Familial non-medullary thyroid cancer: a critical review. J Endocrinol Invest. 2021;44(5):943–50.

Carney JA, Ryan J, Goellner JR. Hyalinizing trabecular adenoma of the thyroid gland. Am J Surg Pathol. 1987;11(8):583–91.

Carney JA, Hirokawa M, Lloyd RV, et al. Hyalinizing trabecular tumors of the thyroid gland are almost all benign. Am J Surg Pathol. 2008;32(12):1877–89.

Castro P, Rebocho AP, Soares RJ, et al. PAX8-PPARgamma rearrangement is frequently detected in the follicular variant of papillary thyroid carcinoma. J Clin Endocrinol Metab. 2006;91(1):213–20.

Chen JH, Faquin WC, Lloyd RV, et al. Clinicopathological and molecular characterization of nine cases of columnar cell variant of papillary thyroid carcinoma. Mod Pathol. 2011;24(5):739–49.

Chou A, Fraser S, Toon CW, et al. A detailed clinicopathologic study of ALK-translocated papillary thyroid carcinoma. Am J Surg Pathol. 2015;39(5):652–9.

Ciampi R, Knauf JA, Kerler R, et al. Oncogenic AKAP9-BRAF fusion is a novel mechanism of MAPK pathway activation in thyroid cancer. J Clin Invest. 2005;115:94–101.

de Kock L, Sabbaghian N, Soglio DB, et al. Exploring the association Between DICER1 mutations and differentiated thyroid carcinoma. J Clin Endocrinol Metab. 2014;99(6):E1072–7.

Deeken-Draisey A, Yang GY, Gao J, et al. Anaplastic thyroid carcinoma: an epidemiologic, histologic, immunohistochemical, and molecular single-institution study. Hum Pathol. 2018;82:140–8.

Dettmer M, Schmitt A, Steinert H, et al. Poorly differentiated oncocytic thyroid carcinoma--diagnostic implications and outcome. Histopathology. 2012;60(7):1045–51.

Dobashi Y, Sugimura H, Sakamoto A, et al. Stepwise participation of p53 gene mutation during dedifferentiation of human thyroid carcinomas. Diagn Mol Pathol. 1994;3:9–14.

Doerfler WR, Nikitski AV, Morariu EM, et al. Molecular alterations in Hürthle cell nodules and preoperative cancer risk. Endocr Relat Cancer. 2021;28(5):301–9.

Dwight T, Thoppe SR, Foukakis T, et al. Involvement of the PAX8/peroxisome proliferator-activated receptor gamma rearrangement in follicular thyroid tumors. J Clin Endocrinol Metab. 2003;88:4440–5.

Elisei R, Cosci B, Romei C, et al. Prognostic significance of somatic RET oncogene mutations in sporadic medullary thyroid cancer: a 10-year follow-up study. J Clin Endocrinol Metab. 2008;93(3):682–7.

Eng C, Smith DP, Mulligan LM, et al. Point mutation within the tyrosine kinase domain of the RET proto-oncogene in multiple endocrine neoplasia type 2B and related sporadic tumours. Hum Mol Genet. 1994;3:237–47.

Eng C, Clayton D, Schuffenecker I, et al. The relationship between specific RET proto-oncogene mutations and disease phenotype in multiple endocrine neoplasia type 2. International RET mutation consortium analysis. JAMA. 1996;276:1575–9.

Erickson LA, Jalal SM, Goellner JR, et al. Analysis of Hurthle cell neoplasms of the thyroid by interphase fluorescence in situ hybridization. Am J Surg Pathol. 2001;25(7):911–7.

Erickson LA, Vrana JA, Theis J, et al. Analysis of amyloid in medullary thyroid carcinoma by mass spectrometry-based proteomic analysis. Endocr Pathol. 2015;26(4):291–5.

Fagin JA, Matsuo K, Karmakar A, et al. High prevalence of mutations of the p53 gene in poorly differentiated human thyroid carcinomas. J Clin Invest. 1993;91:179–84.

French CA, Alexander EK, Cibas ES, et al. Genetic and biological subgroups of low-stage follicular thyroid cancer. Am J Pathol. 2003;162(4):1053–60.

Ganly I, McFadden DG. short review: genomic alterations in Hürthle cell carcinoma. Thyroid. 2019 Apr;29(4):471–9.

Ganly I, Wang L, Tuttle RM, et al. Invasion rather than nuclear features correlates with outcome in encapsulated follicular tumors: further evidence for the reclassification of the encapsulated papillary thyroid carcinoma follicular variant. Hum Pathol. 2015;46(5):657–64.

Ganly I, Makarov V, Deraje S, et al. Integrated genomic analysis of Hürthle cell cancer reveals oncogenic drivers, recurrent mitochondrial mutations, and unique chromosomal landscapes. Cancer Cell. 2018;34(2):256–270.e5.

Gasparre G, Porcelli AM, Bonora E, et al. Disruptive mitochondrial DNA mutations in complex I subunits are markers of oncocytic phenotype in thyroid tumors. Proc Natl Acad Sci U S A. 2007;104(21):9001–6.

Ghossein RA, Leboeuf R, Patel KN, et al. Tall cell variant of papillary thyroid carcinoma without extrathyroid extension: biologic behavior and clinical implications. Thyroid. 2007;17(7):655–61.

Giordano TJ, Kuick R, Thomas DG, et al. Molecular classification of papillary thyroid carcinoma: distinct BRAF, RAS, and RET/PTC mutation-specific gene expression profiles discovered by DNA microarray analysis. Oncogene. 2005;24(44):6646–56.

Gnemmi V, Renaud F, Do Cao C, et al. Poorly differentiated thyroid carcinomas: application of the Turin proposal provides prognostic results similar to those from the assessment of high-grade features. Histopathology. 2014;64(2):263–73.

Gopal RK, Kübler K, Calvo SE, et al. Widespread chromosomal losses and mitochondrial DNA alterations as genetic drivers in Hürthle cell carcinoma. Cancer Cell. 2018;34(2):242–255.e5.

Gowrishankar S, Pai SA, Carney JA. Hyalinizing trabecular carcinoma of the thyroid gland. Histopathology. 2008 Mar;52(4):529–31.

Gucer H, Caliskan S, Kefeli M, et al. Do you know the details of your PAX8 antibody? Monoclonal PAX8 (MRQ-50) is not expressed in a series of 45 medullary thyroid carcinomas. Endocr Pathol. 2020;31(1):33–8.

Guilmette J, Nosé V. Hereditary and familial thyroid tumours. Histopathology. 2018;72(1):70–81.

Hara H, Fulton N, Yashiro T, et al. N-ras mutation: an independent prognostic factor for aggressiveness of papillary thyroid carcinoma. Surgery. 1994;116:1010–6.

Harach HR, Williams GT, Williams ED. Familial adenomatous polyposis associated thyroid carcinoma: a distinct type of follicular cell neoplasm. Histopathology. 1994;25(6):549–61.

Haugen BR, Sherman SI. Evolving approaches to patients with advanced differentiated thyroid cancer. Endocr Rev. 2013;34(3):439–55.

Hirokawa M, Carney JA. Cell membrane and cytoplasmic staining for MIB-1 in hyalinizing trabecular adenoma of the thyroid gland. Am J Surg Pathol. 2000;24(4):575–8.

Ito Y, Hirokawa M, Masuoka H, et al. Prognostic factors of minimally invasive follicular thyroid carcinoma: extensive vascular invasion significantly affects patient prognosis. Endocr J. 2013;60(5):637–42.

Karunamurthy A, Panebianco F. J Hsiao S, et al. Prevalence and phenotypic correlations of EIF1AX mutations in thyroid nodules. Endocr Relat Cancer. 2016;23(4):295–301.

Kazaure HS, Roman SA, Sosa JA. Medullary thyroid microcarcinoma: a population-level analysis of 310 patients. Cancer. 2012;118(3):620–7.

Kim TH, Kim YE, Ahn S, et al. TERT promoter mutations and long-term survival in patients with thyroid cancer. Endocr Relat Cancer. 2016;23(10):813–23.

Kim TH, Lee M, Kwon AY, et al. Molecular genotyping of the noninvasive encapsulated follicular variant of papillary thyroid carcinoma. Histopathology. 2018;72(4):648–61.

Kimura ET, Nikiforova MN, Zhu Z, et al. High prevalence of BRAF mutations in thyroid cancer: genetic evidence for constitutive activation of the RET/PTC-RAS-BRAF signaling pathway in papillary thyroid carcinoma. Cancer Res. 2003;63:1454–7.

Kraus C, Liehr T, Hulsken J, et al. Localization of the human β-catenin gene (CTNNB1) to 3p21: a region implicated in tumor development. Genomics. 1994;23:272–4.

Kunstman JW, Juhlin CC, Goh G, et al. Characterization of the mutational landscape of anaplastic thyroid cancer via whole-exome sequencing. Hum Mol Genet. 2015;24(8):2318–29.

Lai WA, Hang JF, Liu CY, et al. PAX8 expression in anaplastic thyroid carcinoma is less than those reported in early studies: a multi-institutional study of 182 cases using the monoclonal antibody MRQ-50. Virchows Arch. 2020a;476(3):431–7.

Lai WA, Liu CY, Lin SY, et al. Characterization of driver mutations in anaplastic thyroid carcinoma identifies RAS and PIK3CA mutations as negative survival predictors. Cancers (Basel). 2020b;12(7):1973.

Lam AK. Squamous cell carcinoma of thyroid: a unique type of cancer in World Health Organization Classification. Endocr Relat Cancer. 2020;27(6):R177–92.

Lam AK, Fridman M. Characteristics of cribriform morular variant of papillary thyroid carcinoma in post-Chernobyl affected region. Hum Pathol. 2018;74:170–7.

Lam AK, Montone KT, Nolan KA, et al. Ret oncogene activation in human thyroid neoplasms is restricted to the papillary cancer subtype. J Clin Invest. 1992;29:565–8.

Landa I, Ibrahimpasic T, Boucai L, et al. Genomic and transcriptomic hallmarks of poorly differentiated and anaplastic thyroid cancers. J Clin Invest. 2016;126(3):1052–66.

Lauper JM, Krause A, Vaughan TL, et al. Spectrum and risk of neoplasia in Werner syndrome: a systematic review. PLoS One. 2013;8(4):e59709.

Laury AR, Bongiovanni M, Tille JC, et al. Thyroid pathology in PTEN-hamartoma tumor syndrome: characteristic findings of a distinct entity. Thyroid. 2011;21(2):135–44.

Leeman-Neill RJ, Brenner AV, Little MP, et al. RET/PTC and PAX8/PPARγ chromosomal rearrangements in post-Chernobyl thyroid cancer and their association with iodine-131 radiation dose and other characteristics. Cancer. 2013;119(10):1792–9.

Leonardo E, Volante M, Barbareschi M, et al. Cell membrane reactivity of MIB-1 antibody to Ki67 in human tumors: fact or artifact? Appl Immunohistochem Mol Morphol. 2007;15(2):220–3.

Li M, Carcangiu ML, Rosai J. Abnormal intracellular and extracellular distribution of basement membrane material in papillary carcinoma and hyalinizing trabecular tumors of the thyroid: implication for deregulation of secretory pathways. Hum Pathol. 1997;28(12):1366–72.

Lin B, Ma H, Ma M, et al. The incidence and survival analysis for anaplastic thyroid cancer: a SEER database analysis. Am J Transl Res. 2019;11(9):5888–96.

Liu J, Singh B, Tallini G, et al. Follicular variant of papillary thyroid carcinoma: a clinicopathologic study of a problematic entity. Cancer. 2006;107(6):1255–64.

Liu X, Bishop J, Shan Y, et al. Highly prevalent TERT promoter mutations in aggressive thyroid cancers. Endocr Relat Cancer. 2013;20(4):603–10.

Liu Z, Zeng W, Chen T, et al. A comparison of the clinicopathological features and prognoses of the classical and the tall cell variant of papillary thyroid cancer: a meta-analysis. Oncotarget. 2017;8(4):6222–32.

Lui WO, Zeng L, Rehrmann V, et al. CREB3L2-PPARgamma fusion mutation identifies a thyroid signaling pathway regulated by intramembrane proteolysis. Cancer Res. 2008;68(17):7156–64.

Macerola E, Poma AM, Vignali P, et al. Molecular genetics of follicular-derived thyroid cancer. Cancers (Basel). 2021;13(5):1139.

Maenhaut C, Detours V, Dom G, et al. Gene expression profiles for radiation-induced thyroid cancer. Clin Oncol (R Coll Radiol). 2011;23(4):282–8.

Marchiò C, Da Cruz PA, Gularte-Merida R, et al. PAX8-GLIS3 gene fusion is a pathognomonic genetic alteration of hyalinizing trabecular tumors of the thyroid. Mod Pathol. 2019;32(12):1734–43.

Melo M, da Rocha AG, Vinagre J, et al. TERT promoter mutations are a major indicator of poor outcome in differentiated thyroid carcinomas. J Clin Endocrinol Metab. 2014;99(5):E754–65.

Moses W, Weng J, Kebebew E. Prevalence, clinicoPathologic Features, and somatic genetic mutation profile in familial versus sporadic nonmedullary thyroid cancer. Thyroid. 2011;21(4):367–71.

Moura MM, Cavaco BM, Pinto AE, et al. High prevalence of RAS mutations in RET-negative sporadic medullary thyroid carcinomas. J Clin Endocrinol Metab. 2011;96(5):E863–8.

Nabhan F, Ringel MD. Thyroid nodules and cancer management guidelines: comparisons and controversies. Endocr Relat Cancer. 2017;24(2):R13–26.

Nakamura N, Carney JA, Jin L, et al. RASSF1A and NORE1A methylation and BRAFV600E mutations in thyroid tumors. Lab Invest. 2005;85:1065–75.

Nakata T, Kitamura Y, Shimizu K, et al. Fusion of a novel gene, ELKS, to RET due to translocation t(10;12)(q11;p13) in a papillary thyroid carcinoma. Genes Chromosomes Cancer. 1999;25:97–103.

Namba H, Gutman RA, Matsuo K, et al. H-ras protooncogene mutations in human thyroid neoplasms. J Clin Endocrinol Metab. 1990;71:223–9.

Nath MC, Erickson LA. Aggressive variants of papillary thyroid carcinoma: hobnail, tall cell, columnar, and solid. Adv Anat Pathol. 2018;25(3):172–9.

Ngeow J, Mester J, Rybicki LA, et al. Incidence and clinical characterisics of thyroid cancer in prospective series of individuals with Cowden and Cowden-like syndrome characterized by germline PTEN, SDH, or KLLN alterations. J Clin Endocrinol Metab. 2011;96(12):E2063–71.

Nicolson NG, Murtha TD, Dong W, et al. Comprehensive genetic analysis of follicular thyroid carcinoma predicts prognosis independent of histology. J Clin Endocrinol Metab. 2018;103(7):2640–50.

Nieminen TT, Walker CJ, Olkinuora A, et al. Thyroid carcinomas that occur in familial adenomatous polyposis patients recurrently harbor somatic variants in APC, BRAF, and KTM2D. Thyroid. 2020;30(3):380–8.

Nikiforov YE, Rowland JM, Bove KE, et al. Distinct pattern of ret oncogene rearrangements in morphological variants of radiation-induced and sporadic thyroid papillary carcinomas in children. Cancer Res. 1997;57:1690–4.

Nikiforov YE, Bove KE, Rowland JM, et al. RET/PTC1 and RET/PTC3 rearrangements are associated with different biological behavior of papillary thyroid carcinoma (abstract). Mod Pathol. 2000;13:73A.

Nikiforov YE, Erickson LA, Nikiforova MN, et al. Solid variant of papillary thyroid carcinoma: incidence, clinical-pathologic characteristics, molecular analysis, and biologic behavior. Am J Surg Pathol. 2001;25(12):1478–84.

Nikiforova MN, Nikiforov YE. Molecular genetics of thyroid cancer: implications for diagnosis, treatment and prognosis. Expert Rev Mol Diagn. 2008;8:83–95.

Nikiforova MN, Biddinger PW, Caudill CM, et al. PAX8–PPAR gamma rearrangement in thyroid tumors: RT-PCR and immunohistochemical analyses. Am J Surg Pathol. 2002;26:1016–23.

Nikiforova MN, Kimura ET, Gandhi M, et al. BRAF mutations in thyroid tumors are restricted to papillary carcinomas and anaplastic or poorly differentiated carcinomas arising from papillary carcinomas. J Clin Endocrinol Metab. 2003a;88:5399–404.

Nikiforova MN, Lynch RA, Biddinger PW, et al. RAS point mutations and PAX8–PPAR gamma rearrangement in thyroid tumors: evidence for distinct molecular pathways in thyroid follicular carcinoma. J Clin Endocrinol Metab. 2003b;88:2318–26.

Nikiforova MN, Nikitski AV, Panebianco F, et al. GLIS rearrangement is a genomic hallmark of hyalinizing trabecular tumor of the thyroid gland. Thyroid. 2019;29(2):161–73.

Nozaki Y, Yamamoto H, Iwasaki T, et al. Clinicopathological features and immunohistochemical utility of NTRK-, ALK-, and ROS1-rearranged papillary thyroid carcinomas and anaplastic thyroid carcinomas. Hum Pathol. 2020;106:82–92.

Ohashi R, Kawahara K, Namimatsu S, et al. Clinicopathological significance of a solid component in papillary thyroid carcinoma. Histopathology. 2017;70(5):775–81.

Pacini F, Elisei R, Romei C, et al. RET proto-oncogene mutations in thyroid carcinomas: clinical relevance. J Endocrinol Invest. 2000;23(5):328–38.

Panebianco F, Nikitski AV, Nikiforova MN, et al. Characterization of thyroid cancer driven by known and novel ALK fusions. Endocr Relat Cancer. 2019;26(11):803–14.

Pekova B, Sykorova V, Mastnikova K, et al. NTRK fusion genes in thyroid carcinomas: clinicopathological characteristics and their impacts on prognosis. Cancers (Basel). 2021;13(8):1932.

Podda M, Saba A, Porru F, et al. Follicular thyroid carcinoma: differences in clinical relevance between minimally invasive and widely invasive tumors. World J Surg Oncol. 2015;13:193.

Proietti A, Sartori C, Macerola E, et al. Low frequency of TERT promoter mutations in a series of well-differentiated follicular-patterned thyroid neoplasms. Virchows Arch. 2017;471(6):769–73.

Rivera M, Tuttle RM, Patel S, et al. Encapsulated papillary thyroid carcinoma: a clinico-pathologic study of 106 cases with emphasis on its morphologic subtypes (histologic growth pattern). Thyroid. 2009;19(2):119–27.

Rivera M, Ricarte-Filho J, Knauf J, et al. Molecular genotyping of papillary thyroid carcinoma follicular variant according to its histological subtypes (encapsulated vs infiltrative) reveals distinct BRAF and RAS mutation patterns. Mod Pathol. 2010a;23(9):1191–200.

Rivera M, Ricarte-Filho J, Patel S, et al. Encapsulated thyroid tumors of follicular cell origin with high grade features (high mitotic rate/tumor necrosis): a clinicopathologic and molecular study. Hum Pathol. 2010b;41(2):172–80.

Rothenberg HJ, Goellner JR, Carney JA. Hyalinizing trabecular adenoma of the thyroid gland: recognition and characterization of its cytoplasmic yellow body. Am J Surg Pathol. 1999;23(1):118–25.

Salvatore G, Chiappetta G, Nikiforov YE, et al. Molecular profile of hyalinizing trabecular tumours of the thyroid: high prevalence of RET/PTC rearrangements and absence of B-raf and N-ras point mutations. Eur J Cancer. 2005;41(5):816–21.

Santoro M, Dathan NA, Berlingieri MT, et al. Molecular characterization of RET/PTC3; a novel rearrangement version of the RET protooncogene in a human thyroid papillary carcinoma. Oncogene. 1994;9:509–16.

Sheu SY, Schwertheim S, Worm K, et al. Diffuse sclerosing variant of papillary thyroid carcinoma: lack of BRAF mutation but occurrence of RET/PTC rearrangements. Mod Pathol. 2007;20(7):779–87.

Soares P, Fonseca E, Wynford-Thomas D, et al. Sporadic ret-rearranged papillary carcinoma of the thyroid: a subset of slow growing, less aggressive thyroid neoplasms? J Pathol. 1998;185:71–8.

Sohn SY, Lee JJ, Lee JH. Molecular profile and clinicopathologic features of follicular variant papillary thyroid carcinoma. Pathol Oncol Res. 2020;26(2):927–36.

Song T, Chen L, Zhang H, et al. Multimodal treatment based on thyroidectomy improves survival in patients with metastatic anaplastic thyroid carcinoma: a SEER analysis from 1998 to 2015. Gland Surg. 2020;9(5):1205–13.

Suchy B, Waldmann V, Klugbauer S, et al. Absence of RAS and p53 mutations in thyroid carcinomas of children after Chernobyl in contrast to adult thyroid tumours. Br J Cancer. 1998;77:952–5.

Szabo Yamashita T, Baky FJ, McKenzie TJ, et al. Occurrence and natural history of thyroid cancer in patients with cowden syndrome. Eur Thyroid J. 2020;9(5):243–6.

Teng L, Deng W, Lu J, et al. Hobnail variant of papillary thyroid carcinoma: molecular profiling and comparison to classical papillary thyroid carcinoma, poorly differentiated thyroid carcinoma and anaplastic thyroid carcinoma. Oncotarget. 2017;8(13):22023–33.

Us-Krasovec M, Golouh R. Papillary thyroid carcinoma with exuberant nodular fasciitis-like stroma in a fine needle aspirate. A case report. Acta Cytol. 1999;43(6):1101–4.

Van Hengel J, Nollet F, Berx G, et al. Assignment of the human β-catenin gene (CTNNB1) to 3p22–>p21.3 by fluorescence in situ hybridization. Cytogenet Cell Genet. 1995;70:68–70.

Vanzati A, Mercalli F, Rosai J. The "sprinkling" sign in the follicular variant of papillary thyroid carcinoma: a clue to the recognition of this entity. Arch Pathol Lab Med. 2013;137(12):1707–9.

Villar-Taibo R, Peteiro-González D, Cabezas-Agrícola JM, et al. Aggressiveness of the tall cell variant of papillary thyroid carcinoma is independent of the tumor size and patient age. Oncol Lett. 2017;13(5):3501–7.

Volante M, Collini P, Nikiforov YE, et al. Poorly differentiated thyroid carcinoma: the Turin proposal for the use of uniform diagnostic criteria and an algorithmic diagnostic approach. Am J Surg Pathol. 2007;31(8):1256–64.

Volante M, Lam AK, Papotti M, et al. Molecular pathology of poorly differentiated and anaplastic thyroid cancer: what do pathologists need to know? Endocr Pathol. 2021;32(1):63–76.

Vuong HG, Odate T, Duong UNP, et al. Prognostic importance of solid variant papillary thyroid carcinoma: A systematic review and meta-analysis. Head Neck. 2018a;40(7):1588–97.

Vuong HG, Odate T, Ngo HTT, et al. Clinical significance of RET and RAS mutations in sporadic medullary thyroid carcinoma: a meta-analysis. Endocr Relat Cancer. 2018b;25(6):633–41.

Wells SA, Asa SL, Dralle H, et al. Revised American Thyroid Association guidelines for the management of medullary thyroid carcinoma. Thyroid. 2015;25(6):567–610.

Wenig BM, Thompson LD, Adair CF, et al. Thyroid papillary carcinoma of columnar cell type: a clinicopathologic study of 16 cases. Cancer. 1998;82(4):740–53.

Wong KS, Dong F, Telatar M, et al. Papillary thyroid carcinoma with high-grade features versus poorly differentiated thyroid carcinoma: an analysis of clinicopathologic and molecular features and outcome. Thyroid. 2021;31(6):933–40.

Xing M, Liu R, Liu X, et al. BRAF V600E and TERT promoter mutations cooperatively identify the most aggressive papillary thyroid cancer with highest recurrence. J Clin Oncol. 2014 Sep;32(25):2718–26.

Xu B, Wang L, Tuttle RM, et al. Prognostic impact of extent of vascular invasion in low-grade encapsulated follicular cell-derived thyroid carcinomas: a clinicopathologic study of 276 cases. Hum Pathol. 2015;46(12):1789–98.

Xu B, Fuchs T, Dogan S, et al. Dissecting anaplastic thyroid carcinoma: a comprehensive clinical, histologic, immunophenotypic, and molecular study of 360 cases. Thyroid. 2020;30(10):1505–17.

Yakushina VD, Lerner LV, Lavrov AV. Gene fusions in thyroid cancer. Thyroid. 2018;28(2):158–67.

Yang J, Barletta JA. Anaplastic thyroid carcinoma. Semin Diagn Pathol. 2020;37(5):248–56.

Zhou X, Zheng Z, Chen C, et al. Clinical characteristics and prognostic factors of Hurthle cell carcinoma: a population based study. BMC Cancer. 2020;20(1):407.

Zhu Z, Ciampi R, Nikiforova MN, et al. Prevalence of Ret/Ptc rearrangements in thyroid papillary carcinomas: effects of the detection methods and genetic heterogeneity. J Clin Endocrinol Metab. 2006;91:3603–10.

Parathyroid Tumors

Al-Salameh A, Cadiot G, Calender A, et al. Clinical aspects of multiple endocrine neoplasia type 1. Nat Rev Endocrinol. 2021;17(4):207–24.

Arnold A, Kim HG, Gaz RD, et al. Molecular cloning and chromosomal mapping of DNA rearranged with the parathyroid hormone gene in a parathyroid adenoma. J Clin Invest. 1989;83:2034–40.

Arribas B, Cristobal E, Alcazar JA, et al. p53/MDM2 pathway aberrations in parathyroid tumors: p21(WAF-1) and MDM2 are frequently overexpressed in parathyroid adenomas. Endocr Pathol. 2000;11(3):251–7.

Brewer K, Costa-Guda J, Arnold A. Molecular genetic insights into sporadic primary hyperparathyroidism. Endocr Relat Cancer. 2019;26(2):R53–72.

Burgess JR, Harle RA, Tucker P, et al. Adrenal lesions in a large kindred with multiple endocrine neoplasia type 1. Arch Surg. 1996;131(7):699–702.

Carpten JD, Robbins CM, Villablanca A, et al. HRPT2, encoding parafibromin, is mutated in hyperparathyroidism-jaw tumor syndrome. Nat Genet. 2002;32(4):676–80.

Cetani F, Pardi E, Borsari S, et al. Genetic analyses of the HRPT2 gene in primary hyperparathyroidism: germline and somatic mutations in familial and sporadic parathyroid tumors. J Clin Endocrinol Metab. 2004;89(11):5583–91.

Cetani F, Marcocci C, Torregrossa L, et al. Atypical parathyroid adenomas: challenging lesions in the differential diagnosis of endocrine tumors. Endocr Relat Cancer. 2019;26(7):R441–64.

Chow LS, Erickson LA, Abu-Lebdeh HS, Wermers RA. Parathyroid lipoadenomas: a rare cause of primary hyperparathyroidism. Endocr Pract. 2006;12(2):131–6.

Cinque L, Pugliese F, Clemente C, et al. Rare somatic MEN1 gene pathogenic variant in a patient affected by atypical parathyroid adenoma. Int J Endocrinol. 2020;2020:2080797.

Clarke CN, Katsonis P, Hsu TK, et al. Comprehensive genomic characterization of parathyroid cancer identifies novel candidate driver mutations and core pathways. J Endocr Soc. 2019;3(3):544–59.

Cromer MK, Starker LF, Choi M, et al. Identification of somatic mutations in parathyroid tumors using whole-exome sequencing. J Clin Endocrinol Metab. 2012;97(9):E1774–81.

Cryns VL, Thor A, Xu HJ, et al. Loss of the retinoblastoma tumor-suppressor gene in parathyroid carcinoma. N Engl J Med. 1994a;330(11):757–61.

Cryns VL, Rubio MP, Thor AD, et al. p53 abnormalities in human parathyroid carcinoma. J Clin Endocrinol Metab. 1994b;78(6):1320–4.

Erickson LA, Mete O. Immunohistochemistry in diagnostic parathyroid pathology. Endocr Pathol. 2018;29(2):113–29.

Erickson LA, Jin L, Wollan P, et al. Parathyroid hyperplasia, adenomas, and carcinomas: differential expression of p27Kip1 protein. Am J Surg Pathol. 1999;23(3):288–95.

Erickson LA, Jin L, Papotti M, et al. Oxyphil parathyroid carcinomas: a clinicopathologic and immunohistochemical study of 10 cases. Am J Surg Pathol. 2002;26(3):344–9.

Fernandez-Ranvier GG, Khanafshar E, Tacha D, et al. Defining a molecular phenotype for benign and malignant parathyroid tumors. Cancer. 2009;115(2):334–44.

Fraker DL, Travis WD, Merendino JJ, et al. Locally recurrent parathyroid neoplasms as a cause for recurrent and persistent primary hyperparathyroidism. Ann Surg. 1991;213(1):58–65.

Gill AJ, Clarkson A, Gimm O, et al. Loss of nuclear expression of parafibromin distinguishes parathyroid carcinomas and hyperparathyroidism-jaw tumor (HPT-JT) syndrome-related adenomas from sporadic parathyroid adenomas and hyperplasias. Am J Surg Pathol. 2006;30(9):1140–9.

Gill AJ, Lim G, Cheung VKY, et al. Parafibromin-deficient (HPT-JT Type, CDC73 Mutated) parathyroid tumors demonstrate distinctive morphologic features. Am J Surg Pathol. 2019;43(1):35–46.

Haglund F, Juhlin CC, Brown T, et al. TERT promoter mutations are rare in parathyroid tumors. Endocr Relat Cancer. 2015;22(3):L9–L11.

Haven CJ, van Puijenbroek M, Karperien M, et al. Differential expression of the calcium sensing receptor and combined loss of chromosomes 1q and 11q in parathyroid carcinoma. J Pathol. 2004;202(1):86–94.

Haven CJ, van Puijenbroek M, Tan MH, et al. Identification of MEN1 and HRPT2 somatic mutations in paraffin-embedded (sporadic) parathyroid carcinomas. Clin Endocrinol (Oxf). 2007;67(3):370–6.

Hosny Mohammed K, Siddiqui MT, Willis BC, et al. Parafibromin, APC, and MIB-1 are useful markers for distinguishing parathyroid carcinomas from adenomas. Appl Immunohistochem Mol Morphol. 2017;25(10):731–5.

Howell VM, Haven CJ, Kahnoski K, et al. HRPT2 mutations are associated with malignancy in sporadic parathyroid tumours. J Med Genet. 2003;40(9):657–63.

Hu Y, Zhang X, Wang O, et al. The genomic profile of parathyroid carcinoma based on whole-genome sequencing. Int J Cancer. 2020;147(9):2446–57.

Juhlin CC, Erickson LA. Genomics and epigenomics in parathyroid neoplasia: from bench to surgical pathology practice. Endocr Pathol. 2021;32(1):17–34.

Juhlin CC, Kiss NB, Villablanca A, et al. Frequent promoter hypermethylation of the APC and RASSF1A tumour suppressors in parathyroid tumours. PLoS One. 2010a;5(3):e9472.

Juhlin CC, Nilsson IL, Johansson K, et al. Parafibromin and APC as screening markers for malignant potential in atypical parathyroid adenomas. Endocr Pathol. 2010b;21(3):166–77.

Juhlin CC, Nilsson IL, Lagerstedt-Robinson K, et al. Parafibromin immunostainings of parathyroid tumors in clinical routine: a near-decade experience from a tertiary center. Mod Pathol. 2019;32(8):1082–94.

Juhlin CC, Falhammar H, Zedenius J, et al. Lipoadenoma of the parathyroid gland: characterization of an institutional series spanning 28 years. Endocr Pathol. 2020;31(2):156–65.

Kruijff S, Sidhu SB, Sywak MS, et al. Negative parafibromin staining predicts malignant behavior in atypical parathyroid adenomas. Ann Surg Oncol. 2014;21(2):426–33.

Kutahyalioglu M, Nguyen HT, Kwatampora L, et al. Genetic profiling as a clinical tool in advanced parathyroid carcinoma. J Cancer Res Clin Oncol. 2019;145(8):1977–86.

Lin L, Zhang JH, Panicker LM, et al. The parafibromin tumor suppressor protein inhibits cell proliferation by repression of the c-myc proto-oncogene. Proc Natl Acad Sci U S A. 2008;105(45):17420–5.

Lloyd RV, Carney JA, Ferreiro JA, et al. Immunohistochemical analysis of the cell cycle-associated antigens Ki-67 and retinoblastoma protein in parathyroid carcinomas and adenomas. Endocr Pathol. 1995;6(4):279–87.

Mao X, Wu Y, Yu S, et al. Genetic alteration profiles and clinicopathological associations in atypical parathyroid adenoma. Int J Genomics. 2021;2021:6666257.

Pandya C, Uzilov AV, Bellizzi J, et al. Genomic profiling reveals mutational landscape in parathyroid carcinomas. JCI Insight. 2017;2(6):e92061.

Schneider R, Bartsch-Herzog S, Ramaswamy A, et al. Immunohistochemical expression of e-cadherin in atypical parathyroid adenoma. World J Surg. 2015;39(10):2477–83.

Seabrook AJ, Harris JE, Velosa SB, et al. Multiple endocrine tumors associated with germline MAX mutations: multiple endocrine neoplasia type 5? J Clin Endocrinol Metab. 2021;106(4):1163–82.

Shattuck TM, Välimäki S, Obara T, et al. Somatic and germ-line mutations of the HRPT2 gene in sporadic parathyroid carcinoma. N Engl J Med. 2003;349(18):1722–9.

Silva-Figueroa AM, Bassett R, Christakis I, et al. Using a novel diagnostic nomogram to differentiate malignant from benign parathyroid neoplasms. Endocr Pathol. 2019;30(4):285–96.

Simonds WF, Robbins CM, Agarwal SK, et al. Familial isolated hyperparathyroidism is rarely caused by germline mutation in HRPT2, the gene for the hyperparathyroidism-jaw tumor syndrome. J Clin Endocrinol Metab. 2004;89(1):96–102.

Soong CP, Arnold A. Recurrent ZFX mutations in human sporadic parathyroid adenomas. Oncoscience. 2014;1(5):360–6.

Sulaiman L, Haglund F, Hashemi J, et al. Genome-wide and locus specific alterations in CDC73/HRPT2-mutated parathyroid tumors. PLoS One. 2012;7(9):e46325.

Sulaiman L, Juhlin CC, Nilsson IL, et al. Global and gene-specific promoter methylation analysis in primary hyperparathyroidism. Epigenetics. 2013;8(6):646–55.

Talat N, Schulte KM. Clinical presentation, staging and long-term evolution of parathyroid cancer. Ann Surg Oncol. 2010;17(8):2156–74.

Thakker RV. Genetics of parathyroid tumours. J Intern Med. 2016;280(6):574–83.

Thakker RV, Bouloux P, Wooding C, et al. Association of parathyroid tumors in multiple endocrine neoplasia type 1 with loss of alleles on chromosome 11. N Engl J Med. 1989;321:218–24.

Wei Z, Sun B, Wang ZP, et al. Whole-exome sequencing identifies novel recurrent somatic mutations in sporadic parathyroid adenomas. Endocrinology. 2018;159(8):3061–8.

Williams MD, DeLellis RA, Erickson LA, et al. Pathology data set for reporting parathyroid carcinoma & atypical parathyroid neoplasm: recommendations from the international collaboration on cancer reporting (ICCR). Hum Pathol. 2020;110:73–82.

Wynne AG, van Heerden J, Carney JA, et al. Parathyroid carcinoma: clinical and Pathologic Features in 43 patients. Medicine (Baltimore). 1992;71(4):197–205.

Zhu R, Wang Z, Hu Y. Prognostic role of parafibromin staining and CDC73 mutation in patients with parathyroid carcinoma: a systematic review and meta-analysis based on individual patient data. Clin Endocrinol (Oxf). 2020;92(4):295–302.

Adrenal Cortical Tumors

Assié G, Libé R, Espiard S, et al. ARMC5 mutations in macronodular adrenal hyperplasia with Cushing's syndrome. N Engl J Med. 2013;369(22):2105–14.

Assié G, Letouzé E, Fassnacht M, et al. Integrated genomic characterization of adrenocortical carcinoma. Nat Genet. 2014;46(6):607–12.

Assié G, Jouinot A, Fassnacht M, et al. Value of molecular classification for prognostic assessment of adrenocortical carcinoma. JAMA Oncol. 2019;5(10):1440–7.

Aubert S, Wacrenier A, Leroy X, et al. Weiss system revisited: a clinicopathologic and immunohistochemical study of 49 adrenocortical tumors. Am J Surg Pathol. 2002;26(12):1612–9.

Beuschlein F, Fassnacht M, Assié G, et al. Constitutive activation of PKA catalytic subunit in adrenal Cushing's syndrome. N Engl J Med. 2014;370(11):1019–28.

Bisceglia M, Ludovico O, Di Mattia A, et al. Adrenocortical oncocytic tumors: report of 10 cases and review of the literature. Int J Surg Pathol. 2004;12(3):231–43.

Bourdeau I, Parisien-La Salle S, Lacroix A. Adrenocortical hyperplasia: a multifaceted disease. Best Pract Res Clin Endocrinol Metab. 2020;34(3):101386.

Brondani VB, Lacombe AMF, Mariani BMP, et al. Low protein expression of both *ATRX* and *ZNRF3* as novel negative prognostic markers of adult adrenocortical carcinoma. Int J Mol Sci. 2021;22(3):1238.

Carney JA, Gordon H, Carpenter PC, et al. The complex of myxomas, spotty pigmentation, and endocrine overactivity. Medicine (Baltimore). 1985;64(4):270–83.

de Krijger RR, Papathomas TG. Adrenocortical neoplasia: evolving concepts in tumorigenesis with an emphasis on adrenal cortical carcinoma variants. Virchows Arch. 2012;460(1):9–18.

de Reyniès A, Assié G, Rickman DS, et al. Gene expression profiling reveals a new classification of adrenocortical tumors and identifies molecular predictors of malignancy and survival. J Clin Oncol. 2009;27(7):1108–15.

Di Dalmazi G, Kisker C, Calebiro D, et al. Novel somatic mutations in the catalytic subunit of the protein kinase A as a cause of adrenal Cushing's syndrome: a European multicentric study. J Clin Endocrinol Metab. 2014;99(10):E2093–100.

Domènech M, Grau E, Solanes A, et al. Characteristics of adrenocortical carcinoma associated with lynch syndrome. J Clin Endocrinol Metab. 2021;106(2):318–25.

Duregon E, Fassina A, Volante M, et al. The reticulin algorithm for adrenocortical tumor diagnosis: a multicentric validation study on 245 unpublished cases. Am J Surg Pathol. 2013;37(9):1433–40.

Duregon E, Cappellesso R, Maffeis V, et al. Validation of the prognostic role of the "Helsinki Score" in 225 cases of adrenocortical carcinoma. Hum Pathol. 2017;62:1–7.

Else T, Lerario AM, Everett J, et al. Adrenocortical carcinoma and succinate dehydrogenase gene mutations: an observational case series. Eur J Endocrinol. 2017;177(5):439–44.

Espiard S, Drougat L, Libé R, et al. ARMC5 Mutations in a Large Cohort of Primary Macronodular Adrenal Hyperplasia: Clinical and Functional Consequences. J Clin Endocrinol Metab. 2015;100(6):E926–35.

Espiard S, Knape MJ, Bathon K, et al. Activating PRKACB somatic mutation in cortisol-producing adenomas. JCI Insight. 2018;3(8):e98296.

Faillot S, Foulonneau T, Néou M, et al. Genomic classification of benign adrenocortical lesions. Endocr Relat Cancer. 2021;28(1):79–95.

Gatta-Cherifi B, Chabre O, Murat A, et al. Adrenal involvement in MEN1. Analysis of 715 cases from the Groupe d'etude des Tumeurs Endocrines database. Eur J Endocrinol. 2012;166(2):269–79.

Gaujoux S, Pinson S, Gimenez-Roqueplo AP, et al. Inactivation of the APC gene is constant in adrenocortical tumors from patients with familial adenomatous polyposis but not frequent in sporadic adrenocortical cancers. Clin Cancer Res. 2010;16(21):5133–41.

Giordano TJ, Berney D, de Krijger RR, et al. Data set for reporting of carcinoma of the adrenal cortex: explanations and recommendations of the guidelines from the International Collaboration on Cancer Reporting. Hum Pathol. 2021;110:50–61.

Goh G, Scholl UI, Healy JM, et al. Recurrent activating mutation in PRKACA in cortisol-producing adrenal tumors. Nat Genet. 2014;46(6):613–7.

Guo X, Chen H, Fu H, et al. Hereditary leiomyomatosis and renal cell carcinoma syndrome combined with adrenocortical carcinoma on 18F-FDG PET/CT. Clin Nucl Med. 2017;42(9):692–4.

Henry I, Jeanpierre M, Couillin P, et al. Molecular definition of the 11p15.5 region involved in Beckwith–Wiedemann syndrome and probably in predisposition to adrenocortical carcinoma. Hum Genet. 1989;81(3):273–7.

Hsiao HP, Kirschner LS, Bourdeau I, et al. Clinical and genetic heterogeneity, overlap with other tumor syndromes, and atypical glucocorticoid hormone secretion in adrenocorticotropin-independent macronodular adrenal hyperplasia compared with other adrenocortical tumors. J Clin Endocrinol Metab. 2009;94(8):2930–7.

Jouinot A, Assie G, Libe R, et al. DNA methylation is an independent prognostic marker of survival in adrenocortical cancer. J Clin Endocrinol Metab. 2017;102(3):923–32.

Juhlin CC, Goh G, Healy JM, et al. Whole-exome sequencing characterizes the landscape of somatic mutations and copy number alterations in adrenocortical carcinoma. J Clin Endocrinol Metab. 2015;100(3):E493–502.

Juhlin CC, Bertherat J, Giordano TJ, et al. What did we learn from the molecular biology of adrenal cortical neoplasia? from histopathology to translational genomics. Endocr Pathol. 2021;32(1):102–33.

Kamilaris CDC, Stratakis CA, Hannah-Shmouni F. Molecular genetic and genomic alterations in Cushing's syndrome and primary aldosteronism. Front Endocrinol (Lausanne). 2021;12:632543.

Lenzini L, Rossitto G, Maiolino G, et al. A meta-analysis of somatic KCNJ5 K(+) channel mutations in 1636 patients with an aldosterone-producing adenoma. J Clin Endocrinol Metab. 2015;100(8):E1089–95.

Liu T, Brown TC, Juhlin CC, et al. The activating TERT promoter mutation C228T is recurrent in subsets of adrenal tumors. Endocr Relat Cancer. 2014;21(3):427–34.

Louiset E, Duparc C, Young J, et al. Intraadrenal corticotropin in bilateral macronodular adrenal hyperplasia. N Engl J Med. 2013;369(22):2115–25.

Lowe KM, Young WF, Lyssikatos C, et al. Cushing syndrome in carney complex: clinical, pathologic, and molecular genetic features in the 17 affected mayo clinic patients. Am J Surg Pathol. 2017;41(2):171–81.

Maillet M, Bourdeau I, Lacroix A. Update on primary micronodular bilateral adrenocortical diseases. Curr Opin Endocrinol Diabetes Obes. 2020;27(3):132–9.

Martins-Filho SN, Almeida MQ, Soares I, et al. Clinical impact of pathological features including the Ki-67 labeling index on diagnosis and prognosis of adult and pediatric adrenocortical tumors. Endocr Pathol. 2021;32(2):288–300.

Matyakhina L, Freedman RJ, Bourdeau I, et al. Hereditary leiomyomatosis associated with bilateral, massive, macronodular adrenocortical disease and atypical cushing syndrome: a clinical and molecular genetic investigation. J Clin Endocrinol Metab. 2005;90(6):3773–9.

Menon RK, Ferrau F, Kurzawinski TR, et al. Adrenal cancer in neurofibromatosis type 1: case report and DNA analysis. Endocrinol Diabetes Metab Case Rep. 2014;2014:140074.

Mete O, Asa SL, Giordano TJ, et al. Immunohistochemical biomarkers of adrenal cortical neoplasms. Endocr Pathol. 2018;29(2):137–49.

Nakamura Y, Maekawa T, Felizola SJ, et al. Adrenal CYP11B1/2 expression in primary aldosteronism: immunohistochemical analysis using novel monoclonal antibodies. Mol Cell Endocrinol. 2014;392(1-2):73–9.

Nanba K, Blinder AR, Rege J, et al. Somatic *CACNA1H* mutation as a cause of aldosterone-producing adenoma. Hypertension. 2020;75(3):645–9.

Papathomas TG, Pucci E, Giordano TJ, et al. An international Ki67 reproducibility study in adrenal cortical carcinoma. Am J Surg Pathol. 2016;40(4):569–76.

Papotti M, Volante M, Duregon E, et al. Adrenocortical tumors with myxoid features: a distinct morphologic and phenotypical variant exhibiting malignant behavior. Am J Surg Pathol. 2010;34(7):973–83.

Picard C, Orbach D, Carton M, et al. Revisiting the role of the pathological grading in pediatric adrenal cortical tumors: results from a national cohort study with pathological review. Mod Pathol. 2019;32(4):546–59.

Pilati C, Shinde J, Alexandrov LB, et al. Mutational signature analysis identifies MUTYH deficiency in colorectal cancers and adrenocortical carcinomas. J Pathol. 2017;242(1):10–5.

Rassi-Cruz M, Maria AG, Faucz FR, et al. Phosphodiesterase 2A and 3B variants are associated with primary aldosteronism. Endocr Relat Cancer. 2021;28(1):1–13.

Raymond VM, Everett JN, Furtado LV, et al. Adrenocortical carcinoma is a lynch syndrome-associated cancer. J Clin Oncol. 2013;31(24):3012–8.

Rege J, Nanba K, Blinder AR, et al. Identification of somatic mutations in *CLCN2* in aldosterone-producing adenomas. J Endocr Soc. 2020;4(10):bvaa123.

Scholl UI, Stölting G, Nelson-Williams C, et al. Recurrent gain of function mutation in calcium channel CACNA1H causes early-onset hypertension with primary aldosteronism. Elife. 2015;4:e06315.

Stratakis CA, Boikos SA. Genetics of adrenal tumors associated with Cushing's syndrome: a new classification for bilateral adrenocortical hyperplasias. Nat Clin Pract Endocrinol Metab. 2007;3(11):748–57.

Svahn F, Paulsson JO, Stenman A, et al. TERT promoter hypermethylation is associated with poor prognosis in adrenocortical carcinoma. Int J Mol Med. 2018;42(3):1675–83.

Thiel A, Reis AC, Haase M, et al. PRKACA mutations in cortisol-producing adenomas and adrenal hyperplasia: a single-center study of 60 cases. Eur J Endocrinol. 2015;172(6):677–85.

Tissier F, Cavard C, Groussin L, et al. Mutations of β-catenin in adrenocortical tumors: activation of the Wnt signaling pathway is a frequent event in both benign and malignant adrenocortical tumors. Cancer Res. 2005;65(17):7622–7.

Torres MB, Diggs LP, Wei JS, et al. Ataxia telangiectasia mutated germline pathogenic variant in adrenocortical carcinoma. Cancer Genet. 2021;256–257:21–5.

Vaduva P, Bonnet F, Bertherat J. Molecular basis of primary aldosteronism and adrenal cushing syndrome. J Endocr Soc. 2020;4(9):bvaa075.

Wasserman JD, Novokmet A, Eichler-Jonsson C, et al. Prevalence and functional consequence of TP53 mutations in pediatric adrenocortical carcinoma: a children's oncology group study. J Clin Oncol. 2015;33(6):602–9.

Weiss LM. Comparative histologic study of 43 metastasizing and nonmetastasizing adrenocortical tumors. Am J Surg Pathol. 1984;8(3):163–9.

Weiss LM, Medeiros LJ, Vickery AL. Pathologic Features of prognostic significance in adrenocortical carcinoma. Am J Surg Pathol. 1989;13(3):202–6.

Wieneke JA, Thompson LD, Heffess CS. Adrenal cortical neoplasms in the pediatric population: a clinicopathologic and immunophenotypic analysis of 83 patients. Am J Surg Pathol. 2003;27(7):867–81.

Williams TA, Gomez-Sanchez CE, Rainey WE, et al. International histopathology consensus for unilateral primary aldosteronism. J Clin Endocrinol Metab. 2021;106(1):42–54.

Wright JP, Montgomery KW, Tierney J, et al. Ectopic, retroperitoneal adrenocortical carcinoma in the setting of Lynch syndrome. Fam Cancer. 2018;17(3):381–5.

Wu VC, Wang SM, Chueh SJ, et al. The prevalence of CTNNB1 mutations in primary aldosteronism and consequences for clinical outcomes. Sci Rep. 2017;7:39121.

Zheng S, Cherniack AD, Dewal N, et al. Comprehensive pan-genomic characterization of adrenocortical carcinoma. Cancer Cell. 2016;29(5):723–36.

Zilbermint M, Xekouki P, Faucz FR, et al. Primary aldosteronism and ARMC5 Variants. J Clin Endocrinol Metab. 2015;100(6):E900–9.

Adrenal Medullary Tumors and Paragangliomas

Baysal BE. Hereditary paraganglioma targets diverse paraganglia. J Med Genet. 2002;39:617–22.

Baysal BE. Genomic imprinting and environment in hereditary paraganglioma. Am J Med Genet C Semin Med Genet. 2004;129C:85–90.

Baysal BE, Ferrell RE, Willett-Brozick JE, et al. Mutations in SDHD, a mitochondrial complex II gene, in hereditary paraganglioma. Science. 2000;287:848–51.

Bernardo-Castiñeira C, Valdés N, Sierra MI, et al. SDHC promoter methylation, a novel pathogenic mechanism in parasympathetic paragangliomas. J Clin Endocrinol Metab. 2018;103(1):295–305.

Cascon A, Landa I, Lopez-Jimenez E, et al. Molecular characterization of a common SDHB deletion in paraganglioma patients. J Med Genet. 2008;45:233–8.

Casey R, Neumann HPH, Maher ER. Genetic stratification of inherited and sporadic phaeochromocytoma and paraganglioma: implications for precision medicine. Hum Mol Genet. 2020;29(R2):R128–37.

Crona J, Lamarca A, Ghosal S, et al. Genotype-phenotype correlations in pheochromocytoma and paraganglioma: a systematic review and individual patient meta-analysis. Endocr Relat Cancer. 2019;26(5):539–50.

Favier J, Amar L, Gimenez-Roqueplo AP. Paraganglioma and phaeochromocytoma: from genetics to personalized medicine. Nat Rev Endocrinol. 2015;11(2):101–11.

Favier J, Meatchi T, Robidel E, et al. Carbonic anhydrase 9 immunohistochemistry as a tool to predict or validate germline and somatic VHL mutations in pheochromocytoma and paraganglioma-a retrospective and prospective study. Mod Pathol. 2020;33(1):57–64.

Fishbein L, Leshchiner I, Walter V, et al. Comprehensive molecular characterization of pheochromocytoma and paraganglioma. Cancer Cell. 2017;31(2):181–93.

Gimencz-Roqueplo AP, Favier J, Rustin P, et al. Mutations in the SDHB gene are associated with extra-adrenal and/or malignant phaeochromocytomas. Cancer Res. 2003;63:5615–21.

Hao HX, Khalimonchuk O, Schraders M, et al. SDH5, a gene required for flavination of succinate dehydrogenase, is mutated in paraganglioma. Science. 2009;28(325):1139–42.

Havekes B, Corssmit EP, Jansen JC, et al. Malignant paragangliomas associated with mutations in the succinate dehydrogenase D gene. J Clin Endocrinol Metab. 2007;92:1245–8.

Job S, Draskovic I, Burnichon N, et al. Telomerase activation and ATRX mutations are independent risk factors for metastatic pheochromocytoma and paraganglioma. Clin Cancer Res. 2019;25(2):760–70.

King EE, Dahia PLM. Molecular biology of pheochromocytoma and paragangliomas. In: Lloyd RV, editor. Endocrine pathology: differential diagnosis and molecular advances. 2nd ed. New York, NY: Springer; 2010. p. 297–305.

Korpershoek E, Petri BJ, Post E, et al. Adrenal medullary hyperplasia is a precursor lesion for pheochromocytoma in MEN2 syndrome. Neoplasia. 2014;16(10):868–73.

Korpershoek E, Koffy D, Eussen BH, et al. Complex MAX rearrangement in a family with malignant pheochromocytoma, renal oncocytoma, and erythrocytosis. J Clin Endocrinol Metab. 2016;101(2):453–60.

Latif F, Tory K, Gnarra J, et al. Identification of the von Hippel–Lindau disease tumor suppressor gene. Science. 1993;260:1317–20.

Melicow MM. One hundred cases of pheochromocytoma (107 tumors) at the Columbia–Presbyterian Medical Center, 1926–1976: a clinicopathological analysis. Cancer. 1977;40:1987–2004.

Mete O, Asa SL, Giordano TJ, et al. Immunohistochemical biomarkers of adrenal cortical neoplasms. Endocr Pathol. 2018 Jun;29(2):137–49.

Papathomas TG, Suurd DPD, Pacak K, et al. What have we learned from molecular biology of paragangliomas and pheochromocytomas? Endocr Pathol. 2021;32(1):134–53.

Raue F, Frank-Raue K. Multiple endocrine neoplasia type 2: 2007 update. Horm Res. 2007;68:101–4.

Rindi G, Klimstra DS, Abedi-Ardekani B, et al. A common classification framework for neuroendocrine neoplasms: an International Agency for Research on Cancer (IARC) and World Health Organization (WHO) expert consensus proposal. Mod Pathol. 2018;31(12):1770–86.

Seth R, Ahmed M, Hoschar AP, et al. Cervical sympathetic chain paraganglioma: a report of 2 cases and a literature review. Ear Nose Throat J. 2014;93(3):E22–7.

Skala SL, Dhanasekaran SM, Mehra R. Hereditary leiomyomatosis and renal cell carcinoma syndrome (HLRCC): a contemporary review and practical discussion of the differential diagnosis for HLRCC-associated renal cell carcinoma. Arch Pathol Lab Med. 2018;142(10):1202–15.

Thompson LDR, Gill AJ, Asa SL, et al. Data set for the reporting of pheochromocytoma and paraganglioma: explanations and recommendations of the guidelines from the International Collaboration on Cancer Reporting. Hum Pathol. 2020;110:83–97.

Turchini J, Cheung VKY, Tischler AS, et al. Pathology and genetics of phaeochromocytoma and paraganglioma. Histopathology. 2018 Jan;72(1):97–105.

van Nederveen FH, Gaal J, Favier J, et al. An immunohistochemical procedure to detect patients with paraganglioma and phaeochromocytoma with germline SDHB, SDHC, or SDHD gene mutations: a retrospective and prospective analysis. Lancet Oncol. 2009;10(8):764–71.

Williams MD. Paragangliomas of the head and neck: an overview from diagnosis to genetics. Head Neck Pathol. 2017;11(3):278–87.

Molecular Pathology of Soft Tissue and Bone Tumors

22

Adrian Marino-Enriquez, Alanna J. Church,
Neal I. Lindeman, and Paola Dal Cin

Contents

A. Marino-Enriquez · P. Dal Cin
Department of Pathology, The Center for Advanced Molecular
Diagnostics (CAMD), Brigham and Women's Hospital and
Harvard Medical School, Boston, MA, USA

A. J. Church
Department of Pathology, Boston Children's Hospital and Harvard
Medical School, Boston, MA, USA

N. I. Lindeman (✉)
Department of Pathology and Laboratory Medicine, Weill Cornell
Medicine, New York Presbyterian Hospital, New York, NY, USA
e-mail: nel4003@med.cornell.edu

General Concepts

Definition

- Soft tissue and bone tumors are mesenchymal neoplasms that are clinically, morphologically, and biologically heterogeneous
- Soft tissue and bone tumors are classified primarily by their line of differentiation (e.g., leiomyosarcoma, smooth muscle differentiation; chondrosarcoma, cartilaginous differentiation) rather than by anatomic site of origin
- The line of differentiation is unknown for a significant fraction of soft tissue and bone tumors, either due to lack of a nonneoplastic tissue counterpart (e.g., Ewing sarcoma, synovial sarcoma) or due to the absence of differentiation (i.e., undifferentiated sarcomas)
- Undifferentiated sarcomas represent 10–15% of soft tissue and bone sarcomas and are classified according to their histological appearances (spindle cell, round cell, epithelioid, pleomorphic)

Clinical Features

- Mesenchymal tumors display a wide range of biological behaviors, with a variety of presentations, treatments, and outcomes
  - Tumors range from clinically indolent lesions to aggressive malignancies called sarcomas
  - There is a group of mesenchymal tumors of intermediate biologic potential that tends to recur locally but rarely metastasize
- Benign mesenchymal tumors are very frequent, while sarcomas are rare
  - Soft tissue and bone sarcomas are much less common than carcinomas (according to American Cancer Society estimates for 2021), comprising approximately 1% of adult cancers and 14% of childhood cancers in USA

- Metastatic carcinomas to soft tissues and bone are more common than primary malignant mesenchymal neoplasms
- Primary sarcomas of soft tissues are more common than primary sarcomas of bone
- Making a diagnosis is challenging, often requiring expert clinicopathological evaluation, with the integration of clinical, morphological, immunohistochemical, and molecular data
- Prognosis depends on disease *stage*—an anatomical classifier function of tumor size, local extension, and distant spread; and *grade*—a histopathological classifier function of the tumor type, degree of differentiation, mitotic rate, and amount of necrosis
- In general, sarcomas tend to spread to distant sites via hematogenous rather than lymphatic spread

Basic Principles

- In contrast to carcinoma oncogenesis, which proceeds through the accumulation of a sequence of alterations, sarcoma oncogenesis is more often associated with one or few oncogenic alterations
- The identification of recurrent molecular alterations in sarcomas is contributing to a refined morphology-based classification system, with novel entities and subtypes recognized regularly
- In clinical practice, molecular studies are often used to confirm a suspected diagnosis and are particularly useful for small biopsy specimens
  - Predictive biomarkers are increasingly being described as more targeted therapeutics become available
- Gene fusions are characteristically initiating or early oncogenic events in a subset of mesenchymal tumors (see Gene Fusions and Gene Promiscuity section below)
- Focal somatic copy number alterations may also be strongly associated with some diagnose
  - *MDM2* and *CDK4* amplification in well-differentiated and dedifferentiated liposarcoma and parosteal osteosarcoma
  - *SMARCB1* deletion in rhabdoid tumor and epithelioid sarcoma
  - 13q/*RB1* deletion in atypical spindle cell adipocytic neoplasms, cellular angiofibroma and mammary-type myofibroblastoma
- Sequence variants are less common but can be characteristic
  - *PDGFRB* or *NOTCH3* alterations in myofibroma
  - *KIT* or *PDGFRA* activating mutations in GIST
  - *TSC1/TSC2* inactivating variants in PEComa
  - *IDH1/IDH2* mutations in cartilaginous tumors

- Genetic predisposition syndromes due to germline alterations underly the development of some sarcomas, which may be identified either intentionally or incidentally with sequencing techniques
  - *NF1* alterations in neurofibromatosis type 1
  - *TP53* mutations in Li-Fraumeni syndrome, and
  - *RB1* mutations in hereditary retinoblastoma

Gene Fusions and Gene Promiscuity

- Many benign and malignant bone and mesenchymal lesions are characterized by chromosomal rearrangements resulting in gene fusions
- Gene fusions can be identified using a variety of molecular techniques (Fig. 22.1; General Technical Considerations section, below)
- The development of next-generation sequencing (NGS) approaches has increased our appreciation of the number and diversity of gene fusions implicated in sarcomagenesis
- An increasing number of recurrent gene fusions are being described and are generally considered disease-defining—such as *BCOR* or *CIC* rearrangements, characteristic of novel types of undifferentiated round cell sarcomas as described in the 2020 WHO classification
- Oncogenic gene fusions are heterogeneous and can be categorized according to the function of the partners genes involved
  - Gene fusions involving kinases are the predominant oncogenic drivers in mesenchymal tumors, including *ALK, RET, ROS1, FGFR, BRAF*, and *NTRK1-3*
    - These fusions are diagnostically important and represent therapeutic targets for kinase inhibitors
  - Gene fusions involving transcription factors are another class of fusions that results in the creation of chimeric transcription factors with aberrant transcriptional activity, such as *EWSR1* fusions in Ewing sarcoma and *FOXO1* fusions in rhabdomyosarcoma
  - Inactivating rearrangements commonly involve tumor suppressor genes, such as *TP53* or *RB1* rearrangements that are frequent in osteosarcoma and leiomyosarcoma
- Although some gene fusions are unique to a specific sarcoma type and hence serve as diagnostic biomarkers, others occur in different tumor types, including tumors of different lineages such as carcinomas and hematological malignancies
  - *ETV6::NTRK3* fusions are the main oncogenic driver in infantile fibrosarcoma/congenital mesoblastic nephroma, mammary analog secretory carcinoma, and in some cases of acute myeloid leukemia

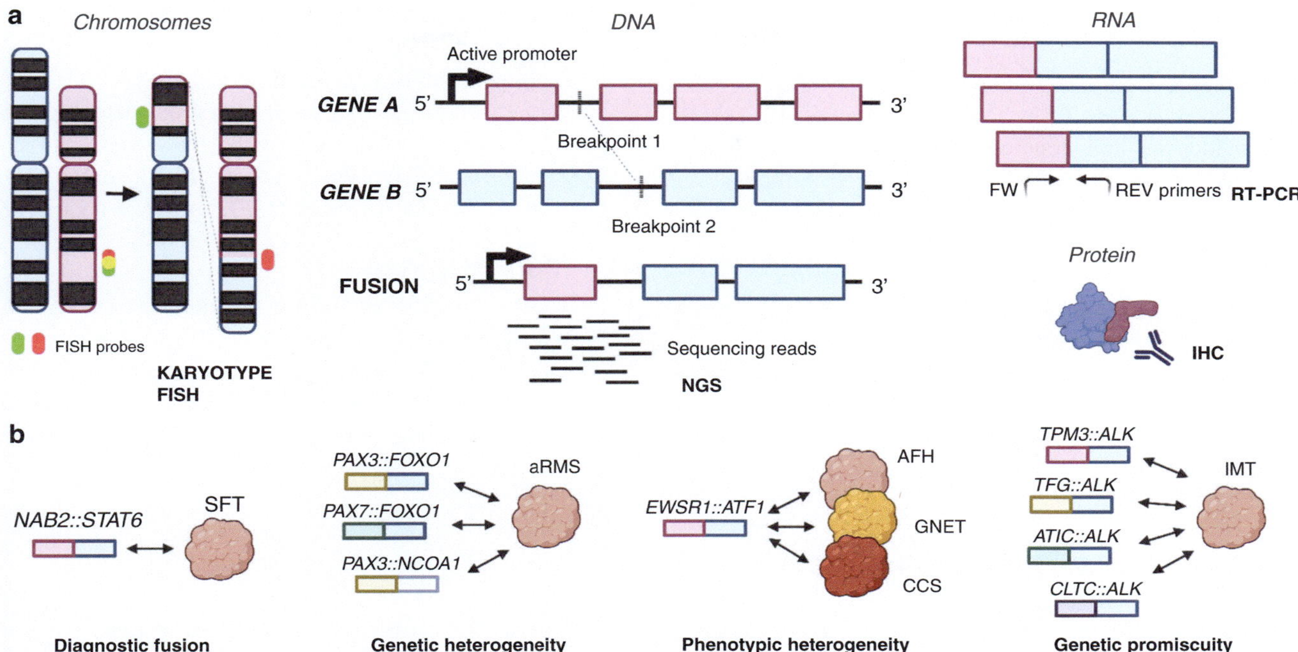

Fig. 22.1 Fusion genes are characteristic of a subset of mesenchymal tumors. (**a**) Diagnostic tools and their application to the identification of gene fusions, illustrated here with an example of a balanced reciprocal chromosomal translocation. Widely available techniques include karyotype and FISH (left), next-generation sequencing (NGS, center), RT-PCR with specific or degenerate primers, and immunohistochemistry (right). (**b**) Examples of genotype/phenotype correlations for gene fusions involved in mesenchymal tumors. Some gene fusions are present in a single tumor type (left, SFT: solitary fibrous tumor); some mesenchymal tumors can be driven by different fusions (genetic heterogeneity; aRMS: alveolar rhabdomyosarcoma); certain fusions may drive different tumor types (phenotypic heterogeneity; AFH, angiomatoid fibrous histiocytoma); GNET: gastrointestinal neuroectodermal tumor; CCS: clear cell sarcoma); some genes form fusions with many different gene partners, demonstrating genetic promiscuity (IMT: inflammatory myofibroblastic sarcoma). Created with BioRender.com

- *EWSR1::ATF1* occurs in angiomatoid fibrous histiocytoma, clear cell sarcoma, gastrointestinal neuroectodermal tumor, clear cell odontogenic carcinoma and hyalinizing clear cell carcinoma of the salivary glands
- Conversely, certain sarcoma types may be driven by different gene fusions, with variable degrees of homology and biological overlap
  - Myxoid liposarcomas may show *EWSR1::DDIT3* or *FUS::DDIT3* fusions
  - Alveolar rhabdomyosarcoma may be driven by *PAX3::FOXO1* or *PAX7::FOXO1* fusions
- Interestingly, some genes are involved in fusions with various fusion partners
  - *EWSR1* can contribute to many different fusions with partners as diverse as *FLI1*, *ATF1*, *WT1*, *NR4A3*, or *DDIT3*, among others
  - Some of these fusions are relatively homologous and occur in tumors that share biological features, while others are very different and occur in clinically and pathologically diverse sarcomas and nonmesenchymal tumors

- "Gene promiscuity" refers to the ability of some genes to contribute to different fusion genes with a variety of fusion partners, although it may also refer to the ability of some fusion genes to drive different tumor types
- Tyrosine kinase genes also pair with a variety of gene partners; importantly, the occurrence of kinase fusions in different tumor types allows for a larger number of patients to benefit from tyrosine kinase inhibitor treatment independent of tumor histology, if the molecular target is present (so-called "histology-agnostic" treatment indications)
  - Treatment of tumors with activated kinase fusions, such as *ALK* or growth factors, such as *PDGFB,* provides significant clinical benefit to sarcoma patients (see Predictive Molecular Alterations section below)
- In practice, the presence or absence of gene fusions in mesenchymal tumors must be interpreted in the context of clinical, morphological and immunohistochemical data to achieve an integrated pathological diagnosis
  - The election of the most appropriate molecular techniques (FISH vs. sequencing techniques, for example) will depend on the differential diagnosis under consideration

General Technical Considerations

Karyotype Analysis

- Provides a whole-genome assessment of large chromosomal abnormalities (>5 Mb), including numerical and structural alterations, at single cell resolution
- Requires fresh tissue
- Requires the establishment of primary cultures
  - In general, soft tissue tumors grow better than bone lesions, and usually require 3–4 days
- Cultured cells are arrested in metaphase, then stained, usually with Giemsa (GTG-banding)
- Admixture of benign stromal cells may lead to overgrow of normal tissue

Fluorescence in situ Hybridization (FISH)

- Robust, widely available technique for identifying gene rearrangements and copy number alterations
- It enables the detection of locus-specific chromosomal abnormalities, including alterations that are cryptic by standard GTG-banding
- Probes hybridize into unique DNA sequences, which vary in size from about 1 to 100 kilobases
  - Up to three fluorochromes (red/orange, green, and aqua) may be used simultaneously
- Requires prior knowledge of a suspected aberration
- Can be used on fresh-frozen tissue, FFPE tissue, cytology specimens and peripheral blood or fluid specimens
  - Interphase FISH can be performed on intact nuclei or tissue sections
  - Metaphase FISH can be performed on metaphase spreads
  - FISH on tissue sections (4–5 μm) is subject to section artifact and requires the analysis of a high number of cells (100–200)
- Poor cell preservation, as well as decalcification process in bone specimens can impair FISH interpretation,
- Various FISH designs may be used, including breakapart, fusion probes, or locus-specific designs to detect alterations of interest
  - Breakapart assay designs use differentially labeled probes (usually red/green) that hybridize to the 5′ and 3′ end of a gene of interest
    - Split signals suggest the presence of a rearrangement
    - The location of the probes is relevant to appropriately interpret unbalanced rearrangements, monosomy, or negative results

- Small insertions or deletions may be missed with breakapart probes
    - Intrachromosomal rearrangements due to paracentric inversions or interstitial deletions that result in fusions of neighboring genes may not be identified with certain FISH probe sets (e.g., inv(X) (p11.2q12)/*TFE3::NONO*, or inv(12) (q13q13)/*NAB2::STAT6*, or del(8)(q13.3q21.1)/ *HEY1::NCOA2*)
  - Breakapart FISH performed on interphase nuclei does not identify the gene partners involved in fusions
  - Dual fusion probe designs can distinguish rearrangement variants and identify both gene partners involved in a fusion, but require separate assays for each partner pair

Single Gene Sequencing Assays

- Progressively replaced by sequencing panels, but still commonly used for validation purposes
  - Widely available and cost-effective
- PCR enables analysis of point mutations, small deletions/ insertions, and molecular variants of translocations with known breakpoints
  - Detection involves gel or capillary electrophoresis and occasionally Sanger sequencing
- Reverse transcription-PCR (RT-PCR) is applied to RNA to detect small molecular variations as DNA PCR
  - Main differences include variable template expression and absence of introns
- Sensitivity may be enhanced by a variety of techniques, including real-time probe detection
- For the detection of translocations, DNA PCR is less useful because breakpoints are usually variable within large introns and require a very large PCR product
  - RT-PCR is preferable because the exons that flank breakpoints are less variable and produce a small enough amplicon for successful analysis (e.g., detection of *HEY1::NCOA2* in mesenchymal chondrosarcoma)
- Digital droplet PCR (ddPCR) is an ultrasensitive approach that uses fluorescence detection to detect sequence variants and copy number alterations within thousands of droplets in an emulsion
  - Applications in sarcomas include the detection of *MYOD1 L122R* in sclerosing rhabdomyosarcoma
  - ddPCR is not routinely used for the detection of fusions at this time, although proof-of-concept designs have been published

Next-Generation Sequencing

- High-throughput sequencing assays are increasingly being integrated into clinical molecular laboratories and offer the opportunity to simultaneously detect multiple types of alterations
- DNA or RNA may be used as the substrate for library preparation
- Targeted panels designed for somatic analyses work with FFPE tissues as they are designed for use with fragmented nucleic acids
- Whole genome and whole transcriptome sequencing assays typically work best with fresh or frozen tissues
- Amplicon library preparation uses PCR and RT-PCR to create sequencing libraries and is often used for smaller panels with a quicker turnaround time
- Hybrid capture library preparation uses baits to capture the regions of interest to create libraries and is typically used for larger panels
- Large panels may interrogate sequence variants, copy number alterations, and rearrangements in one assay
  - Pan-cancer DNA panels may not cover many fusions relevant to mesenchymal tumors
- Multiplex RNA-based sequencing panels are available, with enough sensitivity to be applied to FFPE samples, allowing for the simultaneous detection of multiple fusions; these are particularly useful given the rarity and histologic overlap of many mesenchymal tumors
  - Multiplex RT-PCR panels combine library preparation reactions, which amplify fusion pairings where both partners are known and targeted, e.g., *ETV6::NTRK3*; these would not identify nontargeted gene partners
  - Anchored multiplex PCR library, preparation allows for the identification of multiple fusion partners for a given "anchor" gene; these allow the identification of multiple partner genes, known or unknown
    - By targeting *EWSR1*, an anchored assay could identify and distinguish *EWSR1::FLI1, EWSR1::ERG, EWSR1::ATF1*, and *EWSR1::WT1* fusions, and could identify novel fusion partners
  - Transcriptome seq (RNA-seq) is a costly option that is usually utilized for discovery purposes; sample barcoding and pooling or other cost-reduction strategies may make it a viable option for clinical laboratories that manage large volume of samples
- In-depth review of the technical aspects of NGS is outside of the scope of this chapter; assay design, its chemistry, and bioinformatic analyses require substantial planning and expertise

Chromosomal Microarray

- Chromosomal microarrays (SNP-based or array CGH-based) are the gold standard for copy number evaluation, with a lower limit of detection between 10 kb and 1 Mb depending on the platform
  - They are most accurate in identifying deletions or amplifications involved in creating mesenchymal tumor types (*MDM2* and other oncogenes in well-differentiated/dedifferentiated liposarcoma; *NF1* deletions in malignant peripheral nerve sheath tumor (MPNST); *SMARCB1* deletion in epithelioid sarcoma)
- The use of chromosomal microarrays is limited in the evaluation of mesenchymal tumors due to their inability to detect sequence changes and balanced rearrangements

Methylation Studies

- DNA methylation is an epigenetic modification that is associated with cell type and acquired oncogenic alterations
  - Despite its characteristic heterogeneity and dynamic nature, selection of specific CpG sites allows to generate algorithms and classifiers to reproducibly identify certain tumor types
- Sarcomas can be classified in are 62 methylation classes, according to a methylation-based classifier built using Infinium HumanMethylation450K or EPIC beadchip array methylation profiles, from 1,077 samples representing 54 bone and soft tissue tumor types at the Deutsches Krebsforschungszentrum (DKFZ)
- The DKFZ sarcoma methylation classifier can predict a diagnosis in 55–75% of sarcoma cases, which in 88–91% of cases agree with the accepted histopathological diagnosis
- The current algorithm includes only a subset of sarcomas and performs better for samples with high tumor content (>70%)
- The diagnostic value of this approach is limited at present, given a relatively low rate of prediction and reproducibility; in its current form, methylation profiling can provide supportive evidence to certain diagnoses and may occasionally prompt an unexpected diagnosis, which would require validation

Multimodal Diagnostic Approach

- The diagnosis of soft tissue and bone neoplasms relies on the appropriate integration of clinical, morphological, immunohistochemical and genetic data
  - Soft tissue and bone tumors may be approached with several different modalities of molecular diagnostics, given the variety of alterations that are relevant for different types of sarcomas, as well as the various types of tissue available
  - Most molecular alterations can be identified using different techniques, which should be prioritized according to the differential diagnosis under consideration and the type of sample available (Fig. 22.2)
- Immunohistochemistry (IHC) is a widely used and relatively easy, quick, and inexpensive method of understanding some molecular alterations, typically those that lead to an alteration in the abundance or subcellular localization of protein expression (Table 22.1)
  - Mutation-specific antibodies are being developed for the diagnosis of specific sequence variants (e.g., H3-3A p.G34W, BRAF p.V600E, and IDH1 p. R132H)
- Cytogenetics methods, including karyotype, FISH, and microarray hybridization, are often the ideal method to detect gross molecular alterations that affect large portions of the genome, such as alterations in ploidy, whole chromosomes or chromosome arms, and structural variants
  - FISH analysis is a quick and relatively simple, low-cost, method that can identify structural variants or copy number alterations on paraffin-embedded tissues in a suitable time frame for clinical applications
- Molecular diagnostic methods are increasingly applied to sarcoma samples, particularly NGS, which affords the potential to discover not only small substitutions and insertions/deletions, which can be very difficult to detect by IHC or cytogenetics methods but also larger alterations, including copy number changes and structural variants
  - DNA NGS has limitations for some of these chromosomal alterations, however, and cytogenetics may still be preferred
- RNA-based methods are also increasingly being used, particularly to detect structural variants such as rearrangements, where the breakpoints occur in large introns that exceed the capacity of many DNA-based sequencing assays
  - Fusion-specific single-target RNA methods require a priori knowledge of the specific breakpoints and are utilized mainly for confirmatory purposes

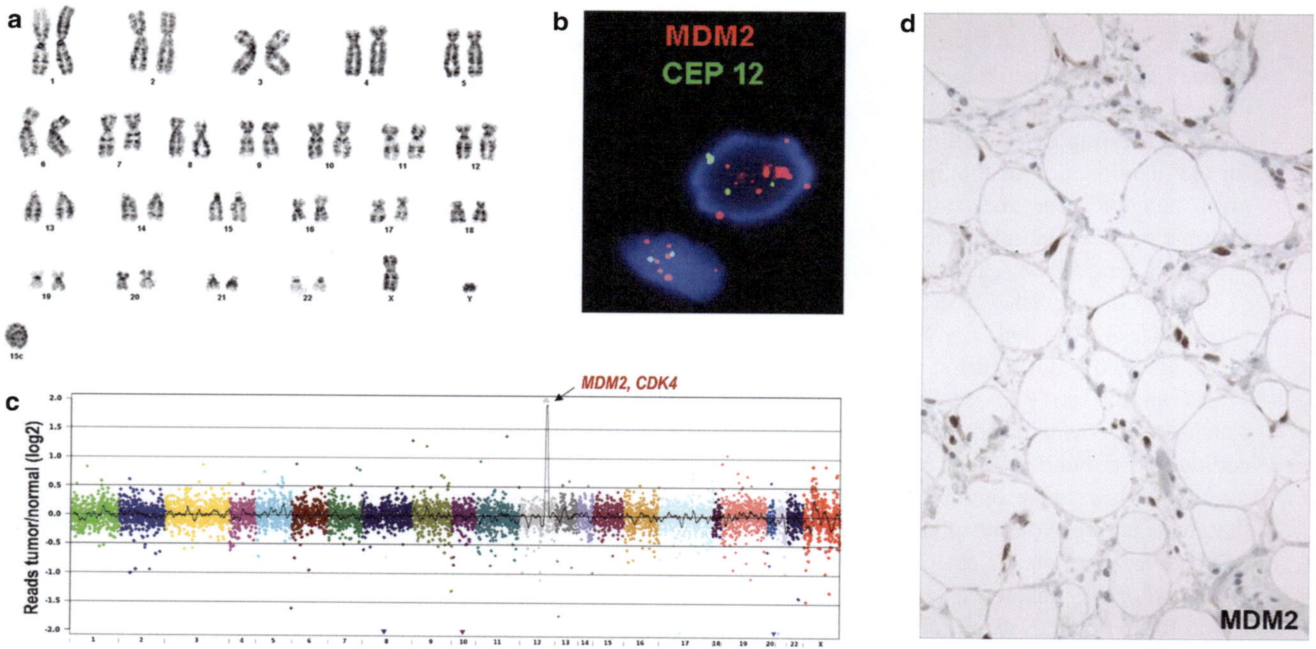

Fig. 22.2 Identification of MDM2 amplification using various diagnostic techniques in a well-differentiated liposarcoma. (**a**) Karyotype demonstrating a ring chromosome as the sole cytogenetic aberration. (**b**) FISH shows multiple red signals corresponding to copies of MDM2. (**c**) NGS-based copy number alteration analysis shows high number of reads mapping to the MDM2 locus, consistent with amplification. (**d**) Immunohistochemistry shows MDM2 overexpression in tumor cell nuclei

Table 22.1 Immunohistochemical correlates of molecular alterations

Immunohistochemical marker	Corresponding molecular alteration	Tumor types	Clinical significance
ALK	*ALK* fusions	*ALK*-rearranged tumors including IMT	Supports diagnosis and treatment
panTRK	*NTRK* fusions	*NTRK*-rearranged tumors including infantile fibrosarcoma and IMT	Supports diagnosis and treatment. Molecular confirmation is recommended
ROS1	*ROS1* fusions	*ROS1*-rearranged tumors including IMT	Supports diagnosis and treatment. Molecular confirmation is recommended
FOSB	*SERPINE1::FOSB*	Pseudomyogenic hemangioendothelioma	Supports diagnosis
CCNB3	*BCOR::CCNB3*	*BCOR::CCNB3*-rearranged round cell sarcoma with	Supports diagnosis
STAT6	*NAB2::STAT6* fusion	Solitary fibrous tumor	Supports diagnosis
DDIT3	*FUS::DDIT3* *EWSR1::DDIT3* fusions	Myxoid liposarcoma	Supports diagnosis
SS18-SSX	*SS18::SSX1* *SS18::SSX2* *SS18::SSX4* fusions	Synovial sarcoma	Supports diagnosis
SDHB (loss)	Loss-of-function alterations in either *SDHA, SDHB, SDHC, SDHD* or *SDHAF2*	GIST (Paraganglioma) (Pheochromocytoma)	Supports a genetic loss-of-function alteration in one of the SDH genes, which may be of somatic or germline origin (further genetic testing is recommended)
SDHA (loss)	Loss-of-function alterations in *SDHA*	GIST (Paraganglioma) (Pheochromocytoma	Supports a genetic loss-of-function alteration in the SDHA gene, which may be of somatic or germline origin (further genetic testing is recommended)
INI1 (loss)	*SMARCB1* loss-of-function alterations	Epithelioid sarcoma	Supports diagnosis. Does not determine somatic or germline origin (further genetic testing is recommended)

Table 22.2 Morphological patterns of soft tissue and bone tumors

Morphological pattern	Tumor type discussed in this chapter
Round cell sarcomas	*BCOR::CCNB3* sarcoma
Spindle cell sarcomas	Infantile fibrosarcoma
Epithelioid cell sarcomas	Epithelioid hemangioendothelioma
Biphasic or mixed sarcomas	Synovial sarcoma
Pleomorphic cell sarcomas	Osteosarcoma
Myxoid sarcomas	Extraskeletal myxoid chondrosarcoma

- RNA-based multiplex sequencing approaches are more readily able to detect a wider variety of structural variants
- RNA methods require a transcript to be expressed, such that loss-of-function structural variants may be missed

Diagnostic Molecular Alterations in Soft Tissue and Bone Tumors

- Molecular studies are frequently performed to confirm or refute a morphological diagnostic impression in certain clinicopathological context, or to resolve a differential diagnosis
  - Less frequently, molecular information is necessary to appropriately classify a neoplasm as one of the increasingly recognized molecularly defined entities

- A morphological, pattern-based approach to soft tissue and bone tumors is crucial for the appropriate interpretation of diagnostic molecular data
- Current morphological classification schemes consider six main morphological categories of soft tissue and bone tumors (Table 22.2); molecular studies can provide useful diagnostic information in each group, as illustrated by the examples below

Round Cell Sarcomas

- Round cell sarcomas are among the most aggressive soft tissue tumors (Table 22.3); they often, but not exclusively, affect children or young adults
- Therapeutic options available for these tumors are limited, but certain therapeutic modalities are very effective against specific entities

Table 22.3 Round cell sarcomas

Diagnosis	Molecular alteration	Preferred molecular method	Alternative molecular methods
Alveolar rhabdomyosarcoma	*PAX3::FOXO1 PAX7::FOXO1*	FISH	RT-PCR, RNA NGS
BCOR-rearranged sarcoma	*BCOR::CCNB3* BCOR ITD	IHC; FISH for BCOR; NGS or PCR for ITD	
CIC-rearranged sarcoma	*CIC::DUX4*	FISH	RNA NGS; IHC
DSRCT	*EWSR1::WT1*	FISH	
Ewing sarcoma	*EWSR1::FLI1* And other EWSR1 fusions	FISH	RT-PCR, IHC, RNA NGS
EWSR1-non ETS round cell sarcoma	*EWSR1::PATZ1 EWSR1::NFATC2*	NGS	FISH
Glomus tumor	*NOTCH* fusions, *BRAF* mutations	NGS	
Mesenchymal chondrosarcoma	*HEY1::NCOA2*	NGS	RT-PCR

- – Proper diagnosis and subtyping have a great clinical impact and are crucial for appropriate patient management
- A subset of round cell sarcomas cannot be unequivocally classified based on clinical, morphologic, and immunophenotypic features
  - – Often resemble Ewing sarcoma morphologically but lack its characteristic immunoprofile and genetic features
  - – The term *undifferentiated round cell sarcoma* provides a useful working category for this heterogeneous group of tumors, which are identified by thoroughly excluding other diagnostic categories
- The application of cytogenetics and molecular genetic techniques allows for the identification of an increasing number of genetically defined subgroups within this category, which represent distinct biological entities (Table 22.3)
- Sequencing panels covering multiple gene fusions are particularly useful for proper classification of undifferentiated round-cell sarcomas

BCOR::CCNB3 sarcoma

- A group of undifferentiated round cell sarcomas show *BCOR* alterations, which may be gene fusions involving *BCOR* (most frequently *BCOR::CCNB3*) or *BCOR* internal tandem duplications
- *BCOR::CCNB3* sarcomas have a predilection for children and males; arise more often in bone than soft tissue, with a predilection for the pelvis, lower limbs, and paraspinal region
- Histologically, sarcomas with *BCOR::CCNB3* are typically composed of primitive small round to ovoid cells arranged in solid sheets or a vague nesting pattern, surrounded by a rich capillary network
- Immunohistochemically, most cases express BCOR and CCNB3 (Fig. 22.3)
- Molecular detection of *BCOR::CCNB3* is possible with various molecular techniques

- – The fusion results from a paracentric inversion on the short arm of chromosome X, which brings together the coding sequences of the two genes, which are normally located ~10Mb apart
- – The proximity of the genes, and the possibility of alternative *BCOR* alterations, determines the design of FISH approaches, which optimally rely on a combined three-color breakapart and bring-together approach: probes flanking *CCNB3* are labeled with one color, while probes flanking centromeric and telomeric regions of *BCOR* are labeled with two different colors
- Sequencing approaches, either targeted DNA or RNA sequencing panels or RT-PCR, can effectively detect *BCOR::CCNB3* fusions
  - – The breakpoint in *BCOR* is invariably located in the terminal end of the coding sequence, either in exon 15 (the last exon) or the 3′UTR
- Of note, structural rearrangements involving *BCOR* have been identified in cases of acute promyelocytic leukemia, endometrial stromal sarcomas, and ossifying fibromyxoid tumors with a variety of breakpoints

Spindle Cell Sarcomas

- Some of the most common mesenchymal tumors (still rare diseases) show spindle cell morphology, such as gastrointestinal stromal tumor (GIST), leiomyosarcoma, malignant peripheral nerve sheath tumor (MPNST), and many dedifferentiated liposarcomas (Table 22.4)
- Management, therapy, and molecular diagnostic approaches vary according to tumor type
- Clinical context, morphological examination and immunohistochemistry usually narrow down the diagnosis to a differential diagnosis that rarely requires molecular studies
  - – Monomorphic tumors with no clear line of differentiation benefit most from unbiased panel-based sequencing approaches

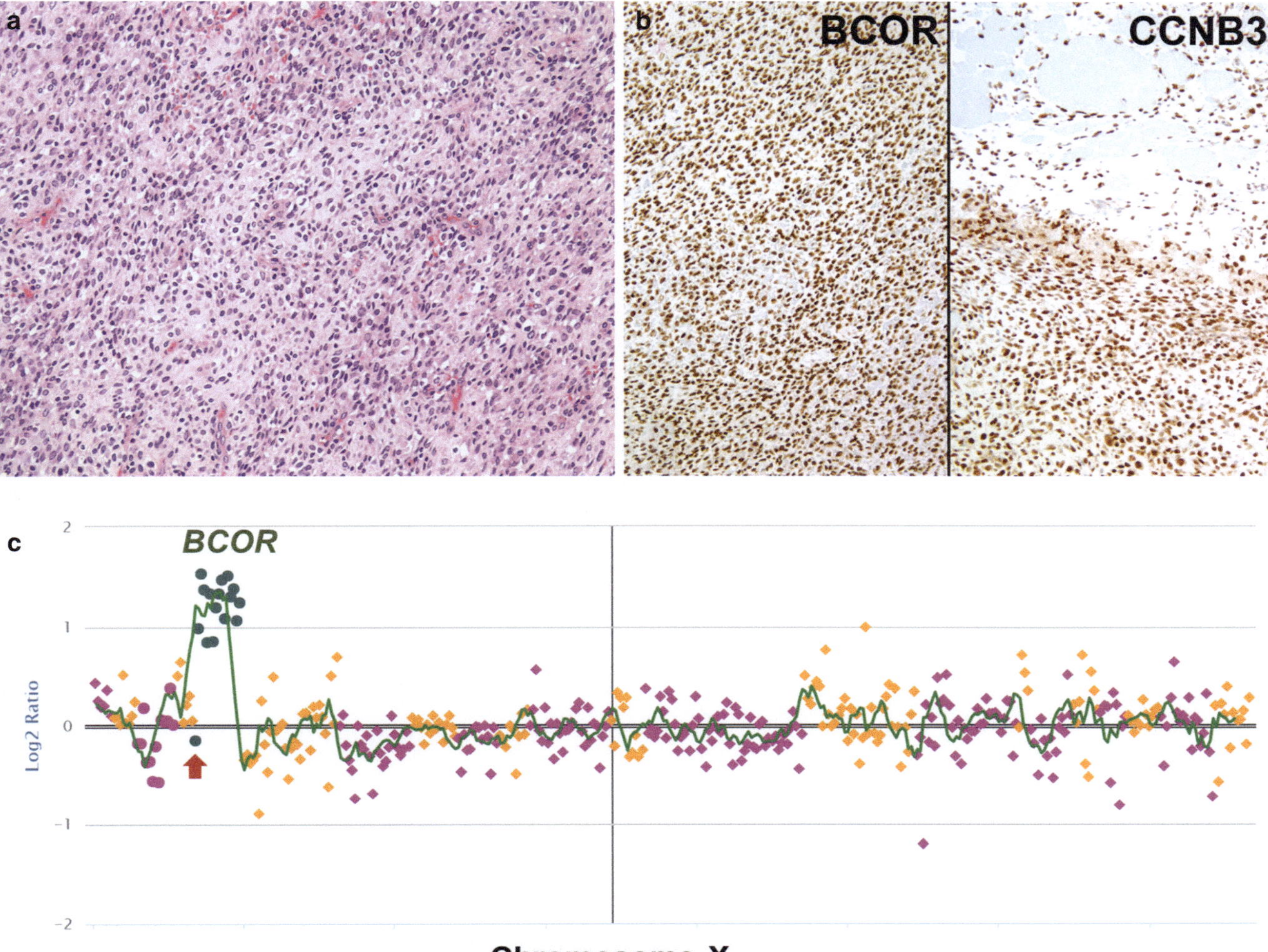

Fig. 22.3 Round cell sarcoma with *BCOR::CCNB3* rearrangement involving the pubic bone of a 20-year-old man. (**a**) Cellular proliferation of relatively monomorphic round cells in a collagenous stroma. (**b**) Tumor cells show diffuse nuclear immunohistochemical expression of BCOR and CCNB3. (**c**) Copy number plot corresponding to chromosome X, from a targeted NGS sequencing panel that includes *BCOR*. Most reads mapping to *BCOR* are overrepresented, indicating copy number gain of exons 1–15. The final segment of the gene (exon 15–5′UTR, red arrow) is not amplified, consistent with a breakpoint in exon 15 of *BCOR*. RNA-based fusion analysis identified a fusion transcript involving exon 15 of *BCOR* and exon 6 of *CCNB3*

- Most GIST is characterized by activating mutations in the receptor tyrosine kinases *KIT* or *PDGFRA*
  - The distribution of *KIT* and *PDGFRA* oncogenic mutations cluster in relatively small regions (exons 9, 11, 13, 17, and 18) and are amenable for detection with targeted sequencing approaches, such as PCR, ddPCR, or amplicon sequencing
  - The specific mutation determines response or resistance to targeted tyrosine kinase inhibitors (see "Predictive Molecular Alterations" section)
- Leiomyosarcoma and MPNST show typical but nondiagnostic molecular alterations
- Dedifferentiated liposarcoma (and its low-grade counterpart, well-differentiated liposarcoma) is characterized by coamplification of *MDM2* and *CDK4*, which results in overexpression of these two proteins that can be detected by IHC and confirmed by FISH
  - The chromosomal substrate of these amplification events are ring chromosomes and/or giant marker chromosomes that can be detected by karyotype
- Synovial sarcoma (sometimes with biphasic morphology, but most often spindle cell monophasic) is characterized by recurrent *SS18::SSX* oncogenic fusions, which can be detected by sequencing techniques (Table 22.4).
  - A traditional RT-PCR approach with two primer sets can identify the most frequent fusions but will miss 5–10% of casesl; DNA or RNA NGS approaches can detect all fusion variants
  - A novel antibody against a fusion-specific epitope detects most fusions

Table 22.4 Spindle cell mesenchymal tumors

Diagnosis	Molecular alteration	Preferred molecular method	Alternative molecular methods
Angiofibroma of soft tissue	*NCOA2* fusions	FISH	NGS
Calcifying aponeurotic fibroma	*FN1::EGF*	FISH	NGS
Clear cell sarcoma	*EWSR1::ATF1/CREB*	FISH	NGS
Dedifferentiated liposarcoma	*MDM2/CDK4* amplification	IHC (lab-dependent) FISH	FISH, MLPA, microarray, NGS
Desmoid fibromatosis	*CTNNB1* mutation	DNA NGS	IHC
DFSP	*COL1A1::PDGFRB*	FISH	NGS
EWSR1-SMAD3 fibroblastic tumor	*EWSR1::SMAD3*	NGS	
Giant cell fibroblastoma	*COL1A1::PDGFB*	FISH	NGS
GIST	*KIT, PDGFRA, SDHA, SDHB, SDHC, SDHD*	NGS	IHC
Infantile fibrosarcoma	*ETV6::NTRK3*	FISH	NGS
Inflammatory myofibroblastic tumor	*ALK* rearrangements	IHC	NGS, FISH
Kaposi's sarcoma	HHV8 infection	IHC	PCR
Low-grade fibromyxoid sarcoma/sclerosing fibroepithelioid sarcoma	*EWSR1/FUS::CREB3L1/2*	IHC	FISH, NGS
MPNST	*NF1, CDKN2A/B, EED, SUZ12* inactivation		
Myofibroblastoma	Monoallelic *RB1* loss	IHC	FISH
Nodular fasciitis	*USP6* rearrangement	FISH	NGS
NTRK-rearranged spindle cell neoplasm	*NTRK* fusions	NGS	
Phosphaturic mesenchymal tumor	*FN1::FGF/FGFR1*	FISH	NGS
Solitary fibrous tumor	*NAB2::STAT6*	IHC	NGS
Spindle cell rhabdomyosarcoma	*VGLL2, NCOA2, CITED2* fusions, *MYOD1* mutations	NGS	FISH, PCR
Synovial sarcoma	*SS18::SSX1, SS18::SSX2, SS18::SSX4*	FISH	NGS, PCR

- A subset of monomorphic spindle cell sarcomas, frequent in children and young adults, are characterized by recurrent kinase fusions that can be identified with molecular studies to prompt treatment with effective molecularly targeted therapies (see Predictive Molecular Alterations section, below)

Infantile Fibrosarcoma

- Also called *congenital fibrosarcoma*, is a locally aggressive and rarely metastasizing sarcoma that typically affects infants
  - Common locations include extremities, followed by trunk
  - Intrabdominal, retroperitoneal, and visceral sites occur less frequently—including the kidney, where the tumor is called *cellular congenital mesoblastic nephroma*
- Morphology
  - Cellular, monomorphic proliferation of ovoid or short spindled cells, arranged in compact sheets or vague fascicles
  - Stroma may be myxoid or collagenous
  - Mitotic rate and necrosis are variable and have no prognostic value

  - The immunohistochemical profile is nonspecific (Fig. 22.4)
- Molecularly, it is characterized by an *ETV6::NTRK3* oncogenic fusion that results from a t(12;15)(p13;q25)
  - Other *NTRK3* fusion partners have been reported, including *EML4::NTRK3* fusions
- Alternative fusions with other tyrosine kinases have been described in small numbers of cases, including *NTRK1, NTRK2, RET, MET, BRAF*, and *RAF1*
- It is unclear to what extent other monomorphic spindle cell tumors with kinase rearrangements, which often occur in older patients and coexpress S100 and CD34, are related to infantile fibrosarcoma or constitute separate clinicopathological entities

Epithelioid Sarcomas

- Mesenchymal tumors with epithelioid morphology are infrequent but include several entities with specific molecular alterations (Table 22.5)
- Most tumors in this group can occur at many anatomic locations, which makes diagnosis challenging and increases the value of molecular diagnostic studies

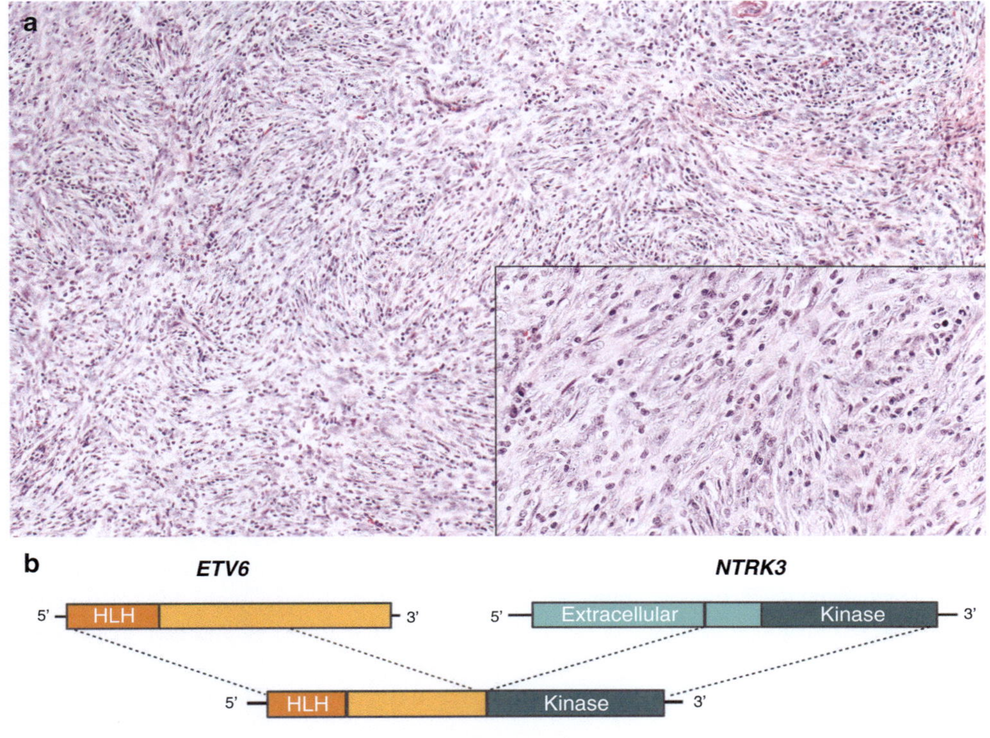

Fig. 22.4 Infantile fibrosarcoma arising in the thigh of a 6-month-old girl. (**a**) Cellular proliferation of densely packed atypical spindle cells with syncytial cytoplasm and ovoid nuclei, arranged in short intersecting fascicles. A lymphocytic inflammatory infiltrate is occasionally present, such as in this case. (**b**) A targeted RNA-based anchored multiplex PCR high-throughput sequencing assay identified an *ETV6::NTRK3* fusion, juxtaposing exon 5 of *ETV6* (NM_001987.4) and exon 15 of *NTRK3* (NM_001243101.1). These breakpoints generate a prototypical chimeric oncogenic kinase that includes the HLH oligomerization domain of ETV6 and the intact kinase domain of NTRK3. Of note, karyotype analysis in this case did not show the characteristic t(12;15), although breakapart FISH demonstrated split *NTRK3* signals. Created with BioRender.com

Table 22.5 Mesenchymal tumors with epithelioid cytomorphology

Diagnosis	Molecular alteration	Preferred molecular method	Alternative molecular methods
Alveolar soft part sarcoma	*ASPCR1::TFE3*	FISH	NGS, IHC
Angiomatoid fibrous histiocytoma	*EWSR1-CREB1*	FISH	NGS
Epithelioid hemangioendothelioma	*WWTR1::CAMTA1, YAP1::TFE3*	IHC	FISH, RT-PCR, NGS
Epithelioid sarcoma	*SMARCB1* loss	IHC	NGS
GIST	*KIT, PDGFRA, SDHA, SDHB, SDHC, SDHD*	NGS	IHC
Malignant rhabdoid tumor	*SMARCB1* loss	IHC	NGS
Myoepithelioma	*EWSR1, FUS, PLAG1* rearrangements	FISH	NGS
Ossifying fibromyxoid tumor	*PHF1, TFE3* rearrangements	FISH	NGS
PEComa	*TFE3* rearrangement *TSC2* loss of function	IHC NGS	NGS
Sclerosing epithelioid fibrosarcoma	*EWSR1:: or FUS:: CREB3L1/2*	IHC (MUC4) FISH	NGS

- Molecular diagnostics may be helpful in distinguishing some of these tumors from undifferentiated carcinomas with overlapping morphology
  - The most appropriate molecular study depends on the differential diagnosis considered and may include mutational profiling or gene fusion detection
- At present, there are not many biology-based therapies available to treat tumors in this group; notable exceptions are mTOR inhibition in PEComa and tyrosine kinase inhibitors in epithelioid GIST treatment

Epithelioid Hemangioendothelioma

- Epithelioid hemangioendothelioma (EHE) is a malignant vascular tumor, often compared to a low-grade angiosarcoma
- Morphologically, it is composed of cords of epithelioid endothelial cells seen in a distinctive myxohyaline stroma (Fig. 22.5)
- Characterized by the recurrent t(1;3) translocation, resulting in a *WWTR1::CAMTA1* fusion gene
- Interestingly, EHE often presents as multifocal lesions, which are all clonally related

- Immunohistochemistry for CAMTA1 can be used as a surrogate marker for the translocation
- Molecular studies are seldom required for diagnosis, but in some cases (particularly in cases involving the bone) may facilitate the diagnosis in small biopsies
- A morphologically distinct variant of EHE with solid architecture and well-formed vascular channels is genetically characterized by *YAP1::TFE3* fusions; these tumors demonstrate strong immunoreactivity for TFE3

Pleomorphic Sarcomas

- Cytological pleomorphism in mesenchymal tumors is associated with aggressive clinical behavior
- Pleomorphic sarcomas are most frequent in older adults (Table 22.6), except for osteosarcoma
- Pleomorphic sarcomas frequently show complex genomes and karyotypes, with loss-of-function alterations in canonical tumor suppressor genes involved in genome maintenance, such as *TP53*, *RB1*, and *ATRX*

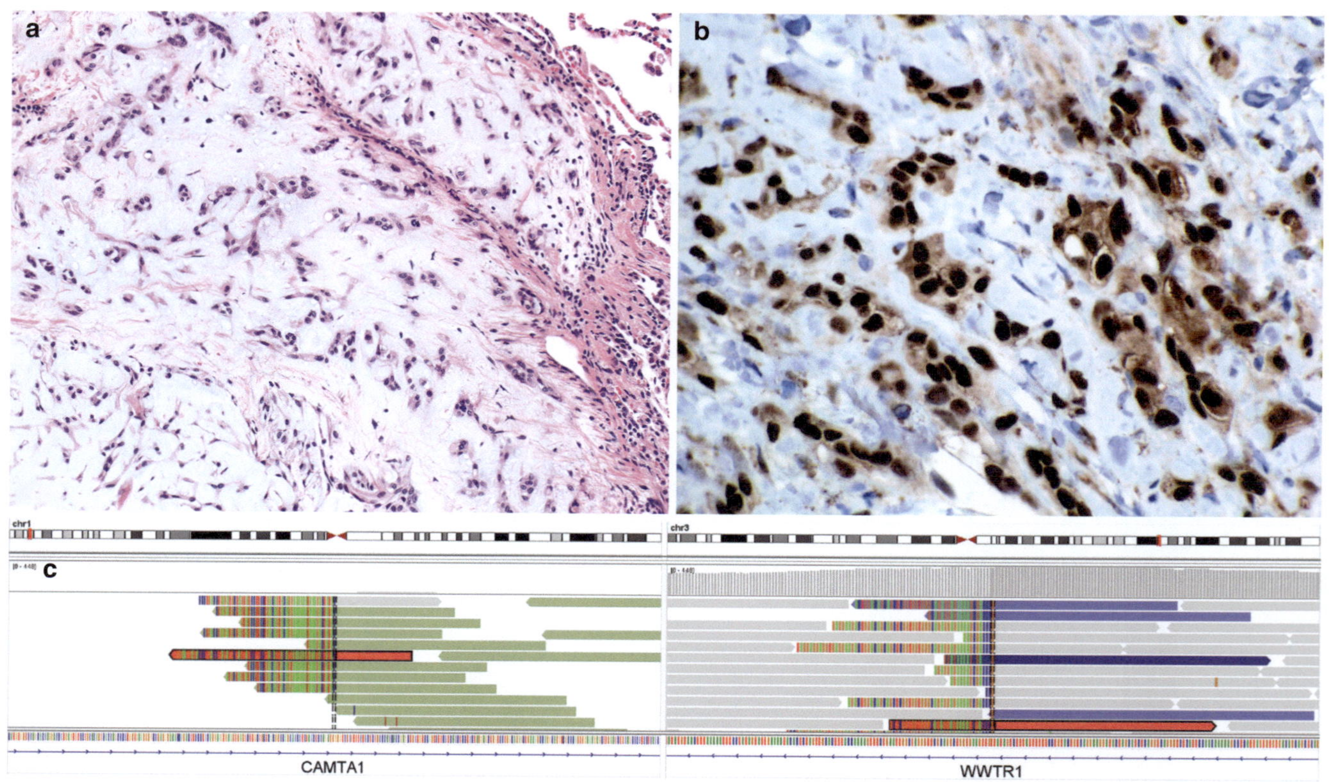

Fig. 22.5 Morphologic appearances (**a**) and immunohistochemical expression of CAMTA1 (**b**) in epithelioid hemangioendothelioma [*IHC image courtesy of Dr. L. Doyle, from Brigham and Women's Hospital and Harvard Medical School*]. (**c**) Next-generation sequencing DNA-based gene panel assay identified reads spanning two breakpoints on chromosomes 1 and 3, mapping to intron 7 of *CAMTA1* and to intron 4 of *WWTR1*, respectively, corresponding to a *CAMTA1::WWTR1* gene fusion

Table 22.6 Pleomorphic mesenchymal tumors

Diagnosis	Molecular alteration	Preferred molecular method
Myxofibrosarcoma	*TP53*, *RB1* inactivation Hippo pathway activation	NGS (limited clinical use)
Osteosarcoma	*TP53*, *RB1* inactivation *MYC* amplification	NGS (limited clinical use)
Pleomorphic leiomyosarcoma	*TP53*, *RB1* inactivation Homologous recombination deficiency	NGS (limited clinical use)
Pleomorphic liposarcoma	*TP53*, *RB1* inactivation Complex genome/ karyotype	NGS (limited clinical use)
Dedifferentiated liposarcoma	*MDM2* and *CDK4* amplification	FISH
Undifferentiated pleomorphic sarcoma	*TP53*, *RB1* inactivation Complex genome/ karyotype	NGS (limited clinical use)

- Genomic instability is a prominent feature, with marked molecular heterogeneity from tumor to tumor and within each tumor
- Molecular studies have limited value in this group of tumors, both for diagnostics and therapeutics
  - Notable exceptions are the identification of germline alterations, and the identification of dedifferentiated variants of mesenchymal tumors with specific underlying molecular alterations (e.g., dedifferentiated solitary fibrous tumor, dedifferentiated GIST, dedifferentiated liposarcoma)
- Even though individual molecular alterations may not have diagnostic value, a given molecular profile or molecular signature may rarely be supportive of certain diagnoses or disclose a therapeutic vulnerability (e.g., microsatellite instability or homologous recombination deficiency)

Osteosarcoma

- Osteosarcoma is a mesenchymal tumor with osteogenic differentiation; most commonly occurs in bone, but primary soft tissue osteogenic sarcomas also occur
- Osteosarcoma is most common in children and young adults
- Histologically, it is defined by the formation of osteoid matrix
  - Histopathological examination is usually sufficient for diagnosis, although osteogenic differentiation may be identified with SATB2 immunohistochemistry
- Acid decalcification may destroy nucleic acids and impede molecular studies
  - For heavily calcified lesions, decalcification with *EDTA* (ethylenediaminetetra acetic acid) is preferred to other methods

- Cytological imprints or processing of a nondecalcified sample dedicated to molecular studies should be considered
- Osteosarcomas show a complex karyotype, with multiple numerical and structural alterations
  - NGS shows a very complex copy number profile with multiple somatic copy number alterations and complex rearrangements involving many chromosomes
  - Mechanistically, these have been associated with whole-genome doubling and chromothripsis (Fig. 22.6)
- *TP53* and *RB1* are inactivated in most cases, often by complex structural rearrangements that may be difficult to identify by conventional sequencing techniques
- Germline pathogenic mutations are observed in up to 28% of patients, most frequently in *TP53*, *RB1*, *RECQL4*, and *CDKN2A*
- Other mesenchymal tumors of bone show nonoverlapping molecular alterations (Table 22.7) that may be useful for diagnosis in limited biopsy material or in cases with discrepant clinical/radiological/morphological data

Myxoid Sarcomas

- A myxoid extracellular matrix is a prominent feature in some sarcoma types (Table 22.8)
- Importantly, some cases of spindle, epithelioid, and pleomorphic sarcomas may show a myxoid matrix focally or diffusely
  - Such cases are rare (such as myxoid SFT or myxoid DFSP), and molecular studies occasionally reveal an unsuspected diagnosis
- Myxoid tumors often display low cellularity, and nucleic acid extraction may yield low concentrations
  - Selection of cellular areas or processing of additional material may be necessary to enable molecular studies

Extraskeletal Myxoid Chondrosarcoma

- Extraskeletal myxoid chondrosarcoma is a rare malignant mesenchymal neoplasm of uncertain differentiation (no evidence of cartilaginous differentiation) that affects middle-aged adults in the deep soft tissue of proximal extremities
- Histologically characterized by a proliferation of uniform cells arranged in cords, clusters, and reticular networks embedded in an abundant myxoid matrix with multilobular architecture
- Molecularly defined by *NR4A3* rearrangements, most often with *EWSR1* as the 5′ partner and occasionally with *TAF15*, as a result of balanced t(9;22) (q22;q12) or t(9;17) (q22;q12) translocations

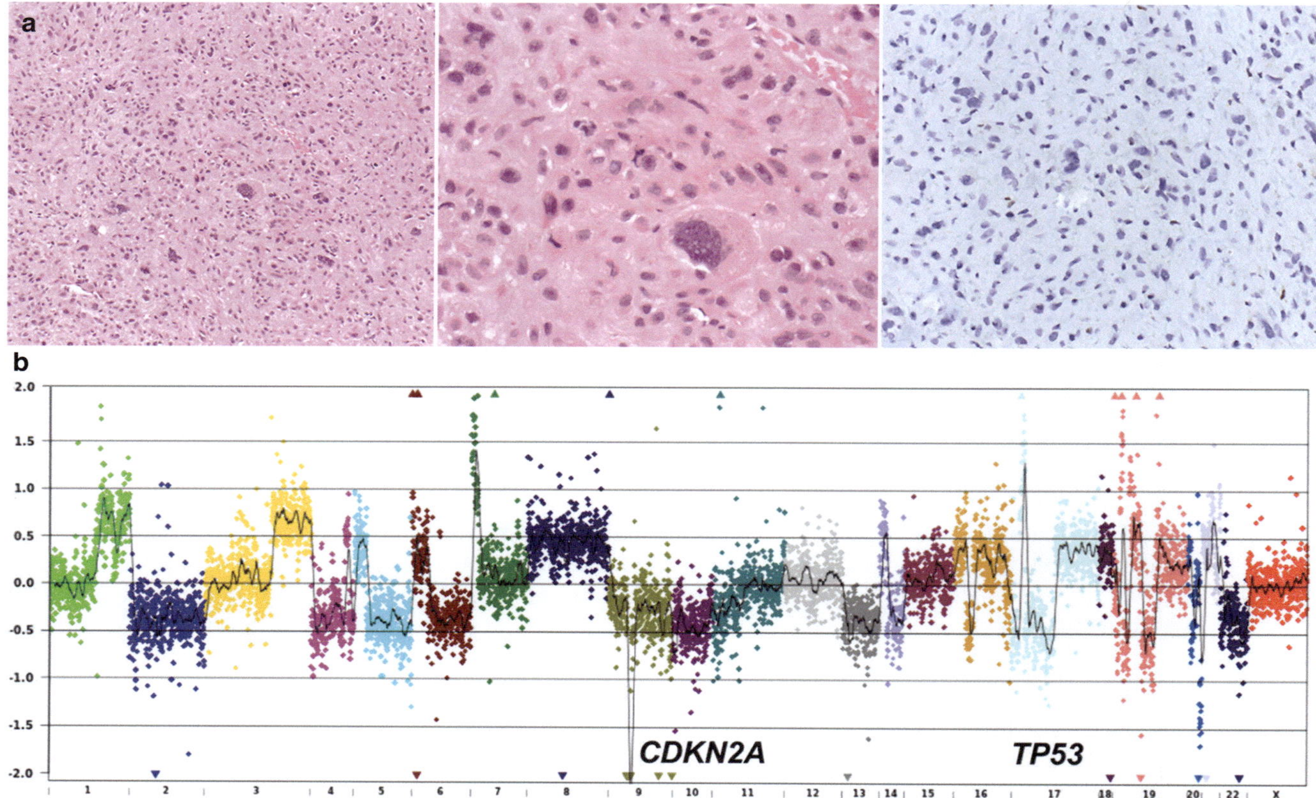

Fig. 22.6 Microscopic and molecular features in osteosarcoma. (**a**) Histopathologic examination of an osteoblastic osteosarcoma of the femur in a 7-year-old girl revealed a proliferation of atypical cells with marked pleomorphism in an eosinophilic extracellular matrix. Immunohistochemical evaluation revealed loss of TP53 expression in tumor cells. A targeted DNA sequencing panel revealed a structural rearrangement disrupting *TP53* in this case. (**b**) Copy number plot of an osteosarcoma of bone showing a complex genomic profile with multiple focal, segmental, and large-scale gains and losses along the targeted genome, including deletion of two copies of *CDKN2A* on chromosome 9, and one copy deletion of *TP53* on chromosome 17p (in this case, a TP53 p.R248Q mutation affected the other allele, although most often biallelic inactivation results from a structural rearrangement

Table 22.7 Selected mesenchymal tumors of bone with known underlying molecular alterations

Diagnosis	Molecular alteration	Preferred molecular method	Alternative molecular methods
Aneurysmal bone cyst	*USP6* rearrangement	FISH	NGS, IHC
Chondroblastoma	*H3F3B/A* p.L36M	PCR	NGS
Chondrosarcoma	*IDH1* or *IDH2* mutations	PCR	IHC for selected mutations, NGS
Enchondroma	*IDH1* or *IDH2* mutations	PCR	IHC for selected mutations, NGS
Fibrous dysplasia	*GNAS* activating mutation	PCR	NGS
Giant cell tumor of bone	*H3A3* codon 34 mutation	IHC	PCR, NGS
Nonossifying fibroma	*KRAS*, *FGFR1* activating mutations	PCR	NGS
Osteoblastoma	*FOS* fusions	NGS	
Osteochondroma	*EXT1* or *EXT2* biallelic loss	NGS	
Osteoid osteoma	*FOS* fusions	NGS	
Paraosteal osteosarcoma	*MDM2* and *CDK4* amplification	IHC	MLPA, FISH, NGS
Synovial chondromatosis	*FN1::ACVR2A*	FISH	NGS

- Variant *TCF12::NR4A3* and *TFG::NR4A3* have been identified in isolated cases

• Demonstration of *NR4A3* rearrangement by FISH or sequencing techniques is useful for the diagnosis of extraskeletal myxoid chondrosarcoma (Fig. 22.7)

Table 22.8 Mesenchymal tumors with a prominent myxoid matrix

Diagnosis	Molecular alteration	Preferred molecular method	Alternative molecular methods
Myxoid liposarcoma	*FUS/EWSR1::DDIT3* *TP53* inactivating mutations *PIK3CA* activating mutations	FISH	NGS, RT-PCR, IHC
Low-grade fibromyxoid sarcoma/sclerosing fibroepithelioid sarcoma	*EWSR1/FUS::CREB3L1/2*	IHC	FISH, NGS
Low-grade myxofibrosarcoma	*TP53*, *RB1* inactivation Hippo pathway activation	NGS (limited clinical use)	
Extraskeletal myxoid chondrosarcoma	*NR43* rearrangement	FISH	NGS

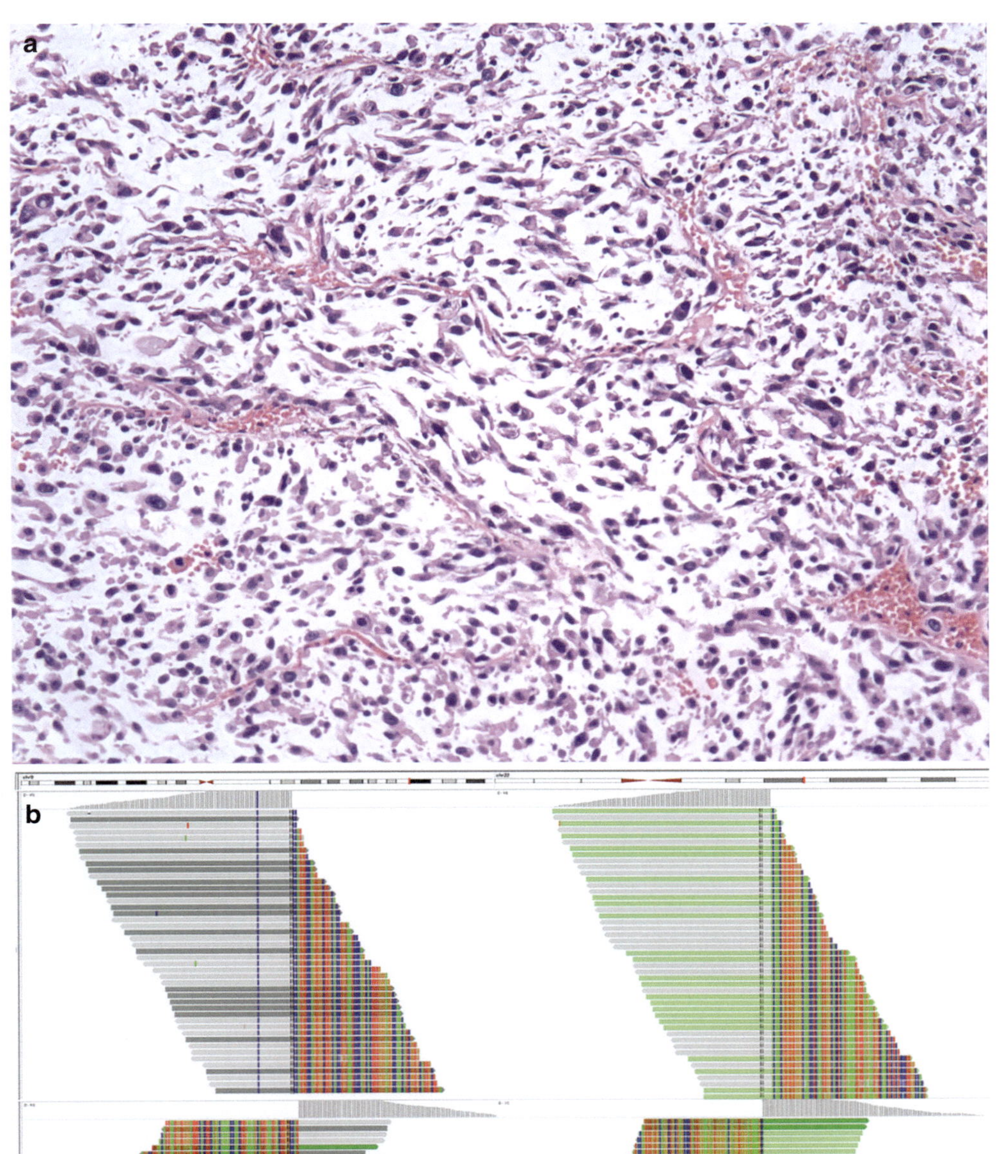

Fig. 22.7 Extraskeletal myxoid chondrosarcoma involving the sacrum of a 63-year-old woman. (**a**) histologically, the tumor is composed of a cellular proliferation of short, atypical spindle cells on a vaguely lobular arrangement with a myxoid matrix. (**b**) A next-generation sequencing DNA-based gene panel assay identified multiple reads spanning two breakpoints on chromosomes 9 and 22, mapping to intron 1 of *NR4A3* and to intron 13 of *EWSR1*, respectively, corresponding to a *EWSR1::NR4A3* gene fusion

Predictive Molecular Alterations in Soft Tissue and Bone Tumors

- Several recurrent molecular alterations in mesenchymal tumors are amenable to therapeutic interventions (Table 22.9)
- Oncogenic kinases are the most prominent group of therapeutic targets in mesenchymal tumors
  - Constitutive activation may be the result of point mutations or small insertions/deletions (such as the *KIT* and *PDGFRA* receptor tyrosine kinases in GIST), gene fusions that result from rearrangements (*ALK, ROS1,* or *RET* in inflammatory fibroblastic tumors) or paracrine activation by secreted growth factors (*PDGFB* in dermatofibrosarcoma protuberans)
- The level of specificity of each tyrosine kinase inhibitor is variable, with some inhibitors targeting a broad spectrum of kinases (sunitinib) and others binding specifically to a few, or a single kinase target
  - The spectrum of mutations covered by each inhibitor varies widely
- Specific inhibitors (of target or even mutation) have increasingly been approved, and knowledge of the specific mutations present in each case by molecular analyses is required to recommend the most appropriate therapeutics

Table 22.9 Therapeutically actionable molecular alterations in mesenchymal tumors

Molecular alteration	Tumor type	Potential therapeutics
KIT or *PDGFRA* mutations	GIST	Tyrosine kinase inhibitors
PDGFB or *PDGFD* gene fusions	DFSP	Tyrosine kinase inhibitors
ALK gene fusions	IMT, IMS	Tyrosine kinase inhibitors
NTRK gene fusions	IMT, spindle cell sarcomas	Tyrosine kinase inhibitors
RET gene fusions	IMT, spindle cell sarcomas	Tyrosine kinase inhibitors
ROS1 gene fusions	IMT, spindle cell sarcomas	Tyrosine kinase inhibitors
COL6A3::CSF1	Tenosynovial giant cell tumor	Tyrosine kinase inhibitors
Homologous recombination deficiency ("BRCAness")	Leiomyosarcoma	PARP inhibitors
Mismatch repair deficiency	Multiple types, rare	Immune checkpoint blockade
TSC1/TSC2 inactivation	PEComa	mTOR inhibitors
PIK3CA activating mutation	Myxoid liposarcoma	PIK3CA inhibitors
IDH1 mutations	Chondrosarcoma	IDH1 inhibitor (ivosidenib)
SMARCB1 inactivation	Epithelioid sarcoma	EZH2 inhibitors
CDK4 amplification	Dedifferentiated liposarcoma	CDK4 inhibitors (in combination)

- The most successful example of biologically targeted therapies in mesenchymal tumors is GIST, with five approved tyrosine kinase inhibitors effective against a range of *KIT* and *PDGFRA* mutations (imatinib, sunitinib, regorafenib, avapritinib, and ripretinib)
  - Other notable targets are *ALK, RET, NTRKn,* and *ROS1* (Table 22.8)
- Resistance to tyrosine kinase inhibitors and clinical progression often occur through secondary mutation of the target
  - Serial molecular studies are required at the time of progression to identify mechanisms of resistance and adjust the therapeutic regimen accordingly
- Molecular signatures reflecting DNA damage repair defects, including mismatch repair defects and homologous recombination deficiency, are identified in subsets of sarcoma cases and may determine susceptibility to checkpoint-blockade immunotherapy and PARP-inhibitor therapy, respectively
- Epigenetic reprogramming due to *SMARCB1* inactivation or *IDH1* mutations lead to therapeutic susceptibility to EZH2 and IDH1 inhibitors
- Cell cycle dysregulation due to *CDK4* amplification or *CDKN2A* inactivation in mesenchymal tumors suggests a role of combination regimens including CDK4/CDK6 inhibitors, although none have been approved based on limited effects in early clinical trials
- Loss of *TSC1* or *TSC2* in PEComas leads to dependence on mTOR signaling and determines sensitivity to mTOR inhibitors, such as rapamycin and similar rapalogs
- By analogy with other tumor types, activation of the PI3K signaling pathway by *PIK3CA* mutation (in myxoid liposarcoma), *AKT1* mutation, or *PTEN* inactivation (in occasional cases of several sarcoma types) may determine consistent sensitivity to inhibitors of the PI3K pathway, although only case reports and exceptional responders have been documented thus far in mesenchymal tumors
- Interpretation of a molecular alteration with potential predictive value in mesenchymal tumors often benefits from a general sense of the overall genomic complexity and molecular context in which the alteration occurs
  - Knowledge of the molecular background, with any genome-wide approach regardless of its resolution (karyotype, NGS, or array), allows for more accurate prediction of response

Suggested Reading

Agaram NP, Chen CL, Zhang L, LaQuaglia MP, Wexler L, Antonescu CR. Recurrent MYOD1 mutations in pediatric and adult sclerosing and spindle cell rhabdomyosarcomas: evidence for a common pathogenesis. Genes Chromosomes Cancer. 2014;53:779–87.
Agaram NP, LaQuaglia MP, Alaggio R, et al. MYOD1-mutant spindle cell and sclerosing rhabdomyosarcoma: an aggressive subtype irre-

spective of age. A reappraisal for molecular classification and risk stratification. Mod Pathol. 2019;32:27–36.

AJCC Cancer Staging Manual. New York, NY: Springer, 2018; p. 507.

Alaggio R, Zhang L, Sung YS, et al. A molecular study of pediatric spindle and sclerosing rhabdomyosarcoma: identification of novel and recurrent VGLL2-related fusions in infantile cases. Am J Surg Pathol. 2016;40:224–35.

Antonescu CR, Katabi N, Zhang L, et al. EWSR1-ATF1 fusion is a novel and consistent finding in hyalinizing clear-cell carcinoma of salivary gland. Genes Chromosomes Cancer. 2011;50:559–70.

Antonescu CR, Sung YS, Chen CL, et al. Novel ZC3H7B-BCOR, MEAF6-PHF1, and EPC1-PHF1 fusions in ossifying fibromyxoid tumors--molecular characterization shows genetic overlap with endometrial stromal sarcoma. Genes Chromosomes Cancer. 2014;53:183–93.

Baranov E, Black MA, Fletcher CDM, Charville GW, Hornick JL. Nuclear expression of DDIT3 distinguishes high-grade myxoid liposarcoma from other round cell sarcomas. Mod Pathol. 2021;34:1367–72.

Baranov E, McBride MJ, Bellizzi AM, et al. A Novel SS18-SSX fusion-specific antibody for the diagnosis of synovial sarcoma. Am J Surg Pathol. 2020;44:922–33.

Barretina J, Taylor BS, Banerji S, et al. Subtype-specific genomic alterations define new targets for soft-tissue sarcoma therapy. Nat Genet. 2010;42:715–21.

Bennett JA, Oliva E. Perivascular epithelioid cell tumors (PEComa) of the gynecologic tract. Genes Chromosomes Cancer. 2021;60:168–79.

Biegel JA, Kalpana G, Knudsen ES, et al. The role of INI1 and the SWI/SNF complex in the development of rhabdoid tumors: meeting summary from the workshop on childhood atypical teratoid/rhabdoid tumors. Cancer Res. 2002;62:323–8.

Board. PPTE. Childhood Soft Tissue Sarcoma Treatment (PDQ®): Health Professional Version. Available from URL: https://www.ncbi.nlm.nih.gov/books/NBK65923/table/CDR0000062934__749/ (accessed September 28, 2021).

Brandao M, Caparica R, Eiger D, de Azambuja E. Biomarkers of response and resistance to PI3K inhibitors in estrogen receptor-positive breast cancer patients and combination therapies involving PI3K inhibitors. Ann Oncol. 2019;30:x27–42.

Butrynski JE, D'Adamo DR, Hornick JL, et al. Crizotinib in ALK-rearranged inflammatory myofibroblastic tumor. N Engl J Med. 2010;363:1727–33.

Capasso M, Montella A, Tirelli M, Maiorino T, Cantalupo S, Iolascon A. Genetic predisposition to solid pediatric cancers. Front Oncol. 2020;10:590033.

Chen BJ, Marino-Enriquez A, Fletcher CD, Hornick JL. Loss of retinoblastoma protein expression in spindle cell/pleomorphic lipomas and cytogenetically related tumors: an immunohistochemical study with diagnostic implications. Am J Surg Pathol. 2012;36:1119–28.

Choi JH, Ro JY. The 2020 WHO classification of tumors of bone: an updated review. Adv Anat Pathol. 2021;28:119–38.

Church AJ. Next generation sequencing. In: Tafe LAM, editor. Genomic medicine: a practical guide. Springer: Cham; 2020.

Church AJ, Calicchio ML, Nardi V, et al. Recurrent EML4-NTRK3 fusions in infantile fibrosarcoma and congenital mesoblastic nephroma suggest a revised testing strategy. Mod Pathol. 2018;31:463–73.

Coffin CM, Hornick JL, Fletcher CD. Inflammatory myofibroblastic tumor: comparison of clinicopathologic, histologic, and immunohistochemical features including ALK expression in atypical and aggressive cases. Am J Surg Pathol. 2007;31:509–20.

Davis JL, Lockwood CM, Stohr B, et al. Expanding the spectrum of pediatric NTRK-rearranged mesenchymal tumors. Am J Surg Pathol. 2019;43:435–45.

Davis JL, Vargas SO, Rudzinski ER, et al. Recurrent RET gene fusions in paediatric spindle mesenchymal neoplasms. Histopathology. 2020;76:1032–41.

Demetri GD, Antonescu CR, Bjerkehagen B, et al. Diagnosis and management of tropomyosin receptor kinase (TRK) fusion sarcomas: expert recommendations from the World Sarcoma Network. Ann Oncol. 2020;31:1506–17.

Dickson MA, Tap WD, Keohan ML, et al. Phase II trial of the CDK4 inhibitor PD0332991 in patients with advanced CDK4-amplified well-differentiated or dedifferentiated liposarcoma. J Clin Oncol. 2013;31:2024–8.

Doyle LA, Nowak JA, Nathenson MJ, et al. Characteristics of mismatch repair deficiency in sarcomas. Mod Pathol. 2019;32:977–87.

Doyle LA, Vivero M, Fletcher CD, Mertens F, Hornick JL. Nuclear expression of STAT6 distinguishes solitary fibrous tumor from histologic mimics. Mod Pathol. 2014;27:390–5.

Drilon A, Laetsch TW, Kummar S, et al. Efficacy of larotrectinib in TRK fusion-positive cancers in adults and children. N Engl J Med. 2018;378:731–9.

Fisher C. The diversity of soft tissue tumours with EWSR1 gene rearrangements: a review. Histopathology. 2014;64:134–50.

Flucke U, van Noesel MM, Wijnen M, et al. TFG-MET fusion in an infantile spindle cell sarcoma with neural features. Genes Chromosomes Cancer. 2017;56:663–7.

Forrest SJ, Al-Ibraheemi A, Doan D, et al. Genomic and immunologic characterization of INI1-deficient pediatric cancers. Clin Cancer Res. 2020;26:2882–90.

Garcia EP, Minkovsky A, Jia Y, et al. Validation of OncoPanel: a targeted next-generation sequencing assay for the detection of somatic variants in cancer. Arch Pathol Lab Med. 2017;141:751–8.

Gounder M, Schoffski P, Jones RL, et al. Tazemetostat in advanced epithelioid sarcoma with loss of INI1/SMARCB1: an international, open-label, phase 2 basket study. Lancet Oncol. 2020;21:1423–32.

Hao X, Billings SD, Wu F, et al. Dermatofibrosarcoma protuberans: update on the diagnosis and treatment. J Clin Med. 2020;9

Hechtman JF, Benayed R, Hyman DM, et al. Pan-Trk immunohistochemistry is an efficient and reliable screen for the detection of NTRK fusions. Am J Surg Pathol. 2017;41:1547–51.

Heinrich MC, Corless CL, Demetri GD, et al. Kinase mutations and imatinib response in patients with metastatic gastrointestinal stromal tumor. J Clin Oncol. 2003;21:4342–9.

International classification of diseases for oncology (ICD-O). Available from URL: https://apps.who.int/iris/handle/10665/96612 (accessed May 20, 2021).

Jo VY. EWSR1 fusions: Ewing sarcoma and beyond. Cancer Cytopathol. 2020;128:229–31.

Kao YC, Fletcher CDM, Alaggio R, et al. Recurrent BRAF gene fusions in a subset of pediatric spindle cell sarcomas: expanding the genetic spectrum of tumors with overlapping features with infantile fibrosarcoma. Am J Surg Pathol. 2018;42:28–38.

Kao YC, Sung YS, Zhang L, et al. EWSR1 fusions with CREB family transcription factors define a novel myxoid mesenchymal tumor with predilection for intracranial location. Am J Surg Pathol. 2017;41:482–90.

Koelsche C, Schrimpf D, Stichel D, et al. Sarcoma classification by DNA methylation profiling. Nat Commun. 2021;12:498.

Koo SC, Janeway KA, Harris MH, et al. A distinctive genomic and immunohistochemical profile for NOTCH3 and PDGFRB in myofibroma with diagnostic and therapeutic implications. Int J Surg Pathol. 2020;28:128–37.

Krystel-Whittemore M, Taylor MS, Rivera M, et al. Novel and established EWSR1 gene fusions and associations identified by next-generation sequencing and fluorescence in-situ hybridization. Hum Pathol. 2019;93:65–73.

Lam SW, Kostine M, de Miranda N, et al. Mismatch repair deficiency is rare in bone and soft tissue tumors. Histopathology. 2021;79:509–20.

Lanocha AA, Zdziarska B. T-cell lymphoblastic leukemia with t(11;22) (q24;q12) and EWSR1 rearrangement. Blood. 2017;129:393.

Lin JJ, Riely GJ, Shaw AT. Targeting ALK: precision medicine takes on drug resistance. Cancer Discov. 2017;7:137–55.

Lutsik P, Baude A, Mancarella D, et al. Globally altered epigenetic landscape and delayed osteogenic differentiation in H3.3-G34W-mutant giant cell tumor of bone. Nat Commun. 2020;11:5414.

McArthur GA, Demetri GD, van Oosterom A, et al. Molecular and clinical analysis of locally advanced dermatofibrosarcoma protuberans treated with imatinib: Imatinib Target Exploration Consortium Study B2225. J Clin Oncol. 2005;23:866–73.

McNeil C. NCI-MATCH launch highlights new trial design in precision-medicine era. J Natl Cancer Inst. 2015;107

Mirabello L, Zhu B, Koster R, et al. Frequency of pathogenic germline variants in cancer-susceptibility genes in patients with osteosarcoma. JAMA Oncol. 2020;6:724–34.

Oliveira G, Polonia A, Cameselle-Teijeiro JM, et al. EWSR1 rearrangement is a frequent event in papillary thyroid carcinoma and in carcinoma of the thyroid with Ewing family tumor elements (CEFTE). Virchows Arch. 2017;470:517–25.

Panagopoulos I, Thorsen J, Gorunova L, et al. Fusion of the ZC3H7B and BCOR genes in endometrial stromal sarcomas carrying an X;22-translocation. Genes Chromosomes Cancer. 2013;52:610–8.

Perry JA, Kiezun A, Tonzi P, et al. Complementary genomic approaches highlight the PI3K/mTOR pathway as a common vulnerability in osteosarcoma. Proc Natl Acad Sci U S A. 2014;111:E5564–73.

Rekhi B, Upadhyay P, Ramteke MP, Dutt A. MYOD1 (L122R) mutations are associated with spindle cell and sclerosing rhabdomyosarcomas with aggressive clinical outcomes. Mod Pathol. 2016;29:1532–40.

Riggi N, Suva ML, Stamenkovic I. Ewing's Sarcoma. N Engl J Med. 2021;384:154–64.

Rosenbaum E, Jonsson P, Seier K, et al. Clinical outcome of leiomyosarcomas with somatic alteration in homologous recombination pathway genes. JCO Precis Oncol. 2020;4 https://doi.org/10.1200/PO.20.00122.

Rossi S, Szuhai K, Ijszenga M, et al. EWSR1-CREB1 and EWSR1-ATF1 fusion genes in angiomatoid fibrous histiocytoma. Clin Cancer Res. 2007;13:7322–8.

Schneider K, Zelley K, Nichols KE, Garber J. Li-Fraumeni syndrome. In: Adam MP, Ardinger HH, Pagon RA, et al., editors. GeneReviews((R)). Seattle, WA, 1993.

Shukla NN, Patel JA, Magnan H, et al. Plasma DNA-based molecular diagnosis, prognostication, and monitoring of patients with EWSR1 fusion-positive sarcomas. JCO Precis Oncol. 2017;2017 https://doi.org/10.1200/PO.16.00028.

Siegel RL, Miller KD, Fuchs HE, Jemal A. Cancer statistics, 2021. CA Cancer J Clin. 2021;71:7–33.

Sumegi J, Streblow R, Frayer RW, et al. Recurrent t(2;2) and t(2;8) translocations in rhabdomyosarcoma without the canonical PAX-FOXO1 fuse PAX3 to members of the nuclear receptor transcriptional coactivator family. Genes Chromosomes Cancer. 2010;49:224–36.

Suurmeijer AJH, Dickson BC, Swanson D, et al. A novel group of spindle cell tumors defined by S100 and CD34 co-expression shows recurrent fusions involving RAF1, BRAF, and NTRK1/2 genes. Genes Chromosomes Cancer. 2018;57:611–21.

Suurmeijer AJ, Dickson BC, Swanson D, et al. The histologic spectrum of soft tissue spindle cell tumors with NTRK3 gene rearrangements. Genes Chromosomes Cancer. 2019;58:739–46.

Tap WD, Gelderblom H, Palmerini E, et al. Pexidartinib versus placebo for advanced tenosynovial giant cell tumour (ENLIVEN): a randomised phase 3 trial. Lancet. 2019;394:478–87.

Tap WD, Villalobos VM, Cote GM, et al. Phase I study of the mutant IDH1 inhibitor ivosidenib: safety and clinical activity in patients with advanced chondrosarcoma. J Clin Oncol. 2020;38:1693–701.

Thway K. Well-differentiated liposarcoma and dedifferentiated liposarcoma: an updated review. Semin Diagn Pathol. 2019;36:112–21.

Thway K, Fisher C. Mesenchymal tumors with EWSR1 gene rearrangements. Surg Pathol Clin. 2019;12:165–90.

Wachter F, Al-Ibraheemi A, Trissal MC, et al. Molecular characterization of inflammatory tumors facilitates initiation of effective therapy. Pediatrics. 2021;148:e2021050990.

Wang WL, Mayordomo E, Zhang W, et al. Detection and characterization of EWSR1/ATF1 and EWSR1/CREB1 chimeric transcripts in clear cell sarcoma (melanoma of soft parts). Mod Pathol. 2009;22:1201–9.

Wang L, Motoi T, Khanin R, et al. Identification of a novel, recurrent HEY1-NCOA2 fusion in mesenchymal chondrosarcoma based on a genome-wide screen of exon-level expression data. Genes Chromosomes Cancer. 2012;51:127–39.

Wegert J, Vokuhl C, Collord G, et al. Recurrent intragenic rearrangements of EGFR and BRAF in soft tissue tumors of infants. Nat Commun. 2018;9:2378.

Yamamoto H, Yoshida A, Taguchi K, et al. ALK, ROS1 and NTRK3 gene rearrangements in inflammatory myofibroblastic tumours. Histopathology. 2016;69:72–83.

Zheng Z, Liebers M, Zhelyazkova B, et al. Anchored multiplex PCR for targeted next-generation sequencing. Nat Med. 2014;20:1479–84.

Jennifer A. Cotter and Eyas M. Hattab

Contents

Central Nervous System Tumors

Introduction

- While morphologic and immunohistochemical evaluations remain the gold standard in the diagnosis of most central nervous system (CNS) neoplasms, more than ever before, molecular techniques supplement and increasingly define tumor classification

J. A. Cotter
Department of Pathology and Laboratory Medicine, Children's Hospital Los Angeles, Keck School of Medicine of the University of Southern California, Los Angeles, CA, USA
e-mail: jcotter@chla.usc.edu

E. M. Hattab (✉)
Department of Pathology and Laboratory Medicine, University of Louisville School of Medicine, Louisville, KY, USA
e-mail: eyas.hattab@louisville.edu

- Predictive gene mutations, gene expression profiles, structural rearrangements, and DNA methylation profiles have been identified for many CNS tumors in recent years and will likely continue to drive future classifications
- Identification of certain genetic alterations that influence the survival or therapeutic responsiveness of some CNS neoplasms has added a new dimension to the significance of ancillary molecular testing in these tumors
- Among the most utilized techniques in the genetic characterization of CNS neoplasms are fluorescence in situ hybridization (FISH), comparative genomic hybridization, and next-generation sequencing (NGS)
- DNA methylation array can be a useful adjunct for classifying more difficult or unusual cases; a methylation profile aligned with an established entity is an accepted criterion for establishing a diagnosis in several entities
- Gene expression profiling and DNA methylation array can allow more precise subgrouping within specific types of CNS tumors (e.g., medulloblastoma and atypical teratoid/rhabdoid tumor [ATRT])

Glial Tumors

Introduction

- Glial neoplasms encompass a heterogeneous group of tumors, all of which have been historically considered to have phenotypic similarity to the glial cells of the normal brain (astrocytes, oligodendrocytes, ependymal cells)
- Gliomas are the most common category of primary CNS tumor and demonstrate variable expression of GFAP
- There are two major groupings of gliomas based on age and growth pattern
  - Adult-type vs. pediatric-type: Despite significant morphologic overlap, there is overwhelming molecular evidence to suggest that pediatric gliomas differ significantly from their adult counterparts and therefore

L. Cheng et al. (eds.), *Molecular Surgical Pathology*, https://doi.org/10.1007/978-3-031-35118-1_23

constitute distinct entities (WHO CNS5); of note, occasionally, cases of the opposite type may occur in patients outside of the expected age range

- Diffuse vs circumscribed (localized): Molecular profiling has shown that diffuse gliomas are characterized by distinct genetic pathways and alterations, further enforcing the morphologic differences between the two groups

- Based on the information above, the most practical approach to working up gliomas starts with determining age group and evaluating whether a glioma is diffuse or circumscribed; neurofilament immunohistochemistry is a useful way to evaluate the extent of infiltration of a tumor into the surrounding brain tissue

Adult-Type Diffuse Gliomas

- Adult-type diffuse gliomas can be separated into two major categories based on their IDH mutation status
- Frequent somatic mutations of *IDH1* (isocitrate dehydrogenase 1) or *IDH2* are observed in diffuse gliomas, particularly in younger adult patients
- *IDH1/2* mutations occur in 60–100% of grades 2–3 diffuse gliomas but are rare in grade 4 tumors and in the pediatric setting
- The IDH proteins catalyze the oxidative decarboxylation of isocitrate to alpha-ketoglutarate, resulting in effects on epigenetic state (i.e., hypermethylation) and genetic regulation
- *IDH1* mutations at codon 132 are by far the most common, accounting for over 90% of IDH mutations
  - These are characterized by a base-pair exchange of guanine to adenine (G395A), resulting in a substitution of the amino acid arginine by histidine (R132H)
  - Other reported *IDH1* mutations are uncommon (<5% each), and include R132C, R132L, R132S, and R132G
- IDH variants outside codons 132 and 172 that may appear on some NGS reports are not necessarily pathologic and should not be confused with the above-described *IDH1/2* mutations that define this group of diffuse gliomas
- Immunohistochemistry for IDH1 R132H protein is commercially available and works well on formalin-fixed, paraffin-embedded samples
- Genetic analysis of IDH1/2 genes by sequencing is recommended for diffuse gliomas when immunohistochemistry for IDH1 R132H is negative in the following scenarios:
  - All diffuse gliomas with grade 2 or 3 morphology, regardless of age
  - Diffuse gliomas with grade 4 morphology in patients younger than 55 years
- Certain molecular alterations (e.g., H3K27-mutant glioma) are mutually exclusive with IDH mutations

- Among adult gliomas, *IDH1/2* mutations correlate with younger age at diagnosis, the presence of *TP53* mutation, combined 1p/19q deletion, and *MGMT* promoter hypermethylation

Adult-Type Diffuse Gliomas, IDH Mutant

Astrocytoma, IDH Mutant
Definition
- Diffuse astrocytic glioma harboring a pathogenic variant in either *IDH1* or *IDH2*
- Histologic grade may range from 2 to 4

Clinical Features
- May occur anywhere in the neuraxis, and imaging features vary depending on the tumor grade, with grade 2 tumors often lacking contrast enhancement, edema, and necrosis, which are more commonly seen in grade 3 or 4 tumors
- Occur in adult patients (median age, 38), with a male predominance; IDH-mutant astrocytomas are uncommon in patients over 55

Pathologic Features
- Astrocytoma, IDH mutant, is a diffuse glioma classically characterized by fibrillary tumor cells with elongate hyperchromatic nuclei but may occur with different cytology, such as gemistocytic or protoplasmic
- Demonstrates an infiltrative growth pattern that is often manifested by perineuronal satellitosis, subpial condensation, and perivascular accumulation (secondary structures of Scherer)
- Grading relies on the identification of increased mitotic activity, necrosis, and/or microvascular proliferation
  - A tumor lacking all three is considered grade 2
  - A tumor with mitotic figures but lacking necrosis and microvascular proliferation is considered grade 3
  - A tumor with necrosis or microvascular proliferation is assigned grade 4
- The term "glioblastoma, IDH mutant" has been dropped in the 2021 WHO classification in favor of astrocytoma, IDH mutant, WHO grade 4; this is in direct recognition that despite the morphologic similarities, glioblastoma, IDH wildtype, is a distinct biologic and genetic entity from astrocytoma, IDH mutant, WHO grade 4
- Almost invariably expresses GFAP and OLIG2

Genetic and Biomarker Findings
- The presence of *ATRX* and/or *TP53* mutations in an IDH mutant diffuse glioma is diagnostic of astrocytoma, IDH mutant

Table 23.1 Familial syndromes

Syndrome	Gene involved	CNS tumors associated
Neurofibromatosis, type 1	*NF1*	Pilocytic astrocytoma, other gliomas, MPNST
Neurofibromatosis, type 2	*NF2*	Meningioma, ependymoma, schwannoma
Schwannomatosis	*SMARCB1*	Cranial or spinal nerve schwannoma
Von Hippel–Lindau syndrome	*VHL*	Hemangioblastoma
Tuberous sclerosis	*TSC1, TSC2*	Subependymal giant cell astrocytoma, cortical tuber
Li–Fraumeni syndrome	*TP53*	High-grade glioma, sarcoma, medulloblastoma
Constitutional mismatch repair deficiency	*MSH6, MSH2, MLH1, MLH2, PMS2, POLE1*	High-grade glioma
Rhabdoid tumor predisposition syndrome	*SMARCB1, SMARCA4* (rarely)	Atypical teratoid/rhabdoid tumor
Nevoid basal cell carcinoma syndrome	*PTCH1, PTCH2, SUFU*	Medulloblastoma, SHH-activated
Carney complex	*PRKAR1A*	Malignant melanotic nerve sheath tumor

- Somatic mutation of the tumor suppressor gene *TP53* and LOH of 17p are common findings (i.e., loss of a wildtype *TP53* allele)
    - Astrocytoma, IDH mutant, is the most common brain tumor associated with Li–Fraumeni syndrome (germline *TP53* mutation, see Table 23.1)
- Over 90% harbor *ATRX* alterations, which are highly associated with the alternate lengthening of telomeres, a mechanism of maintaining telomeres employed by cancer cells
- Loss of nuclear protein expression by ATRX immunohistochemistry is a highly sensitive and specific surrogate marker for *ATRX* mutations
    - The presence of internal positive control in blood vessels and native glia guards against potential misinterpretation
- Astrocytomas, IDH mutant, with loss of immunoexpression of ATRX and/or extensive p53 nuclear immunoreactivity can be diagnosed as such on a histologic basis alone
- Intact ATRX immunoexpression or weak or absent p53 immunoreactivity in an IDH-mutant glioma should prompt further investigation including evaluation for 1p/19q codeletion (see Oligodendroglioma, IDH mutant section below)

Oligodendroglioma, IDH Mutant and 1p/19q Codeleted

Definition

- Diffuse oligodendroglial tumor with a pathogenic variant in either *IDH1* or *IDH2* and 1p/19q combined deletion
- Histologic grade ranges from 2 to 3

Clinical Features

- Occur in adults (median age, 41–47) with a slight male predominance; they frequently present with seizures
- Rare in children and the very old
- The subcortical white matter of the cerebral hemispheres (mostly frontal) is preferentially affected
- Exceptionally rare in the cerebellum, brainstem, or spinal cord
- Patients may present with focal neurological deficits, including seizure disorder, or more generalized symptoms because of increased intracranial pressure
- Radiographically, appear as demarcated lesions in the cortex or subcortical white matter with no significant enhancement or peritumoral edema; calcification is frequent
- Contrast enhancement and prominent peritumoral edema usually indicate higher grade

Pathologic Features

- Oligodendrogliomas are diffuse gliomas that usually show characteristic morphology, including a rich network of thin-walled, branching capillaries (chicken-wire vasculature) (Fig. 23.1)
- Similar to other diffuse gliomas, cortical invasion is essentially a constant feature, often producing the so-called secondary structures of Scherer:
    - Perineuronal satellitosis
    - Perivascular condensation
    - Subpial accumulation
- The uniform tumor cells are characterized by round to oval nuclear morphology with smooth contour and indistinct chromatin pattern; mitotic figures may be scattered; and the cells may acquire perinuclear clearing secondary to a formalin-fixation artifact (fried-egg appearance)
- Presence of marked cytologic atypia, frequent mitotic figures (more than 6 per 10 high-power fields [hpf]), and microvascular proliferation has been reported as indicators of more aggressive behavior
    - Unlike diffuse astrocytomas, the presence of necrosis or vascular proliferation in oligodendroglioma does not confer a grade 4 designation

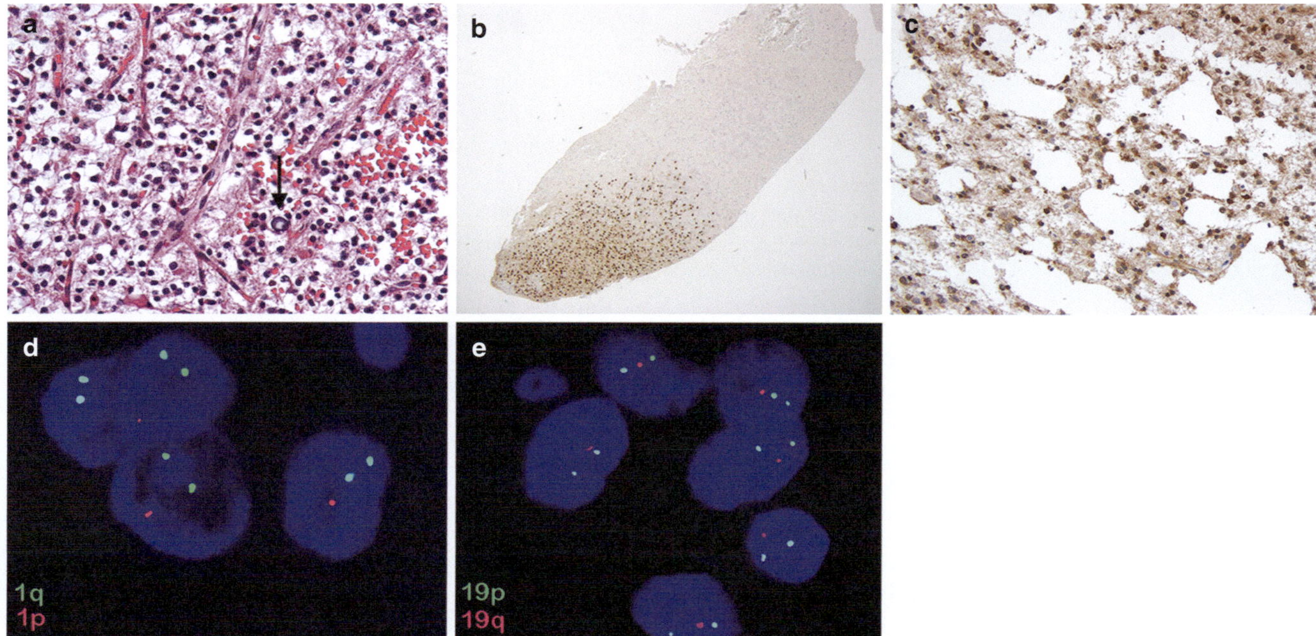

Fig. 23.1 Oligodendroglioma. (**a**) This WHO grade 2 oligodendroglioma shows a moderately cellular tumor comprised of uniform cells with the characteristic "fried-egg" appearance and round nuclei. Note the branching capillaries (chicken-wire) and focal calcification (arrow). (**b**) IDH1-R132H immunohistochemistry can be of exceptional value in identifying the infiltrative edge of oligodendroglioma as in this example. (**c**) While IDH1-R132H staining pattern is cytoplasmic, a variable nuclear immunoreactivity is usually detectable. (**d**, **e**) Dual color FISH assays showing loss of 1p and 19q, respectively

- Even in the presence of "classic" morphologic features of oligodendroglioma, the diagnosis of oligodendroglioma should not be made on histologic grounds alone and should always be followed by appropriate molecular workup

Genetic Features

- Tumors with *IDH1/2* mutation and 1p/19q codeletion are defined as oligodendrogliomas; this codeletion may be identified by FISH or copy number analysis
- *IDH1* or *IDH2* mutation must be present along with whole arm deletions of 1p and 19q
  - While still overall rare, *IDH2* mutations are more common in oligodendrogliomas (4–5%) than in astrocytomas (1%)
- Partial arm deletions of 1p and 19q are not significant and should not be interpreted as supportive of the diagnosis
- Common co-occurring alterations include *TERT* promoter mutations and mutations of *CIC, FUBP1,* and *NOTCH1*
  - *TERT* promoter mutations are less often observed in adolescent oligodendrogliomas than in adult cases
- *CDKN2A* and/or *CDKN2B* homozygous deletion is associated with shorter survival and may serve as a molecular marker of CNS WHO grade 3 tumors

Adult-Type Diffuse Gliomas, IDH Wildtype

- The 2021 WHO Classification (CNS5) eliminated the entity of astrocytoma, IDH wildtype, WHO grades 2–4, in favor of glioblastoma, IDH wildtype, WHO grade 4
- This followed years of mounting molecular evidence that the vast majority of diffuse astrocytomas that are IDH wildtype (regardless of grade) show molecular features characteristic of glioblastoma, IDH wildtype, and behave poorly
- The elimination of astrocytoma, IDH wildtype, grades 2/3, essentially forces the pathologist to consider one of two options:
  - Perform molecular analysis for more accurate classification
  - Issue a "diffuse astrocytoma, not otherwise specified (NOS)" diagnosis with an appropriate explanatory comment; care should be taken to exclude other types of glioma

Glioblastoma, IDH Wildtype

Definition

- A diffuse astrocytic glioma lacking *IDH1/2* mutation and H3 mutation
- Glioblastoma can now be defined either histologically (diffuse glioma with vascular proliferation and/or necro-

sis) or genetically (any diffuse glioma lacking IDH and H3 mutations and exhibiting *EGFR* amplification, *TERT* promoter mutation, or gains of chromosome 7 and loss of chromosome 10)

Clinical Features
- Occurs mostly in adults in sixth to eighth decade of life
- Male:female ratio of 1.5:1
- Preferentially involves the cerebral hemispheres
- Typically demonstrates ring-like enhancement on post-contrast MRI
- Most patients die within 15–18 months

Pathologic Features
- Histologically, glioblastoma, IDH wildtype, is highly cellular with marked cytologic anaplasia and contains foci of vascular proliferation and/or necrosis, whether palisading or geographic (Fig. 23.2)
- Morphology is markedly heterogeneous (hence the prior name "glioblastoma multiforme") and may include small undifferentiated cells, large pleomorphic bizarre cells, spindle cells, or epithelioid cells
- Diffuse gliomas lacking microvascular proliferation and necrosis may be diagnosed as glioblastoma, IDH wildtype, if they are shown to have one of the following poor prognostic molecular features (molecularly defined glioblastoma):
  - *EGFR* amplification
  - Gain of chromosome 7 and loss of chromosome 10 (+7/−10)
  - *TERT* promoter mutation
- As such, WHO grading of glioblastoma has now expanded from the historical histologic features of vascular prolif-

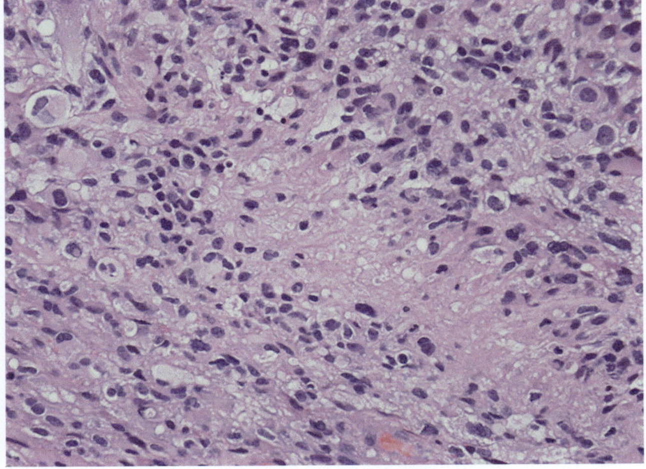

Fig. 23.2 Glioblastoma. A hypercellular glial neoplasm with large hyperchromatic irregular nuclei. Note the high degree of nuclear pleomorphism and the area of palisading necrosis toward the center

eration and/or necrosis to include the three molecular features described above; thus, the presence of any one of the five criteria is sufficient for a WHO grade 4 designation of an IDH/H3-wildtype diffuse glioma

Genetic Features
- Frequent molecular alterations include:
  - *EGFR* amplification (~40%), especially associated with small cell histology
  - *PTEN* mutation/deletion (40%)
  - Mutations of PI3K pathway genes (*PIK3CA* and *PIK3R1*) (25–30%)
  - *TERT* promoter mutations
  - *TP53* mutations, especially associated with giant cell histology
  - Homozygous deletion of *p16* (*CDKN2A*) (~40%) and *p14*^ARF
  - *CDK4* amplification
  - *RB1* mutation/homozygous deletion
  - *MDM2/MDM4* gene amplification
  - Whole chromosome 7 gain and whole chromosome 10 loss
    - Less commonly, 7q gain and/or 10q loss

Pediatric-Type Diffuse Gliomas

Diffuse Gliomas, H3 Defined
- Histone proteins organize DNA into structural subunits and have a key role in the maintenance of gene expression or repression
- Modifications to histone proteins commonly occur on the histone "tail" or N-terminal portion and include acetylation or methylation changes
- Histone mutations in known H3 histone genes, *H3-3A, H3C2, or H3C3*, occur in diffuse gliomas with distinct clinical presentations and affect specific critical amino acids within the N-terminal portion of the histone gene products
- H3 mutations in diffuse gliomas are commonly associated with inactivating mutations of *ATRX*
  - The finding of such an *ATRX* mutation, or loss of ATRX immunoexpression, in an IDH-wildtype glioma should prompt consideration for an H3-mutant glioma

Diffuse Midline Glioma, H3K27-Altered
Definition
- Diffuse glioma occurring in midline sites of the CNS with loss of H3K27me3 harboring somatic K27M mutations in one of the H3 genes, *H3-3A, H3C2,* or *H3C3*, aberrant overexpression of EZHIP, or an *EGFR* mutation

Clinical Features
- Typically arise in thalamus, pons, or spinal cord

- Less frequent sites include the pineal region and cerebellum
- Most frequently encountered in the pediatric setting but may occur in adults
- Usually peaks at about 7–8 years with H3.1 or H3.2 K27–mutant subtypes occurring in slightly younger patients
- CNS WHO grade 4

Pathologic Features
- Four diffuse midline glioma subtypes are recognized:
  - H3.3 K27 mutant (either K27M or K27I)
  - H3.1 or H3.2 K27 mutant
  - H3 wildtype with EZHIP overexpression
  - *EGFR* mutant
- Diffuse pattern of growth is a requirement for the diagnosis, as some circumscribed midline gliomas (e.g., pilo-cytic astrocytomas, subependymomas, and gangliogliomas) can also harbor H3K27M mutation and posterior fossa group A ependymoma may also show loss of K27me3 or EZHIP overexpression
- In addition to a diffuse growth pattern, most diffuse midline gliomas are characterized by small, monomorphic cells with frequent mitotic figures; however, apart from the infiltrative pattern, no specific morphologic requirements are necessary for the diagnosis or for satisfying WHO grade 4 designation (Fig. 23.3)
- Most diffuse midline gliomas express OLIG2, SOX10, and S100 except for *EGFR*-mutant subtype
- Immunohistochemistry for H3K27M-mutant specific protein is a useful surrogate marker for the mutation and when combined with loss of nuclear H3K27me3, offers a greater sensitivity and specificity

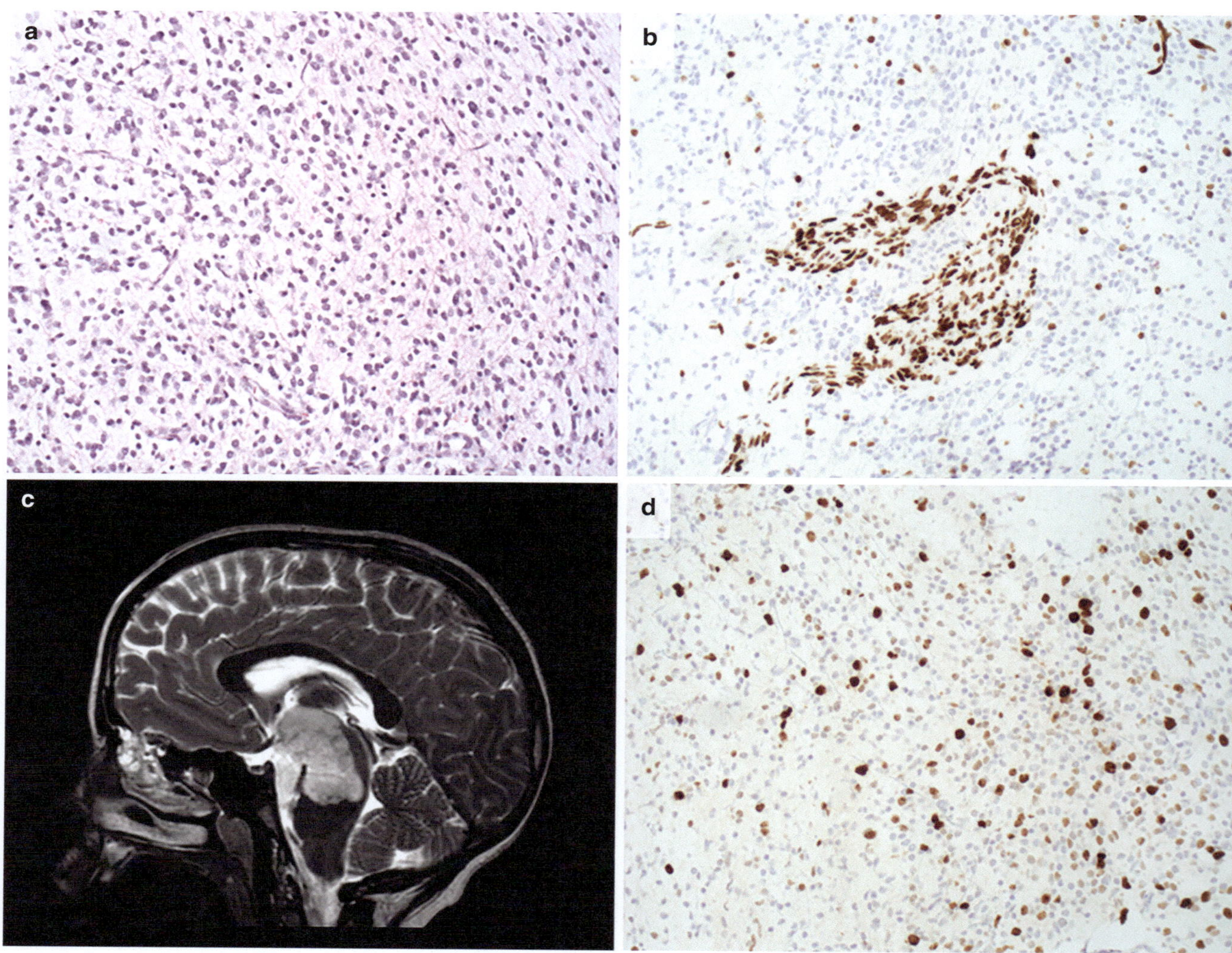

Fig. 23.3 Diffuse midline glioma, H3K27-altered. (**a**) This diffuse glioma is composed of monomorphic ovoid cells infiltrating brainstem structures. (**b**) H3K27me3 immunohistochemistry demonstrates loss of nuclear expression in tumor cells, while expression is retained in endothelial cells as well as in native glial and neuronal cells. (**c**) This tumor involved the pons and midbrain and was found to harbor an H3C2 mutation. (**d**) Ki-67 labeling index is often elevated, but this feature is not required for the diagnosis

- Loss of nuclear immunoreactivity for H3K27me3 is particularly helpful finding in other H3K27-altered diffuse midline gliomas that are not usually detected by K27M immunohistochemistry (e.g., those with EZHIP overexpression or K27I mutation)
- EZHIP (CXorf67) antibodies are commercially available
- Approximately 15% of cases show loss of ATRX nuclear expression
- About one-half show p53 overexpression
- K27 mutation (or other alteration associated with loss of H3K27 trimethylation) is associated with a very poor prognosis and assigned WHO grade 4

Genetic Features

- The most common H3 gene mutation in this tumor is in *H3-3A*
- K27M mutation impairs methylation of its site in the N-terminal region of the histone protein, resulting in the release of repression of transcription (activation)
- While one-third of K27M mutant pediatric tumors have mutations of *H3C2* or *H3C3*, these are rare in adult K27M tumors

Diffuse Hemispheric Glioma, H3 G34 Mutant

Definition

- Diffuse, hemispheric (nonmidline) glioma that harbors *H3-3A* G34R or G34V mutations

Clinical Features

- Occurs in the cerebral hemispheres of adolescent and young adult patients
- Male predominance (1.4:1)
- CNS WHO grade 4

Pathologic Features

- Frequently has the appearance of glioblastoma, including palisading necrosis and vascular proliferation, but may also exhibit an embryonal appearance
- Secondary structures are often prominent
- OLIG2 is typically negative, ATRX lost, and p53 overexpressed
- Given the embryonal-like primitive appearance in some cases and OLIG2 negativity, these tumors may be mistaken for CNS embryonal tumors
  - Secondary structures of Scherer are a clue to the diagnosis of this diffuse glioma (Fig. 23.4)

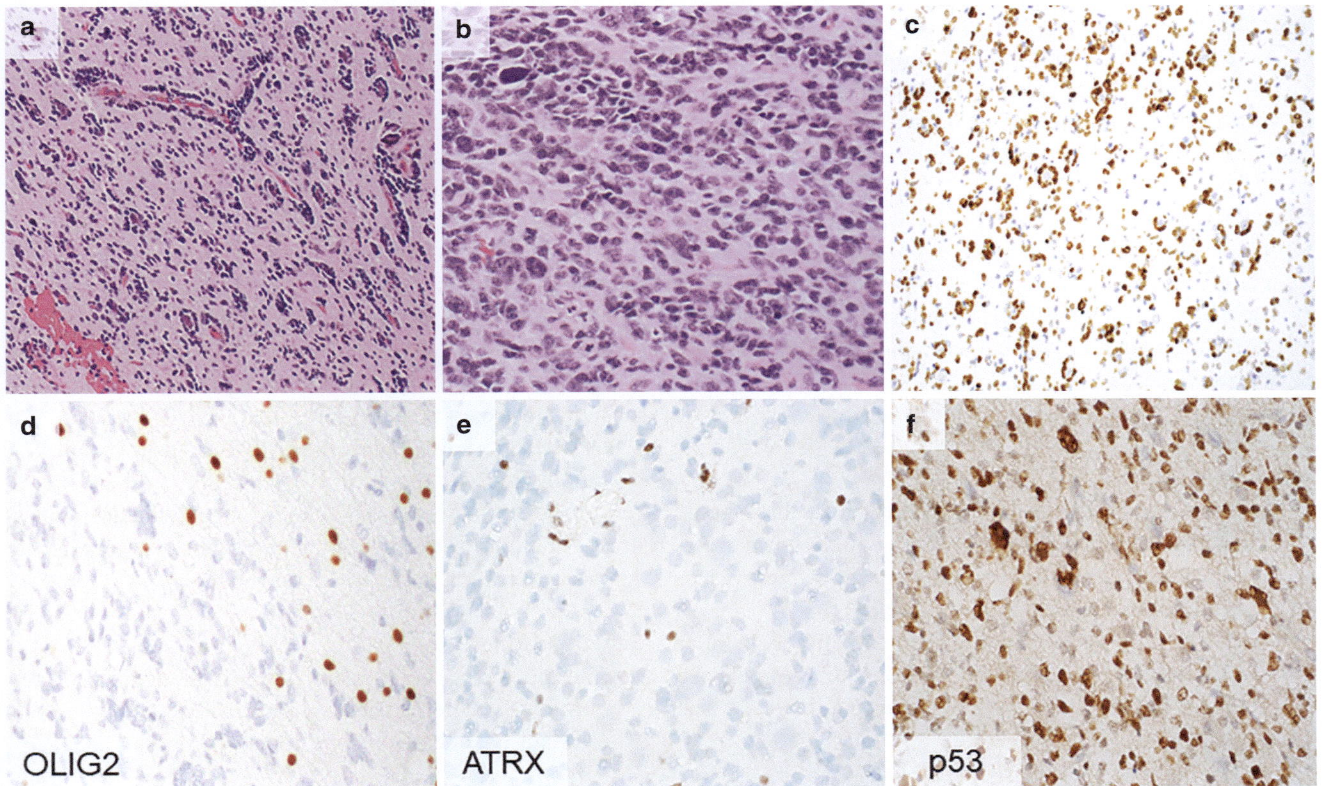

Fig. 23.4 Diffuse cerebral glioma, G34 mutant. A diffuse glioma with cortical infiltration (**a**) and areas of solid growth (**b**) sometimes resembling an embryonal tumor. (**c**) G34R-specific immunohistochemistry is a surrogate marker for the underlying mutation in H3-3A. Lack of OLIG2 expression (**d**), ATRX expression (**e**), and overexpression of p53 protein (**f**) are common supportive findings

- Ganglion cell components and pleomorphic xanthoastrocytoma histology have been described
- Immunohistochemistry for G34R or G34V–mutant proteins is available, but concordance with sequencing results is not well-established
  - Sequencing-based approaches are suggested for hemispheric tumors with OLIG2 loss, ATRX loss, and/or p53 overexpression by immunohistochemistry

Genetic Features

- *H3-3A* G34 mutations are associated with loss of trimethylation of H3 K36, probably resulting in the release of repression of transcription (activation)
- While currently considered gliomas, these tumors probably originate in interneuron precursor cells, and the G34 alteration impairs neuronal differentiation; *PDGFRA* signaling is an essential oncogenic mechanism in these tumors
- *ATRX* mutations are frequent in this category (~95%), and *TP53* mutations commonly co-occur (~90%)
- H3.3 G34 mutations are mutually exclusive with IDH and H3K27 mutations
- MGMT is often hypermethylated

Diffuse Pediatric-Type High-Grade Glioma, H3 Wildtype, and IDH Wildtype

Definition

- A pediatric high-grade diffuse glioma lacking H3 and IDH mutations, occurring in children, adolescents, and young adults, and belonging to one of several specific molecular and epigenetic categories

Clinical Features

- Occurs throughout the CNS, with the majority being supratentorial
- These are typically observed in children and young adults, but detailed epidemiological and clinical features are yet to be well-defined
- CNS WHO grade 4

Pathologic Features

- Tumors with the histologic appearance of a high-grade glioma should show a diffuse growth pattern in this category, frequent mitotic figures, and may show microvascular proliferation or necrosis
- The MYCN subtype tumors are frequently highly cellular and resemble CNS embryonal tumors; they often have circumscribed borders with adjacent brain tissue (Fig. 23.5)

Genetic Features

- The category of H3 wildtype, IDH wildtype, tumors encompasses a heterogeneous group of tumors
- Three molecular subtypes are currently recognized
  - Diffuse pediatric-type high-grade glioma (pHGG) RTK2 (*EGFR* amplifications)
  - pHGG RTK1 (*PDGFRA* amplifications and *TERT* promoter mutations)
  - pHGG MYCN (*MYCN* amplifications)
- High-grade gliomas arising after therapeutic radiation, and those associated with germline mismatch repair deficiency, typically show molecular features similar to this category, with predominance of the pHGG RTK1 molecular subtype
- Gliomas arising in the context of constitutional mismatch repair deficiency syndrome, Lynch syndrome, or Li–Fraumeni should be recognized as distinct from this entity even though they share nearly identical molecular features (see Table 23.1)

Infant-Type Hemispheric Glioma

Definition

- A type of cellular glioma arising in the hemispheres of a very young patient, characteristically harboring one of several RTK gene alterations

Clinical Features

- Occur mostly in infants, usually under 12 months of age
- Arise in the cerebral hemispheres, with a large, frequently cystic and heterogeneous radiologic appearance
- Frequent superficial and leptomeningeal involvement
- CNS WHO grade not assigned

Pathologic Features

- Histologic features are not clearly defined, but most are high-grade (Fig. 23.6)
- Spindle cell regions are commonly present, and gemistocytic areas can be seen
- GFAP expression is typical, and diffuse growth pattern is variably present
- Some extent of overlap is acknowledged with desmoplastic infantile astrocytoma/ganglioglioma, so molecular characterization is key to the diagnosis

Genetic Features

- Four molecular subtypes are currently recognized
  - Infant-type hemispheric glioma, *NTRK*-altered
  - Infant-type hemispheric glioma, *ROS1*-altered

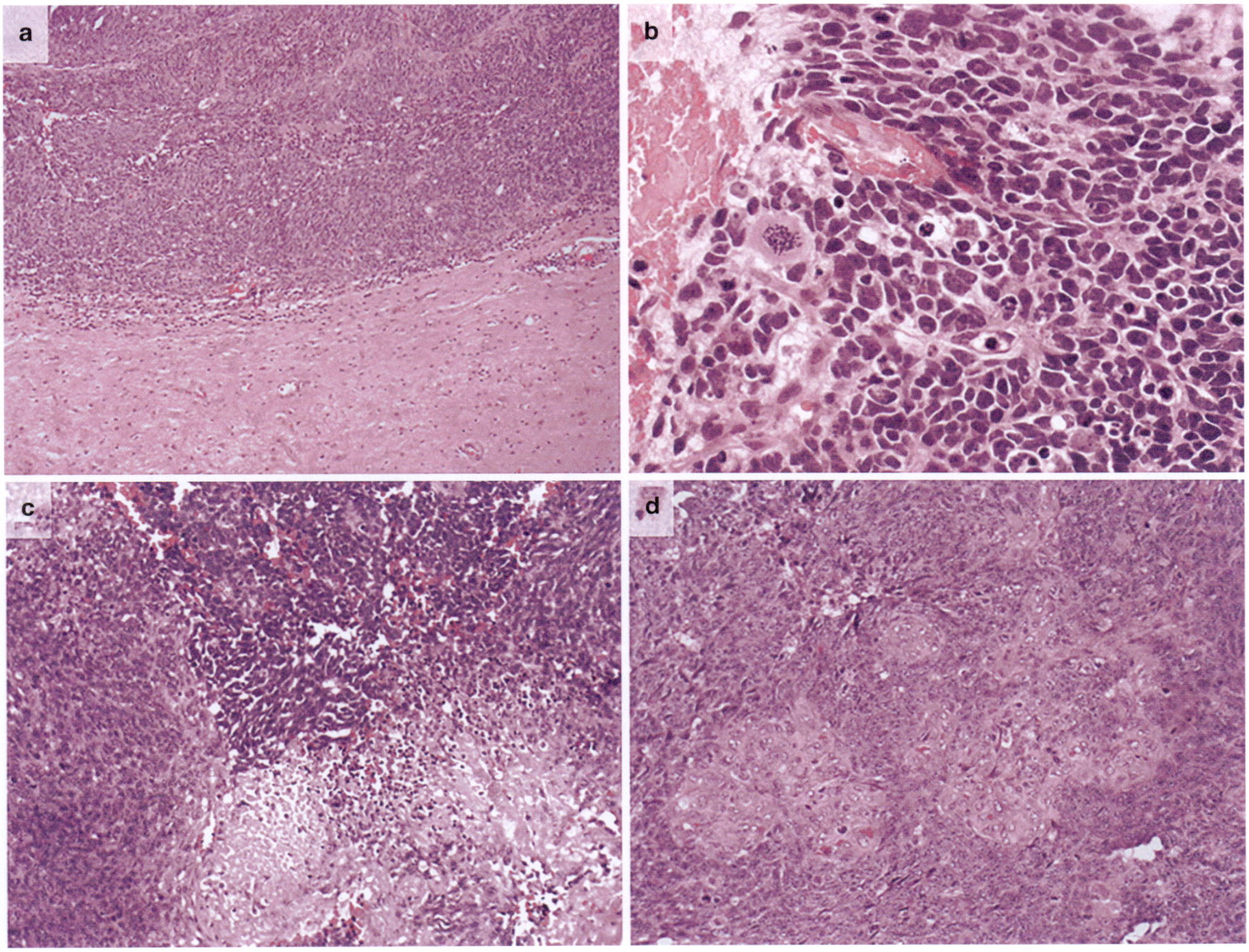

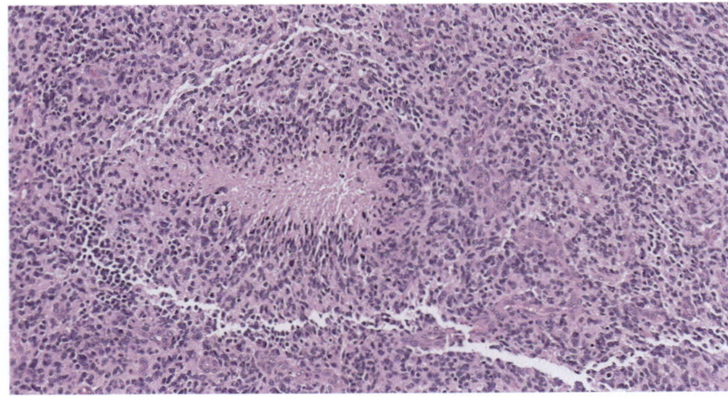

Fig. 23.5 Diffuse pediatric-type high-grade glioma, H3 wildtype and IDH wildtype, MYCN subtype. (**a**) This category of gliomas often features high cellularity and may show circumscribed borders with adjacent brain tissue. (**b**) Nuclear atypia, frequent mitotic figures, and apoptotic bodies are present. (**c**) Necrosis and (**d**) vascular proliferation are observed but, in some cases, these features may be absent

Fig. 23.6 Infantile hemispheric glioma. High-grade histology is seen in most cases, including microvascular proliferation and necrosis. Despite its resemblance to adult glioblastoma, infantile hemispheric glioma is associated with a specific set of molecular driver alterations affecting tyrosine kinase genes *ALK, ROS1, MET*, and *NTRK*. The tumor shown here harbored a TPM3::NTRK fusion

- Infant-type hemispheric glioma, *ALK*-altered
- Infant-type hemispheric glioma, *MET*-altered
- This category of tumors encompasses several types of structural genetic rearrangement that share a general DNA methylation profile and clinical presentation
- RTK fusions are present in 60–80% of cases and serve as potential therapeutic targets

Pediatric-Type Diffuse Low-Grade Gliomas

Diffuse Low-Grade Glioma, MAPK Pathway-Altered

Definition
- Glial tumor with diffuse pattern of growth may show astrocytic or oligodendroglial morphology

- Harbors an alteration involving the MAPK pathway protein; either an internal tandem duplication or mutation in the tyrosine kinase domain of *FGFR1* or *BRAF* p.V600E
- IDH wildtype and H3 wildtype and lacks CDKN2A homozygous deletion

Clinical Features
- Typically occur in childhood, often in the cerebral hemispheres
- Epilepsy may be a presenting feature
- Heterogeneously enhancing masses with cystic elements
- Prognosis appears to follow a low-grade course, but the outcome depends on many factors, and a CNS WHO grade is yet to be assigned

Pathologic Features
- Monotonous cytologic appearance and subtle infiltrative growth pattern
- Histology is typically low-grade lacking necrosis, microvascular proliferation, or brisk mitotic activity
- Nodules or aggregates of tumor cells may be present
- OLIG2 is typically positive; CD34 is sparse; BRAF p. V600E immunoreactivity restricted to the *BRAF* p.V600E-mutant subtype
- While it is too early to associate a specific morphology with a genetic alteration, there is a tendency for *FGFR1*-altered tumors to show an oligodendroglial appearance and a nodularity akin to that of DNT

Genetic Features
- Three genetic subtypes recognized
  - Diffuse low-grade glioma, *FGFR1* tyrosine kinase domain-duplicated
  - Diffuse low-grade glioma, *FGFR1*-mutant
  - Diffuse low-grade glioma, *BRAF* p.V600E-mutant
- Other rare alterations are also encountered
- These subtypes result in MAPK pathway activation
- Methylation classification does not currently assign a specific cluster or class to these tumors
  - Methylation profiling maybe useful in excluding other overlapping entities

Diffuse Astrocytoma, MYB- or MYBL1-Altered
Definition
- A rare glial tumor with diffuse pattern of growth, harboring a pathogenic alteration in the *MYB* or *MYBL1* gene
- CNS WHO grade 1

Clinical Features
- Children, adolescents, and young adults
- Predominantly a cerebral tumor involving cortex and subcortex

- Long-term, drug-resistant epilepsy is the most common presenting symptom, usually from childhood
- MRI features include lack of contrast enhancement and lack of diffusion restriction suggestive of low cellularity; occasionally, cystic
- Most patients become seizure-free following surgical resection, and the overall outcome appears consistent with the benign nature of CNS grade 1 tumors

Pathologic Features
- Monotonous appearing glial cells, occasionally with a vague perivascular condensation (histologic overlap with angiocentric glioma, discussed below)
- Infiltration maybe so subtle, making the tumor difficult to recognize
- Absent or low mitotic activity and absent necrosis and vascular proliferation
  - OLIG2 and CD34 negative
  - Very low Ki-67 labeling index

Genetic Features
- Two genetic subtypes
  - Diffuse astrocytoma, *MYB*-altered
  - Diffuse astrocytoma, *MYBL1*-altered
- Structural rearrangement in the above proteins results in truncation and overexpression
- Reported fusion partners include *PCDHGA1*, *MMP16*, and *MAML2*
- Of note, *MYB::QKI* fusion is associated with angiocentric glioma and is not typically seen in this entity
- IDH and H3 wildtype

Polymorphous Low-Grade Neuroepithelial Tumor of the Young
Definition
- Polymorphous low-grade neuroepithelial tumor of the young (PLNTY) is a diffuse low-grade tumor usually occurring in the cortex and subcortical white matter in young people (second to third decades of life) and characterized by oligo-like components, calcification, CD34 immunoreactivity, and MAPK pathway abnormalities
- CNS WHO grade 1

Clinical Features
- Patients usually present with refractory seizures
- Tumors are usually in the cerebrum with predilection for temporal lobes
- May have cystic and solid appearance on MRI and calcification on CT
- Outcome generally favorable with resection eliminating or improving seizures

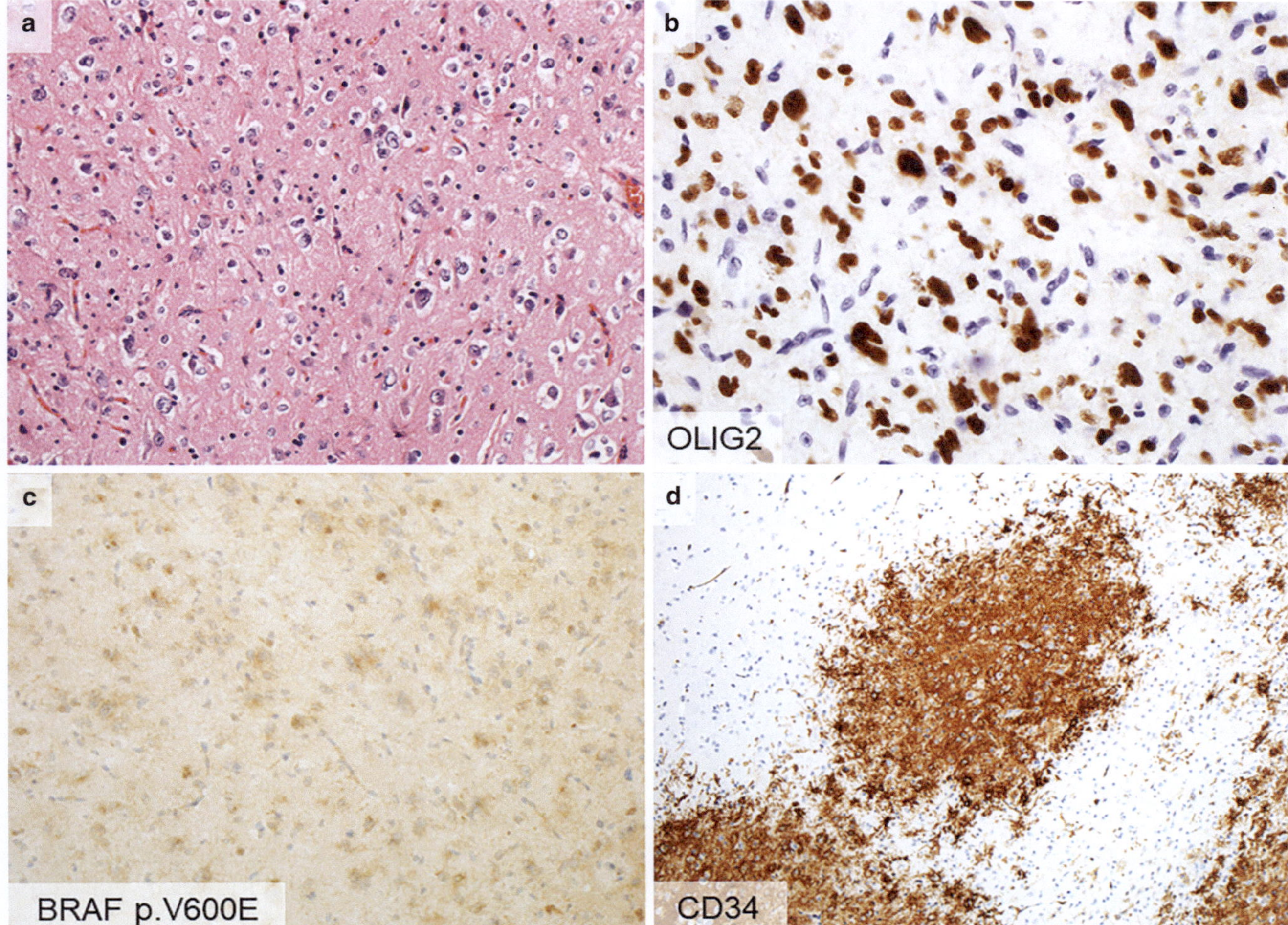

Fig. 23.7 Polymorphous low-grade neuroepithelial tumor of the young (PLNTY). (**a**) H&E shows areas of hypercellularity containing moderately pleomorphic oligodendroglioma-like cells with perinuclear haloes. Few isolated intervening neurons are present. There is often proliferation of thin capillaries, and microcalcification. (**b**) Immunoreactivity is seen in OLIG2. (**c**) BRAF V600E shows patchy cytoplasmic immunoreactivity of unknown significance; this pattern should prompt molecular testing for BRAF p.V600E. (**d**) CD34 shows strong immunoreactivity; this is considered an essential diagnostic feature of PLNTY

Pathologic Features
- Growth patten may be clearly diffuse or appear solid; both are reported
- Frequent calcifications, often confluent
- Oligodendroglioma-like appearance is common but usually mixed with other heterogeneous cytomorphologies (e.g., astrocytic, spindled, pleomorphic) (Fig. 23.7)
- OLIG2 and CD34 positive; low Ki-67 labeling index

Genetic Features
- Like other diffuse low-grade gliomas, show MAPK pathway activation
  - *BRAF* V600E
  - *FGFR2/3* fusions
- IDH and H3 wildtype

Angiocentric Glioma
Definition
- Angiocentric glioma is a rare, slow growing diffuse glioma
- Perivascular arrangement of bland bipolar cells
- Identified in children and young adults
- CNS WHO grade 1

Clinical Features
- Patients usually present with intractable partial seizures
- Tumors are usually in the cortex but do occur in the brainstem
- Imaging shows well-circumscribed tumors with T2 and FLAIR hyperintensity and no contrast enhancement
- Very favorable outcome; gross total resection usually curative

Pathologic Features

- Histologically, angiocentric glioma is characterized by uniform, spindle-shaped, cells with oval to elongated nuclei and speckled chromatin, and pink, tapering cytoplasm
- Tumor cells are intimately associated with vessels
  - Often oriented parallel to the vessels
  - May form streaming arrays in perivascular space with either single or multilayered cells
- Mitotic figures are generally absent with a low Ki-67 labeling index
- OLIG2 is usually negative
- Tumor cells are strongly GFAP-positive and show dot-like EMA staining pattern, like ependymoma

Genetic Features

- Consistent finding is rearrangement or copy number alteration involving *MYB*
- *MYB* activation may drive MAPK signaling
- *MYB::QKI* is the most common fusion (87%)
- Rare cases have been shown to also harbor *BRAF* p.V600E mutation
- Using methylation studies, angiocentric gliomas cluster with or close to other *MYB*-altered pediatric tumors

Circumscribed Astrocytic Gliomas

Pilocytic Astrocytoma

Definition

- Circumscribed glioma frequently occurs in pediatric patients, most commonly in the cerebellum, but supratentorial and spinal cord occurrences are also well-established
- Indolent clinical course warrants a designation of WHO grade 1

Clinical Features

- Usually occur in the first two decades of life, representing the most common pediatric brain tumor
- Cerebellar examples commonly appear as a cyst with a brightly enhancing mural nodule
- Optic pathway gliomas commonly meet the criteria for this diagnosis

Pathologic Features

- Variable loose myxoid areas and compact fibrillar areas (Fig. 23.8)
- Oligodendroglioma-like cytology is a frequent focal finding
- Involvement in the leptomeningeal space is common

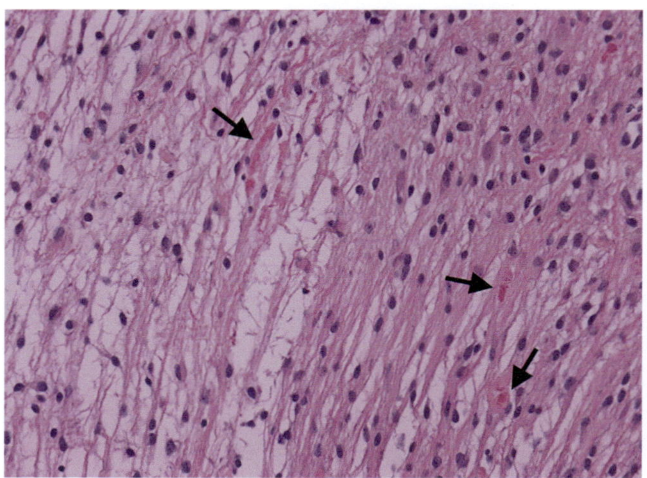

Fig. 23.8 Pilocytic astrocytoma. The characteristic biphasic pattern of relatively compact eosinophilic component made up of bipolar cells with elongated nuclei (right side) and loose hypocellular component with more stellate-appearing tumor cells (left side). Note the scattered Rosenthal fibers (arrows)

- Cyst formation is a common feature of these tumors allowing for a characteristic radiographic appearance
  - Cyst wall is often associated with dense glomeruloid vascular proliferation
- Rosenthal fibers and eosinophilic granular bodies are frequently present
- While these are understood as circumscribed tumors, foci of infiltration are not infrequent, particularly in the cerebellar cortex, where effacement of the cortical architecture can be seen
- Pilomyxoid histologic pattern is an established variant of pilocytic astrocytoma and is more commonly observed in younger patients, often in the hypothalamic region

Genetic Features

- The most common genetic alteration (>80%) observed in pilocytic astrocytomas is fusion of the *KIAA1549* and *BRAF* genes through tandem duplication at 7q34, resulting in constitutive *BRAF* activation
- *BRAF* V600E mutation is observed in a subset of tumors, usually associated with extracerebellar sites
- Gain of chromosome 7 is associated with more aggressive clinical behavior
- Pilocytic astrocytoma is the most common histologic type of glioma in patients with NF1 (see Table 23.1)
  - Optic pathway gliomas in NF1 patients are histologically pilocytic astrocytomas
  - NF1-associated low-grade gliomas usually do not harbor genetic alterations other than biallelic inactivation of *NF1*

High-Grade Astrocytoma with Piloid Features

- Recently described, rare tumors usually occurring in the cerebellum with high-grade histologic features combined with features of pilocytic astrocytoma (piloid features)
- Median age, 40 years; probably rare in pediatric setting
- Molecular testing, specifically DNA methylation profiling, is required for the diagnosis
- Loss of ATRX expression (reflecting underlying *ATRX* mutation) is common, and *CDKN2A/B* homozygous deletion is observed in majority of cases

Pleomorphic Xanthoastrocytoma

Definition

- Circumscribed glioma of the cortex frequently occurs in young adults presenting with seizures
- WHO grade ranges from 2 to 3

Clinical Features

- Children and young adults (first to third decades)
- Almost invariably cerebral, predominantly involving the superficial temporal lobe with frequent meningeal involvement
- Most patients present with seizures
- Frequently shows radiographic cyst formation with an enhancing mural nodule

Pathologic Features

- Grade 2 is reserved for cases lacking significant mitotic activity or other high-grade features
  - Grade 3 is assigned if mitotic figures exceed 5/10 hpf or if necrosis and microvascular proliferation are present
- Eosinophilic granular bodies are common, as are perivascular lymphocyte aggregates, microcalcifications, and regions of xanthomatous change
- Cytologic features of the astrocytic cells are markedly atypical with giant and bizarre tumor cells, even in cases with very low proliferative activity (Fig. 23.9)
- Spindle cell areas or mesenchymal areas are commonly encountered
  - Intercellular reticulin deposition is not always present but can be a helpful supporting feature

Genetic Features

- Recurrent somatic alterations include *CDKN2A/B* loss (94%) and *BRAF* V600E mutation (76%)
- Cases lacking *BRAF* V600E typically have alternate driver mutations affecting genes in the RAS pathway (*BRAF RAF1,* and *NF1*)

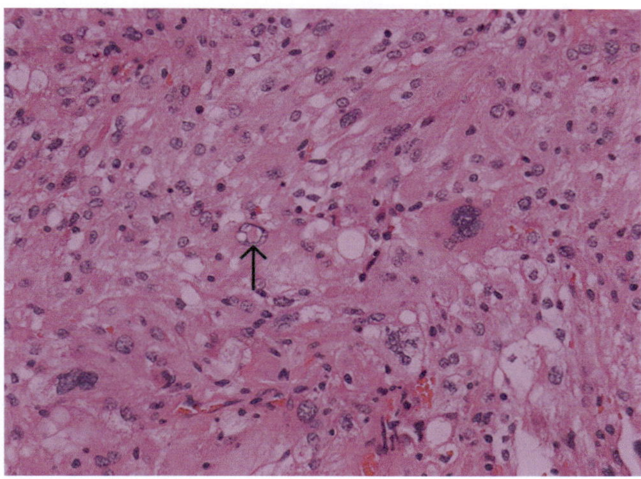

Fig. 23.9 Pleomorphic xanthoastrocytoma. Large bizarre-shaped astrocytes, one with nuclear pseudoinclusion (arrow), are scattered in a background of low-grade astrocytes

Astroblastoma, MN1-Altered

- Astroblastoma is a rare glial tumor of uncertain origin
- Usually manifests in children and young adults as a well-circumscribed, contrast-enhancing solid or cystic hemispheric mass
- Histologically, it combines ependymal features (perivascular pseudorosettes) with astrocytic differentiation (GFAP-positive cells with broad, nontapering processes)
- Epithelioid and rhabdoid cytology may be seen
- Vascular hyalinization is prominent and, in some cases, obscures the neoplastic cells
- Alterations involving *MN1* are recurrent, including *MN1::BEND2* fusion
- *MN1*-altered tumors can be identified by DNA methylation profiling

Chordoid Glioma

- This is a rare, low-grade (WHO grade 2) glioma of the third ventricle of adults that is histologically characterized by cords and clusters of epithelioid, GFAP-positive cells embedded in a mucin-rich background containing lymphoplasmacytic infiltrate
- GFAP and TTF1 are usually positive
- *PRKCA* p.D463H is seen in the majority of chordoid gliomas studied

Subependymal Giant Cell Astrocytoma

- The most common CNS tumor associated with tuberous sclerosis (see Table 23.1)
  - Tumors have biallelic inactivation of *TSC1* or *TSC2*

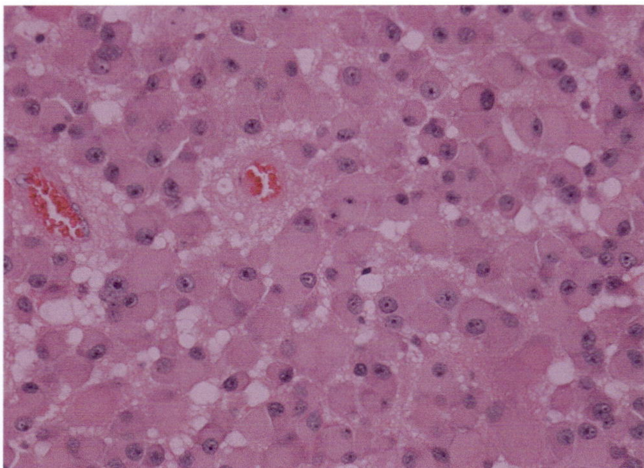

Fig. 23.10 Subependymal giant cell astrocytoma. Sheets of loosely arranged, relatively uniform, round/epithelioid, tumor cells exhibiting large round vesicular nuclei with nucleoli and granular eosinophilic cytoplasm

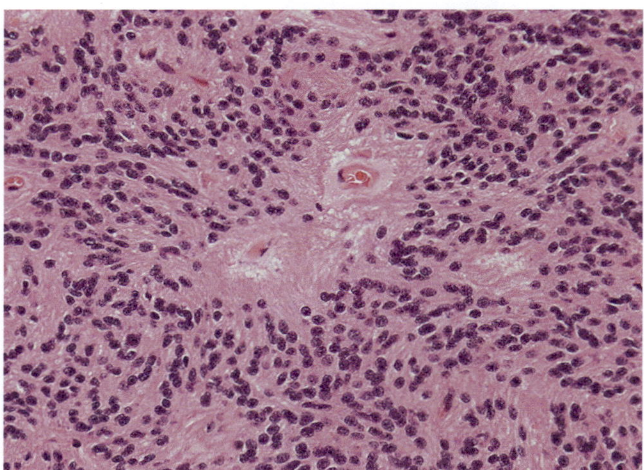

Fig. 23.11 Ependymoma. This WHO grade 2 ependymoma shows a moderately cellular tumor consisting of monomorphic population of cells with oval nuclei. Note the characteristic perivascular pseudorosettes toward the center of the image

- Classic location is in the region of the foramen of Monro, which commonly blocks cerebral spinal fluid (CSF) flow
- Usually arises before age 25
- Histologically, composed of gemistocyte-like or large polygonal cells (Fig. 23.10)
  - Immunoreactivity for S100 and TTF1
  - GFAP, synaptophysin, and NeuN expression are variable

Ependymal Tumors

Introduction

- A category of tumors defined by ependymal-like protein expression and ultrastructure; these occur throughout the neuraxis
- Ependymomas are characterized by moderate cellularity, monotonous cytology, and frequent arrangement of tumor cells in characteristic patterns: perivascular pseudorosettes, ependymal rosettes, and ependymal canals (Fig. 23.11)
- Immunohistochemically, ependymomas are commonly positive for EMA with a pattern of punctate, paranuclear cytoplasmic dot-like expression
  - EMA also highlights the apical surfaces of ependymal canals or rosettes, if present
- GFAP is often accentuated in the perivascular pseudorosettes of ependymomas
- OLIG2 is usually negative or only focally positive in ependymoma cells

- Grading of ependymomas (2 vs 3) has been proposed based on mitotic activity as well as the presence of high-grade features such as necrosis and microvascular proliferation
- Ependymoma molecular genetics suggest that distinct ependymoma entities segregate by anatomic site, with ependymomas of the supratentorial compartment, posterior fossa, and spinal cord each harboring distinct molecular signatures

Posterior Fossa Ependymoma

- Posterior fossa ependymomas are subdivided by methylation status into group A (PF-A) or group B (PF-B)
  - PF-A tumors occur in younger patients and have more aggressive clinical behavior including higher frequency of clinical recurrence
  - PF-B tumors occur in slightly older patients and have a more benign clinical course comparatively
- PF-A and PF-B ependymomas can be readily identified by H3K27me3 immunohistochemistry: H3K27me3 is retained in PF-B tumors and lost in PF-A tumors
- Genetically, PF-A tumors are often "silent" with balanced genomes and few to no somatic variants of clinical significance; when present, 1q gain and 6q loss are associated with shorter survival
- PF-B tumors may be molecularly heterogeneous, and 1q gain does not seem to be a robust marker of poor prognosis in this group

Supratentorial Ependymoma

- Supratentorial ependymomas are localized to the cerebral hemispheres and are more frequent in the frontal and parietal lobes. They account for approximately one-third of all ependymomas
- Molecular testing for supratentorial ependymomas may identify fusions involving the *C11orf95 (ZFTA)* or *YAP1* genes. FISH, NGS, or RNA sequencing can identify these alterations
- 17–30% of supratentorial ependymomas will not show a specific gene fusion and may be diagnosed as supratentorial ependymoma, not elsewhere classified (NEC)
- *ZFTA-RELA* fusion-positive supratentorial ependymomas show activation of NFkB signaling
  - Homozygous loss of *CDKN2A* is also present in a subset of these tumors and may be a predictor of poor overall survival
  - *RELA* fusion ependymomas often show L1CAM immunoreactivity, making this a helpful surrogate immunomarker
- *YAP1-MAMLD1* fusion-positive supratentorial ependymomas occur in young children, girls more often than boys
  - *YAP1-MAMLD1* fusion is thought to be the oncogenic driver, interacting with the transcription factors *TEAD1-4* and *NFI*
  - *YAP1* fusion ependymomas do not show L1CAM immunoexpression

Spinal Ependymoma

- NF2 is associated with spinal ependymomas, especially multiple spinal ependymomas (see Table 23.1)
  - These usually occur in the cervical spinal cord or cervicomedullary region
- An *MYCN*-amplified group of spinal cord ependymomas has been identified and is aggressive compared to other spinal ependymomas
  - H3 K27-altered diffuse midline glioma should be considered and ruled out with immunohistochemistry if necessary

Myxopapillary Ependymoma

- Myxopapillary ependymomas are low-grade but frequently metastatic tumors, classically occurring in the cauda equina region
- Tumor cells are arranged around hyaline vessels with perivascular myxoid material
  - Ki-67 labeling index is very low

Subependymoma

- Subependymomas are indolent (WHO grade 1) tumors usually occurring in the ventricular system and are characterized by regions of low cellularity with fibrillar stroma and occasional aggregates of oval tumor cells
- Subependymoma-like areas can occur in classic ependymomas, and vice versa, ependymoma-like regions can be seen in subependymomas
  - Mixed ependymoma-subependymoma is of unclear grade and significance

Glioneuronal and Neuronal Tumors

Introduction

- Glioneuronal and neuronal tumors encompass a heterogeneous group of CNS tumors characterized by varying degrees of neuronal differentiation in some or all the tumor cells
- Gangliocytomas and neurocytomas demonstrate more extensive differentiation as evidenced by expression of neuron-associated proteins, including NeuN, synaptophysin, and neurofilaments
- Other glioneuronal tumors may have irregularly distributed neuronal antigen expression or cells with neuronal cytomorphology, often accompanied by a second population of glial neoplastic cells
- 14 types of glioneuronal tumors are currently recognized by the WHO, and still, some glioneuronal tumors may not neatly fit into the prescribed categories (i.e., glioneuronal tumors, NEC)
- Generally, these tumors exhibit low rates of proliferation and are clinically indolent
- Frequent cortical locations and slow growth rates contribute to the common presentation of treatment-resistant epilepsy
- Across all categories, *BRAF* alterations and *FGFR1* alterations are recurrent
- The most encountered glioneuronal and neuronal tumors are described in more detail in this section

Ganglioglioma

Definition

- An indolent neoplasm composed of ganglionic and glial cells, in combination

Clinical Features

- Frequently presents with seizures
- Localized to the temporal lobes but can occur anywhere in the neuraxis
- Mineralization is common and may be observed by neuroimaging studies

Pathologic Features

- Ganglion cells may be unevenly distributed in the tumor, and can be recognized by their large size, irregular clustering, or occasional binucleation (Fig. 23.12)
- Ganglion cells typically express neurofilament and synaptophysin and have a weaker expression of NeuN than normal neurons
- Ganglion cell tumors lacking a glial component are more appropriately considered gangliocytomas

Genetic Features

- *BRAF* alterations are common, including *BRAF* V600E and *KIAA1549::BRAF* fusion; other MAPK signaling pathway gene alterations have also been reported
- Mutant *BRAF* has been shown to associate with neuronal cells

Dysembryoplastic Neuroepithelial Tumor

- Dysembryoplastic neuroepithelial tumor (DNT or DNET) presents with seizures due to intracortical location, often in temporal lobe
- Growth pattern is often nodular, observable by neuroimaging and microscopically

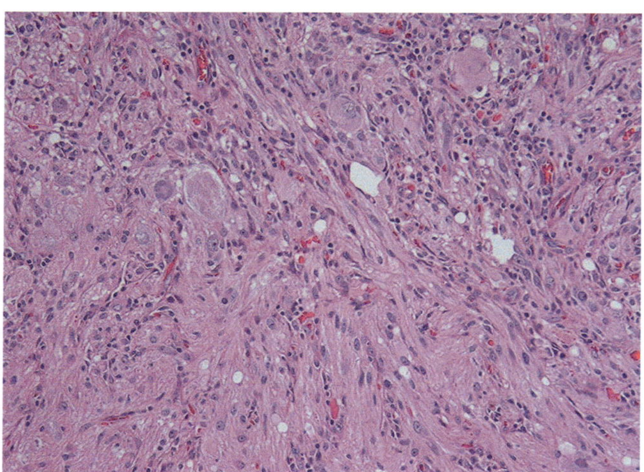

Fig. 23.12 Ganglioglioma. Large dysmorphic neurons are scattered in a background of haphazardly arranged low-grade astrocytes. Note that the tumor is peppered with perivascular lymphocytes

- "Specific glioneuronal element" is the term applied to the combination of round, monotonous, oligodendrocyte-like cells organized around neuronal cells centered in microcysts of mucinous stroma, a common and pathognomonic feature of DNT
- *FGFR1* mutations are common

Desmoplastic Infantile Astrocytoma/ Ganglioglioma

- Occurs as circumscribed desmoplastic lesions of the cerebral hemispheres in very young patients (under age 2 years)
- Superficial cortical location is common, as is meningeal involvement, which may suggest an extraaxial mass by neuroimaging
- A glial component expressing GFAP is seen with extensive reticulin-rich desmoplasia
  - A commingled ganglion cell component may or may not be present
- Foci of primitive or undifferentiated cells are commonly seen, but overall clinical behavior is low-grade
- There is a known histologic overlap with infantile hemispheric gliomas harboring *NTRK/ALK/ROS/MET* fusions; thus, molecular genetic characterization is key to a precise diagnosis
- MAPK pathway gene alterations, including *BRAF* or *CRAF (RAF1)* mutation or fusion, are recurrent variants in this tumor category

Diffuse Glioneuronal Tumor with Oligodendroglioma-Like Features and Nuclear Clusters

- Diffuse glioneuronal tumor with oligodendroglioma-like features and nuclear clusters (DGONC) is a rare, usually supratentorial, tumor with multinucleated cells and nuclear clusters, occurring in pediatric patients
- DNA methylation profiles of these tumors appear distinct and monosomy 14 is common, suggesting this is a distinct entity
- Put forth as a provisional entity in WHO CNS5, awaiting further characterization

Papillary Glioneuronal Tumor

- Papillary glioneuronal tumor (PGNT) is a rare, benign (WHO grade 1) tumor predominantly identified in young adults with seizures
- Characterized by a well-circumscribed, solid to cystic, contrast-enhancing mass in the cerebral hemispheres (most frequent in temporal lobe)

- Shows an obvious papillary/pseudopapillary architecture at low power and is made up of two distinct cell types surrounding a central fibrovascular core
  - An innermost layer of small, cuboidal GFAP-positive, cells with eosinophilic cytoplasm and round nuclei
  - An outer layer of larger clear, synaptophysin-positive, cells with neurocytic or ganglioid appearance
- Mitotic figures are rare or absent, and the Ki-67 labeling index is very low
- *PRKCA* gene fusions are recurrent; the *PRKCA* gene product is involved in the MAPK signaling pathway

Rosette-Forming Glioneuronal Tumor

- Rosette-forming glioneuronal tumor (RGNT) is another rare, benign (WHO grade 1) tumor found exclusively in the posterior fossa of children and young adults
- Often occupies a large portion of the fourth ventricle
- Radiographically, the tumor is a circumscribed, solid mass with heterogeneous enhancement
- Characterized by biphasic histology with clearly defined and spatially separate neurocytic and glial components
  - The neurocytic component shows a homogeneous population of small, clear cells in rosettes with large central spaces or papillae around central vessels and delicate processes forming a neuropil matrix adjacent to tumor cells and extending to central vessels
  - The glial component consists of solid pattern of highly fibrillary cells resembling pilocytic astrocytoma but may have an oligodendroglioma-like appearance with microcysts
  - Both neurocytic and glial components are histologically low grade with rare mitotic figures and low Ki-67 labeling index
- *FGFR1* kinase domain hotspot mutations are common, and co-occurring *PIK3CA* or *PIK3R1* mutations are often seen
- RGNT comprises a distinct DNA methylation class

Myxoid Glioneuronal Tumor

- Myxoid glioneuronal tumor (MGNT) is a very rare, benign (WHO grade 1) tumor most commonly arising in the midline in the region of the septum pellucidum, but can occur in the periventricular region more generally
- Composed of oligodendrocyte-like tumor cells in myxoid background, very similar to DNT
- Recurrent mutations of *PDGFRA* p.K385 characterize this entity

Diffuse Leptomeningeal Glioneuronal Tumor

- Diffuse leptomeningeal glioneuronal tumor (DLGNT) typically involving the leptomeninges of pediatric patients
  - Spinal cord location is common, and in this location, lesions may be primarily parenchymal
- Neoplastic cells are oligodendrocyte-like and myxoid background is present
- WHO grade is not currently assigned for this entity; clinical courses reported are similar to WHO grades 2–3
- *KIAA1549::BRAF* fusion accompanied by 1p deletion (or 1p/19q codeletion) is the molecular signature

Multinodular and Vacuolated Neuronal Tumor

- Multinodular and vacuolated neuronal tumor (MVNT) consists of benign monomorphic neuronal lesions of the gray/white junction, usually in the temporal or frontal lobes of adult patients
- May be incidentally discovered or associated with seizures
- MRI appearance of clustered hyperintense nodules of the deep cortex is considered characteristic
- Tumor cells are positive for OLIG2 and SOX2 and partly positive for MAP2 and synaptophysin; Ki-67 labeling index is very low (<1%)
- Previously debated whether this represented a malformation or a neoplasm, somatic RAS/RAF/MAPK pathway gene alterations suggest these lesions are neoplastic

Central Neurocytoma

- A low-grade (WHO grade 2), well-circumscribed neuronal neoplasm of young adults, preferentially situated in the lateral ventricles near the foramen of Monro
- Histologically, central neurocytoma is comprised of nests of highly uniform round cells resembling those of oligodendroglioma
  - Tumor cells, which are embedded in a delicate fibrillar background, are characterized by a finely speckled chromatin pattern and infrequent mitotic figs
  - A network of fine capillaries and calcifications are frequent findings
  - Tumor cells are intensely immunoreactive for synaptophysin and NeuN but usually negative for GFAP and chromogranin
- No consistent genetic alterations reported
  - Gains on chromosomes 2p, 7, 10q, 18q, and 13q were recorded in about 20% of cases
  - 1p/19q deletions are lacking

Extraventricular Neurocytoma

- A rare tumor of adult patients, histologically similar to central neurocytoma, but arising outside the ventricular system
- Like central neurocytoma, considered WHO grade 2
- *FGFR1::TACC1* fusions are frequent

Embryonal Tumors

Introduction

- Refers to a group of tumors characterized by primitive or undifferentiated histology, formerly considered "primitive neuroectodermal tumors" or "PNET," a term no longer recommended for CNS neoplasms
- Medulloblastoma is four times as common as the other embryonal tumor subtypes
- While medulloblastomas arise exclusively in the posterior fossa, other embryonal tumors occur throughout the neuroaxis
- Most commonly occur before age 5 years
- Molecular characterization is necessary for accurate classification of embryonal tumors with methylation profiling constituting the basis of their revised classification
  - Surrogate immunohistochemistry serves as a useful screening and diagnostic tool
- The association of certain embryonal tumors with inherited tumor syndromes calls for more exhaustive genetic testing and counseling

Medulloblastoma

Definition

- Medulloblastoma is a highly malignant (WHO grade 4), though exquisitely radiosensitive, primitive tumor of the cerebellum or dorsal brainstem, with a tendency to spread along the CSF pathway
- Historically referred to as "cerebellar neuroblastoma"
- Negative for LIN28A (a marker of embryonal tumors with multilayered rosettes) and shows retained INI-1 nuclear expression (a feature lost in ATRT)
- Modern classification requires an integrated approach combining histologic type and molecular/genetic type to produce optimal prognostic and predictive information
- Biologically heterogenous, as reflected by molecular subgroups representing a wide range of risk stratification
- Clinical staging M0–M4 is required

Clinical Features

- Most commonly occurs in children, with over 70% of cases occurring in individuals younger than 16 years and the peaks of presentation at 3 and 7 years
- A quarter of cases are seen in adults (mostly ages 21–40)
- Male predominance with M:F of 1.7:1
- Located in the posterior fossa
- Can occur in the setting of several inherited tumor syndromes (see below)
- Clinical presentation includes cerebellar signs and symptoms and manifestations of CSF flow obstruction
- Radiographically, classic medulloblastomas appear as central, homogeneously contrast-enhancing, solid cerebellar lesions
- Leptomeningeal involvement is common

Histopathologic Patterns

- Classic medulloblastoma (Fig. 23.13)
  - Most common morphologic pattern (~70–80%); any age
  - Made up of sheets of back-to-back, relatively uniform small cells with intensely hyperchromatic nuclei and scant cytoplasm
  - Nuclei are round to oval or carrot-shaped and show abundant apoptotic bodies and frequent mitotic figures

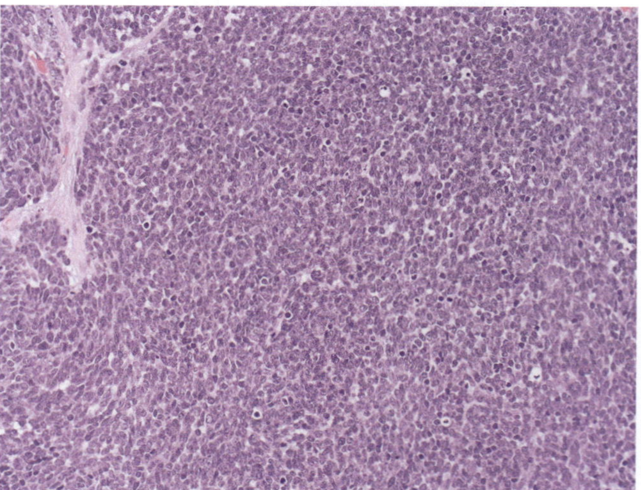

Fig. 23.13 Medulloblastoma, classic histology. Classic medulloblastomas are composed of sheets of closely packed tumor cells with frequent mitotic figures and apoptotic bodies. Classic medulloblastomas are defined by lack of nodularity/desmoplasia and lack of cytologic anaplasia

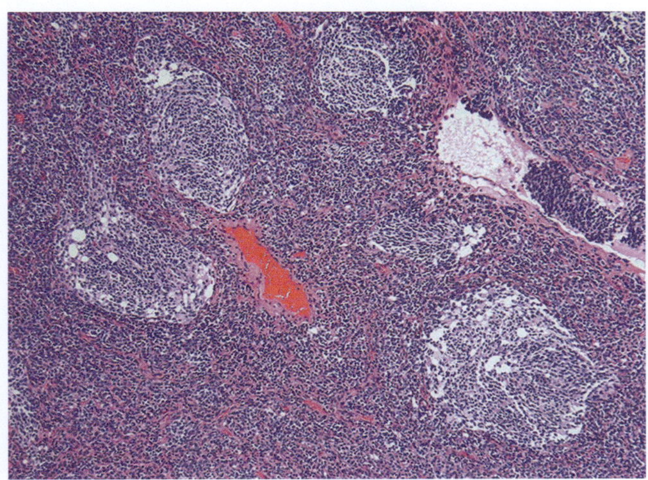

Fig. 23.14 Medulloblastoma, desmoplastic/nodular. This nodular medulloblastoma shows neuropil-rich islands embedded in a highly cellular, desmoplastic background. The cells within the nodules are better differentiated than the hyperchromatic and mitotic, carrot-shaped nuclei of the internodular areas

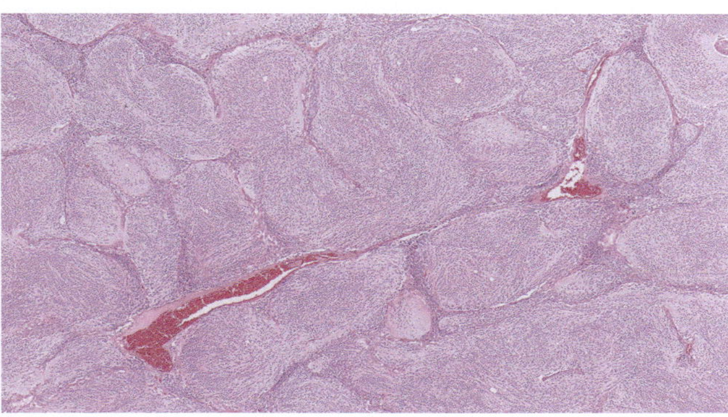

Fig. 23.15 Medulloblastoma with extensive nodularity. These tumors are marked by closely apposed nodules of differentiated cells with minimal internodular areas. These frequently occur in infants and are associated with a good prognosis

- Most often undifferentiated or shows neuronal differentiation, but may occasionally demonstrate glial differentiation
- Leptomeningeal spread in ~40% of cases
- Immunohistochemistry
    Most are immunoreactive for synaptophysin and NeuN but not universally
    GFAP expression is rare
- Any molecular type but nearly all WNT-activated medulloblastomas show classic morphology
• Desmoplastic/nodular medulloblastoma (Fig. 23.14)
- Approximately 20% of medulloblastomas; most common in children <3 years
- Characterized by a highly distinctive biphasic pattern of fibrillar, reticulin-free "pale" nodules embedded in a highly cellular, desmoplastic, and reticulin-rich background
- Tumor cells within the nodules tend to be larger, rounder, and far less anaplastic looking compared with the tightly packed and highly mitotic cells of the internodular areas
- Immunohistochemistry
    Nodules are typically intensely synaptophysin-positive and show much lower proliferative activity than internodular areas
    GAB1 and YAP1 expression, predominantly in internodular areas, is typical in SHH-activated tumors
- Most hemispheric medulloblastomas belong to the desmoplastic/nodular type, especially in adults
- Leptomeningeal spread in ~20% of cases

- Nearly all are associated with the SHH molecular group; mostly the SHH-1 and SHH-2 subgroups
- Along with medulloblastomas with extensive nodularity, comprise most medulloblastomas occurring in the nevoid basal cell carcinoma syndrome
- Genetic counseling and germline mutation testing recommended
• Medulloblastoma with extensive nodularity (Fig. 23.15)
- Comprise <5% of all medulloblastomas; most commonly in children <3 years
- Usually seen in infants and is characterized by an expanded lobular architecture with large areas of neuropil-like tissue harboring small round cells often in a streaming pattern, and markedly reduced internodular areas
- Immunohistochemistry: Similar pattern to desmoplastic/nodular pattern with more intense synaptophysin and NeuN staining
- Usually located in vermis
- Leptomeningeal spread is rare
- Imaging: Enhancing "bunch of grapes" appearance
- May be pathogenetically related to desmoplastic/nodular type
- Like desmoplastic/nodular medulloblastoma, most belong to the SHH-activated group
- Along with desmoplastic/nodular medulloblastomas, they comprise most medulloblastomas occurring in the nevoid basal cell carcinoma syndrome
- Genetic counseling and germline mutation testing recommended
• Large cell/anaplastic medulloblastoma (Fig. 23.16)
- Account for approximately 10% of medulloblastomas
- A rare variant characterized by sheets and lobules of large, round, markedly pleomorphic cells with prominent nucleoli, frequent, often atypical, mitotic figures, and apoptosis

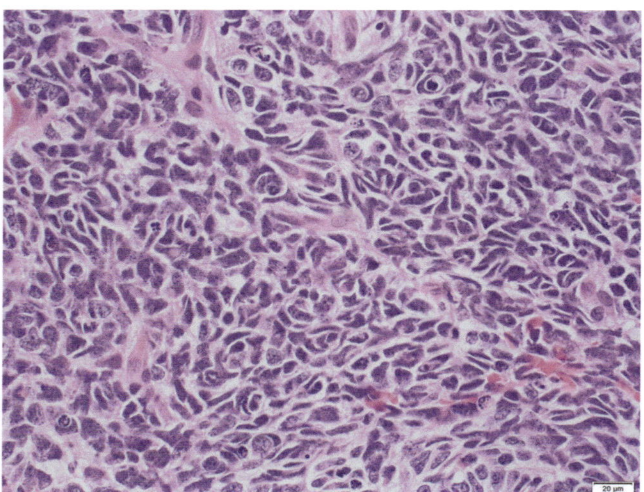

Fig. 23.16 Large cell/anaplastic medulloblastoma. In this histologic type of medulloblastoma, common findings include nuclear pleomorphism, cell–cell wrapping, and a high mitotic rate. The rarer large cell type features nuclear enlargement with prominence of nucleoli. Both are associated with more aggressive clinical behavior

- Nuclear molding and cell–cell wrapping are often seen
- Often cerebellar midline/4th ventricle localization
- Imaging: heterogeneous contrast enhancement
- Leptomeningeal spread in ~60–70% of cases
- Most belong to the SHH-3 subgroup, the non-WNT/non-SHH subgroup 2 or *TP53*-mutant groups
- Majority is sporadic
- Medulloblastoma with myogenic differentiation (medullomyoblastoma)
  - Any of the above variants that show skeletal muscle differentiation
- Medulloblastoma with melanotic differentiation (melanocytic medulloblastoma)
  - Any of the above variants that demonstrate melanocytic differentiation

Molecular Groups

- Molecular subgroups of medulloblastoma are recurrently demonstrated by gene expression profiling and DNA methylation profiling
- Subgroup assignment should ideally be reached based on two independent validated analytical methods (e.g., immunohistochemistry and an orthogonal molecular test)
- Medulloblastoma, WNT-activated
  - 10% of medulloblastomas
  - Older children and adults; rare in infants
  - Slight female predominance

- Cells of origin: Lower rhombic lip and dorsal brainstem, therefore usually central location within the fourth ventricle
- Enhance brightly on imaging
- Leptomeningeal spread is exceptional
- Classic histologic subtype, rarely large cell/anaplastic
- Very good prognosis in children, with overall survival exceeding 90%
- Outcome not as favorable in adults
- Activating mutations in the Wingless (*WNT*) pathway effector *CTNNB1*
- Associated with monosomy 6 in most cases (WNT-α) subgroup
- Immunohistochemistry
  - Nuclear translocation of β-catenin
  - Positive for YAP1
  - Negative for GAB1
- Invariably sporadic; rarely germline *APC* mutations
- Medulloblastoma, SHH-activated
  - Includes two major WHO-recognized molecular groups
    - Medulloblastoma, SHH-activated and *TP53*-wildtype
    - Medulloblastoma, SHH-activated and *TP53*-mutant
  - 30% of medulloblastomas
  - Bimodal age distribution: Infants/children less than 3 years or older teenagers and adults
  - Usually hemispheric location (children and young adults) or vermis (infants)
  - Solid, enhancing on imaging
  - Desmoplastic/nodular classic, or large cell/anaplastic
  - Uncommonly metastatic; prognosis in infants good, others intermediate
  - Associated with loss of chromosome 9
  - Intermediate prognosis between WNT-activated and group 3
  - Poor prognosis associated with *TP53* mutation, *MYCN*, and Gli2 alterations as well as large cell/anaplastic morphology
  - Immunohistochemistry
    - GAB1 and YAP1 positive
    - Negative for nuclear β-catenin
    - p53 is helpful in distinguishing between *TP53*-mutant and *TP53*-wildtype but sequencing is preferred
  - Four provisional *SHH* molecular subgroups recognized (SHH-1, SHH-2, SHH-3, and SHH-4)
  - Germline mutations (*PTCH1, SUFU, TP53, ELP1, PALB2, BRCA2,* and *GPR161*) present in approximately 40% of cases (see Table 23.1)

- o Germline analysis and genetic counseling are recommended
  - Most adult tumors exhibit *TERT* promoter mutations, with most children showing hypermethylation without mutation
  - Cells of origin: Granule cell precursors (external granular layer), therefore usually hemispheric locations
- Medulloblastoma, SHH-activated and *TP53*-wildtype
  - Associated with nevoid basal cell carcinoma syndrome (Gorlin syndrome)
  - All nevoid basal cell carcinoma syndrome-associated medulloblastomas belong to the SHH molecular subgroup
    - o Most due to inactivating germline mutation in *PTCH1*
    - o Rarely due to *SUFU* or *PTCH2* mutations
  - Most are desmoplastic/nodular or medulloblastomas with extensive nodularity, rarely classic or large cell/anaplastic
- Medulloblastoma, SHH-activated and *TP53*-mutant
  - Comprise approximately 10–20% of all SHH-activated medulloblastoma
  - Children aged 4–17 years
  - Approximately 70% show large cell/anaplastic phenotype
  - Immunohistochemistry: Strong, widespread p53 nuclear expression, or very rarely, null pattern of staining
  - Germline *TP53* point mutations (Li–Fraumeni syndrome) associated with SHH-activated group with over one-half SHH-activated and *TP53*-mutant medulloblastomas exhibiting germline *TP53* alterations
  - Isochromosome 17p deletion and *TP53* LOH are characteristic
  - Almost always belong to SHH-3 subgroup
- Non-WNT/non-SHH (group 3/group 4)
  - 60% of medulloblastomas do not fall into the above two categories and are not associated with a particular cell signaling pathway; they comprise eight provisional molecular subgroups
  - Group 3 and group 4 tumors cannot currently be reliably separated by immunohistochemistry, but can be distinguished by gene expression or methylation profiling
  - Classic or large cell/anaplastic histology
  - Frequently metastatic at the time of presentation
  - Imaging: enhanced inhomogeneously
  - Intermediate prognosis
  - Immunohistochemistry

- o GAB1 and YAP1 negative
  - o Negative for nuclear β-catenin
  - Rarely associated with hereditary tumor syndromes (*CREBBP, PALB2,* and *BRCA2*)
  - *MYC* amplification/overexpression seen in subgroup 3 tumors is a poor prognostic feature
  - Isochromosome 17q common

Cancer Predisposition Syndromes and Genes Associated with Medulloblastoma

- Li–Fraumeni syndrome; *TP53* (17p13.1)
- Nevoid basal cell carcinoma syndrome (Gorlin syndrome)
  - SHH-associated genes: *ELP1* and *PTCH1,* less commonly *PTCH2, SUFU*
- Turcot syndrome; *APC* (5q21–q22) and β-catenin mutations
  - Associated with *WNT* pathway activation
- *BRCA2, PALB2*
  - Observed in SHH, group 3, and group 4 tumors
- *CREBBP*; Rubinstein-Taybi syndrome
- *NBN* (*NBS1*); Nijmegen breakage syndrome

Prognosis and Risk Stratification

- Very high risk (<50% survival)
  - SHH with *TP53* mutation
  - Group 3 with metastases at presentation
- High risk (50–75% survival)
  - SHH with metastasis or *MYCN* amplification
  - Group 4 with metastases at presentation
- Standard risk (75–90% survival)
  - SHH lacking *TP53* mutation, *MYCN* amplification, and metastases
  - Group 3 lacking *MYC* amplification and metastases
  - Group 4 lacking chromosome 11 loss and metastases
- Low risk (>90% survival)
  - WNT in patients under 16 years of age
  - Group 4 with chromosome 11 loss and lacking metastases
- Unknown risk
  - WNT with metastases
  - Group 3 with *MYC* amplification and without metastasis
  - Large cell/anaplastic histology in group 3/group 4
  - Isochromosome 17q in group 3
  - Melanotic or myoblastic differentiation

Atypical Teratoid/Rhabdoid Tumor

Definition

- A highly malignant (WHO grade 4) embryonal CNS tumor of infants and very young children with primitive histology, often showing rhabdoid features, and exhibiting biallelic inactivation of *SMARCB1* (aka *hSNF5, INI1, or BAF47*) or rarely of *SMARCA4* (*BRG1*)

Clinical Features

- Rare overall, but most common CNS embryonal tumor in patients under 12 months of age; vast majority of patients are under 2 years of age
- Over 95% of cases are intracranial, and the majority in the posterior fossa
- Spinal examples are rare
- Occurs rarely in adults, usually in the cerebral hemispheres and sellar region
- Slight male predominance
- Clinical and radiographic features are like those of medulloblastoma and other CNS embryonal tumors
- CSF pathway seeding is seen in about one-third of patients at presentation
- Historically thought to have dismal outcome, but more recent studies have shown improved survival with chemoradiation

Pathologic Features

- Architecturally complex, made up of a mixture of large, somewhat rhabdoid cells with reniform nuclei, prominent nucleoli and abundant eosinophilic cytoplasm, and smaller cells resembling those of medulloblastoma (Fig. 23.17)
- Tight fascicles of small spindle cells giving the tumor a mesenchymal appearance may be present
- Some tumor cells may show cytoplasmic vacuolation and/or distinct cell–cell borders
- Mitotic figures, necrosis, hemorrhage, and calcifications are common
- Complex immunohistochemical profile, often expressing EMA, GFAP, actin, and cytokeratins reflecting their primitive neuroectodermal, mesenchymal, and epithelial features
- Nearly all demonstrate the absence of the nuclear immunohistochemical expression of SMARCB1/INI1 protein (Fig. 23.18)

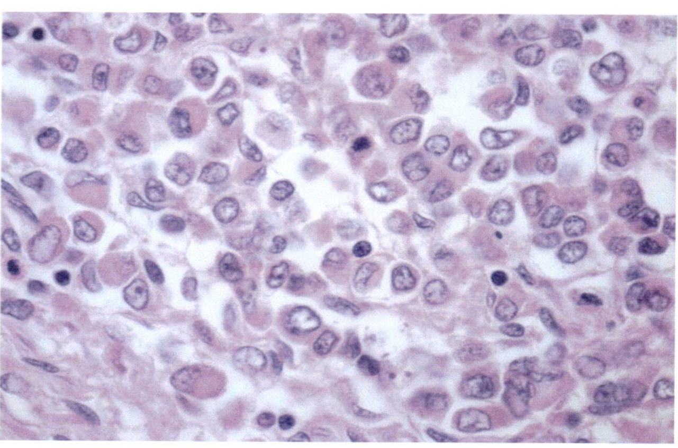

Fig. 23.17 Atypical teratoid rhabdoid tumor. ATRT are characterized by high cellularity, and occasionally contain cells with rhabdoid features, including eccentrically placed eosinophilic cytoplasm. Nuclei with vesicular chromatin and prominent nucleoli are common

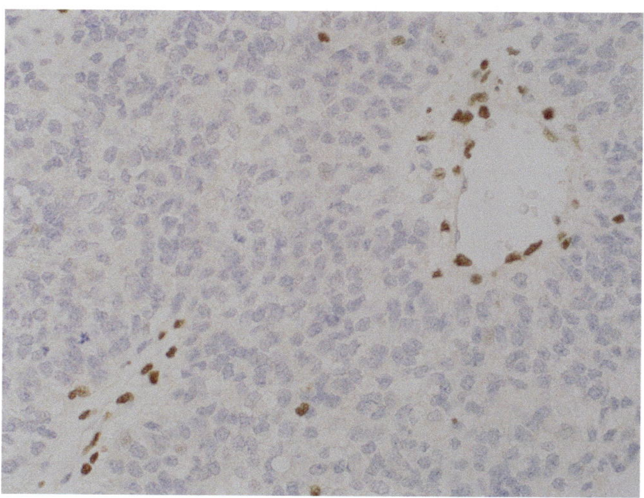

Fig. 23.18 Atypical teratoid rhabdoid tumor, loss of INI1/SMARCB1 expression. ATRT are defined by biallelic loss of SMARCB1 (or rarely SMARCA4) function; thus, all ATRT should show immunonegativity for INI1/SMARCB1 or BRG1/SMARCA4. Nuclear expression of INI1/SMARCB1 is retained in the endothelial cells in the tumor shown

- CNS embryonal tumors with loss of nuclear SMARCB1/INI1 expression are designated ATRT regardless of morphology
  - One exception seems to be the loss of SMARCB1 expression in cribriform neuroepithelial tumor (CRINET), which should display a characteristic cribriform architecture
  - Awareness of SMARCB1 loss in poorly differentiated chordomas and in schwannomatosis-associated schwannomas guards against erroneous diagnosis of ATRT

Genetic Features

- Hallmark is loss of function of *SMARCB1* locus at 22q11.2, most commonly caused by deletion
- Very rare cases may show loss of an alternate gene, *SMARCA4* (0.5–2% of ATRT)
- Germline *SMARCB1* mutations/deletions are present in one-third of ATRT patients, representing the rhabdoid tumor predisposition syndrome 1 (*SMARCB1*) or 2 (*SMARCA4*) (see Table 23.1)
- Gene expression and DNA methylation profiling suggest three subgroups: MYC, SHH, and TYR
  - Some suggestion of a relationship or overlap between the MYC subtype and extra-CNS rhabdoid tumors

Cribriform Neuroepithelial Tumor

- Cribriform neuroepithelial tumor (CRINET) is a provisional WHO entity defined as a nonrhabdoid neuroepithelial tumor characterized by cribriform strands and ribbons and exhibiting loss of nuclear SMARCB1/INI1 expression
- Clusters with ATRT-TYR molecular subgroup on methylation profiling suggesting a close relationship
- Immunohistochemistry: Expresses EMA, tyrosinase, synaptophysin, and vimentin
- Some were previously diagnosed as choroid plexus carcinoma, particularly given its localization in the vicinity of the fourth, third, and lateral ventricles
- Limited information on outcome suggests significantly longer survival than ATRT-TYR

Embryonal Tumor with Multilayered Rosettes

- Embryonal tumor with multilayered rosettes (ETMR) is a rare, highly malignant (WHO grade 4) embryonal tumor typically occurring in patients younger than 24 months and characterized by C19MC alteration or rarely *DICER1* mutation (Fig. 23.19)
- Supratentorial location is more common, but can occur anywhere along the neuraxis
- Three morphologic patterns are recognized; all share the common feature of multilayered rosettes
  - Embryonal tumor with abundant neuropil
  - Ependymoblastoma
  - Medulloepithelioma
- Two molecular subgroups: ETMR, C19MC-altered and ETMR, *DICER1*-mutated
  - It should be noted, however, that about one-half C19MC-altered ETMRs may concurrently harbor heterozygous *DICER1* mutations

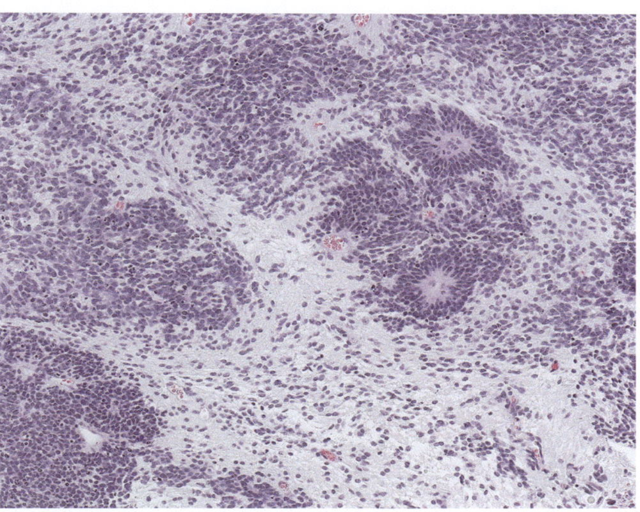

Fig. 23.19 Embryonal tumor with multilayered rosettes. These occur in a variety of histologic patterns, but all share the presence of multilayered rosettes: the concentric arrangement of tumor cells around a central lumen often containing granular debris

- Recurrent genetic alteration is amplification of chromosome 19 microRNA cluster (C19MC)
  - C19MC amplification may be identified by FISH or copy number analysis
  - Diffuse positivity with the surrogate immunomarker LIN28A strongly suggests the presence of C19MC amplification
  - Other CNS tumors that may demonstrate LIN28A staining include occasional gliomas, ATRT, and germ cell tumors
- *DICER1*-mutant ETMR that are not C19MC-altered should be investigated for *DICER1* germline alteration
- Invariably aggressive tumor with poor outcome

CNS Neuroblastoma, *FOXR2*-Activated

- CNS neuroblastoma, FOXR2-activated (CNS NB-FOXR2), is a rare supratentorial embryonal tumor (WHO grade 4) of the cerebral hemispheres, characterized by activation of the transcription factor *FOXR2* by structural rearrangements
- Comprise most embryonal tumors that are not medulloblastoma, ATRT, ETMR, or pineoblastoma
- Similar to most embryonal tumors, it is a disease of childhood
- Variable degrees of neuroblastic and/or neurocytic differentiation can be seen (Fig. 23.20)
- Clinical behavior is overall unclear due to the rarity of cases

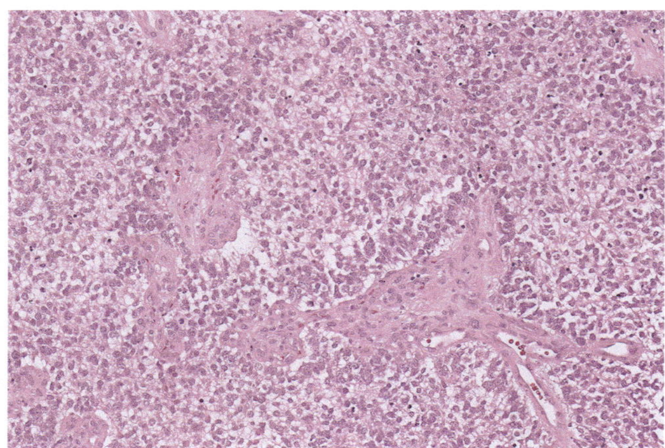

Fig. 23.20 CNS neuroblastoma, *FOXR2*-activated. This entity frequently demonstrates some degree of neurocytic cytology and can contain foci of microvascular proliferation. Gain of 1q by copy number analysis can be a helpful supporting feature

- OLIG2 is typically expressed in tumor cells, with synaptophysin expression largely limited to areas of neurocytic or ganglion cell differentiation
- *FOXR2* is overexpressed, but a variety of structural rearrangements may contribute, making sequencing-based approaches of limited value
  - *FOXR2* expression may also be seen in rare medulloblastomas and a subset of high-grade gliomas
- Gain of 1q is recurrent

CNS Tumor with *BCOR* Internal Tandem Duplication

- Rare malignant (WHO grade 4) embryonal tumor, characterized by *BCOR* internal tandem duplication
- May occur supratentorially or infratentorially
- Well-demarcated on imaging, often with central cystic component
- Solid growth pattern with glioma-like fibrillarity or myxoid/microcystic background with a rich branching capillary network (Fig. 23.21)
- Ependymoma-like perivascular pseudorosettes that are GFAP-negative are characteristic
- OLIG2 expression is variable
- PCR can identify the hallmark feature, *BCOR* internal tandem duplication
- Alternate *BCOR* alterations have been described, including mutations, deletions, or fusions
- DNA methylation profiling suggests this is a distinct class of tumors

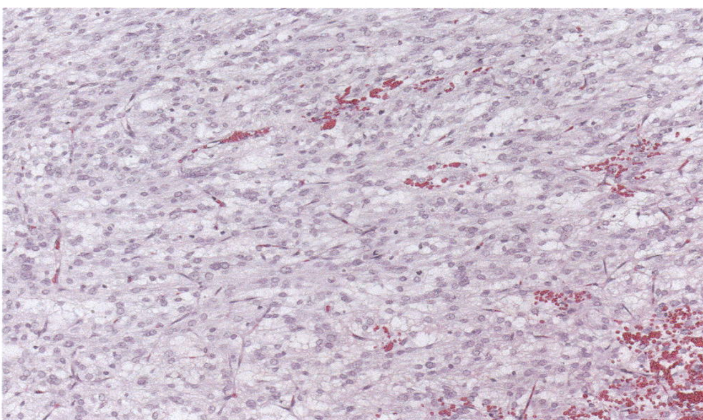

Fig. 23.21 CNS tumor with *BCOR* internal tandem duplication. This tumor with *BCOR* internal tandem duplication shows glioma-like fibrillarity and contained several microcystic areas

Choroid Plexus Tumors

Definition

- Rare CNS tumors derived from choroid plexus epithelium

Clinical Features

- Choroid plexus tumors are distributed in proportion to the density of normal choroid plexus tissue, with about one-half occurring in the lateral ventricle and about one-half in the fourth ventricle
  - Pediatric cases more commonly occur supratentorially, and adult cases are more commonly infratentorial
- Choroid plexus tumors comprise a spectrum including papilloma (WHO grade 1), atypical papilloma (WHO grade 2), and carcinoma (WHO grade 3)
- Atypical papillomas have a fivefold increased risk of recurrence compared to conventional papillomas
- 80% of choroid plexus carcinomas occur in children
- 20% of choroid plexus carcinomas present with disseminated disease

Pathologic Features

- Choroid plexus papillomas resemble normal choroid plexus with a simple papillary arrangement and smooth apical surface
- Atypical papillomas are distinguished by the presence of mitotic figures (>2 per 10 hpf) and occasionally by focal necrosis or solid growth/loss of papillary architecture
- Choroid plexus carcinomas exhibit frank malignant features, including cellular pleomorphism, frequent mitotic figures, and foci of necrosis

Genetic Features

- Choroid plexus papillomas exhibit hyperdiploidy, whereas carcinomas show markedly aberrant copy number profiles with complex patterns of losses and gains
- Choroid plexus carcinomas frequently demonstrate loss of function of the *TP53* tumor suppressor gene and are often associated with Li–Fraumeni syndrome (germline *TP53* mutation) (see Table 23.1)
- The strong association of choroid plexus carcinoma with Li–Fraumeni syndrome has led to recommendations for genetic counseling and consideration of germline *TP53* testing for choroid plexus carcinoma patients
- Methylation profiling suggests adult choroid plexus papillomas are distinct from pediatric examples
- Adult tumors more commonly harbor *TERT* promoter mutations, which associate with shorter progression-free survival

Pineal Parenchymal Tumors

- Neoplasms arising from the pineal parenchymal cells
- Spectrum of tumors from benign (pineocytoma, WHO grade 1) to malignant (pineoblastoma, WHO grade 4)
- Pineocytomas display mature cells arranged in sheets which overall resemble normal pinealocytes, with round nuclei and an open chromatin pattern
- Pineoblastomas have an undifferentiated embryonal appearance with frequent mitotic figures; CSF dissemination is frequent and 5-year survival is under 60%
- Pineoblastomas can occur in conjunction with retinoblastomas
- Three pineoblastoma subgroups have been proposed through molecular analysis: FOXR2, MYC, and RB
- *DICER1* and *DROSHA* mutations are recurrent variants in pineoblastoma

Meningioma

Definition

- Very common, relatively benign (usually WHO grade 1), slowly growing, well-circumscribed tumor of the meninges and dura, thought to arise from the arachnoidal cap cells
- While majority are WHO grade 1, grade can range from 1 to 3

Clinical Features

- Primarily adults, with a peak incidence in the sixth and seventh decades
- Exceedingly rare in children and the very old, but a higher incidence of atypical forms
  - 2.5% of childhood CNS tumors
- Female predominance; more pronounced for spinal meningiomas
- May occur at any location within the CNS; most commonly over the cerebral convexities
- Clinical presentation highly dependent on location; may be incidental finding
- Radiographically, appears as a well-circumscribed, isointense, homogeneously contrast-enhancing, dural-based mass with a "dural tail" sign in most the cases

Pathologic Features (Fig. 23.22)

- Grossly appears as discrete, firm, or rubbery mass with broad dural attachment and a characteristically lobular cut surface
- Typical meningiomas are made up of tight whorls, lobules, or bundles of uniform, bland spindled cells with oval nuclei characterized by a delicate chromatin pattern with occasional central clearing, and infrequent mitotic figures
- Calcification in the form of psammoma bodies is common
- Meningiomas are almost always EMA-positive

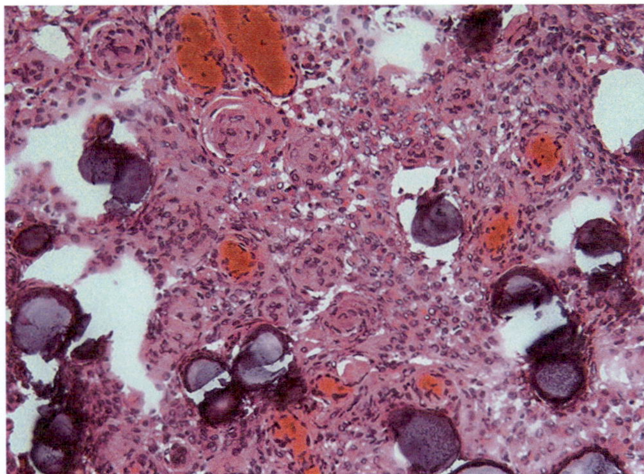

Fig. 23.22 Meningioma. A classic meningioma shows tight whorls and sheets of uniform spindled cells with bland vesicular nuclei. Calcifications in the form of psammoma bodies are abundant

- Histological variants that do not influence the clinical outcome (WHO grade 1)
  - Meningothelial (syncytial)
  - Transitional
  - Fibrous
  - Microcystic
  - Secretory
  - Psammomatous
  - Angiomatous
  - Lymphoplasmacyte-rich
  - Metaplastic
- Histologic variants with less favorable outcome
  - Clear cell (WHO grade 2)
  - Chordoid (WHO grade 2)
  - Papillary
  - Rhabdoid
  - Note that while historically considered higher grade, current recommendations state that papillary and rhabdoid types should be assigned grade based on overall grading criteria, not based on histologic type alone
- WHO CNS5 grading recommendations
  - Grade 2, any of the following:
    - ≥4 mitotic figures/10 hpf
    - Brain invasion
    - Clear cell or chordoid type
    - At least three of the following "soft" criteria
      Increased cellularity
      Small cell change
      Prominent nucleoli
      Sheeting/loss of whorl formation
      Spontaneous necrosis
  - Grade 3, any of the following:
    20 or more mitotic figures/10 hpf
    Frank anaplasia
    TERT promoter mutation
    CDKN2A/B homozygous deletion

Genetic Features

- Chromosome 22 loss; usually in the form of monosomy 22 is the most frequent chromosomal abnormality, encountered in about 50–70% of meningiomas
- *NF2* gene mutations are common (~70% of sporadic meningiomas) and involve fibrous and transitional meningiomas most frequently
  - *NF2* gene mutations are much less frequent (~25%) in meningothelial meningiomas
- *NF2* gene mutations are as frequent in atypical/anaplastic meningiomas as they are in grade 1 meningiomas, suggesting a role for *NF2* inactivation in the formation, rather than the progression of meningioma

- Meningiomas are one of the most common neoplasms to arise in the setting of NF2 and are often multiple (see Table 23.1)
- Higher grade meningiomas show more complex copy number alterations involving chromosomes 1p, 6q, 9q, 10q, 14q, 17p, and 18q, with deletions of 1p, 10q, and 14q being most frequent
- Losses of *CDKN2A/B* region are a poor prognostic feature
- Clear-cell meningiomas are associated with germline *SMARCE1* mutations
- *BAP1* germline mutations are associated with rhabdoid meningiomas as well as other tumors, including melanomas and mesotheliomas

Craniopharyngioma

- A histologically benign (WHO grade 1) epithelial neoplasm of the sellar region that is thought to arise from Rathke pouch remnants
- Two histological and molecular subtypes are recognized
  - Adamantinomatous (associated with activating *CTNNB1* mutations)
    - More common in children and is usually partially cystic
  - Papillary (associated with *BRAF* V600E mutations)
    - Occurs predominantly in adults in the region of the third ventricle
- Compared with their papillary counterpart, adamantinomatous craniopharyngiomas are additionally characterized by the presence of wet keratin, calcifications, and cholesterol clefts (Fig. 23.23)

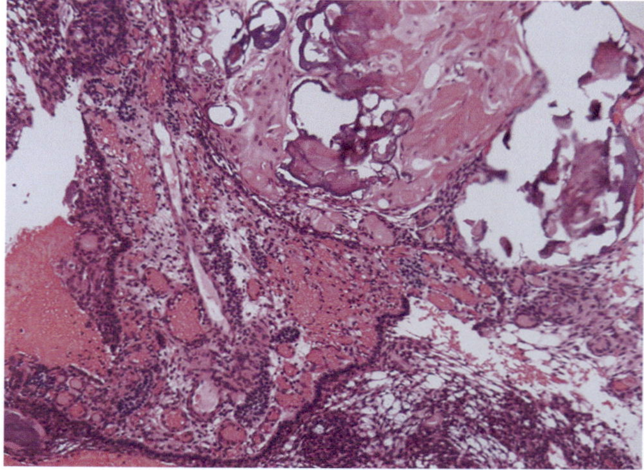

Fig. 23.23 Craniopharyngioma. Adamantinomatous craniopharyngioma is characterized by palisaded squamous epithelium, wet keratin, and calcification

Hemangioblastoma

Definition

- A benign (WHO grade 1), richly vascular tumor of uncertain histogenesis

Clinical Features

- Sporadic cases occur in adults while those associated with von Hippel–Lindau disease (VHL) tend to involve younger patients
- Sporadic cases are largely limited to the cerebellum, but those associated with VHL syndrome may be multiple and additionally manifest in the brainstem and spinal cord
- Most show the characteristic radiographic appearance of a cyst with a contrast-enhancing mural nodule; "flow voids" may be encountered

Pathologic Features

- These are discrete neoplasms made up of a variable mixture of small capillaries and large vacuolated or lipidized stromal cells
- Stromal cell nuclei may show hyperchromasia and nuclear pleomorphism, but mitotic figures are infrequent
- Often cystic but occasionally solid
- Mast cells may be a diagnostically helpful finding

Genetic Features

- Approximately 25% occur in the setting of VHL syndrome (see Table 23.1)
- A minority of sporadic hemangioblastomas show mutations or deletions of the *VHL* gene
- EGFR, VEGF, and VEGF receptors are expressed at high levels in the stromal cells

Solitary Fibrous Tumor

- Fibroblastic tumor, usually dural-based, harboring 12q13 inversion that results in *NAB2::STAT6* fusion
- WHO grade ranges from 1 to 3
- Usually occur in the fifth to seventh decades of life, exceedingly rare in children
- Histologically appear as spindle cell lesions with characteristic branching, staghorn blood vessels
- CD34 is diffusely positive in grade 1 solitary fibrous tumors, and ALDH1 positivity is present in the majority of solitary fibrous tumors (while usually negative in meningiomas)

- *STAT6* nuclear immunoreactivity is a sensitive and specific surrogate marker
- Mitotic count and presence of necrosis correspond to a poorer prognosis
    - WHO grade 1: <5 mitotic figures per 10 hpf
    - WHO grade 2: 5 or more mitotic figures per 10 hpf without necrosis
    - WHO grade 3: 5 or more mitotic figures per 10 hpf with necrosis
- *TERT* promoter mutations are present in a subset of solitary fibrous tumors but do not indicate worse prognosis

Germ Cell Tumors

Definition

- Germ cell tumors of the CNS are rare, preferentially midline tumors of children and young adults that show similar characteristics to those arising in the gonads

Clinical Features

- Children and young adolescents are most affected
- Midline intracranial structures are most commonly involved (vast majority occur in the pineal and suprasellar regions), although they have been reported throughout the CNS
- Pineal region tumors show male predominance
- Pineal region tumors usually present with signs and symptoms of increased intracranial pressure, while those of the suprasellar region may manifest due to visual symptoms or disturbances along the hypothalamic–hypophyseal axis (e.g., diabetes insipidus)

Pathologic Features

- Morphologically, identical to their gonadal and other extragonadal counterparts
- Germinomas are by far the most common subtype (Fig. 23.24)

Genetic Features

- Although it is more common among testicular or mediastinal germ cell tumors, gain of chromosome 12p may be seen in CNS germinomas, often manifesting as isochromosome 12p (Fig. 23.17b)
- Patients with Klinefelter syndrome and Down syndrome appear to be more susceptible to develop germ cell tumors, including those of the CNS, than the average individual

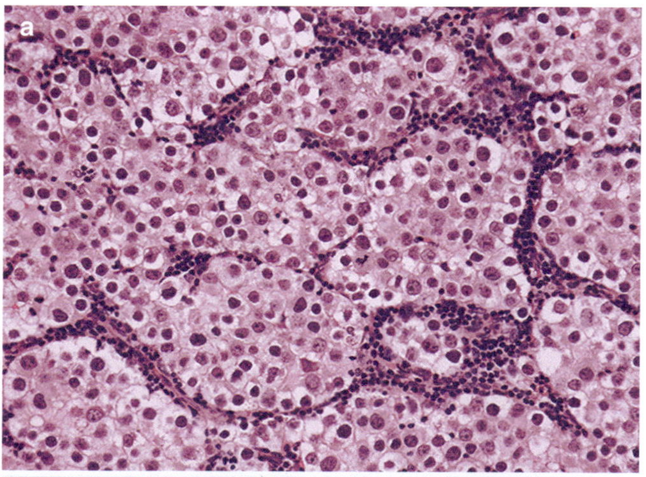

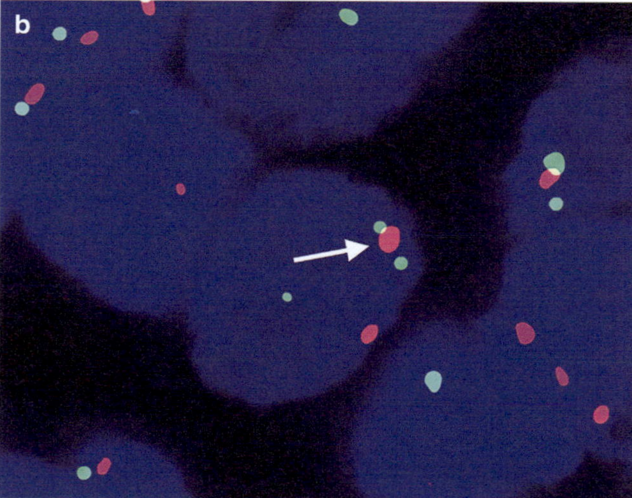

Fig. 23.24 Germinoma. (**a**) Nests of polygonal tumor cells characterized by clear cytoplasm and large round nuclei with prominent nucleoli are decorated by thin fibrous septa-rich in lymphocytes. (**b**) Isochromosome 12p is a common finding in CNS germinoma (two green 12p signals closely juxtaposed to one red centromeric probe signal, arrow)

Peripheral Nerve Sheath Tumors

Schwannoma

Definition

- A benign (WHO grade 1) slowly growing tumor that occurs throughout the peripheral nervous system but often involves the vestibular division of the eighth cranial nerve; only schwannomas of the CNS are discussed next

Clinical Features

- Bilateral vestibular schwannomas are a hallmark of NF2 (see Table 23.1)
- The vestibular division of the eighth cranial nerve is the most common intracranial location
  - Rare intracerebral and intramedullary examples have been reported
- Patients may present with tinnitus, hearing difficulties, or facial paresthesias
- Radiographically, appear as well-circumscribed, often cystic, homogenously contrast-enhancing masses

Pathologic Features

- Encapsulated tumors of moderate to low cellularity
- Biphasic architecture
  - Antoni A areas: Compact, elongated cells arranged in alternating fascicles, sometimes forming distinctive nuclear palisades (Verocay bodies)
  - Antoni B areas: Loosely textured, less cellular, areas with more stellate-looking cells
- Thick hyalinized blood vessels with hemosiderin-laden macrophages
- Aggregates of lipid-laden cells
- Degenerative atypia in the form of nuclear pleomorphism is common, but mitotic figures are infrequent
- Prominent pericellular reticulin
- Intensely and diffusely immunoreactive for S100
- Histologic variants include cellular, plexiform, and melanotic schwannomas

Genetic Features

- Most schwannomas are sporadic
- Schwannomas may arise in the setting of NF2 or schwannomatosis (multiple peripheral schwannomas)
- Inactivating mutations of *NF2* are identified in about 60% of schwannomas
- Some schwannomas may show loss of chromosome 22q in the absence of detectable *NF2* mutations
- Loss of the immunohistochemical expression of the *NF2* product (merlin/schwannomin) is identified in most schwannomas regardless of *NF2* mutations

Neurofibroma

- Benign peripheral nerve tumor associated with biallelic loss of *NF1*

- Arise in the skin, or in deep peripheral nerves
- Spindle cell appearance with "shredded carrot" appearance, abundant collagen, and scattered mast cells
- Usually sporadic but can occur in individuals with NF1 syndrome
  - Multiple neurofibromas or plexiform pattern is suspicious for NF1
- Atypical neurofibromatous neoplasm of uncertain biologic potential is often associated with *CDKN2A/B* loss

Malignant Melanotic Nerve Sheath Tumor

- Peripheral nerve neoplasm with Schwann cell features and melanocytic features
- Melanin pigment may obscure underlying tumor histology
- Express S100, SOX10, and melanocytic markers
- May occur sporadically or in association with Carney complex, a very rare autosomal dominant syndrome characterized by myxomas, endocrinopathy, and skin lesions, associated with *PRKAR1A* germline pathogenic variants
- Clinical behavior is aggressive, and metastasis is common

Malignant Peripheral Nerve Sheath Tumor

- Malignant spindle cell sarcoma arising from peripheral nerve
- 50% are associated with NF1
  - Can arise from preexisting neurofibromas
- Spindle cell histology with areas of high cellularity, numerous mitotic figures, and foci of necrosis
- Subset of cases show divergent differentiation, e.g., rhabdomyosarcomatous ("malignant triton tumor"), cartilaginous, or epithelial
- SOX10 and S100 are frequently negative; loss of H3K27me3 is seen in majority of cases
- *NF1* inactivation, *CDKN2A/B* loss, and complex copy number profiles are seen

Paraganglioma of the Cauda Equina (Cauda Equina Neuroendocrine Tumor)

- Paragangliomas of the cauda equina are histologically identical to their systemic counterparts
- Usually sporadic, unlike their systemic counterparts which are commonly associated with inherited tumor predisposition syndromes
- DNA methylation profiling suggests these are distinct from paragangliomas of other sites

Nonneoplastic Brain Disorders

Introduction

- The ravages of advanced age affect both mind and body, and nowhere do we recognize this more so than with the degenerative diseases of the nervous system
- Combined now with the gross and microscopic changes that have been known for >50 years for many of the neurodegenerative ailments are a vast and rapidly expanding array of genetic and molecular data that have transformed how we look at and classify these diseases
- Many of these disorders have both sporadic and familial forms, providing clues to the pathogenetic basis of these diseases
- Some of these diseases have unique populations of cells affected and unique cytopathic changes, although overlapping neuropathological features are also common
- Neuropathological examination at autopsy is still the gold standard for diagnosing most neurodegenerative diseases
- Investigation of new biomarkers of neurodegenerative disease hopes to recognize these diseases early to allow for earlier treatments
- Described in this section are some of the more prevalent and better studied, although unfortunately, still not fully understood, or treatable disorders with an emphasis on the essential molecular pathological changes involved

General Molecular/Cellular Mechanisms of Neurodegeneration

Protein Aggregation and Transport Dysfunction

- Misfolding and aggregation of proteins is a hallmark of many neurodegenerative diseases, although it is still not known in many cases whether these phenomena are central to the pathogenesis, secondary injury, or even protective to the cell
- Aggregates that form are multimeric but often have a predominant protein and tend to affect certain neuronal or glial cells in different diseases (Table 23.2)
- Aggregates can form intracellularly in cytosol, nucleus, or neuritic processes with diverse cellular pathological consequences, often including cell death
- Soluble oligomers and multimers released in the extracellular interstitial spaces may also be toxic, theoretically causing regional pathology
- Molecular motors within neuritic processes consist of a network of cytoskeletal elements known as kinesins and dyneins
- When aggregates are found in neuritic processes, the major pathological defect is to these molecular motors

Table 23.2 Neurodegenerative diseases and their associated protein aggregate pathology

Neurologic disease	Primary cell type affected	Primary anatomical region affected	Protein aggregation
Alzheimer disease	Neurons	Neocortex, hippocampus, entorhinal cortex	Neurofibrillary (tau) tangles and amyloid plaques
Parkinson disease	Neurons	Substantia nigra, nucleus basalis, locus ceruleus	Lewy bodies (SNCA)
Lewy body disease	Neurons	Substantia nigra, neocortex	Cortical Lewy bodies (SNCA)
Amyotrophic lateral sclerosis	Upper and lower motor neurons	Motor cortex, lower motor neurons anterior horn of spinal cord	Bunina bodies, skein-like aggregates, hyaline bodies (TDP-43, ubiquitin)
Corticobasal degeneration	Neurons and glia	Basal ganglia and cerebral cortex	Cytoplasmic inclusions (tau)
Multisystem atrophy	Glia and neurons (less)	Brainstem, midbrain, cerebellum, striatum	Cytoplasmic inclusions (SNCA)
Prion diseases	Neurons	Variable and diffuse depending on subtype	Amyloid and prion plaques
Frontotemporal dementias	Neurons	Hippocampus, and frontal and temporal cortices	Tau tangles/aggregates (Pick bodies in Pick disease)
Huntington disease	Neurons	Caudate, putamen, and cerebral cortex	Polyglutamine tract inclusions in neurites and nuclei
Progressive supranuclear palsy	Neurons and glia	Globus pallidus, midbrain, pons, subthalamic nucleus	Globose neurofibrillary tangles (neurons) (tau) Coiled bodies (oligodendrocytes) (tau) Tufted and thorn (astrocytes) (tau)

and retrograde and anterograde transport, essential processes that allow the movement of organelles and molecules crucial to cell function along the processes

- The histological finding when neuritic transport is disrupted is spheroid formation (also termed axonal swellings or axonal bulbs)

Mitochondria Dysfunction

- Mitochondria are the energy producing (in the form of oxidative phosphorylation and ATP production) organelles of all cells in the nervous system
- Mitochondrial proteins are encoded by both nuclear DNA and mitochondrial DNA. Mutations in genes from both sources become more numerous with age and have been suggested to play a role in organelle dysfunction and neurodegenerative diseases
- Damaged mitochondria lead to increased accumulation of oxidative molecules including reactive oxygen species, which can injure other organelles and induce apoptosis and/or necrosis
- High levels of oxidants can also be very deleterious to mitochondria inducing a state called mitochondrial permeability transition, with an uncoupling of oxidative phosphorylation often leading to cell death

Neuroinflammation

- An emerging field of inquiry thought to play a part in many neurodegenerative diseases, although unknown whether it serves both a primary and secondary role

- Astrocytes and microglia serve as the local resident cells in the CNS-mediating inflammation
- Circulating immune cells (B cells, T cells, and monocytes/macrophages) and autoantibodies may also play a yet to be defined role in neurodegenerative processes; their importance in multiple sclerosis, infectious disease, and trauma are better established
- The major cytokines and chemokines involved include tumor necrosis factor-α, granulocyte macrophage colony stimulating factor (GM-CSF), interleukin (IL)-1α, IL-1β, IL-2, IL-4, IL-6, interferon-γ, IL-10, IL-12, IL-18, tumor growth factor-β, macrophage inflammatory protein (MIP-1), macrophage chemotactic protein (MCP-1)
- Inflammatory mediators may alter the blood–brain barrier, synapse function, apoptosis, edema, protein aggregation, mitochondrial function, and cell-to-cell communication (glial and/or neuronal)

Survival vs. Apoptosis Factors

- Neurons are continuously signaled throughout life by autocrine, paracrine, and endocrine factors to varying degrees, which promote survival and prevent programmed cell death apoptosis
- Polypeptide factors known as neurotrophins influence survival, differentiation, proliferation, and apoptosis of both neuronal and glial cells
- Examples include nerve growth factor, brain-derived neurotrophic factor, glial-derived neurotrophic factor, ciliary neurotrophic factor, insulin-like growth factor
- Secreted neurotrophins can be taken up at nerve terminals and retrogradely transported to the cell body to exert their

effect or act as secreted ligands with action on specific cell surface receptors, which activate second messenger cascades

- Pathological conditions such as hypoxia/ischemia, electrolyte abnormalities, trauma, neuroinflammation, toxic exposures, and genetic abnormalities can all induce apoptosis through a variety of pathways in the central and peripheral nervous systems (CNS/PNS)

Alzheimer Disease

Clinical/Epidemiology

- Alzheimer disease (AD) is the most common of the neurodegenerative diseases with an increasing incidence every decade of life
- Slight female over male prevalence
- Worldwide disease affecting all races
- The prevalence roughly doubles every 5 years, starting from a level of 1% for the 60–64-year-old population and reaching 40% or more for some 85–89-year-old cohorts
- Disturbances in recent memory formation, difficulties conducting activities of daily living, language impairment, deficits in spatial ability and orientation, and alterations in mood and behavior are the clinical hallmarks of AD
- AD progresses with much variability in individuals with an average of 1–3 years of early symptoms before diagnosis, 1–3 years from diagnosis until need for more intensive care from family or nursing home, and then 1–3 years before death

Gross and Histological Neuropathology

- Brains show variable cerebral atrophy, with hippocampi and adjacent temporal cortices showing consistent diminution in size
- Frontal and parietal lobes also often show diffuse atrophy with narrowed gyri and widened sulci, and occipital lobes are less affected
- Coronal slices invariably display thinned cortical gray matter strips and enlarged ventricular system underscoring the widespread neuronal loss
- The microscopic hallmarks of this disease are amyloid plaques, neurofibrillary tangles, neuronal loss, and reactive glial changes (Fig. 23.25)
- AD plaques can be visualized with Congo red stain and are composed of Aβ protein deposits forming amyloid surrounded by degenerating neuritic processes
- Most AD brains also show amyloid angiopathy

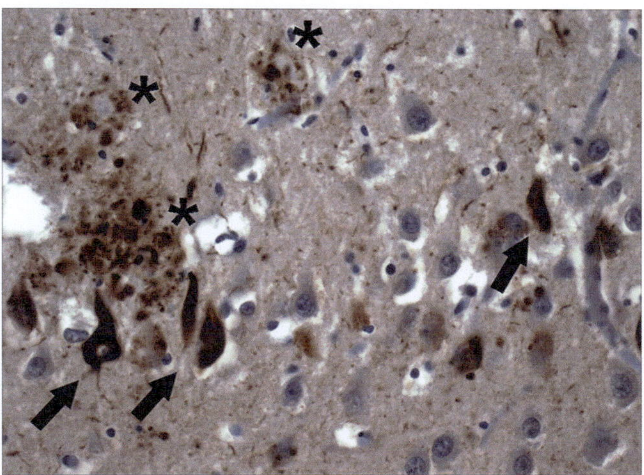

Fig. 23.25 Histological hallmarks of Alzheimer disease. Immunohistochemistry against tau protein highlights neurofibrillary tangles (arrows) and dystrophic neurites within Alzheimer type plaques (asterisks)

Table 23.3 Genes associated with Alzheimer disease

Gene	Chromosome	Protein	Disease onset; transmission
APP	21q21	Amyloid β-(A4) precursor protein	Early-onset; autosomal dominant
PSEN1	14q24.3	Presenilin 1	Early-onset; autosomal dominant
PSEN2	1q31–q42	Presenilin 2	Early-onset; autosomal dominant
APOE (risk factor gene)	19q13.32	Apolipoprotein E	Late-onset; sporadic increased risk of AD (ε4 allele) decreased risk AD (ε2 allele)

- Neurofibrillary tangles are intracellular inclusions of a variety of shapes composed of several proteins with the microtubule-associated protein tau being the predominant; they are best seen with silver stains such as Bodian or Bielschowsky or immunohistochemistry against tau

Genes Associated with Sporadic and Familial AD (Table 23.3)

- Approximately 75% of AD cases are thought to be sporadic, 25% hereditary, with the latter group showing either early or late onsets
- Mutations in presenilin ([PSEN], *PSEN1*, *PSEN2*), or *APP* account for the majority of early-onset familial AD, but there exists a wide spectrum of mutations in these genes as well as variable clinical and pathologic presentations
- Individuals with trisomy 21 who survive beyond 45 years of age nearly all develop AD changes

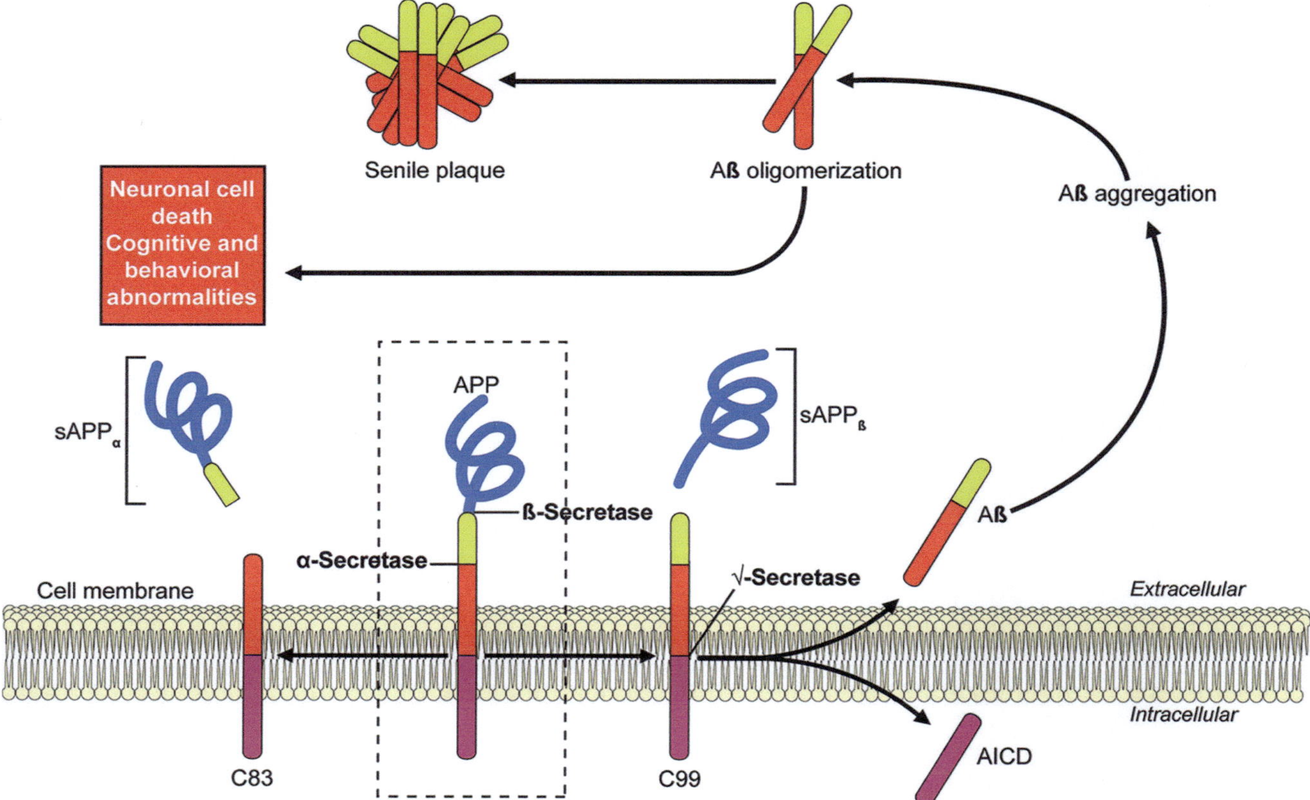

Fig. 23.26 APP processing and Aβ accumulation. Mature APP (center, inside dashed box) is metabolized by two competing pathways, the α-secretase pathway that generates sAPPα and C83 and the β-secretase pathway that generates sAPPβ and C99. Some β-secretase cleavage is displaced by ten amino acid residues and generates sAPPβ′ and C89. All carboxyterminal fragments (C83, C99) are substrates for γ-secretase, generating the APP intracellular domain and, respectively, Aβ, among others. Aβ aggregates into small multimers (dimers, trimers, etc.) known as oligomers. Oligomers appear to be the most potent neurotoxins, while the end-stage senile plaque is relatively inert (adapted from Gandy 2005)

- A genetic susceptibility gene for APOE protein has been identified in sporadic AD
- APOE has three alleles: APOE ε2, APOE ε3, and APOE ε4. Individuals that produce the ε4 form are at a greater risk for developing AD, while those that have the ε2 have decreased risk
- Polymorphisms in the *APOE* isoforms may also confer variable susceptibility
- Genetic testing for the APOE alleles is not generally done outside of the research setting, as the beneficial/detrimental effects are not absolute
- Investigators are actively looking for other risk factor-associated genes using linkage analysis

Possible Molecular Mechanisms of Pathogenesis of AD

- Controversy over the relative importance of the plaques vs. tangles in the pathogenesis of AD has existed for many years in the neuroscience research community of "tauists and βAPPtists"
- Plaques are extracellular deposits of fibrils and amorphous aggregates of amyloid β peptide (Fig. 23.26)
- Neurofibrillary tangles are intracellular fibrillar aggregates of tau that exhibit hyperphosphorylation and oxidative changes
- Mutations in *APP*, *PSEN1*, or *PSEN2* lead to accumulations of atypical Aβ peptide, which makes up the major protein component of amyloid plaques
- Tau mutations have been identified in frontotemporal dementias (*see* below in the section on "Tauopathies"), a group of disorders that are thought to exist in a spectrum with AD
- The "taucentric" and "amyloidocentric" viewpoints of AD neurodegeneration may converge at the level of PSEN
  - *PSEN* mutations lead to both tau tangles and amyloid plaques
  - PSEN is a core component of the γ-secretase responsible for the accurate cleavage of the Aβ peptide

- Aβ aggregates in the form of oligomers and fibrils in the interstitial space may be toxic to neurons
- Aβ aggregates may also form in vessels called amyloid angiopathy
- Prior head trauma predisposes to AD, with the proposed mechanism being a "sensitization" of glia and the neuro-immune response, although the exact molecular triggers are not known

Diagnosis

- An increased number of plaques and tangles combined with the appropriate clinical course are used to make the postmortem diagnosis of AD
- Most pathologists will follow diagnostic guidelines of the NIA–Reagan and Consortium to Establish a Registry (NIA–CERAD) for AD criteria
- New slightly modified criteria from the NIA–Reagan/CERAD criteria were proposed in 2012 and recommended an "ABC" staging protocol for the neuropathologic changes of AD, based on three morphologic characteristics of the disease: A is for amyloid, B is for Braak neurofibrillary tangle staging protocol, and C is for the Consortium to Establish a Registry for AD neuritic plaque scoring system
- Premortem diagnosis of sporadic AD relies on clinical history combined with clinical, cognitive, and laboratory examinations to help rule in AD and rule out other forms of dementias
- New clinical diagnostic criteria recognize early stages of disease with a preclinical onset stage recognized by positive biomarkers ("signature" biomarkers still be researched) and a mild cognitive impairment (MCI) stage that precede AD
- Genetic analysis in familial AD with appropriate genetic counseling is available for *PSEN1*, but because of the small number of families with mutations in *PSEN2* and *APP*, testing for these genes is currently only done in research labs
- Imaging modalities to look for hydrocephalus and generalized cortical and hippocampal atrophy may complement the clinical diagnosis
- "Functional" imaging studies such as positron emission tomography, functional magnetic resonance imaging, and radioligands used for identifying AD pathology are under investigation

Parkinson Disease

Clinical/Epidemiology

- Parkinson disease (PD) affects approximately 1% of population over 65

- Disease onset for sporadic PD varies from 20 to 80 years of age but is most common from 55 to 65; familial forms may occur much early
- Found throughout the world but with variable prevalence in different races and countries
- Slightly higher prevalence in males over females
- Hallmark clinical findings are bradykinesia, resting tremor, rigidity, and postural instability
- Some patients also have overlapping symptoms (and pathology) with AD
- Autonomic, cognitive, and psychiatric disturbances affect some patients

Gross and Histologic Neuropathology

- Pallor of substantia nigra and locus ceruleus grossly
- Lewy bodies (LB): round and eosinophilic cytoplasmic inclusions surrounded by a pale halo, found within the neurons of the substantia nigra (Fig. 23.27), locus ceruleus, nucleus basalis of Meynert, medulla, and thalamus (less frequently in numerous other nuclei)
- Lewy neurites (LN): dystrophic neurites found in similar distribution but also in amygdale and hippocampus
- LB and LN immunostain positively with antibodies against α-synuclein (SNCA) and ubiquitin
- Loss of dopaminergic nigrostriatal neurons and noradrenergic neurons in the substantia nigra and locus ceruleus, respectively, leads to the clinical manifestations of PD

Associated Genes

- Monogenic linkages to PD have only been discovered in the last 10 years, although recognition of familial genetic component has been known for much longer

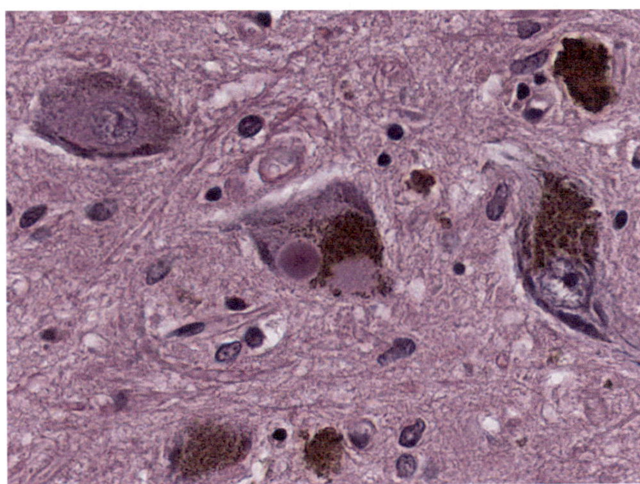

Fig. 23.27 Neuromelanin-containing neurons of the substantia nigra with center neuron displaying a Lewy body

Table 23.4 Genes associated with Parkinson's disease

PARK locus	Gene	Map position	Clinical phenotype	Pathology
PARK1/4	*SNCA*	4q22	Parkinsonism with common dementia	Lewy bodies
PARK2	*PARK2*	6q25–q27	Early-onset, slowly progressing parkinsonism	Lewy bodies
PARK3	Unknown	2p13	Late-onset parkinsonism	Lewy bodies
PARK5	*UCHL1*	4p13	Late-onset parkinsonism	Unknown
PARK6	*PINK1*	1p35–p36	Early-onset, slowly progressing parkinsonism	One case exhibiting Lewy bodies
PARK7	*DJ1*	1p36	Early-onset parkinsonism	Unknown
PARK8	*LRRK2*	12q12	Late-onset parkinsonism	Lewy bodies (usually)
PARK9	*ATP13A2*	1p36	Early-onset parkinsonism with Kufor–Rakeb syndrome	Unknown
PARK10	Unknown	1p32	Unclear	Unknown
PARK11	*GIGYF2*	2q36–q37	Late-onset parkinsonism	Unknown
PARK12	Unknown	Xq	Unclear	Unknown
PARK13	*OMI/HTRA2*	2p13	Unclear	Unknown
PARK14	*PLA2G6*	22q13.1	Parkinsonism with additional features	Lewy bodies
PARK15	*FBX07*	22q12–q13	Early-onset parkinsonism	Unknown
PARK16	Unknown	1q32	Late-onset parkinsonism	Unknown
FTDP-17	*MAPT*	17q21.1	Dementia, sometimes parkinsonism	Neurofibrillary tangles
SCA2	*ATXN2*	12q24.1	Usually ataxia, sometimes parkinsonism	Unknown
SCA3	*ATXN3*	14q21	Usually ataxia, sometimes parkinsonism	Unknown
Gaucher locus	*GBA*	1q21	Late-onset parkinsonism	Lewy bodies

Adapted from Martin et al. (2011)

- A diverse set of ten or more genes is now known to lead to dopaminergic neuron degeneration and PD or related disorders (Table 23.4)
- *SNCA* was the first gene identified with mutations associated with familial PD; polymorphisms in the promoter region may be associated with increased risk of sporadic PD
- 10% of all early-onset familial PD cases are due to a wide variety of mutations in *PARK2*, which encodes for a protein important in the ubiquitination pathway

Possible Molecular Mechanisms of Pathogenesis

- Mitochondrial dysfunction and oxidative stress are thought to be important contributors to neuronal death in PD
- Complex I deficits are found in mitochondria from sporadic and familial PD patients
- Pesticides and toxins such as 1-methyl-4-phenyl-1,2,3,6-tetrohydropyridine (MPTP) can affect complex I function and have been implicated in PD
- MPTP is a contaminant in synthetic opiate production that was accidentally injected by a group of individuals who then developed PD-like syndrome
  - SNCA
    - The physiologic role of SNCA is unknown, but it is associated with lipid rafts, which may be important in synaptic vesicles and synapse formation and function
    - SNCA fibrils are important components of LB and LN, and mutations in the gene promote increased fibrillar aggregations
    - Overexpression of normal SNCA in sporadic AD may occur and increase LB/LN formation
    - Oxidative damage may play a role in the aggregation of SNCA in sporadic PD, and mutant SNCA can interact with mitochondria to increase sensitivity to mitochondrial toxins
    - SNCA may interact with tau or amyloidogenic proteins to increase aggregation
    - The important role of SNCA in both forms of familial PD as well as all sporadic PD place PD as the most prominent diseases known as "synucleinopathies"
    - Dementia with LB, also called diffuse Lewy body disease, is also a synucleinopathy associated with cognitive decline, hallucinations, and parkinsonism and neuropathology shows SNCA aggregates in classical LB and diffuse cortical LB
    - Multiple system atrophy is a sporadic, adult-onset neurodegenerative disease with unknown cause, clinically characterized by variable parkinsonism cerebellar and pyramidal signs, and autonomic failure; histologically characterized by SNCA-positive glial cytoplasmic inclusions
  - Parkin (PARK2)
    - PARK2 functions as an E3 ubiquitin protein ligase import for ubiquitination of proteins targeted for proteosomal degradation (Fig. 23.28)

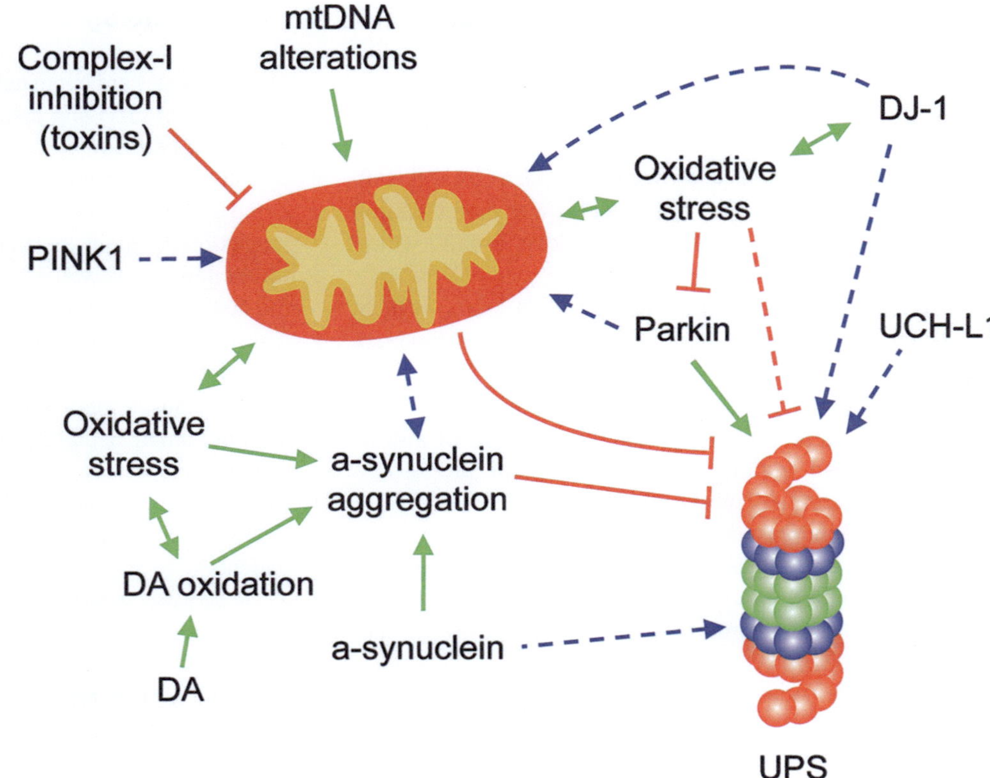

Fig. 23.28 Common pathways underlying PD pathogenesis. Mutations in five genes encoding *SNCA*, *PARK2*, *UCH-L1*, *PINK1*, and *DJ1* are associated with familial forms of PD through pathogenic pathways that may commonly lead to deficits in mitochondrial and ubiquitin–proteosomal system function. Mitochondrial and ubiquitin–proteosomal system dysfunction, oxidative stress, and SNCA aggregation ultimately contribute to the demise of DA neurons in PD. Red lines indicate inhibitory effects, green arrows depict defined relationships between components or systems, and blue-dashed arrows indicate proposed or putative relationships (adapted from Moore DJ et al. 2005, from Annual Reviews)

- o PARK2 may also function as a neuroprotectant molecule by interacting with mitochondria and preventing apoptosis
- o Mutations cause a loss of function of this ubiquitin–proteosomal system and accumulation of neurotoxic proteins and/or loss of the neuroprotectant function
- o PARK2- and PTEN-induced kinase-1 (PINK1) have been shown to act in a common genetic pathway
- – Ubiquitin carboxyl-terminal hydrolase L1 (UCH-L1)
  - o UCH-L1 is a highly abundant, neuron-specific protein that belongs to a family of deubiquitinating enzymes that are responsible for hydrolyzing polymeric ubiquitin chains to free ubiquitin monomers
  - o It may also function as a ubiquitin-protein ligase
  - o UCH-L1 can be found in LB in sporadic PD What a mutated UCH-L1 directly contributes to PD is not known
- – PINK1
  - o PINK1 physiologic function is not currently known
  - o PINK1 has a mitochondrial targeting sequence and a conserved domain similar to the calcium-calmodulin kinase family
  - o Mutations are thought to cause a loss of function in the putative kinase activity leading to mitochondrial dysfunction and PD

- – DJ1
  - o Ubiquitously expressed protein in neurons and glia belonging to the DJ1/ThiJ/PfpI superfamily
  - o DJ1 does not colocalize in LB but is found to be associated with several neurodegenerative tauopathies and with SNCA-positive glial inclusions in multiple system atrophy
  - o Insoluble forms are increased in the brains of sporadic PD patients
  - o Physiologic function of DJ1 is unclear, but it may function as an antioxidant protein or as a sensor of oxidative stress
  - o DJ1 may be a component of the ubiquitin–proteosomal system and may confer protection by functioning as a molecular chaperone or protease to refold or promote degradation of misfolded proteins

Diagnosis

- • Clinical diagnosis relies on history, observation of clinical manifestations, and initial responsiveness to dopaminergic agonist therapy
- • Neuropathological examination at autopsy is required to confirm diagnosis

Amyotrophic Lateral Sclerosis/Motor Neuron Disease

Clinical

- Amyotrophic lateral sclerosis (ALS) is a disease with patients presenting both upper and lower motor neuron signs, which leads to paralysis and death usually within 2–5 years
- Generally, affects older individuals (50–70 years) with an annual incidence of 1–2/100,100 and overall lifetime risk of 1/800
- Slight male-to-female preponderance (1.3:1–1.6:1)
- There is patient-to-patient variability in terms of muscle areas affected initially and the pattern of progressive spread to eventually most muscle groups
- Some patients present with prominent bulbar symptoms secondary to early and more extensive loss of cranial nerve motor neurons
- Upper motor neuron signs include clonus and hyperreflexia
- Lower motor neuron signs include muscle atrophy, weakness, and fasciculations
- The cause is unknown with 90% of cases being sporadic and 10% familial
- Some epidemiological studies suggest the incidence is increasing

Gross and Histological Neuropathology

- Gross changes are not usually noted in the brain, although in long-term surviving patient's, atrophy of the precentral gyrus can be seen
- Spinal cord anterior motor nerve roots are notably thinned compared with posterior sensory nerve roots
- Depletion of upper (corticospinal, Betz cells) and lower motor neurons as well as cranial nerve motor neurons are the histological hallmarks of ALS
- The lateral and anterior medial corticospinal tracts are depleted (lateral sclerosis)
- Skeletal muscle deprived of innervation shows grouped atrophy, small acutely angulated fibers, and fiber-type grouping
- Skein-like inclusions and Bunina bodies are cytoplasmic inclusions that are often found in motor neurons of ALS patients
- Reactive astrocytes and microglia are found in the anterior horns of the spinal cord and motor cortex of the cerebrum

Associated Genes and Possible Molecular Mechanisms of Pathogenesis

- Familial ALS represents about 10% of cases (Table 23.5), and within this group there is significant phenotypic and genotypic heterogeneity

Table 23.5 Genes associated with amyotrophic lateral sclerosis (ALS)

Reported FALS/MND loci	Gene	Chromosomal location
Adult-onset dominant typical ALS		
ALS1[a]	*SOD1*	21q22.1
ALS3		18q21
ALS6	*FUS*	16p11.2
ALS7		20p13
ALS9	*ANG*	14q11.2
ALS10	*TARDBP*	1p36.22
ALS11	*FIG4*	6p21
ALS13	*ATXN2*	12q24.12
Adult-onset dominant atypical ALS		
ALS with/without frontotemporal dementia	*MAPT*	9q21–q22
	C9ORF72	9
	UBQLN2	Xp11.21
ALS14	*VCP*	9p13.3
ALS with dementia/parkinsonism	*MAPT*	17q21.1
Progressive lower motor neuron disease	*DCTN1*	2p13
ALS8	*VAPB*	20q13.32
ALS12	*OPTN*	10p13
Juvenile-onset dominant ALS		
ALS4	*SETX*	9q34.13
Juvenile-onset recessive ALS		
ALS2	*ALS2*	2q33.1
ALS5		15q15.1–q21.1

FALS Familial form of ALS, *MND* Motor neuron disease

[a] Note that both dominant and recessive linked-*SOD1* mutations have been reported (modified from Gros-Louise et al. 2006; Lill et al. 2011)

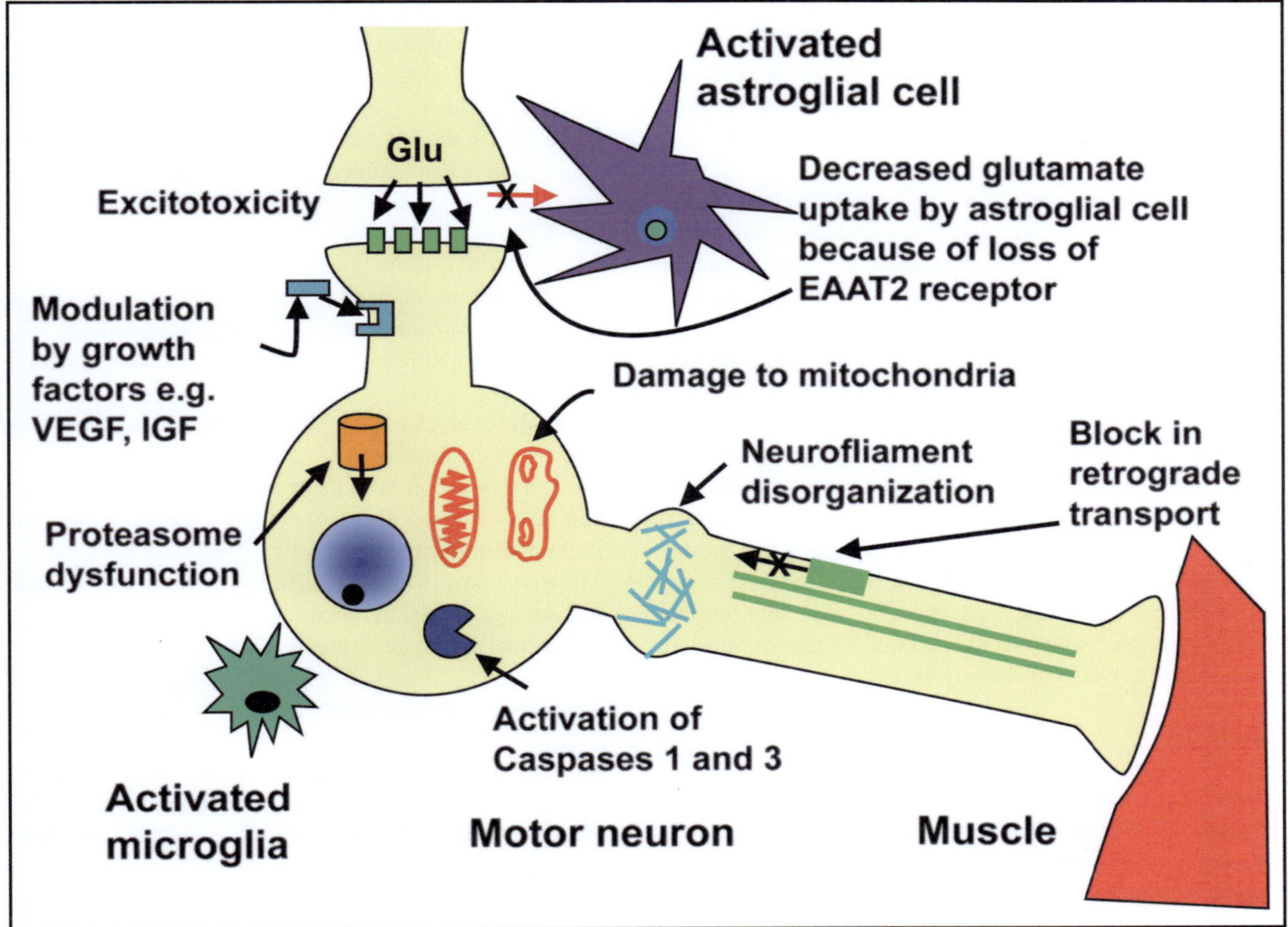

Fig. 23.29 Molecular and cellular processes possibly implicated in the pathogenesis of ALS (adapted from Bruijn et al. 2004; from Annual Reviews)

- While genetic loci have been identified for many of the familial cases of ALS, the function of these genes and their relationship to ALS pathogenesis is unknown
- The search is on for modifier genes and polymorphisms in sporadic ALS
- Small studies of sporadic ALS patients have identified possible altered genes/polymorphisms that need to be confirmed in larger studies: *VEGF*, *EAAT2*, *GRIA2*, *CNTF*, *SMN1*, *SMN2*, *APOE*, and *NEFH*
- Epigenetic and genetic causes have been postulated to bring about several cytotoxic phenotypes, although direct links are not well-established (Fig. 23.29)
  - SOD1
    - o Copper-zinc superoxide dismutase-1 was the first gene identified for familial ALS, and mutations account for 20% of familial cases
    - o Ubiquitously expressed cytoplasmic protein that detoxifies the reactive molecule superoxide to oxy-
    gen and H_2O_2, which can then be cleared by catalase and glutathione peroxidase
    - o >100 mutations have been identified, but the phenotypic variability, even in families with the same mutation, suggests that other genes and/or environmental factors are important
    - o Rodent transgenic models with human *SOD1* mutations display similar clinical and pathological features and have been important research tools
    - o The exact mechanism by which mutated *SOD1* causes ALS is not known despite its discovery >13 years ago
    - o A dominant negative toxic gain of function mechanism is thought to be involved in *SOD1* pathogenesis
    - o Recent animal studies have suggested that mutant *SOD1* expression must be in both glial as well as neuronal cells for the ALS-like disease to manifest

- Hypothesized mechanisms for *SOD1* toxicity include excitotoxicity, oxidative stress, mitochondrial dysfunction, inflammation, axonal transport defect, and/or toxic aggregation, with likely some combination of the above
- TDP-43
  - Transactivating region (TAR) DNA-binding protein 43, a recently identified protein that aggregates with ubiquitin and other proteins in frontotemporal dementias and many cases of ALS, both sporadic and some familial
  - Abnormal aggregation occurs both in the presence and absence of mutation in the gene
  - Interestingly, reports show that abnormal cell localization to the cytoplasm of affected neuronal and glial cells is found with all sporadic ALS cases and some familial cases but not with mutant *SOD1* familial cases, suggesting possible different modes of degeneration in familial and sporadic ALS
  - Normal physiological function and role in the pathophysiology of ALS is not fully understood and an active area of investigation

Diagnosis

- Patients are diagnosed with ALS based on the E1 Escorial criteria that categorizes patients as clinically definite, probable, or possible for ALS based on degree of upper and lower motor neuron findings
- Less commonly muscle biopsies are performed (often to rule out other diseases)
- *SOD1* genetic testing is available and should be considered in the setting of familial ALS with appropriate genetic counseling
- Autopsy findings confirm the diagnosis

Tauopathies

General Comments

- Tau is a microtubule-associated protein that binds microtubules and promotes microtubule assembly, essential components of the cytoskeleton
- Tau is abundantly expressed in the CNS and exists in six isoforms created by alternative mRNA splicing of a single gene
- The different isoforms have either three or four microtubule-binding repeat sequences with similar ratios of either three or four repeat isoforms normally expressed; varying this ratio appears to confer susceptibility to some neurodegenerative disease later in life

- Tau is also a major component of atypical protein aggregates found in paired helical filaments in AD and in several other neurodegenerative diseases collectively thought of as tauopathies
- Mutations in the tau gene and/or altered phosphorylation states have been identified in many of these diseases

FTDP-17T

- Frontotemporal dementia and parkinsonism linked to chromosome 17 associated with tau gene mutations
- Adult-onset, slowly progressive neurodegenerative disease
- Clinically characterized by variable cognitive, behavioral, and motor dysfunction
- Diffuse deposition of tau aggregates in neurons and glia can be identified with silver stains and tau immunohistochemistry
- Mutations have been identified in multiple sites (exonic and intronic) of the tau gene
- Autosomal dominant transmission

Progressive Supranuclear Palsy

- Multisystem degeneration characterized by symptoms of parkinsonism and supranuclear ophthalmoplegia
- Slowly progressive disease affecting middle- and late-aged individuals
- No established genetic or epigenetic etiologies, and most (if not all) cases are sporadic
- Polymorphisms in the tau gene have been identified, and sporadic cases are associated with a homozygous H1 haplotype and overexpression of the 4-repeat tau isoform
- Neurofibrillary tangles, neuropil threads, and glial fibrillary tangles can be identified in multiple areas but tend to concentrate in substantia nigra, basal ganglia, subthalamic nucleus, and brainstem
- Kinase/phosphatase dysregulation may be involved in the pathogenesis of progressive supranuclear palsy

Corticobasal Degeneration

- Adult-onset neurodegenerative disease with focal pathology and corresponding clinical phenotype of primary aphasia, dementia, visual inattention, or rapidly progressive mutism
- Neuropathology consists of focal cortical or deep gray matter degeneration with tau-positive neurons and glia
- Most cases are sporadic; several familial cases have been reported, although they share overlapping pathology and genetics with tau mutations similar to *FTDP-17*

- Sporadic cases are associated with a homozygous H1 haplotype and overexpression of the 4-repeat tau isoform
- The histological hallmark of corticobasal degeneration is swollen or "balloon" neurons in the affected area
- Balloon neurons, neuropil threads, as well as occasional other neurons and glia contain tau-positive aggregates within their cytoplasm

Pick Disease

- Frontotemporal degeneration with three clinical patterns: behavioral syndrome referred to as frontotemporal dementia, progressive nonfluent aphasia, and semantic dementia
- Majority of cases are sporadic
- Familial cases with defined tau mutations have been reported and called "atypical Pick disease" but may be better recognized as some other tauopathy distinct from Pick
- The sporadic form is not related to H1 or H2 haplotypes and appears to be a 3-repeat tau disorder
- Gross neuropathology shows marked variable atrophy ("knife-edge" atrophy) of the frontal, temporal, and parietal lobes with consistently preserved precentral gyrus and posterior two-thirds of the superior temporal gyrus (Fig. 23.30)
- Pick bodies, round cytoplasmic fibrillar inclusions, are consistently found in the fascia dentata of the hippocam-

pus and less common in other nuclei and cortical neurons
- Pick bodies are highlighted with silver stains or tau or ubiquitin immunohistochemistry
- Extensive neuronal loss, gliosis, and occasionally ballooned neurons are seen in affected areas

Trinucleotide Repeat Diseases

General Comments

- These disorders are caused by expanding triplet repeat of nucleotides and affect primarily the CNS/PNS (Table 23.6)
- Generally, the larger the number of repeats the more severe the disease
- "Anticipation," a key genetic feature in these diseases, is the earlier age of onset in successive generations within a family
- A variety of modes of inheritance exist, including AR, AD, X-linked
- Pathogenesis is not well understood, but likely involves loss of function and/or toxic gains of function of affected proteins

Huntington Disease

Clinical

- Huntington disease (HD) is an autosomal dominant transmitted disease with near complete penetrance characterized by chorea and progressive cognitive and behavioral disorders
- Mean age of onset: 40 years
- Prevalence of 5–10/100,000

Gross and Histologic Neuropathology

- Classic gross appearance of the brain on coronal section is widened lateral ventricles and atrophied, flattened caudate and putamen nuclei
- Advanced cases show global brain atrophy
- Neuronal loss and reactive astrocytosis in affected areas
- Polyglutamine expanded repeats cause an accumulation of the abnormal protein and formation of nuclear inclusions within cells of the striatum and cortex
- Atrophy and neuronal loss correlate with severity of clinical disease

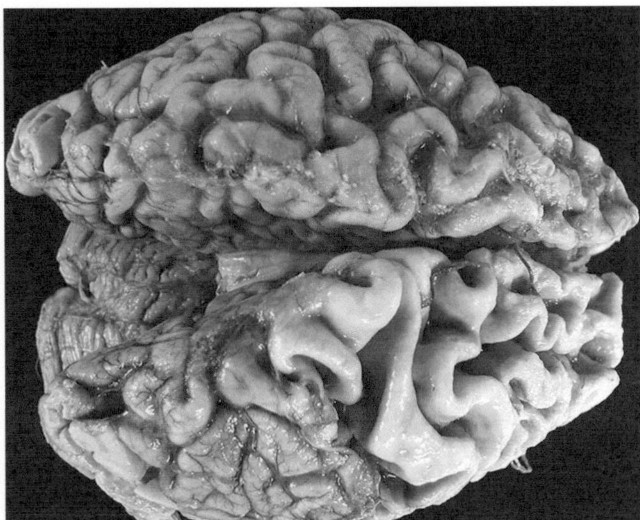

Fig. 23.30 Gross image of superior surface of a brain from a patient with Pick-type frontotemporal dementia. Note the widely spaced, so-called "knife-edge" gyri in the frontal lobes

Table 23.6 Triplet repeat disorders affecting the nervous system

Disease	Symptoms	Gene	Locus	Protein
Noncoding repeats				
Friedreich ataxia	Ataxia, weakness, sensory loss	*FXN*	9q13–q21.1	Frataxin
Fragile X syndrome A	Intellectual disability	*FMR1*	Xq27.3	Fragile X messenger ribonucleoprotein 2
Fragile X syndrome E	Intellectual disability	*FMR2*	Xq28	Fragile X messenger ribonucleoprotein 2
Dystrophia myotonica 1	Weakness, myotonia	*DMPK*	19q13	Dystrophia myotonica protein kinase
Spinocerebellar ataxia 8	Ataxia	Antisense to *KLHL1*	13q21	Undetermined
Spinocerebellar ataxia 12	Ataxia	*PPP2R2B*	5q31–q33	Regulatory subunit of the protein phosphatase PP2A
Huntington disease-like 2	Chorea, dementia	*JPH3*	16q24.3	Junctophilin 3
Polyglutamine disorders				
Spinal and bulbar muscular atrophy	Weakness	*AR*	Xq13–q21	Androgen receptor
Huntington disease	Chorea, dementia	*IT15*	*4p16.3*	Huntingtin
Dentatorubral-pallidoluysian atrophy	Ataxia, myoclonic epilepsy, dementia	*DRPLA*	12p13.31	Atrophin 1
Spinocerebellar ataxia 1	Ataxia	*SCA1*	6p23	Ataxin 1
Spinocerebellar ataxia 2	Ataxia	*SCA2*	12q24.1	Ataxin 2
Spinocerebellar ataxia 3 (Machado–Joseph disease)	Ataxia	*SCA3/MJD*	14q32.1	Ataxin 3
Spinocerebellar ataxia 6	Ataxia	*CACNA1A*	19p13	α-la voltage-dependent calcium channel subunit
Spinocerebellar ataxia 7	Ataxia	*SCA7*	3p12–p13	Ataxin 7
Spinocerebellar ataxia 17	Ataxia	*TBP*	6q27	TATA box binding protein
Polyalanine disorders				
Oculopharyngeal dystrophy	Weakness	*PABPN1*	14q11.2–q13	Poly(A)-binding protein 21
Congenital central hypoventilation syndrome	Respiratory difficulties	*PHOX2B*	4p12	Paired-like homeobox 2B
Infantile spasms	Intellectual disability, epilepsy	*ARX*	Xp22.13	Aristaless-related homeobox, X-linked
Synpolydactyly	Limb malformation	*HOXD13*	2q31–q32	Homeobox D13

Adapted from Di Prospero and Fischbeck (2005)

Associated Genes

- Autosomal dominant form of disease (90% cases; 10% occur de novo) caused by expanded CAG repeats (>36) coding for polyglutamines in the gene, *HTT* (*IT15*), product huntingtin

Possible Molecular Mechanisms of Pathogenesis

- Normal functions of HD are thought to include roles in transport, transcription, and neurogenesis
- The abnormal mutated form of huntingtin with increased glutamines in the amino terminus can form protein aggregates with other proteins including ubiquitin
- How the inclusions lead to cellular dysfunction and degeneration is not known
- Energy depletion, increased apoptosis, impairment of the proteasome–ubiquitin system, oxidative stress, and excitotoxicity of susceptible neurons have all been hypothesized to be involved in the pathogenesis of HD

Diagnosis

- Based on a thorough clinical history, physical and cognitive examination
- CT scanning to look for striatal atrophy can be helpful
- Genetic testing for the mutations of *HTT* can be done in affected individuals and family members
- Genetic counseling is essential
- Fetal genetic testing and in vitro fertilization with preimplantation screening can be performed at some medical centers

Friedrich Ataxia

Clinical

- Friedrich ataxia (FA) is the most common inherited ataxia with worldwide distribution
- European prevalence of 1/29,000 and carrier status of 1:85

- Autosomal recessive
- 85% have onset before age 20; late adult onset is also seen less commonly
- Slow progress with loss of ambulation in about 5–15 years
- Patients have loss of deep sensation and deep tendon reflexes and cerebellar signs including an ataxic speech disorder
- Cardiomyopathy is frequent and can lead to cardiac failure
- Adult-onset diabetes in 10–32%

Gross and Histological Neuropathology

- Spinal cord and dorsal roots are atrophic
- Loss of large myelinated axons and dorsal root ganglion cells
- Degeneration of dorsal column tracts and spinocerebellar tracts
- Cerebellum shows white matter gliosis and dentate nucleus degeneration
- Polyglutamine expanded repeats cause an accumulation of the abnormal protein and formation of nuclear inclusions within cells of the striatum and cortex

Associated Genes

- Homozygous GAA-repeat expansion within the first intron of *FXN*
- Repeats in FA can range from 67 to 1700 (normal 6–34)

Possible Molecular Mechanisms of Pathogenesis

- Frataxin is a mitochondrial protein involved in iron metabolism
- Increased repeats lead to decreased frataxin levels and iron accumulation
- Increased iron is thought to contribute to increased oxidative stress, decreased oxidative phosphorylation, and reduced activity of mitochondrial enzyme complexes containing iron-sulfur clusters

Diagnosis

- Based on satisfying major and minor clinical criteria with emphasis on early-onset, progressive ataxia, and sensory dysfunction
- Genetic testing is widely available to support clinical diagnosis; especially helpful in atypical presentations

Acknowledgments We would like to thank Dr. Brent Harris for his contributions to the previous editions of this chapter.

Suggested Reading

Central Nervous System Tumors

Andreiuolo F, Lisner T, Zlocha J, et al. H3F3A-G34R mutant high grade neuroepithelial neoplasms with glial and dysplastic ganglion cell components. Acta Neuropathol Commun. 2019;7(1):78.

Appay R, Dehais C, Maurage CA, et al. CDKN2A homozygous deletion is a strong adverse prognosis factor in diffuse malignant IDH-mutant gliomas. Neuro-Oncol. 2019;21(12):1519–28.

Baroni L, Sundaresan L, Heled A, et al. Ultra high-risk PFA ependymoma is characterized by loss of chromosome 6q. Neuro-Oncol. 2021;23(8):1360–70.

Belakhoua SM, Rodriguez FJ. Diagnostic pathology of tumors of peripheral nerve. Neurosurgery. 2021;88(3):443–56.

Bitar M, Danish SF, Rosenblum MK. A newly diagnosed case of polymorphous low-grade neuroepithelial tumor of the young. Clin Neuropathol. 2018;37(4):178–81.

Blessing MM, Blackburn PR, Krishnan C, et al. Desmoplastic infantile ganglioglioma: a MAPK pathway-driven and microglia/macrophage-rich neuroepithelial tumor. J Neuropathol Exp Neurol. 2019;78(11):1011–21.

Brat DJ, Aldape K, Bridge JA, et al. Molecular biomarker testing for the diagnosis of diffuse gliomas. Arch Pathol Lab Med. 2022;146(5):547–74.

Brat DJ, Aldape K, Colman H, et al. cIMPACT-NOW Update 3: Recommended diagnostic criteria for "Diffuse astrocytic glioma, IDH-wildtype, with molecular features of glioblastoma, WHO grade IV". Acta Neuropathol (Berl). 2018;136(5):805–10.

Cancer Genome Atlas Research Network, Brat DJ, Verhaak RGW, et al. Comprehensive, integrative genomic analysis of diffuse lower-grade gliomas. N Engl J Med. 2015;372(26):2481–98.

Cavalli FMG, Hübner JM, Sharma T, et al. Heterogeneity within the PF-EPN-B ependymoma subgroup. Acta Neuropathol (Berl). 2018;136(2):227–37.

Chan V, Marro A, Findlay JM, Schmitt LM, Das S. A systematic review of atypical teratoid rhabdoid tumor in adults. Front Oncol. 2018;8:567.

Chen CCL, Deshmukh S, Jessa S, et al. Histone H3.3G34-Mutant interneuron progenitors co-opt PDGFRA for gliomagenesis. Cell. 2020;183(6):1617–1633.e22.

Chun HJE, Johann PD, Milne K, et al. Identification and analyses of extra-cranial and cranial rhabdoid tumor molecular subgroups reveal tumors with cytotoxic T cell infiltration. Cell Rep. 2019;29(8):2338–2354.e7.

Clarke M, Mackay A, Ismer B, et al. Infant high-grade gliomas comprise multiple subgroups characterized by novel targetable gene fusions and favorable outcomes. Cancer Discov. 2020;10(7):942–63.

D'Aronco L, Rouleau C, Gayden T, et al. Brainstem angiocentric gliomas with MYB-QKI rearrangements. Acta Neuropathol (Berl). 2017;134(4):667–9.

Deng MY, Sill M, Chiang J, et al. Molecularly defined diffuse leptomeningeal glioneuronal tumor (DLGNT) comprises two subgroups with distinct clinical and genetic features. Acta Neuropathol (Berl). 2018;136(2):239–53.

Ellison DW, Dalton J, Kocak M, et al. Medulloblastoma: Clinicopathological correlates of SHH, WNT, and non-SHH/WNT molecular subgroups. Acta Neuropathol (Berl). 2011;121(3):381–96.

Ellison DW, Hawkins C, Jones DTW, et al. cIMPACT-NOW update 4: diffuse gliomas characterized by MYB, MYBL1, or FGFR1 alterations or BRAFV600E mutation. Acta Neuropathol (Berl). 2019;137(4):683–7.

Ellison DW, Kocak M, Dalton J, et al. Definition of disease-risk stratification groups in childhood medulloblastoma using combined clinical, pathologic, and molecular variables. J Clin Oncol. 2011;29(11):1400–7.

Ellison DW, Onilude OE, Lindsey JC, et al. beta-Catenin status predicts a favorable outcome in childhood medulloblastoma: the United Kingdom Children's Cancer Study Group Brain Tumour Committee. J Clin Oncol. 2005;23(31):7951–7.

Fisher MJ, Jones DTW, Li Y, et al. Integrated molecular and clinical analysis of low-grade gliomas in children with neurofibromatosis type 1 (NF1). Acta Neuropathol (Berl). 2021;141(4):605–17.

Fritchie K, Jensch K, Moskalev EA, et al. The impact of histopathology and NAB2-STAT6 fusion subtype in classification and grading of meningeal solitary fibrous tumor/hemangiopericytoma. Acta Neuropathol (Berl). 2019;137(2):307–19.

Giangaspero F, Perilongo G, Fondelli MP, et al. Medulloblastoma with extensive nodularity: A variant with favorable prognosis. J Neurosurg. 1999;91(6):971–7.

Giangaspero F, Rigobello L, Badiali M, et al. Large-cell medulloblastomas. A distinct variant with highly aggressive behavior. Am J Surg Pathol. 1992;16(7):687–93.

Gianno F, Antonelli M, Minasi S, et al. Correlation between immunohistochemistry and sequencing in H3G34-mutant gliomas. Am J Surg Pathol. 2020;45(2):200–4.

Gross AM, Wolters PL, Dombi E, et al. Selumetinib in children with inoperable plexiform neurofibromas. N Engl J Med. 2020;382(15):1430–42.

Guerreiro Stucklin AS, Ryall S, Fukuoka K, et al. Alterations in ALK/ROS1/NTRK/MET drive a group of infantile hemispheric gliomas. Nat Commun. 2019;10(1):4343.

Guo C, Pirozzi CJ, Lopez GY, Yan H. Isocitrate dehydrogenase mutations in gliomas: Mechanisms, biomarkers and therapeutic target. Curr Opin Neurol. 2011;24(6):648–52.

Haberler C, Laggner U, Slavc I, et al. Immunohistochemical analysis of INI1 protein in malignant pediatric CNS tumors: Lack of INI1 in atypical teratoid/rhabdoid tumors and in a fraction of primitive neuroectodermal tumors without rhabdoid phenotype. Am J Surg Pathol. 2006;30(11):1462–8.

Hartmann C, Meyer J, Balss J, et al. Type and frequency of IDH1 and IDH2 mutations are related to astrocytic and oligodendroglial differentiation and age: a study of 1,010 diffuse gliomas. Acta Neuropathol (Berl). 2009;118(4):469–74.

Hasselblatt M, Gesk S, Oyen F, et al. Nonsense mutation and inactivation of SMARCA4 (BRG1) in an atypical teratoid/rhabdoid tumor showing retained SMARCB1 (INI1) expression. Am J Surg Pathol. 2011;35(6):933–5.

Hasselblatt M, Thomas C, Hovestadt V, et al. Poorly differentiated chordoma with SMARCB1/INI1 loss: A distinct molecular entity with dismal prognosis. Acta Neuropathol. 2016;132(1):149–51.

Holdhof D, Johann PD, Spohn M, et al. Atypical teratoid/rhabdoid tumors (ATRTs) with SMARCA4 mutation are molecularly distinct from SMARCB1-deficient cases. Acta Neuropathol. 2021;141(2):291–301.

Holsten T, Lubieniecki F, Spohn M, et al. Detailed clinical and histopathological description of 8 cases of molecularly defined CNS neuroblastomas. J Neuropathol Exp Neurol. 2021;80(1):52–9.

Horbinski C, Miller CR, Perry A. Gone FISHing: Clinical lessons learned in brain tumor molecular diagnostics over the last decade. Brain Pathol. 2011;21(1):57–73.

Hou Y, Pinheiro J, Sahm F, et al. Papillary glioneuronal tumor (PGNT) exhibits a characteristic methylation profile and fusions involving PRKCA. Acta Neuropathol (Berl). 2019;137(5):837–46.

Hulsebos TJ, Plomp AS, Wolterman RA, Robanus-Maandag EC, Baas F, Wesseling P. Germline mutation of INI1/SMARCB1 in familial schwannomatosis. Am J Hum Genet. 2007;80(4):805–10.

Huse JT, Snuderl M, Jones DTW, et al. Polymorphous low-grade neuroepithelial tumor of the young (PLNTY): An epileptogenic neoplasm with oligodendroglioma-like components, aberrant CD34 expression, and genetic alterations involving the MAP kinase pathway. Acta Neuropathol (Berl). 2017;133(3):417–29.

Johann PD, Hovestadt V, Thomas C, et al. Cribriform neuroepithelial tumor: molecular characterization of a SMARCB1-deficient non-rhabdoid tumor with favorable long-term outcome. Brain Pathol. 2017;27(4):411–8.

Johnson DR, Giannini C, Jenkins RB, Kim DK, Kaufmann TJ. Plenty of calcification: Imaging characterization of polymorphous low-grade neuroepithelial tumor of the young. Neuroradiology. 2019;61(11):1327–32.

Jones DT, Kocialkowski S, Liu L, et al. Tandem duplication producing a novel oncogenic BRAF fusion gene defines the majority of pilocytic astrocytomas. Cancer Res. 2008;68(21):8673–7.

Jünger ST, Andreiuolo F, Mynarek M, et al. CDKN2A deletion in supratentorial ependymoma with RELA alteration indicates a dismal prognosis: A retrospective analysis of the HIT ependymoma trial cohort. Acta Neuropathol (Berl). 2020;140(3):405–7.

Kleihues P, Schäuble B, zur Hausen A, Estève J, Ohgaki H. Tumors associated with p53 germline mutations: a synopsis of 91 families. Am J Pathol. 1997;150(1):1–13.

Kloosterhof NK, Bralten LB, Dubbink HJ, French PJ, van den Bent MJ. Isocitrate dehydrogenase-1 mutations: a fundamentally new understanding of diffuse glioma? Lancet Oncol. 2011;12(1):83–91.

Koelsche C, Wöhrer A, Jeibmann A, et al. Mutant BRAF V600E protein in ganglioglioma is predominantly expressed by neuronal tumor cells. Acta Neuropathol (Berl). 2013;125(6):891–900.

Kool M, Jones DT, Jager N, et al. Genome sequencing of SHH medulloblastoma predicts genotype-related response to smoothened inhibition. Cancer Cell. 2014;25(3):393–405.

Korshunov A, Remke M, Werft W, et al. Adult and pediatric medulloblastomas are genetically distinct and require different algorithms for molecular risk stratification. J Clin Oncol. 2010;28(18):3054–60.

Korshunov A, Ryzhova M, Jones DTW, et al. LIN28A immunoreactivity is a potent diagnostic marker of embryonal tumor with multilayered rosettes (ETMR). Acta Neuropathol (Berl). 2012;124(6):875–81.

Korshunov A, Sycheva R, Golanov A. Recurrent cytogenetic aberrations in central neurocytomas and their biological relevance. Acta Neuropathol (Berl). 2007;113(3):303–12.

Lafay-Cousin L, Hawkins C, Carret AS, et al. Central nervous system atypical teratoid rhabdoid tumours: the Canadian Paediatric Brain Tumour Consortium experience. Eur J Cancer. 2012;48(3):353–9.

Lee J, Putnam AR, Chesier SH, et al. Oligodendrogliomas, IDH-mutant and 1p/19q-codeleted, arising during teenage years often lack TERT promoter mutation that is typical of their adult counterparts. Acta Neuropathol Commun. 2018;6(1):95.

Louis DN, Ellison DW, Brat DJ, et al. cIMPACT-NOW: a practical summary of diagnostic points from Round 1 updates. Brain Pathol. 2019;29(4):469–72.

Louis DN, Perry A, Wesseling P, et al. The 2021 WHO classification of tumors of the central nervous system: a summary. Neuro Oncol. 2021;23(8):1231–51.

Lucas CHG, Gupta R, Doo P, et al. Comprehensive analysis of diverse low-grade neuroepithelial tumors with FGFR1 alterations reveals a distinct molecular signature of rosette-forming glioneuronal tumor. Acta Neuropathol Commun. 2020;8(1):151.

Lucas CHG, Villanueva-Meyer JE, Whipple N, et al. Myxoid glioneuronal tumor, PDGFRA p.K385-mutant: clinical, radiologic, and histopathologic features. Brain Pathol Zurich Switz. 2020;30(3):479–94.

Macagno N, Vogels R, Appay R, et al. Grading of meningeal solitary fibrous tumors/hemangiopericytomas: analysis of the prognostic value of the Marseille Grading System in a cohort of 132 patients. Brain Pathol Zurich Switz. 2019;29(1):18–27.

Northcott PA, Korshunov A, Witt H, et al. Medulloblastoma comprises four distinct molecular variants. J Clin Oncol. 2011;29(11):1408–14.

Northcott PA, Robinson GW, Kratz CP, et al. Medulloblastoma. Nat Rev Dis Primers. 2019;5(1):11.

Pajtler KW, Wei Y, Okonechnikov K, et al. YAP1 subgroup supratentorial ependymoma requires TEAD and nuclear factor I-mediated transcriptional programmes for tumorigenesis. Nat Commun. 2019;10:3914.

Parsons DW, Jones S, Zhang X, et al. An integrated genomic analysis of human glioblastoma multiforme. Science. 2008;321(5897):1807.

Pekmezci M, Stevers M, Phillips JJ, et al. Multinodular and vacuolating neuronal tumor of the cerebrum is a clonal neoplasm defined by genetic alterations that activate the MAP kinase signaling pathway. Acta Neuropathol (Berl). 2018;135(3):485–8.

Prieto-Granada CN, Wiesner T, Messina JL, Jungbluth AA, Chi P, Antonescu CR. Loss of H3K27me3 expression is a highly sensitive marker for sporadic and radiation-induced MPNST. Am J Surg Pathol. 2016;40(4):479–89.

Qaddoumi I, Orisme W, Wen J, et al. Genetic alterations in uncommon low-grade neuroepithelial tumors: BRAF, FGFR1, and MYB mutations occur at high frequency and align with morphology. Acta Neuropathol. 2016;131(6):833–45.

Ramani B, Gupta R, Wu J, et al. The immunohistochemical, DNA methylation, and chromosomal copy number profile of cauda equina paraganglioma is distinct from extra-spinal paraganglioma. Acta Neuropathol (Berl). 2020;140(6):907–17.

Ramaswamy V, Remke M, Bouffet E, et al. Risk stratification of childhood medulloblastoma in the molecular era: the current consensus. Acta Neuropathol (Berl). 2016;131(6):821–31.

Rausch T, Jones DT, Zapatka M, et al. Genome sequencing of pediatric medulloblastoma links catastrophic DNA rearrangements with TP53 mutations. Cell. 2012;148(1-2):59–71.

Reddy AT, Strother DR, Judkins AR, et al. Efficacy of high-dose chemotherapy and three-dimensional conformal radiation for atypical teratoid/rhabdoid tumor: a report from the Children's Oncology Group trial ACNS0333. J Clin Oncol. 2020;38(11):1175–85.

Reuss DE, Mamatjan Y, Schrimpf D, et al. IDH mutant diffuse and anaplastic astrocytomas have similar age at presentation and little difference in survival: A grading problem for WHO. Acta Neuropathol (Berl). 2015;129(6):867–73.

Reuss DE, Sahm F, Schrimpf D, et al. ATRX and IDH1-R132H immunohistochemistry with subsequent copy number analysis and IDH sequencing as a basis for an "integrated" diagnostic approach for adult astrocytoma, oligodendroglioma and glioblastoma. Acta Neuropathol (Berl). 2015;129(1):133–46.

Rivera B, Gayden T, Carrot-Zhang J, et al. Germline and somatic FGFR1 abnormalities in dysembryoplastic neuroepithelial tumors. Acta Neuropathol (Berl). 2016;131(6):847–63.

Roberts RO, Lynch CF, Jones MP, Hart MN. Medulloblastoma: a population-based study of 532 cases. J Neuropathol Exp Neurol. 1991;50(2):134–44.

Roth JJ, Fierst TM, Waanders AJ, Yimei L, Biegel JA, Santi M. Whole chromosome 7 gain predicts higher risk of recurrence in pediatric pilocytic astrocytomas independently from KIAA1549-BRAF fusion status. J Neuropathol Exp Neurol. 2016;75(4):306–15.

Ryall S, Zapotocky M, Fukuoka K, et al. Integrated molecular and clinical analysis of 1,000 pediatric low-grade gliomas. Cancer Cell. 2020;37(4):569–583.e5.

Sasaki S, Tomomasa R, Nobusawa S, et al. Anaplastic pleomorphic xanthoastrocytoma associated with an H3G34 mutation: a case report with review of literature. Brain Tumor Pathol. 2019;36(4):169–73.

Schindler G, Capper D, Meyer J, et al. Analysis of BRAF V600E mutation in 1,320 nervous system tumors reveals high mutation frequencies in pleomorphic xanthoastrocytoma, ganglioglioma and extra-cerebellar pilocytic astrocytoma. Acta Neuropathol. 2011;121(3):397–405.

Schwalbe EC, Lindsey JC, Nakjang S, et al. Novel molecular subgroups for clinical classification and outcome prediction in childhood medulloblastoma: a cohort study. Lancet Oncol. 2017;18(7):958–71.

Spence T, Sin-Chan P, Picard D, et al. CNS-PNETs with C19MC amplification and/or LIN28 expression comprise a distinct histogenetic diagnostic and therapeutic entity. Acta Neuropathol. 2014;128(2):291–303.

Stone TJ, Keeley A, Virasami A, et al. Comprehensive molecular characterisation of epilepsy-associated glioneuronal tumours. Acta Neuropathol (Berl). 2018;135(1):115–29.

Sturm D, Orr BA, Toprak UH, et al. New brain tumor entities emerge from molecular classification of CNS-PNETs. Cell. 2016;164(5):1060–72.

Taylor MD, Northcott PA, Korshunov A, et al. Molecular subgroups of medulloblastoma: The current consensus. Acta Neuropathol. 2012;123(4):465–72.

Thom M, Blümcke I, Aronica E. Long-term epilepsy-associated tumors. Brain Pathol. 2012;22(3):350–79.

Thom M, Liu J, Bongaarts A, et al. Multinodular and vacuolating neuronal tumors in epilepsy: dysplasia or neoplasia? Brain Pathol Zurich Switz. 2018;28(2):155–71.

Thomas C, Soschinski P, Zwaig M, et al. The genetic landscape of choroid plexus tumors in children and adults. Neuro-Oncol. 2021;23(4):650–60.

Tiwari N, Tamrazi B, Robison N, Krieger M, Ji J, Tian D. Unusual radiological and histological presentation of a diffuse leptomeningeal glioneuronal tumor (DLGNT) in a 13-year-old girl. Childs Nerv Syst ChNS Off J Int Soc Pediatr Neurosurg. 2019;35(9):1609–14.

Torre M, Vasudevaraja V, Serrano J, et al. Molecular and clinico-pathologic features of gliomas harboring NTRK fusions. Acta Neuropathol Commun. 2020;8(1):107.

Vaubel R, Zschernack V, Tran QT, et al. Biology and grading of pleomorphic xanthoastrocytoma-what have we learned about it? Brain Pathol Zurich Switz. 2021;31(1):20–32.

Waszak SM, Northcott PA, Buchhalter I, et al. Spectrum and prevalence of genetic predisposition in medulloblastoma: a retrospective genetic study and prospective validation in a clinical trial cohort. Lancet Oncol. 2018;19(6):785–98.

Waszak SM, Robinson GW, Gudenas BL, et al. Germline Elongator mutations in Sonic Hedgehog medulloblastoma. Nature. 2020;580(7803):396–401.

Weingart MF, Roth JJ, Hutt-Cabezas M, et al. Disrupting LIN28 in atypical teratoid rhabdoid tumors reveals the importance of the mitogen activated protein kinase pathway as a therapeutic target. Oncotarget. 2015;6(5):3165–77.

Witt H, Mack SC, Ryzhova M, et al. Delineation of two clinically and molecularly distinct subgroups of posterior fossa ependymoma. Cancer Cell. 2011;20(2):143–57.

Wolff JEA, Sajedi M, Brant R, Coppes MJ, Egeler RM. Choroid plexus tumours. Br J Cancer. 2002;87(10):1086–91.

Yin XL, Pang JCS, Hui ABY, Ng HK. Detection of chromosomal imbalances in central neurocytomas by using comparative genomic hybridization. J Neurosurg. 2000;93(1):77–81.

Yip S, Butterfield YS, Morozova O, et al. Concurrent CIC mutations, IDH mutations, and 1p/19q loss distinguish oligodendrogliomas from other cancers. J Pathol. 2012;226(1):7–16.

Zetterling M, Berhane L, Alafuzoff I, Jakola AS, Smits A. Prognostic markers for survival in patients with oligodendroglial tumors; a single-institution review of 214 cases. PloS One. 2017;12(11):e0188419.

Zhang J, Wu G, Miller CP, et al. Whole-genome sequencing identifies genetic alterations in pediatric low-grade gliomas. Nat Genet. 2013;45(6):602–12.

Zhukova N, Ramaswamy V, Remke M, et al. Subgroup-specific prognostic implications of TP53 mutation in medulloblastoma. J Clin Oncol. 2013;31(23):2927–35.

Nonneoplastic Brain Disorders

Armstrong RA, Lantos PL, Cairns NJ. Overlap between neurodegenerative disorders. Neuropathology. 2005;25:111–24.

Beal MF. Mitochondria take center stage in aging and neurodegeneration. Ann Neurol. 2005;58:495–505.

Bertram L, Hampel H. The role of genetics for biomarker development in neurodegeneration. Prog Neurobiol. 2011;95:501–4.

Bruijn LI, Miller TM, Cleveland DW. Unraveling the mechanisms involved in motor neuron degeneration in ALS. Annu Rev Neurosci. 2004;27:723–49.

Cameron B, Landreth GE. Inflammation, microglia, and Alzheimer's disease. Neurobiol Dis. 2010;37:503–9.

Cooper AJL, Blass JP. Trinucleotide-expansion diseases. Adv Neurobiol. 2011;1:319–58.

Costanza A, Weber K, Gandy S, Bouras C, Hof PR, Giannakopoulos P, et al. Review: contact sport-related chronic traumatic encephalopathy in the elderly: clinical expression and structural substrates. Neuropathol Appl Neurobiol. 2011;37:570–84.

Da Cruz S, Cleveland DW. Understanding the role of TDP-43 and FUS/TLS in ALS and beyond. Curr Opin Neurobiol. 2011;21:904–19.

Di Prospero NA, Fischbeck KH. Therapeutics development for triplet repeat expansion diseases. Nat Rev Genet. 2005;6:756–65.

Dickson DW, Weller RO. Neurodegeneration: the molecular pathology of dementia and movement disorders. 2nd ed. International Society of Neuropathology: Basel; 2011.

Gandy S. The role of cerebral amyloid beta accumulation in common forms of Alzheimer disease. J Clin Invest. 2005;115:1121–9. Review

Goldman JS, Rademakers R, Huey ED, et al. An algorithm for genetic testing of frontotemporal lobar degeneration. Neurology. 2011;76:475–83.

Gray F, De Girolami U, Poirier J. Escourolle & Poirier manual of basic neuropathology. 4th ed. Philadelphia: Butterworth-Heinemann; 2004.

Gros-Louis F, Gaspar C, Rouleau GA. Genetics of familial and sporadic amyotrophic lateral sclerosis. Biochim Biophys Acta. 1762;2006:956–72. Epub 2006 Feb 10. Review

Holtzman DM, Morris JC, Goate AM. Alzheimer's disease: the challenge of the second century. Sci Transl Med. 2011;3:77sr1.

Hyman BT, Phelps CH, Beach TG, et al. National Institute on Aging-Alzheimer's Association guidelines for the neuropathologic assessment of Alzheimer's disease. Alzheimers Dement. 2012;8:1–13.

Ittner LM, Gotz J. Amyloid-β and tau—a toxic pas de deux in Alzheimer's disease. Nat Rev Neurosci. 2010;12:67–72.

Josephs KA, Petersen RC, Knopman DS, et al. Clinicopathologic analysis of frontotemporal and corticobasal degenerations and PSP. Neurology. 2006;66:41–8.

Kouri N, Whitwell JL, Josephs KA, Rademakers R, Dickson DW. Corticobasal degeneration: a pathologically distinct 4R tauopathy. Nat Rev Neurol. 2011;7:263–72.

Lill CM, Abel O, Bertram L, Al-Chalabi A. Keeping up with genetic discoveries in amyotrophic lateral sclerosis: the ALSoD and ALSGene databases. Amyotroph Lateral Scler. 2011;12:238–49.

Marcolino D. Friedreich's ataxia: past, present and future. Brain Res Rev. 2011;67:311–30.

Martin I, Dawson VL, Dawson TM. Recent advances in the genetics of Parkinson's disease. Annu Rev Genomics Hum Genet. 2011;12:301–25.

Mayeux R. Early Alzheimer's disease. N Engl J Med. 2010;362:2194–201.

McCombe PA, Henderson RD. The role of immune and inflammatory mechanisms in ALS. Curr Mol Med. 2011;11:246–54.

Moore DJ, West AB, Dawson VL, Dawson TM. Molecular pathophysiology of Parkinson's disease. Annu Rev Neurosci. 2005;28:57–87.

Roy S, Zhang B, Lee VM, Trojanowski JQ. Axonal transport defects: a common theme in neurodegenerative diseases. Acta Neuropathol (Berl). 2005;109:5–13.

Seelaar H, Rohrer JD, Pijnenburg YAL, Fox NC, van Swieten JC. Clinical, genetic and pathological heterogeneity of frontotemporal dementia: a review. J Neurol Neurosurg Psychiatry. 2011;82:476–86.

Shulman JM, De Jager PL, Feany MB. Parkinson's disease: genetics and pathogenesis. Annu Rev Pathol. 2011;6:193–222.

Molecular Pathology of Lymphoma

Phillip D. Michaels

Contents

Introduction

Types of Lymphoma

- Multiple lineages of lymphocytes exist, giving rise to lymphoid neoplasms, which include clonal diseases of B-cells and T/NK-cells
- Mature B-cell and T/NK-cell neoplasms can occur at various stages of differentiation; however, some lack a specific analog to normal differentiation or exhibit lineage heterogeneity, thus making a classification based solely upon their normal counterpart not possible

Clinical Features

- Mature B-cell and T/NK-cell lymphomas are a relatively common cancer within the US
  - Estimated incidence of 19.6 per 100,000 men and women per year
  - Death rate is 5.4 per 100,000 men and women per year
- The most common staging classification for lymphoma is the Lugano modification of the Ann Arbor staging system
- Multiple prognostic tools are often employed to assist in therapy choice and clinical trial enrollment (i.e., revised International Prognostic Index (R-IPI) in diffuse large B-cell lymphoma)
- Laboratory and clinical parameters can be utilized in risk stratification and include:
  - Lactate dehydrogenase (LDH)
  - Computed tomography (CT)
  - Positron emission tomography (PET)
  - Bone marrow biopsy
  - Lumbar puncture

P. D. Michaels (✉)
Department of Pathology, Brigham and Women's Hospital and
Harvard Medical School, Boston, MA, USA
e-mail: pdmichaels@bwh.harvard.edu

L. Cheng et al. (eds.), *Molecular Surgical Pathology*, https://doi.org/10.1007/978-3-031-35118-1_24

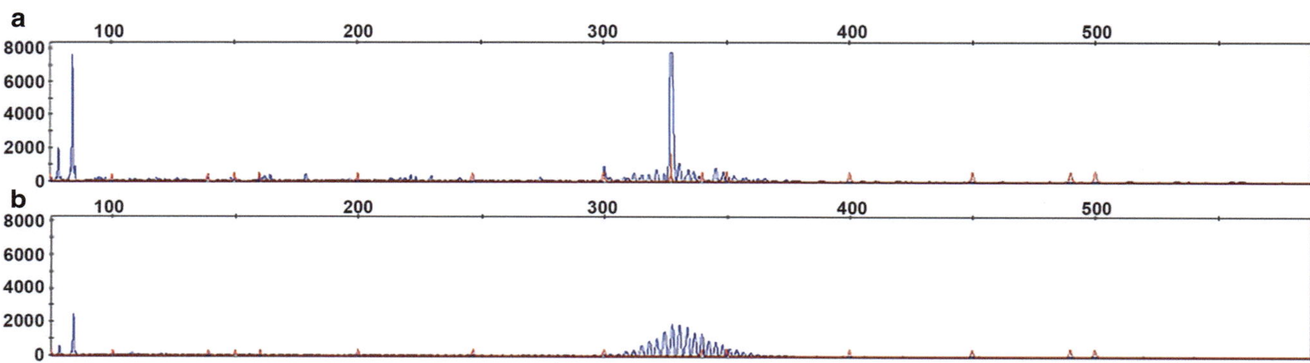

Fig. 24.1 Representative images of capillary electrophoresis tracings for *IGH* gene rearrangement studies including (**a**) clonal pattern and (**b**) polyclonal pattern

Testing Employed

- Conventional cytogenetics is a commonly utilized test that includes a karyotype and fluorescence in situ hybridization (FISH)
  - FISH can be utilized on direct preparations of cells to detect common chromosomal aberrations in lymphomas
  - FISH is also extremely useful for detecting cryptic findings that cannot be seen by karyogram
- As B-cells and T-cells mature, they undergo immunoglobulin (*IG*) and T-cell receptor (*TCR*) gene rearrangement, respectively. For diagnostically challenging cases, assessing for clonality can be a useful adjunct in determining a neoplastic process
- During maturation, the *IG* and *TCR* genes undergo rearrangements to create a diverse repertoire.
  - For IG genes, this includes rearrangements of *IGH*, *IGK*, and *IGL*
  - For TCR genes, this includes *TCRG* and *TCRB*
- Pitfalls exist for assessing clonality in the context of the differentiation of the lymphocyte.
  - For example, in *IGH* gene rearrangement studies, pre-germinal center (pre-GC) B-cell neoplasms express unmutated V_H-region genes, thus having a higher detection rate, while those that are of germinal center (GC) or postgerminal center (post-GC) derivation undergo somatic hypermutation, which leads to a lower rate of clonality detection
- Assessment for IG and TCR gene rearrangements has been utilized by polymerase chain reaction (PCR) ampli-

fication of the variable (V) and joining (J) region of antigen receptor genes (Fig. 24.1)
  - Recently, next-generation sequencing (NGS) assessment for clonal sequences of the IG or TCR amplification products has become more commonplace, particularly in the realm of minimal residual disease (MRD), and may supplant standard PCR clonality assays in the future
- NGS panels evaluating genes associated with lymphoid neoplasms are increasingly being utilized for assessing recurring gene mutations in lymphomas and may provide diagnostic, theranostic, and prognostic information
- Specialty molecular assays are available and provide additional diagnostic and theranostic information in certain contexts. These assays include:
  - Gene expression profiling
  - Gene fusion analysis (RNA sequencing)
  - Chromosomal microarray
    - Array comparative genomic hybridization (aCGH)
    - Oligonucleotide-SNP microarray

B-Cell Lymphomas

Chronic Lymphocytic Leukemia/Small Lymphocytic Lymphoma

- Chronic lymphocytic leukemia/small lymphocytic lymphoma (CLL/SLL) is a mature B-cell lymphoma that often presents with a peripheral lymphocytosis; however, a nodal-based lymphoma may also present

- *IGHV* mutational status is used to assist in prognostication
  - To differentiate between unmutated and mutated cases, if more than 2% of deviation from the germline V_H sequence is present, it is deemed mutated, whereas if less than 2% of the sequences are mutated, then it is considered unmutated
  - Unmutated cases have a worse median survival when compared to those with mutated V_H sequences
- Conventional cytogenetics play a significant role in prognostication
- Approximately 50% of CLL/SLL will have an abnormal karyotype; however, FISH is more often helpful in identifying recurrent chromosomal aberrations
- Common chromosomal aberrations identified in CLL/SLL include:
  - Trisomy 12
    - Often correlates with an abnormal immunophenotype and a more aggressive clinical course
  - Deletion of 13q14 (miR-15a/miR16-1)
  - Deletion of 11q23 (*ATM*)
    - Associated with bulky lymphadenopathy and a poor prognosis
  - Deletion of 17p13 (*TP53*)
    - Associated with a more aggressive clinical course and poor prognosis
- Assessment for mutations in *NOTCH1*, *SF3B1*, *BIRC3*, and *TP53* by targeted NGS has provided prognostic information for patient's undergoing first-line therapy
  - *BIRC3* or *TP53* mutations often have a more aggressive clinical course and may require more intensive therapy
  - *NOTCH1* and *SF3B1* mutations may be overcome with an aggressive treatment approach

Hairy Cell Leukemia

- Hairy cell leukemia (HCL) is a mature B-cell leukemia that often has hairy cells (which can be rare in number)
- Clinical presentation is characterized by monocytopenia and other cytopenias as well as splenomegaly
- A common molecular feature seen in HCL is *BRAF* p.V600E
  - Serves as a helpful diagnostic tool for differentiating between other B-cell leukemias with similar clinical and morphologic features

Nodal Marginal Zone Lymphoma

- Nodal marginal zone lymphoma (NMZL) is a mature B-cell lymphoma composed of primarily small B-cells that arises in the lymph node and may involve the bone marrow

  - When limited to the lymph node, the main differential diagnosis includes follicular lymphoma
  - When involving the bone marrow, the main differential diagnosis includes lymphoplasmacytic lymphoma
- The most common chromosomal aberrations in NMZL include:
  - Trisomy 3
  - Trisomy 7
  - Trisomy 18
- Gene mutations in NMZL include:
  - *MLL2* (35% of cases)
    - Enriched in NMZL
  - *PTPRD* (20% of cases)
    - Unique to NMZL
  - *NOTCH2* (20% of cases)
  - *KLF2* (15–20% of cases)
  - *MYD88* (5% of cases)

Extranodal Marginal Zone Lymphoma of Mucosa-Associated Lymphoid Tissue (MALT Lymphoma)

- Extranodal marginal zone lymphoma of mucosa-associated lymphoid tissue (MALT lymphoma) is a mature B-cell lymphoma that arises in extranodal organs because of infectious agents or autoimmune diseases. They may display lymphoepithelial lesions as well as an expanded lymphoid population with a marginal zone pattern
- 30–40% of MALT lymphomas display a recurrent chromosomal rearrangement (Table 24.1)
- NF-kB activation can be seen in 20% of MALT lymphomas due to inactivation of *TNFAIP3* on chromosome 6p23; however, this has been identified in other marginal zone lymphomas

Splenic Marginal Zone Lymphoma

- Splenic marginal zone lymphoma (SMZL) is a small B-cell lymphoma that presents with splenic involvement as well as invariably involving the bone marrow and peripheral blood, often characterized by polar villi

Table 24.1 Translocations identified in MALT lymphoma

Translocation	Frequency (%)	Site
t(11;18) (q21;q21)/*BIRC3-MALT1*	20–30	Lung, stomach, ocular adnexa
t(14;18) (q32;q21)/*IGH-MALT1*	10–15	Ocular adnexa, skin, liver
t(3;14) (p14.1;q32)/*IGH-FOXP1*	10	Thyroid, ocular adnexa, skin
t(1;14) (p22;q32)/*IGH-BCL10*	5	Lung, small intestine

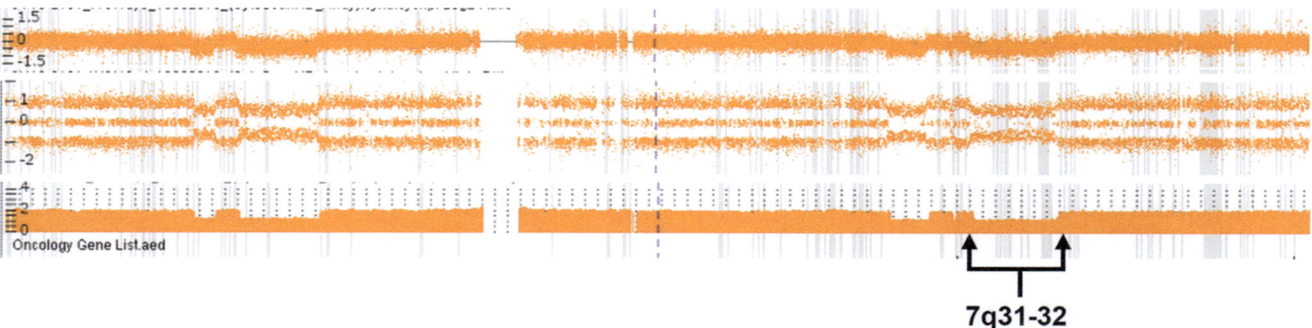

Fig. 24.2 Representative microarray analysis for a splenic marginal zone lymphoma demonstrating deletion of 7q31-32 (courtesy of Dr. C. Bryke, Beth Israel Deaconess Medical Center)

- A subset of SMZL is associated with hepatitis C virus (HCV) infection, which is thought to contribute to lymphomagenesis by mixed cryoglobulinemia, which is characterized by a monoclonal IgM with rheumatoid factor activation
- Karyotypic analysis reveals chromosomal aberrations in greater than 70% of cases, with more than 50% having complex karyotypes
  - The most common chromosomal abnormalities include:
    - Deletion of 7q31-32 (40% of cases)
      - This deletion can be difficult to detect by karyogram and FISH, at which point chromosomal microarray may be help for the minimally deleted region (Fig. 24.2)
  - Other less common chromosomal changes include:
    - Trisomy 3q
    - Deletion of 6q
    - Trisomy 12
    - Translocations of chromosomes:
      - 8q
      - 1q
      - 14q
  - Chromosomal changes that confer a poor prognosis include a complex karyotype (2 or greater chromosomal aberrations), 14q alterations, and deletion of *TP53*
- Molecular pathogenesis of splenic marginal zone lymphoma demonstrates some overlap with other marginal zone lymphomas, particularly in the activation of the NF-kB pathway by copy number changes or mutations which is observed in approximately one-third of cases
- Activation of NOTCH signaling by *NOTCH2* gain-of-function mutations has been identified in 25% of cases

Lymphoplasmacytic Lymphoma

- Lymphoplasmacytic lymphoma (LPL) is a primarily bone marrow-based small B-cell lymphoma that displays a spectrum of mature B-cells with plasmacytoid differentiation
  - A major subset of LPLs displays a monoclonal IgM with bone marrow involvement, which is termed Waldenstrom macroglobulinemia (WM)
- A helpful molecular assay for differentiating LPL from other small mature B-cell lymphomas is analysis for *MYD88*, with p.L265P being the most common variant (identified in greater than 90% of cases)
  - *MYD88* mutations are less common in non-IgM-secreting lymphoplasmacytic lymphoma, with approximately 40–45% of cases carrying the mutation
- MYD88 variants lead to activation of the Toll-like receptors and IL-1b, resulting in activation of the NF-kB pathway promoting cell proliferation
- Additional mutations identified are warts, hypogammaglobulinemia, infection, and myelokathexis (WHIM) syndrome-like mutations in CXCR4 that are identified in addition to MYD88 mutations
  - *CXCR4* variants have implications for disease progression and resistance to therapy
- Cytogenetic findings identified in LPL include:
  - Deletion of 6q (40–50% of cases)
  - Trisomy 18
  - Deletion of 13q
  - Deletion of 17p (*TP53*)
  - Trisomy 4
  - Trisomy 12
  - Deletion of 11q (*ATM*)

Plasma Cell Myeloma

- Plasma cell myeloma (PCM) is often preceded by monoclonal gammopathy of undetermined significance (MGUS) and is identified on bone marrow biopsies as an increase in plasma cells with a monoclonal protein and osteolytic bone lesion (Table 24.2)
- Conventional cytogenetic analysis is a mainstay for the assessment of plasma cell myeloma
- Within the laboratory, bone marrow aspirates often undergo magnetic cell separation to enrich plasma cells, which yields better chromosomal analysis
- As plasma cells traditionally have a low proliferative index, stimulation with IL-4 is often utilized
- Within hyperdiploid PCM, the NF-kB pathway is often activated
- For nonhyperdiploid PCM, secondary genetic events include activation of:
  - NRAS
  - KRAS
  - BRAF
  - FGFR3
- Myc dysregulation and *TP53* inactivation aid in disease progression (Table 24.3)

Mantle Cell Lymphoma

- Mantle cell lymphoma (MCL) is a CD5-positive B-cell lymphoma that often expresses cyclin D1 and SOX11
- Present with a variable clinical course, including significant lymphadenopathy

Table 24.2 Plasma cell myeloma diagnostic criteria

>10% clonal plasma cells in bone marrow
Or
Biopsy-proven plasmacytoma (medullary or extramedullary)
AND
End-organ damage because of PC neoplasm
Or
Biomarker of malignancy
• 60% or greater BM plasma cells
• FLC ratio > 100:1.
• >1 focal bone lesion on MRI.

PC plasma cell, *BM* bone marrow, *FLC* free light chain, *MRI* magnetic resonance imaging

- Morphologically, the lymph node is effaced by small to intermediate-sized lymphocytes with prominent hyalinized blood vessels and histiocytes with eosinophilic cytoplasm (Fig. 24.3a–c)
  - There are two particularly aggressive types of mantle cell lymphoma:
    - Blastoid variant
    - Pleomorphic variant
  - A prognostic feature identified in routine histology is the proliferative index
- The characteristic genetic lesion identified in most MCL cases is the t(11;14)(q13;q32)/*IGH-CCND1* (Fig. 24.3d)
  - It involves the juxtaposition of the *IGH* enhancer element to *CCND1* leads to cyclin-D1 overexpression, Rb phosphorylation, and release of E2F, thus promoting cell cycle progression from G1 to S phase (Fig. 24.4)
- However, a subset of mantle cell lymphoma cases fails to express cyclin D1 or harbor the *IGH-CCND1* translocation
  - In these cases, the most common genetic aberration is a translocation of *CCND2* with light chains (*IGK* or *IGL*)
- Other genetic aberrations include mutations in:
  - TP53
    - Often accompanied by 17p13 deletion leading to biallelic inactivation of *TP53*
  - ATM
  - CCND1
  - Less common mutations: WHSC1, MLL2, BIRC3, MEF2B, and TLR2
  - Rarely, NOTCH2 variants have been identified and appear to be exclusively seen in the blastoid or pleomorphic variants

Follicular Lymphoma

- Follicular lymphoma (FL) is a B-cell lymphoma of germinal center origin and is the second most common mature B-cell lymphoma in the United States and Europe
- There is variability in the clinical course of follicular lymphoma due to different histologic grades, stages, subtypes, and progression to an aggressive B-cell lymphoma (which occurs in 25% of cases)
- Follicular lymphoma subtypes include:

Table 24.3 Common plasma cell myeloma chromosomal aberrations

Hyperdiploid	Nonhyperdiploid				Other abnormalities
gain of odd number chromosomes	cyclin Ds	*FGFR3/MMSET*		MAFs	Gain of 1q21
3, 5, 7, 9, 11, 15, 19, and 21	11q13/*CCND1*	4p16		16p23/c-maf	del(13)(q14)
	12p13/*CCND2*			20q12/*MAF-B*	Trisomy 21
	6p21/*CCND3*			8q24/*MAF-A*	Trisomies 3, 5
					del(17)(p13)/*TP53*

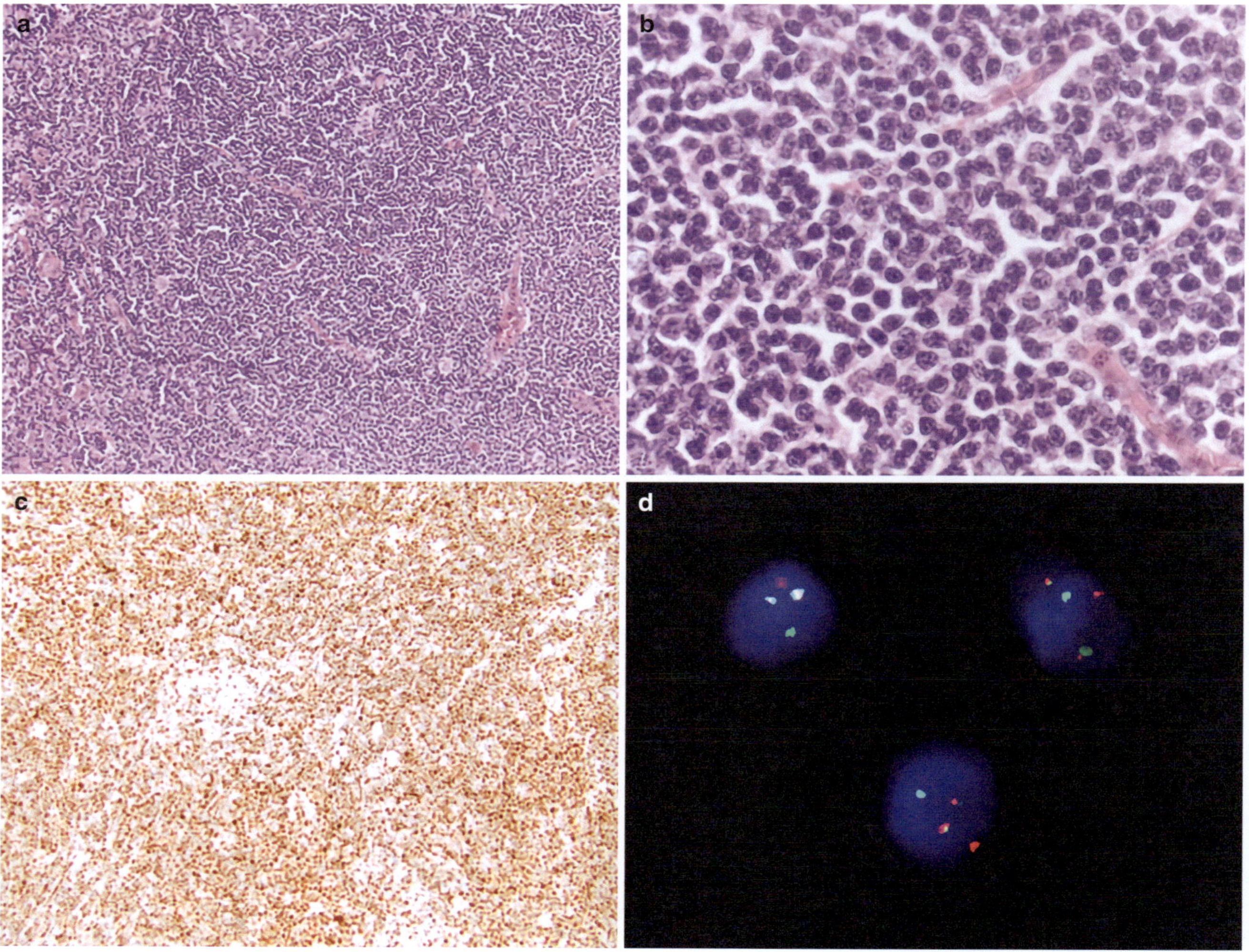

Fig. 24.3 (**a**) Mantle cell lymphoma with prominent hyalinized vessels and histiocytes (200×) (**b**) Morphologic features of mantle cell lymphoma cells include intermediate-sized lymphocytes with irregular nuclear contours, dispersed chromatin, and inconspicuous nucleoli (1000×) (**c**) Immunohistochemistry for CCND1 demonstrating positivity within lymphoma cells (**d**) FISH analysis using an *IGH-CCND1* dual fusion probe in which two fusions are present indicating the presence of t(11;14)(q13;q32)/*IGH-CCND1* (courtesy of Dr. C Bryke, BIDMC)

- Primary cutaneous follicle center lymphoma
- Pediatric-type follicular lymphoma
- Primary duodenal follicular lymphoma
- Primary testicular follicular lymphoma
- Grading of follicular lymphoma is based enumeration of centroblasts in 10 high power fields (HPF)
  - Grade 1 and grade 2 display similar outcomes and are conventionally lumped together as grade 1–2
    - Grade 1: 0–5 centroblasts/HPF
    - Grade 2: 6–15 centroblasts/HPF
  - Grade 3 portends a more aggressive clinical course and is further stratified by the presence of centrocytes within a centroblastic population (Fig. 24.5a, b)
    - Grade 3A: >15 centroblasts/HPF with residual centrocytes
    - Grade 3B: Sheet of centroblasts with no residual centrocytes
- Architectural pattern should be established; however, this may be challenging on core biopsies
  - If grade 3 features are observed in a diffuse area, by definition, the diagnosis is diffuse large B-cell lymphoma
  - Follicular: >75% follicular
  - Follicular and diffuse: 25–75% follicular
  - Diffuse: <25% follicular
- The mainstay of therapy is traditionally R-CHOP (rituximab, cyclophosphamide, doxorubicin, vincristine, and prednisone). However, clinical trials for targeted therapies are ongoing for relapsed/refractory FL

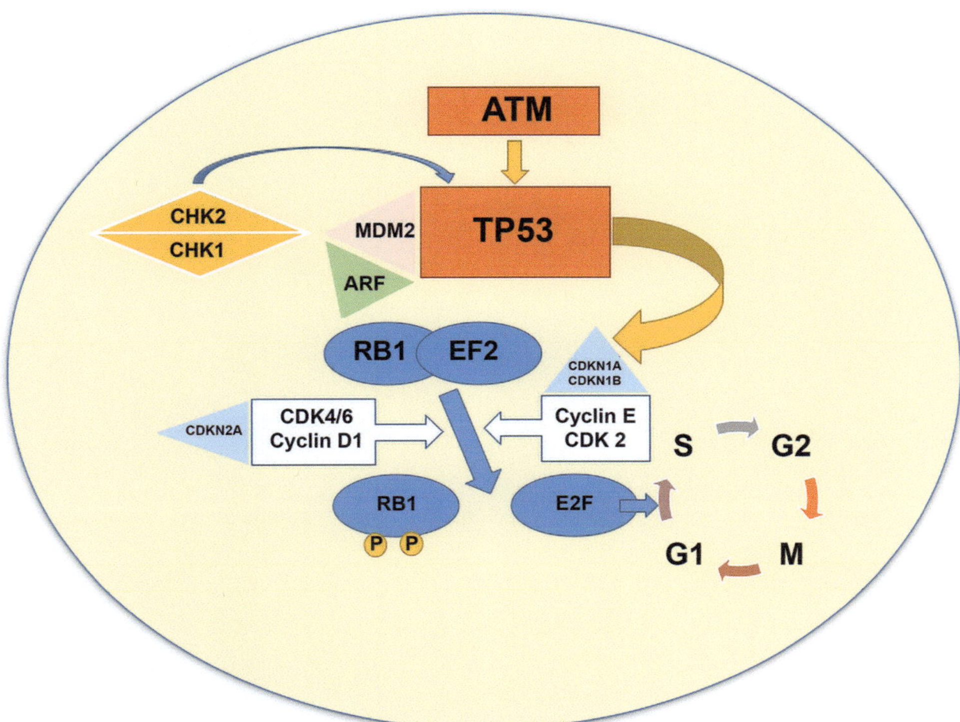

Fig. 24.4 Schematic of cell cycle and DNA damage repair mechanisms present in mantle cell lymphoma (MCL). Cyclin D1 and CDK4/6 complex induce phosphorylation of RB1, leading to E2F, allowing for the progression of the cell cycle from G1 to S phase. The cyclin D1 and CDK4/6 complex is inhibited by CDKN2A. Additionally, Cyclin E and CDK 2 lead to cell cycle progression, which is inhibited by CDKN1A and CDKN1B. For DNA damage repair, TP53 is stabilized by ARF, which inhibits the activity of MDM2, a negative regulator of TP53. ATM is necessary for the activation of TP53 when DNA is damaged. CHK1 and CHK2 are involved in the phosphorylation of TP53, leading to stabilization and decreased degradation. Deleterious events in ATM, TP53, CDKN2A, or overexpression of MDM2 is often observed in MCL

- BCL2 is an antiapoptotic protein that confers cell survival, thus by immunohistochemistry, BCL2 is often positive in the neoplastic follicles of follicular lymphoma (Fig. 24.5c)
- The hallmark chromosomal aberration of FL is t(14;18)(q32;q21)/*IGH-BCL2*, which juxtaposes the *IGH* enhancer element with *BCL2* with multiple breakpoints occurring in BCL2 (Fig. 24.5d)
  - Major breakpoint region (MBR): 50–70% of cases
  - Minor cluster region (mcr): 5–15% of cases
  - Intermediate cluster region (icr): ~13% of cases
- *IGH-BCL2* alone is insufficient to induce lymphomagenesis, and multiple other genetic hits must occur for progression to lymphoma
- Approximately 15% of follicular lymphomas display a rearrangement of *BCL6* on chromosome 3q27, which is more common in grade 3 FL
- Multiple gene mutations occur in follicular lymphoma, with seven being considered prognostic in the modified Follicular Lymphoma International Prognostic Index (m7-FLIPI)
- The most common mutations occur in the chromatin modification group, with *KMT2D* being identified in 80% of cases
- Other genes involved in epigenetic regulation include:
  - CREBBP
  - EZH2
  - EP300

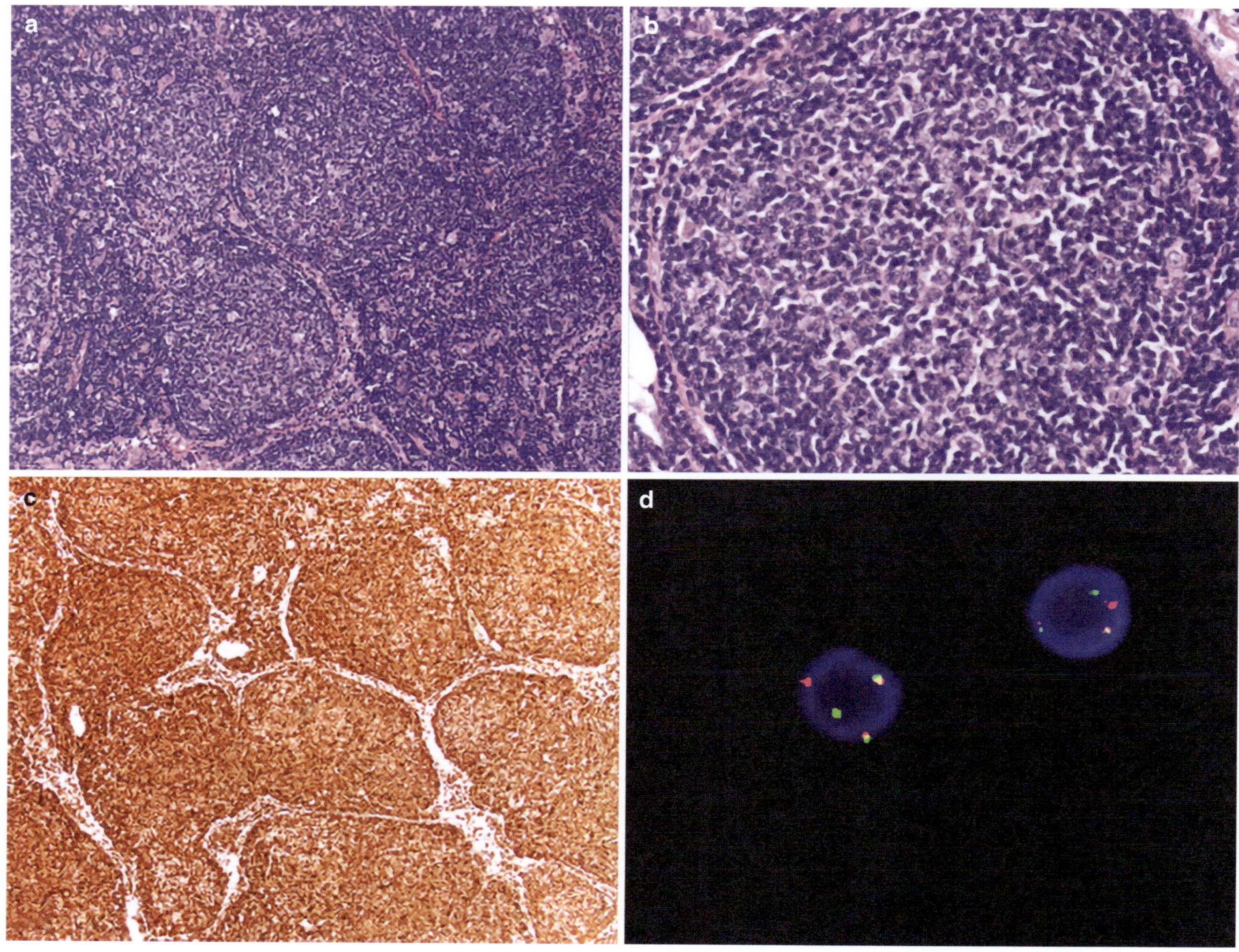

Fig. 24.5 (**a**) Follicular lymphoma with a follicular pattern (200×) (**b**) Neoplastic follicle with centroblasts enumerating >15/HPF (**c**) Immunohistochemistry for BCL2 demonstrating positivity within neoplastic follicles (**d**) *IGH-BCL2* dual fusion probe set demonstrating two fusions, indicating the presence of t(14;18)(q32;q21)/*IGH-BCL2* (courtesy of Dr. C. Bryke, BIDMC)

- Copy number alterations in follicular lymphoma include:
  - Gains: 7, 18, X
  - Losses: 1p36, 6q, 17p13, 9p21.3, 1p16 (LOH)
    - Poor prognosis: Loss 17p13, loss 9p21.3, loss 6q

Diffuse Large B-Cell Lymphoma, NOS

- Diffuse large B-cell lymphoma, NOS (DLBCL), is the most common mature B-cell lymphoma in adults and comprises 30–40% of new non-Hodgkin lymphoma cases
- Diffuse large B-cell lymphoma is lymphoma that displays a diffuse proliferation of large lymphocytes with variable morphologic features
- Morphologic types include:
  - Centroblastic variant (85% of cases)
  - Immunoblastic variant (10% of cases)
  - Anaplastic variant (3% of cases)
- There is heterogeneity in presentation and outcomes; however, with the advent of determining cell-of-origin (COO) by gene expression profiling, two primary groups of diffuse large B-cell lymphoma have been identified: germinal center B-cell-like (GCB) and activated B-cell-like (ABC)
- Determination of COO has provided prognostic information with GCB having a 5-year overall survival of 60% while the ABC group has a 5-year overall survival rate of 35%
- As gene expression profiling is not a routine part of the evaluation of lymphomas, multiple immunohistochemical surrogates have been identified, with the Hans algorithm being the most used

Table 24.4 Common somatic mutations in diffuse large B-cell lymphoma, NOS

Gene	Pathway	Cluster	COO
TP53	Tumor suppressor	C2	NA
KMT2D	Chromatin modifiers	C3	GCB
CREBBP			
EP300			
EZH2			
TNFRSF14			
MEF2B	Transcription factor		
PTEN	Signaling		
MTOR			
IRF8	Immune modulators		
CIITA			
CD58	Immune modulators	C4	GCB
CD70			
CD83			
CARD11	NF-kB signaling		
RHOA	RHOA signaling		
GNA13			
SGK1			
STAT3	JAK/STAT signaling		
KRAS	RAS pathway		
BRAF			
CD79B	BCR signaling	C5	ABC
MYD88	NF-kB signaling		
PIM1	Cell signaling		
ETV6	Transcription		
TBL1XR1			
PRDM1			
CDKN2A	Cell cycle		
BTG1			
BCL2	Apoptosis		
BCL6	Transcription factor	C1	ABC
NOTCH2	NOTCH signaling pathway		
SPEN			
DTX1			
B2M	Immune modulator		
FAS	NF-kB signaling		
TNFAIP3 (A20)			
BCL10			
PRKCB			

Cluster adapted from Chapuy B et al. 2018; COO cell-of-origin

- In reporting the COO by immunohistochemistry, it is important to document the algorithm utilized in the final interpretation
- Newer genomic profiling studies have further stratified diffuse large B-cell lymphoma into subgroups, particularly within the context of GCB and ABC types (Table 24.4)
- Cytogenetic analysis reveals multiple chromosomal aberrations are seen, including gene rearrangements and copy number changes (Tables 24.5 and 24.6)
- Other large B-cell lymphomas exist under the auspices of the 2017 WHO revised 4th ed. and display significant phenotypic and genotypic heterogeneity (Table 24.7)

Table 24.5 Frequent chromosomal rearrangements in diffuse large B-cell lymphoma, NOS

Gene(s)	Frequency (%)	Prognosis
3q27/*BCL6*	20–45	More favorable
t(14;18)/*IGH-BCL2*	21	Worse
1p22/*BCL10*	18	NA
8q24/*MYC*	16	Worse
9p24.1/*PDL1,PDL2*	5	More favorable

Table 24.6 Frequent copy number changes in diffuse large B-cell lymphoma, NOS

Locus	Gene(s) altered	Frequency (%)
17p13	Deletion *TP53*	24
13q14	Deletion *RB1* and miR-15/16	7–14
9p21	Deletion *CDKN2A/B*	35–50
18q21	Gain of *BCL2*	5
10q23	Loss of *PTEN*	8
2p16	Gain/amp of *REL*	15
18p	Gain	23

Table 24.7 Other large B-cell lymphoma subtypes and associated genetic lesions

Lymphoma type	Gene(s) involved	Genetic aberration
Primary diffuse large B-cell lymphoma of the CNS	*BCL6*	Somatic variant or rearrangement
	CD79B	Somatic variants
	CARD11	
	MYD88	
	PRDM1	
Primary cutaneous diffuse large B-cell lymphoma, leg type	*IGH-MYC*	Translocation
	IGH-BCL6	
	BCL2	Amplification
	MALT1	
	CDKN2A/B	Loss
	MYD88	Somatic variants
	CARD11	
	CD79B	
	TNFAIP3	
Diffuse large B-cell lymphoma associated with chronic inflammation	*TP53*	Somatic variant
	MYC	Amplification
	TNFAIP3	
Primary mediastinal (thymic) large B-cell lymphoma	*CIITA*	Somatic variant or rearrangement
	PDL1	Rearrangement (often with *CIITA*)
	PDL2	
	JAK2/PDL1/PDL2	Amplification
	REL	Gains
	BCL11A	
	TNFAIP3	Somatic variant
	SOCS1	
	STAT6	
	PTPN1	
	BCL6	
ALK-positive large B-cell lymphoma	*CLTC-ALK*	Translocation
Plasmablastic lymphoma	*MYC*	Translocation (often with IG genes)

Burkitt Lymphoma

- Burkitt lymphoma (BL) is an aggressive B-cell lymphoma that has three unique clinical subtypes
- Endemic variant of BL is prevalent among younger children, presents most commonly in equatorial Africa, and is Epstein-Barr virus (EBV) associated in greater than 95% of cases
- Sporadic BL is found more frequently in adolescents and young adults with less frequent association with EBV (approximately 5–30% of cases)
- Immunodeficiency-associated BL is seen most commonly in association with human immunodeficiency virus (HIV), and in 25–40% of cases, clonal EBV is detected
- The hallmark genetic aberration of BL is *MYC* dysregulation
  - Most commonly, the *IGH* locus is translocated to *MYC*, juxtaposing the *IGH* enhancer element and leading to MYC overexpression
  - *IGH-MYC* translocations are seen in 80% of all Burkitt lymphoma while the remaining cases involve *MYC* and the light chains
- A provisional entity of "Burkitt-like lymphoma with 11q aberration" has been described
- These cases resemble BL phenotypically; however, they lack *MYC* rearrangements and have a characteristic chromosome 11q alteration
  - Proximal gains at 11q23.2-23.3 and terminal losses at 11q24.1-ter
- These lymphomas typically have a complex abnormal karyotype, which is not commonly seen in BL

High-Grade B-Cell Lymphoma

- High-grade B-cell lymphoma encompasses two unique situations
  - An aggressive B-cell lymphoma that harbors an *MYC* and *BCL2* and/or *BCL6* rearrangement, colloquially known as a "double-hit" or "triple-hit" lymphoma
    - These lymphomas typically display the morphologic features reminiscent of diffuse large B-cell lymphoma; however, less commonly, they can display blastoid features
  - An aggressive B-cell lymphoma that displays blastoid features or morphologic features between that of Burkitt lymphoma and diffuse large B-cell lymphoma but does not harbor a "double-hit" or "triple-hit"
- High-grade B-cell lymphoma with *MYC* and *BCL2* and/or *BCL6* rearrangements are conventionally detected by FISH analysis
  - In 65% of cases, *MYC* is juxtaposed to the IG genes, usually *IGH,* and, less commonly, the light chains. In the remaining cases, MYC has a non-IG gene partner, which includes 9p13 and 3q27/*BCL6*
- High-grade B-cell lymphoma, NOS has had relatively few systematic studies evaluating its genomic profile

- It lacks an *MYC* and *BCL2* and/or *BCL6* rearrangement; however, some studies reveal an *MYC* rearrangement in 20–35% of cases and rarely with *BCL2* amplification

---

T/NK-Cell Lymphomas

T-Cell Prolymphocytic Leukemia

- T-cell prolymphocytic leukemia (T-PLL) is an aggressive mature T-cell lymphoma composed of small to intermediate-sized lymphocytes that present with a profound lymphocytosis and significant bone marrow involvement
- Additional sites of involvement include lymph nodes, liver, and spleen with skin being less common
- By immunophenotypic analysis, most cases are CD4 positive (60% of cases), while less commonly CD4 and CD8 positive (25%) and CD8 positive (15%)
- Although prognosis is poor, the best responses to therapy have been observed in alemtuzumab (anti-CD52)
  - If remission is achieved following immunotherapy, autologous or allogeneic stem cell transplant is often considered
- The lymphomagenesis of T-PLL frequently involves the juxtaposition of the enhancer element of *TRA* locus with *TCL1*, which encodes an oncoprotein and becomes overexpressed in 80% of cases (Fig. 24.6)
- Less commonly, t(X;14)(q28;q11.2) involving *MTCP1*, an oncogene, and the *TRA* locus is identified
- Additional cytogenetic features include:
  - Loss of *ATM* (11q22-23)
  - Abnormalities of chromosome 8 (70–80% of cases)
  - *TP53* (17p13.1) gene deletion
  - Deletion of *CDKN1B* on chromosome 12p13
- Whole-exome, whole-genome sequencing, and targeted sequencing studies of T-PLL have revealed recurrent alterations in the JAK/STAT signaling pathway
  - Mutations include:
    - *JAK3* (30–42% of cases)
    - *JAK1* (8% of cases)
    - *IL2RG* (2% of cases)
    - *STAT5B* (21–36% of cases)
    - *TP53* (often with biallelic inactivation)
    - *ATM* (70–73% of cases)
    - *EZH2*
    - *BCOR*

T-Cell Large Granular Lymphocytic Leukemia

- T-cell large granular lymphocytic leukemia (T-LGLL) is a chronic and indolent cytotoxic T-cell disorder that commonly involves the peripheral blood, bone marrow, liver, and spleen

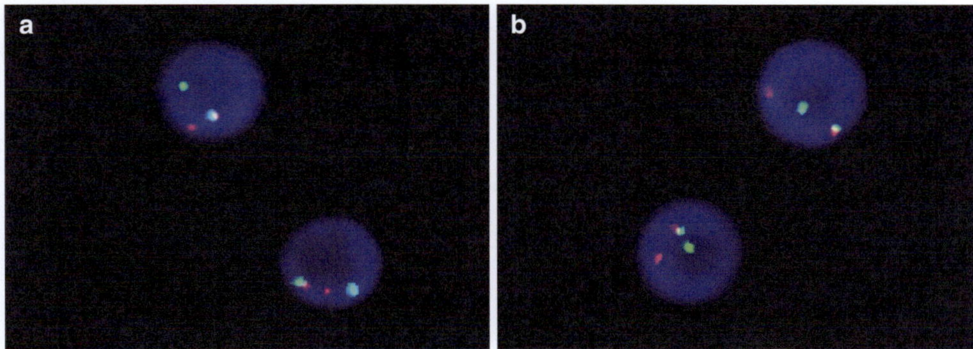

Fig. 24.6 Representative FISH images of break-apart probes for (**a**) *TCL1* and (**b**) *TRA/D*, indicating the presence of rearrangements often identified in T-cell prolymphocytic leukemia (courtesy of Dr. C. Bryke, BIDMC)

Table 24.8 Frequent somatic genetic alterations in adult T-cell leukemia/lymphoma

Pathway	Gene	Type(s) of aberration
TCR and NF-KB signaling	*PLCG1*	Nonsynonymous mutation
	PRKCB	Nonsynonymous mutation
	CARD11	Nonsynonymous mutation and gain
	CD28	Nonsynonymous mutation and gain
	VAV1	Nonsynonymous mutation
	RHOA	Nonsynonymous mutation
	FYN	Nonsynonymous and indel mutations
	IRF4	Nonsynonymous mutation and gain
	CBLB	Nonsynonymous mutation and loss
	TNFAIP3	Indel mutations and loss
	CSNK1A1	Multiple types
GPCR (T-cell trafficking)	*CCR4*	Indel and nonsense mutations
	CCR7	Nonsense mutations
	GPR183	Indel mutations and loss
Other signaling pathways	*STAT3*	Nonsynonymous mutations
	CSNK2A1	Multiple types
	CSNK2B	Nonsense mutations
	NOTCH1	Indel and nonsense mutations
	ATXN1	Indel mutations and loss
Transcriptional regulation	*BCL11B*	Gain
	GATA3	Indel mutations and loss
	CEBPA	Loss
	PRDM1	Loss
	IKZF2	Loss
	ZNF638	Nonsynonymous and indel mutations
	TBL1XR1	Nonsynonymous mutations and loss
	IRF2BP2	Multiple types
	HNRNPA2B1	Multiple types
Immune surveillance	*CD274*	Gain
	PDCD1	Loss
	CD58	Indel mutations and loss
	B2M	Indel mutations
	HLA-A	Loss
	HLA-B	Loss
	FAS	Multiple types
Epigenetic regulation	*TET2*	Multiple types
	EP300	Nonsynonymous and indel mutations
	SETD2	Nonsynonymous and indel mutations
	ARID2	Loss
Telomere regulation and DNA repair	*TP53*	Multiple types and loss
	CDKN2A	Loss
	POT1	Multiple types
Fragile sites	*NRXN3*	Loss

- It is characterized by a persistent (>6 months) and clonal population of large granular lymphocytes (LGLs)
- A common presentation is neutropenia and anemia, often in the setting of pure red cell aplasia
- T-LGLL has an established relationship with autoimmune disorders, particularly rheumatoid arthritis
- Additional features that provide a diagnostic challenge in cases of suspected T-LGLL are in patients with an allogeneic stem cell transplant, as LGLs often expand in lymphocyte reconstitution
- Clonal expansion of LGLs also occurs in association with low-grade B-cell lymphomas, particularly CLL/SLL and HCL
- *STAT3* variants have been identified in approximately one-third of cases, with the SH2 (Src homology 2) domain the most common site of involvement, particularly codons Y640 and D661
- *STAT5B* variants in the SH2 domain have been identified with the N642H mutation likely conferring a more aggressive clinical course

Adult T-Cell Leukemia/Lymphoma

- Adult T-cell leukemia/lymphoma (ATLL) is an aggressive mature T-cell lymphoma driven by human T-cell leukemia virus type 1 Human T-lymphotropic virus type 1 (Table 24.8)

- The disease is often disseminated at presentation and by morphologic examination, the lymphoma cells are highly pleomorphic
- Common sites of involvement include widespread lymph node involvement and peripheral blood dissemination, as well as the bone marrow, spleen, skin, lungs, liver, gastrointestinal tract, and CNS
- Although HTLV-1 is associated with ATLL, it alone is insufficient to cause lymphoma and only 2.5% of those infected with HTLV-1 have ATLL
- There are four categories involving the clinical spectrum of ATLL
  - Smoldering
  - Chronic
  - Acute
  - Lymphomatous
- The acute and lymphomatous forms have a more aggressive clinical course while the smoldering and chronic forms have a more protracted course, although 25% of smoldering and chronic cases progress to an acute phase
- The HTLV-1 Tax protein plays a critical role in leukemogenesis and activates a variety of signaling pathways (Fig. 24.7)
- Through RNA sequencing analysis, fusions were identified that effect TCR signaling and other key pathways necessary for lymphocyte function
  - They include *CTLA4-CD28* and *ICOS-CD28*, which lead to continuous and enhanced CD28 costimulatory signaling and have served as a potential target for anti-CTLA4 immunotherapy

Extranodal NK/T-Cell Lymphoma, Nasal Type

- Extranodal NK/T-cell lymphoma, nasal type (ENKTL) is a primarily extranodal lymphoma that is either of the NK-cell or T-cell lineage and is characterized by EBV infection and local destruction
- It is most prevalent in individuals of Asian descent and those that are indigenous to Mexico, Central and South America
- Morphologic features demonstrate vascular damage with an angiodestructive pattern, necrosis, and by immunohistochemistry, a cytotoxic phenotype is present
- *TCR* and *IG* gene rearrangements are often in the germline configuration, with 10–40% of cases showing a clonal *TCR* gene rearrangement, most often associated with cases demonstrating a cytotoxic T-cell immunophenotype
- Multiple copy number changes have been identified; however, no specific chromosomal change has been implicated in ENKTL (Table 24.9)
- Many types of mutations have been reported throughout different cohorts assessed by Sanger sequencing, next-generation sequencing, and methylation studies (Table 24.10)

Fig. 24.7 Venn diagram indicating the overlap of somatic variants within adult T-cell leukemia/lymphoma and those involved in the Tax protein interactome (adapted from Kataoka et al. 2015)

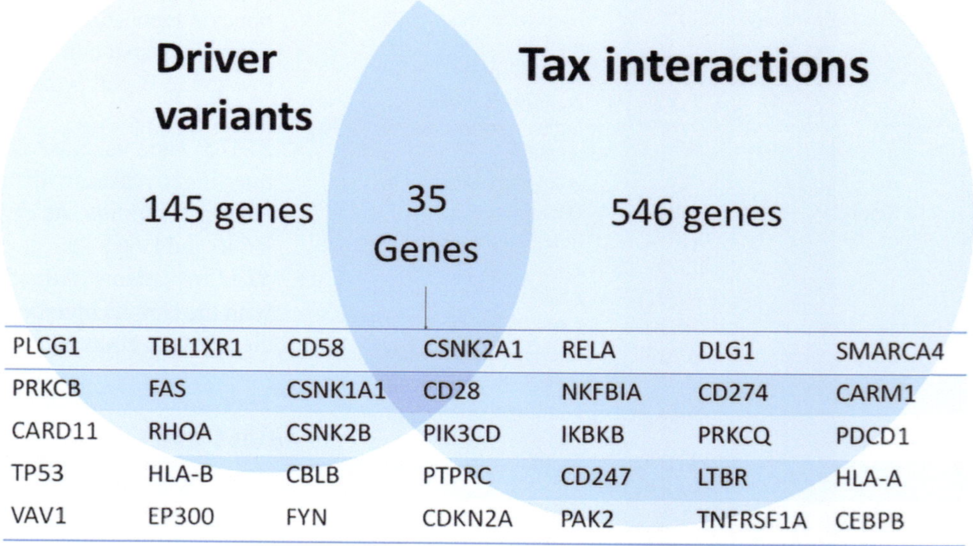

PLCG1	TBL1XR1	CD58	CSNK2A1	RELA	DLG1	SMARCA4
PRKCB	FAS	CSNK1A1	CD28	NKFBIA	CD274	CARM1
CARD11	RHOA	CSNK2B	PIK3CD	IKBKB	PRKCQ	PDCD1
TP53	HLA-B	CBLB	PTPRC	CD247	LTBR	HLA-A
VAV1	EP300	FYN	CDKN2A	PAK2	TNFRSF1A	CEBPB

Table 24.9 Common copy number alterations in extranodal NK/T-cell lymphoma, nasal type

Locus	Copy number alteration
2q	Gain
1p36.23-36.33	Loss
6q16.1-27	Loss
4q12	Loss
5q34-35.3	Loss
7q21.3-22.1	Loss
11q22.3-23.3	Loss
15q11.2-14	Loss

Table 24.10 Common mutations in extranodal NK/T-cell lymphoma, nasal type

Pathway	Gene	Description of variant
Ribosomal assembly	*DDX3X*	Nonsense or missense leading to decrease of RNA unwinding
JAK/STAT signaling	*JAK3*	Gain-of-function variant
	STAT3	Gain-of-function variant
	STAT5B	Gain-of-function variant
	PTPRK	Promoter hypermethylation or monoallelic deletion
Wnt/B-catenin signaling	*CTNNB1*	Exon 3 variants leading to pathway activation
SCF/c-kit signaling	*KIT*	Gain-of-function variant
Tumor suppressor	*TP53*	Loss of function variant
	MGA	
	PRDM1	Deletion and promoter hypermethylation
	ATG5	
	AIM1	
	FOXO3	
	HACE1	
Oncogenes	*KRAS*	Gain-of-function variant
	MYC	
Epigenetic modifiers	*KMT2D*	Loss of function variant
	ARID1A	
	EP300	
	ASXL3	
	BCOR	
Cell cycle regulators	*CDKN2A*	Hypermethylation
	CDKN2B	
	CDKN1A	
Apoptosis	*FAS*	Deletion

Angioimmunoblastic T-Cell Lymphoma and Other T-Cell Lymphomas of T Follicular Helper Cell Origin

- Angioimmunoblastic T-cell lymphoma (AITL) and a subset of peripheral T-cell lymphomas (PTCLs) share a T follicular helper (TFH) cell phenotype and are the neoplastic counterpart of physiologic TFH cells
- TFH markers include CD10, CXCL13, BCL6, ICOS, CXCR5, SAP, MAF, and CD200 (most cases)
- Gene expression signature of TFH cells are identified in AITL and PTCLs that demonstrate a TFH phenotype

Table 24.11 Morphologic patterns of angioimmunoblastic T-cell lymphoma

Pattern	Description
1	Neoplastic cells surround hyperplastic follicles with an attenuated mantle zone
2	Follicles remain but show regressive changes, with neoplastic cells more visible
3	Architecture is effaced with an expanded paracortex with focal regressive follicles still present at the periphery

Table 24.12 Common somatic variants in angioimmunoblastic T-cell lymphoma

Pathway	Gene	Description	Frequency (%)
Signal transduction	*RHOA*	Hot spot missense variant p.G17V	60–70
T-cell signaling	*CD28*	Hot spot missense variants p.D124 and P.T195	5–10
	PLCG1	p.S345F variant	5–10
	FYN	Kinase domain variants	5–10
Epigenetic regulation	*TET2*	Loss of function	50–80
	DNMT3A	Loss of function	20–30
	IDH2	p.R172 hot spot variant appears specific for AITL; less commonly p.R140	20–30

- AITL is a systemic disease that is characterized by lymph node involvement by polymorphous lymphocytes with pale cytoplasm, prominent high endothelial venules (HEVs), and follicular dendritic cell (FDC) meshwork, and a reactive cellular infiltrate. Additionally, most cases (80–95%) are associated with an EBV-positive B-cell immunoblast proliferation in the paracortex
- There are three morphologic patterns of AITL recognized (Table 24.11)
- Common chromosomal aberrations detected by conventional cytogenetics have been reported in up to 90% of cases and include:
  - Trisomies of chromosomes 3, 5, and 21
  - Gain of chromosome X
  - Loss of chromosome 6q
- aCGH reveals additional findings such as recurrent gains of 22q, 19, and 11q13 and loss of 13q
- Structural variants have also been identified in AITL and include *CTLA4-CD28* (58% cases) and rarely *ITK-SYK*, which may confer sensitivity to SYK inhibitors
- PTCLs of TFH cell origin also share many similarities with AITL in their genomic profile, including frequent mutations in *TET2*, *DNMT3A*, and *RHOA* (Tables 24.12).

Peripheral T-Cell Lymphoma, NOS

- Peripheral T-cell lymphoma, NOS (PTCL, NOS) is a heterogeneous group of T-cell lymphomas that do not correspond to a specific category previously mentioned

Table 24.13 Common copy number alterations in peripheral T-cell lymphoma, NOS

Locus	Copy number alteration	Gene(s) implicated
2p15-p16	Amplification	*REL*
3q	Loss	*MDS1*
4q28-q31	Gain	
4q34-qtel	Gain	
7p	Gain	
7q	Gain	
8q24.11	Gain	*MYC*
9p23	Gain	
9p21.3	Loss	*MTAP1, CDKN2A, CDKN2B*
13q	Loss	
17q11-q25	Gain	
19q13.43	Gain	*ZFP28*
22q	Gain	

Table 24.14 Somatic mutations in peripheral T-cell lymphoma, NOS

Pathway	Gene
Epigenetic mediators	*MLL2 (KMT2D)*
	TET2
	KDM6A
	ARID1B
	DNMT3A
	CREBBP
	MLL (KMT2A)
	TET1
	ARID2
Cell signaling	*TNFAIP3*
	APC
	CHD8
	ZAP70
	NF1
	TNFRSF14
	TRAF3
	RHOA
	FYN
	VAV1
Tumor suppressor	*TP53*
	FOXO1
	BCORL1
	ATM
T-cell costimulation	*CD28*
Miscellaneous	*PCLO*
	SETBP1
	CIITA
	FBXW7
	COL6A3
	LRRK1

- PTCLs are traditionally aggressive nodal and/or extranodal T-cell lymphomas that usually present at an advanced stage
- Common sites of involvement, aside from lymph nodes, include bone marrow, liver, spleen, skin, gastrointestinal tract, lungs, and less commonly, CNS
- Morphologic characterization is variable, but most often, a diffuse or expanded paracortex is identified with a cytologic spectrum of polymorphous lymphocytes to that of monomorphic lymphocytes. The lymphocytes are intermediate to large with irregular nuclear contours, vesicular chromatin, and prominent nucleoli. In contrast to AITL, the proliferation of HEVs and/or FDC meshwork is not observed nor are open marginal sinuses. Epithelioid histiocytes may be present, and, if particularly abundant, are consistent with the lympho-epithelioid variant of PTCL
- Gene expression profiling within PTCL, NOS has identified two characteristic groups based on the expression of TBX21 and GATA3. Although both are transcription factors that serve as regulators of T helper (Th) cells, the former allows for skewing into the Th1 cell while the latter toward Th2 cell differentiation
- Next-generation sequencing studies have revealed multiple mutations in PTCL, NOS; however, studies have been limited to rather small cohorts (Tables 24.13 and 24.14)

Anaplastic Large Cell Lymphoma

- Anaplastic large cell lymphoma (ALCL) is further defined by its ALK rearrangement status as either ALCL, ALK-positive, or ALCL, ALK-negative
  - Additionally, two characteristic extranodal sites are described as diagnostic entities:
    - Primary cutaneous ALCL
    - Breast implant-associated ALCL

- ALCL, ALK-positive is a CD30-positive T-cell lymphoma that is composed of large, pleomorphic lymphocytes that typically have abundant cytoplasm and carry the characteristic *ALK* rearrangement (Fig. 24.8)
- ALCL, ALK-positive, accounts for 3% of adult non-Hodgkin and 10–20% of childhood lymphomas
- Survival among ALCL, ALK-positive is more favorable than its ALK-negative counterpart with a long-term survival approaching 80%
- Frequent involvement of both nodal and extranodal sites is seen. Rare cases limited to the skin have been identified, and careful history and staging are necessary to exclude systemic involvement by ALCL, ALK-positive
- ALK encodes a receptor tyrosine kinase, and through multiple types of chromosomal rearrangements, activation of the catalytic domain results in the activation of multiple signaling pathways, including RAS-ERK, JAK/STAT, and PIK-AKT pathways
- There are many characteristic rearrangements with *ALK* in ALCL as well as different immunohistochemical staining patterns (Table 24.15)
- ALCL, ALK-negative is a CD30-positive T-cell lymphoma that morphologically is similar to that of ALCL,

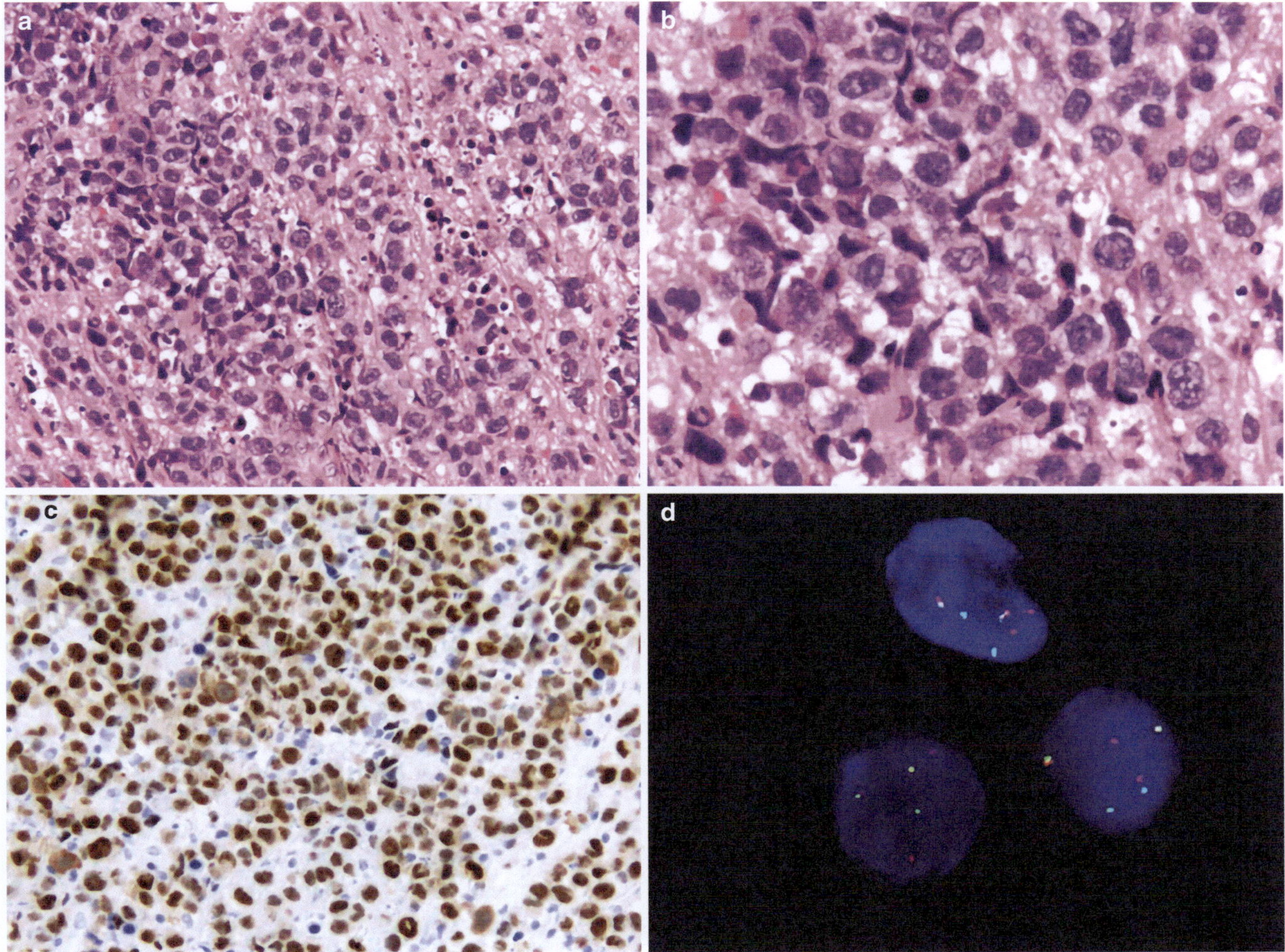

Fig. 24.8 (**a**) Anaplastic appearing lymphoma cells with abundant karyorrhexis and frequent mitotic figures, which is seen frequently in anaplastic large cell lymphoma (200×) (**b**) Pleomorphic lymphocytes with apoptosis, as seen in anaplastic large cell lymphoma (**c**) Immunohistochemistry for ALK demonstrating nuclear positivity with occasional cytoplasmic positivity (**d**) FISH image of *ALK* break-apart probe indicating the presence of an *ALK* rearrangement. Of note, anaplastic large cell lymphomas often demonstrate a tetraploid karyotype, as seen by the extra signals (courtesy of Dr. C. Bryke, BIDMC)

ALK-positive; however, it lacks ALK immunostaining as well as the *ALK* rearrangement
- It often demonstrates clinical differences from ALCL, ALK-positive by its aggressive course and older patient age
- ALCL, ALK-negative localizes to lymph nodes and extra-nodal tissues including the bone, soft tissue, and skin. If a cutaneous lesion is identified, it is important to distinguish cutaneous involvement by a systemic ALCL, ALK-negative, from that of a primary cutaneous ALCL
- The distinction of ALCL, ALK-negative from PTCL, NOS is not always clear; however, strong and uniform expression of CD30 is important as a distinguishing feature from PTCL, NOS, in which a subset of cases will show heterogeneous staining of CD30

- Although multiple studies have identified differences between ALCL, ALK-positive and ALCL, ALK-negative copy number alterations (CNAs), additional changes identified in ALCL, ALK-negative include losses of 6q21, which encompasses the *PRDM1* gene, and 17p13.1 (*TP53*) (Tables 24.16 and 24.17).
- Somatic variants have been identified in ALCL, ALK-negative, that primarily include activation of the JAK/STAT signaling pathway
  - *JAK1* mutations often occur in the kinase domain at p.G1097D/S
  - *STAT3* activating mutations occur in exon 12 of the SH2 domain
    - These mutations appear to be limited only to ALCL, ALK-negative cases and not PTCL, NOS

Table 24.15 Partner genes and immunohistochemical staining pattern of anaplastic large cell lymphoma, ALK-positive

Chromosomal rearrangement	ALK staining pattern	Frequency (%)
t(2;5) (p23;q35.1)/*NPM1-ALK*	Nuclear, diffuse cytoplasmic	84
t(1;2) (q25;p23)/*TPM3-ALK*	Diffuse cytoplasmic with peripheral intensification	13
Inv(2) (p23;q35)/*ATIC-ALK*	Diffuse cytoplasmic	1
t(2;3) (p23;q21)/*TFG-ALK*	Diffuse cytoplasmic	<1
t(2;17) (p23;q23)/*CLTC-ALK*	Diffuse cytoplasmic	<1
t(2;X) (p23;q11–12)/*MSN-ALK*	Membrane staining	<1
t(2;22) (q23;p11.2)/*MYH9-ALK*	Diffuse cytoplasmic	<1
t(2;19) (p23;q13.1)/*TPM4-ALK*	Diffuse cytoplasmic	<1
t(2;17) (p23;q25)/*ALO17-ALK*	Diffuse cytoplasmic	<1

Table 24.16 Common copy number alterations in anaplastic large cell lymphoma

Locus	Copy number alteration
4q13-q21	Loss
11q14	Loss
13q	Loss
7	Gain
17p	Gain
17q24-qter	Gain
6q	Gain

Table 24.17 Chromosomal rearrangements identified in anaplastic large cell lymphoma, ALK-negative

Chromosomal rearrangement	Description	Frequency (%)
t(6;7) (p25.3;q32.2)/*DUSP22-FRA7H*	Decreased DUSP22 expression	20–45
ROS1	Partners include *NFkB2*, *NPM*, or *NCOR2*; constitutive kinase activity and activated JAK/STAT signaling	11
TYK2	Partners include *NFkB2* or *PABPC4*; constitutive kinase activity and activated JAK/STAT signaling	11
inv(3) (q26.3q28)/*TP63-TBL1XR1*	Encodes p63 fusion product that inhibits TP53 pathway	8

Table 24.18 Chromosomal rearrangements in primary cutaneous anaplastic large cell lymphoma

Chromosomal rearrangement	Frequency (%)
6p25.3/*DUSP22-IRF4*	25
t(5;19)(q355;p13)/*NPM1-TYK2*	13

Table 24.19 Common copy number alterations in primary cutaneous anaplastic large cell lymphoma

Copy number alteration	Gene(s) involved	Frequency (%)
1p36.3 loss	*TNFRSF14*	30
6q27 loss		40
7p11.1-7q11.2 loss		45
8p23 loss		45
13q34 loss	*ARHGEF7, TFDP1*	45
16p11.2-16p11.1 loss	*TP53TG3*	45
20p11.1-20q11.2	*REM1*	30

tion involves seroma development around breast implants that contain large, pleomorphic tumor cells as well as "hallmark" cells that can be seen in other ALCLs as well as infiltration into the pericapsular fibrous tissue

- The time to development of BIA-ALCL is approximately 10 years; however, this is highly variable
- Immunophenotypic analysis reveals diffuse CD30 expression, incomplete expression of pan-T-cell antigens, and negative for ALK
- Conventional cytogenetic analysis often reveals a complex karyotype and somatic variants include activating mutations in *JAK1* and *STAT3*
- Primary cutaneous anaplastic large cell lymphoma (c-ALCL) is a T-cell lymphoma of the skin that is composed of large, pleomorphic to anaplastic lymphocytes in which the majority (greater than 75%) of neoplastic cells express CD30. A thorough clinical and staging examination should occur to rule out cutaneous involvement by a systemic ALCL, which carries a different prognosis. Most cases are negative for the *ALK* rearrangement; however, reports of rare instances of ALCL expressing ALK that is limited to the skin have been identified.
- Chromosomal aberrations have been identified in primary c-ALCL, which include recurrent rearrangements as well as frequent copy number changes (Tables 24.18 and 24.19)

Mycosis Fungoides

- Mycosis fungoides (MF) is the most common type of cutaneous T-cell lymphoma that is characterized by epidermotropism and infiltration of small to intermediate-sized lymphocytes with cerebriform nuclear contours

- A new provision entity within the 2017 WHO, revised 4th edition, is breast implant-associated anaplastic large cell lymphoma (BIA-ALCL)
- Although rare, with approximately 1 case per 500,000–3,000,000 women with implants, the clinical presenta-

- MF often has an indolent clinical course that is limited to the skin; however, extracutaneous dissemination can occur in later stages
- The conventional clinical progression of MF is from years to decades of patches that become infiltrated plaques. Lastly, the tumor stage occurs often with a combination of patches and plaques
- According to the International Society for Cutaneous Lymphoma (ISCL) and the European Organization for Research and Treatment of Cancer (EORTC), peripheral blood examination for Sezary cells as well as testing for *TCR* gene rearrangements is recommended for appropriate staging
- Frequent chromosomal aberrations are seen in MF as well as somatic variants that activate *STAT3* and render *PTEN* and *CDKN2A* inactive (Table 24.20)

Sezary Syndrome

- Sezary syndrome (SS) is characterized by the triad of erythroderma, generalized lymphadenopathy, and the presence of clonally related T-cells within the skin, lymph nodes, and peripheral blood. Although MF can present with erythroderma and circulating Sezary cells, it is a clinically and diagnostically distinct entity with different behavior
- SS is much rarer compared to MF and accounts for <5% of cutaneous T-cell lymphomas and is by definition, a disseminated disease
- Immunophenotypic analysis reveals an abundance of CD4 positive/CD7 negative T-cells that comprise greater than 30% or a CD4 positive/CD26 negative T-cell population that comprises greater than 40%. Immunohistochemical findings show expression of PD1 in most cases

Table 24.20 Copy number alterations in mycosis fungoides (tumor stage)

Copy number alteration	Gene(s) involved	Frequency (%)
1q25-q31 gain	*RGS1, RGS2*	35
7p22-p11.2 gain	*RAC1, OCM*	50
7q21 gain	*HGF*	60
7q31 gain	*DOCK4, MET, ING3*	50
9p21	*MTAP, CDKN2A, CDKN2B, DMRTA1*	30
9q21-q22 loss	*CCRK*	35
10p11.2 loss	*MAP3K8, ZEB1, ITGB1*	30
10q26 loss	*MGMT*	40
13q34 loss	*TFDP1*	40
16q21-22 loss	*CMTM1, CKLF, E2F4, CBFB*	30
16q23–24 loss	*BCAR1, WWOX, CDH13*	35
17q12 gain	*PPIP4K2B, MMP28, GRB7, ERBB2*	30

Table 24.21 Copy number alterations in Sezary syndrome

Copy number alteration	Gene(s) involved	Frequency (%)
8q23-24.3 gain	*MYC*	41
10p12-p11.2 loss	*RAB18, MAP3K8, ZEB1, ITGB1*	41
10p11.2 loss	*NRP1*	41
10q22-24 loss	*PTEN*	47
10q25-26 loss	*FGFR2, CASP7, DMBT1*	41
11q22-23 loss	*DDX10*	30
17p13-q11.1 loss	*TP53*	47
17q23-24 gain	*ICAM, MAP3K3*	35

Table 24.22 Common somatic mutations in Sezary syndrome

Gene	Description
T-cell receptor signaling	*PLCG1*
NF-kB activation	*CARD11*
JAK/STAT signaling	*JAK1*
	JAK3
	STAT3
	STAT5B
RAS homologue gene family	*RHOA*
DNA and histone modifiers	*ARID1A*
	ARID5B
	DNMT3A
	DNMT3B
	MLL2 (KMT2D)
	NCOR1
	SETD1A
	SETD1B
	SETD6
	TET1
	TET2
	ZEB1
Tumor suppressor	*TP53*

- Multiple pathways are recurrently mutated in SS and include JAK/STAT signaling, T-cell receptor signaling, NF-kB pathway, epigenetic modifiers, and tumor suppressors (Tables 24.21 and 24.22)

Intestinal T-Cell Lymphoma

- Intestinal T-cell lymphoma is a unique category of mature T-cell lymphomas that are characterized by the intestine being the primary disease site, although this cannot always be elucidated in disseminated cases
- Under the WHO 2017, revised 4th edition, there are currently three categories of intestinal T-cell lymphoma
  - Enteropathy-associated T-cell lymphoma (formerly known as type I EATL)
  - Monomorphic epitheliotropic intestinal T-cell lymphoma (MEITL)

- A category in which criteria are not met for either EATL or MEITL, termed intestinal T-cell lymphoma, NOS

- All forms of intestinal T-cell lymphoma are clinically aggressive and most frequently occur in adults. They are negative for EBV, which is important in distinguishing features from ENKTL, which can present within the intestine

- EATL is the most common primary intestinal T-cell lymphoma in the Western world. Among patients with celiac disease, it is identified in 0.22–1.9 cases per 100,000

- Risk factors for developing EATL include homozygosity for the HLA-DQ2 allele and age (greater than 50 years old)

- EATL is a lymphoma of intraepithelial T-cells that primarily occurs in individuals with celiac disease and is often localized to the jejunum and ileum

- Lymphoma cells are intermediate to large with variable amounts of pleomorphism and background chronic inflammation. Other features indicative of celiac disease are present within, including villous atrophy, crypt hyperplasia, and increased intraepithelial lymphocytes

- Immunophenotypic analysis reveals the neoplastic lymphocytes are often negative for CD5, CD4, and CD8 with expression of cytotoxic markers; however, variability in the immunophenotype is frequently observed

- Copy number alterations in EATL include:
  - Gains of 1q, 7q, 8q24 (involving *MYC*), and 9q34
  - Losses of 8p, 9p (*CDKN2A/B*), and 17p12-13.2 (*TP53*)

- There is a genetic susceptibility to EATL, with more than 90% of EATL patients demonstrating HLA-DQ2.5 heterodimers, which is from *HLA-DQA1*05* and *HLA-DQB1*02* alleles identified in either *cis* or *trans*

- MEITL is a primary intestinal T-cell lymphoma that has no known association with celiac disease. It is much more common in Asian countries as well as those of Hispanic origin. Localization often occurs within the small intestine, with the jejunum being the most commonly affected

- Neoplastic T-cells are monomorphic, intermediate in size, contain pale cytoplasm, and display round and regular nuclear contours with inconspicuous nucleoli. A hallmark feature is the prominent epitheliotropism. Unlike EATL, a chronic inflammatory background is absent

- Multiple variants activating the JAK/STAT signaling pathway as well as the G-protein-coupled receptor (GPCR) signaling pathway have been described:
  - *STAT5B* (63% of cases)
  - *JAK3* (35% of cases)
  - *GNAI2* (24% of cases)
  - *SETD2* (90% of cases in one study) (Tables 24.23 and 24.24)

Table 24.23 Identified somatic mutations in enteropathy-associated T-cell lymphoma

Enteropathy-associated T-cell lymphoma	
Pathway	Gene
Chromatin modifiers	*SETD2*
	TET2
JAK/STAT signaling	*STAT5B*
	JAK1
	STAT3
	SOCS1
	JAK3
RAS family	*NRAS*
	KRAS
DNA damage	*TP53*

Table 24.24 Copy number alterations in monomorphic epitheliotropic intestinal T-cell lymphoma

Copy number alteration	Gene(s) involved
8q24.2 gain	*MYC*
9q34.3 gain	
7q22.1-q23.1 gain	*MET*
7p14.1 loss	
16q12.1-q24.2 loss	*CDH1, CYLD*
11q22.3-q23.2 LOH	*ATM, SDHD*

Hepatosplenic T-Cell Lymphoma

- Hepatosplenic T-cell lymphoma (HSTL) is an aggressive T-cell lymphoma primarily involving the liver and spleen without significant lymphadenopathy and frequent bone marrow involvement

- HSTL is identified in adolescents and young adults that arise in the setting of chronic immune suppression, most often for solid organ transplant. Cases in children on azathioprine and infliximab for Crohn's disease have been reported as well

- Morphologically, the malignant T-cells are monotonous, with intermediate-sized lymphocytes containing pale cytoplasm and dispersed nuclear chromatin with inconspicuous nucleoli

- The lymphocytes involve the cords and sinuses of the spleen, the sinusoids of the liver, and demonstrate an intrasinusoidal pattern within the bone marrow

- Immunophenotypically, the neoplastic T-cells are often CD4 and CD5 negative, with CD8 being variably expressed. The malignancy cells are often gamma delta T-cell receptor-positive; however, rare cases of alpha beta type have been reported

- Conventional cytogenetic analysis frequently reveals isochromosome 7q with variable FISH patterns identified when disease progression is evident

- The critical region leading to 7q amplification has been identified as 7q22.
- Trisomy 8 is seen in a subset of cases
- Similar to intestinal T-cell lymphomas, activation of the JAK/STAT signaling pathway through missense mutations in *STAT3* and *STAT5B* is seen in 42% of cases
  - More recently, variants in chromatin modifiers have been discovered and include *SETD2*, *INO80*, and *ARID1B*

Suggested Reading

Alizadeh AA, Eisen MB, Davis RE, et al. Distinct types of diffuse large B-cell lymphoma identified by gene expression profiling. Nature. 2000 Feb 3;403(6769):503–11. https://doi.org/10.1038/35000501.

Al-Toma A, Goerres MS, Meijer JW, et al. Human leukocyte antigen-DQ2 homozygosity and the development of refractory celiac disease and enteropathy-associated T-cell lymphoma. Clin Gastroenterol Hepatol. 2006 Mar;4(3):315–9. https://doi.org/10.1016/j.cgh.2005.12.011.

Ansell SM, Akasaka T, McPhail E, et al. t(X;14)(p11;q32) in MALT lymphoma involving GPR34 reveals a role for GPR34 in tumor cell growth. Blood. 2012;120(19):3949–57. https://doi.org/10.1182/blood-2011-11-389908. Epub 2012 Sep 10

Arcaini L, Besson C, Frigeni M, et al. Interferon-free antiviral treatment in B-cell lymphoproliferative disorders associated with hepatitis C virus infection. Blood. 2016;128(21):2527–32. https://doi.org/10.1182/blood-2016-05-714667. Epub 2016 Sep 7

Arcaini L, Zibellini S, Boveri E, et al. The BRAF V600E mutation in hairy cell leukemia and other mature B-cell neoplasms. Blood. 2012;119(1):188–91. https://doi.org/10.1182/blood-2011-08-368209. Epub 2011 Nov 9

Attygalle AD, Feldman AL, Dogan A. ITK/SYK translocation in angioimmunoblastic T-cell lymphoma. Am J Surg Pathol. 2013;37(9):1456–7. https://doi.org/10.1097/PAS.0b013e3182991415.

Au WY, Ma SY, Chim CS, Choy C, Loong F, Lie AK, Lam CC, Leung AY, Tse E, Yau CC, Liang R, Kwong YL. Clinicopathologic features and treatment outcome of mature T-cell and natural killer-cell lymphomas diagnosed according to the World Health Organization classification scheme: a single center experience of 10 years. Ann Oncol. 2005;16(2):206–14. https://doi.org/10.1093/annonc/mdi037.

Aukema SM, Kreuz M, Kohler CW, et al. Biological characterization of adult MYC-translocation-positive mature B-cell lymphomas other than molecular Burkitt lymphoma. Haematologica. 2014;99(4):726–35. https://doi.org/10.3324/haematol.2013.091827. Epub 2013 Oct 31

Beà S, Valdés-Mas R, Navarro A, et al. Landscape of somatic mutations and clonal evolution in mantle cell lymphoma. Proc Natl Acad Sci U S A. 2013;110(45):18250–5. https://doi.org/10.1073/pnas.1314608110. Epub 2013 Oct 21

Beck RC, Kim AS, Goswami RS, et al. Molecular/cytogenetic education for hematopathology fellows. Am J Clin Pathol. 2020;154(2):149–77. https://doi.org/10.1093/ajcp/aqaa038.

Bellanger D, Jacquemin V, Chopin M, et al. Recurrent JAK1 and JAK3 somatic mutations in T-cell prolymphocytic leukemia. Leukemia. 2014;28(2):417–9. https://doi.org/10.1038/leu.2013.271. Epub 2013 Sep 19

Boi M, Zucca E, Inghirami G, et al. Advances in understanding the pathogenesis of systemic anaplastic large cell lymphomas. Br J Haematol. 2015;168(6):771–83. https://doi.org/10.1111/bjh.13265. Epub 2015 Jan 6

Bouchekioua A, Scourzic L, de Wever O, et al. JAK3 deregulation by activating mutations confers invasive growth advantage in extranodal nasal-type natural killer cell lymphoma. Leukemia. 2014;28(2):338–48. https://doi.org/10.1038/leu.2013.157. Epub 2013 May 21

Bouska A, Zhang W, Gong Q, et al. Combined copy number and mutation analysis identifies oncogenic pathways associated with transformation of follicular lymphoma. Leukemia. 2017;31(1):83–91. https://doi.org/10.1038/leu.2016.175. Epub 2016 Jun 16

Brito-Babapulle V, Hamoudi R, Matutes E, et al. p53 allele deletion and protein accumulation occurs in the absence of p53 gene mutation in T-prolymphocytic leukaemia and Sezary syndrome. Br J Haematol. 2000;110(1):180–7. https://doi.org/10.1046/j.1365-2141.2000.02174.x.

Brody GS, Deapen D, Taylor CR, et al. Anaplastic large cell lymphoma occurring in women with breast implants: analysis of 173 cases. Plast Reconstr Surg. 2015;135(3):695–705. https://doi.org/10.1097/PRS.0000000000001033. Erratum in: Plast Reconstr Surg. 2015;136(2):426.

Brüggemann M, Kotrová M, Knecht H, et al. Standardized next-generation sequencing of immunoglobulin and T-cell receptor gene recombinations for MRD marker identification in acute lymphoblastic leukaemia; a EuroClonality-NGS validation study. Leukemia. 2019;33(9):2241–53. https://doi.org/10.1038/s41375-019-0496-7. Epub 2019 Jun 26

Bunn PA Jr, Schechter GP, Jaffe E, et al. Clinical course of retrovirus-associated adult T-cell lymphoma in the United States. N Engl J Med. 1983;309(5):257–64. https://doi.org/10.1056/NEJM198308043090501.

Chapuy B, Stewart C, Dunford AJ, et al. Molecular subtypes of diffuse large B cell lymphoma are associated with distinct pathogenic mechanisms and outcomes. Nat Med. 2018;24(5):679–90. https://doi.org/10.1038/s41591-018-0016-8. Epub 2018 Apr 30. Erratum in: Nat Med. 2018;24(8):1292. Erratum in: Nat Med. 2018 Aug;24(8):1290–1.

Chen Y-W, Guo T, Shen L, et al. Receptor-type tyrosine-protein phosphatase κ directly targets STAT3 activation for tumor suppression in nasal NK/T-cell lymphoma. Blood. 2015;125(10):1589–600. https://doi.org/10.1182/blood-2014-07-588970.

Chesi M, Bergsagel PL. Advances in the pathogenesis and diagnosis of multiple myeloma. Int J Lab Hematol. 2015;37(Suppl 1):108–14. https://doi.org/10.1111/ijlh.12360.

Chiaretti S, Zini G, Bassan R. Diagnosis and subclassification of acute lymphoblastic leukemia. Mediterr J Hematol Infect Dis. 2014;6(1):e2014073. https://doi.org/10.4084/MJHID.2014.073.

Choi J, Goh G, Walradt T, et al. Genomic landscape of cutaneous T cell lymphoma. Nat Genet. 2015;47(9):1011–9. https://doi.org/10.1038/ng.3356. Epub 2015 Jul 20

Corrao G, Corazza GR, Bagnardi V, et al. Mortality in patients with coeliac disease and their relatives: a cohort study. Lancet. 2001;358(9279):356–61. https://doi.org/10.1016/s0140-6736(01)05554-4.

Crescenzo R, Abate F, Lasorsa E, et al. Convergent mutations and kinase fusions lead to oncogenic STAT3 activation in anaplastic large cell lymphoma. Cancer Cell. 2015;27(4):516–32. https://doi.org/10.1016/j.ccell.2015.03.006. Erratum in: Cancer Cell. 2015;27(5):744.

Evans PA, Pott C, Groenen PJ, et al. Significantly improved PCR-based clonality testing in B-cell malignancies by use of multiple immunoglobulin gene targets. Report of the BIOMED-2 Concerted Action BHM4-CT98-3936. Leukemia. 2007;21(2):207–14. https://doi.org/10.1038/sj.leu.2404479. Epub 2006 Dec 14

Falini B, Pileri S, Zinzani PL, et al. ALK+ lymphoma: clinico-pathological findings and outcome. Blood. 1999;93(8):2697–706.

Feldman AL, Dogan A, Smith DI, et al. Discovery of recurrent t(6;7) (p25.3;q32.3) translocations in ALK-negative anaplastic large cell lymphomas by massively parallel genomic sequencing. Blood. 2011;117(3):915–9. https://doi.org/10.1182/blood-2010-08-303305. Epub 2010 Oct 28

Fujiwara SI, Yamashita Y, Nakamura N, et al. High-resolution analysis of chromosome copy number alterations in angioimmunoblastic T-cell lymphoma and peripheral T-cell lymphoma, unspecified, with single nucleotide polymorphism-typing microarrays. Leukemia. 2008;22(10):1891–8. https://doi.org/10.1038/leu.2008.191. Epub 2008 Jul 17

Gladden AB, Woolery R, Aggarwal P, et al. Expression of constitutively nuclear cyclin D1 in murine lymphocytes induces B-cell lymphoma. Oncogene. 2006;25(7):998–1007. https://doi.org/10.1038/sj.onc.1209147.

Guo Y, Karube K, Kawano R, et al. Bcl2-negative follicular lymphomas frequently have Bcl6 translocation and/or Bcl6 or p53 expression. Pathol Int. 2007;57(3):148–52. https://doi.org/10.1111/j.1440-1827.2006.02072.x.

Gurth M, Bernard V, Bernd HW, et al. Nodal marginal zone lymphoma: mutation status analyses of CD79A, CD79B, and MYD88 reveal no specific recurrent lesions. Leuk Lymphoma. 2017;58(4):979–81. https://doi.org/10.1080/10428194.2016.1213836. Epub 2016 Aug 11

Hamblin TJ, Davis Z, Gardiner A, et al. Unmutated Ig V(H) genes are associated with a more aggressive form of chronic lymphocytic leukemia. Blood. 1999;94(6):1848–54.

Hamilton-Dutoit SJ, Rea D, Raphael M, et al. Epstein-Barr virus-latent gene expression and tumor cell phenotype in acquired immunodeficiency syndrome-related non-Hodgkin's lymphoma. Correlation of lymphoma phenotype with three distinct patterns of viral latency. Am J Pathol. 1993;143(4):1072–85.

Hans CP, Weisenburger DD, Greiner TC, et al. Confirmation of the molecular classification of diffuse large B-cell lymphoma by immunohistochemistry using a tissue microarray. Blood. 2004;103(1):275–82. https://doi.org/10.1182/blood-2003-05-1545. Epub 2003 Sep 22

Hartmann S, Gesk S, Scholtysik R, et al. High resolution SNP array genomic profiling of peripheral T cell lymphomas, not otherwise specified, identifies a subgroup with chromosomal aberrations affecting the REL locus. Br J Haematol. 2010;148(3):402–12. https://doi.org/10.1111/j.1365-2141.2009.07956.x. Epub 2009 Oct 22

Hongyo T, Hoshida Y, Nakatsuka S, et al. p53, K-ras, c-kit and beta-catenin gene mutations in sinonasal NK/T-cell lymphoma in Korea and Japan. Oncol Rep. 2005;13(2):265–71.

Howell WM, Leung ST, Jones DB, et al. HLA-DRB, -DQA, and -DQB polymorphism in celiac disease and enteropathy-associated T-cell lymphoma. Common features and additional risk factors for malignancy. Hum Immunol. 1995;43(1):29–37. https://doi.org/10.1016/0198-8859(94)00130-i.

https://seer.cancer.gov/statfacts/html/nhl.html

Huang Y, de Leval L, Gaulard P. Molecular underpinning of extranodal NK/T-cell lymphoma. Best Pract Res Clin Haematol. 2013;26(1):57–74. https://doi.org/10.1016/j.beha.2013.04.006. Epub 2013 May 23

Hunter ZR, Xu L, Yang G, et al. The genomic landscape of Waldenstrom macroglobulinemia is characterized by highly recurring MYD88 and WHIM-like CXCR4 mutations, and small somatic deletions associated with B-cell lymphomagenesis. Blood. 2014;123(11):1637–46. https://doi.org/10.1182/blood-2013-09-525808. Epub 2013 Dec 23

Iqbal J, Kucuk C, Deleeuw RJ, et al. Genomic analyses reveal global functional alterations that promote tumor growth and novel tumor suppressor genes in natural killer-cell malignancies. Leukemia. 2009;23(6):1139–51. https://doi.org/10.1038/leu.2009.3. Epub 2009 Feb 5

Iqbal J, Wright G, Wang C, et al. Gene expression signatures delineate biological and prognostic subgroups in peripheral T-cell lymphoma.

Blood. 2014;123(19):2915–23. https://doi.org/10.1182/blood-2013-11-536359. Epub 2014 Mar 14

Jajosky AN, Havens NP, Sadri N, et al. Clinical utility of targeted next-generation sequencing in the evaluation of low-grade lymphoproliferative disorders. Am J Clin Pathol. 2021;156(3):433–44. https://doi.org/10.1093/ajcp/aqaa255.

Jenderny J, Schmidt W, Kochhan L. Chromosome aberrations identified by cytogenetic analysis of the first two clones of cultured amniotic fluid cells compared with QF-PCR results. Cytogenet Genome Res. 2014;142(4):239–44. https://doi.org/10.1159/000362524. Epub 2014 May 16

Jerez A, Clemente MJ, Makishima H, et al. STAT3 mutations unify the pathogenesis of chronic lymphoproliferative disorders of NK cells and T-cell large granular lymphocyte leukemia. Blood. 2012;120(15):3048–57. https://doi.org/10.1182/blood-2012-06-435297. Epub 2012 Aug 2

Kataoka K, Nagata Y, Kitanaka A, et al. Integrated molecular analysis of adult T cell leukemia/lymphoma. Nat Genet. 2015;47(11):1304–15. https://doi.org/10.1038/ng.3415. Epub 2015 Oct 5

Kawamata N, Inagaki N, Mizumura S, et al. Methylation status analysis of cell cycle regulatory genes (p16INK4A, p15INK4B, p21Waf1/Cip1, p27Kip1 and p73) in natural killer cell disorders. Eur J Haematol. 2005;74(5):424–9. https://doi.org/10.1111/j.1600-0609.2005.00417.x.

Kiel MJ, Sahasrabuddhe AA, Rolland DCM, et al. Genomic analyses reveal recurrent mutations in epigenetic modifiers and the JAK-STAT pathway in Sézary syndrome. Nat Commun. 2015;6:8470. https://doi.org/10.1038/ncomms9470.

Kiel MJ, Velusamy T, Rolland D, et al. Integrated genomic sequencing reveals mutational landscape of T-cell prolymphocytic leukemia. Blood. 2014;124(9):1460–72. https://doi.org/10.1182/blood-2014-03-559542. Epub 2014 May 13

King RL, Gonsalves WI, Ansell SM, et al. Lymphoplasmacytic lymphoma with a non-IGM paraprotein shows clinical and pathologic heterogeneity and may harbor MYD88 L265P mutations. Am J Clin Pathol. 2016;145(6):843–51. https://doi.org/10.1093/ajcp/aqw072. Epub 2016 Jun 21

Ko YH, Ree HJ, Kim WS, et al. Clinicopathologic and genotypic study of extranodal nasal-type natural killer/T-cell lymphoma and natural killer precursor lymphoma among Koreans. Cancer. 2000;89(10):2106–16. https://doi.org/10.1002/1097-0142(20001115)89:10<2106::aid-cncr11>3.0.co;2-g.

Kramer MH, Hermans J, Wijburg E, et al. Clinical relevance of BCL2, BCL6, and MYC rearrangements in diffuse large B-cell lymphoma. Blood. 1998;92(9):3152–62.

Küçük C, Jiang B, Hu X, et al. Activating mutations of STAT5B and STAT3 in lymphomas derived from γδ-T or NK cells. Nat Commun. 2015;6:6025. https://doi.org/10.1038/ncomms7025.

Laharanne E, Oumouhou N, Bonnet F, et al. Genome-wide analysis of cutaneous T-cell lymphomas identifies three clinically relevant classes. J Invest Dermatol. 2010;130(6):1707–18. https://doi.org/10.1038/jid.2010.8. Epub 2010 Feb 4

Lamy T, Loughran TP Jr. Current concepts: large granular lymphocyte leukemia. Blood Rev. 1999;13(4):230–40. https://doi.org/10.1054/blre.1999.0118.

Lee SH, Kim JS, Kim J, et al. A highly recurrent novel missense mutation in CD28 among angioimmunoblastic T-cell lymphoma patients. Haematologica. 2015;100(12):e505–7. https://doi.org/10.3324/haematol.2015.133074. Epub 2015 Sep 24

Lemonnier F, Couronné L, Parrens M, et al. Recurrent TET2 mutations in peripheral T-cell lymphomas correlate with TFH-like features and adverse clinical parameters. Blood. 2012;120(7):1466–9. https://doi.org/10.1182/blood-2012-02-408542. Epub 2012 Jul 3

Li S, Feng X, Li T, et al. Extranodal NK/T-cell lymphoma, nasal type: a report of 73 cases at MD Anderson Cancer Center.

Am J Surg Pathol. 2013;37(1):14–23. https://doi.org/10.1097/PAS.0b013e31826731b5.

Li S, Seegmiller AC, Lin P, et al. B-cell lymphomas with concurrent MYC and BCL2 abnormalities other than translocations behave similarly to MYC/BCL2 double-hit lymphomas. Mod Pathol. 2015;28(2):208–17. https://doi.org/10.1038/modpathol.2014.95. Epub 2014 Aug 8

Liu H, Bench AJ, Bacon CM, et al. A practical strategy for the routine use of BIOMED-2 PCR assays for detection of B- and T-cell clonality in diagnostic haematopathology. Br J Haematol. 2007;138(1):31–43. https://doi.org/10.1111/j.1365-2141.2007.06618.x.

Macon WR, Levy NB, Kurtin PJ, et al. Hepatosplenic alphabeta T-cell lymphomas: a report of 14 cases and comparison with hepatosplenic gammadelta T-cell lymphomas. Am J Surg Pathol. 2001;25(3):285–96. https://doi.org/10.1097/00000478-200103000-00002.

Malamut G, Chandesris O, Verkarre V, et al. Enteropathy associated T cell lymphoma in celiac disease: a large retrospective study. Dig Liver Dis. 2013;45(5):377–84. https://doi.org/10.1016/j.dld.2012.12.001. Epub 2013 Jan 10

Manso R, Rodríguez-Pinilla SM, González-Rincón J, et al. Recurrent presence of the PLCG1 S345F mutation in nodal peripheral T-cell lymphomas. Haematologica. 2015;100(1):e25–7. https://doi.org/10.3324/haematol.2014.113696. Epub 2014 Oct 10

Martinez A, Pittaluga S, Villamor N, et al. Clonal T-cell populations and increased risk for cytotoxic T-cell lymphomas in B-CLL patients: clinicopathologic observations and molecular analysis. Am J Surg Pathol. 2004;28(7):849–58. https://doi.org/10.1097/00000478-200407000-00002.

Masuko K, Kato S, Hagihara M, et al. Stable clonal expansion of T cells induced by bone marrow transplantation. Blood. 1996;87(2):789–99.

Matutes E, Brito-Babapulle V, Swansbury J, et al. Clinical and laboratory features of 78 cases of T-prolymphocytic leukemia. Blood. 1991;78(12):3269–74.

McKinney M, Moffitt AB, Gaulard P, et al. The Genetic Basis of Hepatosplenic T-cell Lymphoma. Cancer Discov. 2017;7(4):369–79. https://doi.org/10.1158/2159-8290.CD-16-0330. Epub 2017 Jan 25

Moffitt AB, Ondrejka SL, McKinney M, et al. Enteropathy-associated T cell lymphoma subtypes are characterized by loss of function of SETD2. J Exp Med. 2017;214(5):1371–86. https://doi.org/10.1084/jem.20160894. Epub 2017 Apr 19

Montes-Mojarro IA, Chen BJ, et al. Mutational profile and EBV strains of extranodal NK/T-cell lymphoma, nasal type in Latin America. Mod Pathol. 2020;33(5):781–91. https://doi.org/10.1038/s41379-019-0415-5. Epub 2019 Dec 10

Nairismägi ML, Tan J, Lim JQ, et al. JAK-STAT and G-protein-coupled receptor signaling pathways are frequently altered in epitheliotropic intestinal T-cell lymphoma. Leukemia. 2016;30(6):1311–9. https://doi.org/10.1038/leu.2016.13. Epub 2016 Feb 8

Nakagawa M, Nakagawa-Oshiro A, Karnan S, et al. Array comparative genomic hybridization analysis of PTCL-U reveals a distinct subgroup with genetic alterations similar to lymphoma-type adult T-cell leukemia/lymphoma. Clin Cancer Res. 2009;15(1):30–8. https://doi.org/10.1158/1078-0432.CCR-08-1808.

Nakashima Y, Tagawa H, Suzuki R, et al. Genome-wide array-based comparative genomic hybridization of natural killer cell lymphoma/leukemia: different genomic alteration patterns of aggressive NK-cell leukemia and extranodal Nk/T-cell lymphoma, nasal type. Genes Chromosomes Cancer. 2005;44(3):247–55. https://doi.org/10.1002/gcc.20245.

Novak U, Rinaldi A, Kwee I, et al. The NF-{kappa}B negative regulator TNFAIP3 (A20) is inactivated by somatic mutations and genomic deletions in marginal zone lymphomas. Blood. 2009;113(20):4918–21. https://doi.org/10.1182/blood-2008-08-174110. Epub 2009 Mar 3

Olsen E, Vonderheid E, Pimpinelli N, et al. Revisions to the staging and classification of mycosis fungoides and Sezary syndrome: a proposal of the International Society for Cutaneous Lymphomas (ISCL) and the cutaneous lymphoma task force of the European Organization of Research and Treatment of Cancer (EORTC). Blood. 2007;110(6):1713–22. https://doi.org/10.1182/blood-2007-03-055749. Epub 2007 May 31. Erratum in: Blood. 2008;111(9):4830.

Oschlies I, Lisfeld J, Lamant L, et al. ALK-positive anaplastic large cell lymphoma limited to the skin: clinical, histopathological and molecular analysis of 6 pediatric cases. A report from the ALCL99 study. Haematologica. 2013;98(1):50–6. https://doi.org/10.3324/haematol.2012.065664. Epub 2012 Jul 6

Palomero T, Couronné L, Khiabanian H, et al. Recurrent mutations in epigenetic regulators, RHOA and FYN kinase in peripheral T cell lymphomas. Nat Genet. 2014;46(2):166–70. https://doi.org/10.1038/ng.2873. Epub 2014 Jan 12

Parrilla Castellar ER, Jaffe ES, Said JW, et al. ALK-negative anaplastic large cell lymphoma is a genetically heterogeneous disease with widely disparate clinical outcomes. Blood. 2014;124(9):1473–80. https://doi.org/10.1182/blood-2014-04-571091. Epub 2014 Jun 3

Pastore A, Jurinovic V, Kridel R, et al. Integration of gene mutations in risk prognostication for patients receiving first-line immunochemotherapy for follicular lymphoma: a retrospective analysis of a prospective clinical trial and validation in a population-based registry. Lancet Oncol. 2015;16(9):1111–22. https://doi.org/10.1016/S1470-2045(15)00169-2. Epub 2015 Aug 6

Pham-Ledard A, Prochazkova-Carlotti M, Laharanne E, et al. IRF4 gene rearrangements define a subgroup of CD30-positive cutaneous T-cell lymphoma: a study of 54 cases. J Invest Dermatol. 2010;130(3):816–25. https://doi.org/10.1038/jid.2009.314. Epub 2009 Oct 8

Piccaluga PP, Agostinelli C, Califano A, et al. Gene expression analysis of angioimmunoblastic lymphoma indicates derivation from T follicular helper cells and vascular endothelial growth factor deregulation. Cancer Res. 2007a;67(22):10703–10. https://doi.org/10.1158/0008-5472.CAN-07-1708.

Piccaluga PP, Agostinelli C, Califano A, et al. Gene expression analysis of peripheral T cell lymphoma, unspecified, reveals distinct profiles and new potential therapeutic targets. J Clin Invest. 2007b;117(3):823–34. https://doi.org/10.1172/JCI26833. Epub 2007 Feb 15

Piccaluga PP, Fuligni F, De Leo A, et al. Molecular profiling improves classification and prognostication of nodal peripheral T-cell lymphomas: results of a phase III diagnostic accuracy study. J Clin Oncol. 2013;31(24):3019–25. https://doi.org/10.1200/JCO.2012.42.5611. Epub 2013 Jul 15

Pienkowska-Grela B, Rymkiewicz G, Grygalewicz B, et al. Partial trisomy 11, dup(11)(q23q13), as a defect characterizing lymphomas with Burkitt pathomorphology without MYC gene rearrangement. Med Oncol. 2011;28(4):1589–95. https://doi.org/10.1007/s12032-010-9614-0. Epub 2010 Jul 27

Piva R, Agnelli L, Pellegrino E, et al. Gene expression profiling uncovers molecular classifiers for the recognition of anaplastic large-cell lymphoma within peripheral T-cell neoplasms. J Clin Oncol. 2010;28(9):1583–90. https://doi.org/10.1200/JCO.2008.20.9759. Epub 2010 Feb 16

Rajala HL, Eldfors S, Kuusanmäki H, et al. Discovery of somatic STAT5b mutations in large granular lymphocytic leukemia. Blood. 2013;121(22):4541–50. https://doi.org/10.1182/blood-2012-12-474577. Epub 2013 Apr 17

Rajkumar SV, Dimopoulos MA, Palumbo A, et al. International Myeloma Working Group updated criteria for the diagnosis of multiple myeloma. Lancet Oncol. 2014;15(12):e538–48. https://doi.org/10.1016/S1470-2045(14)70442-5. Epub 2014 Oct 26

Roberti A, Dobay MP, Bisig B, et al. Type II enteropathy-associated T-cell lymphoma features a unique genomic profile with highly recurrent SETD2 alterations. Nat Commun. 2016;7:12602. https://doi.org/10.1038/ncomms12602.

Rodriguez-Justo M, Attygalle AD, Munson P, et al. Angioimmunoblastic T-cell lymphoma with hyperplastic germinal centres: a neoplasia with origin in the outer zone of the germinal centre? Clinicopathological and immunohistochemical study of 10 cases with follicular T-cell markers. Mod Pathol. 2009;22(6):753–61. https://doi.org/10.1038/modpathol.2009.12. Epub 2009 Mar 27

Rosenwald A, Wright G, Chan WC, et al. The use of molecular profiling to predict survival after chemotherapy for diffuse large-B-cell lymphoma. N Engl J Med. 2002;346(25):1937–47. https://doi.org/10.1056/NEJMoa012914.

Salaverria I, Beà S, Lopez-Guillermo A, et al. Genomic profiling reveals different genetic aberrations in systemic ALK-positive and ALK-negative anaplastic large cell lymphomas. Br J Haematol. 2008;140(5):516–26. https://doi.org/10.1111/j.1365-2141.2007.06924.x.

Salaverria I, Martin-Guerrero I, Wagener R, et al. A recurrent 11q aberration pattern characterizes a subset of MYC-negative high-grade B-cell lymphomas resembling Burkitt lymphoma. Blood. 2014;123(8):1187–98. https://doi.org/10.1182/blood-2013-06-507996. Epub 2014 Jan 7

Salido M, Baró C, Oscier D, et al. Cytogenetic aberrations and their prognostic value in a series of 330 splenic marginal zone B-cell lymphomas: a multicenter study of the Splenic B-Cell Lymphoma Group. Blood. 2010;116(9):1479–88. https://doi.org/10.1182/blood-2010-02-267476. Epub 2010 May 17

Scarisbrick JJ, Hodak E, Bagot M, et al. Blood classification and blood response criteria in mycosis fungoides and Sézary syndrome using flow cytometry: recommendations from the EORTC cutaneous lymphoma task force. Eur J Cancer. 2018;93:47–56. https://doi.org/10.1016/j.ejca.2018.01.076. Epub 2018 Feb 21

Schatz JH, Horwitz SM, Teruya-Feldstein J, et al. Targeted mutational profiling of peripheral T-cell lymphoma not otherwise specified highlights new mechanisms in a heterogeneous pathogenesis. Leukemia. 2015;29(1):237–41. https://doi.org/10.1038/leu.2014.261. Epub 2014 Sep 3

Schmitz R, Wright GW, Huang DW, et al. Genetics and Pathogenesis of Diffuse Large B-Cell Lymphoma. N Engl J Med. 2018;378(15):1396–407. https://doi.org/10.1056/NEJMoa1801445.

Scott DW, Wright GW, Williams PM, et al. Determining cell-of-origin subtypes of diffuse large B-cell lymphoma using gene expression in formalin-fixed paraffin-embedded tissue. Blood. 2014;123(8):1214–7. https://doi.org/10.1182/blood-2013-11-536433. Epub 2014 Jan 7

Simpson HM, Khan RZ, Song C, et al. Concurrent Mutations in ATM and Genes Associated with Common γ Chain Signaling in Peripheral T Cell Lymphoma. PLoS One. 2015;10(11):e0141906. https://doi.org/10.1371/journal.pone.0141906.

Sommer VH, Clemmensen OJ, Nielsen O, et al. In vivo activation of STAT3 in cutaneous T-cell lymphoma. Evidence for an antiapoptotic function of STAT3. Leukemia. 2004;18(7):1288–95. https://doi.org/10.1038/sj.leu.2403385.

Spina V, Khiabanian H, Messina M, et al. The genetics of nodal marginal zone lymphoma. Blood. 2016;128(10):1362–73. https://doi.org/10.1182/blood-2016-02-696757. Epub 2016 Jun 22

Stein H, Foss HD, Dürkop H, et al. CD30(+) anaplastic large cell lymphoma: a review of its histopathologic, genetic, and clinical features. Blood. 2000;96(12):3681–95.

Stengel A, Kern W, Zenger M, et al. Genetic characterization of T-PLL reveals two major biologic subgroups and JAK3 mutations as prognostic marker. Genes Chromosomes Cancer. 2016;55(1):82–94. https://doi.org/10.1002/gcc.22313. Epub 2015 Oct 23

Swerdlow SH, International Agency for Research on Cancer, World Health Organization. WHO classification of tumours of haematopoietic and lymphoid tissues. In: World Health Organization classification of tumours. Revised 4th ed. Lyon: International Agency for Research on Cancer, 2017.

Thériault C, Galoin S, Valmary S, et al. PCR analysis of immunoglobulin heavy chain (IgH) and TcR-gamma chain gene rearrangements in the diagnosis of lymphoproliferative disorders: results of a study of 525 cases. Mod Pathol. 2000;13(12):1269–79. https://doi.org/10.1038/modpathol.3880232.

Thorns C, Bastian B, Pinkel D, et al. Chromosomal aberrations in angioimmunoblastic T-cell lymphoma and peripheral T-cell lymphoma unspecified: A matrix-based CGH approach. Genes Chromosomes Cancer. 2007;46(1):37–44. https://doi.org/10.1002/gcc.20386.

Tibiletti MG, Martin V, Bernasconi B, et al. BCL2, BCL6, MYC, MALT 1, and BCL10 rearrangements in nodal diffuse large B-cell lymphomas: a multicenter evaluation of a new set of fluorescent in situ hybridization probes and correlation with clinical outcome. Hum Pathol. 2009;40(5):645–52. https://doi.org/10.1016/j.humpath.2008.06.032. Epub 2009 Jan 13

Vega F, Medeiros LJ, Gaulard P. Hepatosplenic and other gammadelta T-cell lymphomas. Am J Clin Pathol. 2007;127(6):869–80. https://doi.org/10.1309/LRKX8CE7GVPCR1FT.

Velusamy T, Kiel MJ, Sahasrabuddhe AA, et al. A novel recurrent NPM1-TYK2 gene fusion in cutaneous CD30-positive lymphoproliferative disorders. Blood. 2014;124(25):3768–71. https://doi.org/10.1182/blood-2014-07-588434. Epub 2014 Oct 27

Wada DA, Law ME, Hsi ED, et al. Specificity of IRF4 translocations for primary cutaneous anaplastic large cell lymphoma: a multicenter study of 204 skin biopsies. Mod Pathol. 2011;24(4):596–605. https://doi.org/10.1038/modpathol.2010.225. Epub 2010 Dec 17

Weinberg OK, Ai WZ, Mariappan MR, et al. "Minor" BCL2 breakpoints in follicular lymphoma: frequency and correlation with grade and disease presentation in 236 cases. J Mol Diagn. 2007;9(4):530–7. https://doi.org/10.2353/jmoldx.2007.070038. Epub 2007 Jul 25

Willemze R, Jaffe ES, Burg G, et al. WHO-EORTC classification for cutaneous lymphomas. Blood. 2005;105(10):3768–85. https://doi.org/10.1182/blood-2004-09-3502. Epub 2005 Feb 3

Wlodarska I, Martin-Garcia N, Achten R, et al. Fluorescence in situ hybridization study of chromosome 7 aberrations in hepatosplenic T-cell lymphoma: isochromosome 7q as a common abnormality accumulating in forms with features of cytologic progression. Genes Chromosomes Cancer. 2002;33(3):243–51. https://doi.org/10.1002/gcc.10021.

Yoo HY, Kim P, Kim WS, et al. Author reply to Comment on: Frequent CTLA4-CD28 gene fusion in diverse types of T-cell lymphoma, by Yoo et al. Haematologica. 2016;101(6):e271. https://doi.org/10.3324/haematol.2016.148015. Epub 2016 May 31

Molecular Pathology of Leukemia

25

Clayton E. Kibler and Devon S. Chabot-Richards

Contents

Introduction

- Leukemias are a group of malignant neoplasms of hematopoietic stem cells caused by molecular alterations that disrupt cell differentiation and proliferation
  - The abnormal hematopoietic cells can accumulate in the bone marrow and blood and interfere with growth of normal hematopoietic precursors leading to cytopenias including anemia, thrombocytopenia, and neutropenia
  - Neoplastic cells may also involve tissue sites such as soft tissue, bone, organs, and skin
- Acute leukemias involve proliferation of immature hematopoietic cells and progress rapidly, while chronic leukemias are characterized by increased abnormal mature leukocytes and progress slowly
  - Leukemias are also broadly classified based on cell line of origin as myeloid, lymphoid, or ambiguous lineage
- Leukemias are common malignancies in all age groups, accounting for 3% of new cancer cases overall (28% in children, 13% in adolescents) and 4% of cancer deaths in the US
  - The most common types of leukemia among adults are chronic lymphocytic leukemia (38%) and acute myeloid leukemia (31%); acute lymphoblastic leukemia is most common in children and adolescents (74% of cases)
- Molecular and cytogenetic assays play a key role in the classification, risk stratification, monitoring, and direction of treatment including targeted therapy of leukemia

C. E. Kibler · D. S. Chabot-Richards (✉)
Department of Pathology, University of New Mexico,
Albuquerque, NM, USA
e-mail: ckibler@salud.unm.edu;
DChabot-Richards@salud.unm.edu

Hereditary Leukemia Syndromes

- Although most cases of leukemia are sporadic, leukemia, particularly of myeloid origin can rarely occur due to inherited or de novo germline mutations with specific associated genetic and clinical features
  - Myeloid neoplasms with germline predisposition may initially present with myeloid neoplasm or with an antecedent condition or organ dysfunction, e.g., thrombocytopenia, bone marrow failure syndromes, and telomere biology disorders
    - Patients with preexisting platelet disorders or organ dysfunction must be closely monitored due to the increased risk of developing a myeloid neoplasm
    - Monitoring may include bone marrow biopsy
  - Often present in children and young adults, but can present in older individuals depending on the specific gene mutation
  - Family history can be key to diagnosis
  - With some disorders, congenital or physical abnormalities may be present
  - Diagnosis of these disorders is important for proper long-term clinical management and family counseling
  - Specialized molecular testing is often necessary for diagnosis
    - Specific evaluation for germline DNA mutations is needed
      - Sample types that can be used for germline DNA evaluation include skin fibroblast culture (gold standard and free of leukocyte contamination), hair, nails, saliva, and buccal mucosa swabs
    - Many families have unique mutations in the genes of interest that may not be detected by targeted sequencing panels or be reported as a variant of uncertain significance due to lack of evidence as a pathogenic alteration in the literature
      - Whole-exome sequencing can identify novel variants

Myeloid Neoplasms with Germline Predisposition

Myeloid Neoplasms with Germline Predisposition Without a Preexisting Disorder or Organ Dysfunction

- These patients present with acute myeloid leukemia (AML) or myelodysplastic syndrome with no preceding blood count abnormalities or organ dysfunction

Acute Myeloid Leukemia with Germline CEBPA Mutation

- Pathogenesis (Fig. 25.1)
  - *CEBPA* 19q13.1 encodes a transcription factor involved in granulocyte differentiation
  - Caused by biallelic *CEBPA* mutations, with the inheritance of a germline mutation typically in the 5′ end of the gene (corresponding to the N terminus) and an acquired somatic mutation in the 3′ end of the gene (corresponding to the C terminus)
- The presence of biallelic *CEBPA* mutations should prompt evaluation for a germline mutation to differentiate between AML with somatic mutations and AML with germline *CEBPA* mutation
  - In around 10% of cases of AML with biallelic *CEBPA* mutation, a germline mutation is present
- Typically presents as AML in children or young adults

Myeloid Neoplasms with Germline DDX41 Mutation

- Pathogenesis
  - *DDX41* 5q35.3 encodes a likely RNA helicase involved in posttranscriptional gene expression, pre-mRNA splicing, cell proliferation, and differentiation
  - Caused by inherited *DDX41* mutations, often biallelic with one germline and one somatic mutation
- Presents as MDS or AML in older adults

Myeloid Neoplasms with Germline Predisposition and Preexisting Platelet Disorders

- These patients have abnormalities in platelet number and function and an increased risk of developing MDS, AML, and occasionally lymphoid neoplasms

Myeloid Neoplasms with Germline RUNX1 Mutation

- *RUNX1* 21q22.12 encodes a subunit of core binding factor involved in transcriptional regulation of hematopoiesis
- Caused by monoallelic germline *RUNX1* mutations

Myeloid Neoplasms with Germline ANKRD26 Mutation, also Known as Thrombocytopenia 2

- Autosomal dominant disorder caused by germline mutations in *ANKRD26* 10p12.1 usually within the 5′ untranslated region that results in increased gene transcription and signaling through the MPL pathway and impaired platelet production

Myeloid Neoplasms with Germline ETV6 Mutation, also Known as Thrombocytopenia 5

- Autosomal dominant disorder caused by missense germline mutations in *ETV6* 12p13.2, which encodes a transcription factor involved in hematopoiesis

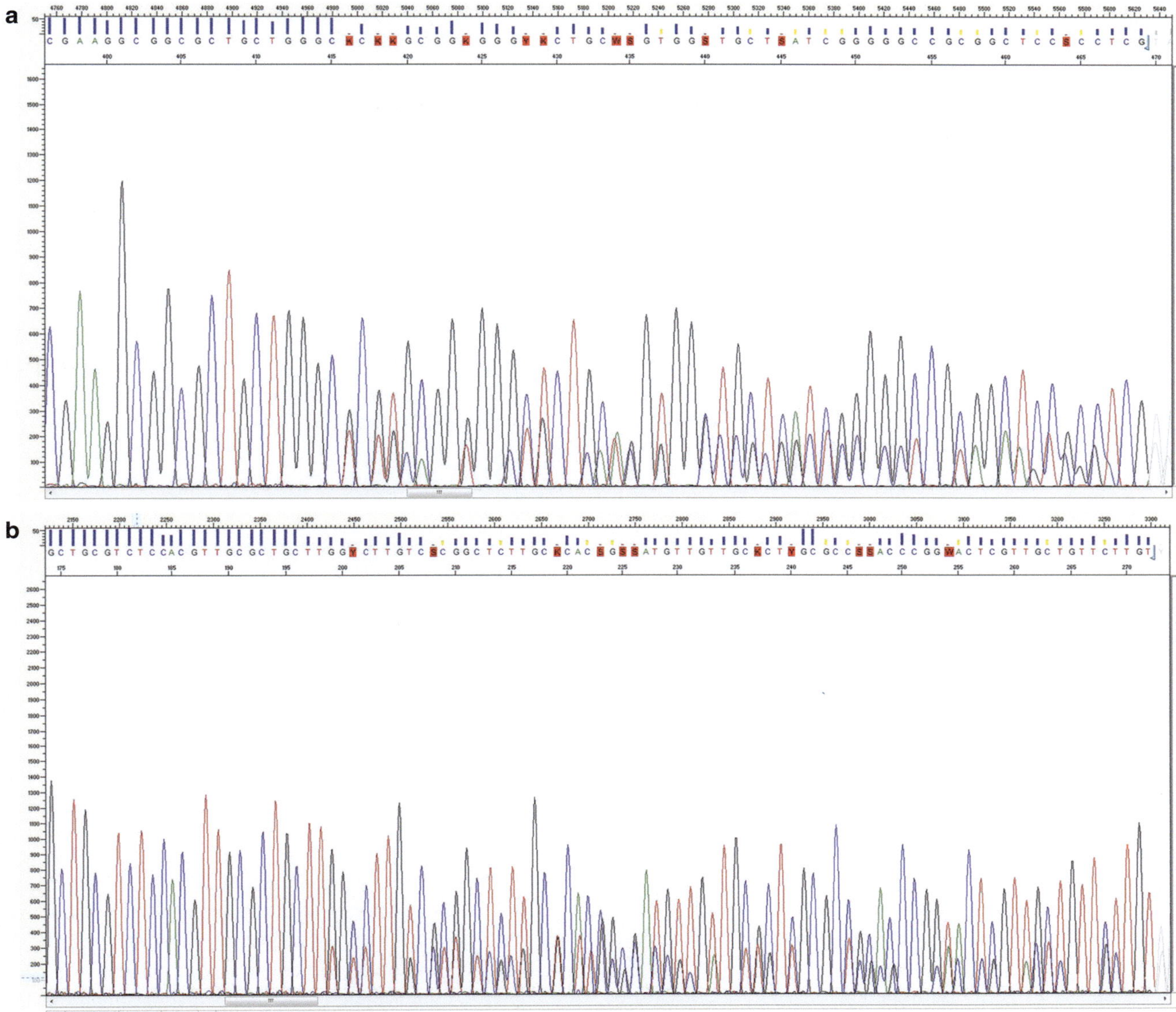

Fig. 25.1 Sanger sequencing of *CEBPA* shows segments with mixed bases in the 3′ end (C terminus) (**a**) and 5′ end (N terminus) (**b**) segments of the gene, consistent with biallelic *CEBPA* mutation

Myeloid Neoplasms with Germline Predisposition and Other Organ Dysfunction

- These patients develop various manifestations of organ dysfunction and have an increased risk of MDS, AML, or ALL

Myeloid Neoplasms with Germline GATA2 Mutation

- This disorder was previously divided into multiple distinct syndromes including MonoMAC syndrome, dendritic cell, monocyte, B- and NK-lymphoid (DCML) deficiency, familial MDS/AML, and Emberger syndrome
  - Now recognized as a single disorder with protean manifestations including infections and lymphedema
- Pathogenesis

- *GATA2* 3q21.3 encodes a zinc-finger transcription factor involved in hematopoietic stem cell differentiation
- Caused by monoallelic germline *GATA2* mutations resulting in loss of function and haploinsufficiency

Myeloid Neoplasms Associated with Bone Marrow Failure Syndromes

- Includes Schwachman–Diamond syndrome, Diamond–Blackfan anemia, and severe congenital neutropenia
- Patients usually present in childhood with bone marrow failure or systemic manifestations such as limb abnormalities and short stature in Fanconi anemia, or pancreatic dysfunction in Schwachman–Diamond syndrome

Myeloid Neoplasms Associated with Telomere Biology Disorders

- Characterized by abnormal telomere maintenance and predisposition to MDS and AML
- Dyskeratosis congenita is the prototypical disorder
  - Manifestations include nail dystrophy, abnormal reticular skin pigmentation, and oral leukoplakia
- Syndromes due to mutations in *TERC* and *TERT* lack mucocutaneous findings and predispose patients to various solid tumors in addition to MDS/AML
  - *TERC* encodes RNA components of telomerase
  - *TERT* encodes reverse transcriptase component of telomerase

Juvenile Myelomonocytic Leukemia (JMML) Associated with Neurofibromatosis, Noonan Syndrome, Or Noonan Syndrome-Like Disorders

- JMML is characterized by the proliferation of granulocytic and monocytic lineages with increased peripheral blood monocyte count $\geq 1 \times 10^9$/L, blast percentage in peripheral blood and bone marrow <20%, splenomegaly, and no Philadelphia (Ph) chromosome or *BCR::ABL1* fusion
  - Patients often have leukocytosis, a shift toward immature granulocytes, thrombocytopenia, and anemia with nucleated red blood cells
- 15% of cases occur in infants with Noonan syndrome, and 10% occur in children with neurofibromatosis type 1
- 90% of patients have mutations of RAS pathway genes, e.g., *PTPN11, SOS-1, KRAS,* and *RAF*
- Patients should be tested for somatic or germline mutation in *PTPN11, KRAS,* or *NRAS* and germline mutation of *CBL* or *NF1*
- Patients with germline mutations in *PTPN11, KRAS,* or *NRAS* may have transient abnormal myelopoiesis of Noonan syndrome, which can resolve without treatment
- Somatic mutations in these genes are associated with aggressive disease course

Myeloid Neoplasms Associated with Down Syndrome

Transient Abnormal Myelopoiesis Associated with Down Syndrome (DS-TAM)

- Epidemiology
  - Occurs in up to 30% of newborns with Down syndrome
  - May not be diagnosed in all cases
- Clinical features
  - Occurs in neonates
  - Thrombocytopenia and less commonly other cytopenias

- May have leukocytosis and increased blasts in peripheral blood
- Hepatosplenomegaly, jaundice, ascites, respiratory distress, bleeding, pericardial effusions, pleural effusions, and sometimes severe clinical complications
- Usually undergoes spontaneous remission by age 3 months
  - Rarely patients may die or require treatment
  - Rare patients develop myeloid leukemia associated with Down syndrome
- Genetic features
  - Associated with acquired *GATA1* mutation
    - Transcription factor involved in normal erythropoiesis and megakaryopoiesis
    - Located at Xp11.23

Myeloid leukemia Associated with Down Syndrome (ML-DS)

- Epidemiology
  - 1–2% of children with Down syndrome develop AML by age 5 years
  - Acute megakaryoblastic leukemia accounts for $\geq 50\%$ of cases
  - 20–30% of patients have a history of TAM, usually 1–3 years prior
- Sites of involvement
  - Involves peripheral blood, bone marrow and almost always has extramedullary involvement such as spleen and liver
- Histomorphology
  - In leukemic phase, blasts are usually present in peripheral blood as well as bone marrow and have distinctive features of megakaryoblasts, including round to slightly irregular nuclei, moderate basophilic cytoplasm, cytoplasmic blebs, and coarse basophilic cytoplasmic granules
  - Dyserythropoiesis and dysgranulopoiesis may also be present
- Genetic testing
  - In addition to trisomy 21, somatic *GATA1* mutations in exon 2 (or rarely exon 3) occur in all cases of DS-TAM and ML-DS
    - *GATA1* is essential for terminal erythroid and megakaryocyte maturation
    - ML-DS is clonally linked to precursor TAM
  - Other genetic abnormalities that may occur include trisomy 8, monosomy 7, and mutations of *CTCF, EZH2, KANSL1, JAK2, JAK3, MPL, SH2B3,* and RAS pathway genes
- Patients respond well to chemotherapy with cytarabine
- Very favorable prognosis compared to children with AML without Down syndrome

Myeloid Leukemias

Introduction

- Epidemiology
  - Acute myeloid leukemia (AML) accounts for 31% of cases of adult leukemia
  - AML accounts for 15–20% of cases of pediatric leukemia
- Sites of disease involvement
  - Involves peripheral blood and bone marrow
  - Extramedullary involvement of acute myeloid leukemia is termed myeloid sarcoma and occurs in 3–9% of cases at diagnosis
    Incidence varies with diagnostic subtype
- General histomorphology
  - Increased myeloid blasts
    Commonly intermediate to large with round to oval nuclei, prominent nucleoli, and blue gray cytoplasm
    May have irregular nuclear contours
    May have cytoplasmic granulation
  - Depending on subtype, it may include other immature myeloid cells such as monoblasts, promonocytes, and promyelocytes
- Cytogenetic findings
  - In addition to specific abnormalities in certain diagnoses, several nonspecific myeloid neoplasm-associated abnormalities have been identified
    Include trisomy 8, loss of chromosome 7 or 7q, deletion of 5q, deletion of 20q, and loss of chromosome 17 or 17p
  - Many myeloid leukemias have normal karyotype
- Molecular findings
  - Gene mutations are common in myeloid leukemia and include tumor suppressor genes like *TP53;* DNA methylation-related genes such as *TET2, IDH1, IDH2,* and *DNMT3A;* signaling pathway genes like *FLT3, KIT, KRAS,* and *NRAS*; chromatin-modifying genes including *ASXL1* and *EZH2*; transcription factors such as *CEBPA* and *RUNX1*; Cohesin-complex genes like *STAG2*; and spliceosome complex genes such as *SRSF2* and *U2AF1*

Acute Myeloid Leukemia with Recurrent Genetic Abnormalities

- AML with recurrent genetic abnormalities generally occur due to chromosomal rearrangements that generate a fusion gene encoding a chimeric protein that is required, but usually not sufficient for leukemogenesis
  - In the multihit model, both class 1 mutations that increase cellular proliferation and class 2 mutations that impair cellular differentiation and maturation are required for leukemogenesis
- These types of AML do not require a blast cell count of ≥20% for diagnosis

Acute Myeloid Leukemia with t(8;21)(q22;q22.1); *RUNX1::RUNX1T1*

- AML with t(8;21)(q21;q22.1) and inv(16)(p13q22)/t(16;16)(p13;q22) are also known as core binding factor leukemias due to having cytogenetic rearrangements that disrupt genes encoding subunits of core binding factor (CBF)
  - Core binding factor (CBF) is a heterodimeric protein complex involved in transcriptional regulation of hematopoiesis
- Pathogenesis
  - *RUNX1* 21q22.12 encodes CBFα, the DNA-binding subunit of CBF
  - *RUNX1T1* 8q21.3 encodes CBFβ, which normally increases the DNA-binding affinity of *RUNX1*
  - The t(8;21)(q22;q22.1) translocation results in the production of an abnormal *RUNX1::RUNX1T1* fusion protein that includes the DNA-binding domain of *RUNX1* and most of *RUNX1T1* except for the C-terminal transactivation domain
  - This fusion protein binds promoter regions of *RUNX1* target genes and blocks transcription, disrupting hematopoiesis
- Epidemiology
  - Accounts for 7% of adult AML and 10–15% of pediatric AML cases and is more common in younger patients
- Sites of disease involvement
  - Extramedullary involvement is rare but can be associated with gingival hyperplasia, cutaneous involvement, or splenomegaly
- Histomorphology
  - Increased myeloblasts in peripheral blood and/or bone marrow with abundant basophilic cytoplasm and characteristic single long, thin Auer rods with tapered ends
  - Granulocytes with maturation, salmon-colored granules, and dysplastic nuclear features
  - May express aberrant B-cell antigens including CD19, CD79a, and PAX-5
  - Mast cell disease may be present
- Genetic testing
  - Conventional karyotype readily detects t(8;21)
  - FISH demonstrates *RUNX1::RUNX1T1* fusion gene
  - RT-PCR is useful for disease monitoring, as >1 log increase of transcript levels or <3 log reduction at remission is associated with relapse
  - 70% of cases show additional chromosome abnormalities, e.g., loss of a sex chromosome or del(9q)

- *KIT* mutations, including D816V, present in up to 50% of cases
  - ○ Associated with coexisting systemic mastocytosis
- Other common gene mutations include *FLT3, NRAS, KRAS, ASXL1, ASXL2*, and *ZBTB7A*

Acute Myeloid Leukemia with inv(16)(p13.1;q22) or t(16;16)(p13.1;q22); *CBFB::MYH11*

- Pathogenesis (Fig. 25.2)
  - *CBFB* 16q22 encodes CBFβ, which normally increases the DNA-binding affinity of CBFα
  - *MYH11* 16p13.1 encodes smooth muscle myosin heavy chain, involved in muscle cell contraction
  - The inv(16)(p13.1q22) inversion or, less commonly, t(16;16)(p13.1;q22) translocation results in the production of an abnormal *CBFB::MYH11* fusion protein
  - This fusion protein sequesters normal CBFα in cytoplasm and functions as a transcriptional repressor of CBF target genes, disrupting hematopoiesis
- Epidemiology
  - Accounts for 8% of adult AML and 6–12% of pediatric AML cases
  - Median age: 40–45 years
- Sites of disease involvement
  - Extramedullary involvement is common, involving skin (especially scalp), gingiva, or central nervous system (CNS)
- Histomorphology
  - Associated with monocytic differentiation
  - Increased blasts and blast equivalents (myeloblasts, monoblasts, promonocytes)
  - Abnormal eosinophils with mixed eosinophilic and basophilic granules
  - Hypercellular marrow with myelomonocytic predominance, may not show effacement by monotonous sheets of blasts
- Genetic testing
  - *CBFB::MYH11* fusion can be detected with FISH for diagnosis or RT-PCR, especially for residual disease monitoring
  - Inversion 16 is subtle and may be overlooked on karyotype
  - 40% of cases show additional chromosome abnormalities, e.g., gains of chromosomes 22, 8, or 21, del(7q), trisomy 22
  - Secondary gene mutations are common (>90% of cases) and include mutations of *KIT, NRAS, KRAS, FLT3,* and less commonly *ASXL2*

Acute Promyelocytic Leukemia with t(15;17)(q24;q21); *PML::RARA*

- Pathogenesis
  - *PML* 15q24 encodes a transcription factor and tumor suppressor that regulates apoptosis
  - *RARA* 17q21 encodes nuclear retinoic acid receptor alpha, which regulates the transcription of genes involved in differentiation, apoptosis, and granulopoiesis
  - The t(15;17)(q24;q21) translocation results in the production of an abnormal *PML::RARA* fusion protein
    - ○ RARA breakpoint occurs in intron 2
    - ○ PML breakpoint is variable
      - ◆ Bcr1 (long form) in intron 6 in 50–60% of cases
      - ◆ Bcr2 (variable form) in exon 6 in 5% of cases
      - ◆ Bcr3 (short form) in intron 3 in 35–45% of cases
  - This fusion protein binds to nuclear corepressor (NCOR)-histone deacetylase complex, enhances histone deacetylase function, inhibits gene transcription, and prevents differentiation of myeloid cells
  - 1–2% of cases of acute promyelocytic leukemia (APL) have variant *RARA* rearrangement with an alternate translocation partner, including *ZBTB16* 11q23.2 (most common), *NUMA1* 11q13.4, *NPM1* 5q35.1, or *STAT5B* 17q21.2
  - *FLT3* internal tandem duplication (ITD) mutations are detected in 40% of cases
- Epidemiology
  - Accounts for 5–8% of all AML cases
  - Subset of cases are therapy-related and occur due to chemotherapy or radiation for a prior malignancy
- Sites of disease involvement
  - Extramedullary disease is rare, with CNS and skin as the most common sites
  - Can present with life-threatening disseminated intravascular coagulation
    - ○ Rapid diagnosis and initiation of all trans-retinoic acid (ATRA) therapy critical to prevent early death due to coagulopathy
- Histomorphology (Fig. 25.3)
  - Increased blasts with bilobed or "sliding plates" nuclei and abnormal promyelocytes
  - Hypergranular variant: more common, low WBC count, marked thrombocytopenia, abnormal promyelocytes with abundant large granules, blasts occasionally contain numerous Auer rods
  - Microgranular variant: less common, high WBC count, blasts lack cytoplasmic granules, morphologically similar to acute monocytic leukemia, rare Auer rods

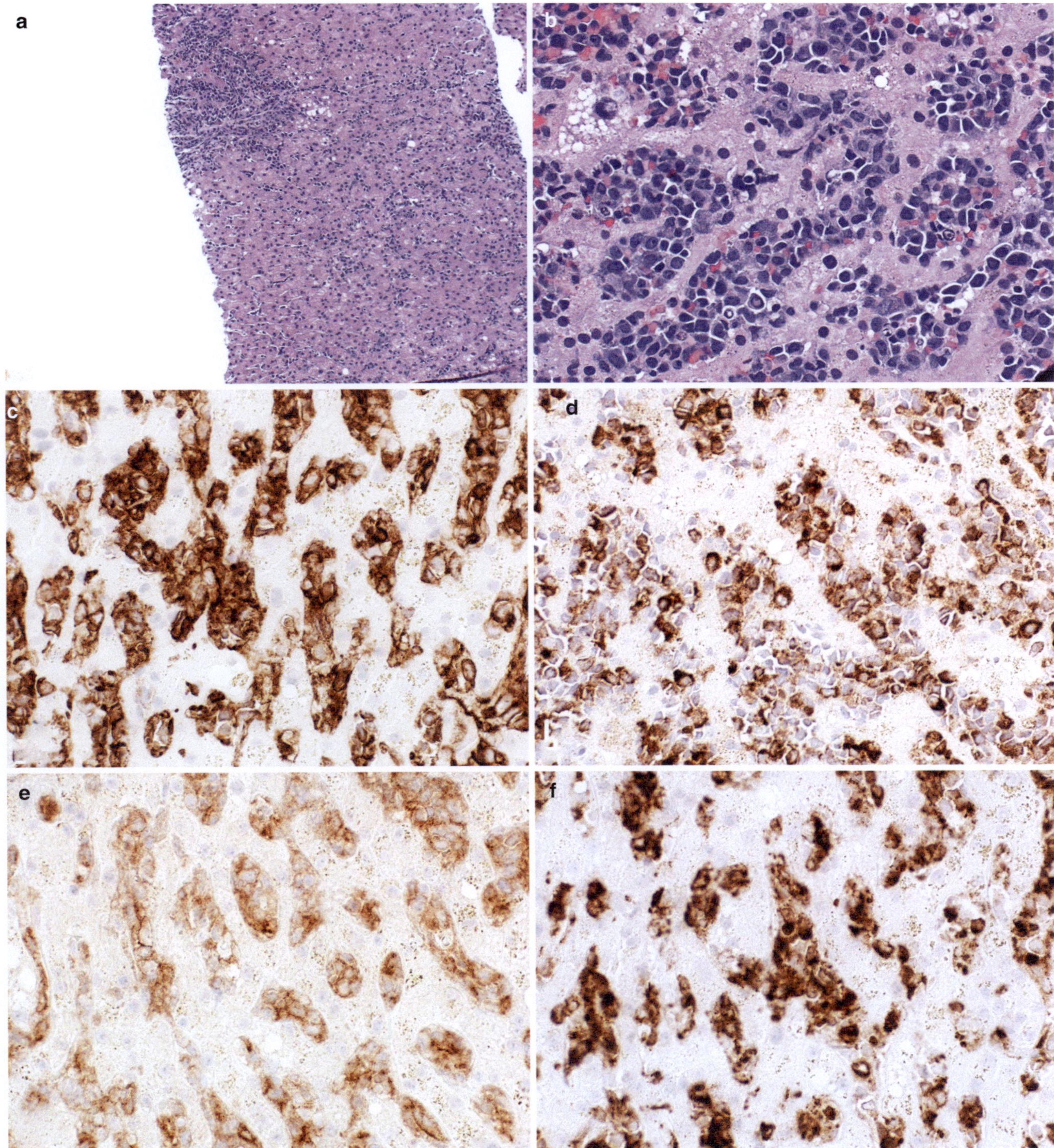

Fig. 25.2 Hepatic involvement by AML with inv.(16)(p13.1;q22). (**a**) Large, atypical cells infiltrate the hepatic sinusoids. (**b**) Higher magnification shows large, irregularly shaped blasts with fine nuclear chromatin and prominent nucleoli. (**c**) The cells are positive with CD45, indicating hematopoietic origin. CD45 staining may not always be positive in acute leukemias. The cells are positive for myeloid lineage-associated antigens myeloperoxidase (**d**) and CD33 (**e**). (**f**) Staining with lysozyme indicates monocytic differentiation

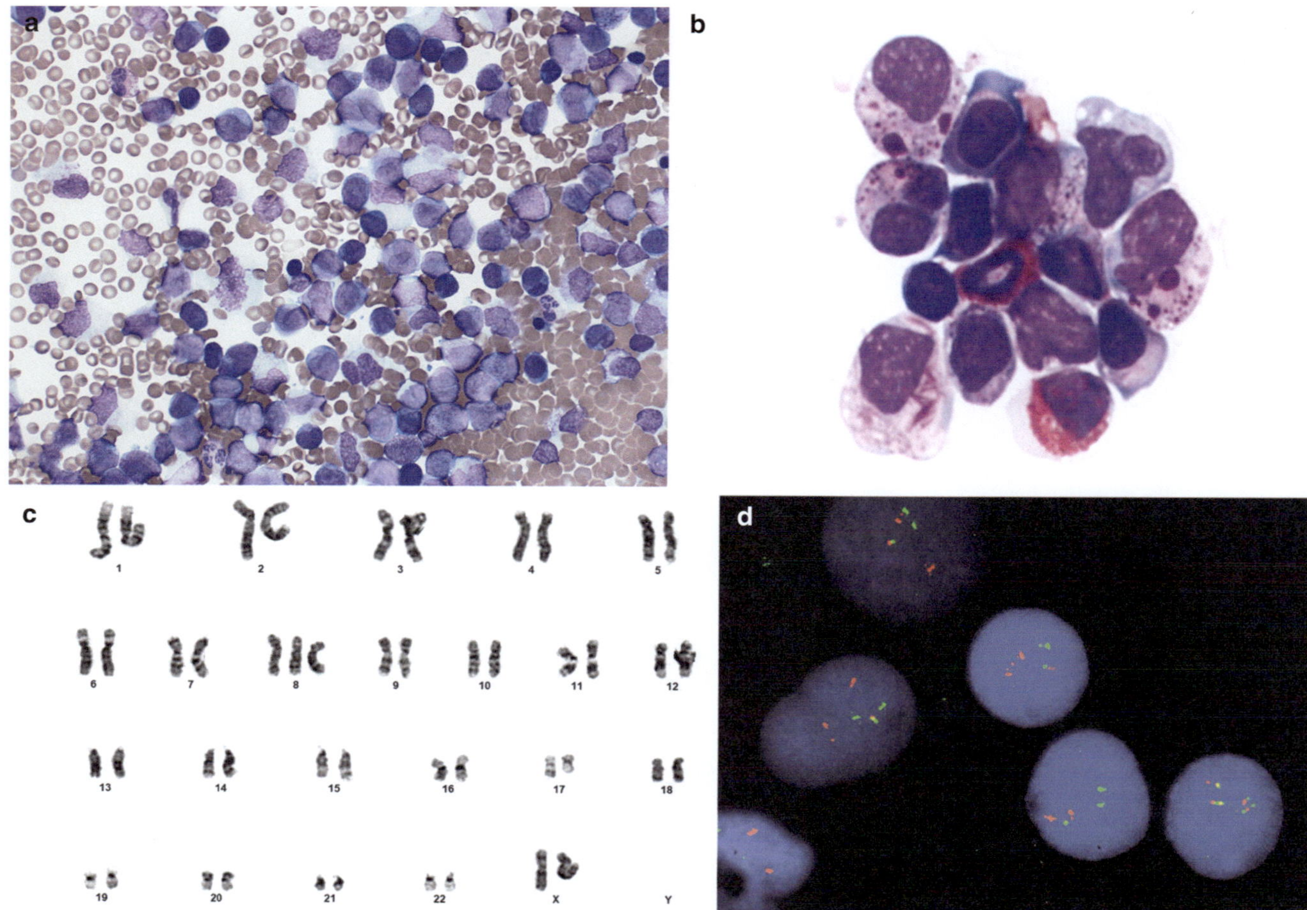

Fig. 25.3 (**a**) Neoplastic promyelocytes in APL showing large size, irregular nuclear contours with bilobed forms, and variably prominent nucleoli. (**b**) A subset of cells may have heavy cytoplasmic granulation and multiple Auer rods, particularly in the hypergranular variant. (**c**) Conventional karyotype from a patient with APL shows a complex translocation involving chromosomes 15, 17, and 19. (**d**) Dual-color, dual-fusion FISH for PML at 15q24.1 (orange) and RARA at 17q21.1 (green) in this case shows fusion signals, confirming t(15;17)

- Genetic testing (Fig. 25.4)
    - Conventional cytogenetics detects t(15;17) in 90% of cases as well as complex variant translocations
    - FISH can detect *PML::RARA* fusion gene and most cryptic rearrangements
    - RT-PCR is useful for the assessment of treatment response and detection of minimal residual disease or early relapse
        - Typically utilizes primers amplifying different breakpoints in PML: bcr1, 2, and 3
        - Can be used for the diagnosis of rare cases unidentified by conventional cytogenetics and FISH
        - Quantitation requires comparison to standard curve
    - 40% of cases have secondary cytogenetic abnormalities, e.g., gain of chromosome 8
    - 30–40% of cases have *FLT3* mutations including *FLT3*-internal tandem duplication and *FLT3*-tyrosine kinase domain mutations

Acute Myeloid Leukemia with t(9;11) (p21.3;q23.3); *MLLT3::KMT2A* or other *KMT2A* rearrangement

- Pathogenesis
    - *KMT2A* 11q23.3 encodes lysine methyltransferase 2A, a DNA-binding transcriptional coactivator protein involved in embryogenesis and hematopoiesis
        - Translocations are associated with both AML and lymphoblastic leukemias
    - *MLLT3* 9p21.3 encodes MLLT3 super elongation complex subunit, a protein involved in the regulation of transcription
    - The t(9;11)(p21.3;q23.3) translocation results in the production of an abnormal *MLLT3:KMT2A* fusion protein
    - This fusion protein impairs cell differentiation and maturation by overriding the normal regulation of transcription of genes associated with early hematopoiesis

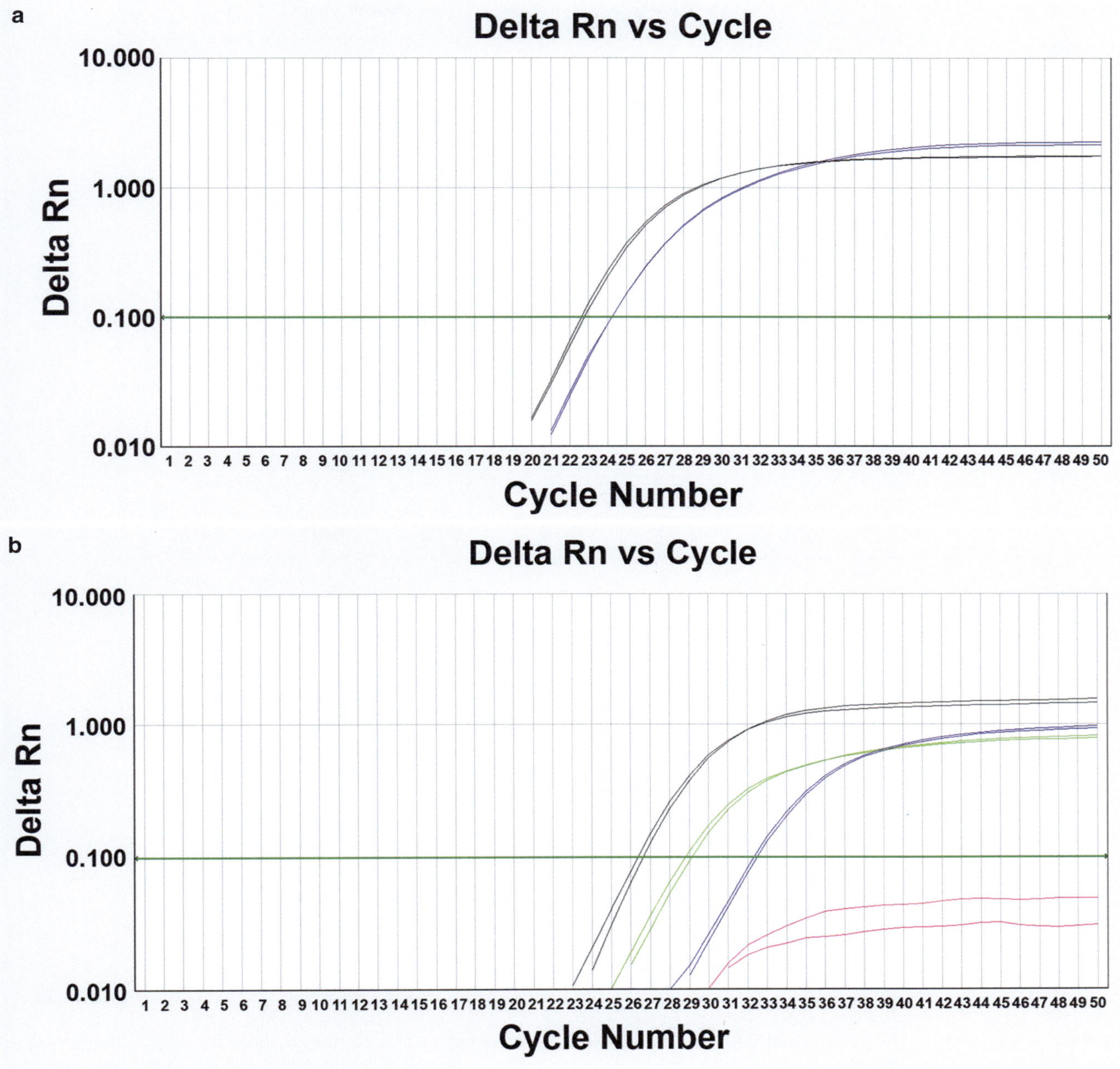

Fig. 25.4 Quantitative RT-PCR positive for the short form (bcr 3) transcript (blue curve) (**a**) and the long form (bcr 1) transcript (green curve) (**b**). In both, amplification of *ABL1* (black curve) serves as a control

- AML with *KMT2A* and other non-*MLLT3* translocations can occur but classified separately from AML with t(9;11)(p21.3;q23.3)
- Epidemiology
  - Accounts for 9–12% of pediatric AML, and 1–2% of adult AML cases
- Sites of disease involvement
  - Extramedullary disease is common, most commonly in skin and CNS
- Can present with disseminated intravascular coagulation
- Histomorphology
  - Increased blasts, usually monoblasts and promonocytes, with hypercellular bone marrow
  - Typically, no significant dysplasia, basophilia, or eosinophilia
- Genetic testing
  - Karyotype readily detects the t(9;11) translocation

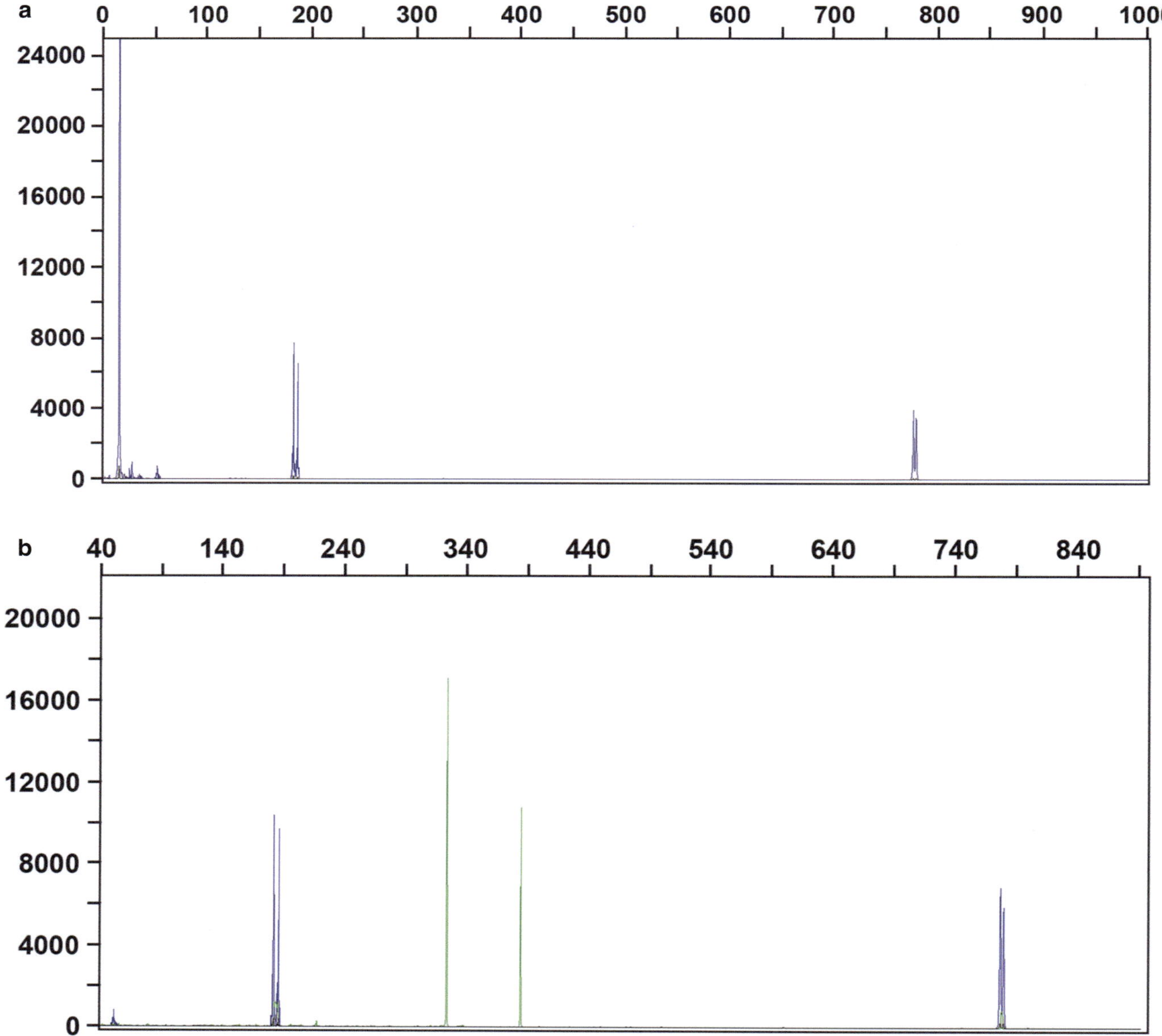

Fig. 25.5 Sizing assays can be used to identify *NPM1* mutations. (**a**) Here, the *NPM1* gene (blue) creates a peak at 186 base pairs, while the 4 base pair insertion causes a second peak at 190 base pairs. (**b**) *NPM1* mutations can occur together with *FLT3* internal tandem duplications, which can also be identified by sizing assays. The *FLT3* gene (green) creates a peak at 329 base pairs, with the larger internal tandem duplication creating a second, longer transcript

- FISH with break-apart probe detects *KMT2A* gene rearrangement
- RT-PCR is useful for disease monitoring and detecting early relapse
- Secondary cytogenetic abnormalities are common including gain of chromosome 8
- More commonly mutated genes include *KRAS, NRAS, PTPN11, FLT3*-TKD, *RUNX1, TET2, PLCG2,* and *ZRSR2*
- 40% of cases show overexpression of *MECOM*

Acute Myeloid Leukemia with t(6;9)(p23;q34.1); *DEK:NUP214*

- Pathogenesis
  - *DEK* 6p22.3 encodes a chromatin remodeling protein that regulates stem cell gene transcription
  - *NUP214* 9q34.13 encodes a nucleoporin involved in nucleocytoplasmic transport
  - The t(6;9)(p23;q34.1) translocation results in the production of an abnormal *DEK::NUP214* fusion protein that may induce proliferation through mTOR pathway

- Epidemiology
  - Represents 1% of all AML cases
- Histomorphology
  - Basophilia, i.e., basophil count ≥2%, occurs in around half of the cases
  - Marrow shows multilineage dysplasia, usually including granulocytic and erythroid
  - 50–60% of cases have TdT expression in myeloid blasts
- Genetic testing
  - Karyotype can detect t(6;9) translocation except in cryptic cases
  - FISH and/or RNA sequencing can detect cryptic translocation
  - A complex karyotype is present in some cases
  - *FLT3*-ITD mutations are very common (70–80% of cases)
  - Additional mutations may include *NRAS, KRAS, FAT1, NOTCH1*, and *PRDM1*

Acute Myeloid Leukemia with inv(3)(q21.3q26.2) or t(3;3)(q21.3;q26.2); *GATA2, MECOM*

- Pathogenesis
  - *GATA2* 3q21.3 encodes a zinc-finger transcription factor that regulates development and proliferation of hematopoietic and endocrine cell lines
  - *MECOM* 3q26.2 encodes zinc-finger transcription factor involved in hematopoiesis, apoptosis, cell differentiation, and proliferation
  - The inv(3)(q21.3q26.2) inversion or t(3;3)(q21.3;q26.2) translocation causes aberrant *MECOM* overexpression and *GATA2* haploinsufficiency
- Epidemiology
  - Accounts for 1–2% of all AML, usually in adults
- Histomorphology
  - Increased myeloblasts, most commonly with morphologies of AML without maturation, myelomonocytic, or megakaryoblastic
  - Increased dysplastic megakaryocytes with unilobed or bilobed nuclei and multilineage dysplasia
- Genetic testing
  - *MECOM* rearrangements involving the long arm of chromosome 3 may be cryptic on routine karyotype but detectable by FISH
  - Additional karyotypic abnormalities are common, e.g., monosomy 7, del(5q), and complex karyotypes
  - Secondary gene mutations are found in 98% of cases, including mutations of RAS or receptor tyrosine kinase signaling pathway genes (e.g., *NRAS, PTPN11, FLT3, KRAS, NF1, CBL*, and *KIT*), *GATA2, RUNX1*, and *SF3B1*

Acute Myeloid Leukemia (Megakaryoblastic) with t(1;22)(p13.3;q13.1); *RBM15::MRTFA*

- Pathogenesis
  - *RBM15* 1p13.3 encodes a protein involved in RNA splicing, stability, and transcription
  - *MRTFA (MKL1)* 22q13.1-q13.2 encodes myocardin-related transcription factor A, a transcriptional coactivator that interacts with serum response factor (SRF)
  - The t(1;22)(p13.3;q13.1) translocation produces a chimeric protein that modulates chromatin organization, differentiation, and extracellular signaling pathways
- Epidemiology
  - Accounts for <1% of all AML
  - Most commonly occurs in infants and young children, with median age of diagnosis of 4 months
- Sites of disease involvement
  - Hepatosplenomegaly is common at diagnosis
- Histomorphology
  - Unlike other entities of AML with recurrent genetic abnormalities, this entity requires ≥20% blasts for diagnosis
    - If <20% blasts, first exclude bone marrow fibrosis causing a falsely low blast count, then monitor for development of definitive evidence of AML, e.g., extramedullary disease or myeloid sarcoma
  - Megakaryoblasts with background myelofibrosis and micromegakaryocytes without multilineage dysplasia in bone marrow
- Genetic testing
  - The t(1;22) translocation can be detected by karyotype and is often the only karyotypic abnormality
  - RT-PCR can detect the *RBM15::MRTFA* fusion transcript and may be useful when bone marrow biopsy is inadequate due to marrow fibrosis (dry tap) and for minimal residual diseases monitoring

Acute Myeloid Leukemia with *BCR::ABL1*

- Pathogenesis
  - *BCR* 22q11.2 encodes a protein that may act as a GTPase-activating protein (GAP)
  - *ABL1* 9q34.1 encodes a nonreceptor tyrosine kinase involved in cell differentiation, division, adhesion, and stress response
  - The t(9;22)(q34.1;q11.2) translocation produces the oncogenic BCR-ABL1 fusion protein
  - Different fusion proteins occur depending on the *BCR* breakpoint
    - The most common fusion protein in AML with *BCR::ABL1* is p210 (b2a2 or b3a2), with a minority of cases having p190 (e1a2)
- Epidemiology
  - Accounts for <1% of all AMLs and <1% of all *BCR::ABL1*-positive leukemias

- Sites of disease involvement
  - Less frequently presents with splenomegaly compared to myeloid blast transformation of chronic myeloid leukemia (CML)
- Histomorphology
  - Resembles blast phase of CML, except basophilia and dwarf megakaryocytes less common
- Genetic testing
  - Conventional cytogenetics detects the t(9;22) translocation in 95% of cases and can identify variant Philadelphia chromosome
  - FISH detects the *BCR::ABL1* fusion gene in >95% of cases including most cryptic rearrangements with greater sensitivity than conventional cytogenetics
  - Quantitative RT-PCR is the most sensitive assay for *BCR::ABL1* transcripts useful for monitoring treatment response, MRD, and early relapse
  - Usually, secondary cytogenetic abnormalities are present such as loss of chromosome 7, gain of chromosome 8, and complex karyotypes
  - Associated gene alterations include loss of *IKZF1* and *CDKN2A*, cryptic deletions within IGH and TRG genes, and mutations of *NPM1* and *FLT3*-ITD

Acute Myeloid Leukemia with Mutated *NPM1*

- Pathogenesis
  - *NPM1* encodes nucleophosmin, which is involved in nucleocytoplasmic transport of proteins
  - *NPM1* mutations are most commonly 4 bp insertions that result in frameshift and alter the nuclear localization motif, leading to abnormal concentration of the protein in cytoplasm
  - Occur in exon 12
- Epidemiology
  - *NPM1* mutations occur in 5% of pediatric AML and 30% of adult AML cases
- Sites of disease involvement
  - May have extramedullary involvement, including gingiva, lymph nodes, or skin
- Histomorphology
  - Varied morphology, most commonly monocytic or myelomonocytic differentiation
  - Blasts often lack CD34
  - Multilineage dysplasia present in up to a quarter of de novo cases
- Genetic testing (Fig. 25.5)
  - *NPM1* mutations typically alter the length of the gene and can be detected by PCR sizing assay
  - Secondary mutations are common including *FLT3, DNMT3A, IDH1, KRAS, NRAS,* and cohesin-complex genes
  - Demonstrates a distinct gene expression profile with upregulation of HOX genes and a unique microRNA signature

- 5–15% of cases have chromosomal abnormalities, such as gain of chromosome 8 and del(9q)

Acute Myeloid Leukemia with In-Frame bZIP *CEBPA* mutation

- Pathogenesis
  - *CEBPA* encodes a transcription factor involved in granulocytic differentiation
  - Mutations in the C-terminus basic leucine zipper interfere with DNA binding and dimerization
  - N terminus mutations are not required for diagnosis
- Epidemiology
  - *CEBPA* mutations occur in 4–9% of pediatric and young adult AML cases, with lower frequency in older adults
- Histomorphology
  - Most cases have features of AML with or without maturation, may have monocytic differentiation
  - May have multilineage dysplasia
    - Presence of *CEBPA* mutations precludes a diagnosis of AML with myelodysplasia-related changes
- Genetic testing
  - *CEBPA* mutations can be detected by gene sequencing
  - Secondary *GATA2* and *FLT3*-ITD mutations may occur
  - <30% of cases have chromosomal abnormalities, such as del(9q) and rarely del(11q)
  - Screening for germline mutation should be considered in cases with biallelic mutations

Acute Myeloid Leukemia with Mutated *RUNX1*

- Pathogenesis
  - *RUNX1* encodes CBFα, a key transcription factor in hematopoiesis
  - Most *RUNX1* mutations occur in the runt homology domain (exons 3-5) or transactivating domain (exons 6–8)
- Epidemiology
  - *RUNX1* mutations occur in 4–16% of AML cases, with higher frequency in older adults
- Histomorphology
  - No specific morphologic features
- Genetic testing
  - Cooperating mutations often occur such as *ASXL1, KMT2A* partial tandem duplication, *FLT3*-ITD, *IDH1* R132, *IDH2* R140 and R172, *SRSF2, EZH2,* and *STAG2* mutations
  - Karyotypic abnormalities may occur including trisomies 8 and 13

Acute Myeloid Leukemia with Myelodysplasia-related Changes (AML-MRC)

- Required criteria for diagnosis:
  - AML with myelodysplasia related gene mutations requires mutation in *ASXL1*, *BCOR*, *EZH2*, *RUNX1*, *SF3B1*, *SRSF2*, *STAG2*, *U2AF1*, or *ZRSR2*
  - AML with myelodysplasia related cytogenetic abnormalities requires a complex karyotype (>3 unrelated clonal abnormalities, del(5q)/t(5q)/add(5q), -7/del(7q), +8, del(12p)/t(12p)/add(12p), i(17q), -17/add(17p), or del (17p), del(20q), and/or idic(X)(q13)
  - AML with mutated *TP53* requires pathogenic *TP53* mutation with variant allele frequency >10%
  - ≥20% blasts in peripheral blood or bone marrow
  - No history of cytotoxic or radiation therapy for an unrelated disease
  - No AML-defining recurrent cytogenetic abnormality
- Pathogenesis
  - Pathogenesis not well understood but similar mechanisms as MDS
  - Often associated with MDS-related unbalanced or rarely balanced cytogenetic abnormalities
- Epidemiology
  - Occurs mainly in elderly patients, rare in children
  - Represents up to 48% of adult AML cases
- Histomorphology
  - Erythroid dysplasia, e.g., multinucleation, abnormal nuclear lobation, megaloblastic changes, ring sideroblasts
  - Granulocytic dysplasia, e.g., hypogranular cytoplasm, nuclear hypo- or hypersegmentation, pseudo-Pelger-Huet nuclei, megaloblastic changes
  - Megakaryocytic dysplasia, e.g., clustering of megakaryocytes, hypolobated nuclei, multiple discrete "pawn ball" nuclei
  - ≥20% myeloblasts that may contain Auer rods, monoblasts, promonocytes, or megakaryoblasts
- Genetic testing
  - Similar chromosome abnormalities as those found in MDS, most commonly complex karyotypes, loss of chromosome 7, del(7q), del(5q), and unbalanced 5q translocations
  - Mutation testing is required for diagnosis of AML with MDS related gene mutations

Therapy-Related Myeloid Neoplasms (t-MN)

- Myeloid neoplasms that occur as a complication of prior cytotoxic and/or radiation therapy
  - Implicated therapies include alkylating agents, topoisomerase II inhibitors, large-field ionizing radiation therapy, antimetabolites, and antitubulin agents

- Includes therapy-related acute myeloid leukemia (t-AML), therapy-related myelodysplasia (t-MDS), and therapy-related myelodysplastic/myeloproliferative neoplasm (t-MDS/MPN)
- Pathogenesis
  - Cytotoxic or radiation therapy induce molecular genetic changes, especially in individuals with heritable mutations in DNA repair or drug metabolism genes
  - Alkylating agents associated with monosomy of chromosome 5 (del(5q)) and 7 (del(7q))
  - Topoisomerase II inhibitors associated with *KMT2A* gene rearrangements and recurrent translocations e.g., t(8;21), t(15;17) and inv(16)
- Epidemiology
  - Accounts for 10–20% of newly diagnosed AML/MDS
- Histomorphology
  - Alkylating agents associated with variable marrow cellularity, multilineage dysplasia, increased small megakaryocytes with hypolobated nuclei, variable blast count, fibrosis common
  - Topoisomerase II inhibitors associated with hypercellular marrow, features of acute monoblastic/monocytic/myelomonocytic leukemia, 11q23 translocations, t(15;17), t(8;21), inv16
- Genetic testing
  - >90% of cases have an abnormal karyotype, with cytogenetic findings correlating with the latent period between the causative therapy and onset of the leukemia
    - o Long latent period is associated with unbalanced chromosomal aberrations including partial loss of 5q, loss of chromosome 7, and del(7q)
      - ♦ Loss of 5q is associated with a complex karyotype and additional chromosomal abnormalities (e.g., del(13q), del(20q), del(11q), del(3p), loss of 17p or chromosome 17, loss of chromosome 18 or 21, or gain of chromosome 8) as well as mutation or loss of *TP53*
    - o Cases with a short latent period often have balanced translocations such as t(9;11)(p21.3;q23.3), t(11;19)(q23.3;p13.1), t(8;21)(q22;q22.1), t(3;21)(q26.2;q22.1), t(15;17)(q24.1;q21.1), and inv(16)(p13.1q22)
  - Gene mutations that frequently occur include *TP53*, which occur in 50% of cases and confer poor prognosis, *TET2*, *PTPN11*, *IDH1/2*, *NRAS,* and *FLT3*

Myeloid Sarcoma

- Tumor mass of proliferating myeloblasts at an extramedullary site, i.e., outside bone marrow

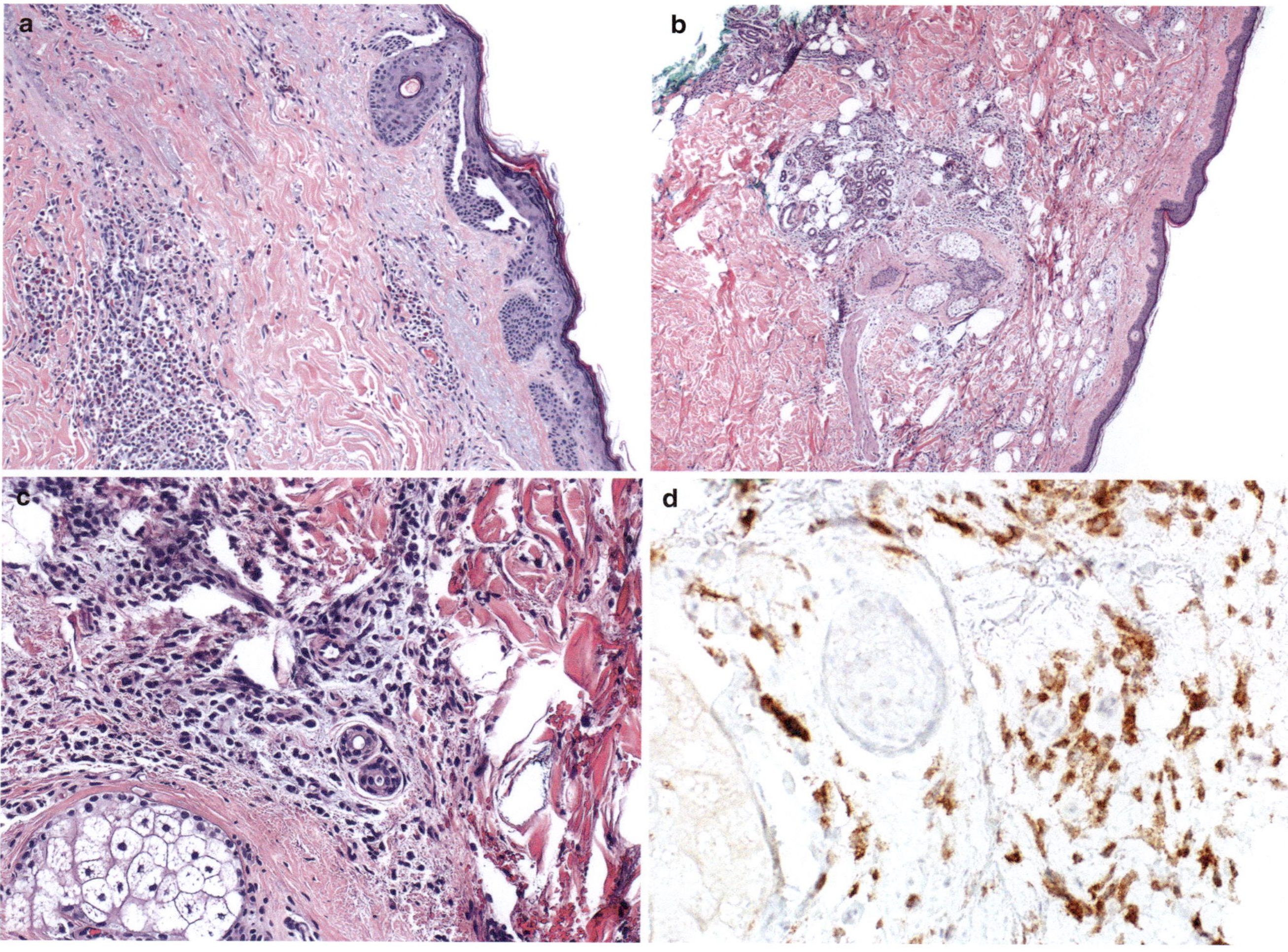

Fig. 25.6 (**a**) Dermal involvement in AML with a dense, tumor like infiltrate of myeloid blasts showing large size and irregular nuclei with prominent nucleoli. (**b**) A more subtle case shows periadnexal infiltration by atypical cells. (**c**) Higher magnification shows irregular nuclear contours and variable cytoplasm. (**d**) Positive staining with CD33 indicates myeloid lineage

- Any body site can be involved including skin (most common), lymph node, gastrointestinal tract, soft tissue, testis
  - Skin involvement termed "leukemia cutis"
- Pathogenesis
  - Involves unique homing and cell–cell interaction proteins including cytokine receptors CCR5, CXCR4, CXCR7, and CX3CR1, and matrix metalloproteinases MMP-9 and MMP-2
- Epidemiology
  - Relatively rare, with male-to-female ratio of 1.2:1 and median age of 56 years
- Histomorphology (Fig. 25.6)
  - Myeloblasts, promyelocytes, myelocytes, or metamyelocytes, potentially with myelomonocytic or monocytic features in diffuse or single-file pattern effacing normal tissue architecture
- Genetic testing
  - 55% of cases demonstrate chromosomal abnormalities, including monosomy 7, trisomy 8, *KMT2A* rearrangement, inv(16), trisomy 4, monosomy 16, loss of 16q, 5q, or 20q, trisomy 11, t(9;11), t(8;17), t(8;16), t(8;21), inv(16), and t(1;11)
  - RAS pathway mutations are common
  - 16% of cases carry *NPM1* mutations, and <15% carry *FLT3*-ITD
    - *NPM1* mutation more common with skin involvement

Lymphocytic Leukemias

B-Lymphoblastic Leukemia (B-ALL)/Lymphoma (B-LBL)

- Defined as neoplasm of lymphoblasts committed to B-cell lineage
  - B-ALL arises in either a hematopoietic stem cell or B-cell progenitor
  - Blood and bone marrow involvement is extensive in B-ALL, absent to minimal in B-LBL
- Epidemiology
  - 80–85% of acute lymphoblastic leukemia (ALL) cases are B-ALL
  - 75% of cases present before age 6 with a slight male predominance
- Histomorphology (Fig. 25.7)
  - Variable WBC count with lymphoblasts with high N:C ratio, finely dispersed nuclear chromatin, and distinct nucleoli in peripheral blood
  - Bone marrow usually extensively replaced by uniform lymphoblasts with patchy to diffuse preserved hematopoiesis, rarely with necrosis
  - Most types of B-ALL do not have specific features to distinguish them from other types of ALL
- Genetic testing
  - Nearly all B-ALL cases have clonal immunoglobulin heavy chain (IGH) rearrangements
  - Up to 70% of cases have T-cell receptor (TCR) gene rearrangements
  - Recurrent cytogenetic abnormalities define specific subtypes of B-ALL in the following section

- Additional cytogenetic abnormalities include del(6q), del(9p), del(12p), and t(17;19)(q22;p13.3), but are either not prognostically important or too rare to be included in the classification
- Many recurrent genetic alterations can occur, including *PAX5,* RAS family oncogenes, and *IKZF1*

B-Lymphoblastic Leukemia/Lymphoma with Recurrent Genetic Abnormalities

B-Lymphoblastic Leukemia/Lymphoma with t(9;22) (q34.1;q11.2); *BCR::ABL1*

- Pathogenesis (Fig. 25.8)
  - *BCR* 22q11.2 encodes a protein that may act as a GTPase-activating protein (GAP)
  - *ABL1* 9q34.1 encodes a nonreceptor tyrosine kinase involved in cell differentiation, division, adhesion, and stress response
  - The t(9;22)(q34.1;q11.2) translocation produces the oncogenic BCR::ABL1 fusion protein
  - Different fusion proteins occur depending on the *BCR* breakpoint
    - Fusions in B-ALL with *BCR::ABL1* include p210 (found in half of adult cases) and p190 (found in most childhood and half of adult cases)
- Epidemiology
  - Accounts for 25% of adult ALL and 2–4% of childhood ALL cases
- Genetic testing
  - *BCR::ABL1* rearrangement can be detected by standard karyotyping, FISH with a break-apart probe, or PCR

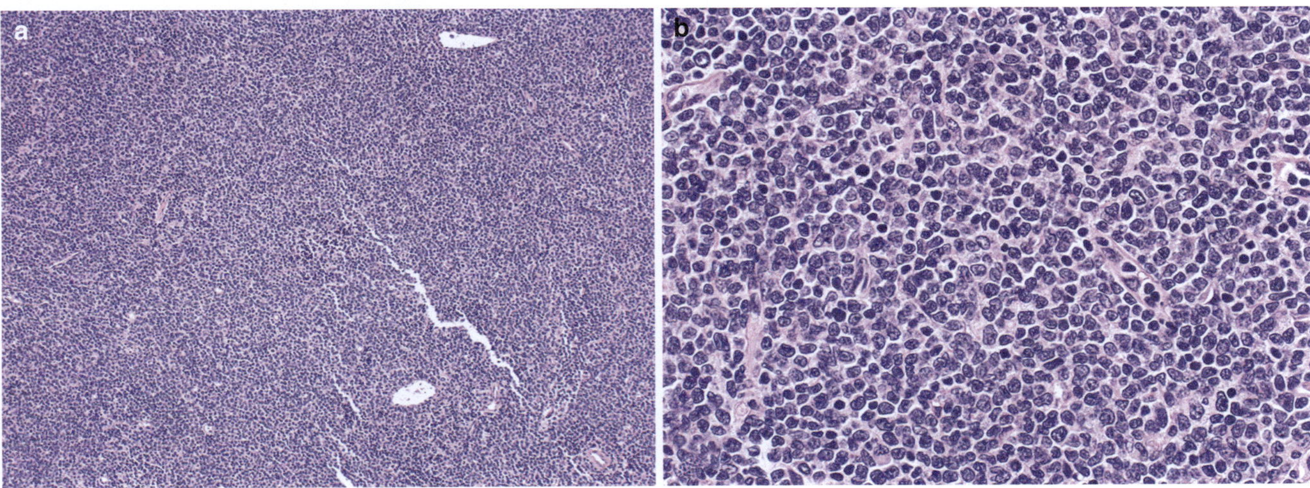

Fig. 25.7 (**a**) Salivary gland excision showing complete effacement by B-ALL with intermediate-sized blasts with high nuclear to cytoplasmic ratios. (**b**) Higher magnification shows irregular nuclear contours with variably prominent nucleoli

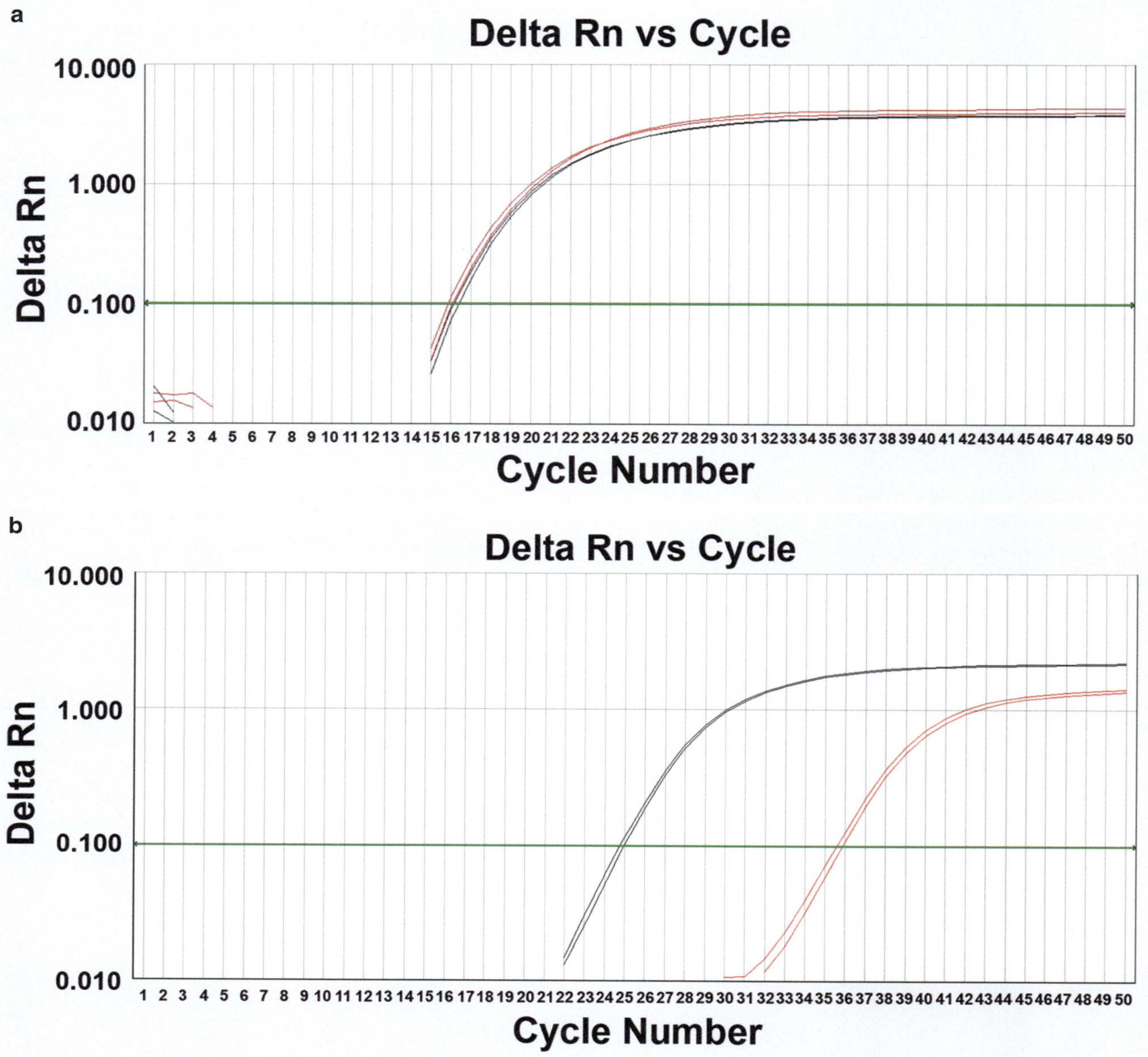

Fig. 25.8 (**a**) RT-PCR with primers for p190 (red) shows a high transcript level at diagnosis of B-ALL with t(9;22) compared to the *ABL1* control (black). (**b**) During treatment, RT-PCR can be used to monitor response with decreasing levels of fusion transcript compared to the control

- Quantitative RT-PCR is the most sensitive assay for *BCR::ABL1* transcripts and is used for monitoring treatment response, MRD, and early relapse
- Secondary genetic abnormalities can occur, rarely including other recurrent B-ALL abnormalities

B-Lymphoblastic Leukemia/Lymphoma with t(v;11q23.3); KMT2A-Rearranged

- Pathogenesis
  - *KMT2A* 11q23.3 encodes lysine methyltransferase 2A, a DNA-binding transcriptional coactivator protein involved in embryogenesis and hematopoiesis
  - *KMT2A* rearrangement with one of >100 fusion partners, such as *AFF1* 4q21, *MLLT1* 19p13, *MLLT3* 9p21.3, leads to production of an oncogenic fusion protein
- Epidemiology
  - Accounts for 70% of new ALL diagnoses in infants <1 year old
  - Less common in older children but becomes increasingly common into adulthood
- Genetic testing
  - *BCR::ABL1* rearrangement can be detected by standard karyotyping, FISH with a break-apart probe, or PCR for specific major translocation partners

- Frequently associated with overexpression of *FLT3*
- Pediatric cases have an overall low mutational burden relative to adults but often present with *PI3K-RAS* pathway mutations
- Common secondary mutations in adult cases include *CDKN2A, IKZF1, KRAS,* and *CREBBP*

B-Lymphoblastic Leukemia/Lymphoma with t(12;21) (p13.2;q22.1); *ETV6:RUNX1*

- Pathogenesis
  - *ETV6* 12p13.2 encodes a transcription factor
  - *RUNX1* 21q22.12 encodes a transcription activator for hematopoietic-specific genes
  - The *ETV6::RUNX1* fusion protein resulting from translocation acts in a dominant negative fashion to inhibit normal RUNX1 function
- Epidemiology
  - Accounts for 25% of child B-ALL cases (excluding infants), rare in adults
- Genetic testing
  - The *ETV6::RUNX1* fusion is virtually undetectable by conventional cytogenetics but can be detected by FISH or qRT-PCR

B-Lymphoblastic Leukemia/Lymphoma with Hyperdiploidy

- Pathogenesis
  - Blasts contain >50 chromosomes, usually <66 and without translocations or other structural alterations
- Epidemiology
  - Accounts for 25% of child B-ALL cases (excluding infants)
  - Decreasing frequency with age, accounting for 7–8% of adult B-ALL cases
- Genetic testing
  - Hyperdiploidy can be detected by conventional karyotype, FISH, or flow cytometric DNA index
  - The chromosomes that most commonly have extra copies include 21, X, 14, and 4
    - Chromosomes 1, 2, and 3 are least common

B-Lymphoblastic Leukemia/Lymphoma with Hypodiploidy

- Pathogenesis
  - Blasts contain <46 chromosomes
  - Subtypes include near-haploid ALL (23–29 chromosomes), low hypodiploid (33–39), high hypodiploid (40–43), and near-diploid (44–45)
- Epidemiology
  - Accounts for 5% of ALL cases
  - Occurs in children and adults, except for near-haploid ALL (children only)
- Genetic testing

- All cases show loss of ≥1 chromosome and sometimes nonspecific structural chromosomal abnormalities
- Karyotype, FISH, and flow cytometric DNA index can detect hypodiploidy
  - In cases of near-haploid or low hypodiploid ALL with endoreduplication, standard karyotyping may falsely indicate a near-diploid or hyperdiploid karyotype
- Near-haploid ALL often has RAS or receptor tyrosine kinase mutations, low hypodiploid ALL usually has loss-of-function *TP53* and/or *RB1* mutations, and high hypodiploid ALL is not associated with any specific gene alterations

B-Lymphoblastic Leukemia/Lymphoma with t(5;14) (q31.1;q32.1); *IGH/IL3*

- Pathogenesis
  - *IL3* 5q31.1 encodes interleukin-3, a growth factor involved in proliferation and differentiation of hematopoietic cells
  - IGH genes 14q32.33 encode immunoglobulin heavy chain
  - The fusion protein leads to overexpression of IL3
- Epidemiology
  - Accounts for <1% of ALL cases
  - Occurs in children and adults
- Histomorphology
  - Reactive eosinophilia in addition to increased lymphoblasts
- Genetic testing
  - The translocation can be detected by karyotype and FISH, but specific FISH probes are not widely available

B-Lymphoblastic Leukemia/Lymphoma with t(1;19) (q23;p13.3); *TCF3::PBX1*

- Pathogenesis
  - *TCF3* 19p13.3 encodes a transcription factor that binds an immunoglobulin enhancer
  - *PBX1* 1q23.3 encodes pre-B-cell leukemia homeobox 1, a transcription factor
  - The *TCF3::PBX* fusion protein transforms cells by constitutively activating PBX-regulated genes
- Epidemiology
  - Accounts for 6% of B-ALL cases
  - Less common in adults
- Genetic testing
  - The t(1;19) translocation may be accompanied by loss of the derivative chromosome 1 leading to an unbalanced translocation
  - ALL cases with the following findings should not be confused with this entity:

- o *TCF3* rearrangement due to t(17;19)(q22;p13) translocation between *TCF3* and *HLF*
- o Hyperdiploid B-ALL with t(1;19) on karyotype not involving *TCF3* or *PBX1*

B-Lymphoblastic Leukemia/Lymphoma, *BCR::ABL1*–Like (Ph-Like ALL)

- Pathogenesis
  - Lymphoblasts lack the *BCR::ABL1* translocation but have gene expression pattern similar to ALL with *BCR::ABL1*
  - Caused by various chromosomal rearrangements and mutations, 80% of cases with cytokine receptor or kinase-activating alterations associated with activation of ABL or JAK-STAT signaling pathways
    - o Cytokine receptor-like factor 2 (*CRLF2*) rearrangements (50% of cases) that commonly carry mutations of *JAK2* or less commonly *JAK1, KRAS, CRLF2, IKZF1, PAX5, ASXL1, ARID2, LRP1B, ITPKB*
    - o *EPOR* rearrangements with partner genes such as *IGH, IGK, LAIR1,* or *THADA*
    - o *JAK2* rearrangements
    - o *ABL* class alterations including *ABL1, ABL2, CSF1R, PDGFRB,* and rarely *PDGFRA* fusions
    - o Mutations of *KRAS, NRAS, NF1, PTPN11, FLT3, NTRK3, LYN, IKZF1,* and *CDKN2A/B* genes
- Epidemiology
  - Accounts for 10–25% of ALL cases
  - Less common in standard-risk ALL in children, progressively higher frequency in high-risk ALL in children to adults
- Genetic testing (Fig. 25.9)
  - Diagnostic methods include gene expression profiling (gold standard), low-density array (LDA) card, flow cytometry to assess CRLF2 overexpression, FISH panel, or whole-exome next-generation sequencing (NGS) and RNA sequencing

B-Lymphoblastic Leukemia/Lymphoma with iAMP21

- Pathogenesis
  - Lymphoblasts with amplification of a portion of chromosome 21 leading to ≥5 copies of *RUNX1* gene
- Epidemiology

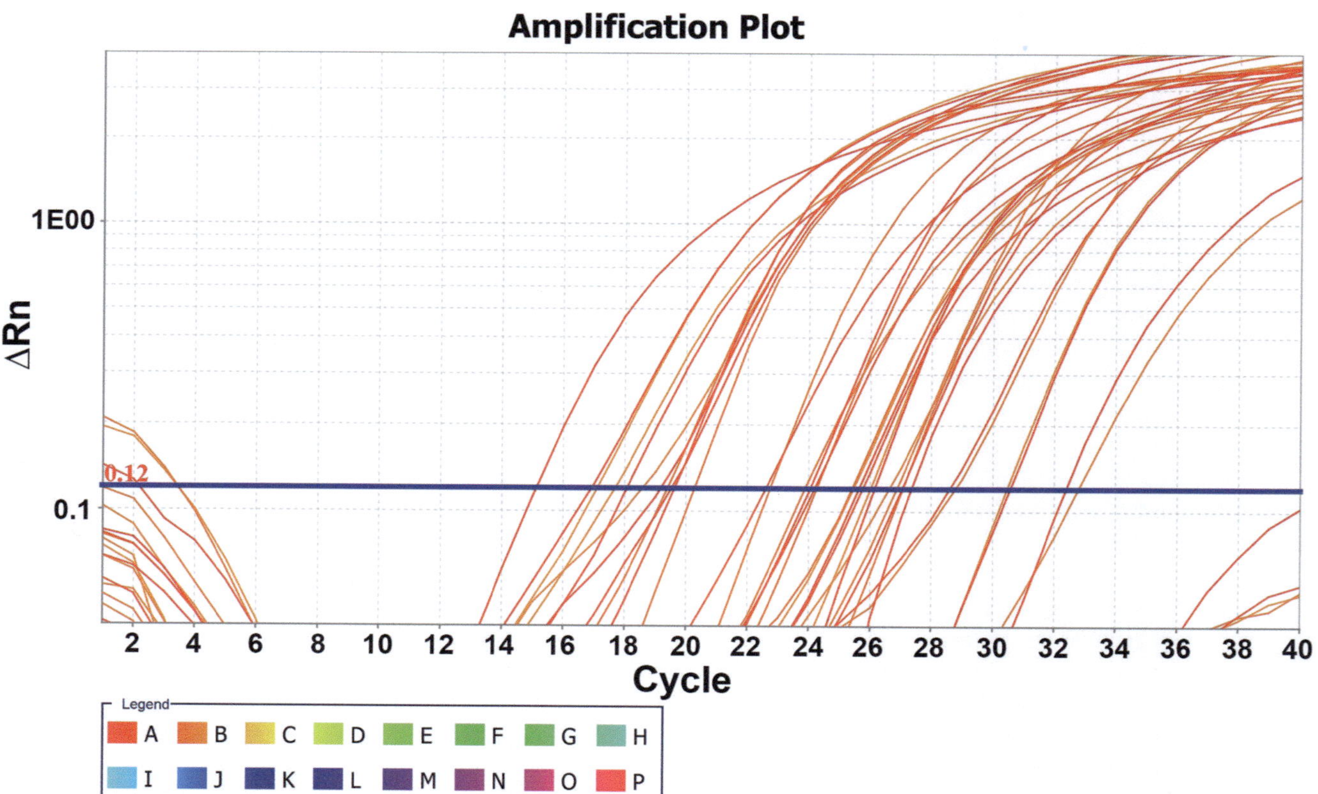

Fig. 25.9 Low-density array card for gene expression profiling in *BCR::ABL1*-like ALL identifies cases based on the amplification of multiple transcripts associated with this entity

- Accounts for 2% of B-ALL cases
- Occurs primarily in children, especially older children who present with low WBC count
- Genetic testing
  - Recognized by FISH for *ETV6::RUNX1* translocation as the *RUNX1* region is consistently amplified
  - 80% of cases have other chromosomal abnormalities, including gain of the X chromosome and chromosome 7 abnormalities
  - Also associated with *RB1* and *ETV6* deletions and *CRLF2* rearrangements

Emerging Subtypes

- Recently, molecular studies have revealed distinct oncogenic subtypes within previously unclassified pediatric B-ALL cases, including *DUX4*-rearranged (accounting for 4% of pediatric B-ALL cases), *ETV6::RUNX1*-like (3%), *ZNF3B4*-rearranged (1%), and *MEF2D*-rearranged (0.5%)

Hairy Cell Leukemia (HCL)

- Hairy cell leukemia is an indolent neoplasm of small mature B cells with characteristic cytologic and immunophenotypic features, propensity for infiltrating splenic red pulp, and bone marrow reticulin fibrosis leading to inaspirability or "dry tap"
- Pathogenesis
  - Derived from *BRAF* V600E-mutant mature memory B cells
    - Rare cases without *BRAF* mutation have *MAP2K1* mutation
  - Most cases (>85%) demonstrate somatic hypermutation of IGHV genes consistent with postgerminal center stage of maturation
  - Neoplastic hairy cells express constitutively activated integrin receptors and matrix metalloproteinase inhibitors that allow homing to bone marrow, splenic red pulp, and hepatic sinusoids
  - Expression of specific cytokines linked to reticulin fibrosis (fibronectin and TGFβ), inhibition of normal hematopoiesis (TGFβ), and prolonged cell survival (TNF, IL6, and Bcl-2)
- Epidemiology
  - Accounts for 2% of lymphoid leukemias
  - Median age of 58 years, rare in younger adults and especially children
  - Higher incidence in men (4:1 male-to-female ratio) and white compared to black populations
- Sites of involvement
  - Hairy cells are found in the bone marrow and spleen, and in small numbers in peripheral blood

- Occasionally infiltrate liver, lymph nodes, and skin, rarely causing prominent abdominal lymphadenopathy
- Histomorphology (Fig. 25.10)
  - Hairy cells have oval nuclei and abundant cytoplasm with "hairy" projections
  - >20% of HCL cases demonstrate marked hypocellularity
  - CD20 staining can highlight inconspicuous hairy cell infiltrates
    - Other useful immunohistochemical stains include Annexin A1, CD123, TRAP, T-bet, CD200, CD103, and BRAF
  - Characteristic flow cytometry phenotype: high side scatter and bright coexpression of CD20, CD22, CD11c, CD103, CD25, and CD200 with moderate CD123 expression
  - Hairy cell leukemia variant (HCLv) lacks classic HCL antigens such as CD25, CD200, CD123, and annexin-A1, lacks *BRAF* V600E mutation, and may have *MAP2K1* mutation
- Genetic testing
  - *BRAF* V600E mutation and somatic mutations of IGHV genes by PCR detected in 90% of cases
  - Chromosomal abnormalities are uncommon and include abnormal number of chromosomes 5 and 7 or translocations

Other Mature B-Cell Leukemias

- Mature or chronic B-cell neoplasms may present with peripheral blood involvement
  - Examples include mantle cell lymphoma, marginal zone lymphoma, follicular lymphoma, and diffuse large B-cell lymphoma
- May represent secondary involvement of the bone marrow by a primary tissue lymphoma
- These conditions are discussed further in the lymphoma chapter

T lymphoblastic Leukemia (T-ALL)/Lymphoma (T-LBL)

- Defined as neoplasm of lymphoblasts committed to T-cell lineage
  - T-ALL arises from a T-cell progenitor and has significant blood and bone marrow involvement
  - T-LBL arises from a thymic lymphocyte and has minimal or absent blood and marrow involvement
- Epidemiology

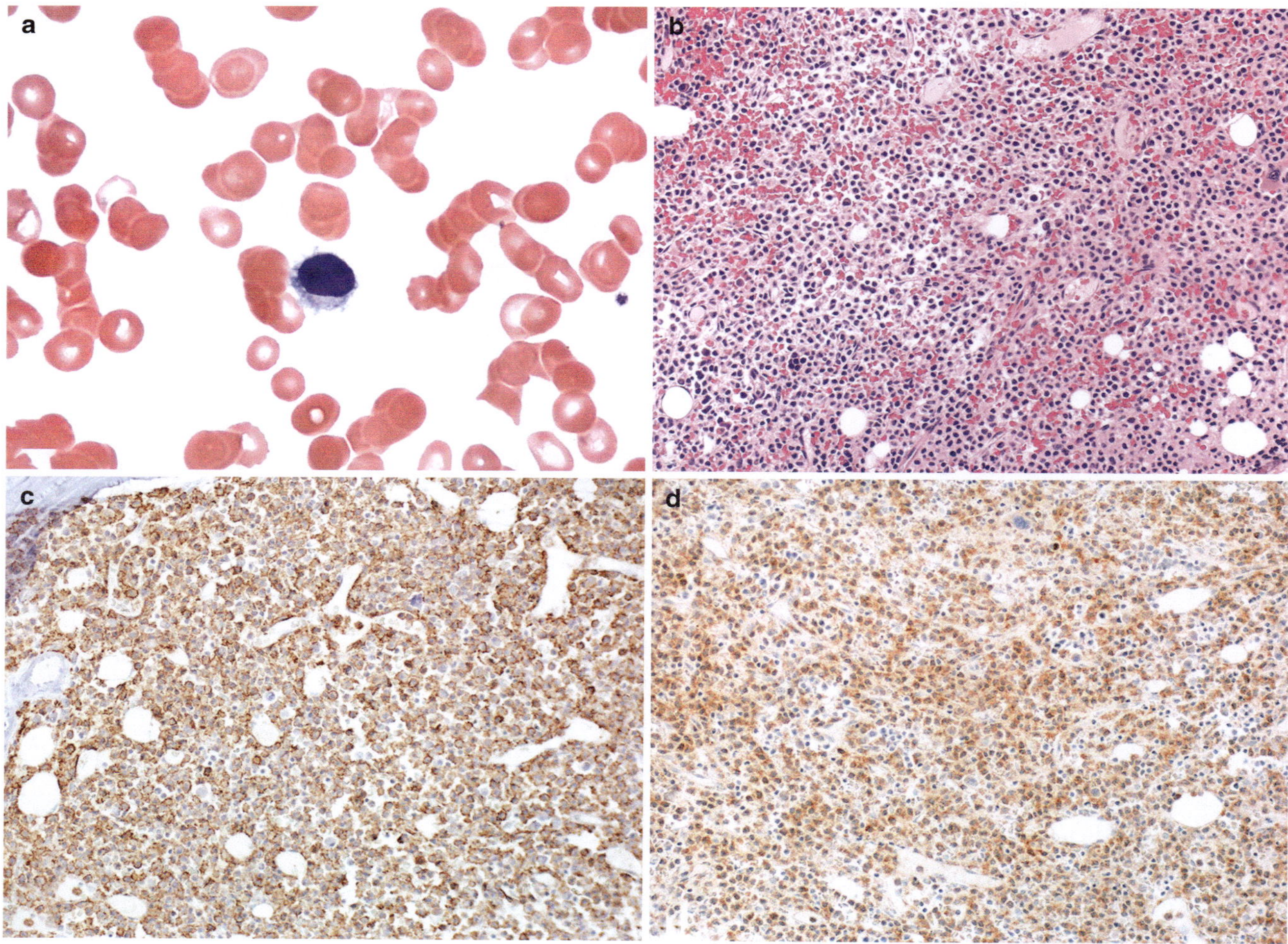

Fig. 25.10 (**a**) Neoplastic lymphocytes in HCL show small size, mature nuclear features, and circumferential cytoplasmic projections. (**b**) Bone marrow involvement in HCL can show variable degrees of infiltration by cells with abundant pink cytoplasm. The cells are brightly positive with CD20 (**c**) and for BRAF (**d**)

- T-ALL accounts for 15% of child ALL and 25% of adult ALL cases
- T-LBL accounts for 85% of all lymphoblastic lymphomas
- Both T-ALL and T-LBL are more common in adolescent males
- Histomorphology (Fig. 25.11)
  - T-ALL
    - Leukocytosis with lymphoblasts with high N:C ratio, finely dispersed nuclear chromatin and distinct nucleoli in peripheral blood, sheets of lymphoblasts in bone marrow with relatively preserved trilineage hematopoiesis
    - May also have tissue involvement
  - T-LBL
    - Most commonly presents as a mediastinal or thymic mass
      - Other nodal and extranodal sites including skin, CNS, and testis may be involved
    - Absent to minimal involvement in peripheral blood, blasts account for <25% of nucleated cells in bone marrow
- Genetic testing
  - T-ALL/LBL virtually always shows clonal rearrangements of T-cell receptor (TR) genes
  - 20% of cases also have IGH gene rearrangements
  - 50–70% of cases have an abnormal karyotype
  - Recurrent cytogenetic abnormalities include translocations involving TR alpha and delta 14q11.2, TR beta 7q35, and TR gamma 7p14-15 loci, *TLX1, TLX3, MYC, TAL1, LMO1, LMO2, LYL1, KMT2A,* t(10;11)(p13;q14) with *PICALM-MLLT10* fusion, and del(9p)
    - The translocations may be undetectable by karyotyping, requiring molecular genetic studies

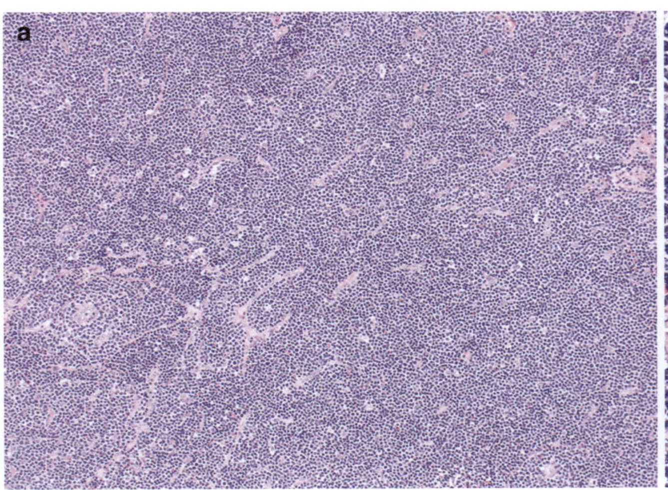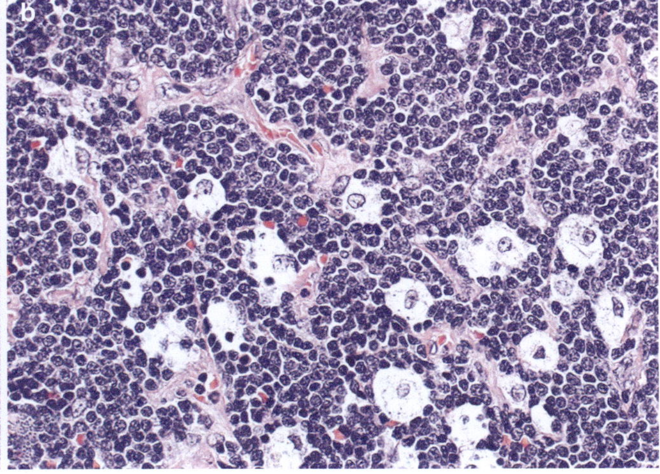

Fig. 25.11 (**a**) Excision of a mediastinal mass consistent with T-LBL shows sheets of intermediate cells with a high nuclear to cytoplasmic ratio. A higher power view shows lymphoblasts with scant cytoplasm and variably prominent nucleoli. (**b**) Numerous intermixed tingible body macrophages are present

- Recurrent gene mutations include *NOTCH1* and *FBXW7* mutations that lead to increased NOTCH1 protein activity

Early T-Cell Precursor Lymphoblastic Leukemia (ETP-ALL)

- Defined as T-ALL meeting all the following criteria:
  - Expression of CD7 (± cytoplasmic CD3, CD2, and CD4)
  - No CD1a or CD8 expression
  - No or weak CD5 expression (<75% of blasts positive)
  - Expression (≥25% of blasts positive) of ≥1 of the following myeloid- or stem cell-associated markers: CD117, CD34, HLA-DR, CD13, CD33, CD11b, and/or CD65
  - Negative for myeloperoxidase
- Epidemiology
  - Accounts for 12–15% of pediatric and 5–10% of adult ALL cases
- Genetic testing
  - Most cases show clonal T-cell receptor gene rearrangements by PCR
  - Gene expression profile similar to normal early thymocyte precursors
  - Overexpressed genes include *CD44, CD34, KIT, GATA2, CEBPA,* and *LYL1*
  - More common mutations include *FLT3,* RAS family genes, *DNMT3A, IDH1,* and *IDH2*

Mature T- and NK-Cell Leukemias

T-Cell Prolymphocytic Leukemia

- Epidemiology
  - Accounts for 2% of mature lymphocytic leukemias in adults, rare in individuals <30 years of age
- Sites of involvement
  - Leukemic T cells are seen in the peripheral blood, bone marrow, lymph nodes, spleen, liver, and occasionally skin
- Histomorphology
  - Often pronounced lymphocytosis in peripheral blood
  - Blasts may have irregular nuclei, cytoplasmic protrusions or blebs, cerebriform appearance in cerebriform variant, or small cell size in small-cell variant
- Genetic testing
  - Clonal rearrangements in T-cell receptor beta and gamma genes (TRB and TRG) are present
  - Chromosomal abnormalities involving *TCL1* and *MTCP1* genes are common, with 80% of cases have inv(14)(q11q32), 10% have t(14;14)(q11;q32), less commonly t(X;14)(q28;q11)
  - 70–80% of cases have chromosome 8 abnormalities including idic(8)(p11), t(8;8)(p11-12;q12), and trisomy 8q
  - Molecular and FISH studies may show mutations or deletion of the *ATM* gene at 11q23
  - Recurrent mutations can occur including in JAK/STAT signaling pathway genes such as *JAK3, STAT5B,* and *JAK1*

T-Cell Large Granular Lymphocytic Leukemia

- Epidemiology
  - Accounts for 2–3% of mature small lymphocytic leukemias, with 73% of cases occurring in individuals aged 45–75 years
- Sites of involvement
  - Primarily involves peripheral blood, bone marrow, liver, and spleen
- Histomorphology
  - Generally, $>2 \times 10^9$/L large granular lymphocytes in peripheral blood with condensed chromatin and abundant pale cytoplasm with azurophilic granules
  - Variable bone marrow involvement, typically in interstitial or intrasinusoidal pattern
- Genetic testing
  - TR gene rearrangements present
  - 40% of cases have *STAT3* mutations

Chronic Lymphoproliferative Disorder of NK Cells (CLPD-NK)

- Epidemiology
  - Rare, occurs predominantly in adults
- Histomorphology
  - Increased mature large granular lymphocytes in peripheral blood and to a lesser extent bone marrow
- Genetic testing
  - 30% of cases have *STAT3* mutations
  - Only method to detect clonal NK-cell population is skewed ratio of X-chromosome inactivation in NK cells in female patients

Aggressive NK-Cell Leukemia

- Epidemiology
  - Rare, most common in young to middle-aged adults
- Histomorphology
  - Sparse to abundant circulating blastic to mature-appearing lymphocytes with sparse cytoplasmic granules
  - Variable patchy and interstitial bone marrow infiltration, associated with secondary hemophagocytosis
- Genetic testing
  - No clonal TR rearrangements or recurrent cytogenetic abnormalities

Adult T-Cell Leukemia/Lymphoma

- Epidemiology
  - Caused by infection with human T-lymphotropic virus type 1 (HTLV-1)
  - Largely restricted to endemic HTLV-1 regions (Japan, Caribbean basin, Iran, Africa), developing in 2–5% of HTLV-1 carriers
  - Occurs only in adults with male-to-female ratio of 1.5:1
- Sites of involvement
  - Involves lymph nodes, peripheral blood, spleen, extranodal sites such as skin
- Histomorphology
  - Peripheral blood shows leukocytosis in acute form, leukemic cells with distinctive multilobed nuclei, and variable eosinophilia
  - Sparse to moderate patchy bone marrow infiltrates
- Genetic testing
  - Clonal TR gene rearrangements and monoclonal integration of HTLV-1 present
  - Recurrent mutations include *CCR4, PLCG1, PRKCB, VAV1, IRF4, FYN, CARD11,* and *STAT3*

Mycosis Fungoides (MF)/Sézary Syndrome (SS)

- MF and SS have overlapping histomorphologic features, but SS presents with abrupt-onset erythroderma (rarely only pruritis), generalized lymphadenopathy, and peripheral blood involvement, whereas MF initially presents with slowly evolving skin lesions without blood involvement
- Epidemiology
  - MF is the most common cutaneous T-cell lymphoma (CTCL), while SS accounts for only 5% of CTCL
  - Both MF and SS tend to occur in adults and predominate in males (1.6:1)
- Histomorphology
  - Peripheral blood shows variable lymphocytosis and eosinophilia, medium to large lymphocytes with distinctly cerebriform nuclei with condensed chromatin
  - Bone marrow may be involved, most often with sparse interstitial infiltrates of lymphocytes with cerebriform nuclei
- Genetic testing
  - Clonal TR gene rearrangements are present
  - Complex numerical and structural chromosomal abnormalities are common including losses of 1p, 6q, and 10q, gain of 8q, and isochromosome 17q
  - Recurrent gene mutations include *PLCG1, CD28, TNFRSF1B, ARID1A,* JAK/STAT pathway genes, *DNMT3A, TP53,* and *CDKN2A*

Acute Leukemias of Ambiguous Lineage

Acute Undifferentiated Leukemia

- Defined as acute leukemia with ≥20% blasts in peripheral blood or bone marrow expressing no markers considered specific for either lymphoid or myeloid lineage, including CD3, MPO, CD22, CD79a, or CD19
- Epidemiology: very rare
- Histomorphology
  - Increased blasts with no Auer rods and nonspecific morphologic features
- Genetic testing
  - Recurrent translocations associated with myeloid or lymphoid leukemias are not present by cytogenetic or FISH analysis
  - May have nonspecific mutations in *BAALC, ERG,* or *MN1*

Mixed Phenotype Acute Leukemia (MPAL)

- Defined as acute leukemia with ≥20% blasts in peripheral blood or bone marrow with expression of markers considered specific for 2 or more lineages among myeloid/monocytic (MPO, NSE, CD11c, CD14, CD64, and lysozyme), T cell (CD3), or B cell (CD19, CD79a, CD22, and CD10)
- Epidemiology
  - Accounts for 2–3% of adult acute leukemias
- Histomorphology
  - May demonstrate 1 or 2 distinct blast populations with lymphoblast or myeloblast cytology

MixedPhenotype Acute Leukemia with t(9;22)(q34.1;q11.2); *BCR::ABL1*

- Epidemiology
  - Most common type of MPAL with recurrent genetic abnormality
  - Accounts for <1% of all acute leukemias, more commonly in adults
- Histomorphology
  - Most commonly demonstrates B-cell and myeloid blasts, but occasionally T-cell and myeloid blasts
- Genetic testing
  - t(9;22) detected by conventional cytogenetics or *BCR::ABL1* fusion gene detected by FISH or PCR
  - p190 fusion transcript more common than the p210 transcript
  - Additional cytogenetic abnormalities and complex karyotypes are common

MixedPhenotype Acute Leukemia with t(v;11q23.3); *KMT2A*-Rearranged

- Epidemiology
  - More common in children and infants than adults
- Histomorphology
  - Usually demonstrates a dimorphic blast population with monoblast and lymphoblast cytology
- Genetic testing
  - *KMT2A* rearrangements involving *AFF1* 4q21.3 or t(9;11) and t(11;19) translocations detected by standard karyotype, FISH with break-apart probe, or PCR
  - Nonspecific secondary cytogenetic or molecular abnormalities may occur

MixedPhenotype Acute Leukemia, B/Myeloid

- Epidemiology
  - Accounts for about 1% of all leukemia cases, more commonly in adults
- Histomorphology
  - Usually demonstrates blasts without distinguishing features or a dimorphic population with lymphoblast and myeloblast cytology
- Genetic testing
  - Most cases show nonspecific clonal cytogenetic abnormalities such as del(6p), 12p11.2 abnormalities, del(5q), chromosome 7 abnormalities, near-tetraploidy or other numerical abnormalities, or complex karyotypes
  - More common gene mutations include *ASXL1, TET1/2, IDH1, IDH2, DNMT3A, NOTCH1, ETV6,* and *IZKF1*

Mixed Phenotype Acute Leukemia, T/Myeloid

- Epidemiology
  - Accounts for <1% of all leukemia cases, more common in children compared to B/myeloid MPAL
- Histomorphology
  - Usually demonstrates blasts without distinguishing features or a dimorphic population with lymphoblast and myeloblast cytology
- Genetic testing
  - Most cases have nonspecific clonal chromosomal abnormalities

Mixed Phenotype Acute Leukemia, NOS

- Very rare type of leukemia that is biphenotypic with blasts showing both T-cell and B-cell lineage commitment

Acute Leukemias of Ambiguous Lineage, NOS

- Rare type of leukemia expressing combinations of markers that do not allow classification as acute undifferentiated leukemia or mixed phenotype acute leukemia

Suggested Triage and Workup of Leukemia Specimens

- Diagnostic workup of suspected leukemia specimens typically involves
  - Collection of relevant clinical and imaging findings
  - Complete blood count and differential with peripheral blood smear evaluation
  - Bone marrow evaluation including aspirate smears, trephine core biopsy, trephine touch preparations, and/or marrow clots
  - Collecting sufficient samples for and performing flow cytometric immunophenotyping and conventional cytogenetic analysis (i.e., karyotype) along with appropriate fluorescence in situ hybridization (FISH) and/or molecular genetic testing
    - If there is inadequate bone marrow aspirate or peripheral blood material, immunohistochemical studies may be utilized as an alternative to flow cytometry, and an additional bone marrow core biopsy may be submitted unfixed in tissue culture media for disaggregation for flow cytometry and genetic studies
    - Acceptable sample types for molecular or genetic testing include cryopreserved cells or nucleic acid, nondecalcified formalin-fixed, paraffin-embedded (FFPE) tissue, or unstained marrow aspirate, peripheral blood smears or other involved tissues if the use of the respective material has been validated
    - Decalcified samples are not acceptable specimens for most genetic testing

Flow Cytometry

- Technique
  - The sample must consist of viable (not formalin-fixed) cells from peripheral blood, plasma, bone marrow, body fluids, or solid tissue
    - Tissue samples are first disaggregated to form cell suspension
  - Cells are incubated with fluorochrome-tagged antibodies against antigens of interest
  - Components of the flow cytometer include fluidics, optics, and electronics
    - Fluidics: Pressurized sheath fluid delivers cells single-file through the laser intercept or interrogation point
    - Optics: Lasers excite the fluorescently labeled cells and detectors (photodiodes or photomultiplier tubes along with dichroic mirrors and bandpass filters) collect visible light scatter and fluorescent light emission spectra
    - Electronics: Optical signals from the detectors are converted to digital signals for computer analysis and graphical representation of the data
    - Compensation techniques can help correct for spectral overlap of fluorochromes
  - Specific cell populations can be gated to further analyze their expression of aberrant markers, light chain clonality, and other parameters
- Uses
  - Separates cells based on size (forward scatter) and internal complexity, e.g., cytoplasmic granules or irregular cell contours (side scatter)
  - Assesses presence and degree of expression of antigens on cells of interest
  - Useful for diagnosis of hematolymphoid neoplasms, triaging additional testing, evaluating prognostic markers such as CD49d in CLL, and minimal residual disease monitoring

Fluorescence In Situ Hybridization (FISH)

- Technique
  - Appropriate sample may consist of dividing cells in metaphase or nondividing cells in interphase including fixed cells, paraffin-embedded tissue sections, air-dried smears, touch preparations, or cytospins
  - The technique is performed by dehydrating the slide with graded ethanol series, adding fluorescently labeled probe specific to DNA target sequence of interest, heating to denature double-stranded DNA, incubating to allow probe hybridization, washing away excess probe, and applying mounting medium with DAPI counterstain for visualization of nuclei
  - FISH probe strategies available for diagnosis of hematological malignancies include dual-color/dual-fusion, break-apart, deletion/duplication, and centromeric probes

- o Dual-color/dual-fusion probes are useful when the two genes involved in a translocation are known
  - ♦ Red probes bracket one gene, green probes bracket the other
  - ♦ Presence of two yellow fusion signals indicates translocation
  - ♦ Separate red and green signals indicate no translocation
- o Break-apart probes are useful when only one gene involved in translocation is known
  - ♦ Red and green probes bracket each end of the gene
  - ♦ Separate red and green signals indicate translocation
  - ♦ Yellow fusion signals only indicate no translocation
- Uses
  - Can help determine the diagnosis or prognosis of hematologic malignancies or detect residual disease

Conventional Cytogenetics

- Technique
  - In contrast to FISH, viable cells that can divide in culture are required
  - The technique is performed by allowing cells to divide in culture with specific mitogens as needed, adding colchicine to cause the cells to arrest in metaphase, adding hypotonic solution to cause cells to swell with water and enhance the spreading of chromosomes, fixing cells, dropping cells onto glass slides, staining, and examination by light microscopy
  - Band resolution varies by sample type and preparation, and higher band resolution provides greater sensitivity
- Uses
  - Can detect aneuploidy (gains or losses of whole chromosomes) or translocations
  - Useful to determine diagnosis or prognosis of hematologic malignancies or identify residual disease or clonal evolution

Molecular DNA/RNA Assays

Common Assays

BCR::ABL1
- Background
  - The *BCR::ABL1* fusion gene is generated by t(9;22)(q34.1;q11.2) reciprocal translocation

- The *ABL1* breakpoint typically occurs between exon 1 and 2 (denoted a2), very rarely between exon 2 and 3 (a3)
- The *BCR* breakpoints include major, minor, and micro breakpoint cluster region or M-BCR, m-BCR, and μ-BCR, respectively
  - o In M-BCR, breakpoints occur in exons 12–16 (also known as b1–b5), most commonly between exons 13 and 14 (b2) or 14 and 15 (b3)
    - ♦ Fusion with *ABL1* a2 breakpoint produces b2a2 or b3a2 transcripts which encode p210 fusion protein
    - ♦ M-BCR is present in most cases of CML and AML with BCR-ABL1
  - o In m-BCR, the breakpoint occurs between exons 1 and 2 (e1)
    - ♦ Fusion with *ABL1* a2 breakpoint produces e1a2 transcript which encodes p190 fusion protein
    - ♦ m-BCR is present in most cases of *BCR::ABL1*-positive B-ALL, with a higher incidence in pediatric cases
  - o In μ-BCR, the breakpoint occurs between exons 19 and 20 (e19)
    - ♦ Fusion with *ABL1* a2 breakpoint produces e19a2 transcript which encodes p230 fusion protein
    - ♦ μ-BCR is rarely present in CML
- Technique
  - The quantitative real-time PCR (qRT-PCR) technique involves the isolation of total cellular RNA from the specimen, reverse transcription to cDNA, PCR amplification, and detection of PCR products in real-time with either nonspecific fluorescent double-stranded DNA-binding dye or fluorescently labeled sequence-specific DNA probes, e.g., the TaqMan assay
  - The specimen must be peripheral blood, bone marrow, or other involved sites that is no older than 48 h due to the instability of RNA, and preferably in EDTA transport media
  - The *BCR::ABL1* qRT-PCR assay utilizes primer sets to amplify the different *BCR::ABL1* fusions and an internal control gene such as *ABL1, B2M, GUSB,* etc., as well as serially diluted positive cell line/plasmid controls
  - The level of *BCR::ABL1* transcripts can either be expressed as the ratio of *BCR::ABL1/ABL1*, or on the International Scale (IS), where the level at diagnosis is set as 100% IS
- Uses
  - Detection of *BCR::ABL1* fusion genes by qRT-PCR is useful for diagnosis and monitoring of chronic myelogenous leukemia (CML) and Philadelphia chromosome-positive B-cell acute lymphoblastic leukemia (Ph(+) B-ALL)
  - CML

- Peripheral blood qRT-PCR testing is recommended at presentation and following complete cytogenetic remission
- Major molecular response is defined as *BCR::ABL1/ABL* ratio with ≥3 log reduction from baseline, or an IS of ≤0.1%
  - ALL
    - The preferred specimen type is bone marrow due to low detection sensitivity of B lymphoblast *BCR::ABL1* transcripts in peripheral blood

PML-RARA

- Technique
  - The *PML::RARA* qRT-PCR assay utilizes primer sets to amplify the different *PML::RARA* fusions and an internal control gene such as *ABL1*
  - Testing with bone marrow specimen is more sensitive than peripheral blood
- Uses
  - Detection of *PML::RARA* fusion genes by quantitative real-time PCR (qRT-PCR) is useful for diagnosis and monitoring of acute promyelocytic leukemia (APL)
  - Testing is recommended at diagnosis, at the end of consolidation therapy, every 3 months for 2 years after complete molecular remission, and 2–4 weeks after a positive result to confirm molecular relapse

Ph-Like

- Technique
  - Ph-like ALL is associated with a variety of rearrangements and mutations
  - Commonly used assays for diagnosis of Ph-like ALL include low-density array (LDA) card, Ph-like FISH panel, flow cytometry to assess CRLF2 overexpression associated with *CRLF2* rearrangements (found in 30–50% of Ph-like ALL cases), and RNA or DNA sequencing
  - The LDA card assay utilizes qRT-PCR to identify 8-gene and 15-gene Ph-like ALL signature
  - The FISH panel commonly consists of *CRLF2, ABL1, ABL2, JAK2, PDGFRB, CSF1R,* and *EPOR* breakapart probe sets and is less sensitive than gene expression profiling
- Uses
  - The LDA screen with reflex to FISH, RNA, and/or DNA sequencing can help with treatment stratification (tyrosine kinase inhibitors may be effective for patients with kinase-activating alterations) and is recommended for adults with B-ALL, children with high-risk B-ALL, and can be considered for children with standard-risk B-ALL

FLT3/NPM1

- Background
  - Types of *FLT3* mutations include internal tandem duplication (ITD) insertions of 6–30 base pairs in exons 14 and 15 (most common) and point mutations in the tyrosine kinase domain (TKD) in exon 20
  - The most common type of *NPM1* mutation is 4 base pair insertion in exon 12
- Technique
  - *FLT3* and *NPM1* insertion mutations can be detected by a single PCR sizing assay
    - To perform, isolate genomic DNA, perform PCR with fluorescently labeled primers encompassing the *FLT3* and *NPM1* insertion regions, and run PCR product on a capillary electrophoresis instrument
    - Positive result shows both shorter wild-type peak and longer mutated peaks
  - Other methods to detect *FLT3* and *NPM1* mutations include Sanger sequencing, qRT-PCR, and next-generation sequencing
    - qRT-PCR and NGS methods are sensitive and can be used for MRD testing
- Uses
  - *FLT3*-ITD mutations are common in hematologic malignancies including AML (20–30% of cases), CML (5–10%), and MDS (5–10%)
    - Considered a high-risk abnormality and associated with increased relapse and shorter overall survival
    - Often accompanied by normal karyotype and *NPM1* mutation
    - *FLT3*-ITD testing is recommended for all patients with acute myeloid leukemia
  - *NPM1* mutations occur in 25–35% of adult AML cases
    - Considered a favorable risk abnormality in cases without coexisting *FLT3* mutation
    - Testing for *NPM1* as well as *CEBPA* and *RUNX1* mutations are recommended for AML except for confirmed core binding factor AML, APL, or AML-MRC

IDH1/2 Mutations

- Background
  - IDH mutations occur in 20% of AML cases, with the incidence of *IDH1* and *IDH2* mutations being equivalent and mutually exclusive
    - Associated with older age, higher platelet counts, normal karyotype, and *NPM1* mutation
  - Common *IDH1* mutations include R132H, R132C, and R132G
  - Common *IDH2* mutations include R140Q, R172K, and R140Q
- Technique (Fig. 25.12)

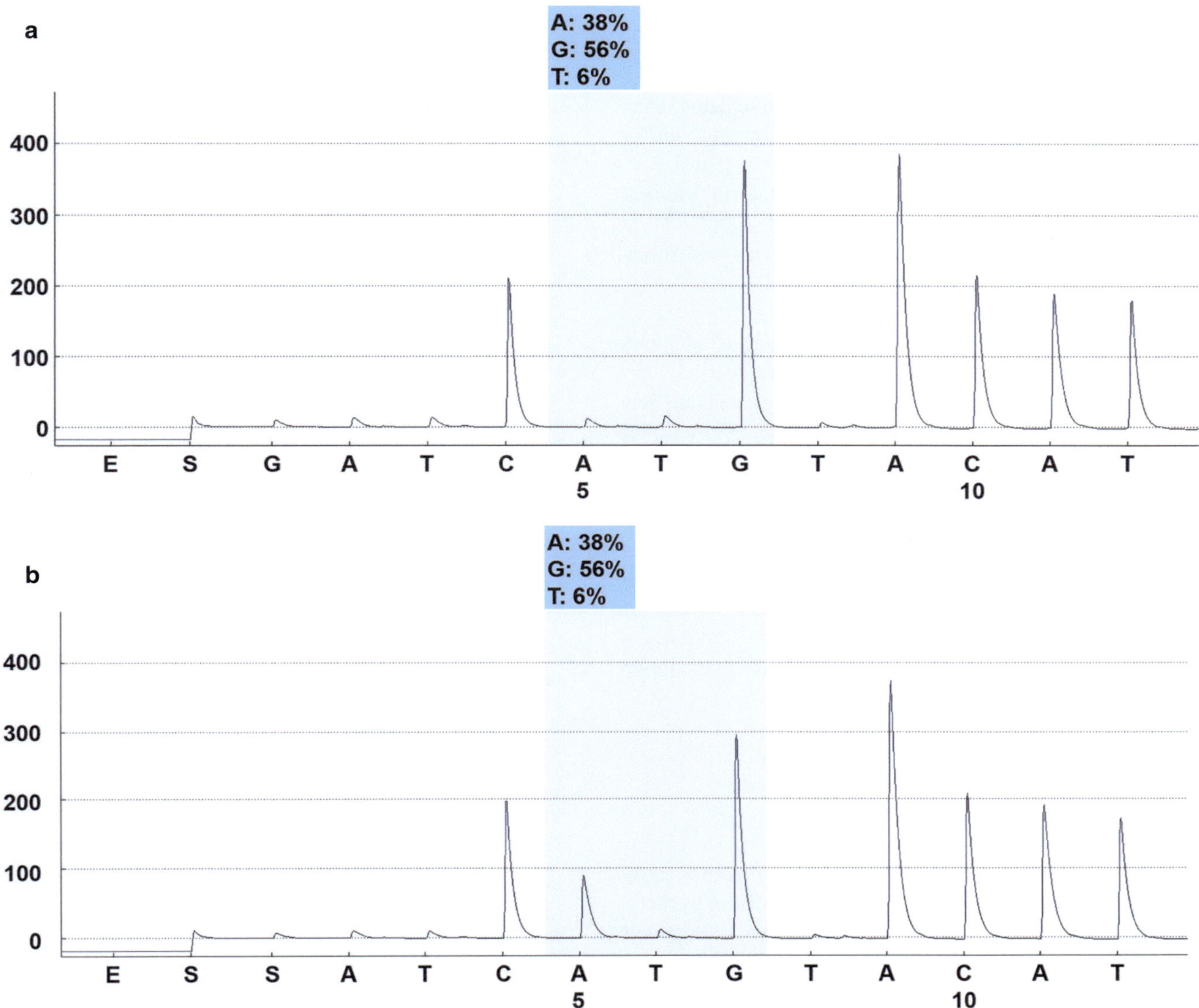

Fig. 25.12 Pyrosequencing for *IDH1* and *IDH2* mutations can rapidly identify treatment targets. (**a**) A nonmutated sequence for codon 140 of *IDH2* reads CGG. (**b**) A sample with an R140Q mutation shows a new A signal and decreased G signal, consistent with CAG

- IDH mutations can be detected by various gene sequencing methods (e.g., whole-exome, massively parallel, Sanger, and pyrosequencing), allelic discrimination assays, high-resolution melting curve analysis, or immunohistochemical staining with antibodies against specific IDH1/2 variants
- Uses
  - The prognostic significance of IDH mutations in AML remains controversial
  - Some patients with IDH mutation have a poor response to traditional chemotherapy
    - Targeted IDH inhibitors enasidenib and ivosidenib are FDA approved for relapsed/refractory AML with *IDH2* and *IDH1* mutation, respectively

Clonality Studies: T- and B-Cell
- Technique
  - Clonality studies evaluate for clonal IG rearrangements primarily to identify B-cell malignancies, and clonal TR gene rearrangements to identify T-cell malignancies
    - However, clonality studies are not entirely lineage-specific as cross-lineage rearrangements can occur in immature or precursor neoplasms
  - Antibody-based methods include flow cytometry, FISH, and immunohistochemistry
    - TR rearrangement flow cytometry assays utilize TCR Vβ or TCR β constant region 1 antibodies
  - Nucleic acid-based methods include multiplex PCR, NGS, and Southern blot hybridization
  - Covered further in the lymphoma chapter

Massively Parallel (Next-Generation) Sequencing Gene Panels

- Many NGS gene panels are commercially variable for different hematologic malignancies, solid tumors, and inherited conditions, and laboratories often create custom panels
- Allow identification of genetic alterations in multiple gene regions simultaneously
  - May be used to identify point mutations, insertions, deletions, fusions, and amplifications
- Technique
  - Sample preparation involves assessment of adequacy and marking areas of tumor for macrodissection
  - DNA or cDNA libraries of specific size range are generated, and target sequences enriched by multiplex PCR amplification
  - DNA fragments are sequenced by sequencing by synthesis (Illumina platforms) or ion semi-conductor-based (Ion Torrent) method
  - Data analysis involves base calling, alignment of reads to reference genome, and identification and annotation of variants
  - A four-tier system is commonly used to report sequence variants
    - Tier I: Strong clinical significance
    - Tier II: Potential clinical significance
    - Tier III: Unknown clinical significance
    - Tier IV: Benign or likely benign
- Uses
  - Useful in AML for diagnosis, risk stratification, treatment planning including predicting response to therapies such as hypomethylating agents, allogeneic transplantation, and FLT3, IDH, and Bcl-2 inhibitors, and response assessment including MRD testing
    - Currently there is no consensus on the genes to include in a myeloid NGS panel, but 2021 National Comprehensive Cancer Network (NCCN) guidelines recommend testing for *ASXL1, c-KIT, FLT3* (ITD and TKD), *NPM1, CEBPA, IDH1, IDH2, RUNX1,* and *TP53* mutations at minimum
  - In ALL, NGS can be used for risk stratification, particularly in *BCR-ABL1*/Ph-negative or Ph-like cases, and MRD assessment
  - In CML, NGS can detect low-level BCR-ABL1 kinase domain mutations and resistance mutations in other genes that may confer resistance to tyrosine kinase inhibitors or predict disease progression

Suggested Reading

Adams J, Nassiri M. Acute promyelocytic leukemia: a review and discussion of variant translocations. Arch Pathol Lab Med. 2015;139(10):1308–13. https://doi.org/10.5858/arpa.2013-0345-RS.

Almond LM, Charalampakis M, Ford SJ, Gourevitch D, Desai A. Myeloid sarcoma: presentation, diagnosis, and treatment. Clin Lymphoma Myeloma Leuk. 2017;17(5):263–7. https://doi.org/10.1016/j.clml.2017.02.027. Epub 2017 Mar 7

American Cancer Society. Cancer Facts & Figs. Atlanta: American Cancer Society; 2021.

Arber DA, Borowitz MJ, Cessna M, et al. Initial diagnostic workup of acute leukemia: guideline from the College of American Pathologists and the American Society of Hematology. Arch Pathol Lab Med. 2017;141(10):1342–93. https://doi.org/10.5858/arpa.2016-0504-CP.

Arber DA, Erba HP. Diagnosis and treatment of patients with acute myeloid leukemia with myelodysplasia-related changes (AML-MRC). Am J Clin Pathol. 2020;154(6):731–41. https://doi.org/10.1093/ajcp/aqaa107.

Aydin C, Cetin Z, Manguoglu AE, et al. Evaluation of ETV6/RUNX1 fusion and additional abnormalities involving ETV6 and/or RUNX1 genes using FISH technique in patients with childhood acute lymphoblastic leukemia. Indian J Hematol Blood Transfus. 2016;32(2):154–61. https://doi.org/10.1007/s12288-015-0557-7.

Bill M, Mrózek K, Kohlschmidt J, Eisfeld AK, Walker CJ, Nicolet D, Papaioannou D, Blachly JS, Orwick S, Carroll AJ, Kolitz JE, Powell BL, Stone RM, de la Chapelle A, Byrd JC, Bloomfield CD. Mutational landscape and clinical outcome of patients with de novo acute myeloid leukemia and rearrangements involving 11q23/KMT2A. Proc Natl Acad Sci U S A. 2020;117(42):26340–6. https://doi.org/10.1073/pnas.2014732117. Epub 2020 Oct 5

Cantor AB. Myeloid proliferations associated with down syndrome. J Hematop. 2015;8(3):169–76. https://doi.org/10.1007/s12308-014-0225-0.

Choi M, et al. RTK-RAS pathway mutation is enriched in myeloid sarcoma. Blood Cancer J. 2018;8(5):43.

Christen F, Hoyer K, Yoshida K, Hou HA, Waldhueter N, Heuser M, Hills RK, Chan W, Hablesreiter R, Blau O, Ochi Y, Klement P, Chou WC, Blau IW, Tang JL, Zemojtel T, Shiraishi Y, Shiozawa Y, Thol F, Ganser A, Löwenberg B, Linch DC, Bullinger L, Valk PJM, Tien HF, Gale RE, Ogawa S, Damm F. Genomic landscape and clonal evolution of acute myeloid leukemia with t(8;21): an international study on 331 patients. Blood. 2019 Mar 7;133(10):1140–51. https://doi.org/10.1182/blood-2018-05-852822. Epub 2019 Jan 4

Coffman B, Chabot-Richards D. Diagnosis of variant translocations in acute promyelocytic leukemia. Adv Mol Pathol. 2021;4(1):37–48. https://www.advancesinmolecularpathology.com/article/S2589-4080(21)00002-8/pdf.

Conant JL, et al. BCR-ABL1-like B-lymphoblastic leukemia/lymphoma: review of the entity and detection methodologies. Int J Lab Hematol. 2019;41 Suppl 1:126–30.

Czuchlewski DR, et al. Myeloid neoplasms with germline predisposition: a new provisional entity within the world health organization classification. Surg Pathol Clin. 2016;9(1):165–76.

Dogan A, Morice WG. Bone marrow histopathology in peripheral T-cell lymphomas. Br J Haematol. 2004;127(2):140–54. https://doi.org/10.1111/j.1365-2141.2004.05144.x.

Forgione MO, McClure BJ, Eadie LN, Yeung DT, White DL. KMT2A rearranged acute lymphoblastic leukaemia: Unravelling the genomic complexity and heterogeneity of this high-risk disease. Cancer Lett. 2020 Jan 28;469:410–8. https://doi.org/10.1016/j.canlet.2019.11.005. Epub 2019 Nov 6

Ganzel C, Douer D. Extramedullary disease in APL: a real phenomenon to contend with or not? Best Pract Res Clin Haematol. 2014;27(1):63–8. https://doi.org/10.1016/j.beha.2014.04.001. Epub 2014 Apr 12

Geyer JT. Myeloid neoplasms with germline predisposition. Pathobiology. 2019;86(1):53–61. https://doi.org/10.1159/000490311. Epub 2018 Jul 26

Gonzales PR, Mikhail FM. Diagnostic and prognostic utility of fluorescence in situ hybridization (FISH) analysis in acute myeloid leukemia. CurrHematolMalig Rep. 2017;12(6):568–73. https://doi.org/10.1007/s11899-017-0426-6.

Graham SJ, Sharpe RW, Steinberg SM, Cotelingam JD, Sausville EA, Foss FM. Prognostic implications of a bone marrow histopathologic classification system in mycosis fungoides and the Sézary syndrome. Cancer. 1993;72(3):726–34. https://doi.org/10.1002/1097-0142(19930801)72:3<726::aid-cncr2820720316>3.0.co;2-p.

Gröschel S, Sanders MA, Hoogenboezem R, et al. Mutational spectrum of myeloid malignancies with inv(3)/t(3;3) reveals a predominant involvement of RAS/RTK signaling pathways. Blood. 2015;125(1):133–9. https://doi.org/10.1182/blood-2014-07-591461.

Gupta AK, Meena JP, Chopra A, Tanwar P, Seth R. Juvenile myelomonocytic leukemia-A comprehensive review and recent advances in management. Am J Blood Res. 2021;11(1):1-21.Published 2021 Feb 15.

Hoelzer D, Bassan R, Dombret H, Fielding A, Ribera JM, Buske C, ESMO Guidelines Committee. Acute lymphoblastic leukaemia in adult patients: ESMO Clinical Practice Guidelines for diagnosis, treatment and follow-up. Ann Oncol. 2016;27(suppl 5):v69–82. https://doi.org/10.1093/annonc/mdw025. Epub 2016 Apr 7

Horna P, Olteanu H, Jevremovic D, Otteson GE, Corley H, Ding W, Parikh SA, Shah MV, Morice WG, Shi M. Single-antibody evaluation of T-cell receptor β constant chain monotypia by flow cytometry facilitates the diagnosis of T-cell large granular lymphocytic leukemia. Am J Clin Pathol. 2021;156(1):139–48. https://doi.org/10.1093/ajcp/aqaa214.

Jain S, Abraham A. BCR-ABL1-like B-acute lymphoblastic leukemia/lymphoma: a comprehensive review. Arch Pathol Lab Med. 2020;144(2):150–5. https://doi.org/10.5858/arpa.2019-0194-RA. Epub 2019 Oct 23

Janeway CA Jr, Travers P, Walport M, et al. Immunobiology: the immune system in health and disease. 5th ed. New York: Garland Science; 2001. Available from: https://www.ncbi.nlm.nih.gov/books/NBK27098/

Leisch M, Jansko B, Zaborsky N, Greil R, Pleyer L. Next generation sequencing in AML-on the way to becoming a new standard for treatment initiation and/or modulation? Cancers (Basel). 2019;11(2):252. https://doi.org/10.3390/cancers11020252. Published 2019 Feb 21

Li Y, Wang J, Wei H, Mi Y, Wang Y. Favorable outcome of allogenic stem cell transplant in patients with acute myeloid leukemia carrying t(6;9)(p22;q34)/DEK-NUP214 rearrangement.

Blood. 2019;134(Supplement_1):5103. https://doi.org/10.1182/blood-2019-126476.

Lilljebjörn H, Fioretos T. New oncogenic subtypes in pediatric B-cell precursor acute lymphoblastic leukemia. Blood. 2017;130(12):1395–401. https://doi.org/10.1182/blood-2017-05-742643. Epub 2017 Aug 4

Liu X, Gong Y. Isocitrate dehydrogenase inhibitors in acute myeloid leukemia. Biomark Res. 2019;7:22. https://doi.org/10.1186/s40364-019-0173-z.

Maitre E, et al. Hairy cell leukemia: 2020 update on diagnosis, risk stratification, and treatment. Am J Hematol. 2019;94(12):1413–22.

Mast KJ, et al. Pathologic features of down syndrome myelodysplastic syndrome and acute myeloid leukemia: a report from the Children's Oncology Group Protocol AAML0431. Arch Pathol Lab Med. 2020;144(4):466–72.

McKinnon KM. Flow cytometry: an overview. Curr Protoc Immunol. 2018;120:5.1.1–5.1.11. Published 2018 Feb 21. https://doi.org/10.1002/cpim.40.

McNerney ME, Godley LA, Le Beau MM. Therapy-related myeloid neoplasms: when genetics and environment collide. Nat Rev Cancer. 2017;17(9):513–27. https://doi.org/10.1038/nrc.2017.60.

Paterson AL, Liu H, ElDaly H. The role and contribution of clonality studies in the diagnosis of lymphoproliferative disorders. Eur J Haematol. 2019;102(6):472–8. https://doi.org/10.1111/ejh.13228. Epub 2019 Mar 28

Qasrawi A, et al. Prognostic impact of Philadelphia chromosome in mixed phenotype acute leukemia (MPAL): a cancer registry analysis on real-world outcome. Am J Hematol; 2020. https://doi.org/10.1002/ajh.25873. ePub

Rogers HJ, et al. Comparison of therapy-related and de novo core binding factor acute myeloid leukemia: a bone marrow pathology group study. Am J Hematol. 2020;95(7):799–808.

Sibaud V, Beylot-Barry M, Thiébaut R, Parrens M, Vergier B, Delaunay M, Beylot C, Chêne G, Ferrer J, de Mascarel A, Dubus P, Merlio JP. Bone marrow histopathologic and molecular staging in epidermotropic T-cell lymphomas. Am J Clin Pathol. 2003;119(3):414–23. https://doi.org/10.1309/qh6xlrf3mvuf2m8m.

Sinha C, Cunningham LC, Liu PP. Core binding factor acute myeloid leukemia: new prognostic categories and therapeutic opportunities. Semin Hematol. 2015;52(3):215–22. https://doi.org/10.1053/j.seminhematol.2015.04.002.

Swerdlow SH, Campo E, Harris NL, Jaffa ES, Pileri SA, Stein H, Thiele J (eds). WHO classification of tumours of haematopoietic and lymphoid tissues (Revised 4th edition). Lyon: lARC; 2017.

Index